W9-CFL-501

Principles of Organic Synthesis

NY Times
Aug 24, 1963

The Educational Low-Priced Books Scheme is funded by the Overseas Development Administration as part of the British Government overseas aid programme. It makes available low-priced, unabridged editions of British publishers' textbooks to students in developing countries. Below is a list of some other books on chemistry published under the ELBS imprint.

Atkins
Physical Chemistry
Oxford University Press

Emsley
The Elements
Oxford University Press

Furniss *et al*. (revisers)
Vogel's Textbook of Practical Organic Chemistry
Longman

Jeffery *et al*. (revisers)
Vogel's Textbook of Quantitative Chemical Analysis
Longman

Lee
Concise Inorganic Chemistry
Chapman & Hall

Liptrot
Modern Inorganic Chemnistry
Collins Educational

Liptrot, Thompson and Walker
Modern Physical Chemistry
Collins Educational

Norman and Waddington
Modern Organic Chemistry
Collins Educational

Sharpe
Inorganic Chemistry
Longman

Shriver, Atkins and Langford
Inorganic Chemistry
Oxford University Press

Principles of Organic Synthesis

Third Edition

R. O. C. NORMAN KBE FRS

Rector, Exeter College,
University of Oxford, UK

J. M. COXON FRSNZ

Professor of Chemistry,
University of Canterbury, New Zealand

EL BS with Chapman & Hall

Educational Low-Priced Books Scheme
funded by the British Government

Chapman & Hall
2–6 Boundary Row, London SE1 8HN

© 1993, R. O. C. Norman

All rights reserved. No part of this work covered by the
copyright hereon may be reproduced or used in any
form or by any means – graphic, electronic, or
mechanical, including photocopying, recording, taping,
or information storage or retrieval systems – without
written permission of the publishers.

First edition 1968 (Methuen & Co. Ltd)
Second edition 1978
Third edition 1993
Reprinted 1995

ELBS edition first published in 1993
Reprinted 1995

ISBN 0 7514 0195 1

Typeset in 10/12pt Times New Roman by
Best-set Typesetter Ltd, Hong Kong

Printed in Great Britain at the Alden Press, Oxford

Preface

Organic chemistry is one of the most rapidly developing of the sciences. Each year, there are new applications of organic compounds, for example, in medicine, in agriculture, and in the new technologies such as optoelectronics and superconductors; tens of thousands of new compounds are synthesized or isolated from natural sources; and new synthetic methods and reagents are introduced. It might seem to the student that ever more needs to be learned.

Fortunately that is not so. The profound increase in our understanding of the pathways by which organic compounds react – their mechanisms of reaction – has provided a relatively simple superstructure on which the vast array of the facts of the subject can be hung. The mechanistic principles are relatively few, and yet they account for the enormous range of reactions of organic compounds.

The purpose of this book is to show how these principles can be applied both to acquiring a knowledge of organic synthetic processes and to planning the construction of organic compounds. It is designed for those who have had no more than a brief introduction to organic chemistry. Nor is it intended to be comprehensive; for example, the vast body of evidence on which reaction mechanisms are based is not included, nor are experimental details given. The object has been to convey a broad understanding rather than to produce a reference text.

The book is in two parts. In Part I, reaction mechanism is set in its wider context of the basic principles and concepts that underlie chemical reactions: chemical thermodynamics, structural theory, theories of reaction kinetics, mechanism itself, and stereochemistry. In Part II, these principles and concepts are applied to the formation of particular types of bonds, groupings, and compounds: for example, how small molecules can be built on to give larger ones by the construction of new carbon–carbon bonds, or how one functional group can be transformed into another. The final chapter in Part II describes the planning and detailed execution of the multi-step syntheses of several complex naturally occurring compounds.

There have been numerous important developments since the second edition of this book was published, and every chapter has been brought up to date to include them. Among the more notable are: the stereochemical control of reactions, reflecting the need to synthesize efficiently compounds with several asymmetric centres; the use of organotransition-metal reagents, leading to a new chapter; the exploitation of lithium diisopropylamide (LDA) as a base in the formation of new C—C bonds; free-radical reactions for the synthesis of C—C bonds; uses of organosilicon compounds; and subtle new protective and condensing reagents in polypeptide synthesis.

The text is extensively cross-referenced and there is a detailed index.

R. O. C. NORMAN

J. M. COXON

Contents

Nomenclature and conventions

Some readers of this book will have taken introductory courses in organic chemistry in which the systematic nomenclature of the International Union of Pure and Applied Chemistry (IUPAC) was used; others will be used to the more common names of organic compounds. We have continued the practice of the previous editions of using common names for most simple compounds because this is still the language used by professional chemists – for example, in papers in the literature describing their research results – so that early familiarity with it is important. We have largely followed the conventions of the *Journal of Organic Chemistry*.

Organic compounds are frequently expressed on paper by a formula such as

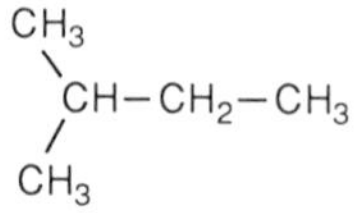

$$CH_3$$
$$CH-CH_2-CH_3$$
$$CH_3$$

for 2-methylbutane. This can be cumbersome for all but small molecules, and we have generally used a convention in which a C—C bond is represented by a line; when more than one is present in an aliphatic chain, a zigzag line is used except for alkynes, which are linear; and hydrogen atoms attached to carbon are not shown. Examples are:

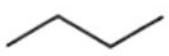

butane 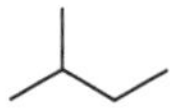   2-methylbutane 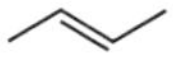   *trans*-2-butene

propyne 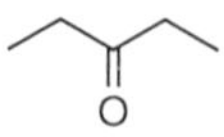   3-pentanone 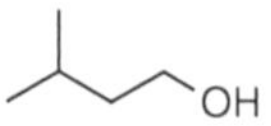   3-methyl-1-butanol

In Part II especially, we have also used abbreviations for groups such as Et (ethyl) and Bu^t (tertiary butyl, or 2-methyl-2-propyl).

In reaction schemes, when two or more reagents are present they are connected by hyphens, and when a compound is present which acts only as a solvent it follows an oblique; for example

$$MeI - EtO^- / EtOH$$

indicates that the reactant is treated with methyl iodide and ethoxide ion in ethanol solvent. Notice the difference between

$$\xrightarrow{\text{NH}_2^- / \text{NH}_3} \qquad \text{and} \qquad \xrightarrow{\text{Na - NH}_3}$$

where ammonia is involved only as solvent in the former but as both solvent and reagent (the source of protons in reduction by sodium) in the latter.

The following are the more frequently used common names with their usual representation.

Aliphatic hydrocarbons

ethylene	$CH_2=CH_2$
propylene	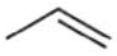
acetylene	 $H-\!\!\equiv\!\!-H$

Alcohols

isopropanol	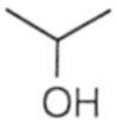
t-butanol	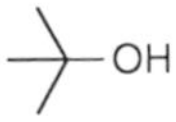
ethylene glycol	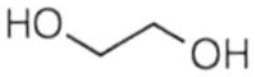
pinacol	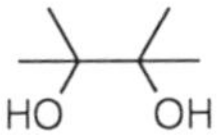

Aldehydes

formaldehyde	$CH_2=O$
acetaldehyde	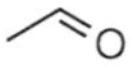
propionaldehyde	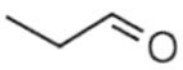

Ketones

acetone	
pinacolone	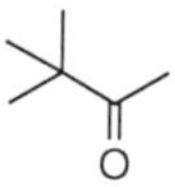

Acids

formic acid	HCO_2H
acetic acid	$AcOH$
propionic acid	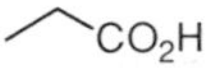 CO_2H
butyric acid	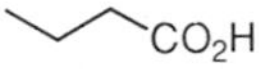 CO_2H
acrylic acid	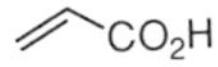 CO_2H
oxalic acid	HO_2C-CO_2H
malonic acid	 $HO_2C \quad CO_2H$
succinic acid	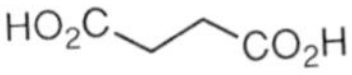 $HO_2C \quad CO_2H$
glutaric acid	 $HO_2C \quad CO_2H$
adipic acid	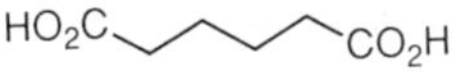 $HO_2C \quad CO_2H$
maleic acid	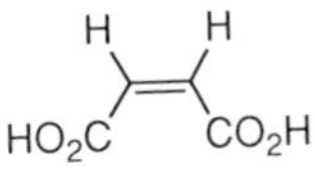
fumaric acid	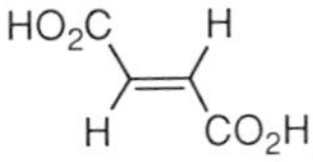
glycolic acid	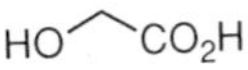 $HO \quad CO_2H$
lactic acid	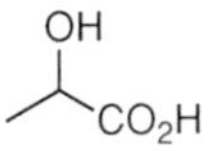
tartaric acid	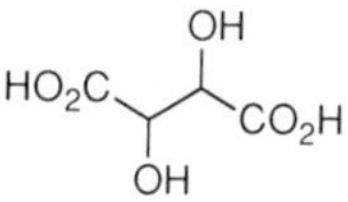

Amines

methylamine	$MeNH_2$
dimethylamine	Me_2NH
trimethylamine	Me_3N
triethylamine	Et_3N

Other aliphatic compounds

diethyl ether	Et_2O
chloroform	$CHCl_3$
carbon tetrachloride	CCl_4
ethyl orthoformate	$HC(OEt)_3$
acetonitrile	$MeCN$
acrylonitrile	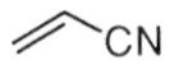
diglyme	$(MeOCH_2CH_2)_2O$

acetoacetic ester

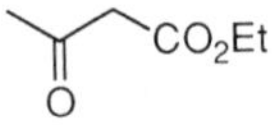

malonic ester EtO_2C CO_2Et

Aromatic compounds

acetophenone

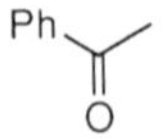

aniline $PhNH_2$

anisole	$PhOMe$
benzaldehyde	$PhCHO$
benzoic acid	$PhCO_2H$
benzophenone	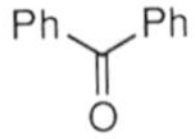
catechol	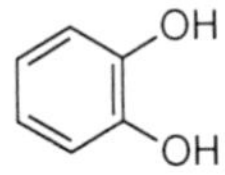
cinnamic acid	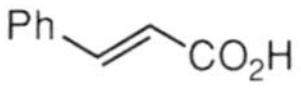
cumene	$PhCHMe_2$
mesitylene	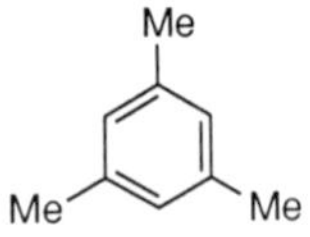
phenol	$PhOH$
phthalic acid	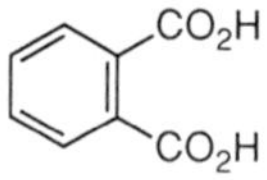
salicylic acid	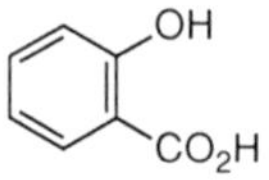
stilbene	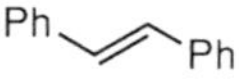
styrene	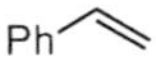
toluene	$PhMe$
xylene	$+$ *m*- and *p*- isomers

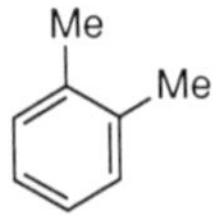

Common names for groups

methyl, Me CH_3

ethyl, Et	CH_3CH_2	vinyl	$CH_2=CH$
isopropyl, Pr^i	$(CH_3)_2CH$	acetyl, Ac	CH_3CO
butyl, Bu	$CH_3CH_2CH_2CH_2$	phenyl, Ph	C_6H_5
isobutyl	$(CH_3)_2CHCH_2$	tolyl	$CH_3-C_6H_4$
t-butyl, Bu^t	$(CH_3)_3C$	tosyl	$4\text{-}CH_3-C_6H_4-SO_2O$
neopentyl	$(CH_3)_3CCH_2$		

Part I

Introduction to Part I

The five chapters which comprise Part I are concerned with the principles that govern organic reactions.

In chapters 1–3, the principles of chemical thermodynamics, the theory of molecular structure, and the principles of reaction kinetics are briefly summarized (textbooks on physical chemistry should be consulted for detailed treatments); the emphasis in these chapters is on the application of the principles and theories to the structures and reactions of organic compounds.

Thermodynamic principles show that a reaction will 'go' – that is have an equilibrium constant greater than unity – only if the products have a lower free-energy content than the reactants. The free energy of a species is related to its enthalpy, which is determined essentially by the strengths of the bonds it contains, and to its entropy, which is a measure of its degree of disorder: a low free energy corresponds to strong bonding forces and a high degree of disorder. From thermodynamic considerations there follows, for example, an understanding of why it is possible to reduce acetylene to ethylene at room temperature and to carry out the reverse reaction at high temperatures, why small changes in structure can have a very large effect on the position of dynamic equilibrium between isomers, and why some ring-closure reactions occur so much more efficiently than equivalent bimolecular reactions.

In the second chapter, the strengths of bonds and the shapes of organic compounds are related to the theories of molecular structure. Delocalized bonds are emphasized; for example, the different chemical behaviour of benzene and ethylene is related to the very large stabilization energy in benzene which arises from delocalization. Other properties of organic compounds that are of importance in synthesis, such as the acidities of C—H bonds in various environments, also follow from structural theory.

That there should be a negative free-energy change is in practice a necessary but not a sufficient condition for a reaction to occur, for the *rate* at which it takes place may be negligible. Thermodynamic considerations alone indicate that hydrocarbons should not coexist with air, for the free-energy change involved in their oxidation to carbon dioxide and water is significantly negative; in practice, however, their rates of combustion at ordinary temperatures are negligible. The third chapter sets out the theories of reaction kinetics and the effects of temperature on rate, and then introduces correlations of the rates of specific types of reaction with structure.

The planning of syntheses is helped considerably by an understanding of the mechanisms by which reactions occur. It must be emphasized that mechanisms are *theories* and not *facts* of the subject: they have been deduced from experimental observations and in some instances they transpire to be incorrect or at least in need of refinement. One should say, 'the mechanism is thought to be', rather than 'the mechanism is'. Nonetheless, the current mechanistic theories, which are surveyed in the fourth chapter, not only provide an intellectually satisfying and unifying picture of the complexity of reactions, but also enable predictions to be made, with increasing assurance as the degree of rationalization of the subject increases, of the effects which structural modifications will have on the course of a reaction.

Stereochemistry – the study of the spatial relationships of atoms and bonds – would in the past have been a natural adjunct of the study of molecular structure. It is now as important to considerations of chemical dynamics as to those of chemical statics, and follows naturally, in the last chapter, the study of kinetics and mechanism. Indeed, it is closely intertwined with mechanism: many naturally occurring compounds have a complex and highly specific stereochemistry and it has only been through an understanding of the stereoelectronic principles of reactions that their syntheses have been successfully planned and executed.

Chemical thermodynamics 1

1.1
Equilibrium

All chemical reactions are in principle reversible: reactants and products eventually reach *equilibrium*. In some cases, such as the esterification of an acid by an alcohol,

$$RCO_2H \quad + \quad R'OH \quad \rightleftharpoons \quad RCO_2R' \quad + \quad H_2O$$

the position of the equilibrium is quite closely balanced between reactants and products, whereas in other cases the equilibrium constant is either very high or very low, so that the reaction goes essentially to completion in one direction or the other (given the appropriate conditions). From the point of view of devising an organic synthesis it is necessary to know whether the position of equilibrium will favour the desired product. The factors which determine the equilibrium constant of a reaction and its variation with changes in conditions follow from the principles of thermodynamics.

1.2
Free energy

Chemical systems are subject to opposing influences: the tendency to minimize their enthalpy (heat content), H, is opposed by the tendency to maximize their entropy (degree of disorder), S. Chemical reactions are usually carried out at constant pressure and under these conditions the compromise between the two trends is determined by the value of the function $(H - TS)$: this function tends to decrease, and the compromise situation – equilibrium – corresponds to its minimum value. It is convenient to define a new function, G, the Gibbs free energy, as $G = H - TS$. A process will occur spontaneously if, as a result, G decreases and it will continue until G reaches a minimum.

The principles of thermodynamics lead to an equation that relates the equilibrium constant, K, to the change in free energy accompanying a reaction. If the sums of the free energies of reactants and products in their standard states (i.e. for 1 mole at a pressure of 1 bar) at temperature T are G_1^0 and G_2^0 respectively, then the change in free energy on reaction, ΔG^0, is $G_2^0 - G_1^0$ and is given by

$$RT \ln K = -\Delta G^0$$

where R is the gas constant ($8.314\,\mathrm{J\,K^{-1}\,mol^{-1}}$).

A knowledge of the standard free energies of reactants and products thus enables us to predict whether a given set of reactants is likely to yield a desired set of products. If ΔG^0 is negative, the equilibrium constant (greater than 1) is favourable to the formation of the products, whereas if ΔG^0 is positive it is correspondingly unfavourable.

Standard free energies have been measured for a number of organic and inorganic compounds. For example, those of ethylene, ethane, and hydrogen in the gaseous state at 25°C are 68, -33, and $0\,\text{kJ}\,\text{mol}^{-1}$ respectively. (By convention, the most stable allotrope of an *element* in its standard state at 25°C is assigned a free-energy value of zero. The scale of free energies is therefore an arbitrary one, but this is immaterial since we are always concerned with *differences* in free energies.) Therefore the reaction

$$C_2H_6 \quad \rightleftharpoons \quad C_2H_4 \quad + \quad H_2$$

has $\Delta G^0 = +101\,\text{kJ}\,\text{mol}^{-1}$, so that we cannot obtain ethylene from ethane in significant amount under these conditions although we may expect to obtain ethane from ethylene. Notice, however, that even though a free-energy change may be favourable, the *rate* of formation of the product may be too slow for the reaction to be practicable, or the reaction may take a different course (e.g. ethylene might react with two molecules of hydrogen to give two molecules of methane, which is also a thermodynamically favourable process). Fulfilment of the thermodynamic criterion is therefore a *necessary* but not a *sufficient* condition for a reaction to occur. The factors which control rates of reaction and the pathways which reactions take are discussed in subsequent chapters.

Thermodynamic data also give no information about the *reasons* why the free energies of compounds have particular values. Ethylene, for example, has a positive standard free energy of formation, whereas that for ethane is negative; that is, ethylene is unstable with respect to carbon and hydrogen and ethane is stable. These are empirical *facts*; the reasons underlying them lie in the realm of *theory*, namely, the theory of chemical structure and bonding, which is discussed in the next chapter.

1.3 Bond energies

Since $G^0 = H^0 - TS^0$ and $RT \ln K = -\Delta G^0$, it follows that

$$RT \ln K = -\Delta H^0 + T\Delta S^0$$

Therefore a knowledge of the differences between the enthalpies and entropies of reactants and products gives information about whether a reaction will 'go'. This is generally a more useful approach than considering free-energy differences, for three reasons: free energies are not known in most cases; enthalpy differences – the heats of reactions – can be calculated with reasonable precision from those of reactions which have been measured; and entropy differences can usually be inferred to the accuracy required (section 1.4) (in the conditions of most organic reactions the $T\Delta S^0$ term is in any case much smaller than the ΔH^0 term).

The calculation of enthalpy differences is based on the first law of thermo-dynamics: if the heats of the reactions $A \rightarrow B$ and $B \rightarrow C$ are respectively x and y, then that of the reaction $A \rightarrow C$ is $(x + y)$, since otherwise it would be possible to construct a cycle of reactions that generated energy, in contravention of the first law. The calculations are facilitated by constructing a table of *bond energies*, taking as the reference point the heat of formation of molecules from their constituent atoms. For example, the bond energy of the C—H bond is defined as one-quarter of the heat of the reaction

$$CH_4 \text{ (g)} \longrightarrow C \text{ (g)} + 4 H \text{ (g)}$$

since four C—H bonds are broken in this process. It is not possible to determine this experimentally, but it can be calculated since it is the sum of the heats of the following reactions which have been measured*:

$$CH_4 \text{ (g)} + 2 O_2 \text{ (g)} \longrightarrow CO_2 \text{ (g)} + 2 H_2O \text{ (l)}$$

$$CO_2 \text{ (g)} \longrightarrow C \text{ (s)} + O_2 \text{ (g)}$$

$$2 H_2O \text{ (l)} \longrightarrow 2 H_2 \text{ (g)} + O_2 \text{ (g)}$$

$$2 H_2 \text{ (g)} \longrightarrow 4 H \text{ (g)}$$

$$C \text{ (s)} \longrightarrow C \text{ (g)}$$

The importance of bond energies is that, to a close approximation, they are constant for a particular bond in different structural environments, so that, given the bond energies 414, 357, and 462 kJ mol^{-1} for C—H, C—O, and O—H bonds respectively, the total bond energy of methanol that contains three C—H, one C—O, and one O—H bond, is calculated to be 2061 kJ mol^{-1}, in fair agreement with the experimental value (1985 kJ mol^{-1}) derived from the heat of combustion of methanol.

Bond energies for a number of commonly occurring structural units are given in Table 1.1.

Bond energies quoted above are *average* bond energies and it is often more useful to know the energy required to break a particular bond in a compound, i.e. the *bond dissociation energy*. Although the O—H bond energy is, by definition, one-half the heat of formation of water, in fact a greater quantity of energy is required to break the first O—H bond in water ($H_2O \rightarrow HO + H$; $\Delta H = 491$ kJ mol^{-1}) than the second ($HO \rightarrow O + H$; $\Delta H = 433$ kJ mol^{-1}). The bond dissociation energy is therefore greater than the bond energy and this is commonly the case, except for diatomic molecules for which the values are necessarily identical. Some typical bond dissociation energies are set out in Table 1.2.

---

* The heat of the last reaction, the atomization of carbon, is necessary because the heat of formation of carbon dioxide is measured for solid carbon whereas bond energies refer to atoms, i.e. 'gaseous' carbon.

Table 1.1 Bond energies (kJ mol^{-1}) at 25°C

H—H	435	C—C	347	C—O	357
H—F	560	C=C	610	C=O^a	694
H—Cl	428	C≡C	836	C=O^b	736
H—Br	364	N--N	163	C=O^c	748
H—I	297	N=N	418	H—C	414
F—F	150	N≡N	940	H—N	389
Cl—Cl	238	C—N	305	H—O	462
Br—Br	188	C=N	614	O—O	155
I—I	150	C≡N	890	S—S	251

^{a} In formaldehyde. ^{b} In other aldehydes. ^{c} In ketones.

Table 1.2 Bond dissociation energies (kJ mol^{-1})

H—CH$_3$	426	H—OH	491	CH$_3$—F	447
H—CH$_2$CH$_3$	401	H—NH$_2$	426	CH$_3$—Cl	339
H—CH(CH$_3$)$_2$	385	CH$_3$—OH	376	CH$_3$—Br	280
H—C(CH$_3$)$_3$	372	CH$_3$—NH$_2$	334	CH$_3$—I	226
H—CH$_2$Ph	322	CH$_3$—CH$_3$	347	CH$_3$—NO$_2$	238

The premise on which the use of Table 1.1 is based, that bond energies are constant for a particular bond in different environments, requires closer examination. If it were strictly true, the heats of combustion of the three isomeric pentanes, for example, each of which contains four C—C and 12 C—H bonds, would be the same. In fact, they are different: CH$_3$(CH$_2$)$_3$CH$_3$, 3533; (CH$_3$)$_2$CHCH$_2$CH$_3$, 3525; (CH$_3$)$_3$CCH$_3$, 3513 kJ mol^{-1}. Nevertheless, the percentage differences are so small in this and many similar cases that bond-energy data are of considerable use. However, there are certain structural environments in which appreciable differences between observed and predicted bond energies occur. One example, in Table 1.1, concerns the carbonyl group, whose bond energy is nearly 10% greater in ketones than in formaldehyde, and is greater still (804 kJ mol^{-1}) in carbon dioxide.

A more general deviation occurs with *conjugated* compounds, that is, compounds possessing alternating single and double bonds. For example, consider the following hydrogenations:

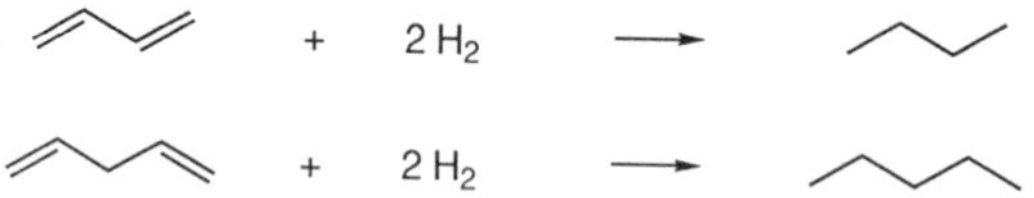

+ 2 H$_2$ ⟶ $\Delta H^0 = -238 \cdot 7\,\mathrm{kJ\,mol}^{-1}$

+ 2 H$_2$ ⟶ $\Delta H^0 = -254 \cdot 2\,\mathrm{kJ\,mol}^{-1}$

If bond energies were strictly additive, the heats of hydrogenation in the above reactions should be the same since, in each, two C=C double bonds are converted into single bonds, two H—H bonds are broken, and four C—H bonds are formed. The lower value for 1,3-butadiene shows that this is a more stable compound than predicted by examination of 1,4-pentadiene, and this proves to be so for other conjugated systems. We shall refer to the difference between predicted and observed values of heats of hydrogenation as the *stabilization energy*.

The attachment of alkyl groups to an alkene also leads to the introduction of stabilization energy. For example, whereas the heat of hydrogenation of ethylene is 137 kJ mol^{-1}, that of propylene ($CH_3CH{=}CH_2$: one alkyl group) is 126 kJ mol^{-1} and that of *trans*-2-butene ($CH_3CH{=}CHCH_3$: two alkyl groups) is 115 kJ mol^{-1}, i.e. the stabilization energies are 11 and 22 kJ mol^{-1}, respectively.

For those cyclic, conjugated compounds which are aromatic (section 2.4b) heats of hydrogenation are in many cases very considerably less than the predicted values. The stabilization energies of benzene, naphthalene, and anthracene are respectively 150, 255, and 349 kJ mol^{-1}.

An understanding of the origin of stabilization energies follows from current theories of molecular structure (chapter 2). It is necessary here to emphasize that bond-energy data must be used with care but that, nevertheless, conjugated and aromatic systems excepted, they provide a useful working basis for predictive purposes.

1.4 Entropies

Molecules undergo three types of motion: translation, rotation, and vibration. For each, in accord with quantum theory, only certain *energy states* occur. The entropy of a compound is related to the number of ways in which the molecules can be distributed amongst these energy states; the greater the number, the larger is the degree of disorder, i.e. the larger is the entropy. The total entropy is the sum of the individual contributions:

$$S_{tr} + S_{rot} + S_{vib}$$

The distribution of molecules amongst the energy states is governed by the *Boltzmann distribution function*: the number of molecules in an excited state, n_e, as a fraction of the number in the lowest, or ground, state, n_l, is given by $n_e/n_l = \exp(-\varepsilon/kT)$ where ε is the energy difference and k is the Boltzmann constant. It follows that the larger the energy difference is, the smaller is the population of an excited state and so the smaller is the contribution to the entropy. In the extreme case, the first excited state could be so much higher in energy than the ground state that all the molecules would be in the latter: this would correspond to complete order and therefore to $S = 0$. Conversely, the more closely spaced the energies of the states, the more heavily populated the excited states will be: the number of permutations possible for the arrangement of the molecules increases and the entropy rises*.

The translational energy levels of a molecule are very closely spaced, so that the higher levels are well populated and S_{tr} is relatively large; for example, for

---

* If, for example, the spacing of the levels is such that, of six particles, one is likely to be in the excited state, then, since any one of the six may be that particle, there are six ways in which the system can be arranged; if two are likely to be in the excited state, the number of possibilities is 30; and so on.

methane $S_{tr}^0 = 144\,\mathrm{J\,K^{-1}\,mol^{-1}}$ at 25°C. Rotational energy levels are less closely spaced and S_{rot} is correspondingly smaller; for example, $S_{rot}^0 = 32\,\mathrm{J\,K^{-1}\,mol^{-1}}$ for methane at 25°C.

There are two forms of vibrational motion: stretching and bending vibrations*. The energy levels for stretching vibrations are usually widely spaced, so that the vibrational contribution to the entropy is negligible. The energy levels for bending vibrations are closer together and the contribution to the entropy, particularly for a molecule which has a large number of bending modes, is significant.

Many polyatomic molecules possess entropy by virtue of their undergoing *internal rotations* which correspond to bending vibrations. For example, the methyl groups in ethane, CH_3—CH_3, rotate with respect to each other, so that the molecule can adopt any of a number of possible *conformations*. In one of the two shown in the projection diagram below, the hydrogen atoms on adjacent carbons eclipse each other, and in the other they are fully staggered (p. 27). Two conformations of butane, showing possible arrangements of the four carbon atoms, are also drawn. The conformations are not necessarily equally energetic: in butane, for example, the first of the below two structures is of lower energy content than the second, largely because of the repulsive forces between the two methyl groups in the latter (p. 28). Nevertheless, the energy differences are normally small, so that there is a contribution to the entropy of the molecule. The magnitude of conformational entropy increases with the number of available conformations and hence with the length of the chain of atoms.

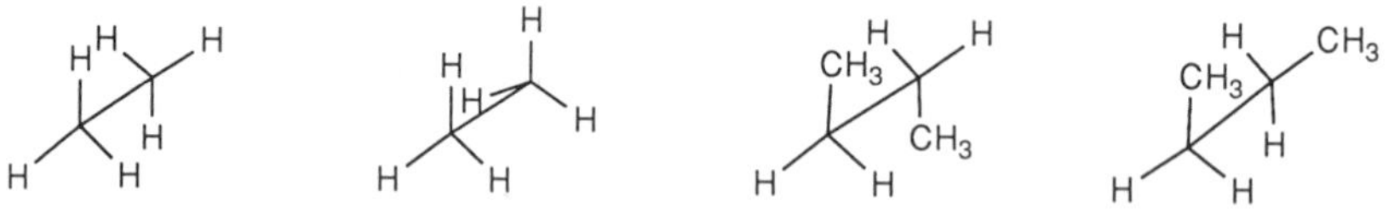

The most significant aspects of the above discussion are, first, the importance of the translational entropy and, second, the significance of the conformational entropy.

(*i*) In a reaction which results in an increase in the number of species, such as the fragmentation $A \rightarrow B + C$, there is a considerable increase in entropy because of the gain of three degrees of translational freedom. (There are also changes in the rotational and vibrational contributions to the entropy but these are normally much smaller and can be neglected for the present argument.) On the other hand, such reactions may result in a decrease in the number of bonds and, associated with this, an increase in the enthalpy of the system. The simplest example of such a reaction is the dissociation of the hydrogen molecule, $H_2 \rightarrow 2H \cdot$ ($\Delta H^0 = +435\,\mathrm{kJ\,mol^{-1}}$). Even when there is no overall change in the number of bonds, there may still be an increase in the enthalpy of the system; an example

---

* A non-linear molecule containing n atoms ($n > 2$) has $3n - 6$ vibrational modes of which $n - 1$ are stretching modes and $2n - 5$ are bending modes, together with three translational and three rotational degrees of freedom.

is the dehydrogenation of ethylene, $CH_2{=}CH_2 \rightarrow CH{\equiv}CH + H_2$ ($\Delta H^0 = +167\,kJ\,mol^{-1}$). In these cases, the enthalpy and entropy terms are opposed and whether ΔG^0 is negative or positive (and hence whether K is greater or less than unity) depends on whether the gain in translational entropy is larger or smaller than the increase in enthalpy.

For the dissociation of the hydrogen molecule at 25°C, $\Delta S_{tr} \sim 100\,J\,K^{-1}\,mol^{-1}$, and for the dehydrogenation of ethylene, $\Delta S_{tr} \sim 117\,J\,K^{-1}\,mol^{-1}$. The contributions from the increase in translational freedom to the decrease in free energy during the forward reaction ($=T\Delta S^0$) at 25°C are therefore ~30 and $35\,kJ\,mol^{-1}$, respectively. In each case, these values are far smaller than the values for the increase in enthalpy, so that ΔG^0 is large and positive and equilibrium lies essentially completely to the left. In general, this is true of such fragmentations unless ΔH^0 is less than about $40\,kJ\,mol^{-1}$.

At much higher temperatures, however, this situation may be reversed. For example, for the dehydrogenation of ethylene at 1000°C, $\Delta S_{tr} \sim 150\,J\,K^{-1}\,mol^{-1}$, so that $T\Delta S^0 \sim 190\,kJ\,mol^{-1}$. The free-energy change is now significantly negative and equilibrium lies almost completely to the right so, whereas it is possible to reduce acetylene at or near room temperature to obtain ethylene, it is also possible to dehydrogenate ethylene at very high temperatures in order to produce acetylene and this process has been used industrially.

In contrast to fragmentations, the entropy changes which accompany reactions in which there is no change in the number of species, such as $A + B \rightarrow C + D$, are usually small: for a typical case, the esterification of acetic acid by methanol, ΔS_{tr}^0 is $-4{\cdot}6\,J\,K^{-1}\,mol^{-1}$ at 25°C, equivalent to a contribution of only $1{\cdot}4\,kJ\,mol^{-1}$ to the free-energy change at this temperature. The equilibrium constant in such reactions tends, therefore, to be governed over a wide temperature range by the enthalpy change. However, exceptions to this generalization occur when, particularly in equilibria involving ions, one or more of the species is *solvated*; entropy changes associated with the 'freezing' of solvent molecules become important.

(*ii*)   Consider the equilibrium between 1-hexene and cyclohexane,

for which $K = 6 \times 10^9$ at 25°C. The enthalpy change is favourable: ΔH^0 from bond-energy data is $-84\,kJ\,mol^{-1}$, in close agreement with the experimental value of $-82\,kJ\,mol^{-1}$. If this factor alone were involved, K would be $\sim10^{14}$. The entropy change is, however, unfavourable because the number of conformations of 1-hexene (which arise from rotations about C—C bonds) is much greater than that for cyclohexane, where rotations are restricted by the ring structure. It is this factor which accounts for almost the entire entropy change of $-86{\cdot}5\,J\,K^{-1}\,mol^{-1}$.

For the formation of the six-membered ring above, the entropy change, though markedly negative, does not offset the favourable enthalpy change: ΔG^0 is $-52\,kJ\,mol^{-1}$. Smaller rings, however, are *strained* because the bond angles are distorted from their natural values (section 2.4a) so that the ΔH term is less favourable. Although less internal freedom is lost on ring closure (for the smaller open chain compounds have fewer possible conformations), the free-energy change is less favourable. This is also true of the formation of larger rings, for although the ring strain is very small (for 7- to 11-membered rings) or zero (for larger rings) (section 2.4a), the loss of internal freedom increases with increase in ring size.

**1.5
Further
applications of
thermodynamic
principles**   (*a*)   *Tautomerism*

In principle, it might be possible to establish equilibrium between any two isomers, e.g.

In practice, it is common to find that conditions are not available for attainment of the equilibrium: each isomer may be isolated and is stable with respect to the other indefinitely. However, there are many systems in which isomers exist in dynamic equilibrium; their interconversion involves the detachment of a group from one position and its re-attachment at another. The isomers are then known as *tautomers* and the phenomenon as tautomerism.

The most frequently encountered tautomeric systems are those in which the tautomers differ in the position of a hydrogen atom. For example, 2,4-pentanedione consists of a mixture of keto and enol forms:

keto tautomer   enol tautomer

Its solution in hexane consists of 8% of the keto tautomer and 92% of the enol tautomer and it exhibits the reactions typical of both the carbonyl group, $C{=}O$, and the enol group, $C{=}C{-}OH$. On the other hand, acetone, whose tautomeric equilibrium is also between keto and enol forms,

contains only about $10^{-5}\%$ of the enol.

The entropy changes in these reactions are negligible and the reason for the difference between them lies in the enthalpy terms. The equilibrium constant for the enolization of acetone, $c.$ 10^{-7}, corresponds to an enthalpy change in favour of the keto form* of $c.$ $39\,kJ\,mol^{-1}$, and this is consistent with the bond-energy data in Table 1.1 which show that replacement of C=O, C—C, and C—H bonds in the keto form by C—O, C=C, and O—H bonds in the enol is strongly endothermic. The enthalpy change for enolization of the dione, for which K is $c.$ 11, is $c.$ $6.6\,kJ\,mol^{-1}$ in favour of the enol form: relatively, therefore, the enol form of the dione is about $45\,kJ\,mol^{-1}$ more stable than the enol of acetone.

This difference can be accounted for by two factors which increase the bonding of the enol form of the dione compared with that in the enol form of acetone. First, the C=C bond in the former is conjugated to C=O, corresponding to a stabilizing interaction of about $17\,kJ\,mol^{-1}$. Second, its hydroxylic hydrogen is *hydrogen-bonded* to the carbonyl group (p. 34),

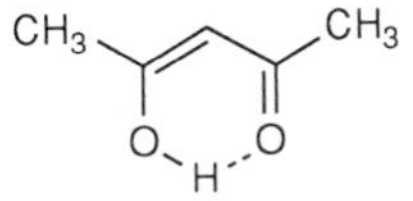

and the strength of this bond is about $25\,kJ\,mol^{-1}$. These two effects provide extra bonding of $42\,kJ\,mol^{-1}$, in close agreement with the experimental value.[†]

In contrast, phenol exists essentially entirely in the enol form:

This is because the enol possesses the stabilization energy of the aromatic ring ($c.$ $150\,kJ\,mol^{-1}$) whereas the keto tautomer, which is conjugated but not aromatic, has a very much smaller stabilization energy ($c.$ $20\,kJ\,mol^{-1}$): the enthalpy change is therefore strongly in favour of the enol.

The situation is different when a second phenolic group is introduced, *meta* to the first,

---

* If $\Delta S^0 = 0$, $RT\ln K = -\Delta H^0$; $RT = c.$ $2.4\,kJ\,mol^{-1}$ at room temperature, whence $\Delta H^0 = c.$ $39\,kJ\,mol^{-1}$.
† In water, 2,4-pentanedione consists of 84% of keto and 16% of enol tautomer. The keto tautomer is now stabilized by hydrogen-bonding of its two carbonyl-oxygen atoms with the solvent, effectively nullifying one of the stabilizing influences in the enol tautomer.

Simple calculation leads to a ΔH^0 value close to zero, essentially because two strongly bonded carbonyl groups are present in the keto structure to offset the aromatic stabilization energy of the enol. Consistent with this, *m*-dihydroxybenzene has properties characteristic of both a phenol and a ketone: for example, it undergoes the rapid electrophilic substitutions such as bromination which are characteristic of phenols (p. 386) and it is reduced by sodium amalgam to 1,3-cyclohexanedione in the manner characteristic of αβ-unsaturated carbonyl compounds (p. 638). An extension of the argument rationalizes the behaviour of 1,3,5-trihydroxybenzene, which is more fully ketonized.

Many other systems of general structure X=Y—Z—H are tautomeric: X=Y—Z—H ⇌ H—X—Y=Z. Those most frequently met are the following:

carbon triad	$\underset{H}{C}-\overset{C}{}\equiv C$	⇌	$C=\overset{C}{}-\underset{H}{C}$
e.g. [*]	$PhCH_2CH{=}CH_2$	⇌	$PhCH{=}CHCH_3$
azomethine	$\underset{H}{C}-\overset{N}{}{\equiv}C$	⇌	$C{=}\overset{N}{}-\underset{H}{C}$
e.g.	$ArCH_2N{=}CHAr'$	⇌	$ArCH{=}NCH_2Ar'$
nitro–aci-nitro	$\underset{H}{C}-\overset{O^-}{\underset{}{N^+}}{=}O$	⇌	$C{=}\overset{O^-}{\underset{H}{N^+}}-O$
e.g.	CH_3-NO_2	⇌	$CH_2{=}N\!\!\begin{smallmatrix}O^-\\OH\end{smallmatrix}$
nitroso–oxime	$\underset{H}{C}-\overset{N}{}{\equiv}O$	⇌	$C{=}\overset{N}{}-\underset{H}{O}$
e.g.	CH_3-NO	⇌	$CH_2{=}N\!\!\begin{smallmatrix}\\O\\H\end{smallmatrix}$
diazoamino	$\underset{H}{N}-\overset{N}{}{\equiv}N$	⇌	$N{=}\overset{N}{}-\underset{H}{N}$
e.g.	$ArNHN{=}NAr'$	⇌	$ArN{=}NNHAr'$

[*] Equilibrium, attained by heating with a strong base, lies to the right because the product possesses stabilization energy owing to the conjugation between the C=C bond and the aromatic ring.

Tautomerism in which the tautomers differ in the position of a hydrogen atom is referred to as *prototropy*, for the mechanisms of interconversion involve the removal of a proton from one position and the gain of a proton at the other (pp. 145, 146). There are also systems in which the migrating group is of anionic type, such as hydroxyl, and this form of tautomerism is referred to as *anionotropy*. For example, alcohols (a) and (b) form an equilibrium mixture when either is treated with dilute sulfuric acid at about 100°C:*

| (a) | (b) |
| 30% | 70% |

As with prototropic systems, the equilibrium constant in anionotropic systems is strongly dependent upon the structures of the tautomers. For example, although in the anionotropic system above the equilibrium is relatively balanced between the tautomers, in the equilibrium between (c) and (d),

equilibrium is so strongly in favour of the latter tautomer that the former is not detectable in the equilibrated mixture.

The position of equilibrium is dictated mainly by the enthalpy term, for the entropy change in a reaction $A \rightleftharpoons B$ is small. The more highly conjugated of two tautomers is therefore the predominant one at equilibrium, for it possesses the larger stabilization energy: the conjugated alcohol (d) has a lower enthalpy than

---

* The acid acts by protonating the hydroxyl group in small degree; water leaves the molecule to form a *delocalized* cation (p. 110) which it can re-attack at either of two carbon atoms. Repeated reactions of this type lead to the equilibrium mixture:

1,2-Disubstituted alkenes such as (b) exist in isomeric forms,

Both are formed in these equilibria, but the *trans*-isomer is generally the more stable and predominates. For simplicity, only the *trans*-isomer is shown in this and later examples.

the non-conjugated alcohol (c). Neither (a) nor (b) is conjugated, but in the latter the C=C bond is attached to two alkyl substituents whereas in the former it is attached to one. This leads to a *small* enthalpy difference (p. 34). (The difference in free energies corresponding to an equilibrium of 70% (b) and 30% (a) is only $2\,\mathrm{kJ\,mol^{-1}}$ at 25°C.)

The rates of interconversion in anionotropic systems vary widely, tending to be greater under given conditions when more highly conjugated systems are involved. This can be important in synthesis, e.g. an attempt to convert alcohol (c) into an ester in acid-catalyzed conditions would give the ester of (d):

(b)   Ring closure

Organic reactions frequently lead to the formation of cyclic compounds from open-chain (acyclic) compounds. We shall consider two general classes of ring-closure reactions: (i) A → B and (ii) A → B + C.

The comparison of a ring closure of type A → B with an analogous inter-molecular reaction is instructive. For example, acetaldehyde exists in aqueous solution in equilibrium with its hydrate:

$$CH_3CH{=}O \quad + \quad H_2O \quad \rightleftharpoons \quad CH_3CH(OH)_2$$

The equilibrium lies to the left, despite a favourable decrease in enthalpy, because of the unfavourable entropy change ($-68\cdot5\,\mathrm{J\,K^{-1}\,mol^{-1}}$, equivalent to a contribution of $20\cdot4\,\mathrm{kJ\,mol^{-1}}$ to ΔG^0 at 25°C) which results mainly from the loss of translational freedom. For the analogous equilibrium between the acyclic and cyclic forms of 5-hydroxypentanal, however, the equilibrium lies well to the right ($K \sim 16$):

This is because ΔS_{tr}^0 is zero and the principal source of the entropy change is the loss of internal freedom which, for six-membered cyclization, is less than the loss of translational freedom for the corresponding intermolecular reaction.

The equilibrium between 4-hydroxybutanal $HO{-}(CH_2)_3{-}CHO$, and its cyclic form is slightly less favourable ($K = 8$) because, although the entropy factor is more favourable than for formation of the six-membered ring, the five-membered ring is slightly strained (p. 28) so that the enthalpy factor is less favourable. The equilibrium constants for the corresponding three- and four-membered rings are negligible because ring strain is considerable* and those for rings with more than six atoms decrease rapidly because the entropy factor becomes increasingly unfavourable (for rings containing 7–11 members, there is also some internal strain). For example, $K = 0.2$ and 0.1, respectively, for the seven- and eight-membered rings.

(c) Unstable compounds

The use of the words *stability* and *instability* often gives rise to confusion. The thermodynamic definitions are clear: the stability of a compound refers to the standard free energy of its formation; a compound may be stable or unstable with respect to its elements. This use of stability must be distinguished from that relating to other reactions, for example a compound may be unstable to heat (e.g. a peroxide), to water (e.g. an acetal in the presence of acid) or to air (e.g. a drying-oil). We shall always refer to the conditions in which a compound is unstable and here draw attention to the thermal stability of organic compounds.

Most organic compounds are stable to heat to temperatures of over 200°C, and since reactions are usually carried out below this temperature the decomposition of one of the reactants does not usually present a problem. Thermal instability within this temperature range is, however, associated with certain structural groups. The extent of decomposition at a particular temperature and within a given time does not depend directly on the thermodynamic properties of the material, but is determined by the *rate* of decomposition. Nevertheless, the rate is usually related to the thermodynamic properties (see section 3.6) and in general weaker bonds undergo faster bond-cleavage than stronger bonds. For example, the extent of decomposition of methane ($D_{C-H} = 426\,\text{kJ}\,\text{mol}^{-1}$) is negligible at 100°C, whereas that of dibenzoyl peroxide ($C_6H_5COO-OCOC_6H_5 \rightarrow 2C_6H_5CO_2\cdot$; $D_{O-O} = 130\,\text{kJ}\,\text{mol}^{-1}$) is about 50% after 30 minutes.

Such instability may be usefully applied. For example, certain organic reactions may be initiated by the introduction of a reactive species such as the benzoyloxy radical (section 17.1) and these reactions can be brought about by adding dibenzoyl peroxide and heating to a temperature at which the rate of generation of benzoyloxy radicals brings about the initiated reaction at a practicable rate.

Other commonly occurring weak bonds are those of the halogens. Many compounds containing C—H bonds may be chlorinated or brominated† by being

---

* α-Hydroxyaldehydes form *dimers* (six-membered ring) which revert to the monomers in aqueous solution:

$$2 \quad \text{HO}\diagup\text{CHO} \quad \rightleftharpoons \quad \text{(dimer structure)}$$

† Fluorine is not normally introduced in this way since the reactions are so strongly exothermic (because of the very weak bond energy of F—F as compared with H—F and C—F) as to be violent.

heated with the halogen, reaction occurring through the mediation of halogen atoms, e.g.

$$CH_4 \ + \ Cl_2 \ \xrightarrow{\text{via Cl·}} \ CH_3Cl \ + \ HCl$$

Again, however the thermodynamic criterion for a reaction to occur ultimately dictates the issue. In the chlorination of methane, ΔS^0 is very small and ΔH^0 ($-100\,kJ\,mol^{-1}$; cf. Tables 1.1 and 1.2) is dominant in the free-energy change. For the iodination of methane, however, $\Delta H^0 = +53\,kJ\,mol^{-1}$ and the free-energy change is unfavourable so that despite the ease with which suitable conditions for iodination can be established (i.e. the generation of iodine atoms), reaction does not occur.

Finally, certain compounds are unstable to heat not because they contain any intrinsically weak bonds but because decomposition can lead to the formation of a strongly bonded molecule. For example, azobisisobutyronitrile decomposes fairly rapidly below 100°C because of the favourable enthalpy change in the process owing mainly to the formation of the strongly bonded molecular nitrogen:

Such compounds are also useful as initiators of free-radical reactions (section 17.1).

Problems 1. The bond energies (25°C) of C—C, C=C, C—H, and H—H are, respectively, 347, 610, 414, and 435 kJ mol⁻¹. Calculate the enthalpy of the reaction,

$$CH_2{=}CH_2 \ + \ H_2 \ \longrightarrow \ CH_3{-}CH_3$$

In practice, the reduction of ethylene to ethane can be carried out readily at room temperature. Under what conditions might it be possible to carry out the reverse reaction?

2. Estimate ΔH^0 (from the data in Tables 1.1 and 1.2) for the following reactions:

(a) $CH_4 \ + \ Cl_2 \ \rightleftharpoons \ CH_3Cl \ + \ HCl$

(b) $C_2H_5OH \ \rightleftharpoons \ C_2H_4 \ + \ H_2O$

(c) $CH_3CHO \ + \ H_2O \ \rightleftharpoons \ CH_3CH(OH)_2$

In which cases would you expect ΔS to contribute significantly to ΔG? Which are likely to have favourable equilibrium constants for the forward reaction at room temperature? What would be the effect on the equilibrium constants of increasing the temperature?

3. The entropy changes for the formation of ethyl chloride by (a) the chlorination of ethane ($C_2H_6 + Cl_2 \rightarrow C_2H_5Cl + HCl$), and (b) the addition of hydrogen chloride to ethylene ($CH_2{=}CH_2 + HCl \rightarrow C_2H_5Cl$), are (a) $+2$ and (b) $-130\,J\,K^{-1}\,mol^{-1}$. Comment.

4. For which of the following reactions would you expect K to be <1 at room temperature? (Assume that the entropy changes in the reactions of ethylene are all $-130\,J\,K^{-1}\,mol^{-1}$ and state any other assumptions you make.)

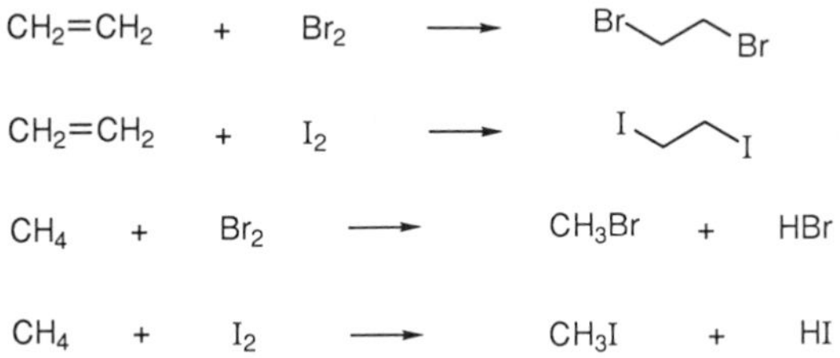

$$CH_2{=}CH_2 \quad + \quad Br_2 \quad \longrightarrow \quad Br{\diagup\!\diagdown\!\diagup}\,Br$$

$$CH_2{=}CH_2 \quad + \quad I_2 \quad \longrightarrow \quad I{\diagdown\!\diagup\!\diagdown}\,I$$

$$CH_4 \quad + \quad Br_2 \quad \longrightarrow \quad CH_3Br \quad + \quad HBr$$

$$CH_4 \quad + \quad I_2 \quad \longrightarrow \quad CH_3I \quad + \quad HI$$

What effect would raising the temperature have?

2 Molecular structure

2.1
Bonding

The principles of thermodynamics relate the concentrations of chemical species in equilibrium to the enthalpies and entropies of those species. Bond energies, closely related to enthalpies, have precise values which may be measured, but thermodynamic principles give no information about the *origin* of these bond energies. It is the purpose of this chapter to show how the current theories of molecular structure lead to an understanding of the strengths of bonds and other physical properties of organic compounds.

2.2
The covalent bond

The two most commonly employed methods for describing the structures of organic compounds, each of which is related to the theories of wave mechanics, are the molecular-orbital (MO) and valence-bond (VB) methods.

(a) The molecular-orbital method

In this treatment, bonding is described as arising from the overlap of atomic orbitals (AOs) of the atoms involved, leading to the formation of molecular orbitals (MOs). The interaction of one AO on one atom with one on another gives two MOs, of which one is a *bonding orbital* (i.e. the energy is less than that of the separated atoms) and the other is an *antibonding* orbital (i.e. the energy is greater than that of the separated atoms). Like an AO, an MO can contain two electrons, so that if the interacting AOs are each associated with two electrons, both MOs are occupied and there is no resulting bonding. However, if each AO is associated with one electron, only the lower-energy, bonding MO is occupied and the greater the extent of overlap of these AOs, the stronger is the resulting bond. It is therefore necessary to consider the interaction of singly occupied orbitals – that is, unpaired electrons – and their geometrical relationship which determines how effectively they overlap.

Atomic carbon has two electrons in its $1s$ orbital, two in its $2s$ orbital, and unpaired electrons in two of its $2p$ orbitals. It would therefore be expected to form only two bonds, whereas it is tetravalent in almost all stable organic compounds; the few stable tervalent carbon compounds (free radicals) owe their stability to unusual structural features (p. 523) and divalent carbon species exist only as short-lived intermediates in certain reactions (e.g. p. 101).

The tetravalence of carbon may be understood in terms of the concept of *hybridization*. The difference in energies of the $2s$ and $2p$ orbitals of the carbon atom is approximately $400\,kJ\,mol^{-1}$, so for the expenditure of this amount of energy a $2s$ electron could be promoted to the empty $2p$ orbital giving four unpaired electrons and hence four covalent bonds. This proves to be thermo-dynamically practicable, for the formation of two extra covalent bonds more than offsets the excitation energy. However, an atom which forms bonds by using three unpaired p electrons and one unpaired s electron would yield a molecule lacking spherical symmetry because three of the bonds would be different from the fourth, whereas the saturated compound methane possesses spherical symmetry, the four bonds being directed towards the corners of a regular tetrahedron of which carbon is the centre. The only way in which four such equivalent bonds can be formed is by the 'mixing' of the three $2p$ orbitals and the one $2s$ orbital; the new orbitals are combinations of these and are described as sp^3-hybridized orbitals.

Moreover, as well as providing equivalent bonds, the hybridized orbitals are strongly directional, as shown in Figure 2.1. The resulting overlap with the AOs of four atoms which are suitably placed is greater than that which could be achieved by the three directional p orbitals and the non-directional s orbital.

Figure 2.1   The formation of an sp^3-hybrid atomic orbital.[*]

There are two other types of hybridization of particular importance. In one the $2s$ orbital is mixed with two of the $2p$ orbitals, giving three sp^2-hybridized orbitals which are coplanar and at $120°$ to each other. The remaining $2p$ orbital is perpendicular to this plane. In this state, two carbon atoms are able to form a double bond, one part of which arises from the overlap of two sp^2 orbitals along the internuclear axis (σ-bond) and the other from the lateral overlap of the two p orbitals (π-bond), as shown for ethylene in Figure 2.2.[†]

---

[*] Opposite phases of the wave function are represented in this book by plain and hatched areas of the orbital's contour (in some texts, + and − signs are used). They can be thought of as corresponding to the peaks and troughs of a wave: when two of like phase overlap they reinforce each other and give a bonding orbital, whereas the interaction of opposite phases is destructive and antibonding.
[†] For clarity of presentation, the overlap of p orbitals to form π-bonds will not normally be shown; instead, a typical representation will be:

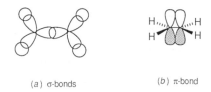

(a) σ-bonds   (b) π-bond

Figure 2.2   Orbital structure of ethylene. (a) Viewed perpendicularly to the plane of the sp^2 orbitals. (b) Viewed along the plane of the sp^2 orbitals.

A number of properties of compounds containing carbon–carbon double bonds follow from this description. First, because the bond strength is related to the extent of overlap of the AOs, there is a tendency for the lateral interaction of the p orbitals to bring the nuclei closer together than in singly bonded molecules, the extra bonding offsetting the extra internuclear repulsion, e.g. the C=C bond length in ethylene is 0.134 nm whereas the C—C bond length in ethane is 0.154 nm. Second, since lateral p-orbital overlap is not so effective as the sp^2 overlap along the internuclear axis, the π-bond is weaker than the σ-bond. Consequently, the bond-energy of C=C is less than twice that of C—C (p. 8). Third, p-orbital overlap is reduced by rotation of one carbon atom relative to the other around the internuclear axis, falling to zero when the angle of rotation is 90°. Since this results in reduction of the bond energy, there is a strong tendency for π-bonded systems to resist rotation. This gives rise to the existence of geometrical isomers, for the energy required to convert, e.g. *cis*-2-butene (1) into *trans*-2-butene (2) is far greater than is available at ordinary temperatures. Similarly, in a bridgehead alkene such as (3) the constrained geometry of the ring system would not allow effective p-orbital interaction and the thermodynamic stability of the compound would be correspondingly low. In fact, this compound cannot be obtained.

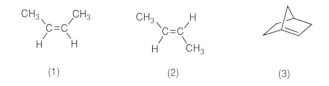

(1)   (2)   (3)

Nitrogen, like carbon, can form double bonds by promotion of one $2s$ electron followed by sp^2-hybridization. Four electrons then occupy the three hybridized orbitals, one of the orbitals containing a pair of electrons and the other two orbitals each containing a single electron. The fifth electron occupies a p orbital perpendicular to the plane of the sp^2 hybrids. σ-Bonds may then be formed by the singly occupied sp^2 orbitals and a π-bond may be formed by the p orbital, so that, for example acetoxime has the structure (4).

$$CH_3 \diagdown \atop CH_3 \diagup C=N \diagdown OH$$

(4)

As in doubly bound carbon, the geometrical requirement for the lateral overlap of *p* orbitals leads to coplanar systems around the double bond and resistance to rotation. Geometrical isomerism, as in unsymmetrical oximes (5) and (6) and azo-compounds (7) and (8), therefore occurs.

$$R \diagdown \atop R' \diagup C=N \diagdown OH \qquad R \diagdown \atop R' \diagup C=N \diagup OH \qquad R \diagdown \atop N=N \diagdown R \qquad R \diagdown N=N \diagup R$$

(5)   (6)   (7)   (8)

The third important type of hybridized orbital is obtained by the mixing of one $2s$ and one $2p$ orbital. Two sp-hybridized orbitals are formed which point in opposite directions, and the remaining $2p$ orbitals are each singly occupied and are available for forming two π-bonds. The simplest example of a compound formed in this way is acetylene, which is constructed as follows:

$$p_y \quad p_z \quad p_y \quad p_z$$

(b)   The valence-bond method and the concept of resonance

In this method, an unpaired electron on one atom is paired with one on another and the possible dispositions of the bonding pair are considered. For example, for hydrogen chloride three such dispositions are possible corresponding to the bonding pair being symmetrically located (H:Cl), on chlorine (H^+Cl^-), and on hydrogen (H^-Cl^+). The actual structure lies between these three and can be regarded as a weighted average of them, the weighting depending on the relative stabilities of the three. Since chlorine is so much more electronegative than hydrogen, H^+Cl^- is a more stable structure than H^-Cl^+ and the latter can be ignored. Hydrogen chloride is then described as a *resonance hybrid* of H:Cl and H^+Cl^- and is written as

$$H-Cl \quad \longleftrightarrow \quad H^+ \ Cl^-$$

The individual structures that contribute to the resonance hybrid are described as *resonance structures* (or sometimes as canonical structures).

A most important aspect of the concept of resonance, justified by wave-mechanical calculations, is that the actual structure is of lower energy (i.e. is more stable) than any of the individual resonance structures.* This proves to be of particular significance in understanding the structures and properties of conjugated systems (section 2.3).

2.3
Delocalization

In both the MO and VB treatments of simple organic compounds, the bonds are regarded as *localized*. In MO language, they are formed by the pairing of electrons derived from two AOs by the appropriate overlapping of these AOs. The characteristics of a particular bond – its length and its strength – are determined solely by the natures of these AOs and are independent of other electronic features of the molecule. This is justified by the fact that a bond of given type has the same properties in a wide variety of structural environments. For example, a bond formed by two sp^3-hybridized carbon atoms is always 0.154 nm long with a strength of about 347 kJ mol^{-1}.

The situation is different in *conjugated* molecules such as butadiene, CH_2=CH—CH=CH_2. First, the heat of hydrogenation of butadiene (238.7 kJ mol^{-1}) is less than twice that of 1-butene (126.7 kJ mol^{-1}), so that the compound is more stable by 14.7 kJ mol^{-1} than expected from the principle of additivity of bond energies. Secondly, the central C—C bond is 0.146 nm long, compared with 0.154 nm for the C—C bond length in saturated compounds. The constancy of bond properties, which holds so well for saturated and simple unsaturated compounds and for which there is wave-mechanical justification, breaks down for conjugated systems.

A new concept, again justified wave-mechanically, is introduced: namely, delocalization. In the MO method, butadiene is treated in the usual way insofar as its σ-bonds are concerned, but the four $2p$ orbitals interact laterally over the whole length of the molecule to establish four *delocalized* π-MOs of which the two of lowest energy are occupied (Figure 2.3). The like-phase overlap of the p orbitals on C-1 and C-2, and on C-3 and C-4, in both π_1 and π_2 provides π-bonding between these pairs. The interaction between the p orbitals on C-2 and C-3 is bonding in π_1 but antibonding in π_2. However, the overlap is greater in π_1 because the orbitals are larger, so that overall there is a degree of π-bonding between C-2 and C-3, consistent with the extra stability of the diene and the shorter length of the bond between C-2 and C-3 compared with that in an alkane.

VB theory approaches the problem by developing the concept of resonance. The ionic structures (b) and (c) and the non-polar structure (d) (in which the

---

* In wave-mechanical language, the wave function for the bonding pair, ψ, is given by

$$\psi = \psi_{\text{covalent}} + \lambda\psi_{\text{ionic}}$$

where λ represents the relative importance of the wave function for the ionic structure, H^+Cl^-.

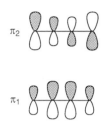

π_2

π_1

Figure 2.3 Occupied π-MOs in butadiene.

single electrons on the terminal carbons are formally paired) contribute to the resonance hybrid:

$$CH_2{=}CH{-}CH{=}CH_2 \quad \overset{+}{C}H_2{-}CH{=}CH{-}\overset{-}{C}H_2 \quad \overset{-}{C}H_2{-}CH{=}CH{-}\overset{+}{C}H_2 \quad \overset{\cdot}{C}H_2{-}CH{=}CH{-}\overset{\cdot}{C}H_2$$

(a)   (b)   (c)   (d)

Since the structures (b), (c), and (d) contain one fewer covalent bond than (a), and the formation of (b) and (c) requires the expenditure of energy to separate unlike charges, the energies of (b), (c), and (d) are higher than that of (a) and the structures are of less importance in the hybrid. However, because each of the three describes the central C—C bond as a double bond, the complete wave function corresponds to there being some double-bond character in this bond, and its shorter length than that of C—C in a saturated molecule is accounted for. Moreover, inclusion of resonance structures (b), (c), and (d) leads to a lower energy than that corresponding to (a) alone. The value of the decrease in energy achieved by inclusion of the structures (b–d) is defined as the *resonance energy* or, perhaps more satisfactorily, the *stabilization energy* of the molecule.*

In its description of molecules like butadiene, the MO method is conceptually more satisfying than the VB method which refers to structures that have no real existence. However, the VB method has the advantage of being easier to express graphically when considering the reactions of organic compounds. It is widely used in this book. Nonetheless, it is easily misapplied and care must be exercised in its use. The following principles provide general guidance to its application.

(*i*) Resonance is only significant when the contributing structures are of comparable energy. This is true, for example, of the two Kekulé structures for

---

* The extra stability of conjugated compounds compared with non-conjugated compounds, as measured, for example, by heats of reduction, is not entirely attributable to delocalization. We would expect that the strength of a C—C bond would be dependent on the state of hybridization of each carbon atom. Then, since in the reduction of 1-butene an sp^2–sp^3 C—C bond is converted into an sp^3–sp^3 C—C bond whereas in the reduction of butadiene an sp^2–sp^2 C—C bond is converted into an sp^3–sp^3C—C bond, there could be a difference in the heats of hydrogenation of butadiene as compared with that of two molecules of 1-butene irrespective of there being a special stabilization associated with the π orbital system. Likewise, the smaller length of the single C—C bond in butadiene than that in saturated compounds could, at least in part, result from the fact that sp^2-hybridized carbon has a smaller covalent radius than sp^3-hybridized carbon.

benzene, (9) and (10), which have the same energy: the resulting stabilization energy is considerable (p. 36). It is not true, however, of the structures (11) and (12) for ethylene for which the high-energy structure (12) may be neglected; that is, inclusion of (12) has a negligible effect on the calculated energy.

(9)   (10)   (11)   (12)

(*ii*)   The nuclei in each of the resonance structures must be in the same relative positions. For example, the structure (14) does not contribute to the structure of isobutylene (13) but is instead the isomeric compound, *trans*-2-butene.

(13)   (14)   (15)

(*iii*)   Only those structures should be included in which no atom is associated with more than the maximum number of electrons that its valence shell can accommodate. For example, $\bar{RO}=CHR$ is not a possible structure in the description of the anion from an ether, $RO-\bar{C}HR$, since ten electrons are associated with oxygen, whose maximum is eight; but $\bar{RS}=CHR$ *is* a possible structure in the description of the corresponding sulfur anion, because sulfur, having available $3d$ orbitals, can accommodate more than eight electrons in its valence shell. However, $\overset{+}{RO}=CHR$, corresponding to eight electrons in oxygen's valence shell, is an allowed resonance structure in the description of the carbocation $RO-\overset{+}{C}HR$.

(*iv*)   Electron pairing must be conserved. The structure (15) does not contribute to the structure of butadiene but is a representation of an (electronically) excited state of this molecule in which the complete pairing of the ground state has been lost as the result of a spin-inversion.

2.4
Applications of structural theory

(*a*)   *Molecular structure and shape*

The origin of directed valencies has already been briefly discussed in terms of the shapes of atomic orbitals and hybridized orbitals. We shall now describe the bonding and the geometry of the more important groups and systems in organic chemistry in the same way.

(i) Saturated compounds.   Methane has a tetrahedral structure because the four sp^3-hybridized orbitals of carbon are directed towards the corners of a regular tetrahedron. Substitution of one hydrogen atom by, say, a chlorine atom to give CH_3Cl leads to a slight distortion of the tetrahedron because the repulsive forces between pairs of atoms, and pairs of electron-pairs, are no longer the same. This holds also for CH_2Cl_2 and $CHCl_3$, but with CCl_4 the regular tetrahedral structure is again established.

Substitution of one hydrogen in methane by a *group* of atoms leads to a different effect. In ethane, for example, not only is there a slight distortion from tetrahedral angles, but also different *conformations* are made possible by the axial symmetry of the C—C σ-bond. Two of these, the *staggered* conformation (16) and the *eclipsed* conformation (17) are shown. They are perhaps more easily visualized by reference to the *Newman projections* (18) and (19), respectively, which correspond to viewing the molecule from one end along the C—C bond.

(16)   (17)   (18)   (19)

As a result of the interactions between pairs of protons, and pairs of electron-pairs, on the two carbon atoms, the conformations differ in energy. The staggered conformation is the most stable of all the possible conformations and the eclipsed conformation is the least stable: the difference in energy between these two extremes is about $12\,kJ\,mol^{-1}$. Application of the Boltzmann distribution relationship shows that at any instant there are about 100 times as many molecules in the staggered as in the eclipsed conformation. The difference in energies of these two conformations – the *torsional strain* – provides a barrier to rotation about the single bond, but this is so small that, at ordinary temperatures, rotation is extremely rapid.

For more complex molecules a larger number of eclipsed and staggered conformations is possible. Four of those for butane, which differ in arrangement about the central C—C bond, are shown (20–23) and the variation of energy with rotation about this bond is shown in Figure 2.4. Rotation occurs rapidly so that it is not possible to isolate the *anti* and *gauche* conformations as discrete compounds.

Alicyclic compounds with three members are necessarily coplanar and therefore possess not only angular strain (the CCC angle being 60° rather than the preferred tetrahedral angle) but also torsional strain, for the hydrogen atoms on adjacent carbons are eclipsed. For cyclobutane and cyclopentane and their derivatives,

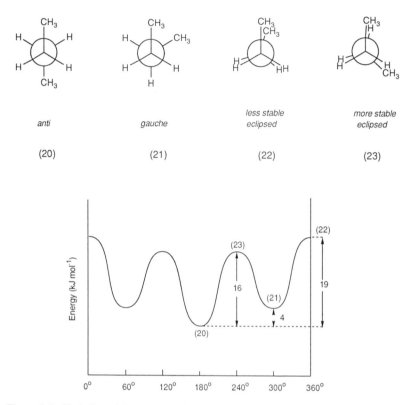

Figure 2.4   Variation of the energy of butane with rotation about the central C—C bond.

the tendency for the CCC angles to be as near to tetrahedral as possible would lead to the carbon skeletons being coplanar, but this situation corresponds to the complete eclipsing of the adjacent hydrogen atoms and it proves to be thermodynamically more favourable for the rings to buckle. For cyclohexane, both angular strain and torsional strain are eliminated by the ring adopting the *chair* conformation (24) which is shown also in the Newman projection (25) (the molecule being viewed along two parallel C—C bonds). Other conformations possess no angular strain but all have torsional strain: the *boat* conformation (26) has four eclipsing interactions (see 27) as well as a repulsive force between the hydrogen atoms at the opposite ends of the boat. Consequently cyclohexane exists predominantly in the chair form.

Seven-membered rings and above can all be constructed with strain-free CCC angles, but for C_7 to C_{11} rings all conformations possess some torsional strain. For rings larger than 11-membered, strain-free conformations again occur.

A measure of the strain in alicyclic rings is given by the heat of combustion of the molecule per methylene group less the value for the strainless cyclohexane. Data for rings possessing three to eleven members are given in Table 2.1.

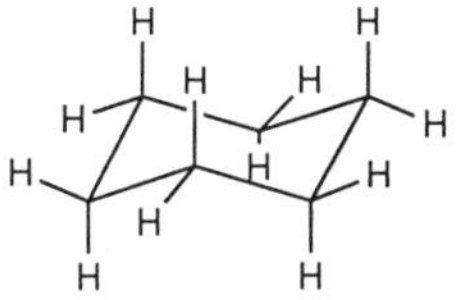

(24)

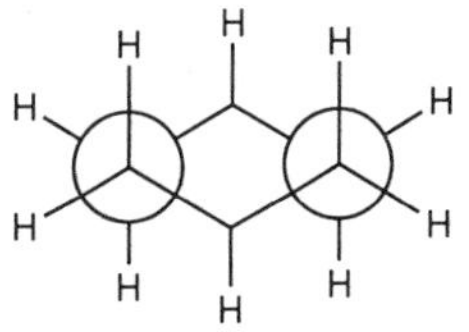

(25)

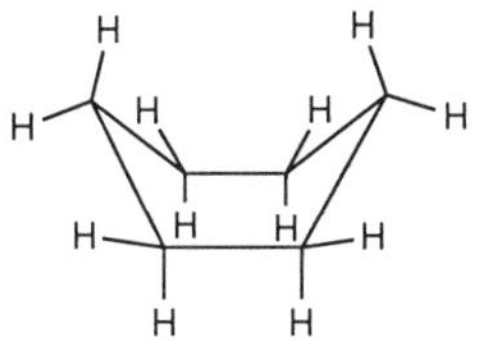

(26)

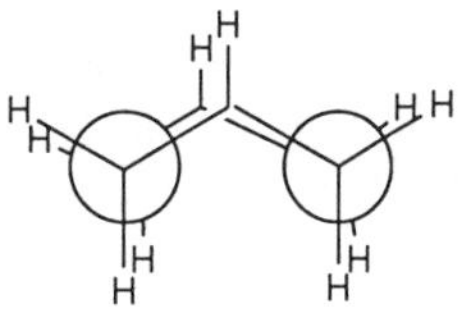

(27)

Table 2.1 Strain in alicyclic rings of varying size

No. of carbons in the ring	Heat of combustion per methylene group $(kJ\,mol^{-1})$	Strain energy $(kJ\,mol^{-1})$
3	696	38
4	685	27
5	663	5
6	658	0
7	662	4
8	663	5
9	664	6
10	663	5
11	662	4

(*ii*) *Dipolar bonds: the inductive effect.*   Whenever two dissimilar atoms are bonded, the bonding pair is not symmetrically placed with respect to the two nuclei because they have different charges. Theoretically, bond polarity is accommodated by the inclusion in the wave function of a parameter which determines the relative probabilities that the bonding electrons are in the vicinity of each nucleus. Experimentally, this asymmetry of bonds is manifested in the occurrence of dipole moments. These are measured by the distance between the centres of positive and negative charge in the molecule multiplied by the size of the charge and have values in the region 10^{-30} coulomb metre (Cm). The Debye unit (D) provides a useful scale of measurement: $1\,D = 3.3 \times 10^{-30}\,Cm$.

It is convenient to refer also to *bond moments*, for, like bond length and bond energy, the bond moment of a particular bond varies little with the structural environment except when conjugative influences are introduced. The vector sum of the individual bond moments in a molecule gives the resultant dipole moment.

Heteronuclear diatomic molecules invariably possess dipole moments. That of hydrogen chloride is 1.03 D, the negative end of the dipole being the chlorine atom. Symmetrical polyatomic molecules have no dipole moment, even though the individual bonds may have large moments, for the vector sum of the bond moments is zero. For example, CH_4 and CCl_4 have zero resultant moments, although both C—H and C—Cl bonds have moments, whereas CH_3Cl, CH_2Cl_2 and $CHCl_3$ have significant moments. Again, p-dichlorobenzene, unlike its *ortho* and *meta* isomers, has a zero moment, although each nuclear carbon is positively polarized, relative to those in benzene, by the electronegative chlorine substituents.

The concept of bond moments can be usefully extended to establish a scale of *inductive* effects. Since methane has no dipole but methyl chloride has a dipole whose negative end is the chlorine atom, chlorine is described as having an electron-withdrawing inductive effect ($-I$). Groups of atoms may also have inductive effects, e.g. in nitromethane carbon is attached to a positively charged nitrogen atom (p. 31), so that the —NO_2 group has a strong $-I$ effect. A procedure has been established for deriving a quantitative scale of inductive effects relative to hydrogen, and the order for some commonly occurring atoms and groups is:

$$(CH_3)_3C < CH_3 < (H) < OCH_3 < CF_3 < Br < Cl < F < CN < NO_2 < \overset{+}{N}(CH_3)_3$$

The inductive effect of an atom or group acts most powerfully on the atom to which it is bonded but atoms further away are affected to some degree. For example, in 1-chlorobutane the electron-pair in the C—Cl bond is displaced towards chlorine. The nucleus of the carbon atom is less strongly shielded than that of the corresponding carbon in butane, so that it is effectively a more electronegative atom. The electron-pair of the adjacent C—C bond is accordingly displaced towards the carbon to which chlorine is attached and, in turn, the second carbon is slightly affected. In this way, the replacement of hydrogen by a more electronegative atom results in electron displacements throughout the molecule, the effect on successive atoms falling off with their increasing distance from the electronegative centre. The effect may be represented

$$\overset{\delta\delta\delta+}{CH_3}\!-\!\overset{\delta\delta+}{CH_2}\!-\!\overset{\delta\delta+}{CH_2}\!-\!\overset{\delta+}{CH_2}\!-\!\overset{\delta-}{Cl}$$

(*iii*) *Unsaturated compounds.* The elements most commonly involved in unsaturated systems are carbon, oxygen, and nitrogen.

In double-bonded groupings, both carbon and nitrogen are sp^2-hybridized. Carbon forms three σ-bonds, by means of its three singly occupied sp^2 orbitals, which are coplanar and at (or close to) 120° to each other and a π-bond by means of its singly occupied p orbital which is perpendicular to this plane. Nitrogen systems are similar save that one of the sp^2 orbitals contains two electrons and constitutes a non-bonding pair. In heteropolar double bonds it is appropriate to include ionic structures in the valence bond descriptions. As a result of the increasing order of electronegativities, carbon < nitrogen < oxygen, resonance structures of the types (28–30) contribute to the hybrid structures of carbonyl compounds, imines, and nitroso compounds, respectively, and these bonds have significant polarity and slightly *less* double-bond character than the homopolar double bonds, C=C and N=N.

The nitro group also contains sp^2-hybridized carbon and may be regarded as derived from the nitroso group by the transference of one electron to oxygen from nitrogen's non-bonding pair followed by the formation of a covalent bond:

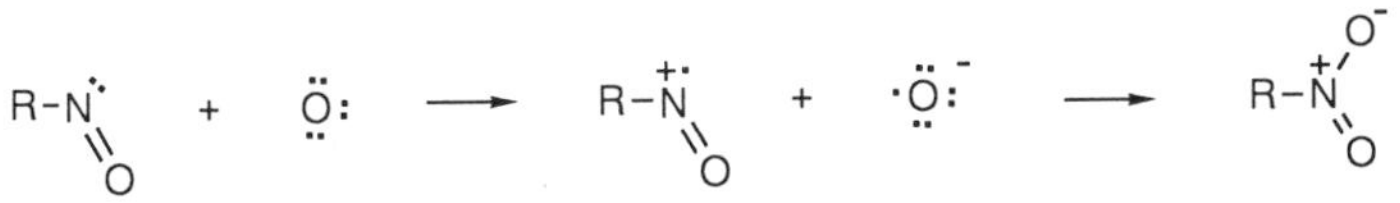

It should be noted that the oxygen atoms now become equivalent, so that a nitro compound is described as a resonance hybrid of the structures (31) and (32).

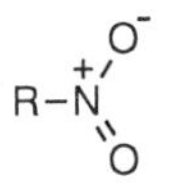

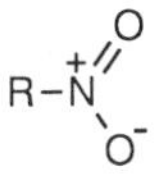

In the MO description, nitrogen forms three coplanar σ-bonds by means of singly occupied sp^2 orbitals and its *p* orbital, which contains a pair of electrons, interacts with the *p* orbitals of each of the two oxygen atoms, as shown in (33), to form three delocalized π-orbitals of which the two of lowest energy are occupied:

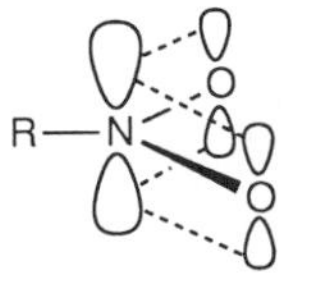

(33)

In triple-bonded groupings, carbon and nitrogen are *sp*-hybridized. Carbon forms two collinear σ-bonds by means of its singly occupied *sp* orbitals and two π-bonds by means of its two singly occupied and mutually perpendicular *p* orbitals. For nitrogen, one of the *sp* orbitals contains a non-bonding electron pair. In the VB description of the C≡N bond it is appropriate to include the ionic structure —C≡N for the bond is significantly polar.

(*iv*) *Conjugated compounds.*   Special properties are associated with systems in which a π-bond is conjugated either with a second π-bond or with an atom which possesses a pair of electrons in a *p* orbital. These are: stabilization energy, single-bond lengths which are shorter than those in non-conjugated compounds, and (in some cases) the modification of dipolar properties.

1. π-Bond–π-bond conjugation.   The simplest example is butadiene, $CH_2{=}CH{—}CH{=}CH_2$ (p. 24). This is a symmetrical molecule, and conjugation does not lead to the appearance of a dipole. In VB language, contributions from the ionic structures $\overset{+}{C}H_2{—}CH{=}CH{—}\overset{-}{C}H_2$ and $\overset{-}{C}H_2{—}CH{=}CH{—}\overset{+}{C}H_2$ are necessarily equal and their dipoles therefore nullify each other. This is not true, however, when π-bonds of different types are in conjugation. For example, in an αβ-unsaturated carbonyl compound such as butenal the contribution of structure (34) is greater than that of (35) because oxygen is significantly more electronegative than carbon: the carbonyl group thus polarizes the C=C bond.

$$CH_3{-}\overset{+}{C}H{-}CH{=}CH{-}O^- \qquad\qquad CH_3{-}\overset{-}{C}H{-}CH{=}CH{-}\overset{+}{O}$$

(34)   (35)

A group such as carbonyl which withdraws electrons from an adjacent group *via* the π-bonding framework is described as having a −*M* effect, the negative sign indicating electron-withdrawal and *M* standing for *mesomeric.** Other groups of −*M* type include ester (36), nitrile (37), and nitro (38); the curved arrows in the representations denote the direction of the mesomeric effect.

$$C{=}C{-}C{=}O \qquad\qquad C{=}C{-}C{\equiv}N \qquad\qquad C{=}C{-}\overset{+}{N}\underset{O}{\overset{O^-}{{<}}}$$

(36)   (37)   (38)

2. π-Bond–p orbital conjugation.   In vinyl chloride (chloroethene) the *p* orbital on the carbon which is attached to chlorine can overlap with both the *p* orbital on the second carbon atom and one of the filled *p* orbitals on chlorine: three delocalized MOs are established of which the two of lowest energy are occupied. Since the *p* orbital on chlorine is initially filled, its participation in the delocalized π-system leads to the partial removal of electrons from chlorine and the appearance of a dipole moment directed from chlorine towards carbon. This is opposed to the dipole established in the C—Cl σ-bond as a result of the −*I* effect of chlorine, with the overall result that the dipole moment of vinyl chloride (1.44 D) is considerably smaller than that of ethyl chloride (2.0 D) in which only the −*I* effect is operative. The capacity of chlorine for donating electrons into a molecular π-system is described as a +*M* effect.

---

*This is the adjective from the noun *mesomerism* ('inbetween-ness') which is sometimes used to describe a resonance hybrid which lies 'between' the individual structures.

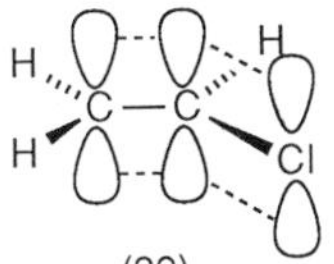

(39)   (40)   (41)

In VB terminology, vinyl chloride is described as a hybrid of the structures (40) and (41), the latter symbolizing the +*M* effect of chlorine. Both descriptions also indicate that the C—Cl bond should be shorter than in a saturated alkyl chloride, as is found.

Other elements with unshared *p*-electrons which take part in forming delocalized π-systems include the other halogens, oxygen, and nitrogen, e.g.

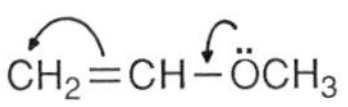

$$CH_2=CH-\ddot{O}CH_3 \qquad CH_2=CH-\ddot{N}(CH_3)_2$$

(42) 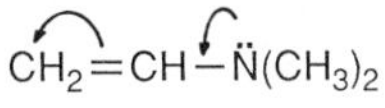   (43)

In each case the substituent has a +*M* effect.

Two further points should be noted. First, bond strength is dependent on the extent of the overlap of the combining atomic orbitals (p. 20), so that in these conjugated systems the more nearly equal in size the *p* orbitals are, the more effective is the π-orbital overlap. Hence fluorine is more effective than chlorine in conjugating with carbon, and oxygen is more effective than sulfur. Secondly, as the nuclear charge in an atom is increased, so also is the hold of the nucleus on the surrounding electrons so that, for comparably sized atoms, the ability to conjugate decreases as the atomic number increases. Hence the order of +*M* effects is —NR$_2$ > —OR > —F.

3. Hyperconjugation.   There is evidence that a C—H bond which is adjacent to a π-bond can take part in a delocalized system. The simplest example is propylene: a C—H bonding orbital in the methyl group which is orthogonal to the plane containing the carbon atoms interacts with the adjacent *p* orbital to form delocalized MOs:

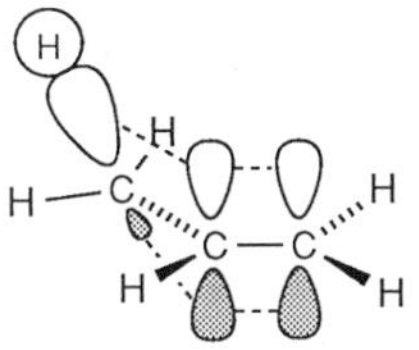

This is termed hyperconjugation.

In VB terminology, propylene is described as a hybrid of structures (44) and (45).

$$
\underset{(44)}{\overset{H}{\underset{\diagdown}{CH_2 - CH = CH_2}}} \quad \longleftrightarrow \quad \underset{(45)}{\overset{H^+}{CH_2 = CH - \bar{C}H_2}}
$$

The evidence for hyperconjugation in this case is that the C—C single bond is slightly shorter than that in ethane and that the heat of hydrogenation ($126\,\mathrm{kJ\,mol^{-1}}$) is less than that of ethylene ($137\,\mathrm{kJ\,mol^{-1}}$). However, it may be that part, if not all, of the measured stabilization energy of $11\,\mathrm{kJ\,mol^{-1}}$ is due to the fact that an sp^2–sp^3 carbon-carbon bond is stronger than an sp^3–sp^3 carbon-carbon bond, while the shortened bond length may result from sp^2-carbon having a smaller covalent radius than sp^3-carbon (cf. footnote, p. 25). Nevertheless, whatever the origin of the stabilization energy, it is a general fact that, amongst a given group of isomeric alkenes, the most stable (unless strain factors are involved) is the one in which the C=C bond is attached to the largest number of alkyl groups. This fact has many consequences in organic synthesis, as, for example, in the direction taken in elimination reactions (p. 97).

Hyperconjugation is thought to play an important part in the stabilization of carbocations (p. 60),*

$$
\underset{R}{\overset{H}{\underset{\diagup}{R^{''''}C - \overset{+}{C}^{''''}\underset{\diagdown R}{R}}}} \quad \longleftrightarrow \quad \overset{H^+}{\underset{R}{R^{''''}C = C^{''''}\underset{\diagdown R}{R}}}
$$

Hyperconjugation involving C—C σ-bonds is believed to be important in some instances (e.g. electrophilic aromatic substitution, p. 349), and the C—Si bond is especially important in the hyperconjugative stabilization of carbocations (p. 480).

(*v*) *Hydrogen-bonding and chelation.* A hydrogen atom which is bonded to an electronegative atom can form a *hydrogen-bond* to a second electronegative atom. The binding results from the polarization of the electron cloud of the second electronegative atom by the small, positively polarized hydrogen atom. Hydrogen bonds are longer than covalent bonds and are much weaker: their strengths lie in the range 10–$40\,\mathrm{kJ\,mol^{-1}}$, which is of the order of one-tenth the strength of covalent bonds. Conventionally, the bonds are represented by dotted lines, e.g.

$$
\underset{H}{\overset{R}{\underset{\diagdown}{O - H \cdots\cdots O}}}\overset{R}{\diagup}
$$

---

*Hyperconjugation is almost certainly more significant in this case than in stabilizing an alkene because the resonance structures for the carbocation contain the same numbers of bonds and are likely to be of more nearly equal energy than the resonance structures (44) and (45). The latter, with one fewer bond, should be a much less important contributor to the hybrid than (44) (cf. the first principle in applying resonance theory, p. 25).

The only elements concerned in hydrogen-bonding are nitrogen, oxygen, fluorine, and, to a lesser extent, chlorine. The bond may be formed both between molecules of the same type, as in water, alcohols, and carboxylic acids, and between molecules of different types, as in the interaction between the proton of an amino group and the oxygen of a carbonyl group:

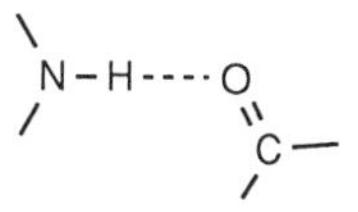

Hydrogen-bonding leads to an increase in intermolecular 'aggregation' forces and is manifested particularly in the boiling point and solubility of the organic compound. The boiling point is raised because energy is required to separate the hydrogen-bonded molecules in their translation to the gaseous state, e.g. ethanol boils more than 100°C higher than dimethyl ether, which has the same molecular formula, and alcohols boil at higher temperatures than the corresponding thiols. Two compounds which can form hydrogen-bonds tend to be more soluble in each other than in those which cannot: alcohols, and to a lesser extent ethers, are more soluble in water than are hydrocarbons of comparable molecular weight.

Intramolecular hydrogen-bonds can also be formed, and are of particular significance when the resulting ring is five- or six-membered. The phenomenon is then described as *chelation* and is illustrated for the enol form of acetoacetic ester (46) and for *o*-nitrophenol (47). Since chelation does not give rise to intermolecular aggregation forces, chelated compounds have 'normal' boiling points. For example, *o*-nitrophenol is much more volatile than its *para*-isomer, for only the latter forms intermolecular hydrogen-bonds.

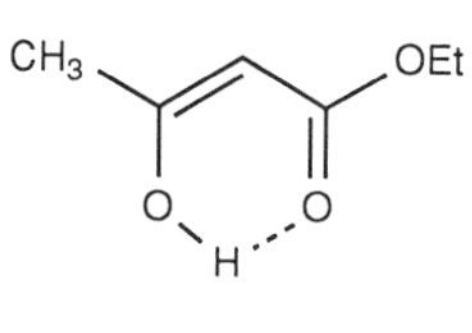

(46) 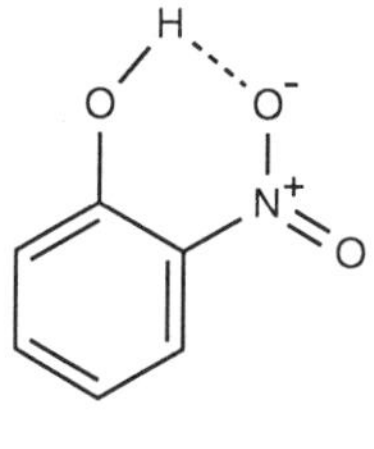　　　　　(47)

(b)　*Aromatic character*

Many cyclic, conjugated compounds possess markedly different physical and chemical properties from those expected by comparison of their structures with acyclic analogues. The simplest example is benzene, which may be regarded as the parent compound of the aromatic series. The only possible classical representations are (48) and (49) but neither alone is adequate, for there is not an alternation of single and double bonds: each C—C bond is the same length

(0.139 nm), rather closer to a normal double bond (0.134 nm) than to a normal single bond (0.154 nm). Moreover, whereas the heat of hydrogenation of the double bond in cyclohexene is 119 kJ mol^{-1}, that for the three double bonds in benzene is 207 kJ mol^{-1}; the latter value is 150 kJ mol^{-1} less than three times the former, so that benzene is more stable by this amount than predicted on the basis of its formal similarity to an alicyclic compound. Finally, benzene does not exhibit the ready addition reactions characteristic of alkenes. These facts are accommodated by both the VB and the MO theories. According to the former, benzene is described as a hybrid of the two structures (48) and (49) (which are known as Kekulé structures) and, since these are of equal energy content, the resulting stabilization energy is considerable.

(48)   (49)   (50)   (51)   (52)

The more energetic Dewar structures (50–52) are much less important, but the inclusion of wave functions corresponding to them results in a slightly greater reduction in the calculated total energy.

The MO treatment describes benzene as containing six sp^2-hybridized carbon atoms, each of which forms σ-bonds with two carbon atoms and one hydrogen atom, the six remaining electrons being in p orbitals each of which overlaps with two neighbours, as shown in (53). The planarity of the ring, which corresponds to strainless CCC angles (120°), allows maximum overlap of these p orbitals. Six delocalized π orbitals are thereby established of which the three of lowest energy (i.e. bonding orbitals) are occupied. The relative energies of these are shown in Figure 2.5. The overall effect is to provide the same degree of π-bonding between each pair of carbon atoms.

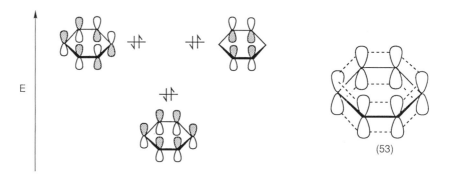

Figure 2.5   Relative energies of the π-orbitals in benzene.

In addition to accounting for the stability of benzene, both the treatments outlined above also embrace the equivalence of the bonds and account for the inertness of benzene to addition, for the addition of, for example, one molecule of bromine would lead to the simple conjugated system (54) and the loss of over 125 kJ of stabilization energy. This results in the reaction being endothermic by about 40 kJ mol^{-1}, whereas the addition of bromine to ethylene is exothermic.

Polycyclic hydrocarbons composed of fused benzene rings are also aromatic. The simplest examples are naphthalene, anthracene, and phenanthrene, which can be represented as the hybrids (55), (56), and (57), respectively. In each, delocalization is achieved by extensive *p*-orbital overlap, and the stabilization energies are greater than that of benzene. In general, for compounds containing equal numbers of benzene rings, that for which the greatest number of Kekulé structures can be drawn has the largest stabilization energy (e.g. phenanthrene > anthracene).

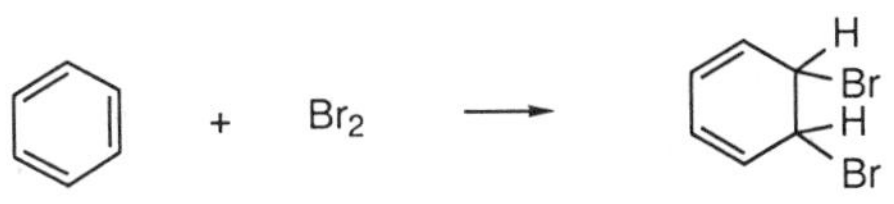

(54)

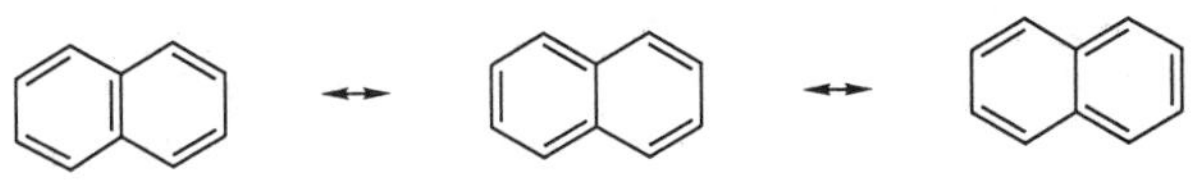

(55) stabilization energy: 255 kJ mol^{-1}

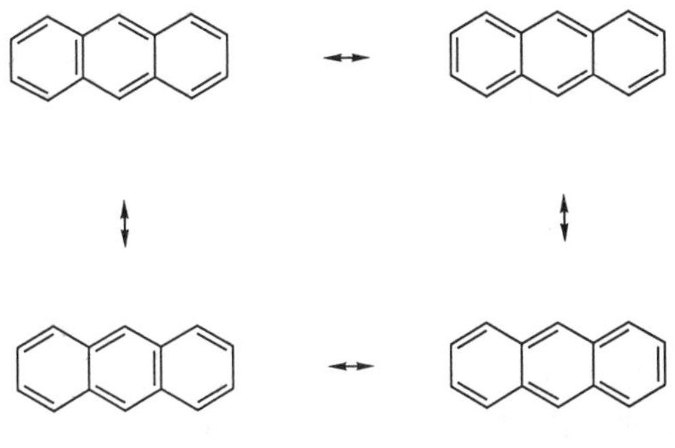

(56)   stabilization energy: 349 kJ mol^{-1}

The bond lengths are not equal in these systems. For instance, the 1,2-bond in naphthalene (0.136 nm) is shorter than the 2,3-bond (0.142 nm). A rough estimate of the relative lengths of the bonds can be obtained by examining the Kekulé structures: for naphthalene, the 1,2-bond is represented as a double bond in two of these structures and as a single bond in the third, while the opposite is the case for the 2,3-bond. Inspection of phenanthrene shows that the 9,10-bond is a double bond in four of the five structures: its length is close to that in an alkene.

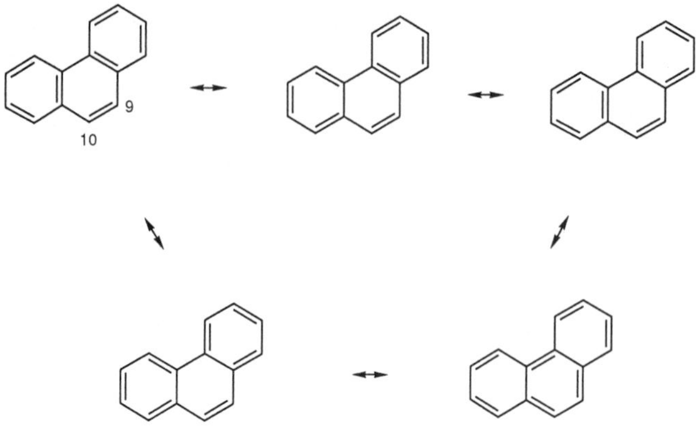

(57) stabilization energy: 380 kJ mol⁻¹

The stabilization energy of a polycyclic hydrocarbon is less than that of the sum of its constituent benzenoid rings and this fact has important consequences in reactivity. For example, addition across the central ring in anthracene gives a compound containing two benzene rings and therefore results in the loss of only $(349 - 2 \times 150)$, i.e. $49\,\text{kJ}\,\text{mol}^{-1}$ of stabilization energy, compared with the loss of nearly $150\,\text{kJ}\,\text{mol}^{-1}$ on addition to benzene. In practice, anthracene undergoes many addition reactions at the 9,10-positions, e.g. bromine gives 9,10-dibromo-9,10-dihydroanthracene (58).

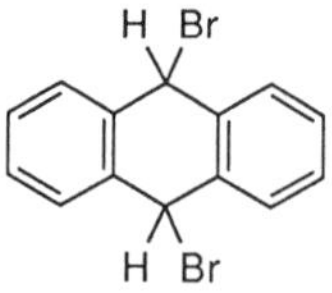

(58)

The properties which are associated with the carbocyclic, six-membered rings of benzene and the polycyclic hydrocarbons, and which result from the extensive *p*-orbital interaction in these planar, conjugated systems, are summarized in the phrase *aromatic character*. To display aromatic properties, however, compounds do not necessarily need to be either entirely carbocyclic or six-membered, and aromatic compounds of different types will now be introduced.

(*i*) sp^2-Hybridized nitrogen may replace a CH group. The nitrogen atom forms two σ-bonds with the adjacent carbon atoms by using two singly occupied sp^2 orbitals, possesses an unshared electron-pair in its third sp^2 orbital, and contributes one *p*-electron to the delocalized π orbitals. The most important examples of such compounds are pyridine (59) and its benzo-derivatives, quinoline (60) and isoquinoline (61), and pyrimidine (62).

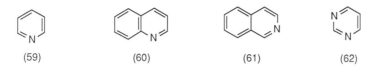

(59)   (60)   (61)   (62)

In VB terms, pyridine may be described as a hybrid of the structures (63) and (64), and, because nitrogen is more electronegative than carbon, it is appropriate to include also, though with reduced weighting, the ionic structures (65–67). As a result, the electron densities at the 2-, 4-, and 6-positions are less than at the carbon atoms in benzene, and this is of importance in connection with the chemical properties of the compound (pp. 359, 397). The same considerations hold for the related systems (60–62).

(63)   (64)   (65)   (66)   (67)

(*ii*)   Two CH groups, each of which provides one *p*-electron for the delocalized π-system, may be replaced by one atom which supplies two *p*-electrons. For example, furan (68) is aromatic because it contains a conjugated, cyclic, and planar system and six π-electrons in delocalized MOs, four of which are provided by carbon atoms and the remaining two by oxygen. The VB representation is in terms of the structures (68–72).

(68)   (69)   (70)   (71)   (72)

The contribution of the ionic structures (69–72) corresponds to there being a negative charge on each of the nuclear carbons although this is small because the dipolar structures are of higher energy, and are therefore less important contributors, than the 'classical' structure (68). Nevertheless, this is of importance in determining the reactivity of the carbon atoms. In addition, again because of the relatively high energy of the ionic structures, the stabilization energy of furan (*c.* 85 kJ mol^{-1}) is much less than that of benzene. As a result, addition reactions occur more easily, for there is less stabilization energy to be lost (e.g. the Diels–Alder reaction, section 9.1).

Pyrrole (73) and thiophen (74) and their benzo-derivatives such as indole (75) are aromatic for the same reason as furan. For thiophen and its derivatives, a new characteristic is introduced: since sulfur has relatively low-lying and unfilled 3*d* orbitals, it may accept π-electrons, so that resonance structures such as (76) contribute to the hybrid. As a result, thiophen is more strongly stabilized than furan and its carbon atoms are less strongly negatively polarized. These characteristics are reflected in its chemistry (e.g. thiophen, unlike furan, does not undergo the Diels–Alder reaction).

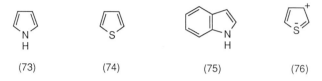

(73)   (74)   (75)   (76)

The cyclopentadienyl anion (78) fulfils the criteria for aromaticity (it is iso-electronic with pyrrole) and proves to be a relatively stable species: the ion is formed from cyclopentadiene (77) on treatment with base in conditions in which acyclic alkenes such as 1,4-pentadiene are unreactive.

(77)   (78)

(*iii*)   An atom which contributes an empty *p* orbital may be introduced into six-membered aromatic systems. For example, cycloheptatrienyl bromide (79) is a salt, unlike aliphatic bromides, and this may be ascribed to the fact that the cycloheptatrienyl (tropylium) cation is aromatic and consequently strongly resonance-stabilized. Again, tropone (80) has markedly different properties from aliphatic ketones, being very high-boiling, strongly polar ($\mu = 4.3\,D$), and lacking ketonic properties. It is evident that dipolar structures such as (81) and (82), which constitute the aromatic tropylium system, provide a better representation of the compound.

(79)   (80)   (81)   (82)

It is necessary to introduce here a final criterion for the occurrence of aromatic character. So far, each of the aromatic systems we have described has contained six π-electrons. Insofar as the simple application of resonance theory is concerned, there is no reason why cyclopentadienyl anions and cycloheptatrienyl cations should be comparatively stable whereas cyclopentadienyl cations and cyclo-heptatrienyl anions should not be; as many equivalent structures can be drawn for $C_5H_5^+$ as for $C_5H_5^-$ and for $C_7H_7^-$ as for $C_7H_7^+$. Yet cyclopentadienyl bro-mide, unlike tropylium bromide, is not ionic, and cycloheptatriene, unlike cyclopentadiene, does not form an anion in basic solution.

The explanation is that only systems which contain $(4n + 2)$ π-electrons (where *n* is an integer) are aromatic and the reason for this lies in the relative energies of the π MOs. For example, cyclobutadiene might exist as a square structure in which the four p AOs combined to give four delocalized π MOs whose relative energies are shown in Figure 2.6a. Since four p electrons are available, two would occupy the lowest MO and the other two would be distributed one each in the degenerate MOs and would have the same spin: two further electrons would be required to complete these MOs in order to provide significant binding energy. A more stable disposition is in fact available: the molecule adopts a rectangular shape (83) in which the four p orbitals form two *localized* π-bonds. Even this is a relatively high-energy structure because of the strain in the CCC angles: it has only a fleeting existence, rapidly dimerizing to a mixture of (84) and (85).

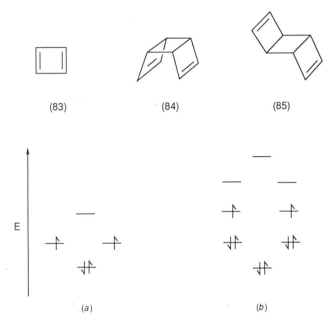

Figure 2.6   Molecular orbitals formed by systems containing (*a*) four p AOs and (*b*) eight p AOs.

For a compound such as cyclooctatetraene in which eight p AOs were able to overlap, the energy levels of the eight delocalized π MOs would be as in Figure 2.6b; there would again be incomplete pairing in the MOs. Instead of its adopting the planar structure (86) necessary for the complete overlapping of its p AOs, cyclooctatetraene exists as the tub-shaped structure (87) in which each p orbital can overlap with only one other, and this gives it the properties of an aliphatic alkene (e.g. ready addition of bromine).

It might be argued that cyclooctatetraene is not aromatic because the angular strain in the necessary planar structure would more than offset the resulting stabilization energy. This thesis is, however, at variance with other observations, such as the comparative stability of the cyclooctatetraene dianion (88) and the cyclopropenyl cation (89); both these systems should be highly strained and their stabilities are evidently associated with their possessing ten and two electrons, respectively, in delocalized MOs which form complete-shell configurations.

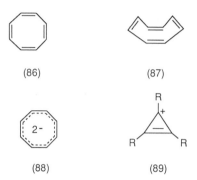

(86)　　　　　　　　(87)

(88)　　　　　　　　(89)

Finally, attempts to make pentalene (90) and heptalene (91) have failed, whereas azulene (92) is an aromatic compound with a stabilization energy of about $170 \, kJ \, mol^{-1}$. Pentalene and heptalene do not contain $(4n + 2)$ π-electrons, but azulene does, and it is especially notable that azulene has a dipole moment, suggesting that the seven-membered ring in some degree donates one π-electron to the five-membered ring so that each ring approximates to a six π-electron system. In valence-bond terms, ionic structures such as (93) are important contributors to the hybrid.

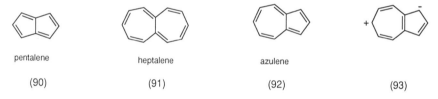

pentalene　　　　　heptalene　　　　　azulene

(90)　　　　　　(91)　　　　　　(92)　　　　　　(93)

The following conclusions may be noted. (1) Cyclic, conjugated, planar compounds which possess $(4n + 2)$ π-electrons tend to be strongly stabilized and are described as aromatic. (2) Such compounds do not have the properties typical of their aliphatic analogues; in particular, addition reactions are rare except when the stabilization energy is relatively small or the product of addition is itself strongly resonance-stabilized. (3) Because of their stability, aromatic compounds are often readily formed both from acyclic compounds by ring-closure and from alicyclic compounds by dehydrogenation or elimination of other groups. Illustrations of conclusions (2) and (3) are numerous throughout this book.

(c)   Acidity and basicity

The customary definitions of acids and bases, due to Brønsted, are that an acid is a species having a tendency to lose a proton and a base is a species having a tendency to add a proton. An alternative definition of each is due to Lewis who defined an acid as a species capable of accepting an electron-pair and a base as a species capable of donating an electron-pair. Since a base can accept a proton only by virtue of being able to donate an electron-pair, all Lewis bases are also Brønsted bases and *vice versa*. However, Lewis's definition of acids embraces compounds which do not have a tendency to lose a proton, e.g. boron trifluoride is a Lewis acid, as illustrated by its reaction with ammonia,

$$BF_3 \quad + \quad :NH_3 \quad \longrightarrow \quad F_3\bar{B}-\overset{+}{N}H_3$$

We shall use the term acid according to the Brønsted definition and shall refer to the compounds embraced only by Lewis's definition as Lewis acids.

Acids and bases are necessarily conjugate entities: that is, if the species HA can donate a proton to the species B, then since the reaction is reversible, the species BH^+ can donate a proton to A^-:

$$HA \quad + \quad B \quad \rightleftharpoons \quad A^- \quad + \quad BH^+$$

The species A^- is referred to as the *conjugate base* of the acid HA, and BH^+ is termed the *conjugate acid* of the base B. Clearly, if HA is a strong acid, A^- must be a weak base. It is, however, more convenient to discuss acidity and basicity separately.

It must also be remembered that the acidity or basicity of a species depends not only on the structure of that species but also on the nature of the solvent. Aniline is a weak base in water (which is a weak proton-donor) but a strong base in sulfuric acid; amide ion is a far stronger base in water than in liquid ammonia (for water is a stronger proton-donor than ammonia). Unless specified, the reference solvent is taken to be water.

(*i*) *Acids.* The most important elements in organic systems from which protons are donated are oxygen, sulfur, nitrogen, and carbon. The acidities of the groups —O—H, —S—H, $\diagdown$N—H, and $\diagup$C—H vary widely with the structure of the remainder of the molecule, but one principle is of pre-eminent importance in determining acidity: any factor which stabilizes the anion of an acid relative to the acid itself increases the strength of the acid. This follows immediately from the thermodynamic principles governing an equilibrium (section 1.5) and is implicit throughout this discussion.

1. Acidity of O—H groups. Ethanol (pK 16)* is a very weak acid compared with acetic acid (pK 4.8). Two factors are responsible. First, the carbonyl group is electron-attracting compared with the methylene group, so that the negative charge is better accommodated in $CH_3—CO—O^-$ than in $CH_3—CH_2—O^-$.

---

* The pK value of an acid is defined as $-\log_{10} K$ where K is the equilibrium constant for its ionization.

Secondly, both acetic acid and its anion are resonance-stabilized, being hybrids of (94) and (95), and (96) and (97), respectively. However, the stabilization energy of the ion, in which the resonance structures are of equal energy, is greater than that of the acid, for which the dipolar structure (95) is of high energy compared with (94). Consequently, there is an increase in stabilization energy when acetic acid ionizes whereas there is none in the ionization of ethanol. These two factors are responsible for a difference of about $60\,kJ\,mol^{-1}$ in the free energies of ionization.

(94)   (95)   (96)   (97)

The discussion of acetic acid has revealed two of the principal factors which govern acidity: the inductive and mesomeric effects. Further illustration of the operation of these effects is given in the following examples.

(1)   The strengths of the mono-halogen-substituted acetic acids increase with the increasing $-I$ effect of the halogen (I < Br ~ Cl < F):

$X-CH_2-CO_2H$	X =	H	F	Cl	Br	I
pK		4.75	2.66	2.86	2.86	3.12

(2)   An increase in the number of $-I$ substituents further promotes acidity:

	$ClCH_2-CO_2H$	Cl_2CH-CO_2H	Cl_3C-CO_2H
pK	2.86	1.30	0.65

(3)   The acid-strengthening influence of a group of $-I$ type is reduced as the group is moved further from the acid centre:

	$ClCH_2-CO_2H$	$ClCH_2-CH_2-CO_2H$	$ClCH_2-CH_2-CH_2-CO_2H$
pK	2.86	4.08	4.52

(4)   Alkyl groups are acid-weakening to an extent depending on their $+I$ effects: $(CH_3)_3C > CH_3-CH_2 > CH_3 > H$:

	$H-CO_2H$	CH_3-CO_2H	$CH_3-CH_2-CO_2H$	$(CH_3)_3C-CO_2H$
pK	3.77	4.75	4.88	5.05

(5)   Phenol (pK 10) is more acidic than ethanol because the negative charge in its anion is delocalized over the aromatic ring, as symbolized by the contributions of the structures (100–102):

(98)   (99)   (100)   (101)   (102)

(6) The introduction of a substituent of electron-attracting type into the nucleus of phenol has an acid-strengthening effect and is greater when the substituent is *ortho* or *para* to the phenolic group than when it is *meta* for, in the former case, it interacts directly with a negatively polarized carbon atom (see structures 100–102). For example, the greater acidity of *p*-nitrophenol (pK 7.1) than phenol may be attributed to the large contribution to the anion of the structure (103); the effect of the nitro group in *m*-nitrophenol (pK 8.35) is not directly transmitted to the oxyanion (see structure 104). Groups of electron-releasing type reduce the acidity of phenol in the same way.

(103)   (104)

(7) In a conjugated acid, both the position of a substituent and the relative importance of its inductive and mesomeric effects are important in determining acidity. Consider, for example, the acid-strengths of the following benzoic acids:

X	pK
H	4.2
Cl	3.8
OCH$_3$	4.1

Y	pK
H	4.2
Cl	4.0
OCH$_3$	4.5

Both *m*-Cl and *m*-OCH$_3$ have an acid-strengthening influence because of the electron-withdrawing inductive effects of the groups, that of chlorine being greater than that of methoxyl. In the *para*-substituted acids, the $-I$ effects are opposed by the acid-weakening $+M$ effects of the substituents, as symbolized by the contributions of the resonance structures (105) and (106). For chlorine, the $-I$ effect is more powerful than the $+M$ effect so that, whereas *p*-chlorobenzoic acid is weaker than its *meta*-isomer, it is stronger than benzoic acid. For methoxyl, on the other hand, the $-I$ effect is less powerful than the $+M$ effect so that *p*-methoxybenzoic acid is a weaker acid than both its *meta*-isomer and benzoic acid:*

(105)   (106)

---

* The differences in pK values quoted above are numerically small, but it should be remembered that the scale is logarithmic: a difference in pK of 0.3 corresponds to a factor of 2 in the dissociation constants.

Similar considerations serve to rationalize the acid strengths (and hence the Hammett σ-values, p. 63) of other substituted benzoic acids.

(8) Sulfonic acids are very much stronger than carboxylic acids because of the strong −I effect of the sulfone group (—SO_2—) and the greater delocalization of charge in the sulfonate anion (107–110) than in the carboxylate anion. Benzenesulfonic acid and its derivatives are useful as strong acid catalysts in organic synthesis as alternatives to sulfuric acid for, while being strong acids, they do not bring about the side reactions (e.g. oxidation and sulfonation) characteristic of sulfuric acid.

$$
\begin{array}{cccc}
\underset{\overset{\|}{O}}{\overset{O}{\|}}\quad & & & \\
R\!-\!S\!-\!O^- & R\!-\!S\!=\!O & R\!-\!S\!=\!O & R\!-\!S\!\overset{2+}{\underline{\ }}O^- \\
(107) & (108) & (109) & (110)
\end{array}
$$

2. Acidity of S—H groups.   Thiols, RSH, are stronger acids than the corresponding alcohols. This can be understood by considering the ionization of each as occurring in two steps:

$$
\begin{array}{ccccccc}
RS\!-\!H & \longrightarrow & RS\cdot & + & H\cdot & \longrightarrow & RS^- & + & H^+ \\
RO\!-\!H & \longrightarrow & RO\cdot & + & H\cdot & \longrightarrow & RO^- & + & H^+
\end{array}
$$

Although the second step is energetically more favourable for RO· than RS· because oxygen is more electronegative than sulfur, the first step is more favourable for the thiol because the S—H bond ($340\,\text{kJ mol}^{-1}$) is considerably weaker than the O—H bond ($462\,\text{kJ mol}^{-1}$) and it is the latter factor which dominates in the values for the free energies of ionization.

3. Acidity of N—H groups.   The N—H bond is intrinsically less acidic than the O—H bond because nitrogen is less electronegative than oxygen. Carboxamides are very feebly acidic compared with carboxylic acids, e.g. they are not dissolved by sodium hydroxide solution. However, just as sulfonic acids are much stronger than carboxylic acids, so sulfonamides, RSO_2NH_2 and RSO_2NHR', are much more strongly acidic than carboxamides, giving delocalized anions represented by structures (111–113). Although they are not acidic enough to liberate carbon dioxide from sodium hydrogen carbonate, they are dissolved by sodium hydroxide solution. Use may be made of this in separating primary and secondary amines: the mixture is treated with a sulfonyl chloride to give a mixture of the sulfonamides, $R'SO_2NHR$ and $R'SO_2NR_2$, of which only the former has an acidic proton and can be extracted into sodium hydroxide solution.

$$
\begin{array}{ccc}
R\!-\!S\!-\!\overset{-}{N}R' & R\!-\!S\!=\!NR' & R\!-\!S\!=\!NR' \\
(111) & (112) & (113)
\end{array}
$$

Two carbonyl groups bonded to NH yield an effect comparable with that of one sulfone group: imides, such as succinimide (114) do not react with sodium hydrogen carbonate but are soluble in sodium hydroxide solution. The delocalized structure of the anion from an imide is represented as (115–117). Use is made of the acidity of imides in a method for forming C—N bonds (p. 303).

(114)   (115)   (116)   (117)

4. Acidity of C—H groups. The sp^3 C—H bond is less acidic than the N—H bond because carbon is less electronegative than nitrogen. As would be expected from the earlier discussion, the acidity is increased when the CH group is attached to an increasing number of carbonyl groups, for the negative charge of the conjugate base is then increasingly effectively delocalized, as shown for the ion from $(CH_3CO)_3CH$:

(118)   (119)   (120)   (121)

The pK values of CH_3COCH_2—H, $(CH_3CO)_2CH$—H, and $(CH_3CO)_3C$—H are respectively 20, 9, and 6.

Carbonyl is one of a number of groups of $-M$ type which have a marked acid-promoting influence: the more important, in decreasing order of effectiveness, are —NO_2, —CHO, —COR, —CN, and —CO_2R. For example, the acidity of nitromethane (pK 10.2) derives from the delocalization of the charge on the anion on to the oxygen atoms of the nitro group:

(122)   (123)

The acidity of C—H bonds is of great significance in organic chemistry because the anions formed by ionization of the bond are intermediates in many synthetic processes (chapter 7). The following structural situations are of particular importance.

(1) The C—H bond is adjacent to *one* group of $-M$ type; pK values lie in the range 10–20. Reactions involving anions derived from compounds containing

these structures are the aldol reaction $\left(\diagdown\diagup\!\!\text{CH—CHO and }\diagdown\diagup\!\!\text{CH—COR, p. 209}\right)$,

the Claisen condensation $\left(\diagdown\diagup\!\!\text{CH—CO}_2\text{R, p. 226}\right)$, and the Thorpe reaction

$\left(\diagdown\diagup\!\!\text{CH—CN, p. 229}\right)$.

(2)   The C—H bond is adjacent to *two* groups of $-M$ type. Typical examples are the keto-ester $CH_3COCH_2CO_2Et$ and the diester $CH_2(CO_2Et)_2$, whose uses are described in section 7.4. The pK-values of such compounds are in the range 4–12.

(3)   The C—H bond is part of cyclopentadiene or a derivative. The compounds are acidic because the derived anion, having six π-electrons and fulfilling the criteria for aromaticity, is strongly resonance-stabilized (p. 40).

(4)   The C—H bond in the haloforms. This is acidic because the conjugate base is stabilized both by the inductive effects of three halogen atoms and by charge-delocalization (e.g. structures 124–127) as the halogens (other than fluorine) have unfilled and relatively low-lying d orbitals. These anions, by loss of a halide ion, form carbenes, which are intermediates in many reactions (p. 101).

(124)   (125)   (126)   (127)

(5)   The C—H bond in acetylene itself and mono-substituted acetylenes, $RC\equiv CH$. This is very much more acidic than that in an alkane or an alkene, e.g. acetylene has pK 25 and forms the acetylide ion when treated with amide ion in liquid ammonia, whereas methane (pK c. 45) is unreactive. The reason is that an s-electron is on average nearer the nucleus than a p-electron, so that it is more strongly held by that nucleus; the electrons in a bond formed by sp-hybridized carbon are more strongly held by the carbon nucleus than those in bonds formed by carbon in the sp^2 or sp^3 states because of the greater proportion of s character of the electrons in the first case. sp-Hybridized carbon is therefore more electro-negative than carbon in other hybridized states and the charge on the acetylene anion is more readily accommodated.

(*ii*)   *Bases*.   The common bases include both anions, e.g. H_2N^- and EtO^-, and neutral molecules containing at least one unshared pair of electrons, e.g. NH_3 and EtOH. For the former group, the basicity is weakened by any factor which stabilizes the negative charge of the anion, while for the latter, the basicity is increased by any factor which stabilizes the positive charge on the conjugate acid of the base. Anionic bases are far stronger than their neutral analogues (e.g. $H_2N^- \gg NH_3$).

Amongst anionic bases, the charge is normally associated with oxygen, nitrogen or carbon. Since the electronegativities of these elements fall in the order $O > N > C$, the order of basicities is $R_3C^- > R_2N^- > RO^-$. For example, amide ion is a far stronger base than hydroxide ion, and methide ion (CH_3^-) is of such high energy that it does not exist in organic media. However, Ph_3C^-, in which the negative charge is delocalized by the aromatic rings, is more stable and is used (as sodium triphenylmethyl) in certain reactions which require a particularly powerful base (see p. 227). The order of basicities of oxyanions, $(CH_3)_3CO^- > CH_3O^- > PhO^- > CH_3CO_2^-$, follows immediately from the principles discussed above which govern acidity: t-butoxide ion is a stronger base than methoxide ion because the three electron-releasing methyl groups in the former destabilize the negative charge; phenoxide ion is a weaker base than an alkoxide ion because the charge is delocalized over the aromatic ring; and acetate ion is still weaker because the charge delocalization by oxygen is even more effective.

Nitrogen is the most important basic element in uncharged bases. Typical alkylamines have pK values in the region 9–11 (where pK refers to the conjugate acid formed by the base, e.g. CH_3—NH_3^+ has pK 10.6). The conjugation of the amino group to a carbon–carbon double bond or an aromatic ring reduces the basicity considerably because the resonance stabilization in the base (represented, in the case of aniline (pK 4.6), by the contribution of structures such as 128) is lost on protonation. The protonation of pyrrole and related compounds, in which the unshared electron-pair on nitrogen completes the aromatic sextet, results in the complete loss of the aromatic stabilization energy, and these compounds are essentially non-basic.

(128)   (129)   (130)

Amides are much less basic than amines because the resonance energy of the delocalized system (129–130) is lost on protonation of the nitrogen atom.

Nitriles are only very weakly basic. The explanation corresponds to that given for the greater acidity of alkynes than alkanes (p. 48): namely, since the sp-hybridized nitrogen in a nitrile is a more electronegative species than the nitrogen in an amine, $RC{\equiv}NH^+$ is more acidic than RNH_3^+. Imines and heterocyclic aromatic compounds such as pyridine (pK 5.2) occupy an intermediate position. Amidines, however, are much stronger bases than both imines and amines. This is because, although both the base (131) and its conjugate acid (132) are resonance stabilized, the stabilization energy of the latter, whose principal resonance structures are equivalent, is greater than that of the former, in which one of the corresponding structures is dipolar and of high energy. The stability of the conjugate acid formed by guanidine (133) serves to make this as strong a base as the alkali-metal hydroxides.

(131)   (132)

(133)

The oxygen atom in neutral molecules is only weakly basic. Nevertheless, many reactions of alcohols, aldehydes, ketones, and esters occur *via* the conjugate acids formed by these compounds and the solubility in sulfuric acid of oxygen-containing compounds such as esters, ethers, and nitro compounds is due to the formation of oxonium ion salts in this medium (e.g. R_2OH^+ HSO_4^-). Both aliphatic and aromatic C=C bonds are also weakly basic, e.g. benzenoid compounds form salts such as (134) with hydrogen chloride in the presence of aluminium trichloride. The basicity of polycyclic aromatic hydrocarbons increases with increase in the number of carbon atoms which can delocalize the positive charge of the conjugate acid.

(134)

Problems

1. Draw orbital representations of the following compounds: ethylene; allene (CH_2=C=CH_2); 1,3-butadiene; nitromethane; acrylonitrile (CH_2=CHCN); hydrazine (NH_2NH_2).

2. The stabilization energies of conjugated compounds are usually obtained by comparing either the heats of hydrogenation or the heats of combustion of these compounds with those of appropriate non-conjugated compounds. Why does the former method give the more reliable values?

    Would you expect biphenyl (Ph—Ph) to have a greater stabilization energy than twice that of benzene? If so, by approximately how much?

    Given that the heat of hydrogenation of styrene (PhCH=CH_2) is $-326 \, \text{kJ mol}^{-1}$, calculate the stabilization energy of styrene (see Table 1.1 for bond-energy values).

3. Arrange each of the following groups in decreasing order of acid-strength:
    (*i*) HCO_2H, CH_3CO_2H, $ClCH_2CO_2H$, FCH_2CO_2H.
    (*ii*) Phenol, *m*- and *p*-chlorophenol, *m*- and *p*-cresol.

(*iii*) Benzoic acid, *m*- and *p*-nitrobenzoic acid, *m*- and *p*-methoxybenzoic acid.

(*iv*) 1,4-Pentadiene and cyclopentadiene.

(*v*)

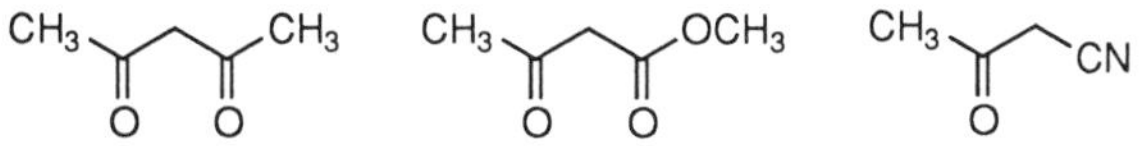

(*vi*)

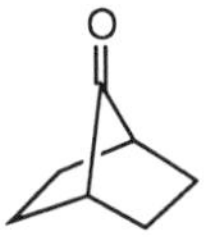

and

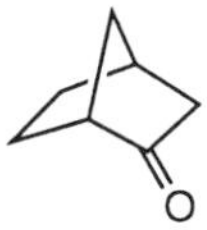

(*vii*)   CHF$_3$ and   CHCl$_3$

(*viii*)

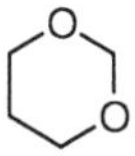

and

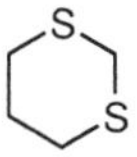

4. Arrange each of the following groups in decreasing order of base-strength:
 (*i*) Ammonia, aniline, *m*- and *p*-nitroaniline.
 (*ii*) Ethoxide ion, t-butoxide ion, acetate ion, and phenoxide ion.
 (*iii*) Pyrrole and pyrrolidine (tetrahydropyrrole).
 (*iv*) Pyridine and piperidine (hexahydropyridine).

5. How can you account for the following:
 (*i*) The dipole moment of 1,2-dichloroethane increases as the temperature is raised.
 (*ii*) The dipole moment of *p*-nitroaniline (6.2 D) is larger than the sum of the values for nitrobenzene (3.98 D) and aniline (1.53 D).
 (*iii*) The dipole moment of propenal (CH_2=CHCHO; 3.04 D) is greater than that of propionaldehyde (CH_3CH_2CHO; 2.73 D).
 (*iv*) Picric acid (2,4,6-trinitrophenol) liberates carbon dioxide from aqueous sodium carbonate, but phenol does not.
 (*v*) *NN*-Dimethylation triples the basicity of aniline but increases the basicity of 2,4,6-trinitroaniline by 40 000-fold.
 (*vi*) The bond dissociation energy of the PhCH$_2$—H bond (322 kJ mol^{-1}) is considerably smaller than that of the CH$_3$—H bond (426 kJ mol^{-1}).
 (*vii*) The boiling point of ethanol is very much higher than that of its isomer, dimethyl ether.
 (*viii*) Boron trifluoride and aluminium trichloride are Lewis acids.

3 Chemical kinetics

3.1 Rates of reaction

It is possible to deduce from thermodynamic data whether or not a particular reaction can *in principle* yield a particular set of products in significant amounts: if the reaction A → B is accompanied by a decrease in free energy, it is possible in principle to convert A largely into B. However, thermodynamic data do not give information about the *rates* of reactions, and it is quite possible that the percentage conversion of A into B will be negligible even after many years. For example, the reaction between hydrogen and oxygen to give water is accompanied by a large negative free energy change but nevertheless the rate at which water is formed is insignificant at room temperature, although it is very rapid indeed at high temperatures. Again, the reaction between acetaldehyde and hydrogen cyanide to give the cyanohydrin,

is accompanied by a favourable free-energy change but is very slow unless a small quantity of a base is added.

In effect, the thermodynamic criterion for a reaction to proceed is a necessary but not a sufficient one; metastable equilibria are common, as in the case of the hydrogen-oxygen reaction. It is necessary that other conditions should be met which result in equilibrium being attained at a practicable rate. It is the purpose of this chapter to set out the principles which govern the rates of reactions.

3.2 The orders of reaction

Many organic reactions follow simple kinetic equations. For example, the hydrolysis of ethyl bromide by sodium hydroxide,

$$CH_3CH_2Br + NaOH \longrightarrow CH_3CH_2OH + NaBr$$

follows *second-order* kinetics:

$$d[CH_3CH_2OH]/dt = k[CH_3CH_2Br][OH^-]$$

This and other evidence led to the accepted mechanism for this reaction in which as the bromine, in the form of bromide ion, leaves the carbon atom to which it was bonded, so the hydroxide ion approaches and forms a bond to that carbon atom (p. 102). However, the rate constant, k, is far smaller than the frequency of collisions between the reactant molecules. This is almost invariably the case, and the reason for it is discussed in sections 3.4 and 3.5.

There are also first-order reactions, such as the solvolysis of t-butyl chloride,

$$(CH_3)_3C-Cl \quad + \quad H_2O \quad \longrightarrow \quad (CH_3)_3C-OH \quad + \quad HCl$$

in which the slow or *rate-determining* step is the ionization of the chloro-compound, followed by the rapid reaction of the organic cation with water:

$$(CH_3)_3C-Cl \quad \xrightarrow{\text{slow}} \quad (CH_3)_3C^+ \quad + \quad Cl^-$$

$$(CH_3)_3C^+ \quad + \quad H_2O \quad \longrightarrow \quad (CH_3)_3C-OH \quad + \quad H^+$$

Reactions which have complex mechanisms can have complex kinetics, including non-integral orders with respect to one or more of the reactants. An example is the decomposition of acetaldehyde at high temperatures:

$$CH_3CHO \quad \longrightarrow \quad CH_4 \quad + \quad CO$$

for which

$$d[CH_4]/dt = k[CH_3CHO]^{3/2}$$

These kinetics are consistent with the mechanism

$$CH_3CHO \quad \xrightarrow{k_1} \quad \cdot CH_3 \quad + \quad \cdot CHO$$

$$CH_3CHO \quad + \quad \cdot CH_3 \quad \xrightarrow{k_2} \quad CH_3\dot{C}O \quad + \quad CH_4$$

$$CH_3\dot{C}O \quad \xrightarrow{k_3} \quad \cdot CH_3 \quad + \quad CO$$

$$2 \cdot CH_3 \quad \xrightarrow{k_4} \quad C_2H_6$$

Reaction is initiated by the fragmentation of a molecule of the aldehyde. The methyl radical which is formed abstracts hydrogen from a second molecule of the aldehyde, giving methane and an acetyl radical; the latter rapidly gives carbon monoxide and a new methyl radical which reacts with a third molecule of aldehyde. The chain continues to be propagated in this way until two methyl radicals meet and dimerize. Since the concentration of methyl radicals is low, the probability of occurrence of the dimerization step compared with the chain-propagating step is small, so that a relatively small number of radicals from the initiation step can yield a large quantity of methane. It can be shown that the kinetics should then be given by

$$d[CH_4]/dt = k_2 \left[\frac{k_1}{k_4}\right]^{1/2} [CH_3CHO]^{3/2}$$

which is identical with the observed kinetics, where $k = k_2[k_1/k_4]^{1/2}$.

Finally, there are reactions where a catalyst is involved. For example, for the acid-catalyzed bromination of acetone,

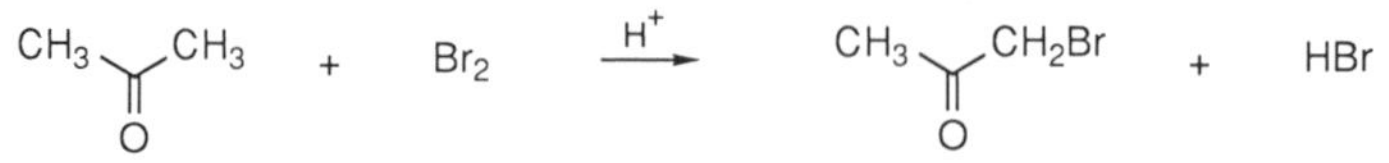

the rate of consumption of bromine is independent of the concentration of bromine until this is very low, whereas it is dependent on the concentration both of acetone and of the protons in solution:

$$-d[Br_2]/dt = k[CH_3COCH_3][H^+]$$

It follows that the rate of the reaction is determined by a step in which bromine plays no part. This step might be the acid-catalyzed enolization of acetone,

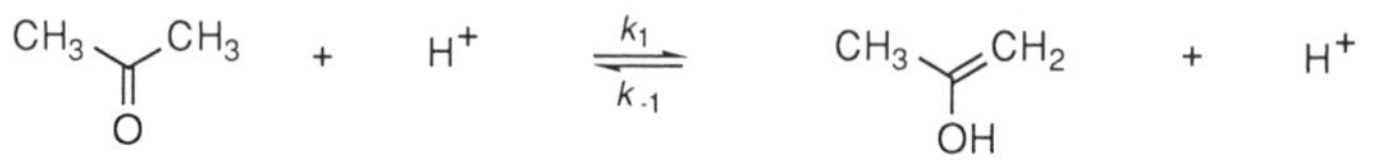

the enol then reacting with bromine as follows:

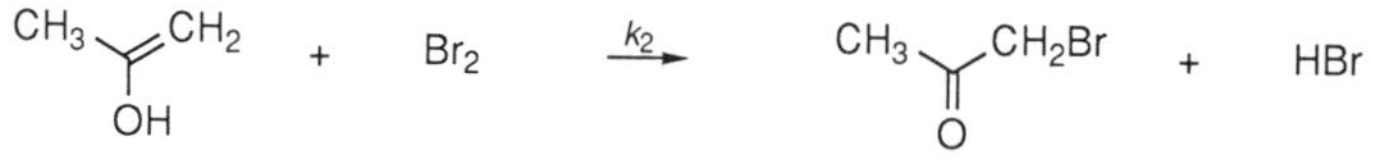

Now, if the rate of formation of the enol is slow compared with the rates of both reactions which destroy it, the enol can be present in only very small concentrations: it will grow in concentration from its initial zero value until it reaches a small *steady-state* concentration when it is destroyed as fast as it is formed, i.e.

$$k_1[CH_3COCH_3][H^+] = k_{-1}[enol][H^+] + k_2[enol][Br_2]$$

i.e. $$[enol] = k_1[CH_3COCH_3][H^+]/(k_{-1}[H^+] + k_2[Br_2])$$

It follows that the rate of loss of bromine is given by

$$
\begin{aligned}
-d[Br_2]/dt &= k_2[enol][Br_2] \\
&= k_1k_2[CH_3COCH_3][H^+][Br_2]/(k_{-1}[H^+] + k_2[Br_2]) \\
&= k_1[CH_3COCH_3][H^+]
\end{aligned}
$$

providing that $k_2 \gg k_{-1}$ and that $[Br_2]$ does not fall to too low a value. This is the experimentally observed result, and other evidence supports the mechanism that has been deduced.

The *steady-state* treatment applied here is of value in understanding a variety of reactions in which a reactive, short-lived intermediate is involved.

For many reactions, a plot of the logarithm of the rate constant against the reciprocal of the absolute temperature is approximately linear with negative gradient (i.e. the reaction is faster at higher temperatures): $\ln k = B - C/T$, where B and C are constants. Differentiation gives $d(\ln k)/dT = C/T^2$ which is of the same form as the equation obtained by differentiating the expression for the equilibrium constant of a reaction (p. 6): $d(\ln K/dt) = \Delta H/RT^2$.

Moreover, the equilibrium constant of a reaction is equal to the ratio of the rate constants for the forward and reverse steps of the process and Arrhenius (1889), therefore, suggested that the constant C above should be replaced by $\Delta E/R$, where ΔE is an energy term in the expression for the rate of the reaction analogous to the enthalpy term ΔH in the equilibrium. The expression $d(\ln k)/dT = \Delta E/RT^2$ gives, on integration, the *Arrhenius equation*,

$$k = Ae^{-\Delta E/RT}$$

where A, like ΔE, is a constant for the reaction concerned. ΔE represents a critical energy which the molecules must possess in order for reaction to occur: for a given value of A, the rate constant decreases as ΔE becomes larger.

The significance of A and ΔE are discussed in the two succeeding sections. For the moment it should be emphasized that the fact that a reaction may have a favourable free-energy change does not imply that reaction will occur at other than an infinitesimal rate; it is necessary also that the values of A and $\Delta E/RT$ should be such that k is in the range of what may be termed practicable rate constants.

3.3
The effect of temperature on reaction rates

3.4
Collision theory

The simplest interpretation of the Arrhenius equation for a bimolecular reaction is that, for reaction to occur, the two reactants must collide and the total energy possessed by them must be at least equal to ΔE: the rate constant should be given by the frequency of collisions times the fraction of the collisions which involve suitably activated molecules, $e^{-\Delta E/RT}$. It ought then to be possible to equate A with Z, the collision number (i.e. the number of collisions per second when there is only one molecule of reactant per unit volume*). We shall first examine the validity of this proposal.

Values of Z lie in the region of $10^{11}\,dm^3\,mol^{-1}\,s^{-1}$, and for many simple gas-phase reactions and for some reactions in solution the experimental value of A is of this order. For these cases, equating A with Z is justified. There are, however, many reactions for which A is considerably smaller than Z: the reaction between butadiene and propenal, for which $A \sim 10^6\,dm^3\,mol^{-1}\,s^{-1}$ is an example

---

* The expression for Z contains a temperature term, e.g. for collisions between molecules of the same type which have molecular weight M and diameter σ, $Z = 4\sigma^2(\pi RT/M)^{1/2}$. However, the effect of changes in T on the value of Z is negligible compared with the effect on the value of the exponential term, unless ΔE is very small.

In general, collision theory fails to account for values of A which are substantially less than Z and this is because it does not allow for the fact that two molecules possessing the necessary activation may, on collision, be unsuitably oriented with respect to each other for reaction to occur. A factor P, representing the *probability* that a collision between suitably activated molecules will result in reaction, is therefore introduced:

$$k = PZe^{-\Delta E/RT}$$

It is not immediately obvious how collision theory can be applied to first-order reactions, such as the solvolysis of t-butyl chloride. The problem was solved by Lindemann, who suggested that the necessary activation energy is acquired by the reactant molecule as the result of its collision with other molecules. The activated molecule can then either undergo reaction or dissipate its extra energy in further collisions. Then, for a reaction $A \rightarrow B$,

$$A + A \underset{k_{-1}}{\overset{k_1}{\rightleftharpoons}} A + A^*$$

$$A^* \overset{k_2}{\rightarrow} B$$

where A^* represents A in the suitably activated state. A small concentration of A^* is thereby maintained, and the steady-state treatment gives

$$d[B]/dt = k_1k_2[A]^2/(k_{-1}[A] + k_2)$$

Hence, if $k_{-1}[A] \gg k_2$, $d[B]/dt = k_1k_2[A]/k_{-1}$, i.e. first-order kinetics should be followed. At very low values of [A], $d[B]/dt = k_1[A]^2$, i.e. second-order kinetics should be followed. At intermediate values of [A] the kinetics should not be described by either the first- or the second-order equations. Where these predictions have been tested, they have been found to hold.

As with bimolecular reactions, it is sometimes difficult to interpret the value of A in unimolecular reactions. In many, A is of the order of $10^{11}\,dm^3\,mol^{-1}\,s^{-1}$ and may be equated with the collision frequency, but for others it is less than Z by several powers of ten. The most satisfactory explanation is that when a complex molecule possesses the critical activation energy, the chances may be small that this energy is in that area of the molecule at which reaction is to occur. Again, therefore, a probability factor, P, is necessary.

The interpretation of E in collision theory is more satisfactory than that of A. Consider the reaction between a hydrogen molecule and a hydrogen atom, $H_2 + H \rightarrow H + H_2$. As the atom approaches the molecule, the existing H—H bond begins to break and the new H—H bond begins to form. This corresponds to a total energy for the system which is greater than that of the original atom and molecule in isolation, and the total energy continues to rise until a point is

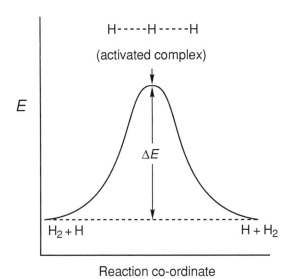

Figure 3.1   Energy profile for the reaction $H_2 + H \cdot \rightarrow H \cdot + H_2$.

reached at which the hydrogen atom which is being transferred is symmetrically placed between the other two hydrogens, as represent by H----H----H. That is, the bonding energy of the two *partial* bonds in H----H----H is less than that of one *full* bond. Thereafter, the energy of the system falls again until the new system of hydrogen atom and hydrogen molecule is obtained. The energy changes during the reaction are shown in the *energy profile*, Figure 3.1, as a function of the progress made in the formation of the new H—H bond.

There is in effect an energy barrier between the two pairs of reactants, and ΔE is the energy necessary for the reactants to surmount the barrier, being described as the *activation energy* of the reaction. The structure corresponding to the highest point on the energy profile is termed the *activated complex* and in this case corresponds to the linear arrangement of three hydrogen atoms which are joined by two partial bonds. For this particular reaction, the difference in energies of the activated complex and the reactants has been calculated theoretically and the value agrees closely with the experimental result ($36\,\mathrm{kJ\,mol^{-1}}$), providing justification for this interpretation of ΔE in the Arrhenius equation. It should also be noted that calculations show that the energy of the activated complex in this reaction is minimal when the complex is linear: that is, the hydrogen atom most easily reacts by approaching the hydrogen molecule along the molecular axis.

It can be appreciated from this discussion why the hydrogen atom–molecule reaction consists of a one-step bimolecular process rather than the alternative two-step process in which the hydrogen molecule first decomposes to two hydrogen atoms one of which then combines with the third hydrogen atom. The latter path would have an activation energy of at least $435\,\mathrm{kJ\,mol^{-1}}$ (the dissociation energy of H_2), whereas in the bimolecular path the energy evolved as the new H—H bond is formed in effect helps to bring about the dissociation of the original H—H bond.

3.5
Transition-state theory

An alternative approach to the collision theory is based on the application of thermodynamic principles to the activated complex. Consider a reaction $A + B \rightarrow C$ in which the activated complex is represented as AB^* and whose energy profile is shown in Figure 3.2. Transition-state theory treats AB^* as a normal chemical species one of whose vibrations is replaced by a translational degree of freedom: a loose bond between A and B, instead of undergoing a stretching-and-contracting vibrational motion, flies apart either to give A and B or to give C. It can be shown that the frequency, v, with which this happens is given by $v = \mathbf{k}T/h$ where $\mathbf{k}$ is the Boltzmann constant and h is Planck's constant. The activated complex, or transition state, is regarded as being in equilibrium with the reactants, i.e.

$$\frac{[AB^*]}{[A][B]} = K^{\neq}$$

where the symbol $\neq$ signifies that this is not a conventional stable equilibrium. The rate at which AB^* is transformed into C is given by the product of the concentration of AB^* and the frequency with which AB^* breaks down to products, so that the rate of loss of A is given by

$$-d[A]/dt = d[C]/dt = [AB^*]\mathbf{k}T/h$$

Since $-d[A]/dt = k_2[A][B]$, where k_2 is the bimolecular rate constant,

$$k_2 = \mathbf{k}T/h \frac{[AB^*]}{[A][B]} = \frac{\mathbf{k}T}{h}K^{\neq}$$

Hence the rate constant* is related to the equilibrium constant, $K^{\neq}$. Now, $-RT\ln K^{\neq} = \Delta G^{\neq} = \Delta H^{\neq} - T\Delta S^{\neq}$, where $\Delta G^{\neq}$, $\Delta H^{\neq}$, and $\Delta S^{\neq}$ are the differences in free energies, enthalpies, and entropies, respectively, of AB^* and $A + B$. It follows that

$$k_2 = \frac{\mathbf{k}T}{h}e^{-\Delta G^{\neq}/RT} = \frac{\mathbf{k}T}{h}e^{-\Delta H^{\neq}/RT}e^{\Delta S^{\neq}/R}$$

The term $\Delta H^{\neq}$, the *enthalpy of activation*, corresponds closely to the activation energy term, ΔE, in the Arrhenius equation (for liquids and solids, $\Delta E = \Delta H^{\neq} + RT$). The term $(\mathbf{k}T/h)e^{\Delta S^{\neq}/R}$ corresponds to the A factor; $\Delta S^{\neq}$ is known as the *entropy of activation*. One important merit of transition-state theory is that it rationalizes the probability factor, P. A reaction in which the transition state is

---

* Strictly, k_2 should be multiplied by a factor (the transmission coefficient) which is the probability that AB^* will dissociate into products instead of back into the reactants. In most cases this factor is close to unity.

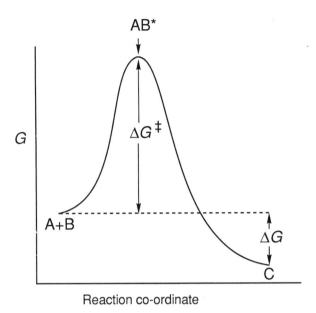

Figure 3.2   Energy profile for A + B → C.

highly organized will have a large, negative $\Delta S^{\neq}$, corresponding to a small value for P. Examples of the dependence of $\Delta S^{\neq}$ on the reaction type are given in the following section.

3.6
Applications of kinetic principles

(*a*)   *Structure, rates and equilibria*

The addition of hydrogen chloride to propylene could in principle give two products:

$$CH_3CH=CH_2 \quad + \quad HCl \quad \longrightarrow \quad CH_3CH_2-\underset{\underset{Cl}{|}}{CH_2} \quad \text{and} \quad CH_3\underset{\underset{Cl}{|}}{CH}-CH_3$$

In practice the latter product predominates, and in general the orientation in the addition of a compound HX to an unsymmetrical alkene is described by *Markovnikov's rule* (1870) which states that 'the hydrogen atom of HX adds to that carbon which bears the greater number of hydrogens'. The basis of the rule is as follows.

The rate-determining step in the reaction is the addition of a proton to one of the carbon atoms of the double bond, giving a carbocation* which then reacts rapidly with the anion X^- from HX. One of two such carbocations might be formed, e.g.

$$CH_3CH{=}CH_2 \; + \; H^+$$

$$CH_3{-}CH_2{-}\overset{+}{C}H_2 \xrightarrow{\;Cl^-\;} CH_3{-}CH_2{-}CH_2Cl$$

$$CH_3{-}\overset{+}{C}H{-}CH_3 \xrightarrow{\;Cl^-\;} CH_3{-}\underset{\underset{Cl}{|}}{C}H{-}CH_3$$

Of these two intermediate ions, the secondary ion, $CH_3{-}\overset{+}{C}H{-}CH_3$, is thermodynamically the more stable for two reasons: the concentration of positive charge at the central carbon atom is reduced by the two electron-releasing $(+I)$ methyl groups attached to it (p. 30), and the charge is also delocalized on to both these groups by the hyperconjugative effect (p. 25). Now, the rates of formation of the two ions are not dependent on their stabilities but rather on the free energies of the *transition states* which precede them. The detailed structures of the transition states are not known; the new C—H bond is partly formed at the transition state, but to an unknown degree. However, it is known that the carbocations are of relatively high free energies compared with the reactants, as represented in Figure 3.3, and there is therefore a smaller change in energy when the transition states are transformed into the intermediate carbocations than when they revert to the reactants. Consequently, it is reasonable to conclude that there is a smaller reorganization of the molecular geometry in passage from transition state to intermediate than in passage from transition state to reactants; that is, the transition states somewhat resemble the carbocations to which they give rise. Therefore, the two transition states can be represented as

$$CH_3{-}CH\overset{\delta+}{\cdots}CH_2 \quad \text{and} \quad CH_3{-}\overset{\delta+}{CH}\cdots CH_2$$

$$\begin{array}{c} \vdots \\ H \\ \vdots \\ \overset{..}{Cl}\,^{\delta-} \end{array} \qquad \begin{array}{c} \vdots \\ H \\ \vdots \\ \overset{..}{Cl}\,^{\delta-} \end{array}$$

corresponding to there being a proportion of the unit positive charge which is ultimately possessed by the carbocations on the appropriate carbon atoms of the

---

* These ions were formerly called *carbonium ions*. However, it is now considered appropriate to reserve the termination *-onium* for ions whose central atom has a complete electron octet, e.g.

$$R_4N^+ \qquad\qquad R_3O^+ \qquad\qquad R_3S^+$$

ammonium   oxonium   sulfonium

whereas these carbon cations have a sextet in the carbon valence shell. The name *carbenium ion* has been suggested for them, but it is not widely used. The description *carbocation* will be used here.

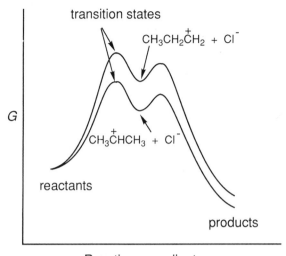

Figure 3.3 Energy profiles for the addition of HCl to propylene.

transition states. It follows that the factors which determine the relative stabilities of the carbocations are also effective in determining the relative stabilities of the transition states, so that the more stable of the two ions is formed the faster and corresponds to a lower maximum on the energy profile (Figure 3.3).

This discussion is generalized by *Hammond's postulate* which states that if a reaction step, A + B → C, is strongly endothermic, the transition state resembles the product C, whereas if it is exothermic, the transition state resembles the reactants.

Hammond's postulate provides a useful working principle, for many reactions occur *via* intermediates whose structures are known and which are of relatively high energy content. In such cases, it is helpful to regard the intermediate as a model for the transition state, for conclusions can then be drawn about the effect which a change in the structure of a reactant is likely to have on the rate of the reaction. Some examples are illustrative.

(1)   Alkenes substituted with electron-withdrawing groups undergo addition of HX in the opposite manner to that of unsubstituted alkenes, e.g.

$$\text{CO}_2\text{Et} \quad + \quad \text{HCl} \quad \longrightarrow \quad \text{Cl} \diagdown \diagup \text{CO}_2\text{Et}$$

This is because the intermediate $\overset{+}{\text{CH}}_2$—CH_2—CO_2Et is of lower energy than CH_3—$\overset{+}{\text{CH}}$—CO_2Et, in which the positive charge is *adjacent* to the strongly electron-withdrawing ester group.

(2)   The rates of addition of HX to three typical alkenes decrease in the order, $(\text{CH}_3)_2\text{C}{=}\text{CH}_2 > \text{CH}_3$—$\text{CH}{=}\text{CH}_2 > \text{CH}_2{=}\text{CH}_2$. This is because the relative stabilities of the carbocations formed in the rate-determining step are,

$$CH_3\underset{CH_3}{\overset{CH_3}{\underset{|}{\overset{|}{C}}}}-CH_3 \quad > \quad CH_3-\overset{+}{CH}-CH_3 \quad > \quad \overset{+}{CH}_2-CH_3$$

which in turn follows from the fact that the stability is increased by the electron-releasing methyl group, three such groups being more effective than two, and two more effective than one.

(3) Compounds in which a C—H bond is adjacent to one or more groups of −M type are acidic (p. 47) and undergo a number of base-catalyzed reactions such as bromination. Typical values of pK and the first-order rate constants for bromination at the same C—H position are:

CH₃CO⟍	EtO₂C⟍ ⟋CO₂Et	CH₃CO⟍ ⟋CO₂Et	CH₃CO⟍ ⟋COCH₃
H	H	H	H

pK	20	13.3	10.7	9
$k(s^{-1})$	5×10^{-10}	2×10^{-5}	10^{-3}	2×10^{-2}

The more acidic the C—H group, the faster is the rate at which bromination occurs. This is because bromination, and the other base-catalyzed reactions, occur *via* the corresponding anions, e.g. CH_3—CO—$\bar{C}H_2$ from acetone, and it is the formation of the anions which constitutes the rate-determining step:

$$CH_3\overset{O}{\underset{\|}{C}}CH_3 \quad + \quad OH^- \quad \xrightarrow{\text{slow}} \quad CH_3\overset{O}{\underset{\|}{C}}\bar{C}H_2 \quad + \quad H_2O$$

$$CH_3\overset{O}{\underset{\|}{C}}\bar{C}H_2 \quad + \quad Br_2 \quad \xrightarrow[\text{-Br}^-]{\text{fast}} \quad CH_3\overset{O}{\underset{\|}{C}}CH_2Br \quad \longrightarrow \quad \text{(further bromination)}$$

Hence any factor which stabilizes the anion, and thereby increases the acidity of the C—H bond, stabilizes the transition state which precedes the anion and enhances the rate of reaction. Some exceptions to this generalization are known; e.g. the pK of nitromethane is 10.2, so that it is a slightly stronger acid than $CH_3COCH_2CO_2Et$ (pK 10.7), but its rate of ionization (5×10^{-8} per second) is much less. It must be emphasized, therefore, that this rate-equilibrium correlation is not quantitative and should be used only as a helpful guide.

(4) Quantitative correlations of rate and equilibria do, however, occur. The second-order rate constants for the alkaline hydrolysis of the benzoic esters R—C_6H_4—CO_2Et (in aqueous acetone at 25°C), where R is a *meta* or *para* substituent on the benzene nucleus, together with the pK values for the corresponding acids (in water at 25°C) are:

R	*p*-CH₃	*m*-CH₃	H	*p*-Cl	*m*-Cl	*m*-NO₂	*p*-NO₂
$10^3 k/\text{dm}^3\,\text{mol}^{-1}\,\text{s}^{-1}$	2.3	3.5	4.9	21.2	36.3	310	510
pK	4.37	4.27	4.20	3.98	3.83	3.49	3.42

A plot of $\log k$ against pK is linear:

$$\log k = -\rho pK + A$$

where ρ and A are constants. Since the point for the unsubstituted compound is fitted by the straight line, $\log k_0 = -\rho pK_0 + A$, so that

$$\log(k/k_0) = \rho(pK_0 - pK)$$

Since $\log k \propto \Delta G^{\neq}$ and $pK \propto \Delta G$, where $\Delta G^{\neq}$ and ΔG are the free energies of activation and ionization, respectively, there is a *linear free energy relationship* between the hydrolysis rates and the acid strengths. This correlation has the following basis. The rate-determining step of the hydrolysis is the addition of hydroxide ion to the carbonyl group of the ester:

The transition state has some of the character of the resulting intermediate and can be represented as

In the passage to the transition state, therefore, the electron density in the vicinity of the ester group is increased, just as it is in the ionization of the corresponding acid ($ArCO_2H \rightarrow ArCO_2^-$). A substituent which withdraws electrons from the nuclear carbon adjacent to the functional centre (e.g. m-NO_2) stabilizes both the transition state of the hydrolysis and the product of the ionization, resulting in a reduction of the free energy of activation of the former (increasing k relative to k_0) and in the free energy of ionization of the latter (decreasing pK relative to pK_0). An electron-releasing group (e.g. p-CH_3) has the opposite effect.

Analogous correlations are obtained between $\log k$ for reactions of other derivatives of benzoic acid, such as the acid-catalyzed hydrolysis of benzamides, and the pK values of the corresponding benzoic acids. The magnitude of ρ varies with the reaction, and its sign (positive or negative) depends on whether the reaction rate is increased or decreased, respectively, by the withdrawal of electron density from the functional centre. It is convenient to define ($pK_0 - pK$) as σ, which is a constant for a given substituent. Then

$$\log(k/k_0) = \sigma\rho$$

which is known as the *Hammett equation*. Typical σ values are:

Substituent		CH_3	F	Cl	NO_2	OCH_3
σ value	*meta* position	−0.07	0.34	0.37	0.71	0.11
	para position	−0.17	0.06	0.23	0.78	−0.27

The order of effects measured by σ values follows qualitatively from the principles which govern the polar properties of substituents and the strengths of acids. As a typical example, the σ value of m-NO_2 is less than that of p-NO_2

because the $-I$ effect of this group (p. 30) is reinforced when the substituent is in the *para* position by its $-M$ effect (p. 32).

Linear free-energy relationships do not hold for *ortho*-substituents or in aliphatic systems because steric effects become important: a substituent can physically impede the approach of the reagent to the functional centre. This can be of considerable significance, e.g. ethyl 2,6-dimethylbenzoate is effectively inert to base-catalyzed hydrolysis. Many other examples of *steric hindrance* will be described in later chapters.

(b)   The effect of solvent

The rate of a reaction in solution is almost always dependent, and often very strongly dependent, on the nature of the solvent.

Consider again the solvolysis of t-butyl chloride, the rate-determining step of which is the formation of the t-butyl cation (p. 53):

$$(CH_3)_3C-Cl \xrightarrow{\text{slow}} (CH_3)_3C^+ + Cl^-$$

At the transition state, the carbon–chlorine bond is partially broken and the covalent bonding pair, which is ultimately associated completely with chlorine, has been considerably displaced towards chlorine

$$(CH_3)_3\overset{\delta+}{C}----\overset{\delta-}{Cl}$$

so that there is a separation of unlike charges in passage from reactant to transition state.

Two characteristics of the solvent play a part in determining the relative free energies of reactant and transition state and therefore the rate of reaction. First, energy is needed to separate the unlike charges, and the amount of energy decreases as the dielectric constant of the solvent increases.* Consequently, the reaction rate increases with the dielectric constant. The dielectric constants of commonly used solvents are:

Water	79	Ethanol	24
Dimethyl sulfoxide, $(CH_3)_2SO$ (DMSO)	49	Acetone	21
Dimethylformamide, $HCON(CH_3)_2$ (DMF)	37	Acetic acid	6
Methanol	33	Diethyl ether	4
Hexamethylphosphoric triamide $((CH_3)_2N)_3PO$ (HMPA)	30	Benzene	2

Second, the *solvating power* of the solvent is important. The transition state can be stabilized by solvation of both the developing positive and the developing negative ions. This occurs with water and alcoholic solvents: the developing carbocation is solvated by the electron-rich hydroxylic oxygen and the developing

---

*The dielectric constant of a medium is equal to the attractive force between opposite charges in a vacuum relative to that in the medium.

chloride ion is solvated by hydrogen-bonding to the hydroxylic hydrogen, which is positively polarized by the electronegative oxygen atom:

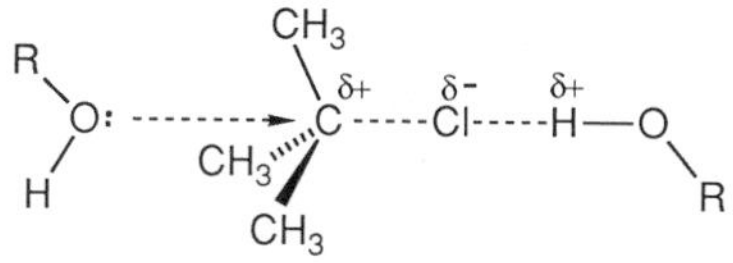

The solvents DMSO, DMF, and HMPA, although strongly polar as measured by their dielectric constants, are less effective solvents than alcohols because, while they can stabilize cations through interaction with their negatively polarized oxygen atoms, they do not contain positively polarized hydrogen atoms for hydrogen-bonding to anions. They are described as *polar aprotic* solvents.

In a second class of reaction, e.g.

$$HO^- \quad + \quad (CH_3)_3S^+ \quad \longrightarrow \quad HO-CH_3 \quad + \quad (CH_3)_2S$$

there is a partial neutralization of unlike charges in passage to the transition state,

$$\overset{\delta-}{HO}\cdots\cdots CH_3\cdots\cdots\overset{\delta+}{S}(CH_3)_2$$

transition state

and these reactions occur more slowly as the dielectric constant and solvating power of the solvent are increased.

In a third class of reactions there is a slight dispersal of charge in the rate-determining step, e.g. the alkaline hydrolysis of methyl iodide:

$$HO^- \quad + \quad CH_3-I \quad \longrightarrow \quad \left[\overset{\delta-}{HO}\cdots\cdots CH_3\cdots\cdots\overset{\delta-}{I} \right]^- \quad \longrightarrow \quad HO-CH_3 \quad + \quad I^-$$

transition state

The reactant hydroxide ion, in which the charge is more concentrated, is more effectively solvated and stabilized than the transition state, so that the reaction occurs more slowly as the anion-solvating power of the solvent is increased. An extreme example of this type is provided by the reaction of bromobenzene with an alkoxide ion:

$$Ph-Br \quad + \quad {}^-OR \quad \longrightarrow \quad Ph-OR \quad + \quad Br^-$$

The formation of the transition state,

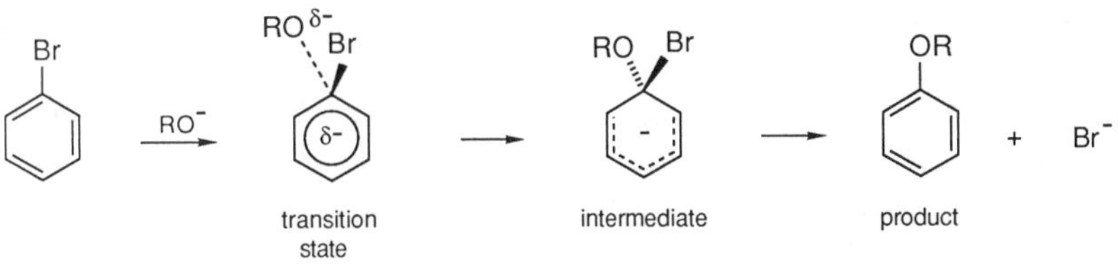

involves the dispersal of charge, so that the reaction is slower in a solvent which solvates anions strongly, such as an alcohol, than in a non-solvating medium. In practice, reaction with t-butoxide ion occurs about nine powers of ten faster in dimethyl sulfoxide, which has little ability to stabilize $(CH_3)_3C—O^-$, than in t-butanol, where the rate is negligible.

In general, reactions where the activation process involves the separation of unlike charges occur more rapidly as the polarity of the medium is increased, whereas those involving the partial neutralization of charge or the dispersal of charge occur more slowly. Many synthetic processes are of ionic type, as will be seen in the course of the text and it is necessary, therefore, to pay careful attention to the choice of the solvent in planning an organic synthesis.

(c) Ring closure

Whereas the esterification of acetic acid by methanol is very slow in the absence of an acid catalyst, the lactonization of 4-hydroxybutyric acid,

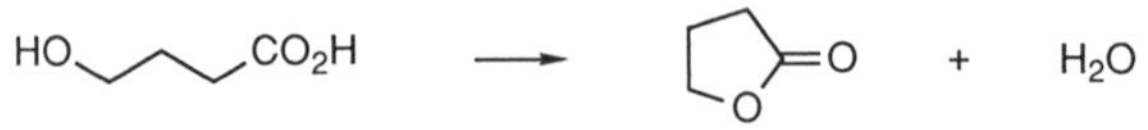

occurs essentially spontaneously. The reason for the difference becomes apparent when the transition states for the processes are considered. In each case, the rate-determining step involves the formation of the new C—O bond and at the transition states this bond is partially formed:

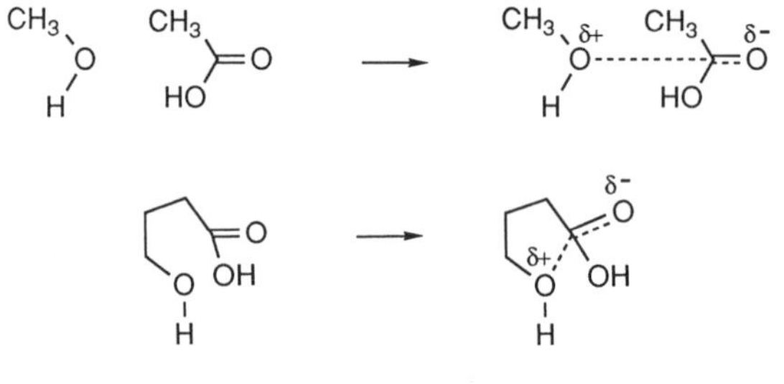

In the esterification, passage to the transition state necessitates two molecules coming together to form one species so that translational freedom is lost and $\Delta S_{tr}^{\neq}$ is correspondingly large and negative. On the other hand, in the lactonization only internal, or vibrational, freedom is lost, and the $\Delta S_{vib}^{\neq}$ term is much smaller than the $\Delta S_{tr}^{\neq}$ term for the intermolecular analogue, though still negative. The $\Delta H^{\neq}$ terms are similar in magnitude, for the same types of bond-forming and bond-breaking processes occur in each case and there is comparatively little strain in the lactone ring which is being formed. Consequently $\Delta G^{\neq}$ is smaller for lactonization and this process occurs faster than esterification.

This discussion both reveals the significance of the *PZ* factor of the Arrhenius equation in these reactions and shows that the factors which determine the relative rates of the two reactions are essentially the same as those which determine the relative equilibrium constants (cf. p. 16).

The similarity between the factors controlling rates and those controlling equilibria can be taken further. For example, in an intramolecular reaction leading to a four-membered ring, $\Delta S^{\neq}$ is less negative than in one leading to a five-membered ring for the acyclic reactant in the former case has fewer conformations than that in the latter, so that there is less loss of internal freedom ($\Delta S^{\neq}_{vib}$) in the formation of the transition state. However, $\Delta H^{\neq}$ is larger for the four-membered ring because of the strain in the cyclic transition state and this factor outweighs the $T\Delta S^{\neq}$ term. Hence $\Delta G^{\neq}$ is larger for the formation of a four-membered ring, and ring closure occurs more slowly (e.g. β-hydroxy-acids, unlike γ-hydroxy-acids, do not lactonize spontaneously).

At first sight surprisingly, three-membered rings are usually formed faster than four-membered rings. This is because the greater strain (leading to a higher $\Delta H^{\neq}$) in the transition state for the former is more than offset by the greater probability (less negative $\Delta S^{\neq}$) that the ends of the chain shall come together for bond formation.

The entropy term is less favourable for the formation of a six-membered than a five-membered ring and, unless there is a significant strain factor involved in the formation of the five-membered ring, this is formed the faster. For larger sized rings the entropy term becomes increasingly unfavourable and there is in addition a small amount of torsional strain in the cyclic transition states for seven- to twelve-membered rings.

The combined effects of the enthalpy and entropy factors are illustrated for the cyclization of some bromo-amines:

$$\begin{matrix} (CH_2)_{n-2} \\ \diagup \quad \diagdown \\ H_2N \qquad CH_2Br \end{matrix} \longrightarrow \begin{matrix} (CH_2)_{n-2} \\ \diagup \quad \diagdown \\ HN \rule{1cm}{0.4pt} \end{matrix}$$

n	3	4	5	6	7	10	15
$k_{relative}$	0.12	0.002	100	1.7	0.03	10^{-8}	10^{-4}

Cyclizations of bifunctional compounds compete with the corresponding intermolecular reaction between two molecules of the compound. In typical reaction conditions, the formation of five- and six-membered rings is strongly favoured over the intermolecular process, and indeed such cyclizations often occur in milder conditions than analogous intermolecular reactions, as in the lactonization of 4-hydroxybutyric acid described above. The formation of three-, four-, and seven-membered rings competes less successfully and yields of the cyclic products are usually low. Rings with more than seven members are usually not formed in significant amounts. It should, however, be remembered that the competition between the intra- and inter-molecular reactions depends not only on the ring size but also on the concentration of the reactant. Since two molecules are necessary for the intermolecular process and only one for cyclization, high dilutions favour cyclization, and this principle has been employed in a technique for synthesizing large rings (p. 229).

(d)   *Thermodynamic versus kinetic control*

In most reactions which can proceed by two or more pathways each of which gives a different set of products, the products isolated are those derived from the pathway of lowest free energy of activation, regardless of whether this path results in the greatest decrease in the free energy of the system. These reactions are described as being *kinetically controlled*.

If, however, the reaction conditions are suitable for equilibrium to be established between the reactants and the kinetically controlled products, a different set of products, formed more slowly but corresponding to a lower free energy for the system, can in some instances be isolated. Such reactions are described as being *thermodynamically controlled*.

Consider the reaction of naphthalene with concentrated sulfuric acid. Two monosulfonated products are in principle obtainable:

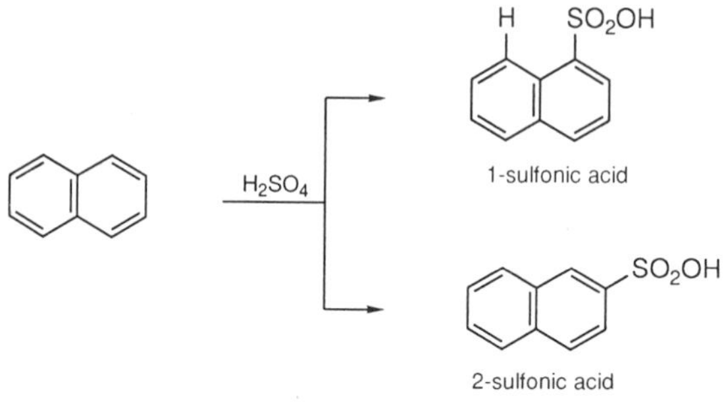

The 1-derivative is formed the faster of the two but the 2-derivative is thermodynamically the more stable (at least in part because of the repulsive forces between the sulfonic acid group and the hydrogen, shown, in the 1-isomer). The situation can be represented schematically as in Figure 3.4.

At low temperatures (*c.* 80°C), sulfonation at the 1-position occurs fairly rapidly whereas that the 2-position is very slow. The free energy of activation for the desulfonation of the 1-sulfonic acid is such that in these conditions this product is essentially inert and is therefore isolated. At higher temperatures (*c.* 160°C), desulfonation of the 1-sulfonic acid becomes important and equilibrium is fairly rapidly established between this product and the reactants. The rate of formation of the 2-sulfonic acid is now also greater, so that gradually most of the naphthalene is converted into the 2-derivative and this becomes the major product.

Problems   1.   It is said in some elementary textbooks that 'reaction rates double when the temperature rises by 10°C.' What is the activation energy of such reactions? (Assume that the statement refers to reactions occurring at or near room temperature.)

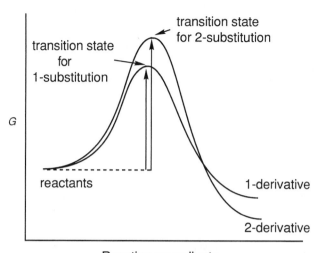

Figure 3.4 Free energy profile for the sulfonation of naphthalene*.

2. The second-order rate constant for the alkaline hydrolysis of ethyl phenyl-acetate ($PhCH_2CO_2Et$) in aqueous acetone varies with temperature as follows:

$T(°C)$	10	25	40	55
$10^2\,k(dm^3\,mol^{-1}\,s^{-1})$	1.74	4.40	10.4	23.6

Calculate the activation energy, ΔE, and the pre-exponential factor, PZ, for the reaction.

3. Two reactions have, respectively, $A = 10^9\,dm^3\,mol^{-1}s^{-1}$, $\Delta E = 60\,kJ\,mol^{-1}$, and $A = 10^{10}\,dm^3\,mol^{-1}s^{-1}$, $\Delta E = 70\,kJ\,mol^{-1}$. At what temperature would the rate constants be equal?

4. The kinetics for the solvolysis of Ph_2CHCl in aqueous acetone are given by

$$-d[Ph_2CHCl]/dt = k[Ph_2CHCl]/(1 + k'[Cl^-])$$

Suggest an explanation.

5. Phenylnitromethane, $PhCH_2NO_2$, is a liquid which dissolves in sodium hydroxide solution. Acidification of the solution precipitates the tautomer,

$$PhCH{=}\overset{\overset{\displaystyle O^-}{|}}{\underset{}{N^+}}{-}OH$$

as a solid, but this then slowly reverts to the original liquid. Explain the chemistry of the formation of the tautomer and draw an energy profile consistent with the observations.

---

* For simplicity, intermediates are not shown.

6. The Cannizzaro reaction on benzaldehyde is thought to have the following mechanism:

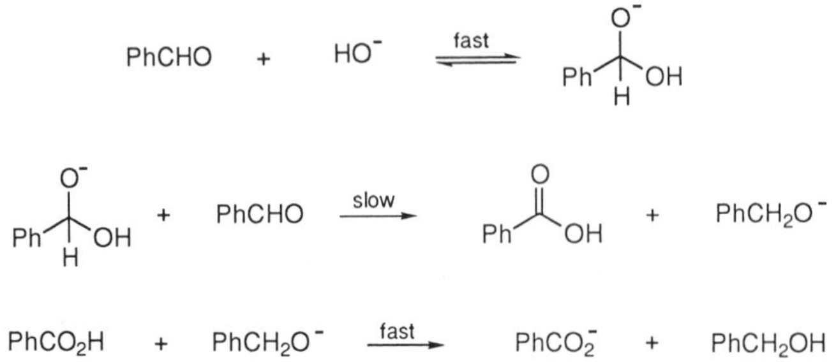

$$PhCHO + HO^- \xrightleftharpoons{\text{fast}} Ph\underset{H}{\overset{O^-}{\cancel{C}}}OH$$

$$Ph\underset{H}{\overset{O^-}{\cancel{C}}}OH + PhCHO \xrightarrow{\text{slow}} Ph\overset{O}{\cancel{C}}OH + PhCH_2O^-$$

$$PhCO_2H + PhCH_2O^- \xrightarrow{\text{fast}} PhCO_2^- + PhCH_2OH$$

What kinetics would you expect the reaction to follow?

7. Derive the kinetic expression given on p. 54 for the decarbonylation of acetaldehyde.

8. The following mechanistic scheme has been postulated for the reaction between iodine and an organic compound, HA, in the presence of hydroxide ions in aqueous solution:

$$HA + OH^- \underset{k_{-1}}{\overset{k_1}{\rightleftharpoons}} A^- + H_2O$$
$$A^- + I_2 \xrightarrow{k_2} IA + I^-$$

Assuming that the concentration of A^- is at all times very much smaller than that of HA and IA, derive an equation for the rate of change of [HA] in terms of the rate constants and the concentrations of HA, OH^-, and I_2 (OH^- and I_2 being present in excess).

9. In a reaction sequence

$$A \underset{k_{-1}}{\overset{k_1}{\rightleftharpoons}} B \xrightarrow{k_2} C$$

C has a lower free energy than A, and $k_2 \gg k_{-1} \gg k_1$. Draw an energy profile, and derive the kinetic equation.

Mechanism 4

4.1 Unit processes in organic reactions

Many organic reactions seem at first glance to be highly complex, involving extensive reorganization of the bonds of the reactants. An example is the Skraup synthesis in which quinoline is obtained in yields of up to 90% by heating a mixture of aniline, glycerol, nitrobenzene, and concentrated sulfuric acid:

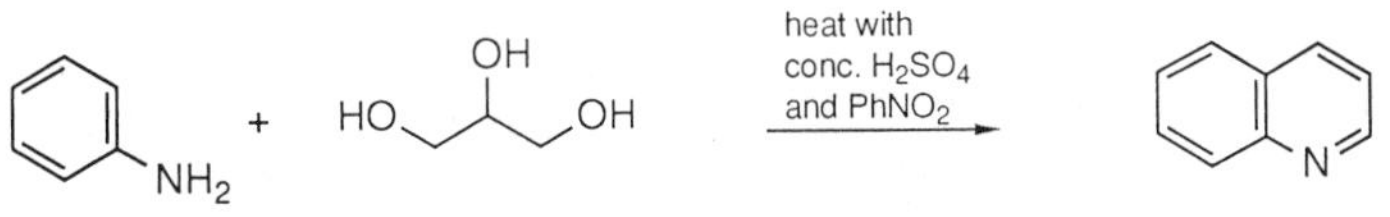

Such complex reactions usually consist of a number of simple steps in each of which a comparatively small reorganization of bonds takes place and that lead to intermediates which undergo further reaction until the final product is reached. The Skraup synthesis, for example, consists of the following steps:

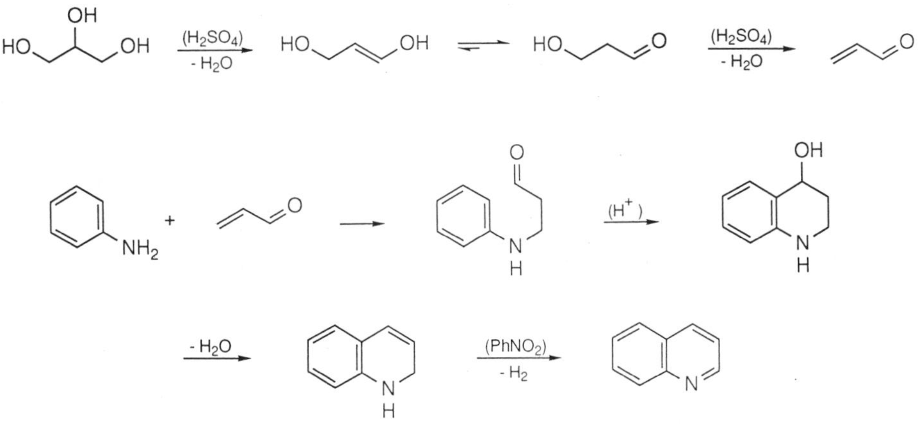

Even reactions which may appear simpler often consist of a number of steps. For instance, the nitration of benzene by a mixture of concentrated nitric and sulfuric acids, whose stoichiometry is represented by

$$PhH \quad + \quad HNO_3 \quad \xrightarrow{(H_2SO_4)} \quad PhNO_2 \quad + \quad H_2O$$

occurs through the following stages:

$$HO-NO_2 \quad + \quad H_2SO_4 \quad \rightleftharpoons \quad H_2\overset{+}{O}-NO_2 \quad + \quad HSO_4^-$$

$$H_2\overset{+}{O}-NO_2 \quad \rightleftharpoons \quad H_2O \quad + \quad NO_2^+$$

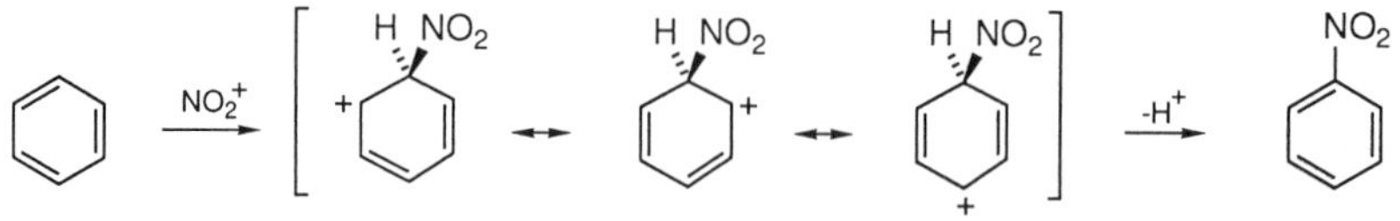

There is, indeed, a comparatively small number of basic processes and each reaction is a combination of these. Five such processes may be recognized: (a) bond-breaking, (b) bond-forming, (c) synchronous bond-breakage and bond-formation, (d) intramolecular migration, and (e) electron-transfer.

(a)   Bond-breaking

A covalent bond A—B may break in one of two ways: either *homolytically* to give A· + B·, or *heterolytically*, to give A⁺ + B⁻ or A⁻ + B⁺.

(i)   *Homolysis*, or homolytic fission, occurs on heating to moderate temperatures those compounds that either contain an intrinsically weak bond, such as O—O, or that, on dissociation, liberate a particularly strongly bonded molecule, such as N_2. Peroxides illustrate the former class, e.g.

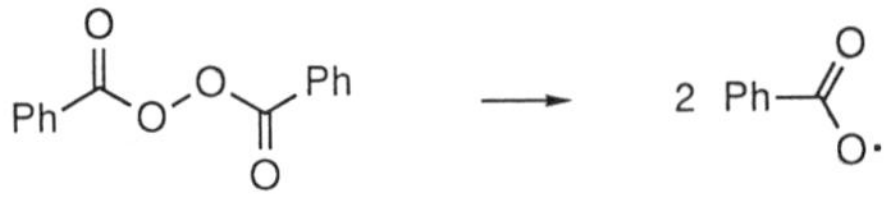

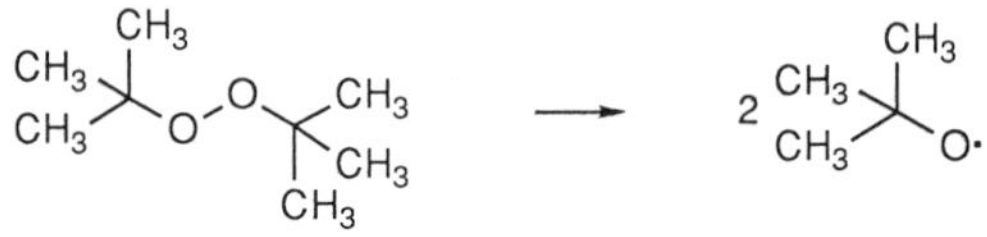

and certain azo compounds illustrate the latter, e.g.

The activation energy for homolysis is usually barely greater than the energy of the bond which breaks, for the activation energy for the reverse reaction is negligible. In turn, the dissociation energy for a bond is affected by the presence of substituents which can stabilize the radicals which are formed by delocalization of the unpaired electron in each. Consequently, the rate of homolysis of a particular type of bond is related to the delocalization energy in the resulting radicals. For example, the faster rate of dissociation of dibenzoyl compared with di-t-butyl peroxide can be related to the delocalization of the unpaired electron in the benzoyloxy radical

$$Ph-C\overset{\displaystyle O}{\underset{\displaystyle O\cdot}{}} \quad \longleftrightarrow \quad Ph-C\overset{\displaystyle O\cdot}{\underset{\displaystyle O}{}}$$

which has no counterpart in the t-butoxy radical.

Homolysis can also be induced by ultraviolet (or sometimes visible) light, providing that the molecule possesses an appropriate light-absorbing group (chromophore). The principles are discussed in detail in chapter 16.

(*ii*) *Heterolysis*, or heterolytic fission, occurs when the species formed are relatively stable. For example, whereas methyl chloride is relatively inert to water, t-butyl chloride is fairly rapidly hydrolyzed, reaction occurring through the t-butyl cation:

$$(CH_3)_3C-Cl \quad \rightleftharpoons \quad (CH_3)_3C^+ \quad + \quad Cl^-$$

$$(CH_3)_3C^+ \quad + \quad H_2O \quad \longrightarrow \quad (CH_3)_3C-\overset{+}{O}H_2 \quad \xrightarrow{-H^+} \quad (CH_3)_3C-OH$$

The difference in behaviour is rationalized as follows. The transition state for the heterolysis involves the partial breakage of the C—Cl bond, the bonding pair departing with chlorine, and may be represented as $(CH_3)_3C^{\delta+}\text{----}Cl^{\delta-}$. The partially positively polarized carbon species is stabilized, relatively to that involved in the heterolysis of methyl chloride $(CH_3^{\delta+}\text{----}Cl^{\delta-})$, by the three methyl substituents, just as the t-butyl cation is stabilized relative to the methyl cation. Consequently the activation energy for the heterolysis of t-butyl chloride is less than that for methyl chloride.

The stability of the departing anion is also important. The order of ease of heterolysis of some t-butyl compounds is

$$(CH_3)_3C-OH \quad < \quad (CH_3)_3C-OAc \quad < \quad (CH_3)_3C-Cl$$

just as the acid strengths lie in the order

$$H-OH \quad < \quad H-OAc \quad < \quad H-Cl$$

Chloride ion is described as a better *leaving group* than acetate, and acetate as better than hydroxide. The capacity of hydroxyl as a leaving group is increased by protonation (H_3O^+ is a far stronger acid than H_2O), and this has synthetic utility: for example, t-butanol is unaffected by chloride ion but is converted into t-butyl chloride by hydrogen chloride:*

---

* This is simply the reverse of the hydrolysis of t-butyl chloride. The direction which the reaction takes depends on the relative concentrations of water and chloride ion.

$$(CH_3)_3C-OH \xrightleftharpoons{HCl} (CH_3)_3C-\overset{+}{O}H_2 + Cl^-$$

$$(CH_3)_3C-\overset{+}{O}H_2 \rightleftharpoons (CH_3)_3C^+ + H_2O$$

$$(CH_3)_3C^+ + Cl^- \longrightarrow (CH_3)_3C-Cl$$

(b) Bond-forming

(i) Two radicals (or atoms) may combine, e.g.

$$2 \cdot CH_3 \longrightarrow CH_3-CH_3$$

$$\cdot CH_3 + Cl\cdot \longrightarrow CH_3-Cl$$

$$2 Cl\cdot \longrightarrow Cl_2$$

These reactions are the reverse of homolysis. They normally have very low activation energies and occur rapidly, although some radicals, particularly those which are sterically congested (section 17.1), are stable in certain conditions.

Many reactions involve free-radical intermediates: the radicals react with neutral molecules giving new radicals which perpetuate a chain reaction. Combinations of radicals such as those above lead to the termination of the chain and are disadvantageous. However, despite the large rate constants for combination, long chains can be propagated because the concentration of radicals is usually small so that their rate of combining (which is proportional to the product of the concentrations of two radicals) does not necessarily compete effectively with the chain-propagating steps in which a radical reacts with a molecule which is present in far higher concentration.

(ii) Oppositely charged ions may combine, e.g.

$$(CH_3)_3C^+ + Cl^- \longrightarrow (CH_3)_3C-Cl$$

This is the reverse of heterolysis. When the positive ion is a carbocation, reaction is almost always rapid and rather unselective with respect to different anions. It should be noted that positive ions which are electronically saturated do not combine with anions, e.g. quaternary ammonium ions such as $(CH_3)_4N^+$ form stable salts in solution, comparable with those formed by alkali-metal cations, since the quaternary nitrogen, unlike the electron-deficient carbon in a carbocation, does not possess a low-lying orbital which can accept two electrons.

(iii) An ion may add to a neutral molecule, e.g.

$$(CH_3)_3C^+ + H_2O \longrightarrow (CH_3)_3C-\overset{+}{O}H_2$$

$$Cl^- + AlCl_3 \longrightarrow AlCl_4^-$$

For addition to a *cation*, the neutral molecule must possess at least one unshared pair of electrons (i.e. it must be a base, the positive ion acting as a Lewis acid), whereas for addition to an *anion* it must possess the ability to accept an electron-pair (i.e. it must be a Lewis acid). It is convenient to define new descriptive terms for these and other species. Cations and electron-deficient molecules which have a tendency to form a bond by accepting an electron-pair from another species are termed *electrophilic* ('electron-seeking'), and anions and molecules with unshared pairs of electrons which have a tendency to form a bond by donating an electron-pair to another species are termed *nucleophilic* ('nucleus-seeking'). These terms are most commonly applied to species reacting at carbon, e.g. in the reaction

$$(CH_3)_3C^+ \quad + \quad OAc^- \quad \longrightarrow \quad (CH_3)_3C-OAc$$

the acetate ion is acting as a nucleophile.

(c) Synchronous bond breakage and bond formation

In the transition state of these reactions, one or more bonds are partially broken and one or more are partially formed. The energy required for bond breakage is partly supplied by the energy evolved in bond formation, so that the activation energy is less than that for the bond-breaking reactions alone.

There are two major categories of synchronous reactions: (*i*) a single bond is broken and a single bond is formed; (*ii*) a double bond is converted into a single bond (or a triple bond into a double bond) while a single bond is formed. Examples of the two are:

(i) CN^- + CH_3-I $\longrightarrow$ CH_3-CN + I^-

(ii) CN^- + (structure: CH_3, $>C=O$, H) $\longrightarrow$ (structure: CH_3, $NC-C-O^-$, H)

These two reactions clearly have the same basis: one bond is cleaved heterolytically while a second bond is formed by nucleophilic attack on carbon. Despite this similarity, it is convenient to discuss the two types of reaction separately.

(*i*) The examples below are classified with respect to the element which undergoes synchronous substitution.

Carbon is substituted by nucleophiles when it is attached to an electronegative group which can depart as a relatively stable anion. Whereas displacement on an alkane, e.g.

$$HO^- \quad + \quad CH_3-H \quad \longrightarrow \quad HO-CH_3 \quad + \quad H^-$$

does not take place because of the high energy of the hydride ion, the reactions of alkyl halides and alkyl sulfates and sulfonates, such as

$$H_3N \quad + \quad CH_3-I \quad \longrightarrow \quad H_3\overset{+}{N}-CH_3 \quad + \quad I^-$$

$$RO^- \quad + \quad CH_3-OSO_2OCH_3 \quad \longrightarrow \quad RO-CH_3 \quad + \quad CH_3OSO_2O^-$$

$$RO^- \quad + \quad CH_3-OSO_2Ar \quad \longrightarrow \quad RO-CH_3 \quad + \quad ArSO_2O^-$$

occur readily. The transition states are represented as, e.g.

$$\overset{\displaystyle H}{\underset{\displaystyle H\ \ H}{RO\overset{\delta-}{\cdots\cdots}C\overset{\delta-}{\cdots\cdots}OSO_2OCH_3}}$$

and a common symbolism for this reaction type, which indicates the electronic movements, is:

$$RO^- \quad CH_3\!-\!OSO_2OCH_3 \longrightarrow RO\!-\!CH_3 \ + \ CH_3OSO_2O^-$$

Note that groups which form less stable anions, such as —OH and —NH_2, are not displaced from carbon, e.g. it is not possible to form an ether from an amine,

$$RO^- \ + \ CH_3\!-\!NH_2 \ \xrightarrow{\ \ \times\ \ } \ RO\!-\!CH_3 \ + \ NH_2^-$$

Carbon is substituted by electrophiles when it is attached to a strongly electropositive group, i.e. a metal, as in the reaction of a Grignard reagent with an acid (p. 188):

$$H^+ \ + \ CH_3\!-\!MgBr \longrightarrow CH_4 \ + \ Mg^{2+} \ + \ Br^-$$

Carbon does not undergo synchronous substitutions with radicals or atoms.

Hydrogen undergoes attack by nucleophiles when it is attached to an electronegative group, e.g.

$$H_2O \ + \ H\!-\!Cl \longrightarrow H_3O^+ \ + \ Cl^-$$

$$HO^- \ + \ H\!-\!CH_2CHO \longrightarrow H_2O \ + \ ^-CH_2CHO$$

These reactions are reversible and are, of course, better described as acid–base equilibria; the principles which govern the position of the equilibrium have been described (p. 42). It should be noted that, whereas reaction is normally very fast when hydrogen is attached to oxygen, nitrogen, or a halogen, abstraction from carbon is often very slow; for example, the first-order rate constants for the ionization of acetic acid and nitromethane in aqueous solution at 25°C are, respectively, about 10^6 and 10^{-8} per second.

Hydrogen is attacked by electrophiles when it is bonded to a strongly electropositive element, as in certain metal hydrides such as lithium aluminium hydride, $LiAlH_4$, which contains the AlH_4^- ion. For example, the electrophilic carbon of the carbonyl group reacts to form a C—H bond (p. 130):

$$H_3\bar{Al}\!-\!H \quad \overset{R}{\underset{R'}{C}}\!=\!O \longrightarrow H\overset{R}{\underset{R'}{\searrow}}O^- \ + \ AlH_3 \longrightarrow H\overset{R}{\underset{R'}{\searrow}}\!-\!O\bar{Al}H_3$$

Hydrogen is attacked by radicals and atoms in a number of different structural environments, e.g.

$$\Delta H \atop (kJ\ mol^{-1})$$

$$Cl\cdot \ + \ H\!-\!CH_3 \longrightarrow HCl \ + \ \cdot CH_3 \qquad -2$$

$$Br\cdot \quad + \quad H{-}CH_3 \quad \longrightarrow \quad HBr \quad + \quad \cdot CH_3 \quad +62$$

$$CH_3CH_2\cdot \quad + \quad H{-}Br \quad \longrightarrow \quad CH_3CH_3 \quad + \quad Br\cdot \quad -37$$

The chief characteristic of these reactions is that they inevitably lead to the formation of a new radical which can itself abstract an atom from another molecule so that chain reactions are propagated. Those abstractions which are exothermic, such as that of the ethyl radical with hydrogen bromide, usually occur rapidly (a rough guide is that the activation energy is about 5% of the dissociation energy of the bond which is broken), whereas those which are endothermic are relatively slow (the activation energy is necessarily at least as great as the reaction enthalpy). Nevertheless, processes which are only moderately endothermic take part in chain reactions. For example, the reaction of the bromine atom with methane constitutes one step in the radical-catalyzed bromination of methane (p. 531). Conversely, reactions of the reverse type, such as that of the ethyl radical with hydrogen bromide, also occur in chain processes such as the radical-catalyzed addition of hydrogen bromide to alkenes (p. 534).

Oxygen and *nitrogen* undergo synchronous substitution only rarely; both elements, as, for example, hydroxide ion and ammonia, instead react primarily as nucleophiles, adding to unsaturated electrophilic centres and substituting at saturated electrophilic centres. Some useful examples are known, however, such as the formation of *N*-oxides, e.g.

and the formation of amines from Grignard reagents and *O*-methylhydroxylamine (p. 196)

$$BrMg{-}R \quad NH_2{-}OCH_3 \quad \longrightarrow \quad R{-}NH_2 \quad + \quad MgBr(OCH_3)$$

The *halogens* chlorine, bromine, and iodine undergo displacement by nucleophiles as in the halogenation of carbanions, e.g.

$$CH_3COCH_2^- \quad Br{-}Br \quad \longrightarrow \quad CH_3COCH_2Br \quad + \quad Br^-$$

The success of the reaction is related to the ability of a halogen atom to depart as an anion.

Halogen molecules also react with radicals, e.g.

$$\cdot CH_3 \quad + \quad Cl_2 \quad \longrightarrow \quad CH_3Cl \quad + \quad Cl\cdot$$

as in the radical-catalyzed halogenation of alkanes (p. 530).

The fluorine molecule reacts so violently with most organic compounds that it has little use as a synthetic reagent. The vigour of reaction is a result of the small bond dissociation energy of F—F together with the very large bond energies of H—F and C—F (p. 8).

(*ii*) Nucleophiles, electrophiles, radicals, and atoms add to unsaturated bonds, of which those most frequently encountered are:

C=C in alkenes and aromatic compounds

C≡C in alkynes

C=O in aldehydes, ketones, amides, acids, esters, anhydrides, and acid halides

C≡N in nitriles

Other unsaturated groupings include C=N (in imines), N=O (in nitroso and nitro compounds), N=N (in azo compounds), and C=S (in thio analogues of carbonyl-containing compounds).

There are marked differences in the ease with which addition occurs. Simple alkenes react readily with electrophiles and radicals, e.g.

$$CH_2=CH_2 \quad + \quad H^+ \quad \longrightarrow \quad CH_3-CH_2^+$$

$$CH_2=CH_2 \quad + \quad Br\cdot \quad \longrightarrow \quad CH_2Br-CH_2^\cdot$$

but do not react with nucleophiles. On the other hand, the carbonyl group reacts with nucleophiles, e.g.

$$R_2C=O \quad + \quad CN^- \quad \longrightarrow \quad NC\overset{R}{\underset{R}{-}}\hspace{-0.5em}{-}O^-$$

this difference from alkenes resulting from the fact that oxygen is better able than carbon to accommodate the negative charge in the adduct. Likewise, the cyano group reacts with nucleophiles to give an adduct in which the charge is accommodated by nitrogen, e.g.

$$R-C≡N \quad + \quad OH^- \quad \longrightarrow \quad \overset{R}{\underset{HO}{>}}{=}N^-$$

Alkynes react with electrophiles less readily, and with nucleophiles more readily, than alkenes. For example, acetylene itself, unlike ethylene, reacts with alkoxide ions in alcoholic solution at high temperatures and pressures:

$$RO^- \quad + \quad H{-}\!\!\equiv\!\!{-}H \quad \longrightarrow \quad \overset{RO}{\underset{H}{>}}{=}\bar{C}H$$

The difference is ascribed to the fact that unsaturated carbon is more electronegative than saturated carbon (cf. the greater acidity of acetylene than ethylene).

An aromatic double bond in benzene reacts less readily than alkenes with all reagents as a result of the loss of aromatic stabilization energy which accompanies addition, e.g.

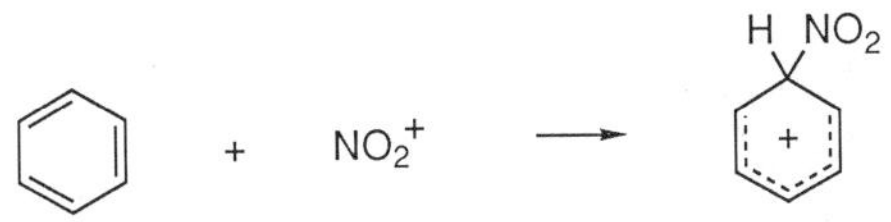

However, the presence of substituents which can stabilize the resulting adduct increases the rate of addition.

(*d*) *Intramolecular migration*

In certain structural situations a group migrates from one atom to another in the same species without becoming 'free' of that species. Three classes of migration may be distinguished.

(1)   The group migrates together with the bonding-pair by which it was attached in the original species. For example, the neopentyl cation rearranges to a tertiary cation by the migration of a methyl group, represented as follows:

The thermodynamic driving force for the reaction lies in the greater stability of a tertiary than a primary carbocation (p. 62).

(2)   The group migrates with one electron of the original bonding-pair, e.g.

The driving force is the greater stability of the tertiary radical, in which the unpaired electron is delocalized over two aromatic rings, than the primary radical.

(3)   There are anionic counterparts of (1), e.g.

the driving force of which lies in the greater capacity of oxygen than of carbon to be anionic. It was at one time thought that in this and related rearrangements, the group migrates with neither of the original bonding electrons, as represented by

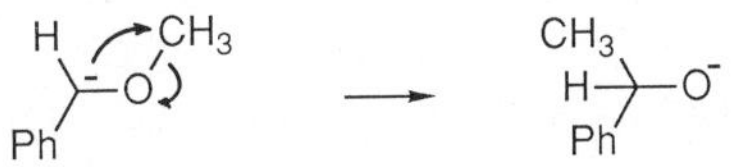

but it is now believed that radical-pairs are involved:

These, and related, migrations occur as unit steps in intramolecular re-arrangements (chapter 14).

(e) Electron-transfer

Species which have a strong tendency to give up one electron (low ionization potential) can react with species with a strong tendency for electron-acceptance (high electron affinity) by electron-transfer. The well known mode of formation of inorganic salts from electropositive metals and electronegative elements or groups has as its organic analogues such reactions as

$$Na\cdot \; + \; Ph_2C{=}O \; \longrightarrow \; Na^+ \; + \; Ph_2\dot{C}{-}O^-$$

$$Fe^{2+} \; + \; RO{-}OH \; \longrightarrow \; Fe^{3+} \; + \; RO\cdot \; + \; OH^-$$

Organic compounds may be electron-acceptors, as in the above examples, or donors, as in

$$Fe^{3+} \; + \; PhO{-}H \; \longrightarrow \; Fe^{2+} \; + \; PhO\cdot \; + \; H^+$$

Donation or acceptance by a neutral organic compound gives rise to a radical which normally undergoes further reaction. For example, the acceptance of an electron by acetone from magnesium gives $(CH_3)_2\dot{C}{-}O^-$ which dimerizes to

Treatment with acid leads to pinacol, $(CH_3)_2C(OH){-}C(CH_3)_2(OH)$ (pinacol reduction, p. 656).

It should be noted that the uptake of electrons by an organic compound constitutes reduction (chapter 20) and the loss of electrons constitutes oxidation (chapter 19).

4.2
Types of
reaction

Combinations of the unit processes described in the previous section give reactions of various types. For example, when ethyl acetate is heated under reflux in ethanol containing sodium ethoxide, the sodium derivative of acetoacetic ester is formed:

$$2\,CH_3CO_2Et \; + \; Na^+OEt^- \; \longrightarrow \; [CH_3COCHCO_2Et]^-Na^+ \; + \; 2\,EtOH$$

The unit processes which combine in the overall reaction are: an acid-base equilibrium (1), bond-formation (2), heterolytic bond-breakage (3), and a second acid-base equilibrium which lies well to the right (4)

$$CH_3CO_2Et \ + \ EtO^- \ \overset{(1)}{\rightleftharpoons} \ \overset{-}{C}H_2CO_2Et \ + \ EtOH$$

$$EtO_2CCH_2^- \ \underset{EtO}{\overset{CH_3}{\diagup}}{=}O \ \overset{(2)}{\rightleftharpoons} \ EtO_2CCH_2{-}\underset{EtO}{\overset{CH_3}{\diagup}}{-}O^- \ \overset{(3)}{\rightleftharpoons} \ EtO_2C{\diagdown}\overset{CH_3}{\diagup} \ + \ EtO^-$$

$$EtO_2C{\diagdown}\overset{CH_3}{\diagup}_{O} \ + \ EtO^- \ \overset{(4)}{\rightleftharpoons} \ EtO_2C{\diagdown}\overset{CH_3}{\diagup}_{O}{}^- \ + \ EtOH$$

It is convenient to employ descriptive terms to overall processes such as this, and the reactions described hereafter are classified as addition, elimination, substitution, condensation, rearrangement, pericyclic reactions, and oxidation–reduction.

Subdivision of these reactions is helpful, and in many cases the reaction can be classified with respect to the nature of the reagent which reacts with a given class of organic compound. For example, additions to unsaturated bonds are divided into those in which the reagent is, respectively, an electrophile, a nucleophile, and a radical or atom. The following are the more important reagents.

(i) Electrophiles

* Hydrogen, in the form of a proton-donor such as H_3O^+ or a carboxylic acid
* Halogens: chlorine, bromine, and iodine
* Nitrogen, as the nitronium ion (NO_2^+), the nitrosonium ion (NO^+), 'carriers' of these ions from which the ions are readily abstracted (e.g. alkyl nitrites, RO—NO), and aromatic diazonium ions (ArN_2^+)
* Sulfur, as sulfur trioxide, usually derived from sulfuric acid
* Oxygen, as hydrogen peroxide, peroxyacids, and ozone
* Carbon, as a carbocation (e.g. $(CH_3)_3C^+$)

(ii) Nucleophiles

* Oxygen, in H_2O, ROH, RCO_2H, and their conjugate bases (OH^-, RO^-, etc.)
* Sulfur, in H_2S and RSH and their conjugate bases, and hydrogen sulfite ion (HSO_3^-)
* Nitrogen, in NH_3 and NH_2^-, amines, and amine derivatives (e.g. NH_2OH)
* Carbon, in cyanide ion (CN^-), acetylide ions ($HC{\equiv}C^-$ and $RC{\equiv}C^-$), carbanions (e.g. $CH_3COCH_2^-$), and organometallic compounds (e.g. $CH_3{-}MgI$)
* Halide ions (F^-, Cl^-, Br^-, I^-)
* Hydrogen, as hydride ion and 'carriers' of this ion (e.g. $LiAlH_4$).

(iii) *Radicals and atoms*

- Halogen atoms ($\cdot$Cl, $\cdot$Br)
- Alkyl and aryl radicals (e.g. $\cdot$CH$_3$, Ph$\cdot$)

4.3
Addition

(a) *Electrophilic addition*

(i) Alkenes. Electrophilic reagents add to alkenes in a two-step process: an organic cation is formed in the first step and reacts with a nucleophile in the second.

The organic cations are not simple carbocations but have *bridged* or otherwise complexed structures. The best understood of the bridged ions is the bromonium ion, which is usually represented as a hybrid, e.g.

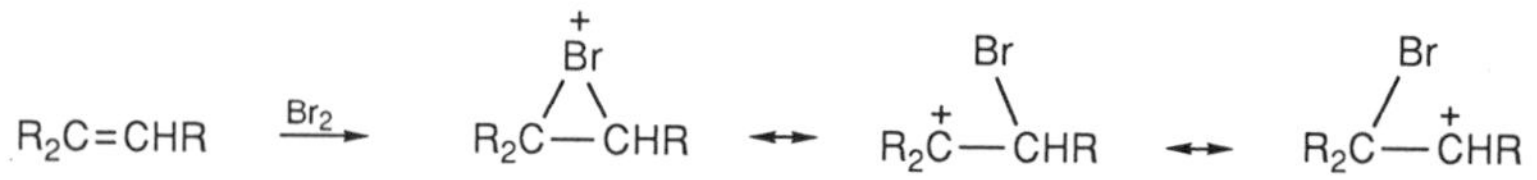

in which the second structure, a tertiary carbocation, is a more important contributor to the hybrid than the third, a secondary carbocation. The chloronium ion is believed to be less strongly bridged; that is, the cyclic structure is of relatively less importance in the hybrid. Adducts with sulfur and selenium electrophiles, e.g.

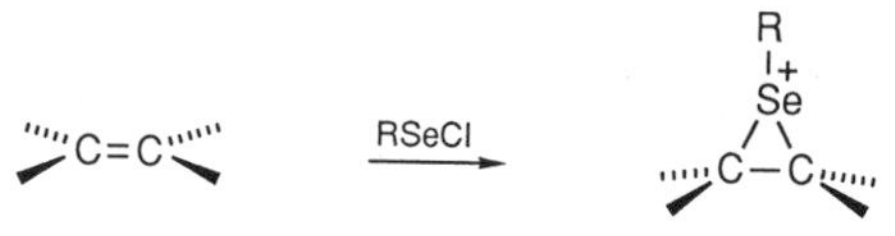

are believed to be more strongly bridged.

Hydrogen halides and other acids do not form bridged ions with alkenes. With one exception, the characteristics of these reactions are consistent with the mediation of carbocations; the exception – their stereoselectivity (p. 86) – can be accounted for by the contribution of a structure of the type

The second step determines both the regioselectivity and the stereoselectivity of the addition. A strongly bridged ion behaves like an epoxide in an S_N2 reaction (p. 590): the nucleophile preferentially attacks the less substituted carbon atom from the opposite side of the bridge, e.g.

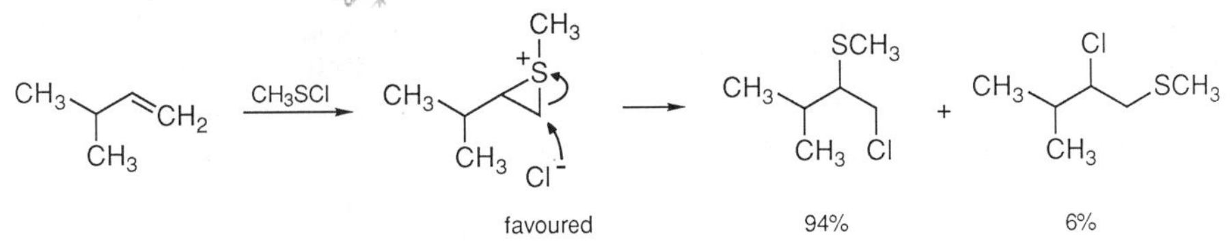

CH3SCl
favoured
94%
6%

Even when approach from the opposite side is hindered, as in addition to norbornene, *anti* addition still occurs:

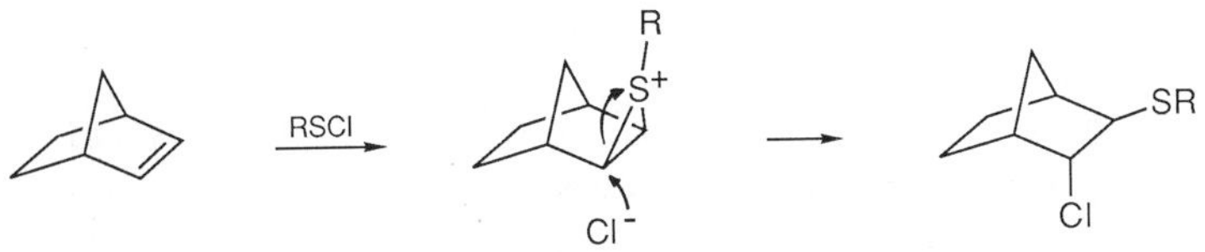

RSCl

In many cases, bridging is strong enough to ensure that the *anti* product is at least the major, if not the exclusive, product, but it is not strong enough to direct attack to the less substituted carbon. Insofar as its regioselectivity is concerned, the intermediate ion behaves as the more stable of the two possible cations (and will, for simplicity, be displayed as such). For example, the addition of bromine to propylene in water, where water competes with bromide ion as nucleophile, gives, as well as the dibromo compound, more of the 2-hydroxy product than its 1-isomer, consistent with the formation of the more stable secondary carbocation:

$$CH_3CH{=}CH_2 \xrightarrow[-Br^-]{Br_2} CH_3\overset{+}{C}H{-}CH_2Br \xrightarrow[-H^+]{H_2O} \underset{\text{major}}{CH_3CH(OH){-}CH_2Br} + \underset{\text{minor}}{CH_3CH(Br){-}CH_2OH}$$

Likewise, the hydrogen in HX (X = Cl, Br, or I) adds mainly to the less substituted carbon atom of an alkene (*Markovnikov's rule*) e.g.

$$(CH_3)_2C{=}CHCH_3 \xrightarrow[-Cl^-]{HCl} (CH_3)_2\overset{+}{C}{-}CH_2CH_3 \xrightarrow{Cl^-} \underset{\text{major product}}{(CH_3)_2CCl{-}CH_2CH_3}$$

Other aspects of electrophilic addition, and further details of regio- and stereoselectivity, are as follows.

(1) Since the formation of the transition state in the addition of a neutral reagent to an alkene involves the separation of unlike charges, e.g.

$$CH_2{=}CH_2 \xrightarrow{HCl} \overset{\delta+}{CH_2}{\cdots}CH_2{\cdots}\overset{\delta+}{H}{\cdots}\overset{\delta-}{Cl} \longrightarrow \overset{+}{C}H_2{-}CH_3 + Cl^-$$

reaction occurs faster in a polar, solvating solvent than in a non-polar one.

(2) The organic reactant acquires positive charge in passage to the transition state, so that electron-releasing substituents increase the rate of addition and

electron-attracting substituents decrease it. Quantitatively, however, there is a marked dependence on the nature of the intermediate cation: for example, as the number of methyl substituents attached to C=C is increased, the rate of addition of bromine in methanol increases regardless of which carbon atom of the double bond is methylated:

$$CH_2=CH_2 \quad CH_3CH=CH_2 \quad \begin{array}{c}(CH_3)_2C=CH_2\\CH_3CH=CHCH_3\end{array} \quad (CH_3)_2C=C(CH_3)_2$$

$k_{relative}$ c.   1 10^2 10^4 10^6

In contrast, the rate of addition of a proton depends on the number of methyl groups at only one of the two carbon atoms:

$$CH_2=CH_2 \quad CH_3CH=CH_2 \quad (CH_3)_2C=CH_2$$
$$CH_3CH=CHCH_3 \quad (CH_3)_2C=C(CH_3)_2$$

$k_{relative}$ c.   1 10^5 10^9

The explanation is that, in the formation of a bridged bromonium ion, it is the total electron-releasing capability that is important, whereas protonation yields a species closer in structure to a carbocation. The stability of the developing carbocation is increased much more by an adjacent methyl group than by a non-adjacent one: $CH_3\overset{+}{C}H-CH_3$ and $CH_3\overset{+}{C}H-CH_2CH_3$ are of similar stability, as are $(CH_3)_2\overset{+}{C}-CH_3$ and $(CH_3)_2\overset{+}{C}-CH_2CH_3$.

(3) Substituents in the alkene of electron-attracting type ($-I$ and/or $-M$) retard the rate of electrophilic addition since they destabilize carbocations and, since they act more strongly when they are bound directly to positively charged carbon than when they are further away from it, they also reverse the regioselectivity, e.g.

$$F_3C-CH=CH_2 + HCl \longrightarrow F_3C-CH_2-\overset{+}{C}H_2 + Cl^- \longrightarrow F_3C-CH_2-CH_2Cl$$

$$O_2N-CH=CH_2 + HCl \longrightarrow O_2N-CH_2-\overset{+}{C}H_2 + Cl^- \longrightarrow O_2N-CH_2-CH_2Cl$$

Halogen substituents appear at first sight to behave anomalously: they retard addition, acting in this sense as electron-attracting groups, but orient in the same way as the electron-releasing alkyl substituents. The reason is that the halogens have opposed inductive ($-I$) and mesomeric ($+M$) effects. Consider the addition of a proton to ethylene, giving $CH_3-\overset{+}{C}H_2$, and to vinyl chloride, giving $CH_3-\overset{+}{C}HCl$ or $\overset{+}{C}H_2-CH_2Cl$. The third of these ions, $\overset{+}{C}H_2-CH_2Cl$, is relatively less stable than the ethyl cation because of the $-I$ effect of chlorine. The second, $CH_3-\overset{+}{C}HCl$, is more strongly destabilized by chlorine's $-I$ effect, since the halogen is nearer the positively charged centre, but it is stabilized by the mesomeric effect,

$$CH_3-\overset{+}{C}H-Cl \longleftrightarrow CH_3-CH=\overset{+}{C}l$$

and this factor serves to make $CH_3-\overset{+}{C}HCl$ more stable than $\overset{+}{C}H_2-CH_2Cl$ although it is relatively less stable than $CH_3-\overset{+}{C}H_2$ (i.e. the $-I$ effect of chlorine outweighs its $+M$ effect). Consequently, the ease of formation of the three ions decreases in the order $CH_3-\overset{+}{C}H_2 > CH_3-\overset{+}{C}HCl > \overset{+}{C}H_2-CH_2Cl$, so that vinyl chloride reacts less rapidly than ethylene but orients in the same manner as propylene. In the other common substituents of $-I$, $+M$ type, alkoxyl and amino, the $+M$ effect is of greater consequence than the $-I$ effect so that not only is the direction of addition in accordance with Markovnikov's rule but also the rate of addition is enhanced relative to the unsubstituted compound. For example, methyl vinyl ether reacts rapidly with hydrogen chloride:

$$CH_3O-CH=CH_2 \ + \ HCl \ \longrightarrow \ CH_3O-CHCl-CH_3$$

(4) The addition of bromine and chlorine to most alkenes is stereospecific: only *anti adduct* is formed, as shown for the reactions of bromine with *cis-* and *trans-*2-butene:

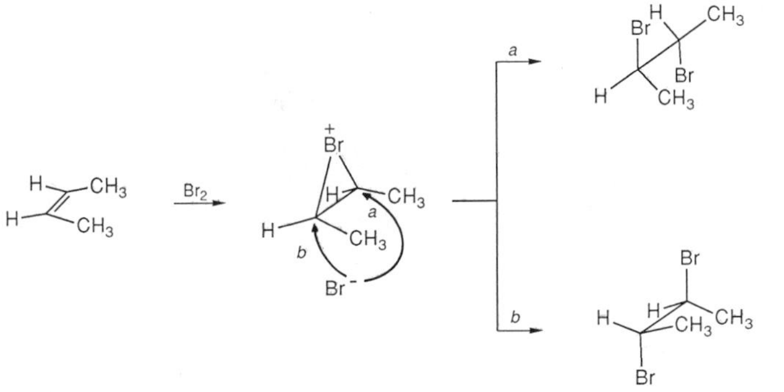

(±)-2,3-dibromobutane

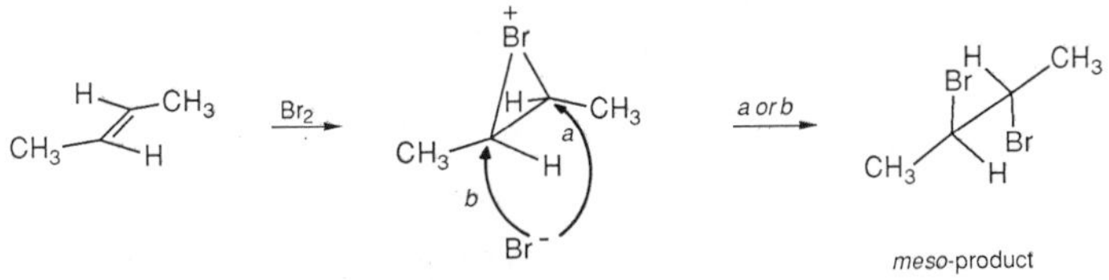

meso-product

However, when the alkene contains at least one aryl group attached to the double bond, the selectivity is less: for example, *cis*-PhCH=CHCH$_3$ gives *anti* and *syn* adducts in the ratio 85:15 with bromine, and rather more of the *syn* adduct with chlorine. This is considered to be because stabilization of the positive charge in the intermediate by its delocalization on to the aromatic ring reduces the effectiveness of bridging and allows rotation to occur, e.g.

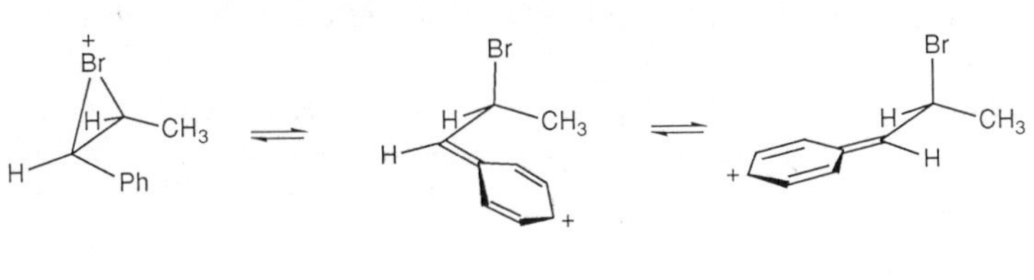

The effect is more pronounced with chlorine, which is a weaker bridging group. The addition of hydrogen halides to alkenes yields mostly *anti* product, consistent with back-side attack on the complex,

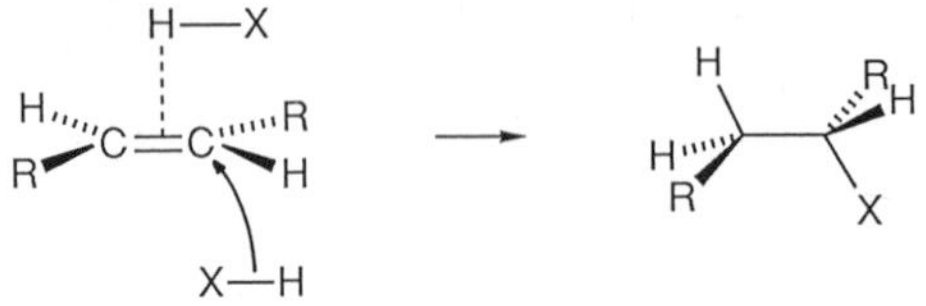

but when an aryl group is present, the *syn* adduct predominates. As with bridging, stabilization of the intermediate by the aryl group reduces the effectiveness of complex formation, and reaction to give *syn* adduct may be dictated by collapse of an ion-pair:

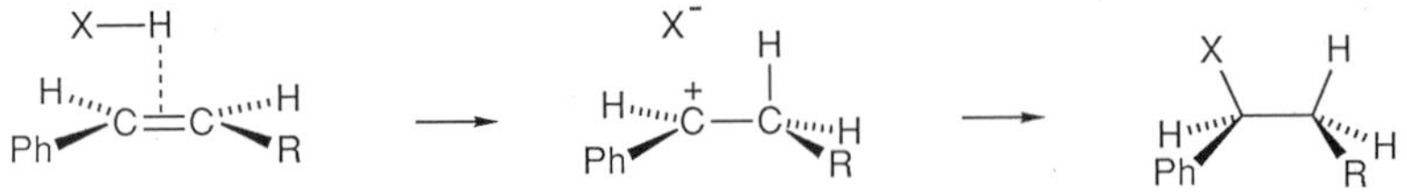

(5)   The intermediate cation can lose a proton in preference to reacting with a nucleophile. This is favoured both when addition is sterically hindered, as in the reaction

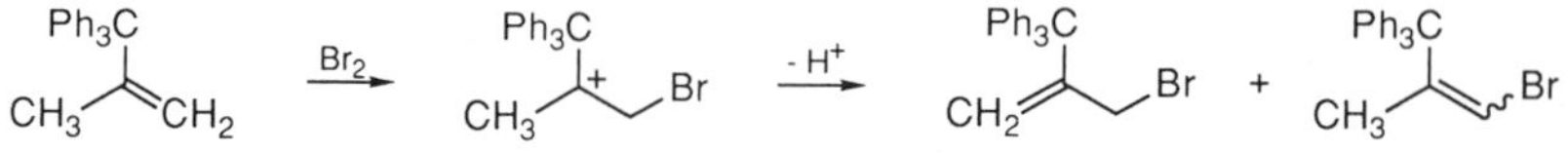

and in addition to enols, e.g.

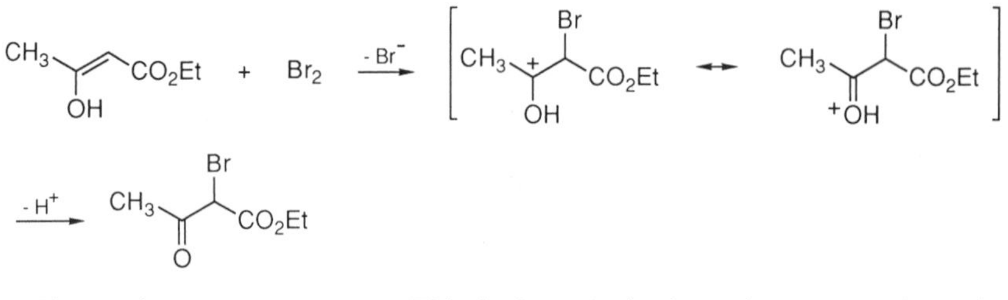

(6)   The cation can rearrange. This is favoured when the new carbocation is significantly the more stable, e.g.

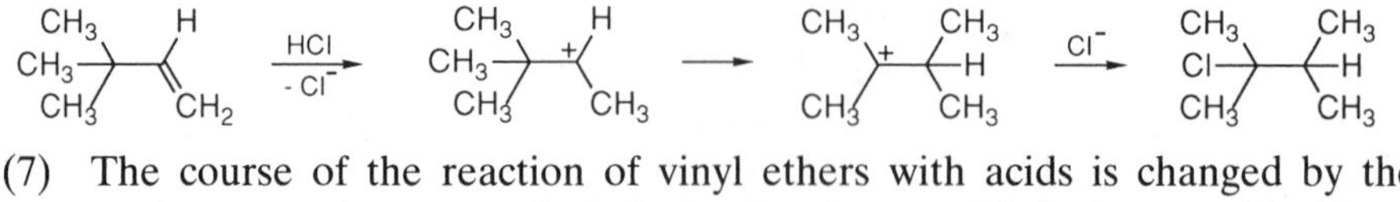

(7)   The course of the reaction of vinyl ethers with acids is changed by the presence of water: whereas methyl vinyl ether reacts with hydrogen chloride to give the adduct, CH_3—$CHCl$—OCH_3 (p. 85), reaction with hydrochloric acid gives acetaldehyde and methanol. Here, the intermediate carbocation reacts with water to give a hemiacetal which is further hydrolyzed (cf. the hydrolysis of acetals, p. 89).

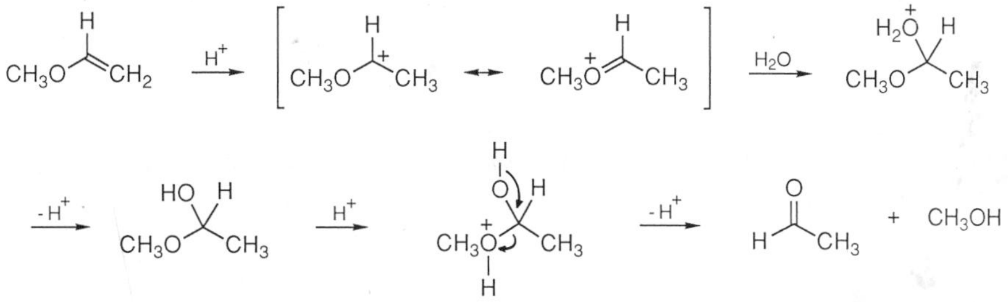

(8)   Addition to a system containing two or more *conjugated* double bonds gives mixtures of products because the charge in the intermediate carbocation is delocalized over two or more carbon atoms each of which can be attacked in the second step, e.g.

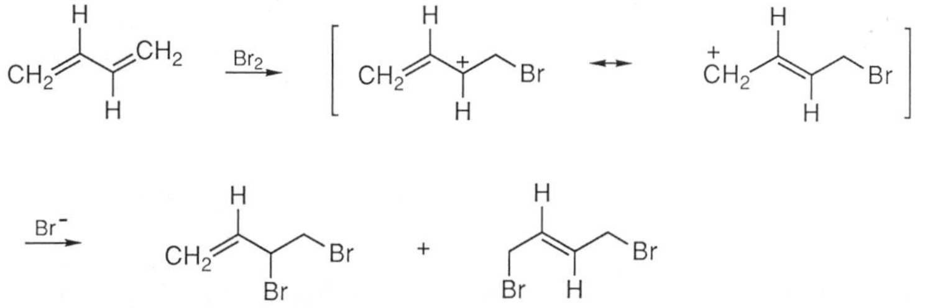

(*ii*)   *Alkynes* are somewhat less reactive than alkenes towards electrophiles (p. 78). Markovnikov's rule is followed, and the other general characteristics of the reactions are similar, although the stereoselectivity is lower. In general, reaction with bromine or chlorine gives mainly *anti* adduct, consistent with a bridged-ion intermediate.

(*iii*)   *Aromatic* carbon–carbon double bonds normally react with electrophiles by substitution rather than addition because addition would result in the loss of the aromatic stabilization energy. If, however, the product of addition is itself strongly stabilized, addition can compete with substitution. For example, anthracene and bromine give the 9,10-dibromo-adduct which possesses the aromatic stabilization energy of two benzene rings:[*]

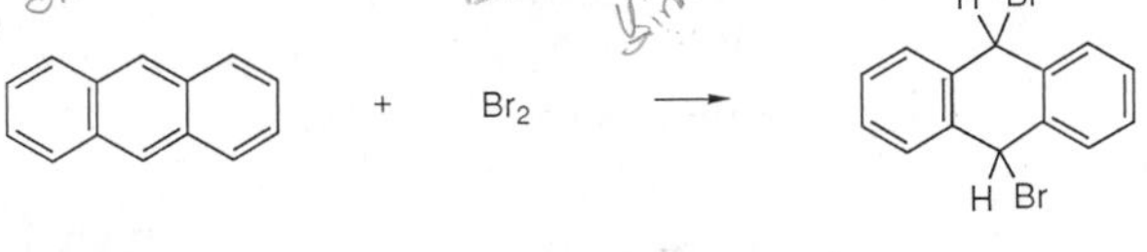

(*iv*)   *Carbonyl groups* do not react with electrophiles such as bromine, and their reactions with proton acids are so readily reversible that the products cannot be isolated, e.g.

[*] The loss of aromatic stabilization energy on 1,4-addition to benzene is $150\,\text{kJ mol}^{-1}$, whereas on 9,10-addition to anthracene it is only $(349 - (2 \times 150)) = 49\,\text{kJ mol}^{-1}$.

$$\underset{R}{\overset{R}{\diagdown}}C=O \quad + \quad HCl \quad \rightleftharpoons \quad \underset{R}{\overset{R}{\diagdown}}\underset{Cl}{\overset{OH}{\diagup}}$$

However, the addition of an acid to the oxygen atom of a carbonyl group promotes the ease of addition of a nucleophile to the carbon atom. These reactions are more conveniently classified as nucleophilic additions (see below).

(b)   Nucleophilic addition

(i)   *Carbonyl groups.*   The main features of the addition of nucleophiles to the carbonyl groups in aldehydes and ketones are as follows.

(1)   The rate-determining step is the addition of the nucleophile, e.g.

$$\underset{R}{\overset{R}{\diagdown}}C=O \quad + \quad CN^- \quad \longrightarrow \quad \underset{R}{\overset{R}{\diagdown}}\underset{CN}{\overset{O^-}{\diagup}}$$

$$\underset{O^-}{\overset{HO}{\diagdown}}S=O \;\; \underset{R}{\overset{R}{\diagdown}}C=O \quad \longrightarrow \quad \underset{O}{\overset{HO}{\diagdown}}\underset{R}{\overset{R}{S}}—O^-$$

Since the carbonyl group acquires a partial negative charge in the transition state of this step, electron-withdrawing groups enhance the rate and electron-releasing groups retard it.

(2)   Addition is completed by the uptake of a proton from the solvent. The complete reaction scheme for the addition of hydrogen cyanide to a ketone in aqueous solution is therefore:

$$\underset{R}{\overset{R}{\diagdown}}C=O \; + \; CN^- \longrightarrow \underset{R}{\overset{R}{\diagdown}}\underset{CN}{\overset{O^-}{\diagup}} \xrightarrow{H_2O} \underset{R}{\overset{R}{\diagdown}}\underset{CN}{\overset{OH}{\diagup}} \; + \; OH^-$$

It is now apparent why this reaction is base-catalyzed: hydrogen cyanide is a rather weak acid and the function of the base is to increase the concentration of the active entity, cyanide ion, the base being regenerated in the second step.

(3)   Bulky groups in the vicinity of the carbonyl group retard the addition, e.g. acetone reacts much faster than diethyl ketone with sodium hydrogen sulfite and gives a much higher yield·of the adduct (56% and 2%, respectively). The carbonyl group in 2,4,6-trimethylbenzaldehyde is so strongly hindered that the compound fails to react with the hydrogen sulfite ion.

(4)   Aromatic aldehydes and ketones react less rapidly than their aliphatic analogues. This is a result of the fact that, in passage to the transition state, the stabilization due to the conjugation between the carbonyl double bond and the aromatic ring is destroyed. Electron-attracting substituents on the aromatic ring facilitate addition and electron-releasing groups retard it.

(5)   Aliphatic αβ-unsaturated carbonyl compounds are similarly less reactive

than their saturated analogues. Addition usually occurs instead at the C=C bond (p. 91).

(6)   Whereas reactive nucleophiles such as cyanide ion add readily to carbonyl, less reactive nucleophiles such as water often react slowly unless the activity of the carbonyl group is increased by hydrogen-bonding between the oxygen atom and an acid HA, e.g.

Hydrate formation is readily reversible and hydrates can only be isolated from compounds which contain strongly electron-withdrawing substituents at the carbonyl group, such as —CCl_3 or a second carbonyl group.* However, aldehydes react with alcohols in the presence of acid to give isolable acetals, e.g.

a hemiacetal

an acetal

The intermediate hemiacetal is not isolable[†] but undergoes acid-catalyzed elimination of water to give a new cation; here, water acts as a good leaving-group and cation formation is aided by the electron-releasing ability of the alkoxy-substituent (cf. the ready S_N1 solvolysis of α-halo-ethers; p. 109). It should be noted that the reactions involved in the formation of an acetal are all readily reversible and the position of the resulting equilibrium depends on the relative concentrations of the alcohol and water: whereas acetals are formed from aldehydes by treatment

---

* This is because an electron-attracting group reduces the stability of the dipolar structure in the hybrid RR′C=O ↔ RR′$\overset{+}{C}$—$\overset{-}{O}$ and thereby lowers the stabilization energy of the carbonyl group; consequently ΔH for hydration becomes more favourable. Conversely, an electron-releasing group strengthens the C=O bond, so that the equilibrium constant for hydration falls in the order formaldehyde > other aldehydes > ketones (cf. bond energy values, p. 8).

[†] See, however, the discussion of 4- and 5-hydroxy-aldehydes, which exist predominantly as hemiacetals in certain cases, p. 16.

with an alcohol in the presence of an *anhydrous* acid, they are readily hydrolyzed to the aldehyde by treatment with *aqueous* acid.

Ketones do not react with monohydric alcohols to give ketals but do so with 1,2-diols to give cyclic ketals, e.g.

This difference represents a further example of the importance of the entropy factor in determining the positions of equilibria: whereas the formation of a ketal from a ketone with two molecules of a monohydric alcohol involves the loss of three degrees of translational freedom, reaction of a ketone with a dihydric alcohol involves no change in the number of degrees of translational freedom.

Both aldehydes and ketones give the thio-analogues of acetals and ketals when treated with thiols in the presence of acid:

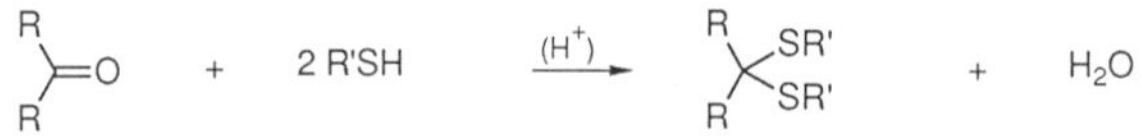

(*ii*) *Other carbonyl groups.* The carbonyl group in derivatives of acids (acid halides, anhydrides, esters, and amides) is also attacked by nucleophiles but reaction is completed by the departure of an electronegative group and not by the addition of a proton, e.g.

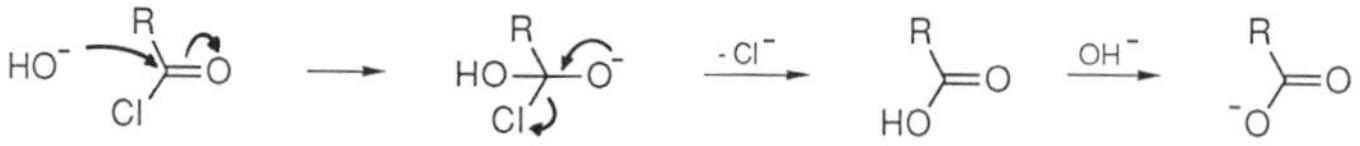

These reactions are therefore classified as substitutions (cf. section 4.5).

(*iii*) *Nitriles.* The carbon–nitrogen triple bond in nitriles is attacked by powerful nucleophiles such as hydroxide ion in water:

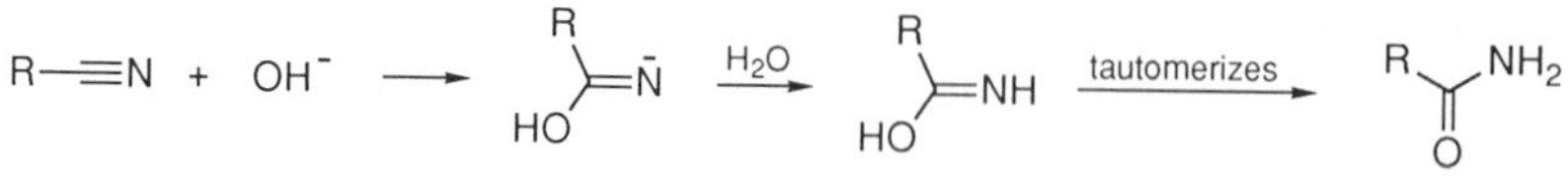

and by weaker nucleophiles such as water and alcohols in the presence of acid

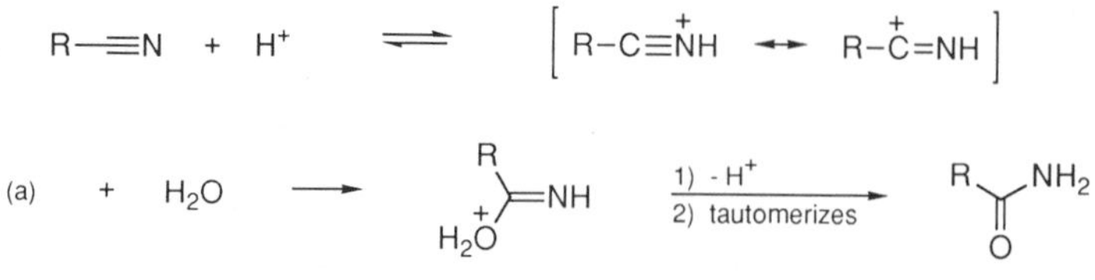

(b)   + R'OH ⟶ [structure: R with =NH, R'—O⁺ with H] ⟶ [structure: R with =NH₂⁺, R'O]

the conjugate
acid of an
imino-ester

(*iv*)   *Alkenes* are not attacked by nucleophiles unless the carbon–carbon double bond is conjugated to a group of −*M* type (p. 32). The more powerful nucleophiles such as carbanions then add to the double bond as follows (Michael addition, section 7.5)

[reaction scheme: R⌒⌒R' with C=O, →CH(CO₂Et)₂→ [bracketed resonance structures with EtO₂C, CO₂Et, O⁻]]

[scheme: →H⁺→ product with R, R', EtO₂C, CO₂Et, O]

[scheme: Ph⌒⌒Ph with O, →KCN / AcOH / EtOH→ Ph⌒⌒Ph with CN, O]

The increased ease of addition compared with that of an unconjugated alkene arises from the stability given to the intermediate anion, and therefore to the preceding transition state, by the delocalization of the charge onto the electronegative oxygen. Reaction is completed by the uptake of a proton from the solvent, which is usually an alcohol.

Less powerful nucleophiles such as alcohols add to these systems only in the presence of an acid, e.g.

[scheme: R⌒⌒O with R', + R"OH →H⁺→ product with R, R"O, H, R', O]

(*v*)   *Alkynes* react with powerful nucleophiles such as alkoxide ion in an alcoholic solvent:

[scheme: R≡H →R'O⁻→ R with =CH⁻, R'O →R'OH→ R with =CH₂, R'O + R'O⁻]

Less powerful nucleophiles require catalysis, and mercury(II) ion is frequently used because of its tendency to complex with, and draw electrons from, the triple bond. For example, water reacts in the presence of mercury(II) sulfate and dilute sulfuric acid, giving an enol which tautomerizes to a ketone (in the case of acetylene itself, to acetaldehyde):

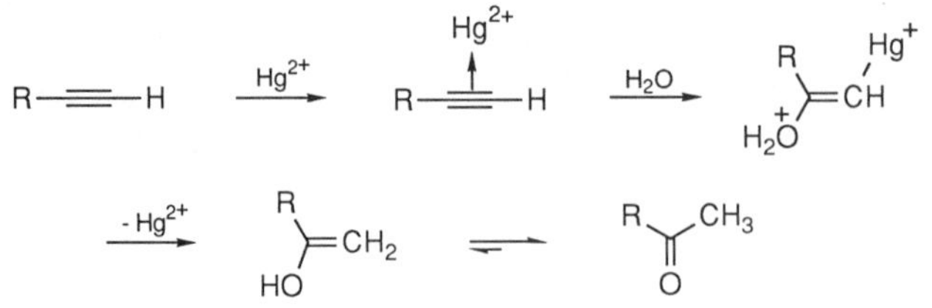

(c) *Radical addition*

Alkenes react readily with free radicals. The synthetically useful reactions are *chain* reactions: they are usually initiated by a peroxide or azobisisobutyronitrile (AIBN) which, on heating or irradiation, yield reactive radicals (R·), e.g.

$$\text{Initiator} \longrightarrow R\cdot$$

$$R\cdot + HBr \longrightarrow RH + Br\cdot$$

propagation
$$Br\cdot + R'CH{=}CH_2 \longrightarrow R'\dot{C}H{-}CH_2Br$$
$$R'\dot{C}H{-}CH_2Br + HBr \longrightarrow R'CH_2{-}CH_2Br + Br\cdot$$

Chains are terminated when two radicals meet, but since the concentration of radicals is usually very small, the probability of two of them colliding is much smaller than that of a radical colliding with a reactant molecule. Providing that the rate constants for the propagation steps are large enough, long chains are propagated.

(*i*) *Regioselectivity*. Radicals add more rapidly to the less substituted carbon atom of an unsymmetrical alkene, regardless of the nature of the substituents. This is partly for steric reasons and partly because all substituents have some stabilizing influence on an adjacent radical centre; examples are delocalization, e.g.

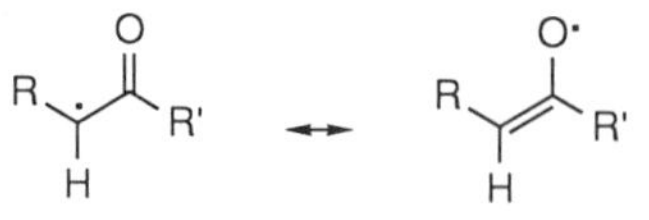

and hyperconjugation, e.g.

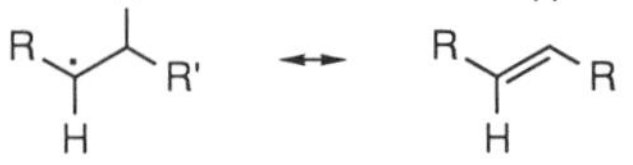

(*ii*) *Stereoselectivity*. At room temperature the reaction shows some stereoselectivity; for example, (*E*)- and (*Z*)-2-bromo-2-butene each give 75% of

(±)-2,3-dibromobutane and 25% of the *meso*-isomer. It is believed that this results from rapid rotation about the new C—C bond in the intermediate radical to give a predominance of the less congested radical, before reaction with hydrogen bromide takes place on the face opposite the new C—Br bond:

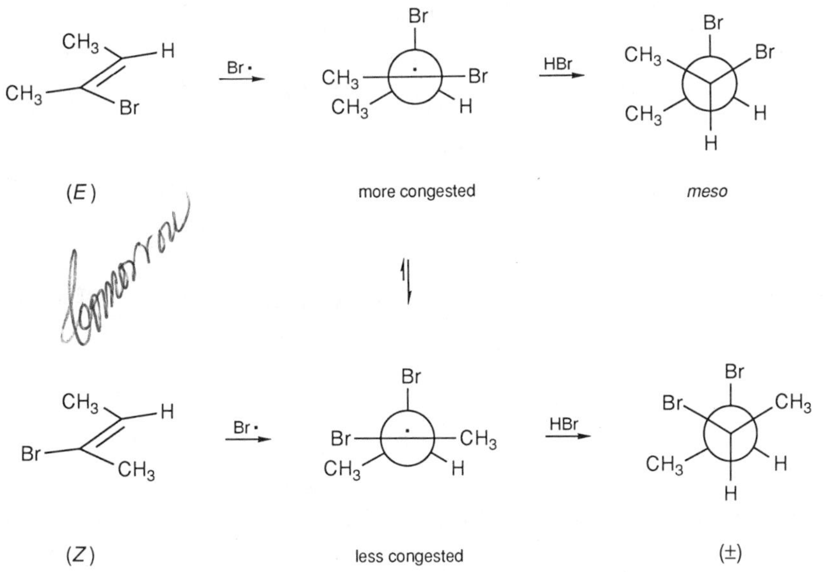

However, at very low temperatures (e.g. −78 °C) the reaction is stereospecific: the (*E*)-alkene gives only *meso*-product and the (*Z*)-alkene gives only (±)-product. It is thought that rotation in the intermediate radical is now relatively slow, possibly because there is a degree of bridging in the radical by bromine (cf. bromonium ions, p. 82).

4.4 Elimination

Elimination reactions are classified under two general headings: (*a*) β-eliminations in which groups on adjacent atoms are eliminated with the formation of an unsaturated bond,

e.g. $$CH_3CH(Br)CH_3 \xrightarrow[-H^+,\ -Br^-]{NaOEt} CH_3CH{=}CH_2$$

and (*b*) α-eliminations, in which two groups are eliminated from the same atom,

e.g. $$CHCl_3 \xrightarrow[-H^+]{OH^-} CCl_3^- \xrightarrow{-Cl^-} :CCl_2$$

In the latter case, unstable species are formed which undergo further reactions.

(a) β-Elimination

The unsaturated bond formed by elimination may be any one of a number of types (C=C, C≡C, C=O, C≡N, etc.). We shall consider in detail only those eliminations leading to C=C unsaturation and, in particular, the case in which one of the two groups eliminated is hydrogen, e.g.

$$
\begin{array}{ccc}
CH_3 & & CH_3 \\
CH_3\!\!-\!\!Br & \xrightarrow[55\,^\circ C]{EtOH} & \diagdown\!\!=\!\!CH_2 \quad + \quad HBr \\
CH_3 & & CH_3
\end{array}
$$

This group of reactions may be further subdivided into those which occur by a bimolecular mechanism (E2) and those which occur by a unimolecular mechanism (E1).

(i) *The bimolecular process.* Here, elimination is facilitated by attack by a base on the hydrogen atom which is to be removed, e.g.

$$
\begin{array}{c}
EtO^- \\
\end{array}
\quad
\begin{array}{c}
H \\
\diagdown \\
C\!-\!C \\
\diagdown \\
Br
\end{array}
\quad\longrightarrow\quad
C\!=\!C \quad + \quad EtOH \quad + \quad Br^-
$$

The other group eliminated must be one capable of departing as a neutral molecule or a relatively stable anion. In the transition state, the new double bond is partially formed and the groups being eliminated are partially removed:

$$
\begin{array}{c}
EtO^{\delta-} \\
\vdots \\
H \\
\vdots \\
C\!\!=\!\!\!=\!\!C \\
\vdots \\
Br^{\delta-}
\end{array}
$$

The extent to which these three bond changes have occurred at the transition state would be expected to depend on the structure of the organic compound and the strength of the base. In the generalized case, the transition state can be described as a hybrid of the structures (a)–(d) (B: = base),

$$
\begin{array}{cccc}
B: & B^+ & B^+ & B: \\
\diagdown & \diagdown & \diagdown & \diagdown \\
H & H & H & H \\
\diagdown & \diagdown & \diagdown & \diagdown \\
C\!-\!C & C\!=\!C & \bar{C}\!-\!C & C\!-\!\overset{+}{C} \\
\diagdown & \diagdown & \diagdown & \diagdown \\
X & X^- & X & X^- \\
\\
(a) & (b) & (c) & (d)
\end{array}
$$

amongst which (c) would be expected to be relatively more important when X^- is a poor leaving group and/or the negatively charged carbon is adjacent to an electron-attracting group, and (d) should be relatively more important when X^- is a very good leaving group and B: is a weak base. This is consistent with observation:

in particular, whereas the transition state appears to have alkene-like character (importance of b) when X^- is a good leaving group (e.g. Br^-) and a reasonably strong base is used, it is better described by (c) when the leaving group is a poor one (e.g. $N(CH_3)_3$) (Hofmann elimination, p. 98). At the other extreme, a transition state of type (d) can lead to a carbocation intermediate (E1 elimination).

The following are the principal features of the reaction.

(1)   The reaction rate increases with increasing strength of the base, e.g.

$$CH_3CO_2^- \ < \ HO^- \ < \ EtO^- \ < \ Me_3CO^- \ < \ NH_2^-$$

(2)   The rate increases with increase in the capacity of the second eliminated group to depart with the covalent bonding-pair. Most leaving groups are present as neutral substituents which depart as anions and, amongst substituents in which the same element is bonded to carbon, leaving-group ability parallels the stability of the anion. For example, alcohols and ethers do not undergo E2 elimination because hydroxide and alkoxide ions are relatively high energy species, whereas a sulfonate undergoes elimination readily,

$$\longrightarrow \qquad BH^+ \ + \ C{=}C \ + \ RSO_2O^-$$

because of the stability of the sulfonate anion (the conjugate base of a strong acid). There is therefore a general tendency for the ease of departure of a group to be related to the stability of the group as an anion. However, when different elements are being compared, a second factor, the dissociation energy of the C—X bond which is broken on the departure of X, is also concerned. The ease of the E2 reaction on alkyl halides is —I > —Br > —Cl ≫ —F, for, although the electronegativities of the groups lie in the opposite order, the bond strengths are in the order C—I < C—Br < C—Cl < C—F and this factor is here dominant, so that iodide is the best leaving group of the series and fluoride is the worst.

(3)   Amongst alkyl groups, the order of reactivity is tertiary > secondary > primary, e.g.

$$(CH_3)_3C{-}Br \ \xrightarrow{\ E:\ } \ (CH_3)_2C{=}CH_2 \quad \text{faster than}$$

$$(CH_3)_2CH{-}Br \ \xrightarrow{\ B:\ } \ CH_3CH{=}CH_2 \quad \text{faster than}$$

$$CH_3CH_2{-}Br \ \xrightarrow{\ B:\ } \ CH_2{=}CH_2$$

The reason is that the thermodynamic stability of an alkene increases as the C=C double bond is bonded to an increasing number of alkyl groups (p. 34) and this stabilizing influence in the product of the eliminations is reflected in some degree in the preceding transition state in which the π-bond is partly formed.

(4)   Elimination occurs more readily when the new double bond comes into conjugation with existing unsaturated bonds because the stabilization energy

due to the conjugation in the products is partly developed at the transition state. For example, elimination of hydrogen bromide occurs more readily from $CH_2=CH—CH_2—CH_2Br$, to give $CH_2=CH—CH=CH_2$, than from 2-bromobutane. An existing unsaturated group of $-M$ type facilitates elimination more strongly when the proton to be eliminated is attached to the α-carbon atom than when it is attached to the β-carbon. For example, base-catalyzed elimination occurs faster from β-bromopropionaldehyde than from α-bromopropionaldehyde:

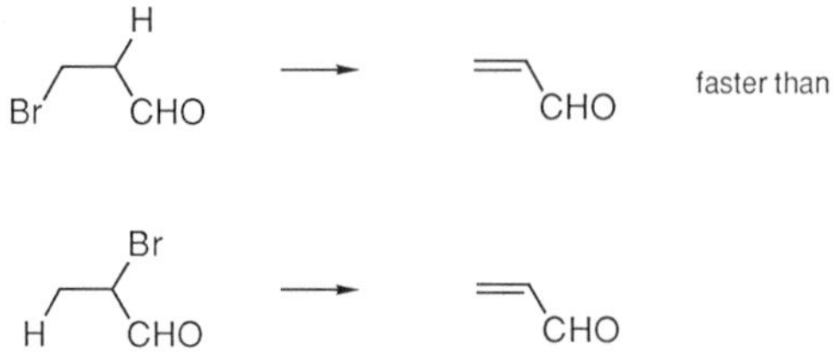

This is because the $-M$ group stabilizes structures such as

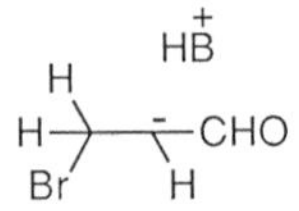

which contribute to the transition state structure (B = base), thereby lowering the activation energy. An activating effect of this type is responsible for the ready base-catalyzed elimination of water from β-hydroxy carbonyl, nitrile, or nitro compounds, whereas simple alcohols are inert to base-catalyzed elimination. In these cases, there is evidence for the occurrence of an enolate as an intermediate, e.g.

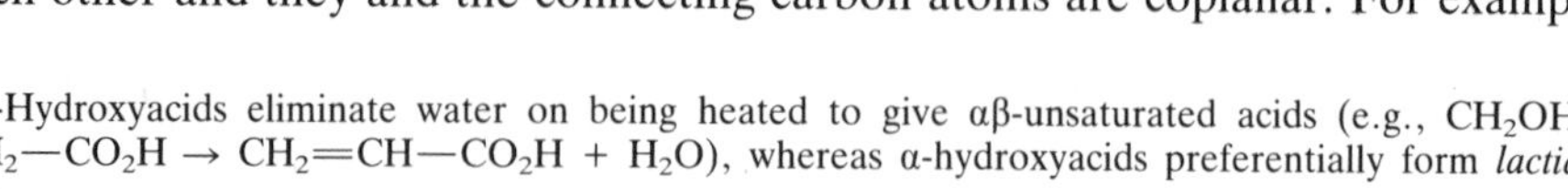

This has synthetic consequences, for several base-catalyzed reactions lead to such β-hydroxy-substituted compounds from which the corresponding αβ-unsaturated ones can therefore readily be obtained (see especially section 7.2).*

(5) The reaction occurs fastest when the two eliminated groups are *anti* to each other and they and the connecting carbon atoms are coplanar. For example,

---

* β-Hydroxyacids eliminate water on being heated to give αβ-unsaturated acids (e.g., $CH_2OH—CH_2—CO_2H \rightarrow CH_2=CH—CO_2H + H_2O$), whereas α-hydroxyacids preferentially form *lactides*, e.g.

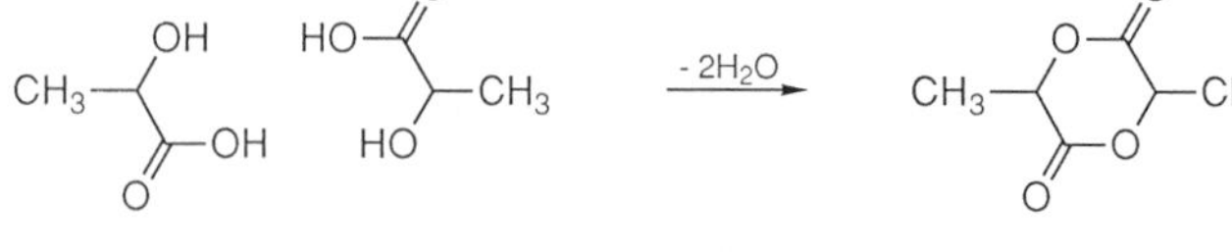

The γ- and δ-hydroxy compounds form *lactones* (p. 66).

whereas the compound (1) readily eliminates toluene-*p*-sulfonic acid, its isomer (2), in which there is no β-hydrogen *anti* to the toluene-*p*-sulfonate group, is inert even to strong bases (see also p. 171).

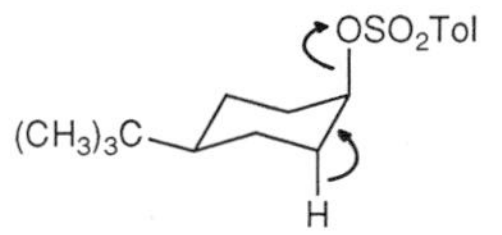

(1) 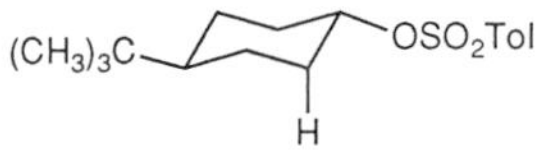   (2)

This steric requirement owes its origin to the tendency for maximization of bonding in the transition state. As the two leaving groups depart, the carbon orbitals involved in their bonding are changing from sp^3 to p; with an antiperiplanar arrangement, these developing p orbitals have some ability to overlap laterally, that is, to provide some bonding energy:

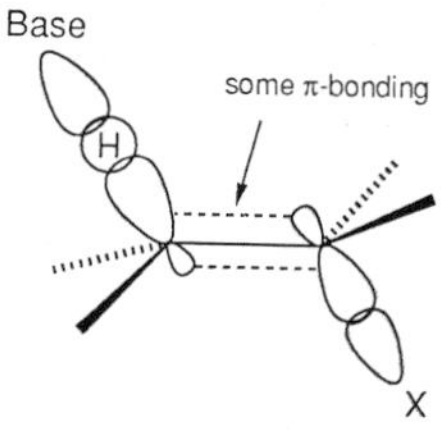

The partial formation of the alkene structure at the transition state is reflected also in the formation of more of the less sterically congested *trans*-alkene when geometrical isomers are possible, e.g.

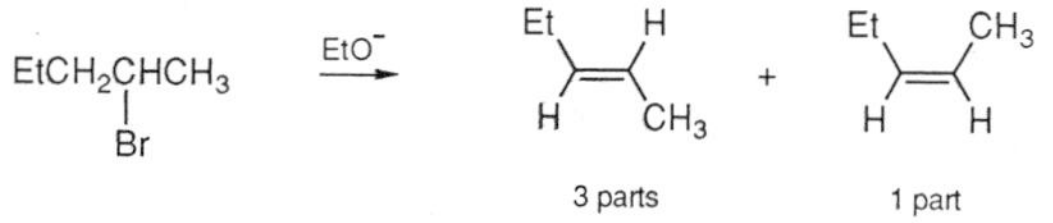

(6)   A compound which can eliminate in each of two ways to give different alkenes normally gives as the major product the more highly substituted one, e.g.

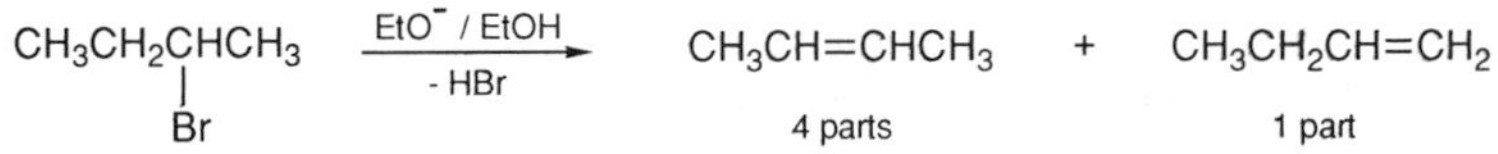

This generalization is known as the *Saytzeff rule* and has the same underlying principle as described in (3), namely, that the more substituted of the two alkenes is the more stable and this difference is reflected in the rate-determining transition states.

Exceptions to the Saytzeff rule occur in two circumstances. First, when the proton to be removed for Saytzeff elimination is in the sterically more hindered environment, the use of a base whose basic centre is also sterically hindered can lead to the predominance of the less substituted alkene. The following data are illustrative.

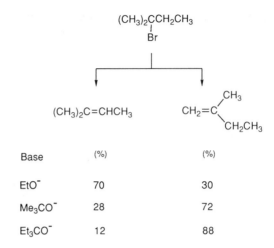

Base	(%)	(%)
EtO⁻	70	30
Me₃CO⁻	28	72
Et₃CO⁻	12	88

Second, elimination from quaternary ammonium ions (Hofmann elimination) usually gives mainly the less substituted alkene (*Hofmann rule*), e.g.

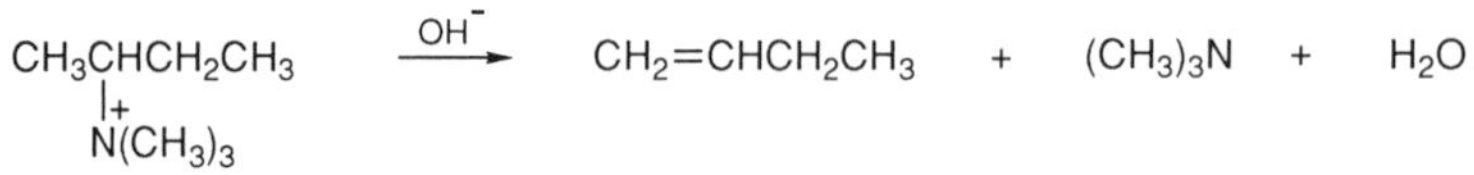

$$CH_3CHCH_2CH_3 \quad \xrightarrow{\text{OH}^-} \quad CH_2=CHCH_2CH_3 \quad + \quad (CH_3)_3N \quad + \quad H_2O$$
$$\overset{|+}{N(CH_3)_3}$$

The difference between these eliminations and those which follow the Saytzeff rule is attributed to the relatively poor leaving-group property of an amine. As a result, the C—H bond is relatively more fully broken at the transition state, that is, the structures

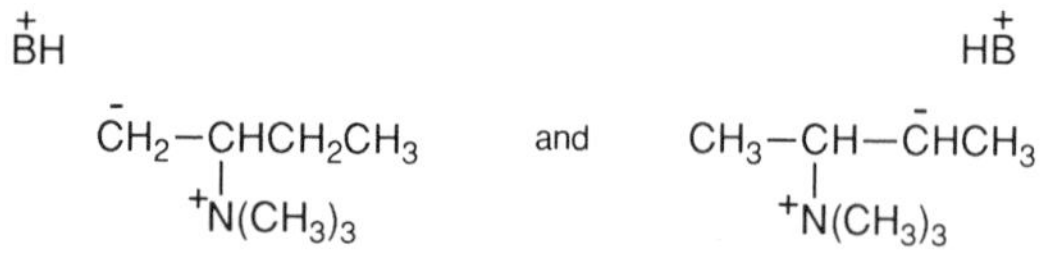

$$\overset{+}{B}H \qquad\qquad\qquad\qquad \overset{+}{H}\overset{+}{B}$$
$$\overset{-}{C}H_2-CHCH_2CH_3 \quad \text{and} \quad CH_3-CH-\overset{-}{C}HCH_3$$
$$\overset{|}{{}^+N(CH_3)_3} \qquad\qquad\qquad \overset{|}{{}^+N(CH_3)_3}$$

are more important contributors to the corresponding transition states than in elimination from a halide, so that the transition states are more 'carbanion-like' and less 'alkene-like'. Since an alkyl substituent destabilizes an adjacent negative charge, the first of the above structures is the more stable and this causes the less substituted alkene to be the major product.

(7) The S_N2 reaction always competes with the E2 reaction. The proportion of elimination is determined by the natures of the base and the alkyl group; the underlying theory is discussed below (section 4.5).

(8) Elimination does not occur to give 'bridgehead' alkenes in which the geometry of the ring-system precludes the close attainment to coplanarity required for *p*-orbital overlap in the alkene (*Bredt's rule*). For example,

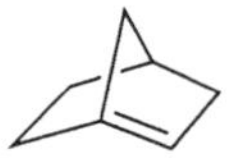

has never been obtained.

(*ii*) *The unimolecular process.* Elimination can also occur without the participation of a base. The reaction occurs in two stages, the first step, unimolecular heterolysis, being rate-determining:

$$\underset{X}{\overset{H}{\underset{|}{\overset{|}{C}}}}-C \xrightarrow[-X^-]{slow} \overset{H}{\underset{}{C}}-\overset{+}{C} \xrightarrow[-H^+]{fast} C=C$$

The characteristics of this process (E1) are as follows.

(1)   The order of reactivity of alkyl groups is tertiary > secondary > primary. This is because the rate-determining step is the formation of a carbocation and the stabilities of these ions (and the transition states preceding them) increase in the order primary < secondary < tertiary (p. 61). The rate of elimination from primary alkyl groups is usually negligible.

(2)   The effects on the rate of the nature of the leaving group are the same as in the E2 process.

(3)   The direction of elimination from the carbocation usually follows the Saytzeff rule (p. 97), e.g.

$$\underset{CH_3CH_2}{\overset{CH_3}{CH_3}}\!\!\!>\!\!-Cl \longrightarrow \underset{CH_3}{\overset{CH_3}{}}\!\!>\!\!\overset{+}{-}CH_2CH_3 \longrightarrow \underset{CH_3}{\overset{CH_3}{}}\!\!=\!\!\underset{CH_3}{\overset{H}{}} + \underset{CH_3CH_2}{\overset{CH_3}{}}\!\!=\!\!\underset{H}{\overset{H}{}}$$

$$ \text{4 parts} \text{1 part}$$

However, when the product of Saytzeff elimination is the more sterically compressed, the less substituted alkene may be formed, e.g.

$$(CH_3)_3CCH_2\!\!-\!\!\overset{+}{\underset{CH_3}{\overset{CH_3}{}}}\!\!< \longrightarrow \underset{CH_3}{\overset{(CH_3)_3CCH_2}{}}\!\!>\!\!=\!\!CH_2$$

because the alternative product,

$$\underset{H}{\overset{CH_3}{CH_3\!-\!}}\!\!\underset{}{\overset{CH_3}{<}}\!\!\underset{CH_3}{\overset{CH_3}{>}}$$

possesses significant repulsive forces between the hydrogens on the t-butyl group and those on the methyl group *cis* to it.

(4)   The carbocation may not only eliminate but may also add a nucleophilic species to give a substitution product (S_N1 reaction). The E1 and S_N1 reactions are therefore competitive (section 4.5). Further, the carbocation may rearrange (section 4.7).

(5)   Alcohols undergo E1 elimination under the influence of a strong acid, water being the leaving group:

$$\underset{OH}{\overset{H}{\underset{|}{C}}}-C \underset{}{\overset{H^+}{\rightleftharpoons}} \underset{\overset{+}{O}H_2}{\overset{H}{\underset{|}{C}}}-C \xrightarrow{-H_2O} \overset{H}{C}-\overset{+}{C} \xrightarrow{-H^+} C=C$$

With tertiary alcohols, reaction occurs in aqueous sulfuric acid with a co-solvent such as tetrahydrofuran at about 50°C, making it a useful process, but primary and secondary alcohols require treatment with 95% sulfuric acid at about 150 and 100°C, respectively, and these conditions are too severe for many organic compounds.

(6)   Lewis acids, such as silver ion, increase the reactivity of halides in E1 reactions, e.g.

$$R-Cl \quad + \quad Ag^+ \quad \longrightarrow \quad R^+ \quad + \quad AgCl$$

(*iii*)   *Elimination to give C≡C bonds.*   Base-catalyzed elimination from alkenyl derivatives,

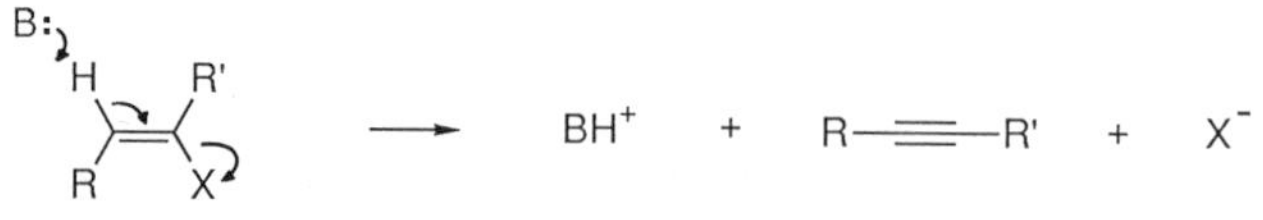

$$\longrightarrow \quad BH^+ \quad + \quad R-\!\!\!\equiv\!\!\!-R' \quad + \quad X^-$$

is similar in principle to the E2 reaction for the formation of alkenes. Strong bases are necessary, and amide ion is commonly used.

The product of elimination is usually a terminal alkyne, e.g.

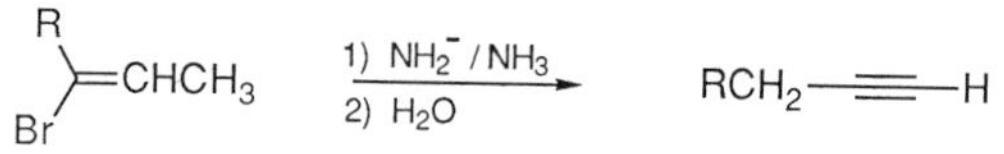

This is because the first product formed undergoes a series of prototropic shifts, catalyzed by the strong base:

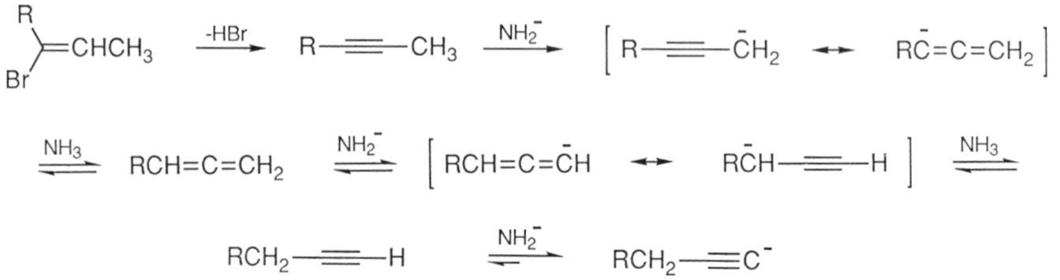

The thermodynamic driving force is the stability of the acetylenic anion which, on the addition of water, gives the terminal alkyne.

(*iv*)   *Elimination to give C=O bonds.*   Base-catalyzed elimination can give carbonyl compounds, as in the case of nitrate esters:

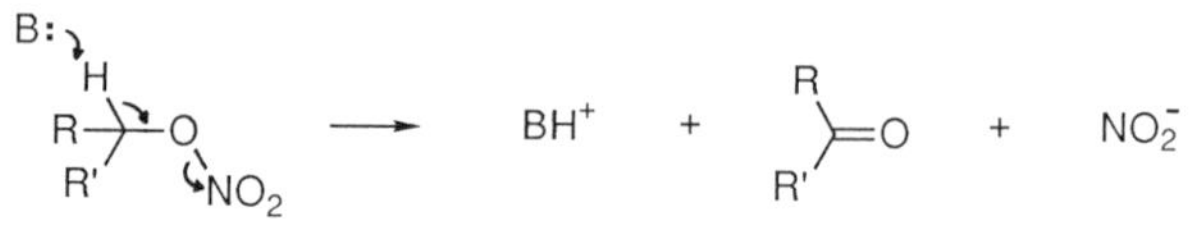

(*v*)   *Elimination to give C≡N bonds.*   Derivatives of aldoximes undergo base-catalyzed elimination to give nitriles:

$$\text{B:} \quad \overset{H}{\underset{R}{\underset{\underset{OSO_2Ar}{\big|}}{\big\rangle}}}=N \quad \longrightarrow \quad BH^+ \quad + \quad R-\!\!\!\equiv\!\!N \quad + \quad ArSO_2O^-$$

As in the E2 reaction, the rate depends on the capacity of the leaving group to depart with the bonding-pair (toluene-*p*-sulfonates are therefore suitable derivatives for elimination) and an *anti* geometry is required.

(*b*)   α-*Elimination*

The haloforms undergo base-catalyzed α-elimination with the formation of *carbenes*. The reaction normally occurs in two stages, the second of which is rate-determining, e.g.

$$CHCl_3 \quad + \quad OH^- \quad \rightleftharpoons \quad CCl_3^- \quad + \quad H_2O$$

$$CCl_3^- \quad \longrightarrow \quad :CCl_2 \quad + \quad Cl^-$$

The carbanion intermediates owe their stability in part to the ability of all the halogens except fluorine to accommodate the negative charge in *d* orbitals, e.g.

$$\overset{Cl}{\underset{Cl}{\big\rangle}}{}^-\!\!-Cl \quad \longleftrightarrow \quad \overset{Cl}{\underset{Cl}{\big\rangle}}=\bar{Cl} \quad \longleftrightarrow \quad \text{etc}$$

The resulting order of reactivity in the first step is $CHI_3 > CHBr_3 > CHCl_3 \gg CHF_3$.

The stability of the carbenes is determined by the ability of the halogens to supply *p*-electrons to the electron-deficient carbon, e.g.

$$F\overset{\displaystyle\cdot\cdot}{\frown}F \quad \longleftrightarrow \quad F\overset{\displaystyle\cdot}{\frown}\overset{+}{F} \quad \longleftrightarrow \quad \overset{+}{F}\overset{\displaystyle\cdot}{\frown}F$$

Since the $+M$ effects of the halogens lie in the order $F > Cl > Br > I$, the ease of the second step in carbene formation is greatest when the haloform contains fluorine atoms. Indeed, the three mixed haloforms, CHF_2Cl, CHF_2Br, and CHF_2I, undergo *concerted* elimination to carbenes,

$$HO^- \curvearrowright H\overset{\displaystyle F}{\underset{\displaystyle F}{\overset{\frown}{C}}}\!\!-Br \quad \longrightarrow \quad H_2O \quad + \quad :CF_2 \quad + \quad Br^-$$

Dihalocarbenes are reactive species which cannot be isolated. In the normal conditions for their formation (using hydroxide ion in a hydroxylic medium) they are hydrolyzed to acids,

$$HO^- \quad + \quad :CCl_2 \quad \longrightarrow \quad HO-\bar{C}Cl_2 \quad \xrightarrow{H_2O} \quad HO-CHCl_2 \quad \xrightarrow{\text{hydrolysis}}$$

$$HCO_2H \xrightarrow{\ OH^-\ } HCO_2^-$$

The addition to the reaction medium of a more strongly nucleophilic species than hydroxide ion (e.g. RS^-) diverts the carbene from hydrolysis. The addition of an alkene leads to cyclopropane derivatives whose formation occurs in a stereo-specific manner (p. 164), e.g.

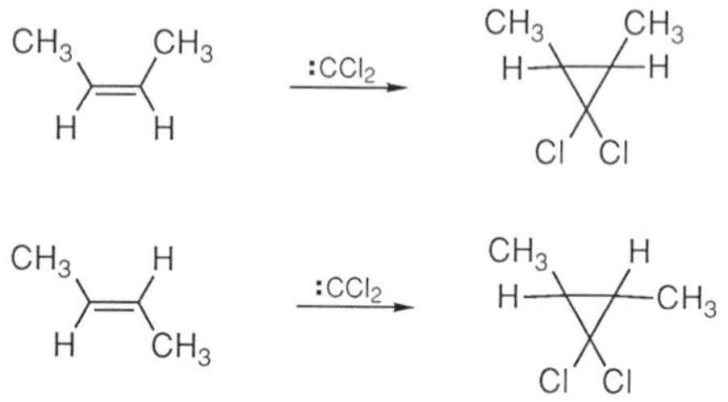

Carbenes also react with phenoxide ions (Reimer–Tiemann reaction, p. 374) and with primary amines (carbylamine reaction, p. 304).

Methylene itself, $:CH_2$, is of higher energy content than the dihalocarbenes and is correspondingly more difficult to obtain, more reactive, and less selective. It may be prepared by ultraviolet irradiation of diazomethane:

$$CH_2N_2 \xrightarrow{\ hv\ } :CH_2 + N_2$$

Its reactivity is illustrated by its reactions with cyclohexene, which include both addition to the double bond and insertion in the saturated and unsaturated C—H bonds,

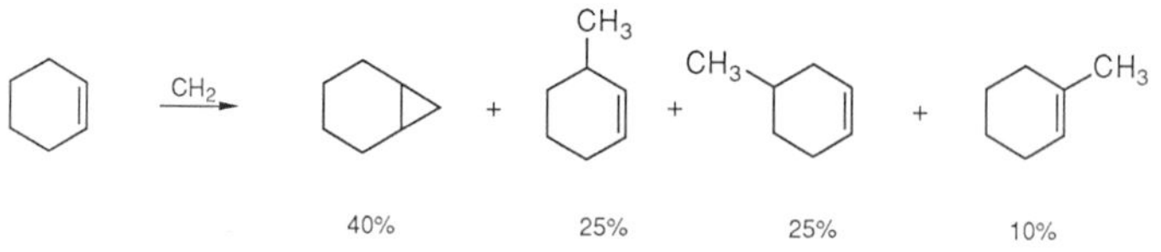

4.5
Substitution   One group attached to carbon may be replaced by another group in one of three ways: (*a*) by synchronous substitution, (*b*) by elimination followed by addition, and (*c*) by addition followed by elimination. The last applies only to unsaturated carbon.

(*a*) *Synchronous substitution*

The reagent may be a nucleophile (S_N2 reaction) or an electrophile (S_E2 reaction). Atoms and radicals do not substitute directly at carbon.

(*i*) *The S_N2 reaction.*   The reaction may be represented in general by

$$Nu^- \frown C\overset{\frown}{-}X \longrightarrow Nu-C \quad + \quad X^-$$

where Nu^- is a nucleophile and X^- is a leaving group.* The following are the principal characteristics of the process.

(1)   The relative reactivities of different leaving groups are the same as in the E2 reaction (p. 95), e.g. $I > Br > Cl \gg F$. As in that process, hydroxy, alkoxy, and amino groups are not displaced as the corresponding anions, since these are of too high an energy content, so that alcohols, ethers, and amines are inert to nucleophiles (but see electrophilic catalysis, p. 106). Sulfates and sulfonates are particularly reactive since the leaving group is, in each case, the anion of a strong acid. This makes dimethyl sulfate a useful methylating agent for alcohols in basic solution:

$$ROH \quad + \quad OH^- \rightleftharpoons RO^- \quad + \quad H_2O$$

$$RO^- \quad + \quad CH_3-OSO_2OCH_3 \longrightarrow RO-CH_3 \quad + \quad {}^-OSO_2OCH_3$$

(2)   The carbon atom at which substitution occurs undergoes inversion of its configuration (Walden inversion, p. 160) because the nucleophile approaches along a line diametrically opposite the bond to the leaving group, e.g.

This corresponds to a transition state in which the carbon undergoing substitution is sp^2-hybridized and possesses a p orbital perpendicular to the plane of the bonds which overlaps in part with an orbital of the leaving group and in part with that of the incoming nucleophile:

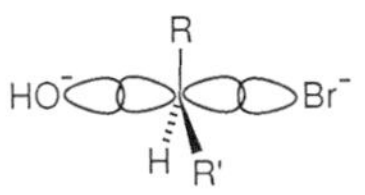

---

*Nucleophiles may be negatively charged, as shown above (e.g. OH^-), or neutral. In the latter case a salt is formed which is either the final product,

$$Nu: \frown C\overset{\frown}{-}X \longrightarrow \overset{+}{Nu}-C \quad + \quad X^-$$

or, if the nucleophilic atom is attached to hydrogen, an intermediate,

$$HNu: \frown C\overset{\frown}{-}X \xrightarrow{-X^-} H\overset{+}{Nu}-C \quad X^- \xrightarrow{-H^+} Nu-C \quad + \quad HX$$

Examples of reactions of neutral nucleophiles are:

$$(CH_3)_3N \quad + \quad CH_3-Br \longrightarrow (CH_3)_3\overset{+}{N}-CH_3 \quad Br^-$$

$$(CH_3)_2NH \quad + \quad CH_3-Br \longrightarrow \underset{\underset{H}{|}}{(CH_3)_2\overset{+}{N}-CH_3} \quad Br^- \longrightarrow (CH_3)_3N \quad + \quad HBr$$

(3)   The nucleophiles may be any of those listed on p. 81. Typical examples of reactions on alkyl halides, RX, are:

Nucleophile	Product
OH^-	Alcohols, R—OH
$R'O^-$	Ethers, R—OR'
$R'S^-$	Thioethers (sulfides), R—SR'
$R'CO_2^-$	Esters, R—OCOR'
$R'C\equiv C^-$	Alkynes, R—C$\equiv$CR'
CN^-	Nitriles, R—CN
NH_3	Amines, R—NH$_2$
R'_3N	Quaternary ammonium salts, $R'_3RN^+X^-$

A nucleophile is also a base, so that elimination (E2) competes with substitution. Just as in the E2 reaction the rate of elimination increases with the strength of the base, so in the S_N2 reaction the rate increases with the power of the nucleophile. Since a species is both base and nucleophile by virtue of its possessing the same characteristic – an unshared electron-pair – it might be expected that nucleophilic power would parallel basic strength. This is true, however, only in two general circumstances: first, when a common nucleophilic atom is involved (e.g. anions such as HO^- and RO^- are both stronger bases and more powerful nucleophiles than their conjugate acids, H_2O and ROH); and second, when elements in the same *period* are involved (e.g. ammonia is both a stronger base and a more powerful nucleophile than water). It is not necessarily true in other circumstances, particularly when elements in the same *group* are compared: for example, a thiol anion, RS^-, although less basic than its oxygen analogue, RO^-, is more strongly nucleophilic so that the $S_N2/E2$ ratio is higher for reactions with thiol anions than for those with alkoxide anions.

The main reason for this reversal of basicity and nucleophilicity can be understood by reference to the factors which determine the relative strengths of acids in a given medium, namely, the dissociation energy of the bond which breaks on ionization, and the electron affinity of the group from which the proton is lost. For thiols and alcohols, these factors are in opposition. S—H bonds ($340\,kJ\,mol^{-1}$) are weaker than O—H bonds ($462\,kJ\,mol^{-1}$), but RS· is less electronegative than RO·. However, the very large difference in bond energies dominates the situation, with the result that thiols are stronger acids than the analogous alcohols (i.e. thiol anions are weaker bases than alkoxide anions). Nucleophilicity is also determined, in part,* by the strength of the bond formed by the nucleophile, in this case with a carbon atom, and by the ease with which the nucleophile can release an electron to form a bonding-pair. For thiols and alcohols, these factors are again in opposition, but the bond-strength factor (C—S, 272; C—O, $357\,kJ\,mol^{-1}$) is in this instance of less importance than the

---

*It must be remembered that nucleophilicity refers to a *kinetic* phenomenon (the rate at which a species reacts at carbon) whereas basicity refers to an *equilibrium*. In particular, the processes discussed here have not occurred completely at the transition state for the substitution but are completed in the equilibrium.

electronegativity factor, with the result that RS^- is more strongly nucleophilic than RO^-.

At least one other factor is of importance in determining nucleophilicity. A species whose nucleophilic centre is attached to an atom possessing an unshared pair of electrons (e.g. NH_2—NH_2) is more strongly nucleophilic than predicted by consideration of its basicity. The reason is not clear.

(4)   The order of reactivity of alkyl groups is methyl > primary > secondary > tertiary, at least in part because of the increased steric hindrance to the approach of the reagent as the carbon atom is more heavily substituted. As an example, the relative reactivities of the following alkyl bromides towards iodide ion in acetone are:

CH_3—Br	CH_3CH_2—Br	$(CH_3)_2CH$—Br	$(CH_3)_3C$—Br
145	1	8×10^{-3}	5×10^{-4}

The order of reactivities in the E2 reaction is the opposite of this (p. 95), so that it follows that the S_N2/E2 ratio is greatest for a primary halide and least for a tertiary halide. The following results for the reactions of alkyl bromides with ethoxide ion in ethanol at 55°C are typical:*

CH_3CH_2—Br $\longrightarrow$ CH_3CH_2—OEt + CH_2=CH_2
90%   10%

It should be noted that tertiary halides rarely give significant yields in S_N2 reactions, e.g. t-butyl cyanide cannot be obtained from t-butyl chloride and cyanide ion; the product is $(CH_3)_2C$=CH_2.

(5)   Steric hindrance is also pronounced when the β-carbon atom is increasingly substituted by alkyl groups. For the reaction with iodide ion in acetone, relative reactivities of alkyl bromides are:

CH_3–CH_2–Br	CH_3CH_2–CH_2–Br	$(CH_3)_2CH$–CH_2–Br	$(CH_3)_3C$–CH_2–Br
1	0.8	4×10^{-2}	10^{-5}

*The S_N2/E2 ratio decreases with increase in temperature because the E2 reaction has the smaller (less negative) activation entropy.

Inspection of models shows that, for neopentyl bromide, approach of the nucleophile along the line of the C—Br bond is inevitably impeded by a methyl group, whatever geometry is established by rotation about the single bonds.

(6) Suitable solvents include both hydroxylic compounds such as methanol and ethanol and polar aprotic solvents, particularly DMSO, DMF, HMPA, and acetone. With anionic nucleophiles, reaction rates are much greater in the aprotic solvents owing to the lack of stabilization of the nucleophile by hydrogen bonding (p. 64), e.g.

$$CH_3(CH_2)_2CH_2Br \quad + \quad N_3^- \quad \longrightarrow \quad CH_3(CH_2)_2CH_2N_3 \quad + \quad Br^-$$

solvent	EtOH	DMSO	DMF	HMPA
k relative	1	1.3×10^3	3×10^3	2×10^5

This stabilization becomes less important as the size of the nucleophile increases, so that the ratio of rate constants for reactions in aprotic and hydroxylic solvents is less for, say, MeS^- than MeO^-. A further manifestation of this trend is that, amongst halide ions, the order of nucleophilicities is $Cl^- > Br^- > I^-$ in an aprotic solvent, but $I^- > Br^- > Cl^-$ in a hydroxylic medium. In spite of the greater rates of reaction in aprotic solvents, S_N2 reactions are mostly conducted in hydroxylic solvents unless the rate is impracticably small or an alternative reaction occurs, partly because they are cheaper and partly because they dissolve a wide range of both organic compounds and inorganic salts. With neutral reactants, where passage to the transition state involves the generation of opposite charges, e.g.

$$(CH_3)_3N: \frown CH_3 \frown I \longrightarrow \left[(CH_3)_3 \overset{\delta+}{N} ----- CH_3 ------\overset{\delta-}{I} \right] \longrightarrow (CH_3)_4N^+ \ I^-$$

transition state

rates are greater in solvents such as alcohols which can solvate both positive and negative species than in aprotic solvents which can only solvate the former.

(7) Reaction rates are increased by electrophilic catalysis. For alkyl halides, silver ion and mercury (II) ion are suitable, since they form strong bonds with halide ions. For alcohols, strong acids are effective, for example, primary alcohols form bromides with hot concentrated hydrobromic acid,

$$RCH_2OH \overset{H^+}{\rightleftharpoons} RCH_2\overset{+}{O}H_2 \overset{Br^-}{\underset{-H_2O}{\longrightarrow}} RCH_2-Br$$

Ethers are cleaved by hydrogen iodide, e.g.

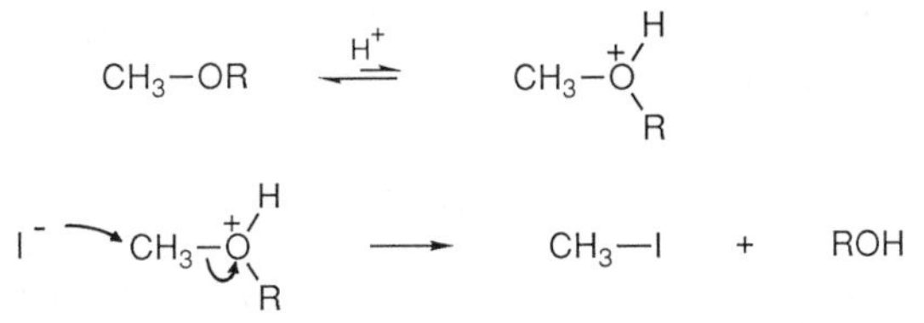

$$CH_3-OR \overset{H^+}{\rightleftharpoons} CH_3-\overset{+}{\underset{R}{O}}{}^H$$

$$I^- \overset{}{\frown} CH_3-\overset{+}{\underset{R}{O}}{}^H \longrightarrow CH_3-I + ROH$$

However, strong acids are too disruptive for sensitive organic compounds, and the Lewis acid boron tribromide, which also provides the nucleophile (Br^-), is more suitable:

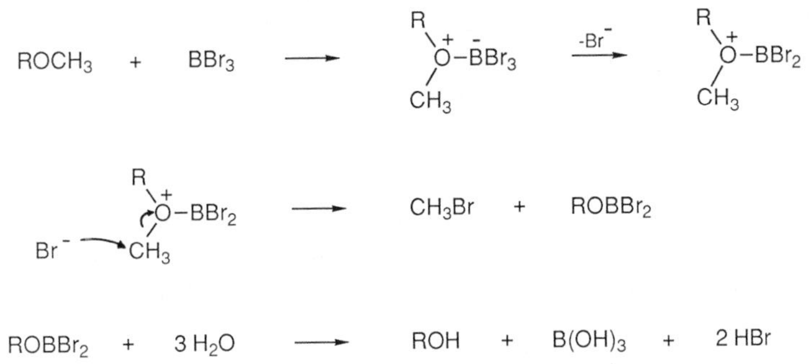

$$ROCH_3 + BBr_3 \longrightarrow \overset{R}{\underset{CH_3}{\overset{\backslash +}{O}-\bar{B}Br_3}} \overset{-Br^-}{\longrightarrow} \overset{R}{\underset{CH_3}{\overset{\backslash +}{O}-BBr_2}}$$

$$\overset{R}{\underset{Br^- \frown CH_3}{\overset{\backslash +}{O}-BBr_2}} \longrightarrow CH_3Br + ROBBr_2$$

$$ROBBr_2 + 3 H_2O \longrightarrow ROH + B(OH)_3 + 2 HBr$$

(8) When both a weak nucleophile and a poor leaving group are involved, reaction rates can be increased by the addition of iodide ion as a nucleophilic catalyst. For example, the displacement

$$\text{(pyridine)} N: \overset{R}{\frown} CH_2 \overset{\frown}{-}OAc \longrightarrow \text{(pyridine)} \overset{+}{N}-CH_2R + AcO^-$$

is effectively catalyzed because iodide ion is both a stronger nucleophile than pyridine and a better leaving group than acetate:

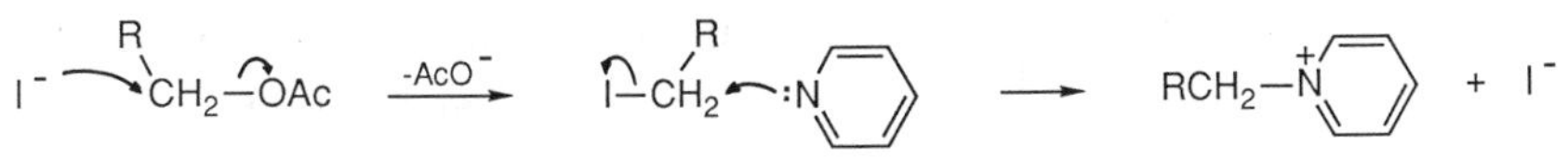

$$I^- \overset{R}{\frown} CH_2 \overset{\frown}{-} OAc \overset{-AcO^-}{\longrightarrow} I \overset{R}{\frown} CH_2 \frown :N \text{(pyridine)} \longrightarrow RCH_2 -\overset{+}{N} \text{(pyridine)} + I^-$$

(9) Another technique for aiding nucleophilic reactivity is to add a crown ether such as 18-crown-6,

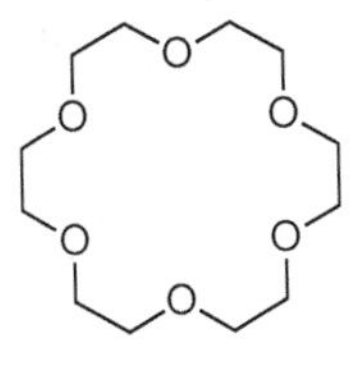

which specifically solvates cations such as Na^+ and K^+. For example, KF is insoluble in benzene and unreactive to organic halides. However, it dissolves in benzene in the presence of 18-crown-6 and the fluoride ion, being itself essentially unsolvated, acts as a good nucleophile, e.g.

$$RCH_2-Br \quad + \quad KF \quad \xrightarrow[PhH]{18\text{-}crown\text{-}6} \quad RCH_2-F \quad + \quad KBr$$

(10) Allylic and benzylic compounds and α-carbonyl-substituted compounds such as $BrCH_2COCH_3$ and $BrCH_2CO_2Et$ react far more rapidly than the corresponding alkyl compounds. It is believed that this is because the transition state is stabilized by interaction of the p orbital on the carbon undergoing substitution with the adjacent π-system. α-Alkoxy compounds are also very reactive, e.g. CH_3OCH_2Cl reacts about 5000 times faster than CH_3Cl with iodide ion in acetone. In this case, the stabilizing effect in the transition state is thought to be interaction of the p orbital on the carbon undergoing substitution with a (filled) oxygen $2p$ orbital.

(ii) The S_N2' reaction. Allylic compounds can undergo substitution with a double-bond shift (S_N2' reaction) in addition to the S_N2 reaction. The S_N2' reaction usually occurs only when the S_N2 process is sterically hindered, e.g.

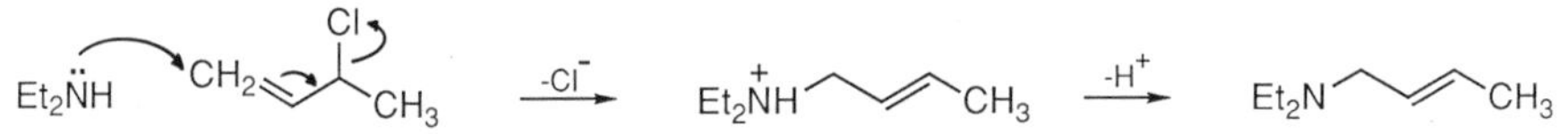

(iii) The S_E2 reaction. Carbon undergoes bimolecular electrophilic substitution (S_E2 reaction) when it is attached to strongly electropositive atoms, i.e. metals. A typical reaction is that of an organomercury compound with bromine,

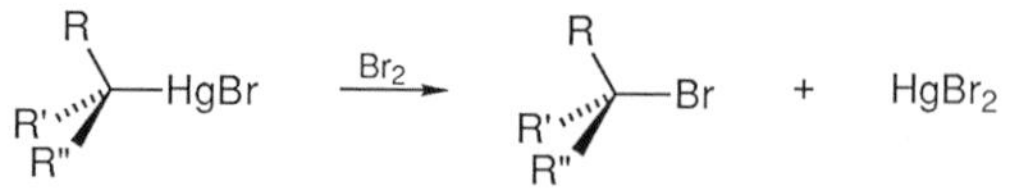

The reaction occurs with *retention* of configuration at carbon, as shown above, in contrast to the *inversion* which is characteristic of the S_N2 reaction.

(iv) The S_Ni reaction. Alcohols which possess an aromatic substituent react with thionyl chloride to give the corresponding chlorides with retention of the configuration of the hydroxyl-bearing carbon atom, e.g.

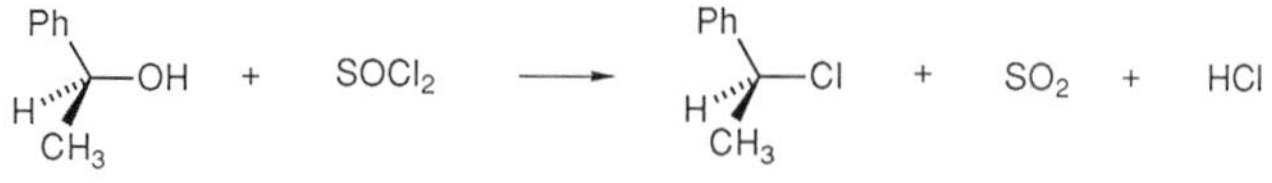

This is described as an S_Ni reaction (substitution, nucleophilic, internal). Reaction occurs through the chlorosulfite which collapses, with the elimination of sulfur dioxide, to give an ion-pair and thence the chloride:

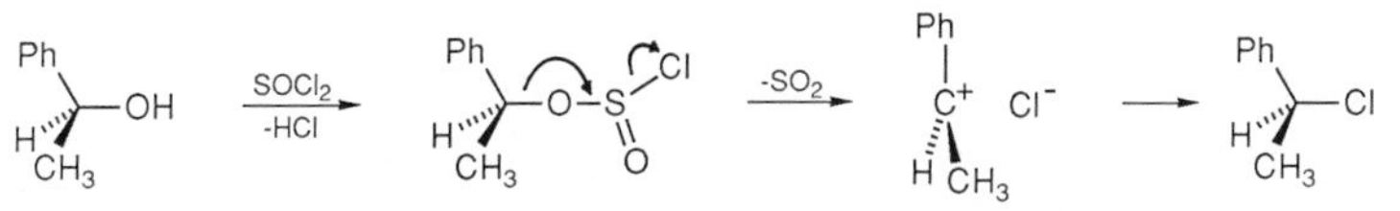

The role of the aromatic substituent is probably to stabilize the cationic part of the ion-pair by delocalizing the positive charge.

In the presence of pyridine, however, inversion of configuration occurs, probably because the pyridine removes a proton from the hydrogen chloride generated in the first step and the resulting chloride then reacts with the chlorosulfite in the S_N2 manner:

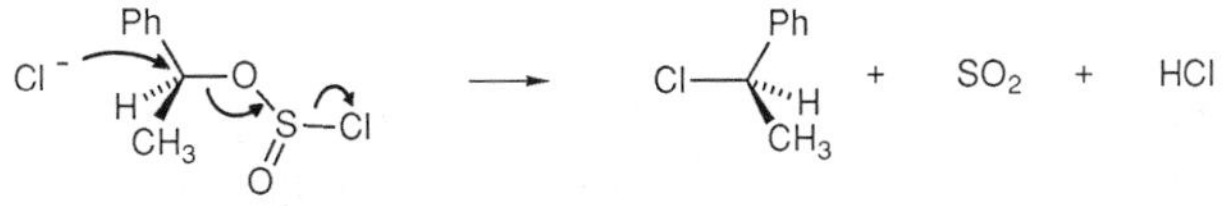

(b)   Elimination followed by addition

When a carbon atom is attached to a group which has a strong capacity for departure with the bonding electron-pair, a unimolecular solvolysis (S_N1 reaction) can occur, e.g.

$$(CH_3)_3C-Cl \longrightarrow (CH_3)_3C^+ + Cl^-$$

$$(CH_3)_3C^+ + H_2O \longrightarrow (CH_3)_3C-\overset{+}{O}H_2 \xrightarrow{-H^+} (CH_3)_3C-OH$$

The S_N1 process is related to the S_N2 reaction in the same way as E1 elimination is related to E2 elimination. The following are the characteristics of the reaction.

(1)   Reaction is facilitated by substituents which stabilize the carbocation, i.e. groups of $+I$ and/or $+M$ type. Among alkyl halides the order of reactivity is tertiary > secondary > primary, which is the opposite of that in the S_N2 reaction. Ether groups typify those of $+M$ type, e.g. methyl chloromethyl ether is rapidly hydrolyzed in water, reaction occurring through a delocalized carbocation-oxonium ion (note that the first product, a hemiacetal, is further hydrolyzed, p. 86)

$$CH_3O-CH_2-Cl \xrightarrow{-Cl^-} \left[CH_3O-\overset{+}{C}H_2 \longleftrightarrow CH_3\overset{+}{O}=CH_2 \right] \xrightarrow{H_2O}$$

$$CH_3O-CH_2-\overset{+}{O}H_2 \xrightarrow{-H^+} CH_3O-CH_2-OH \longrightarrow CH_2=O + CH_3OH$$

(2)   The relative reactivities of leaving groups are as in the E1 reaction. The range of reactivities is enormous: for example, relative rates for the solvolysis of $PhCHCH_3-X$ in 80% aqueous ethanol at $75\,°C$ are approximately:

Leaving group	$CF_3SO_2O^-$	Tosylate$^-$	I^-	Br^-	Cl^-	AcO^-
$k_{relative}$	10^{14}	10^{10}	10^8	10^7	10^6	1

(3) The carbon atom at which substitution occurs does not maintain its configuration. This is a result of the fact that a carbocation forms three coplanar bonds and can be attacked from either side, so that an optically active compound gives a mixture of products:

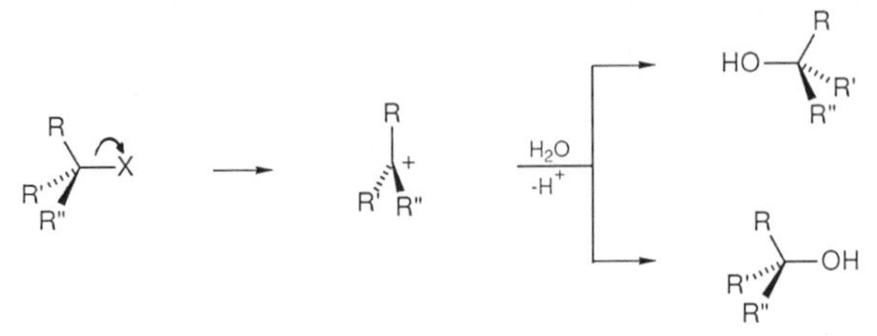

Attack would appear to be equally probable from both sides of the ion, but the two products are not normally formed in equal amounts: the major product has the opposite configuration to that of the reactant (i.e. *inversion* predominates over *retention*). It is believed that when the carbocation and X⁻ are formed, they exist as an ion-pair for a short time during which X⁻ shields the side of the carbon atom to which it was attached so that the incoming nucleophile more easily approaches from the other side. The less stable the carbocation, the more likely it is to be attacked before X⁻ separates from it. For example, inversion of configuration is much more extensive in the solvolysis of C_6H_{13}—$CHCH_3$—Cl compared with Ph—$CHCH_3$—Cl, in which the positive charge on the carbocation is delocalized on to the aromatic ring.

(4) E1 elimination competes with S_N1 solvolysis, e.g.

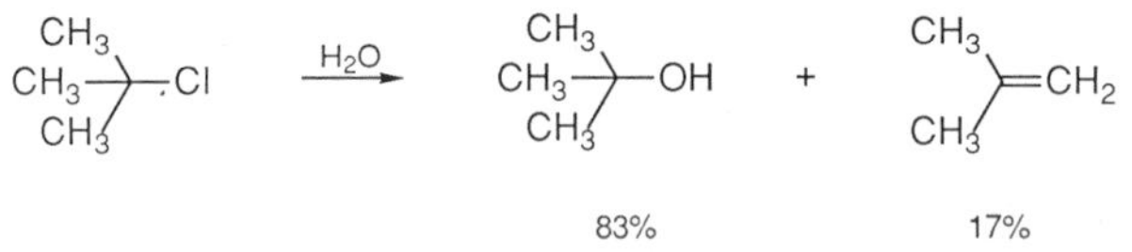

Elimination relative to substitution is favoured by more highly alkylated compounds since the alkene product then has more alkyl substituents (p. 34). The $E1/S_N1$ ratio is, however, independent of the nature of the leaving group because competition between the two processes occurs only after the carbocation has been formed.

(5) Allylic systems give mixtures of products, e.g.

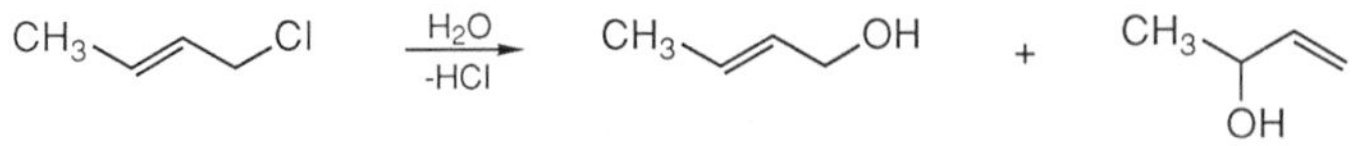

This is a result of the fact that the intermediate carbocation is delocalized and can react with the nucleophile (water) at each of two carbon atoms:

(6)   The reaction rate can be increased by the presence of electron-rich substituents which are stereochemically suited to interact with the carbon atom undergoing substitution. For example, the β-chlorosulfide, $C_2H_5SCH_2CH_2Cl$, is hydrolyzed in aqueous dioxan 10 000-times faster than its ether analogue, $C_2H_5OCH_2CH_2Cl$. This has been ascribed to *neighbouring-group participation* by the sulfur atom:

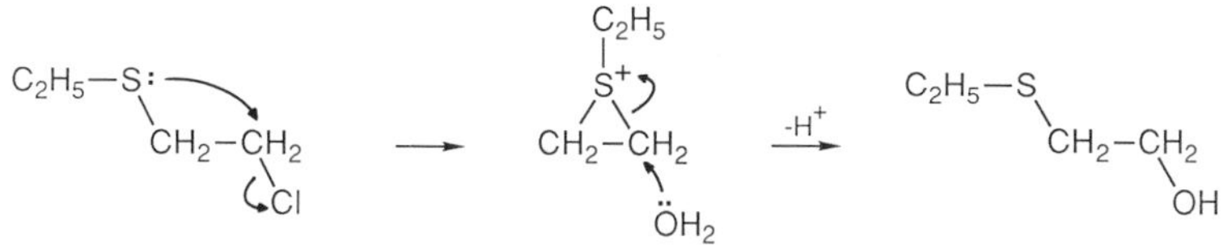

Instead of there being a relatively slow one-step reaction, two relatively rapid reactions occur. In the first, sulfur acts as an internal nucleophile. This step is favoured as compared with the intermolecular reaction with water because of the much more favourable entropy of activation for the intramolecular process (p. 65) and it is favoured as compared with the corresponding reaction of the oxygen analogue because sulfur is a more powerful nucleophile than oxygen (p. 104). In the second step, the three-membered ring is opened by attack by water. Other atoms and groups which enhance S_N2 rates in this way include nitrogen in amines, oxygen in carboxylate and alkoxide ions, and aromatic rings (p. 122). Participation is only effective when the interaction involves three-, five-, and six-membered rings (cf. p. 67). Participation by the oxygen in carboxylates and alkoxides and by nitrogen in primary or secondary amines can lead to cyclic compounds if these are stable in the reaction conditions. For example, 2-chloroethanol with alkali gives ethylene oxide,

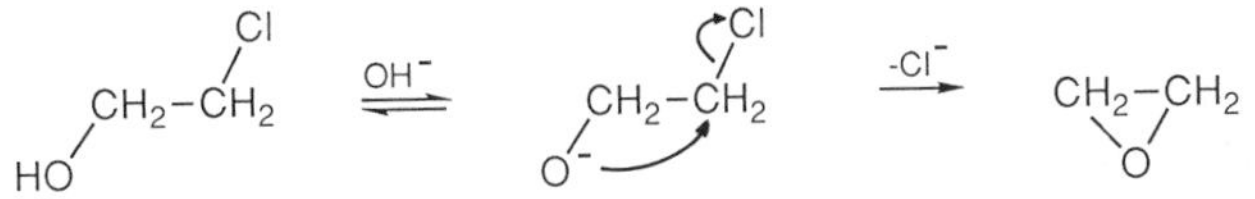

and pentamethylenediamine hydrochloride gives piperidine on being heated,

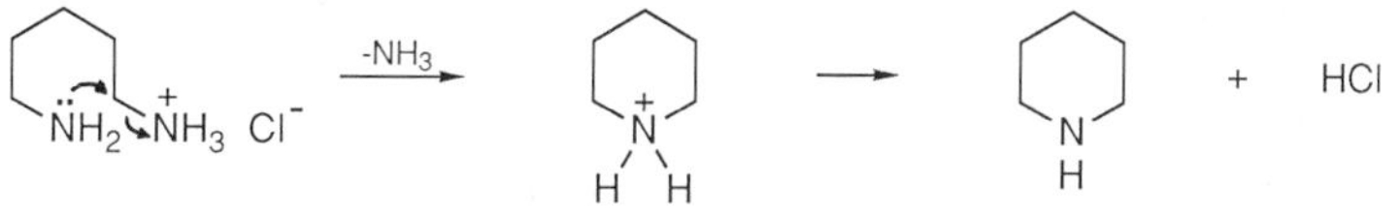

Neighbouring-group participation can also lead to rearrangement (chapter 14).

(7)   Bridgehead halides are essentially inert under S_N1 conditions, for example, the relative rates of solvolysis at 25°C in ethanol are:

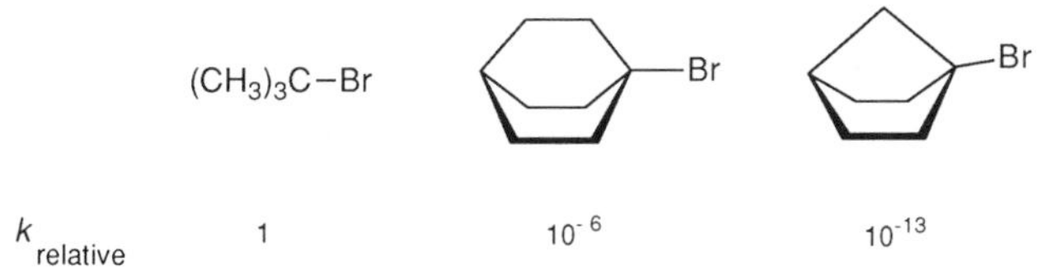

The reason is that the intermediate carbocation cannot attain the coplanarity needed for optimum formation of its three sp^2 bonds and the transition state is correspondingly distorted from its optimum configuration.

(8)   Alkenyl, alkynyl, and aryl halides are inert. However, aromatic diazonium ions, where the leaving group is the very strongly bonded nitrogen molecule, are reactive (p. 411):

$$Ar-\overset{+}{N}{\equiv}N \longrightarrow Ar^+ + N{\equiv}N$$

$$Ar^+ + H_2O \longrightarrow ArOH + H^+$$

(*i*)   *Summary of factors controlling substitution and elimination by unimolecular and bimolecular mechanisms.*   Unimolecular reactions are favoured by:

- Very good leaving groups (e.g. sulfonates such as p-CH_3—C_6H_4—SO_2O or, better, CF_3—SO_2O)
- Acid catalysis to aid the departure of the leaving group (e.g. silver ion for alkyl halides and protons for alcohols)
- Attachment of carbocation-stabilizing groups (usually alkyl, aryl, or alkoxyl) to the carbon atom undergoing reaction; tertiary > secondary > primary
- Polar solvent, and one capable of solvating both the developing carbocation and the anionic leaving group
- Presence of only weak bases and nucleophiles (e.g. solely an alcoholic solvent).

Bimolecular reactions are favoured by:

- Lack of carbocation-stabilizing substituents
- Presence of a strong base and/or a good nucleophile
- Aprotic solvents for anionic nucleophiles

Both E1:S_N1 and E2:S_N2 ratios increase with increased substitution by alkyl groups or conjugating substituents in the incipient alkene.

The E2:S_N2 ratio is also increased by:

- Use of a strong base and also by a bulky one (e.g. $(CH_3)_3C$—O^-) which can abstract a proton from the periphery of the molecule but is hindered as a nucleophile
- Increased alkylation at the site of reaction, causing increased hindrance to substitution.

The S_N2:E2 ratio is increased by:

- Use of a nucleophile which is a relatively weak base (e.g. RS^-, I^-)
- Use of an aprotic solvent
- Reduced substitution at carbon.

(ii) Summary of differences between primary, secondary, and tertiary alkyl halides.

Primary halides:

- Do not normally undergo unimolecular reactions
- Undergo S_N2 in preference to E2 reactions even with quite strong bases such as EtO^-; but with strong bases which are also hindered, the E2 reaction predominates, e.g.

$$CH_3(CH_2)_2CH_2Br \quad \begin{cases} \xrightarrow{EtO^-} & CH_3(CH_2)_2CH_2OEt \quad + \quad CH_3CH_2CH=CH_2 \\ & \qquad\quad 90\% \qquad\qquad\qquad\quad 10\% \\ \\ \xrightarrow{(CH_3)_3CO^-} & CH_3(CH_2)_2CH_2OC(CH_3)_3 \quad + \quad CH_3CH_2CH=CH_2 \\ & \qquad\quad 15\% \qquad\qquad\qquad\qquad 85\% \end{cases}$$

Secondary halides:

- Undergo S_N1 and E1 reactions in polar solvents in the absence of strong bases or good nucleophiles
- Undergo the E2 reaction in the presence of strong bases (e.g. EtO^-)
- Undergo the S_N2 reaction with good nucleophiles (e.g. RS^-, I^-) in polar aprotic solvents.

Tertiary halides:

- Undergo S_N1 and E1 reactions in polar, solvating solvents, e.g.

$$(CH_3)_3C-Br \xrightarrow{EtOH} (CH_3)_3C-OEt \quad + \quad (CH_3)_2C=CH_2$$
$$\qquad\qquad\qquad\qquad\quad 80\% \qquad\qquad\qquad 20\%$$

- Undergo E2 reactions in the presence of strong bases, e.g.

$$(CH_3)_3C-Br \xrightarrow{EtO^-} (CH_3)_2C=CH_2 \quad + \quad EtOH \quad + \quad Br^-$$

- Do not undergo S_N2 reactions.

(c) Addition followed by elimination

Unsaturated carbon in suitable environments undergoes substitution *via* a two-step process consisting of addition followed by elimination. The more important classes of reaction are nucleophilic substitution at carbonyl groups and nucleophilic, electrophilic, and free-radical substitution at aromatic carbon.

(i) Nucleophilic substitution at carbonyl groups. Derivatives of carboxylic acids in which the carbonyl group is adjacent to an electronegative substituent are susceptible to substitution *via* the addition–elimination mechanism, e.g.

$$\underset{Cl}{\overset{R}{>}}\!\!=\!O \;+\; OH^- \;\xrightarrow{\text{slow}}\; HO\overset{R}{-}\!\!\underset{Cl}{>}\!\!-O^- \;\xrightarrow[-Cl^-]{\text{fast}}\; \underset{HO}{\overset{R}{>}}\!\!=\!O \;\underset{}{\overset{OH^-}{\rightleftharpoons}}\; RCO_2^-$$

It should be noted that the carbonyl group in these environments differs from that in aldehydes and ketones only by virtue of the fact that the anionic intermediate derived by addition can eliminate a group as an anion (Cl$^-$, in the above example) whereas the corresponding anionic adducts derived from aldehydes and ketones cannot (H$^-$ and R$^-$ being of too high an energy content); the latter group of intermediates react instead with a proton (p. 88). Further, the reaction of an acid halide with a nucleophile differs from the S$_N$2 reaction of an alkyl halide only in that the former gives an intermediate whereas the latter cannot. The factors governing the ease of reaction (nature of the group R, the leaving group, and the nucleophile) are essentially the same in the two types of reaction, although acyl derivatives are in general more reactive than the corresponding alkyl derivatives. The following points are notable.

(1)   Amongst the commoner acyl derivatives, the order of reactivity is acid halide > anhydride > ester > amide. Acid halides react rapidly with the weak nucleophile water, anhydrides react slowly, and esters and amides are essentially inert (but see below). The order is the result of two electronic factors which operate in the same sense: first, the $-I$ effects of the halogens, oxygen, and nitrogen decrease in that order, so that the charge on the intermediate

$$Nu\overset{R}{-}\!\!\underset{X}{>}\!\!-O^-$$

is best accommodated when X = halogen and least well when X = NH$_2$; and second, the $+M$ effects of these elements, which stabilize the reactant,

$$\underset{O}{\overset{R}{\|}}\!\!\diagdown\!X \;\longleftrightarrow\; \underset{O^-}{\overset{R}{}}\!\!\diagdown\!\overset{+}{X}$$

fall in the order N > O > halogen (p. 33), so that the greatest amount of stabilization energy is lost in passage to the transition state when X is nitrogen and the least when X is halogen.

(2)   Since passage to the transition state involves the uptake of negative charge,* the rates of reaction are increased by the presence of electron-attracting substituents and decreased by electron-releasing substituents. For example, ethyl *m*-

---

* This is the B$_{AC}$2 mechanism of ester hydrolysis: B signifies that it is in basic conditions, AC that the ester bond broken is that between the acyl group and oxygen, and 2 that it is bimolecular. Hydrolysis can also be catalyzed by acids, the alkyl-oxygen bond (AL) can be cleaved (i.e. in S$_N$1- and S$_N$2-type reactions), and cleavage can occur without participation of the nucleophile (hydroxide ion or water), so that there are in principle eight mechanisms for hydrolysis. All but two (B$_{AC}$1 and A$_{AL}$2) are known, and the commonest are B$_{AC}$2 and A$_{AC}$2. Examples of A$_{AC}$2, A$_{AC}$1, A$_{AL}$1, and B$_{AL}$2 reactions are given later.

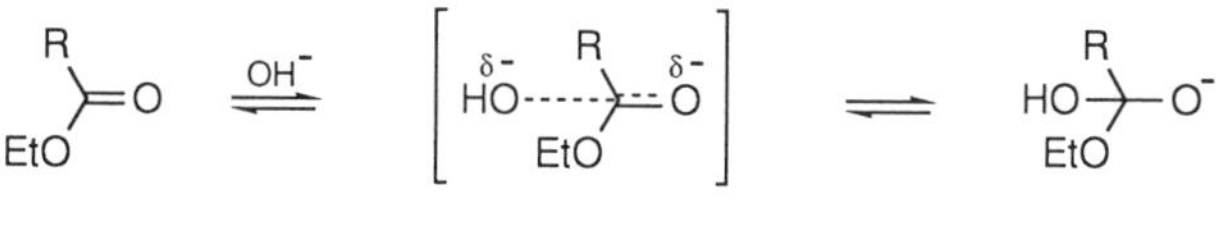

transition state

nitrobenzoate is hydrolyzed by hydroxide ion more than ten times as rapidly as ethyl benzoate (see also p. 63). The polar effect of one ester group on the other in diethyl oxalate ($EtO_2C—CO_2Et$) is great enough to cause this ester to be hydrolyzed even by water, towards which ethyl acetate is inert.

(3)   When the leaving group forms a stable anion, equilibrium lies well to the right, e.g.

$$RCOCl + H_2O \rightleftharpoons RCO_2H + H^+ + Cl^-$$

When it does not, an unfavourable equilibrium in the substitution may be followed by a second and strongly favourable one so that there is effectively complete conversion into products, e.g.

$$RCO_2Et + OH^- \rightleftharpoons RCO_2H + EtO^- \rightleftharpoons RCO_2^- + EtOH$$

Neither of these conditions applies to transesterification,

$$RCO_2R' + R''O^- \rightleftharpoons RCO_2R'' + R'O^-$$

but the amount of RCO_2R'' can be increased by carrying out the reaction in a considerable excess of the corresponding alcohol $R''OH$ so that $R'O^-$ is largely removed.

(4)   The reactions are subject to electrophilic catalysis. For example, although carboxylic acids are unreactive to alcohols alone, they can usually be esterified in the presence of a small quantity of concentrated sulfuric acid or about 3% by weight (of the alcohol) of hydrogen chloride:

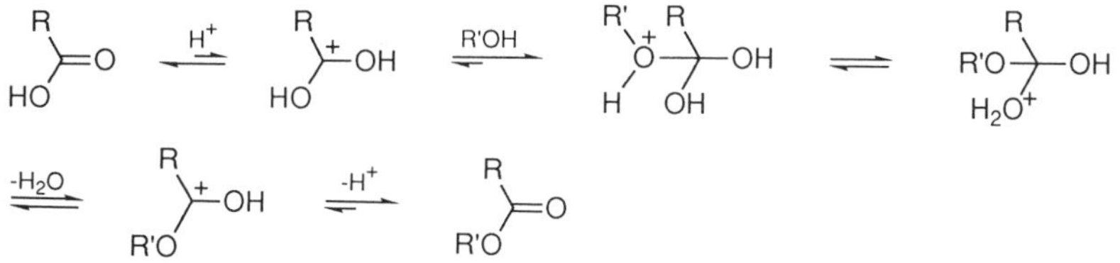

Since the equilibrium constant is about one, it is usual to carry out the reaction either in a large excess of the alcohol or by azeotropic distillation with toluene to remove the water as it is formed. Conversely, esters can be hydrolyzed by aqueous acid ($A_{AC}2$). However, this procedure, leading to an equilibrium, is not so efficient as the base-catalyzed method of hydrolysis which goes essentially to completion and it is chosen only when the ester contains another functional group which is susceptible to an unwanted reaction with base.

(5)   Carboxylic acids have one property which places them in a separate class from their derivatives: nucleophiles which are of moderate basic power convert them into their conjugate bases, RCO_2^-, which are inert to all but the most

powerful nucleophiles. For example, whereas esters react with hydrazine to give acid hydrazides, acids form only salts. Reactions may be effected with nucleophiles such as alcohols by employing an acid catalyst as described above, but nucleophiles which are stronger bases, e.g. ammonia, do not react in acidic conditions because the nucleophile is itself more or less fully protonated and therefore deactivated. The carboxylate ion is attacked by the AlH_4^- ion (p. 658) and the exceptionally powerful nucleophilic organic moiety in organolithium compounds (p. 199):

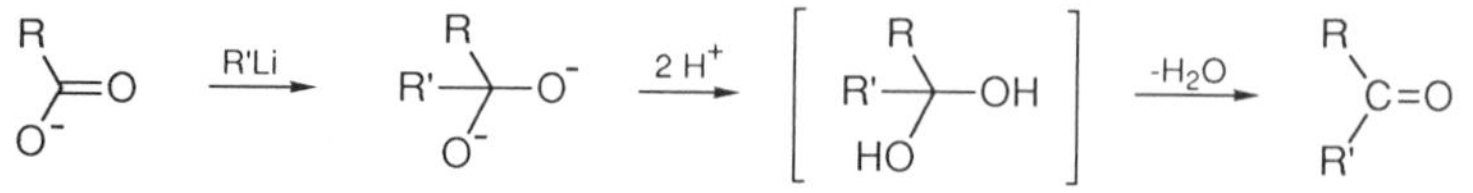

However, acids can be converted into strong acylating agents by reaction with a carbodiimide such as dicyclohexylcarbodiimide, e.g.

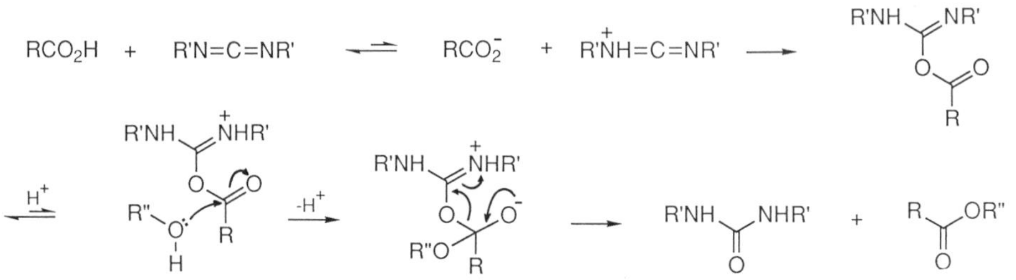

The corresponding reaction with amines is of particular importance in the synthesis of polypeptides (p. 338).

(6) Substitution at carbonyl groups is very markedly subject to steric hindrance. For instance, tertiary acids such as $(CH_3)_3C—CO_2H$, unlike primary and secondary acids, cannot be esterified by an alcohol in the presence of acid nor can their esters be hydrolyzed by treatment with hydroxide ion or aqueous acid.

Several procedures are available in cases of steric hindrance.

Hindered acids, like unhindered ones, can be converted into methyl esters with diazomethane (p. 437):

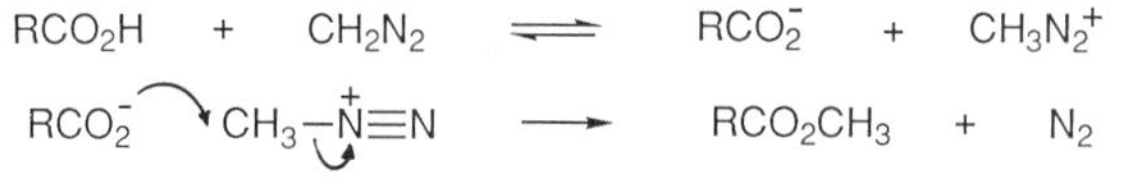

Hindered aromatic acids can be esterified by dissolving them in concentrated sulfuric acid and pouring the solution into the alcohol, e.g.

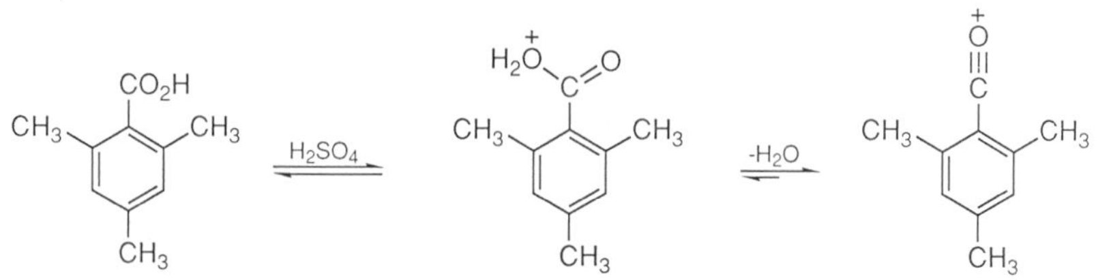

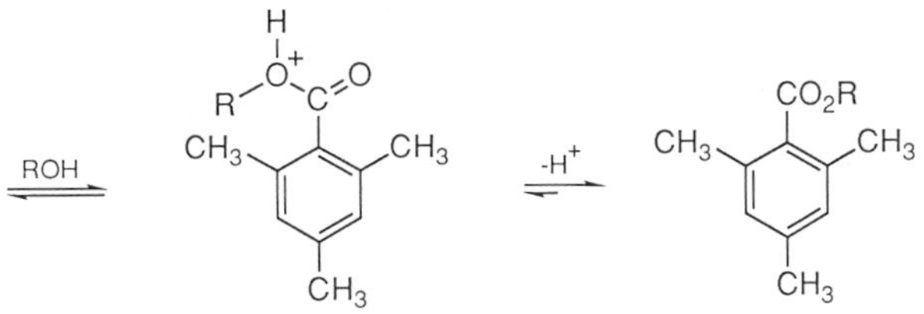

The driving force in the formation of the acylium ion lies partly in the delocalization of its charge on to the aromatic ring and partly in the relief of steric congestion.

Hindered t-butyl esters, as well as unhindered ones, can be hydrolyzed by acid:

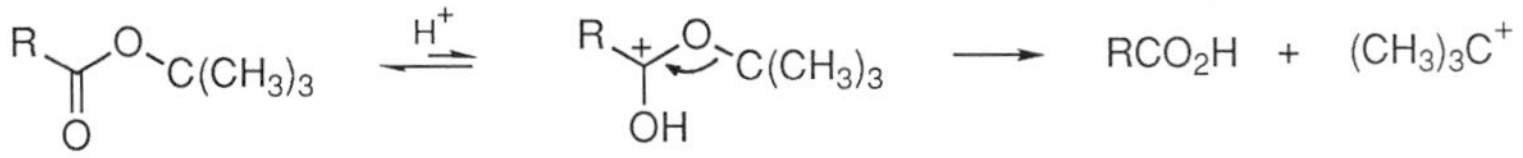

$$(CH_3)_3C^+ \; + \; H_2O \; \xrightarrow{\; -H^+ \;} \; (CH_3)_3C-OH$$

The success of this $A_{AL}1$ reaction lies in the stability of a tertiary carbocation. It is a widely used process in polypeptide synthesis, where trifluoroacetic acid is usually employed as the catalyst (p. 333).

Hindered esters of primary alcohols or methanol can be cleaved by treatment with a strong nucleophile such as a thiol anion in a polar aprotic solvent:*

Hindered aromatic esters can be hydrolyzed by the reverse of their method of formation: the solution of the ester in concentrated sulfuric acid is poured into water ($A_{AC}1$ reaction).

When it is the alcohol that is sterically congested, reactions with acid chlorides and anhydrides can be speeded by up to about 10 000-times by the addition of 4-dimethylaminopyridine (DMAP) as a nucleophilic catalyst. This is both a strong nucleophile owing to the $+M$ effect of the dimethylamino group,

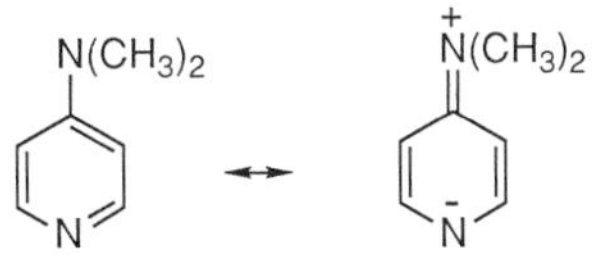

and a good leaving group:

---

* The only established case of hydrolysis by S_N2 reaction of this sort ($B_{AL}2$) is the reaction of β-propiolactone with water,

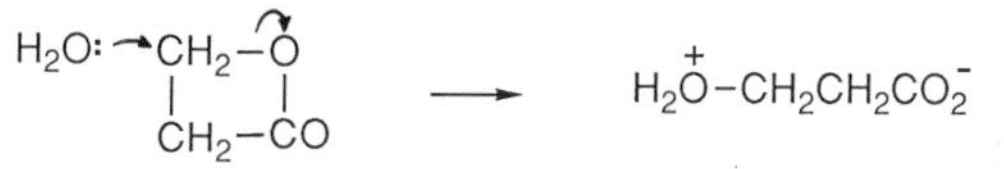

which is facilitated by the relief of strain as the ring opens in passage to the transition state.

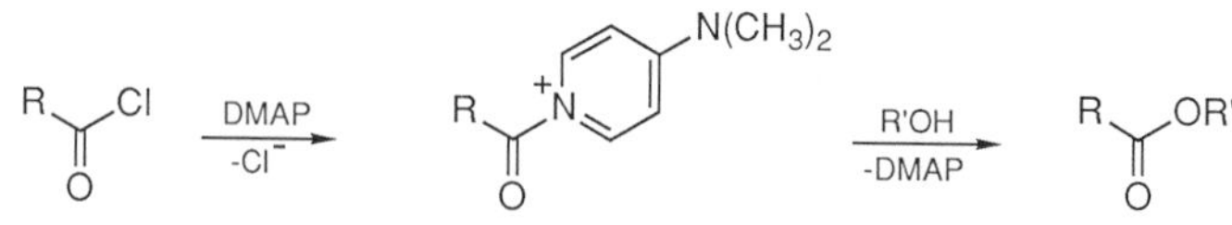

This is a particularly useful method for esterifying tertiary alcohols.

(*ii*) *Nucleophilic substitution at aromatic carbon.* Benzene and its halogen-derivatives, like ethylene and alkenyl halides, are inert to nucleophiles in normal laboratory conditions. More vigorous conditions lead to substitution, e.g.

$$PhCl \xrightarrow[\text{350 °C (pressure)}]{\text{NaOH}} PhOH$$

However, the introduction of groups of $-M$ type, *ortho* or *para* with respect to the leaving group, causes significant increases in rate. For example, *p*-nitrochlorobenzene is hydrolyzed on boiling with sodium hydroxide solution and 2,4,6-trinitrochlorobenzene (picryl chloride) is hydrolyzed even more readily.

The reason is that these reactions occur by addition of the nucleophile to the aromatic ring, giving an anionic intermediate from which elimination occurs:

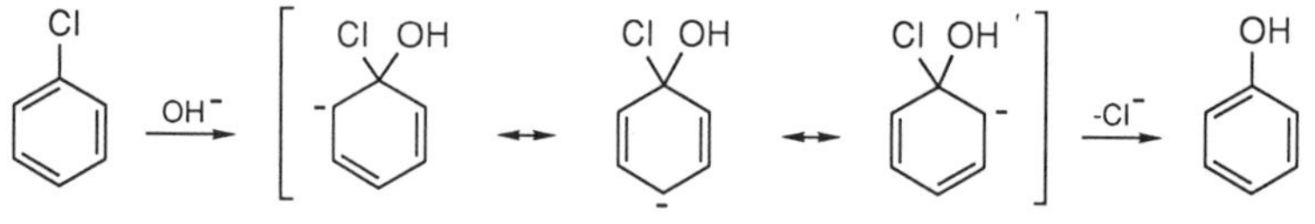

The intermediate is of high energy content relative to the reactants, for the loss of aromatic stabilization energy is far greater than the delocalization energy of the anion, but the activation energy is reduced by the attachment of $-M$ groups at *ortho* or *para* positions because the charge is then further delocalized over electronegative atoms, e.g.

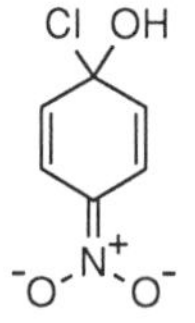

Note that a *meta* substituent can act only through its inductive effect, so that *m*-nitrochlorobenzene is less reactive than its *para* isomer.

An entirely different mechanism for nucleophilic aromatic substitution applies to reactions of the halobenzenes with amide ion in liquid ammonia. Here the power of amide ion as a base takes precedence over its nucleophilic property and elimination occurs to give an unstable *benzyne*. The benzyne then reacts with amide ion.

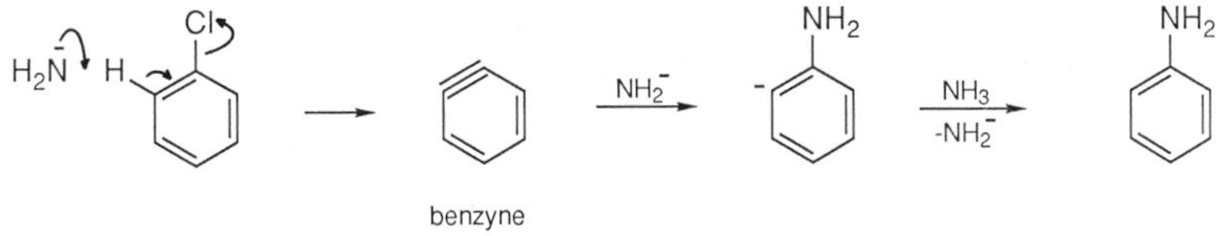

benzyne

These reactions are discussed in chapter 12.

(*iii*)   *Electrophilic substitution at aromatic carbon.*   Aromatic compounds react with electrophiles by the addition–elimination mechanism illustrated by the nitration of benzene:

$$HNO_3 + 2 H_2SO_4 \rightleftharpoons NO_2^+ + H_3O^+ + 2 HSO_4^-$$

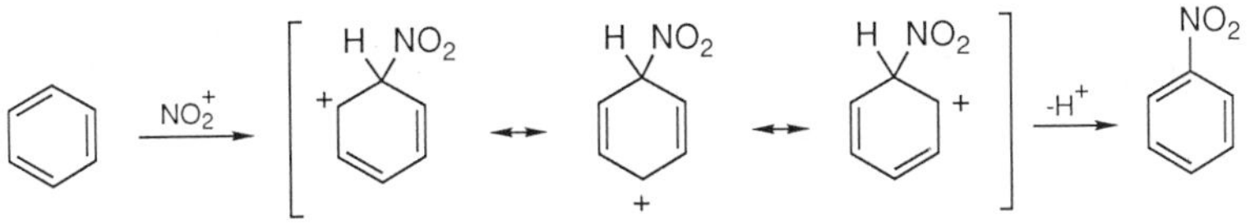

The activation energy is lowered by the presence of an electron-releasing substituent on the benzene ring, the effect being greater when the substituent is *ortho* or *para* to the entering reagent (cf. nucleophilic aromatic substitution, above). Hence substituents of $+I$ and/or $+M$ type direct reagents to the *ortho* and *para* positions and activate these positions relative to those in benzene, e.g.

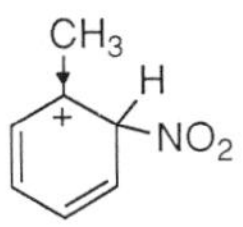

and

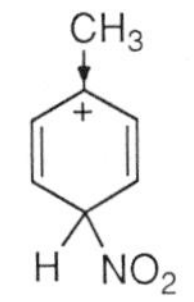

are of lower energy content than

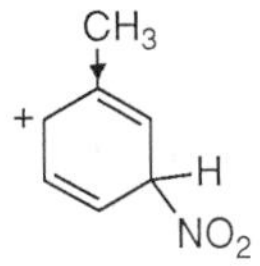

and all three are stabilized relative to

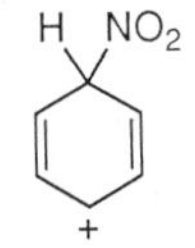

Conversely, electron-attracting substituents, which destabilize the intermediate cation with respect to the reactant (more strongly when the reagent enters the *ortho* or *para* positions than the *meta* position), deactivate the molecule and cause substitution to occur predominantly at the *meta* position, e.g.

$PhNO_2 \xrightarrow{\text{nitration}}$ *m*-dinitrobenzene (90%) + small proportions of the *o*- and *p*-isomers

These reactions are discussed fully in chapter 11.

(*iv*)   *Free-radical substitution at aromatic carbon.*   Like electrophiles and nucleophiles, free radicals and atoms react with aromatic compounds by the addition–elimination mechanism. An example is the formation of biphenyl by the thermal decomposition of dibenzoyl peroxide in benzene:

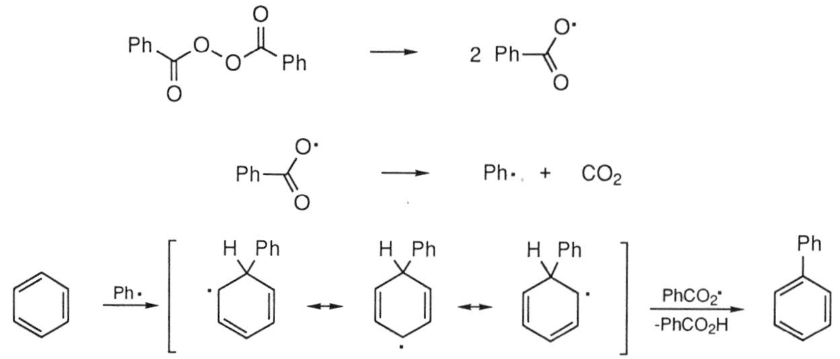

Substituents in the *ortho* or *para* positions are able to increase the delocalization of the intermediate radical, e.g.

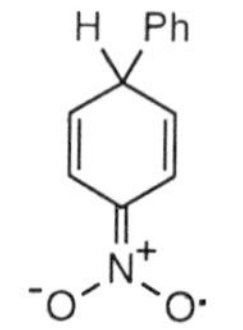

and consequently activate these positions to substitution. However, compared with the powerful stabilizing influences in ionic reactions, the effects in free-radical substitution are small. For example, benzene is about 10^5 times as reactive as nitrobenzene in nitration and about one-third as reactive in free-radical phenylation.

4.6 Condensation

The term condensation is applied to those reactions in which a small molecule such as water or an alcohol is eliminated between two reactants, as in the *Claisen condensation* (p. 226), e.g.

$$2\ CH_3CO_2Et \xrightarrow{(EtO^-)} CH_3COCH_2CO_2Et\ +\ EtOH$$

It should be emphasized that a condensation is not a reaction of a special mechanistic type but consists of a combination of the reaction types so far described. A key step in the Claisen condensation, for example, is a nucleophilic substitution at the carbonyl group of an ester,

The term is sometimes used more widely, to embrace reactions in which C—C bonds are formed without elimination, e.g.

$$2 \ CH_3CH_2CHO \quad \xrightarrow[10\,°C]{NaOH\,/\,H_2O} \quad \underset{HO \qquad CHO}{\overset{CH_3CH_2 \qquad CH_3}{H{\rightarrow}\!\!<\!\!<{-}H}}$$

However, it is more helpful to use the restrictive definition and to refer to this type of reaction as an *addition* (aldol addition) to distinguish it from the condensation which occurs at higher temperatures (p. 211), e.g.

$$2 \ CH_3CH_2CHO \quad \xrightarrow[80\,°C]{NaOH\,/\,H_2O} \quad \underset{H \qquad CHO}{\overset{CH_3CH_2 \qquad CH_3}{>\!\!=\!\!<}} \quad + \quad H_2O$$

4.7
Rearrangement

Rearrangement reactions fall into two categories: (*a*) those in which the migrating group is never fully detached from the system in which it migrates (intramolecular) and (*b*) those in which it becomes completely detached and is later re-attached (intermolecular).

(*a*)   *Intramolecular rearrangements*

Some of the systems in which migration occurs have been described (p. 79). The principles involved will be discussed with reference to the rearrangements of carbocations, for these are of wide synthetic application (chapter 14).

The simplest example is the solvolysis of neopentyl bromide. In a polar solvent and with a Lewis acid to promote it, S_N1 heterolysis occurs to give the neopentyl cation, a methyl group migrates to give the isomeric tertiary carbocation, and this undergoes part-elimination and part-addition:

The chief characteristics are the following.

(1)   The thermodynamic driving force for the migration is the increased stability of the tertiary compared with the primary carbocation. This step is rapid, for it competes successfully with the expected rapid attack of the solvent on the initially formed carbocation.

(2)   Other reactions leading to carbocations also result in rearrangement. For example, the addition of hydrogen iodide to t-butylethylene gives mainly a rearranged product:

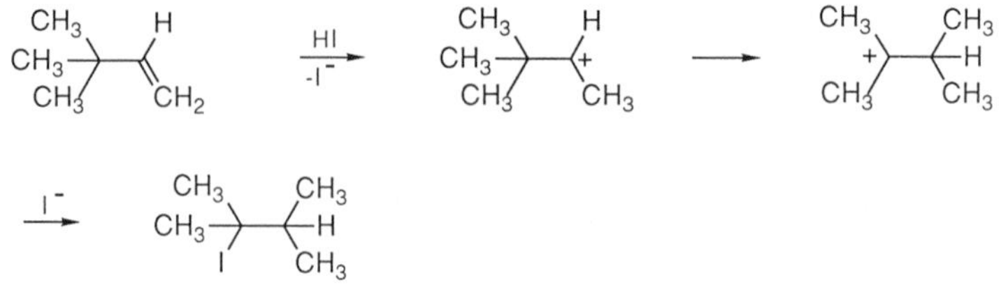

The possibility that rearrangement will occur must always be considered when S$_N$1 and E1 reactions and electrophilic addition to double bonds are involved.

(3)   In compounds in which two or three different groups are attached to the β-carbon, that group migrates which is best able to supply electrons to the carbocation, e.g. phenyl migrates in preference to methyl (see below).

(4)   Aryl groups on the β-carbon not only have a much stronger tendency than alkyl groups to migrate, but also speed reaction by participating in the rate-determining step (cf. p. 111). For example PhC(CH$_3$)$_2$—CH$_2$Cl undergoes solvolytic rearrangement thousands of times faster than neopentyl chloride because the rate-determining step in the former case involves the formation not of the high-energy primary carbocation PhC(CH$_3$)$_2$—CH$_2$$^+$ but of the delocalized ('bridged') phenonium ion:

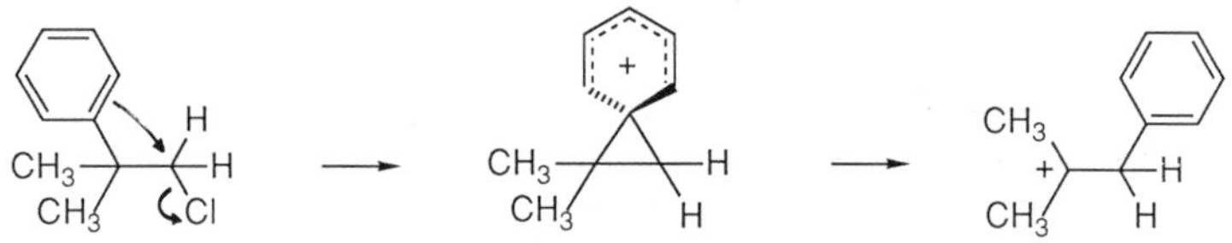

phenonium ion

As expected for a reaction involving the formation of an ion in which positive charge is delocalized over an aromatic system, the rate of reaction is increased by the presence of electron-releasing groups in the aromatic ring and retarded by electron-withdrawing groups.

(b)   Intermolecular rearrangements

These reactions are strictly not representative of a new mechanistic type because they are combinations of the processes described above. For example, the re-arrangement of *N*-chloroacetanilide to *o*- and *p*-chloroacetanilide, catalyzed by hydrochloric acid, consists of the formation of chlorine by a displacement process followed by the electrophilic substitution of acetanilide by chlorine:

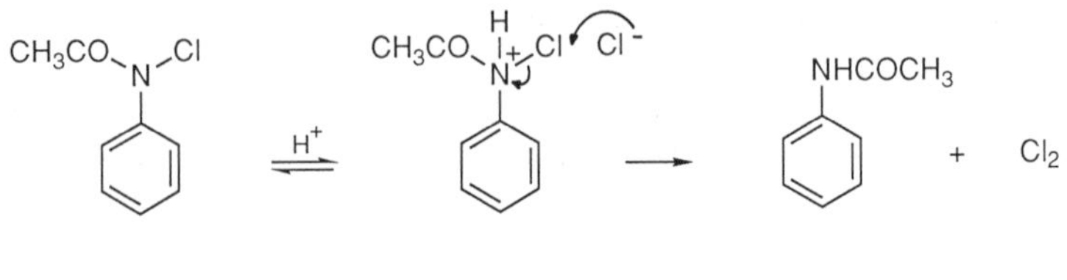

$$\longrightarrow$$

NHCOCH$_3$, Cl + NHCOCH$_3$, Cl + HCl

Prototropic (p. 14) and anionotropic (p. 15) shifts are also intermolecular rearrangements (see also the S_N1 solvolysis of allyl halides, p. 110).

A pericyclic reaction is one involving a concerted bond reorganization in which the essential bonding changes occur within a cyclic array of the participating atomic centres. An example is the reaction between butadiene and propenal which occurs on heating:

CHO heat CHO

It must be emphasized that it is not possible to determine the exact manner in which electron reorganization takes place: in this example the arrows could equally well have been directed in the opposite sense. The important feature is that neither ions nor radicals are formed as intermediates.

Pericyclic reactions can be subdivided into (a) cycloadditions, (b) electrocyclic reactions, (c) cheletropic reactions, (d) sigmatropic reactions, and (e) ene-reactions.

(a)   Cycloadditions

The reaction above is an example of a cycloaddition.

(b)   Electrocyclic reactions

These are intramolecular pericyclic reactions which involve either the formation of a ring, with the generation of one new σ-bond and the consumption of one π-bond, or the converse. An example is the thermal interconversion of 1,3,5-hexatriene and 1,2-dihydrobenzene:

(c) *Cheletropic reactions*

These are processes in which two σ-bonds which terminate at a single atom are made or broken in a concerted reaction, for example:

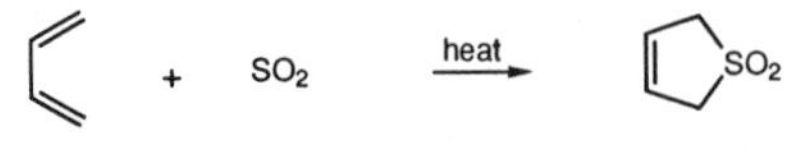

(d) *Sigmatropic rearrangements*

In these, an atom or a group migrates within a π-electron system without change in the number of σ- or π-bonds. An example is the *Claisen rearrangement*, e.g.

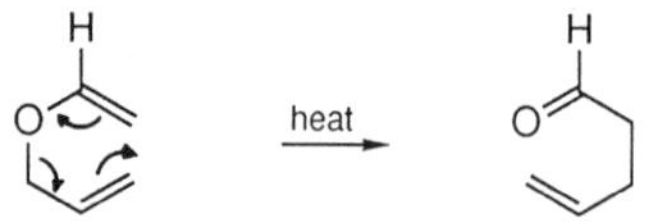

(e) *The ene-reaction*

This is the reaction of an allylic compound with an alkene:

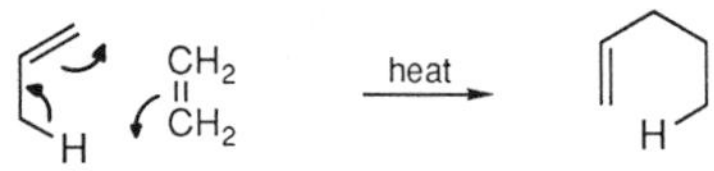

(f) *Mechanisms of pericyclic reactions*

The mechanisms of pericyclic reactions, including in particular their stereo-chemical outcome, followed from the recognition that they have orbital-symmetry requirements. The theory has proved as successful in explaining why certain reactions which appear reasonable on paper do *not* occur as in accounting for the courses of those that do. It has been developed in several forms, of which the frontier orbital treatment is the one used here. It is applied below to explaining why two molecules of an alkene do not form a cyclobutane and why a conjugated diene reacts with an alkene to form a six-membered ring. In chapter 9 the theory is developed to encompass the other types of pericyclic reactions.

(i) *Frontier orbitals and orbital symmetry.*   Just as the outer shell of electrons of an atom is regarded as especially significant in determining the chemistry of that atom, so it is reasonable that, for a molecule, it is the highest occupied molecular orbital (HOMO) which is the key to determining reactivity. This is termed the *frontier orbital*.

Since the ground state of almost all molecules has a pair of electrons in the HOMO, bonding interaction between two molecules* cannot involve only the HOMO of each for this would lead to an orbital occupancy greater than two, in contravention of Pauli's principle. The HOMO of one reactant needs therefore to interact with an unoccupied MO of the second. Now, since the bonding interaction between two orbitals increases as the energies of the two become more nearly equal, it is expected that the HOMO of one reactant should interact efficiently with the lowest unoccupied molecular orbital (LUMO) of the second. As usual, the two orbitals must have the same phase. Consider ethylene. The HOMO is that which results from overlap of carbon $2p$ atomic orbitals of like phase (p. 21); the LUMO is constructed from these two orbitals with opposite phases:

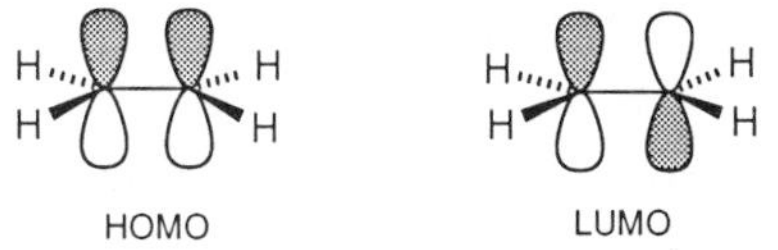

HOMO   LUMO

The HOMO and the LUMO differ in symmetry with respect to reflection in a plane which bisects the alkene perpendicularly to the plane of the molecule: the HOMO is symmetric with respect to this mirror plane and the LUMO is correspondingly antisymmetric.

Consider now the hypothetical cycloaddition of two molecules of ethylene to give cyclobutane:

$$2 \ CH_2=CH_2 \longrightarrow \begin{array}{c} CH_2\text{-}CH_2 \\ | \quad\ | \\ CH_2\text{-}CH_2 \end{array}$$

If the molecules were to approach each other with their molecular planes parallel, the LUMO of one molecule and the HOMO of the other would interact as follows:

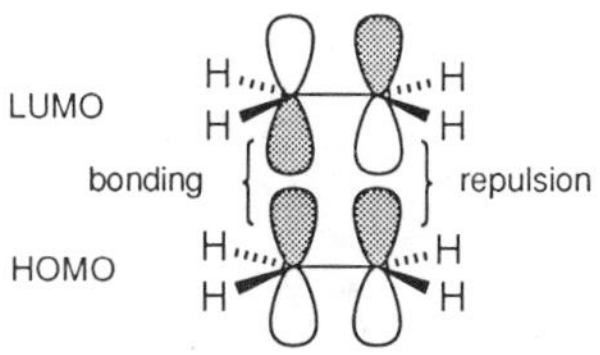

A concerted reaction cannot occur: it is described as *symmetry-forbidden*. Only one of the two new C—C bonds could be formed and the bonding energy associated with this, when offset against the loss of π-bonding of two ethylene molecules, serves to make this (two-step) process of such high activation energy that it is effectively impossible.

---

* This discussion also applies when reaction occurs between two centres in the same molecule.

The approach of the molecules towards each other described above is such that one surface of the π-orbital of one molecule interacts with one surface of the π-orbital of the other. The interaction is described as *suprafacial* with respect to each reactant. However, there is in principle an alternative geometry of approach, namely one in which the molecular planes are perpendicular, so that one surface of the π-orbital of one molecule interacts with *both* surfaces of the π-orbital of the other:

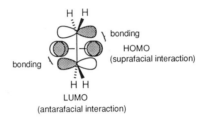

The interaction is described as *suprafacial* with respect to the former component and *antarafacial* with respect to the latter. In principle, bonding interaction could occur between the two pairs of lobes of like phase, but for this to be significant, the further alkene molecule would need to twist about its original π-bond. The energy required for this is evidently so much greater than that which would be gained through the partial formation of two new C—C bonds at the transition state that the reaction does not occur. That is, the process, though symmetry-allowed, is sterically inaccessible.

Consider next the reaction between an alkene and a 1,3-diene. The π-molecular orbitals of the diene are formed by combination of the carbon $2p$ orbitals:*

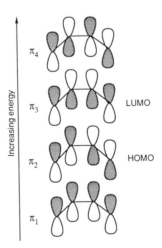

* In constructing π-MOs for dienes and polyenes, it is useful to remember that the lowest energy MO has no nodes, the next has one, the next two, and so on. In this case, the nodes are at the middle of the central C—C bond (π_2), the middles of the two terminal bonds (π_3), and the middles of all three bonds (π_4).

It is apparent that the symmetries of the HOMO of the diene and the LUMO of ethylene are such that, when the reactants approach each other with their molecular planes parallel, two new C—C bonds can be formed at the same time:

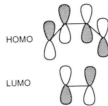

Consequently, two new C—C bonds are partially formed at the transition state and when the bonding energy associated with these is set against that lost as two C—C π-bonds are partially destroyed, the activation energy for this concerted reaction is relatively low.

Frontier-orbital theory is straightforward to apply and successfully rationalizes the outcome of a wide variety of pericyclic reactions. However, its inherent limitation should be recognized: the outcome of a reaction is determined by the characteristics of its transition state, whereas frontier-orbital theory is concerned only with the initial interaction of the reactants.

4.9 Oxidation–reduction

As strictly defined, a compound or group is described as undergoing *oxidation* when electrons are wholly or partly removed from it. For example, the methyl group is oxidized when methane is converted by bromine into methyl bromide, because the electron-pair in the C—Br bond is less under the control of the carbon atom than the pair in the original C—H bond since bromine is more electronegative than hydrogen. It is more convenient, however, to employ a narrower definition: oxidation throughout this book is used to cover reactions in which an electron is completely lost from an organic compound, e.g.

$$PhO^- \xrightarrow{Fe^{3+}} PhO\cdot$$

oxygen is gained, e.g.

$$RCHO \xrightarrow{[O]} RCO_2H$$

or hydrogen is lost, e.g.

$$RCH_2OH \xrightarrow{-[2H]} RCHO$$

Reduction is used to describe the converse reactions.

Oxidation and reduction are complementary in that in any system in which one species is oxidized, another is reduced. The term used is customarily that appropriate to the reaction undergone by the organic compound concerned, e.g. the reaction of an alcohol with dichromate is described as an oxidation, although of course the dichromate ion is reduced. In some cases, however, this usage is unsatisfactory, as in the *Cannizzaro reaction* in which the disproportionation of two molecules of certain aldehydes leads to the oxidation of one and the reduction of the other:

$$2 \text{ PhCHO} \xrightarrow{\text{OH}^-} \text{PhCO}_2\text{H} + \text{PhCH}_2\text{OH}$$

(a) Oxidation

Oxidation is normally brought about in one of the following ways.

(1)   By the effective loss of two electrons to a metal ion which is capable of a two-electron reduction, reaction occurring within an intermediate compound, e.g.

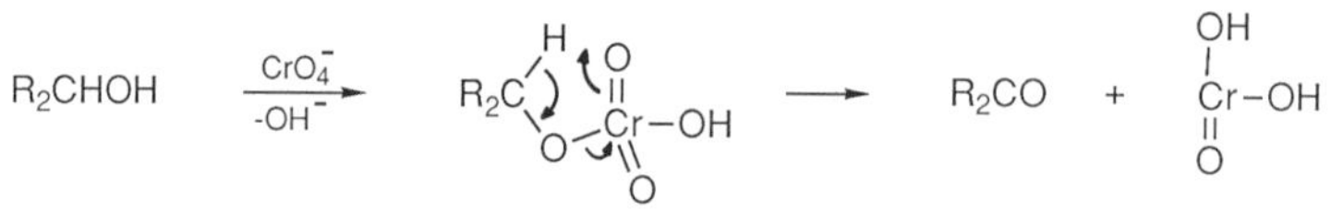

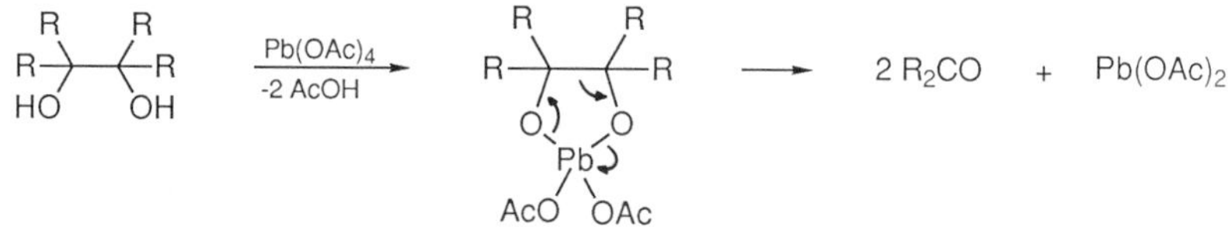

(2)   By removal of an electron, as in the oxidation of phenols by $\text{Fe(CN)}_6{}^{3-}$ (p. 597), e.g.

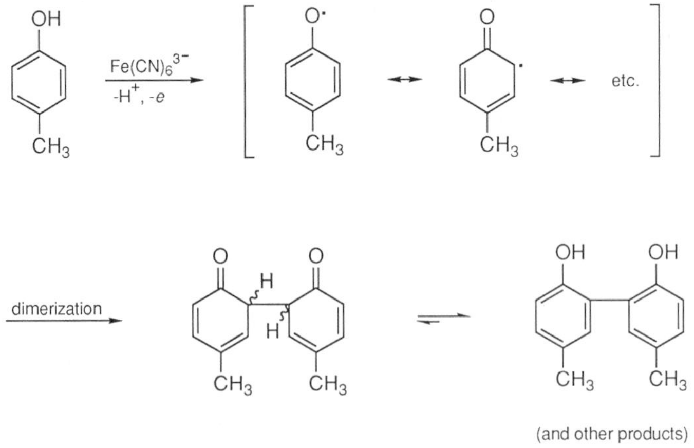

(and other products)

The requirement for the oxidizing agent is that it should be capable of one-electron reduction (here, Fe(III) to Fe(II)) characterized by a suitable redox potential. The requirement for the organic compound is that it should give a relatively stable radical on oxidation (achieved in this case by the delocalization of the unpaired electron over the benzene ring).

(3)   By removal of a hydrogen atom, as in the radical-catalyzed autoxidation of aldehydes (p. 557),

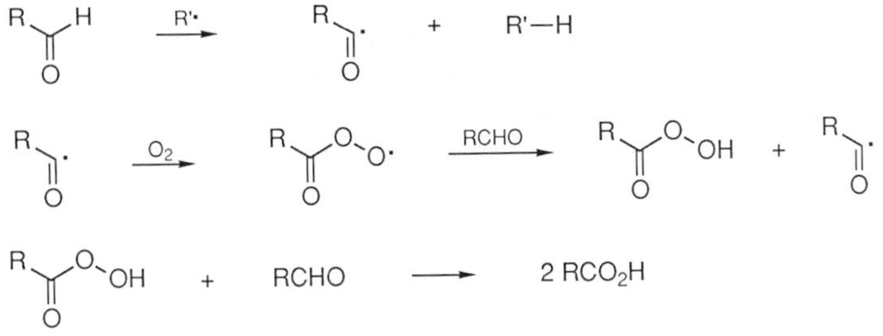

(4)   By removal of hydride ion, as in the *Cannizzaro reaction* (p. 654), e.g.

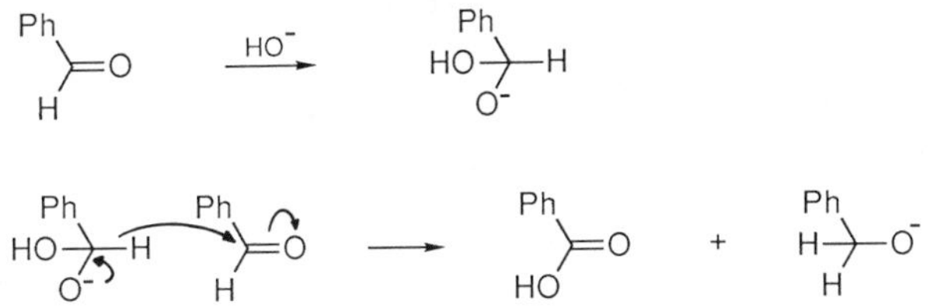

(5)   By the insertion of oxygen, as in the epoxidation of an alkene by a peroxyacid (p. 588):

(6)   By catalytic dehydrogenation, as in the palladium-catalyzed conversion of cyclohexane into benzene (p. 606).

(*b*)   *Reduction*

The commoner mechanisms of reduction are the following:

(1)   By addition of an electron, as in the formation of pinacols,

$$R_2C{=}O \xrightarrow{\ e\ } R_2\dot{C}{-}O^-$$

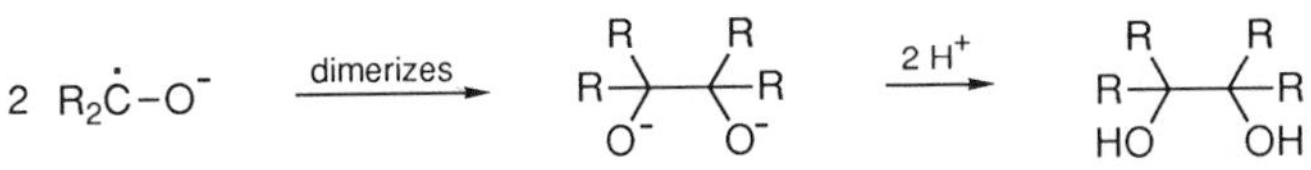

Two electrons may be transferred, as in the reduction of alkynes by sodium in liquid ammonia:

$$RC{\equiv}CR \xrightarrow{\ e\ } \left[RC{=}CR\right]^{\overline{\cdot}} \xrightarrow{\ H^+\ } \left[R\overset{\cdot}{C}{=}CHR\right]$$

$$\xrightarrow{\ e\ } \left[R\overset{\bar{}}{C}{=}CHR\right] \xrightarrow{\ H^+\ } \begin{matrix} R \\ \diagup \\ H \end{matrix}{>}{=}{<}\begin{matrix} H \\ \diagdown \\ R \end{matrix}$$

The requirement of the reducing agent is that it should have a strong tendency to donate an electron, e.g. electropositive metals such as sodium, and transition-metal ions in low valency states such as Cr(II).

(2)   By addition of hydride ion, usually from a complex metal hydride (p. 651), e.g.

$$H_3\bar{Al}{-}H \quad \overset{R}{\underset{R'}{>}}{=}O \longrightarrow H{-}\overset{R}{\underset{R'}{>}}{-}O^- \ + \ AlH_3$$

Reaction may occur *via* a cyclic transition state, as in the *Meerwein–Ponndorf–Verley* reduction of aldehydes and ketones (p. 654):

$$(CH_3)_2C\underset{H}{\overset{O}{\diagdown}}Al(OCH(CH_3)_2)_2 \quad \underset{{-}(CH_3)_2CO}{\rightleftharpoons} \quad H{-}\overset{R}{\underset{R}{>}}{-}OAl(OCH(CH_3)_2)_2$$

(3)   By catalytic hydrogenation, as in the reduction of alkenes over Raney nickel (a form of nickel in a very finely divided state):

$$\overset{R}{\diagdown}\!\!/\!\!\overset{R'}{\diagup} \xrightarrow{\ H_2\text{-}Ni\ } \overset{R}{\underset{H}{>}}{-}{-}{<}\overset{R'}{\underset{H}{}}$$

These reactions occur in a stereoselective manner, giving mainly the *cis*-dihydro adduct.

Problems   1.   Arrange the following in decreasing order of their reactivity towards an electrophilic reagent (e.g. H^+):

$$CH_2{=}CH\overset{+}{N}(CH_3)_3 \qquad CH_2{=}CH_2 \qquad CH_2{=}CHCH_3$$

$$CH_2{=}CHOCH_3 \qquad CH_2{=}CHBr \qquad CH_2{=}CHNO_2$$

2.   What reactivity differences would you expect between the compounds in each of the following pairs:

(i)   CH$_3$—CH=CH—CO$_2$Et   and   CH$_2$=CH—CH$_2$—CO$_2$Et

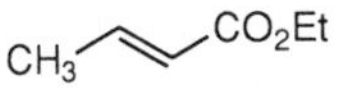

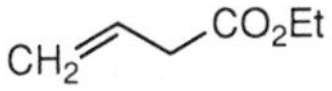

(ii)   EtOEt   and   CH$_2$=CHOEt

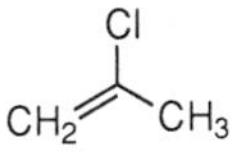

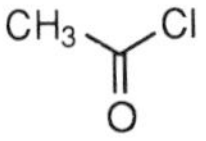

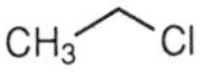

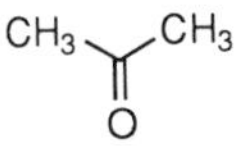

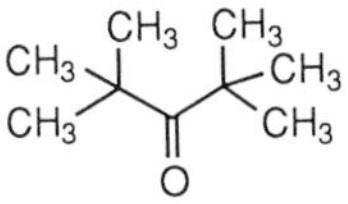

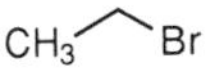

(iii)   CH$_2$=CH—CH$_2$—Cl   and   CH$_2$=CH—CHCl—CH$_3$

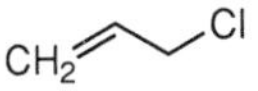

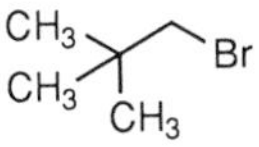

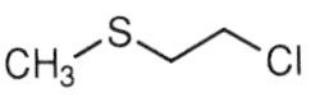

(iv)   CH$_3$—CO—Cl   and   CH$_3$—CH$_2$—Cl

(v)   CH$_3$—CO—CH$_3$ and   (CH$_3$)$_3$C—CO—C(CH$_3$)$_3$

(vi)   CH$_3$—CH$_2$—Br   and   (CH$_3$)$_3$C—Br

(vii)   CH$_3$—S—CH$_2$—CH$_2$—Cl   and   CH$_3$—S—CH$_2$—CH$_2$—CH$_2$—Cl

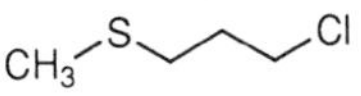

3. Account for the following:

   (i) Whereas t-butyl chloride almost instantly gives a precipitate with alcoholic silver nitrate, the chloride (I) is inert, even on prolonged boiling.

(I)

   (ii) Acetals are stable to bases but are readily hydrolyzed by acids.
   (iii) When the solvent polarity is increased, the rate of the S$_N$2 reaction,

$$HO^- + CH_3OSO_2Ph \rightarrow CH_3OH + PhSO_2O^-$$

is slightly reduced but that of the S$_N$2 reaction,

$$Et_3N + EtI \rightarrow Et_4N^+I^-$$

is greatly increased.

   (iv) The alkaline hydrolysis of ethyl bromide is catalyzed by iodide ion.
   (v) The dehydration of CH$_3$—CH(OH)—C(CH$_3$)$_3$ with concentrated sulfuric acid gives tetramethylethylene.
   (vi) Simple β-keto-acids, RCOCH$_2$CO$_2$H, readily decarboxylate on being heated, but the compound (II) is stable.

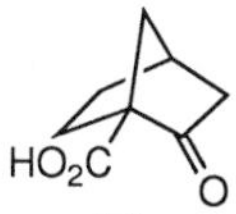

(II)

(*vii*)   The isomeric bromo-ethers (III) and (IV) undergo solvolysis in acetic acid to give the same mixture of products.

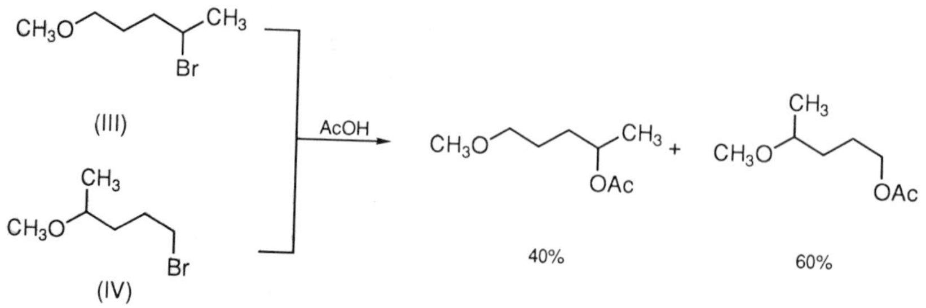

4.   What would you expect to be the major products of the following reactions:

(*i*)

(*ii*)

(*iii*)

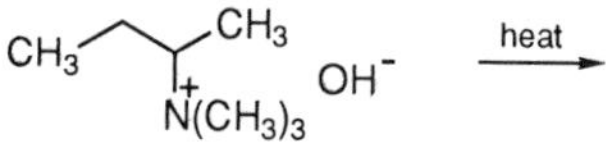

(*iv*)

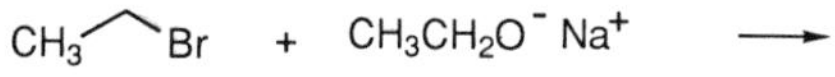

(*v*)

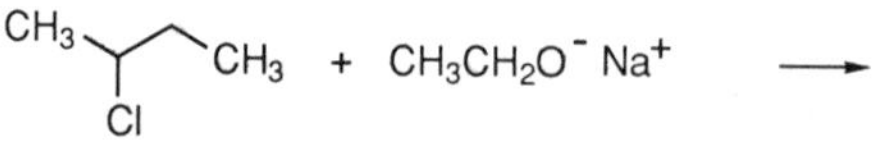

(*vi*)

(*vii*)

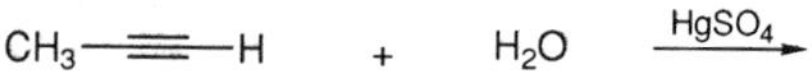

(*viii*)

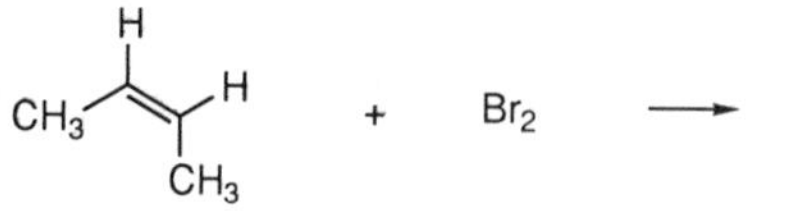

(*ix*)

(*x*)

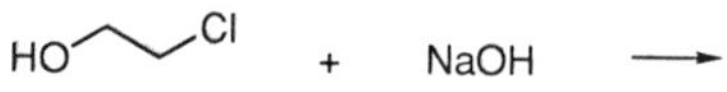

(*xi*) $CH_3\diagup\!\!\diagdown\!\!\diagup\!\!\diagdown CO_2H$ $\xrightarrow{\text{heat}}$

5. Indicate schematically the mechanisms of the following reactions:
    (*i*)   The acid-catalyzed hydrolysis of an amide, $RCONH_2$.
    (*ii*) The base-catalyzed hydrolysis of an ester, RCO_2CH_3.
    (*iii*) The acid-catalyzed bromination of acetone.
    (*iv*) The base-catalyzed self-condensation of ethyl acetate.

6. What would be the products if the following hydrolyses were carried out in water labelled with ^{18}O?

$$CH_3CO_2C(CH_3)_3 \quad \xrightarrow{H^+}$$

$$CH_3CO_2CH_3 \quad \xrightarrow{OH^-}$$

5 Stereochemistry

(*a*) *Enantiomers*

A compound whose structure is such that it is not superimposable on its mirror image exists in two forms, known as *enantiomers*. Lactic acid is an example: the two forms may be represented two-dimensionally as (1) and (2)

Other compounds of the general type *Cabcd* also exist in enantiomeric forms; they are described as *chiral*, from the Greek *cheir* for 'hand' (since the non-superimposability of enantiomers is like that of a left and a right hand) and the carbon atom with four different substituents is described as *asymmetric* or a *stereocentre*.

The physical and chemical properties of enantiomers are identical in a symmetrical environment but differ in a dissymmetric environment. In particular, enantiomers rotate the plane of plane-polarized light in opposite directions, although to the same extent per mole in the same conditions: they are described as being *optically active*. For example, lactic acid (1) is laevorotatory in aqueous solution and can be described as $(-)$-lactic acid; the enantiomer (2) is dextrorotatory and can be described as $(+)$-lactic acid.

A dissymmetric environment can also be provided by one enantiomer of another enantiomeric pair: $(+)$ and $(-)$ enantiomers usually react at different rates with another enantiomer (e.g. p. 150) and give different products (diastereomers, p. 140).

The phenomenon of enantiomers is known as optical isomerism and its importance stems from the fact that many naturally occurring compounds are optically active. Lactic acid, for example, occurs in sour milk as the dextrorotatory enantiomer and in muscle tissue as the laevorotatory enantiomer.

The criterion of non-superimposability of a structure and its mirror-image is a necessary and sufficient one for the existence of enantiomers. An alternative approach for deciding whether a particular structure is capable of existing in optically active forms is to examine the symmetry of the molecule: a structure lacking an alternating axis of symmetry is not superimposable on its mirror image and therefore represents an optically active species.* It is common practice to check whether the compound possesses a plane of symmetry (a one-fold alternating axis) or a centre of symmetry (a two-fold alternating axis). For example, *cis*-1,3-dichlorocyclobutane (3) cannot exist in optically active forms because it has two planes of symmetry and compound (4) cannot because it has a centre of symmetry.

(3)   (4)   (5)

Caution must be exercised if this approach is adopted; for example, the spiro-compound (5) possesses neither a centre nor a plane of symmetry, but it cannot exist in enantiomeric forms because it possesses a four-fold alternating axis of symmetry. (The reader should show that (5) is superimposable on its mirror image.)

(b) Projection diagrams

Since stereochemistry is a three-dimensional science which requires two-dimensional representation on paper, a number of conventions for representing structures has been developed. The earliest was Fischer's projection formula, illustrated for compounds of the type $Cabcd$, such as lactic acid. The molecule is first oriented so that the atom C is in the plane of the paper; two substituents, e.g. a and d, are arranged at the top and bottom relative to C, each below the plane of the paper and inclined equally to it; and the other two substituents, b and c, are arranged to the left and right of C, each above the plane of the paper and inclined equally to it. The resulting structure (6) is then projected onto the plane to give (7): a molecule shown in Fischer projection as (7) has the structure (6).

(6)   (7)   (8)

The Fischer projection has certain disadvantages. It is easily misused; for example, it must be remembered that (7) and (8) are not representations of the same

---

* An object possesses an n-fold alternating axis of symmetry if rotation of $360°/n$ about an axis followed by reflection in a plane perpendicular to that axis brings the object into a position indistinguishable from its original one.

compound but represent enantiomers. Further, the representation of compounds containing more than one asymmetric carbon atom is inadequate, e.g. the projection (9) for (−)-threose implies an eclipsing relationship of the groups attached to the two asymmetric atoms, whereas staggered conformations are more stable (cf. the conformational preferences of butane, p. 27). Moreover, the realistic presentation of many of the reactions of such compounds necessitates the representation of staggered structures (section 5.3).

$$\begin{array}{c} CHO \\ | \\ HO-C-H \\ | \\ H-C-OH \\ | \\ CH_2OH \end{array}$$

(9)

There are three conventions which avoid these difficulties. In one, the 'saw-horse' representation, the bond between the carbon atoms is drawn diagonally, implying that it runs downwards through the plane of the paper and is slightly elongated for clarity. The substituents on each of the carbons are then projected on to the plane of the paper and can be represented as staggered or eclipsed, e.g. for (−)-threose:

two of the staggered conformations of (−)-threose

an eclipsed conformation of (−)-threose

In the second, the Newman projection, the molecule is viewed along the bond joining the asymmetric carbon atoms and these atoms are represented as superimposed circles, only one circle being drawn. The remaining bonds and the substituents are then projected into the plane of the paper, the bonds to the nearer carbon being drawn to the centre of the circle and those to the further carbon being drawn only to the perimeter. The projections for (−)-threose which correspond to those above in the sawhorse representation are (10)–(12).

(10)   (11)   (12)

In the third convention, the longest chain, or backbone, of the molecule is drawn as a zig-zag in the plane of the paper, and substituents in front and behind the plane are drawn with wedged lines and dashed lines, respectively, e.g.

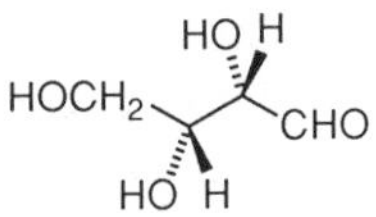

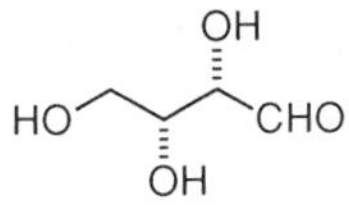

(–)-threose

This method is particularly suitable for chain compounds which possess multiple asymmetric carbon atoms.

(c) Conventions for describing chirality

The most widely used convention for describing the relative positions of the atoms or groups attached to an asymmetric carbon atom is based upon their atomic mass. First, the atoms attached to the stereocentre are listed in order of their decreasing atomic mass (e.g. $Cl > O > C > D > H$). If two or more of these atoms are the same, the atomic numbers of the atoms bonded to them define the precedence (e.g. $CH_2F > CH(OCH_3)_2 > CH_3$). If an atom forms a double or triple bond, it is treated as though each such bonded atom is duplicated or triplicated*, e.g.

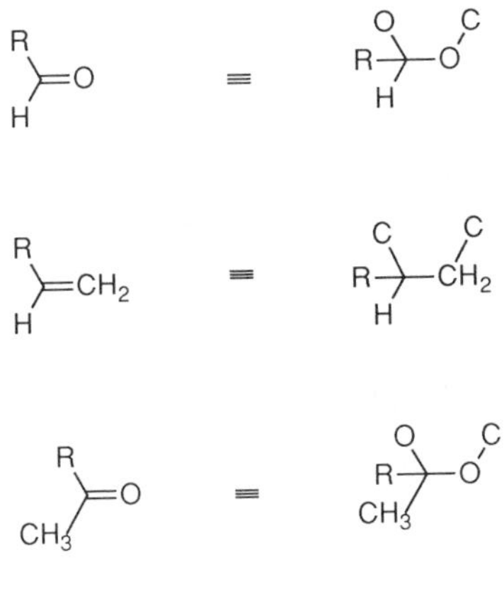

* This order of precedence is also applied in a convention for naming geometrical isomers. If the two substituents on the same side of a $C=C$ bond are of higher precedence, the prefix Z (German *zusammen*, together) is used; if they are on opposite sides, E (German *entgegen*, across), e.g.

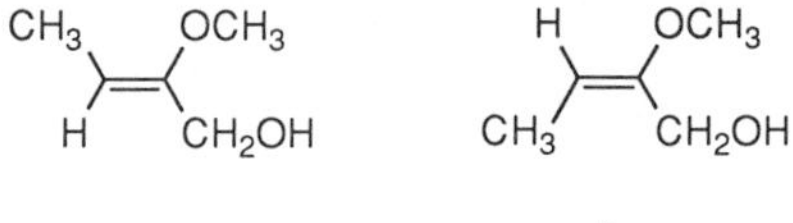

This has much wider generality than *cis,trans* nomenclature, which is applicable only if at least one pair of substituents on adjacent unsaturated carbon atoms are the same.

Then the asymmetric carbon atom is viewed from the side opposite the substituent of lowest priority. If the other groups appear in the order of decreasing priority in a clockwise direction, the compound is given the prefix R (Latin *rectus*, right); if anticlockwise, S (Latin *sinister*, left). For example, for glyceraldehyde, HOCH$_2$—CH(OH)—CHO, the priority order is OH > CHO > CH$_2$OH > H; viewing from the direction opposite the hydrogen atom gives, in a Newman projection in which the hydrogen is 'hidden',

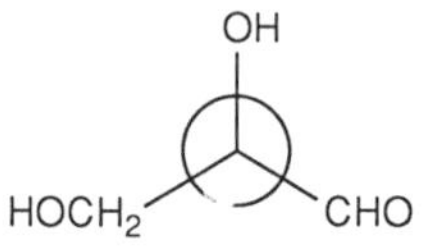

(R)-glyceraldehyde 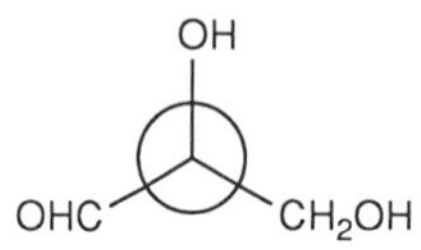   (S)-glyceraldehyde

Another convention employs the prefix D for (+)-glyceraldehyde and compounds having the same configuration that are derived from it, and the prefix L for compounds having the mirror-image configuration, e.g.

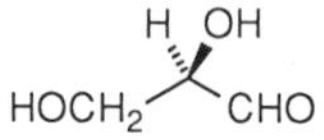

D-(+)-glyceraldehyde 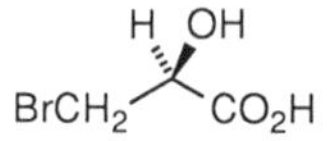   D-(−)-bromolactic acid

This convention is now usually employed only in naming members of two groups of compounds: sugars and α-amino acids. Sugars that have the same configuration as D-glyceraldehyde at the stereocentre most distant from the carbonyl group are termed D-sugars,* e.g.

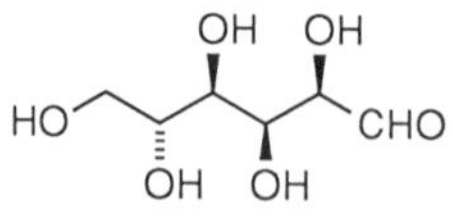

D-(+)-glucose

All the optically active α-amino acids of which naturally occurring peptides and proteins are constituted are formally related to L-glyceraldehyde, NH$_2$ replacing OH, and are named as L-amino-acids, e.g.

---

* The logic of this is that the D-sugars have been synthesized from D-glyceraldehyde by reactions which preserve the latter's stereochemistry and introduce a new carbonyl group further from its stereocentre, e.g.

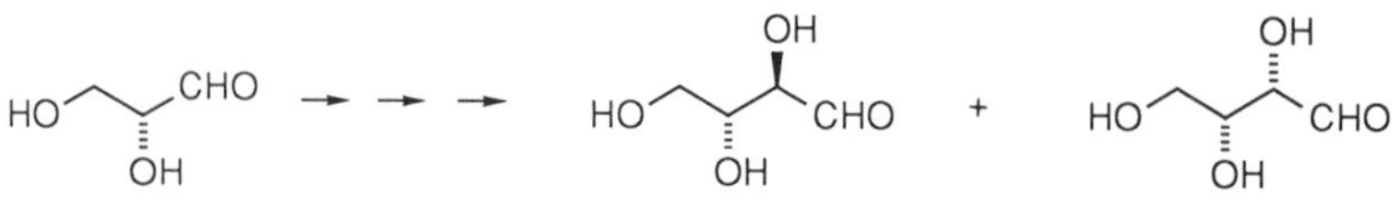

D-erythrose   D-threose

L-serine

L-phenylalanine

(d) Structural situations which give rise to optical isomerism

In addition to compounds of the type $Cabcd$ which possess one asymmetric carbon atom and fulfil the conditions necessary for the occurrence of enantiomeric pairs, there are several other structural situations which lead to the occurrence of optical isomerism. Some of these structures contain one or more stereocentres and others owe their optical activity to the presence of other dissymmetric features.

(i) *Compounds possessing a stereocentre other than carbon.* Since the valencies of nitrogen in an amine are directed approximately towards three corners of a tetrahedron with the lone-pair being directed towards the fourth corner, it might be expected that an amine of the type $Nabc$ would exist in enantiomeric forms: the mirror images (13) and (14) are not superimposable. However, the separation of such acyclic amines into optically active isomers has never been achieved

(13)   (14)

because the isomers are too rapidly interconvertible by a flapping mechanism,

i.e.   (13)   ⇌   (14)

However, the planar transition state for flapping of some aziridines, e.g.

(15)   (16)

is sufficiently strained that it has been possible to isolate the individual enantiomers.

Ternary phosphorus, arsenic, and antimony compounds are configurationally more stable than their nitrogen analogues and several have been resolved, e.g. PMeEtPh, whose enantiomeric forms have the structures (17) and (18). Since quaternary ammonium ions cannot flap in the manner of amines, those of the type $Nabcd^+$ exist as enantiomeric pairs some of which have been separated, e.g. (19) and (20). Phosphonium and arsonium salts have also been resolved.

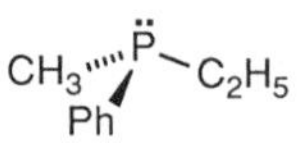

(17) 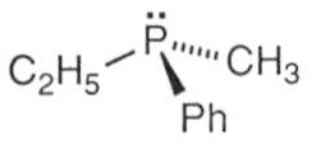   (18) 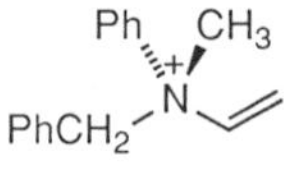   (19) 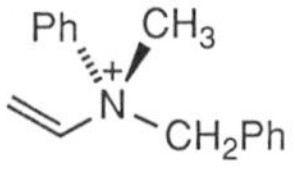   (20)

Optically active sulfonium salts, sulfoxides, and silicon and germanium analogues of hydrocarbons have also been obtained, e.g.

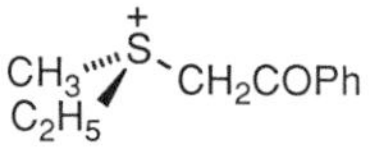

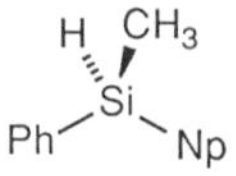

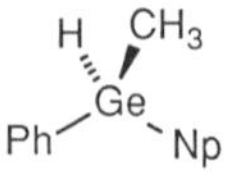

(Np = 1-naphthyl)

(ii)   Compounds possessing two or more asymmetric carbon atoms.   Compounds with two asymmetric carbon atoms in which at least one substituent is not common to both carbons occur in four optically isomeric forms, e.g. the four isomers of structure $CH(OH)(CH_2OH)—CH(OH)(CHO)$ are:

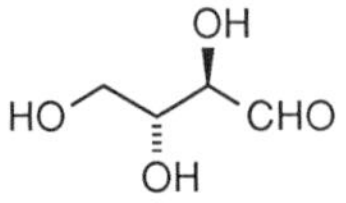

(−)-erythrose 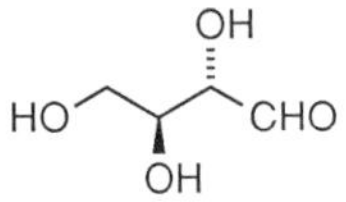   (+)-erythrose 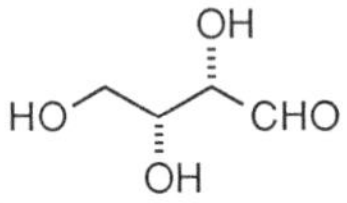   (−)-threose 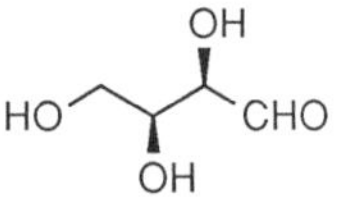   (+)-threose

In general, a compound possessing n distinct asymmetric carbon atoms exists in 2^n optically active forms.

It should be noted that, whereas (+)- and (−)-erythrose and (+)- and (−)-threose are mirror-image pairs and therefore have identical properties in a symmetric environment, neither of the erythroses bears a mirror-image relationship to either of the threoses. These and other stereoisomers which are not enantiomers are described as *diastereomers* (or diastereoisomers) and, unlike enantiomers, they differ in physical and chemical properties.

Compounds which contain two asymmetric carbon atoms and are of the type *Cabc–Cabc* exist in only three isomeric forms. Two of these are non-superimposable mirror images of each other and are optically active and the third, a diastereomer of the first two, contains a plane of symmetry, is superimposable on its mirror image, and is not optically active, e.g.

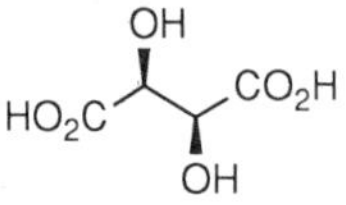

(S,S)-(−)-tartaric	(R,R)-(+)-tartaric	(R,S)-tartaric	(24)
acid	acid	acid	
(21)	(22)	(23)	plane of symmetry perpendicular to the paper

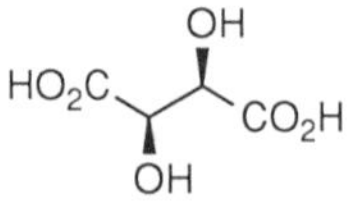

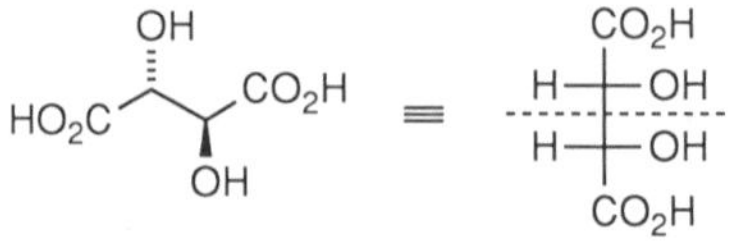

The inactive (R,S)-diastereomer is usually described as a *meso* form. As with other examples of diastereomers, the properties of *meso* forms are different from those of the isomeric mirror-image pairs: for example, *meso*-tartaric acid melts at a lower temperature (140°C) than the (R,R)- and (S,S)-isomers (170°C), and is less dense, less soluble in water, and a weaker acid.

It is simple to extend this discussion to compounds which contain more than two asymmetric carbon atoms. For example, the aldohexoses, CH_2OH—$(CHOH)_4$—CHO, possess four such atoms, each of which is distinct from the remainder, and there are therefore 2^4 or 16 stereoisomers all of which are optically active. On the other hand, some of the derived saccharic acids, HO_2C—$(CHOH)_4$—CO_2H, have symmetry properties which make them optically inactive: ten stereoisomers exist of which eight are optically active (four pairs of enantiomers) and two, (25) and (26), are *meso* forms.

```
       CO2H                    CO2H
   H —— OH               H  ——  OH
   H —— OH               HO ——  H
   H —— OH               HO ——  H
   H —— OH               H  ——  OH
       CO2H                    CO2H

      (25)                     (26)
```

(*iii*) *Allenes.* An *sp*-hybridized carbon atom possesses one electron in each of two mutually perpendicular *p* orbitals. When it is joined to two sp^2-hybridized carbon atoms, as in an allene, two mutually perpendicular π-bonds are formed and consequently the π-bonds to the sp^2-carbons are in perpendicular planes (27). Allenes of the type $abC\!=\!C\!=\!Cab$ ($a \neq b$) are therefore not superimposable on their mirror images and, despite the absence of any asymmetric atoms, exist as enantiomers (28) and (29). Several optically active compounds have been obtained (e.g. a = phenyl, b = 1-naphthyl).

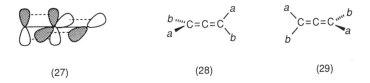

(27)   (28)   (29)

(*iv*) *Alkylidenecycloalkanes.* The replacement of one double bond in an allene by a ring does not alter the basic geometry of the system and appropriately substituted compounds exist in optically active forms, e.g. (30). Related compounds in which sp^2-carbon is replaced by nitrogen, e.g. (31), have also been obtained as optical isomers.

(30)   (31)   (32)

(v) *Spirans.* The replacement of both double bonds in an allene by ring systems gives a spiran; appropriately substituted compounds have been obtained in optically active forms, e.g. (32).

(vi) *Biphenyls.* The two conformations, (33) and (34), of *meso*-2,3-dichloro-butane are non-superimposable mirror-images of each other. However, it is not possible to isolate optically active forms because rotation about the central C—C bond occurs rapidly and results in the interconversion of the two structures. On the other hand, when the barrier to rotation about a C—C bond exceeds about $80 \, kJ \, mol^{-1}$, rotation is slow enough at room temperature for the isolation of optically active isomers to become practicable. This situation holds for certain biphenyls of the general structure (35).

(33)   (34)   (35)

Optical isomerism in biphenyls is possible because the conformation in which the benzenoid rings are coplanar and which possesses a plane of symmetry is strained with respect to non-coplanar conformations such as (35) as a result of the steric repulsions between pairs of *ortho*-substituents on the adjacent rings. The substituents X and Y require to be fairly bulky for the occurrence of the necessary barrier to rotation and their presence also ensures that the molecule has the necessary lack of symmetry for optical activity. Examples of compounds of this type which have been obtained in optically active form are biphenyl-2,2'-disulfonic acid (X = Y = SO₂OH) and the diamine (X = Y = N(CH₃)₂).

A given energy barrier between a biphenyl and its enantiomer is surmounted more rapidly at higher temperatures. Hence one enantiomer is converted, with first-order kinetics, into a mixture of the enantiomers at a rate which increases with temperature, and the optical activity of the species falls eventually to zero when equilibrium is reached. This process – racemization (p. 145) – occurs less rapidly as the sizes of X and Y are increased. The introduction of one or two more *ortho*-substituents in place of hydrogen atoms therefore reduces the rate of racemization to an extent depending on the sizes of the groups, e.g. racemization of the compounds (36) occurs at 118°C with a half-life of 91 minutes (Z = OCH₃), 125 minutes (Z = NO₂), and 179 minutes (Z = CH₃).

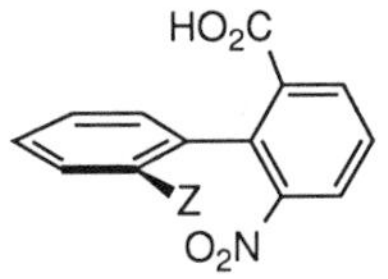

(36) 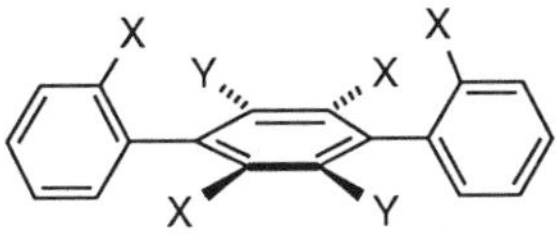   (37)

Optical isomerism occurs also in suitably substituted polyphenyls but the stereochemistry is more complex since both *meso* forms and geometrical isomerism are also possible. For example, compounds of the type (37) exist in three stereoisomeric forms: there are enantiomeric *cis* isomers, one of which is (37), and a *trans* isomer (in which the X groups on the end rings are on opposite sides) which, having a centre of symmetry, is an optically inactive (*meso*) form.

There are other compounds which are optically active due to restricted rotation about single bonds, of which the following are examples:

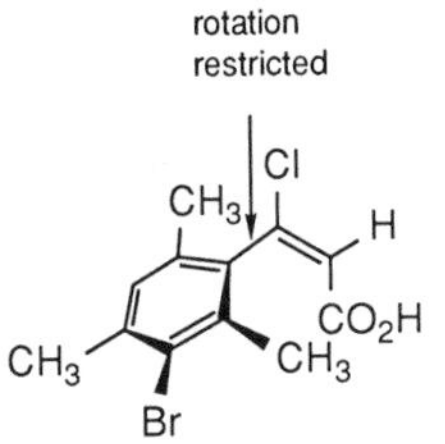

(38) 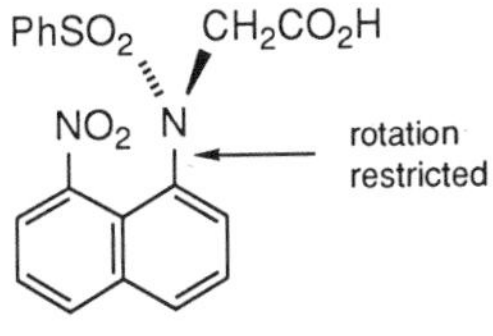   (39)

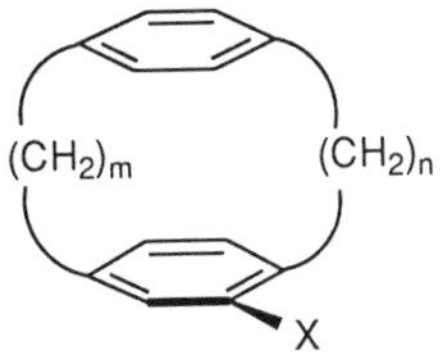

(40)

Compounds of the type (40) ('paracyclophanes') have been resolved when *m* and *n* are fairly small (e.g. when *m* = 3, *n* = 4, and X = CO_2H) for then the benzenoid rings cannot rotate at a significant rate. When the connecting methylene chains are longer, however, rotation is so fast that the enantiomers rapidly equilibrate.

(e)   *Absolute configuration*

The complete structure of a chiral molecule is not elucidated until the absolute configuration of the compound – the actual arrangement of the atoms in space – is known. The first determination of absolute configuration, by an X-ray method (1951), showed that sodium rubidium (+)-tartrate has the absolute (R,R) configuration (41).

$$RbO_2C \qquad \overset{H \; OH}{\underset{H \; OH}{\diagup}} \qquad CO_2Na$$

(41)

The absolute configurations of other optically active compounds can be obtained from that of the (+)-tartrate by correlative methods. The simplest of these is to convert one optically active compound into another without breaking any bond to the relevant stereocentre. The absolute configurations of reactant and product are then necessarily the same, so that if one is known the other follows. For example, the known sodium rubidium (R,R)-(+)-tartrate can be converted in this way successively into (R)-(+)-malic acid, (R)-(+)-isoserine, (S)-(−)-bromolactic acid, and (R)-(−)-lactic acid, all of whose absolute configurations are thereby established.

(R,R)-(+)-tartaric acid   (R)-(+)-malic acid   (R)-(+)-isoserine

(S)-(−)-bromolactic acid   (R)-(−)-lactic acid

It must be emphasized that there is no simple relationship between absolute configuration and the sign of rotation of an optically active compound; a structural alteration in which the absolute configuration of the chiral centre is maintained can result in a change of the direction of rotation, as in the above conversion of (R)-(+)-isoserine into (S)-(−)-bromolactic acid.

(f) Racemates and racemization

The optical activities of equal numbers of molecules of enantiomers nullify each other. Such an assembly is known as a *racemate* and is often symbolized by (±).

A racemate can crystallize in one of three ways. When each enantiomer has a greater affinity for molecules of its own kind than for those of the other, the enantiomers tend to crystallize separately to give a *racemic mixture* of two types of crystals. When each has a greater affinity for molecules of the other enantiomer, the crystal grows by the laying down of (R)- and (S)-molecules alternately, to give a *racemic compound*. When there is little difference between the affinities of one enantiomer for molecules of its own type and for those of the other, the arrangement in the solid is random and a *racemic solid solution* is obtained.

Racemates can be obtained in three ways: by mixing equal amounts of the enantiomers, by synthesis, and by racemization.

(*i*) *Synthesis.*   The synthesis of an asymmetric compound from symmetric reactants necessarily produces a racemic modification. Consider, for example, the addition of hydrogen cyanide to acetaldehyde. If the approach of cyanide ion to the aldehyde in the conformation (42) is as shown, the product is (R)-lactonitrile, but it is statistically as probable that the approach will be to the mirror-image conformation (43), giving (S)-lactonitrile. It is apparent that any particular pathway to the (R)-product corresponds to an equally probable pathway to the (S)-product, so that a racemate is produced.

(42)

(43)

(*ii*) *Racemization.*   Racemization, defined as the production of a racemate from one enantiomer, is normally an undesirable process; that is, when

it is necessary to synthesize an optically active compound, it is necessary to avoid steps which result in racemization.

Racemization can occur in a number of ways, one or more of which may apply to a particular optically active compound. The commoner mechanisms are as follows.

(1) *By rotation about a single bond.* The biphenyls and related compounds, in which optical activity results from the restriction of rotation about a single bond, racemize when enough thermal energy is present for the energy barrier between the enantiomers to be surmounted at a practicable rate (p. 142).

(2) *Via an anion.* Since carbanions RR'R''C$^-$, like amines, are not optically

stable, an enantiomer of the type RR'R"C—X racemizes in an environment in which X is reversibly removed without its bonding pair of electrons. The group X is most commonly hydrogen, and if one or more of the other substituents is of the type which stabilizes a carbanion, racemization can occur in basic conditions. For example, enantiomers containing a carbonyl group adjacent to the asymmetric carbon racemize in this way:

Other groups of $-M$ type, such as nitro and ester (p. 32), have the same effect as carbonyl.

(3) *Via a cation.* Carbocations are coplanar at the tervalent carbon atom, so that if the ion RR'R"C$^+$ is formed from an enantiomer RR'R"C—X by the reversible removal of X with its bonding pair, racemization occurs. The group X is usually a halogen, and its removal can be aided by a Lewis acid, e.g.

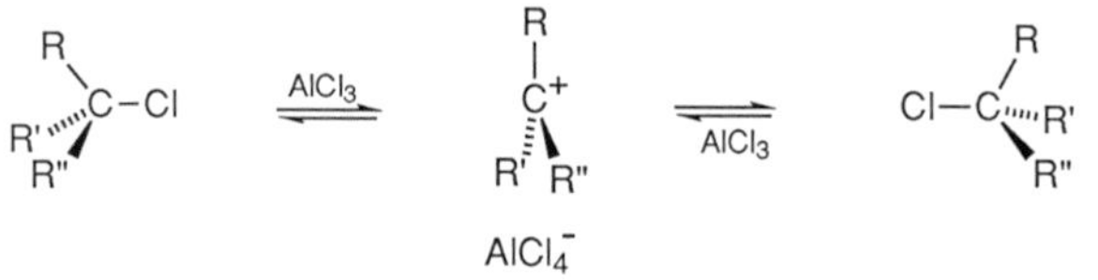

(4) *Via reversibly formed, inactive intermediates.* This mechanism is similar to methods (2) and (3) save that the intermediate is a relatively stable compound as compared with the transient carbanions and carbocations. For example, optically active ketones containing an α-CH group racemize when treated with acid which catalyzes the formation of the optically inactive enol tautomer:

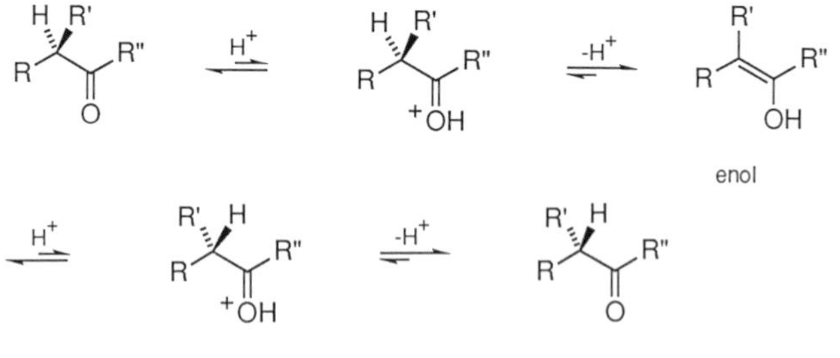

This method is complementary to base-catalyzed racemization.

(5) *By S$_N$2 reaction.* The racemization of optically active halides, in which the halogen is attached to the asymmetric carbon, occurs in the presence of the corresponding halide ion. This is a result of the fact that the S$_N$2 reaction causes inversion of configuration (p. 160).

Repetition of this process leads to the racemate.

(g)   Epimerization

Epimerization is said to occur when there is a change in configuration at one stereocentre in a compound which possesses more than one such centre. It results in the formation of a diastereomer, not the enantiomer, of the starting material. The mechanisms of epimerization parallel those of racemization.

(h)   Resolution

The resolution of a racemate consists of separating the enantiomers and isolating them in a pure state. (In synthesis, it is common practice to attempt to obtain only the required enantiomer in pure form.) There are two widely used approaches to resolution: *via* diastereomers, and by biochemical methods.

(i)   Via diastereomers.   The reaction of each of a pair of enantiomers with an optically active compound gives two diastereomers. Since diastereomers do not have identical properties, it is usually possible to separate them: crystallization is most commonly employed. Provided that the reaction leading to the diastereomers can be reversed, resolution can be achieved.

The criteria for a satisfactory resolution by this method are that the diastereomers should be easily formed, well crystalline, and easily broken up. These conditions are met by employing salts: if the required enantiomers are acids or bases, or can be converted readily into them, treatment with an optically active base or acid, respectively, gives the diastereomeric salts and, after separation, the required enantiomers can be obtained by treating each diastereomer with mineral acid to isolate an organic acid, or with base to isolate an organic base. Many optically active acids and bases which are suitable as resolving agents occur naturally, e.g. (+)-tartaric acid and the alkaloids quinine and brucine.

Enantiomers which are neither acids nor bases are usually converted into acidic derivatives for resolution. For example, alcohols can be treated with phthalic anhydride to give the acid phthalate esters (44) which, after separation, are reconverted into the enantiomeric alcohols by hydrolysis with sodium hydroxide or reduction with lithium aluminium hydride (p. 658). Carbonyl compounds can be resolved by treatment with the semicarbazide (45), resolution of the resulting enantiomeric acids (46) with an optically active base, and reconversion by hydrolysis. Amino acids, which are amphoteric, are usually first converted into their N-acetyl derivatives which are resolved as acids and reconverted by hydrolysis.

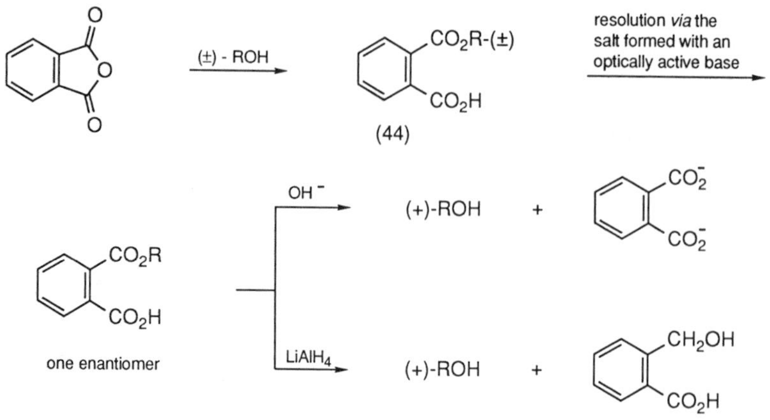

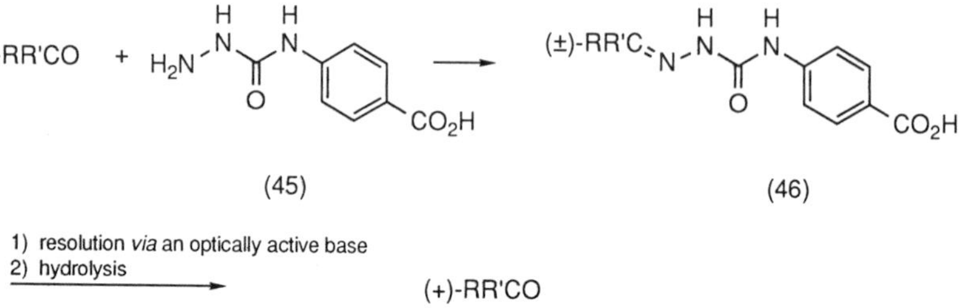

(44)

one enantiomer

$(\pm)$-RR'CO +

(45)

(46)

1) resolution *via* an optically active base
2) hydrolysis
───────────────────→   (+)-RR'CO

Compounds which cannot be converted readily into acidic or basic derivatives may be obtained in optically active form by synthesis from an optically active precursor. There are, however, some special, rather than general, methods of resolution *via* diastereomers.

First, the diastereomeric molecular complexes formed with a chiral reagent may be employed. For example, the optically active fluorenone derivative (47), itself obtained by resolution as an acid, can be used to resolve derivatives of aromatic compounds such as naphthalenes which complex with it; the separated diastereomers are usually reconverted readily into the reactants by heating or chromatography. A related method employs the inclusion complexes (clathrates) formed by certain chiral compounds.

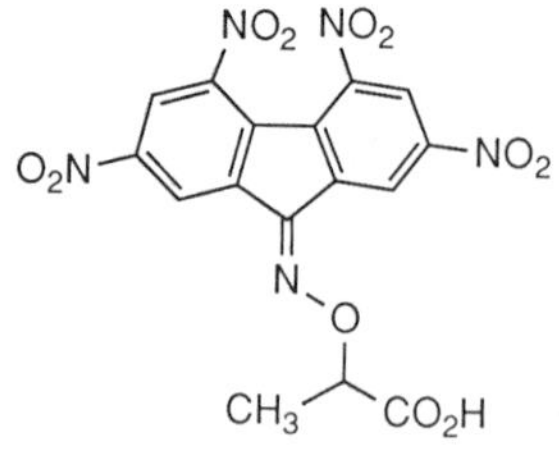

(47)

Second, in some cases enantiomers can be separated by chromatography on an optically active support such as silica in which surface hydroxyl groups have been derivatized by treatment with (R)-phenylglycine. Diastereomeric adsorbates are formed which have different stabilities; the enantiomer forming the less stable adsorbate is eluted first.

(*ii*) *By biochemical methods.* These methods differ from those in which diastereomers are formed and separated in employing the differences in *rates* of reactions of each of a pair of enantiomers with, or in the presence of, an asymmetric material of natural origin. The asymmetric compounds used include both enzymes and living organisms such as moulds. Their use is special rather than general, but where they are successful they are usually very efficient. The following are examples.

(1) Whereas the laboratory reaction between acetaldehyde and hydrogen cyanide gives optically inactive lactonitrile (p. 145), the presence of the enzyme emulsin gives a nearly optically pure product.

(2) The mould *Penicillium glaucum* preferentially destroys the (+)-isomer of racemic ammonium tartrate, leaving the pure laevorotatory salt.

(3) The reaction of racemic acetylphenylalanine with *p*-toluidine catalyzed by the enzyme papain gives the *p*-toluidide of acetyl-L-phenylalanine and leaves unchanged D-phenylalanine:

These three examples are typical of methods which are described respectively as asymmetric synthesis, asymmetric destruction, and asymmetric kinetic resolution.

(*iii*) *Other methods.* A variety of less general methods has been used for resolution. The earliest ever employed was the hand-picking of the asymmetric crystals of the enantiomeric sodium ammonium tartrates, but this method is limited because it is applicable only to racemic mixtures (p. 144), it requires that the crystals be easily distinguishable by eye, and it is excessively tedious.

Another method is to seed a saturated solution of the enantiomers with a crystal of one form which can induce the crystallization of the same enantiomer. If a crystal of one form is not available, it is sometimes possible to induce selective crystallization by seeding with a crystal of an optically active form of another molecule. In other cases one enantiomer may crystallize spontaneously from a supersaturated solution.

Finally, resolution has been achieved in one instance by inducing a photo-chemical reaction with circularly polarized light: irradiation of racemic $CH_3CH(N_3)CON(CH_3)_2$ preferentially destroys one enantiomer, depending on the direction of polarization of the light, and leaves some of the reactant enriched in the other enantiomer.

(*i*)   *Diastereoselective synthesis*

Many naturally occurring compounds contain several stereocentres, as do many other structures that are important synthetic targets (for example, compounds that are thought likely to have valuable biological properties). In order to maximize the overall yield in the synthesis of such compounds, it is important to employ reactions for the generation of each stereocentre that give as high a yield as possible of the desired stereoisomer.

A reaction between a symmetric organic compound and a symmetric reagent which creates a stereocentre inevitably yields equal amounts of the two enantiomers (p. 145). However, if one of the components is asymmetric, two diastereomers are formed in different amounts.*

This is the basis of diastereoselective synthesis and it can be illustrated by the addition of a nucleophile Nu^- to the carbonyl group of a chiral compound R—CO—CLMS, where L, M, and S represent, respectively, the largest, medium-sized, and smallest of the substitutents (other than the keto group) on the asymmetric carbon. Prediction of the stereochemistry of the major product of the reaction can be made by applying *Cram's rule*, which was originally based on experimental observations and has now been rationalized, with the aid of theoretical calculations, in the *Felkin-Anh model*, as follows. The lowest-energy transition state corresponds to a conformation of the carbonyl compound in which, in Newman projection, the C—L bond is perpendicular to the carbonyl group. The nucleophile approaches the carbonyl carbon in a plane perpendicular to that of the —CO— fragment, from the side opposite the C—L bond and at an obtuse angle with C=O, corresponding approximately to the tetrahedral angle of Nu—C—O in the product. There are two such conformations, (48) and (50), and the former is of lower energy because there is less steric interference between the nucleophile and the smallest group, S, than between the nucleophile and the larger group, M. The major product is therefore (49). Likewise, the enantiomeric ketone gives mainly the enantiomer of (49).

---

*When a stereocentre is established during a reaction between two symmetric components, en-antiomeric transition states are involved which, except that they are non-superimposable mirror images, are structurally identical and therefore of the same stability. However, if one of the components is asymmetric, the transition states are diastereomeric and therefore can differ in energy, whence rate differences arise.

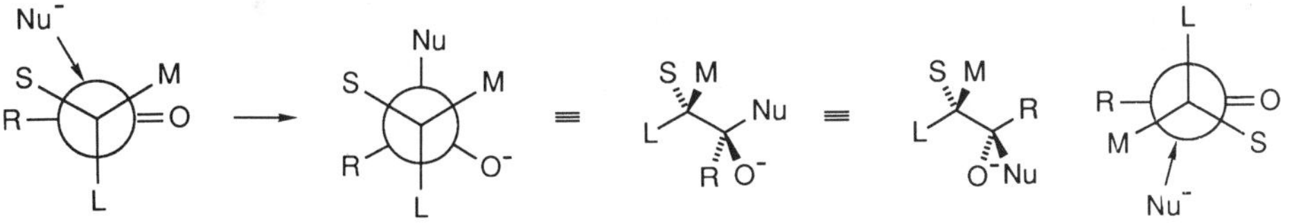

For example, the reduction of (R)-PhCHMe—COMe (in which the sizes of the groups decrease in the order Ph > CH$_3$ > H) with lithium aluminium hydride gives three times as much of the product predicted by Cram's rule as its diastereomer:*

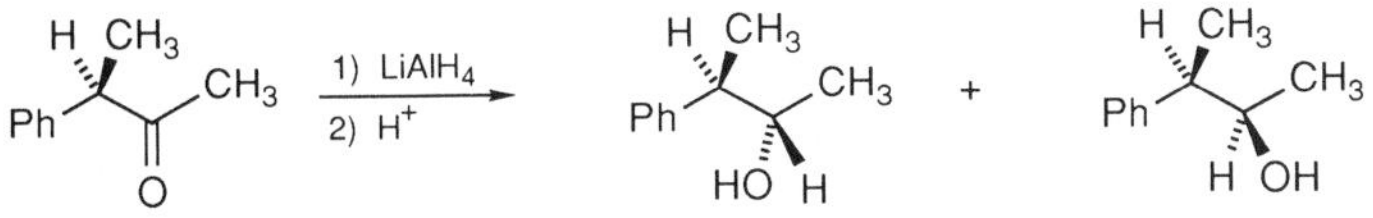

3 parts   1 part

There are two common circumstances when Cram's rule is not necessarily followed. When the reagent is associated with a metal ion and one of the substituents on the asymmetric carbon is an alkoxy or other complexing group, the preferred transition state corresponds to conformation (51); and when one of the substituents is strongly electronegative (e.g. bromine), it corresponds to (52) as a result of the tendency of the negatively polarized oxygen and bromine atoms to be as far apart as possible.

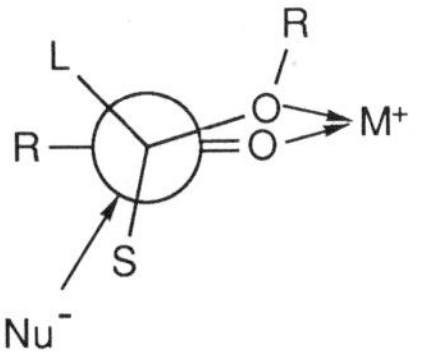

(51) 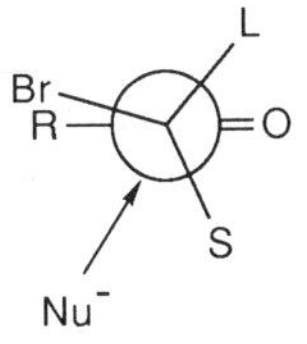   (52)

This approach to stereoselectivity has two limitations: the necessary stereocentre may not be present in the organic reactant and in most cases the degree of selectivity is usually not high. It is therefore generally better to employ an asymmetric reagent: in particular, for reduction of C=O and for certain additions to C=C, asymmetric reagents are commercially available which, because of the bulkiness associated with their reactive sites, are highly enantioselective (e.g. pp. 487, 653).

* It is convenient to define the *diastereomeric excess* (d.e.) in the reaction as the difference in yields of the two diastereomers as a fraction of their total yield, ×100%. In this case, the d.e. = (3 − 1)/4 × 100 = 50%.

**5.2
The
stereochemistry
of cyclic
compounds**

The shapes of cycloalkanes containing three to six carbon atoms were discussed briefly in chapter 2 (p. 28) and these compounds are here considered in the wider context of their stereochemical properties.

(a)   Types of stereoisomerism in cycloalkanes and their derivatives

(i)   Cyclopropanes.   The carbon atoms in cyclopropanes must inevitably lie in a plane. A monosubstituted (53) and a 1,1-disubstituted (54) cyclopropane exist in only one form, for the structures possess a plane of symmetry. A 1,2-disubstituted compound exists in four forms: a pair of *cis* enantiomers (55) and a pair of *trans*

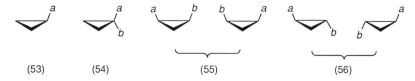

enantiomers (56). If the substituents are the same ($a = b$) the *cis* form possesses a plane of symmetry and is therefore a *meso* type.

(ii)   Cyclobutanes.   The CCC-angles would be least strained if cyclobutanes adopted a coplanar-ring structure, but this structure corresponds to maximum steric repulsion (torsional strain) between substituents on adjacent carbon atoms. As a result of these opposing factors, the ring buckles slightly from the plane.

As with cyclopropanes, monosubstituted and 1,1-disubstituted cyclobutanes exist in only one form whereas 1,2-disubstituted cyclobutanes exist as diastereomeric pairs of enantiomers (57, 58) (or a pair of enantiomers and a *meso* form if $a = b$). However, both the *cis* and the *trans* 1,3-disubstituted derivatives (59) and (60) possess a plane of symmetry so that there are only two isomers and neither is optically active. (This is true of any disubstituted cycloalkane containing an even number of carbon atoms where the substituents are on the opposite sides of the ring.)

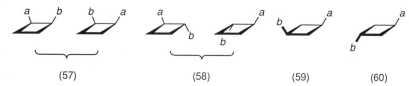

(iii)   Cyclopentanes.   The opposing angular and torsional strain factors which apply to cyclobutanes apply also to cyclopentanes, and these compounds adopt an 'envelope' structure.

'envelope' conformation
of cyclopentane

The 1,2- and 1,3-disubstituted compounds exist in four forms, there being two diastereomeric (*cis* and *trans*) pairs of enantiomers (cf. cyclopropanes).

(*iv*) *Cyclohexanes*.   The cyclohexane system is by far the most commonly occurring of the cycloalkanes in natural products, doubtless because of its stability and the (related) ease of its formation (p. 28).

Unlike the 3- to 5-membered rings, cyclohexanes can adopt a conformation which is free both of angular strain and of torsional strain: this is the *chair* conformation (61) shown also in Newman projection (63). By co-ordinated rotations about its single bonds, the molecule can flip into the mirror-image chair conformation (62), and since the energy barrier between these forms is only 42 kJ mol^{-1}, flipping occurs rapidly at room temperature (k c. 10^4 s^{-1}). In between the chair structures there are two other notable conformations: the skew-boat conformation (64), which is 23 kJ mol^{-1} less stable than the chair and the boat conformation (65) (Newman projection (66)), which is 30 kJ mol^{-1} less stable. It should be noted that the last is free of angular strain but possesses considerable torsional strain as a result of four eclipsing interactions (*xx*, etc.) and repulsion between the two groups *y* (the so-called bowsprit-flagpole interaction). Although the skew-boat form is not strainless, the torsional strain forces are considerably less than those in the boat form and the structure is (7 kJ mol^{-1}) more stable. The complete energy profile is shown in Figure 5.1; the chair conformations constitute over 99.9% of the mixture of isomers.

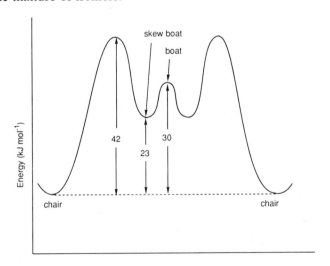

Figure 5.1   Energy profile and conformations of cyclohexane.

Axial and equatorial bonds.   The twelve C—H bonds in the chair structure of cyclohexane are of two types (67). Six of the bonds are parallel to the three-fold axis of symmetry of the molecule, i.e. they are represented by vertical lines in the

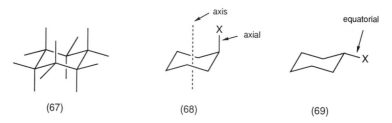

(61)   (62)   (63)

(64)   (65)   (66)

plane of the paper, and are described as *axial*, and the remaining six bonds are inclined at an angle of 109° 28′ to the three-fold axis and are described as *equatorial*. In derivatives of cyclohexane, the substituents which are in one or other of these two situations are described as axial and equatorial substituents, respectively.

(67)   (68)   (69)

Monosubstituted cyclohexanes. The substituent in a monosubstituted cyclohexane may be either axial (68) or equatorial (69). These structures are isomeric, but it is not possible to isolate two isomers – *conformational isomers* – because the rate of interconversion of the two, through ring flipping, is too rapid and an equilibrium is maintained. The two isomers normally have different energies, so that the equilibrium constant is not unity. The difference arises because there are repulsive forces between the axial substituent and the axial hydrogen atoms on C_3 and C_5 which are not present when the substituent is equatorial and which destabilize the axially substituted compound. The difference in stability increases with the size of the substituent, e.g. for methyl, phenyl, and t-butyl, ΔG^0 is c. 7·1, 11·3, and 23·4 kJ mol^{-1}, so that the equilibrium constants are about 20, 100, and 10^4, respectively, in favour of the equatorially substituted conformational isomer*.

---

*From $RT \ln K = -\Delta G^0$. Typical relationships between K and $-\Delta G^0$ at room temperature are (approximately):

$-\Delta G^0$/(kJ)	3	6	11	17	23
K		3	10	10^2	10^3	10^4

Disubstituted cyclohexanes. There are three geometrically different relation-ships between the substituents, X, in a 1,2-disubstituted cyclohexane of the formula $C_6H_{10}X_2$: each substituent may be axial (70), each may be equatorial (72), or one may be axial and the other equatorial (74). The three structures are shown together with their mirror images.

None of the structures is immediately superimposable upon its mirror image, and three pairs of enantiomers might be expected. In practice, however, there are fewer stereoisomers as a result of the rapid interconversions of chair confor-mations: (70) is in equilibrium with (73) (as can be seen by converting (70) into

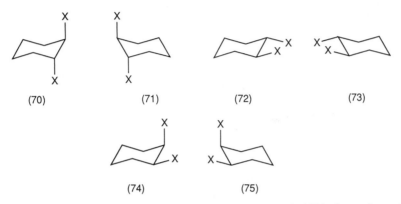

the alternative chair conformation and rotating this through 120° about the axis of symmetry) and likewise (71) is in equilibrium with (72). These four structures therefore constitute one pair of enantiomers. The remaining pair are superimposable by chair flapping, and therefore constitute a (±)-pair which are unresolvable because they interconvert too rapidly.

Largely for historical reasons* the structures (70)–(73) are described as *trans* and the two which constitute the (±)-pair are described as *cis*. Clearly, the relationship between the X groups in (70) and (71) is a *trans* one, but it is not so in the conformational isomers (73) and (72). It is useful to retain the terminology however, and *trans* substituents in cyclohexanes (and other rings) are defined as substituents which are on opposite sides of the ring when the ring is moulded into a planar form; *cis* substituents are those which are correspondingly on the same side of the ring. The *trans* isomer is usually more stable when the substituents are

---

* It was originally thought that cyclohexanes have planar rings. The relationship between the isomers above was therefore imagined to be similar to that of the geometrical isomers of alkenes, i.e.

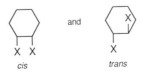

The non-resolvable *cis* isomer is now seen to be the conformational isomer with one axial and one equatorial substituent, and the resolvable *trans* isomer to be the conformational isomer with both substituents either axial or equatorial.

both equatorial (cf. monosubstituted cyclohexanes), e.g. for $X = CH_3$, ΔG^0 is about $11 \cdot 3\,kJ\,mol^{-1}$ so that only about one molecule in a hundred is present in the diaxial form. The *trans* isomer is more stable than the *cis* isomer, in which one substituent is necessarily in the unfavourable axial situation, e.g. for $X = CH_3$, ΔG^0 is about $7 \cdot 5\,kJ\,mol^{-1}$.

This discussion of the 1,2-disubstituted cyclohexanes can readily be extended to the 1,3- and 1,4-analogues. The 1,3-compounds exist in three discrete stereoisomeric forms: an enantiomeric *trans* pair in which one substituent is axial and the other equatorial, (76) and (77), and a *cis* isomer which has a plane of symmetry (not resolvable) and which consists of interconvertible conformational isomers having both substituents respectively equatorial (78) and axial (79), the

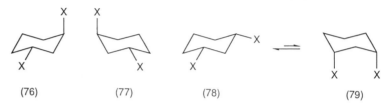

(76)   (77)   (78)   (79)

former predominating. The *cis* isomer (78) is in this case the more stable, e.g. for $X = CH_3$, by $7 \cdot 5\,kJ\,mol^{-1}$.

The 1,4-compounds exist in only two forms: a *trans* compound, with both substituents either equatorial (80) or axial (81); and a *cis* compound, with one axial and one equatorial substituent (82). Both *cis* and *trans* compounds possess a plane of symmetry and are optically inactive and the *trans* isomer is the more stable.

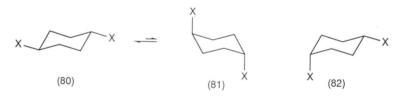

(80)   (81)   (82)

(*v*) *Cyclohexene.* Since the four carbon atoms in the fragment $C—C=C—C$ are necessarily coplanar, cyclohexene cannot adopt either a chair or a boat form. Instead, its preferred conformation is the half-chair (83).

(83)   (84)

(*vi*) *Cyclohexanone.* Compared with cyclohexane, the ring in cyclohexanone is buckled very slightly from the chair structure in order to accommodate the trigonal carbon atom whose optimal CCC angle is 120°. As shown in (84), the equatorial hydrogen atoms on the α-carbon atom are then nearly eclipsed by the carbonyl oxygen, so that equatorial substituents are somewhat destabilized by steric repulsions. In some cases this results in the greater stability of the axially substituted conformational isomer (e.g. 2-bromocyclohexanone) and in others, of the boat-shaped conformation [e.g. *cis*-2,4-di-t-butylcyclohexanone (85)].

(85)

(*vii*)   *Larger rings*.   Heat of combustion data show that cycloalkanes containing 7- to 11-membered rings are all slightly strained (p. 29). This is because the conformations in which the CCC angles are tetrahedral do not correspond to the optimal staggered arrangement of hydrogen atoms on adjacent carbons. Indeed, to reduce the strain due to this partial eclipsing, the rings deform slightly and an optimum situation is achieved in which the total strain (angular + torsional) is minimized. Rings containing 12 or more carbon atoms can, however, adopt conformations in which there is neither angular nor torsional strain.

Detailed information about the preferred conformations of cycloalkanes containing 7 or more members is not complete. It is, however, known that cycloheptanone preferentially adopts the somewhat deformed chair structure (86).

(86)

(*b*)   *Fused rings*

A fused ring system is one in which adjacent rings have two atoms in common. By far the most important is the decalin system, of which there are two isomers; the preferred conformations of each are shown.

| *cis*-decalin | *trans*-decalin |
| (87) | (88) |

In the *trans* isomer each ring is fused to the other by equatorial bonds, whereas in the *cis* isomer one ring is joined to the other by an axial bond. Consequently, in line with the greater stability of *trans*- than *cis*-1,2-dimethylcyclohexane (p. 155), the *trans* isomer is the more stable (by $11\,kJ\,mol^{-1}$).

trans-Decalin is incapable of undergoing ring flipping, whereas the *cis* compound interconverts rapidly with the conformational isomer in which the rings shown in (87) are in the alternative chair forms. The *trans* compound possesses a centre of symmetry and is therefore optically inactive. The *cis* compound is asymmetric in both conformations, which are non-superimposable mirror images of each other, but since ring flipping between them is rapid the compound is a non-resolvable ($\pm$)-pair.

The 9-methyldecalin system occurs in many natural products (e.g. cholesterol). It differs from the decalin system in that the difference in stabilities between the *cis* and *trans* forms is smaller, as a result of the fact that in the *trans* compound (89) the methyl group is sterically compressed by four hydrogen atoms and in the *cis* compound (90) by only two. Depending on the presence of other groups in substituted decalins, either the *cis* or the *trans* isomer may be the more stable.

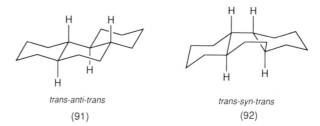

(89)   (90)

Polycyclic systems composed of fused six-membered rings also occur commonly, particularly the perhydrophenanthrene system. Perhydrophenanthrene itself exists in ten stereoisomeric forms (four enantiomeric pairs and two *meso* isomers); the most stable isomer (*trans-anti-trans*) and the least stable (*trans-syn-trans*, in which the central ring is in the boat conformation) are shown.*

trans-anti-trans
(91)

trans-syn-trans
(92)

(c)   Bridged rings

A bridged ring system is one in which adjacent rings have more than two atoms in common. Many such systems occur naturally: examples are derivatives of bornane (93) in terpenes, and tropane (94) in alkaloids.

---

* The prefixes *cis* and *trans* are used to denote the stereochemistry at the bonds of fusion between rings, and *syn* and *anti* refer to the orientation of the terminal rings with respect to each other.

(93)   (94)

Many other bridged-ring systems have been synthesized, among them, nor-bornadiene, nortricyclane, 'Dewar benzene' (bicyclo[2.2.0]hexadiene), barrelene (bicyclo[2.2.2]octatriene), cubane, and adamantane. With the exception of the last, each is strained in some degree and 'Dewar benzene' is particularly unstable, but in adamantane the four rings each adopt the strainless chair conformation and the compound is the most stable isomer of molecular formula $C_{10}H_{16}$. Urotropin (hexamethylenetetramine) (p. 307) is a heterocyclic analogue of adamantane in which nitrogen atoms replace CH at the four bridgeheads.

norbornadiene	nortricyclane	'Dewar benzene'
(95)	(96)	(97)

barrelene	cubane	adamantane
(98)	(99)	(100)

**5.3
Stereochemistry
and reactivity**

Reactivity is related to stereochemistry in two ways. First, many reactions have specific requirements for the relative positions of the atoms or groups that are involved. For example, in E2 eliminations the four atoms involved need to attain a coplanar relationship and preferably one in which the eliminated groups are *anti* (the so-called *antiperiplanar* arrangement, p. 97). Second, the steric requirements of non-reacting groups help to determine reactivity; for example, S_N2 displace-ments on neopentyl bromide, $(CH_3)_3C—CH_2\,Br$, occur very slowly because the three methyl groups hinder the approach of the nucleophile (p. 106). These characteristics are termed respectively *stereoelectronic* and *steric* factors.

Two levels of stereochemical control can be recognized. A reaction may be *stereospecific*, that is, stereoisomerically different starting materials give stereoisomerically different products, or it may be *stereoselective*, that is, of two or more possible stereoisomeric products, one is formed in predominance.

The stereochemical factors controlling reactivity and the stereospecificity and stereoselectivity of reactions will be discussed with respect first to acyclic and then to cyclic compounds.

(a)   Acyclic compounds

(i) S_N2 reactions.   These are stereospecific, involving inversion of configuration at the reaction centre:

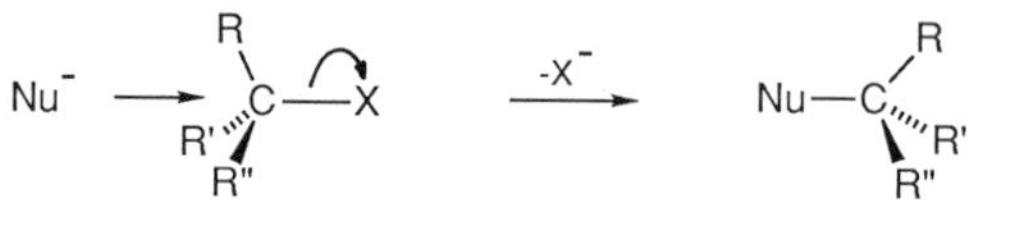

(101)

If inversion of configuration is impossible, S_N2 displacements do not occur, e.g. (102) is inert to hydroxide ion

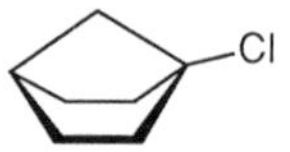

(102)

S_N2 reactions are also susceptible to steric hindrance (p. 105).

(ii) S_N1 reactions.   Unlike their bimolecular analogues, S_N1 reactions are not stereospecific. This is because the intermediate carbocation is coplanar and can be attacked from both sides to give a mixture of the two possible products:

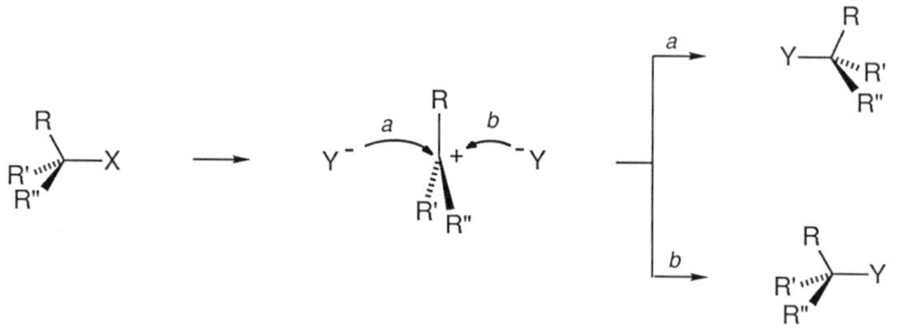

In practice, the product usually contains more of the enantiomer in which the original configuration has been inverted (p. 110).

(iii)   E2 eliminations.   The stereoelectronic requirement is that the four atoms that take part in the elimination should be able to attain a coplanar arrangement. Of the two possibilities (syn and anti), elimination occurs much more rapidly in the antiperiplanar conformation:

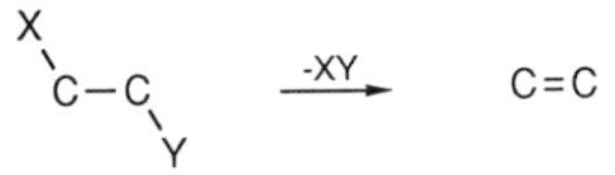

$$\overset{X}{\underset{Y}{\overset{|}{C}-\overset{|}{C}}} \xrightarrow{\ -XY\ } \quad C=C$$

Consider the elimination of hydrogen bromide, induced by hydroxide ion, from the diastereomers of PhCHBr—CHPhCH$_3$. Reaction occurs through the *anti* conformations (103) and (104) respectively, with the result that the former diastereomer gives specifically (E)-alkene and the latter gives specifically (Z)-alkene:

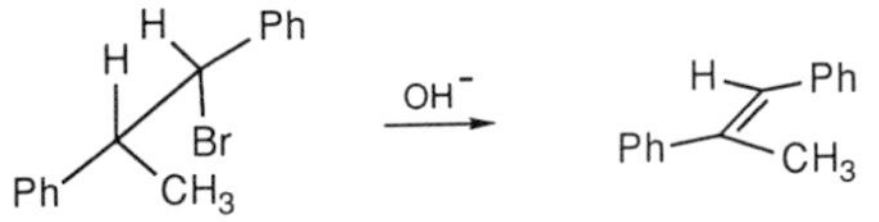

(103)

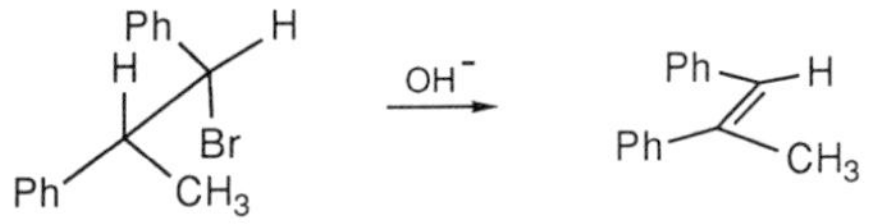

(104)

Similarly, the elimination of bromine, induced by iodide ion, gives specifically *trans*-2-butene from *meso*-2,3-dibromobutane and specifically *cis*-2-butene from both (R,R)- and (S,S)-2,3-dibromobutane:

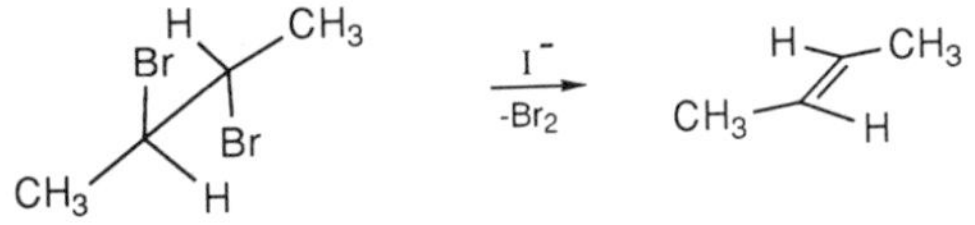

meso

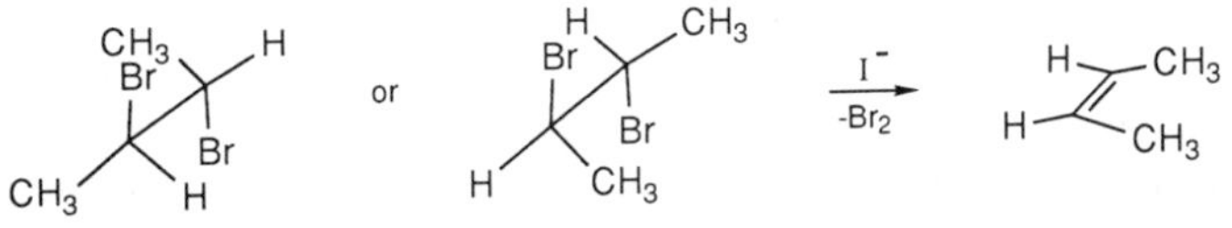

or

In addition to the fact that each of the above reactions is stereospecific, the rate of formation of the *trans* product is greater in each case than that of the *cis* product. This is because the repulsive forces between the nearby aromatic rings or methyl groups which are present in the *cis* isomers, and serve to make them less stable than the *trans* isomers, are present in some degree in the transition states in which the *cis* products are being developed.

When the *anti* conformation cannot be attained for geometrical reasons or when it would be very sterically congested, a slower *syn* elimination occurs. This is most commonly encountered with elimination from quaternary ammonium ions.

These are examples of stereospecificity in elimination reactions; there are also situations in which a particular compound reacts stereoselectively to give a pre-dominance of one of two or more products. For example, the base-induced elimination of hydrogen chloride from 1,2-diphenyl-1-chloroethane can occur through each of two *anti* conformations to give *cis*- and *trans*-stilbene respectively, the latter in predominance:

A priori, either or both of two factors might be responsible for the pre-ponderance of the *trans* isomer: the conformation of the reactant leading to the major product might be more heavily populated than that leading to the minor product and the transition state for the formation of the major product might be of lower energy than that leading to the minor product. Now, the activation energy of the elimination is large compared with the rotational energy barrier which separates the two significant conformations and, in these conditions, a general proposition – the *Curtin–Hammett principle* – applies: the relative pro-portions of the products are not determined by the relative populations of the ground state conformations but depend only on the relative energies of the transition states of the various processes. In the present example, *trans*-stilbene predominates over *cis*-stilbene because the transition state which leads to it is of lower energy; this in turn follows because of the repulsive forces between the two aromatic rings in the transition state leading to *cis*-stilbene (see above).

The basis of the Curtin–Hammett principle is as follows. Suppose the two conformers, A and B, have free energies G_1 and G_2 and the transition states for the formation of the products from each have free energies $G_1^{\neq}$ and $G_2^{\neq}$ (Figure 5.2). The ratio of the products, P_1 and P_2, is equal to the ratio of their rates of formation, i.e. $[P_1]/[P_2] = (dP_1/dt)/(dP_2/dt) = k_1[A][X]/k_2[B][X] = k_1[A]/k_2[B]$ (where $[X]$ symbolizes the concentration of all other species in the rate equation). From transition-state theory (p. 58),

$$k_1 = (kT/h)e^{-\Delta G_1^{\neq}/RT} \text{ and } k_2 = (kT/h)e^{-\Delta G_2^{\neq}/RT}$$

where $\Delta G_1^{\neq} = G_1^{\neq} - G_1$ and $\Delta G_2^{\neq} = G_2^{\neq} - G_2$. In addition, the conformers A and B are in equilibrium, i.e. $[B]/[A] = K = e^{-\Delta G/RT}$ where $\Delta G = G_2 - G_1$. It follows that

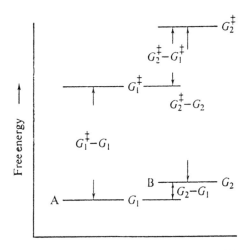

Figure 5.2   Energies of reactants and transition states.

$$[P_1]/[P_2] = \frac{e^{-\Delta G_1^*/RT}}{e^{-\Delta G_2^*/RT}} \cdot e^{\Delta G/RT} = e^{(\Delta G + \Delta G_2^* - \Delta G_1^*)/RT}$$
$$= e^{(G_2^* - G_1^*)/RT}$$

which is dependent only on the difference in free energies of the two transition states.

(iv)   Pyrolytic eliminations.   Certain eliminations which occur on heating, and without the need for an added reagent, take place through cyclic transition states in which the eliminated groups are *syn* (p. 296), e.g. xanthate pyrolysis:

(v)   El elimination.   Unlike E2 eliminations, E1 eliminations are not stereo-specific. Essentially, this is because the intermediate carbocation can undergo rotation about the bonds to the tervalent carbon atom before a proton is eliminated (however, more subtle factors are involved in determining the precise ratios of the possible products), e.g.

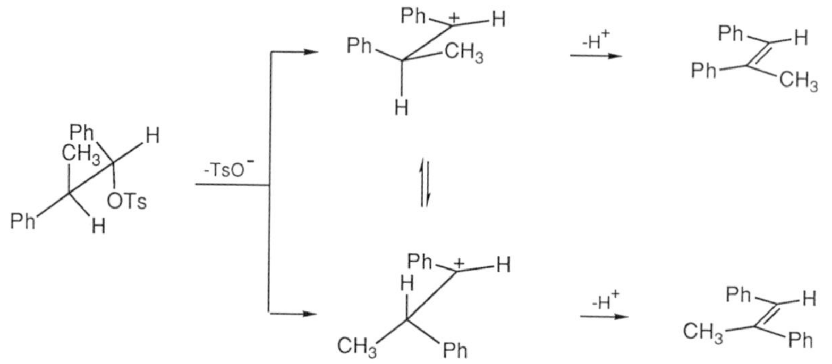

(vi)   Electrophilic addition to alkenes.   This has been described in detail (p. 85). In summary, *anti* addition occurs exclusively when the electrophile forms a strongly bridged intermediate cation (RSCl, RSeCl), and also when it forms a less strongly bridged cation provided that there is no aryl-conjugated substituent (Br$_2$, Cl$_2$). *Anti* addition predominates in the addition of hydrogen halides to alkenes with no aryl substituent, whereas *syn* addition is predominant with aryl-conjugated alkenes.

(vii)   Molecular addition to alkenes.   A number of additions to alkenes occur in a concerted manner to give *cis* products. A typical example is the *Diels–Alder reaction*, e.g.

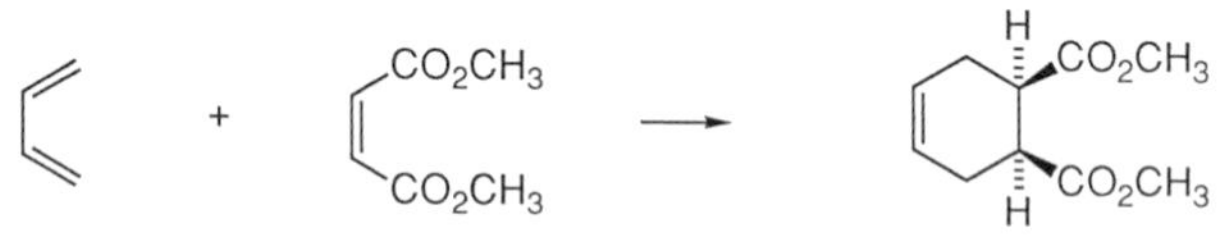

The full stereochemical details of Diels–Alder addition are described later (p. 272).

Other *syn* additions to alkenes include hydroxylation by osmium tetroxide (p. 591) and by potassium permanganate (p. 592), catalytic hydrogenation (p. 633), reduction by diimide (p. 634), and the formation of alcohols *via* hydroboration (p. 487).

(viii)   Addition to alkynes.   Electrophilic addition gives mainly *trans*-products, e.g.

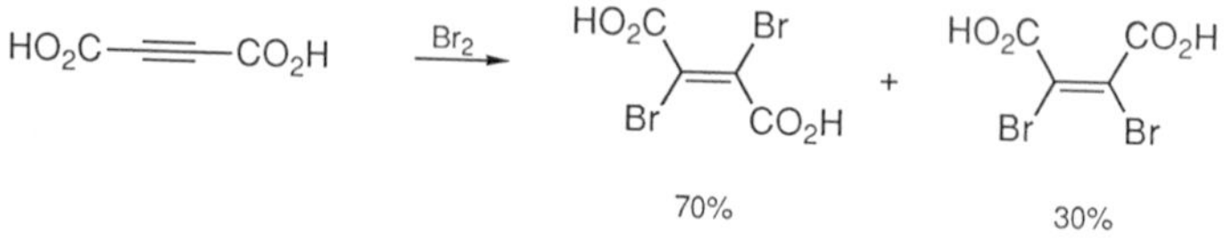

Nucleophilic addition also occurs in the *anti* manner, e.g.

$$Ph\text{---}\equiv\text{---}H \xrightarrow{CH_3O^- \cdot CH_3OH} \quad \begin{array}{cc} Ph & OCH_3 \\ \diagdown & \diagup \\ & \\ H & H \end{array}$$

Catalytic hydrogenation gives mainly the *cis* alkene (p. 639), whereas chemical reduction with sodium in liquid ammonia gives mainly the *trans* alkene (p. 639), as does the reduction of α-acetylenic alcohols by lithium aluminium hydride (p. 640).

(*ix*) *Addition to carbonyl groups.* Addition of a symmetric reagent to the carbonyl group of an optically inactive aldehyde or ketone necessarily occurs with neither stereospecificity nor stereoselectivity (p. 145). If, however, the carbonyl group is part of a stereosystem, addition can occur with some degree of selectivity to give a predominance of one diastereomer (p. 150).

(*x*) *Intramolecular rearrangements.* Rearrangements of electrophilic type (section 14.1) display, in appropriate cases, three stereochemical properties: the migrating group retains its configuration, there is inversion of configuration at the migration origin, and there is also inversion of configuration at the migration terminus.

The first property is shown in those reactions in which an asymmetric alkyl group migrates, as in the following Hofmann rearrangement (p. 442):

(105)

Manifestation of the second property is rare; an example is on p. 433. The third property is shown in reactions of the pinacol type (p. 434), e.g.

(106)

(*xi*) *Neighbouring-group participation.* Participation by neighbouring groups can result not only in an enhancement of the rate of a reaction (p. 111) but also in the alteration of the normal stereochemical course. This is particularly significant in substitutions; the following are examples.

(1)   The optically active 3-bromo-2-butanol (107) reacts with hydrogen bromide to give (±)-2,3-dibromobutane, whereas the diastereomeric (108) gives *meso*-2,3-dibromobutane. If reaction occurred through 'open' carbocations, each compound would give a mixture of (±) and *meso* products; however, the neighbouring

bromine atom fixes the stereochemistry of the intermediate ion, as shown, in a way similar to that in which the *anti* addition of bromine to alkenes is controlled (p. 85).

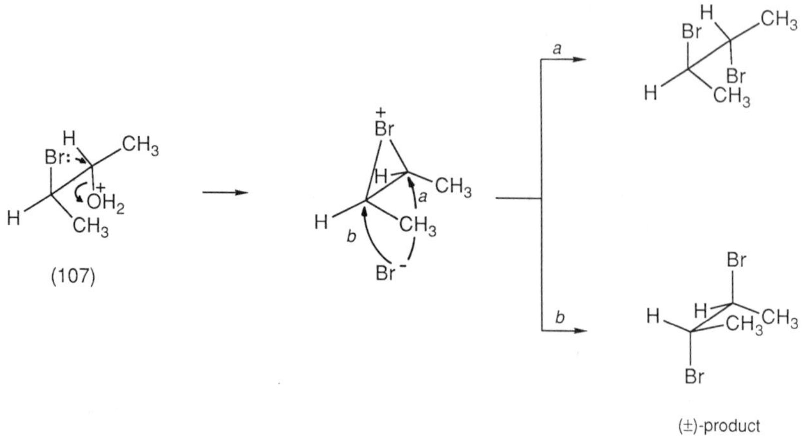

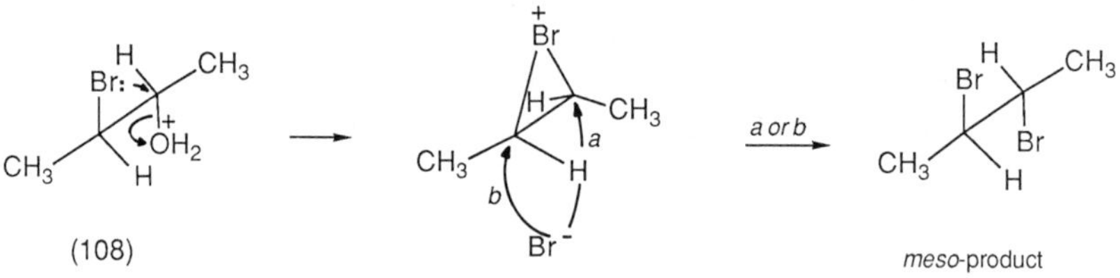

(107)

(108)

(±)-product

meso-product

(2)   The phenyl group participates similarly in the solvolysis of the diastereomeric toluene-*p*-sulfonates (109) and (110) in acetic acid:

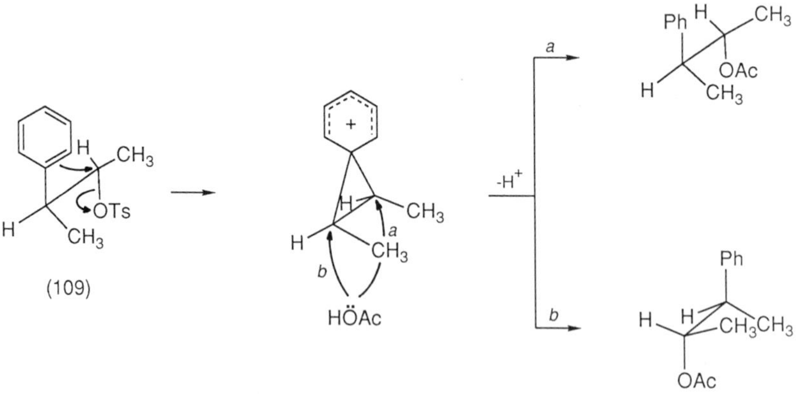

(109)

HÖAc

(3)   The carboxylate group in the α-bromopropionate ion (111) directs the stereochemistry of the unimolecular solvolysis below by preventing the attack of solvent from the side opposite to the departing bromide ion, with the result that the configuration of the asymmetric carbon atom is retained:

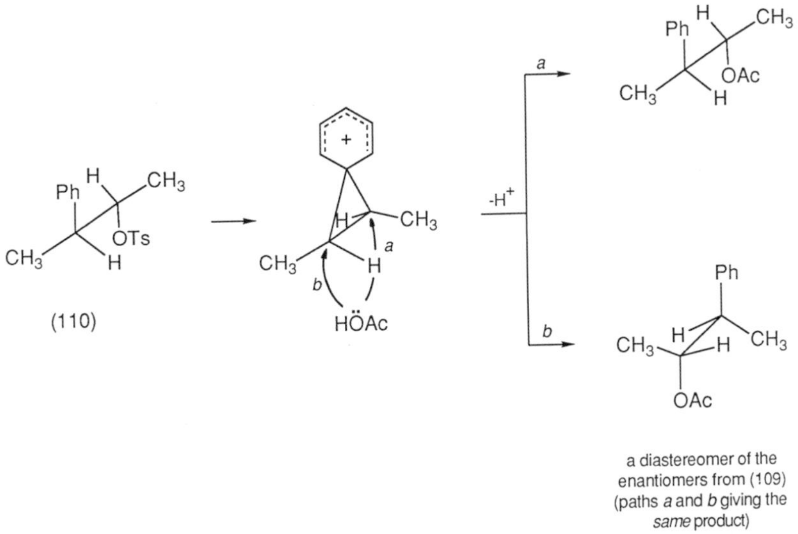
a
b
-H+
H
Ph
CH3
OTs
OAc
HÖAc
(110)
a diastereomer of the
enantiomers from (109)
(paths a and b giving the
same product)

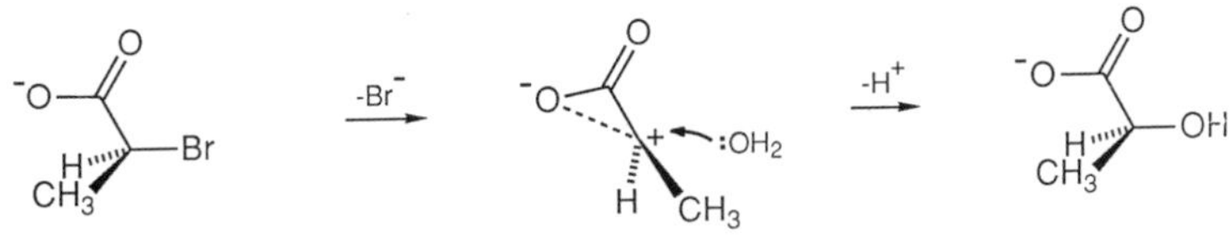
-Br-
-H+
:OH2
Br
OH
(111)

(*b*) *Cyclic compounds*

With the exception of the final section, the following discussion concerns only cyclohexane systems. This is because the six-membered ring system is the most widely occurring in nature and has been investigated the most extensively. It should, however, be clear that many of the propositions apply in a general sense to other ring systems.

(*i*) *Equilibria.* Monosubstituted cyclohexanes exist as equilibrium systems in which axially and equatorially substituted compounds interconvert rapidly through the flipping of the ring, although one or other of the conformational isomers usually predominates (p. 154). However, disubstituted and more complex cyclohexanes can exist in stereoisomeric forms which are not normally interconvertible, whether or not the ring is able to flip. For example, *cis*- and *trans*-menthane exist as separate compounds even though ring flipping occurs in each; and *cis*- and *trans*-decalin exist as separate compounds because ring flipping cannot occur to interconvert the two (p. 157).

cis-menthane

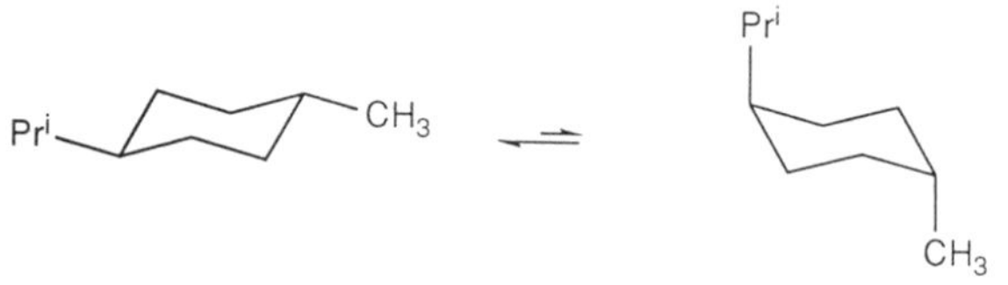

trans-menthane

There are, however, structural situations which do allow interconversion of such stereoisomers in certain conditions. For example, the incorporation of a carbonyl group in menthane gives *cis* and *trans* isomers which are interconvertible in both acidic and basic conditions: the *trans* isomer, having both alkyl substituents equatorial, predominates.

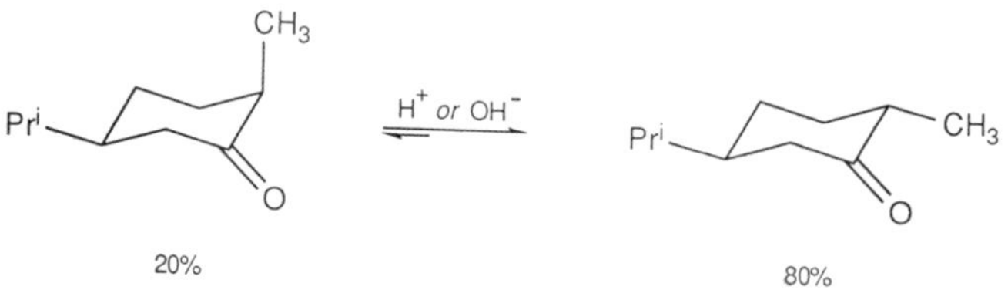

Interconversion occurs, in acid conditions, through the enol tautomer of the ketone (p. 146) and, in basic conditions, through the enolate (p. 146). A second example is the interconversion of the decalin derivatives (112) and (113) by the action of ethoxide ion; an enolate is again the intermediate. Equilibrium lies to the right because (113) has its ester group in the equatorial position.

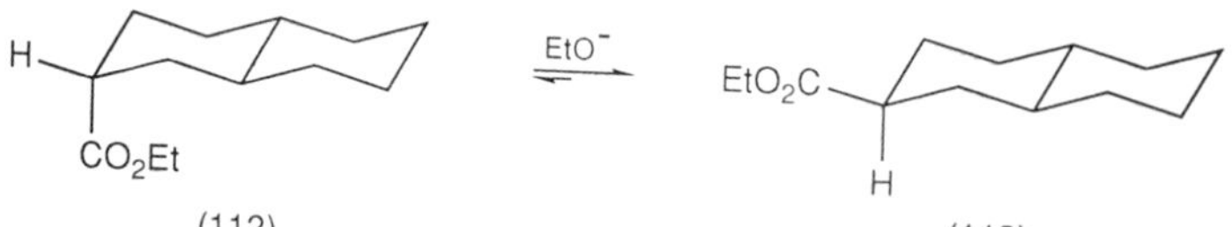

It will be clear from these examples that interconversion of this type can, in general, be brought about by the methods suitable for the racemization of optically active compounds (p. 146).

Interconversion can also be brought about if the ring system readily undergoes ring opening and ring closure. The best known example is glucose: α-D-glucose and β-D-glucose are interconverted fairly rapidly in solution through the open-chain tautomer (114). The β form predominates (β:α ~ 2:1) because all the substituents are in equatorial positions.

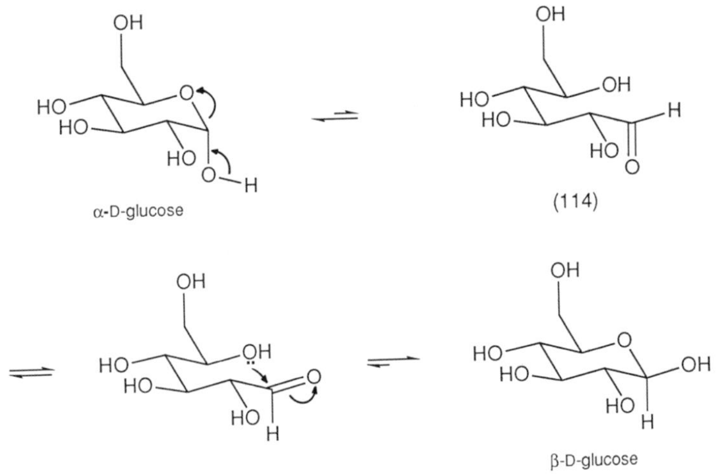

α-D-glucose   (114)

β-D-glucose

When crystalline α-D-glucose is dissolved in water, the initial rotation of the solution, corresponding to $[\alpha]_D^{20} = 111°$, falls gradually to an equilibrium value of 52.5° (the β form has $[\alpha]_D^{20} = 19°$). The phenomenon is known as *mutarotation*.

(*ii*) *Substitutions.* Equatorially and axially substituted conformational isomers normally react at different rates. Which isomer reacts the faster depends on whether the steric requirements of the transition state are less or greater than those of the reactants.

The first case is illustrated by the S_N1 solvolysis of *cis-* and *trans*-4-t-butyl-cyclohexyl toluene-*p*-sulfonate.* The *cis* isomer is the less stable, having the ester group in the axial conformation, but the transition states differ less in stability because the substituent has become somewhat removed from the ring system. The situation is that in Figure 5.3, from which it is clear that the activation energy should be smaller for the *cis* compound. In practice, solvolysis of the *cis* isomer is faster by a factor of about 3 and is described as being *sterically assisted*.

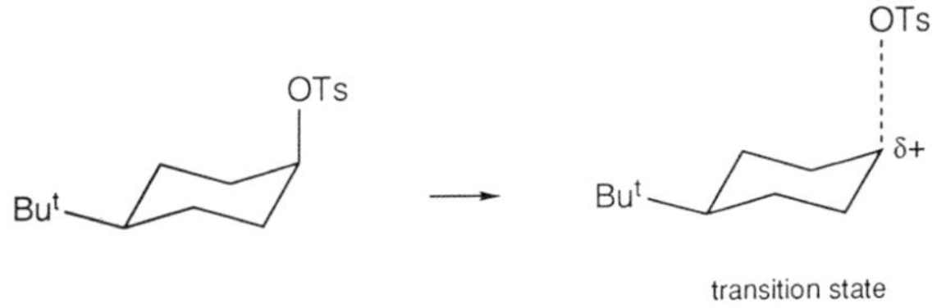

transition state

---

* The large size of the t-butyl group results in the conformational isomer in which the group is axial being of such high energy content relative to that in which it is equatorial that the latter isomer constitutes essentially 100% of the material. The discussion would, of course, apply to other systems of this sort or ones in which the conformation was rigidly held, such as the *trans*-decalin system.

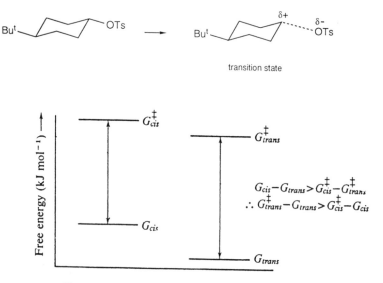

Figure 5.3   Energies of reactants and transition states.

The second case is illustrated by the hydrolysis of the *cis* and *trans* esters (115) and (116). Here, the *cis* compound is again the less stable, but the difference in stabilities of the transition states is greater than that of the reactants because the addition of hydroxide ion further increases the molecular crowding. Consequently the *trans* isomer reacts the more rapidly, by a factor of about 20.

These two cases are examples of the more general proposition that, if passage to the transition state is associated with a decrease in steric crowding, the axial

substituent will react faster, whereas if it is associated with an increase in crowding, the equatorial substituent will react faster.

(*iii*) *E2 eliminations.*   The stereoelectronic requirement is a periplanar arrangement, preferably *anti*, of the atoms involved. This requirement can affect both rates and products. For example, neomenthyl chloride undergoes base-catalyzed dehydrochlorination readily, giving the expected 3-menthene (Saytzeff rule, p. 97); reaction occurs through the favoured conformation in which both alkyl groups are equatorial. The diastereomeric menthyl chloride, on the other hand, reacts at only one-twohundredth the rate and gives 2-menthene; elimination occurs through the unfavourable conformation in which all three substituents are axial and in which the proton for Saytzeff elimination is not in the necessary *anti* position:

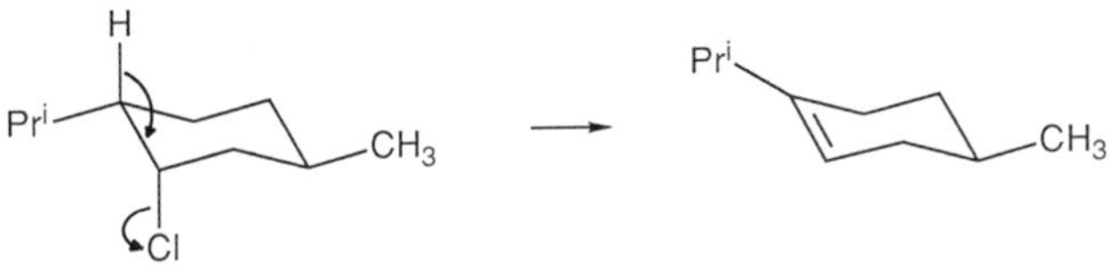

neomenthyl chloride

3-menthene

menthyl chloride

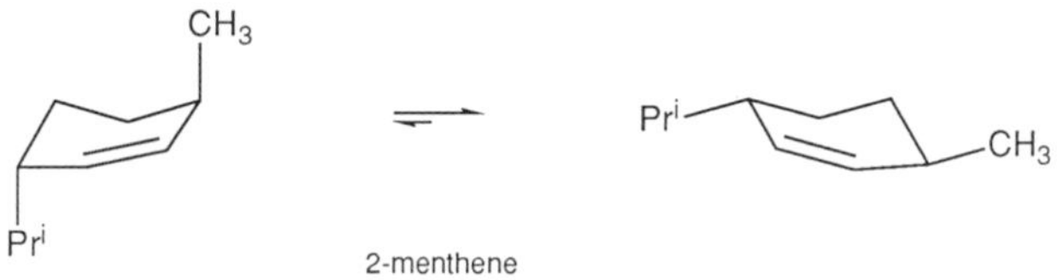

2-menthene

It can be readily appreciated why β-hexachlorocyclohexane undergoes dehydro-chlorination more slowly (by several powers of ten) than any of its diastereomers.

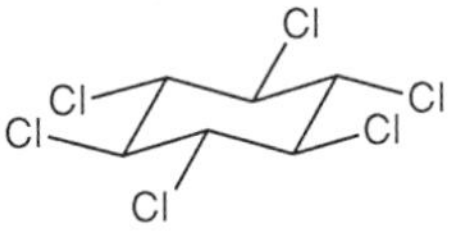

β-hexachlorocyclohexane

When the achievement of the periplanar conformation is impossible, rearrange-ment may precede elimination, e.g.

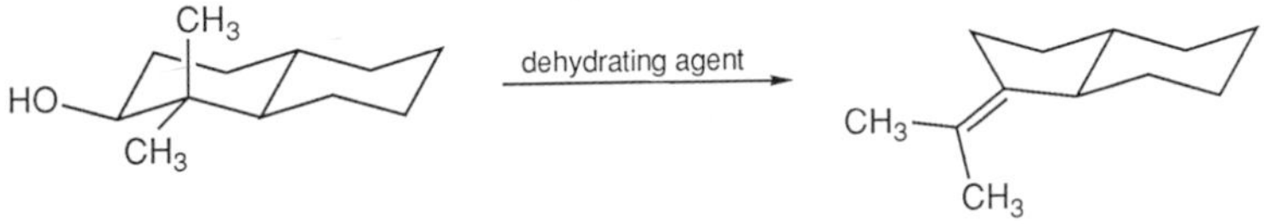

dehydrating agent

(Note that the requirement of such eliminations – inversion at the migration terminus – is met, p. 165).

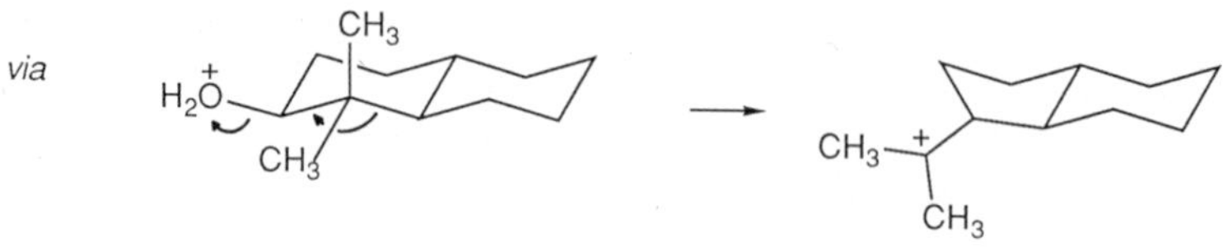

via

(*iv*) *Fragmentations.* 1,3-Disubstituted cyclohexanes which possess a hydroxyl substituent and a good leaving group such as toluene-*p*-sulfonate undergo base-induced fragmentation, providing that the leaving group can attain the equatorial position:

That is, the stereoelectronic requirement is that the leaving group and the C—C bond which breaks should be in the *anti* conformation.

Use can be made of this reaction in the formation of *trans*-cyclodecane derivatives:

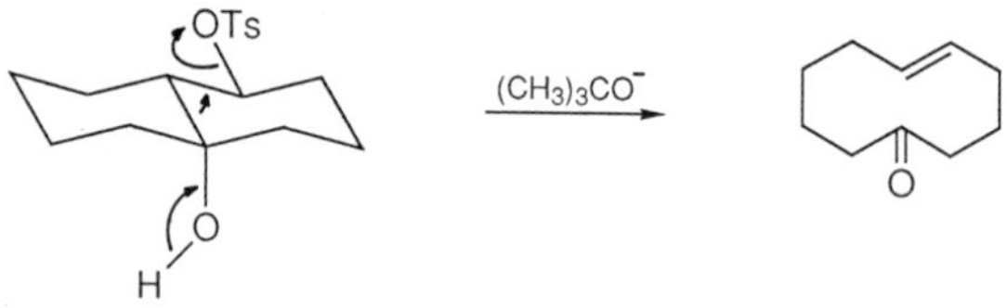

(*v*) *Electrophilic addition to alkenes.* Additions which occur *via* bridged ions to give *anti* products (p. 83) form diaxial compounds which, if ring flipping is possible, can convert predominantly into the di-equatorial products, e.g.

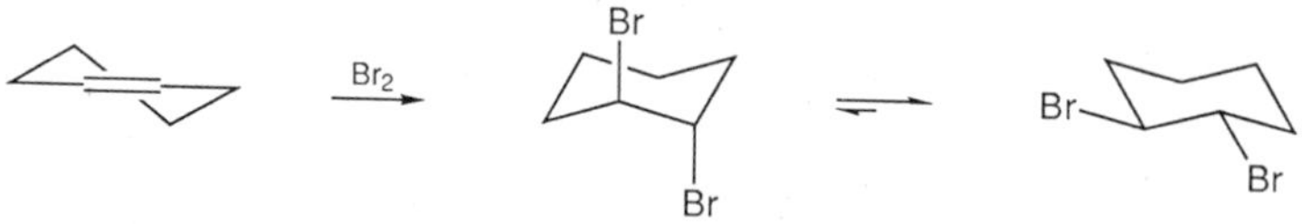

(ring flipping possible)

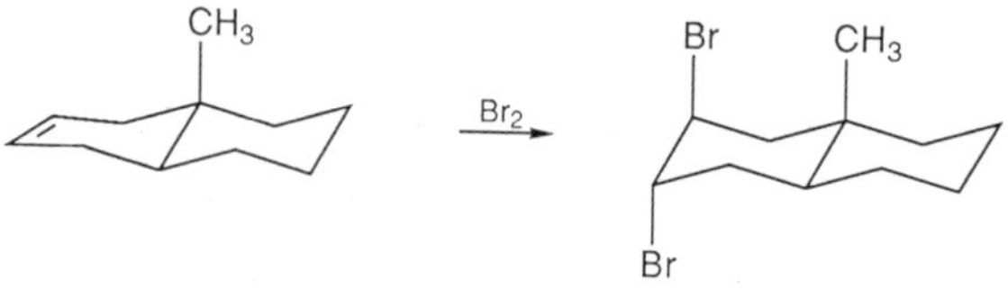

(ring flipping not possible)

(*vi*) *Intramolecular reactions*.   The stereoelectronic requirement – the *anti* conformation – can lead to different behaviour by diastereomers. For example, treatment of *trans*-2-aminocyclohexanol with nitrous acid gives only cyclopentanecarboxaldehyde, by rearrangement:

However, the *cis* isomer gives a mixture of the ring-contracted aldehyde and cyclohexanone. In this case, alternative *anti* conformations are available for rearrangement, one involving ring contraction and the other hydride migration:

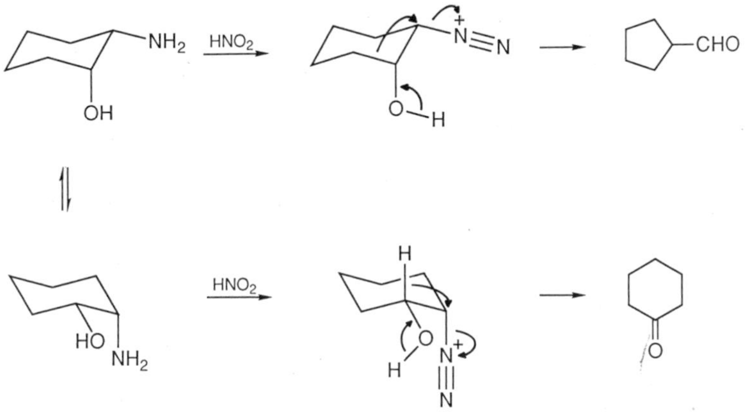

A slightly different situation holds for *cis*- and *trans*-2-chlorocyclohexanol. The oxyanion formed by the *trans* isomer with base is suitably sited to displace chloride ion intramolecularly by back-side attack; reaction occurs rapidly to give the epoxide. The *cis* isomer cannot achieve the necessary conformation for displacement and instead a slower reaction occurs in which a hydride ion migrates to give cyclohexanone:

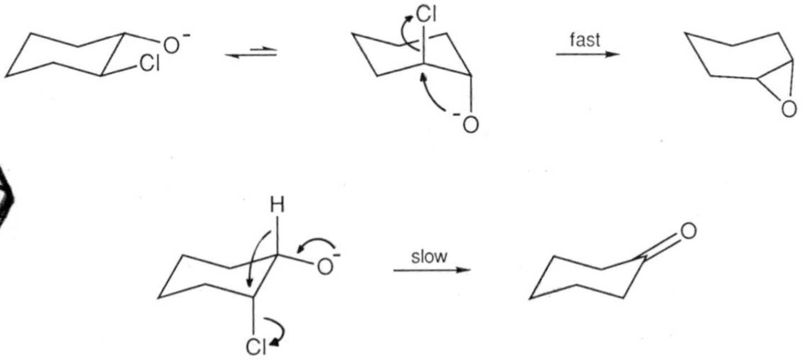

Just as epoxides are formed from diaxial chloro-alcohols as in the example above, so also they yield diaxial products on ring opening: a nucleophile attacks from the opposite side to the quasi-equatorial bond from carbon to oxygen, e.g.

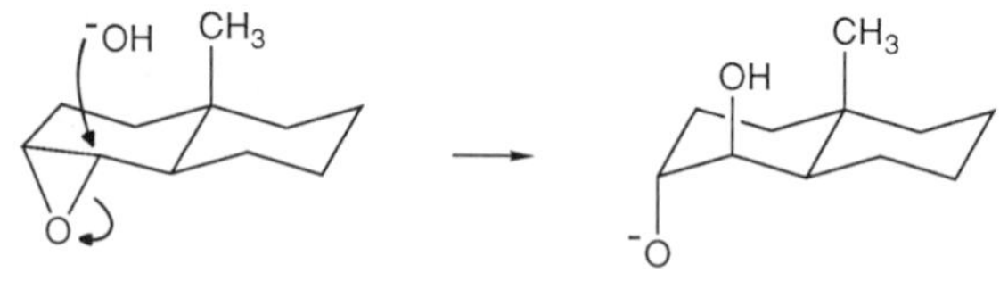

(vii)   Cyclohexanone and related compounds.   Cyclohexanone reacts with borohydride ion to give cyclohexanol faster than cyclopentanone. This is because
cyclohexanone possesses torsional strain, owing to the eclipsing of its oxygen
atom by two adjacent hydrogens, which is relieved in passage to saturated
derivatives, whereas the torsional strain in the five-membered system is increased
in passage to saturated derivatives.

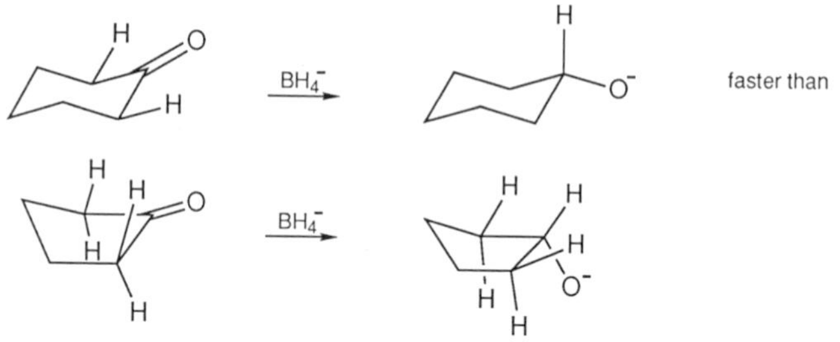

Conversely, cyclohexyl derivatives are less reactive in S_N1 and S_N2 reactions
than their cyclopentyl analogues, for passage to the transition state in the former
case is accompanied by an increase in strain and in the latter by a decrease, e.g.

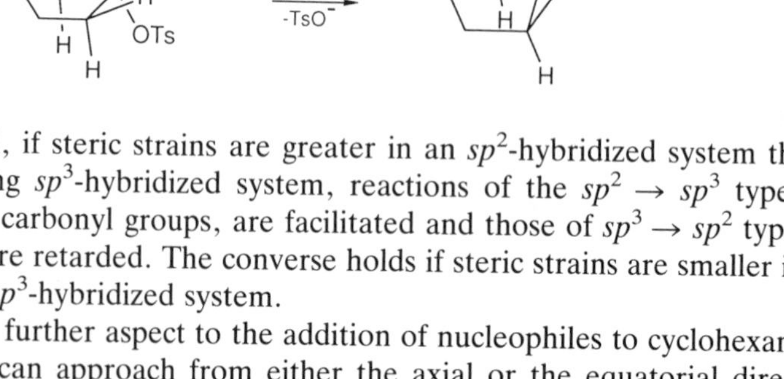

In general, if steric strains are greater in an sp^2-hybridized system than in the
corresponding sp^3-hybridized system, reactions of the $sp^2 \rightarrow sp^3$ type, such as
additions to carbonyl groups, are facilitated and those of $sp^3 \rightarrow sp^2$ type, such as
solvolyses, are retarded. The converse holds if steric strains are smaller in the sp^2-
than in the sp^3-hybridized system.

There is a further aspect to the addition of nucleophiles to cyclohexanones: the
nucleophile can approach from either the axial or the equatorial direction, the
carbonyl oxygen moving into an equatorial or axial position, respectively. Axial

approach is generally favoured* but is the more sterically hindered by the hydrogen atoms or other substituents on C-3 or C-5. Which route predominates depends on the balance between the two effects; for example, with 4-t-butylcyclohexanone (which exists essentially entirely with the t-butyl group equatorial, p. 169), axial approach dominates with a small nucleophile such as lithium aluminium hydride (p. 653) but equatorial approach dominates with Grignard reagents (p. 192).

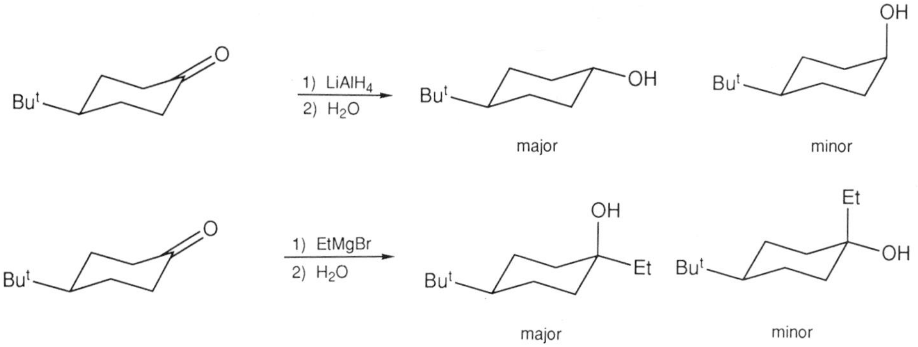

1. In what stereoisomeric forms would you expect the following compounds to exist?   **Problems**

(*a*)   CH₃CH—CHCH₃

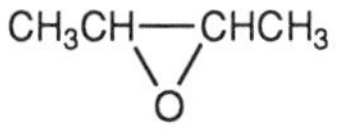

(*b*)   (CH₃)₂C=C=C=C(CH₃)₂

(*c*)   (C₂H₅)(CH₃)C=C=C(CH₃)(C₂H₅)

(*d*)   PhCH(OH)——≡—CH(OH)CH₃

(*e*)

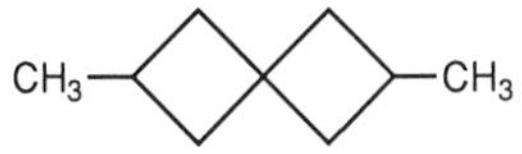

(*f*)

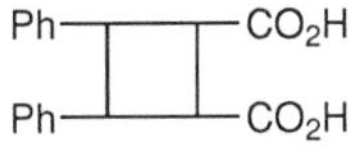

(*g*)

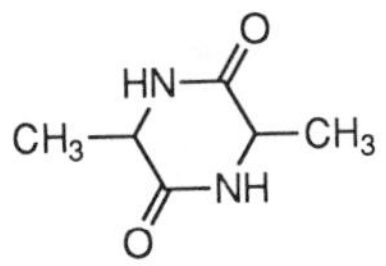

(*h*)

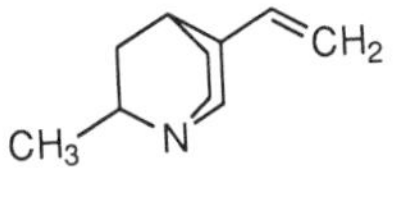

(*i*)

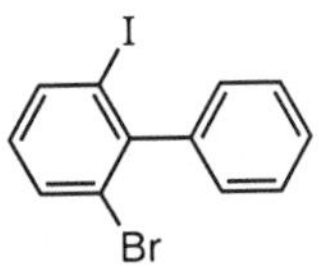

(*j*)

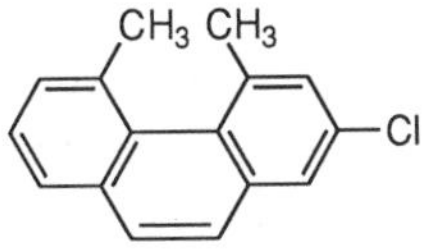

---

* The reason is not certain. One suggestion is that, when the reagent approaches from the equatorial direction, the C—O bond, which in cyclohexanone is not quite eclipsed by the C₂—H and C₆—H bonds, has to pass through a fully eclipsed position on its way to the transition state (i.e. as it moves into the axial position): torsional strain is thereby generated.

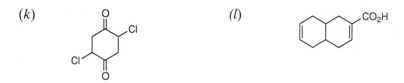

(k)

(l)

2. Draw Newman projection diagrams of the following structures, which are shown as Fischer projections. Which compound is optically inactive?

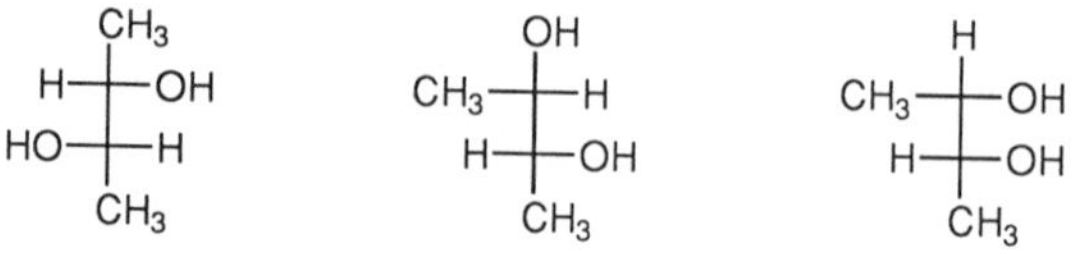

$$
\begin{array}{c}
CH_3 \\
H\!-\!\!\!-\!OH \\
HO\!-\!\!\!-\!H \\
CH_3
\end{array}
\qquad
\begin{array}{c}
OH \\
CH_3\!-\!\!\!-\!H \\
H\!-\!\!\!-\!OH \\
CH_3
\end{array}
\qquad
\begin{array}{c}
H \\
CH_3\!-\!\!\!-\!OH \\
H\!-\!\!\!-\!OH \\
CH_3
\end{array}
$$

3. Which of the following pairs of diastereomers are epimers, and which of the pairs of epimers could be readily interconverted?

(a)

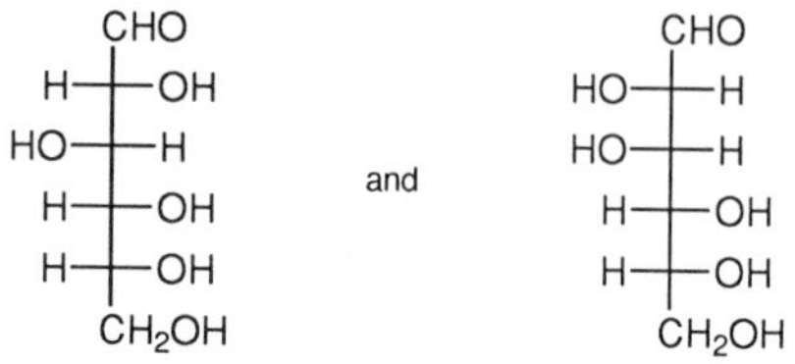

$$
\begin{array}{c}
CHO \\
H\!-\!\!\!-\!OH \\
HO\!-\!\!\!-\!H \\
H\!-\!\!\!-\!OH \\
H\!-\!\!\!-\!OH \\
CH_2OH
\end{array}
\qquad \text{and} \qquad
\begin{array}{c}
CHO \\
HO\!-\!\!\!-\!H \\
HO\!-\!\!\!-\!H \\
H\!-\!\!\!-\!OH \\
H\!-\!\!\!-\!OH \\
CH_2OH
\end{array}
$$

(b)

$$
\begin{array}{c}
CHO \\
H\!-\!\!\!-\!OH \\
H\!-\!\!\!-\!OH \\
H\!-\!\!\!-\!OH \\
CH_2OH
\end{array}
\qquad \text{and} \qquad
\begin{array}{c}
CHO \\
HO\!-\!\!\!-\!H \\
H\!-\!\!\!-\!OH \\
HO\!-\!\!\!-\!H \\
CH_2OH
\end{array}
$$

(c)

and

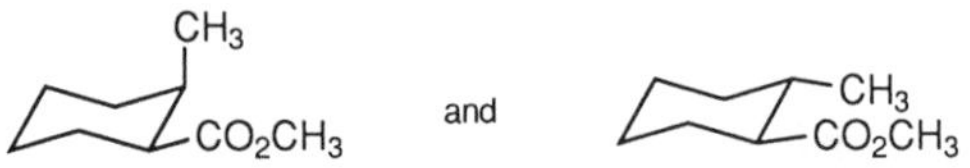

(d)

and

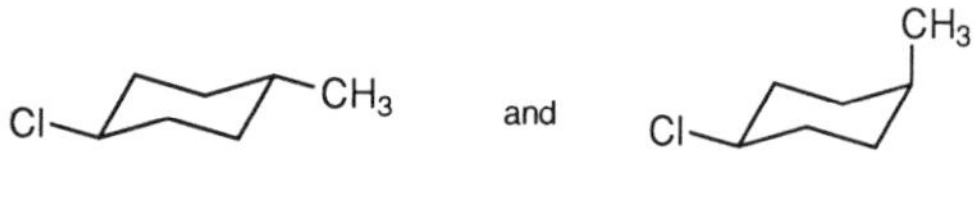

4. What products would you expect from the following reactions?
    (i)   The dehydration of 1,1,3-triphenyl-3-p-chlorophenylpropan-2-ol.
    (ii) The treatment of (I) with (a) OH⁻ and (b) HBr.

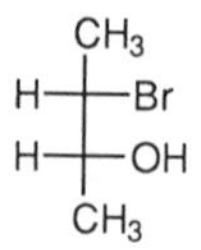

$$
\begin{array}{c}
CH_3 \\
H\!-\!\!\!-\!Br \\
H\!-\!\!\!-\!OH \\
CH_3
\end{array}
$$

(I)

(*iii*)   The elimination of bromine from *meso*-1,2-dibromo-1,2-diphenyl-ethane by treatment with iodide ion.

(*iv*)   The treatment of $CH_3CHXCH_2CH_3$ with base, (*a*) when X = Cl, (*b*) when X = $\overset{+}{N}Me_3$.

5.   Account for the following:

(*i*)   When an optically active sample of 2-iodobutane is treated with radioactive iodide ion in solution, the initial rate of racemization is twice the initial rate of uptake of radioactivity.

(*ii*)   (*R*)-α-Bromopropionic acid gives (*S*)-lactic acid with very concentrated alkali, but (*R*)-lactic acid with dilute alkali.

(*iii*)   (*R*)-2-Butyl acetate is hydrolyzed in basic conditions to give (*R*)-2-butyl alcohol; the alcohol, with hydrochloric acid, gives a mixture of (*R*)- and (*S*)-2-butyl chloride.

(*iv*)   The base-induced elimination of hydrogen chloride occurs more rapidly from *cis*- than from *trans*-1-chloro-4-methylcyclohexane.

(*v*)   Whereas tne diequatorial and diaxial conformations of *trans*-1,2-dimethylcyclohexane occur in the ratio of about 99:1 at room temperature, those of *trans*-1,2-dibromocyclohexane occur in about equal amounts.

Part II

Introduction to Part II

Whereas Part I was concerned with the principles of organic reactions, Part II is concerned with their practice; the structure of each chapter, with the exception of the last, is designed to answer the question, 'How may one go about constructing a particular type of bond in a given environment?'. Experimental details are not included, but many of the general reactions described are illustrated by examples taken from *Organic Syntheses*, Collective Volumes 1–7. These describe carefully tested experimental procedures and are referred to by inclusion in the text of the appropriate collective volume number in square brackets.

It will have become apparent from the earlier chapters that, in a large number of instances, the construction of a bond necessitates the generation, from a relatively stable compound, of a reactive species which is able to form a bond with a second compound. Chapters 6–8, each of which is concerned with the synthesis of aliphatic carbon–carbon bonds, all illustrate this principle. In the first of the three chapters the method adopted is to bond one of the two carbon atoms to a metal; then, because of the electropositive character of metals, the carbon atom is negatively polarized and able to react with a second carbon atom which is attached to an electronegative group and is therefore positively polarized. In the second of the chapters the method involves the use of basic media in order to generate a carbanion from a C—H-containing group of a suitable type. This, like the negatively polarized carbon in a C—metal bond, reacts with a carbon atom which is in an electronegative environment. In the last of the three chapters the opposite approach is adopted: acidic conditions are used to generate a carbocation which then reacts with a carbon atom which is in a relatively electron-rich environment.

A different principle is employed in chapter 9. Here, two molecules are brought together in such a way that two new bonds are formed in a concerted manner, to give a cyclic product; variants of this method include intramolecular cyclization and rearrangement. In most, but not all, cases new carbon–carbon bonds are formed.

The methods for forming aliphatic carbon–nitrogen bonds (chapter 10) are related to the methods of chapters 7 and 8 for forming carbon–carbon bonds. One approach, the more widely applicable, employs a species containing a nucleophilic nitrogen atom to form a bond to an electronegatively substituted carbon atom; the other involves the generation of a positively charged nitrogen species which reacts at an electron-rich carbon atom.

The stability of aromatic nuclei necessitates the use of certain different methods for bonding to aromatic carbon compared with those appropriate in the aliphatic series. In general, reagents of both electrophilic and nucleophilic type are employed, the former class being of wider applicability; they are described in chapters 11 and 12, respectively. Aromatic chemistry is completed by a survey of the reactions of diazonium ions (chapter 13): amino substituents may be replaced, *via* diazotization, by a wide variety of other groupings. Since the amino group can usually be introduced readily into aromatic rings through nitration followed by reduction, these reactions are of considerable use.

Intramolecular rearrangements (chapter 14), in which the skeleton of a molecule is transformed by the migration of a group to an adjacent atom, are characteristic of a large number of structural situations. Many, especially those which involve electron-deficient intermediates, may be usefully employed in synthesis; it is perhaps equally necessary to recognize the need, when planning a synthesis, to guard against the possibility that an intermediate will be formed which will undergo an unwanted rearrangement.

Reagents in which the key atom is phosphorus, sulfur, silicon, or boron provide a relatively recent addition to the synthetic chemist's armoury. They are now recognized as of major importance for their versatility and, in several cases, for their specificity. They are grouped together in chapter 15.

As well as carbocations and carbanions, two other classes of highly reactive carbon-containing species provide the basis for a variety of synthetic methods. One comprises electronically excited molecules, which can be formed by the ultraviolet or visible irradiation of unsaturated or aromatic compounds. These take part in a diversity of reactions and find especial use in the construction of complex cyclic compounds, often of a very strained kind (chapter 16). The other class are free carbon radicals. These can be generated in several ways and are capable of reacting with a number of organic groupings. Although their reactions are less commonly employed in synthesis than those of carbocations and carbanions, largely because free radicals are far less selective entities, there is nevertheless a diversity of processes in which they are usefully engaged (chapter 17).

The reactions of complexes in which a transition metal is bound to one or more organic ligands are now being recognized for their considerable synthetic potential. The usefulness of seven transition metals is described in chapter 18 and it is likely that this will be one of the most active areas of development over the next few years.

Almost any multi-stage synthesis involves steps in which the oxidation or the reduction of particular groups is required, and there have been greater recent advances in the control of these than of any other fundamental processes. In particular, highly selective methods have been developed which make it possible to oxidize (or reduce) at a required position in a molecule without affecting other oxidizable (or reducible) groups (chapters 19 and 20).

The construction of heterocyclic systems (chapter 21) utilizes only those reactions which have already been discussed. The subject is, however, suitably

treated separately for two reasons: first, ring-closure reactions which lead to five-
or six-membered rings occur in milder conditions than analogous reactions in
acyclic chemistry; second, mechanistically complex processes which might, in
acyclic chemistry, give correspondingly complex mixtures frequently give particular aromatic heterocycles essentially specifically, the driving force being the
development of the associated aromatic stabilization energy.

The final chapter describes and discusses the syntheses of a number of interesting
naturally occurring compounds. These have been chosen, above all, as illustrating
the fine control which can now be exercised in complex synthetic operations as a
result of our understanding of the electronic and stereoelectronic principles of
organic chemistry. It is to be hoped that students will not attempt to *learn* these
syntheses; their planning and their individual stages are dissected in such a
manner that the student should both develop some insight into the planning of
such syntheses and appreciate their artistic and scientific aspects.

6 Formation of carbon–carbon bonds: organometallic reagents

**6.1
Principles**

All the non-metallic elements to which carbon is commonly bonded are more electronegative than carbon. As a consequence, the carbon atom is positively polarized and, if the atom or group attached to it has a marked capacity for accepting an electron-pair, the carbon is susceptible to attack by nucleophiles, e.g.

$$HO^- \curvearrowright CH_3 \curvearrowleft I \longrightarrow HO-CH_3 \quad + \quad I^-$$

$$H_3N: \curvearrowright \overset{CH_3}{\underset{CH_3}{\diagdown}}\!\!=\!\!O \longrightarrow H_3\overset{+}{N}\!\!-\!\!\overset{CH_3}{\underset{CH_3}{\diagdown}}\!\!-\!\!O^-$$

Conversely, carbon bonded to an electropositive element – a metal – is negatively polarized and, because a metal can release a bonding electron-pair to form a cation, the carbon atom is susceptible to attack by electrophiles. The counterparts to the nucleophilic reactions above are:

$$I\!\!-\!\!CH_3 \quad CH_3\!\!-\!\!M \longrightarrow CH_3\!\!-\!\!CH_3 \quad + \quad M^+ \quad + \quad I^-$$

$$O\!\!=\!\!\overset{CH_3}{\underset{CH_3}{\diagup}} \quad CH_3\!\!-\!\!M \longrightarrow {}^-O\!\!-\!\!\overset{CH_3}{\underset{CH_3}{\diagdown}}\!\!-\!\!CH_3 \quad + \quad M^+$$

where M is a monovalent metal and methyl iodide and acetone are regarded as electrophilic reagents attacking the negatively polarized carbon in CH_3—M.

Organometallic compounds are enormously versatile reagents.* Their reactivity depends on the nature of the metal atom, there being a steady decrease in reactivity with decrease in the electropositive character of the metal and, by appropriate choice of the metal, a wide range of syntheses may be carried

---

*The term organometallic compound is here used in the restricted sense of describing compounds which contain a carbon–metal bond.

out. This, together with the ease with which organometallic compounds can be obtained, gives the reagents extensive applications.

(a) Preparation

The following are the general methods of preparation of organometallic compounds. They are discussed in more detail in the later sections concerned with individual metals, together with their limitations and the methods of choice in particular instances.

(1) *From metals and organic halides.* The simplest method of preparation is by treatment of the metal with an organic halide in an unreactive solvent, e.g.

$$R-Br \quad + \quad 2\,Li \quad \longrightarrow \quad R-Li \quad + \quad LiBr$$

(2) *Metal-halogen exchange.* The exchange reaction between an organometallic compound and an organic halide, e.g.

$$R-Br \quad + \quad R'-Li \quad \rightleftharpoons \quad R-Li \quad + \quad R'-Br$$

is an equilibrium which favours the organometallic compound in which the metal is attached to the more electronegative carbon.* For example, since aromatic (sp^2) carbon is more electronegative than aliphatic (sp^3) carbon, aryl-metal compounds can be prepared from alkyl-metal compounds, e.g.

$$Ph-Br \quad + \quad Bu-Li \quad \rightleftharpoons \quad Ph-Li \quad + \quad Bu-Br$$

(3) *Metal-metal exchange.* Organometallic compounds react with metallic salts by interchanging metals. The equilibrium favours formation of the organometallic compound which contains the less electropositive metal, e.g.

$$2\,R-Li \quad + \quad CdCl_2 \quad \rightleftharpoons \quad R_2Cd \quad + \quad 2\,LiCl$$

(4) *Metallation of hydrocarbons.* A C—H bond which is significantly acidic (i.e. which gives rise to a reasonably stable carbanion) reacts with an organometallic reagent in which the carbon is less electronegative to give the corresponding C-metal derivative, e.g.

$$H-\!\!\equiv\!\!-H \quad + \quad C_2H_5MgBr \quad \longrightarrow \quad H-\!\!\equiv\!\!-MgBr \quad + \quad C_2H_6$$

(b) Structure and reactivity

The organic compounds of sodium and potassium are salt-like compounds which are insoluble in non-polar solvents. Those of less electropositive elements such as

---

* The following is a simple way of understanding this. The carbon–metal bond has considerable ionic character and may be represented as a hybrid of structures such as CH_3——Li (purely covalent) and CH_3^- Li^+ (purely ionic). The more electronegative the carbon, the more important the ionic structure is and the more strongly bonded is the compound.

magnesium are essentially covalent and are soluble in, for example diethyl ether. There is, however, no sharp distinction between the two groups and it is possible to classify C—metal bonds according to their percentage ionic character. Values for the commoner bonds are as follows:

Metal	K	Na	Li	Mg	Cd
% Ionic character	51	47	43	35	15

The reactivity of organometallics can be correlated with their ionic character. Compounds of sodium and potassium are by far the most reactive, for example they are spontaneously inflammable in air, whereas organomagnesium compounds react with oxygen but far less vigorously. Lithium compounds are more reactive than magnesium compounds, e.g. unlike the latter, they react with carboxylate ions (p. 199); and cadmium compounds are less reactive than magnesium compounds, e.g. they do not react with ketones.

6.2 Organo-magnesium compounds (Grignard reagents)

Organomagnesium compounds, known as Grignard reagents after their discoverer, are prepared and used in solution in an ether, in which they exist as co-ordination complexes and dimers, e.g.

(a) Preparation

The standard method for the preparation of a Grignard reagent is the reaction of an organic halide in dry diethyl ether with magnesium metal (as turnings, granules or powder). The reagents are not isolated from solution but are used directly for the required synthesis.

$$RX \quad + \quad Mg \quad \longrightarrow \quad RMgX$$

The group R may be alkyl, aryl, or alkenyl, although alkenyl halides require special conditions. Amongst the halides, the order of reactivity is I > Br > Cl >> F; organomagnesium fluorides have not been prepared. The choice amongst chloride, bromide, and iodide is based on the accessibility of the halide and in general chlorides are used because they are cheaper to obtain. Since methyl chloride and methyl bromide are gases at room temperature, methyl iodide (b.p. 43°C) is generally used to make the methyl Grignard and likewise ethyl bromide is preferred to ethyl chloride for the ethyl Grignard. Grignard reagents will be referred to in this discussion as RMgX, where the nature of X is unspecified.

Alkyl and aryl halides, except aryl chlorides, normally react readily with magnesium in diethyl ether solution provided that water is rigorously excluded; aryl chlorides react in tetrahydrofuran. The magnesium is covered with a solution in diethyl ether of about 10% of the halide to be used, the exothermic reaction is normally initiated by stirring, and the remainder of the halide in ether solution is then added at such a rate that the ether is maintained at reflux by the heat evolved in the reaction. With some halides it is necessary to add a crystal of iodine or a small quantity of another (preformed) Grignard such as methylmagnesium iodide in order to initiate reaction. Grignard reagents react with oxygen, and it is helpful, although not usually essential, to exclude air. During the formation of the Grignard reagent, air is largely kept out by the blanket of ether vapour above the refluxing solution and the reagent is usually used straightaway.

Allyl halides react readily in diethyl ether, but since they also react readily with Grignard reagents (p. 190) it is important to keep the concentration of the halide to a minimum in the presence of the Grignard reagent. This is done by the slow addition of a dilute solution of the halide to a large excess of a vigorously stirred suspension of magnesium in ether.

Vinyl halides do not react in diethyl ether but do so in tetrahydrofuran. Reaction requires longer times and higher temperatures than with alkyl or aryl halides, e.g.

$$\diagup\!\!\diagup^{Cl} \quad \xrightarrow[\text{9 hours at 50 °C}]{\text{Mg / THF}} \quad \diagup\!\!\diagup^{MgCl}$$

Grignard reagents are also frequently prepared by the metallation procedure (p. 185), using a preformed Grignard reagent. The method is only suitable when the atom attached to magnesium in the compound to be prepared is markedly more electronegative than that in the preformed Grignard compound so that the equilibrium,

$$RH \quad + \quad R'MgX \quad \rightleftharpoons \quad RMgX \quad + \quad R'H$$

lies well to the right. Alkynyl Grignards are normally prepared by this method, e.g.

$$R-\!\!\equiv\!\!-H \quad + \quad CH_3CH_2MgBr \quad \longrightarrow \quad R-\!\!\equiv\!\!-MgBr \quad + \quad C_2H_6$$

and so are those from other acidic hydrocarbons such as cyclopentadiene,

(b)   Reactivity

Grignard reagents react with all organic functional groups except tertiary amines, C=C and aromatic double bonds, C≡C bonds, and ethers,* and they cannot therefore be prepared from compounds containing groupings other than these. Therefore while their reactivity is usefully applied in synthesis, it also provides a limiting factor. In general, the compounds are of more use in the synthesis of small and simple molecules than for polyfunctional molecules, and it is to be noted that they are rarely used in the syntheses of complex naturally occurring compounds (chapter 22) where highly specific reactions are required.

A useful working guide to the mode of reaction of Grignard reagents is that the direction of reaction is such that the magnesium atom is transferred to a more electronegative atom, as in the metallation procedure for their synthesis. All OH- and NH-containing compounds react by replacement of hydrogen,[†] e.g.

$$ROH \quad + \quad R'MgX \quad \longrightarrow \quad RO\text{-}MgX \quad + \quad R'H$$

$$R_2NH \quad + \quad R'MgX \quad \longrightarrow \quad R_2N\text{-}MgX \quad + \quad R'H$$

and in their reactions at carbon centres the magnesium is similarly transferred to oxygen or nitrogen when one of these elements is present, e.g.

$$R_2C{=}O \quad + \quad R'MgX \quad \longrightarrow \quad R_2R'C\text{-}O\text{-}MgX$$

(i)   Allyl and benzyl Grignard reagents.   Allyl Grignard reagents exist in solution as mixtures of isomers in rapid equilibrium, e.g.

However, the major product of the reaction of such a mixture with an electrophile is often one which is apparently derived from the minor constituent of the mixture. For example, in the above equilibrium the first isomer, containing the more highly substituted double bond, is the major component, but the product obtained on reaction is usually that corresponding to the minor component. The reason is that reaction of the major constituent can result in a shift of the double bond,

(E^+ = an electrophilic centre, such as the carbon atom in a carbonyl compound)

---

* They react with *strained* cyclic ethers, p. 194.
[†] The reaction of a Grignard reagents with heavy water (deuterium oxide) may be used to introduce deuterium into organic compounds:

$$RMgX \quad + \quad D_2O \quad \longrightarrow \quad RD \quad + \quad MgX(OD)$$

and this can be aided by the occurrence of the reaction through a six-membered cyclic transition state,

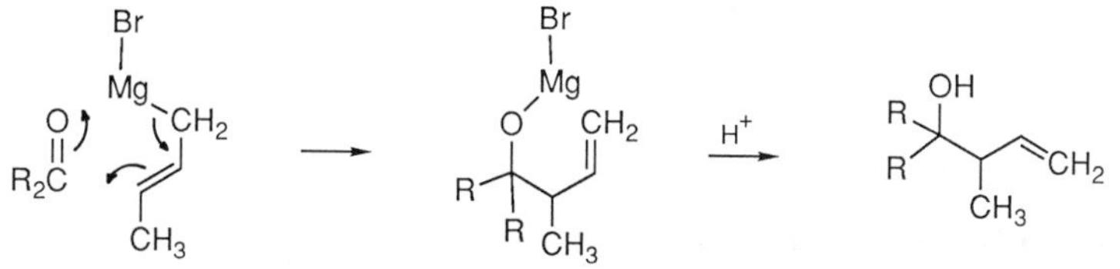

However, if the carbonyl group is hindered, reaction tends to occur at the less substituted carbon atom of the Grignard reagent.

Benzylic Grignards can also react in this way, e.g.

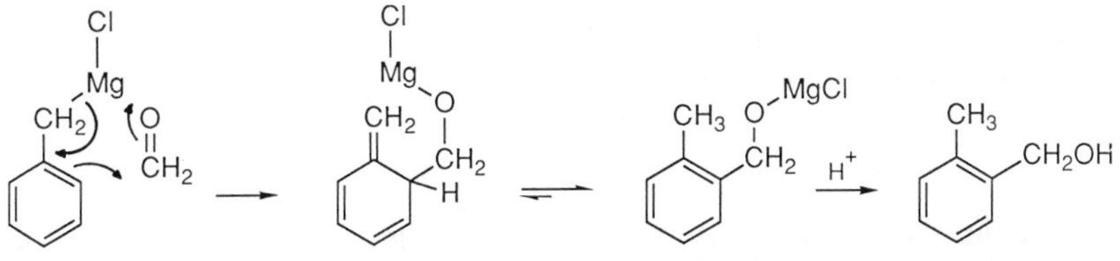

Even when reaction at the further carbon of the allylic-type Grignard cannot be aided by formation of a suitable cyclic transition state, it can still occur, and this property is usefully employed in the synthesis of pyrrole and indole derivatives. Pyrrole reacts with Grignard reagents at its NH group to give a N—Mg derivative which reacts with electrophiles at the 2-carbon atom, e.g.

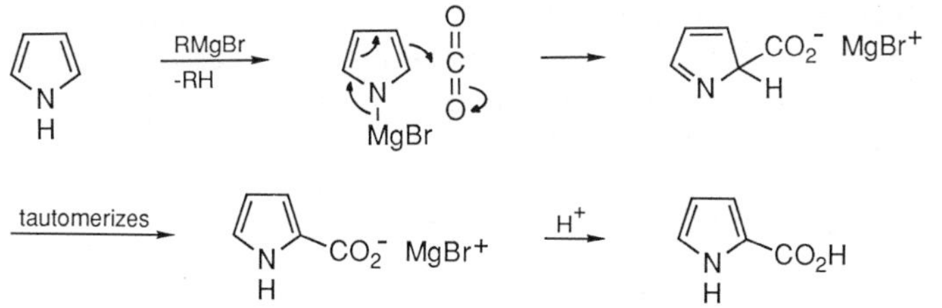

The Grignard reagents from indole behave analogously in that reaction occurs at the 3-position, as in the synthesis of the plant-growth hormone, heteroauxin (indole-3-acetic acid).*

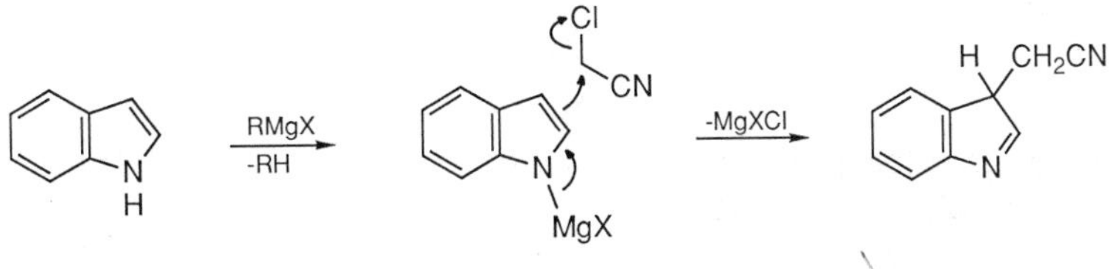

*The reasons why pyrrole and indole react with electrophiles chiefly at their 2- and 3-positions, respectively, are discussed later.

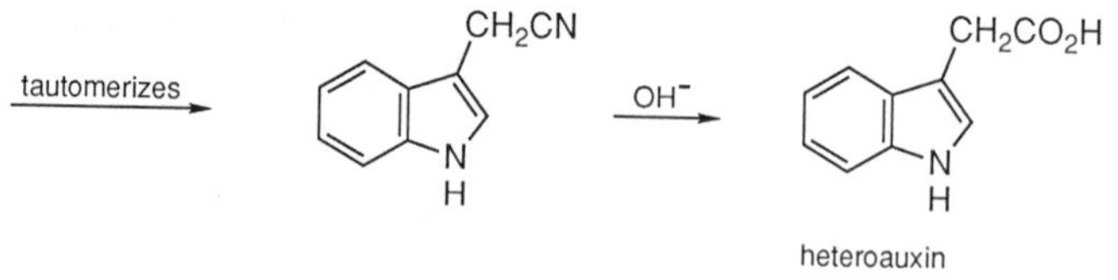

heteroauxin

(c) Formation of carbon–carbon bonds

The reactions of Grignard reagents at carbon atoms in various environments are classified here with reference to the type of compound which is obtained.

(i) *Hydrocarbons.* Grignard reagents react with alkyl halides and related compounds in the S_N2 manner, e.g.

The reactions with saturated halides are slow and the yields poor, but allyl and benzyl halides (which are more reactive than alkyl halides in S_N2 reactions) react efficiently, e.g.

94%

Alkyl compounds containing better leaving groups than the halides, such as alkyl sulfates and sulfonates, react in much higher yields than the alkyl halides. For example, propylbenzene is obtained in 70–75% yield from benzyl chloride and diethyl sulfate [1],

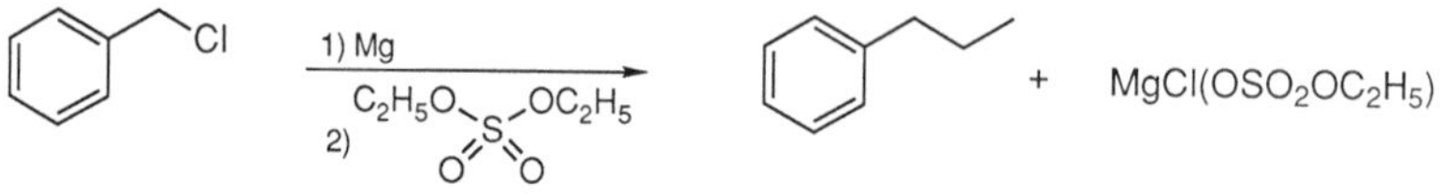

pentylbenzene is obtained in 50–60% yield from benzyl chloride and butyl toluene-*p*-sulfonate [2],

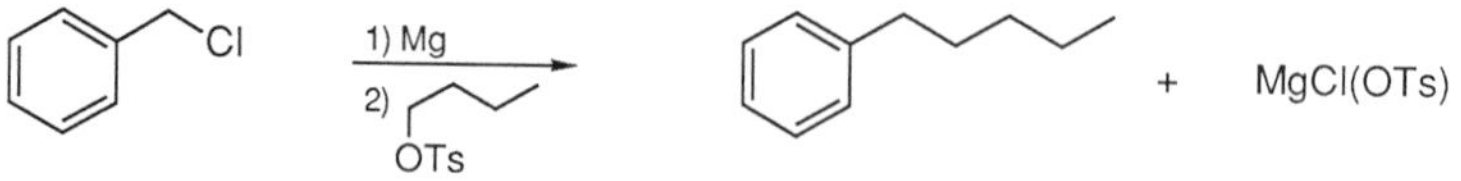

and isodurene may be prepared in up to 60% yield from mesityl bromide and dimethyl sulfate [2].

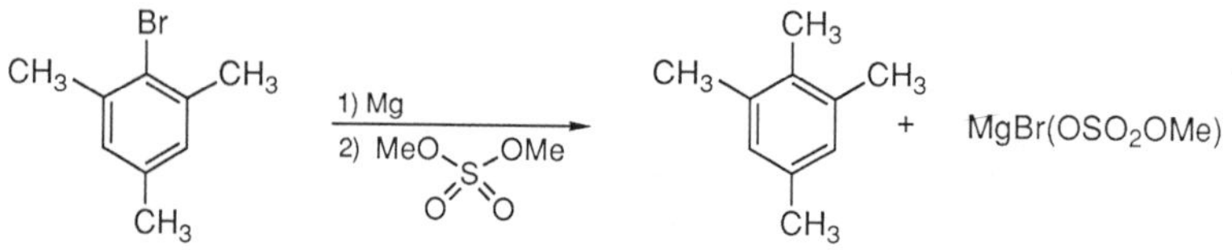

(*ii*) *Alcohols.* Grignard reagents react at the carbonyl groups of aldehydes and ketones to give the magnesium derivatives of alcohols which are converted into alcohols by treatment with acid:

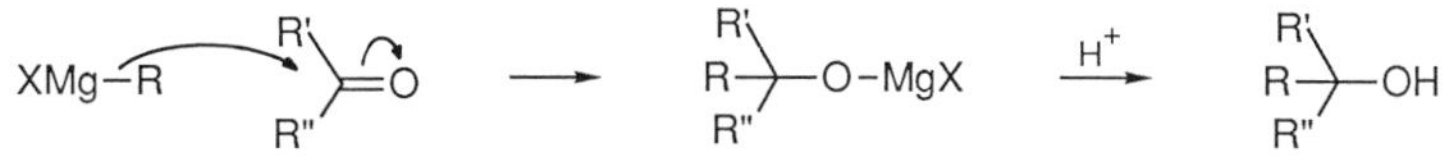

Formaldehyde (R' = R″ = H) gives primary alcohols, other aldehydes give secondary alcohols, and ketones give tertiary alcohols. For example, the Grignard reagent from cyclohexyl chloride reacts with formaldehyde to give cyclohexylmethanol in 65% yield [1]:

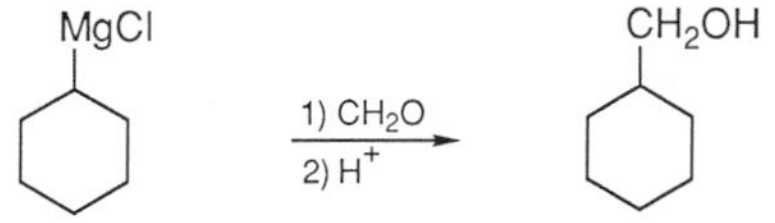

When the alcohol is sensitive to acid (e.g. tertiary alcohols, which are readily dehydrated) it is advisable to decompose the magnesium salt of the alcohol with aqueous ammonium chloride; basic magnesium salts are precipitated and the alcohol remains in the ether layer.

Stereoselectivity. The reaction of a Grignard reagent with a carbonyl group can establish a stereocentre. If the reactants are symmetric, equal amounts of the two enantiomers are formed, e.g.

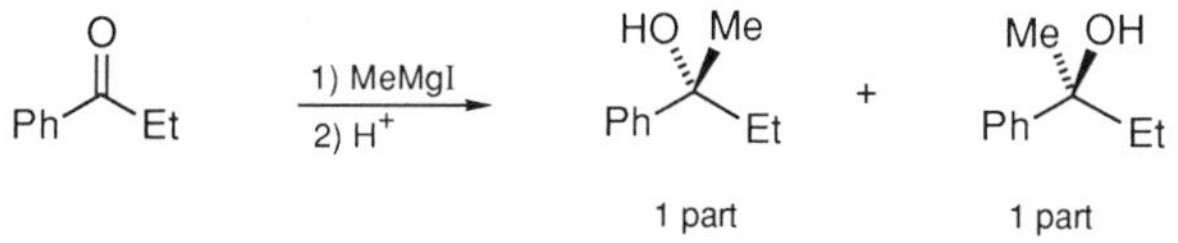

since reaction is equally likely on either face of the carbonyl group. However, if one of the reactants is itself asymmetric, there is a predominance of one of the two possible diastereomers. The results are in accord with Cram's rule (p. 150): for example,

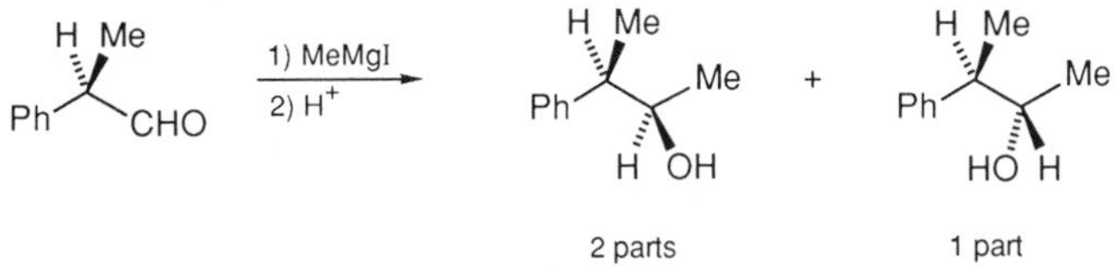

consistent with the preferred direction of attack:

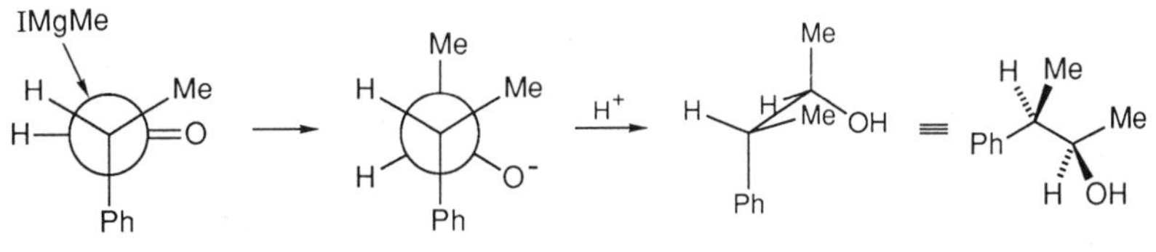

The stereoselectivity in reactions with cyclohexanones is governed by steric hindrance, approach of the reagent being mainly from the equatorial direction (p. 175), e.g.

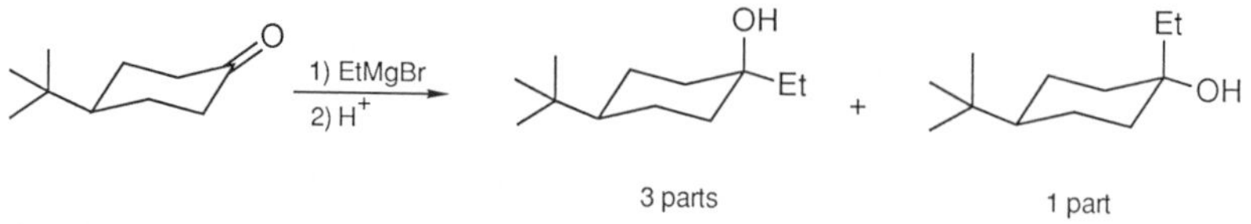

3 parts   1 part

With an axial substituent at the 3-position, as in 3,3,5-trimethylcyclohexanone, the stereoselectivity is more prononounced, e.g.

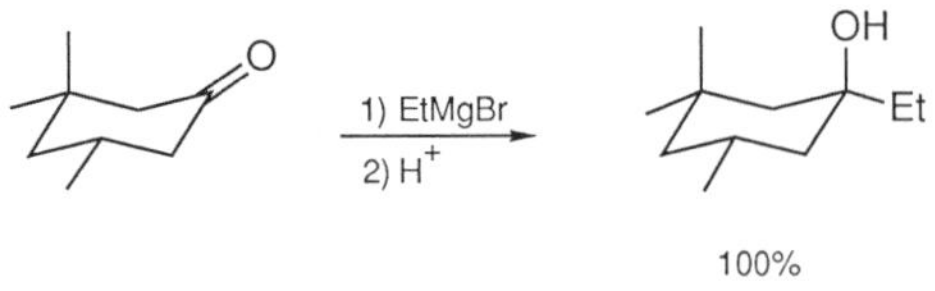

100%

Limitations.   The reactions with aldehydes and ketones usually give good yields of alcohols except with ketones which contain bulky groups. For example, di-t-butyl ketone does not give tertiary alcohols with Grignard reagents. If the ketone contains somewhat less bulky groups but the Grignard reagent contains a branched alkyl group, yields are again low, e.g. whereas diisopropyl ketone reacts with methylmagnesium bromide to give the tertiary alcohol in 95% yield,

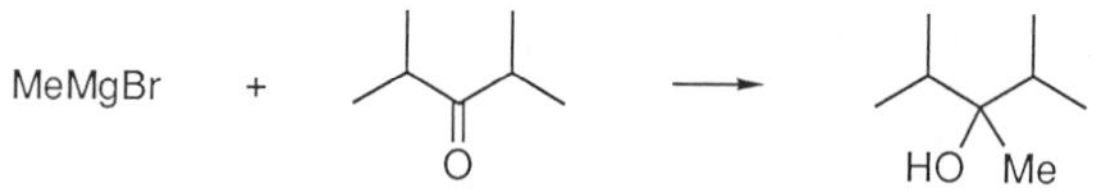

it fails to give tertiary alcohols with isopropyl and t-butyl Grignard reagents.

In these cases one or both of two side reactions take place. If the ketone has at least one hydrogen atom on one of its two α-carbon atoms, *enolization* can occur: the Grignard reagent acts as a base rather than as a nucleophile and abstracts an activated hydrogen, giving the *enolate*. Treatment with acid regenerates the ketone.

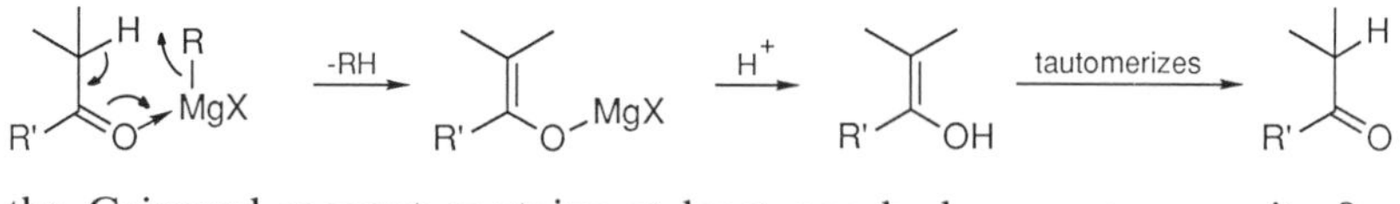

If the Grignard reagent contains at least one hydrogen atom on its β-carbon atom, *reduction* can occur by hydride-ion transfer within a six-membered cyclic transition state (cf. Meerwein–Ponndorf–Verley reduction, p. 654).

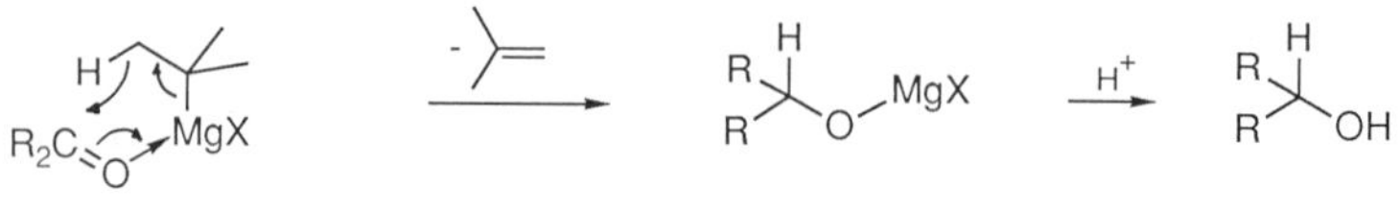

When each of these structural features is present, enolization and reduction are competitive, e.g. diisopropyl ketone and t-butylmagnesium bromide give 35% of the enolate and 65% of the secondary alcohol.

αβ-Unsaturated ketones.   Whereas Grignard reagents react with αβ-unsaturated aldehydes by addition to C=O, e.g.

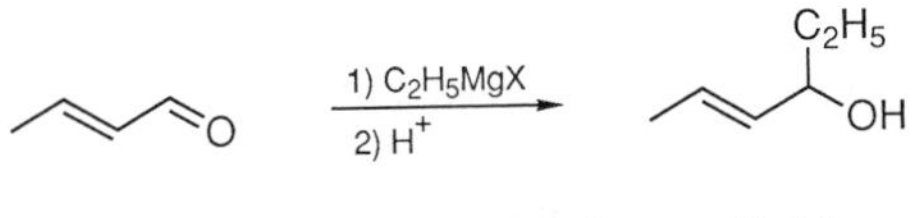

those prepared from normal commercially available magnesium react with αβ-unsaturated ketones both by addition to C=O and by addition to C=C, e.g.

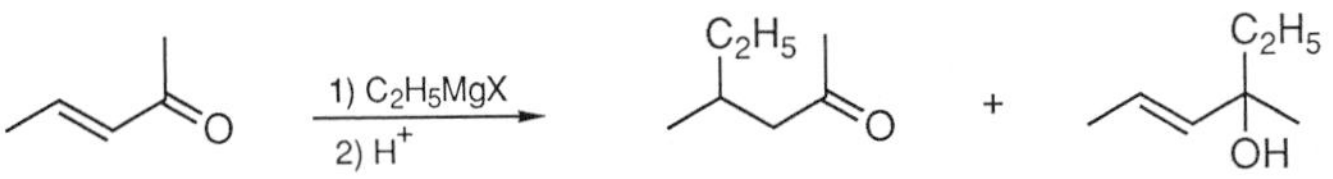

The latter type of reaction (described as 1,4-addition, reflecting the formation of an enolate intermediate, as shown below) competes increasingly effectively with addition to C=O (1,2-addition) as the sizes of the substituents at the carbonyl groups are increased.

It used to be thought that 1,4-addition occurred in the same way as in other additions of nucleophilic carbon to C=C—C=O (e.g. Michael reaction, p. 240), for example,

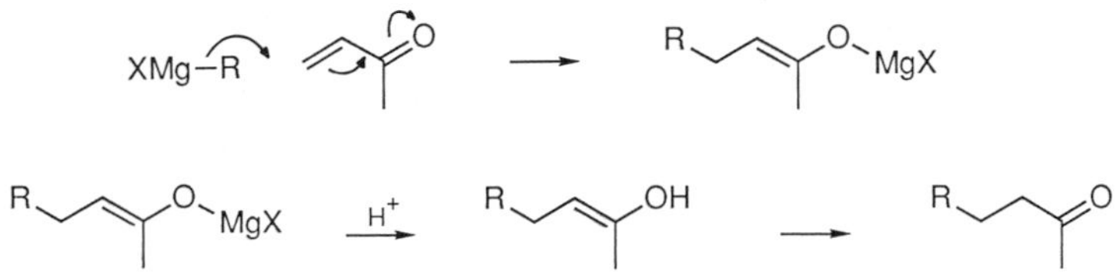

but this is now known not be so: when the magnesium used to prepare the Grignard reagent is freed of traces of transition metal impurities such as copper, only 1,2-addition occurs. Moreover, 1,4-addition is promoted compared with 1,2-addition by the addition of a catalytic quantity of a copper(I) compound such as CuBr. Although it is not known for certain how the copper acts, one possibility is that it forms an organocuprate with the Grignard reagent; this adds in the 1,4-manner to give the same enolate as above, and is regenerated by reaction with further Grignard reagent. The mode of 1,4-addition is considered further on p. 201.

Alternative methods for synthesizing alcohols.   Acid chlorides react with Grignard reagents to give ketones which react further to give tertiary alcohols:

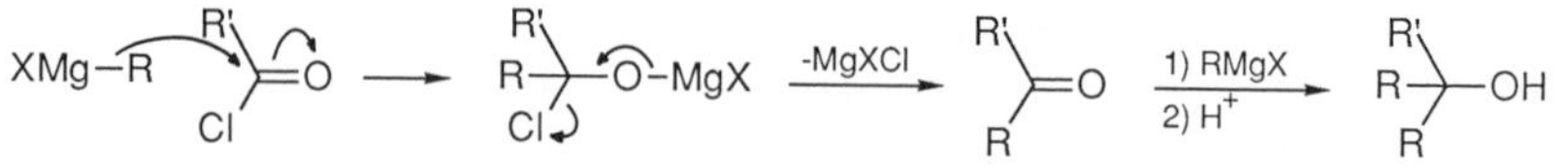

Esters react analogously to acid chlorides, e.g. phenylmagnesium bromide and ethyl benzoate give triphenylmethanol in 90% yield [1]:

$$2\,PhMgBr + PhCO_2Et \xrightarrow{-MgBr(OEt)} Ph_3C{-}OMgBr \xrightarrow{H^+} Ph_3C{-}OH$$

Whereas acyclic ethers and cyclic ethers containing essentially strainless rings (e.g. tetrahydrofuran) do not react with Grignard reagents, the strained small-ring cyclic ethers do so, for the strain is relieved during ring opening. For example, butylmagnesium bromide and ethylene oxide give 1-hexanol in 60% yield [1]:

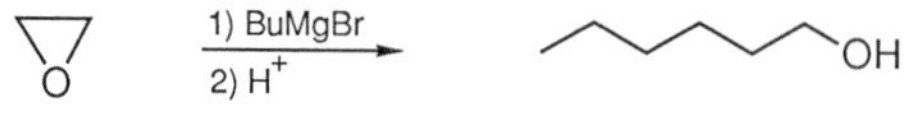

Trimethylene oxide reacts similarly. These procedures provide a rapid means of extending carbon chains by two and three atoms, respectively.

(*iii*) *Aldehydes*. The reaction of a Grignard reagent with ethyl orthoformate gives an acetal which is converted by mild acid hydrolysis into the aldehyde. It is believed that the acetal ionizes, aided by magnesium acting as a Lewis acid and by the carbocation-stabilizing ability of the two remaining alkoxy groups; reaction of this cation with further Grignard reagent gives the acetal:

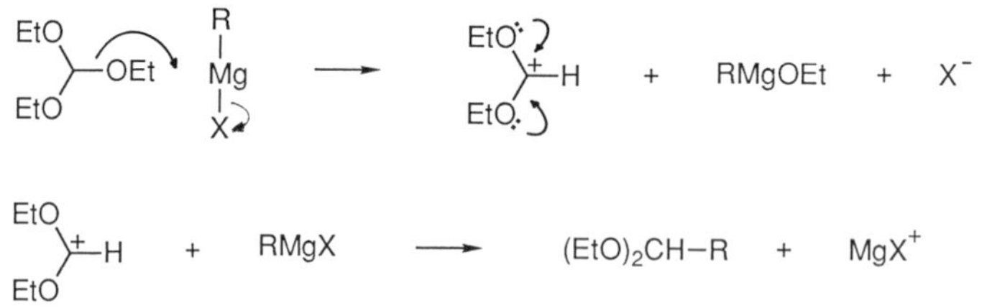

For example, 1-bromopentane can be converted into hexanal in up to 50% yield [2].

(*iv*) *Ketones*. Three methods are available. In one, Grignard reagents add to the triple bond in nitriles to give magnesium derivatives which are unreactive to further addition and on hydrolysis give ketones *via* the unstable ketimines:

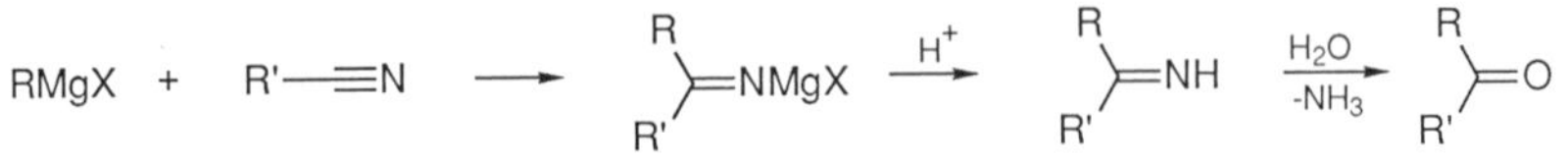

Second, *NN*-disubstituted amides react with Grignard reagents to give magnesium derivatives which yield ketones with acid:

$$RMgX + \underset{R''_2N}{\overset{R'_1}{\diagdown}}C{=}O \longrightarrow R{-}\underset{R''_2N}{\overset{R'_1}{\diagup}}OMgX \xrightarrow{H^+} \underset{R}{\overset{R'_1}{\diagdown}}C{=}O + MgX^+ + R''_2NH$$

Third, since acid chlorides are more reactive than ketones towards nucleophiles, their reactions with Grignard reagents can be used to give ketones provided that only one equivalent of Grignard is used and with the THF as solvent at $-78°C$, e.g.

$$C_6H_{11}MgBr \xrightarrow[\text{2) H}^+]{\text{1) PhCOCl}} C_6H_{11}\overset{Ph}{\underset{O}{\diagup}}$$

93%

(*v*) *Carboxylic acids*. Grignard reagents add to carbon dioxide to give salts of carboxylic acids from which the free acids are generated by treatment with mineral acid:

$$XMg{-}R \overset{\overset{O}{\underset{\parallel}{C}}}{\underset{\parallel}{O}} \longrightarrow R{-}\overset{O^-}{\underset{O}{\diagup}} MgX^+ \xrightarrow[-MgX^+]{H^+} R{-}CO_2H$$

(Grignard reagents, unlike organolithium compounds (p. 199) are not sufficiently reactive to add to the resonance-stabilized carboxylate ion.)

Reaction may be carried out either by pouring an ether solution of the Grignard reagent on to solid carbon dioxide ('dry ice') or by passing gaseous carbon dioxide from a cylinder into the Grignard solution. For example, using the latter method trimethylacetic acid can be obtained in 70% yield from t-butyl chloride [1].

$$Me_3C{-}Cl \xrightarrow{Mg} Me_3C{-}MgCl \xrightarrow{CO_2} Me_3C{-}CO_2^- MgCl^+ \xrightarrow{H^+} Me_3C{-}CO_2H$$

(*d*) *Reaction at elements other than carbon*

Grignard reagents may be used to attach various other elements to carbon. The following types of compounds can be obtained.

(1) *Hydroperoxides*. The slow addition of a Grignard reagent at low temperatures to ether through which oxygen is bubbling gives the magnesium derivative of a hydroperoxide from which the hydroperoxide is obtained with acid. t-Butyl hydroperoxide can be obtained in 90% yield in this way:

$$Me_3C{-}MgX \xrightarrow{O_2} Me_3C{\diagup}O{\diagdown}O{\diagup}MgX \xrightarrow{H^+} Me_3C{\diagup}O{\diagdown}OH$$

(2) *Thiols*. Reaction with sulfur leads to thiols:

$$RMgX + S \longrightarrow R{-}S{-}MgX \xrightarrow{H^+} RSH$$

(3) *Sulfinic acids*. Sulfur dioxide reacts analogously to carbon dioxide (p. 195):

$$RMgX \quad + \quad SO_2 \quad \longrightarrow \quad R-S\overset{O^-}{\underset{O}{\diagdown\!\!/}} \; MgX^+ \quad \xrightarrow{H^+} \quad R-S\overset{OH}{\underset{O}{\diagdown\!\!/}}$$

a sulphinic acid

(4) *Iodides*. The reaction with iodine,

$$XMg\!-\!R \qquad I\!-\!I \quad \longrightarrow \quad R\!-\!I \quad + \quad MgXI$$

provides a useful method for preparing iodides when the standard methods are unsuccessful. For example, iodides are commonly prepared from chlorides by treatment with sodium iodide in acetone (S_N2 displacement), but this method fails for the highly hindered neopentyl chloride (cf. p. 105). However, the Grignard reagent may be formed from neopentyl chloride and it reacts with iodine to give neopentyl iodide in good yield:

$$Me_3C-CH_2Cl \xrightarrow{Mg} Me_3C-CH_2MgCl \xrightarrow[-MgClI]{I_2} Me_3C-CH_2I$$

(5) *Amines*. The reagent employed is *O*-methylhydroxylamine:

$$XMg\!-\!R \qquad NH_2\!-\!OCH_3 \quad \longrightarrow \quad R\!-\!NH_2 \quad + \quad MgX(OCH_3)$$

This provides a useful method for t-alkyl amines such as $(CH_3)_3C-NH_2$, as these are not obtainable from S_N2 reactions between t-alkyl halides and ammonia.

(6) *Derivatives of phosphorus, boron, and silicon.*

$$3\,RMgX \quad + \quad PCl_3 \quad \longrightarrow \quad R_3P \quad + \quad 3\,MgXCl$$

$$3\,RMgX \quad + \quad BCl_3 \quad \longrightarrow \quad R_3B \quad + \quad 3\,MgXCl$$

$$4\,RMgX \quad + \quad SiCl_4 \quad \longrightarrow \quad R_4Si \quad + \quad 4\,MgXCl$$

In the last case it is possible to isolate the intermediate silanes, $RSiCl_3$, R_2SiCl_2, and R_3SiCl, by using calculated quantities of the Grignard reagent.

6.3
Organosodium compounds

Organosodium compounds react in the same way as Grignard reagents but far more vigorously. In addition, they react with ethers by S_N2 displacement and must therefore be prepared in hydrocarbon solvents. Even so, their reaction with the halides from which they are prepared (*Wurtz* coupling),

$$Na\!-\!R \quad + \quad R\!-\!X \quad \longrightarrow \quad R\!-\!R \quad + \quad NaX$$

is so much more rapid than that of the corresponding Grignard reagents that special techniques are required to obtain them. These difficulties, combined with their spontaneous flammability in air, have restricted the synthetic value of the compounds, particularly since the advent of organolithium compounds.

Organolithium compounds are somewhat less reactive than their sodium analogues but more reactive than Grignard reagents. The fact that they undergo some reactions of which Grignard reagents are incapable gives them their special uses in synthesis.

(a) *Preparation*

Like Grignard reagents, organolithium compounds can often be made by treating an organic halide in ether with lithium metal:

$$RX \quad + \quad 2\,Li \quad \longrightarrow \quad RLi \quad + \quad LiX$$

Because of the reactivity of organolithiums with oxygen, the reaction is carried out in an atmosphere of dry nitrogen or (better) argon. Organolithium compounds are more reactive than Grignard reagents towards alkyl halides and it is advantageous to cool the reaction mixture to about $-10°C$ to minimize the extent of the Wurtz coupling reaction,

$$RLi \quad + \quad RX \quad \longrightarrow \quad R{-}R \quad + \quad LiX$$

However, aryl halides are much less reactive towards nucleophiles and the synthesis of aryllithium compounds can be carried out at the boiling point of the solvent.

Lithium metal does not always react well with aryl and vinyl halides and the corresponding lithium compounds are then conveniently prepared by the metal-halogen exchange reaction using, e.g. preformed butyllithium:

$$RBr \quad + \quad BuLi \quad \longrightarrow \quad RLi \quad + \quad BuBr$$

The metallation reaction is suitable for the preparation of the lithium derivatives of comparatively acidic hydrocarbons, e.g.

Many organolithium reagents are now commercially available.

(b)   Reactions

The reactions of organolithium compounds in the main parallel those of Grignard reagents and attention is directed here to those reactions which either only the lithium compounds undergo, or in which they are the more effective reagents.

(1)   Lithium compounds are less readily prevented by steric hindrance from reacting at carbonyl groups. For example, whereas isopropylmagnesium bromide does not add to diisopropyl ketone but brings about enolization and reduction (p. 193), isopropyllithium adds successfully:

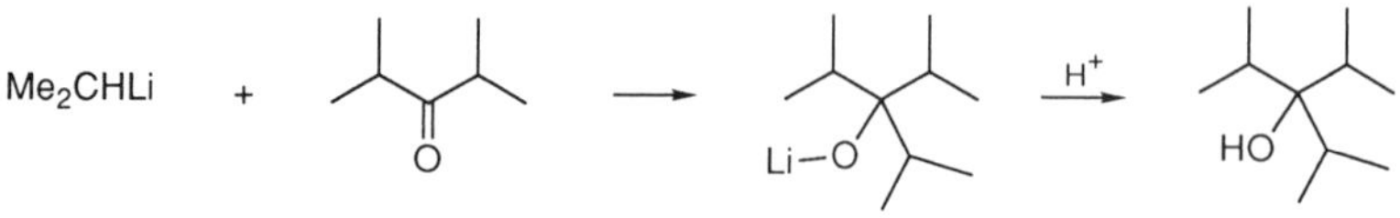

(2)   Whereas Grignard reagents often react with αβ-unsaturated ketones predominantly by 1,4-addition (p. 193), lithium compounds react predominantly by 1,2-addition, e.g.

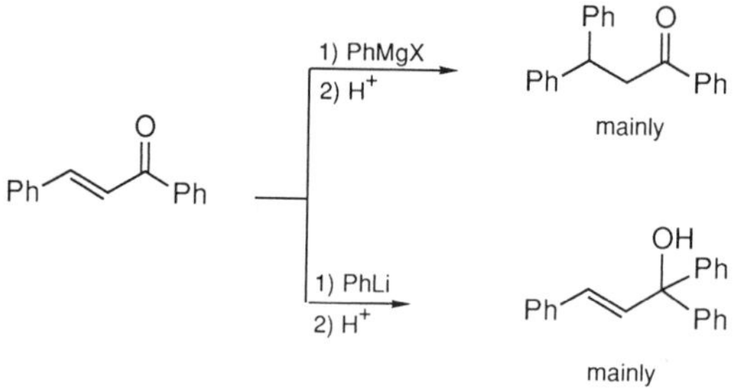

(3)   Organolithium compounds react more efficiently with alkyl halides and Wurtz coupling reactions can be carried out in good yield. An interesting example is the synthesis of optically active 9,10-dihydro-3:4,5:6-dibenzophenanthrene* from optically active 1,1'-binaphthyl-2,2'-dicarboxylic acid:

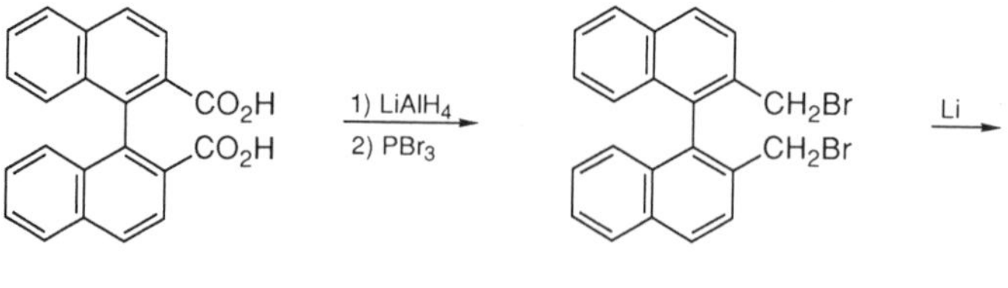

---

* This was the first synthesis (1950) of an optically active compound of this type. Repulsion between the *peri*-hydrogen atoms forces the lower left-hand benzenoid ring to lie somewhat above or somewhat below the upper left-hand ring, giving the two enantiomers. The molecule thus has a spiral shape.

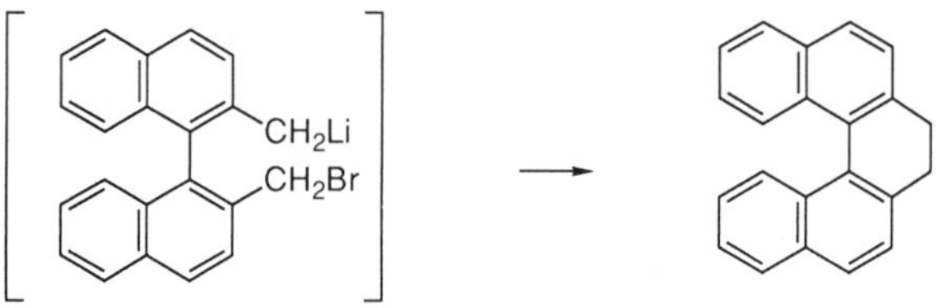

9,10-dihydro-3:4,5:6-dibenzophenanthrene

(4)   Carbon dioxide, which reacts with Grignard reagents to give carboxylic acids, reacts with organolithium compounds to give ketones:

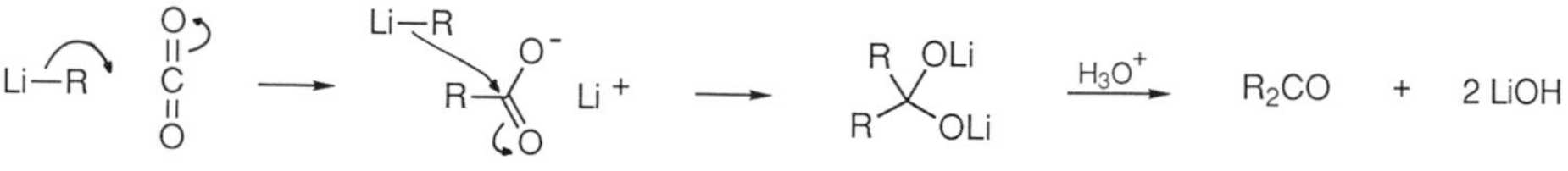

The difference arises from the fact that the lithium compounds are more strongly nucleophilic than Grignard reagents and are able to react with the intermediate resonance-stabilized carboxylate anion.

Similarly, carboxylic acids may be converted into ketones:

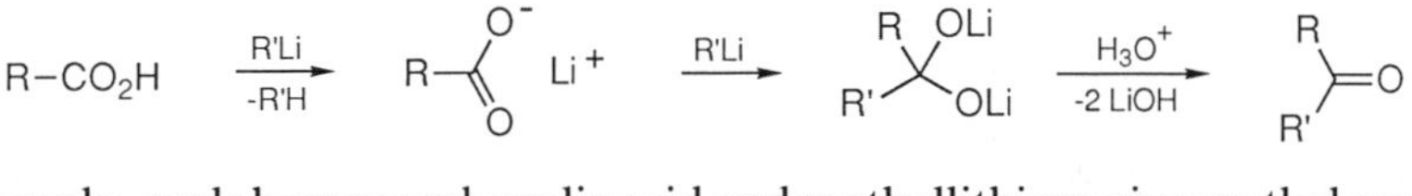

For example, cyclohexanecarboxylic acid and methyllithium give methyl cyclohexyl ketone in up to 94% yield [5]:

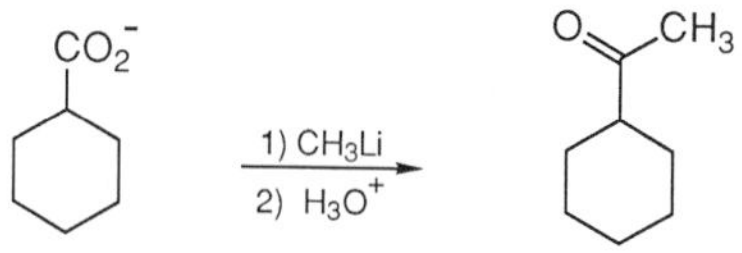

(5)   Organolithium compounds, unlike Grignard reagents, react with C=C bonds. Primary alkyllithiums and simple alkenes such as ethylene react only at high pressures (100–500 atmospheres) and give mixtures of compounds of different chain lengths,

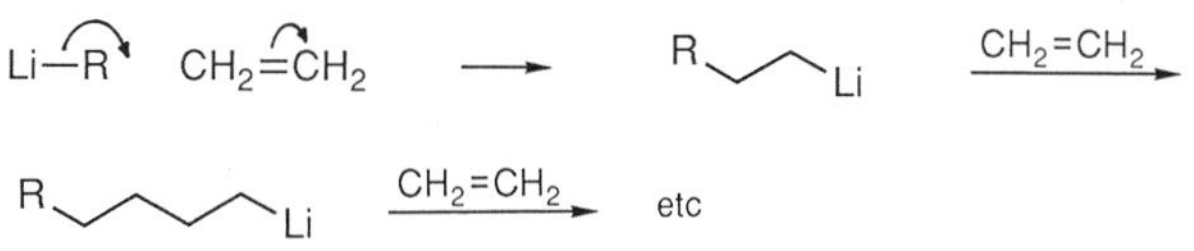

but secondary and tertiary alkyllithiums react efficiently with ethylene at low temperatures to give monomeric products, e.g.

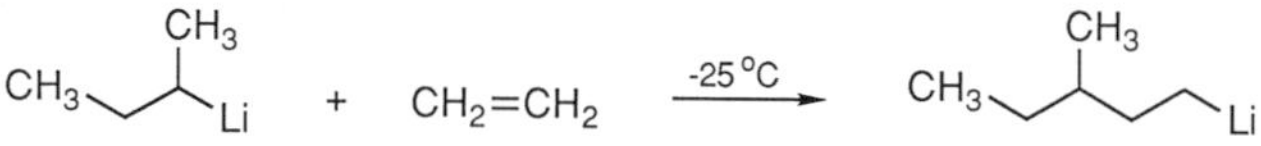

Higher temperatures are required with other alkenes and secondary or tertiary alkyllithiums and further reactions render these processes of little use, but conjugated alkenes react with these and with primary alkyllithiums under mild conditions to give monomeric products, e.g.

(6)   Unlike Grignard reagents, lithium compounds are sufficiently nucleophilic to react at the nuclear carbon atoms of aromatic systems which are activated to nucleophiles (see chapter 12 for a fuller discussion). For example, phenyllithium reacts with pyridine at 110°C to give an adduct which is decomposed by water to 2-phenylpyridine (40–49%) [2]:

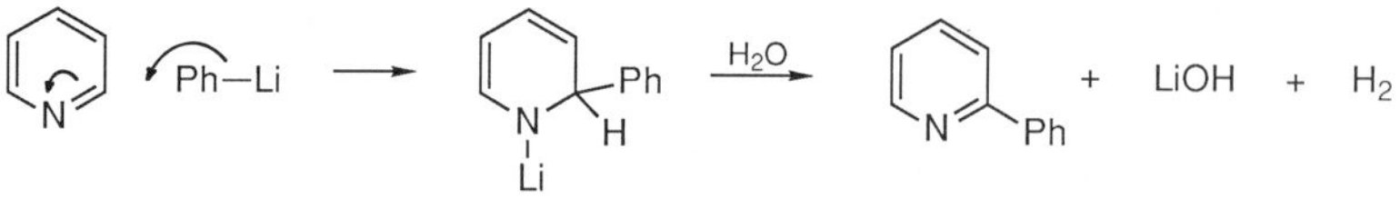

6.5 Organocopper compounds

(a)   Preparation

Two types of copper(I) compound are of value in synthesis: organocopper compounds, RCu, and lithium organocuprates, R_2CuLi, where R is alkyl (primary, secondary, or tertiary), alkenyl, or aryl. They can be prepared from the corresponding lithium compounds (p. 197) with copper(I) iodide in an aprotic solvent such as an ether and in an inert atmosphere:

$$RLi \quad + \quad CuI \quad \longrightarrow \quad RCu \quad + \quad LiI$$

$$2\,RLi \quad + \quad CuI \quad \longrightarrow \quad R_2CuLi \quad + \quad LiI$$

Alkenyl compounds of two types can also be made from alkynes. Acetylene itself reacts with either an organocopper compound or an organocuprate, with almost complete *syn* stereoselectivity,

Alkynes of the type $RCH_2C\equiv CH$ react with the complex of RCu and $MgBr_2$ (though not with R_2CuLi) with both regioselectivity (the copper atom becoming attached to the less substituted carbon atom) and *syn* stereoselectivity, e.g.

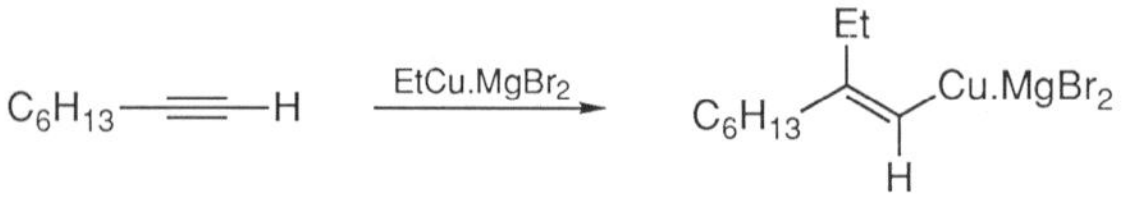

(b)   *Reactions*

(*i*)   *With αβ-unsaturated carbonyl compounds.*   Reaction takes place specifically in the 1,4-manner, e.g.

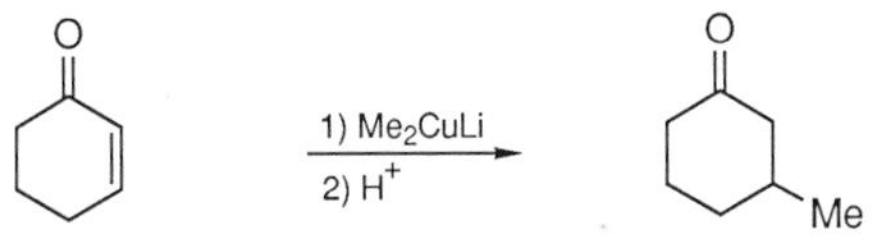

in contrast to organolithium compounds (p. 198) and Grignard reagents from purified magnesium (p. 193). A mechanism which has been suggested involves an intermediate adduct in which copper has a formal oxidation state of +3 and which undergoes a *reductive elimination* (p. 563), e.g.

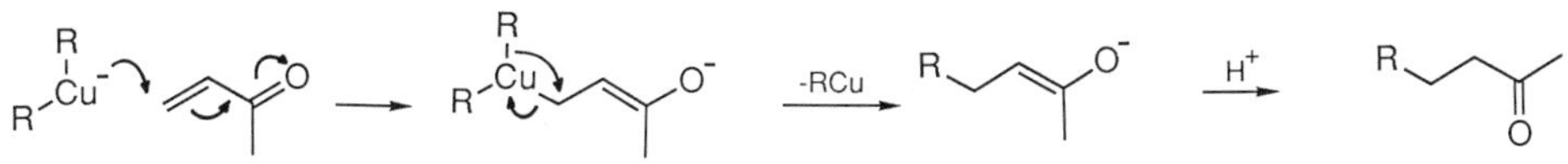

but it remains unclear why 1,4-addition should occur so much more readily than 1,2-addition.

The organocuprates react more rapidly and usually give better yields than organocopper compounds and are the reagents of choice. However, as the equation above shows, one of the R groups in R_2CuLi is 'wasted' as RH, and if this group if difficult to synthesize or particularly expensive to obtain, it is better to employ a mixed organocuprate in which one of the groups attached to copper is l-pentynyl:

Alkynyl groups are not transferred from copper to αβ-unsaturated carbonyl compounds so that the R group in the mixed organocuprate is used efficiently for this purpose; the by-product, l-pentyne, can be reconverted into its copper derivative with a Cu(I) salt.

1,4-Addition of organocopper compounds and organocuprates also occurs to C≡C conjugated to a carbonyl group and to C=C and C≡C conjugated to —CO_2R or —CN. However, isolated carbonyl groups react very slowly and alternative methods are used.

(*ii*)   *With acid halides.*   These form ketones, e.g.

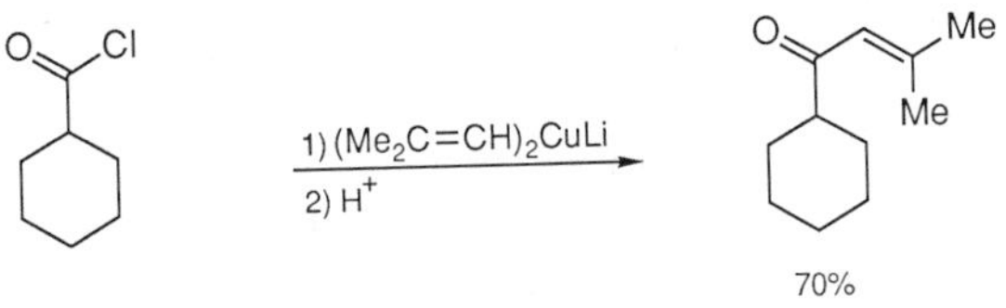

70%

(*iii*) *With epoxides.* These form alcohols by reaction, with inversion of configuration, at the less substituted carbon atom of the epoxide, as in the S_N2 process, e.g.

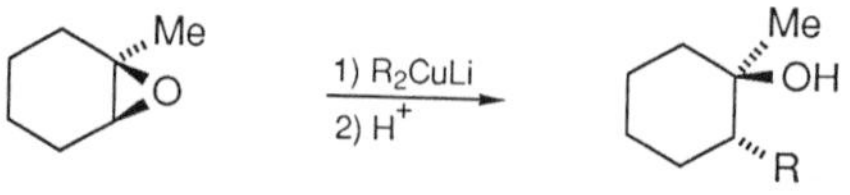

(*iv*) *With halides.* Organocuprates react with alkyl (primary or secondary), alkenyl, and aryl bromides and iodides, but not chlorides, by displacement of halide ion in a process known as a *coupling* reaction. Except with secondary alkyl compounds and the coupling of an alkenylcuprate with an alkenyl halide, yields are generally high, e.g.

$$C_{10}H_{21}Br \xrightarrow{Bu_2CuLi} C_{14}H_{30}$$

1-bromodecane → tetradecane 80%

80%

The yields with secondary alkyl halides are improved by employing a cyano-complexed cuprate:

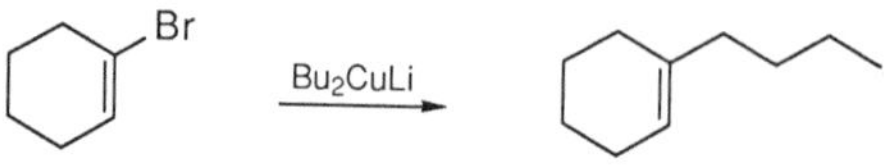

$$2\,RLi + 1/2\,Cu_2(CN)_2 \longrightarrow R_2Cu(CN)Li_2$$

The yields of alkenyl-alkenyl coupled products are improved by including $ZnBr_2$ and a catalytic quantity of $Pd(PPh_3)_4$, e.g.

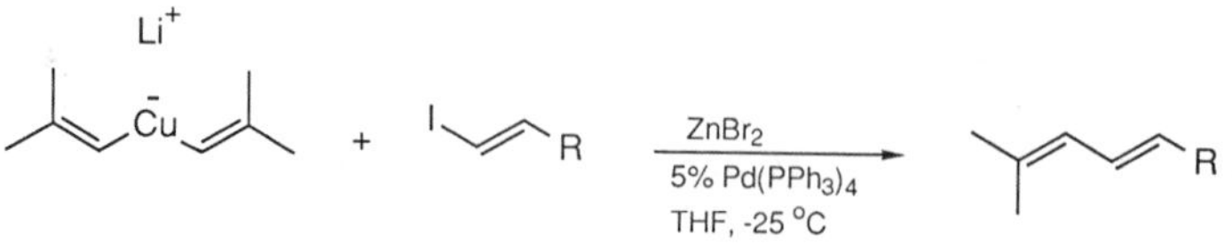

Displacement of the halide occurs with almost complete stereoselectivity: the yield of (*Z*)-isomer in the example above is less than 0.5%. Since alkenylcuprates can be made with a high degree of stereoselectivity from alkynes of the type $RCH_2C{\equiv}CH$ (p. 200), this enables two double bonds to be introduced in a stereochemically predetermined way.

The mechanism of coupling is not clear. Alkyl bromides react with inversion of configuration, suggestive of an S_N2 reaction. However, optically active alkyl iodides yield racemic mixtures, and the fact that alkenyl halides react also argues against an S_N2 process. It is possible that, in these cases, electron-transfer processes are involved.

Cadmium alkyls and aryls may be obtained by the metal-metal exchange reaction (p. 185), using either a Grignard reagent or an organolithium compound:

$$2\ RMgX\quad +\quad CdCl_2\quad \longrightarrow\quad R_2Cd\quad +\quad 2\ MgXCl$$

$$2\ RLi\quad +\quad CdCl_2\quad \longrightarrow\quad R_2Cd\quad +\quad 2\ LiCl$$

They are far less reactive compounds than both Grignard reagents and organolithium compounds. In particular, they do not react with ketones or esters, although they react with acid chlorides:

$$2\ R'COCl\quad +\quad R_2Cd\quad \longrightarrow\quad 2\ R'COR\quad +\quad CdCl_2$$

This specificity gives organocadmium compounds their main use in synthesis. Although it is possible to obtain ketones from both Grignard reagents and organolithium compounds, it is often necessary to introduce a ketonic function into a molecule which possesses other functions susceptible to attack by magnesium and lithium compounds. The cadmium compound may then be used.

Mercury(II) salts are electrophiles which add to alkenes and substitute in aromatic compounds; the latter reactions are described later (p. 388).

The reaction of an alkene with mercury(II) acetate usually occurs rapidly in aqueous tetrahydrofuran. It is thought that there is bridging in the intermediate cation (p. 82), but the regiochemistry, as with the addition of bromine, is dictated by the tendency for formation of the more stable of the two possible carbocations, e.g.

If the water in the solvent is replaced by an alcohol or a carboxylic acid, the corresponding ether or carboxylic ester is formed, e.g.

The importance of the reactions lies in the ease with which the mercury substituent can be removed by reduction with sodium borohydride,

in a reaction believed to involve radicals,

initiation   R—Hg—OAc $\xrightarrow{\text{NaBH}_4}$ R—Hg—H $\longrightarrow$ R· + ·HgH

propagation $\begin{cases} \text{R· + R—Hg—H} \longrightarrow \text{RH + R—Hg·} \\[1em] \text{R—Hg·} \longrightarrow \text{R· + Hg(0)} \end{cases}$

Yields in each step are usually high, for example, 1-methylcyclohexene forms l-methylcyclohexanol in 90% overall yield,

1) Hg(OAc)$_2$ - THF / H$_2$O
2) NaBH$_4$

This process is an alternative to the hydration of alkenes with sulfuric acid and is the method of choice when another function in the organic compound is sensitive to strong acids. It is also complementary to the hydration of alkenes by the hydroboration route, which occurs with the alternative regioselectivity (p. 484).

Alkynes are hydrated in aqueous sulfuric acid in the presence of a catalytic quantity of mercury(II) sulfate (p. 91), e.g.

$$R\text{—}\!\!\!=\!\!\!\text{—}H \quad \xrightarrow[\text{H}_2\text{SO}_4 \text{ - H}_2\text{O}]{\text{HgSO}_4 \text{ cat.}} \quad \overset{R}{\underset{O}{\bigvee}}\!\!CH_3$$

6.8 Organozinc compounds

Diiodomethane reacts with a zinc-copper alloy to form an unstable organometallic adduct, ICH$_2$ZnI which, in the presence of an alkene, transfers CH$_2$ to form a cyclopropane (*Simmons–Smith reaction*), e.g.

Zinc is also employed in the Reformatsky reaction (p. 217).

Problems

1. How would you obtain each of the following from phenylmagnesium bromide? Ph$_3$C—OH; Ph$_2$C=CH$_2$; PhC(CH$_3$)=CH$_2$; PhCO$_2$H; PhCH$_2$Ph; PhCHO; PhCH$_2$CH$_2$OH; PhD.

2. What products would you expect from the reaction of methylmagnesium iodide with each of the following? CH$_3$CO$_2$Et; CH$_3$CN; ClCO$_2$Et; CH$_3$CO—O—COCH$_3$; ClCH$_2$OCH$_3$; CH$_2$=CH—CH$_2$Br; PhCOCH=CH$_2$; (CH$_3$)$_3$C—CO—C(CH$_3$)$_3$.

3. How would you employ organometallic reagents to make the following?

(*a*) 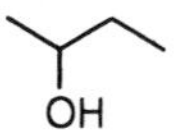
OH

(*b*) 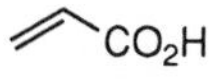
CO2H

(*c*) 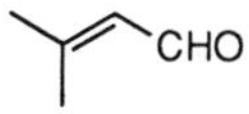
CHO

(*d*) 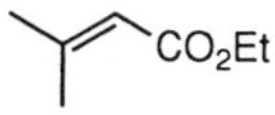
CO2Et

(*e*) 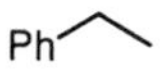
Ph

(*f*) 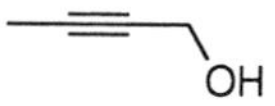
OH

(*g*) 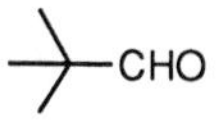
CHO

(*h*) 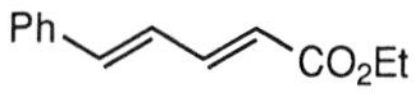
Ph
CO2Et

(*i*) 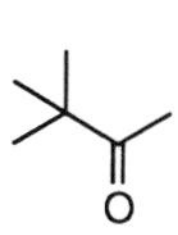
O

(*j*) 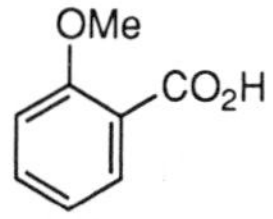
OMe
CO2H

(*k*) 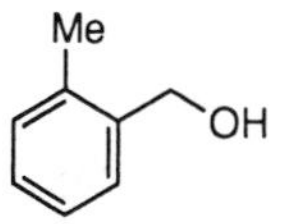
Me
OH

(*l*) 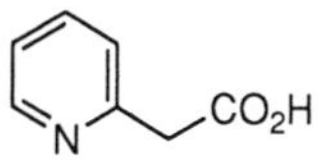
N
CO2H

4. How would you use organocopper reagents in the preparation of the following?

(*a*) 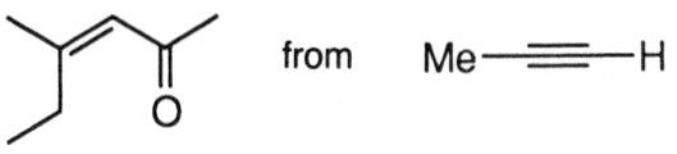
O
from
Me
H

(*b*) 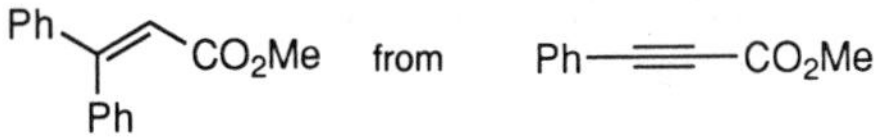
Ph
CO2Me
Ph
from
Ph
CO2Me

(*c*) 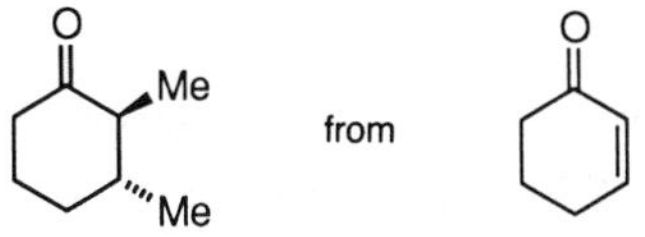
O
Me
Me
from
O

7 Formation of carbon–carbon bonds: base-catalyzed reactions

7.1
Principles The formation of carbon–carbon bonds by base-catalyzed reactions is closely related to their formation from organometallic reagents. In each method negatively polarized carbon reacts with the electrophilic carbon of carbonyl groups, alkyl halides, and related compounds. The difference between the two procedures is that whereas the method of inducing negative polarization in carbon in the latter case is to attach carbon to an electropositive metal, in the former case a base is employed to abstract a proton from a C—H bond to give a *carbanion*.

The scope of base-catalyzed reactions depends on three facts: a wide range of compounds is able to form carbanions, these anions react with electrophilic carbon in a variety of environments, and the basicity of the reagent used to abstract the proton in the generation of the carbanion may be varied widely so as to produce suitable conditions for a particular reaction.

The factors which govern the formation of a carbanion have been briefly discussed (section 2.4). The most general structural requirement is that the C—H bond from which proton abstraction is to occur is adjacent to one or more groups of $-M$ type which can stabilize the anion, e.g.

$$CH_3\diagup\diagdown O \quad \underset{BH^+}{\overset{B}{\rightleftharpoons}} \quad {}^-CH_2\diagup\diagdown O \quad \longleftrightarrow \quad CH_2\diagup\diagdown O^-$$

For example, alkanes do not undergo base-catalyzed reactions because the corresponding anions, being of very high energy content, are formed in negligible concentration. The more common of the activating groups, arranged in approximately decreasing power of activation, are:

$$-NO_2 \quad > \quad -COR \quad > \quad -CN \quad > \quad -CO_2R$$

A hydrogen atom attached to carbon which is bonded to one of these groups (an 'α-hydrogen') is described as being *activated*. The presence of two such substituents adjacent to a C—H bond considerably increases the acidity of that bond (section 2.4), and many useful reactions employ di-activated compounds such as malonic ester, $CH_2(CO_2Et)_2$, and acetoacetic ester, $CH_3COCH_2CO_2Et$.

In the majority of cases, the negative charge is shared with an oxygen atom and these ions are usually referred to as *enolates* (i.e. they are formally derived from enols, $C{=}C{-}OH$). There are also synthetically useful carbanions in which the charge is not delocalized, especially anions from alkynes, $RC{\equiv}C^-$ (section 7.6), and cyanide ion (section 7.7). Enolates, although resonance-stabilized, are nevertheless high-energy species which react with positively polarized carbon. The two classes of reaction of general importance are:

(1)   with carbonyl-containing compounds

(2)   with alkyl halides or sulfonates

The reactions are similar in principle in that each involves the displacement of an electron-pair from the carbon which is attacked on to an electronegative atom. They differ in that only the former reaction is readily reversible (so that many of the reactions described in the ensuing text can be employed in the reverse sense for the cleavage of $C{-}C$ bonds).

The commonly used bases, in order of decreasing base-strength (section 2.4), are:

$$Ph_3C^- > (Me_2CH)_2N^- > EtO^- > HO^- > R_3N$$

The choice of base is determined by the reactivity of the enolate-forming component and is discussed below.

Until fairly recently, most base-catalyzed reactions were carried out with an oxy-anion base such as HO^- or EtO^-. However, for mono-activated compounds where pK is in the range 10–20, the concentration of the enolate is small, e.g.

$$CH_3COCH_3 + OH^- \underset{}{\overset{K=10^{-4}}{\rightleftharpoons}} CH_3COCH_2^- + H_2O$$

pK 20 $\qquad\qquad\qquad\qquad\qquad\qquad\qquad$ pK 16

and reaction occurs relatively slowly. It has therefore become common practice to employ a base whose conjugate acid is significantly stronger than the enolate-forming compound. For example, if the pK of the base's conjugate acid is 4 units greater than that of the carbonyl compound, formation of the enolate is 99.99% complete. A particularly valuable very strong base is $(Me_2CH)_2N^-$ ('LDA'), which has the added advantage of being hindered as a nucleophile. It is prepared, as its lithium salt, by reaction of the amine (pK 40) with butyllithium in an aprotic solvent such as tetrahydrofuran at low temperatures,

$$(Me_2CH)_2NH + BuLi \xrightarrow[0\,°C]{THF} (Me_2CH)_2N^-\ Li^+ + C_4H_{10}$$

lithium diisopropylamide

(LDA)

Two further points should be noted. First, the charge on an enolate is shared by carbon and a second, more electronegative element (usually oxygen, but in the case of activation by -CN, nitrogen). Indeed, the major portion of the charge is associated with the latter, for example, in the hybrid

$$^-CH_2\diagup\diagup O \quad\longleftrightarrow\quad CH_2\diagup\diagup O^-$$

the second resonance structure is the more important contributor. Reaction of the enolate with a carbonyl group or a halo-compound could therefore occur at the more electronegative atom, e.g.

In practice, reaction at carbon normally dominates in each case, probably because the products are thermodynamically more stable than those derived by reaction through oxygen, this difference in stability being reflected in the transition states for the processes. Two exceptions to this generalization – ring closure when the ring formed by reaction at carbon would be significantly strained whereas that formed by reaction at oxygen is not, and alkylation when a particularly reactive electrophile is used – are described later (pp. 239, 233). In contrast to reaction of an enolate with a carbonyl compound or an alkyl halide, reaction with a chlorosilane occurs at oxygen to form a Si—O bond since this is a significantly stronger bond than Si—C, the difference being reflected in transition-state energies. This has important applications (e.g. p. 216).

Secondly, the activated compound may have two active sites, for example, 2-methylcyclohexanone could react through

(a)   (b)

Which provides the major pathway can depend on whether the process is subject to kinetic or thermodynamic control. Enolate (a) is the more stable, probably because of stabilization by the methyl group of the important alkene-like structure (c) (p. 34):

(c)

Therefore, when the two enolates have time to equilibrate before reacting with another reagent (thermodynamic control), reaction occurs mainly as though through (a); an example is alkylation (p. 234). In contrast, reaction with LDA in 1,2-dimethoxyethane at −78°C gives 99% of (b) and 1% of (a) and subsequent reaction is almost entirely by way of (b). This is the result of kinetic control: the CH_2 group is sterically more accessible than the methyl-substituted CH and is attacked the more rapidly; in an aprotic solvent at the very low temperature used the enolate does not revert to reactant at a significant rate so that equilibration between (b) and (a) does not occur.

7.2
Reactions of enolates with aldehydes and ketones

(a) The aldol reaction

The aldol reaction is one of the most commonly used methods for forming new C—C bonds. It is capable of wide variation in the reactants and providing regiochemical control and, in some cases, stereochemical control. It consists of the reaction between two molecules of aldehydes or ketones, which may be the same or different: one molecule is converted into a nucleophile by forming its enolate in basic conditions or its enol in acidic conditions (p. 256) and the second acts as an electrophile. The reaction takes its name from the trivial name of the product of the dimerization of acetaldehyde discovered by Wurtz in 1872:

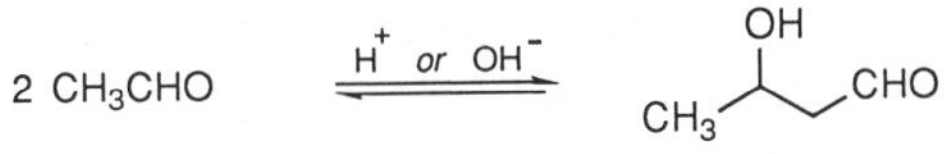

The mechanism of the base-catalyzed reaction is as follows:

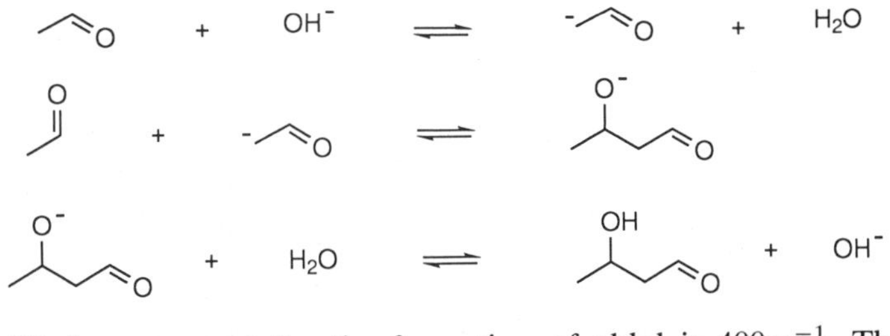

The equilibrium constant for the formation of aldol is $400\,M^{-1}$. The kinetics (rate $= k[CH_3CHO][OH^-]$) show that the first step, generation of the enolate, is rate-determining. In a closely related case, however, the self-condensation of acetone to give diacetone alcohol, the kinetics (rate $= k[CH_3COCH_3]^2[OH^-]$) show that the rate-determining step is the reaction of the enolate with a second molecule of acetone:

The difference is attributed to the fact that the carbonyl group in acetone is less rapidly attacked by nucleophiles than that in acetaldehyde, partly because the electron-releasing character of methyl compared with hydrogen renders the adduct from acetone (and the preceding transition state) less stable, and partly because the carbonyl group in acetone is more hindered. It is a general feature of aldol reactions that the step involving C—C bond-formation is facilitated by electron-attracting groups on the carbonyl component and retarded by electron-releasing groups.

Other general features of the reaction which must be borne in mind in synthetic applications are the following.

(1)   Reaction is readily reversible, and the position of equilibrium is not always favourable to the product. For example, this is true for the formation of diacetone alcohol from acetone ($K = 0.04\,\mathrm{M}^{-1}$) and a special experimental procedure is used to obtain a high yield of the product. Barium hydroxide is placed in the thimble of a Soxhlet extractor over a flask of boiling acetone. As acetone comes into contact with the catalyst a small proportion of diacetone alcohol is formed which is then automatically returned to the boiler by a syphon mechanism. Since this product boils over 100°C higher than acetone, it accumulates in the boiler while acetone continues to reflux, condenses on to the catalyst, and returns to the boiler together with further product. In this way diacetone alcohol may be obtained in 70% yield [1].

(2)   Aldols as such are not always isolated from the condensation. For example, that from acetaldehyde readily forms a cyclic hemiacetal:

On being heated, this trimer is converted back into the aldol which, in hot basic conditions, undergoes dehydration:*

---

* The products of base-catalyzed eliminations such as this, and of the corresponding acid-catalyzed eliminations, are not formed stereospecifically but generally consist mainly of the *trans* isomer (cf. p. 15) and will be shown as such.

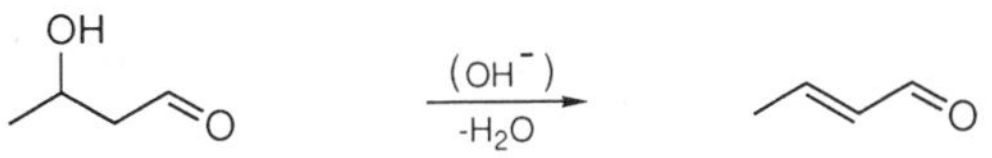

Diacetone alcohol is more stable, but dehydration can be brought about by acid or base:

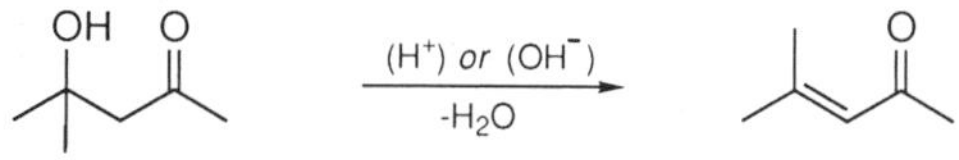

(3) The use of more concentrated alkali can lead directly to the unsaturated carbonyl product which can then undergo further condensations. For instance, acetaldehyde gives resins as a result of successive condensations of the type,

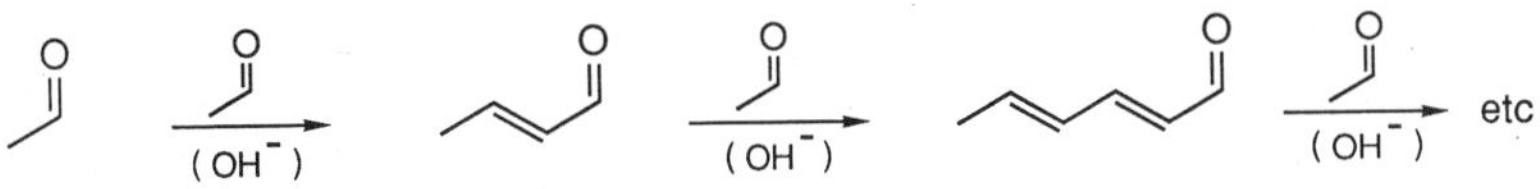

However, it is usually possible to select reaction conditions in which the first dehydration product is formed in good yield, e.g.

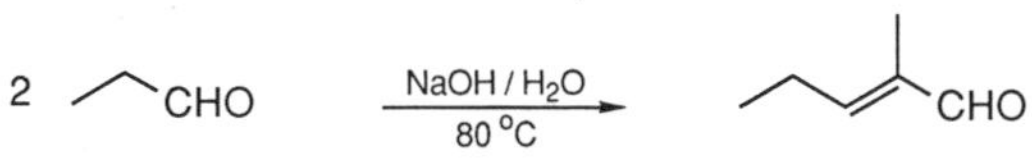

The process is then referred to as the *aldol condensation*, and it is in the formation of such αβ-unsaturated carbonyl compounds that the synthetic utility of the aldol reaction mainly lies.

(*i*) *Unsymmetrical ketones*. Whereas an aldehyde has only one carbon atom activated by the carbonyl group from which proton abstraction may occur, a ketone may have two. For example, butanone might undergo reaction *via* either of the anions $\bar{C}H_2COCH_2CH_3$ and $CH_3CO\bar{C}HCH_3$ leading to

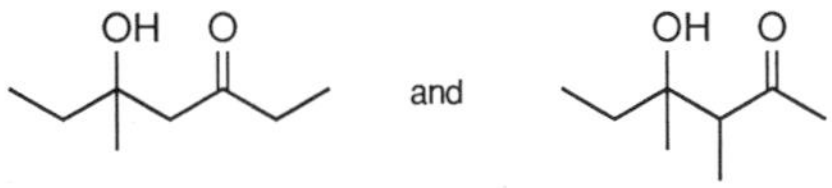

respectively. In practice, the former product predominates, even though a higher concentration of the enolate leading to the latter would be expected under conditions of thermodynamic control (p. 208). This is probably because there is greater steric hindrance to attack on the carbonyl group of butanone by $CH_3CO\bar{C}HCH_3$ than by $\bar{C}H_2COCH_2CH_3$. In general, the enolate component from an unsymmetrical ketone is that derived from the less highly alkylated carbon.

(*ii*) *Mixed reactions*. If each of two aldehydes contains an α-hydrogen atom, aldol reaction can give each of four products: each aldehyde can provide an enolate and each can act as the carbonyl component. Such condensations are only of synthetic value if a required product can easily be separated from the resulting mixture. In general, it is better to employ an alternative strategy: namely, to use a pre-formed enolate (p. 213).

If, however, one of the two components has no α-hydrogen, only two products can be formed. Further, if this component has the more reactive carbonyl group of the two, one product predominates. For example, the action of alkali on a mixture of formaldehyde and acetaldehyde leads, initially, to β-hydroxypropionaldehyde,

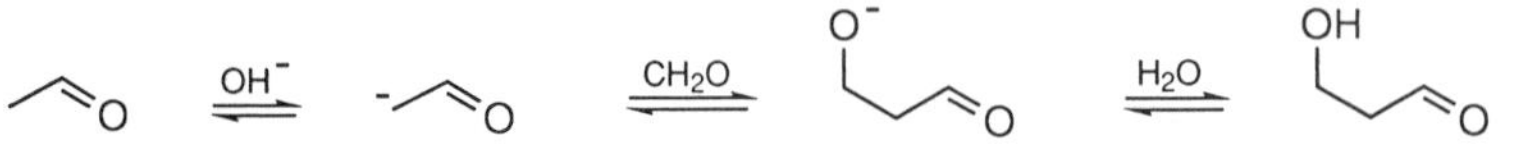

for only acetaldehyde can form an enolate and the carbonyl group in formaldehyde is the more reactive to nucleophilic addition.

In practice, this reaction does not give β-hydroxypropionaldehyde as the final product. If the reaction is carried out in the gas phase at a high temperature (e.g. using sodium silicate as catalyst, at 300°C), dehydration occurs

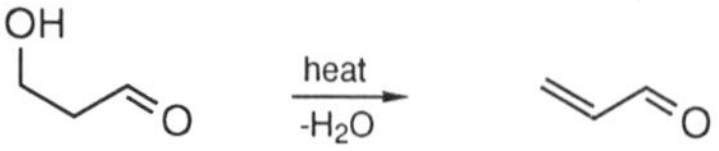

whereas at low temperatures further aldol reactions occur until each of the three α-hydrogens of acetaldehyde has been replaced,

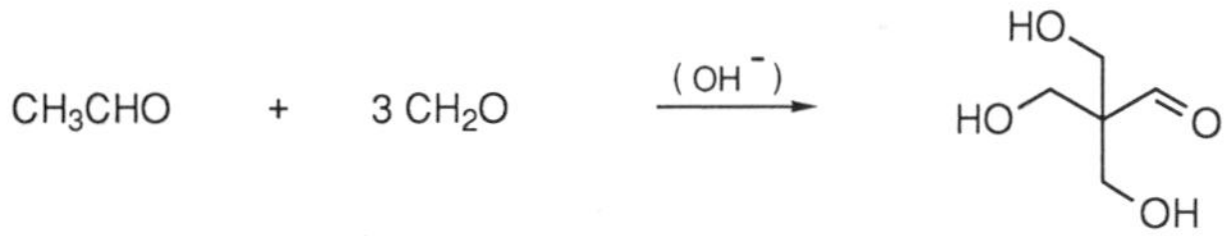

$$CH_3CHO \quad + \quad 3\,CH_2O \quad \xrightarrow{(OH^-)}$$

This product, like formaldehyde, has no active hydrogen and next undergoes the crossed Cannizzaro reaction characteristic of such aldehydes (p. 654):

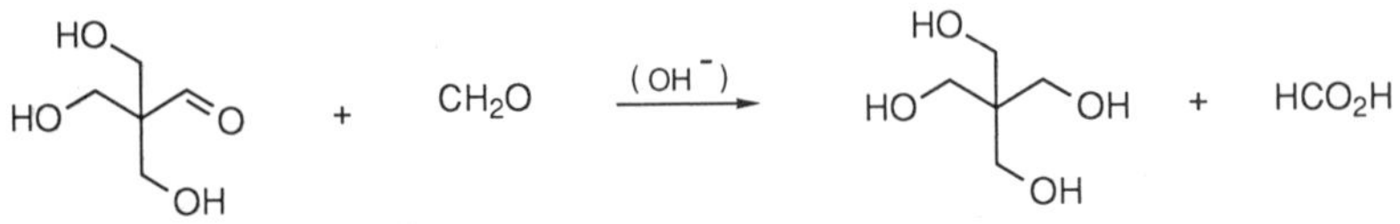

The product, pentaerythritol, can be obtained in about 55% yield in this way using calcium hydroxide as the basic catalyst [1]. This reaction is also carried out industrially using lime as the catalyst; the pentaerythritol is converted into its tetranitrate ester for use as an explosive.

The aldol reaction on a mixture of an aldehyde and a ketone each of which contains active hydrogen may also in principle yield four products but, since the carbonyl group in aldehydes is usually more reactive towards nucleophiles than that in a ketone, only two products are likely to be formed in significant amounts: the dimer of the aldehyde and the product derived by reaction of the enolate from the ketone with the carbonyl of the aldehyde. Moreover, the aldehyde dimer can be minimized by the slow addition of the aldehyde to a mixture of the ketone and the base.

The aldol reaction on a mixture of a ketone and an aldehyde which possesses no active hydrogen usually gives essentially one product. For example, benzalacetone is obtained in about 70% yield from benzaldehyde and excess of acetone in the presence of 10% sodium hydroxide solution [1],

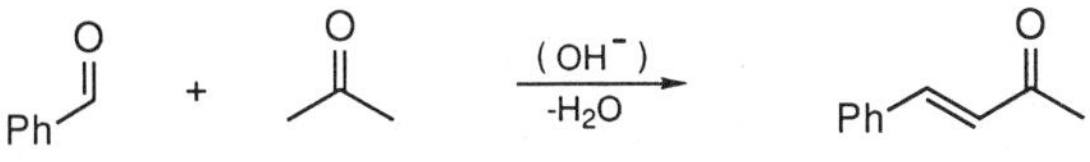

benzalacetone

and, if acetone is not in excess, dibenzalacetone can be obtained in about 80% yield [2]:

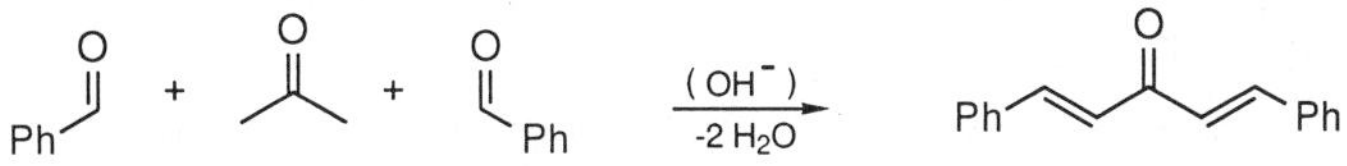

dibenzalacetone

Any ketone which possesses the grouping $-CH_2-CO-CH_2-$ can give an analogous dibenzal derivative and the reaction with benzaldehyde has been employed to reveal the presence of such groupings.

When an α-diketone possessing no active hydrogen is used in place of an aldehyde, cyclic products are formed, e.g.

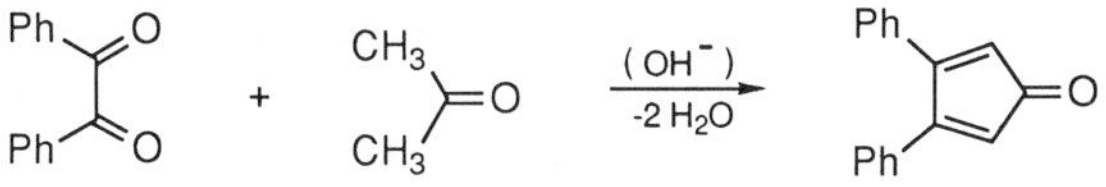

In the above examples, formation of the condensation product from the aldol adduct is promoted by the stabilizing effect of the conjugated phenyl group on the C=C bond which is developing at the transition state.

(iii)   *Use of pre-formed enolates in aldol reactions.*   The disadvantage that more than one product is formed in mixed aldol reactions when each carbonyl compound contains an active hydrogen can be avoided by pre-forming the enolate of one of the components with LDA, e.g.

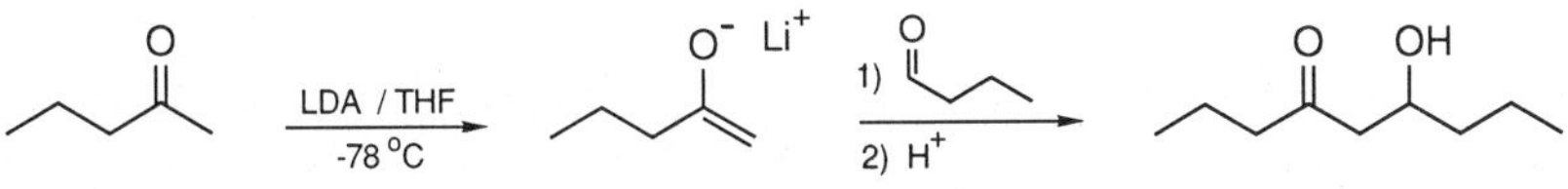

65%

This development, giving regiochemical control, has greatly increased the scope of aldol reactions.

(iv)   *Intramolecular aldol reactions.*   Internal reactions are particularly valuable for generating five- and six-membered cyclic enones, e.g.

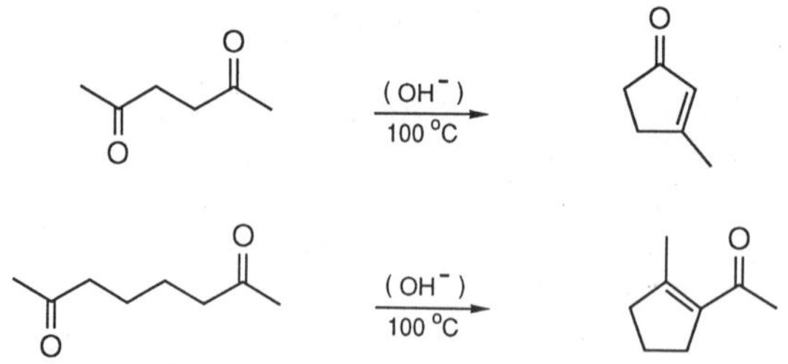

Three- and four-membered rings are not formed in this way owing to their strain (unfavourable enthalpy of activation): the intermolecular reaction is favoured. The second example above shows that five-membered are formed in preference to seven-membered rings and, in general, rings larger than six-membered are not available owing to some strain occurring in their formation (7–11 members) and an increasingly unfavourable entropy of activation (p. 67).

Bicyclic compounds, of value in the synthesis of certain natural products (e.g. p. 765), can also be obtained, e.g.

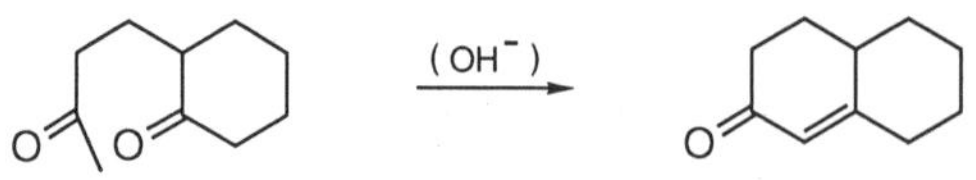

(v) *Diastereoselectivity in aldol reactions.*   An aldol reaction can create two asymmetric centres, e.g.

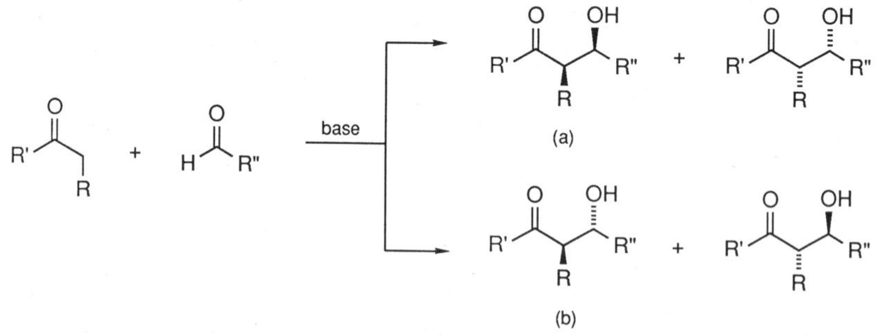

The diastereomers of type (a) are usually referred to as *syn* or *erythro* and those of type (b) as *anti* or *threo*.

Unless one of the substituents is itself asymmetric, a racemic mixture of any pair of enantiomers is inevitably formed. However, the diastereomers are not usually formed in equal amounts; either can predominate depending on the reaction conditions and the nature of the substituents, and in some cases does so essentially exclusively. The results can be rationalized in most instances and the theory that has evolved is of value in predicting and controlling the stereochemical outcome of new reactions. The main tenet of the theory is that the new C—C bond is formed, at least in part, *via* a six-membered cyclic transition state which includes the metal counter-ion. This is coupled with recognizing that the geometrically isomeric forms of an enolate, which are not formed at the same rate and may or may not be able to equilibrate under the reaction conditions, yield cyclic transition states of different stability.

There are two general cases. In one, the (Z)- and (E)-forms of the enolate are in rapid equilibrium, and the product distribution is determined by the relative stabilities of the cyclic transition states (M = metal) (only one enantiomer of each product is shown in this and subsequent examples):

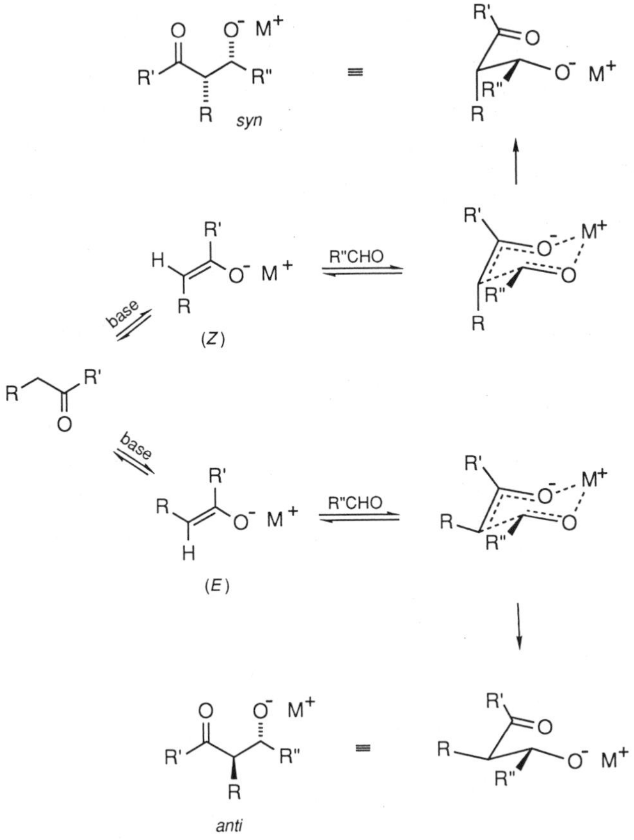

The chair-shaped transition state leading to the *anti* product has both R and R" in (approximately) equatorial positions, whereas in that leading to *syn* product R is in the less stable axial position. The former is therefore of lower energy, leading to a predominance of *anti* product. The conditions for this to occur (thermodynamic control) are the use of a relatively weak base such as hydroxide ion, high temperatures, and long reaction times.

In the other general case, kinetic control is achieved by the use of a very strong base such as LDA, low temperatures, and short reaction times. The (Z)- and (E)-enolates are formed rapidly and irreversibly, and their relative amounts determine the product composition. For a ketone MeCH$_2$COR the (Z)-enolate is normally formed the faster of the two providing that R is reasonably bulky (presumably because the methyl substituent is eclipsed by oxygen in the (Z)-isomer but by the bulkier R group in the (E)-isomer and this strain factor is reflected in the transition states for their formation). The major product is then the *syn* diastereomer. The selectivity can be very high when R is particularly large, such as t-butyl, e.g.

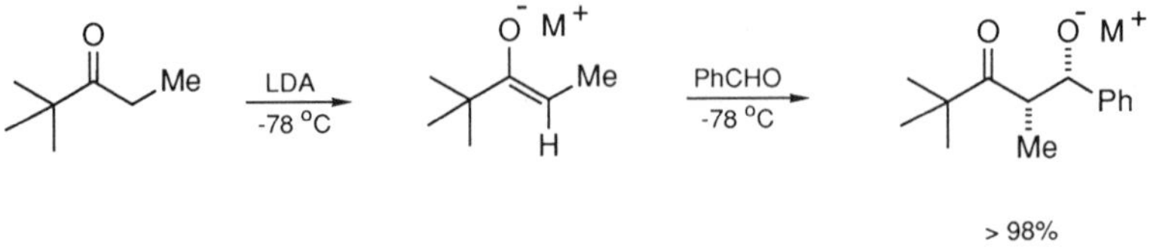

> 98%

but it falls off as the size of R is reduced (e.g. the *syn*:*anti* ratio is about 4:1 when R is isopropyl under these conditions). Moreover, in some cases the (E)-enolate is formed the faster.

There are two techniques for increasing the degree of diastereoselectivity. One is to pre-form the enolate of the desired stereochemistry. Chlorotrimethylsilane is included with the ketone and base: the enolates are trapped as silyl enol ethers, e.g.

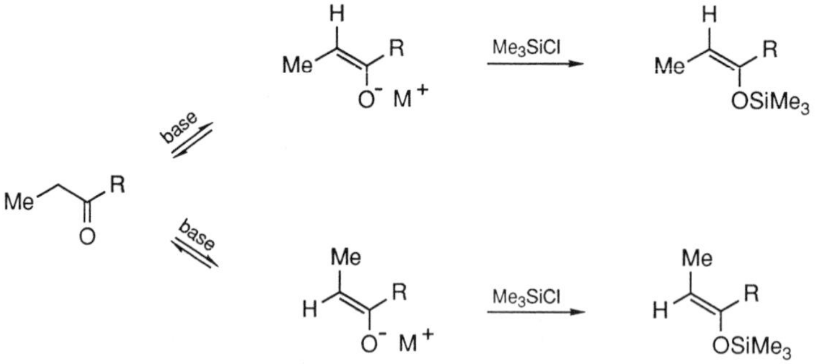

which can be separated by distillation or chromatography and then converted into pure (Z)- or (E)-enolate by treatment with fluoride ion, e.g.

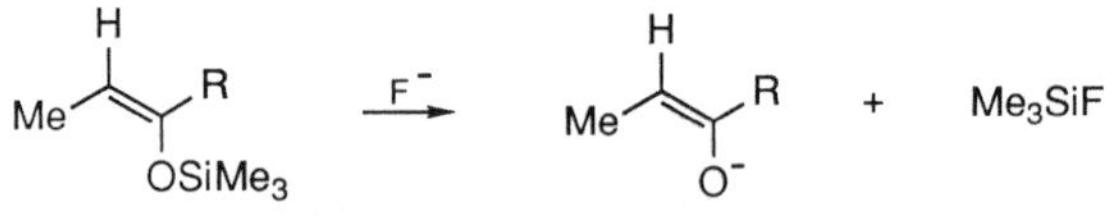

Reaction with the second carbonyl compound then yields mainly *syn* and *anti* products from (Z)- and (E)-enolates, respectively. However, although the *syn*:*anti* ratio from a (Z)-enolate is usually large, the *anti*:*syn* ratio from an (E)-enolate is usually small, for example, for reactions with benzaldehyde:

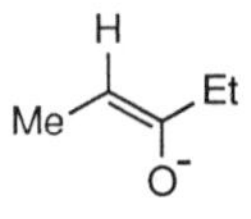

syn : anti = 90 : 10

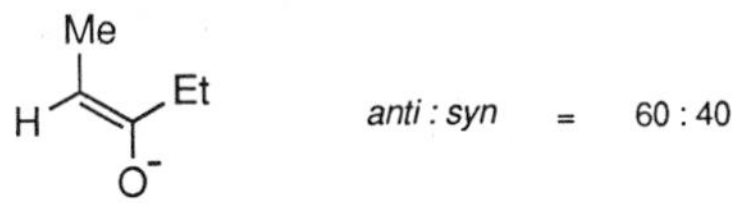

anti : syn   =   60 : 40

The second technique exploits the observation above that, when R in MeCH$_2$COR is very bulky, kinetically controlled conditions give very largely *syn* product: that is, a large R group is introduced which can later be removed, e.g.

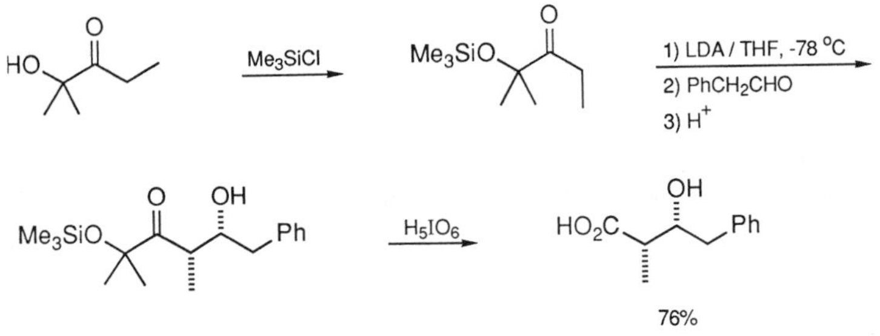

76%

(b)   The Claisen reaction

The base-catalyzed reaction between an ester which contains active hydrogen and an aldehyde which does not is known as the Claisen reaction. For example, ethyl acetate and benzaldehyde react in the presence of sodium ethoxide to give ethyl cinnamate in about 70% yield [1]:

The success of the reaction depends on the fact that the carbonyl in an aldehyde is more reactive towards nucleophiles than one in an ester. Aldehydes which contain an α-hydrogen atom are not suitable components in the Claisen reaction because they preferentially undergo self-condensation.

(c)   The Reformatsky reaction

Although this is not a base-catalyzed reaction, it is included here since it is the reaction of an enolate with an aldehyde or ketone. The enolate is generated from an α-bromo-ester with zinc in diethyl ether, e.g.

and adds to the carbonyl compound to give, after hydrolysis, an aldol-type product

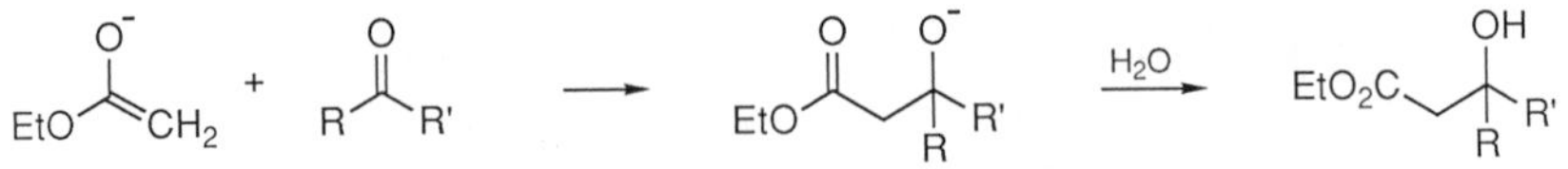

Whereas reactions of enolates with aromatic aldehydes are often followed by base-catalyzed elimination to give αβ-unsaturated compounds, the absence of base in the Reformatsky reaction results in the isolation of the β-hydroxy-ester, e.g.

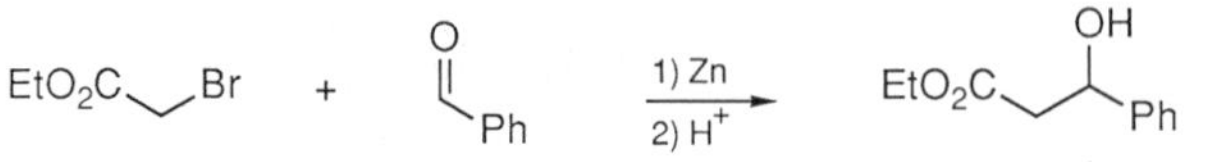

α-Bromoketones cannot be used in the Reformatsky reaction because the zinc enolate preferentially reacts as it is formed with more of the ketone.

(d)   The Perkin reaction

This reaction consists of the condensation of an acid anhydride with an aromatic aldehyde catalyzed by a carboxylate ion. The mechanism is surprisingly complex, although each step is of a well established type. The anhydride first provides the enolate under the influence of the basic carboxylate ion:

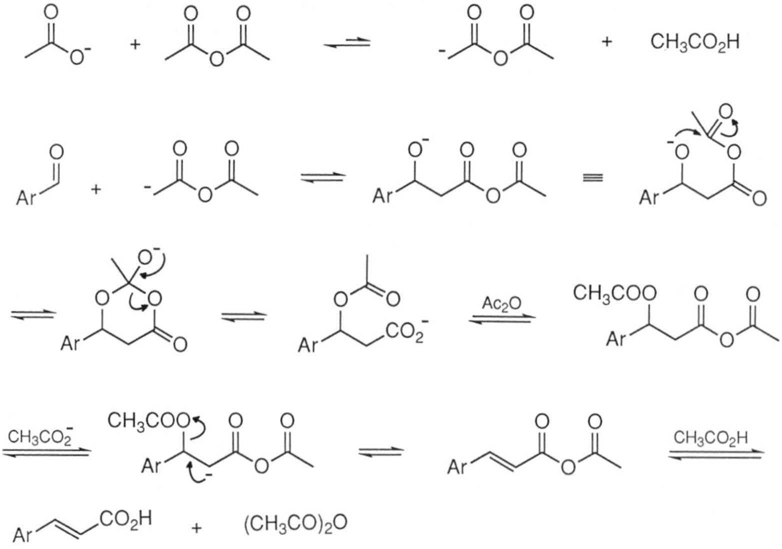

Electron-attracting groups on the aromatic ring increase the efficiency of the reaction, e.g. p-nitrobenzaldehyde gives an 82% yield of p-nitrocinnamic acid. Conversely, electron-releasing groups reduce the efficiency, e.g. p-dimethylaminobenzaldehyde fails to condense. Because of the relatively weak basicity of carboxylate ions, fairly high temperatures are generally required.

Methods based on the Perkin reaction have been used in the synthesis of α-amino acids. The first of these was Erlenmeyer's azlactone synthesis, in which an acylglycine is condensed with an aromatic aldehyde in the presence of acetic anhydride and sodium acetate. The acylglycine is converted into an azlactone by the dehydration of its enol tautomer with acetic anhydride (cf. the ready formation of γ-lactones, p. 66), e.g.

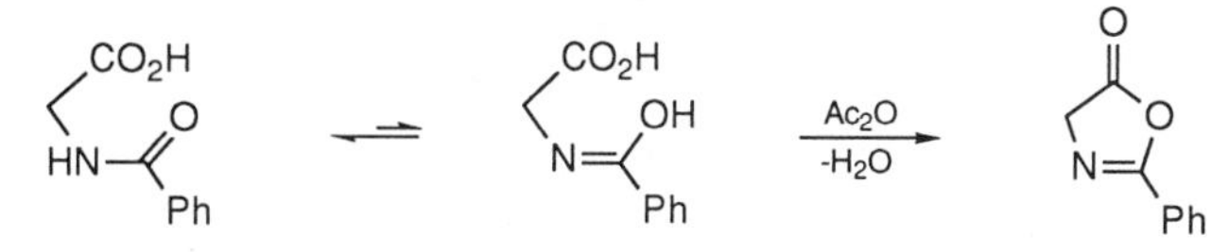

benzoylglycine

The azlactone, whose methylene group is activated by the carbonyl group, undergoes the Perkin condensation with the aldehyde,

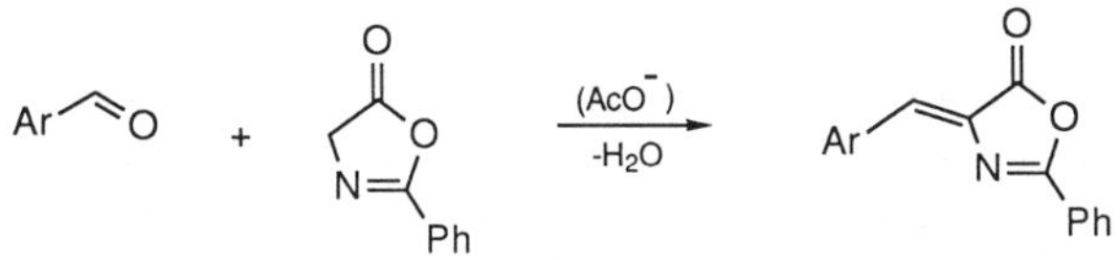

This product is then reduced and hydrolyzed to give the α-amino acid. For example, (±)-phenylalanine may be obtained in about 65% yield by treating the azlactone below with red phosphorus and aqueous hydriodic acid [2]:

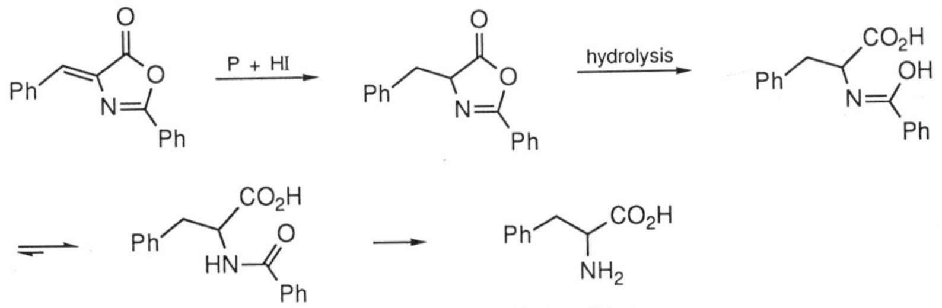

(±)-phenylalanine

The Erlenmeyer reaction proceeds much more readily than the Perkin reaction, e.g. the condensation above is complete after about 1 hour at 100 °C. However, if the acylglycine cannot form an azlactone (e.g. benzoyl-N-methylglycine) condensation occurs much less readily.

Variants of the azlactone synthesis in which analogues of azlactones are used are sometimes advantageous. Hydantoin, thiohydantoin, and rhodanine have each been employed as the enolate-forming component of the condensation.

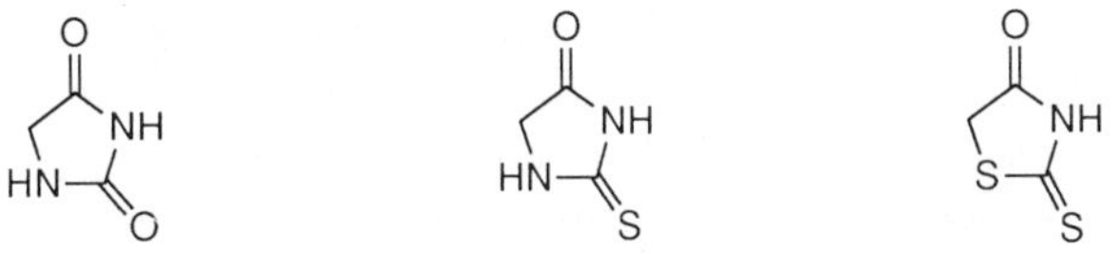

hydantoin   thiohydantoin   rhodanine

By omitting the reduction step, α-keto-acids may also be obtained from azlactones, e.g.

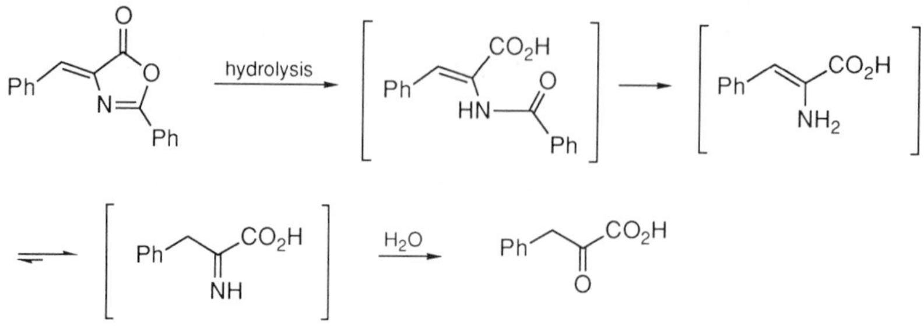

(e)   The Stobbe condensation

Ketones normally react with esters in the presence of base by a mechanism in which the enolate from the ketone displaces alkoxide ion from the ester (Claisen condensation, p. 226). Dialkyl succinates, however, behave differently from other esters in that the enolate from the ester adds to the carbonyl group of the ketone:

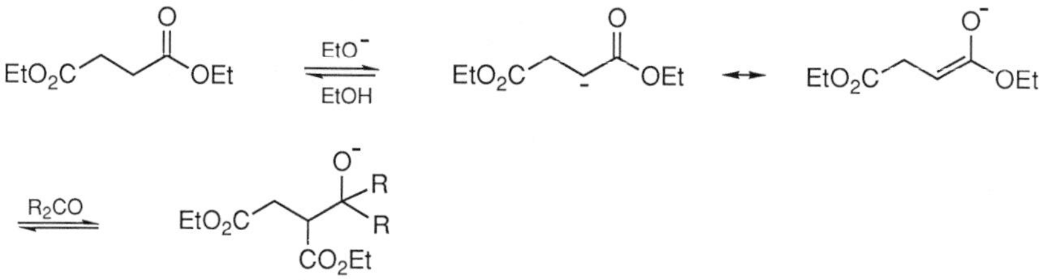

The resulting adduct cyclizes to a γ-lactone which then undergoes base-catalyzed ring opening to give a carboxylate salt:

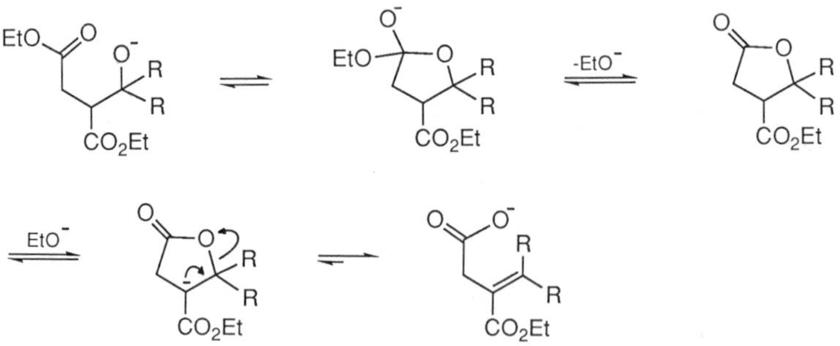

The stability of the carboxylate anion is the basis for the success of the reaction for it results in the equilibrium being favourable to this product, whereas in the reaction of monobasic esters with ketones, where the corresponding situation does not exist, the equilibria established favour the product of the Claisen condensation.

Aldehydes containing α-hydrogen undergo self-condensation in preference to the Stobbe reaction, but those with no α-hydrogen, e.g. benzaldehyde, react successfully.

The Stobbe condensation leads to the attachment of a three-carbon chain to a ketonic carbon atom, whereas the condensations so far described lead to the attachment of a two-carbon chain.

The procedure is usefully employed in the extension of aromatic rings, e.g.

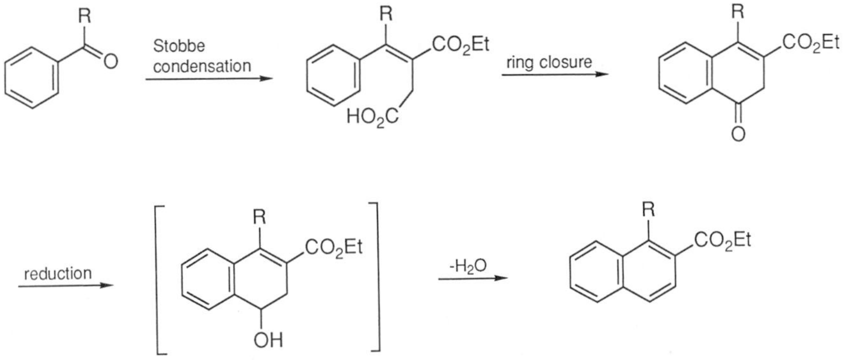

Ring closure is accomplished by the Friedel–Crafts reaction (section 11.3) and the final dehydration occurs spontaneously because of the accompanying gain in aromatic stabilization energy.

(f)   The Darzens reaction

The base-catalyzed condensation between an α-halo-ester and an aldehyde or ketone gives an epoxy-containing ester (glycidic ester):

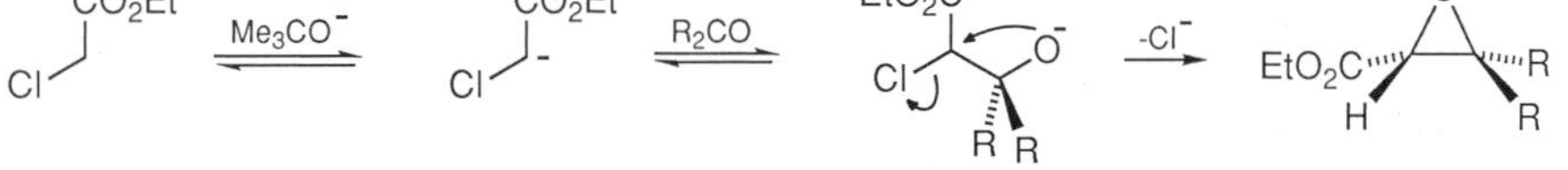

A hindered base such as potassium t-butoxide is used so that S_N2 displacement of the chloride is disfavoured, compared with removal of a proton, by steric hindrance.

The difference between this and the Claisen reaction is that in the former the oxyanion formed by the addition of the enolate to the carbonyl group preferentially displaces halide ion in an intramolecular nucleophilic substitution (cf. p. 111), whereas in the latter the oxyanion removes a proton from the solvent.

The glycidic acids obtained by base-catalyzed hydrolysis of the esters readily undergo a decarboxylative rearrangement in the presence of acids:

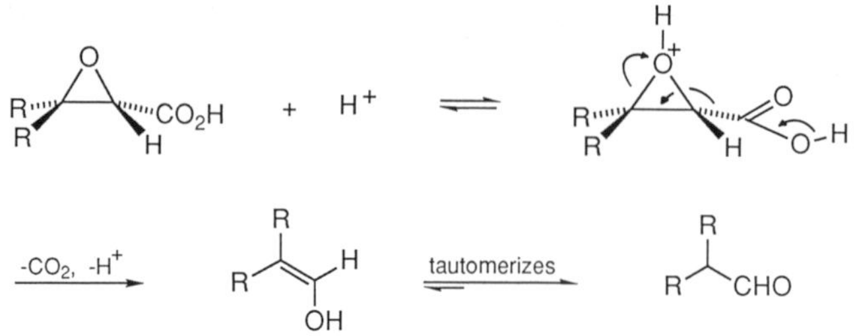

The overall process therefore consists of the addition of *one* carbon atom, as aldehyde, to a carbonyl group,

(g)   Other reactions of aldol type

The reactions of aldehydes and ketones described so far have been with enolates derived from aldehydes, ketones, esters, and anhydrides. In addition, useful reactions may be carried out using carbanions derived from nitro-compounds, nitriles, and potentially aromatic systems.

(i)   Nitro-compounds: Henry reaction. Nitroalkanes with α-CH bonds react readily with aldehydes in the presence of base (usually OH^-) to give β-hydroxynitro-compounds:

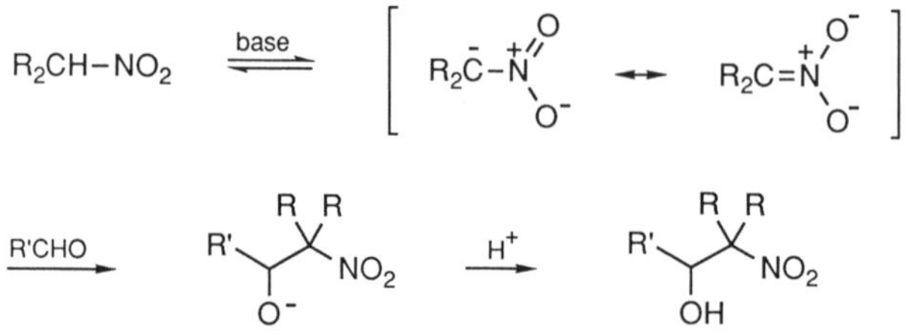

With primary nitroalkanes or nitromethane, further condensations can occur, for example nitromethane and excess of formaldehyde give the trimethylol derivative (cf. the reaction of acetaldehyde and formaldehyde, p. 212):

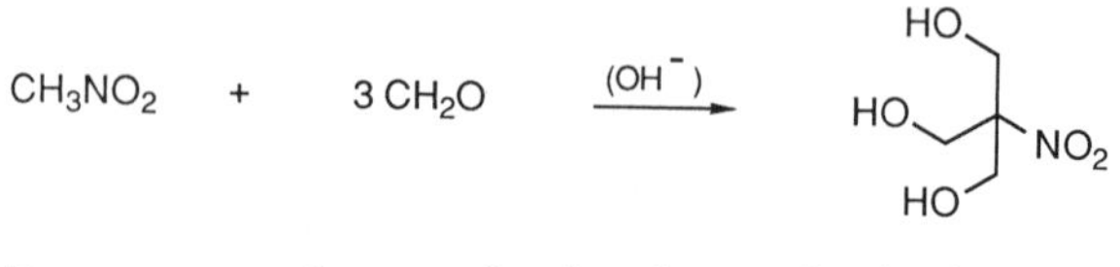

β-Hydroxynitro-compounds are of value in synthesis through their ready conversion into other intermediates:

(1) To β-amino-alcohols, by reduction of the nitro group (e.g. catalytically or with lithium aluminium hydride)
(2) To β-nitroketones, by oxidation (e.g. with pyridinium chlorochromate, p. 609)
(3) To αβ-unsaturated nitro compounds, by dehydration (e.g. with acetic anhydride).

With aromatic aldehydes, the Henry reaction usually leads directly to the αβ-unsaturated nitro-compound, for base-catalyzed dehydration of the initial product occurs readily as a result of the increased conjugation in the product, e.g. benzaldehyde and nitromethane give ω-nitrostyrene in 80% yield [1]:

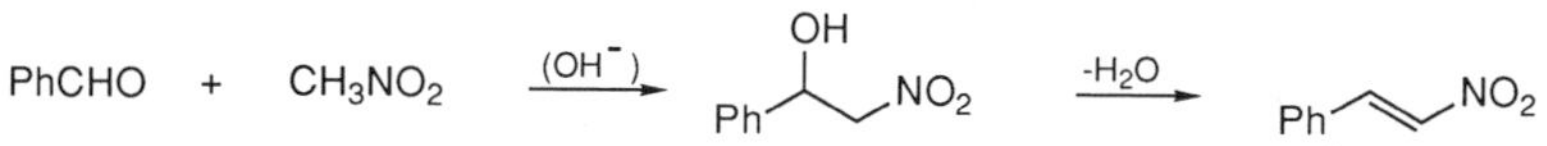
PhCHO
+
CH_3NO_2
(OH^-)
OH
Ph
NO_2
$-H_2O$
Ph
NO_2

ω-Nitrostyrene and its derivatives are useful starting materials in the Bischler–Napieralski synthesis of isoquinolines (p. 706) since the necessary β-arylethylamines may readily be derived from them by reduction.

(*ii*) *Nitriles.* Nitriles containing α-hydrogen atoms behave analogously to nitro compounds, e.g.

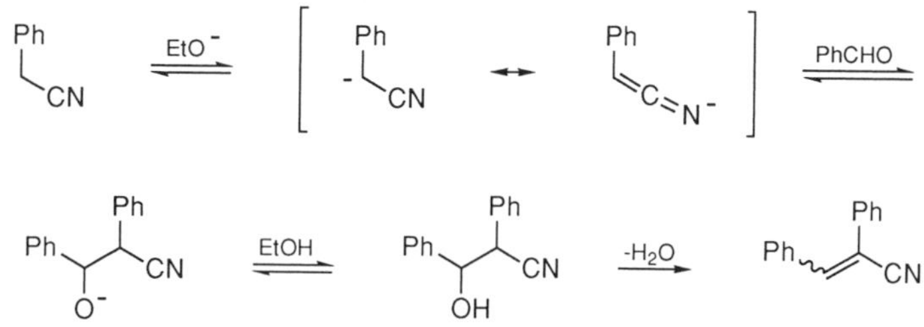
Ph
CN
EtO^-
PhCHO
EtOH
$-H_2O$
OH
O^-

(*iii*) *Potentially aromatic systems.* The anion derived by proton abstraction from cyclopentadiene is an aromatic system containing six π-electrons (p. 40) and the resulting stabilization energy renders cyclopentadiene acidic enough to undergo base-catalyzed condensations, e.g.

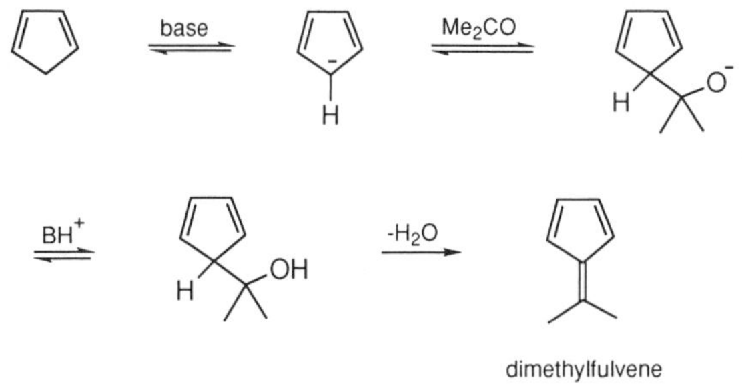
base
Me_2CO
H
O^-
BH^+
OH
$-H_2O$
dimethylfulvene

Related compounds such as indene and fluorene react similarly.

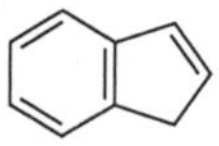

indene 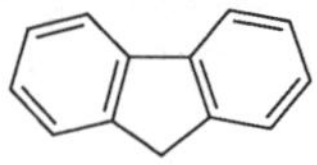   fluorene

(h)   The Knoevenagel reaction

Compounds in which a methylene group is bonded to two groups of $-M$ type, such as malonic ester, have much more acidic C—H bonds than those with only one such group. Consequently a weaker base than those used in the condensations described so far can provide a sufficient concentration of enolate ions, and sufficiently rapidly, for reactions with aldehydes, and in some cases ketones, to occur at a practicable rate. When amines such as piperidine are used as catalysts, the processes are known as Knoevenagel reactions.

Formation of the new C—C bond* is followed by base-catalyzed elimination, e.g.

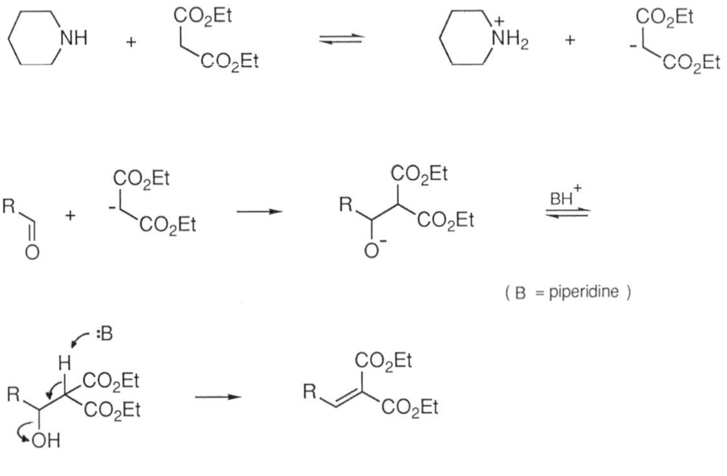

(B = piperidine)

* It is believed that, at least in some cases, it is the imine or the immonium salt formed from the carbonyl compound and the amine which reacts with the enolate, e.g.

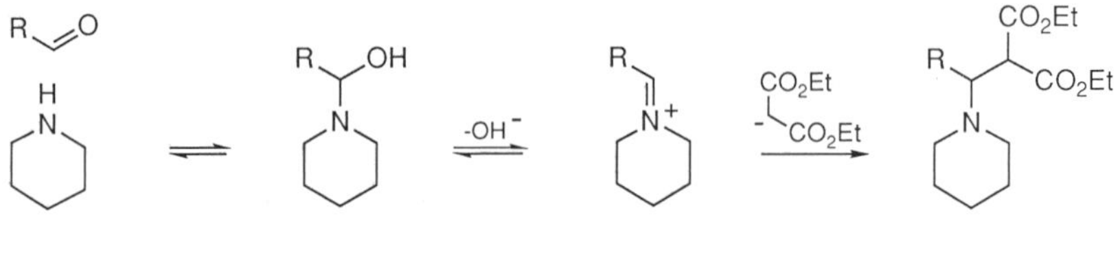

When malonic acid is used, one carboxyl group is eliminated, probably as follows:

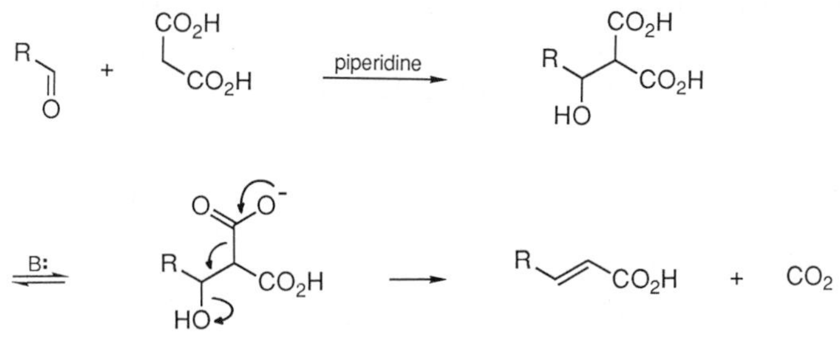

For example, heating under reflux a solution of malonic acid and benzaldehyde in pyridine containing a little piperidine as catalyst gives cinnamic acid in 80% yield:

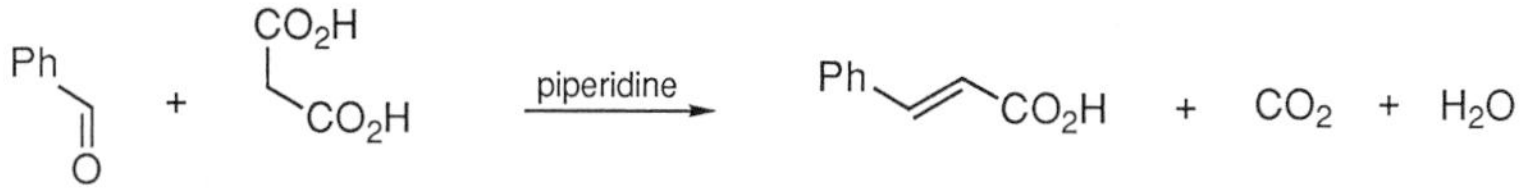

The Knoevenagel reaction is more useful with aromatic than with aliphatic aldehydes. The latter react readily, but the C=C bond in the products, being activated to nucleophiles by the conjugated carbonyl, usually reacts further (Michael addition, section 7.5), e.g.

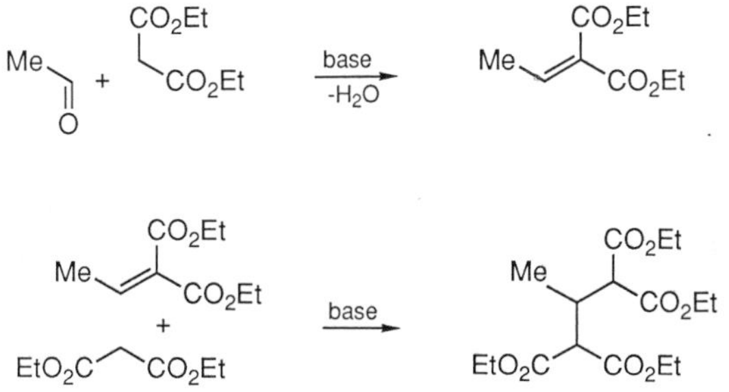

The corresponding addition to the C=C bond of the product from an aromatic aldehyde occurs less readily because of the loss of conjugation to the aromatic system which such addition involves.

Ketones do not undergo the Knoevenagel reaction with malonic acid or its esters, but do so with cyanoacetic acid and its esters which contain more highly activated α-hydrogen.

(i) *Summary*

It is apparent from the preceding discussion that a variety of types of reactions are often available for the synthesis of a particular compound. For example, cinnamic acid may be prepared from benzaldehyde in each of the following ways:

- Claisen reaction (CH₃CO₂Et, EtO⁻; 2 hours at 0–5°C, followed by hydrolysis of ethyl cinnamate): *c.* 70%
- Perkin reaction (Ac₂O, AcO⁻; 5 hours at 180°C): 55%
- Knoevenagel reaction (CH₂(CO₂H)₂, piperidine; 1 hour at 110°C): 80%
- Reformatsky reaction (CH₂BrCO₂Et, Zn then dehydration): 50–60%.

The Perkin reaction requires the most vigorous conditions and usually gives the lowest yield. However, it has the advantage over the Claisen and Reformatsky reactions in that if a nitro- or halogen-substituent is present in the aromatic aldehyde the yield is increased whereas the other two methods cannot be employed. Nevertheless, the Perkin reaction has generally been superseded by the Knoevenagel reaction.

**7.3
Reactions of enolates with esters**

(a)   *The Claisen condensation*

The self-condensation of an ester which contains an α-hydrogen atom is known as the Claisen (ester) condensation. The simplest example is the formation of acetoacetic ester (ethyl 3-oxobutanoate) from ethyl acetate, catalyzed by ethoxide ion:

acetoacetic ester

The reaction is usually carried out by refluxing very dry ethyl acetate over sodium wire. The sodium reacts with the 2–3% of ethanol which is present in commercial ethyl acetate, generating ethoxide ion which then catalyzes the reaction. The product is obtained as the sodium salt of acetoacetic ester from which the free ester is liberated by treatment with acetic acid. Yields of about 30% are obtained [1].

The Claisen condensation differs from the aldol reaction only after the oxyanion has been formed by addition of the enolate to the carbonyl group: in the Claisen reaction this anion eliminates ethoxide ion to give a β-keto-ester, whereas in the aldol reaction the anion gains a proton to give a β-hydroxy-aldehyde or ketone.

Each step in the Claisen condensation is reversible, and acetoacetic ester can be isolated in significant yield only because it is essentially completely removed from equilibrium by conversion into its anion under the influence of ethoxide ion (i.e. acetoacetic ester is a much stronger acid than ethanol). However, in those cases in which the β-keto-ester product does not contain a C—H bond adjacent to both keto and ester groups so that this marked acidic character is absent, the equilibria are unfavourable to reaction when ethoxide ion is the base. For example, ethyl isobutyrate, Me_2CHCO_2Et, fails to give Me_2CH—CO—CMe_2—CO_2Et in these conditions. This problem can be surmounted by using a very much stronger base than ethoxide ion, and the triphenylmethide ion, added as sodium triphenyl-methyl, is conveniently employed. This results in the essentially complete removal of the ethanol formed in the reaction:

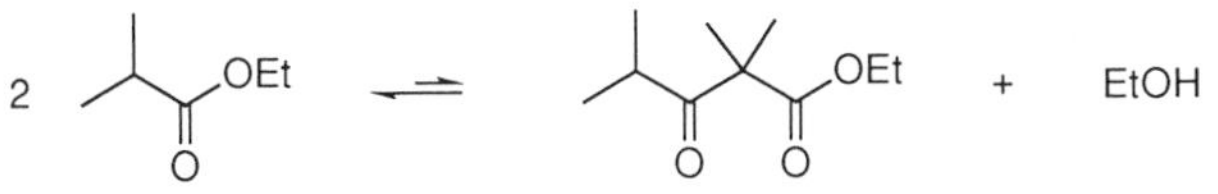

$$EtOH + Ph_3C^- \rightleftharpoons EtO^- + Ph_3CH$$

(i) *Mixed reactions.* As with aldol reactions, the Claisen condensation on a mixture of two esters each of which contains α-hydrogen may yield four products. In general, therefore, this is an unsatisfactory method for obtaining a particular β-keto-ester. Instead, it is best to pre-form the enolate of one component with LDA and to treat it with the acid chloride corresponding to the other component:

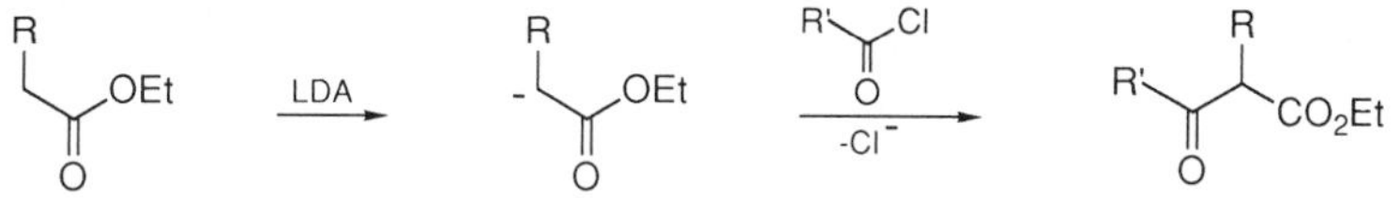

However, mixed reactions are successful when only one of the two esters has α-hydrogen and the other has the more reactive of the two carbonyl groups. Diethyl oxalate and ethyl formate are particularly suitable. The use of the former leads to compounds which are both β-keto-esters and α-keto-esters. Since α-keto-esters decarbonylate on being heated, the method provides a route to β-dibasic-esters, as in the preparation of diethyl phenylmalonate in 80–85% yield [2]:

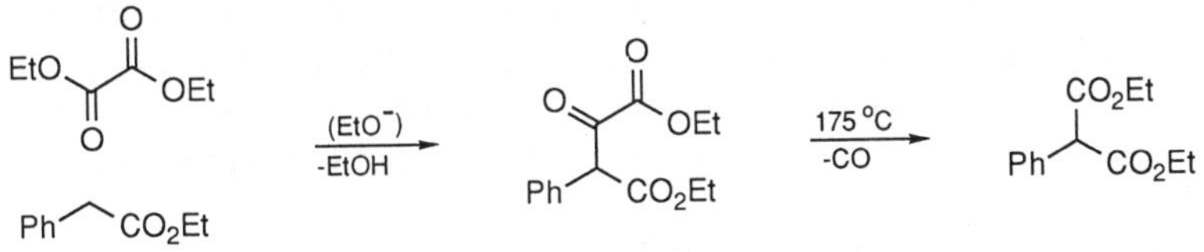

Since halobenzenes are not reactive towards nucleophiles in normal conditions (p. 399) and do not react with malonic ester in the presence of ethoxide, this provides a convenient route to phenylmalonic acid and related compounds.

The products from reactions with diethyl oxalate may also be used to prepare α-keto-acids: hydrolysis of the initially formed dibasic keto-ester gives a dibasic acid which, as a β-keto-acid, is readily decarboxylated by heat:

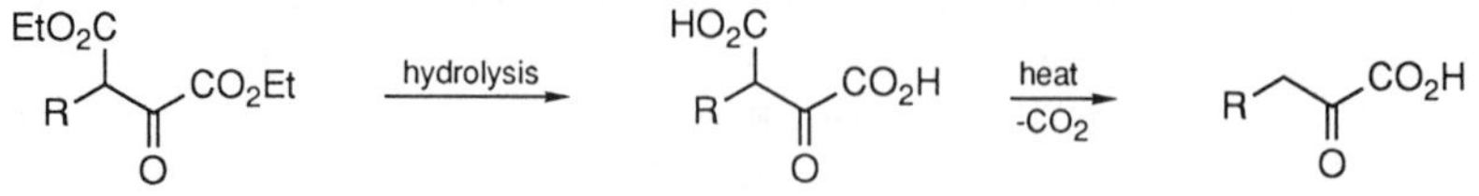

Ethyl formate reacts in Claisen conditions with esters containing α-hydrogen, giving β-aldehydo-esters:

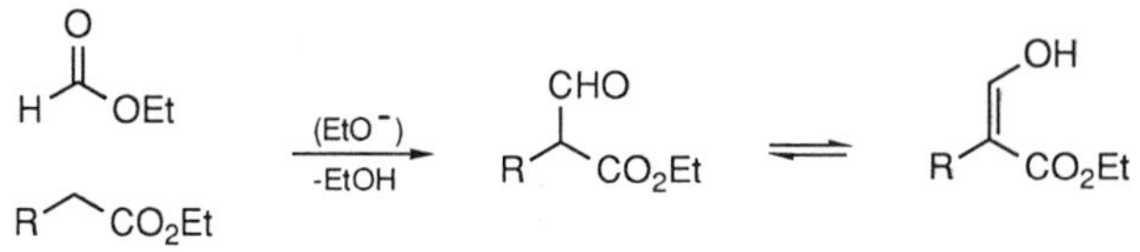

(b)   The Dieckmann reaction

The Claisen condensation on the di-esters of C_6 and C_7 dibasic acids occurs intramolecularly to give five- and six-membered cyclic β-keto-esters, respectively. For example, diethyl adipate in toluene reacts with sodium metal to give the sodium salt of 2-carboethoxycyclopentanone from which the free ester (75–80%) is liberated with acetic acid [2]:

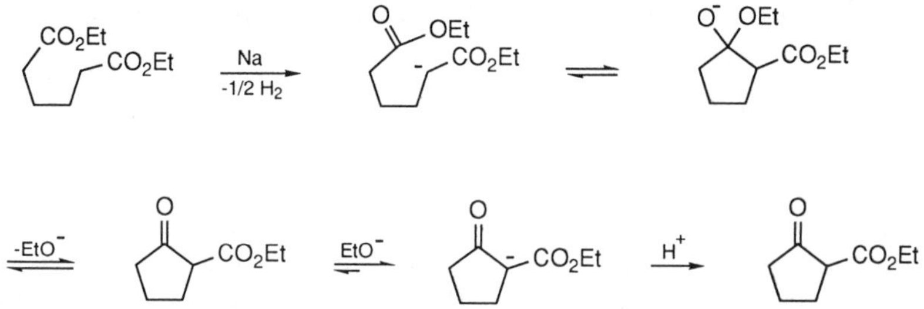

Diethyl pimelate reacts analogously to form 2-carboethoxycyclohexanone in 60% yield.

The corresponding seven- and eight-membered ring compounds are formed in much lower yield, and rings with between nine and 12 members in negligible yield. This is because there is a small degree of strain in these rings and, as the chain length is increased, the entropy of activation becomes increasingly unfavourable, so that the ring closure competes less effectively with intermolecular condensation.

Di-esters of shorter-chain dibasic acids do not undergo the Dieckmann reaction because of the strain which would result in the small rings formed (higher heat of activation). In the case of diethyl succinate, an intermolecular condensation between two molecules is followed by cyclization to the cyclohexanedione system:

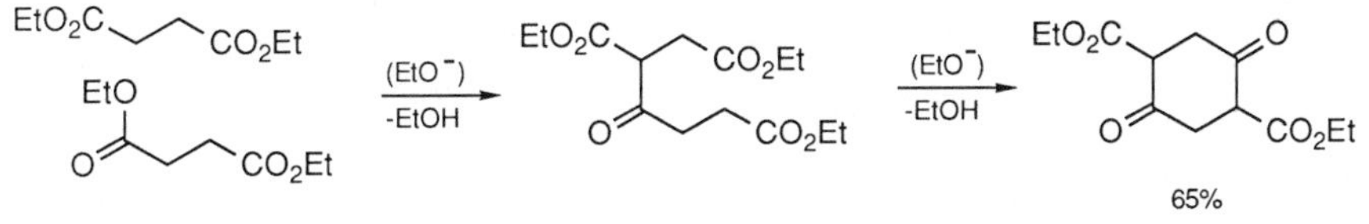

(c) The Thorpe reaction

The cyclization of dinitriles in the presence of base is closely analogous to the Dieckmann reaction. The initial product, a β-imino-nitrile, is readily hydrolyzed to a β-keto-nitrile:

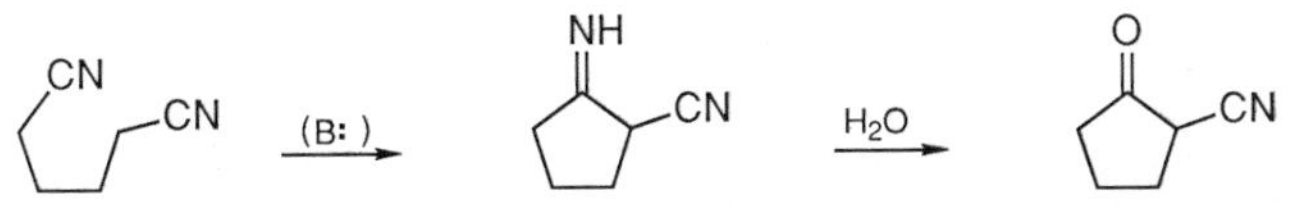

As in the Dieckmann reaction, satisfactory yields are normally obtained only for five- and six-membered rings. However, the problem that arises in trying to make larger rings, that intermolecular reaction competes increasingly favourably with intramolecular reaction, has been overcome by employing an experimental technique devised by Ziegler. The dinitrile is added in very dilute solution in benzene to a solution of the basic catalyst in a large quantity of ether. This reduces considerably the probability of the intermolecular reaction and in this way yields of 95% (seven-ring), 88% (eight-ring), and 60–80% (rings with 14 or more members) have been obtained; yields of 9–13 membered rings are, however, less than 15%. The basic condensing agent must be soluble in ether and highly hindered so as to minimize its reaction as a nucleophile with the nitrile group: lithium salts of amines are commonly used.

(d) Reactions of esters with ketones

Four products might result from a base-catalyzed reaction on a mixture of an ester and a ketone each of which contains α-hydrogen: each of the two enolates might react with each of the two carbonyl groups. In practice, one product, that derived from the enolate of the ketone and the carbonyl of the ester, normally pre-dominates, e.g.

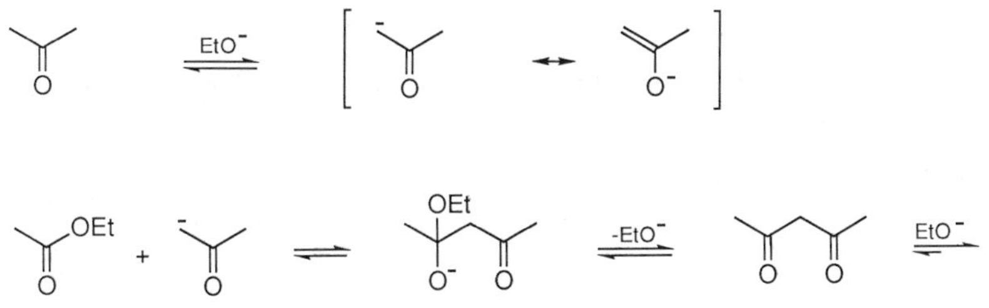

Two factors are responsible. First, the two Claisen products are converted into their anions by ethoxide ion so that the equilibria are favourable to these products but not to the two aldol products (cf. the self-reaction of acetone, p. 210). Secondly, acetone and 2,4-pentanedione are more acidic than ethyl acetate and acetoacetic ester, respectively, so that the formation of 2,4-pentanedione as its anion is the predominant reaction.

This procedure for preparing β-diketones is limited in scope to reactions between acetate esters and methyl ketones because of the reversibility of Claisen condensation. Consider the reaction between 3-pentanone and ethyl acetate. The expected β-diketone,

can yield, in the reverse reactions, not only the initial reactants (cleavage at *a*) but also butanone and ethyl propionate (cleavage at *b*):

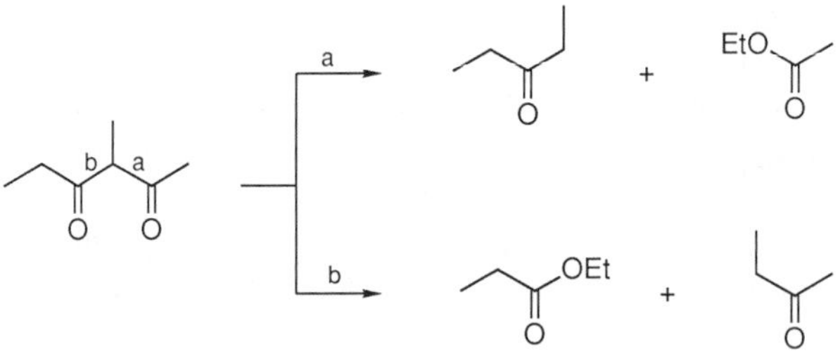

Reactions may now occur involving the reactants produced in *b*, giving rise to a complex mixture of products.

Reaction may also be effected on to aromatic esters. For example, ethyl benzoate and acetophenone react in the presence of sodium ethoxide to give the sodium salt of dibenzoylmethane; the diketone is liberated (60–70% yield) on acidification [3]:

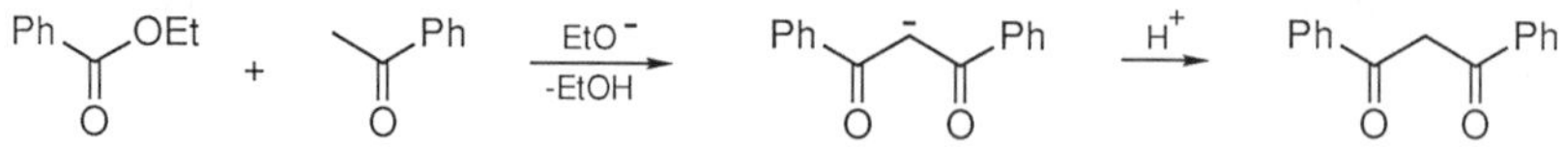

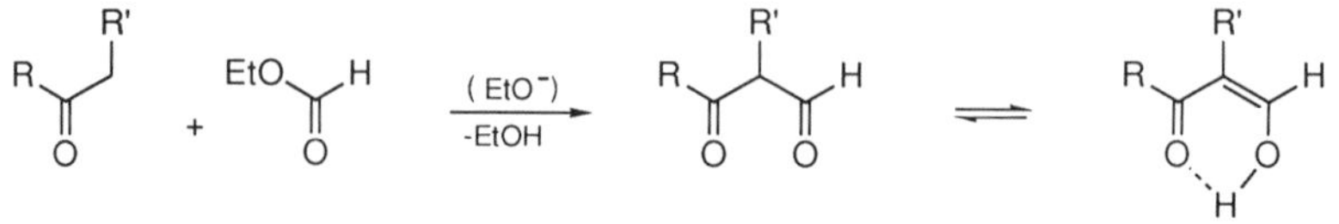

The condensation of ethyl formate with ketones gives β-keto-aldehydes, and this reaction can be used when it is necessary to alkylate a ketone at the less reactive of its two α-CH positions: the more reactive position is blocked by reaction with ethyl formate followed by treatment with N-methylaniline. Alkylation is then effected at the other α-CH position and the blocking substituent is removed by hydrolysis. This technique has been employed in the introduction of angular methyl groups in steroid synthesis, e.g.

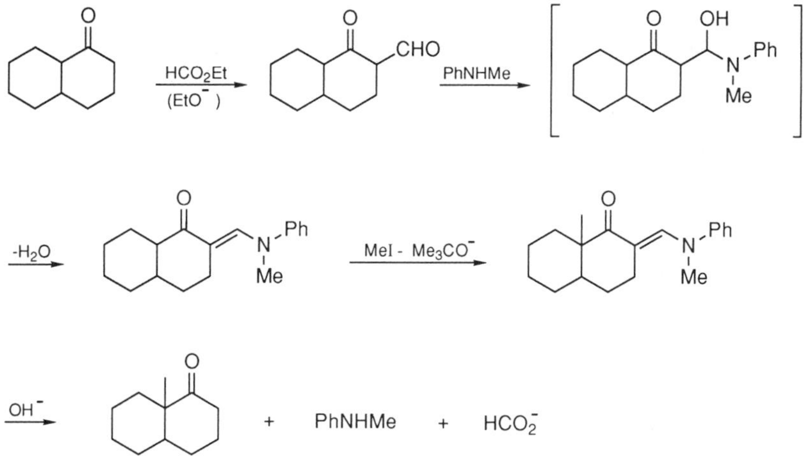

9-methyl-1-decalone

(e) Enamines

Enamines may be prepared from secondary amines and ketones which contain an α-hydrogen atom:*

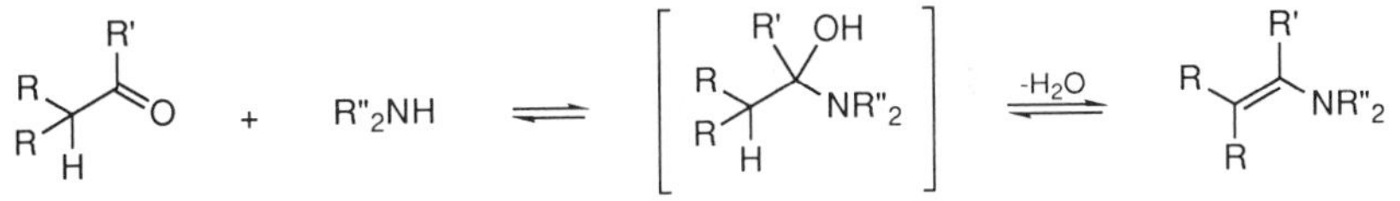

The reaction is driven to the right by removing the water as it is formed, either by azeotropic distillation (e.g. with toluene) or with molecular sieves.

* Those prepared from ammonia or primary amines are unstable with respect to their tautomers,

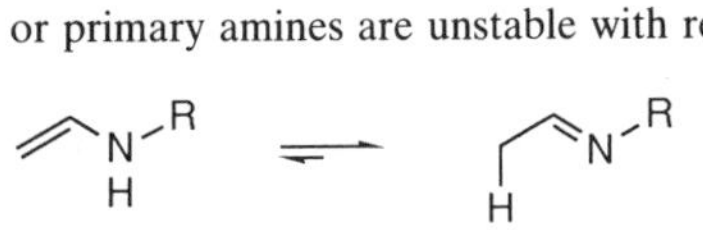

and cannot be isolated (cf. the instability of simple enols with respect to their carbonyl tautomers).

Enamines are closely related to the enolates derived from ketones,

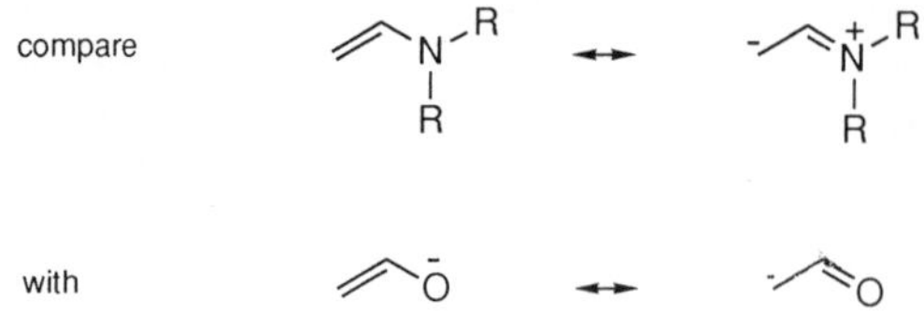

and behave analogously in their reactions with acid chlorides, e.g.

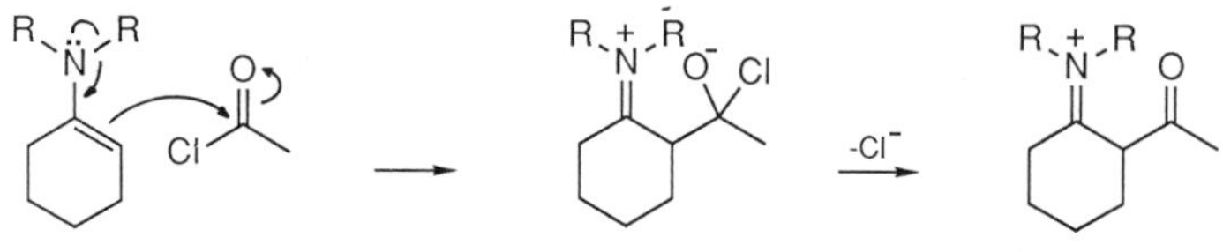

Mild hydrolysis removes the original amine, so that the overall reaction is the formation of a β-diketone:

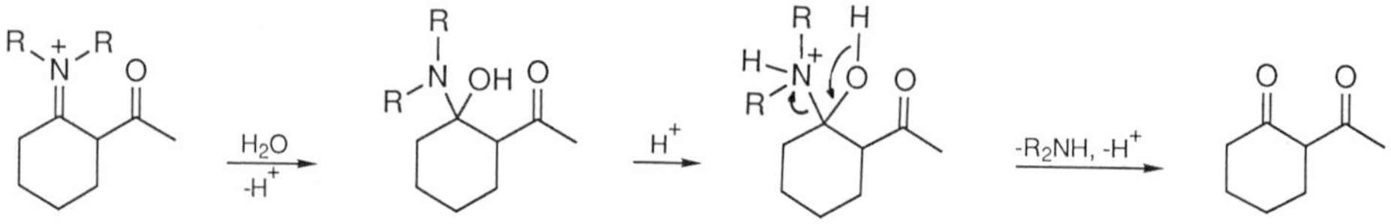

The enamine from an unsymmetrical ketone which can react at two sites is predominantly the less substituted one, e.g.

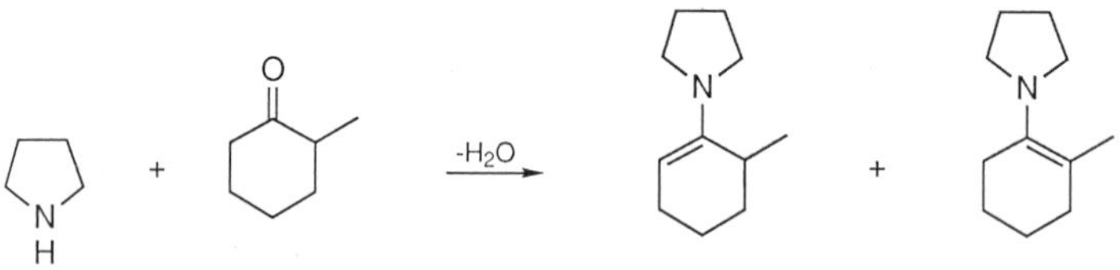

This is because an enamine is stabilized by interaction of the alkene π-system with the unshared electron-pair in a *p* orbital on nitrogen (cf. aniline, p. 49). This requires coplanarity of the bonds on the unsaturated carbon atoms and those to nitrogen, which is possible in (a) but, owing to steric repulsion, is not possible in (b). This can be usefully applied in synthesis.

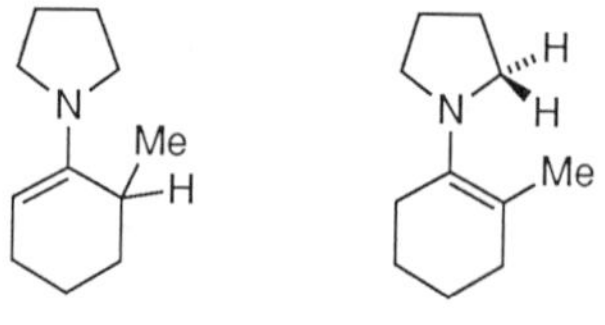

(a)   (b)

In addition to the ability to add to carbonyl groups, enolates, like other nucleo-   **7.4**
philes, displace halide ions from alkyl halides and sulfonates with the formation of   **The alkylation**
C—C bonds,   **of enolates**

$$C^- \frown C-X \longrightarrow C-C + X^-$$

(X = halogen or OSO$_2$R)

Halides were formerly used, but nowadays the toluene-*p*-sulfonate (tosylate) is often employed because it generally reacts more efficiently. Tosylates are readily available from alcohols with toluene-*p*-sulfonyl chloride (TsCl) in the presence of pyridine to absorb the hydrogen chloride:

$$ROH + TsCl \xrightarrow{\text{pyridine}} ROTs + HCl$$

Reaction can also occur at the oxygen atom of the enolate,

$$\diagup\!\!\!\!\diagdown O^- \frown CH_3-X \longrightarrow \diagup\!\!\!\!\diagdown O-CH_3 + X^-$$

but this is normally the less important reaction, e.g.

71%   13%

The reason is the same as that which applies to the aldol and related reactions: the *C*-alkyl product is thermodynamically the more stable of the two, and this is reflected in the stabilities of the transition states. However, with very reactive halogen compounds, such as a 1-chloro-ether ROCH$_2$Cl, reaction at oxygen predominates. This can be rationalized by the Hammond postulate: the transition state occurs earlier in the reaction profile so that the new C—C and C—O bonds are less fully formed and the stabilities of the products are of less importance in determining reactivity than the distribution of charge in the enolate, which is greater at the oxygen atom. *O*-Alkylation also predominates with very good leaving groups in an aprotic solvent like HMPA in which the enolate ion is not solvated and is more reactive (p. 65); this is again because the transition state occurs earlier in the reaction profile.

It is convenient to group these reactions into those in which the enolate is derived from (*a*) a monofunctional compound and (*b*) a bifunctional compound.

(*a*) *Alkylation of monofunctional compounds*

Although compounds containing a hydrogen atom activated by one group of $-M$ type react readily with carbonyl groups in the presence of ethoxide ion, their reactions with alkyl halides in these conditions are inefficient. This is because ethoxide ion is a relatively weak base, so that the enolate is formed only in small concentration and most of the ethoxide ion added remains as such and is able itself to displace on the alkyl halide.

One way of obviating this problem is to use a stronger base so as to generate a larger concentration of enolate, and also one which is sterically hindered from reacting as a nucleophile. The t-butoxide ion is frequently used, as in the conversion of cyclohexanone into 2-methylcyclohexanone:

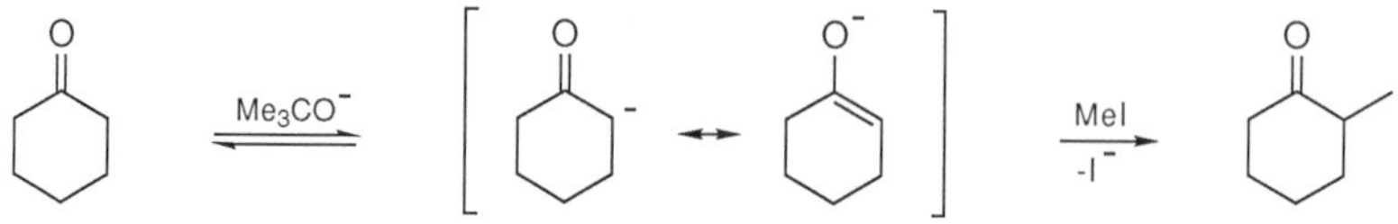

Nowadays, LDA in tetrahydrofuran is commonly used. However, the two methods can yield different products with unsymmetrical ketones. For example, the methylation of 2-methylcyclohexanone with methyl iodide occurs mainly at the 6-position when LDA is used and mainly at the 2-position with t-butoxide:

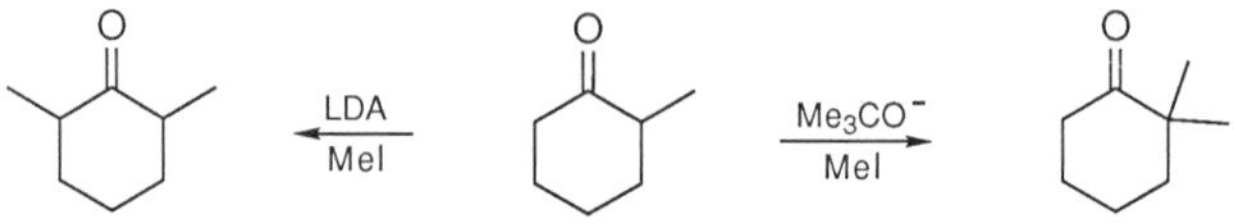

There is kinetic control with LDA: proton abstraction is faster, and essentially irreversible, from the less hindered α-CH position. With t-butoxide, there is thermodynamic control: equilibration occurs between the two enolates and the methyl-substituted one, being the more stable (p. 208), is present in higher concentration. However, when reaction at the more highly substituted position is strongly sterically hindered, alkylation with t-butoxide occurs at the less highly substituted carbon, e.g.

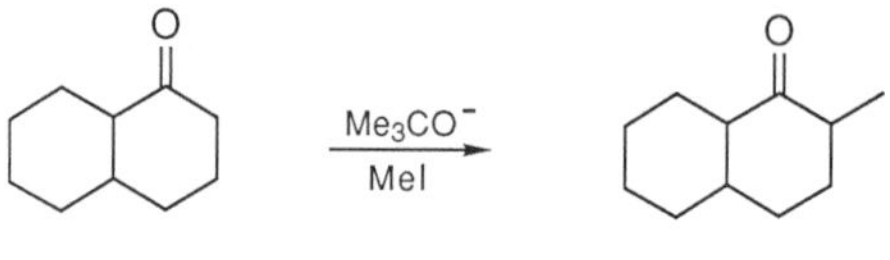

(i) *The halogen compound.* Alkylations of enolates are S_N2 processes and show the characteristics of those reactions which have been described previously (p. 103). These are:

(1) The order of reactivity is I > Br > Cl ≫ F.

(2) Unsaturated and aromatic halides are unreactive except, in the latter case, when there is a strongly electron-withdrawing group, such as nitro, at a 2- or 4-position (p. 400).

(3) Primary alkyl halides undergo substitution in generally good yield, but because enolates are very strong bases, secondary and tertiary halides preferentially undergo E2 elimination (p. 95).

A modification of the method is suitable for tertiary halides: the enolate is converted into its silyl enol ether (p. 275) which is both less nucleophilic and much less basic than the enolate but which effects substitution when the halide is activated with a Lewis acid, e.g.

$$EtMe_2C-Cl \quad + \quad TiCl_4 \quad \rightleftharpoons \quad EtMe_2C^+ \quad + \quad TiCl_5^-$$

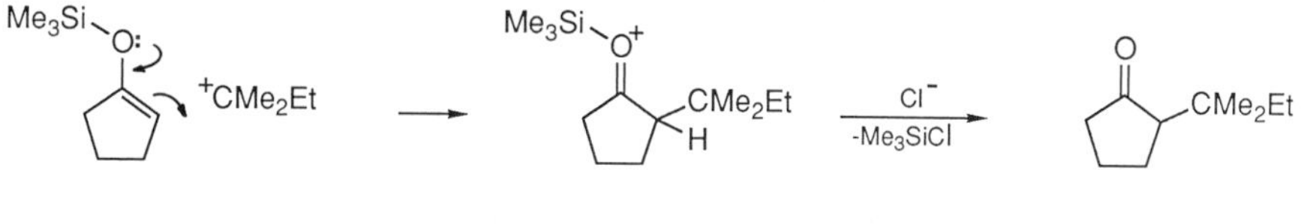

60% [7]

(*ii*) *Monoalkylation v. dialkylation*. The occurrence of at least some dialkylation at an activated CH_2 group is inevitable; as reaction proceeds, an enolate-forming product accumulates at the expense of the enolate-forming reagent. The mono-alkylated product is favoured by use of an excess of the latter over the halide; conversely, if the dialkylated product is required, at least a two molar excess of halide is used.

It is, however, possible to obtain only monalkylated product by pre-forming the enolate with the exact amount of LDA and adding the halide subsequently; no base is then available to react with the product as it is formed.

(*iii*) *Alkylation of αβ- or βγ-unsaturated ketones*. Ketones of these types form enolates in which the charge is shared by two carbons as well as by oxygen:

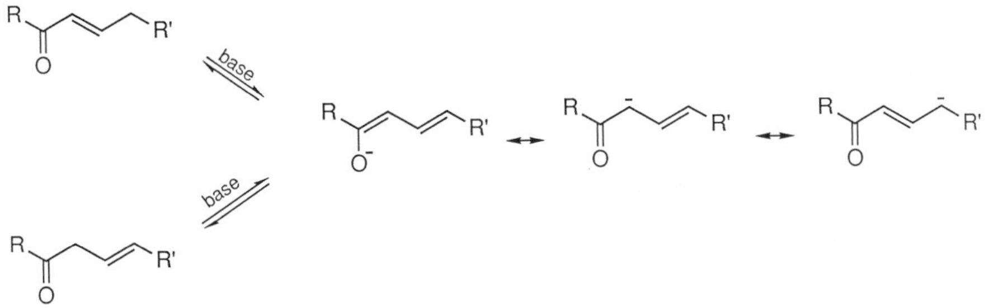

Alkylation could therefore occur at each of these two atoms, but in practice it takes place essentially entirely at the carbon nearer to the oxygen atom, e.g.

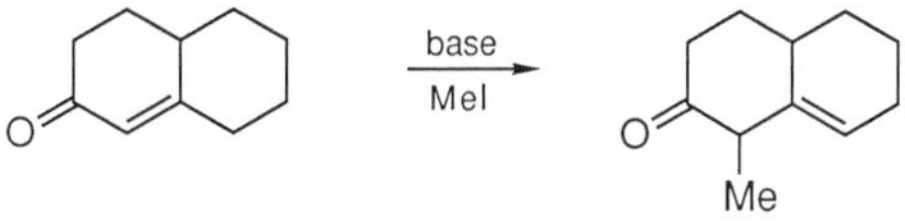

It is thought that this is because this carbon atom in the enolate is the more strongly negatively polarized by the nearby electron-rich oxygen atom.

(b) Alkylation of bifunctional compounds

A C—H bond which is adjacent to two groups of $-M$ type is more acidic than one adjacent to one such group (p. 48) and may be alkylated in milder conditions. In other respects the characteristics of alkylation of bifunctional compounds parallel those for monofunctional compounds; in particular it should be noted that tertiary halides are unsuitable as alkylating agents.

Two particularly important bifunctional compounds in synthetic procedures are malonic ester and acetoacetic ester.

(i) Malonic ester.
Malonic ester may be successfully monoalkylated in the presence of ethoxide ion:

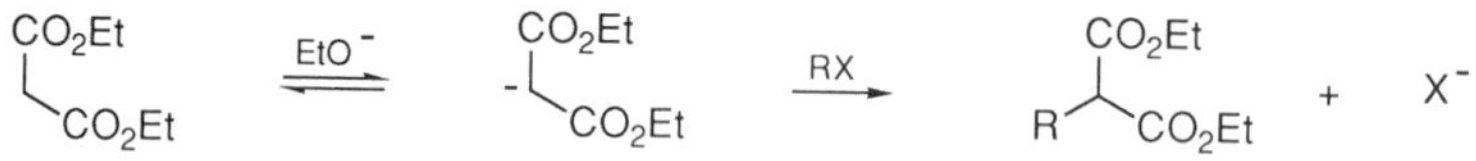

The monoalkylated product still contains an activated α-hydrogen and a second alkyl group may be introduced. However, slightly more vigorous conditions are usually required for the second alkylation, and it is therefore possible to isolate the product of monoalkylation in good yield by using only one equivalent of the alkyl halide. For example, methyl bromide gives methylmalonic ester (diethyl methylmalonate), without external heating, in about 80% yield [2].

The substituted malonic esters may be hydrolyzed to the acids which, having two carboxyl groups on one carbon atom, are readily decarboxylated on being heated (cf. β-keto-acids, p. 296):

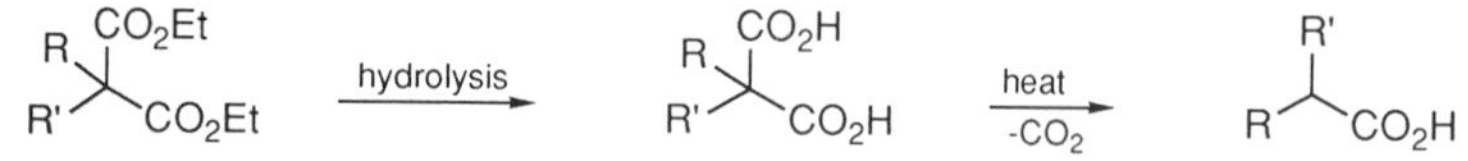

The ease of all these reactions, together with the ready availability of malonic ester, makes this a convenient synthetic method for aliphatic acids of the types RCH_2CO_2H and $RR'CHCO_2H$ from halides containing the groups R and R'. For example, pelargonic acid (nonanoic acid) can be prepared in about 70% yield from 1-bromoheptane [2]:

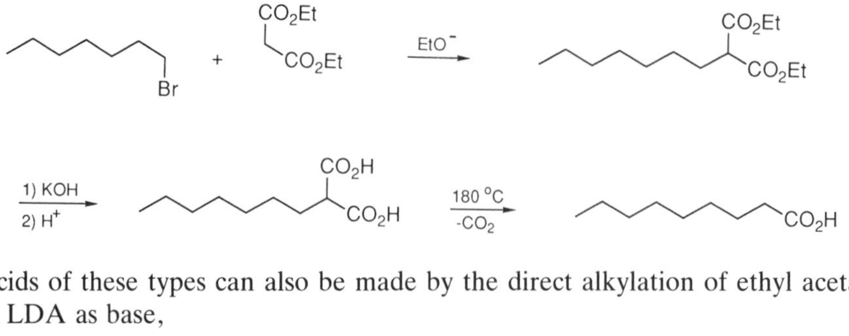

Acids of these types can also be made by the direct alkylation of ethyl acetate with LDA as base, followed by hydrolysis of the ester. The success of this method derives from the fact that the ester is fully converted into the enolate so that the competing Claisen condensation does not occur. Nevertheless, although this procedure is shorter than that *via* malonic ester, the latter is often used in preference owing to the expense of LDA and the need to use special techniques to avoid water or oxygen coming into contact with it (see also p. 239).

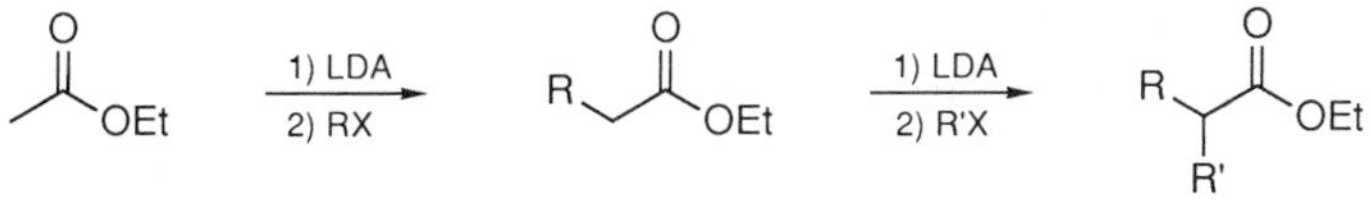

Malonic ester may also be used for the synthesis of three- and four-membered alicyclic compounds from dibromides, e.g.

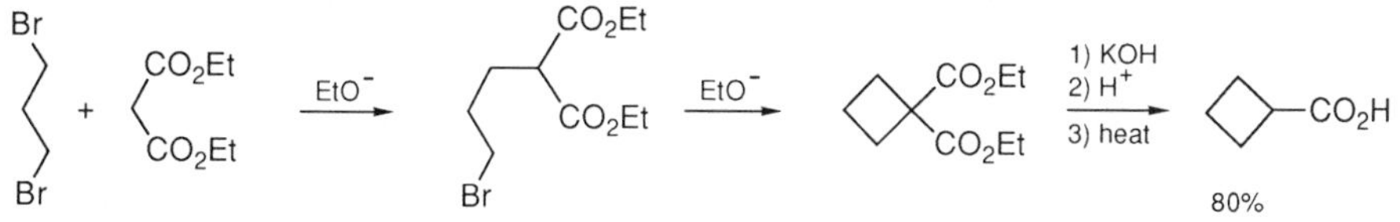

When malonic ester or a monoalkylated derivative is treated with iodine in the presence of a base, reaction occurs to give a tetrabasic ester. A probable mechanism is:

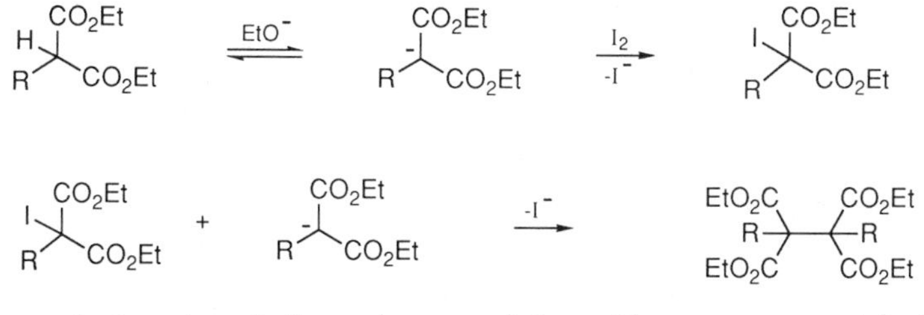

Hydrolysis and decarboxylation give succinic acid or a symmetrical dialkyl derivative:

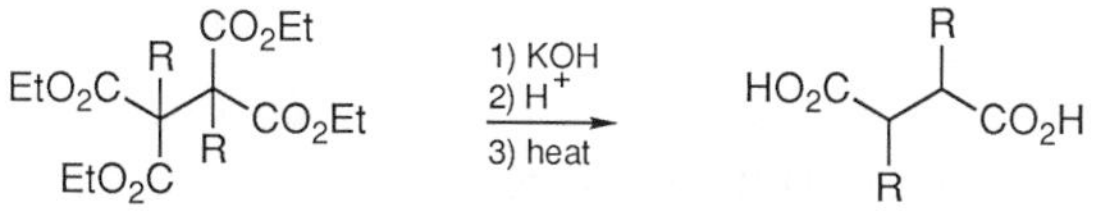

The reaction of malonic ester and its monoalkyl derivatives with bromine or chlorine in the presence of base is different from that with iodine: reaction stops at the first stage and the bromo- or chloro-derivative may be isolated (cf. the readier reaction of nucleophiles with alkyl iodides than with bromides or chlorides, p. 103):

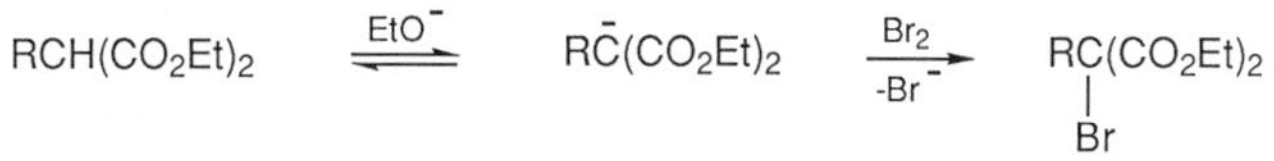

The products are useful intermediates in the synthesis of α-amino acids (p. 326).

(*ii*) *Acetoacetic ester.* Like malonic ester, acetoacetic ester can be mono- and di-alkylated by alkyl halides in the presence of base.

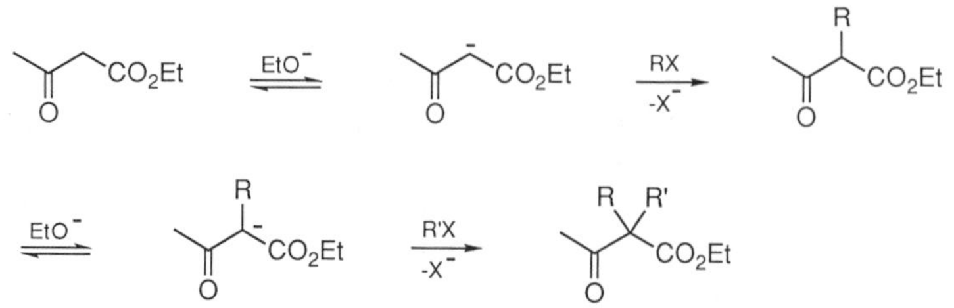

The products undergo two types of hydrolytic cleavage, depending on the conditions. Dilute acid leads to hydrolysis of the ester group and the resulting β-keto-acid loses carbon dioxide (p. 296):

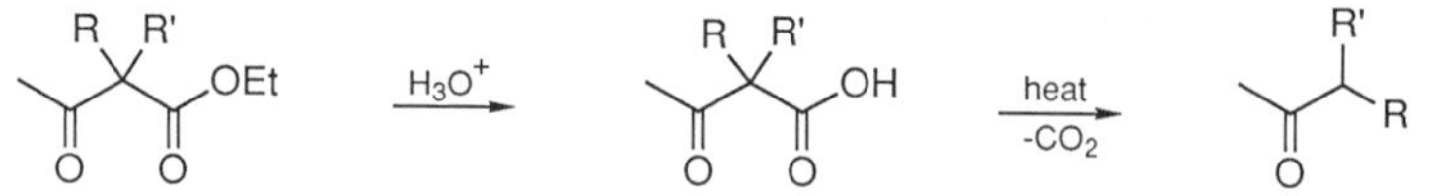

Sodium hydroxide solution, on the other hand, induces scission of a carbon–carbon bond in the reverse manner of the Claisen reaction:

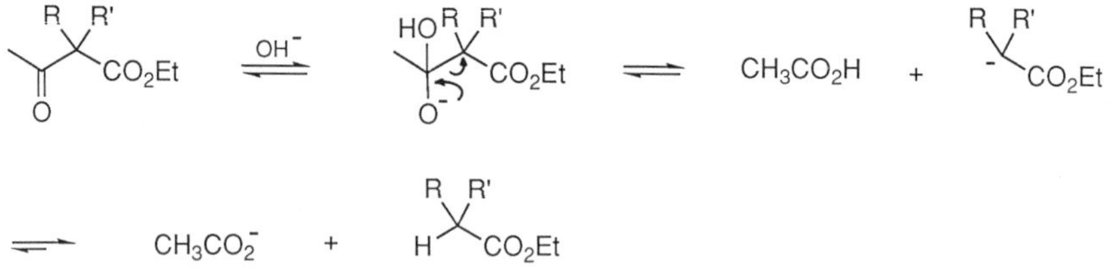

The equilibria are favourable to the products as the result of the stability of the acetate ion.

It is therefore possible to make from acetoacetic ester both methyl ketones of the general type $CH_3COCHRR'$ and acids of the type $CHRR'$—CO_2H. The former is the more useful process, for the compounds obtained in the latter process may be contaminated with by-products derived from the ketonic scission and they are therefore more suitably prepared from malonic ester.

Other β-keto-esters, of the types $RCOCH_2CO_2Et$ and $RCOCHR'CO_2Et$, react analogously, providing routes to ketones of the types $RCOCH_2R'$ and $RCOCHR'R''$. These could also be prepared from other ketones by using LDA, e.g.

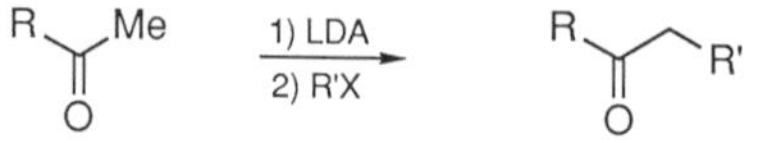

but the longer method *via* the β-keto-ester is still often preferred, for one or more of three reasons: LDA is expensive and difficult to handle; enolates from mono-functional compounds are so strongly basic that elimination occurs with both secondary and tertiary halides and alkylation is only successful with methyl and primary halides (p. 234), whereas the much less basic enolates from keto-esters can be alkylated by secondary halides are well; and LDA may not provide the required regiochemical control. For example, the reaction

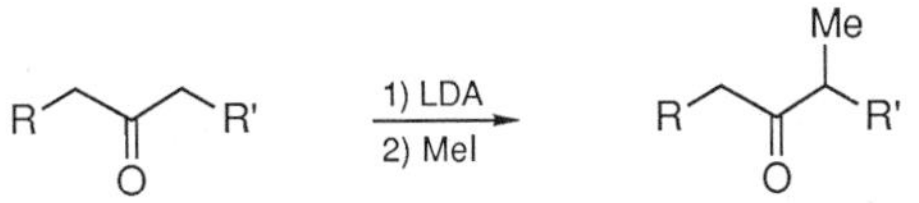

would not be efficient owing to competitive methylation at the alternative CH_2 group; it would be more efficient to use the keto-ester route,

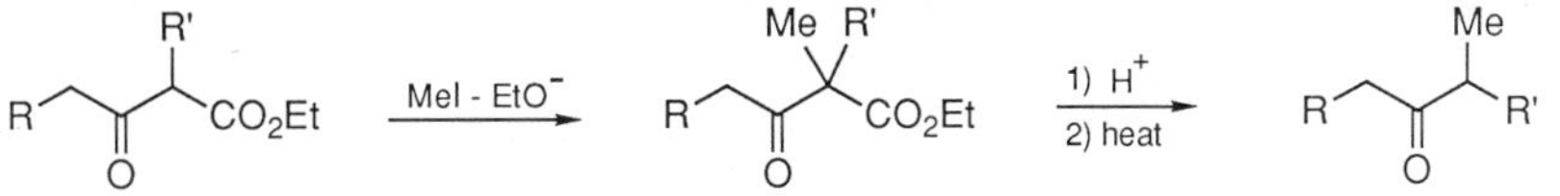

since there is then no alkylation at the less strongly activated CH_2 group.

Like malonic ester, acetoacetic ester and its monoalkyl derivatives undergo a base-catalyzed reaction in the presence of iodine. Hydrolysis and decarboxylation of the products give γ-diketones:

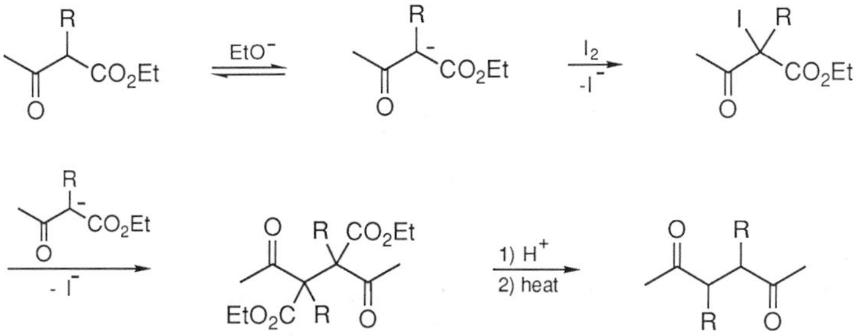

Unlike malonic ester, acetoacetic ester does not form a four-membered alicyclic compound with 1,3-dibromopropane. The first alkylation occurs normally, but the second, involving ring closure, occurs on oxygen instead of carbon for, although the latter is favoured in intermolecular reactions (p. 233), it is here made less favourable by the fact that a strained four-membered ring would be formed whereas O-alkylation gives a six-membered ring:

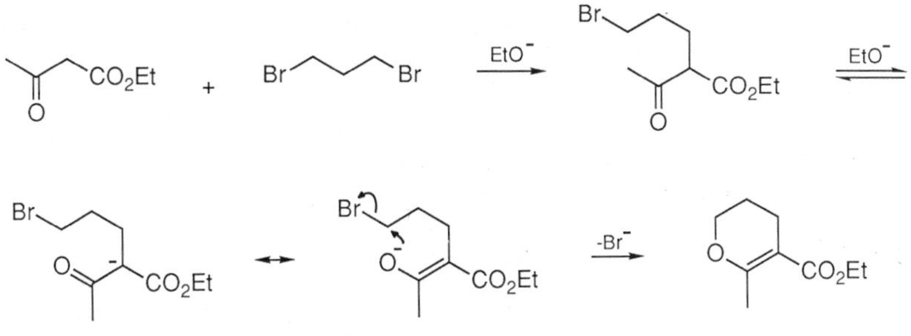

(*iii*) *Alkylation of dianions.* A compound which possesses two activated CH groups of markedly different acidity is alkylated exclusively at the more acidic site when one equivalent of a base is used, e.g.

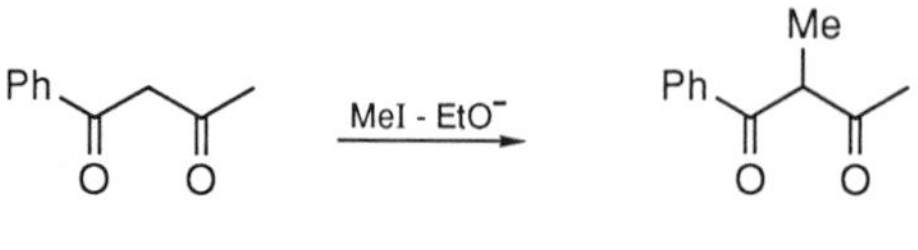

This is because, although the alternative enolate PhCOCH$_2$COCH$_2^-$, stabilized by only one $-M$ group, is the more reactive, it is present in negligible concentration. However, if two equivalents of a very strong base are used, the dianion is formed; mono-alkylation then occurs at the alternative, more basic, carbon atom.

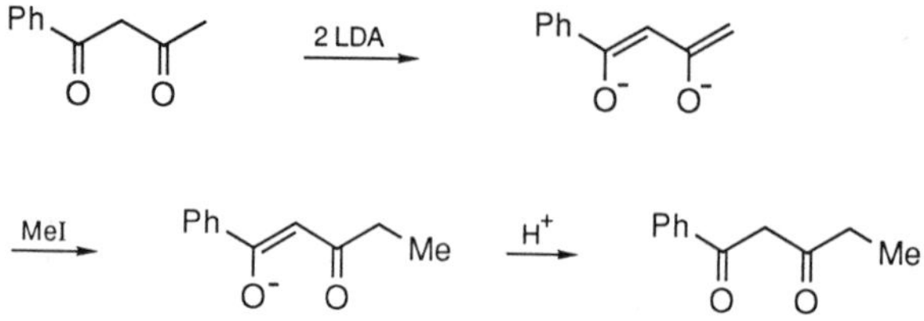

7.5 Addition of enolates to activated alkenes

Although enolates, in common with other nucleophiles, do not react with simple alkenes, they do so if the C=C bond is conjugated to a group of $-M$ type. The reason is that the anion formed by addition, and therefore the preceding transition state, is then stabilized sufficiently by the delocalization of the charge on to an electronegative element for addition to occur at a practicable rate, e.g.

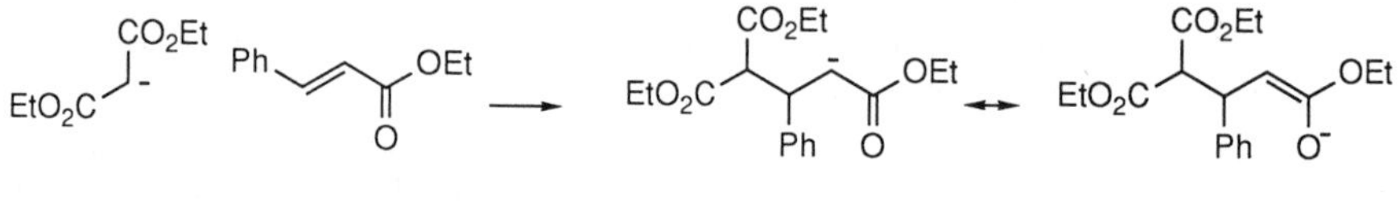

The adduct formed may react with a proton on acidification at either carbon or oxygen but since the resulting tautomers equilibrate rapidly in the presence of acids and the keto-tautomer is the more stable, this is the product which is isolated. The overall reaction between malonic ester and ethyl cinnamate is

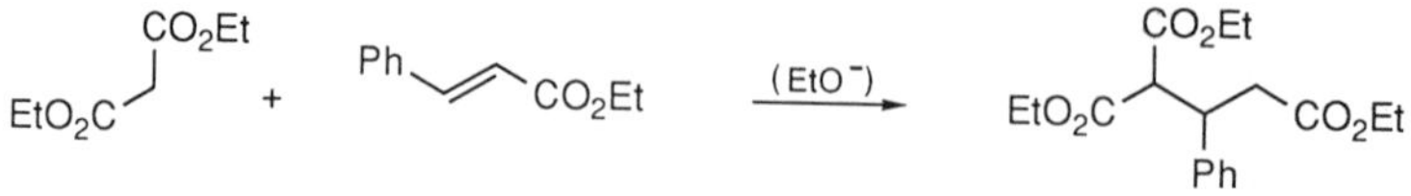

These additions to activated alkenes are usually referred to as *Michael reactions*, a name which was originally applied to those reactions involving the enolates from

acetoacetic and malonic esters. The alkene may be activated by conjugation to carbonyl, ester, nitro, and nitrile groups, and the enolate-forming component may be a bifunctional compound such as malonic ester, or a monofunctional compound such as nitromethane, e.g.

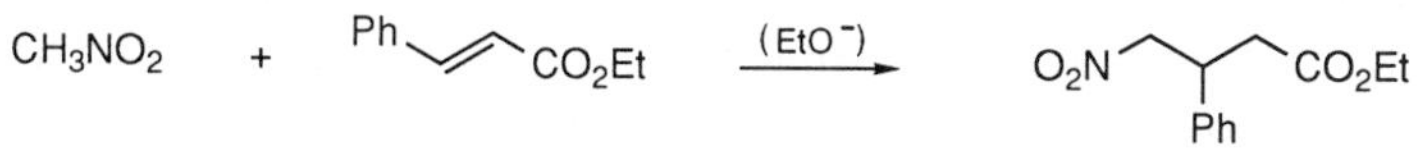

Michael addition may follow spontaneously the reaction of an aliphatic aldehyde with malonic ester (p. 225),

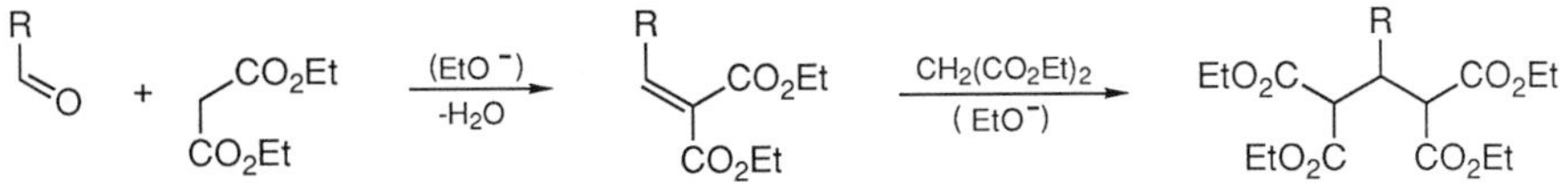

Hydrolysis and decarboxylation of the products give (substituted) glutaric acids:

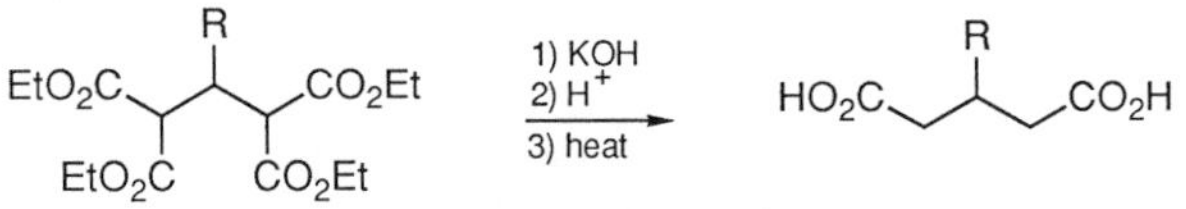

Michael addition to an $\alpha\beta$-unsaturated ketone may be followed by an intramolecular Claisen condensation, e.g. [2]:

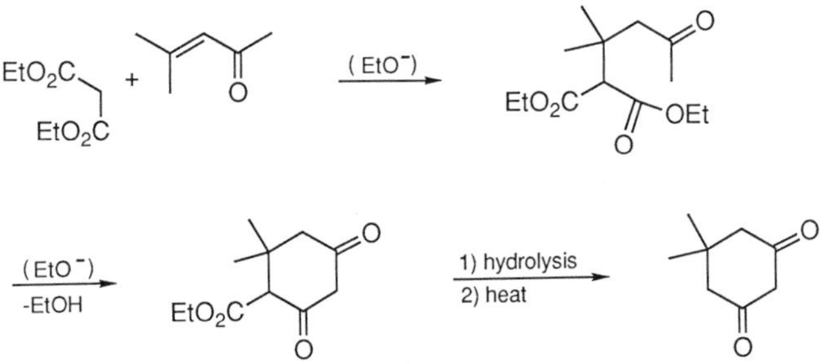

7.6 Reactions involving alkynes

Acetylene and its monosubstituted derivatives are markedly more acidic than alkenes and alkanes and are able to take part in base-catalyzed reactions with both carbonyl-containing compounds and alkyl halides. A strong base is necessary and amide ion in liquid ammonia is commonly used. The typical reactions are:

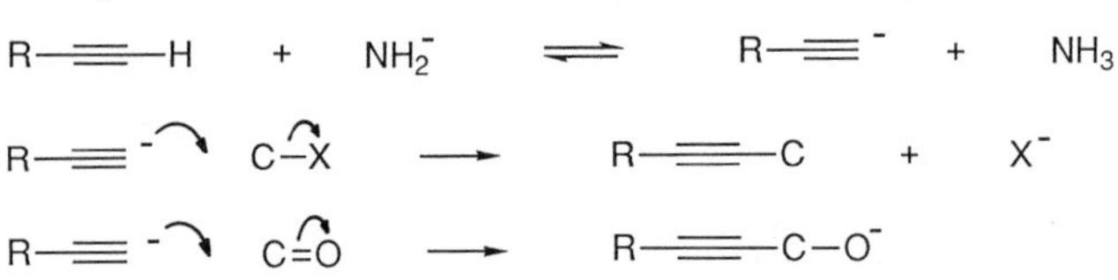

In the usual reaction conditions, sodamide is first formed by treating liquid ammonia with sodium metal in the presence of a catalytic quantity of an iron(III) salt. (In the absence of the iron(III) salt the formation of sodamide, $Na + NH_3 \rightarrow NaNH_2 + \frac{1}{2}H_2$, is very slow and the solution contains solvated electrons which are powerfully reducing, p. 637.) The alkyne is then passed into the ammonia solution, the alkyne anion being formed by the acid-base equilibrium above. Pre-formed sodamide or potassium amide may also be used.

An alternative method is to pass the alkyne into liquid ammonia and then to add sodium at such a rate that no blue colour develops for any length of time. This method has the disadvantage that one-third of the alkyne is reduced to the corresponding alkene, e.g.

$$3 \text{ H}\!-\!\!\equiv\!\!-\!\text{H} \quad + \quad 2\,\text{Na} \quad \longrightarrow \quad 2\,\text{Na}^+ \; {}^-\!\!\equiv\!\!-\!\text{H} \quad + \quad \text{H}_2\text{C}=\text{CH}_2$$

(a)   Reactions with alkyl halides

The reaction is successful only with halides of the type RCH_2CH_2Hal, i.e. those in which there is no branching at the α- or β-carbon atoms. Good yields are then obtained, as in the formation of 1-hexyne from acetylene and butyl bromide (70–77%) [4]:

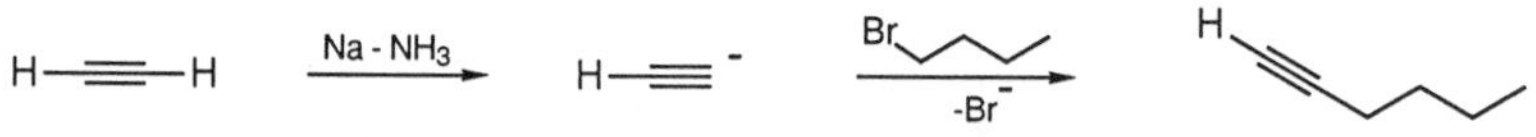

As usual in nucleophilic displacements on halides, the order of reactivity is iodide > bromide > chloride. Use may be made of this fact when selectivity is required; an example occurs in the synthesis of oleic acid:

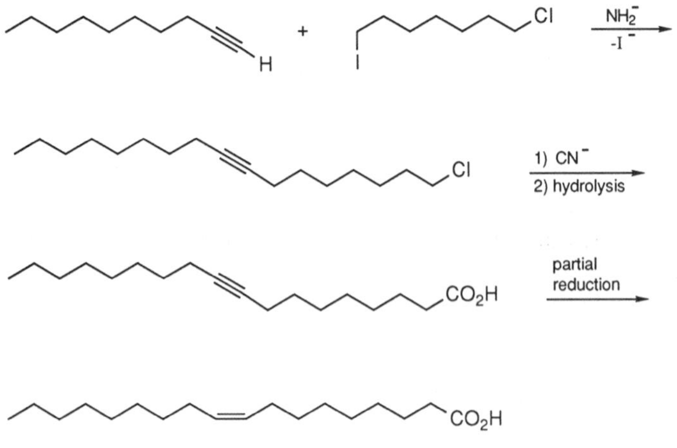

oleic acid

(b)   *Reactions with carbonyl groups*

Alkyne anions react with aldehydes and ketones to form α-acetylenic alcohols, e.g. [3],

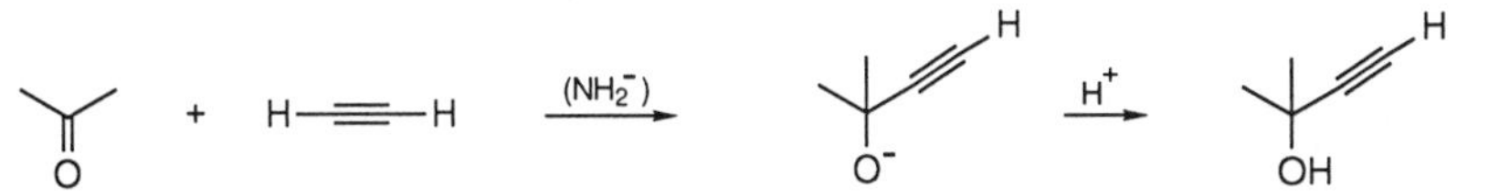

Copper(I) acetylides are made by treating acetylene or its monosubstituted derivatives with an ammonia solution of a copper(I) salt. Those from mono-substituted acetylenes are unreactive towards halides and carbonyl groups but that from acetylene itself reacts readily. Since copper(I) acetylide, unlike sodium acetylide, is not decomposed by water, reactions may be conducted in aqueous solution and this makes for both ease of handling and economy. Many commercial products are derived from the reaction between copper(I) acetylide and two molecules of formaldehyde, including tetrahydrofuran, butadiene (for synthetic rubbers), and adipic acid and hexamethylenediamine (the constituents of 6.6-nylon).

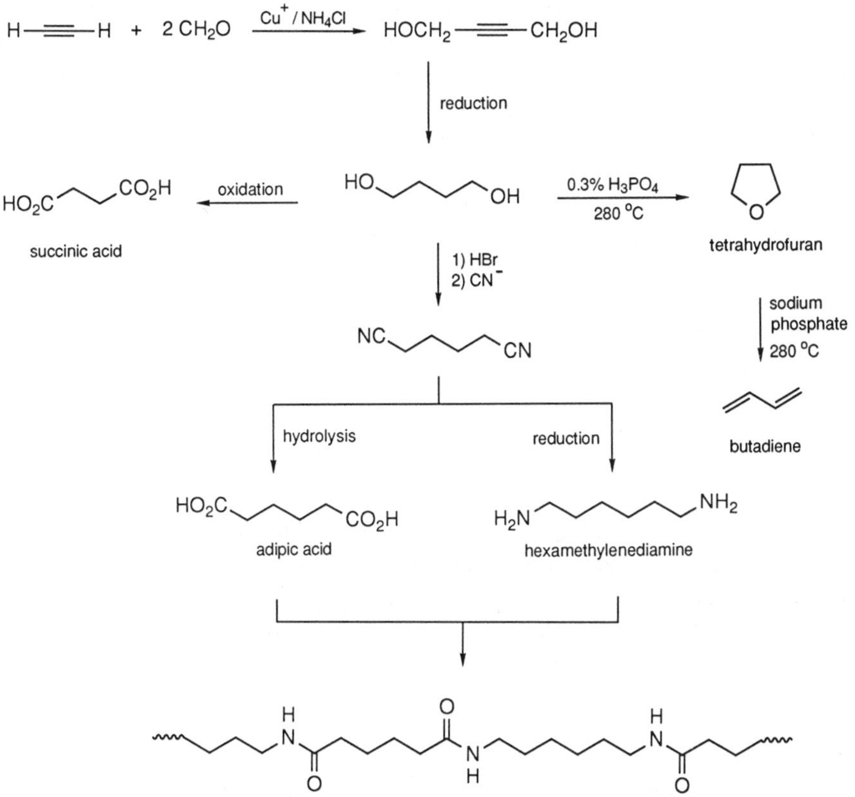

6.6-nylon

7.7 Hydrogen cyanide is isoelectronic with acetylene and, like acetylene, it is a weak
Reactions acid whose anion may be generated by base and is reactive towards alkyl halides
involving and carbonyl groups. It is usually more convenient to introduce the cyanide as
cyanide cyanide ion (e.g. NaCN) rather than as hydrogen cyanide.

(a)   Reactions with alkyl halides and sulfonates

Primary and secondary halides and sulfonates undergo nucleophilic displacement
to give the corresponding nitriles:

$$NC^- \quad C-X \longrightarrow NC-C \quad + \quad X^-$$

Tertiary compounds do not yield nitriles but undergo elimination to give alkenes
(cf. the reactions of other nucleophiles).

These reactions provide a way of extending aliphatic carbon chains by one
carbon atom. The following transformations of the nitrile are useful.

(1)   To carboxylic acids, by hydrolysis. An important example is the synthesis
of malonic acid, and hence malonic ester, in 75–80% yield from chloroacetic acid
[2]. Chloroacetic acid is first converted into its sodium salt (or otherwise the
addition of cyanide ion would liberate hydrogen cyanide), displacement with
cyanide ion gives sodium cyanoacetate and alkaline hydrolysis yields sodium
malonate. The addition of calcium chloride precipitates calcium malonate from
which malonic acid is liberated by treatment with hydrochloric acid.

$$\underset{CO_2H}{\overset{Cl}{|}} \xrightarrow{OH^-} \underset{CO_2^-}{\overset{Cl}{|}} \xrightarrow{CN^-} \underset{N}{\overset{}{}}\diagup\!\!\!\diagup CO_2^- \xrightarrow{OH^-} {}^-O_2C \diagup CO_2^-$$

$$\xrightarrow{Ca^{2+}} {}^-O_2C \diagup CO_2^- \; Ca^{2+} \xrightarrow{2\,HCl} HO_2C \diagup CO_2H \quad + \quad CaCl_2$$

(2)   To amines, by catalytic reduction ($-CN \rightarrow -CH_2NH_2$, p. 661).
(3)   To aldehydes, by Stephen reduction (p. 665).

$$R-\!\!\!\equiv\!\!\!N \xrightarrow[\text{2) H}_2\text{O}]{\text{1) HCl - SnCl}_2} R-CHO$$

Aryl halides do not react readily with cyanide ion unless the aromatic nucleus is
especially activated towards nucleophilic attack (p. 400). For example, 2,4-dinitro-
chlorobenzene gives 2,4-dinitrobenzonitrile readily, but to obtain benzonitrile it is
necessary to treat bromobenzene with anhydrous copper(I) cyanide at 200°C in
the presence of pyridine or quinoline:

$$PhBr \xrightarrow[200\,°C]{CuCN\,/\,pyridine} PhCN$$

In general, it is more satisfactory to prepare aromatic nitriles through aromatic
diazonium salts (p. 413).

(b) Reactions with carbonyl compounds

Hydrogen cyanide adds to aldehydes and ketones to give cyanohydrins (α-hydroxynitriles):

One of the first mechanistic studies in organic chemistry concerned this reaction. Lapworth discovered in 1903–4 that the reaction is catalyzed by base, and he rationalized this by suggesting that the reactive species is the cyanide ion:

$$HCN + OH^- \rightleftharpoons CN^- + H_2O$$

The complete reaction therefore corresponds to those in which other carbanions react with carbonyl compounds: addition of cyanide ion to the carbonyl group gives an oxyanion which, by abstracting a proton from hydrogen cyanide, generates a further cyanide ion:

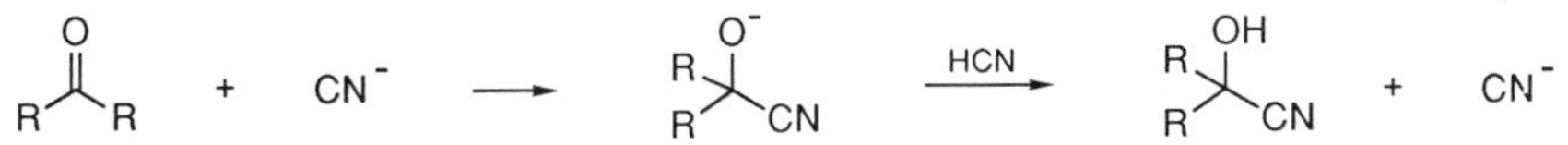

Typical examples of cyanohydrin formation are the preparation of glycolonitrile (c. 76%) from formaldehyde [3],

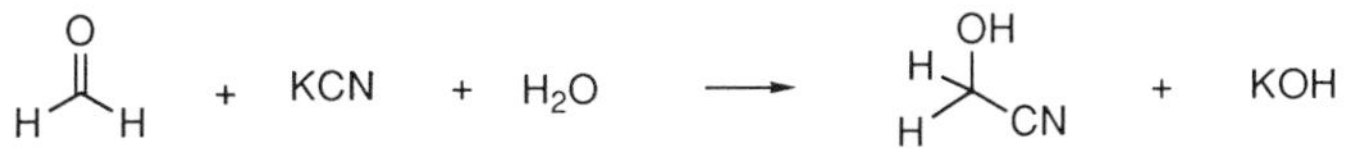

and of acetone cyanohydrin (77%) from acetone in a similar manner [2].

As usual in addition reactions at carbonyl groups, aldehydes are more reactive than ketones, partly beause of the unfavourable inductive effect of the second alkyl group in a ketone and partly because of the increased steric hindrance in addition to ketones. For example, diisopropyl ketone does not react. In the aromatic series, alkyl aryl ketones react unless the alkyl group is particularly bulky or the aromatic ring has large ortho-substituents, but diaryl ketones such as benzophenone are inert. Aromatic aldehydes behave anomalously (see p. 246).

Activated alkenes react with cyanide in the manner of the Michael addition, e.g. [2]:

It should be noted that the C=C bond is reactive enough for addition to occur readily in acidic conditions, a process probably helped by protonation of the carbonyl oxygen (p. 89). The product, which contains a carbonyl group, is inert to further addition in these conditions but if the solution is allowed to become basic a second molecule of hydrogen cyanide is added.

Cyanohydrins are of synthetic value because of their ready conversion by hydrolysis into α-hydroxy-acids or esters. For example, (±)-lactic acid may be obtained from acetaldehyde,

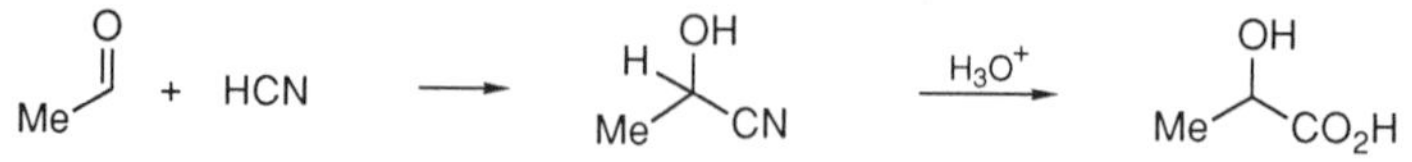

and methyl α-methacrylate (polymerization of which gives Perspex) may be obtained from acetone cyanohydrin by treatment with sulfuric acid in methanol, which brings about both esterification and dehydration:

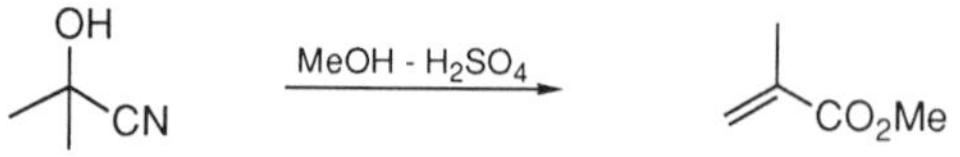

methyl α-methacrylate

The reaction of an aliphatic aldehyde with sodium cyanide in the presence of ammonium chloride gives an α-amino-nitrile (p. 308); hydrolysis gives an α-amino acid (Strecker synthesis, pp. 308, 329), e.g.

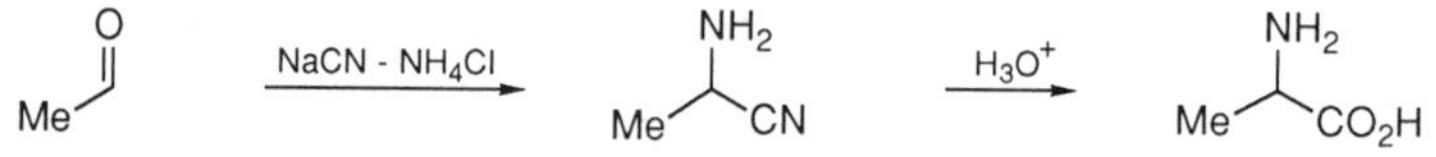

(±)-alanine

(*i*) *Aromatic aldehydes.* These do not form cyanohydrins but instead undergo the *benzoin reaction*. Benzaldehyde and sodium cyanide in ethanol give benzoin itself in about 80% yield [1]:

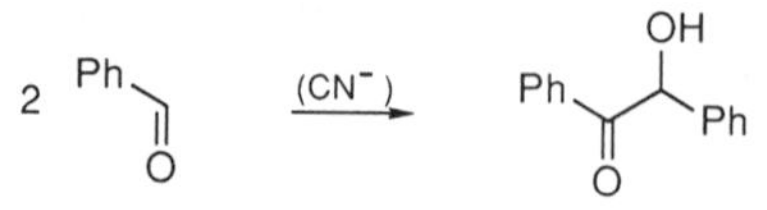

benzoin

Reaction occurs through the cyanide addition product which, by base-abstraction of a proton from the α-carbon, gives a carbanion; this reacts with a second molecule of the aldehyde and hydrogen cyanide is then eliminated.

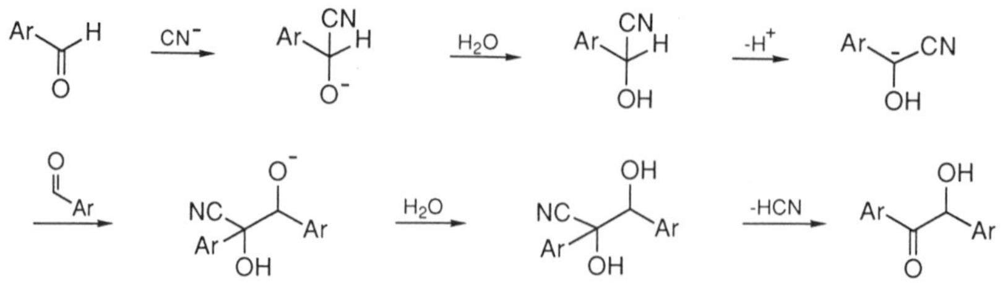

Cyanide ion owes its ability to effect this reaction to two properties: first, it is a reactive nucleophile; second, the cyanide group, by its capacity to delocalize the negative charge on the carbanion, assists the formation of this species. The difference between aliphatic and aromatic aldehydes is ascribed to the fact that the further delocalization of the negative charge over the aromatic ring provides sufficient extra driving force for the reaction to occur.

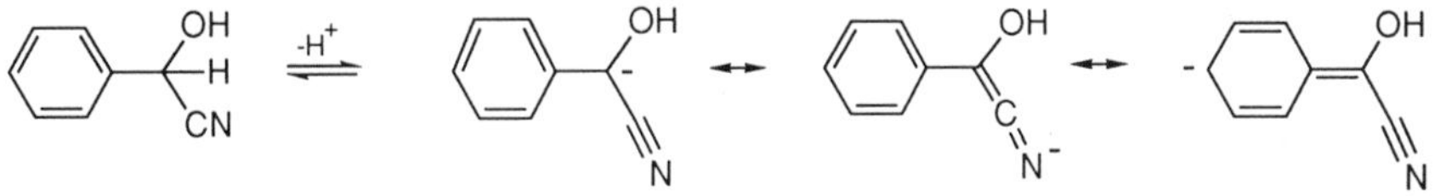

Vitamin B$_1$ (thiamine) brings about similar reactions both in the laboratory and in living organisms. The activity is due to the thiazolium ring whose C$_2$-hydrogen (shown) is acidic enough to exchange with deuterium in heavy water.

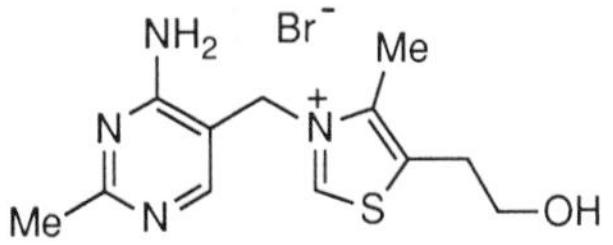

thiamine bromide

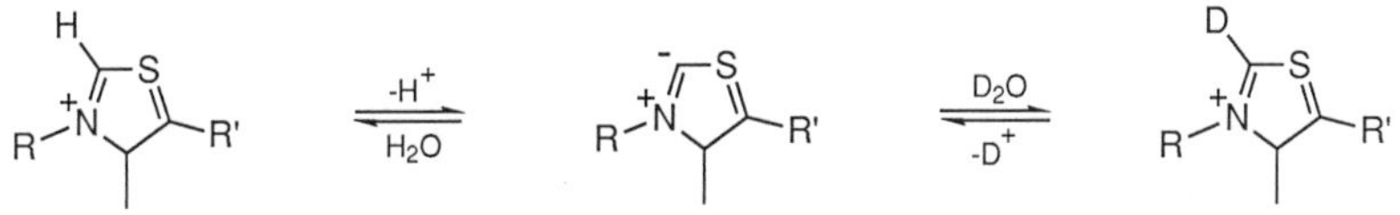

The carbanion intermediate evidently owes its relative stability (compared, say, with that of the anion formed by ionization of the methylene group adjacent to the quaternary nitrogen) to a combination of the presence of an adjacent positive pole and the fact that the charge resides on unsaturated carbon (p. 48). It brings about the benzoin condensation in a manner completely analogous to cyanide ion, to which it bears a close resemblance:

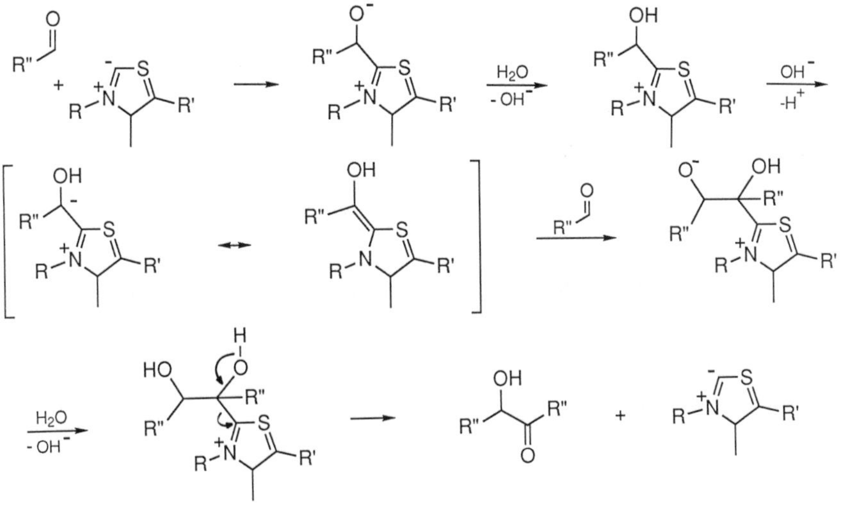

The principle of this method has been exploited in a procedure for effecting aliphatic condensations: the thiazolium cation

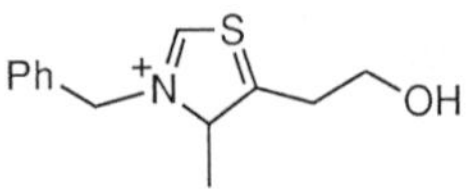

is used. For example, the condensation

occurs in about 70% yield in the presence of the thiazolium cation and Et_3N [7].

Problems   1. How would you employ base-catalyzed reactions in the synthesis of the following compounds?

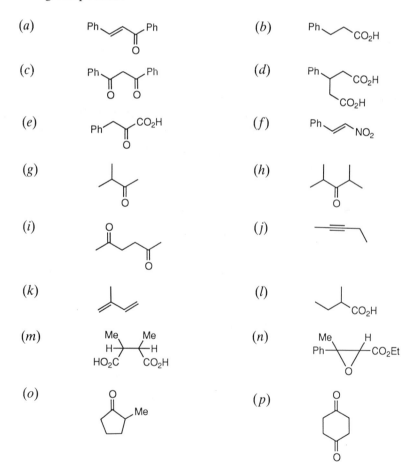

(a)

(b)

(c)

(d)

(e)

(f)

(g)

(h)

(i)

(j)

(k)

(l)

(m)

(n)

(o)

(p)

2. What products would you expect from the following reactions?

(a)

(b)

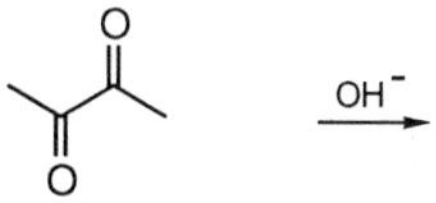

(c)

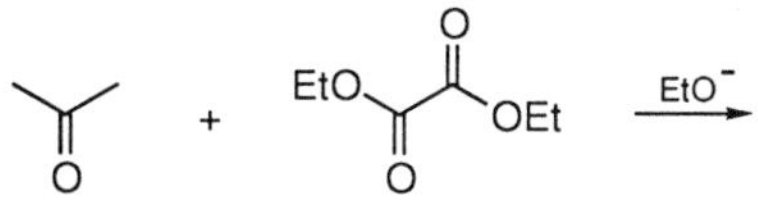

(d)

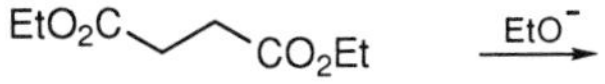

(e)

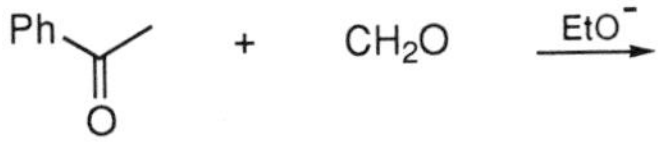

(f)

(g)

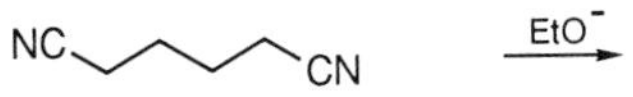

3. Account for the following:
   (i) Amorgst simple reagents, only cyanide ion catalyzes the self-condensation of benzaldehyde.
   (ii) Although diethyl succinate undergoes self-condensation with base, it may be successfully used in the Stobbe condensation to react with aldehydes and ketones.
   (iii) When ketones containing α-hydrogen react with esters, the product is that derived by displacement of alkoxide ion from the ester by the enolate from the ketone. However, when the ester is ethyl chloro-acetate, the product is that derived by addition of the enolate from the ester to the carbonyl group of the ketone.
   (iv) 2,4,6-Trinitrotoluene reacts with benzaldehyde in the presence of pyridine, but toluene does not.
   (v) The thiazolium ion

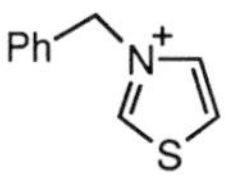

   catalyzes the reaction,

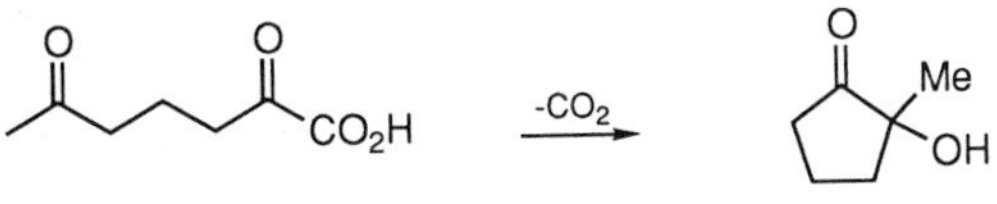

4. How would you employ LDA in the following conversions?

(a)

(b)

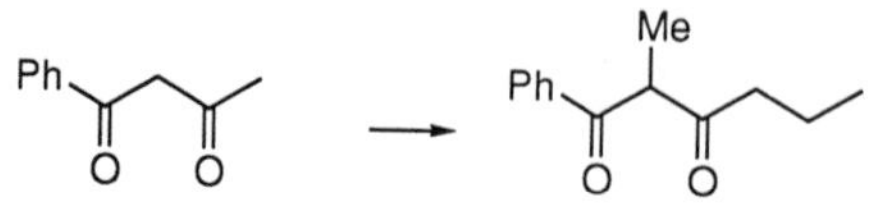

(c)

(d)

(e)

Formation of aliphatic carbon–carbon bonds: acid-catalyzed reactions

8

The principle which is applied in acid-catalyzed reactions is the generation of an electrophilic species, with the aid of an acid, in the presence of a nucleophile with which the electrophile then reacts.

The electrophile may be obtained either from an alkyl or acyl halide by treatment with a Lewis acid, e.g.

$$Me_3C-Cl \quad AlCl_3 \longrightarrow Me_3C^+ \ AlCl_4^-$$

or, more commonly, by the addition of a proton to a double bond. The double bond may be C=C, as in the dimerization of alkenes, e.g.

$$\underset{CH_2}{\diagup} \ + \ H^+ \longrightarrow \underset{CH_3}{\diagup}{}^+$$

or a carbonyl group, as in the self-condensations of aldehydes and ketones, e.g.

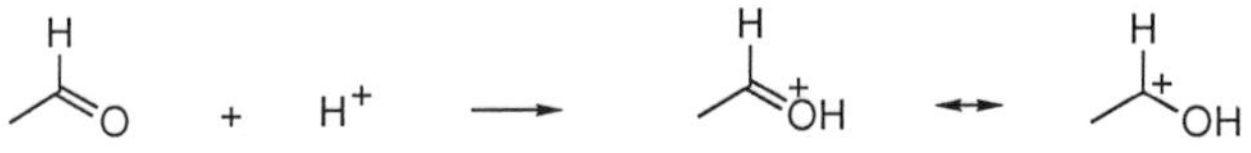

$$\overset{H}{\underset{O}{\diagup}} \ + \ H^+ \longrightarrow \overset{H}{\underset{\overset{+}{O}H}{\diagup}} \longleftrightarrow \overset{H}{\underset{OH}{\diagup}}{}^+$$

The electrophile in the Mannich reaction (section 8.6) is generated from an aldehyde and an amine in the presence of an acid, e.g.

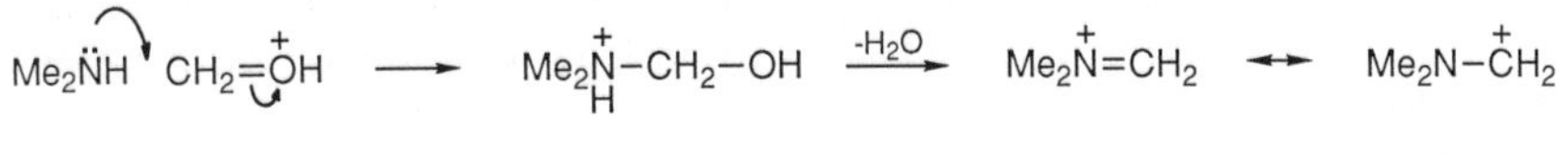

$$Me_2\overset{\cdot\cdot}{N}H \quad CH_2=\overset{+}{O}H \longrightarrow \underset{H}{Me_2\overset{+}{N}-CH_2-OH} \xrightarrow{-H_2O} Me_2\overset{+}{N}=CH_2 \longleftrightarrow Me_2N-\overset{+}{C}H_2$$

The nucleophile may be an alkene, an enol, or, in some applications of the Mannich reaction, a compound of related type such as indole. A typical example is the reaction of ethylene with t-butyl chloride in the presence of aluminium trichloride,

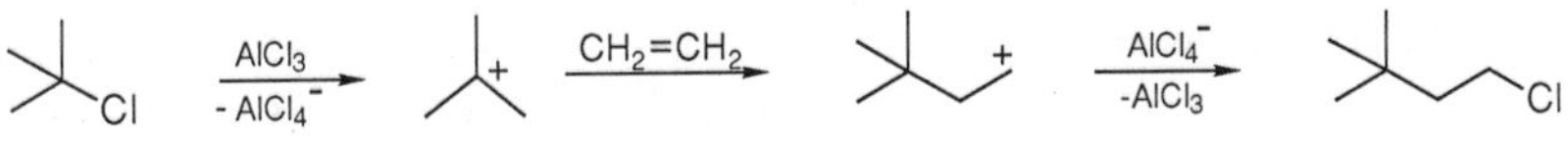

$$\underset{Cl}{\diagdown} \xrightarrow[-AlCl_4^-]{AlCl_3} \diagdown^+ \xrightarrow{CH_2=CH_2} \diagdown^+ \xrightarrow[-AlCl_3]{AlCl_4^-} \underset{Cl}{\diagdown}$$

The ensuing discussion delineates the various combinations of electrophile and nucleophile which are of value in synthesis.

**8.2
The self-
condensation of
alkenes**

The treatment of isobutylene with 60% sulfuric acid gives a mixture of 2,4,4-trimethyl-1-pentene and 2,4,4-trimethyl-2-pentene. Reaction occurs by the protonation of one molecule of the alkene to give a carbocation which adds to the methylene group (Markovnikov's rule) of a second molecule; the new carbocation then eliminates a proton:

4 parts   1 part

The conditions must be carefully controlled. When dilute sulfuric acid is used, the first carbocation reacts preferentially with water to give t-butyl alcohol,

$$Me_3C^+ \xrightarrow{H_2O} Me_3C-\overset{+}{O}H_2 \xrightarrow{-H^+} Me_3C-OH$$

and if more concentrated acid is employed, the second carbocation reacts with a further molecule of isobutylene,

and further polymerization can occur. The concentration of the basic species water must be such that the simple hydration reaction is minimized while sufficient water is present to favour removal of a proton from the dimeric carbocation over further addition of alkene.

This discussion indicates the difficulties necessarily present in effecting the dimerization of an alkene. Nonetheless, the example cited is of great industrial importance: the mixture of pentenes is reduced catalytically to 2,2,4-trimethyl-pentane (isooctane), used as a high-octane fuel.

In a modified procedure, isooctane is obtained directly from isobutylene and concentrated sulfuric acid by carrying out the reaction in the presence of isobutane. Dimerization occurs as above, but the resulting carbocation, instead of eliminating a proton or reacting with more isobutylene, abstracts hydride ion from isobutane:

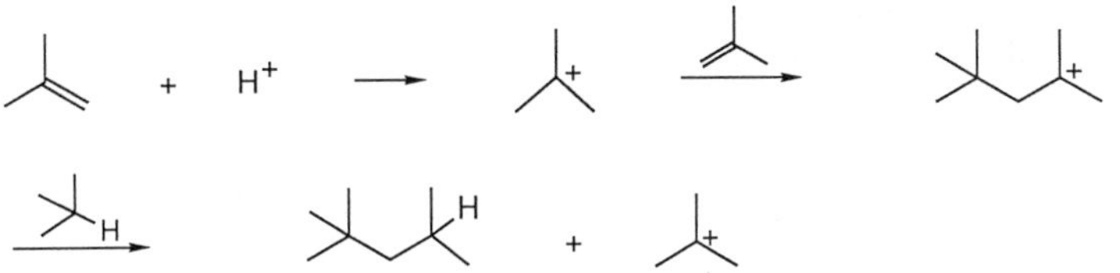

A new t-butyl cation is produced and adds to isobutylene so that a chain reaction is propagated.

Dienes undergo acid-catalyzed cyclization provided that the stereochemically favoured five- or six-membered rings are formed. The example chosen is the conversion of ψ-ionone into α- and β-ionone, the second of which was used in a synthesis of vitamin A:

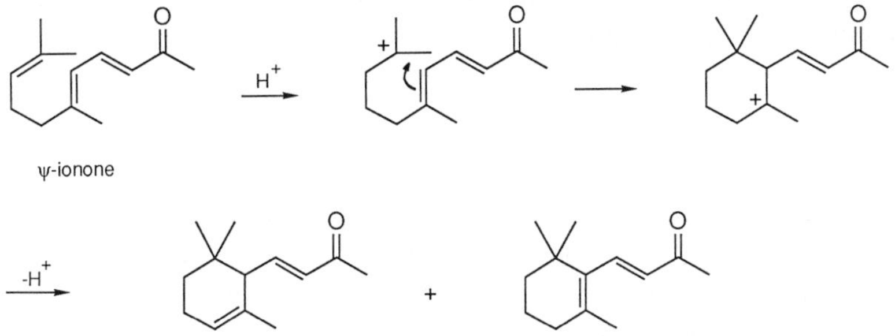

ψ-ionone

α-ionone   +   β-ionone

8.3
Friedel–Crafts
reactions

The names of Friedel and Crafts were originally associated only with the alkylation and acylation of aromatic systems in the presence of Lewis acids (section 11.3), e.g.

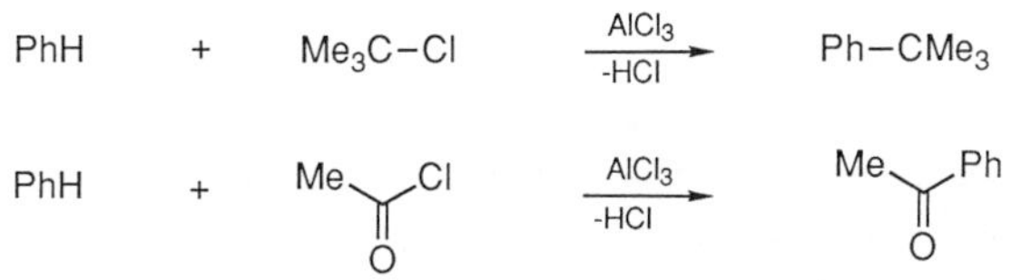

$$\text{PhH} \quad + \quad \text{Me}_3\text{C−Cl} \quad \xrightarrow[\text{−HCl}]{\text{AlCl}_3} \quad \text{Ph−CMe}_3$$

$$\text{PhH} \quad + \quad \underset{\text{O}}{\text{Me}\diagup\diagdown\text{Cl}} \quad \xrightarrow[\text{−HCl}]{\text{AlCl}_3} \quad \underset{\text{O}}{\text{Me}\diagup\diagdown\text{Ph}}$$

These reactions are fully discussed in chapter 11. Analogous processes occur with alkenes, although they are not so widely applicable.

(a) Alkylation

The simplest example of an efficient alkylation is the preparation of neohexyl chloride (75%) from t-butyl chloride and ethylene at about −10°C in the presence of aluminium trichloride:

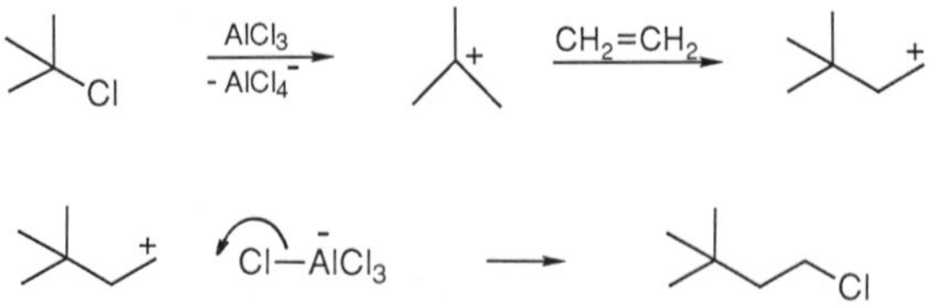

Several side reactions are normally encountered in alkylation. First, alkenes are often isomerized by aluminium trichloride. Second, alkyl halides may rearrange, e.g. propyl chloride gives isopropyl derivatives (cf. p. 364). Finally, the halide produced may react further. These problems do not arise in the example cited because ethylene cannot isomerize, t-butyl halides are not rearranged by Lewis acids, and the primary halide product is far less reactive than the tertiary halide towards aluminium trichloride. It must be emphasized, however, that the synthetic utility of alkylation is restricted.

(b) Acylation

The acylation of an alkene is brought about by an acid chloride or anhydride in the presence of a Lewis acid. As in aromatic Friedel–Crafts reactions (section 11.3), the electrophilic reagent can be either a complex of the acid chloride and Lewis acid or an acylium cation derived from it, e.g.

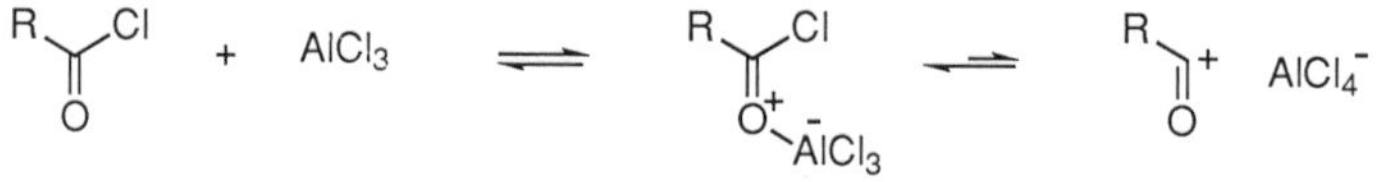

Addition of the electrophile to the alkene can be followed either by uptake of chloride ion to give a β-chloroketone, e.g.

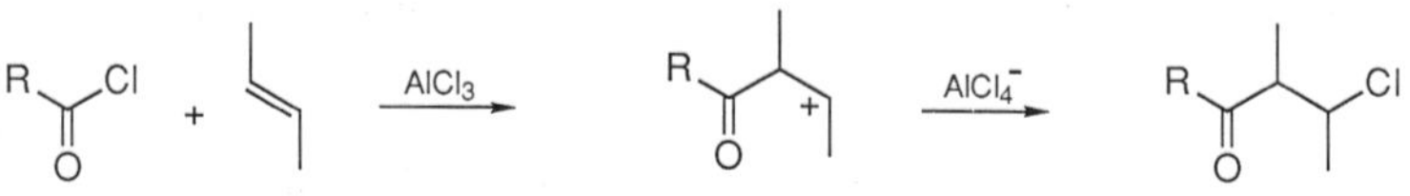

or by elimination, via a 1,5-hydrogen shift (p. 291), to give a βγ-unsaturated ketone. The latter then rearranges by a prototropic shift, catalyzed by the released acid, to the more stable conjugated ketone, e.g.

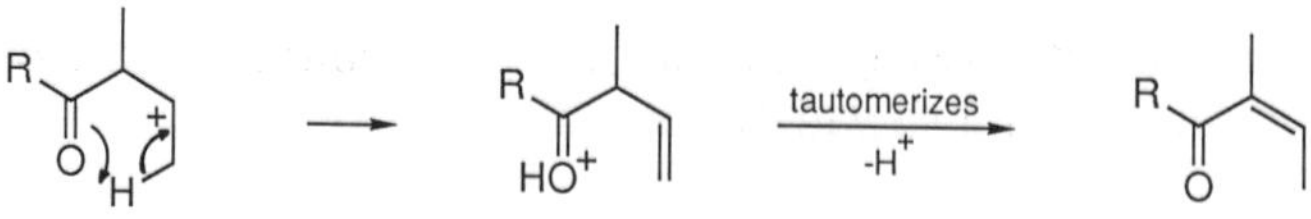

Since the β-chloroketone undergoes ready base-catalyzed elimination, both routes lead to the αβ-unsaturated ketone. If the βγ-unsaturated ketone is the required product, reaction is carried out with a pre-formed acylium salt in the presence of a weak, hindered (i.e. non-nucleophilic) base to take up the acid as it is formed. It is thought than an 'ene' reaction (p. 295) occurs, e.g.

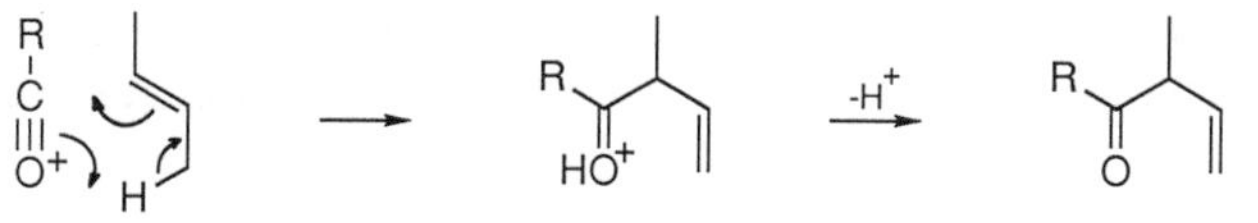

Acylations suffer from the same disadvantage as alkylations in that the alkenes are rearranged by aluminium trichloride and it is therefore usually more satisfactory to use a less vigorous Lewis acid such as tin(IV) chloride. However, the other side reactions ecountered in alkylations do not apply: the acid halides and anhydrides do not rearrange and the products are much less reactive than the starting materials. Acylation is therefore a useful synthetic process, as illustrated by a synthesis of (±)-thioctic acid:*

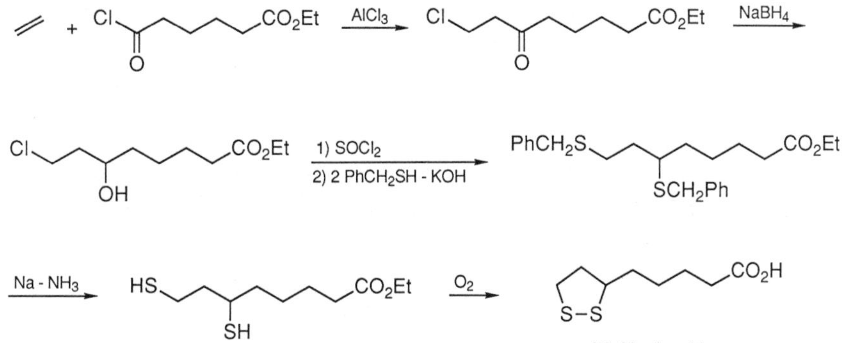

(±)-thioctic acid

8.4
Prins reaction

The treatment of an alkene with formaldehyde in the presence of an acid gives a 1,3-diol together with the cyclic acetal derived from this and a second molecule of formaldehyde:

$$CH_2{=}O \; + \; H^+ \; \rightleftharpoons \; CH_2{=}\overset{+}{O}H \; \longleftrightarrow \; \overset{+}{C}H_2{-}OH$$

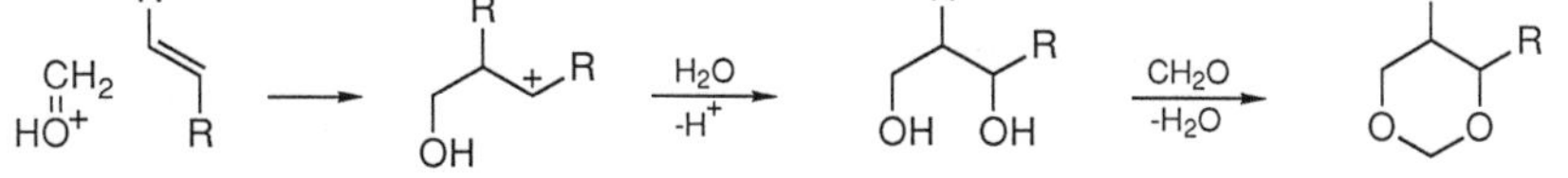

Ethylene itself requires very vigorous conditions for reaction, but alkylated alkenes, which are more reactive towards electrophiles, react fairly readily. Alkenes of the type RCH=CHR give mainly 1,3-diols in low yield, whereas those of the types RCH=CH$_2$ and R$_2$C=CH$_2$ give mainly acetals (1,3-dioxans),

---

* Thioctic acid (lipoic acid) is a cofactor in the physiological oxidative-decarboxylation of pyruvic acid to acetic acid:

$$CH_3COCO_2H \xrightarrow{\;1/2\,O_2\;} CH_3CO_2H \; + \; CO_2$$

often in good yield. For example, 4-phenyl-1,3-dioxan can be isolated in about 80% yield by refluxing a mixture of styrene, 37% formalin, and concentrated sulfuric acid [4]:

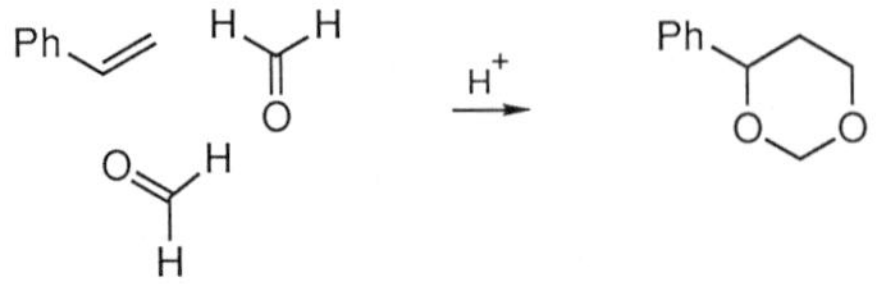

8.5
Reactions of aldehydes and ketones   (*a*) *Self-condensations*

Aldehydes and ketones which are capable of enolization undergo self-condensation when treated with acids. The acid has two functions: first, it enhances the reactivity of the carbonyl group towards the addition of a nucleophile (p. 89), e.g.

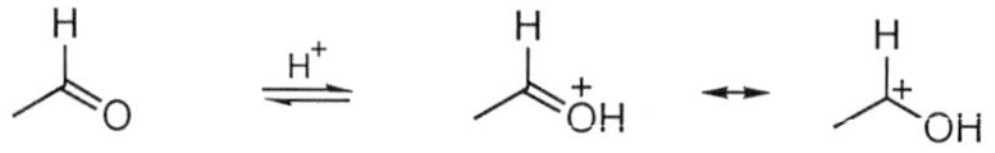

Second, it catalyzes the enolization of the carbonyl compound, e.g.

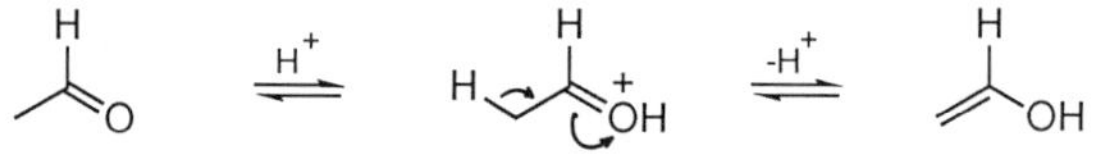

A molecule of the enol then reacts with a molecule of the (activated) carbonyl compound,

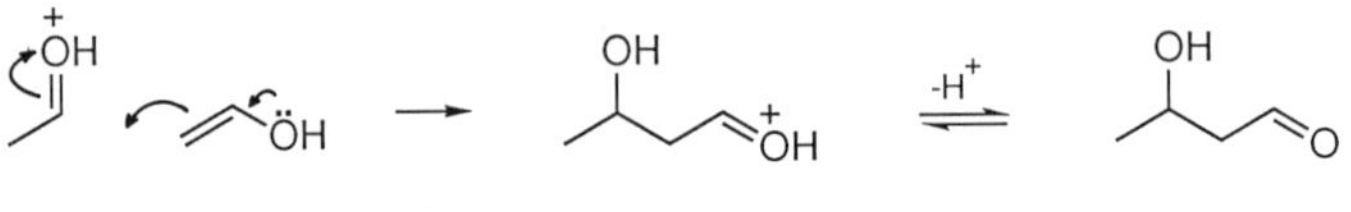

Acid-catalyzed dehydration *via* the enol normally follows:

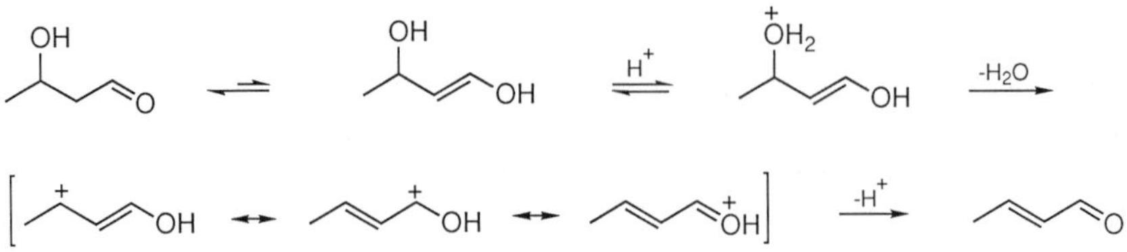

A ketone which possesses at least one hydrogen atom on each of its α-carbon atoms can react further. For example, when acetone is saturated with hydrogen chloride, a mixture of mono- and di-condensation products is obtained:

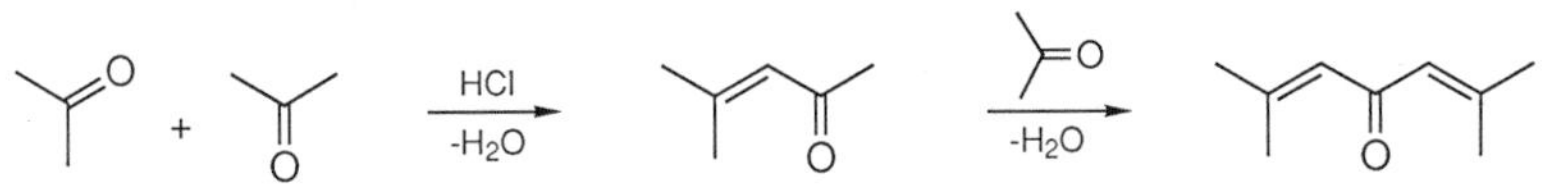

The yields in these acid-catalyzed aldol condensations are usually low, and the base-catalyzed process is normally preferred. However, in some instances the processes give different major products, and the acid-catalyzed method may then be appropriate. For example, the base-catalyzed reaction of benzaldehyde with butanone gives mainly product (a) whereas the acid-catalyzed reaction gives mainly (b):

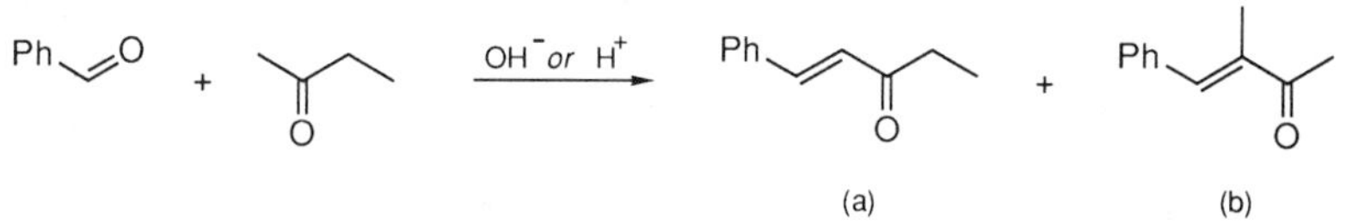

(a)   (b)

The reason is that, in the former, equilibria are established with the two possible aldol adducts and the major product is determined by the ease of dehydration of each to give an aryl-conjugated product; this favours (a) since the enolate (c) is formed faster than the more substituted enolate (d) (p. 208):

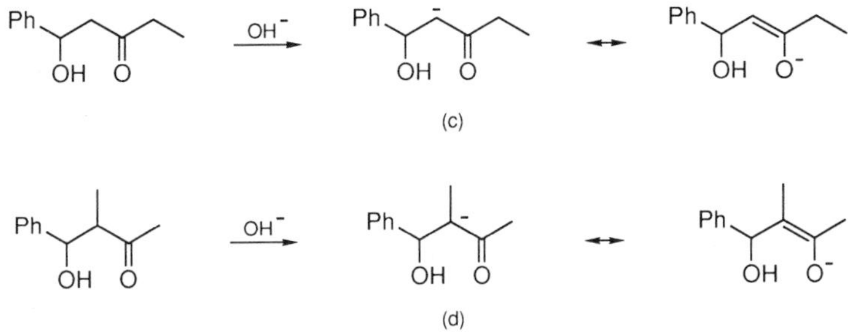

(c)

(d)

In contrast, in acidic conditions the critical step is not dehydration but formation of the new C—C bond, and the regioselectivity is governed by the more rapid formation of the enol (e), which is a 1,2-dialkylated alkene, than (f), which is monoalkylated:

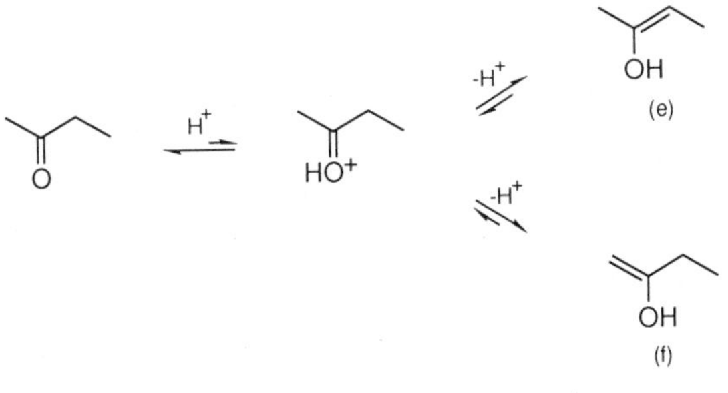

(e)

(f)

Finally, some acids give cyclic polymers with acids, for example, acetaldehyde and a little concentrated sulfuric acid give the cyclic trimer, paraldehyde, and some of the cyclic tetramer, metaldehyde:

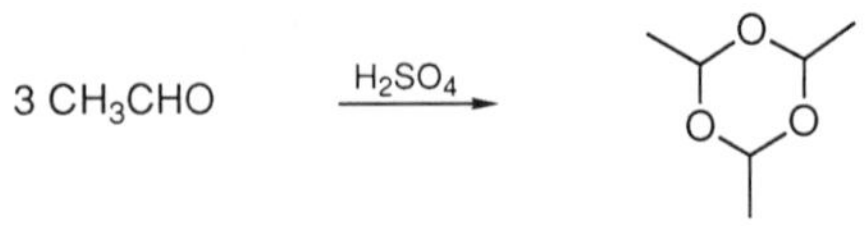

$$3 \ CH_3CHO \xrightarrow{\text{H}_2\text{SO}_4}$$

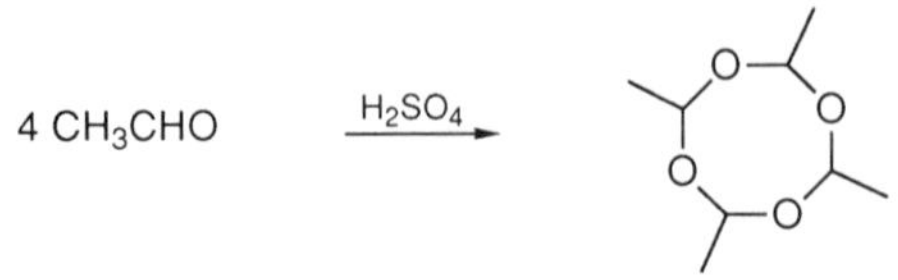

$$4 \ CH_3CHO \xrightarrow{\text{H}_2\text{SO}_4}$$

Polymerization is reversible and the aldehyde may be recovered by warming with dilute acid. Formaldehyde yields a solid polymer, paraformaldehyde, on evaporation of formalin,

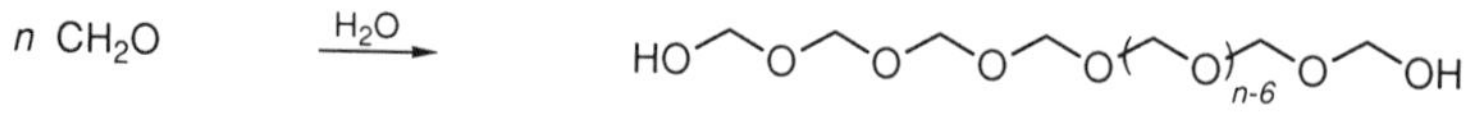

$$n \ CH_2O \xrightarrow{\text{H}_2\text{O}}$$

and treatment with sulfuric acid gives trioxan (trioxymethylene):

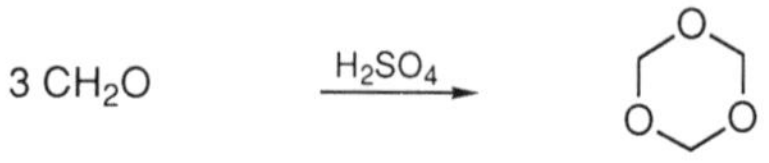

$$3 \ CH_2O \xrightarrow{\text{H}_2\text{SO}_4}$$

Formaldehyde and acetaldehyde are conveniently stored in the form of paraformaldehyde and paraldehyde, respectively; treatment with dilute acid immediately before use or *in situ* generates the free aldehyde. Ketones do not form analogous oxygen-linked polymers, but treatment of acetone with concentrated sulfuric acid gives mesitylene (1,3,5-trimethylbenzene).

(b)   Crossed reactions

As in base-catalyzed reactions, crossed reactions between two carbonyl compounds each of which can enolize are likely to result in a mixture of four products. However, if only one of the two compounds can enolize and the other has the more reactive carbonyl group, a single product may be formed in good yield.

An example is the reaction between acetophenone and salicylaldehyde, catalyzed by anhydrous hydrogen chloride; only the former compound can enolize and the latter has the more reactive carbonyl group. Condensation is followed by acid-catalyzed reactions to give an oxonium salt which is the parent of the anthocyanidin system:*

---

* Oxonium salts normally exist only in solution. The stability of the anthocyanidin cation results from the aromatic character of the six-membered conjugated oxonium ring which contains six π-electrons.

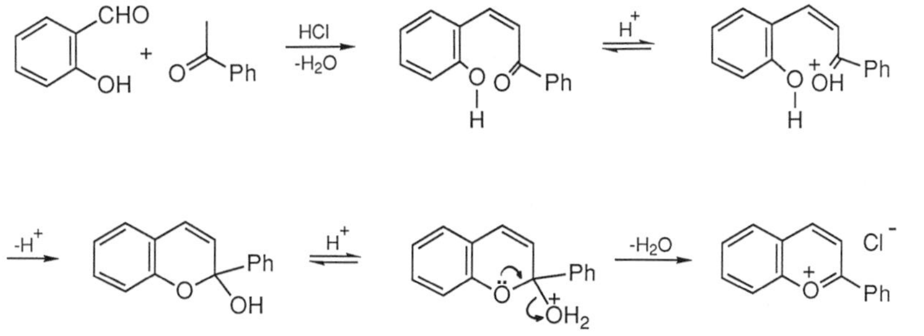

The application of this synthesis to the formation of the naturally occurring anthocyanidins and anthocyanins is described later (p. 714).

(c) *Reactions between ketones and acid chlorides or anhydrides*

Reactions can also be effected between ketones and compounds with very reactive carbonyl groups such as acid chlorides and anhydrides. Aqueous conditions must be avoided to prevent hydrolysis of the chloride or anhydride and it is convenient to use a Lewis acid to catalyze the enolization step, e.g.

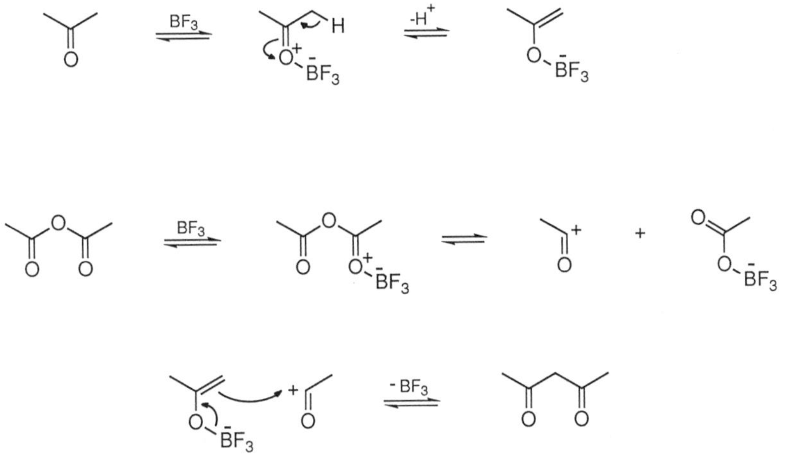

2,4-Pentanedione can be obtained in this way by passing gaseous boron trifluoride into a mixture of acetone and acetic anhydride, adding copper(II) acetate to precipitate the product as the copper(II) derivative of the enol, and reconverting this into the pentanedione with acid [3]. The yield (80%) is considerably higher in this case than in the alternative base-catalyzed process from acetone and ethyl acetate (p. 229).

As in the aldol reaction, the acid- and the base-catalyzed methods give mainly different products with unsymmetrical ketones, e.g.

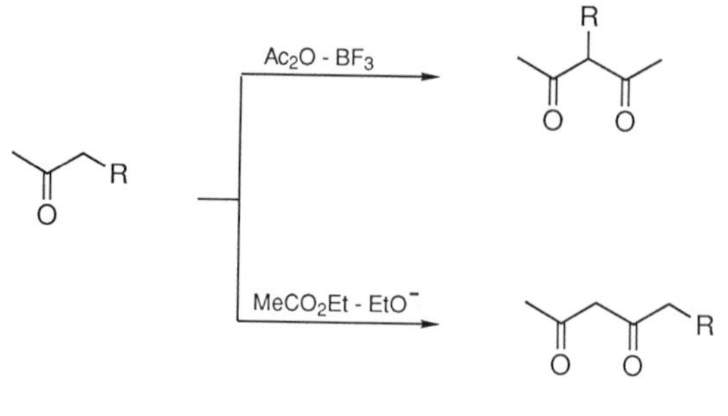

(d) Reactions of 2-methylpyridine and related compounds

2-Methylpyridine reacts with aldehydes in the presence of zinc chloride which catalyzes its conversion into the nitrogen-analogue of an enol, e.g.

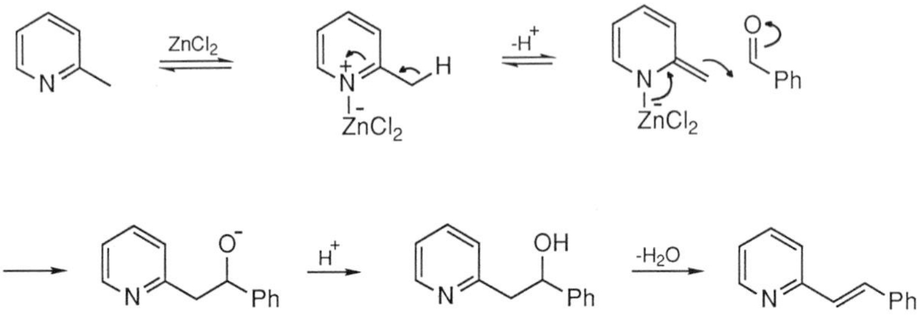

Similar reactions occur with 4-methylpyridine, 2- and 4-methylquinoline, and 1-methylisoquinoline, each of which can undergo the corresponding acid-catalyzed reaction, e.g.

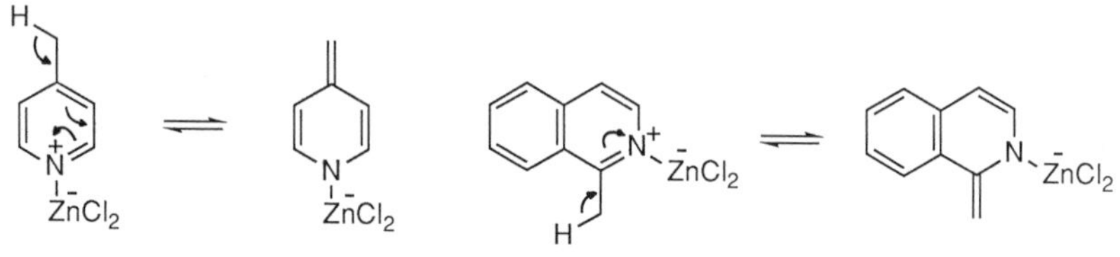

The other methyl derivatives of these heterocyclic compounds are either unable to form the analogous methylene derivatives (e.g. 3-methylpyridine) or fail to do so because the loss of aromatic stabilization energy on enolization is too great, e.g.

(non-benzenoid)

The quaternary salts of the enolizable methyl derivatives of heterocyclic compounds undergo analogous reactions merely on being heated. A particularly useful example is the synthesis of pinacyanol from 2-methylquinoline ethiodide and ethyl orthoformate:

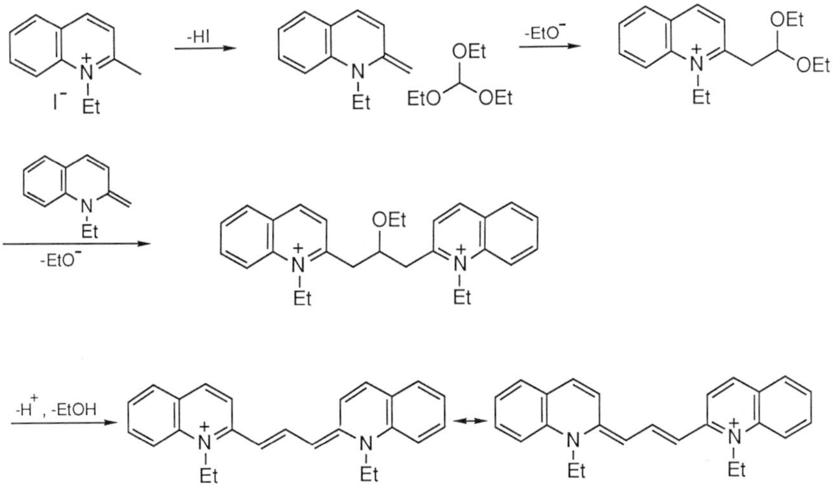

pinacyanol (cation)

Pinacyanol and related compounds made by analogous routes have highly coloured cations and are used as photographic sensitizers.

8.6 Mannich reaction

Compounds which are enolic, or potentially enolic, and also certain alkynes, react with a mixture of an aldehyde (usually formaldehyde) and a primary or secondary amine in the presence of an acid to give, after basification, an aminomethyl derivative. For example, by refluxing a mixture of acetone, diethylamine hydrochloride, paraformaldehyde, methanol, and a little concentrated hydrochloric acid, and treating the product with base, 1-diethylamino-3-butanone can be obtained in up to 70% yield [4]:

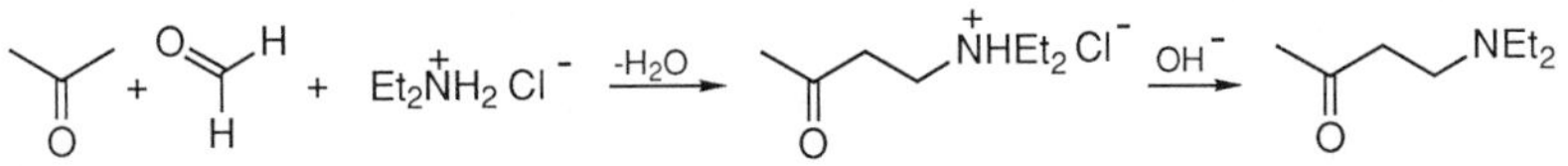

(a) Mechanism

The amine reacts with formaldehyde in the presence of acid to give an adduct which eliminates water to form an electrophile, e.g.

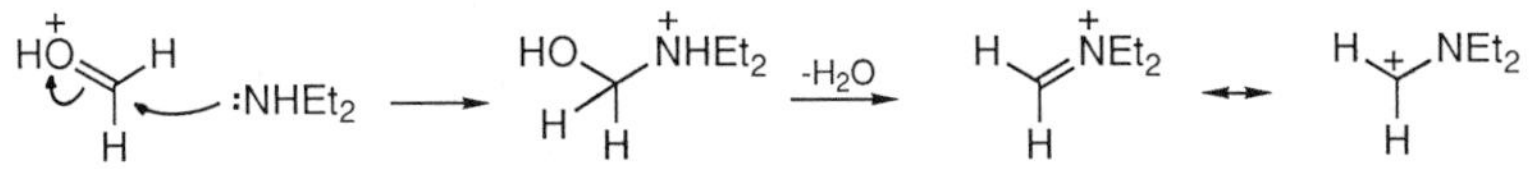

and the acid also catalyzes the conversion of acetone into its enol tautomer,

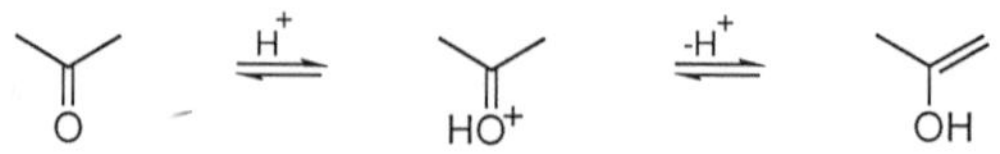

The enol then reacts with the electrophile and the resulting adduct tautomerizes to the amine salt,

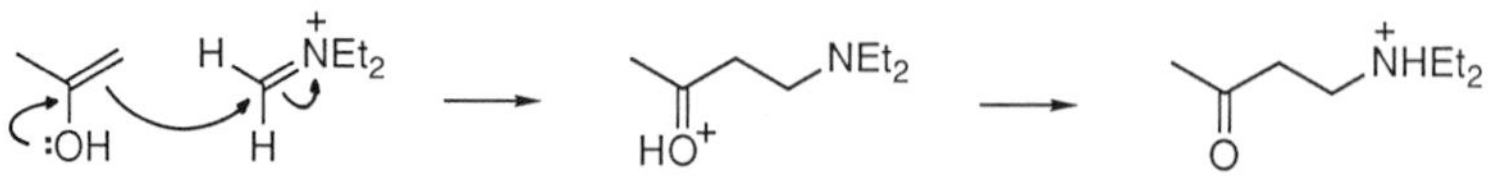

(b) Structural requisites in the reactants

The amine may be primary or secondary. In the former case, the product, a secondary amine, usually reacts further, e.g.

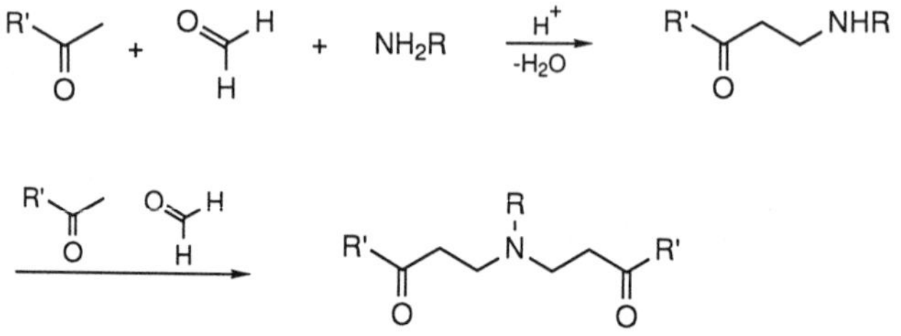

Ammonia may also be employed, usually resulting in three successive reactions, e.g.

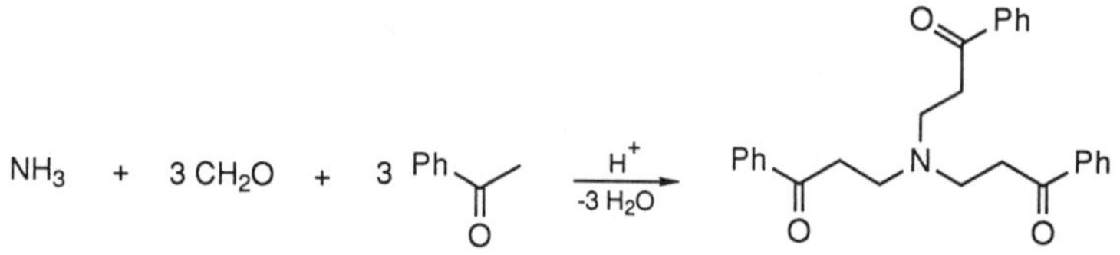

The aldehyde is commonly formaldehyde, but higher aldehydes have been used successfully and several examples are given later.

The third reactant may be an enol, a related compound, or any compound capable of undergoing enolization in the presence of acid. The more important classes of compound are the following; each is illustrated by the reaction of a typical member with formaldehyde and dimethylamine.

- Aldehydes

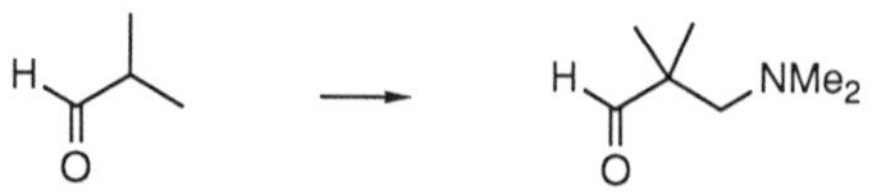

● Ketones

● β-Dibasic acids, β-cyano-acids, β-keto-acids, etc.

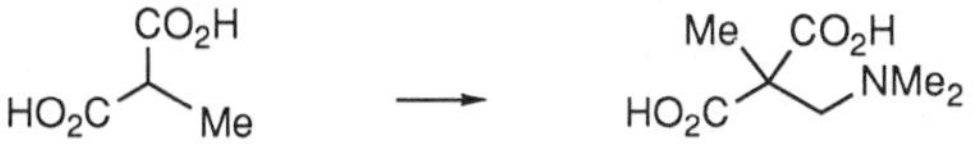

● 2-Methylpyridine and related compounds (see p. 260)

● Phenols,*

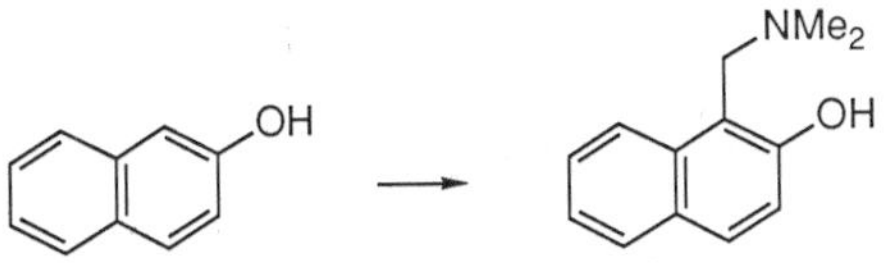

● Furan, pyrrole, indole, and derivatives,*

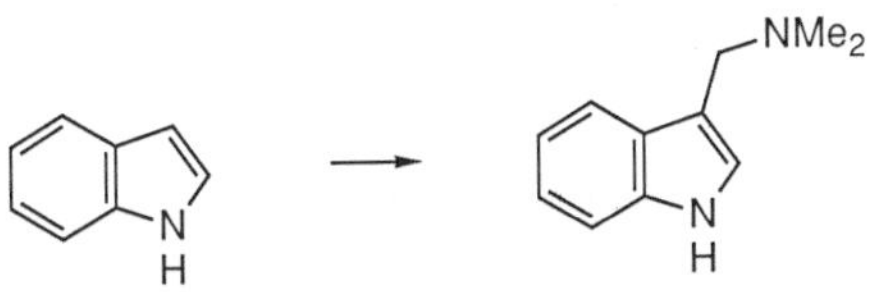

gramine

(These last-named compounds are sensitive to acids (p. 358) and the Mannich reaction must be carried out in a weak acid, usually acetic acid, at a low temperature.)

Phenylacetylene and some of its nuclear-substituted derivatives, although not enolic, also react, e.g.

(*i*) *Pre-formed immonium salts.* The use of a pre-formed immonium salt avoids the need for an acid catalyst and thereby increases the scope of the reaction. The salts can be prepared in general by the reaction,

---

*The mechanisms of reactions of electrophiles with phenols, furan, pyrrole, indole, and related compounds and the reasons for the preferred positions of reactivity in the nucleus, are discussed in chapter 11.

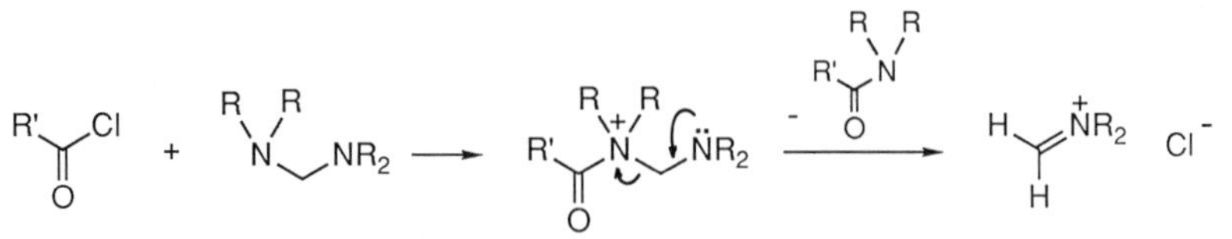

A widely employed reagent is *Eschenmoser's salt*, prepared by heating the quaternary ammonium salt from trimethylamine and diiodomethane at about 150°C in tetrahydrothiophen dioxide:

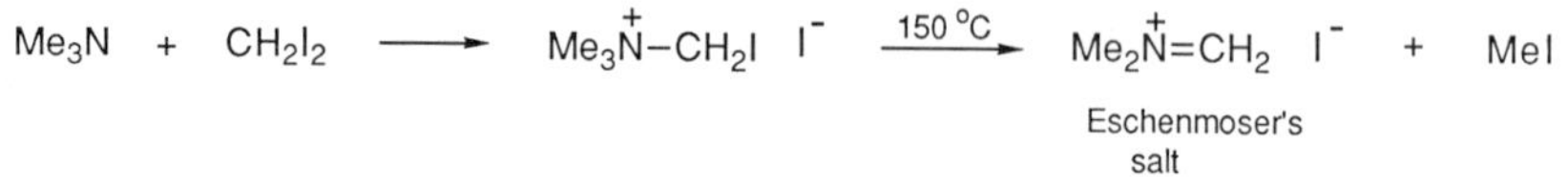

It is believed that reaction occurs as follows:

Compounds which are moderately enolic react readily, e.g.

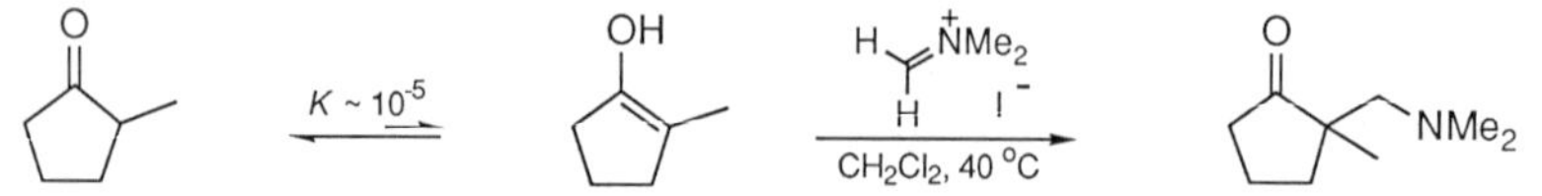

(c) Mannich bases as intermediates in synthesis

The facts that the Mannich reaction is applicable to a wide range of compounds, is usually efficient, and gives products (Mannich bases) which undergo a number of types of transformation, give the process extensive applications in synthesis. The following are the more important uses.

(i) Formation of αβ-unsaturated carbonyl compounds. The Mannich reaction normally gives the hydrochloride of the Mannich base (p. 261). These salts are usually stable at room temperature but those derived from aliphatic compounds undergo elimination on being heated, e.g.

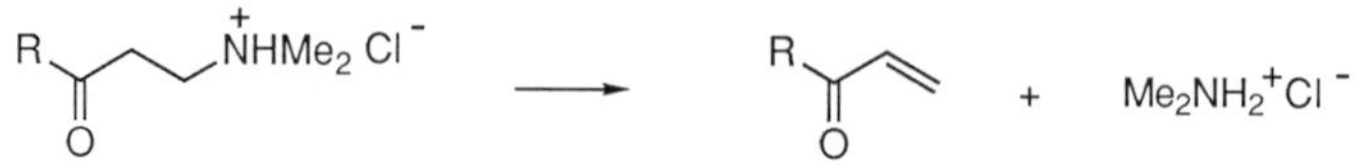

The reaction is similar to the Hofmann elimination of a quaternary ammonium salt (p. 98), but whereas the latter process requires a strong base (hydroxide ion), Mannich base hydrochlorides undergo elimination readily because the double bond which is generated is conjugated with a second unsaturated group.

Eliminations of this type have two uses. First, reduction of the C=C bond (e.g. catalytically, p. 631) gives the next highest homologue of the ketone from which the Mannich base was derived:

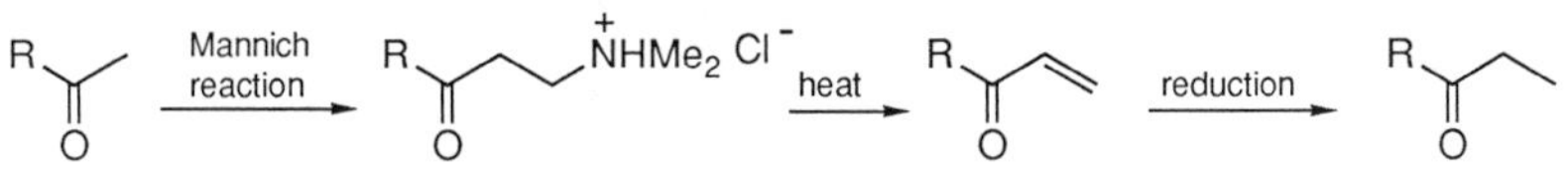

Second, the quaternary salts of Mannich bases are latent sources of αβ-unsaturated carbonyl compounds required for condensation reactions. For example, for a base-catalyzed reaction utilizing butenone, it is better to employ the quaternized Mannich base from acetone, from which the unsaturated ketone is generated *in situ* by the action of base, than to use the free ketone, since this readily polymerizes. The best known example of this application is Robinson's ring extension, used in building up the ring system of steroids.

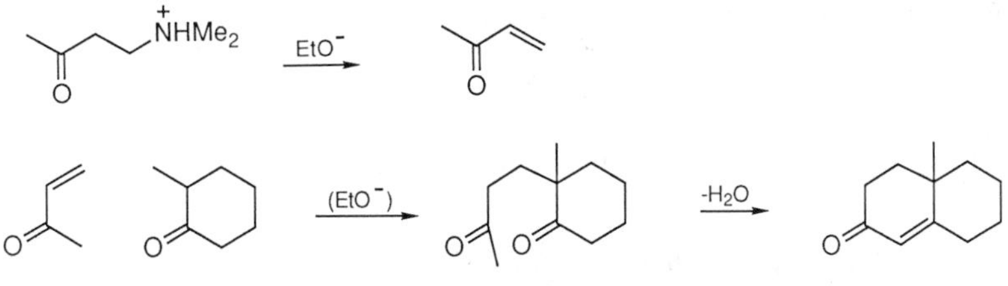

(*ii*)   *Replacement of the amino group.*   The Mannich bases derived from aromatic systems such as phenols and indoles are benzylic-type compounds and, as such, are particularly susceptible to nucleophilic displacements (S_N2 reactions). Advantage can be taken of this to replace the amino group by other functions. For this purpose they are first quaternized, for a tertiary amine is a more labile leaving group than an amide ion.

(1)   An alternative to the synthesis of heteroauxin described previously (p. 189) employs the Mannich base from indole (gramine). Methylation with dimethyl sulfate followed by treatment with cyanide ion and then hydrolysis gives heteroauxin:

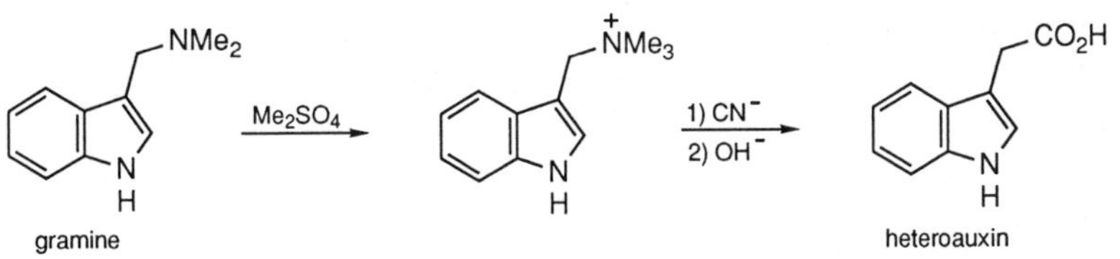

(2)   Gramine is also the starting material in a synthesis of tryptophan, one of the α-amino acids contained in proteins. The quaternized compound is treated with acetamidomalonic ester (p. 328) in the presence of base; the enolate displaces on the quaternary salt to give a compound which is readily hydrolyzed and decarboxylated to tryptophan:

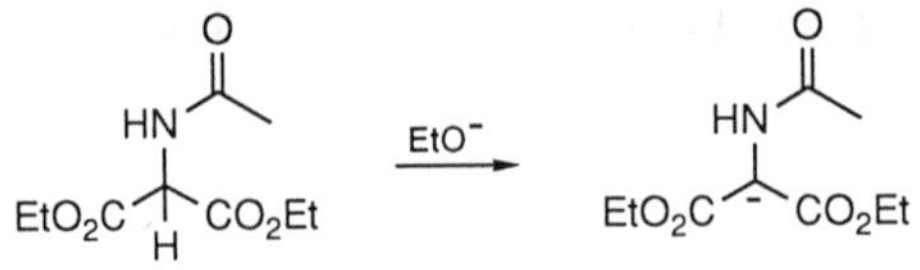

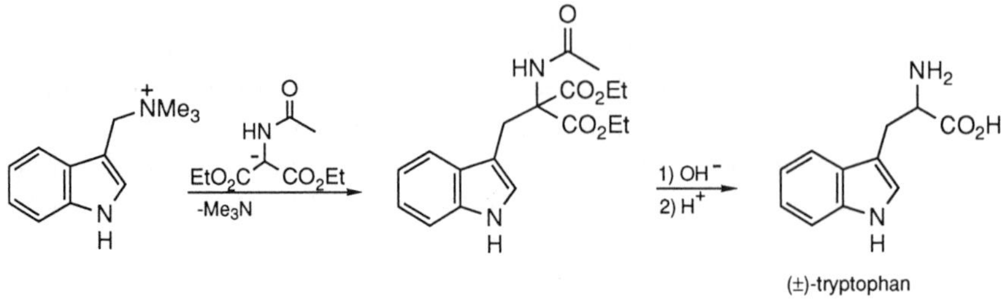

(3)   Benzylic systems are also susceptible to hydrogenolysis,

$$ArCH_2X \xrightarrow{2\,'H'} ArMe \;+\; HX$$

where X is an oxygen or nitrogen function (p. 643).

(*iii*)   *The use of aldehydes other than formaldehyde: alkaloid synthesis.*   A number of elegant syntheses of alkaloids have been based on the Mannich reaction, perhaps the most outstanding being Robinson's (1917) synthesis of tropinone, required for the synthesis of atropine. The following mixture, on standing for several days at pH 5, gave tropinone in 90% yield.*

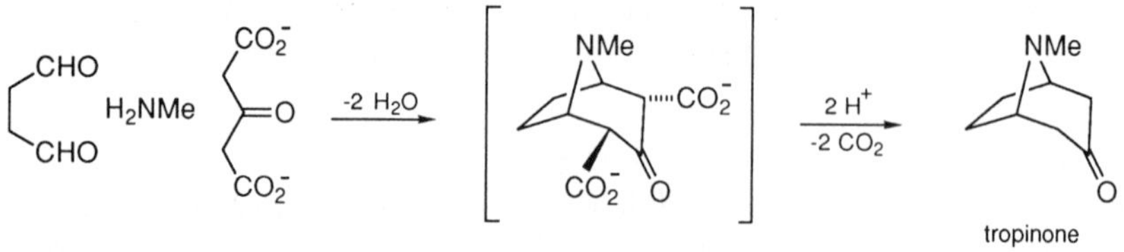

The synthesis consists of two Mannich reactions followed by the spontaneous decarboxylation of the dibasic β-keto-acid. Tropinone can be converted into atropine by reduction of the carbonyl group followed by formation of the ester with tropic acid:

---

* Robinson's synthesis was inspired by his consideration of the possible ways in which alkaloids such as atropine and cocaine may arise naturally. His successful synthesis of atropine in the mild physiological conditions of pH and temperature was the first example of a probable biosynthetic process achieved in the laboratory. It has since become apparent, by the use of radioactive tracer techniques, that Mannich-type reactions play a central role in the formation in plants of the heterocyclic systems of several groups of alkaloids.

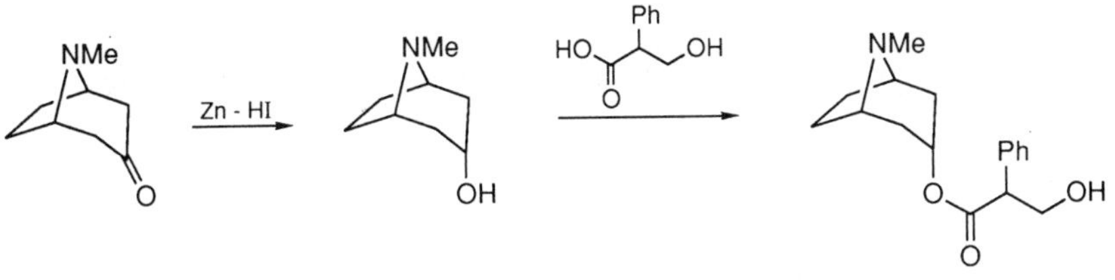

atropine

This synthesis has been adapted to the syntheses of cocaine (from coca leaves),

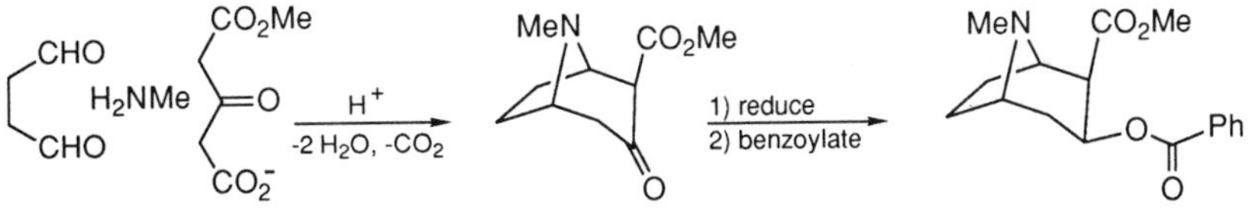

(±)-cocaine

and pseudopelletierine (58–68%) (an alkaloid obtained from the root bark of the pomegranate tree) [4],

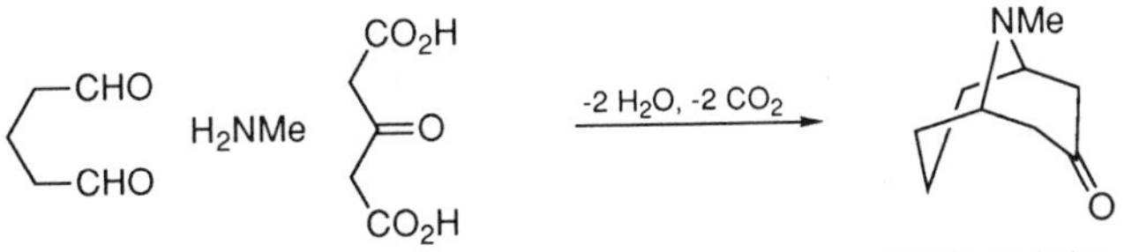

pseudopelletierine

(*iv*) *Local anaesthetics*. Both atropine and cocaine have certain anaesthetic properties, and an examination of a range of such compounds has indicated that the critical arrangement of atoms which is associated with local anaesthetic action is

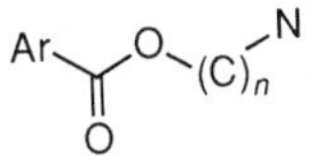

i.e. an amino function connected by a chain of carbon atoms to an alcoholic function which is esterified by an aromatic acid. These structures, where $n = 2$, can be obtained by Mannich reactions on ketones followed by reduction of the carbonyl group and esterification of the resulting alcohol, and compounds whose anaesthetic properties are as good as those of cocaine but which are less toxic have been made in this way. Tutocaine is a typical example:

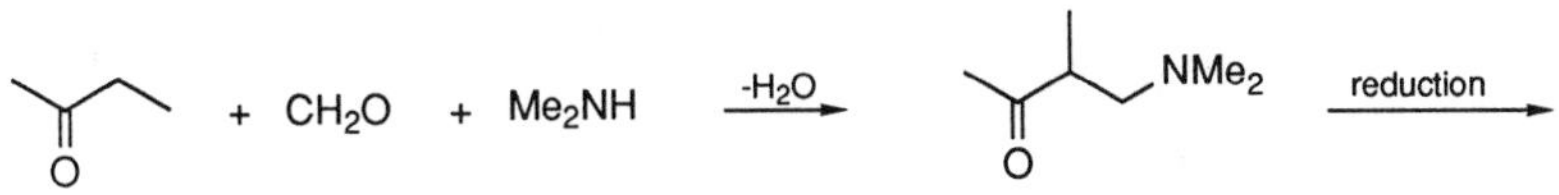

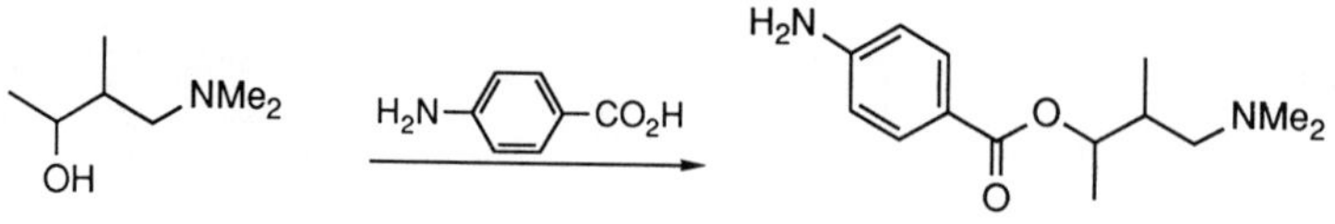

tutocaine

Problems   1.   What products would you expect from the following reactions?:

(a)   +   -Cl $\xrightarrow{\text{AlCl}_3}$

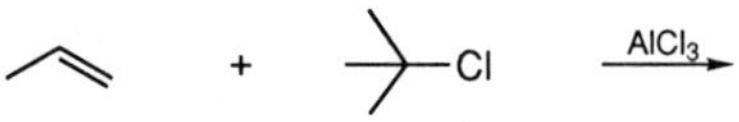

(b)   +   CH$_2$O $\xrightarrow{\text{H}_2\text{O - H}^+}$

(c)   +   CH$_2$O $\xrightarrow{\text{H}^+}$

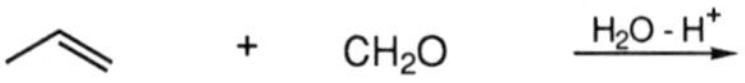

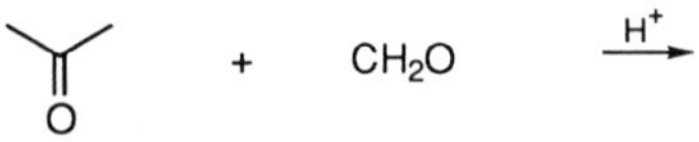

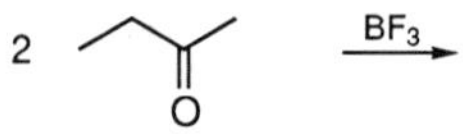

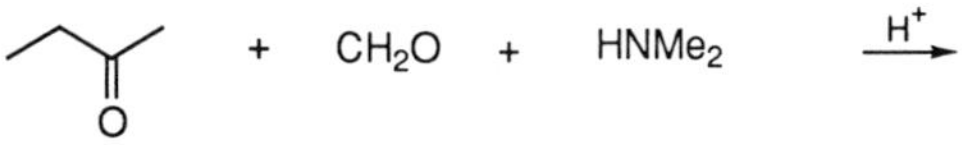

(d)   2 $\xrightarrow{\text{BF}_3}$

(e)   +   CH$_2$O   +   HNMe$_2$ $\xrightarrow{\text{H}^+}$

(f)   + $\xrightarrow{\text{HCl}}$

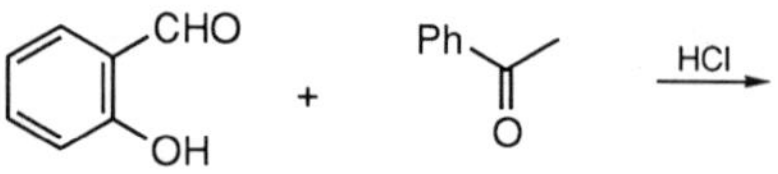

2.   Outline synthetic methods for the following which make use of acid-catalyzed reactions:

(a)   (b)

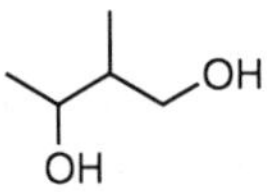

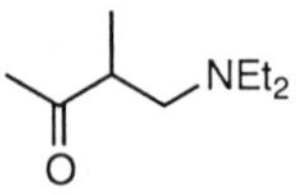

(c) 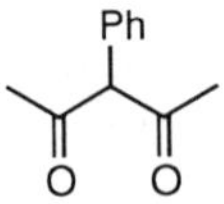   (d)

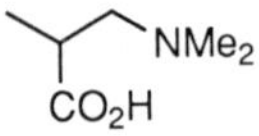

(e) 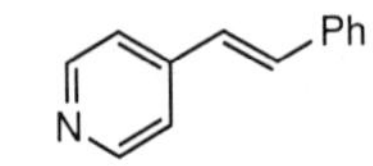   (f)

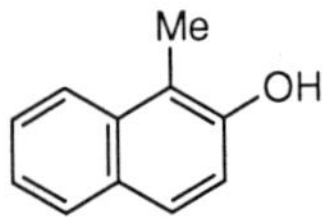

(g)

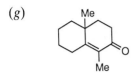

3. Formulate mechanisms for the following reactions:

(i)

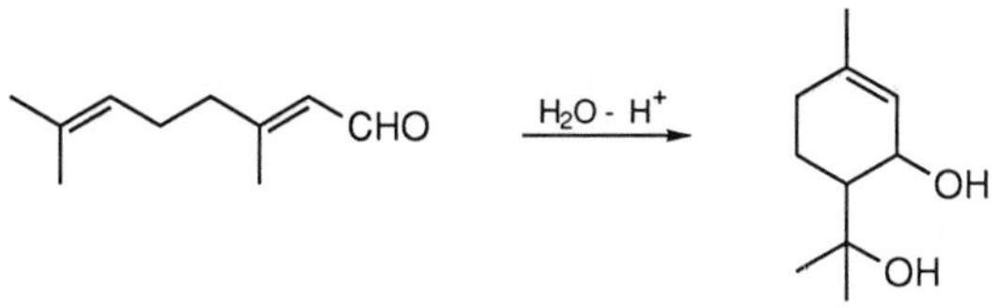

(ii)

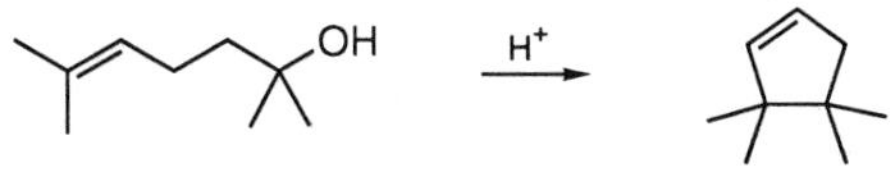

(iii)

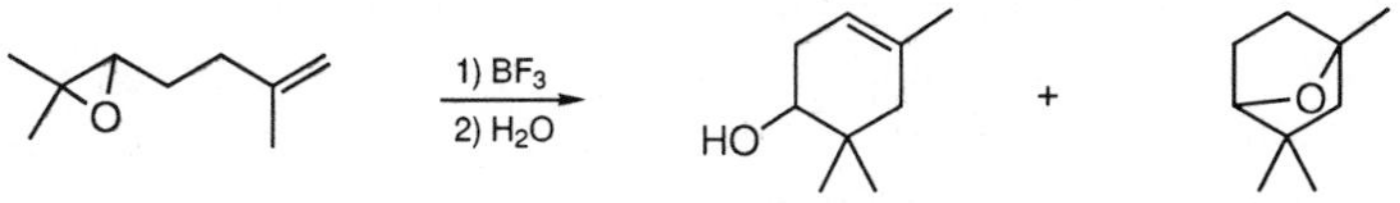

9 Pericyclic reactions

9.1
Cycloadditions (*a*)   *Diels–Alder reaction*

The reaction between butadiene and ethylene

is the simplest example of a general procedure for forming six-membered rings, developed by Diels and Alder, in which a conjugated diene reacts with a compound containing a C=C or C≡C bond (a *dienophile*). The reactions are described as $[_{\pi}4 + {}_{\pi}2]$ cycloadditions, denoting the numbers of interacting π-electrons in each component. They are normally brought about by heating the components alone or in an inert solvent, the temperature required depending on the structures of the reactants.

Diels–Alder reactions have large negative entropies of activation, reflecting not only the loss of translational entropy when two molecules come together but also the high degree of ordering which corresponds to the mutual orientation of four atomic centres. However, the activation enthalpy is often small, so that rates of reaction are large even at moderate temperatures. The reverse reaction can often be brought about at high temperatures (i.e. as the $T\Delta S$ term becomes increasingly important).

A brief account of these reactions in terms of frontier-orbital analysis was given earlier (section 4.8). A more detailed treatment of the interaction between the highest-occupied molecular orbital (HOMO) of one reactant and the lowest-unoccupied molecular orbital (LUMO) of the other will be given here.

(*i*)   *The dienophile.*   Although the Diels–Alder reaction between butadiene and ethylene is successful (in low yield), it is generally unsatisfactory, or fails completely, with other simple alkenes or alkynes. However, when the unsaturated bond of the dienophile is conjugated to a group of $-M$ type, such as carbonyl, nitro, or cyano, reaction occurs under milder conditions and normally gives good yields. It is thought that this is because the substituent lowers the energy of the LUMO of the dienophile, thereby bringing it closer in energy to the HOMO of the diene and increasing the bonding interaction in the transition state. Sometimes greater reactivity can be achieved by carrying out the reaction in the presence of a Lewis acid such as BF_3, which complexes with the $-M$ substituent and so serves to withdraw electrons even more strongly; such reactions can occur at room temperature or below. The following reactions of butadiene with various dienophiles are illustrative:

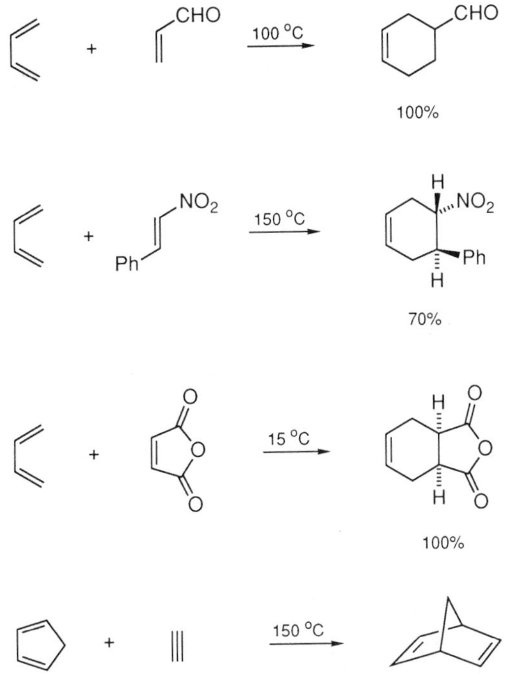

(*ii*) *The diene.* In order for reaction to occur, the diene must be capable of achieving the *s-cis* conformation which the formation of a six-membered ring requires. This is always possible for acyclic dienes but it can result in wide variations in reactivity. For example, *cis*-1-substituted butadienes are less reactive than their *trans*-isomers because the substituent in the former suffers steric crowding in the *s-cis* conformation which raises the activation energy:

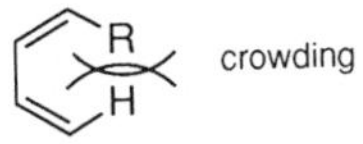

Cyclic dienes react only if they are of *cis* type; for example,

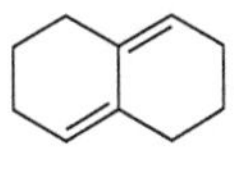

fails to react.

Just as the compatibility of the energies of the dienophile's LUMO and the diene's HOMO and thereby the reactivity, is increased by a $-M$ substituent in the former, so it is also increased by an electron-releasing group in the latter. Conversely, if the diene contains an electron-withdrawing substituent, the dienophile requires an electron-releasing one for ready reaction; in this case, the interaction is between the diene's LUMO and the dienophile's HOMO (e.g. p. 274).

Aromatic compounds are effective dienes providing that the loss of aromatic stabilization energy is not too great. For example, although benzene is inert, anthracene reacts with maleic anhydride at 80°C,

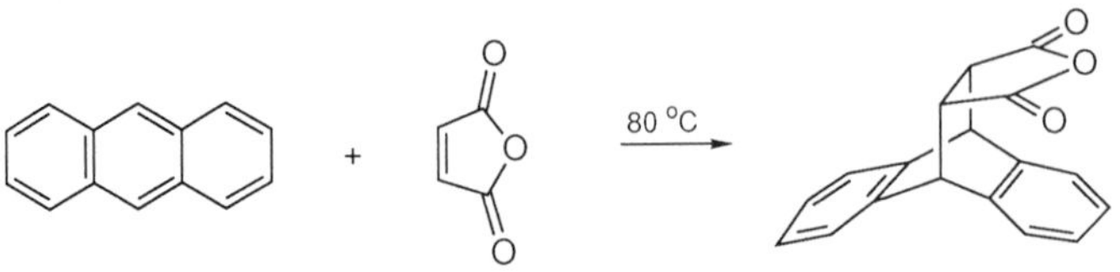

the difference reflecting the fact that whereas a reaction with benzene would result in the loss of $150 \, \text{kJ mol}^{-1}$ of stabilization energy, only $(349 - (2 \times 150))$, i.e. about $50 \, \text{kJ mol}^{-1}$ is lost on addition to anthracene.

Amongst monocyclic heteroaromatic compounds, furan is reactive, e.g.

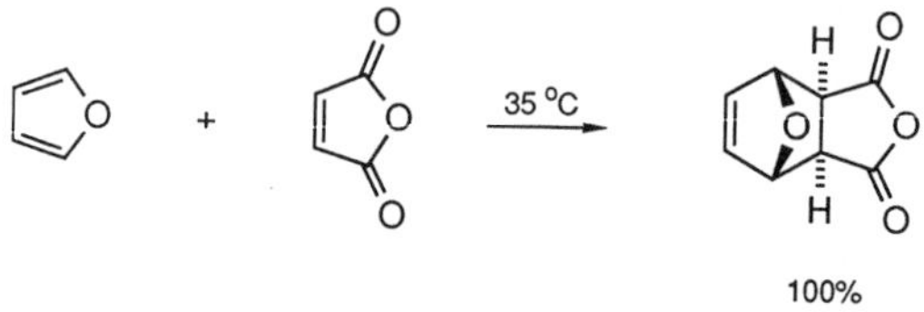

100%

whereas the more stabilized aromatic, thiophen, is inert. Pyrrole reacts with maleic anhydride, but not in the Diels–Alder manner (p. 358).

(*iii*) *Stereochemistry.* The cycloadditions are stereospecific. For example, dimethyl maleate and dimethyl fumarate react with butadiene to give, respectively, *cis*- and *trans*-dimethyl cyclohexene-4,5-dicarboxylates:

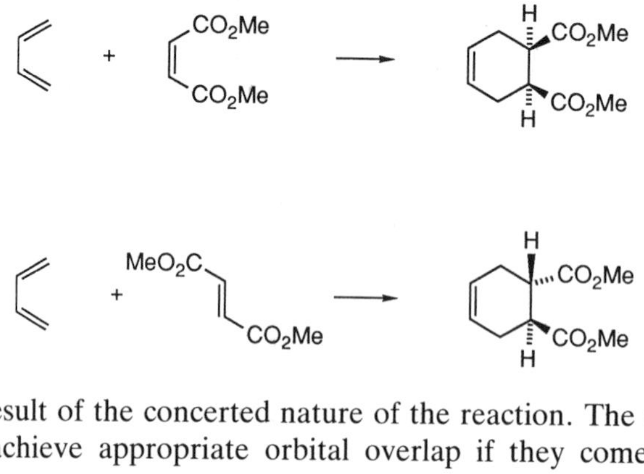

This is a result of the concerted nature of the reaction. The diene and dienophile can only achieve appropriate orbital overlap if they come together with their molecular planes (nearly) parallel; the orbital interaction is suprafacial with respect to both reactants:

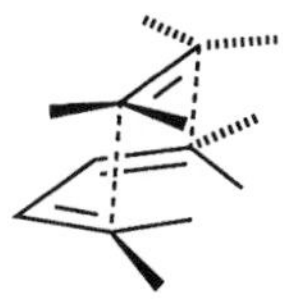

A further stereochemical factor applies with suitably substituted reactants. For example, in the dimerization of cyclopentadiene, where one molecule acts as diene and the other as dienophile, two orientations in the product are possible, each of which arises from suprafacial-suprafacial interaction of the reactants:

*endo** *exo**

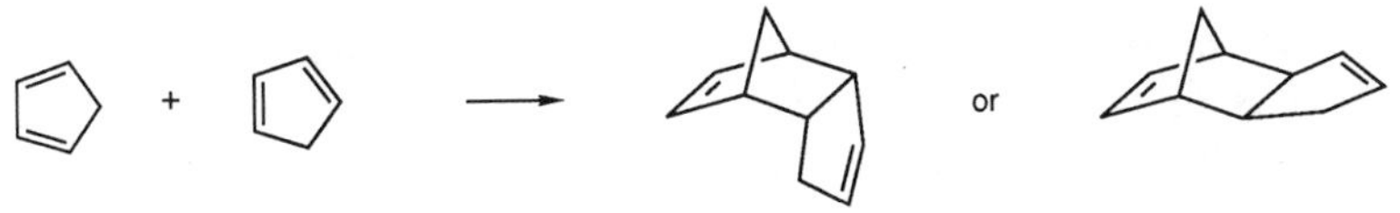

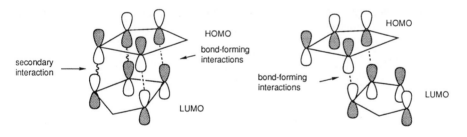

In practice, in this and most other cases the thermodynamically less stable *endo* product predominates. This has been accounted for in terms of a secondary interaction of the HO and LU orbitals in the transition state for *endo* addition which is sterically impossible for *exo* addition:

transition state for *endo* addition

transition state for *exo* addition

This secondary interaction does not lead to bonding in the product, but, because the overlapping orbitals are of like phase, it can stabilize the transition state.

Since, usually, the *exo* product is thermodynamically more stable, it is sometimes possible to form it by the use of conditions under which formation of the kinetically favoured adduct is reversible. The reaction of furan with maleimide is an example:

* The term *endo* indicates that the substituent is directed to the *inside* of the boat-shaped cyclohexene ring; *exo* indicates that the substituent is directed to the *outside* of the ring.

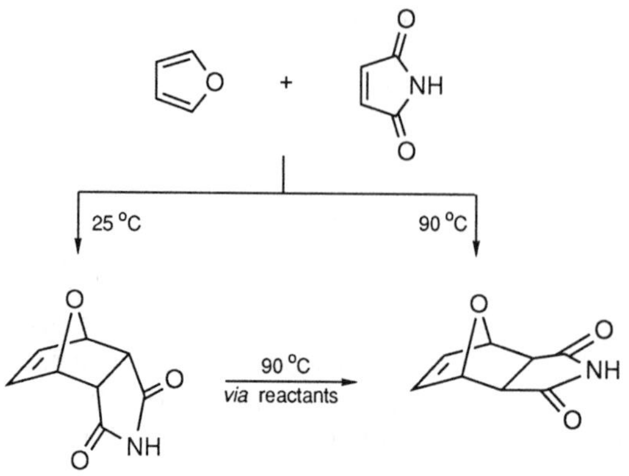

The less stable *endo* isomer is formed the faster and predominates at 25°C, where the reaction is effectively irreversible. At 90°C, this product is in fairly rapid equilibrium with the reactants, and the less rapidly formed but more stable *exo* isomer gradually accumulates.

Maleic anhydride and furan give the *exo*-adduct even at low temperatures (p. 272). This is because, although the *endo*-adduct is formed about 500 times faster than the *exo*-isomer, it reverts to reactants about 10000 times faster and *exo*-product accumulates.

(iv) Regiochemistry. When both diene and dienophile contain substituents, more than one product can be formed, e.g.

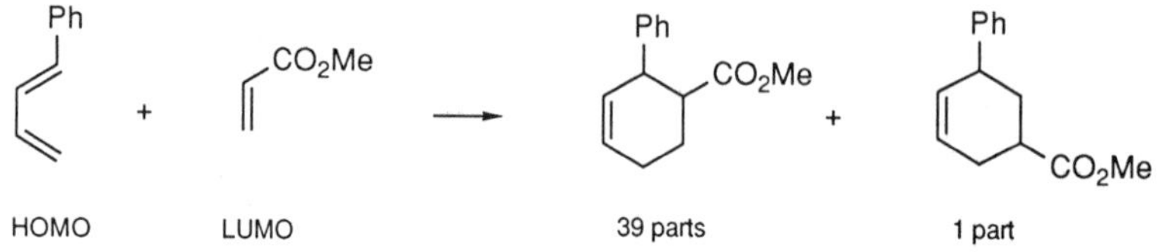

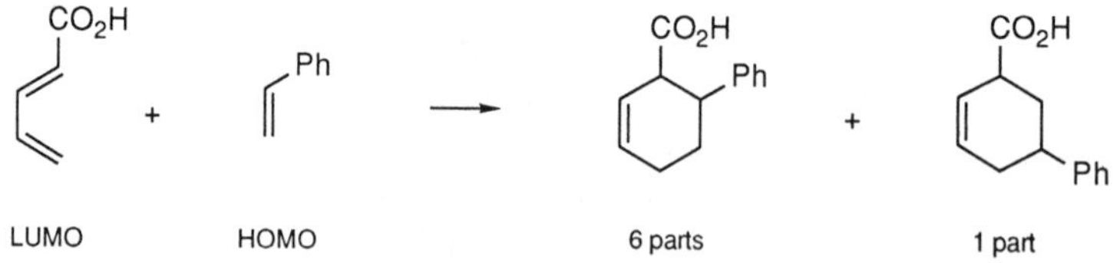

It is believed that 1,2-products predominate in these cases because bonding at the transition state is not only more effective when the HOMO of one reactant and the LUMO of the other are more closely matched in *energy* (whence the pairings noted above and as described earlier) but also when the *sizes* of the orbitals at the reacting termini are the more closely matched, e.g.

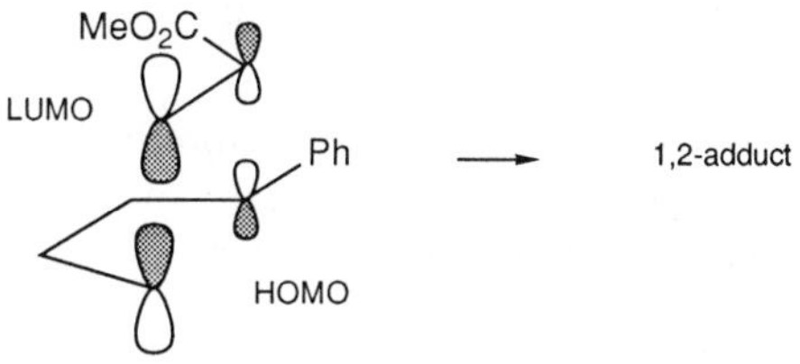

LUMO

MeO$_2$C

Ph

HOMO

⟶   1,2-adduct

is a better match than the alternative that leads to the 1,3-adduct.

Regioselectivity is particularly marked with butadiene containing a 2-OSiMe$_3$ substituent, e.g.

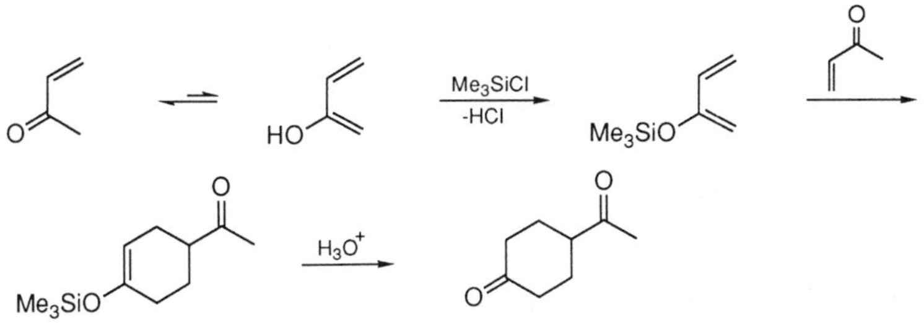

HO

Me$_3$SiCl
-HCl

Me$_3$SiO

Me$_3$SiO

H$_3$O$^+$

A modification of this reaction, employing *Danishefsky's diene*, enables αβ-unsaturated ketones to be formed, e.g.

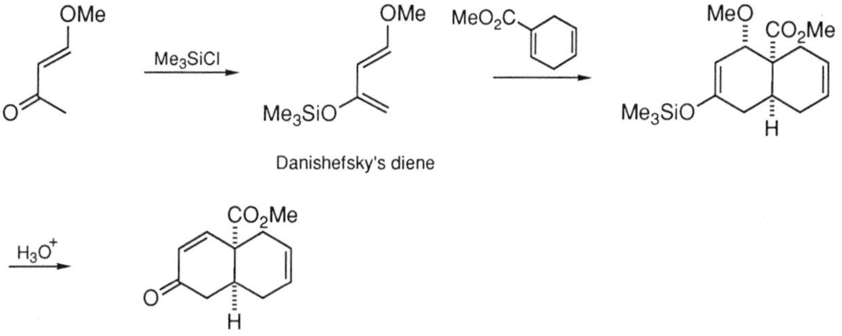

OMe

Me$_3$SiCl

OMe

Me$_3$SiO

Danishefsky's diene

MeO$_2$C

MeO   CO$_2$Me

Me$_3$SiO   H

H$_3$O$^+$

CO$_2$Me

O   H

(v)   Applications.   The Diels–Alder reaction is of very great importance in the synthesis of naturally occurring compounds, partly because of the large number of structures to which it is applicable and partly because of its stereospecificity. By the appropriate choice of starting materials, it is often possible in one step to assemble a product with several groupings in the required stereochemical configuration. Some examples are:

(1)   The reaction

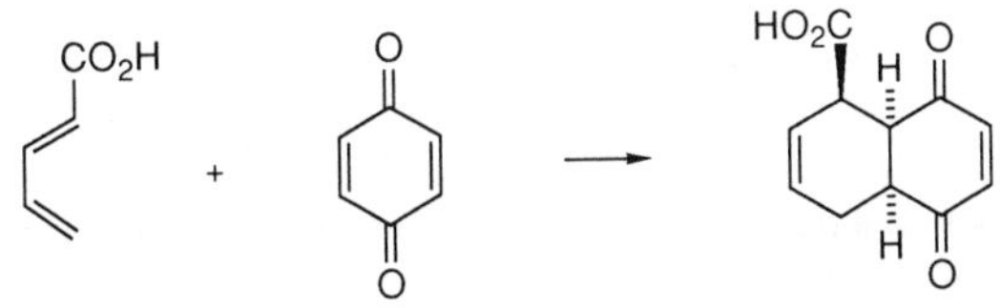

CO$_2$H

+

HO$_2$C   O
H

H   O

was employed in the synthesis of the alkaloid reserpine to assemble three stereocentres (p. 731).

(2)   The reaction between furan and dimethyl acetylenedicarboxylate, in a synthesis of cantharidin (from the body fluid of cantharides beetles such as the Spanish fly). Catalytic hydrogenation of the product gave a new dienophile which

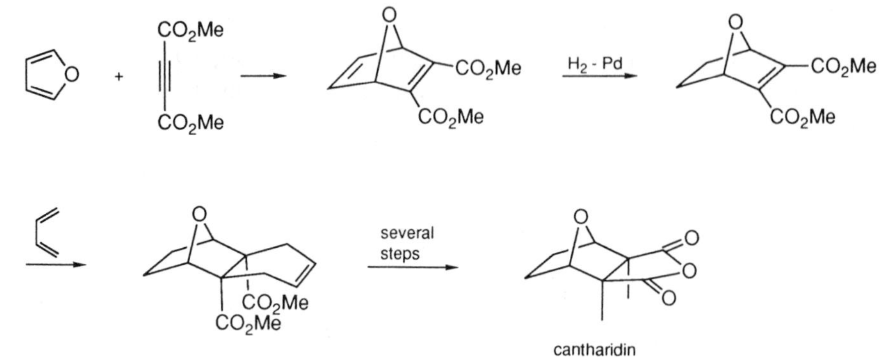

cantharidin

reacted with butadiene in a second Diels–Alder reaction.* The required stereochemistry having been established, the ester groups were converted into methyl and the ring derived from butadiene was oxidized in several steps to the required anhydride.

(3)   Intramolecular Diels–Alder reactions can be used to assemble bicyclic compounds, e.g.

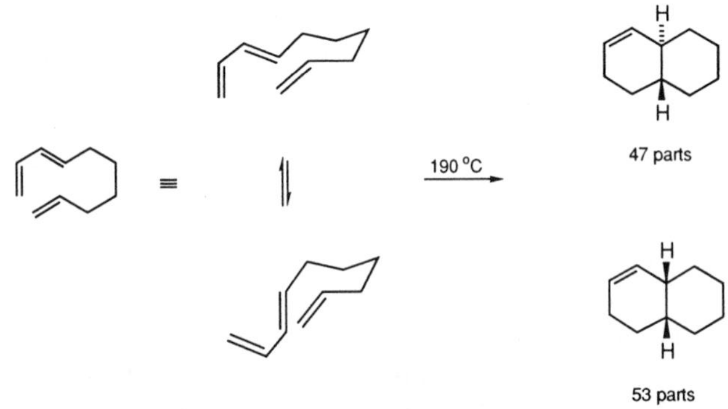

47 parts

53 parts

The activating effect of a carbonyl group conjugated to the dienophile is shown by

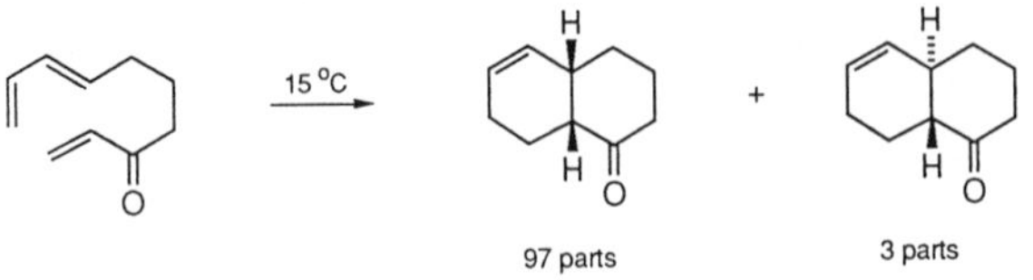

97 parts   3 parts

* The stereochemistry of this reaction results from the fact that addition of butadiene is less hindered by the oxide bridge than by the ethylene bridge.

(*vi*) *Enantioselective synthesis*. Although stereocentres can be created in Diels–Alder reactions (e.g. three in example (1) above), racemic mixtures are obtained unless one or both reactants is asymmetric. An approach being developed to induce enantioselectivity is the use of a chiral catalyst. For example, the optically active alkyldichloroborane (1) complexes with (*E*)-methyl crotonate to give (2). The approach of cyclopentadiene from the rear face of the dienophile is strongly hindered by the naphthyl group and attack from the front gives product in almost optically pure state:

(1) (2)

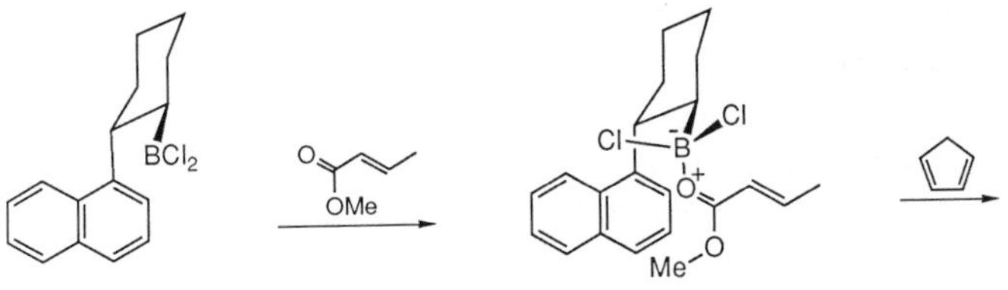

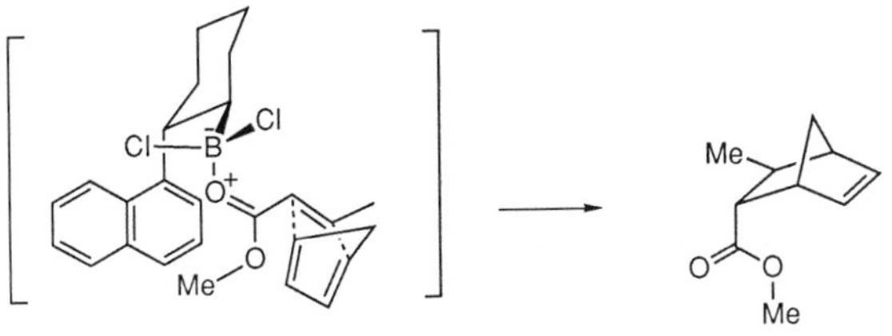

(*b*) *1,3-Dipolar additions*

A large number of types of five-membered ring may be formed by the reaction of '1,3-dipolar' compounds with unsaturated bonds (dipolarophiles) such as C=C, C≡C, C=O, and C≡N. The 1,3-dipolar components are compounds whose representation requires ionic structures which include ones with charges on atoms bearing a 1,3-relationship, e.g.

diazomethane	$CH_2{=}\overset{+}{N}{=}\overset{-}{N}$	↔	$\overset{-}{C}H_2{-}\overset{+}{N}{\equiv}N$
benzonitrile oxide	$Ph{-}C{\equiv}\overset{+}{N}{-}\overset{-}{O}$	↔	$Ph{-}\overset{-}{C}{=}\overset{+}{N}{=}O$

The 1,3-dipole is a structural variant of the diene component in the Diels–Alder reaction; in the dipolar compound, four π-electrons are distributed over three atoms instead of the four in a diene. Moreover, the HOMO and LUMO of a 1,3-dipole are of similar symmetry to those of a diene with respect to the two-fold axis and to the mirror plane which bisects the molecule:

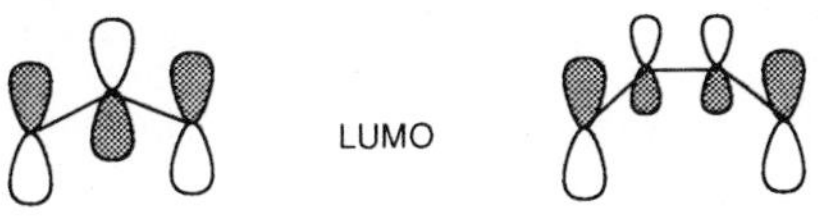

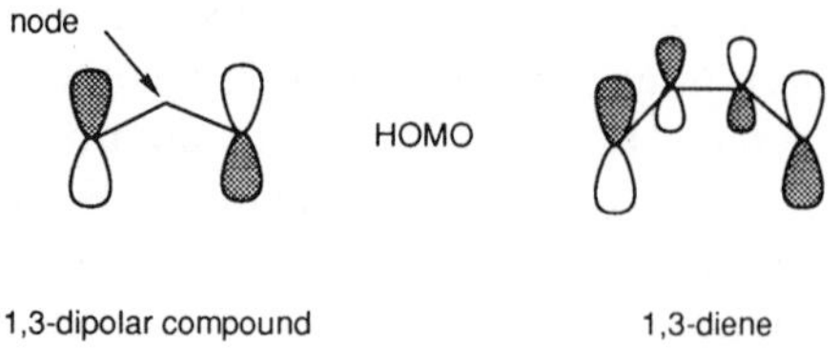

1,3-dipolar compound   1,3-diene

Consequently, concerted cycloaddition to an alkene is symmetry-allowed.

Unlike dienes which are stable, isolable molecules, 1,3-dipoles are often short-lived species which must necessarily be generated *in situ* and trapped as soon as they are formed; even those which are isolable, such as diazoalkanes, $R—\bar{C}H—\overset{+}{N}\equiv N$, and azides, $R—\bar{N}—\overset{+}{N}\equiv N$, often tend to be thermally labile materials, best handled in solution and used immediately after synthesis.

The following reactions illustrate those which are used for preparing 1,3-dipolar compounds *in situ*:

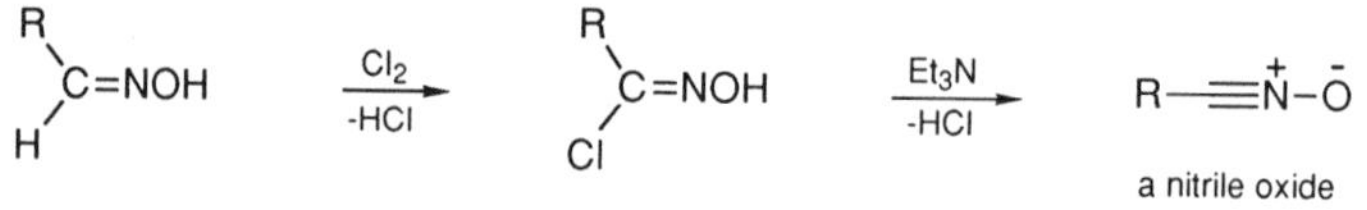

a nitrile oxide

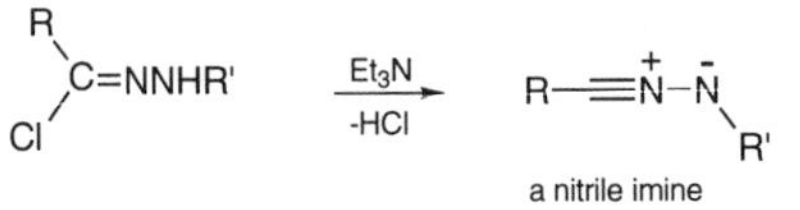

a nitrile imine

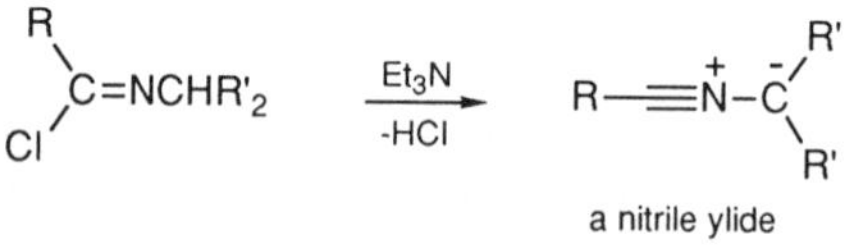

a nitrile ylide

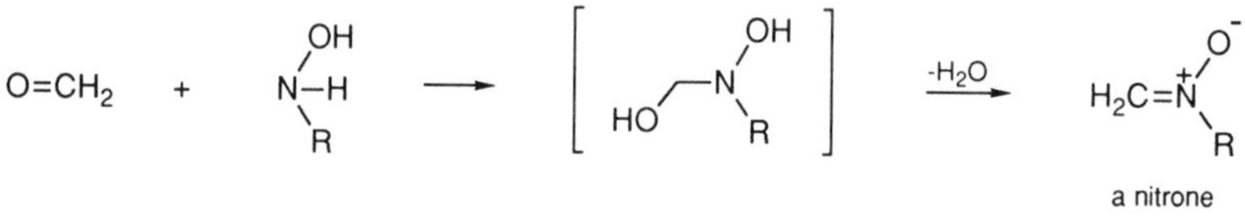

a nitrone

(*i*) *Applications.* A variety of heterocyclic compounds can be constructed by reactions of 1,3-dipolar compounds with alkenes and alkynes. Some examples are:

(1) Diazoalkanes yield pyrazole derivatives, e.g.

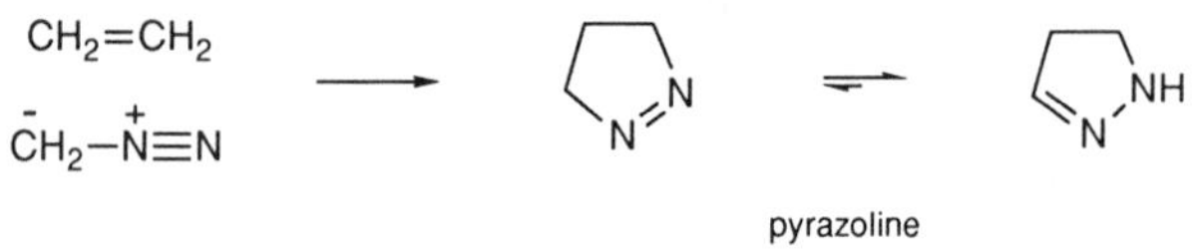

pyrazoline

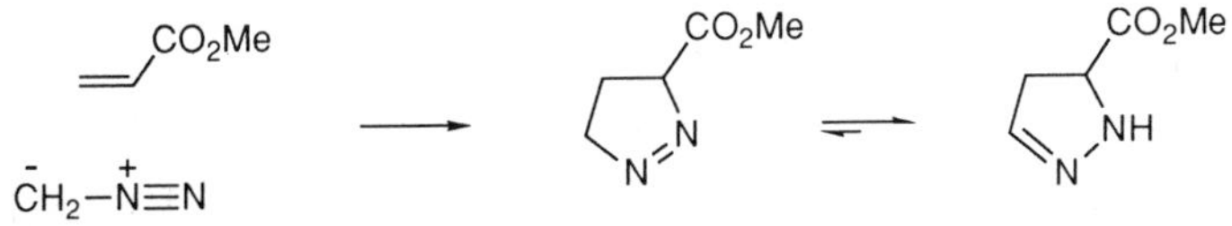

(2) Azides yield triazole derivatives,

(3) Nitrile oxides yield isoxazole derivatives,

(4) Nitrile ylides yield pyrrole derivatives,

The dipolarophile may also be a carbonyl compound or a nitrile. In these cases, there are two possible orientations for the addition, e.g.

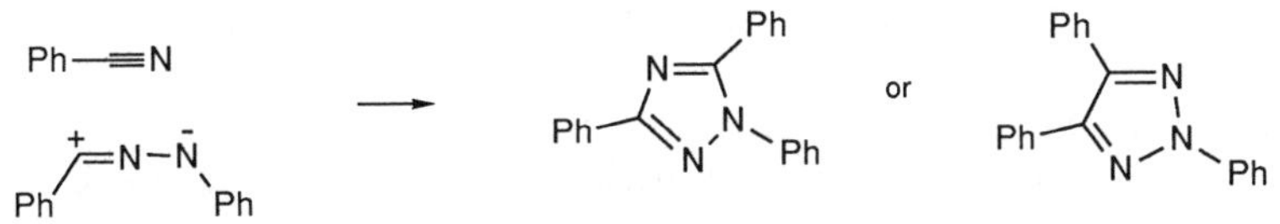

One product usually predominates: in the above example it is the first of the two. This and related results can be accounted for by considering the energies of the bonds being formed. For example, in the first product above, two C—N bonds are formed, whereas in the second, one C—C and one N—N bond are formed; the first combination corresponds to the greater bonding energy and this is reflected in the preceding transition states.

The value of 1,3-dipolar cycloaddition in the synthesis of natural products is now well recognized. An elegant example, the synthesis of (racemic) luciduline (an alkaloid whose dextrorotatory enantiomer occurs in *Lycopodium lucidulum*), made use of both 1,3-dipolar and Diels–Alder cycloaddition:

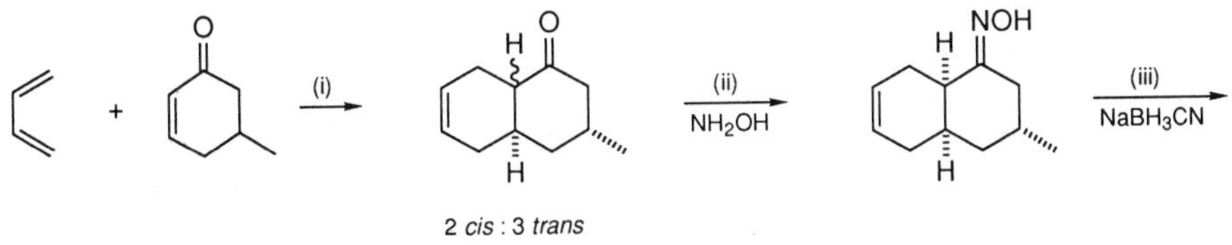

2 *cis* : 3 *trans*

luciduline

The Diels–Alder reaction – step (i) – was carried out at room temperature in the presence of tin(IV) chloride. This results initially in the formation of a *cis*-ring junction and also in a *cis* relationship between the hydrogen atoms at the ring junction and the methyl substituent (the latter because the suprafacial addition of the diene is to the less hindered face of the dienophile). However, the bridgehead hydrogen adjacent to the carbonyl group undergoes acid-catalyzed enolization and the major product was the (unwanted) *trans*-isomer. It was found that the *cis*-isomer reacted the faster of the two with hydroxylamine in step (ii), and by conducting this reaction at high pH so as to effect rapid *cis-trans* isomerization *via* the enolate, it was possible to obtain almost entirely the required *cis*-ring-fused oxime. This was reduced with sodium cyanoborohydride in methanol: this is a hydride-transfer agent (p. 651) which delivered H⁻ from the less hindered side of the oxime, to create a hydroxylamine function stereoselectively as shown. The nitrone, generated in step (iv) with paraformaldehyde in hot toluene, underwent a spontaneous cycloaddition to give a bridged oxazolidine. In step (v) this was methylated with methyl fluorosulfonate and the resulting salt was reduced in step (vi) with lithium aluminium hydride (p. 651). In the final step, the secondary alcohol was oxidized to carbonyl with chromium(VI) oxide in acetone.

(c) [$_\pi$2 + $_\pi$2] Cycloadditions

As described in chapter 4, cycloaddition of alkenes to form cyclobutanes is a symmetry-forbidden process when the geometry of the transition state requires suprafacial-suprafacial interaction of the orbitals.

However, alkenes do combine in the $[_\pi 2 + _\pi 2]$ manner photochemically. The simplest interpretation of this is as follows. When an alkene absorbs a quantum of light, an electron is promoted from the HOMO to the LUMO, so that there are now two singly occupied molecular orbitals (SOMO). The highest occupied orbital now therefore has the symmetry of the LUMO and consequently can interact with the LUMO of a second molecule of the alkene which is in its ground state to give a cyclobutane in its first excited state; this loses energy and returns to the ground state by fluorescence or collisional deactivation.

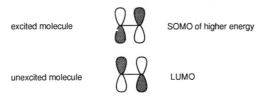

These photochemical reactions are considered in more detail in chapter 16.

The formation of cyclobutanes from alkenes also occurs in two circumstances by routes which are neither photochemical nor concerted, but involve thermally generated diradicals. First, alkenes with radical-stabilizing substituents react on heating to give head-to-head adducts, e.g.

Second, highly strained cycloalkenes dimerize, sometimes spontaneously, e.g.

(*i*) *Cumulenes.* Cumulenes – compounds which contain adjacent double bonds, such as allene ($CH_2{=}C{=}CH_2$), ketene ($CH_2{=}C{=}O$), and isocyanates ($RN{=}C{=}O$) – participate readily in $[_\pi 2 + _\pi 2]$ cycloadditions which have been proved in a number of cases to be concerted. For example, ketene reacts with dienes to give [2 + 2] cycloadducts in a concerted reaction, to the exclusion of the formation of [4 + 2] adducts.

From the earlier analysis of the $[_\pi 2 + _\pi 2]$ cycloaddition of alkenes, $[_\pi 2 + _\pi 2]$ addition for ketenes is forbidden when the geometry of approach is suprafacial-

suprafacial. However, consider the alternative suprafacial-antarafacial interaction. The efficiency of bonding interaction is very low unless one of the components undergoes twisting about its original π-bond, and this is the constraint which makes a transition state derived from this array inaccessible for alkenes. On the other hand, if the component whose π orbital interacts in an antarafacial manner is a cumulene, one of the carbons involved is *sp*-hybridized. This reduces the geometric constraint but, more importantly, makes available another orbital whose participation stabilizes the transition state. This further orbital is the LUMO of the adjacent double bond. This is very low-lying, in ketene especially, and its participation accounts for the preference by ketene of $[_\pi2 + _\pi2]$ over $[_\pi4 + _\pi2]$ cycloaddition.

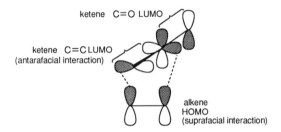

The propensity for $[_\pi2 + _\pi2]$ cycloaddition amongst cumulenes causes them readily to dimerize. Substituted ketenes form cyclobutanediones,

$$R_2C=C=O$$
$$O=C=CR_2$$

but ketene itself dimerizes in the liquid phase to a β-lactone, the $[2 + 2]$ addition involving the carbonyl group of one component:

$$H_2C=C=O$$
$$H_2C=C=O$$

This dimerization forms part of the industrial synthesis of acetoacetic ester; the ketene is obtained by the pyrolysis of acetone, and the dimer is converted into acetoacetic ester by reaction with ethanol, as shown.

Other examples of dimerizations of cumulenes and of $[2 + 2]$ cycloadditions involving cumulenes are the following:

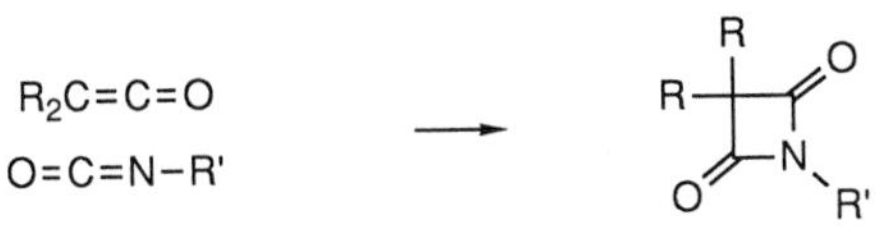

$$S=C=CR_2$$
$$R_2C=C=S$$ ⟶ (four-membered dithietane ring with =CR_2 and R_2C)

$$R_2C=C=O$$
$$O=C=N-R'$$ ⟶ (β-lactam ring, R, R, O, N-R')

$$H_2C=CH-CN$$
$$H_2C=C=CH_2$$ ⟶ (methylenecyclobutane with CN)

Ketenes also undergo $[_\pi2 + _\pi2]$ cycloadditions with ketones, catalyzed by Lewis acids, e.g.

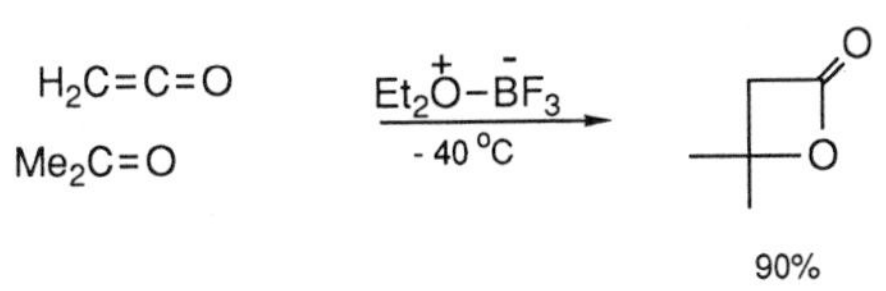

$$H_2C=C=O$$
$$Me_2C=O$$ $\xrightarrow[-40\,°C]{Et_2\overset{+}{O}-\overset{-}{B}F_3}$ (β-lactone)

90%

and Lewis acids can also divert $[_\pi4 + _\pi2]$ cycloadditions to $[_\pi2 + _\pi2]$ reactions, e.g.

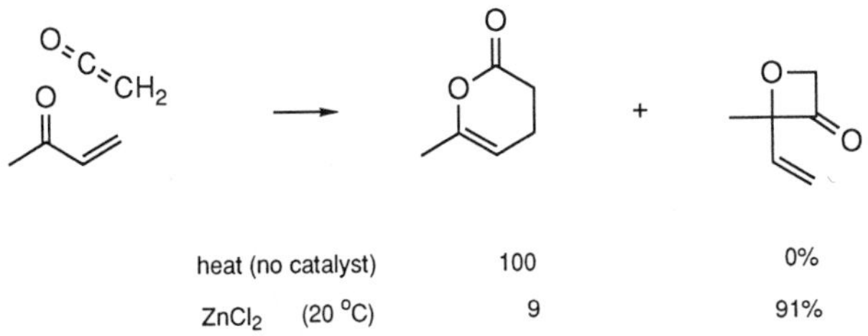

(scheme) ⟶ (dihydropyranone) + (β-lactone)

	100	0%
heat (no catalyst)	100	0%
ZnCl$_2$ (20 °C)	9	91%

Finally, the intramolecular cyclization of diketenes forms the basis of a method for making large-ring ketones, e.g.

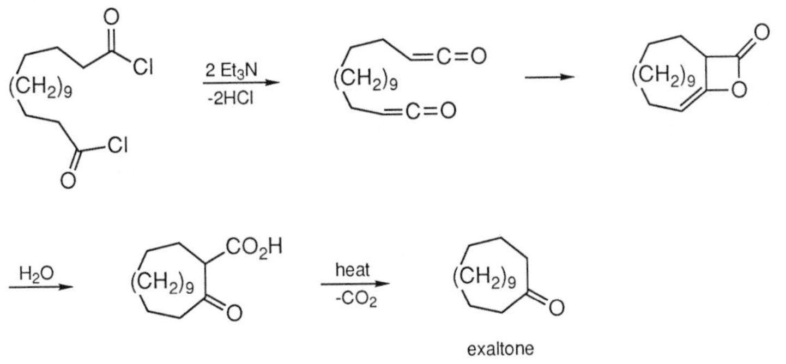

(diacid chloride, (CH$_2$)$_9$) $\xrightarrow[-2HCl]{2\ Et_3N}$ (bis-ketene, (CH$_2$)$_9$) ⟶ (fused ring ketone, (CH$_2$)$_9$)

$\xrightarrow{H_2O}$ (CH$_2$)$_9$ ring with CO$_2$H and O $\xrightarrow[-CO_2]{heat}$ (CH$_2$)$_9$ ring ketone

exaltone

Yields are lower than in the Ziegler process from dinitriles (p. 229) but the starting materials are more readily accessible.

Electrocyclic reactions are those in which either a ring is formed with the generation of a new σ-bond and the loss of a π-bond or a ring is broken with the opposite consequence, e.g.

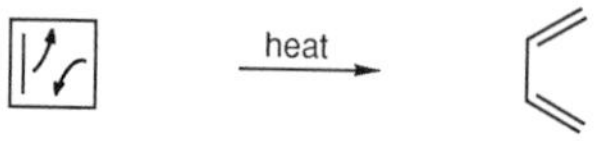

The reactions are stereospecific. For example, *cis*-3,4-dimethylcyclobutene gives solely *cis,trans*-2,4-hexadiene, which is not the thermodynamically most stable isomer:

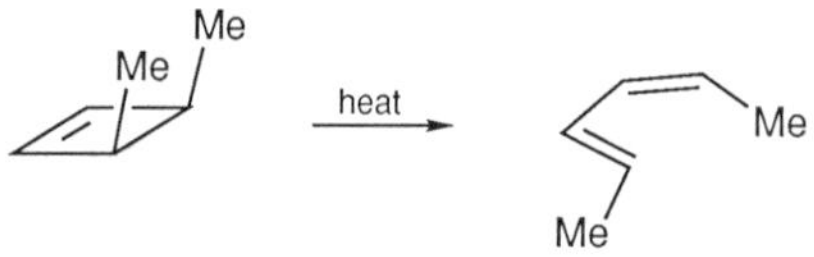

Considerations of orbital symmetry again offer an explanation. In the transition state for ring opening of a cyclobutene, the HOMO of the σ-bond that is undergoing fission interacts with the LUMO of the C=C bond. This can only happen if the σ-bond opens in a *conrotatory* manner:*

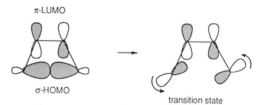

For the reverse reaction – the ring closure – precisely the opposite path is followed (principle of microscopic reversibility) so that conrotatory motion of the termini is likewise required.

Conrotatory ring opening of *trans*-3,4-dimethylcyclobutene could in principle lead to two products, depending upon the sense of the rotation: the *trans,trans*-dimethyl diene and the *cis,cis*-isomer. However, severe steric crowding would result from the conrotatory movement which turns both methyl groups inwards in the formation of the latter compound,

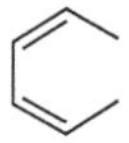

*This also follows if the σ-LUMO and the π-HOMO are considered.

so raising the activation energy relative to that for the *trans,trans*-isomer. Consequently, only the *trans,trans*-compound is formed.

In contrast, strongly electron-withdrawing groups preferentially rotate inward, e.g.

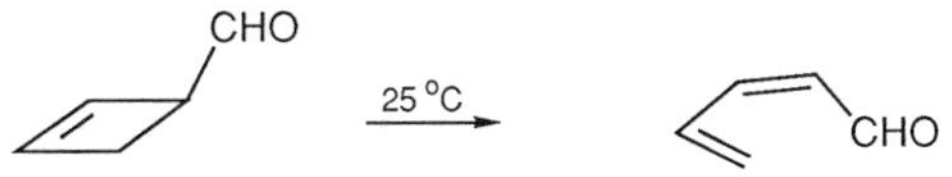

This can be exploited in synthesis, e.g.

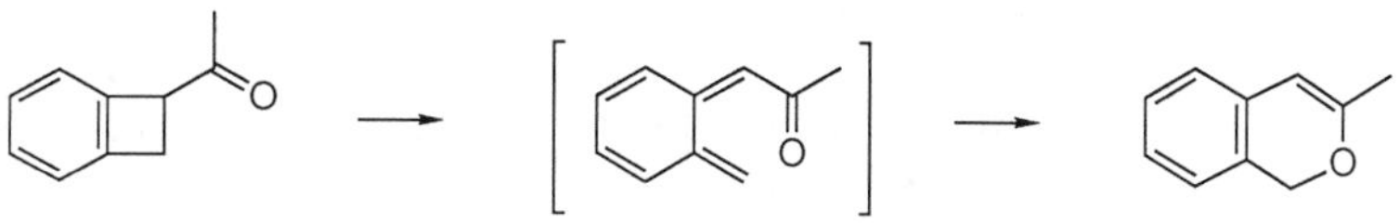

The theory is readily extended to polyenes. For example, the opening of a 1,3-cyclohexadiene, and likewise the ring closure, require *disrotatory* motion:

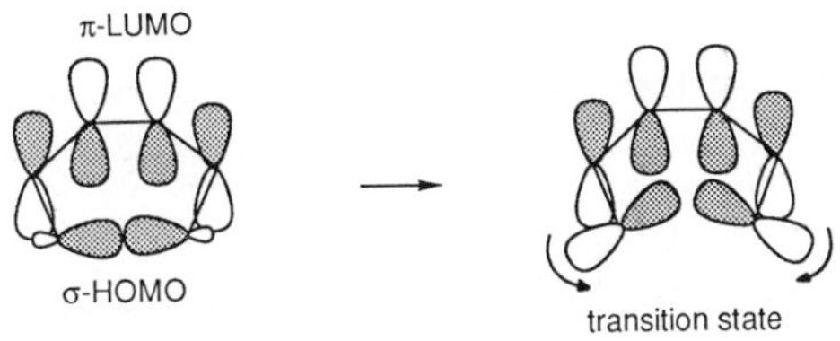

It can be understood why, for example, *trans,cis,trans*-2,4,6-octatriene gives specifically *cis*-5,6-dimethylcyclohexadiene:

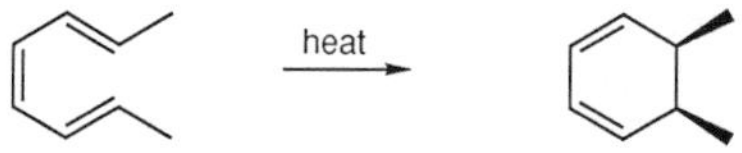

When the ring closures of dienes and trienes are carried out photochemically, the stereospecificity is the opposite to that of the thermal reaction: the symmetry of the SOMO of higher energy governs the reaction and the observed stereochemical course follows, e.g.

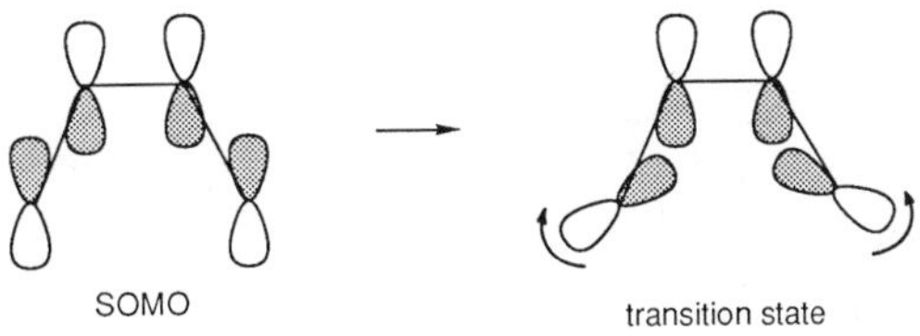

However, it is recognized that this is an oversimplified treatment: a full understanding requires examination of all the π orbitals of polyene and product, as well as of the σ orbital of the bond formed between the termini.

The stereochemistry of electrocyclic reactions is summarized in Table 9.1:

Table 9.1 Direction of motion of the termini in electrocyclic ring closure or the reverse ring opening

Number of π-electrons in ring closure reaction	Rotatory motion	
	Thermal reaction	Photochemical reaction
4	con	dis
6	dis	con

(*i*) *Applications.* The direction taken by an electrocyclic reaction depends on the relative stabilities of the ring and open-chain reactants. With cyclobutenes, the latter is favoured because of the strain in the ring, e.g.

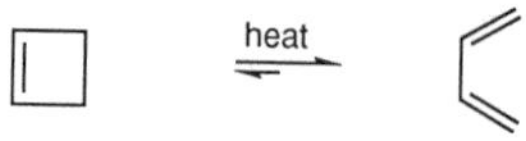

In contrast, with benzocyclobutenes the ring structure is favoured because the loss of aromatic stabilization energy on ring opening outweighs the strain in the ring, e.g.

The latter equilibrium opens up two synthetic routes. First benzocyclobutenes can be synthesized, e.g.

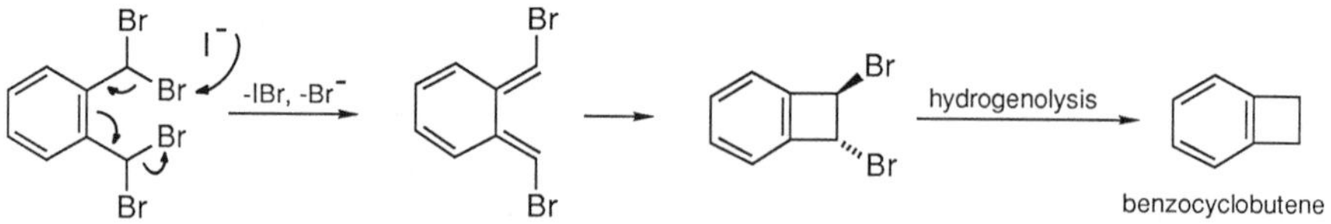

Second, a benzocyclobutene is a masked conjugated diene; a reagent that reacts only with the diene can trap it as it is formed. This principle was applied in an elegant steroid synthesis in which two of the rings were formed with a predetermined *trans* stereochemistry:

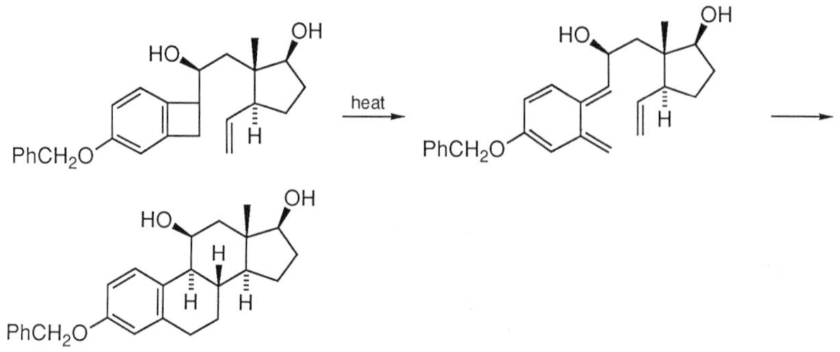

(*ii*)   *Valence tautomerism*.   Electrocyclic equilibria that occur at ambient temperature give rise to the phenomenon of valence tautomerism. For example, cyclooctatetraene is in equilibrium with *cis*-bicyclo[4.2.0]octa-2,4,7-triene, which it forms by the symmetry-allowed disrotatory ring closure:*

9.3 Cheletropic reactions

Cheletropic reactions are ones in which two σ-bonds which terminate at a single atom are made or broken in a concerted reaction, e.g.

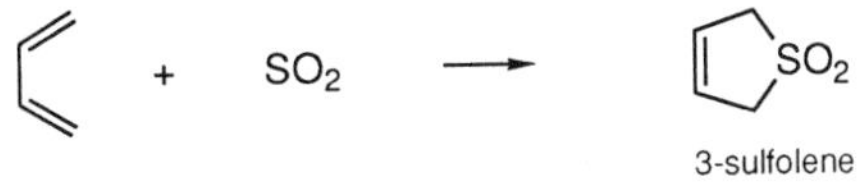

Cheletropic addition is related to Diels–Alder addition in that the 2π-electron system in the dienophile of the latter is replaced by an unshared pair of electrons on a single atom in the former.

(*a*)   *Cheletropic reactions of dienes*

The HOMO of a molecule like sulfur dioxide or carbon monoxide is that which has a lone-pair of electrons in the plane containing the atoms; the LUMO is a *p* orbital perpendicular to this plane:

For a symmetry-allowed cycloaddition of sulfur dioxide to a diene, the SO_2 lies in a plane which bisects the *s-cis* conformation of the diene, and the reaction is suprafacial with respect to both diene and SO_2:

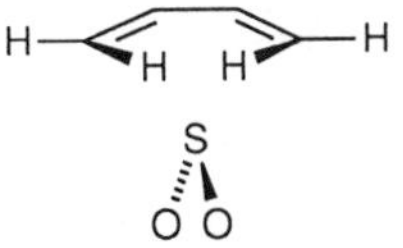

---

* Valence tautomerism can be studied by NMR spectroscopy. If equilibrium occurs slowly with respect to the NMR time scale, the spectrum of each tautomer is observed; if it is fast, a time-averaged spectrum of the two is observed. At intermediate rates, line-broadened spectra are obtained, analysis of which enables the rate of interconversion to be derived.

The appropriate matching of orbitals is:

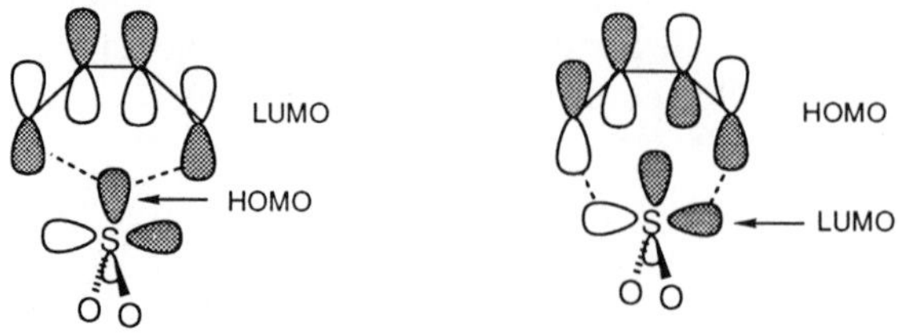

In the transition state, the terminal carbon atoms of the diene must move in the disrotatory manner so that the HOMO of SO_2 can interact with the LUMO of the diene, or the LUMO of SO_2 with the HOMO of the diene. The reality of this prediction of frontier-orbital theory is proved by the fact that *trans,trans*-1,4-disubstituted dienes give specifically the more crowded *cis*-substituted 3-sulfolenes, e.g.

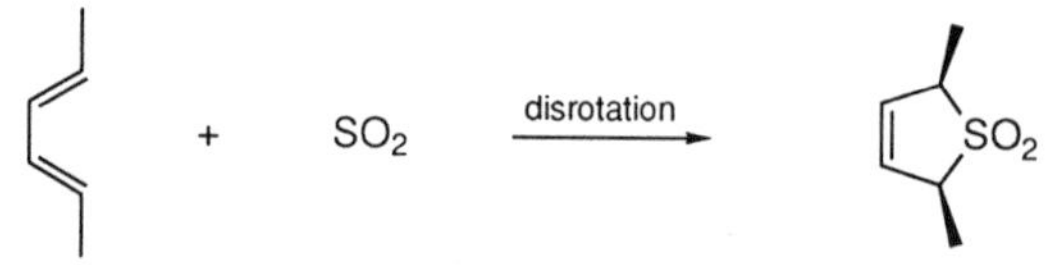

and *cis,trans*-disubstituted dienes give *trans*-substituted 3-sulfolenes.

As with electrocyclic reactions, the opposite stereochemistry is observed if the reaction is photochemical rather than thermochemical:

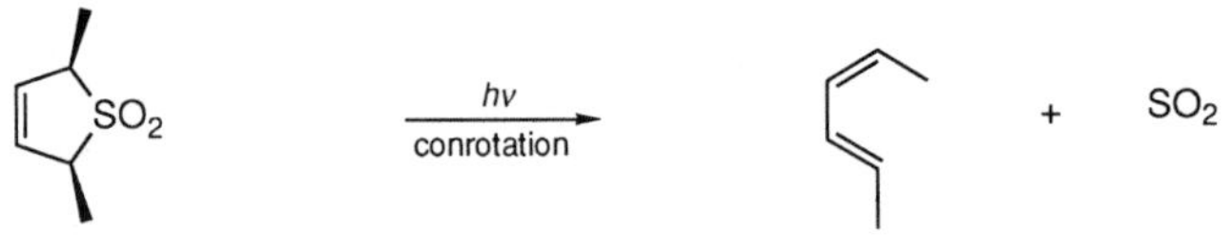

(*i*) *Applications.* The reaction between butadiene and sulfur dioxide and the reverse reaction which occurs at high temperatures, provide a useful method for 'carrying' butadiene, so avoiding the experimental inconvenience of handling gaseous butadiene at the elevated temperatures which may be needed for reaction. For example, when 3-sulfolene is heated in the presence of a dienophile, butadiene is released and immediately trapped in a Diels–Alder reaction.

There are few other reactions where a molecule like sulfur dioxide is sequestered by a diene, but there are others where the reverse process – extrusion of a, usually, stable molecule – occurs. For instance, Diels–Alder adducts of cyclopentadienones extrude carbon monoxide, e.g.

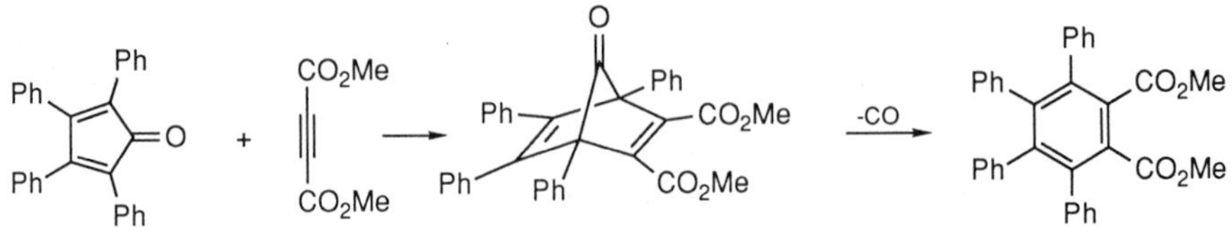

(b) Cheletropic reactions of trienes

2,7-Dihydrothiepin dioxides undergo thermolysis with a high degree of stereospecificity, e.g.

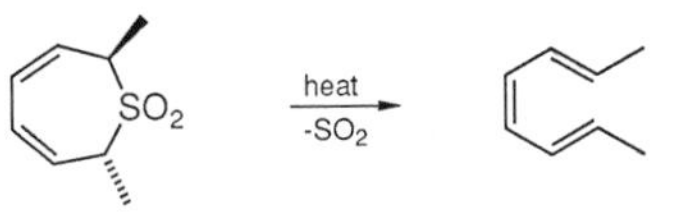

The reaction therefore occurs with a conrotatory motion of the triene termini. If the cheletropic reaction is linear, i.e. if the HOMO of SO_2 reacts suprafacially as it does in the case of dienes, the triene must react antarafacially, which is consistent with the observed conrotation:

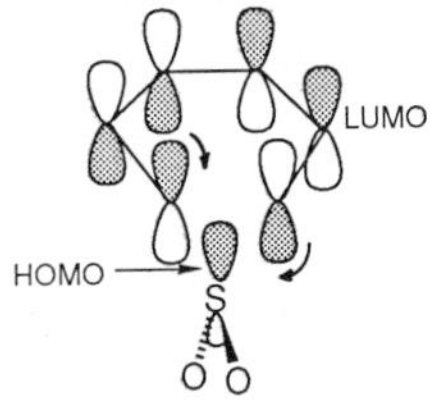

(c) Cheletropic reactions of simple alkenes

Cheletropic reaction of sulfur dioxide with alkenes is symmetry-forbidden if both components interact suprafacially. Antarafacial reaction of a simple alkene is sterically very unlikely (cf. discussion of $[_\pi 2 + _\pi 2]$ cycloaddition of alkenes), so that the reaction is likely to involve the SO_2 antarafacially. It is known in the stereospecific extrusion of SO_2 from episulfones, e.g.

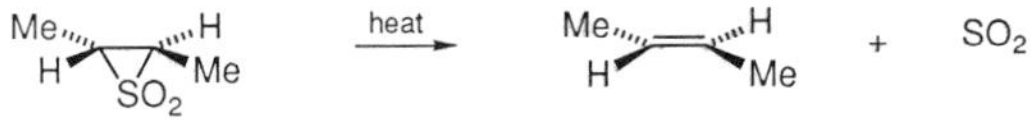

The orbital interaction in the transition state is represented by:

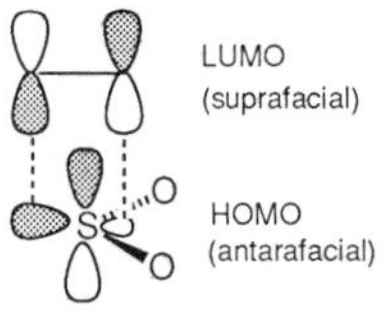

The reactions of singlet carbenes and nitrenes with alkenes are also of this type, e.g.

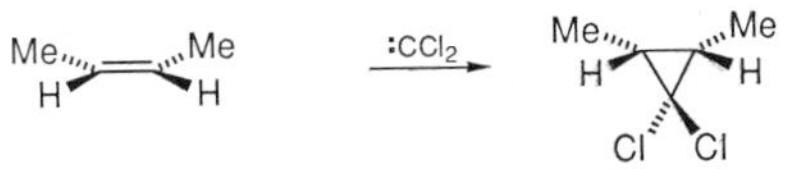

Only singlet carbenes and nitrenes behave in this way: the addition of triplet carbenes or nitrenes to a double bond results in a diradical which is sufficiently long-lived, during the time required for spin-inversion, for stereospecificity to be lost as a result of rotations about bonds.

The development of phase-transfer catalysis has greatly increased the synthetic utility of dichlorocarbene. A phase-transfer catalyst is usually a quaternary ammonium salt where the cation contains large alkyl groups which confer upon it solubility in organic solvents (e.g. $C_{16}H_{33}\overset{+}{N}Me_3$). Such a cation will transport hydroxide as its counter-ion into chloroform from aqueous solution. This greatly facilitates the reaction,

$$HO^- \;\; + \;\; H-CCl_3 \;\; \rightleftharpoons \;\; H_2O \;\; + \;\; {}^-CCl_3$$

which otherwise could only occur at the interface of the organic and aqueous phases. The trichloromethyl carbanion, CCl_3^-, is relatively short-lived, decomposing to give dichlorocarbene,

$$ {}^-CCl_3 \;\; \longrightarrow \;\; Cl^- \;\; + \;\; :CCl_2 $$

which is trapped by the alkene.

**9.4
Sigmatropic
rearrangements** A sigmatropic rearrangement is a pericyclic reaction which involves the migration of a σ-bonded atom or group within a π-electron system; overall, the numbers of π- and σ-bonds remain separately unchanged. The rearrangements can be divided into two classes: (1) those where the group which migrates is bonded through the same atom in both reactant and product, (2) those where the migrating group is bonded through different atoms in the reactant and the product. An example of the former is,

where hydrogen is transferred from the methylene group to the alternative terminus of the diene; the latter is exemplified by the Claisen rearrangement, e.g.

where the allyl group which migrates is bound by different atoms before and after the reaction. The first reaction is thermodynamically favourable because a dialkyl-substituted diene is more stable than a mono-alkylated one (p. 34) and the second reaction reflects the strength of the C=O bond.

These two reactions illustrate, respectively, a [1,5]-shift and a [3,3]-shift. The figures in parentheses denote the numbers of essential interacting centres in the two groups which are formed by breaking the migrating σ-bond.

(a) [1,j]-Sigmatropic rearrangements

The occurrence and stereochemistry of sigmatropic rearrangements can be accounted for, like other pericyclic reactions, in terms of the symmetry of frontier orbitals. Consider the [1,5]-shift of hydrogen as in the first example:

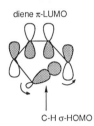

For a maximum of bonding to occur in the transition state when the HOMO (or LUMO) of the C—H σ-bond interacts with the LUMO (or HOMO) of the diene π-system, the hydrogen is transferred suprafacially. Since this arrangement is easily accessible geometrically, the [1,5]-shift of hydrogen in dienes readily occurs thermally.

Owing to the difference in symmetry between the LUMO of a simple alkene and that of a diene, a similar suprafacial [1,3]-transfer of hydrogen in a substituted alkene is symmetry-forbidden. A [1,3]-hydrogen transfer would be allowed if the π-bond were to interact antarafacially,

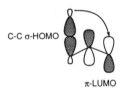

but the geometric constraint that this imposes is too great and concerted [1,3]-transfer of hydrogen is not observed. The stability of the triene,

derives from the fact that concerted thermal isomerization to toluene, which is thermodynamically much the more stable, is a symmetry-forbidden process.

On the other hand, a [1,7]-transfer of hydrogen in a triene, which must also be antarafacial if it is to be symmetry-allowed, is sterically feasible when the system can form a helical structure:

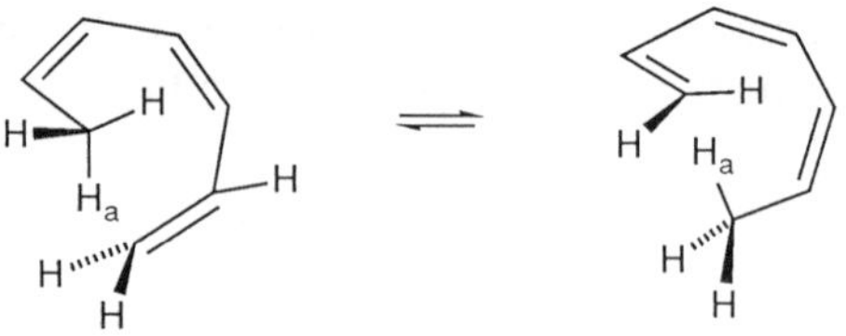

The classic example is the thermal interconversion of vitamin D and precalciferol:

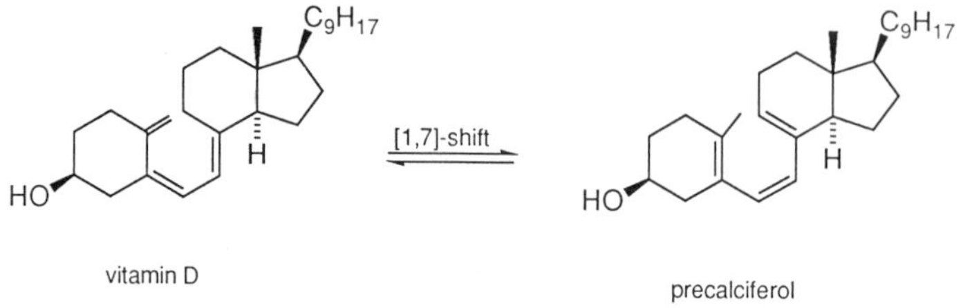

vitamin D

precalciferol

In sigmatropic shifts of hydrogen, only the polyene can participate in an antarafacial manner on account of the spherical symmetry of the hydrogen *s* orbital. With other functions such as alkyl groups, the migrating group can be the antarafacial component and this expands the possibility for reaction. Consider the [1,3]-shift of an alkyl group in an alkene:

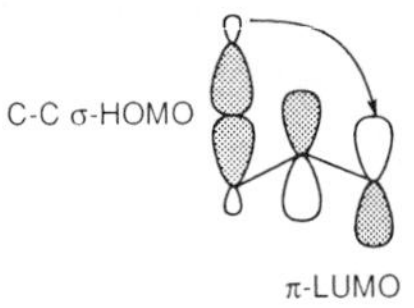

As shown, there is the possibility of an interaction which is antarafacial for the C—C σ-bond and suprafacial for the π-bond. Again, this is likely to require a transition state with a geometrically difficult access, particularly for flexible molecules, but this interaction is known in molecules whose rigid framework builds into the ground state of the reactant much of the ordering which would otherwise have to be achieved at the expense of an unfavourable entropy of activation in a flexible analogue. An example is:

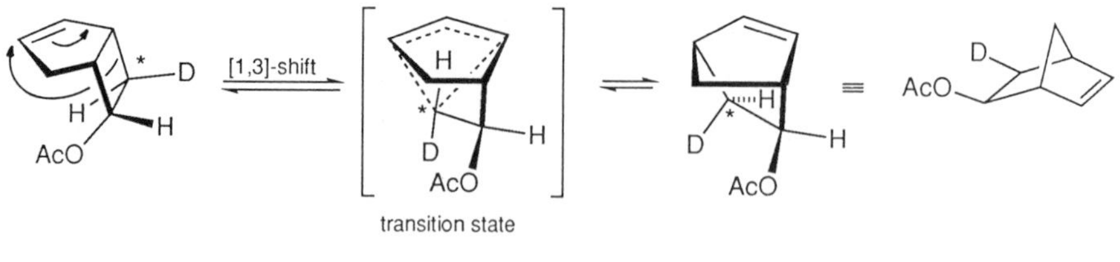

The antarafacial character of the interaction of the alkyl σ-bond is demonstrated by the fact that the configuration of the migrating carbon (marked *) is inverted in the change.

Alkyl migrations occur with carbocations and compounds with a good leaving group, X. Interaction of the HOMO of the migrating σ-bond with the LUMO of the empty *p* orbital or of the C—X σ-bond is consistent with the retention of configuration of the migrating group:

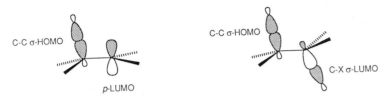

In contrast, the concerted [1,2]-shift in a carbanion is predicted to be symmetry-forbidden.* Such rearrangements do, however, occur but not by a concerted mechanism (p. 447).

(b) [i,j]-Sigmatropic rearrangements

The most frequently encountered sigmatropic rearrangements of this category are [3,3]-shifts for, as the notation implies, the transition state involves the stereochemically favourable six-membered ring.

The *Cope rearrangement* is the [3,3]-sigmatropic rearrangement of 1,5-dienes, e.g.

$$\text{R} \quad \xrightarrow{\text{165 - 185 }^{\circ}\text{C}} \quad \text{R}$$

The stereochemical outcome of the rearrangements of acyclic compounds is consistent with their occurrence through chair-shaped transition states. For example, *meso*-3,4-dimethyl-1,5-hexadiene gives *cis,trans*-2,6-octadiene,

whereas a boat-shaped transition state would give *trans,trans*- or *cis,cis*-product.

The valence tautomers of bullvalene undergo interconversion (between over a million identical forms) by Cope rearrangement, e.g.

---

* This is usually discussed in terms of a treatment of pericyclic reactions which is an alternative to the frontier-orbital method. The transition state for the [1,2]-shift of a carbanion would contain a cyclic array of 4π-electrons. This is an 'anti-aromatic', high-energy configuration, in contrast to the stable 'aromatic' arrangements in the [1,2]-shift of a carbocation (2π-electrons) or a [1,5]- or [3,3]-shift (6π-electrons).

The *Claisen rearrangement* of allyl alkenyl ethers is another example of a [3,3]-sigmatropic rearrangement, differing from the Cope rearrangement in the replacement of a CHR fragment by oxygen but having the same stereochemical outcome, e.g.

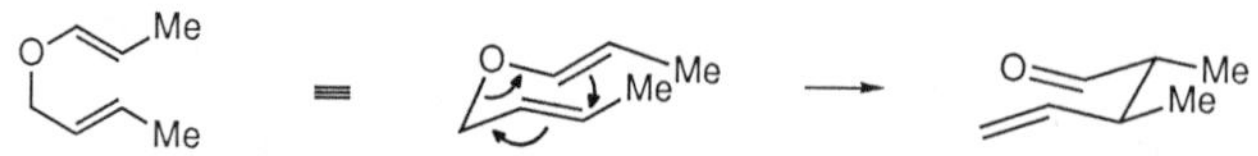

The reaction is widely employed for building up bi- and polyfunctional carbon chains, for example:

(1) β-Keto-esters to γδ-unsaturated ketones, e.g.

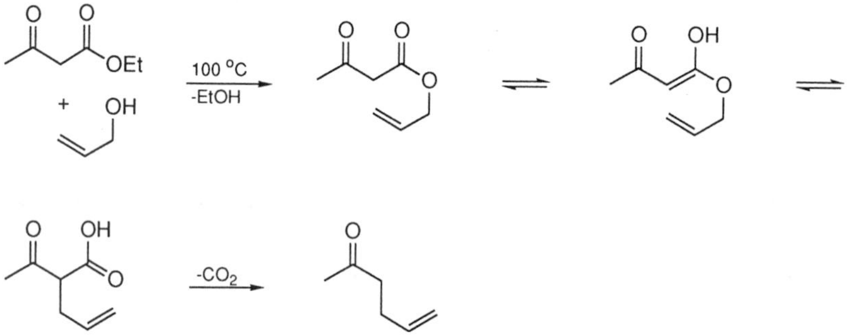

(2) Allyl alcohols to γδ-unsaturated amides (*Eschenmoser reaction*), e.g.

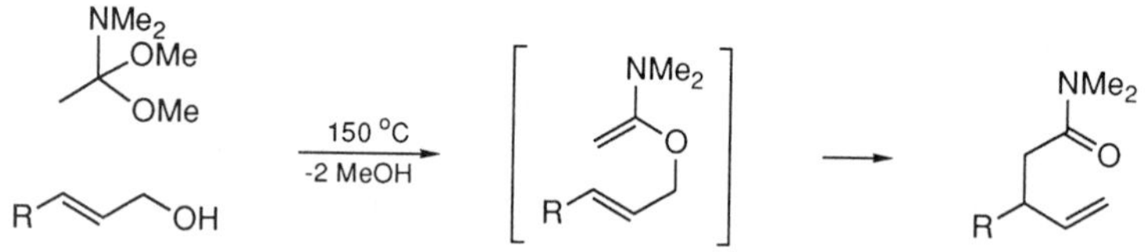

or γδ-unsaturated esters, e.g.

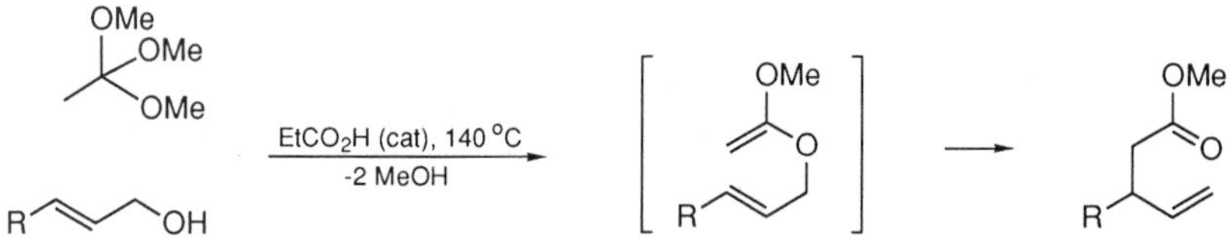

(3) Prenylation, e.g.

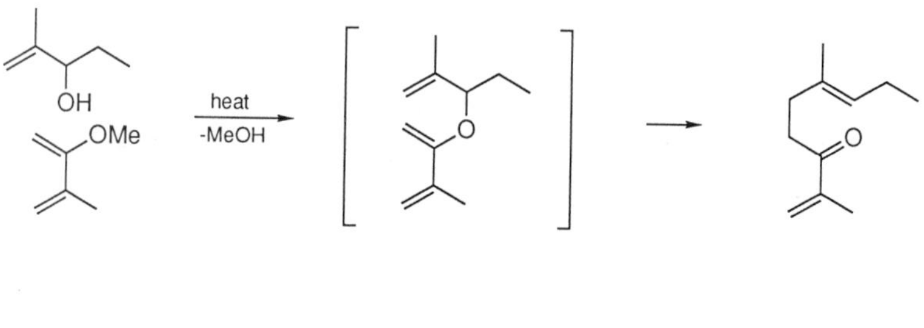

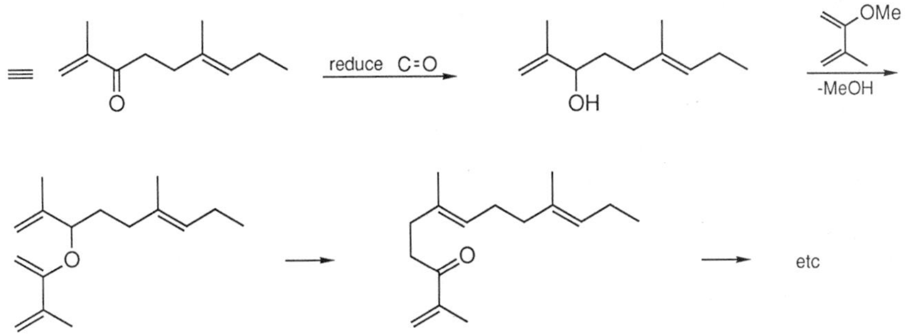

Claisen rearrangements of aryl allyl ethers also occur, e.g.

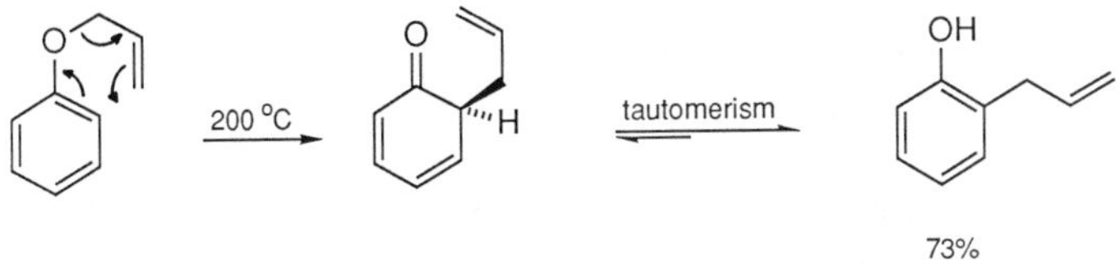

73%

If both *ortho*-positions are substituted, the cyclohexadienone intermediate undergoes a second [3,3]-shift, giving a *p*-allylphenol:

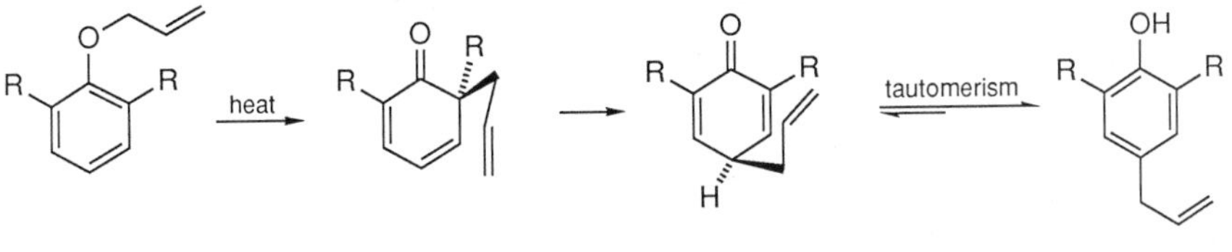

[3,2]-Sigmatropic rearrangements are also well established. The reactants possess a two-atom migrating unit in which one atom has a negative charge; this is equivalent to the two π-electrons of an unsaturated bond in a three-atom unit. An example is the Sommelet rearrangement in which the two-atom unit is a nitrogen ylide, $-\overset{+}{N}R_2-CH_2^-$ (p. 448).

**9.5
The ene-
reaction and
related
reactions**

The ene-reaction is a reaction of an allylic compound with an alkene in a manner which resembles both cycloaddition and a [1,5]-sigmatropic shift of hydrogen, e.g.

For example, allyl alcohols form aldehydes,

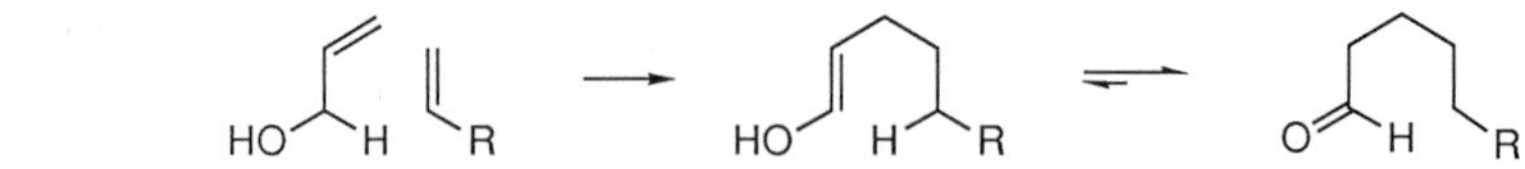

The reactions require higher temperatures than Diels–Alder reactions but occur faster with conjugated enones, especially (as in Diels–Alder reactions) with Lewis-acid catalysis, e.g.

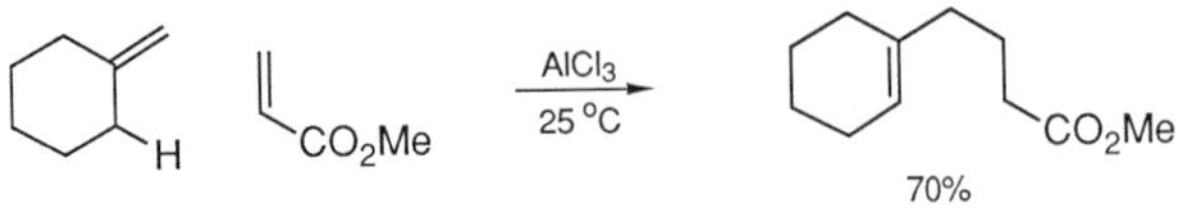

The reverse process also occurs, as in the pyrolysis of esters:

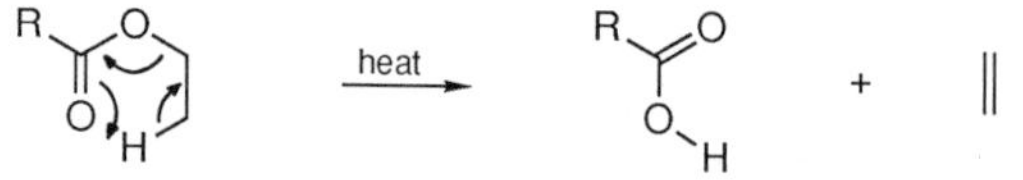

The temperature required for this reaction is high (300–500°C), which restricts its utility. Conditions are much milder, however, for the related *Chugaev reaction* in which a xanthate ester is heated at 100–200°C:

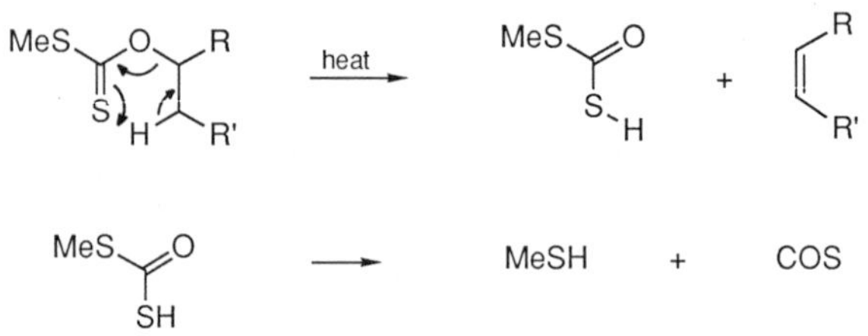

The synthetic value of the Chugaev reaction lies in the stereospecificity which its concertedness confers.

The ready decarboxylation of β-keto-acids has been shown to occur analogously, e.g.

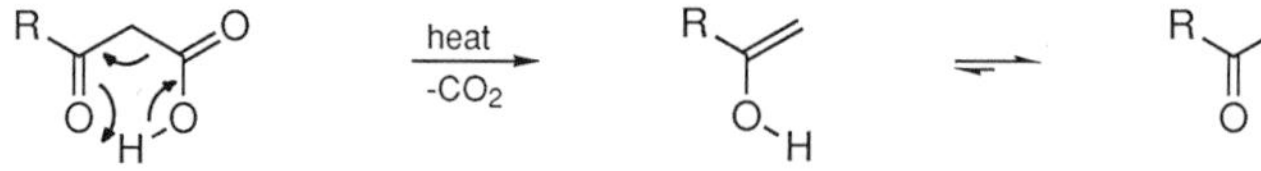

The *Cope reaction* is a related pericyclic reaction of amine oxides which involves a five-membered cyclic transition state. It occurs under mild conditions and is useful in the generation of non-conjugated polyenes, e.g.

Sulfoxides react similarly, and applications of the process are described later (p. 473).

1. Draw the structure of the chief product of each of the following thermal **Problems** reactions:

(a)

(b)

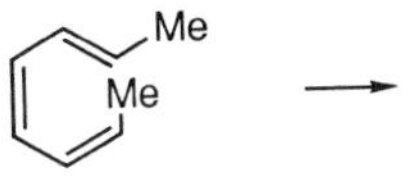

(c)

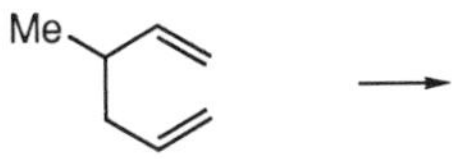

(d)

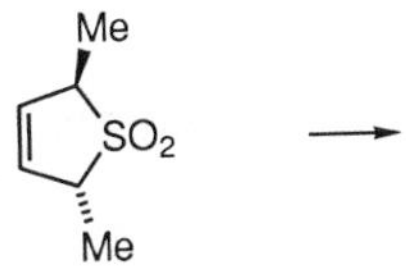

(e)

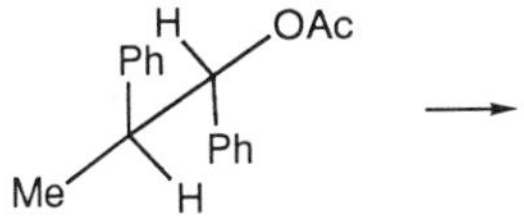

(f)

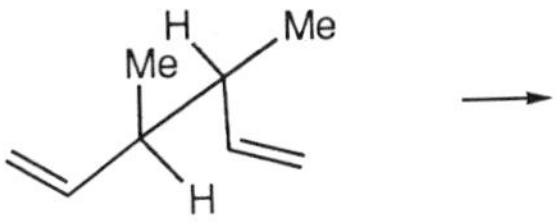

(g)

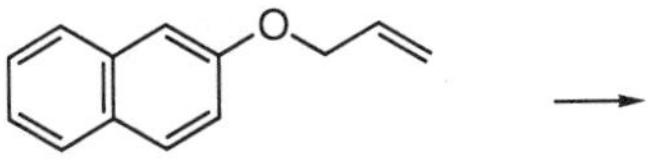

(h)

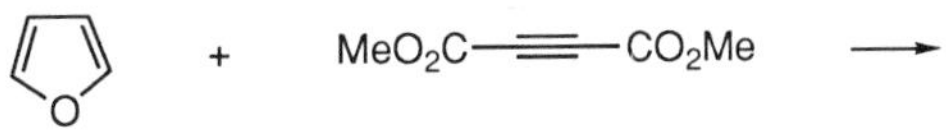

MeO_2C—≡—CO_2Me ⟶

(i)

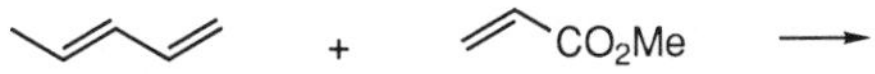

CO_2Me ⟶

(j)

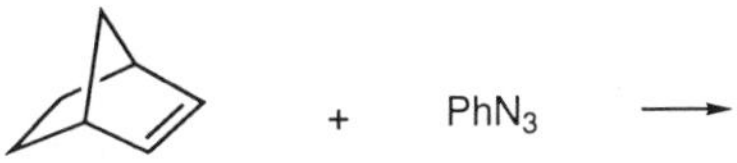

+ PhN_3 ⟶

(k)

Ph   +   Ph—≡$\overset{+}{N}$–$\overset{-}{O}$ ⟶

(l)

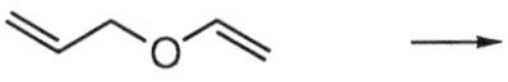

2. How would you employ pericyclic reactions in the synthesis of the following?

(a)

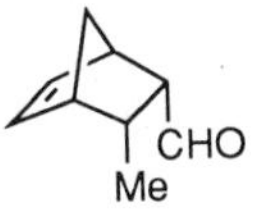

(b)

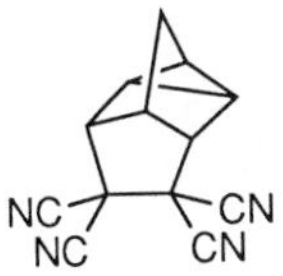

(c)

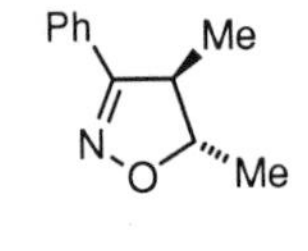

(d)

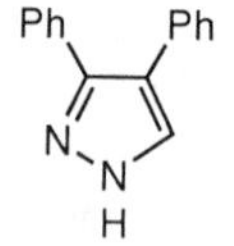

(e)

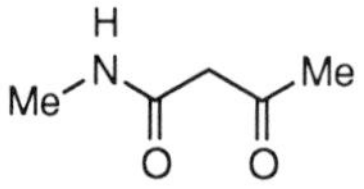

(f)

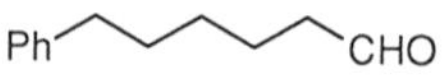

3. Rationalize the following thermal reactions:

(a)

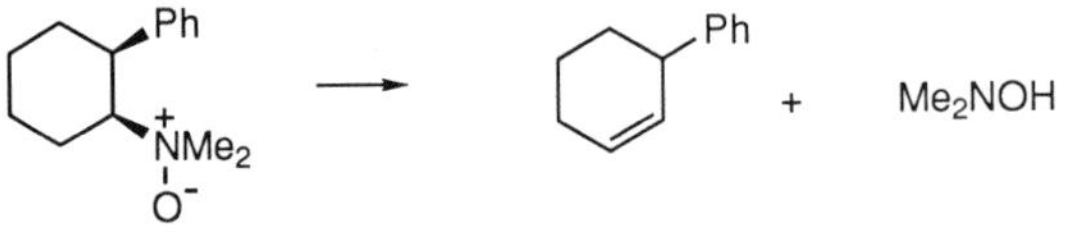

(b)

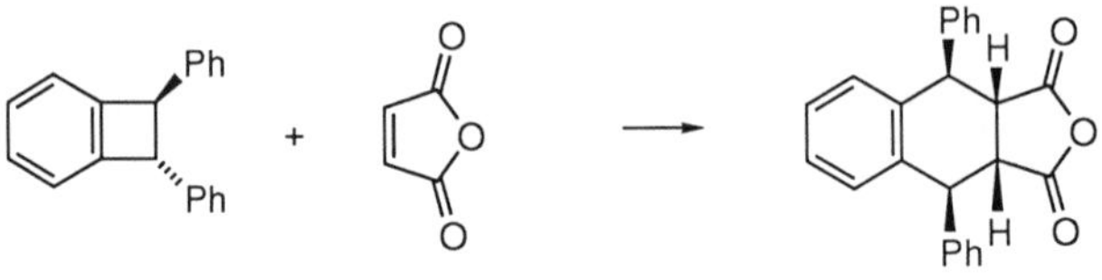

(c)

(d)

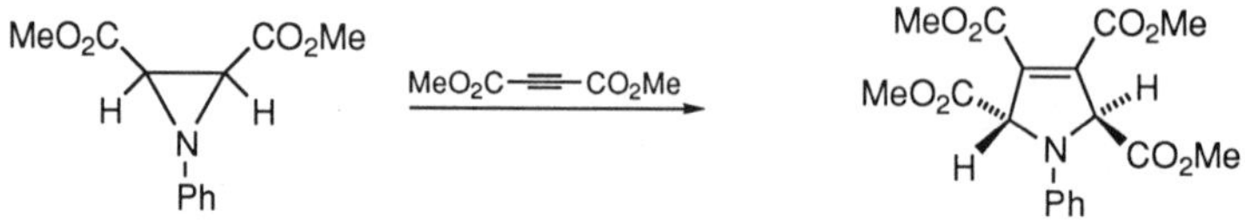

The formation of aliphatic carbon–nitrogen bonds

10

With the exception of a few methods involving free-radical reactions (section 17.4), the methods for forming bonds between nitrogen and aliphatic carbon fall into two categories. In the first, nucleophilic nitrogen reacts with electrophilic carbon and, in the second, electrophilic nitrogen reacts with nucelophilic carbon. The first category is by far the more important.

(a) Nucleophilic nitrogen

A ternary nitrogen atom possesses an unshared pair of electrons and is therefore nucleophilic. Like a carbanion, ternary nitrogen can react both with saturated carbon from which a group can be displaced with the covalent bonding-pair (S_N2 reaction),

$$N: \quad\longrightarrow\quad C{-}X \quad\xrightarrow{\;-X^-\;}\quad \overset{+}{N}{-}C \;+\; X^-$$

and with unsaturated carbon, leading initially to an adduct, e.g.

$$N: \quad\longrightarrow\quad C{=}O \quad\longrightarrow\quad \overset{+}{N}{-}C{-}O^-$$

The products of these reactions depend on the structures of the reactants. In the S_N2 process, if the nitrogen atom is not bonded to hydrogen, a quaternary salt is formed:

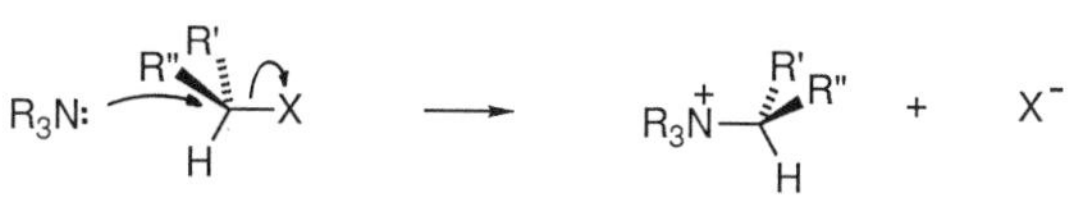

$$R_3N: \quad\longrightarrow\quad \underset{H}{\overset{R''\,R'}{C}}{-}X \quad\longrightarrow\quad R_3\overset{+}{N}{-}\underset{H}{\overset{R'\;R''}{C}} \;+\; X^-$$

but if one or more hydrogen atoms are bonded to nitrogen, a proton may be lost to a basic species (such as a second molecule of the nitrogen compound), giving a new ternary nitrogen species which can react further, e.g.

Reactions at unsaturated centres may be followed by a tautomeric shift, to give an adduct, e.g.

or, if the carbonyl group is attached to an electronegative leaving group, by the elimination of a proton and an anion, e.g.

In the former case, elimination from the adduct commonly ensues in appropriate structural situations, e.g.

Therefore the overall process may consist of substitution at saturated carbon, addition to unsaturated carbon (sometimes followed by elimination), or substitution at unsaturated carbon. Each of these processes has an exact counterpart in the reactions of carbanions (chapter 7).

The commonest nitrogen nucleophiles are ammonia and its derivatives. These may be alkyl- or aryl-derivatives (primary, secondary, or tertiary amines), nitrogen-containing derivatives such as hydrazine, NH_2—NH_2, and related compounds (e.g. $PhNH$—NH_2), and oxygen-containing derivatives such as hydroxylamine, HO—NH_2.

Nitrogen attached to an electron-attracting group of $-M$ type has very little nucleophilic reactivity. The reason is that these systems are resonance-stabilized, e.g.

and the loss of stabilization energy which occurs when the nitrogen atom bonds to carbon renders the transition states for the reactions of such high energy content, relative to the reactants, that reaction is very slow. For example, amides, imides, sulfonamides, and nitramide (NO_2—NH_2) are very weak nucleophiles.

Amide ions (e.g. NH_2^-, present in a solution of sodamide in liquid ammonia) are considerably more nucleophilic than the corresponding amines. This is of particular value in effecting reactions with imides: the grouping —CO—NH—CO— is relatively acidic (p. 47) so that, in basic conditions, the nucleophile (—CO—$\overset{-}{\text{N}}$—CO—) is present and, for example, can react with alkyl halides (pp. 303, 326).

Other nitrogen nucleophiles include azide ion ($\overset{-}{\text{N}}$=$\overset{+}{\text{N}}$=$\overset{-}{\text{N}}$) and nitrite ion ($\overset{-}{\text{O}}$—N=O). The latter is an *ambident* nucleophile, being capable of reacting at either nitrogen or oxygen. Its mode of reaction is determined by the structure of the second reactant and the reaction conditions.

(b)   Electrophilic nitrogen

Nitrogen is electrophilic both in certain cations such as aromatic diazonium ions, ArN$_2^+$ (section 13.4), the nitronium ion, O=$\overset{+}{\text{N}}$=O, and the nitrosonium ion, $\overset{+}{\text{N}}$=O, and in neutral molecules such as alkyl nitrites, RO—N=O, and nitroso compounds, R—N=O. The former class of electrophile is of only limited importance in aliphatic chemistry, although of major importance in aromatic chemistry (section 11.4).

The nitrogen atom in alkyl nitrites and related compounds is electrophilic by virtue of the fact that the addition of a nucleophile gives a relatively stable oxyanion; reaction is completed by the elimination of an anion, e.g.

This reaction is the analogue of that between a nucleophile and a carboxylic ester (p. 114).

The carbon centres to which electrophilic nitrogen bonds include both carbanions and neutral molecules such as enols which are particularly reactive towards electrophiles.

**10.2
Substitution by nucleophilic nitrogen at saturated carbon**

(a)   Reactions of ammonia and amines

The treatment of an alkyl halide with ammonia gives, initially, the conjugate acid of the corresponding primary amine, e.g.

$$\text{H}_3\text{N:} \longrightarrow \text{CH}_3\text{—X} \xrightarrow{\text{-X}^-} \text{H}_3\overset{+}{\text{N}}\text{—CH}_3$$

This then reacts with more ammonia in an acid-base equilibrium,

$$\text{H}_3\overset{+}{\text{N}}\text{—CH}_3 \quad + \quad \text{NH}_3 \quad \rightleftharpoons \quad \text{H}_2\text{N—CH}_3 \quad + \quad \text{NH}_4^+$$

and the primary amine so generated reacts with a second molecule of the halide:

$$MeNH_2 \ + \ MeI \ \longrightarrow \ Me_2\overset{+}{N}H_2 \ + \ I^-$$

Further reactions lead to the tertiary amine and the quaternary ammonium salt.

Consequently a mixture of products is usually formed and the simple alkylation of ammonia is an inefficient process for the preparation of secondary and tertiary amines. However, primary amines can be obtained in good yield by the use of a considerable excess of ammonia, for then the amine formed is a relatively ineffective competitor for reaction with the halide.

Secondary amines can be obtained by treating an alkyl halide with an excess of a primary amine, but this is a wasteful process and it is often more convenient to employ one of the methods for the reductive reaction of a carbonyl compound and a primary amine (p. 669), or to use tertiary aromatic amines as starting materials (p. 404). Tertiary amines are obtained by the alkylation of secondary amines. Quaternary ammonium salts may be made either by the alkylation of tertiary amines or, for quaternary salts of the type R_4N^+, by the treatment of ammonia with a large excess of an alkyl halide.

(*i*) *Halides.* As in other S_N2 processes, aryl and alkenyl halides are inert towards amines and ammonia. Amongst alkyl halides, primary halides react efficiently but secondary halides usually give a significant yield of an elimination product and tertiary halides give entirely the elimination product (i.e. ammonia and amines react preferentially as bases and bring about E2 elimination; cf. p. 105). Tertiary alkyl amines such as $(CH_3)_3C—NH_2$ are usually prepared either by reaction of a Grignard reagent with *O*-methylhydroxylamine (p. 196) or by the *Ritter reaction*. The latter consists of treating a tertiary alcohol, or the corresponding alkene, with concentrated sulfuric acid and a nitrile. A tertiary carbocation is formed which is attacked by the nitrile;* the resulting quaternary ion is decomposed by water to give an amide which, in the acidic conditions, is hydrolyzed to the amine, e.g.

(*ii*) *Gabriel's method.* The problem which arises in attempts to make primary amines by the monoalkylation of ammonia (namely, that further alkylation in

---

*The nitrogen atom of a nitrile is sufficiently nucleophilic to react with a carbocation but not so as to react with the less electrophilic carbon atoms of halides, ketones, etc. The greater nucleophilicity of an amine than a nitrile has a parallel in the greater basicity of the former, p. 49.

some degree is unavoidable) is circumvented by a procedure devised by Gabriel. This is based on the fact that phthalimide, having an acidic N—H group (p. 47), reacts with base to give a nitrogen-containing anion which, as a strong nucleophile, displaces on alkyl halides. Hydrolysis of the resulting compound with alkali gives the primary amine.

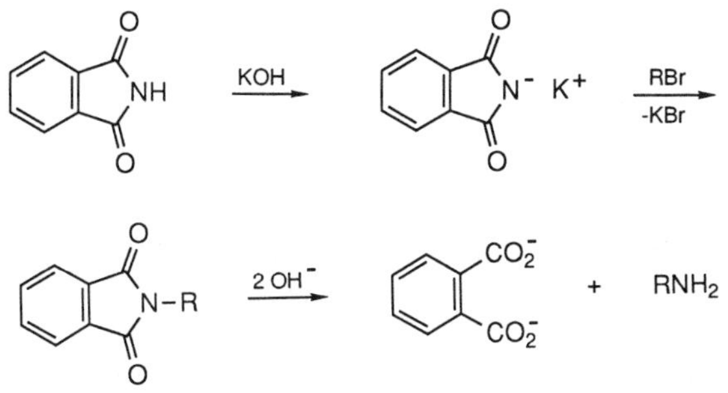

For example, potassium phthalimide and excess of 1,2-dibromoethane react at 180–190°C to give β-phthalimidylethyl bromide in 75% yield [1]; hydrolysis gives β-bromoethylamine.

(*iii*) *Other leaving groups.* As usual in S_N2 reactions, alcohols are relatively inert to displacement by nucleophilic nitrogen. However, reaction can be brought about in vigorous conditions in the presence of a Lewis acid, which aids the departure of hydroxide ion by coordinating with oxygen. For example, methylamine is made industrially by the reaction of methanol with ammonia at 400°C under pressure, in the presence of alumina:

$$\text{MeOH} \quad + \quad \text{NH}_3 \quad \xrightarrow[\text{400 °C}]{\text{Al}_2\text{O}_3} \quad \text{MeNH}_2 \quad + \quad \text{H}_2\text{O}$$

Ethers, like alcohols, are relatively inert, but strained cyclic ethers react. For example, ethylene oxide reacts with secondary amines to give β-hydroxy-tertiary amines,

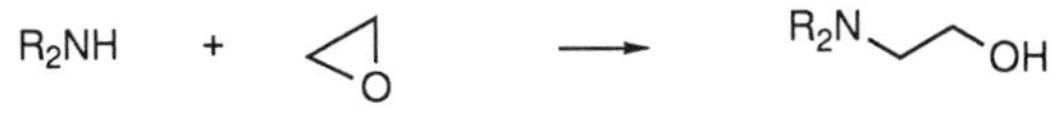

and ammonia reacts with ethylene oxide to give successively β-hydroxy- primary, secondary, and tertiary amines,

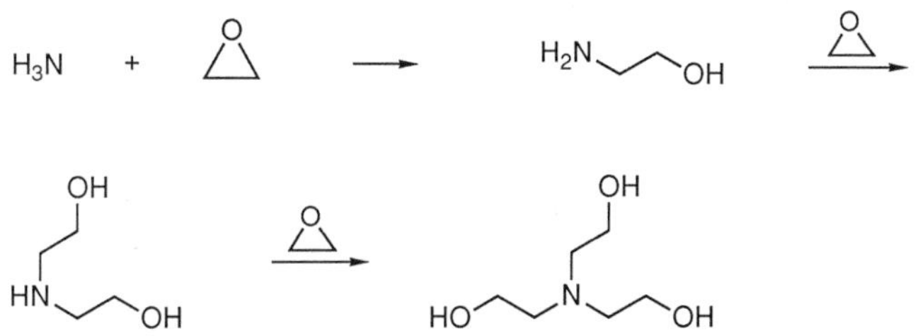

These reactions are analogous to those of small ring cyclic ethers with other nucleophiles, such as hydroxide ion (p. 590) and Grignard reagents (p. 194); in each case reaction is facilitated by the release of ring strain which occurs in passage to the transition state.

A related example is the reaction of ammonia with β-propiolactone to give β-alanine:

β-alanine

Here, the relief of ring strain in the transition state so lowers the free energy of activation as to give this reaction preference over the normal reaction of ammonia with lactones (or esters), i.e. attack at the carbonyl group to give an amide (the hydrolysis of β-propiolactone occurs in a similar manner).

Intramolecular nucleophilic substitution can occur readily to give three-, five-, and six-membered alicyclic rings (but not other ring sizes; cf. p. 67). For example, distillation of (the zwitterionic) ethanolamine hydrogen sulfate with aqueous sodium hydroxide gives ethyleneimine,

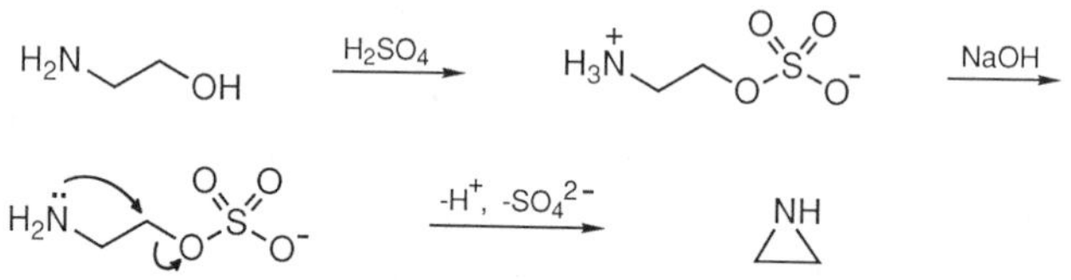

and pyrrolidine and piperidine are formed by heating the salts of tetramethyl-enediamine and pentamethylenediamine, respectively, e.g.

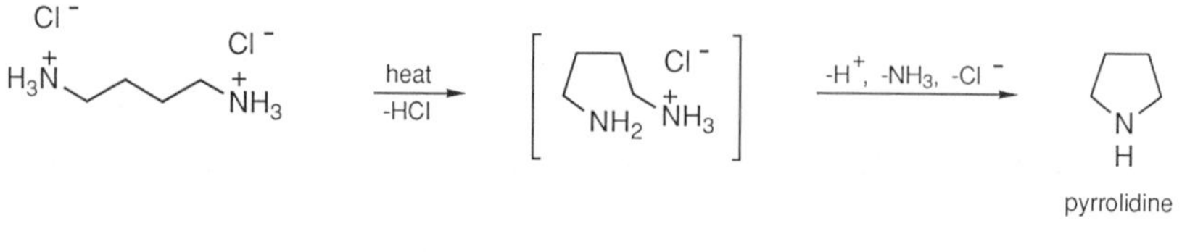

pyrrolidine

(iv) *Carbylamine reaction.* Primary amines react with chloroform and alkali-metal hydroxides to give carbylamines (isonitriles). (The unpleasant odour of these compounds has in the past been used as a test for primary amines.) Reaction probably occurs *via* dichlorocarbene (p. 101).

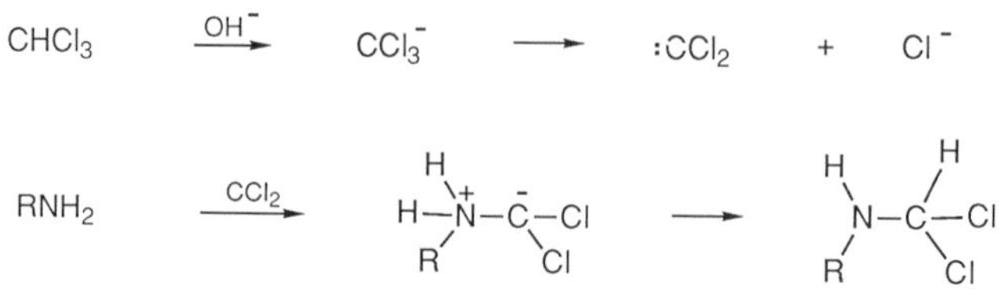

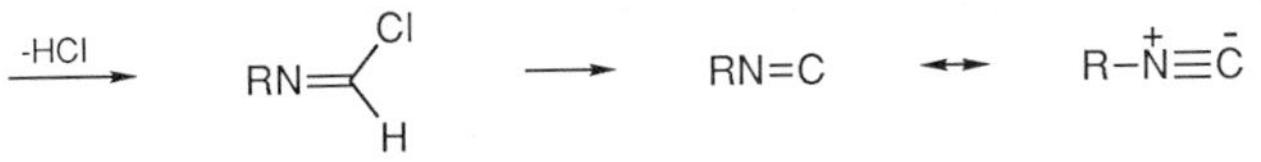

(b) Reactions of other nitrogen nucleophiles

(i) Nitrite. Metal nitrites can react with alkyl halides at both nitrogen and oxygen, giving nitro-compounds and nitrites, respectively.

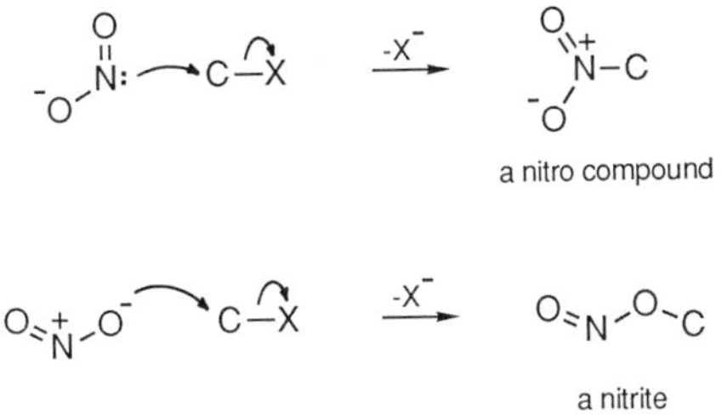

a nitro compound

a nitrite

The proportions of the two products are determined by the structures of the reactants and the reaction conditions. First, with a given halide, silver nitrite suspended in ether gives a higher proportion of nitro-compound than alkali-metal nitrites. Second, with a given nitrite, the proportion of nitro-compound decreases from primary to secondary to tertiary halides, e.g.

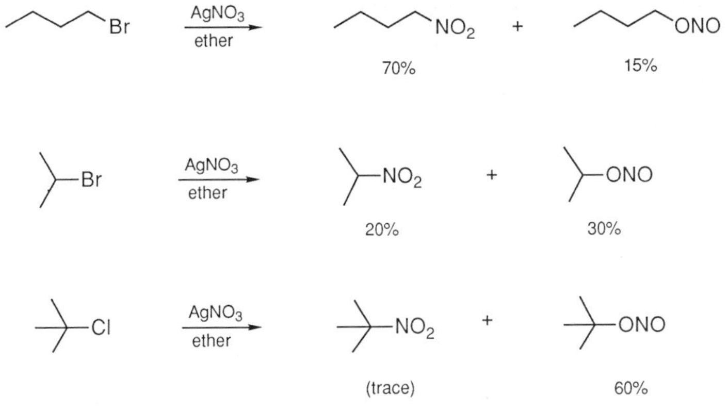

Third, although alkali-metal nitrites normally give very low yields of nitro-compounds, the use of the polar aprotic solvents dimethylformamide and di-methyl sulfoxide markedly increases the yields (to *c.* 50–60%) from primary and secondary alkyl bromides and iodides. However, tertiary halides undergo elimination in these conditions (e.g. t-butyl bromide gives isobutylene, cf. p. 105).

Nitromethane can be obtained conveniently, although not in high yield, by treating sodium chloroacetate with sodium nitrite. The resulting nitro-compound readily undergoes decarboxylation (cf. the analogous β-keto-acids) and nitromethane may be isolated by distillation, the nitrite by-product remaining in solution.

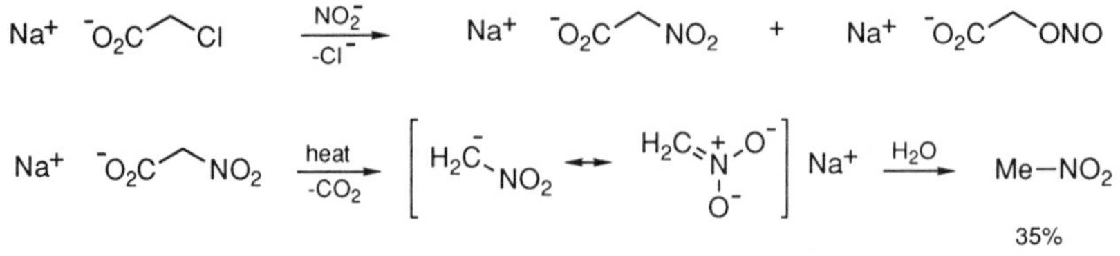

(ii) Azide ion. Halides react with azides such as sodium azide to give alkyl azides. The reaction provides a method for preparing primary amines *via* catalytic reduction of the azide, e.g.

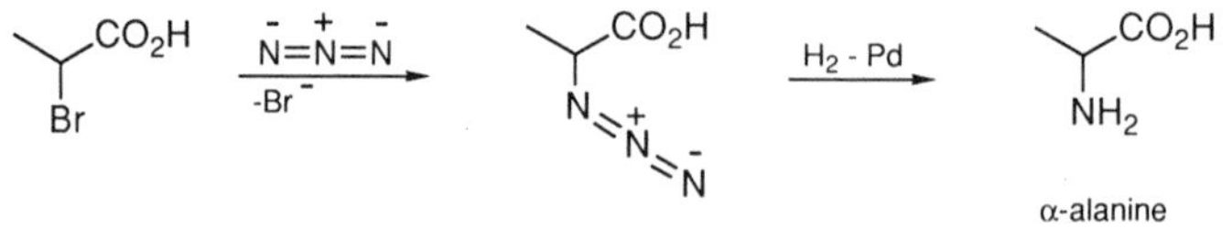

(iii) Hydrazine. Hydrazine reacts with alkyl halides to give alkylhydrazines. The introduction of the first alkyl group increases the nucleophilicity of the alkylated nitrogen (since alkyl groups are electron-releasing relative to hydrogen), so that further alkylation tends to occur, e.g.

$$H_2N-NH_2 \xrightarrow[-HI]{MeI} MeNH-NH_2 \xrightarrow[-HI]{MeI} Me_2N-NH_2$$

Monoalkylhydrazines are obtained more satisfactorily from alkylureas (p. 313) by the Hofmann bromination procedure (p. 441):

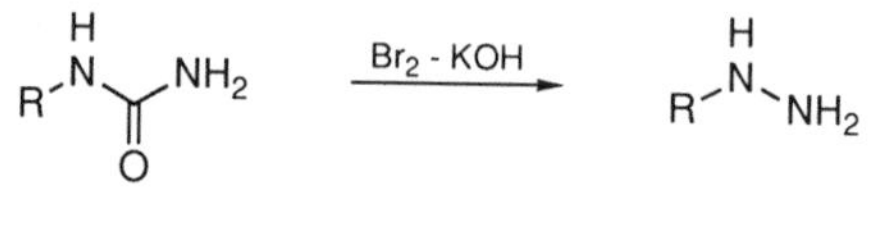

**10.3
Addition of nucleophilic nitrogen to unsaturated carbon**

(a)   Reaction with aldehydes and ketones

(i)   Ammonia.   Ammonia reacts with aldehydes and certain ketones, but the products are usually complex. With aldehydes, the first step is a simple nucleophilic addition,

$$H_3N: \quad \overset{R}{\underset{H}{\diagdown}}{=}O \quad \longrightarrow \quad H_3\overset{+}{N}\overset{R}{\underset{H}{-}}O^- \quad \longrightarrow \quad H_2N\overset{R}{\underset{H}{-}}OH$$

but the products are unstable except when the aldehydic carbon is attached to a strongly electron-attracting group. For example, an ether solution of acetaldehyde absorbs ammonia and gives a white crystalline precipitate which is probably a polymer of the adduct,

$$\overset{Me}{\underset{H}{\diagdown}}{=}O \quad + \quad NH_3 \quad \longrightarrow \quad H_2N\overset{Me}{\underset{H}{-}}OH \quad \longrightarrow \quad polymer$$

but it cannot be obtained pure, whereas chloral gives a stable adduct,[*]

$$\overset{Cl_3C}{\underset{H}{\diagdown}}{=}O \quad + \quad NH_3 \quad \longrightarrow \quad H_2N\overset{Cl_3C}{\underset{H}{-}}OH$$

The fate of the adduct depends on the structure of the aldehyde. Formaldehyde is so reactive towards nucleophiles that the primary addition product reacts further, giving ultimately urotropine (hexamethylenetetramine) whose structure is similar to that of adamantane (p. 159):

$$6 \ H_2C{=}O \quad + \quad 4 \ NH_3 \quad \xrightarrow{via \ HOCH_2NH_2} \quad$$

urotropine

Many aromatic aldehydes (e.g. benzaldehyde) give condensation products whose formation involves dehydration of the initial adduct followed by further reaction:

$$\overset{Ar}{\underset{H}{\diagdown}}{=}O \quad \xrightarrow{NH_3} \quad H_2N\overset{Ar}{\underset{H}{-}}OH \quad \xrightarrow{-H_2O} \quad \overset{Ar}{\underset{H}{\diagdown}}{=}NH \quad \xrightarrow{NH_3}$$

$$\overset{Ar}{\underset{H_2N}{\diagdown}}\overset{}{\underset{}{-}}NH_2 \quad \xrightarrow{2 \ ArCHO} \quad \xrightarrow{-2 \ H_2O}$$

Most other aldehydes give polymers of the corresponding imines, but if the reaction is carried out in the presence of a reagent which can react with the imine, the process can yield useful products. Two examples follow.

---

[*] The stability of the ammonia adducts of aldehydes parallels roughly the stability of the corresponding hydrates (1,1-diols), e.g. chloral forms a stable hydrate whereas that of acetaldehyde exists only in solution (p. 89). Likewise, 1,1-diamines, $RCH(NH_2)_2$, are unstable.

(1) *Strecker synthesis.* The reaction between the aldehyde and ammonia is carried out in the presence of cyanide ion. The dehydration product of the adduct reacts with cyanide to give an α-aminonitrile:

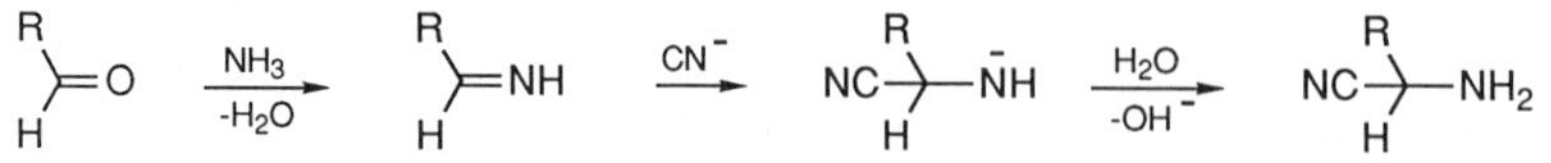

It is convenient to employ aqueous ammonium chloride and sodium cyanide as the reagents, ammonia being generated *in situ* by hydrolysis. Hydrolysis of the α-aminonitrile gives an α-amino acid, and the total process constitutes a useful method of synthesis of these compounds.

(2) *Reductive amination.* When the reaction between an aldehyde and ammonia is carried out in the presence of a reducing agent such as hydrogen on Raney nickel, the imine is reduced as it is formed to a primary amine:

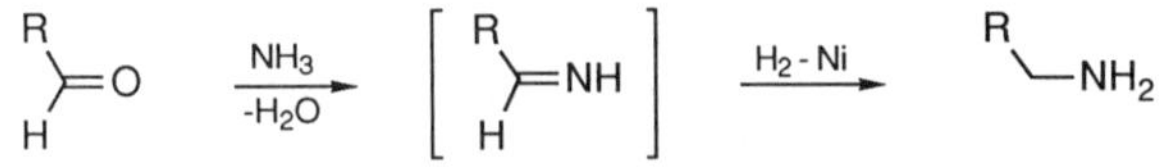

The scope and limitations of reductive reactions are described later (p. 668).

Ketones do not give adducts with ammonia. Reaction at the carbonyl group occurs reversibly, but only leads to products if further reaction can follow so as to give a stable product. Three examples are illustrative.

(1) Acetoacetic ester reacts to give an imine which tautomerizes to a conjugated amine (the nitrogen analogue of the enol form of acetoacetic ester):

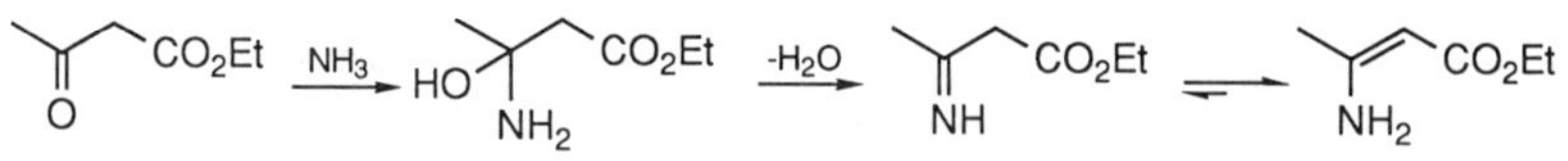

(2) 2,4-Pentanedione reacts at one carbonyl group to give an adduct whose amino group is suitably placed to react intramolecularly at the second carbonyl group. Dehydration occurs to give 2,5-dimethylpyrrole:

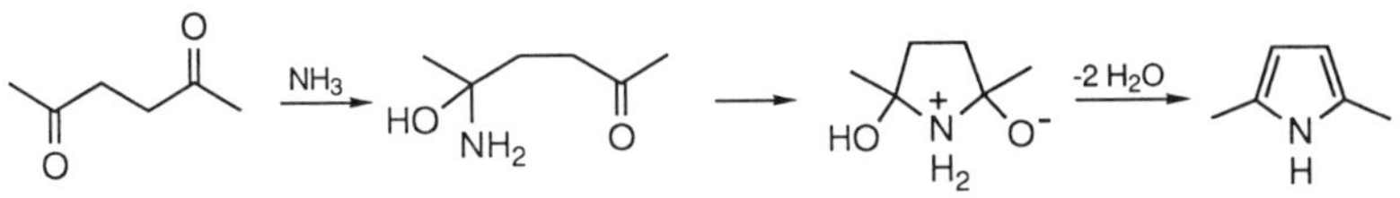

(3) The imine can be trapped by the reductive method described for aldehydes:

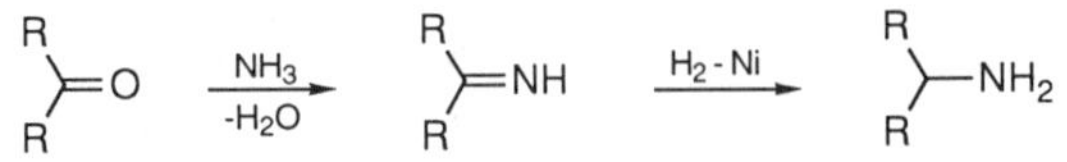

Acetone behaves in a different manner, first undergoing self-condensation under the influence of ammonia which here acts as a base rather than as a nucleophile. Two products are isolable. The first is derived by Michael-type addition of ammonia to the reaction product,

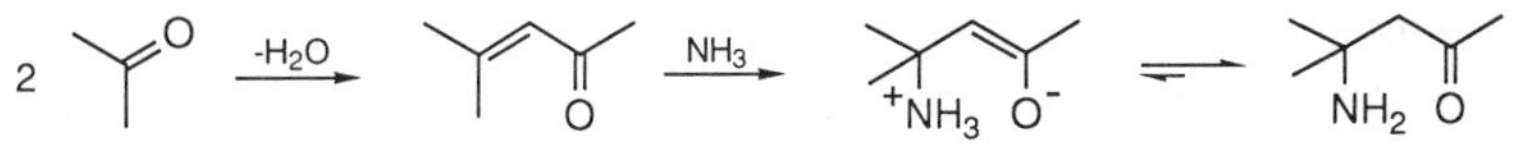

The second product is derived from a similar reaction on the product of further condensation:

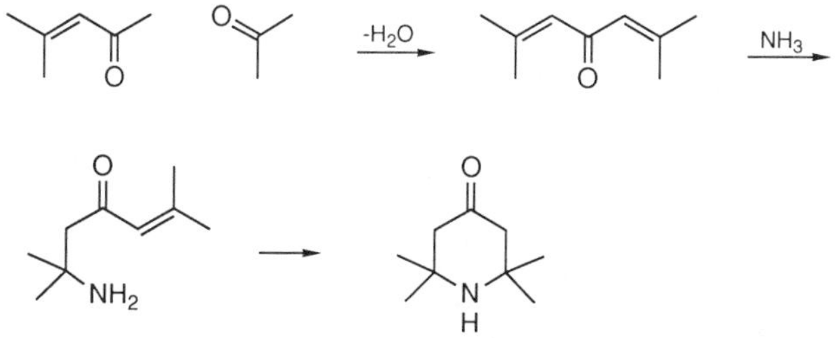

(ii) Primary amines. As with ammonia, aldehydes and ketones react with primary amines initially by addition. Dehydration then occurs to give an imine:

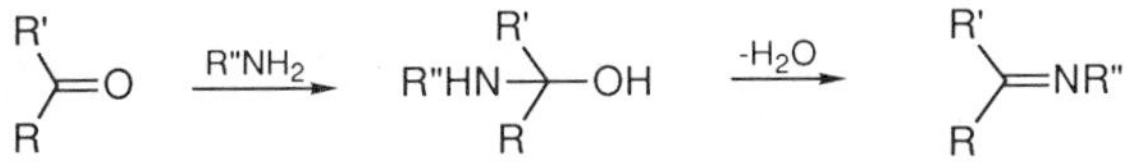

The imines, except for those from aromatic amines, are unstable and tend to polymerize unless they are trapped in a further addition process. For example, reductive amination gives secondary amines,

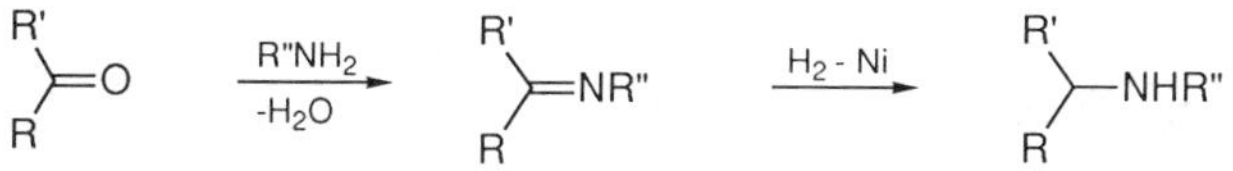

Imines from aromatic amines are usually stable and can be isolated. For example, benzaldehyde and aniline react exothermically to give benzylideneaniline in about 85% yield [1]:

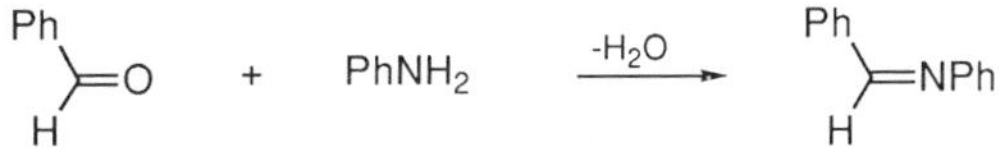

The stability of these imines is evidently associated with the conjugation between the aromatic ring of the original amine and the imino double bond.

α-Dicarbonyl compounds react with aromatic *ortho*-diamines by consecutive addition-elimination processes to give quinoxalines:

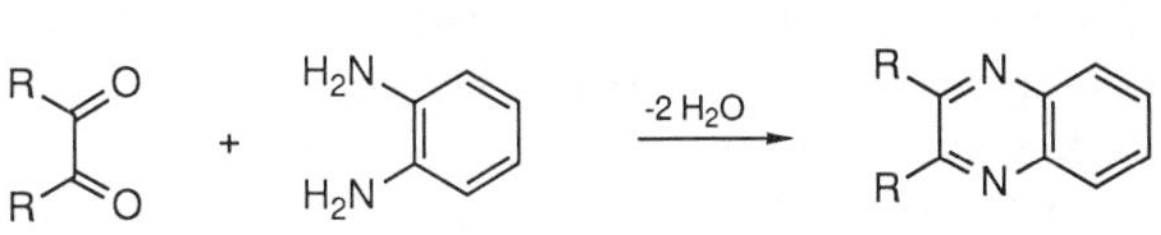

a quinoxaline

(*iii*) *Secondary amines.* Aldehydes and ketones which possess a hydrogen atom attached to an α-carbon atom react with secondary amines to give *enamines*, e.g.

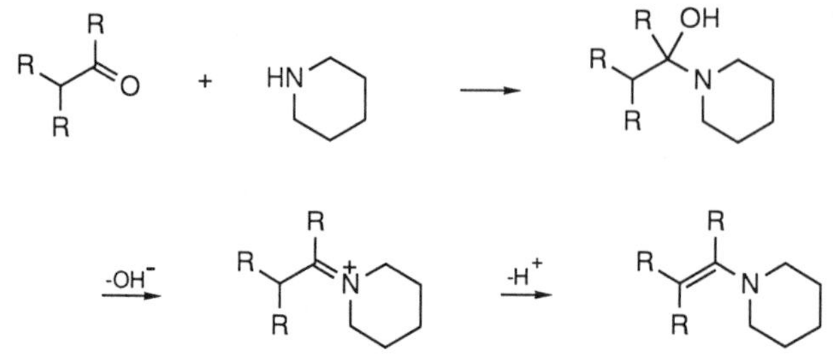

The synthetic uses of enamines are discussed elsewhere (p. 232).

The reductive amination procedure leads to tertiary amines:

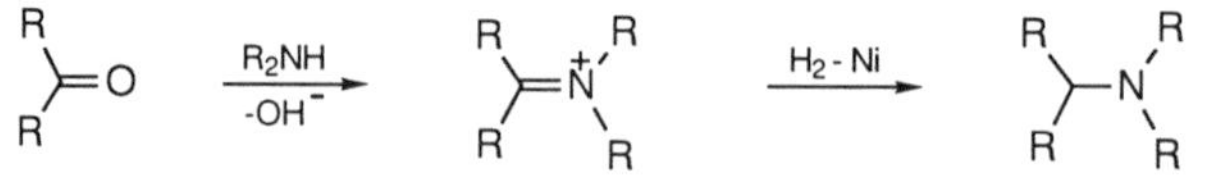

Dianilinoethane reacts with aldehydes by consecutive nucleophilic reactions, the second occurring intramolecularly:

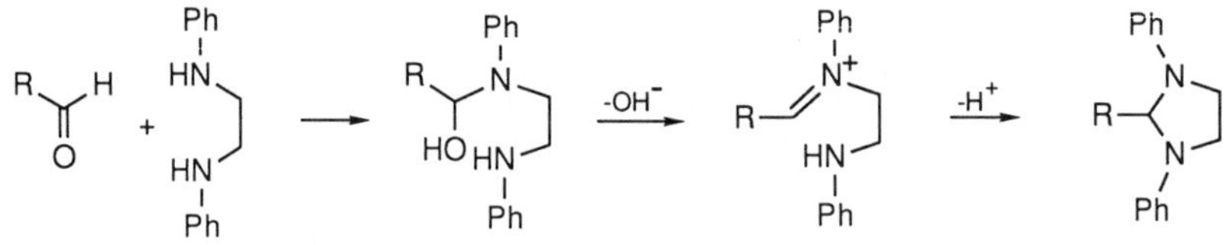

The crystalline products may be used for the characterization of aldehydes.

(*iv*) *Other nitrogen nucleophiles.* Hydroxylamine reacts readily with aldehydes and ketones to give oximes; addition of the nitrogen nucleophile is followed by elimination of water,

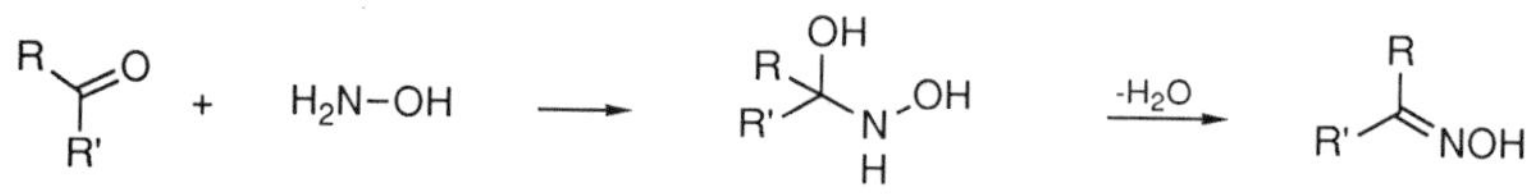

The reaction may be carried out simply by treating a mixture of the carbonyl compound and hydroxylamine hydrochloride with sodium hydroxide: the hydroxylamine reacts as it is released from its salt. In this way, for example, benzophenone oxime may be obtained in over 98% yield [2]. Alternatively, for less reactive carbonyl compounds, the reaction is carried out by adding sodium acetate to the carbonyl compound and hydroxylamine chloride; hydroxylamine is then present in a buffered medium of near-neutral pH. The reason for adopting this procedure is that the reactivity of a carbonyl group towards nucleophiles is increased by the addition of an acid (p. 89) but, if the pH is too low, hydroxylamine is present almost entirely as its unreactive conjugate acid. As a result of these conflicting requirements the reaction has a pH-optimum of about 5.

Oximes are used both for the characterization of aldehydes and ketones, since they are stable, crystalline compounds, and in synthesis; ketoximes give amines *via* the Beckmann rearrangement (p. 444) and aldoximes can be dehydrated by acetic anhydride to nitriles:

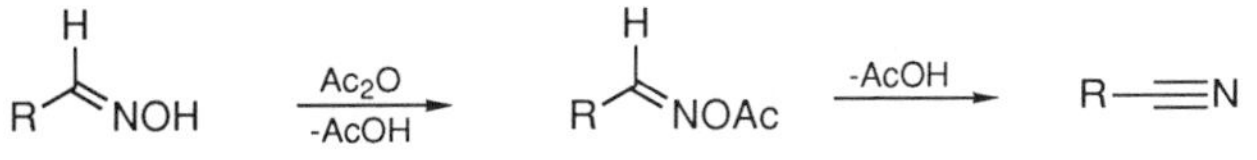

Hydrazine reacts with aldehydes and ketones analogously to hydroxylamine except that the first product, a hydrazone, can react with a second molecule of the carbonyl compound to give an azine, so that a mixture of two products is usually obtained:

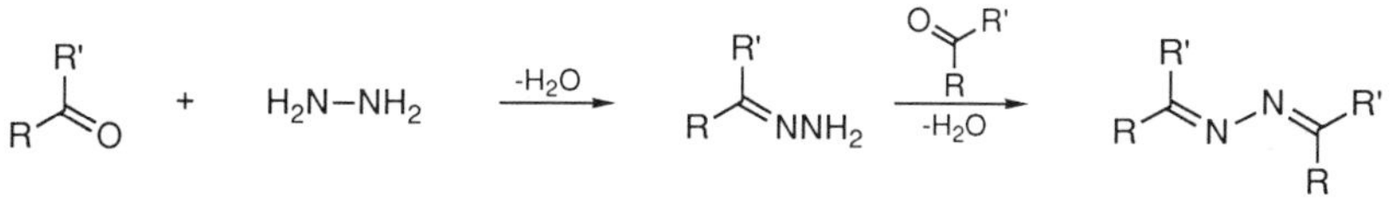

a hydrazone   an azine

In order to obtain derivatives for the characterization of a carbonyl compound it is more satisfactory to use a monosubstituted hydrazine: semicarbazide, phenyl-hydrazine, and 2,4-dinitrophenylhydrazine are commonly chosen. 2,4-Dinitro-phenylhydrazine is usually preferred to phenylhydrazine because it gives products which have higher melting points and are more easily isolated. A convenient small-scale method is to add hydrochloric acid dropwise to a mixture of the carbonyl compound and 2,4-dinitrophenylhydrazine in diethyleneglycol dimethyl ether; the derivative precipitates.

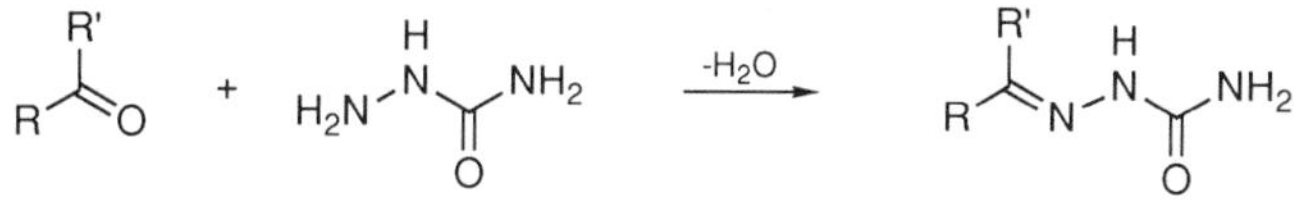

a semicarbazone

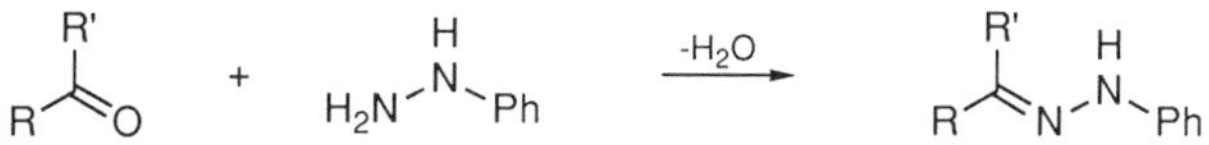

a phenylhydrazone

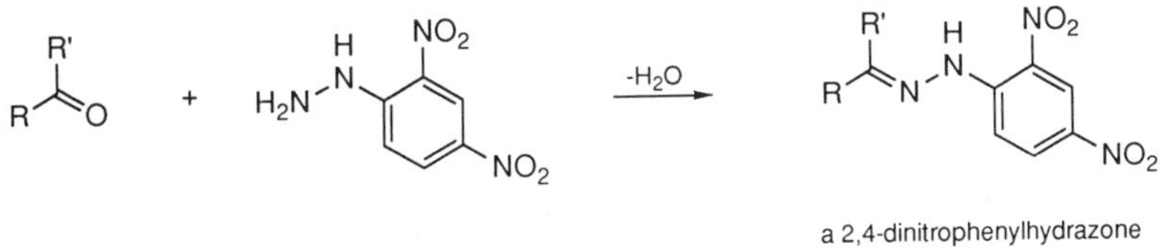

a 2,4-dinitrophenylhydrazone

β-Dicarbonyl compounds and β-keto-esters react in a more complicated manner with hydroxylamine, hydrazine, and substituted hydrazines. The initial addition-elimination process gives a derivative which is stereochemically suited to intra-molecular reaction, e.g.

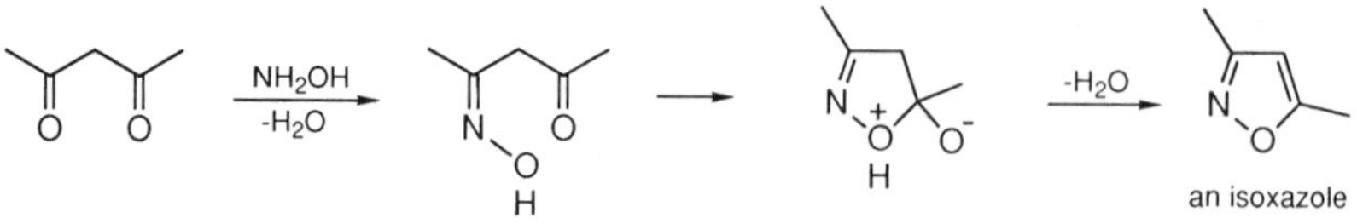

an isoxazole

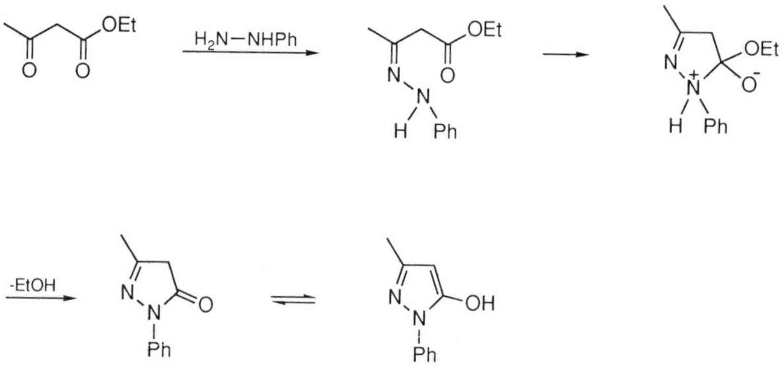

a pyrazolone

These reactions are important in the synthesis of heteroaromatic compounds.

(b) Reaction with other types of unsaturated carbon

(i) *Nitriles*. Amidines may be obtained by heating nitriles with ammonium chloride:

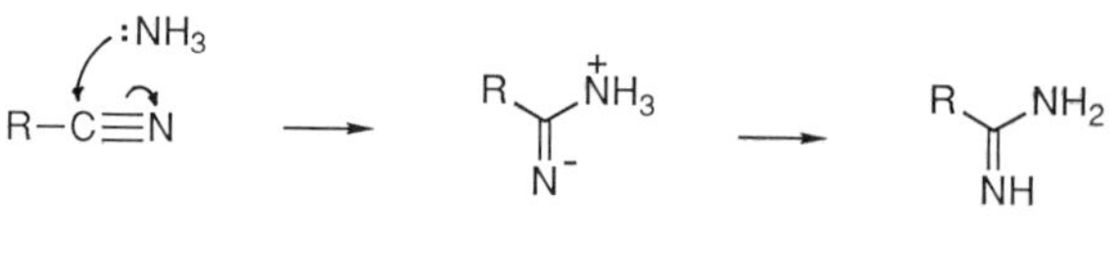

an amidine

Likewise, cyanamide gives guanidine:

guanidine

(*ii*) *Cyanates and isocyanates*.   The action of heat on ammonium cyanate gives urea, reaction probably occurring *via* ammonia and (the unstable) isocyanic acid:

$$NH_4^+ NCO^- \quad \rightleftharpoons \quad HN{=}C{=}O \quad + \quad NH_3$$

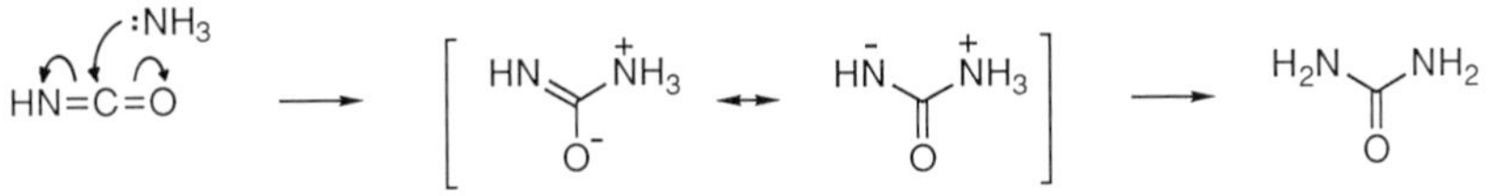

urea

The principle underlying this reaction can be applied to the synthesis of both mono- and disubstituted ureas. Monosubstituted ureas may be obtained by heating an amine with an alkali cyanate, e.g.

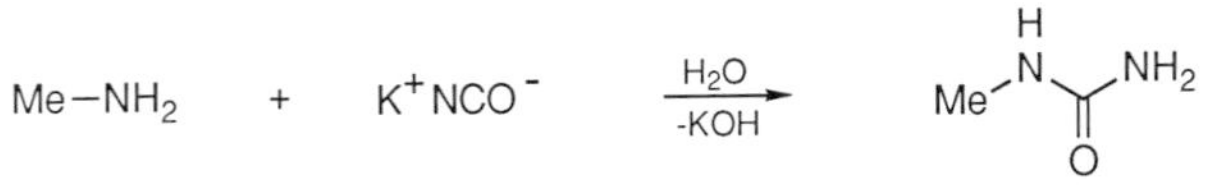

N-methylurea

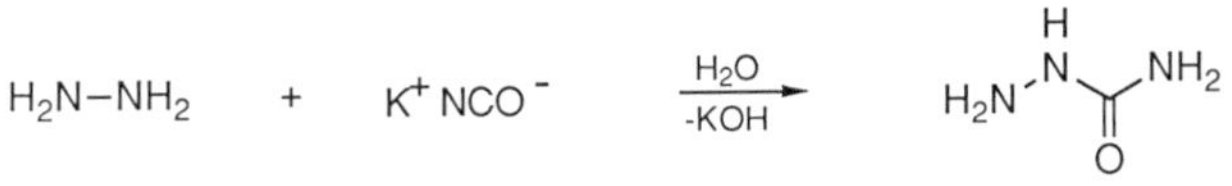

semicarbazide

α-Amino-esters react further to give hydantoins, e.g.

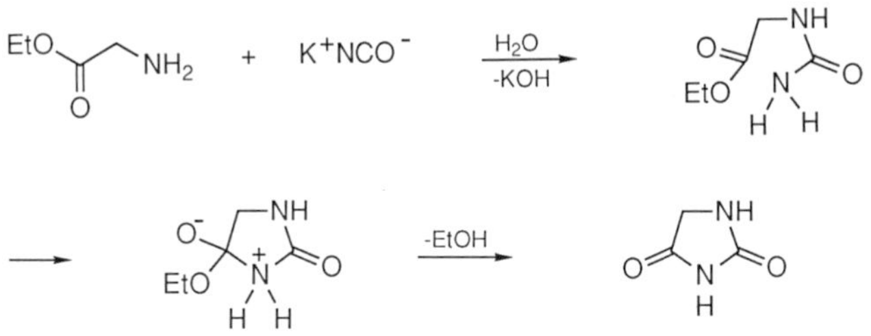

hydantoin

Disubstituted ureas may be prepared by heating organic isocyanates with amines, e.g.

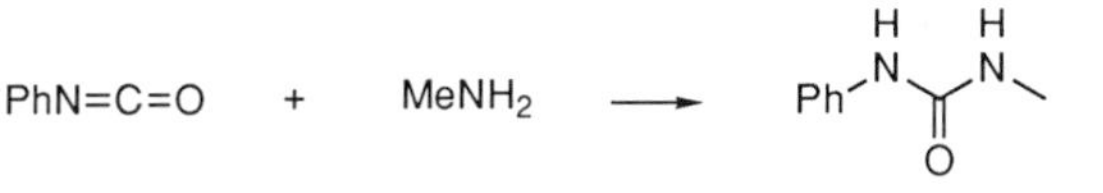

$$PhN{=}C{=}O \quad + \quad MeNH_2 \quad \longrightarrow$$

N-methyl-*N'*-phenylurea

A reaction of this type is used in the formation of polyurethane foams. A diisocyanate is first polymerized with a dihydric alcohol such as ethylene glycol:

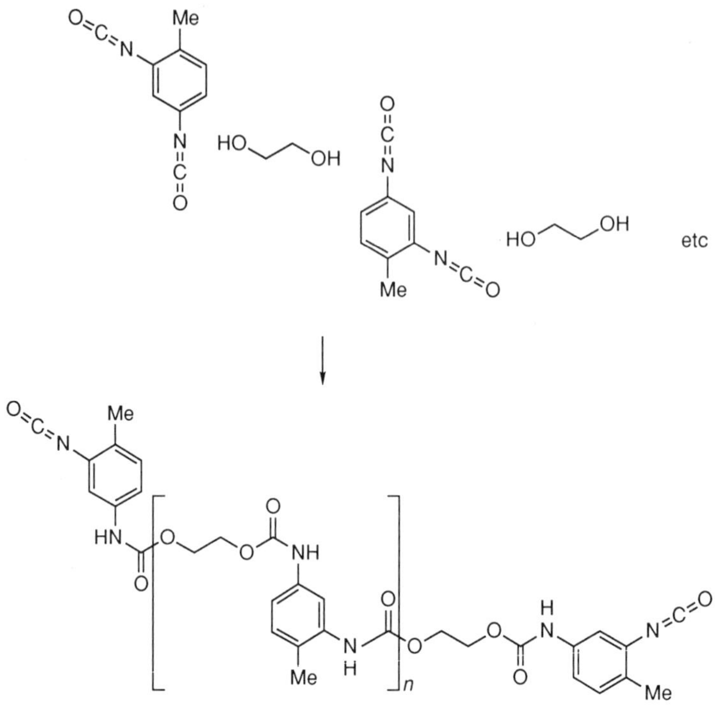

Treatment of the resulting polymer with water converts some of the terminal isocyanate groups into amino groups,

$$\sim\!\!ArN\!\!=\!\!C\!\!=\!\!O \;+\; H_2O \longrightarrow \left[\; \sim\!\!ArNH\!-\!CO_2H \;\right] \longrightarrow \sim\!\!ArNH_2 \;+\; CO_2$$

which react with the remaining isocyanate groups to extend the polymeric chain:

$$\sim\!\!ArNH_2 \;+\; \sim\!\!ArN\!\!=\!\!C\!\!=\!\!O \longrightarrow \sim\!\!ArNH\!-\!\underset{\underset{O}{\|}}{C}\!-\!NHAr\!\sim$$

The carbon dioxide which is liberated in the first step is embedded in the polymer and causes the characteristic foam.

(*iii*) *Thiocyanates and isothiocyanates.* Thiourea is obtained by heating ammonium thiocyanate at 160–170°C (cf. the formation of urea from ammonium cyanate):

$$NH_4^+NCS^- \;\rightleftharpoons\; HN\!\!=\!\!C\!\!=\!\!S \;+\; NH_3$$

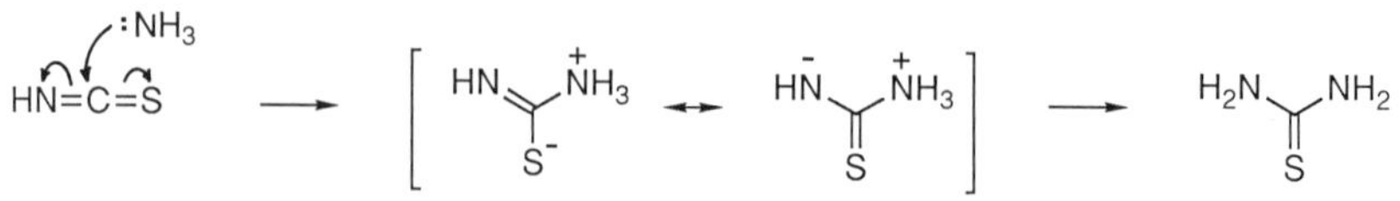

Likewise, organic isothiocyanates react with ammonia to give monosubstituted thioureas, e.g. methyl isothiocyanate gives *N*-methylthiourea in about 75% yield [3]:

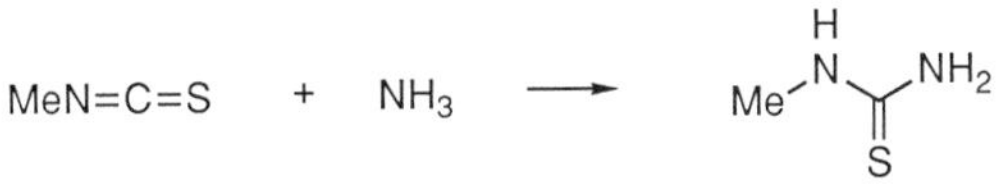

$$MeN=C=S \quad + \quad NH_3 \quad \longrightarrow$$

N-methylthiourea

Reaction with amines gives disubstituted thioureas,

$$RN=C=S \quad + \quad R'NH_2 \quad \longrightarrow$$

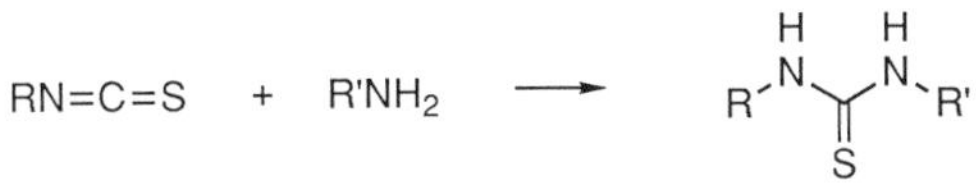

Symmetrically substituted dithioureas are more easily obtained by treating amines with carbon disulfide in refluxing alcohol. Reaction probably occurs by addition of one molecule of the amine to carbon disulfide to give a dithiocarbamic acid; this loses hydrogen sulfide and the resulting isothiocyanate reacts with a second molecule of the amine, e.g.

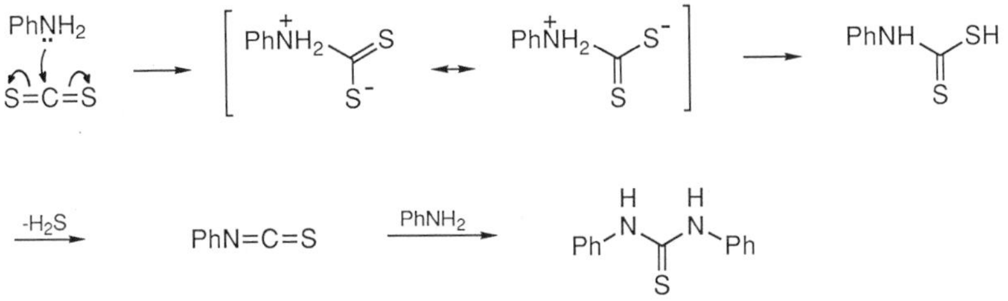

$$\xrightarrow{-H_2S} \qquad PhN=C=S \qquad \xrightarrow{PhNH_2}$$

The last reaction is reversible, and by treatment with acid to remove the amine, isothiocyanates can be obtained:

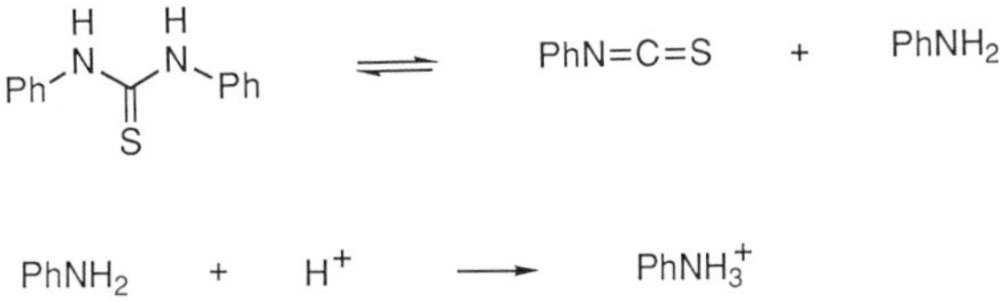

$$\rightleftharpoons \quad PhN=C=S \quad + \quad PhNH_2$$

$$PhNH_2 \quad + \quad H^+ \quad \longrightarrow \quad PhNH_3^+$$

The treatment of disubstituted thioureas with mercury(II) oxide (to remove hydrogen sulfide) has been used to obtain carbodiimides, reagents used in peptide synthesis (p. 338):

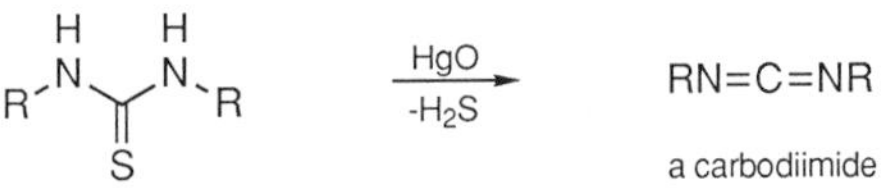

$$\xrightarrow[-H_2S]{HgO} \qquad RN=C=NR$$

a carbodiimide

**10.4
Substitution by
nucleophilic
nitrogen at
unsaturated
carbon**

A carbonyl group attached to a group capable of departing with the covalent bonding-pair is susceptible to substitution by nitrogen nucleophiles, e.g.

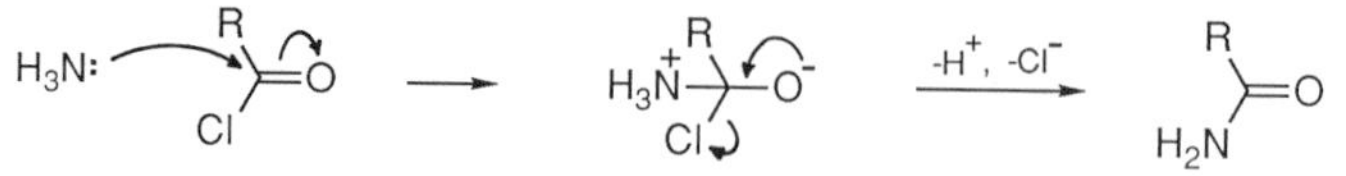

These reactions are analogous to the base-catalyzed hydrolyzes of the carbonyl-containing compounds (p. 114). As in hydrolysis, the order of reactivity of compounds containing different leaving groups, X, is: acid halides > anhydrides > esters > amides, i.e. X = Cl (etc.) > OCOR > OR > NH_2. Amides are effectively inert to displacement (except in suitable intramolecular processes, e.g. formation of succinimide, below). Carboxylic acids react, but vigorous conditions are required because the nitrogen nucleophile, acting as a base, converts the acid almost entirely into the unreactive carboxylate ion.

(a)   Reactions of ammonia and amines

Acids react with ammonia only at high temperatures, e.g. benzoic acid and ammonia, heated in a sealed tube at 200°C, give benzamide:

Amides and substituted amides are more conveniently prepared by treating acid chlorides or anhydrides with ammonia or an amine; reaction can be very vigorous, particularly when acid chlorides are used, and care is needed.

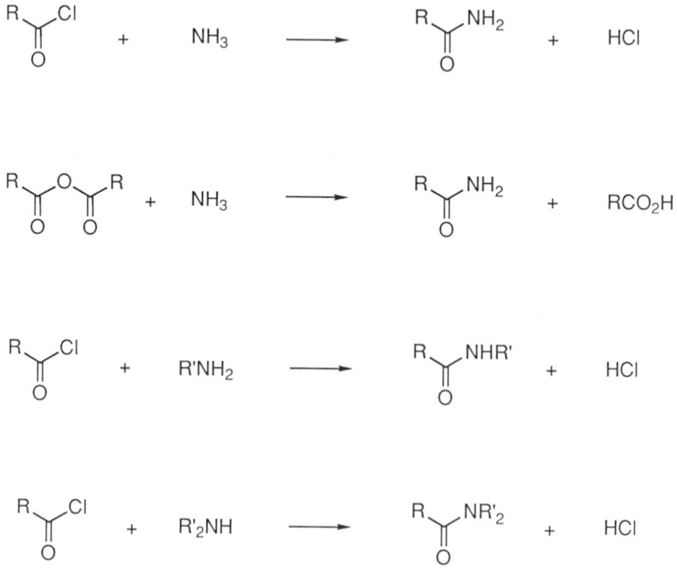

Amides themselves are so weakly nucleophilic that further reaction, e.g. $RCONH_2 + RCOCl \rightarrow RCONHCOR + HCl$, occurs only slowly and the mono-acylated derivative may be isolated in high yield. However, intramolecular displacement can occur if a five- or six-membered ring is thereby formed, e.g. succinimide can be obtained by heating the acyclic diamide of succinic acid:

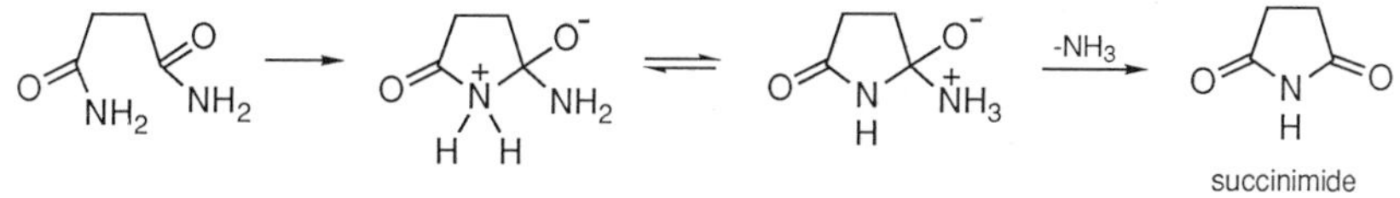

succinimide

In practice, it is easier to obtain succinimide and related cyclic imides directly from the corresponding anhydrides by treatment with ammonia at a high temperature. For example, by heating phthalic anhydride with aqueous ammonia so that the water is gradually distilled and the temperature rises to about 300°C, phthalimide is obtained in over 95% yield [1]:

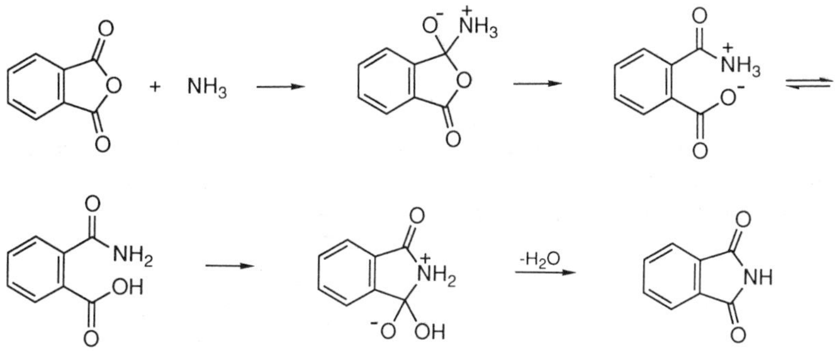

Ethyl chloroformate and phosgene (carbonyl chloride) react with amines to give urethanes and substituted ureas, respectively:

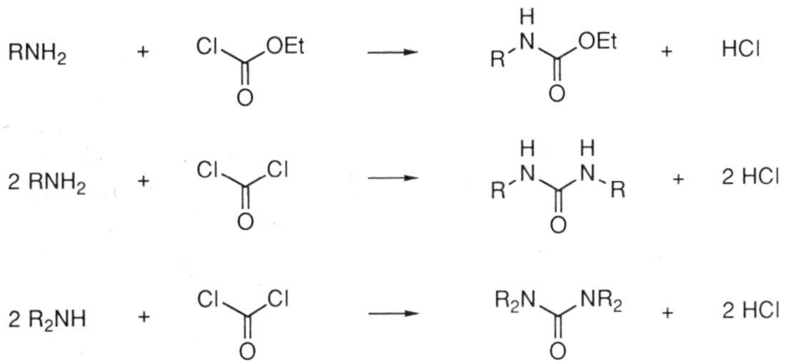

Finally, the grouping, $-C\begin{smallmatrix} OEt \\ NH_2^+ \end{smallmatrix}$ contained in imino-ether hydrochlorides, reacts analogously to the ester grouping. For example, acetamidine hydrochloride can be obtained in about 85% yield from acetonitrile by the acid-catalyzed addition of ethanol followed by treatment of the resulting imino-ether hydro-chloride with ammonia [1]:

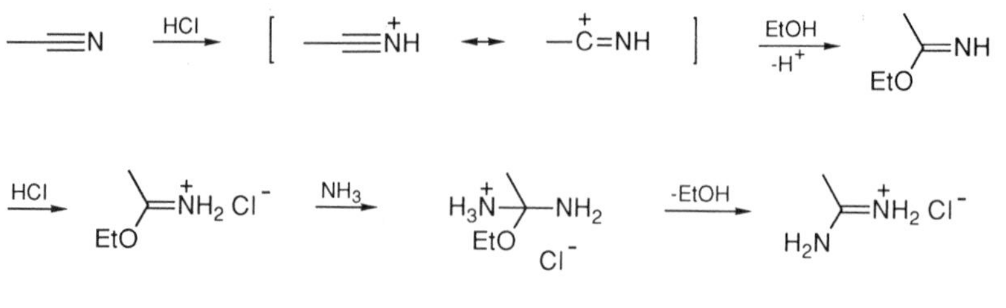

(b)   Reactions of other nitrogen nucleophiles

Hydrazine reacts with acid chlorides, anhydrides, and esters to give acid hydra-zides, e.g.

Introduction of the acyl group reduces the nucleophilicity of both nitrogens, more strongly affecting that to which it is attached; further acylation requires more vigorous conditions and occurs at the unsubstituted nitrogen atom,

Hydroxylamine also reacts with carboxylic acid derivatives giving (tautomeric) hydroxamic acids, e.g.,*

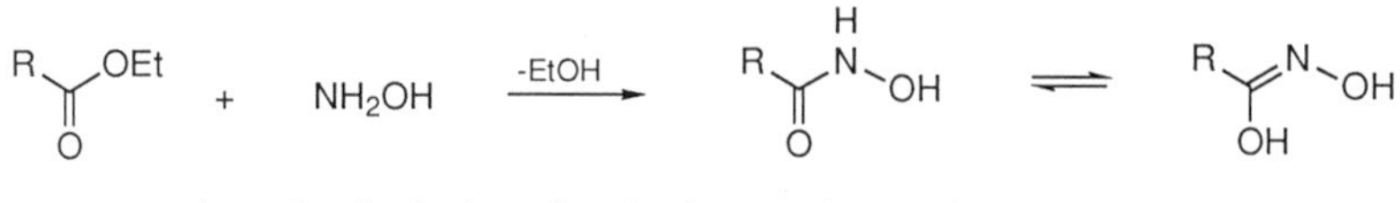

The reverse reaction, hydrolysis of a hydroxamic acid to a carboxylic acid and hydroxylamine, is applied in an industrial process for making carboxylic acids. An aliphatic nitro compound, available through vapour-phase nitration (p. 551), is heated with hydrochloric acid; the acid catalyzes the formation of the tautomer,

---

* Hydroxamic acids give deeply coloured complexes with iron(III) ion, and this provides a method for detecting the ester group in qualitative analysis.

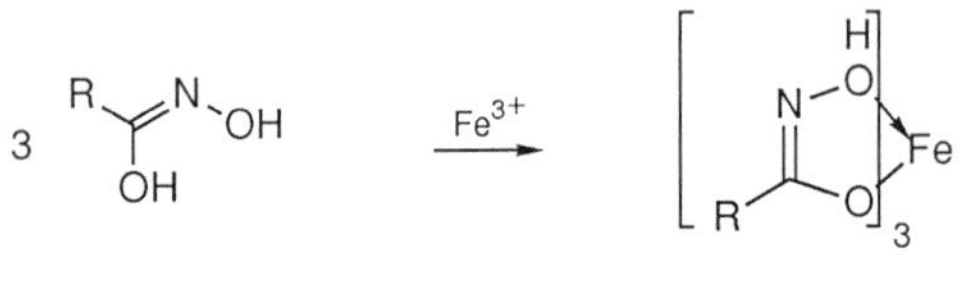

an anionotropic shift *via* the addition and elimination of water and the reverse of hydroxamic acid formation:

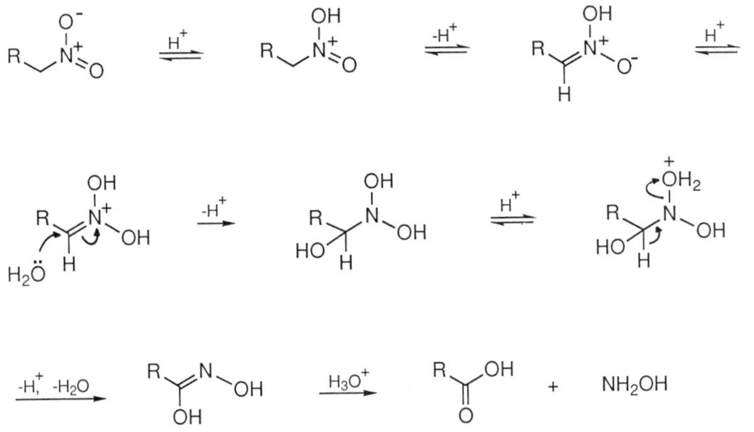

Acid azides, required for the preparation of amines *via* the *Curtius rearrangement* (p. 443), may be made by treating acid chlorides with sodium azide:

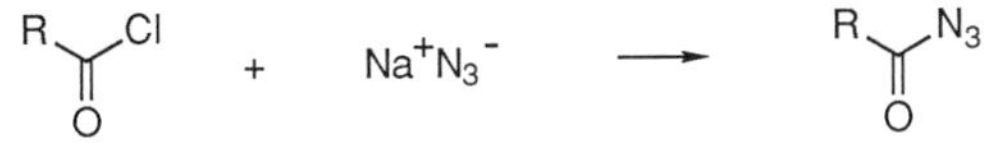

10.5 Reactions of electrophilic nitrogen

Four nitrogen-containing groups may be bonded to aliphatic carbon by procedures involving electrophilic nitrogen. The groups are nitroso (—NO), nitro (—NO$_2$), arylazo (—N=NAr), and arylimino (=NAr). The introduction of the arylazo group is discussed subsequently (section 13.4).

(a) Nitrosation

The nitroso group may be introduced in one of two ways. First, unsaturated carbon which is strongly activated towards electrophiles, such as that in an enol, reacts with nitrous acid or an organic nitrite in an acid-catalyzed process; the electrophile is thought to be the nitrosonium ion, NO$^+$:

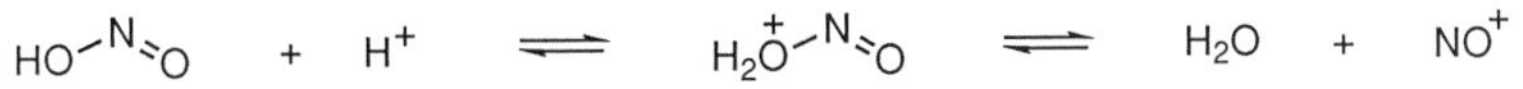

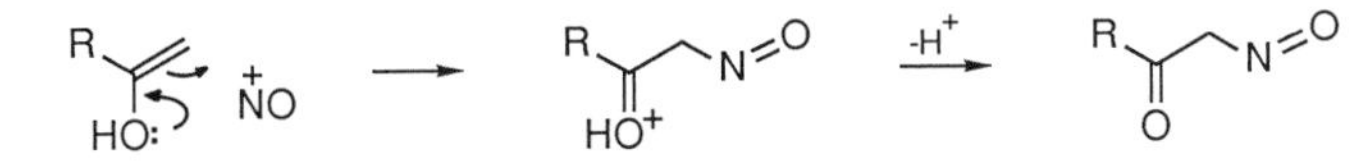

If the product has a hydrogen atom attached to the nitroso-bearing carbon, it tautomerizes to the more stable oxime:

For example, acetoacetic ester, when treated in acetic acid solution with aqueous sodium nitrite, gives the oximino-derivative:

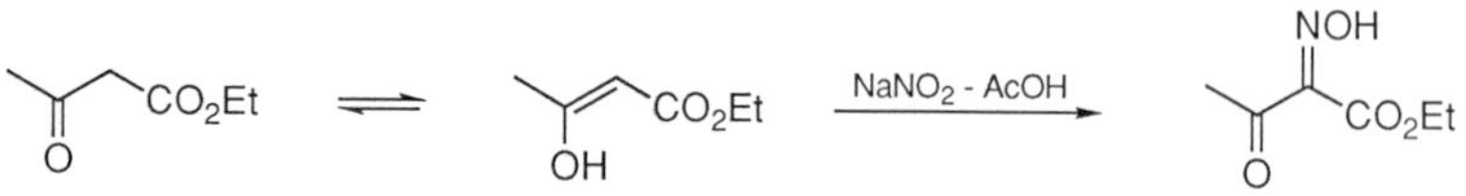

Compounds which are not significantly enolic but are capable of enolization also react, for the enolization is catalyzed by acid. For example, butanone reacts with ethyl nitrite* and hydrochloric acid to give biacetyl monoxime in about 70% yield [2]; note that reaction occurs at the methylene and not the methyl group since the corresponding enol is formed the more readily of the two (cf. p. 257).

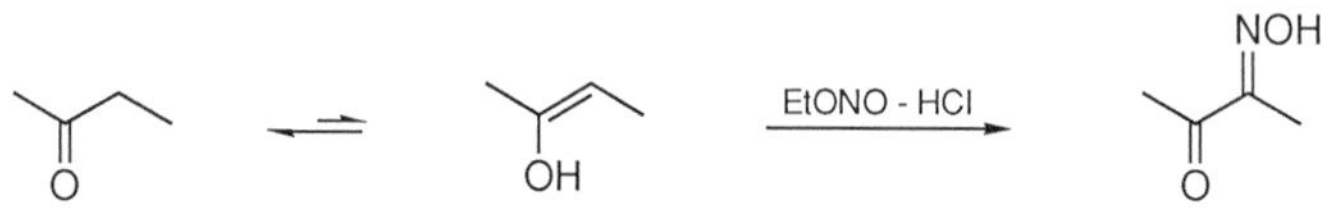

Likewise, nitromethane gives methyl nitrolic acid *via* its tautomer:

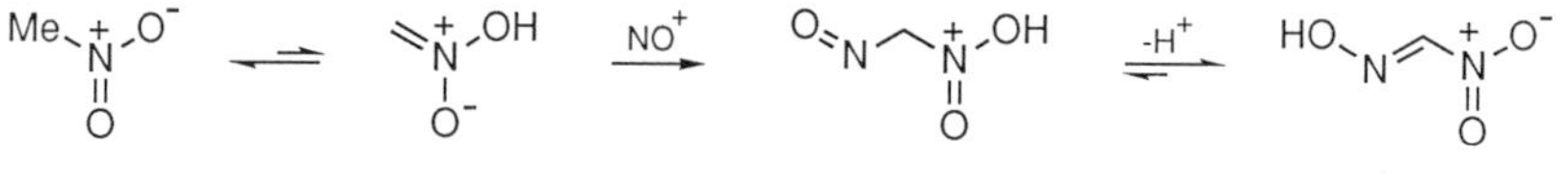

methyl nitrolic acid

In the second method a compound containing a C—H group adjacent to one or more groups of $-M$ type is treated with base (usually ethoxide ion in ethanol) in the presence of an alkyl nitrite. An enolate is generated and reacts with the nitrite in a manner analogous to the Claisen condensation (p. 226):

---

* Ethyl nitrite is prepared by treating an aqueous ethanolic solution of sodium nitrite with sulfuric acid, analogous to the esterification of a carboxylic acid (p. 115):

$$NaNO_2 \quad + \quad H^+ \quad \longrightarrow \quad HO-NO \quad + \quad Na^+$$

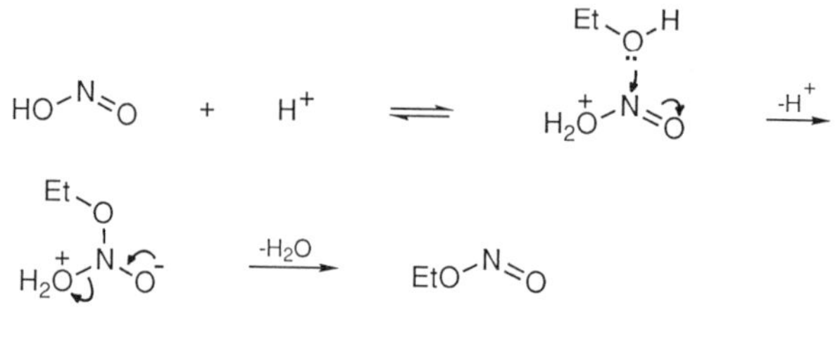

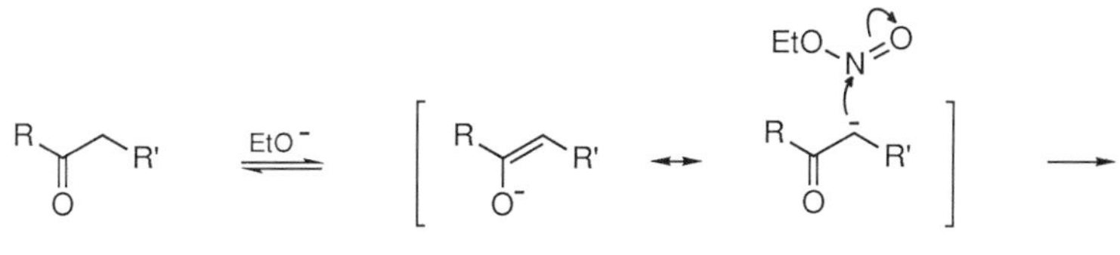

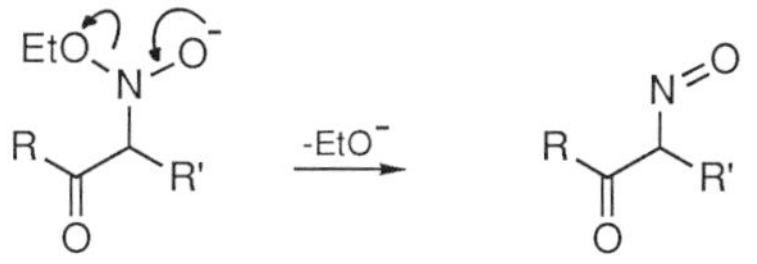

If the product contains a hydrogen atom on the nitroso-bearing carbon, its acidity is such that it is converted essentially completely into its (delocalized) conjugate base (cf. the synthesis of β-keto-esters, p. 226); acidification gives the oxime.

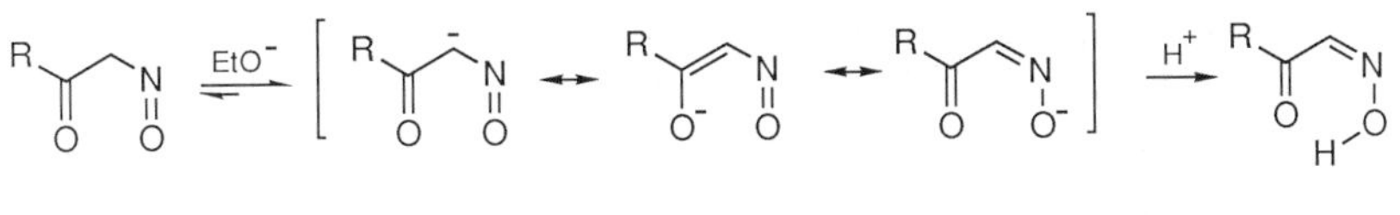

If, however, the nitroso product cannot be removed from the system in this way, a reverse Claisen reaction involving heterolytic carbon–carbon bond fission can take place:

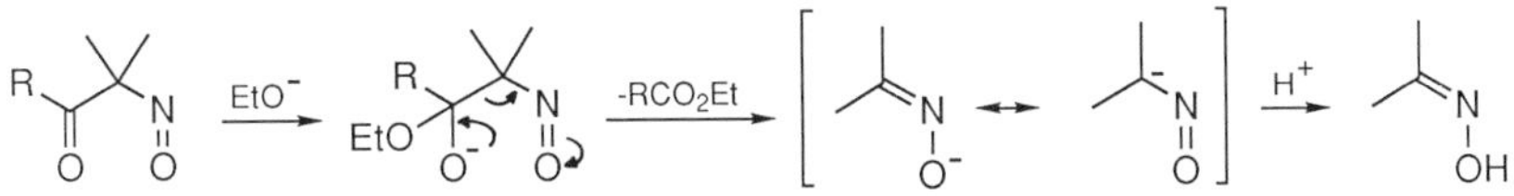

This principle was applied in the synthesis of quinine to effect a required ring opening:

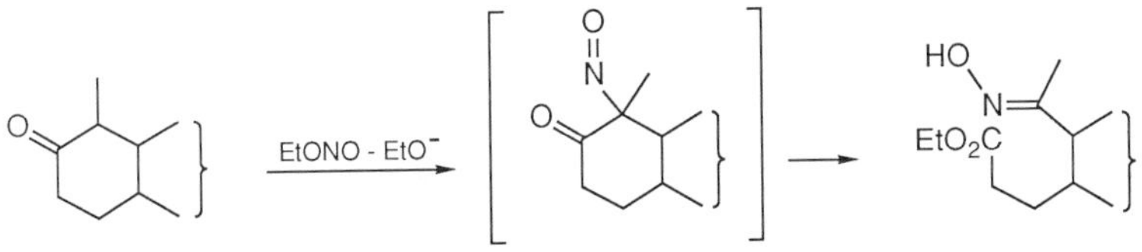

Enols and potentially enolic compounds are also those which are capable of giving enolates, so that the acid- and base-catalyzed methods are alternative procedures. The following examples include both methods.

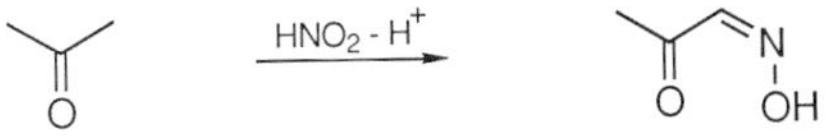

69%

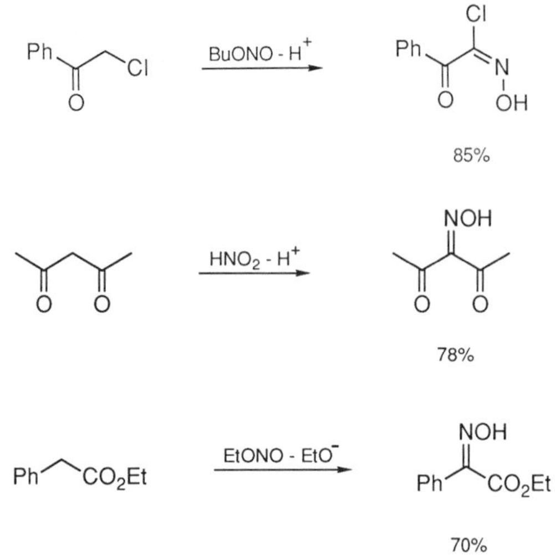

85%

78%

70%

The products of nitrosations have two main synthetic uses. First, reduction leads to amino derivatives. Those derived from ketones are unstable because they readily undergo self-condensation, but several heterocyclic syntheses are successfully carried out by reducing β-keto-oximes in the presence of compounds with which the products react to form ring systems such as pyrroles, e.g.

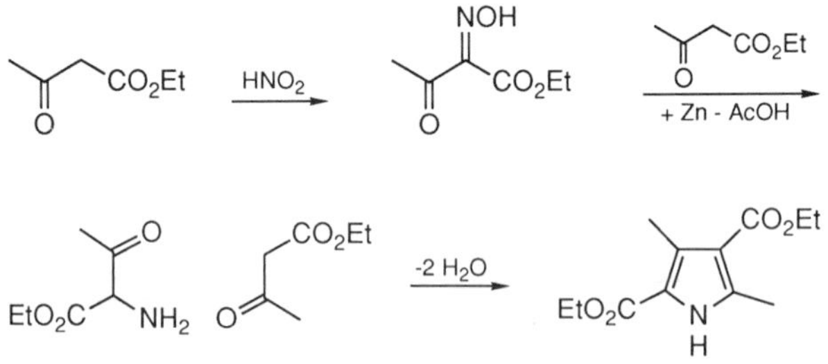

Second, hydrolysis converts the oxime into a carbonyl group, e.g.

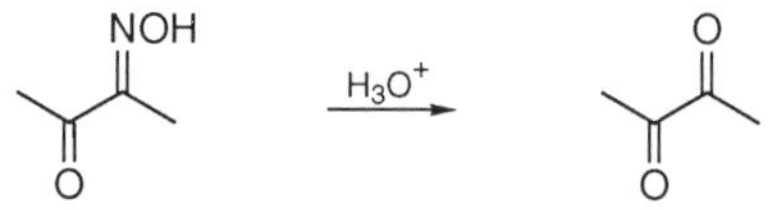

so that the overall process can be used for the transformation, —CO—CH₂— → —CO—CO—.

(b) Nitration

The nitronium ion, NO_2^+, is generated, analogously to the nitrosonium ion, by treating concentrated nitric acid with a powerful acid such as sulfuric acid. This method is widely used for bonding aromatic carbon to the nitro group (p. 376), but it is not suitable for the nitration of aliphatic systems because of the oxidations and degradations which tend to occur in these very vigorous conditions. However, a base-catalyzed procedure analogous to nitrosation may be used: an enolate-forming compound is treated with base in the presence of an organic nitrate. For example, benzyl cyanide and methyl nitrate* react in the presence of ethoxide ion to give a nitro compound which, on alkaline hydrolysis followed by acidification, gives phenylnitromethane in 50–55% yield [2]:

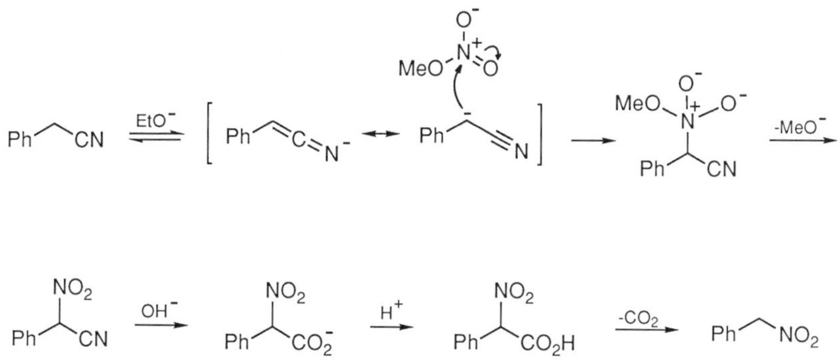

(c) Formation of imines

Enolate-forming compounds react with aromatic nitroso compounds to give imines:[†]

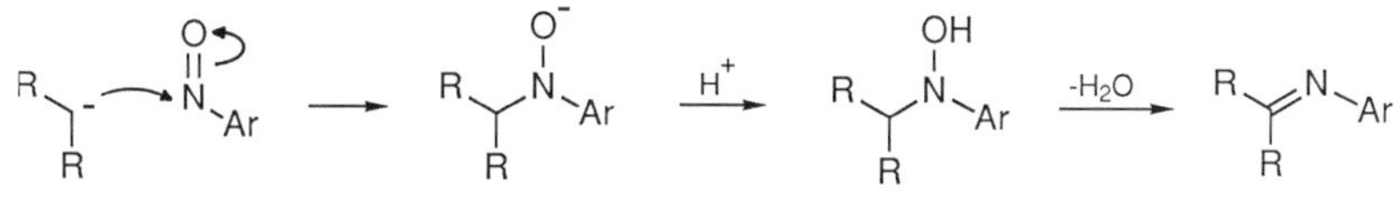

The reaction provides a method for the oxidation of activated methylene groups to carbonyl groups which are released on acidic hydrolysis of the imine. The readily prepared p-nitrosodimethylaniline (p. 380) is usually employed as the nitroso compound. For example, reaction with 2,4-dinitrotoluene, whose methyl group is activated by the ortho and para nitro groups, gives 2,4-dinitrobenzaldehyde:

---

* Methyl nitrate is prepared from methanol and nitric acid in the presence of concentrated sulfuric acid (cf. ethyl nitrite; p. 320). It is not very stable and is used immediately.
† This reaction bears the same relationship to that between an enolate and an alkyl nitrite as the aldol reaction bears to the Claisen condensation.

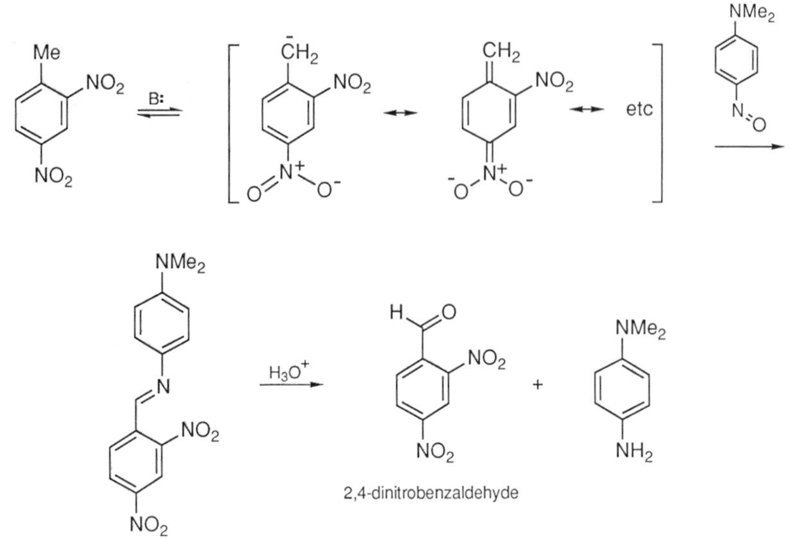

2,4-dinitrobenzaldehyde

10.6 α-Amino acids, peptides, and proteins

Proteins are naturally occurring long-chain polymers, of universal occurrence in living systems, which are derived from α-amino acids linked together by amide bonds:

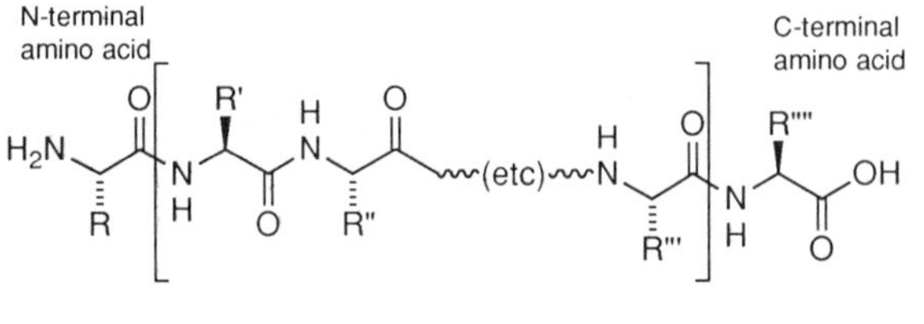

The commonest α-amino acids, R—CH(NH_2)CO_2H, in proteins are: R = H (glycine), CH_3 (alanine), $(CH_3)_2CH$ (valine), $(CH_3)_2CHCH_2$ (leucine), $CH_3CH_2CH(CH_3)$ (isoleucine), $HOCH_2$ (serine), $HSCH_2$ (cysteine), $CH_3SCH_2CH_2$ (methionine), HO_2CCH_2 (aspartic acid), $HO_2CCH_2CH_2$ (glutamic acid), $H_2N—(CH_2)_3—CH_2$ (lysine), $PhCH_2$ (phenylalanine), together with

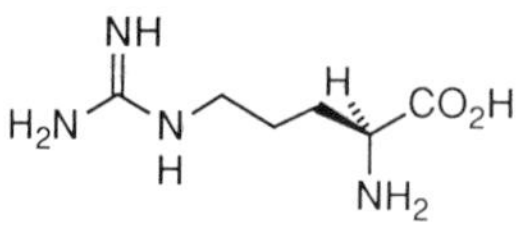

arginine

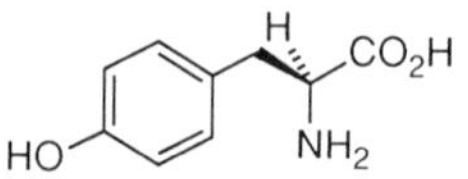

tyrosine

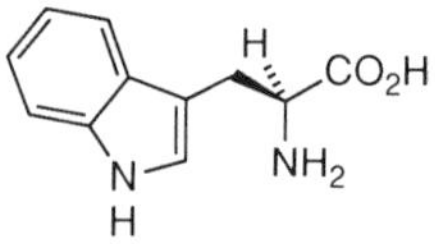

tryptophan

histidine 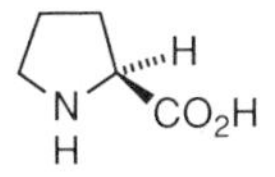   proline 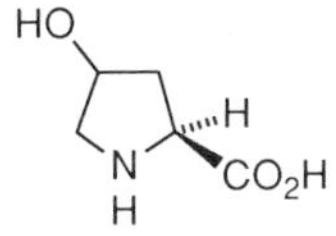   hydroxyproline

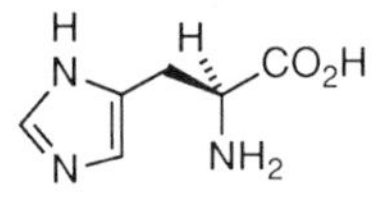

Except in one case (glycine) the carbon atom of the —CHR— fragment in the protein chain is asymmetric and each has the same (L) configuration at this centre:

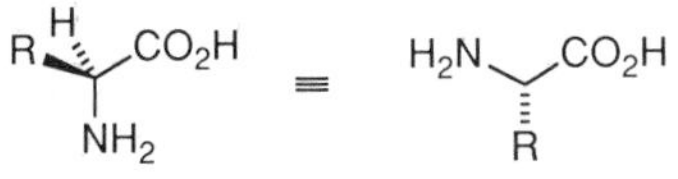

L α-amino acid

Peptides, which also occur naturally, are similar polymeric materials but of smaller chain length. The (arbitrary) distinction between peptides and proteins is that those polymers with molecular weights less than 10 000 are termed peptides and those with higher molecular weights are termed proteins.

The acid-catalyzed hydrolysis of peptides and proteins yields the constituent α-amino acids, and the syntheses of peptides achieved to date have been based on the reverse of this process. The α-amino acids are therefore of considerable importance. They are zwitterionic compounds, $H_3\overset{+}{N}$—CHR—CO_2^-, both in the solid state and in solution, although in the latter conditions the zwitterion is in equilibrium with both its conjugate base and its conjugate acid, the position of equilibrium depending on the acidity of the medium:

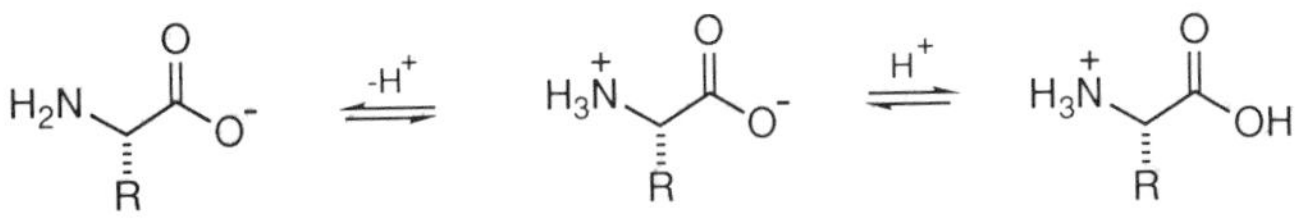

The physical properties of α-amino acids reflect their zwitterionic structure: they are high-melting solids (>200°C), insoluble in non-polar solvents, and soluble in water (the solubility decreasing as the group R is made increasingly non-polar). Since they are normally obtained in aqueous solution together with inorganic salts, the isolation of the more soluble members of the group is difficult. They can be obtained in one of three ways.

(1)   Copper(II) ion is added to precipitate the copper chelate,

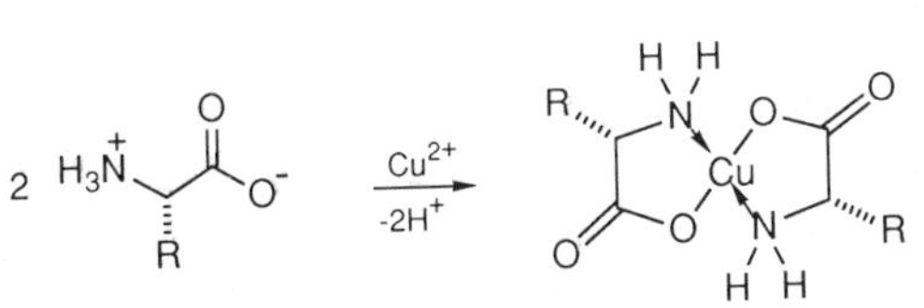

(2)   Hydrochloric acid is added, the solution is evaporated to dryness, and the amino acid hydrochloride is extracted into alcohol (leaving the inorganic salts as

residue). Lead oxide is then added to remove chloride ion, followed by hydrogen sulfide to remove lead ion, and evaporation leaves the amino acid which can be recrystallized from water or aqueous alcohol.

(3) The simplest and most modern method is to separate them from inorganic salts on an ion-exchange resin.

(a) The synthesis of α-amino acids

The following are the more general methods for the synthesis of α-amino acids, illustrated with reference to some of those derived from proteins.

(i) From α-halo-acids.
The simplest method consists of converting a carboxylic acid into its α-bromo-derivative and treating this with ammonia:

(racemic)

The *Hell–Volhard–Zelinsky procedure* is normally employed for the first step. The acid is treated with bromine in the presence of a small quantity of phosphorus, the phosphorus tribromide formed converts the acid into its acid bromide, this undergoes (electrophilic) bromination at the α-position *via* its enol tautomer, and the resulting α-bromo acid bromide exchanges with more of the acid to give α-bromo acid together with more acid bromide for further bromination:

$$2\,P \quad + \quad 3\,Br_2 \quad \longrightarrow \quad 2\,PBr_3$$

The conversion of the α-bromo acid into the α-amino acid may be accomplished with *excess* of ammonia (p. 302), but better yields and purer products are usually obtained by the *Gabriel procedure* (p. 303).

The use of malonic ester considerably increases the versatility of this general method. First, the appropriate alkyl group can be attached by the standard alkylating procedure (p. 236), and second, bromination of the resulting substituted malonic acid occurs readily with bromine alone because these acids, unlike mono-carboxylic acids, are significantly enolic. The synthesis of leucine is illustrative:

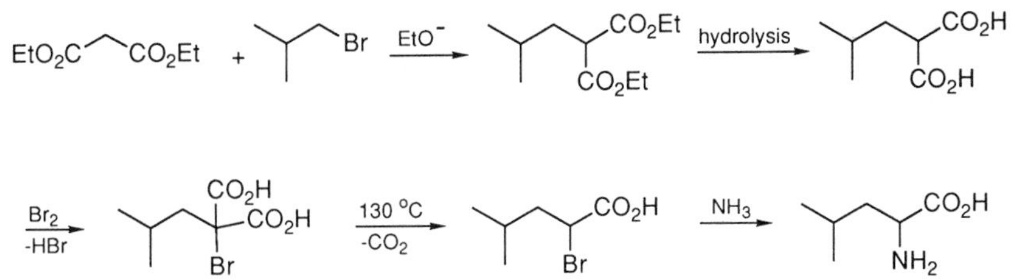

leucine

This approach may be combined with the Gabriel procedure, as in a synthesis of methionine:

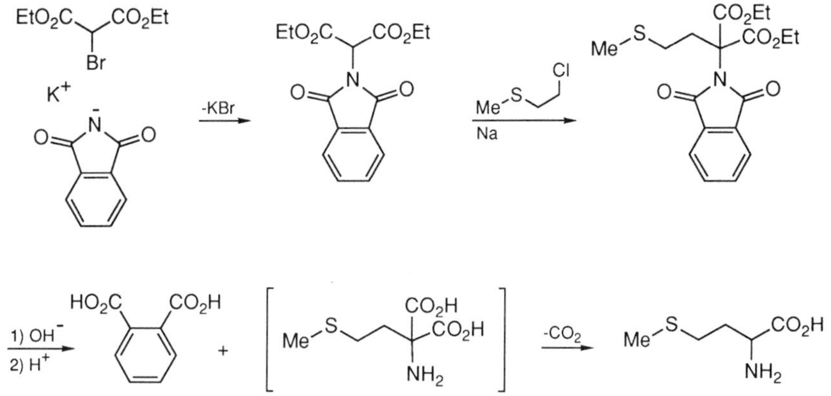

methionine

Application of the same method to the synthesis of cysteine illustrates the use of a protective group: the —SH group in cysteine is held in the form of its benzyl-derivative until the final stage and is then released by hydrogenolysis (p. 644):

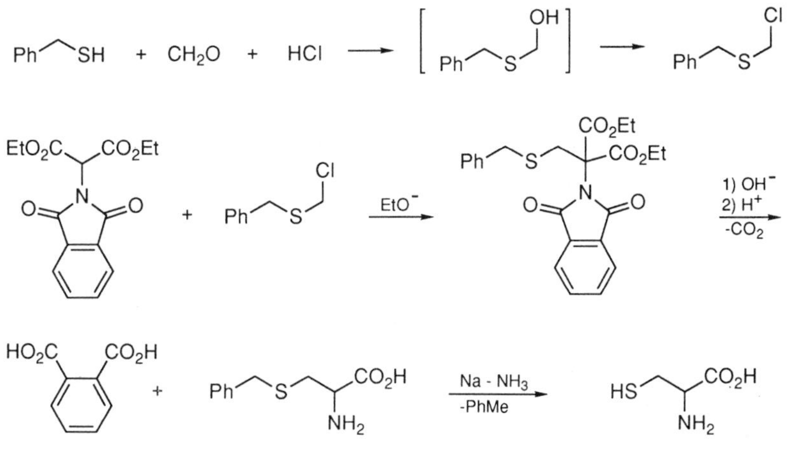

cysteine

The amino group may also be introduced *via* nitrosation at an activated C—H group. For example, alkylation of acetoacetic ester with 2-bromobutane followed by treatment with an alkyl nitrite in the presence of base gives a nitroso-derivative which, being incapable of ionizing, undergoes cleavage (cf. p. 321). Reduction of the resulting oxime gives isoleucine:

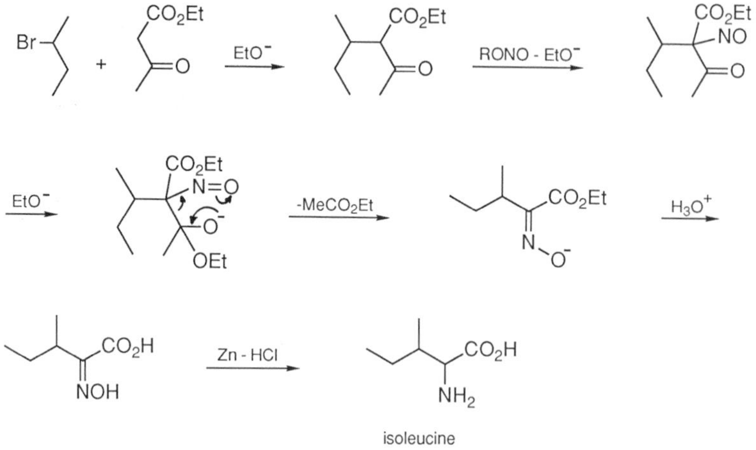

isoleucine

A further modification employs acetamidomalonic ester, i.e. the amino group is introduced in a protected form *before* attachment of the required alkyl residue. Acetamidomalonic ester is obtained by nitrosation of malonic ester (cf. p. 319) followed by reduction in the presence of acetic anhydride. An example of its use occurs in a synthesis of glutamic acid:

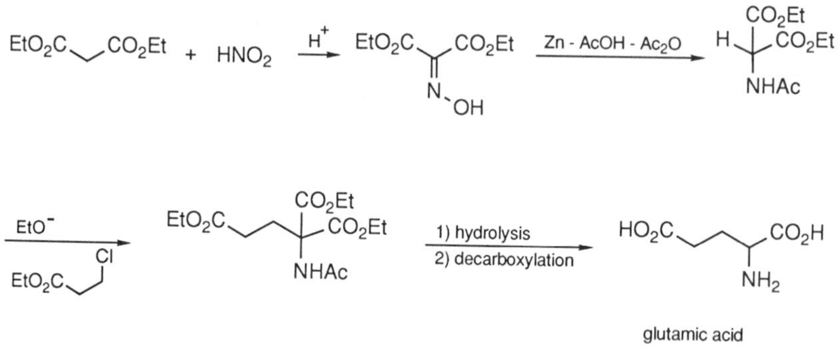

glutamic acid

A variation of this method is used in the synthesis of lysine where the alkyl residue is introduced by Michael addition:

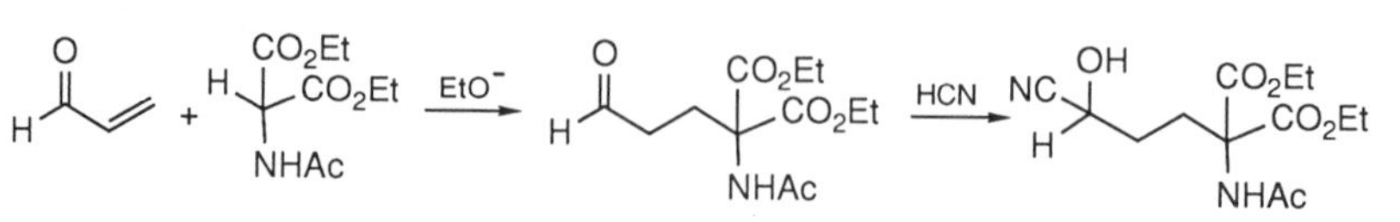

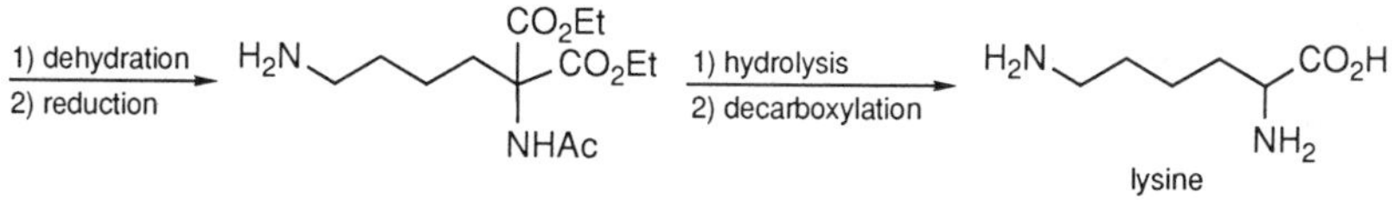

lysine

(*ii*) *Strecker reaction.* This reaction may be carried out either with potassium cyanide and ammonium chloride (p. 308) or with hydrogen cyanide and ammonia; the latter conditions obviate the need to separate the product from potassium chloride. The intermediate α-aminonitrile is conveniently hydrolyzed with sulfuric acid, sulfate ion then being removed from the solution by the addition of barium carbonate.

Methionine has been synthesized by the Strecker procedure:[*]

methionine

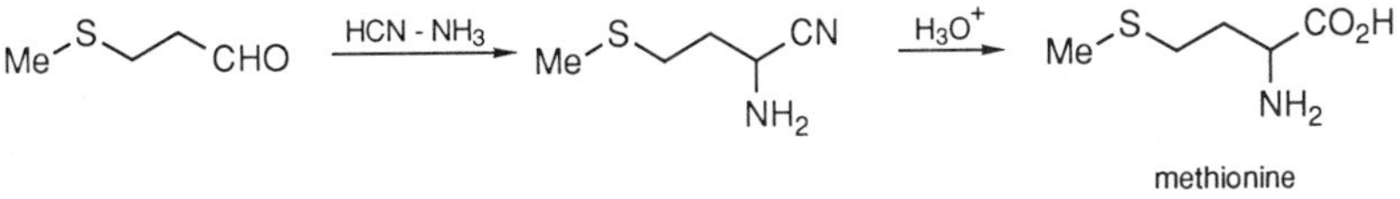

The following synthesis of serine by the Strecker procedure illustrates a method for introducing an alcoholic group in a protected form. The ether linkage is hydrolyzed at the final stage by boiling with hydrobromic acid.

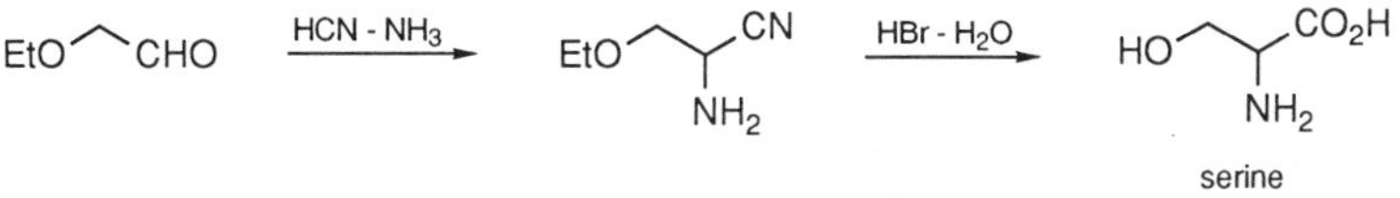

serine

(*iii*) *Curtius method.* Acid azides undergo thermal rearrangement to isocyanates (p. 443). In the presence of water the isocyanate reacts to form an amine *via* the unstable carbamic acid. The required acid azides are readily obtained from malonic ester and its derivatives, e.g.

[*] β-Thiomethylpropionaldehyde may be prepared as follows:

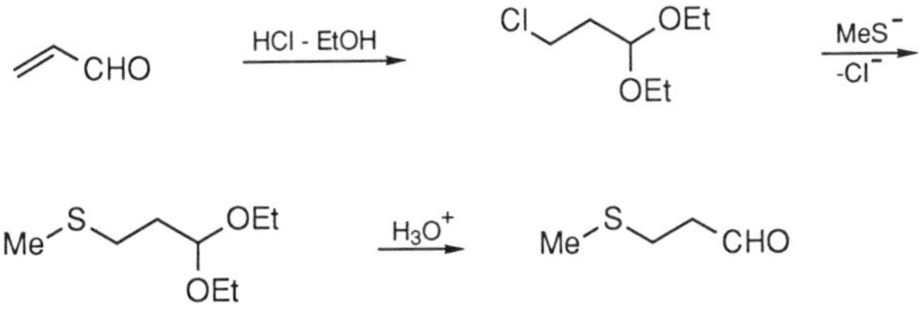

The addition of hydrogen chloride to propenal ('anti-Markovnikov'; see p. 84) is followed by conversion of the aldehyde into its diethyl acetal which protects the compound against base-catalyzed reaction during treatment with the methanethiol anion.

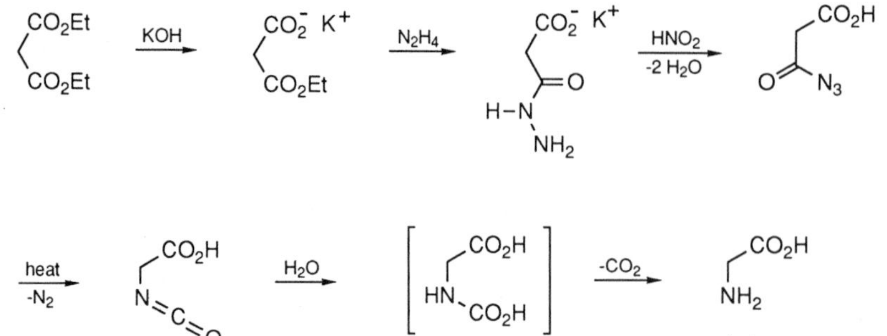

glycine

(iv) Condensation methods. The aromatic-containing α-amino acids are usually prepared by Perkin-type reactions between aromatic aldehydes and the activated methylene groups of hydantoin and related cyclic compounds. The general procedures have been outlined earlier (p. 219), and are suitable for the following conversions:

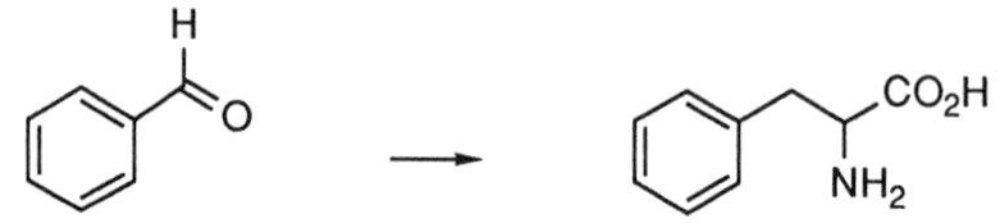

phenylalanine

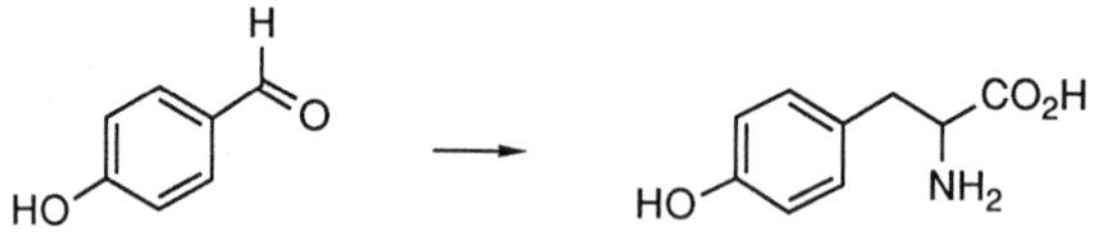

tyrosine

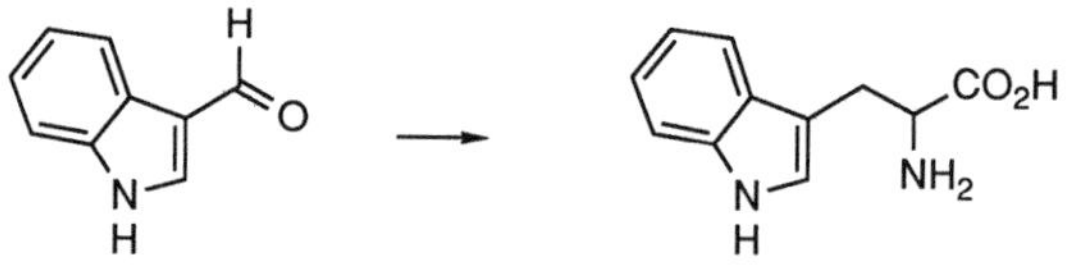

tryptophan

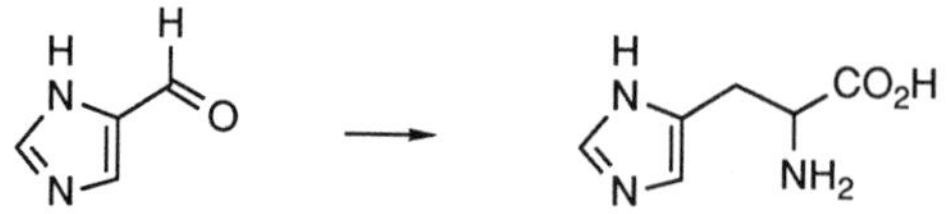

histidine

(b) The synthesis of peptides

The synthesis of peptides from α-amino acids presents three main problems. First, a reaction between two α-amino acids could give each of four products (two by self-condensation and two by crossed reaction). In order to achieve a specific mode of reaction it is necessary to protect the amino group of one reactant and the carboxyl group of the other so that reaction can only occur in one way; after the peptide link has been formed, the protecting groups are removed. Secondly, it is necessary to activate the carboxyl group which is to be bonded to amino in order that the peptide-forming step should take place in mild conditions in which undesirable side reactions do not occur.

Thirdly, it is necessary to prevent racemization of the optically active centres in the peptide units, for the synthesis of a natural peptide requires that all these centres should have the L-configuration. Racemization usually occurs as follows: a basic reagent initiates formation of the oxazolone system,

the optically active centre racemizes *via* the enol form of the oxazolone,

and ring opening of the DL-mixture of oxazolones (by, for example, reaction with the amino group of the amino acid to be bonded) gives DL-peptide.*

Finally, when α-amino acids which contain other functional groups such as —SH are employed, additional protection is necessary.

In the synthesis of peptides, the *step-wise* technique is usually adopted. One amino acid is protected at its carboxyl end with a group Y and the second is protected at its amino end with a group Z and activated at its carboxyl group by conversion into a derivative —COX. Reaction between the two forms the peptide bond. The group Z is now removed and a third amino acid, protected at its amino end and activated at its carboxyl end, is introduced to form the second peptide bond. Repetition of the procedure gives the required peptide.

---

*The problem of racemization is not encountered when the amino acid unit is glycine, which is not asymmetric, or proline or hydroxyproline, which cannot form oxazolones of this kind.

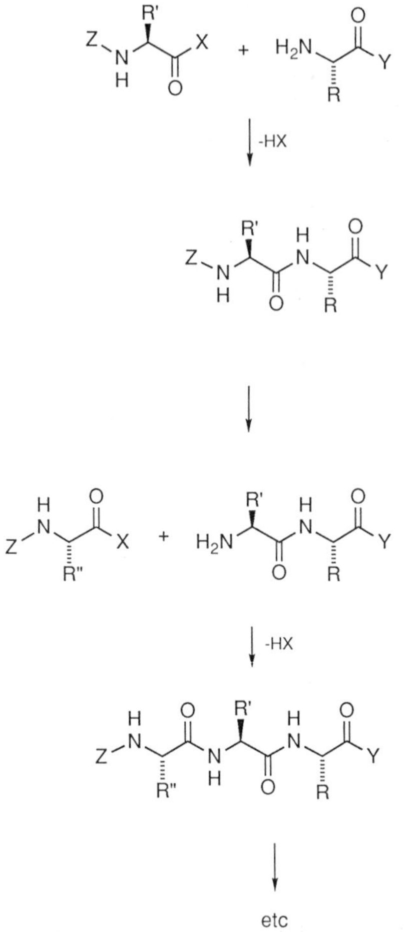

-HX

-HX

etc

The chief advantage of the step-wise procedure is that the likelihood of racemization is minimized because the oxazolone-forming intermediates (i.e. peptide chains terminating in —COX) are not involved. In some syntheses, however, it is desirable to bond two pre-formed peptide units and racemization is then only avoided by using the acid azide method of activation (p. 338).

(*i*) *Methods of protection.* A practicable method of protection must have the following characteristics: the protective group must be capable of introduction in conditions in which side reactions, including racemization, do not occur; it must be inert in the conditions in which the peptide link is formed; and it must be removable in conditions which do not affect other bonds, in particular the peptide bonds.

The carboxyl group is normally protected by converting it with isobutylene in the presence of sulfuric acid into its t-butyl ester:

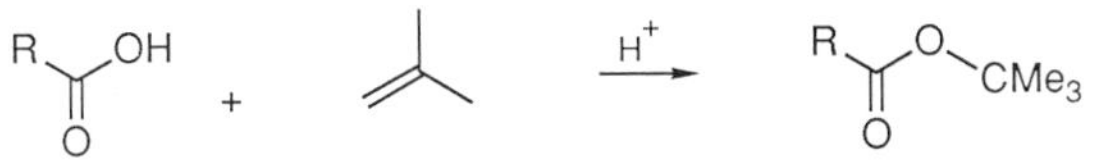

The protecting group may be removed by mild acid hydrolysis *via* the readily formed t-butyl cation:

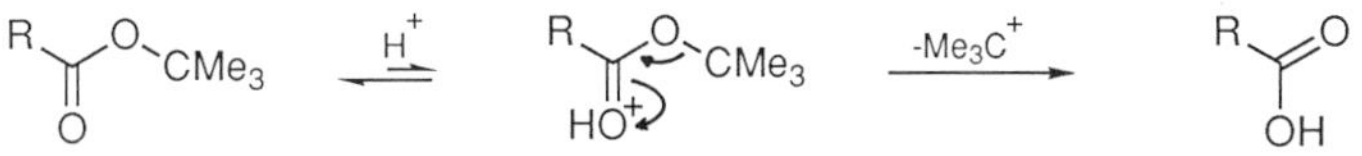

$$Me_3C^+ \quad + \quad H_2O \quad \xrightarrow{-H^+} \quad Me_3C-OH$$

The amino group is commonly protected in organic synthesis by acetylation or benzoylation. Neither method is suitable here because hydrolysis of the amide linkages to the protective groups also cleaves the peptide bonds. Certain acyl-amino groups can, however, be cleaved by methods other than basic hydrolysis, in particular, by catalytic hydrogenolysis and mild acid hydrolysis. Four derivatives are widely employed.

(1) *The benzyloxycarbonyl group.* Benzyl chloroformate (from benzyl alcohol and phosgene) reacts with amino groups to give benzyloxycarbonyl-derivatives:

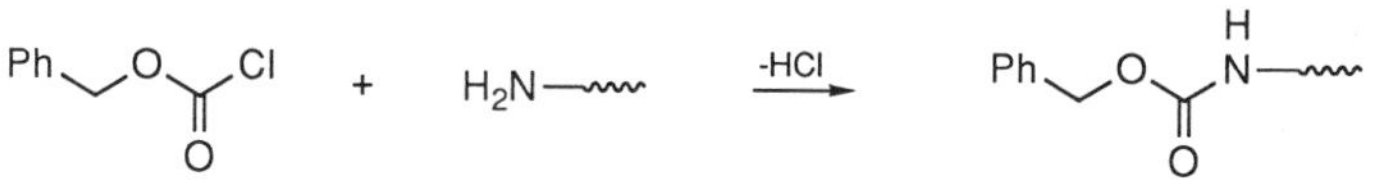

The protecting group may be removed by hydrogenolysis,* carried out with hydrogen on a palladium catalyst,

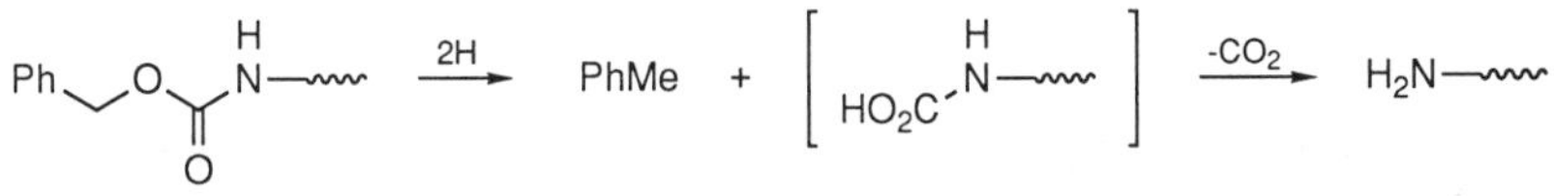

but sulfur-containing compounds (i.e. units derived from cysteine and cystine) poison the catalysts. A suitable alternative reducing agent is sodium in liquid ammonia.

(2) *The t-butyloxycarbonyl group ('Boc').* This was originally introduced by reaction of the amino group with t-butyl azidoformate,

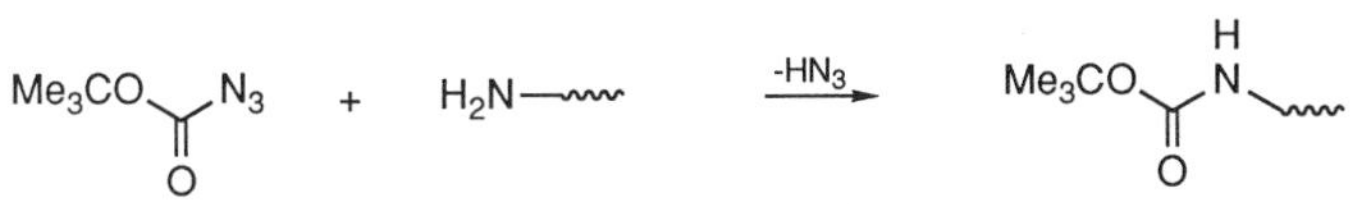

However, the azidoformate is a hazardous compound and it has been replaced by two stable compounds which react readily with the amino group at room temperature: di-t-butyl dicarbonate [7],

---

* Systems of the types $ArCH_2O—$ and $ArCH_2N<$ undergo hydrogenolysis (i.e. hydrogenation with bond fission) in catalytic and other conditions, see p. 643.

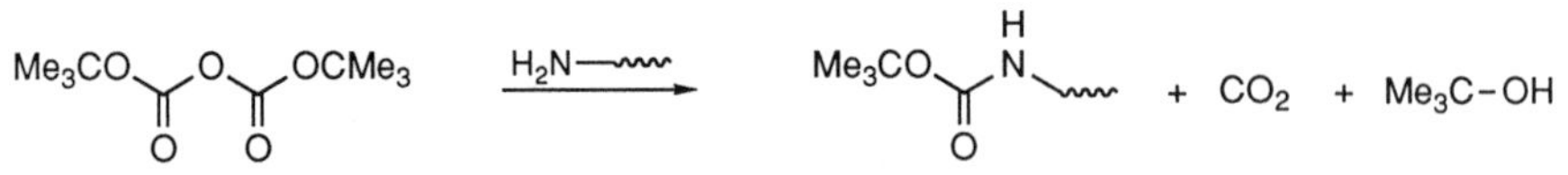

and an oxime derivative,*

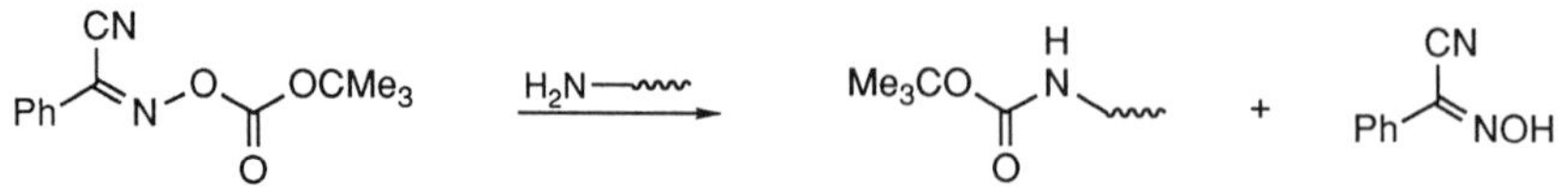

Its high reactivity towards the amino group is due in part to the $-M$ effect of the cyano-group:

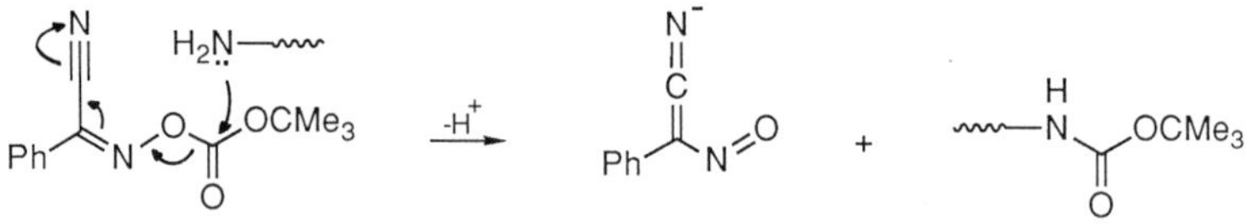

After peptide-bond formation, the t-butyloxycarbonyl group is removed with cold trifluoroacetic acid:

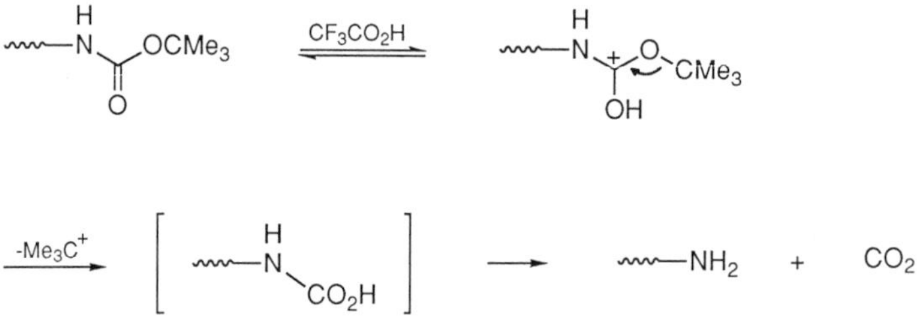

In order to prevent the t-butyl cation which is liberated from reacting with peptide components such as the activated aromatic rings of tyrosine and tryptophan, an efficient carbocation scavenger such as anisole or dimethyl sulfide is included.

(3) *The biphenylyl analogue of Boc ('Bpoc')*. The advantage of this protecting group compared with Boc is that it can be removed under milder acidic conditions – an aqueous solution of the weakly acidic CF_3CH_2OH – as a result of the additional stabilizing influence of the aromatic rings on the leaving carbocation:

* This is prepared from phenylacetonitrile [6]:

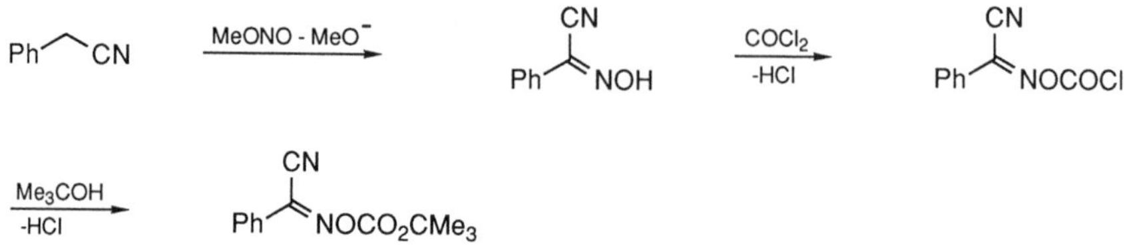

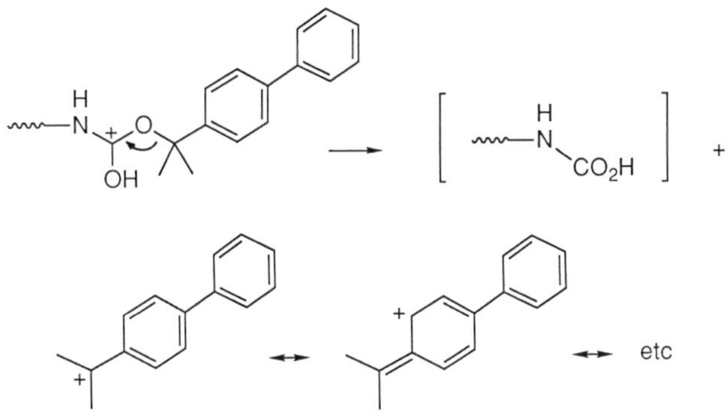

(4)   *The 9-fluorenylmethyloxycarbonyl group ('Fmoc')*. The important characteristic of this protecting group is that it can be removed by treatment with a
weak base, such as piperidine, which does not affect other linkages. The reason
doubtless lies in the aromatic stabilization energy of the 9-fluorenyl carbanion
(p. 40), whether or not this is an intermediate (as shown) or a contributor to the
transition state of an E2 elimination.

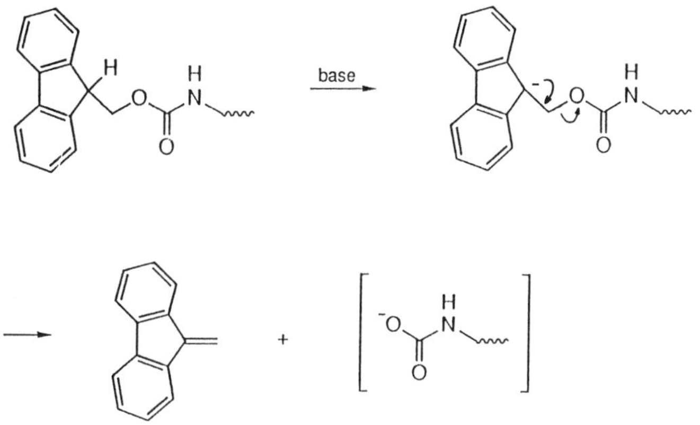

Racemization *via* the oxazolone is not a significant problem with any of the
above four amino-protecting groups.

Additional protection.   In most polypeptide syntheses, one or more amino
acids is involved which contains a reactive group in the side-chain. This must be
protected with a group that is not affected when the α-amino protecting group is
removed. There are two general cases. When the α-amino group is protected by t-
butyloxycarbonyl, which is removed by trifluoroacetic acid, the side-chains are
protected by groups that are unaffected by this but are removable by hydrogenolysis, namely benzyloxycarbonyl for side-chain amino and benzyl for side-
chain hydroxyl, thiol, and carboxyl. When the α-amino group is protected by
benzyloxycarbonyl, which is removed by hydrogenolysis, or by Bpoc or Fmoc,
which are removed by very mild acidic or basic conditions respectively, side-chain

protection is usually by groups that can be removed by trifluoroacetic acid, namely t-butyloxycarbonyl for amino and t-butyl for hydroxyl and carboxyl.

Glutamic acid and lysine present particular problems. The former occurs in peptides and proteins as both α-glutamyl and γ-glutamyl residues

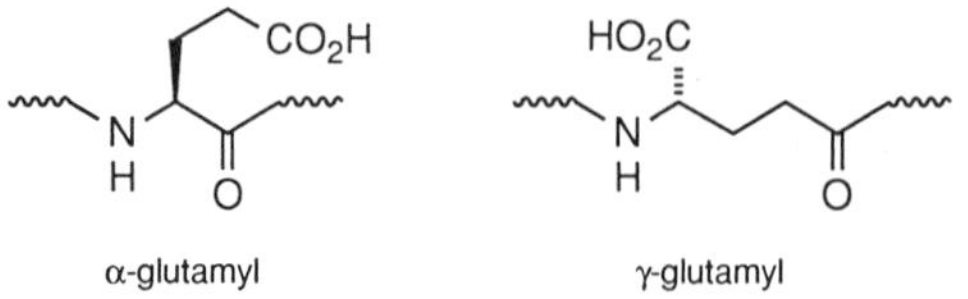

α-glutamyl   γ-glutamyl

and it is therefore necessary to be able to protect each of the two carboxyl groups separately. The γ-carboxyl group can be protected as follows: (1) both carboxyl groups are benzylated, (2) mild acid hydrolysis preferentially regenerates the γ-carboxyl group,* (3) the amino-protecting group Z is introduced, (4) the free acid is esterified with isobutylene, and (5) the α-benzyl group is cleaved by hydrogenolysis:

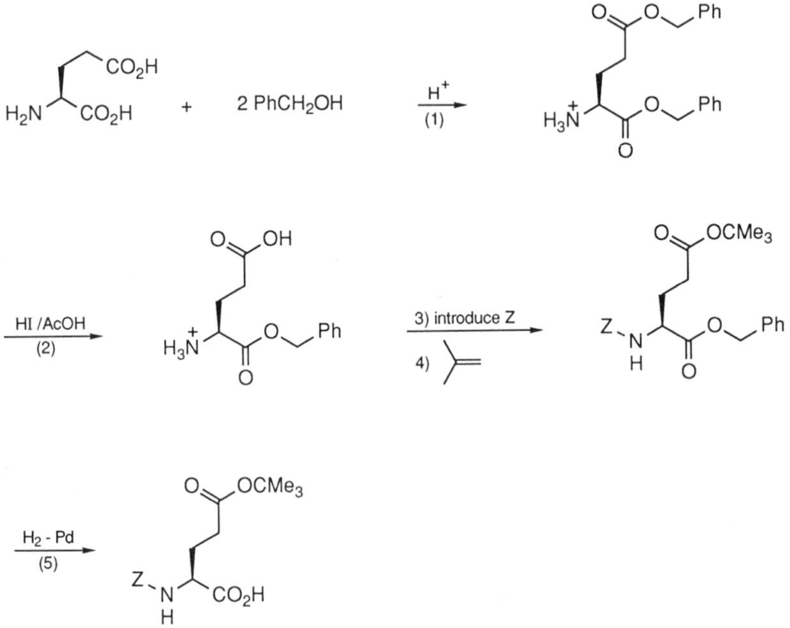

* The acyl-oxygen of the α-carboxyl group is less basic than that of the γ-carboxyl group because it is more affected by the electron-withdrawing positive pole. Consequently acid-catalyzed hydrolysis,

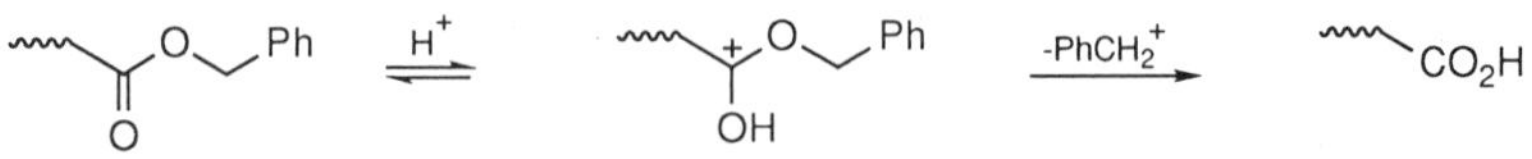

occurs more readily at the γ-ester.

The α-carboxyl group is protected by acid-catalyzed methylation of the γ-carboxyl, t-butylation of the α-carboxyl, and cleavage of the γ-methyl ester with base:*

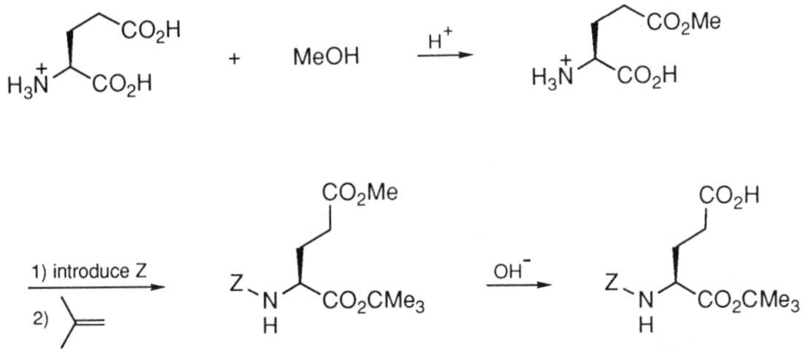

Lysine contains an amino group in the side-chain, although it is always present in peptides and proteins as the α-lysyl group. The ε-amino group may be selectively protected by adding the protecting agent to lysine's copper chelate, e.g.

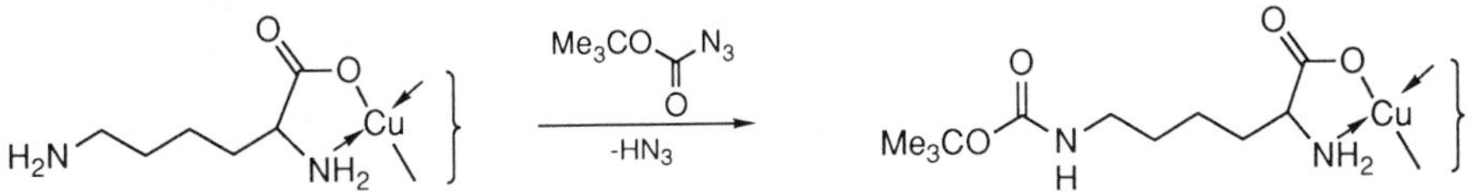

Finally, the guanidino group in arginine can be protected by nitration, or by conducting the synthesis at a pH at which the group is essentially fully protonated (the pK of —NH—C(NH$_2$)=NH is ~12, so that at pH 7 only about one molecule in 10^5 is not protonated).

(ii) *Methods of activation.*   Carboxylic acids react with amines only under very vigorous conditions (cf. ammonia, p. 340). It is consequently necessary to convert the acid into a derivative which is more reactive towards nucleophiles, the requirement being that the group X in the derivative R—CO—X should be a good leaving group. Derivatives of several types have been used.

    (1) *Acid chlorides (X = Cl).* These are exceptionally reactive towards nucleophiles and peptide-bond formation occurs readily but chloride ion is so good a leaving group that N-carboxyanhydrides are formed from benzyloxycarbonyl-protected acid chlorides:

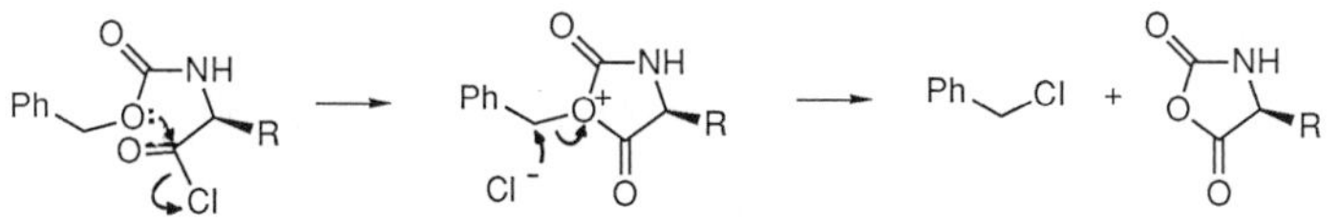

*The readier acid-catalyzed esterification of the γ-carboxyl has the same basis as the readier acid-catalyzed hydrolysis of the γ-ester (preceeding footnote). The selective base-catalyzed hydrolysis occurs because the t-butyl ester is more hindered than the methyl ester.

(2) *Acid azides* ($X = N_3$). Conversion of an acid into its azide can be accomplished under relatively mild conditions (esterification, treatment with hydrazine and then with nitrous acid, p. 443). Azides, though not so reactive as chlorides, are sufficiently reactive for peptide-bond formation to occur smoothly without the competing reaction leading to the N-carboxyanhydride.

(3) *Mixed anhydrides* ($X = OCOR$). Mixed anhydrides are readily formed by displacement by the nucleophilic carboxylate anion on an acid chloride. A typical example is the reaction with ethyl chloroformate, triethylamine being added to generate the carboxylate ion:

Reaction of an amino group with the resulting mixed anhydride occurs smoothly:

(4) *Activated esters* ($X = OR$). Although alkyl esters are not very reactive towards amines, aryl esters, particularly those with electron-attracting substituents, react readily. The reason is that the negative charge in aryloxide anions is delocalized over the aromatic ring and over *ortho-* and *para*-substituents of $-M$ type, e.g.

so that ArO^- is a much better leaving group than RO^-.* As a consequence, the p-nitrophenyl group is widely used for activation of carboxyl:

(5) *Carbodiimide.* Carboxylic acids react readily with amines in the presence of acid and a carbodiimide (usually dicyclohexylcarbodiimide), a disubstituted urea precipitating from solution. The probable mechanism is:†

---

* This is also, of course, the basis for the greater acidity of phenol than an alcohol, and of *p*-nitrophenol than phenol, p. 45.
† The two C=N bonds make the central carbon atom very reactive towards nucleophiles, particularly in the presence of an acid. The resulting adduct, containing the grouping —CO—O—C=NR, is analogous to an anhydride and reacts readily with amino groups.

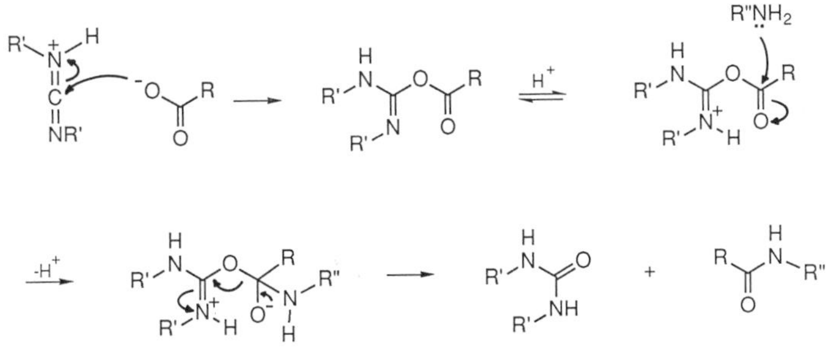

An improved procedure is to add the acid to the carbodiimide in the presence of *N*-hydroxybenzotriazole. The hydroxylic oxygen in the triazole is strongly nucleophilic and reacts with the adduct of acid and diimide to form a type of activated ester; this then reacts readily with the amine:

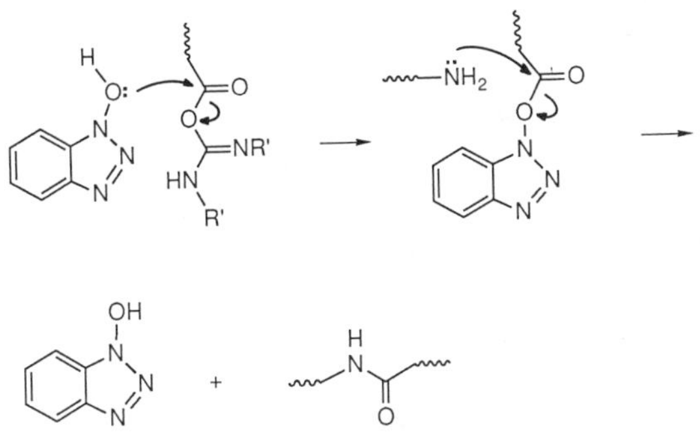

Both the activated ester and the carbodiimide methods are suitable only for the step-wise synthesis of peptides, for carboxyl-terminating peptides are racemized by these procedures.

(iii) Solid-phase peptide synthesis. The step-wise approach to peptide synthesis, when carried out by conventional chemical techniques in which each intermediate is purified before the next step, is of limited value for large peptides because the overall yield is likely to be very low. For example, if 50 synthetic steps were involved, then even if each gave a 90% yield, the overall yield would be only $(0.9^{50} \times 100) = 0.5\%$; for 100 steps, it would be only 0.003%. Since many peptides of medical or biological importance require at least this number of operations, a new approach was called for.

The breakthrough was made by the American, Bruce Merrifield, who introduced solid-phase synthesis. The growing peptide chain is attached by a chemical bond to a solid support in the form of an insoluble polymer. Instead of purification procedures after each step, impurities and unused reagents are simply removed by washing and filtration. The result is that each step is quantitative, or nearly so, and the overall yield is high. Moreover, the steps can be completed comparatively quickly.

An early example was the synthesis of bradykinin (a naturally occurring non-apeptide with important physiological properties). The solid support used was a copolymer of styrene and divinylbenzene. Styrene itself gives a linear polymer, $(CH_2—CHPh)_n$, and the introduction of a small percentage of divinylbenzene yields a polymer with cross-linked chains:

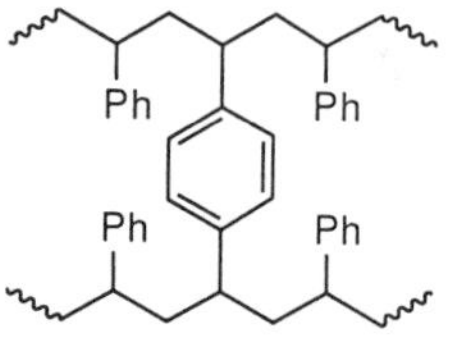

The resulting material has a gel structure with good permeability. About 5% of the benzene rings were then chloromethylated (p. 369) and the first amino acid, arginine, was introduced as t-butyloxycarbonylnitroarginine, its free carboxylate group reacting at the benzyl chloride centres so that the arginine was held to the solid as a benzyl ester (step 1). The t-butyloxycarbonyl group was removed with hydrochloric acid (step 2), triethylamine was added to liberate the amino group (step 3), and the second amino acid, phenylalanine (protected as its t-butyloxycarbonyl-derivative), was introduced, reaction being brought about by dicyclohexylcarbodiimide (step 4). The operations in steps 2–4 were repeated with the appropriate amino acids until the nonapeptide had been formed and the peptide was then detached from the solid by passing hydrogen bromide through a suspension of the solid in trifluoroacetic acid. This acid treatment also removed the final t-butyloxycarbonyl group and the benzyl protecting-group of a serine unit, and catalytic hydrogenation removed the nitro protecting-groups of two arginine units. Bradykinin was isolated in 68% yield.

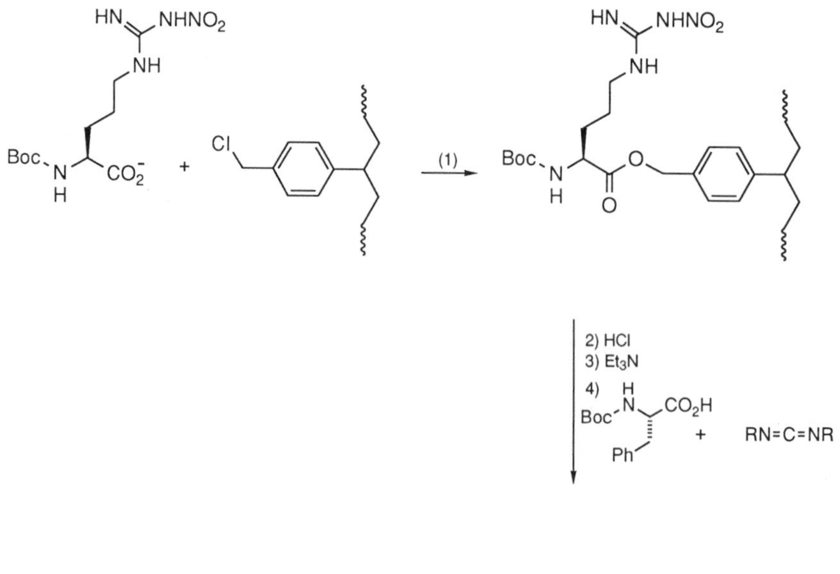

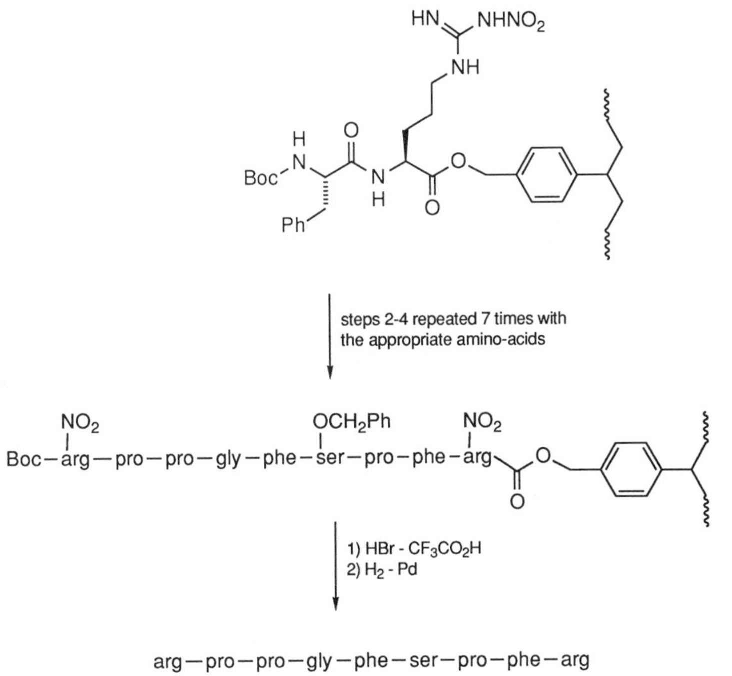

steps 2-4 repeated 7 times with
the appropriate amino-acids

$$\text{Boc}-\overset{\overset{\displaystyle NO_2}{|}}{\text{arg}}-\text{pro}-\text{pro}-\text{gly}-\text{phe}-\overset{\overset{\displaystyle OCH_2Ph}{|}}{\text{ser}}-\text{pro}-\text{phe}-\overset{\overset{\displaystyle NO_2}{|}}{\text{arg}}$$

1) HBr - CF$_3$CO$_2$H
2) H$_2$ - Pd

arg – pro – pro – gly – phe – ser – pro – phe – arg

bradykinin

(Boc = t-butyloxycarbonyl; arg = arginine; pro = proline; gly = glycine;
phe = phenylalanine; ser = serine)

The solid-phase method has now been developed by automation of the chemical operations. A remarkable example of its use is the synthesis of the enzyme ribonuclease. This contains 124 amino acids, and the synthesis required nearly 12 000 automated operations.

There have been a number of developments of Merrifield's method, both in the type of polymer used and in the protection–deprotection regime. For example, in early applications, such as the synthesis of bradykinin, α-amino groups were protected with Boc and side-chains with benzyloxy residues. A complementary regime is now possible: α-amino groups are protected with Fmoc (p. 335) (removed with piperidine after each peptide bond has been formed) and side-chains with t-butyloxy groups, which can be removed at the end with trifluoroacetic acid. Moreover, when the first amino acid is attached to the polymer through a suitable 'linking' group, e.g.

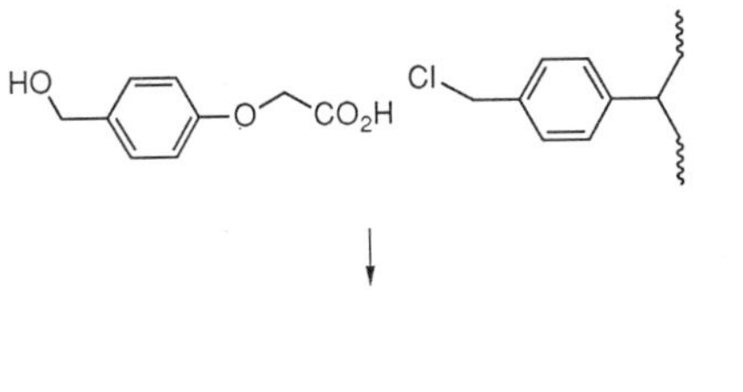

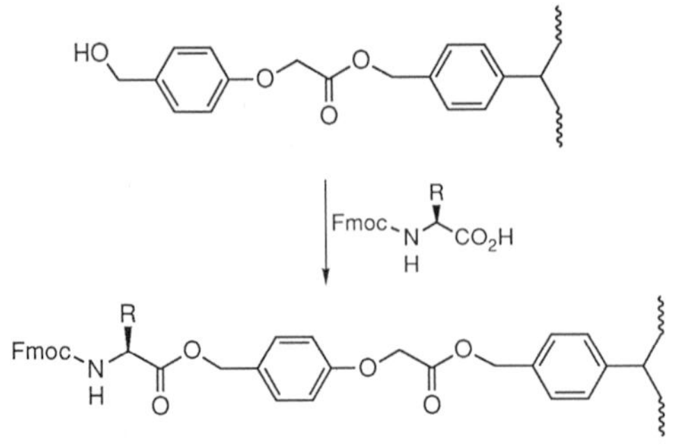

the final peptide can be removed from the support at the same time as the side-chain protectors, as a result of the stabilizing influence of the phenolic oxygen on the intermediate carbocation in the acid cleavage:

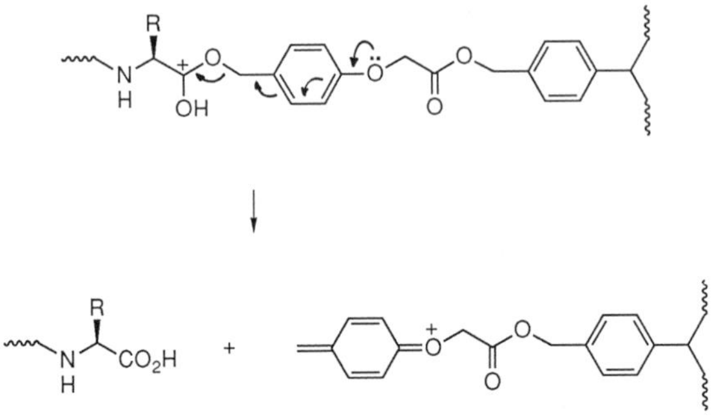

Problems 1. What products may be obtained from the reaction of ammonia with each of the following: methyl iodide; acetyl chloride; formaldehyde; acetaldehyde; chloral; acetone; benzaldehyde; ethylene oxide; 2,4-pentanedione; 2,5-hexanedione; acetonitrile; phenyl isothiocyanate; cyanamide?

2. Illustrate the types of product which can be obtained from the reactions of acetoacetic ester with compounds containing nucleophilic nitrogen.

3. Outline routes to the following compounds:

(a)   Me$_3$N

(b)

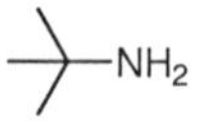

(c)

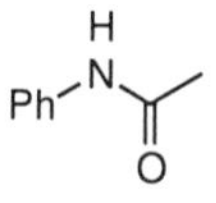

(d)

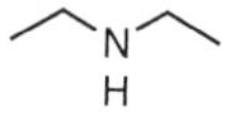

(e)

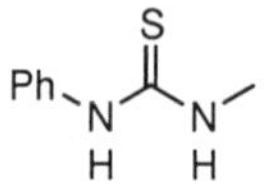

(f)

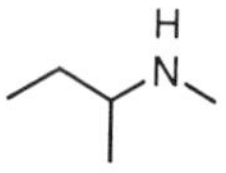

(g)
HS—CH₂—CH(NH₂)—CO₂H

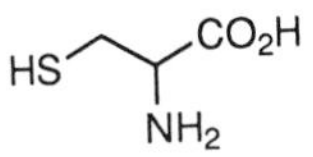

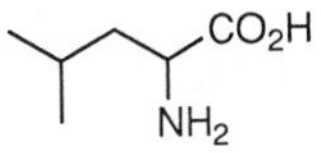

(h)
(CH₃)₂CH—CH₂—CH(NH₂)—CO₂H

(i)
CH₃CH₂—NO₂

(j)
CH₃CH₂—O—N=O

(k)
CH₃CH₂—O—NO₂

(l)
Ph—CH₂—S—C(=NH₂⁺)—NH₂   Cl⁻

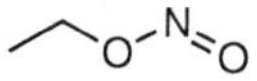

(m)
Ph—C(=NH)—NH₂

(n)
NC—C(CH₃)₂—N=N—C(CH₃)₂—CN

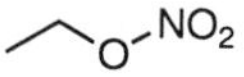

(o)
cyclohexyl—N=C=N—cyclohexyl

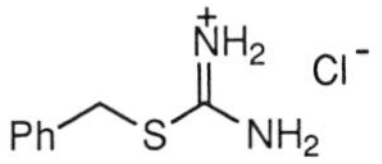

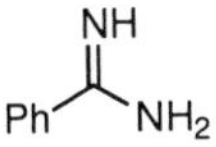

(p)
pyrrolidine-2-CO₂H (N–H)

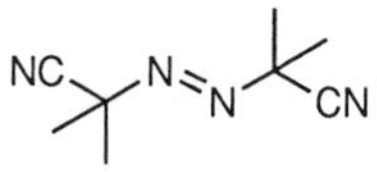

(q)
2,4-diacetyl-3,5-dimethylpyrrole

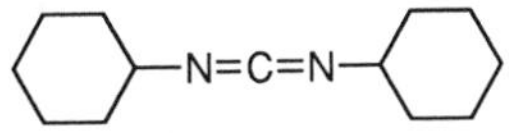

(r)
3,5-diphenyl-1H-pyrazole (Ph, Ph)

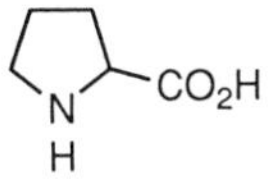

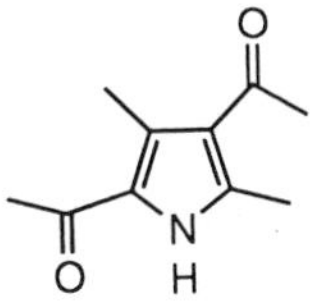

(s)
HO₂C—CH(NH₂)—CH₂—CH₂—C(=O)—NH—CH(CH₃)—C(=O)—NH—CH₂—CO₂H

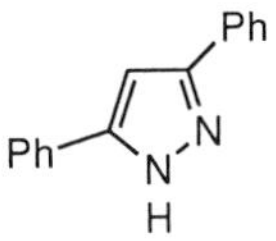

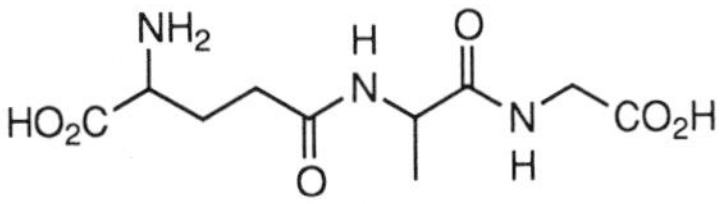

11 Electrophilic aromatic substitution

11.1
The mechanism
of substitution

The substitution of benzene by an electrophilic reagent (E^+) occurs in two stages: the reagent adds to one carbon atom of the nucleus, giving a carbocation in which the positive charge is delocalized over three carbon atoms, and a proton is then eliminated from this adduct:

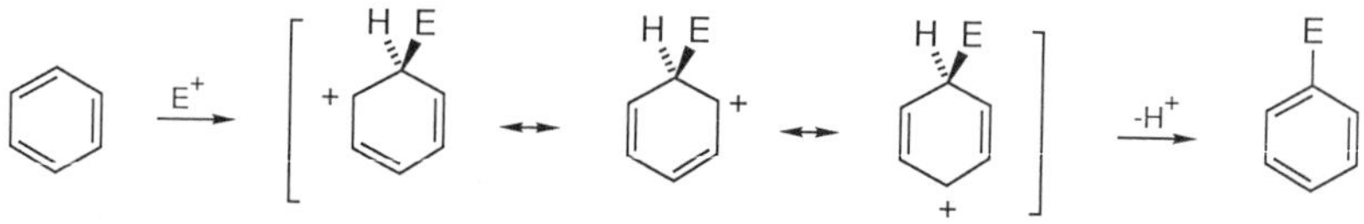

The electrophile may be a charged species, such as the nitronium ion, NO_2^+, which participates in nitration (section 11.4), and the t-butyl cation, $(CH_3)_3C^+$, which participates in Friedel–Crafts t-butylation (section 11.3), or it may be a neutral species which can absorb the pair of electrons provided by the aromatic nucleus. The latter class includes reagents such as sulfur trioxide (in sulfonation, section 11.5) which absorb the electron-pair without bond breakage,

and those such as the halogens in which uptake of the electron-pair leads to bond breakage and the formation of a stable anion,

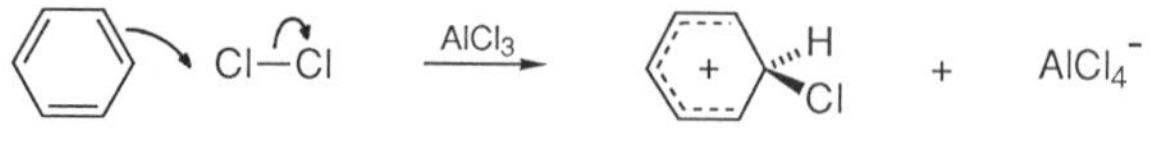

The following are the chief characteristics of these reactions.

(1) The intermediate carbocation adducts are too unstable to be isolated as salts except in special circumstances. For example, benzotrifluoride reacts with nitryl fluoride (NO_2F) and boron trifluoride at low temperatures to give a crystalline product thought to be the salt,

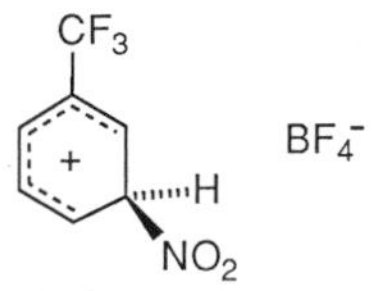

This product is stable only to $-50°C$, above which it decomposes into *m*-nitrobenzotrifluoride, hydrogen fluoride, and boron trifluoride. The relative stability of this adduct is associated with the stability of the fluoroborate anion.

(2) In most instances, the first step in the process is rate-determining, e.g. in the nitration and bromination of benzene. There are some reactions in which the second step (loss of the proton) is rate-determining; sulfonation is the best known example.

(3) The reactions are, with few exceptions, irreversible and the products formed are kinetically controlled (p. 68). Two important exceptions are sulfonation (section 11.5) and Friedel–Crafts alkylation (section 11.3a); the reversibility of these reactions can lead to the formation of the thermodynamically controlled products in appropriate conditions. Use may be made of this fact in synthesis (p. 364), but in some situations it proves to be disadvantageous (p. 390).

(4) Substitution is subject to electrophilic catalysis. For example, benzene reacts with bromine at a negligible rate but in the presence of iron(III) bromide reaction is comparatively fast. The catalyst acts by aiding the removal of bromide ion:

Lewis acids, such as iron(III) bromide in the example above, are customarily used as catalysts, as in halogenation (section 11.6) and Friedel–Crafts alkylation (section 11.3a).

(5) The electrophile may add to a position that is already substituted (a process known as *ipso*-reaction). The carbocation so formed can react in one of three ways:

(*i*) A substitution product can be formed by departure of the substituent as a cation or its equivalent, e.g.

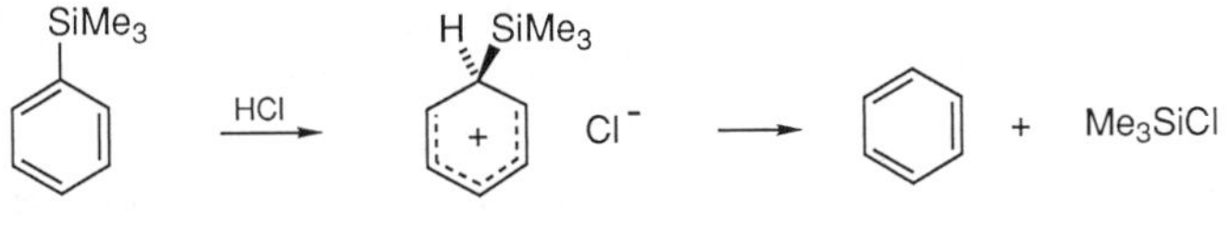

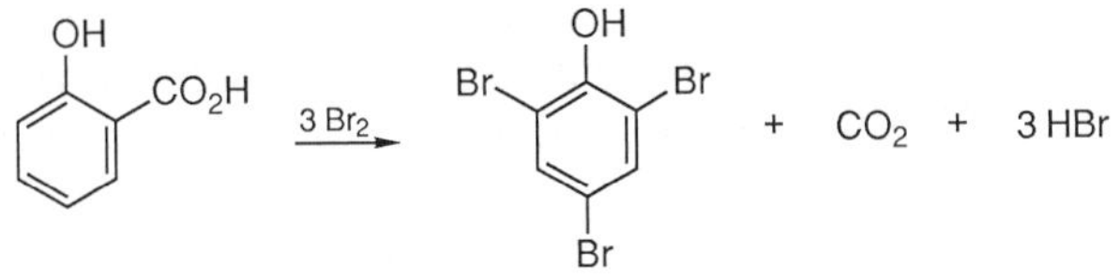

(*ii*)   When the substituent is not capable of departing readily in this way (e.g. methyl), a nucleophile can react to give a diene as product. This is relatively rare, since most electrophilic reactions with aromatic compounds are conducted under conditions in which only very weakly nucleophilic species are present, in which case the carbocation reverts to the reactants. However, nitration with nitric acid in acetic anhydride, where acetic acid is present (p. 376), can give dienes, e.g.

cis and trans

50%

together with normal substitution products.

(*iii*)   Carbocations from phenols can give dienones by loss of the hydroxylic proton, e.g.

Comparison with the reactions between alkenes and electrophiles.   In their reactions with electrophilic reagents, aromatic compounds resemble alkenes in that the first step of the process consists of the addition of the electrophile to sp^2-hybridized carbon with the formation of a carbocation. There are, however, two *general* differences, important exceptions to which are discussed below.

First, whereas the carbocation adduct from an alkene and an electrophile normally reacts with a nucleophile by addition (p. 252), that from an aromatic compound reacts by elimination. The difference arises from the fact that in the latter case elimination regenerates the aromatic system and liberates the associated stabilization energy. In fact, the addition of one mole of hydrogen to benzene is endothermic, whereas the reduction of ethylene is exothermic by about $140 \, kJ \, mol^{-1}$. Aromatic compounds are characterized *in general* by their undergoing substitution, just as alkenes are characterized by their undergoing addition.

Second, aromatic compounds react less rapidly than alkenes with a given electrophile, e.g. whereas benzene is hardly affected by bromine, ethylene reacts instantly. This is because the formation of the intermediate carbocation is accompanied, in the addition to benzene, by the loss of the aromatic stabilization energy; although this loss is offset to some extent by the delocalization energy in the resulting ion (p. 344), it leads to a more endothermic reaction than in the case of ethylene.

Although the causes of the two principal differences between aromatic compounds and alkenes are revealed in the above discussion, two qualifications are necessary. First, although benzene and its simple derivatives are relatively inert to addition, compounds in which two or more benzene rings are fused together are often quite susceptible to addition. For example, anthracene reacts with bromine to give, initially, 9,10-dibromo-9,10-dihydroanthracene,

We have discussed previously the principles which underlie the relative ease of addition across the 9,10-positions of anthracene (p. 38). Second, some substituted benzenes react very rapidly with electrophiles, e.g. acetanilide reacts rapidly with bromine to give mainly *p*-bromoacetanilide, and the reaction of dimethylaniline with bromine is even faster (its activation energy is close to zero). This is because certain substituents are able to stabilize the carbocation intermediate, and hence the preceding transition state, thereby lowering the activation energy; in the case of the bromination of acetanilide, the stabilization is represented by the contribution of the resonance structure,

which symbolizes the delocalization of the positive charge over nitrogen as well as over three of the nuclear carbon atoms.

11.2
Directive and
rate-controlling
factors

The role played by nitrogen in the bromination of acetanilide, described above, illustrates one principle of great importance in electrophilic aromatic substitution: the rate of reaction is strongly dependent upon the nature of the substituent(s) in the aromatic nucleus. Further, the relative ease of substitution at different positions in an aromatic compound is also determined by the nature of the substituent(s): in the case of acetanilide, the order of reactivity at the nuclear carbons is *para* > *ortho* ≫ *meta*. Both the directive effects and the rate-controlling effects of substituents are of great importance in synthetic applications of electrophilic aromatic substitutions. The effects are conveniently discussed under four headings: (*a*) monosubstituted benzenes, (*b*) di- and polysubstituted benzenes, (*c*) bi- and polycyclic hydrocarbons, and (*d*) heteroaromatic compounds.

(a) Monosubstituted benzenes

In the first step of its reaction with an electrophile, the benzene ring provides two electrons to form a new covalent bond with the electrophile, yielding a carbocation. It is possible to represent this ion as a hybrid of three resonance structures (p. 344) but it is not possible to represent adequately the structure of the preceding transition state. At the transition state, the electron-pair which ultimately forms the new covalent bond has been partly transferred from the aromatic ring to the electrophile. The residual aromatic system therefore bears a fraction of the unit positive charge which it bears in the intermediate so that a useful working representation of the transition state is as follows:

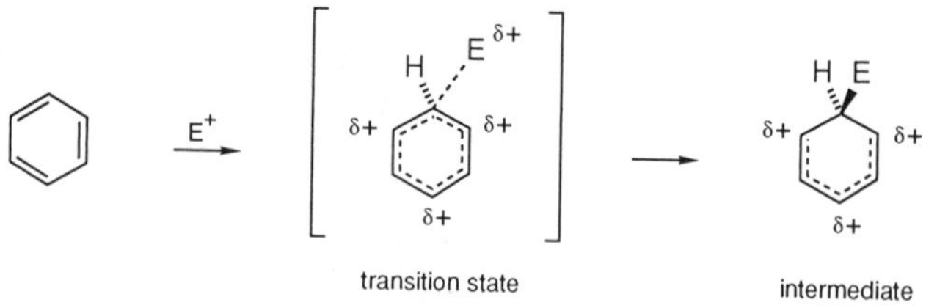

Therefore any factor which stabilizes the intermediate in a particular case also stabilizes the transition state and in practice it is convenient to employ the intermediate as a model for the transition state (cf. Hammond's postulate, p. 61).

In a reaction on a monosubstituted benzene, the intermediates for *ortho*, *meta*, and *para* substitution are, respectively,

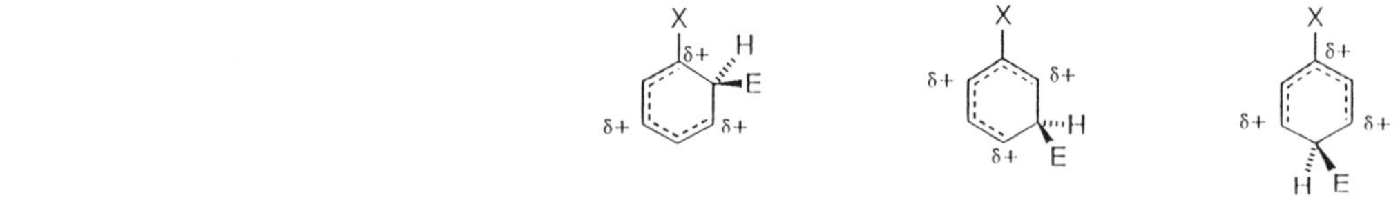

The effect on the stabilities of these ions of groups, X, of different polar types will now be considered.

If X is an electron-attracting substituent of $-I$ type (e.g. $N(CH_3)_3^+$), each of the three ions is less stable than that formed by benzene. The destabilizing effect is greatest when reaction is at the *ortho* or *para* position because the substituent is then adjacent to a carbon atom which bears a portion of the positive charge, as shown by the resonance structures whereas when reaction is at the *meta* position it is further removed from a positively charged carbon atom. Consequently, while all three positions are less reactive than one in benzene, the *ortho* and *para* positions are less reactive than the *meta* position. The substituent is said to be *deactivating* and *meta-directing*.

Substituents of $-I$, $-M$ type (e.g. NO_2, COR, CO_2R, CN) are also deactivating and *meta*-directing. The $-M$ effect serves to increase the deactivation of the ring due to the $-I$ effect; the extra bonding associated with the $-M$ effect, e.g.

is lost on passage to the transition state. Nonetheless, the deactivating influence of these uncharged substituents is less than that of the positively charged $N(CH_3)_3^+$ group.

The $-O^-$ substituent is of $+I$, $+M$ type. The $+I$ effect activates all three ring positions; the *ortho* and *para* positions are more influenced than the *meta* position since the substituent is adjacent to a positively charged carbon atom in the transition state, and the $+M$ effect activates the *ortho* and *para* positions specifically by further delocalization, as represented by the structures,

The substituent is therefore *activating* and *ortho*, *para-directing*.

Alkyl substituents are also of $+I$, $+M$ type and therefore activating and *ortho*, *para*-directing, although the effects are much weaker than with $-O^-$. The $+M$ influence arises from hyperconjugation that can be of C—H or C—C type (p. 34) and represented by structures such as

and

Some substituents have opposed inductive ($-I$) and mesomeric ($+M$) effects, e.g. amino and hydroxyl groups and halogen atoms. Reaction at the *para* (or *ortho*) position then leads to an intermediate in which the positive charge is delocalized onto the substituent,

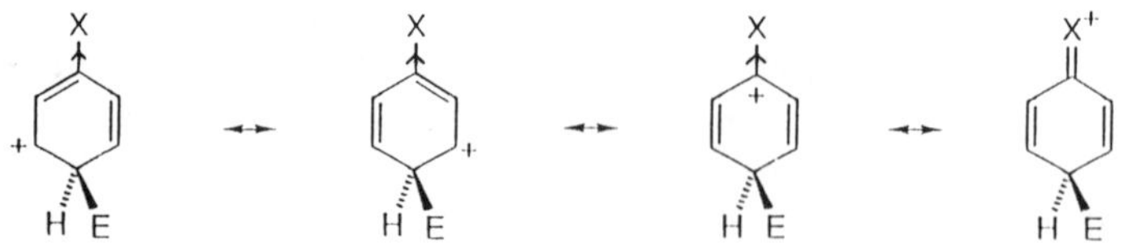

However, this stabilizing influence is offset by the unfavourable inductive effect of the substituent which decreases the stability of the first three structures relative to the corresponding structures from benzene. The resulting effect depends on the relative importance of the $-I$ and $+M$ effects of each group. Addition of the reagent at a *meta* position gives an intermediate which is destabilized relative to that from benzene by the $-I$ effect. Consequently the *meta* position is deactivated, and always less reactive than the *para* (or *ortho*) position, whether or not this is itself deactivated.

The similarity in the mode of operation of the polar effect on the *ortho* and *para* positions leads to the expectation that the ratio of the reactivities at these positions should be 2:1, there being two *ortho* positions and one *para* position. In practice, the ratio is normally different from this, for two main reasons. First, more subtle electronic effects than those apparent from the above discussion seem to be involved and, second, steric hindrance between substituent and reagent can lead to *ortho*:*para* ratios which are considerably less than 2:1.

Individual groups of compounds behave as follows.

(i) Alkylbenzenes.   The nitration of toluene by nitric acid in acetic anhydride at 0°C gives *o*-, *m*-, and *p*-nitrotoluene in the ratios (expressed as percentages) 61·5:1·5:37. The total reactivity of toluene compared with benzene is 27 and combination of this result with the isomer distribution of the nitrotoluenes leads to the following data for the relative reactivities of each nuclear position in toluene compared with one position in benzene: *o*, 50; *m*, 1·3; *p*, 60. (This measure of the reactivity of a particular nuclear carbon in a given reaction is referred to as the *partial rate factor*.)

This result illustrates the directive and activating behaviour of a substituent CH$_3$ of $+I$ type. Other alkylbenzenes behave similarly in giving predominantly the *ortho* and *para* derivatives, as illustrated by the following partial rate factors for the nitration of four compounds in acetic anhydride and the chlorination of two in acetic acid.

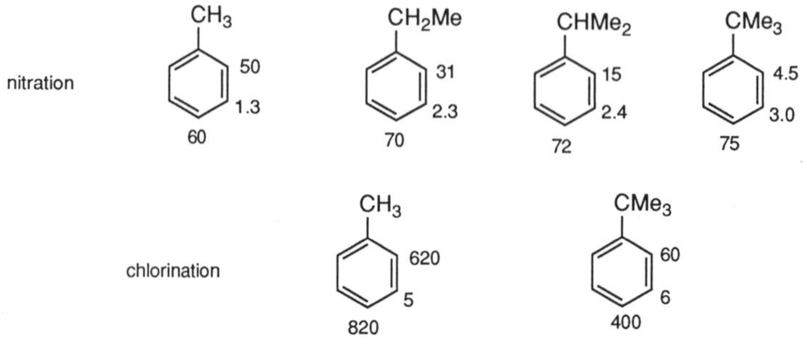

Three points should be noted. First, the *ortho*:*para* ratio falls off sharply as the size of the alkyl group is increased, as a result of steric hindrance to *ortho*-substitution. This imposes limitations in synthesis, e.g. whereas it is possible to obtain *o*-nitrotoluene in about 60% yield, the yield of *o*-nitro-t-butylbenzene is less than 10%.

Second, the absolute values of the partial rate factors for reaction on a given compound depend on the nature of the reagent. This results from variation in the *selectivity* of the reagent. A reagent of such reactivity that it reacted on every collision with an aromatic compound would not discriminate between the three nuclear positions of toluene, or between toluene and benzene: the partial rate factors would all be unity. As the reactivity of the reagent decreases, there is an increasing demand for an electron-pair to be supplied at the relevant nuclear carbon in order that the activation energy barrier can be surmounted. The ease of supply of this pair is determined by the polar character and the position of the substituent, so that the greater the demand by the reagent, the greater will be the differential effects of substituents. Therefore the biggest differences in selectivity between different positions occur with the least reactive electrophiles: in the above case, chlorine is less reactive and more selective than the nitrating agent.

Third, the *para* position in t-butylbenzene is more reactive than that in toluene in nitration but less reactive in chlorination (and many other reactions). It is believed that this is because, although the t-butyl group stabilizes the positively charged transition state more effectively than methyl, it hinders the stabilizing solvation of the positively polarized ring carbon atom adjacent to it more strongly than does the smaller methyl group. When solvation is relatively unimportant, as in nitration in acetic anhydride, the reactivity order is *p*-t-Bu > *p*-Me, but when it is more important, as in chlorination in acetic acid, the opposite order results.

In summary, the greater reactivity of alkylbenzenes compared with benzene enables electrophilic substitutions to be carried out in slightly milder conditions than are necessary for benzene and the directive properties of the alkyl groups are such that *para* derivatives and, in some cases, *ortho* derivatives, can be obtained.

(*ii*) *Benzenes substituted with electron-attracting groups.* These include compounds in which the substituent is a positive pole, such as $PhNMe_3^+$, and those in which it is a neutral unsaturated group of strong dipolar character in which the positive end of the dipole is attached to the benzene ring, e.g.

The substituents are strongly deactivating and *meta*-directing, e.g. nitrobenzene undergoes nitration with a mixture of concentrated nitric and sulfuric acids at only about a hundred-thousandth the rate of benzene, and *m*-dinitrobenzene accounts for about 93% of the dinitrobenzenes formed.

Because of their low nuclear reactivities, these compounds require much more vigorous conditions than benzene and in many cases reaction cannot be brought about (e.g. nitrobenzene does not undergo Friedel–Crafts reactions, section 11.3).

(*iii*) *Benzenes substituted with groups of* −*I,* +*M type.*   These are all *ortho*, *para*-directing, but the ease of reaction (activation or deactivation) varies over a wide range, depending on the substituent, e.g. in the same conditions, the relative rates of chlorination at the *para* positions of *NN*-dimethylaniline and bromo-benzene are about $10^{20}:1$. Consequently, as will be described later, the reaction conditions for the successful substitution of a particular compound in this group vary widely.

1.   *Amino and hydroxyl substituents and related groups.*   For these substituents, the conjugative effect (+*M*) dominates over the inductive effect (−*I*). That is, the contribution made to the transition state by structures of the type

far more than outweighs the deactivating influence due to the inductive effect of nitrogen or oxygen when substitution occurs at the *ortho* or *para* positions. These positions are consequently strongly activated.

The amino group is more strongly activating than hydroxyl, for it has the larger +*M* and the smaller −*I* effect; quaternary cationic nitrogen is more stable than ternary cationic oxygen. However, the —O⁻ substituent is more strongly activating than both —NH₂ and —OH because, first, its +*M* effect is greater and, second, it possesses a +*I* effect. Electrophilic reactions are normally conducted in acidic media in which phenoxide ions are present in negligible concentration compared with phenols, but for reaction of a phenol with the electrophilic aromatic diazonium ions it is necessary to use basic conditions in order to make use of the strong activating influence of −O⁻ (p. 421).

N-Alkyl-derivatives of aniline and *O*-alkyl-derivatives of phenol are of reactivity comparable with aniline and phenol themselves but the acyl-derivatives are much less reactive. This is because the unshared electron-pair on nitrogen or oxygen is already delocalized within the substituent, and is not so readily available for π-orbital overlap in the electron-deficient transition state.

The smaller activating effect of acetamido compared with amino is usefully applied in synthesis, e.g. whereas aniline is so reactive towards bromine that it gives 2,4,6-tribromoaniline essentially instantaneously, acetanilide reacts more slowly and the monobromo-derivatives (principally the *para*, together with some of the *ortho* compound) may be isolated. Further, since aromatic amines are very susceptible to oxidation, it is normally necessary to carry out substitutions on their acyl derivatives and then remove the acyl group by hydrolysis.

2.   *Halogen substituents.*   Fluorine has a weaker $+M$ and a stronger $-I$ effect than oxygen, and for *para* substitution the effects approximately nullify each other so that reaction occurs about as readily as at one carbon in benzene. Both the $+M$ and the $-I$ effects of the halogens fall in the order, $F > Cl > Br > I$ (p. 33), and the resultant effect leads to the order, $F \sim H > Cl \sim Br \sim I$. In many substitutions, chloro-, bromo-, and iodobenzene are about one-tenth as reactive at their *para* positions as benzene is at any one position. The *ortho* positions of all four halobenzenes are less reactive than the *para* positions and the *meta* positions are strongly deactivated.

In summary, substitutions of the halobenzenes require conditions comparable in vigour with those for benzene and yield mainly the *para*-derivatives.

3.   *Biphenyls and styrenes.*   Biphenyl is activated in the *ortho* and *para* positions and weakly deactivated in the *meta* position. The latter result follows from the weak $-I$ effect of sp^2-hybridized carbon (p. 48) and the former from the ability of the phenyl substituent to delocalize the positive charge on the transition states for *ortho* and *para* substitution, e.g. that for *para* substitution is represented as the hybrid,

Consequently, the compound is *ortho*, *para*-directing and of reactivity comparable with toluene.

The introduction of an electron-attracting substituent into biphenyl reduces the ease of reaction and causes substitution to occur at the *ortho* and *para* positions of the unsubstituted ring. Conversely, an electron-releasing substituent increases the reactivity and causes reaction to occur in the substituted ring.

Styrene and its derivatives behave similarly. For example, cinnamic acid is nitrated predominantly in the *ortho* and *para* positions and reacts at about one-tenth the rate of benzene. In effect, the —CH=CH—CO$_2$H group is similar to a halogen atom: the inductive effect of the group ($-I$) reduces the ease of reaction, but the deactivation at the *ortho* and *para* positions is somewhat offset by the extra delocalization of the positive charge on to the aliphatic side-chain:

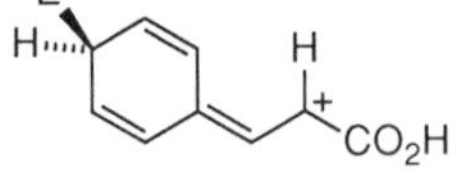

That is, the pair of π-electrons in the alkene bond of cinnamic acid behaves analogously to a pair of *p*-electrons on a halogen substituent.

(*iv*) *Summary*. Table 11.1 summarizes the directive and rate-controlling effects of the more common substituents in electrophilic aromatic substitution.

Table 11.1 Directive and rate-controlling effects of substituents in electrophilic aromatic substitution

Substituent	Polar character	Directive effect	Rate-controlling effect
O$^-$	$+I, +M$	*o, p*	Very powerfully activating
NH$_2$, NHR, NR$_2$ } OH, OR }	$-I, +M$	*o, p*	Powerfully activating
NHCOR, OCOR	$-I, +M$	*o, p*	Activating
Ph	$-I, +M$	*o, p*	Moderately activating
CH$_3$ and other alkyl groups	$+I$	*o, p*	Moderately activating
(H)			–
F	$-I, +M$	*o, p*	Comparable with benzene
Cl, Br, I } CH=CH—CO$_2$H (etc.) }	$-I, +M$	*o, p*	Weakly deactivating
CO$_2$R, CO$_2$H, CHO, COR, CN, } NO$_2$, SO$_2$OH }	$-I, -M$	*m*	Strongly deactivating
NH$_3$$^+$, NR$_3$$^+$	$-I$	*m*	Strongly deactivating

(*b*) *Di- and polysubstituted benzenes*

To a first approximation, two or more substituents affect the ease of reaction at a particular position in benzene independently of each other.* For example, the

---

* A linear free-energy relationship holds approximately. That is, if the introduction of each of two substituents separately alters the free energy of activation for reaction at a particular position by amounts x and y, the presence of both substituents alters the free energy of activation by $(x + y)$. Since $\log k \propto \Delta G^{\neq}$, the partial rate factor for substitution at a particular position in the disubstituted compound is equal to the *product* of the partial rate factors for the appropriate positions of the two monosubstituted compounds.

introduction of successive methyl groups into benzene increases the reactivity and pentamethylbenzene is the most reactive compound in the series.

Use can be made of this principle in synthesis. For instance, the acetamido group activates *ortho* positions and slightly deactivates *meta* positions, whereas the methyl group activates *ortho* less strongly than acetamido but also slightly activates *meta* positions. Consequently, in *p*-acetamidotoluene,

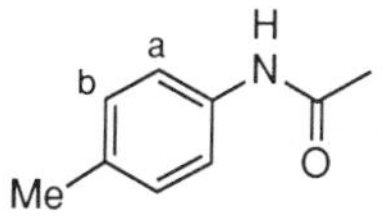

position *a*, which is *ortho* to acetamido and *meta* to methyl, is more reactive than position *b*, which is *meta* to acetamido and *ortho* to methyl. Substitution therefore occurs predominantly at *a*, and *meta*-substituted toluenes, which are obtained in only very low yield by the direct substitution of toluene, may in this way be obtained by removal of the acetamido group (e.g. p. 414).

(c)   Bi- and polycyclic hydrocarbons

The delocalization of the positive charge in the transition states of electrophilic substitutions is increased by the fusion of two or more benzene rings and the polycyclic hydrocarbons are therefore all more reactive than benzene. Directive effects are also introduced.

Substitution in naphthalene is illustrative. The transition states for reaction at the 1- and 2-positions may be represented as follows:

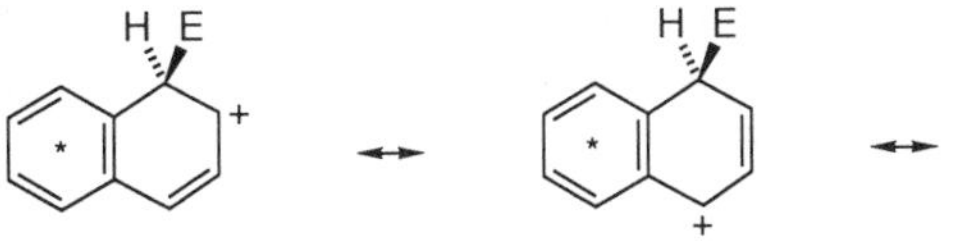

1-substitution:

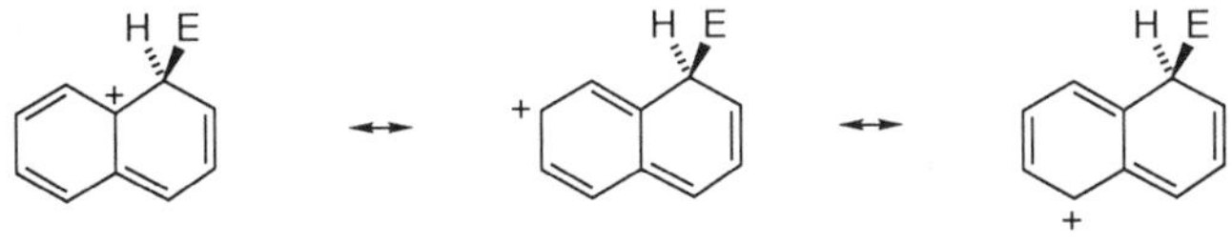

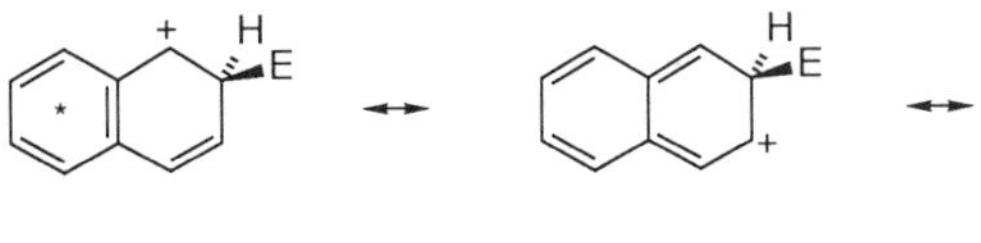

2-substitution:

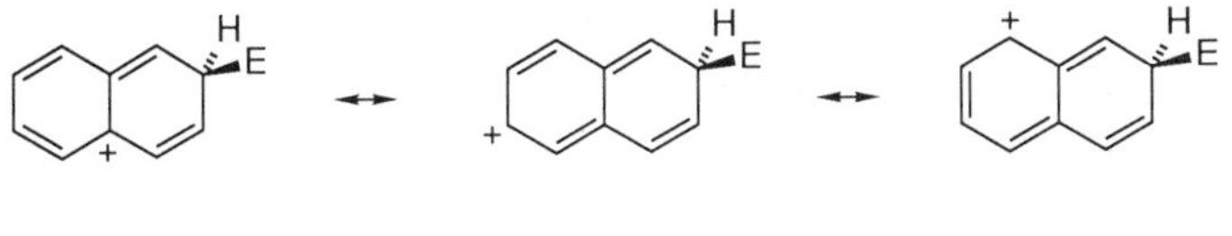

In each of the two transition states, the positive charge is more extensively delocalized than in reaction on benzene, leading to lower activation energies. Further, the three starred structures are of benzenoid type and therefore of lower energy content than the remainder in which the benzenoid nature of the second ring has been interrupted. Since there are two such low-energy contributors for 1-substitution as compared with one for 2-substitution, it is understandable that the 1-position should be the more reactive. In a typical reaction, nitration, the partial rate factors for 1- and 2-substitution are 470 and 50, respectively.

There are, however, two conditions in which substitution at the 2-position predominates. One applies when the reaction is thermodynamically controlled, as in sulfonation at high temperatures (p. 383), because the 1-derivative, in which there is steric repulsion between the substituent and the *peri*-hydrogen atom shown,

is thermodynamically the less stable. The other applies when the reaction is kinetically controlled but the reagent is particularly bulky, as in Friedel–Crafts acetylation in nitrobenzene (p. 368), for reaction at the 1-position is then markedly hindered sterically by the *peri*-hydrogen.

The presence of an electron-attracting group in naphthalene reduces the reactivity and causes substitution to occur in the unsubstituted ring, mainly at the 5- and 8-positions (i.e. the two 1-positions of that ring). An electron-releasing group activates the molecule further and reaction occurs in the substituted ring. If the group is in the 1-position, substitution occurs at the 2- and 4-positions (i.e. *ortho* and *para* to the electron-releasing group), but a 2-substituent directs almost entirely to the 1-position, although the 3-position is also an *ortho* position. The reason is that the stabilization of the transition state which is provided by the substituent is more effective when the appropriate resonance structure is benzenoid (1-substitution) than when it is not (3-substitution), e.g.

The reactivities of polycyclic hydrocarbons follow from the principles described for naphthalene. Examples, including data for partial rate factors in nitration at the most reactive position, are tabulated.

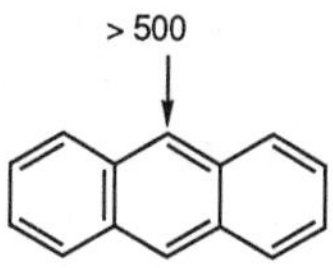

> 500

anthracene

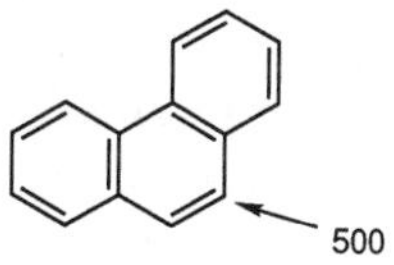

phenanthrene

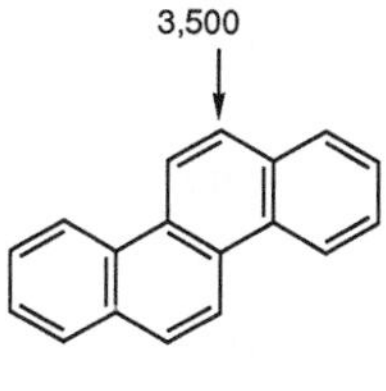

— 500

3,500

chrysene

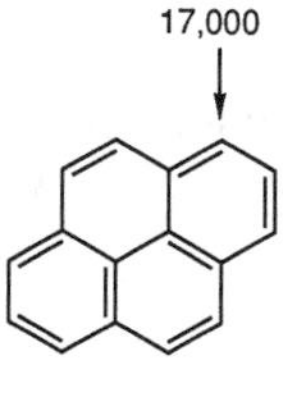

17,000

pyrene

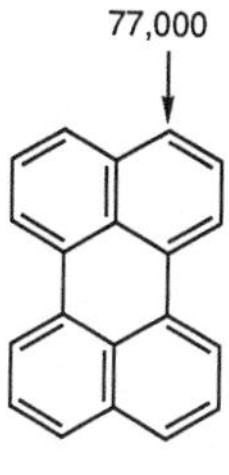

77,000

perylene

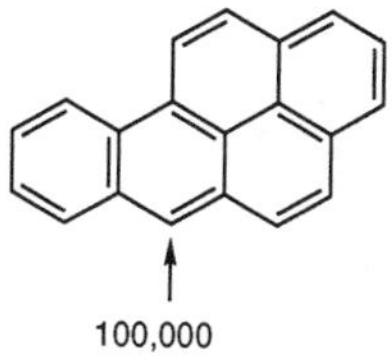

3:4-benzopyrene

100,000

(d) *Heteroaromatic compounds*

(i) *Five-membered rings.*   The principles which govern the electrophilic substi-tutions of this group of heteroaromatic compounds will be illustrated by reference to pyrrole.

Pyrrole is highly reactive at both the 2- and 3-positions. The reason is that the transition state for substitution at each position is strongly stabilized by the accommodation of the positive charge by nitrogen,

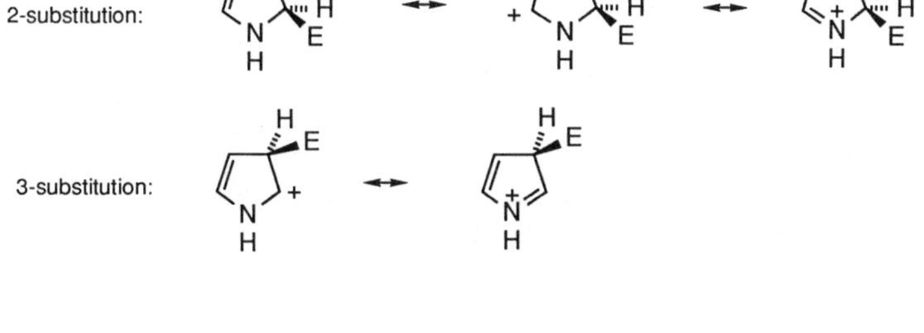

2-substitution:

3-substitution:

in just the way that aniline owes its reactivity to the exocyclic nitrogen (p. 352). 2-Substitution predominates because the positive charge in the transition state is delocalized over a total of three atoms, compared with two for 3-substitution.

The similarity of pyrrole and aniline is particularly apparent in their reactions with bromine: each reacts at all its activated carbon atoms, pyrrole giving tetra-bromopyrrole and aniline giving 2,4,6-tribromoaniline. In fact, pyrrole is even more strongly activated than aniline and should perhaps be compared with the phenoxide ion: each undergoes the Reimer–Tiemann reaction, unlike other benzenoid compounds (p. 374). In addition, pyrrole undergoes Friedel–Crafts acylation in the absence of a catalyst (p. 365) and the Hoesch reaction, characteristic of polyhydric phenols (p. 372).

One difficulty in dealing with pyrrole is that it is readily converted by acid into the trimer,

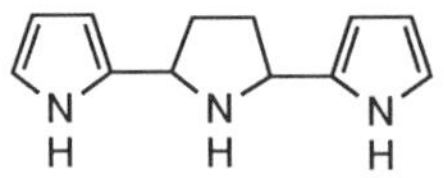

Accordingly, special conditions need to be used to carry out reactions normally involving acids (e.g. sulfonation, p. 382). Pyrroles substituted with electron-attracting groups are less prone to polymerization and can be more conveniently handled.

Furan and thiophen are also activated towards electrophiles and react predominantly at the 2-position. The underlying theory is similar to that for pyrrole, namely, that the heteroatom is able to delocalize the positive charge on the transition state. Since oxygen accommodates a positive charge less readily than nitrogen, furan is less reactive than pyrrole, just as phenol is less reactive than aniline. The $+M$ effect of sulfur is smaller than that of oxygen because the overlap of the differently sized p orbitals of carbon and sulfur is less than in the case of carbon and oxygen so that, understandably, thiophen is less reactive than furan.

A revealing trend of reactivities is shown by the behaviour of these three heterocyclic compounds towards maleic anhydride. Pyrrole is sufficiently reactive towards electrophiles to take part, as a nucleophile, in a Michael addition (p. 240):

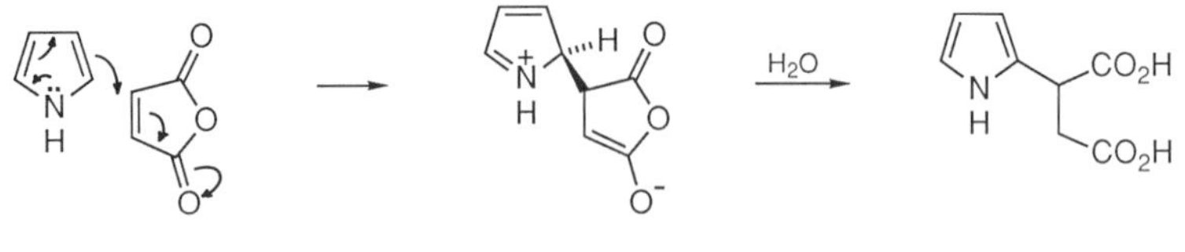

Furan is less reactive towards electrophiles than pyrrole and instead undergoes the Diels–Alder reaction (p. 272), differing in this respect from benzene because less aromatic stabilization energy is lost on 1,4 addition. Thiophen, which is less reactive both to electrophiles and as a conjugated diene, does not react.

The 2:3-benzo-derivatives of pyrrole, furan, and thiophen are activated and react in the hetero-ring. Substitution occurs mainly at the 3-position as a result of the fact that the stabilizing influence of the heteroatom on the transition state is more effective when the appropriate resonance structure is benzenoid (3-substitution) than when the benzenoid system is disrupted (2-substitution), as illustrated for indole:

3-substitution   2-substitution

(*ii*) *Six-membered rings.*   The principles governing the reactivity of these compounds are illustrated by reference to pyridine. The transition states for substitution at the 3- and 4-positions can be represented as the hybrids,

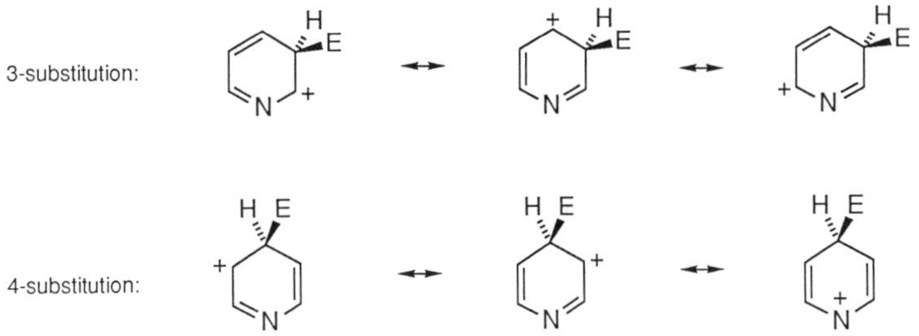

3-substitution:

4-substitution:

In each, the positive charge is less well accommodated than in reactions on benzene because nitrogen is more electronegative than carbon. Hence both the 3- and 4-positions are deactivated, the latter the more strongly because of the high energy of the contributing structure which contains divalent positive nitrogen.* The 2-position resembles the 4-position, as reference to the appropriate resonance structures will show.

Many electrophilic substitutions are conducted in acidic media in which pyridine is present almost entirely as its conjugate acid and this is even less reactive than pyridine itself. Very vigorous conditions are required to bring about reaction, e.g. nitration requires 100% sulfuric acid with sodium and potassium nitrates at 300°C, and even then the yield of 3-nitropyridine is only a few percent.

Quinoline and isoquinoline are also deactivated, though less so than pyridine, and reaction normally occurs in the homocyclic ring at the 5- and 8-positions (cf. the behaviour of naphthalene containing an electron-attracting substituent, p. 356). There are, however, many exceptions.

---

* Careful distinction should be made between the ability of (*a*) divalent nitrogen and (*b*) quaternary nitrogen, to accommodate a positive charge. The latter system, in which nitrogen possesses an octet, is relatively stable and this stability is responsible for the activating effect of nitrogen in aniline and pyrrole.

quinoline   isoquinoline

Pyridazine, pyrimidine, and pyrazine (p. 682), containing two heterocyclic nitrogen atoms, are more strongly deactivated than pyridine and electrophilic substitutions are unsuccessful. Derivatives are obtained either *via* nucleophilic substitution (section 12.2) or by synthesizing a compound which contains a powerfully activating substituent such as hydroxyl or amino and with which electrophiles react (e.g. p. 719).

11.3 Formation of carbon–carbon bonds

A carbon atom which is bonded to an electronegative atom or group, X, is positively polarized and therefore electrophilic, although not normally sufficiently so for it to react with aromatic compounds other than those of exceptionally high reactivity. However, its electrophilicity may be enhanced by the addition of a species which can accept electrons from X and reaction then occurs with less reactive aromatic compounds.

This principle may be applied to the formation of bonds between aromatic and aliphatic carbon atoms in a variety of ways, some of which are of wider applicability than others. The most commonly used processes are Friedel–Crafts alkylation and acylation.

(a) Friedel–Crafts alkylation

Aliphatic compounds which take part in alkylations are halides, alcohols, esters, ethers, alkenes, aldehydes, and ketones. Reactions of the first four classes of compound are normally catalyzed by Lewis acids and those of the last three (the unsaturated compounds) by proton acids.

Halides are the reagents of most frequent choice in alkylation and aluminium trichloride is usually employed as the catalyst. With primary and secondary halides, it is believed that an ion-pair is formed with the Lewis acid, the carbocation part of which then effects substitution,* e.g.

---

* The reaction cannot be an S_N2 substitution on the halide,

since carbocation rearrangements occur (p. 430) and an asymmetric halide gives racemic product; nor is it an S_N1 reaction, since methyl chloride, bromide, and iodide show different regioselectivities with substituted benzenes which would not be the case for a free methyl cation.

$$RCH_2Cl \quad + \quad AlCl_3 \quad \rightleftharpoons \quad \left[RCH_2^+ \, AlCl_4^- \right]$$

ion-pair

Free carbocations are thought to be involved with tertiary halides. For example, benzene, t-butyl chloride, and iron(III) chloride give t-butylbenzene in 80% yield *via* the t-butyl cation:

$$Me_3C^+ \quad + \quad FeCl_4^-$$

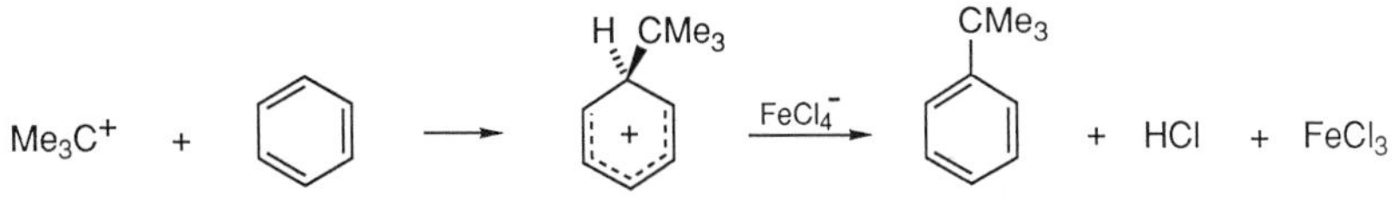

Benzyl halides are very reactive as alkylating agents, just as they are in other nucleophilic displacements, but alkenyl and aryl halides are inert. The use of di- and polyhalides leads to successive alkylations: benzene reacts with dichloromethane in the presence of aluminium trichloride to give diphenylmethane,

$$PhH \quad + \quad CH_2Cl_2 \quad \xrightarrow[\text{-HCl}]{(AlCl_3)} \quad PhCH_2Cl \quad \xrightarrow[\text{-HCl}]{PhH\ (AlCl_3)} \quad PhCH_2Ph$$

with 1,2-dichloroethane to give bibenzyl,

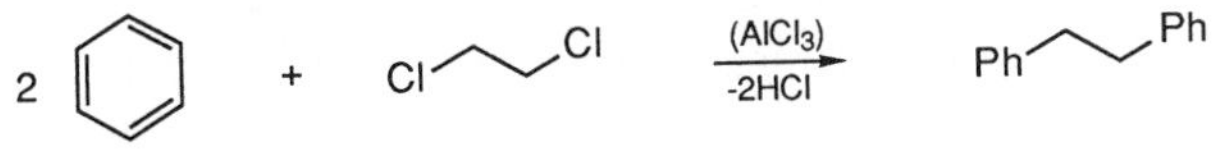

and with carbon tetrachloride to give triphenylmethyl chloride (steric hindrance evidently preventing the displacement of the fourth chlorine),

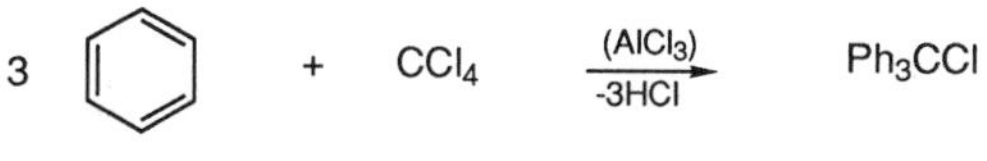

Alcohols, esters, and ethers react analogously to halides.

Alkenes react *via* the carbocations which they form with proton acids:

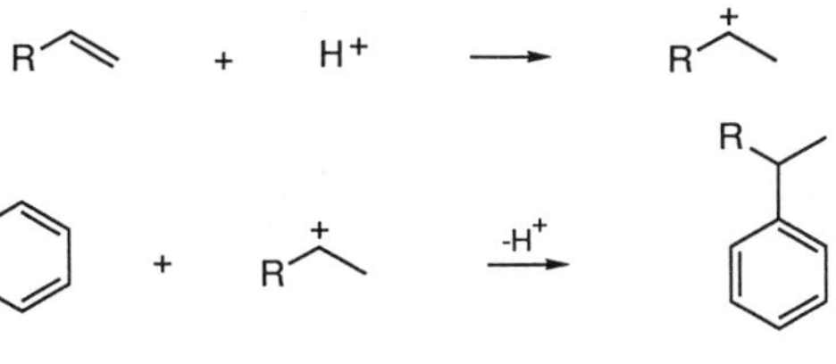

For example, the addition of cyclohexene to benzene in concentrated sulfuric acid at 5–10°C gives cyclohexylbenzene in about 65% yield [2]:

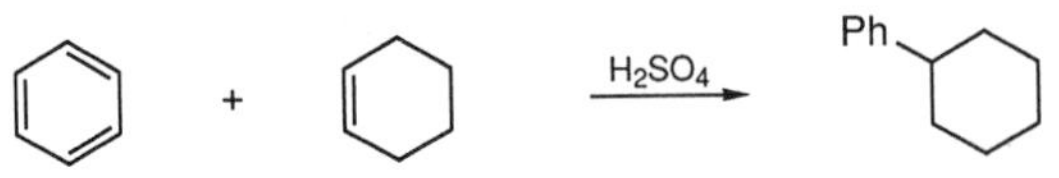

Styrene (required for the production of polystyrene, styrene-butadiene copoly-mer rubbers, etc.) is made industrially by the ethylation of benzene on an aluminium trichloride catalyst followed by dehydrogenation (p. 606):

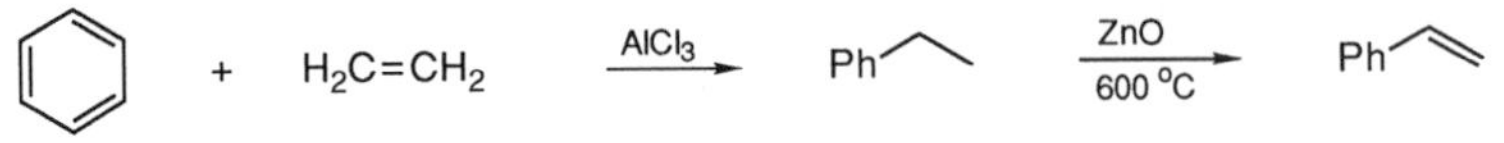

The Bogert–Cook synthesis of phenanthrene and its derivatives makes use of intramolecular alkylation by an alkene, e.g.

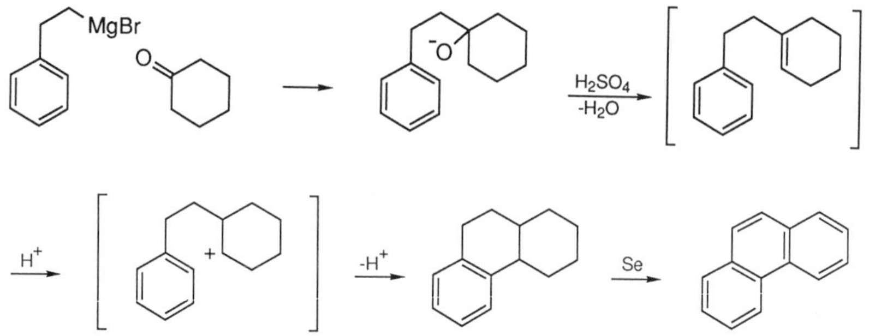

Aldehydes and ketones can also act as alkylating agents in the presence of proton acids but this process does not compete favourably with acid-catalyzed self-condensation of the carbonyl compound (p. 256) except when alkylation is intramolecular and stereochemically favoured. For example, 4-methyl-2-quinolone can be obtained by acid-treatment of the amido-ketone formed from aniline and acetoacetic ester (p. 705):

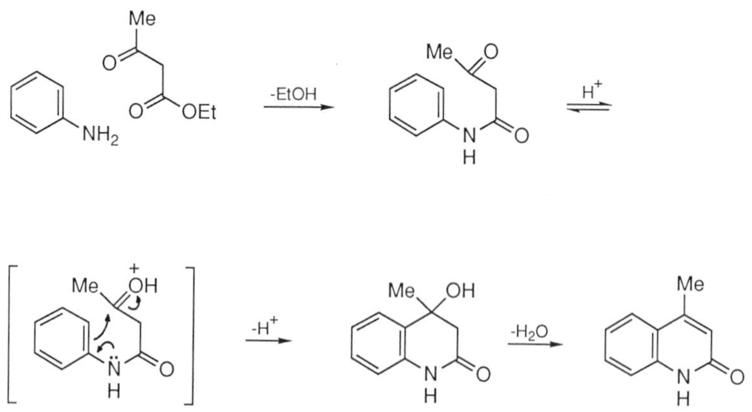

The Skraup synthesis of quinoline also involves acid-catalyzed intramolecular alkylation (p. 704).

(*i*)   *Reactivity of the aromatic compound.*   Aromatic compounds whose nuclear reactivity is comparable with or greater than that of benzene can be successfully alkylated but strongly deactivated compounds do not react. For example, chlorobenzene reacts but nitrobenzene is inert. Phenols do not react satisfactorily because they react with the Lewis acids at oxygen ($ArOH + AlCl_3 \rightarrow ArOAlCl_2 + HCl$) and the resulting compound is usually only sparingly soluble in the reaction medium so that it reacts slowly. Conversion of the phenol into its methyl ether leads to successful alkylations providing that a low temperature is used to minimize acid-catalyzed cleavage of the ether group. Aromatic amines complex strongly with Lewis acids and are not suitable for alkylations.

The Lewis acids used as catalysts differ in activity, the order for the commoner compounds in alkylations with halides being: $AlCl_3 > SbCl_5 > FeCl_3 > SnCl_4 > ZnCl_2$.

The catalyst is chosen with reference to the reactivities of the aromatic compound and the alkylating agent; in general, tertiary halides require milder catalysts than primary halides. It is advisable to employ the least active Lewis acid consistent with the occurrence of alkylation at a practicable rate, for the more active catalysts tend to induce the isomerizations described below.

Alkylations with alcohols and ethers may be catalyzed by either a Lewis acid or a proton acid. Among Lewis acids, boron trifluoride is the reagent of choice because of its strong tendency to complex with oxygen. The proton acids used are hydrogen fluoride, concentrated sulfuric acid, and phosphoric acid; sulfuric acid is often unsuitable because of its capacity for forming sulfonated byproducts.

The reactions of alkenes, aldehydes, and ketones are normally proton-catalyzed.

(*ii*)   *Problems attendant upon alkylation.*   Three problems are encountered in alkylation, each of which considerably reduces the general scope of the process.

(1)   Since alkyl groups are activating in electrophilic substitutions, the product of alkylation is more reactive than the starting material and further alkylation inevitably occurs. For example, the methylation of benzene with methyl chloride in the presence of aluminium chloride gives a mixture containing toluene, the xylenes, the tri- and tetra-methylbenzenes, pentamethylbenzene, and hexamethylbenzene. Careful control of the molar ratio of the reactants and the reaction conditions can yield a particular polyalkylbenzene as the predominant product: for example, a mixture of the tetramethylbenzenes may be obtained by fractionation and durene (1,2,4,5-tetramethylbenzene) may then be isolated by freezing.

In general, in order to obtain a monoalkylated product minimally contaminated by polyalkylated products, an excess of the aromatic compound should be used.

(2)   Many alkyl groups rearrange during alkylation, e.g. benzene and propyl halides give mixtures of propylbenzene and isopropylbenzene. The extent to which isomerization competes with direct alkylation depends on the structure of the alkyl halide, the nature of the catalyst, and the reactivity of the aromatic compound.

First, isomerization is especially common with primary halides, as in the formation of isopropylbenzene from a propyl halide, and is fairly common with secondary halides. These rearrangements are understandable in that primary carbocations are less stable than secondary, and secondary less stable than tertiary. However, in some instances tertiary halides rearrange during alkylation, e.g.

A probable explanation is that, although the tertiary ion is the more stable of the two, it is also the less reactive, so that the faster reaction of the secondary ion dominates, even though this ion is present in smaller concentration. t-Butyl halides do not undergo rearrangement during alkylation, probably because this would involve formation of the highly energetic primary carbocation.

Second, the extent of isomerization is reduced by using a less powerful Lewis acid catalyst. For example, the rearrangement above does not occur when iron(III) chloride is the catalyst; $PhCMe_2CH_2CH_3$ is then the product.*

Third, isomerization becomes less significant as the reactivity of the aromatic compound is increased.

(3)   Alkylation is reversible, so that reaction is thermodynamically controlled. For example, a monosubstituted benzene usually gives mainly the *meta*-alkyl-derivative, since this is thermodynamically the most stable. This principle may be usefully applied, e.g. benzene and excess of ethyl bromide react in the presence of aluminium trichloride to give 1,3,5-triethylbenzene in 87% yield, further substitution probably being sterically impeded.

Just as tertiary alkyl groups are the most readily introduced during alkylation, so they are the most readily removed by the reverse reaction, departing as the relatively stable tertiary carbocations:

---

*This is probably because the weaker catalyst, while polarizing the C-halogen bond sufficiently for alkylation to occur, does not completely break the bond to form the alkyl cation through which isomerization occurs.

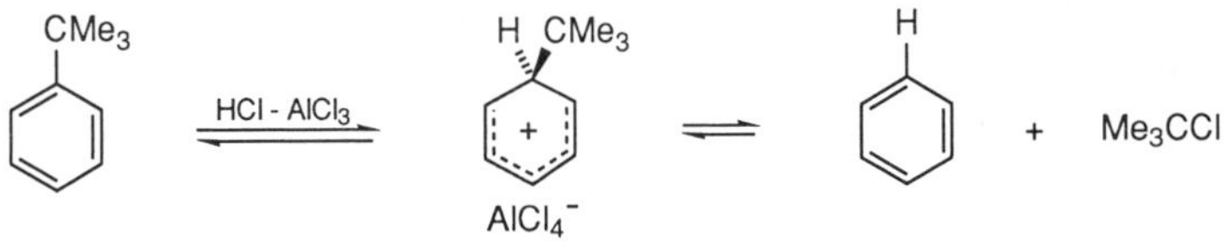

This enables the t-butyl group to be used to protect the most reactive position in a compound in order to effect reaction elsewhere; the t-butyl group is subsequently removed by the addition of an excess of benzene which draws the equilibrium in the desired direction. For example, the Friedel–Crafts acylation of toluene gives largely the *para* acyl-derivative (p. 368). The following scheme enables the *ortho* isomer to be obtained:

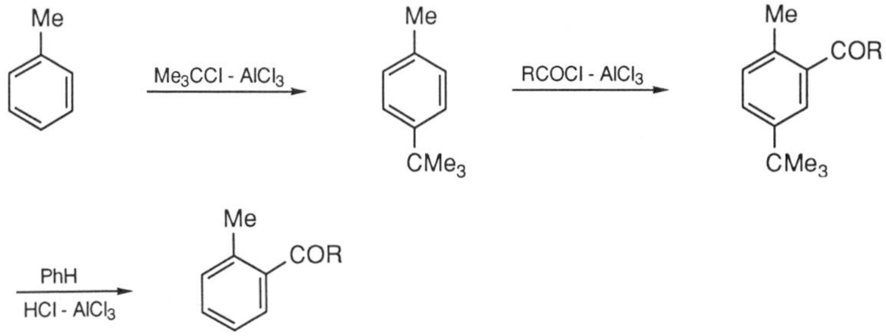

The less easily accessible 1,2,3-trialkylbenzenes may also be prepared by application of this principle. t-Butylation of a *m*-dialkylbenzene gives the 5-t-butyl derivative (thermodynamic control), the desired alkyl group is then introduced (reaction occurring at the 2- rather than the 4-position because of the hindrance due to the t-butyl group), and the t-butyl group is removed by reaction with more of the starting *m*-dialkylbenzene.

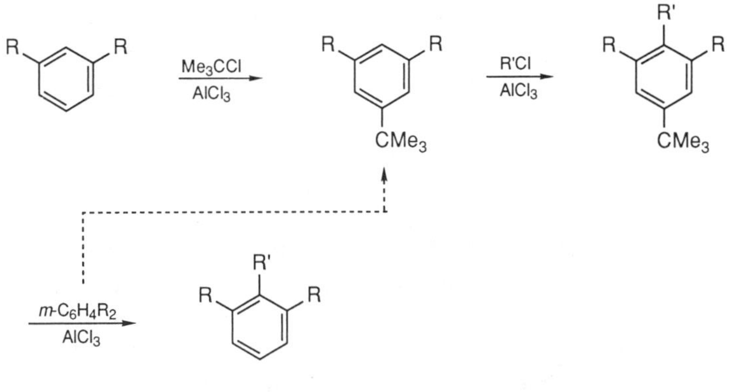

(b)   *Friedel–Crafts acylation*

The acylation of aromatic rings may be brought about by an acid chloride or anhydride in the presence of a Lewis acid or, in some circumstances (p. 369), by a carboxylic acid in the presence of a proton acid.

In the Lewis acid-catalyzed methods, two electrophiles are involved. One is an oxygen-bonded complex, e.g.

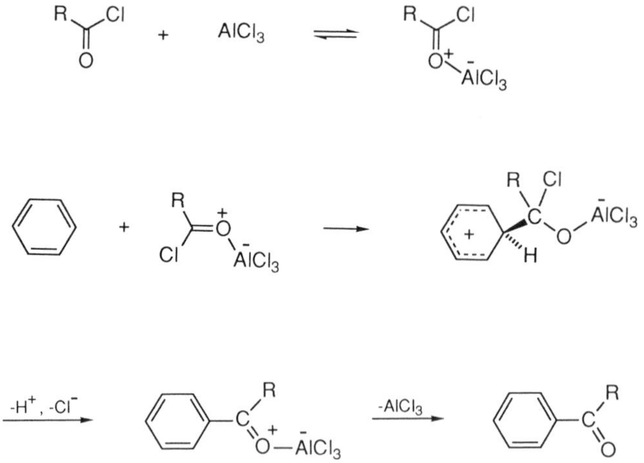

The other is an acylium ion which is formed from the complex in low concentration but is more reactive as an electrophile,

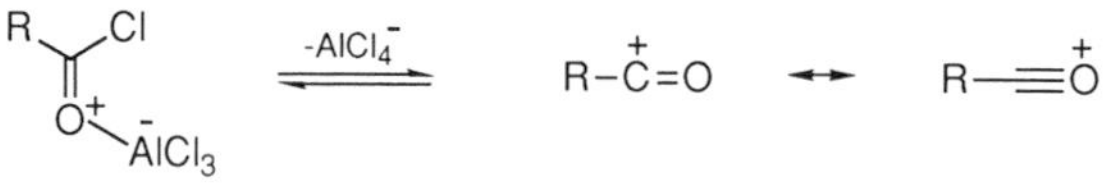

Which of the two is the more effective in a particular case depends on the nature of R (for example, formation of the acylium ion is favoured when R is aromatic, since its positive charge can be delocalized on to the aromatic ring) and the aromatic compound (the acylium ion is relatively the more important electrophile with less reactive compounds).

In a typical procedure, acetic anhydride is added to a refluxing solution of bromobenzene in carbon disulfide containing suspended aluminium trichloride. After removal of the solvent and decomposition of the resulting complex with hydrochloric acid, p-bromoacetophenone is obtained in about 75% yield [1].

The aromatic compounds which may be acylated are in general the same as those which may be alkylated. Benzene and compounds of comparable or greater reactivity undergo the reaction but deactivated molecules such as benzaldehyde, benzonitrile, and nitrobenzene are inert: nitrobenzene is a common solvent for acylation. The difficulty encountered in the alkylation of phenols (p. 363) applies also in acylation and may be surmounted by acylating the methyl ether at a low temperature. An alternative in some instances is to employ concentrated sulfuric acid as the catalyst, as in the preparation of phenolphthalein from phenol and phthalic anhydride:

phenolphthalein

The order of activity of Lewis acids in the benzoylation of toluene and chloro-benzene is: $SbCl_5 > FeCl_3 > AlCl_3 > SnCl_4$. As with alkylation, it is often wise to employ the weakest reagent consistent with successful reaction in order to minimize side-reactions, e.g. thiophen is polymerized during acetylation by acetyl chloride in the presence of aluminium trichloride but gives 2-acetylthiophen in about 80% yield when tin(IV) chloride is used [2].

There are a number of important differences between acylation and alkylation.

(1)   Whereas alkylation does not require stoichiometric quantities of the Lewis acid since this is regenerated in the last stage of the reaction, acylation requires greater than equivalent quantities because the ketone which is formed complexes with the Lewis acid.

(2)   Since acyl groups deactivate aromatic nuclei towards electrophilic sub-stitution, the products of acylation are less reactive than the starting materials and the mono-acylated product is easy to isolate. This makes acylation a more useful procedure than alkylation, and alkyl-derivatives are often more satisfactorily obtained by acylation followed by reduction of carbonyl to methylene than by direct alkylation.

(3)   A further advantage of acylation over alkylation is that the isomerizations and disproportionations which are characteristic of the latter process do not occur in the former. There is, however, one limitation: attempted acylation with derivatives of tertiary acids may lead to alkylation. For example, trimethylacetyl chloride reacts with benzene in the presence of aluminium trichloride to give mainly t-butylbenzene, liberating carbon monoxide:

$$Me_3C-COCl \xrightarrow{AlCl_3} Me_3C-\overset{+}{C}O \xrightarrow{-CO} Me_3C^+ \xrightarrow[-H^+]{PhH} Ph-CMe_3$$

The driving force for the decarbonylation no doubt resides in the relative stability of tertiary carbocations. However, more reactive aromatic compounds can react before decarbonylation occurs, e.g. anisole in the same conditions gives mainly p-$CH_3O-C_6H_4-COCMe_3$.

(4)   The complex of the acylating agent and Lewis acid is evidently very bulky, for *ortho*, *para*-directing monosubstituted benzenes give very little of the *ortho*

product. For example, toluene, which gives nearly 60% of the *ortho*-derivative on nitration, gives hardly any *o*-methylacetophenone on acetylation; the *para*-isomer may be obtained in over 85% yield. When nitrobenzene is used as the solvent, steric hindrance to *ortho* substitution is even more marked, possibly because a solvent molecule takes part in the acylating complex, e.g. whereas the acetylation of naphthalene in carbon disulfide solution gives mainly the expected 1-acetylnaphthalene, reaction in nitrobenzene gives predominantly 2-acetylnaphthalene, providing a convenient route to 2-naphthoic acid by oxidation with hypochlorite (p. 619):

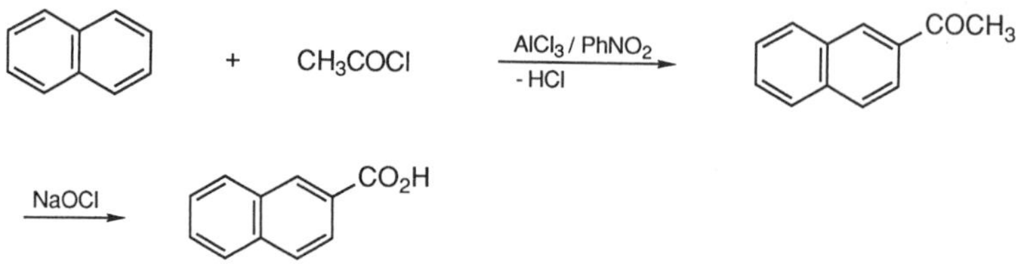

2-naphthoic acid

(*i*) *Cyclizations.* Intramolecular Friedel–Crafts acylations are of particular value in building up cyclic systems; dibasic acid anhydrides are much used in these reactions.

For example, benzene and succinic anhydride in the presence of aluminium trichloride give β-benzoylpropionic acid in 80% yield [2] and reduction of carbonyl to methylene, conversion of the acid group to the acid chloride, and cyclization with aluminium trichloride give α-tetralone:

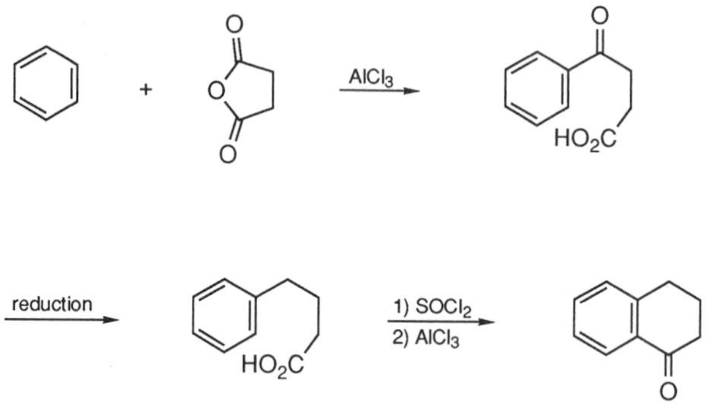

α-tetralone

1-Alkylnaphthalenes may then be obtained by a Grignard reaction followed by dehydration and dehydrogenation:

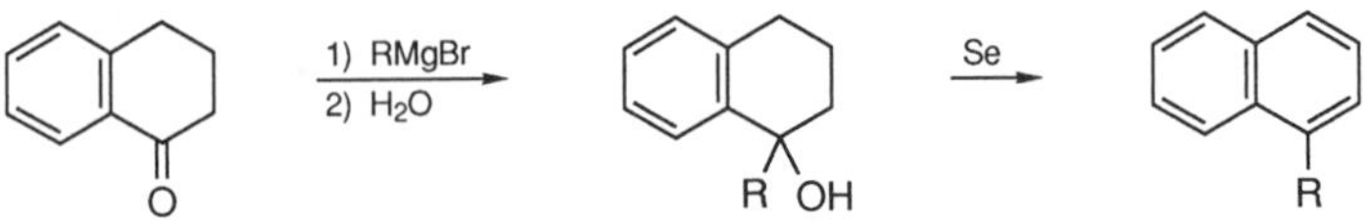

Many variants of this method are possible. By starting with naphthalene, phenanthrene and some of its derivatives may be obtained. By using phthalic anhydride, two new aromatic rings may be built on, as in the synthesis of 1:2-benzanthracene:

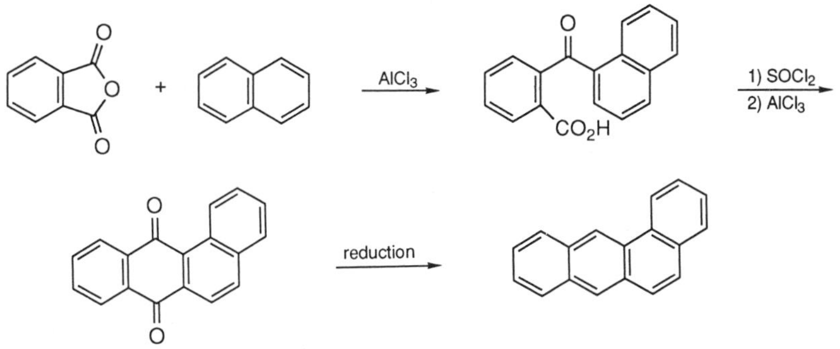

The cyclization step may be carried out either by treating the acid chloride with aluminium trichloride as above or, more conveniently, by treating the acid itself with liquid hydrogen fluoride, concentrated sulfuric acid, or polyphosphoric acid. Sulfuric acid is the least satisfactory reagent since it can lead to sulfonated byproducts.

(c) Chloromethylation

The chloromethyl group, —CH_2Cl, can be introduced into aromatic compounds by treatment with formaldehyde and hydrogen chloride in the presence of an acid. For example, benzyl chloride may be obtained in about 80% yield by passing hydrogen chloride into a suspension of paraformaldehyde and zinc chloride in benzene; the acid first liberates formaldehyde from paraformaldehyde and then takes part in the reaction.

$$PhH \quad + \quad CH_2O \quad + \quad HCl \quad \xrightarrow{(ZnCl_2)} \quad PhCH_2Cl \quad + \quad H_2O$$

Fluoromethylation, bromomethylation, and iodomethylation may be carried out with the appropriate halogen acid.

Studies of the mechanism indicate that the electrophilic entity is the hydroxymethyl cation. This reacts to give an alcoholic product that, in the presence of hydrogen chloride, is converted into the chloromethyl product:

$$CH_2{=}O + H^+ \rightleftharpoons \left[CH_2{=}\overset{+}{O}H \quad \longleftrightarrow \quad \overset{+}{C}H_2{-}OH \right]$$

$$\xrightarrow[-H^+]{PhH} \quad PhCH_2OH \quad \xrightarrow[-H_2O]{HCl} \quad PhCH_2Cl$$

Chloromethylation, unlike Friedel–Crafts reactions, is successful even with quite strongly deactivated nuclei such as that of nitrobenzene, although *m*-dinitrobenzene and pyridine are inert.

Two complications can occur in chloromethylation. First, the chloromethyl product can alkylate another molecule of the aromatic compound in the presence of the acid catalyst, e.g.

$$PhH \xrightarrow[ZnCl_2]{CH_2O - HCl} PhCH_2Cl \xrightarrow[ZnCl_2]{PhH} PhCH_2Ph$$

This secondary reaction is of particular significance when the aromatic compound is strongly activated and for this reason chloromethylation is not a suitable procedure for phenols and anilines.

Second, the chloromethyl group is activating, although less so than methyl because the chlorine substituent in the methyl group reduces the $+I$ effect of that group. It is usually difficult to avoid the occurrence of some further chloromethylation, although this is not nearly so important a problem as it is in Friedel–Crafts alkylation.

The reaction conditions may be varied widely. Anhydrous hydrogen chloride may be replaced by the concentrated aqueous acid; formaldehyde may be introduced as paraformaldehyde or methylal $(CH_2(OCH_3)_2)$; and zinc chloride may be replaced by sulfuric acid or phosphoric acid or omitted altogether in the chloromethylation of very reactive aromatic compounds such as thiophen. In a typical example, a mixture of naphthalene, paraformaldehyde, glacial acetic acid, 85% phosphoric acid, and concentrated hydrochloric acid, heated at 80°C for 6 hours, gives a 75% yield of 1-chloromethylnaphthalene.

The principal value of chloromethylation lies in the ease of displacement of the benzylic chloride by nucleophiles. Conversion into the corresponding alcohols, $ArCH_2OH$, ethers, $ArCH_2OR$, nitriles, $ArCH_2CN$, and amines, $ArCH_2NR_2$, may be accomplished efficiently, and treatment with enolates, such as that from malonic ester, leads to extension of the aliphatic carbon chain, e.g.

$$Ar\diagup Cl \xrightarrow[EtO^-]{CH_2(CO_2Et)_2} Ar\diagup\diagdown{}^{CO_2Et}_{CO_2Et} \xrightarrow[2) - CO_2]{1) \text{ hydrolysis}} Ar\diagup\diagdown CO_2H$$

(d)   Gattermann–Koch formylation

The formyl group, —CHO, may be introduced into aromatic compounds by treatment with carbon monoxide and hydrogen chloride in the presence of a Lewis acid:

$$ArH + CO \xrightarrow{HCl + \text{Lewis acid}} ArCHO$$

The reaction is carried out either under pressure or in the presence of copper(I) chloride.

It was thought at one time that the reaction occurs through the formation of formyl chloride, HCOCl, from carbon monoxide and hydrogen chloride, followed by a Friedel–Crafts acylation catalyzed by the Lewis acid, but formyl chloride has never been obtained and it is now considered probable that the electrophilic species is the formyl cation, $[HC{\equiv}\overset{+}{O} \leftrightarrow H\overset{+}{C}{=}O]$, formed without the mediation of formyl chloride:

$$HCl \;+\; CO \;+\; AlCl_3 \longrightarrow HCO^+ \;+\; AlCl_4^-$$

$$ArH \;+\; HCO^+ \xrightarrow{\;-H^+\;} ArCHO$$

The role of copper(I) chloride may be to aid the reaction between carbon monoxide and hydrogen chloride *via* the complex which it forms with carbon monoxide.

Formylation is unsuccessful with aromatic compounds of lower nuclear reactivity than the halobenzenes so nitrobenzene may be used as solvent. It is also unsuccessful with amines, phenols, and phenol ethers because of the formation of complexes with the Lewis acid. One drawback in the application of the reaction to polyalkylated benzenes is that rearrangements and disproportionations occur: e.g. *p*-xylene gives 2,4-dimethylbenzaldehyde.

In a typical procedure, carbon monoxide and hydrogen chloride are passed into toluene which contains suspended aluminium trichloride and copper(I) chloride. *p*-Tolualdehyde is formed as a complex with the Lewis acid and is isolated in 50% yield by treatment with ice and distillation in steam [2].

(e) Gattermann formylation

This is an alternative to the Gattermann–Koch reaction, employing hydrogen cyanide instead of carbon monoxide. The initial product is an immonium chloride which is converted into the aldehyde with mineral acid:

$$ArH + HCN + HCl \xrightarrow{\text{Lewis acid}} ArCH{=}NH_2^+\; Cl^- \xrightarrow{H_3O^+} ArCHO + NH_4Cl$$

The ionic intermediate, $[HC{\equiv}\overset{+}{N}H \leftrightarrow H\overset{+}{C}{=}NH]$, analogous to the formyl cation, is thought to be the electrophilic entity. The reaction is unsuccessful with deactivated compounds such as nitrobenzene, and compounds of moderate reactivity, such as benzene and the halobenzenes, give only low yields. Yields from more reactive compounds are considerably higher, e.g. anthracene gives the 9-aldehyde in 60% yield.

Unlike the Gattermann–Koch reaction, Gattermann formylation is successful with phenols and phenol ethers, e.g. *p*-anisaldehyde is formed from anisole almost quantitatively in the presence of aluminium trichloride. More reactive nuclei still can be formylated in the presence of the weaker Lewis acid zinc chloride and furan reacts even in the absence of a catalyst to give the 2-aldehyde.

To avoid the use of hydrogen cyanide, it is convenient to use zinc cyanide from which hydrogen cyanide is generated *in situ* by reaction with hydrogen chloride. For example, mesitaldehyde is obtained in over 75% yield by passing hydrogen chloride into a solution of mesitylene in tetrachloroethane in the presence of zinc cyanide, adding aluminium trichloride, and decomposing the resulting immonium hydrochloride with hydrochloric acid [2]:

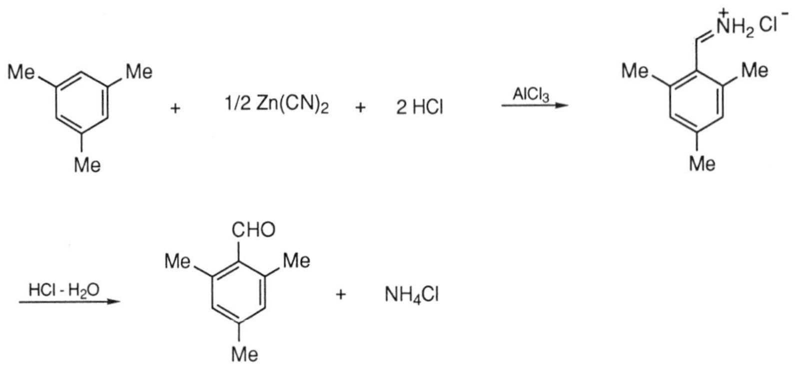

(f) Hoesch acylation

This reaction is an adaptation of Gattermann formylation: the use of an aliphatic nitrile in place of hydrogen cyanide leads to an acyl-derivative of the aromatic compound.

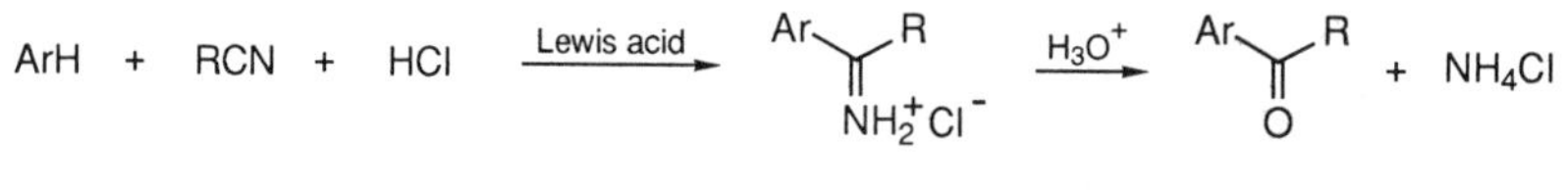

The reaction occurs only with the most highly activated aromatic compounds such as di- and polyhydric phenols. Monohydric phenols react mainly at oxygen to give imido-esters,

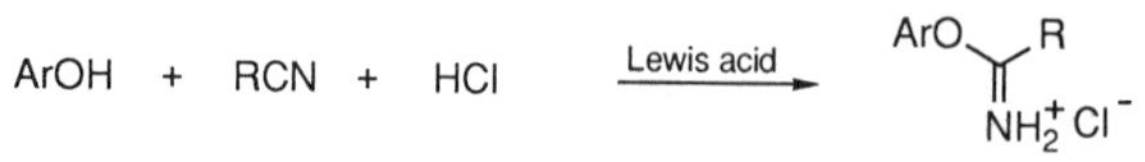

but the combination of two or three hydroxyl groups *meta* to each other so increases the reactivity of the nuclear positions *ortho* or *para* to hydroxyl that nuclear acylation occurs. For example, phloroacetophenone may be obtained in 80% yield by passing hydrogen chloride into a cooled solution of phloroglucinol (1,3,5-trihydroxybenzene) and acetonitrile in ether containing suspended zinc chloride and then hydrolyzing the resulting precipitate of the immonium salt by boiling in aqueous solution [2].

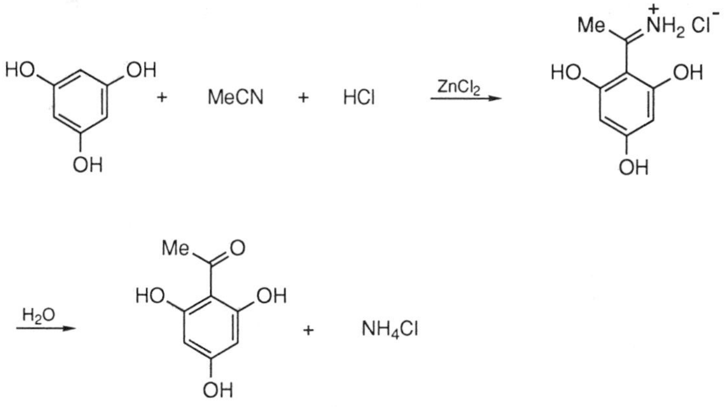

phloroacetophenone

(g) Vilsmeyer formylation

N-Formylamines, from secondary amines and formic acid, formylate aromatic compounds in the presence of phosphorus oxychloride. The mechanism is thought to be as follows:

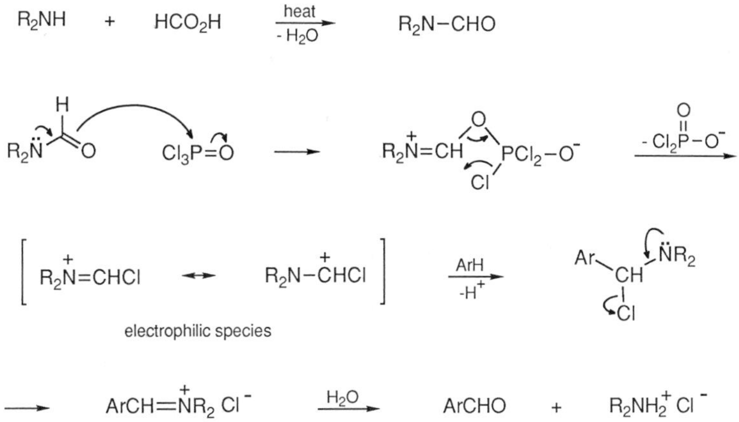

electrophilic species

Only the most reactive aromatic compounds are formylated, e.g. benzene and naphthalene are unreactive but anthracene gives the 9-aldehyde (84%), NN-dimethylaniline gives p-dimethylaminobenzaldehyde (80%), and thiophen gives the 2-aldehyde (70%). The method is particularly effective for compounds such as pyrroles which are not formylated by other procedures, e.g. pyrrole gives the 2-aldehyde in 85% yield and indole gives the 3-aldehyde in 97% yield, each with dimethylformamide [4].

(h) Reimer–Tiemann formylation

The treatment of phenols with chloroform in basic solution gives aldehydes, e.g.

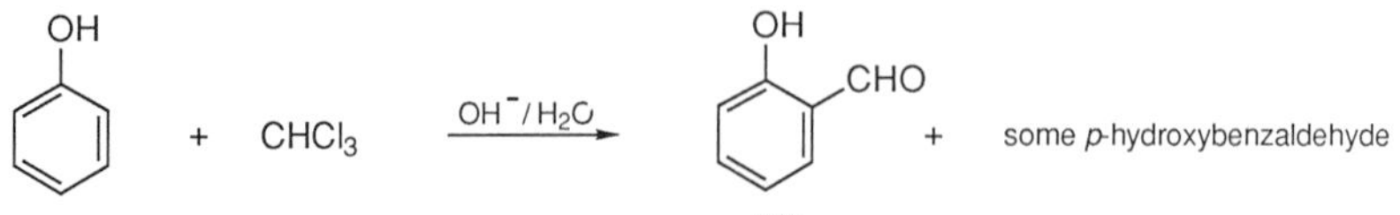

$$\text{OH} \quad + \quad CHCl_3 \quad \xrightarrow{OH^-/H_2O} \quad \text{(salicylaldehyde)} \quad + \quad \text{some } p\text{-hydroxybenzaldehyde}$$

40%

The reaction occurs through dichlorocarbene, which is generated from chloroform and alkali (p. 101) and, being electrophilic, is attacked by the strongly nucleophilic phenoxide ion. Hydrolysis of the benzal chloride follows, and acidification yields the aldehyde.

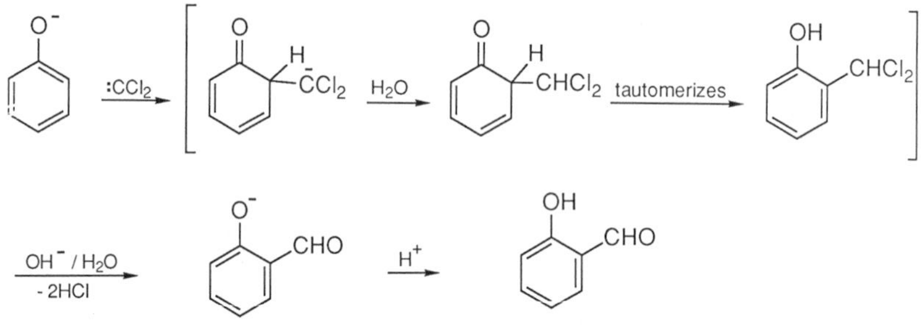

Pyrrole, which resembles phenol in its reactivity to electrophiles (p. 358), undergoes the Reimer–Tiemann reaction, via the strongly nucleophilic pyrrolate anion, giving pyrrole-2-aldehyde, but a second product, 3-chloropyridine, is also obtained. Each product derives from the same intermediate:

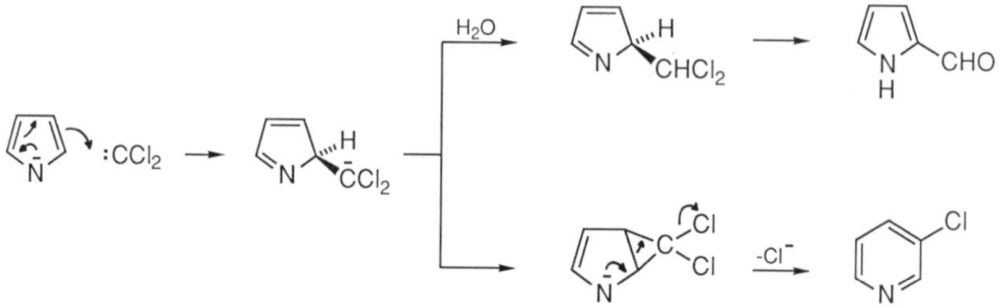

Phenols with blocked *para* positions give cyclohexadienones in which the two chlorine substituents, being in neopentyl-type environments, are resistant to hydrolysis in the basic conditions (p. 105) and remain intact, e.g.

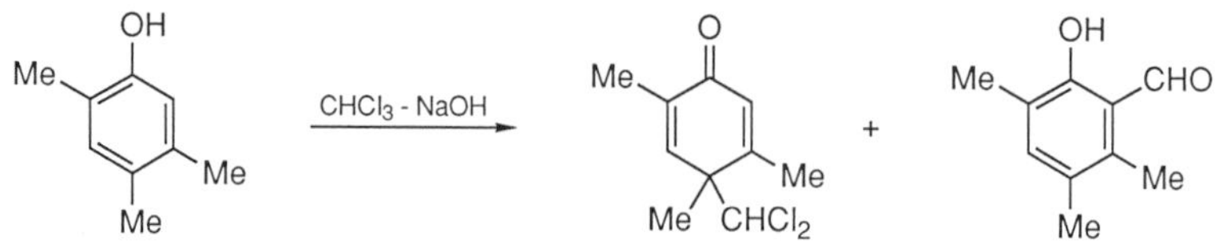

40%

Use may be made of this property for introducing angular methyl groups into decalin derivatives (cf. p. 231): the angular group is introduced as —CHCl₂ and converted into —CH₃ by hydrogenolysis, e.g.

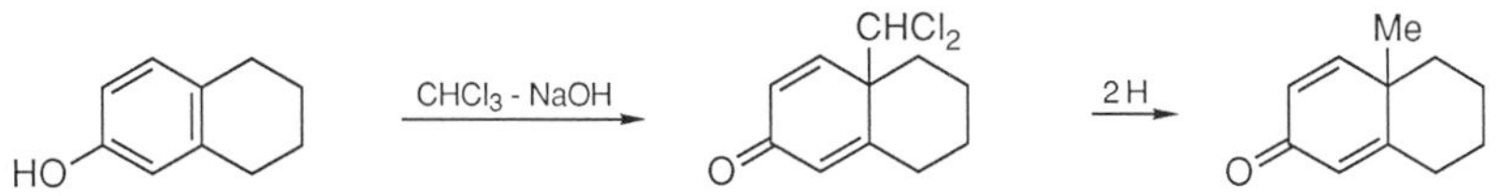

(i) Kolbe–Schmitt carboxylation

Phenoxide ions are reactive enough to add to carbon dioxide, which is a relatively weak electrophile; reaction is carried out under pressure at about 100°C. The *ortho* product predominates, probably because of the stabilizing influence of chelation on the transition state for *ortho* substitution, e.g.

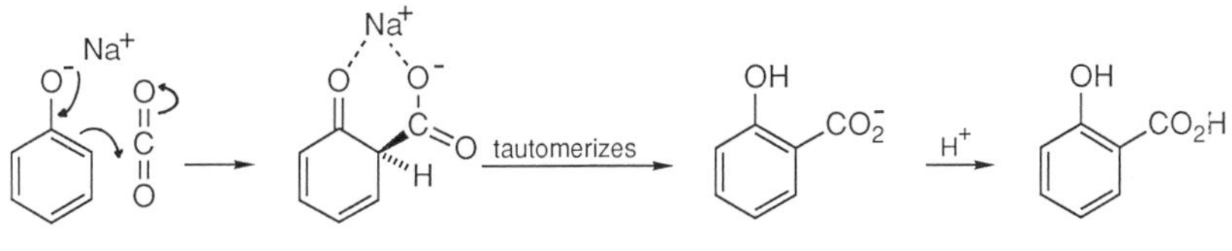

However, the reaction is reversible and at about 240°C the less rapidly formed, but more stable, *para*-isomer predominates.

Pyrrole undergoes a similar reaction, giving pyrrole-2-carboxylic acid, on being heated with ammonium carbonate at 120°C.

(j) The Mannich reaction

This reaction is suitable for bonding aliphatic carbon to the reactive positions of phenols, pyrroles, and indoles (p. 263). For example, pyrrole, formaldehyde, and dimethylamine give 2-dimethylaminomethylpyrrole,

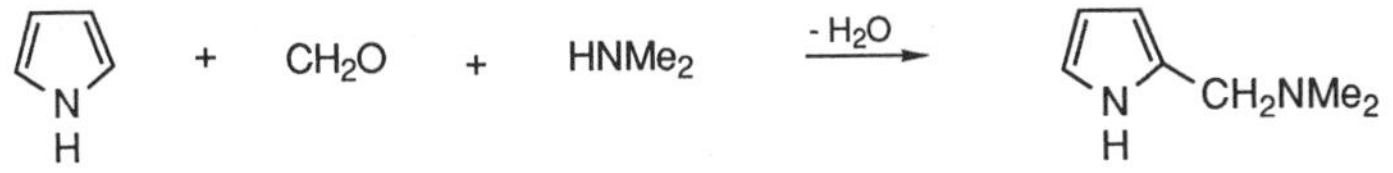

and indole reacts, as usual, at the 3-position to give gramine,

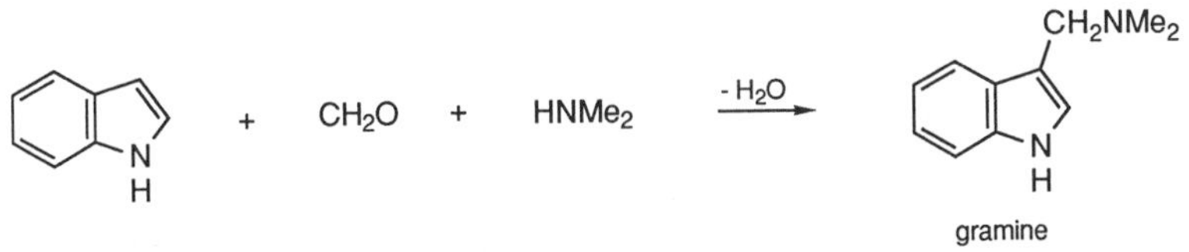

Furan resists Mannich conditions, but 2-methylfuran, whose methyl group increases the reactivity of the nucleus to electrophiles, reacts at the 5-position.

$$\text{Me} \overset{\frown}{\underset{O}{\bigcirc}} + CH_2O + HNMe_2 \xrightarrow{-H_2O} \text{Me} \overset{\frown}{\underset{O}{\bigcirc}} CH_2NMe_2$$

Synthetic uses of the Mannich bases from some of these aromatic compounds have been described previously (p. 264).

11.4 Formation of carbon–nitrogen bonds

(a) Nitration

By far the most commonly employed method for bonding nitrogen to aromatic systems is by nitration. The reason is that a very wide variety of nitration conditions is available, so that suitable procedures may usually be found for nitrating compounds of both very high and very low nuclear reactivity. The more frequently employed methods, arranged in approximately decreasing order of their vigour, are:

- A mixture of concentrated nitric and concentrated sulfuric acid,
- Fuming nitric acid in acetic anhydride,
- Nitric acid in glacial acetic acid,
- Dilute nitric acid.

The mixture of concentrated nitric and sulfuric acids generates the nitronium ion, NO_2^+, and this is the active electrophilic species. Its formation results from protonation of nitric acid by sulfuric acid (i.e. nitric acid acts as a base towards a powerful proton-donor), followed by heterolysis:

$$H_2SO_4 + HO-NO_2 \rightleftharpoons HSO_4^- + H_2\overset{+}{O}-NO_2$$

$$H_2\overset{+}{O}-NO_2 \rightleftharpoons H_2O + NO_2^+$$

The molecule of water formed is essentially completely removed by protonation, so that the equilibria favour the nitronium ion. However, added water can reduce the concentration of nitronium ions and lower the nitrating power of the system.

A solution of nitric acid in acetic anhydride contains a number of species in equilibrium:

$$HO-NO_2 + Ac-OAc \rightleftharpoons HO-Ac + AcO-NO_2$$

$$2\ AcONO_2 \rightleftharpoons AcOAc + O_2N-O-NO_2$$

$$O_2N-O-NO_2 \rightleftharpoons NO_2^+ + NO_3^-$$

It is believed that the nitronium ion is again the principal electrophilic entity.

Dilute nitric acid contains a negligible concentration of the nitronium ion and it acts as a nitrating agent only by virtue of the small concentration of nitrous acid which is normally contained in it. This gives rise to the nitrosonium ion, NO^+, which nitrosates the aromatic ring; the nitroso-compound is oxidized by nitric acid to the nitro-compound which generates more nitrous acid to continue the chain-reaction:

$$ArH \;+\; HNO_2 \;\longrightarrow\; ArNO \;+\; H_2O$$

$$ArNO \;+\; HNO_3 \;\longrightarrow\; ArNO_2 \;+\; HNO_2$$

Consequently, only those compounds (e.g. phenols) which undergo nitrosation can be nitrated with dilute nitric acid.

(*i*) *Benzene.*   Benzene is nitrated satisfactorily by a mixture of concentrated nitric and sulfuric acids ('mixed acids'). Since the nitro group deactivates the aromatic nucleus towards electrophiles, it is relatively easy to prevent di- or polynitration: reaction at 30–40°C yields nitrobenzene rapidly and in about 95% yield. However, dinitration can be brought about by raising the temperature to 90–100°C.

(*ii*) *Moderately activated nuclei.*   Although the alkylbenzenes are more reactive than benzene (e.g. toluene is nitrated at about 25 times the rate of benzene in the same conditions), nitric acid in sulfuric acid is still the most satisfactory method. The *o*- and *p*-nitro-derivatives predominate, but the proportion of the former falls as the alkyl group increases in size (p. 351).

(*iii*) *Strongly activated nuclei.*   Naphthalene is reactive enough to be nitrated by nitric acid in acetic acid, giving mainly 1-nitronaphthalene, and similar conditions give 2-nitrothiophen (70–85%) from thiophen [2] and nitromesitylene (75%) from mesitylene [2].

Phenol, which is considerably more reactive, is nitrated with dilute nitric acid and gives comparable amounts of *o*- and *p*-nitrophenol. Aniline is too readily oxidized for direct nitration to be practicable. The amino group is first protected by acetylation, acetanilide is nitrated, and the acetyl group is removed by hydrolysis. *NN*-Dimethylaniline is less easily oxidized than aniline and can be nitrated with dilute nitric acid.

Furan and pyrrole are polymerized by acidic nitrating systems but nitration in acetic anhydride at low temperatures is successful. Each compound gives mainly the 2-nitro-derivative, that from furan being formed *via* an adduct:

$$HNO_3 \;+\; Ac_2O \;\longrightarrow\; AcO\text{--}NO_2 \;+\; AcOH$$

3-Nitroindole can be obtained by treating indole with ethyl nitrate in ethanol containing sodium ethoxide. Reaction occurs *via* the indolate anion:*

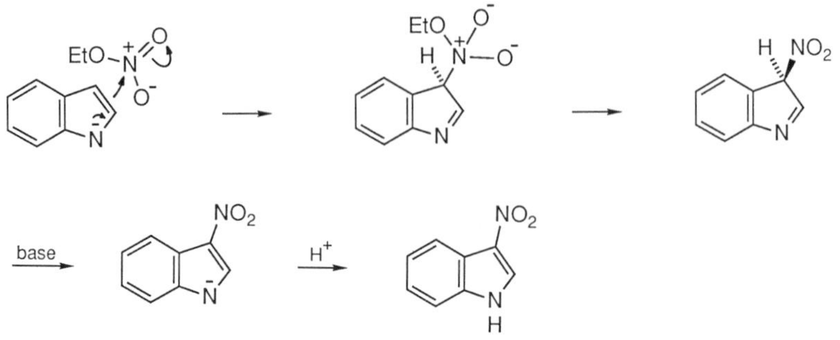

(*iv*) *Moderately deactivated nuclei.* The halobenzenes, which are about one-tenth as reactive as benzene, are nitrated in the same way as benzene. The *p*-nitro-derivative predominates in each case and further nitration gives chiefly the 2,4-dinitro-derivatives. Concentrated nitric and sulfuric acids at or below room temperature are also suitable for the nitration of rather less reactive compounds such as benzaldehyde, acetophenone, and methyl benzoate, each of which gives mainly the *m*-nitro-derivative. For example, nitration of methyl benzoate at 5–15°C gives methyl *m*-nitrobenzoate in 80% yield [1].

The nitration of *NN*-dimethylaniline in concentrated sulfuric acid gives the *m*-nitro-derivative in about 60% yield [3]. This is because in the strongly acidic conditions essentially all the amine is present as the *meta*-directing anilinium ion; in less acidic conditions the *p*-nitro-derivative is the main product.

(*v*) *Strongly deactivated nuclei.* Nitrobenzene is less reactive than the deactivated compounds such as benzaldehyde but the mixed-acid procedure is successful at about 90–100°C: *m*-dinitrobenzene is obtained in about 80% yield. *m*-Dinitrobenzene itself is so strongly deactivated that further nitration requires fuming nitric acid and fuming sulfuric acid at 110°C for several days; 1,3,5-trinitrobenzene is formed. This compound is more satisfactorily obtained from toluene: the activating influence of the methyl group allows 2,4,6-trinitrotoluene to be formed with mixed acid and the methyl group is removed by oxidation and decarboxylation.†

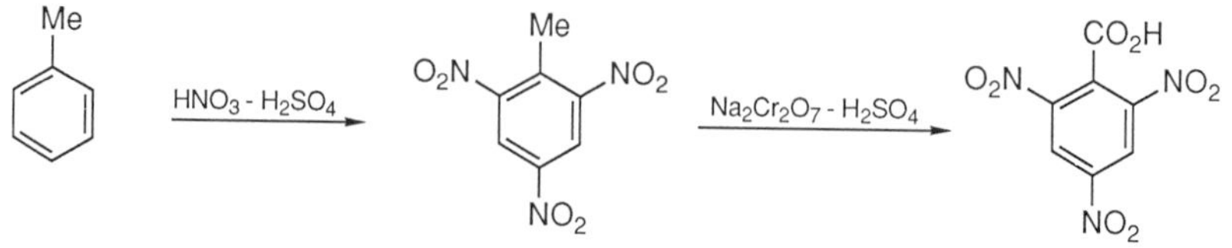

---

* cf. the reaction of enolates with alkyl nitrates, p. 323.
† Decarboxylation occurs readily because of the presence of the electron-attracting nitro groups. Reaction probably occurs *via* the carbanion, see p. 389.

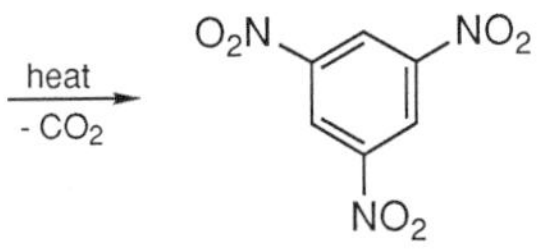

Pyridine is very powerfully deactivated, partly because in the acidic conditions necessary for reaction it is fully protonated (unlike the very weakly basic nitro compounds). Even the use of 100% sulfuric acid and a mixture of sodium and potassium nitrates at 300°C gives only 5% of 3-nitropyridine. 3-Substituted pyridines are better prepared by starting with the naturally occuring nicotinic acid, converting it into 3-aminopyridine *via* Hofmann bromination (p. 442), and making use of the reactivity of the corresponding diazonium salt (p. 410).

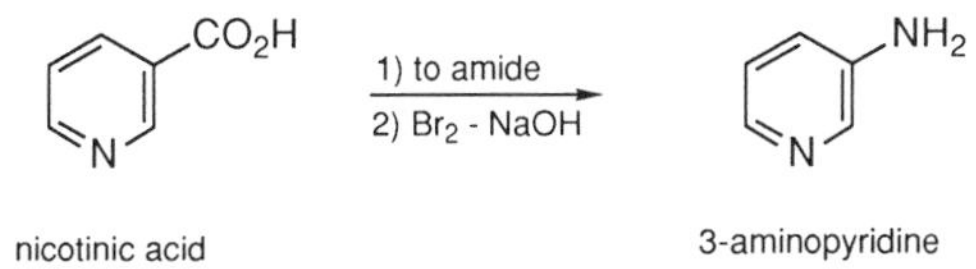

The properties of pyridine and related heterocycles are profoundly modified in the corresponding *N*-oxides; for example, pyridine *N*-oxide, obtained from pyridine by oxidation with 30% aqueous hydrogen peroxide in acetic acid at 70–80°C (see p. 622), undergoes nitration with mixed nitric and sulfuric acids at 100°C to give mainly the 4-nitro-derivative. The explanation is that mesomeric electron-release from the oxide oxygen atom stabilizes the transition state for 4-substitution, as represented by the contribution of the resonance structure *c*:

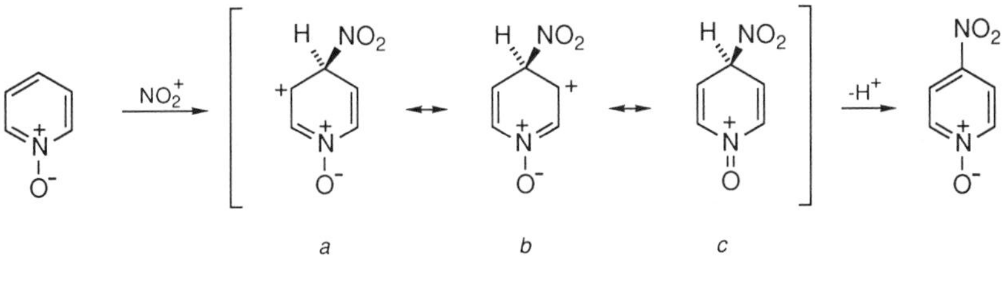

Substitution at the 2-position (but not at the 3-position) should likewise be facilitated, but the activating influence of oxygen at the 2-position is evidently much weaker than at the 4-position, since the 2-nitro-derivative is formed only in very small amount. Since pyridine *N*-oxides are readily deoxygenated (e.g. with phosphorus trichloride), nitration of pyridine *N*-oxide provides a useful route to 4-nitropyridine.

Quinoline and isoquinoline are considerably more reactive than pyridine. The effect of the heteroatom is not strongly felt in the benzenoid ring and nitration in concentrated sulfuric acid at 0°C gives, with quinoline, comparable quantities of the 5- and 8-nitro-derivatives and, with isoquinoline, almost entirely the 5-nitro-derivative.

(b) Nitrosation

Nitrosation is brought about by treatment of an aromatic compound with sodium nitrite and a strong acid. Nitric acid cannot be used since it oxidizes the nitroso product to a nitro compound (p. 622) and sulfuric acid or hydrochloric acid is normally chosen.

The electrophilic entity is the nitrosonium ion, NO^+, that is generated from nitrous acid in a manner analogous to the formation of the nitronium ion from nitric acid.

$$NaNO_2 \quad + \quad H_2SO_4 \quad \longrightarrow \quad HNO_2 \quad + \quad NaHSO_4$$

$$HO{-}NO \quad + \quad H_2SO_4 \quad \rightleftharpoons \quad H_2\overset{+}{O}{-}NO \quad + \quad HSO_4^-$$

$$H_2\overset{+}{O}{-}NO \quad \rightleftharpoons \quad H_2O \quad + \quad NO^+$$

Nitrosation is limited to the very reactive nuclei of phenols and tertiary aromatic amines. Primary aromatic amines undergo diazotization (section 13.1) (but see below) and secondary amines undergo N-nitrosation ($ArNHR + NO^+ \rightarrow ArNR{-}N{=}O + H^+$). In a typical procedure, 2-naphthol in sodium hydroxide solution is treated with sodium nitrite, and concentrated sulfuric acid is then added: 1-nitroso-2-naphthol is obtained in over 90% yield [1].

The treatment of NN-dimethylaniline with sodium nitrite and hydrochloric acid gives p-nitrosodimethylaniline, as its hydrochloride, in 85% yield.

The nitroso compounds formed are easily oxidized to nitro compounds (p. 622) and reduced to amines (p. 671). In the latter context, they find a synthetic application when the normal route to an aromatic amine, via nitration, cannot be employed because of the disruption of a sensitive compound in nitrating conditions. For example, in one synthesis of adenine (6-aminopurine) the last of three amino groups was introduced via nitrosation:*

---

*The primary amino groups in this molecule are not diazotized because the unshared electron-pair on each is withdrawn into the ring by the electron-attracting heteroatoms. The nucleus, however, is sufficiently activated by the *combined* effect of the two amino groups to undergo C-nitrosation.

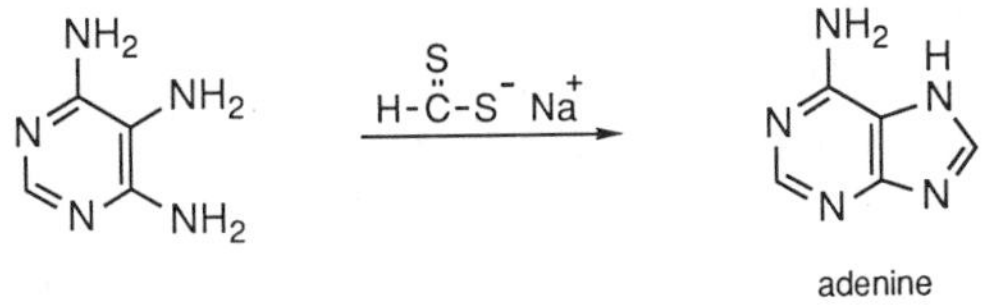

adenine

(c)   Diazonium coupling

The highly reactive nuclei of phenoxide ions, aromatic amines, pyrrole, etc. react with aromatic diazonium ions, ArN_2^+. These reactions are discussed subsequently (section 13.4).

**11.5
Formation of
carbon–sulfur
bonds**

(a)   Sulfonation

As in nitration, the conditions for sulfonation may be varied widely and suitable conditions for the reactions of both strongly activated and strongly deactivated nuclei are available.

The electrophilic sulfonating entity is sulfur trioxide, which is present in concentrated sulfuric acid as the result of the equilibrium,

$$2\ H_2SO_4 \rightleftharpoons SO_3 + H_3O^+ + HSO_4^-$$

Its concentration is reduced by the addition of water but is greater in fuming sulfuric acid. Sulfur trioxide is also 'carried' by pyridine in the form of pyridinium-1-sulfonate (from pyridine and sulfur trioxide),

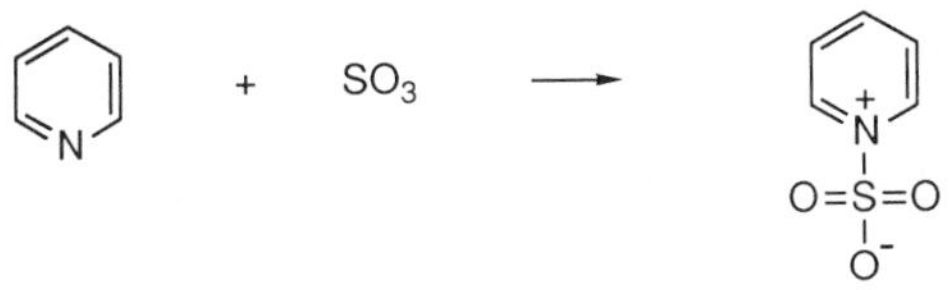

and this compound is a suitable sulfonating agent for compounds which are unstable to acids.

The mechanism of sulfonation is essentially the same as that of other electrophilic substitutions:

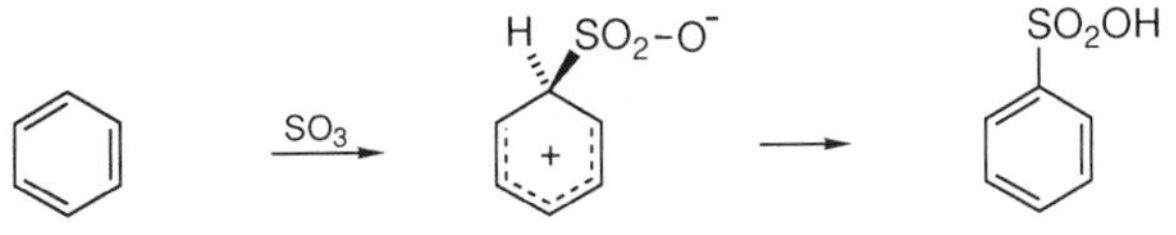

However, unlike most substitutions, sulfonation is readily reversible: the sulfonic acid group is eliminated by heating the compound with dilute sulfuric acid.

Applications of the reverse process are described below.

Benzene is normally sulfonated with 5–20% oleum and these conditions are suitable also for the weakly deactivated halobenzenes. For more reactive compounds, such as the alkylbenzenes and the polycyclic hydrocarbons, sulfuric acid is satisfactory, e.g. in these conditions toluene gives the three toluenesulfonic acids in the approximate proportions: *ortho*, 10: *meta*, 1: *para*, 20.

Pyrrole, furan, and indole, which are decomposed by acids, are successfully sulfonated by pyridinium-1-sulfonate. Reaction occurs at the 2-position in each case (unexpectedly, in the case of indole). The more stable thiophen may be sulfonated with 95% sulfuric acid, but a higher yield (about 90% of thiophen-2-sulfonic acid) is obtained with pyridinium-1-sulfonate.

Aniline reacts with sulfuric acid to form a salt but on strong heating this rearranges, through phenylsulfamic acid, to sulfanilic acid.

Phenol readily undergoes disulfonation to give the 2,4-disulfonic acid.

Strongly deactivated nuclei can be sulfonated with oleum at high temperatures. Benzenesulfonic acid gives the *m*-disulfonic acid with 20–40% oleum at 200°C, nitrobenzene gives *m*-nitrobenzenesulfonic acid similarly, and even pyridine gives a high yield (70%) of pyridine-3-sulfonic acid when sulfonated with oleum in the presence of mercury(II) sulfate at 230°C. As usual, quinoline and isoquinoline are more reactive than pyridine and give, respectively, mainly the 8-sulfonic acid and the 5-sulfonic acid with sulfuric acid at 220°C.

(*i*) *Applications of the reverse of sulfonation.* The reversibility of sulfonation makes possible (1) the use of the sulfonic acid group for *protection* and (2) the formation, in appropriate conditions, of the thermodynamically most stable product.

(1) The use of sulfonate as a protective group is illustrated by the synthesis of *o*-nitroaniline. Acetanilide is sulfonated to give almost exclusively the *p*-sulfonic acid; nitration then occurs *ortho* to the acetamido group (and is accompanied by the hydrolysis of the acetyl group); and the sulfonic acid group is finally eliminated by treatment with dilute sulfuric acid.

(2) The principles of the competition between kinetic and thermodynamic control in sulfonation have been discussed earlier (section 3.7d). Sulfonation at low temperatures gives the kinetically controlled products, e.g. mainly toluene-p-sulfonic acid from toluene and naphthalene-1-sulfonic acid from naphthalene, whereas at about 160°C the more stable toluene-m-sulfonic acid and naphthalene-2-sulfonic acid predominate. This is of particular value in enabling 2-naphthyl derivatives to be obtained through conversion of the 2-sulfonic acid into 2-naphthol (p. 402) and thence into 2-naphthylamine (p. 407), diazotization of which enables other groups to be introduced.

(b) Chlorosulfonation

Aromatic compounds which are neither powerfully deactivated nor unstable to acids react with chlorosulfonic acid, for example benzene at 20–25°C gives benzenesulfonyl chloride in 75% yield [1]:

$$PhH \quad + \quad 2\ ClSO_2OH \quad \longrightarrow \quad PhSO_2Cl \quad + \quad H_2SO_4 \quad + \quad HCl$$

An important example occurs in the synthesis of sulfanilamide (the first of the Sulfa drugs).

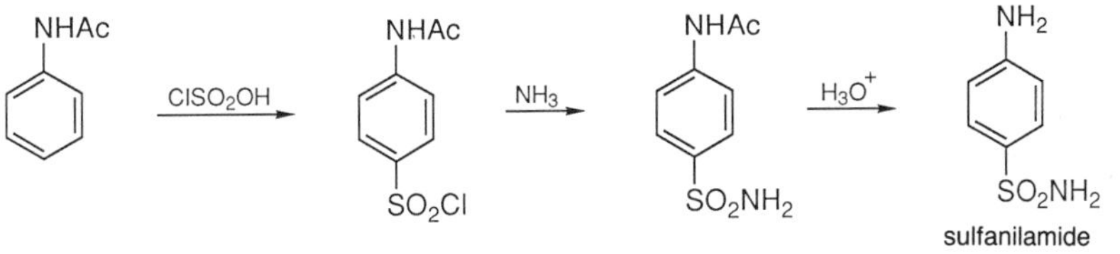

(c) Sulfonylation

Benzene reacts with benzenesulfonyl chloride in the presence of aluminium trichloride to give diphenyl sulfone:

$$PhSO_2Cl \quad + \quad PhH \quad \xrightarrow{\ AlCl_3\ } \quad PhSO_2Ph \quad + \quad HCl$$

The reaction, which is analogous to Friedel–Crafts acylation, has the same limitations as chlorosulfonation.

**11.6
Formation of
carbon–
halogen bonds**

(a) Chlorination

Three general procedures are available for chlorination, the choice of method being based on the reactivity of the aromatic compound.

The mildest conditions involve the use of molecular chlorine, usually in solution in acetic acid or a non-polar solvent such as carbon tetrachloride. The chlorine may be introduced as the gaseous element or may be generated *in situ* from an *N*-chloroamine and hydrogen chloride,

$$R_2N-Cl \quad + \quad HCl \quad \longrightarrow \quad R_2NH \quad + \quad Cl_2$$

Since hydrogen chloride is formed as a result of aromatic chlorination, this procedure for generating chlorine is continuous once started and the concentration of chlorine in solution can be maintained at the desired level according to the amount of hydrogen chloride added initially.

The reactivity of molecular chlorine is increased by the addition of a Lewis acid. Aluminium trichloride and iron(III) chloride are commonly used; they are usually introduced as aluminium amalgam and iron filings, respectively, from which the metal chlorides are generated *in situ*. The function of the Lewis acid is to withdraw electrons from the chlorine molecule thereby increasing its electrophilic character. It should be emphasized that this polarizing influence does not lead to complete ionization (e.g. $Cl_2 + AlCl_3 \rightarrow Cl^+ + AlCl_4^-$), i.e. the reaction should be represented as:

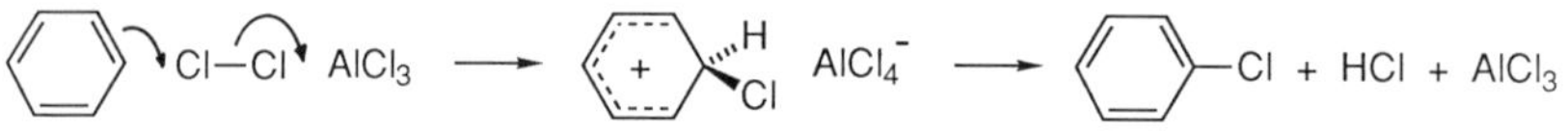

Still more reactive conditions are obtained by using an acidified solution of hypochlorous acid. It has been suggested that the effective electrophile is the chlorinium ion, Cl^+,

$$HO-Cl \quad + \quad H^+ \quad \rightleftharpoons \quad H_2\overset{+}{O}-Cl \quad \rightleftharpoons \quad H_2O \quad + \quad Cl^+$$

but this may be an oversimplification of the mechanism. A similarly reactive form of chlorine is derived from a solution of chlorine in carbon tetrachloride in the presence of silver sulfate.

(b) Bromination

The methods for bromination are closely related to those for chlorination: molecular bromine, added as such, or generated *in situ* from an *N*-bromoamine and acid, may be used; increased electrophilic power may be obtained by using bromine in the presence of a Lewis acid such as iron(III) bromide (i.e. addition of iron filings). A 'positive' brominating entity is present in acidified hypobromous acid and can also be obtained from bromine and silver sulfate in concentrated sulfuric acid.

Bromine itself can catalyze molecular bromination by virtue of its ability to form tribromide ion:

$$ArH \quad Br-Br \quad Br-Br \quad \longrightarrow \quad Ar\overset{+}{H}Br \quad + \quad Br-\overset{-}{Br}-Br$$

A second-order term in bromine therefore occurs in the kinetics of bromination. However, bromine is a much weaker Lewis acid than iron(III) bromide.

Iodine is often used as a catalyst in brominations. It acts by forming iodine bromide which facilitates bromination by removing bromide ion as IBr_2^-:

$$I_2 \; + \; Br_2 \; \rightleftharpoons \; 2\,IBr$$

$$ArH \quad Br\!-\!Br \quad IBr \longrightarrow \overset{+}{Ar}HBr \; + \; IBr_2^-$$

(c) Iodination

Iodine is a relatively weak electrophile which alone reacts with only the most strongly activated aromatic compounds such as phenols and amines. However, less reactive compounds can be iodinated in the presence of nitric acid or peroxyacetic acid; for example, benzene and iodine in the presence of nitric acid give iodobenzene in 86% yield [1].

It used to be thought that iodine and benzene react reversibly to give only a small proportion of iodobenzene,

$$PhH \; + \; I_2 \; \rightleftharpoons \; PhI \; + \; HI$$

and that the function of the additive is to oxidize the hydrogen iodide and so displace the equilibrium to the right. However, it is now realized that this cannot be the explanation, since the rate of the back reaction is very slow. Evidence has been obtained that, in nitric acid, a more reactive iodinating agent than iodine is formed, namely, the protonated form of NO_2I:

$$PhH \; + \; \overset{+}{I}\!-\!\underset{H}{\overset{|}{O}}\!-\!NO \longrightarrow PhI \; + \; HONO \; + \; H^+$$

In peroxyacetic acid, it is possible that the iodinating entity is iodine acetate:

$$I_2 \; + \; CH_3CO_2OH \; \rightleftharpoons \; IOCOCH_3 \; + \; HOI$$

(d) Fluorination

Fluorine is far too reactive a compound to be used as a halogenating agent. Aryl fluorides are prepared *via* aromatic diazonium ions (p. 412).

(e) Applications in synthesis

Highly activated nuclei react readily with molecular halogens and polyhalogenation occurs. For example, phenol and bromine in aqueous solution give a precipitate of a tetrabromodienone which, on washing with $NaHSO_3$ solution, forms 2,4,6-tribromophenol:

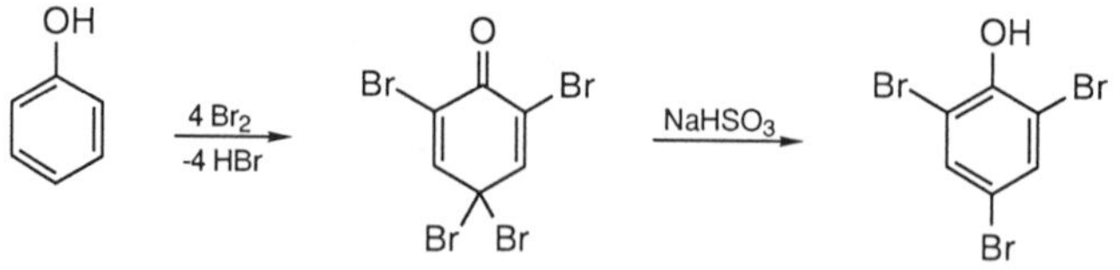

The successive reactions occur through the very reactive phenoxide and sub-stituted phenoxide ions, even though these are present in small concentrations; the deactivating effect of each bromo-substituent is offset by its acid-strengthening influence, so that relatively more of the corresponding phenoxide ion is present as successive bromines are introduced. However, it is possible to control the extent of halogenation by using an acidic medium to suppress ionization. For example, the chlorination of phenol can be controlled to give mainly 2,4-dichlorophenol; the derivative, 2,4-dichlorophenoxyacetic acid, is manufactured as a selective weed killer.

Pyrrole, each of whose four nuclear positions is highly activated, gives tetra-bromo and tetraiodopyrrole with bromine and iodine, respectively, and with chlorine gives a pentachloro-derivative,

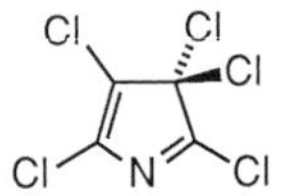

Furan is polymerized at room temperature by the acid liberated during hal-ogenation, but at $-40°C$ chlorination is reasonably efficient: the 2-chloro and 2,5-dichloro products predominate. Thiophen can also be chlorinated at low temperatures and at $-30°C$ gives a mixture of products containing mainly the 2-chloro and 2,5-dichloro derivatives. Iodination in the presence of mercury(II) oxide gives 2-iodothiophen.

Compounds of moderately high nuclear reactivity give monohalo-derivatives with the molecular halogens. For example, bromine in carbon tetrachloride reacts with mesitylene to give bromomesitylene (80%), with naphthalene to give 1-bromonaphthalene (75%), and with acetanilide to give p-bromoacetanilide (80%).

Compounds of the order of reactivity of benzene require a catalyst for halogen-ation. For example, benzene and chlorine give chlorobenzene in the presence of aluminium amalgam, and benzene and bromine give bromobenzene in the presence of iron filings, aluminium amalgam, or iodine.

These conditions are also suitable for quite strongly deactivated compounds, although high temperatures are needed in some cases. For example, nitrobenzene gives a 60% yield of m-bromonitrobenzene when treated with bromine in the presence of iron filings at 135–145°C [1] and even pyridine is brominated at 300°C to give the 3-bromo- and 3,5-dibromo derivatives.

Strongly deactivated compounds can often be successfully halogenated with a source of 'positive' halogen. For example, *m*-dinitrobenzene reacts with acidified hypobromous acid,

$$\text{NO}_2\text{-C}_6\text{H}_4\text{-NO}_2 \xrightarrow{\text{HOBr - H}^+} \text{Br-C}_6\text{H}_3(\text{NO}_2)_2$$

and benzoic acid and iodine in similar conditions give *m*-iodobenzoic acid in 75% yield.

11.7
Other reactions

(a) Hydroxylation

Hydrogen peroxide and peroxyacids in acidic media bring about electrophilic hydroxylation. Proton acids probably act by hydrogen-bonding to one oxygen atom of the peroxy-compound, so increasing the tendency for heterolysis of the O—O bond under the influence of the aromatic compound,

$$\text{ArH} + \text{H–O–O–R} \cdots \overset{+}{\text{H}}\text{–OH}_2 \longrightarrow \overset{+}{\text{ArH}}(\text{OH}) + \text{ROH} + \text{H}_2\text{O}$$

$$\longrightarrow \text{ArOH} + \text{ROH} + \text{H}_3\text{O}^+ \quad (\text{R} = \text{H or acyl})$$

and Lewis acids such as boron trifluoride act by coordination to oxygen,

$$\text{ArH} + \text{H–O–}\overset{+}{\text{O}}\text{–R}(\overset{-}{\text{BF}}_3) \longrightarrow \overset{+}{\text{ArH}}(\text{OH}) + \text{RO–}\overset{-}{\text{BF}}_3$$

$$\longrightarrow \text{ArOH} + \text{ROH} + \text{BF}_3$$

In general, electrophilic hydroxylation is not a satisfactory method for preparing phenols because the hydroxyl group which is introduced powerfully activates the positions in the nucleus which are *ortho* and *para* to it and further hydroxylation occurs readily. The dihydroxy products are then oxidized to quinones and rupture of the aromatic ring may follow.

However, if the positions which are activated by the entering hydroxyl group are substituted, further reaction is less likely to occur. In these circumstances reasonable yields of phenols may be obtained. For example, mesitylene can be converted into mesitol in 88% yield (based on the mesitylene which has reacted) by treatment with trifluoroperoxyacetic acid in the presence of boron trifluoride:

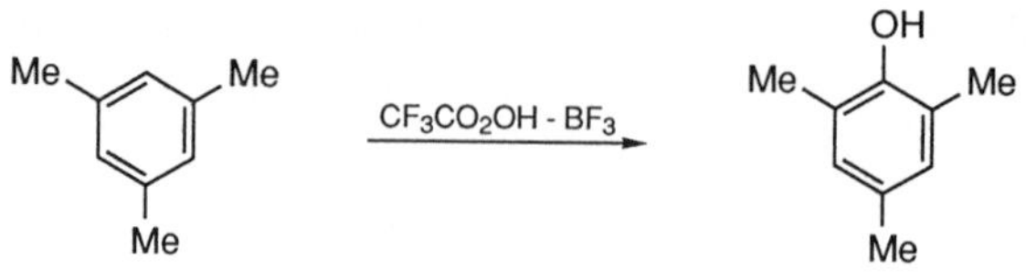

mesitol

(b)   Metallation

Salts of divalent mercury bring about the mercuration of aromatic compounds,*
e.g.

$$PhH \quad + \quad Hg(NO_3)_2 \quad \longrightarrow \quad PhHgNO_3 \quad + \quad HNO_3$$

Yields are low unless air and water are excluded and mercury(II) oxide is added to
remove nitric acid and prevent the reverse reaction.

The mercury derivatives have useful applications. For example, o-iodophenol is
conveniently synthesized by treating phenol with mercury(II) acetate, converting the o-acetoxymercury derivative which is formed into the o-chloromercury-
derivative, and treating this with iodine:

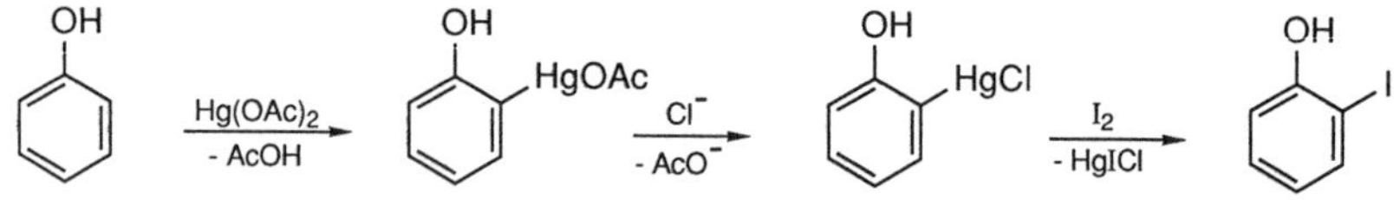

2-Bromo- and 2-iodofuran may be obtained in a similar way from furan:

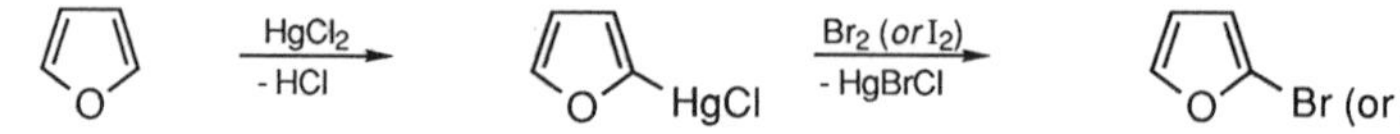

Thallium(III) salts have similar applications. For example, p-xylene and thallium
tris(trifluoroacetate) give the thallium derivative which, with potassium iodide at
0°C, forms the iodo-compound in 80–84% overall yield [6]:

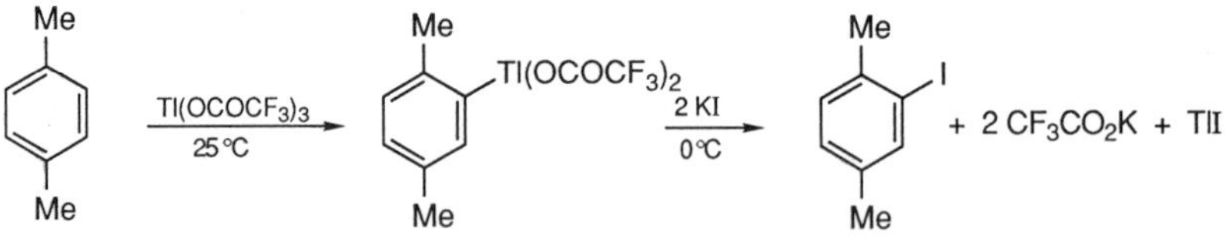

The thallation of monosubstituted benzenes with ortho-, para-directing substituents takes place almost entirely at the para position, probably because of the bulk
of the thallium reagent, and this provides a good method for introducing a p-iodo
group.

---

* Electrophilic metallations by metal salts should not be confused with the metallation of aromatic
compounds with metal, p. 185.

(c)   The displacement of groups other than hydrogen

A large number of electrophilic substitutions are known in which atoms and groups other than hydrogen are displaced from the aromatic ring (ipso-substitution). Few of these have synthetic value, but many can compete as unwanted reactions when attempts are being made to effect other substitutions; particular attention is drawn in the following to these side-reactions.

(i)   Decarboxylation. Powerfully activating substituents lead to the ready displacement of a carboxyl group, as carbon dioxide, from aromatic compounds.* For example, pyrrole- and furan-carboxylic acids decarboxylate on being heated; the reaction can be regarded as involving an internal electrophilic substitution by hydrogen, e.g.

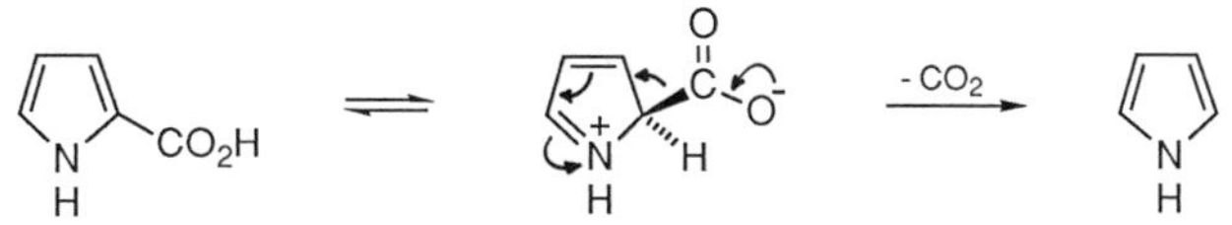

The acid-catalyzed decarboxylation of phenolic acids is similar, e.g.

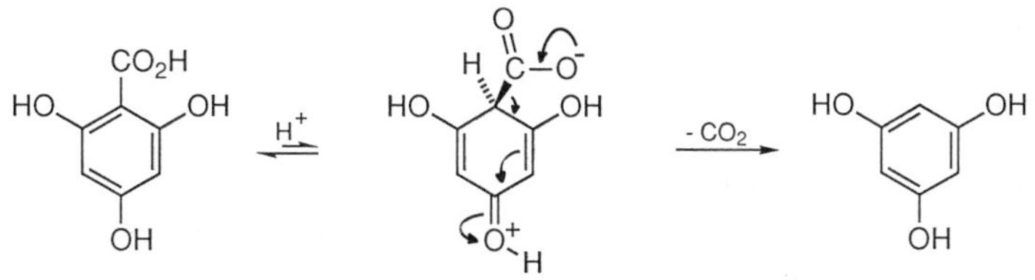

Electrophiles other than the proton can displace carboxyl groups from activated nuclei. For example, salicylic acid and bromine give 2,4,6-tribromophenol:

* These decarboxylations are entirely different from those which occur in aromatic carboxylic acids containing strongly electron-attracting groups. The latter involve the formation of *carbanions* which derive moderate stability from the presence of the electron-attracting group(s). Typical examples are:

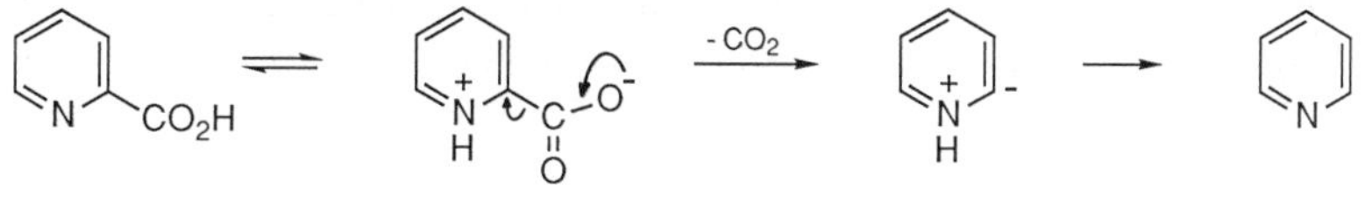

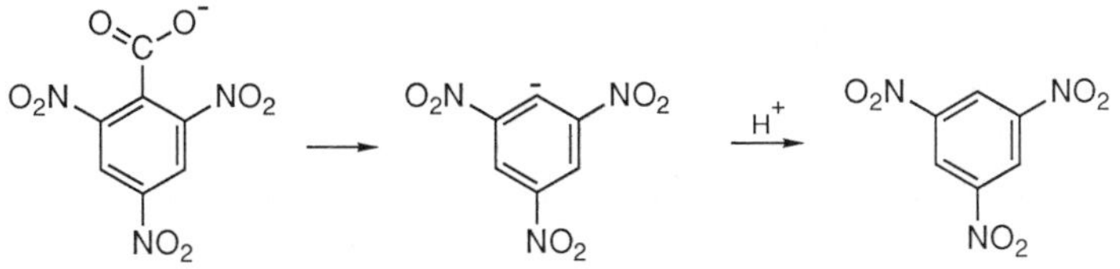

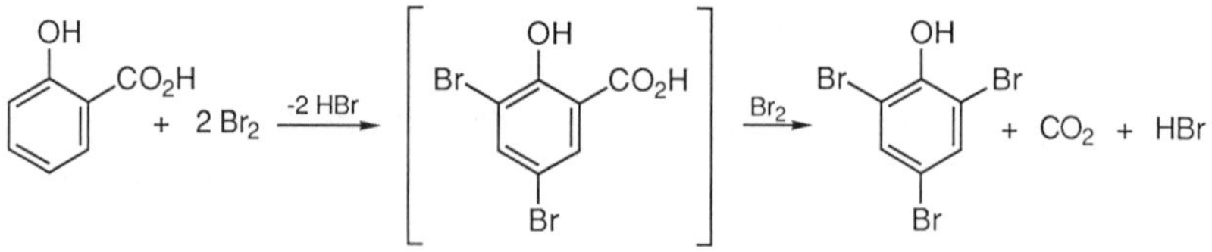

(ii)   Desulfonation.   The sulfonic acid group is readily displaced from aromatic rings by acids (p. 381) and can also be displaced from strongly activated positions (e.g. *ortho* or *para* to hydroxyl or amino) by halogens, e.g.

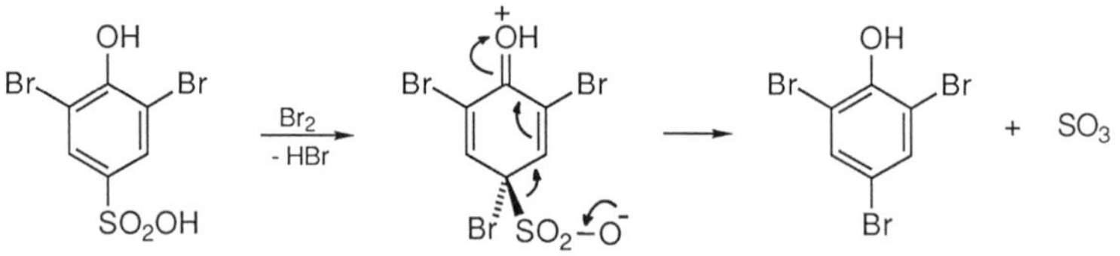

and in nitrating conditions, e.g.

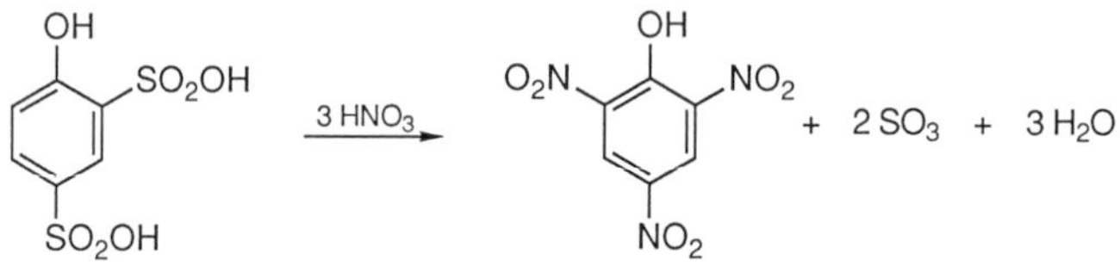

picric acid

The last reaction provides a useful method for preparing picric acid, for attempts to trinitrate phenol itself lead to extensive oxidation. Phenol-2,4-disulfonic acid is readily obtained by sulfonating phenol (p. 382).

(iii)   Dealkylation.   Tertiary alkyl groups are displaced from aromatic rings not only by acids (the reverse of Friedel–Crafts alkylation, p. 364) but also by the halogens. Reaction is especially favourable if the alkyl group is *ortho* or *para* to a strongly activating substituent, but occurs to some extent even in an unactivated situation, e.g. the chlorination and bromination of t-butylbenzene lead to some chlorobenzene and bromobenzene, together with the t-butylhalobenzenes.

This reaction, which can be a disadvantage in halogenations, is apparently limited to tertiary alkyl groups, no doubt because of the comparative stability of tertiary carbocations as leaving groups, e.g.

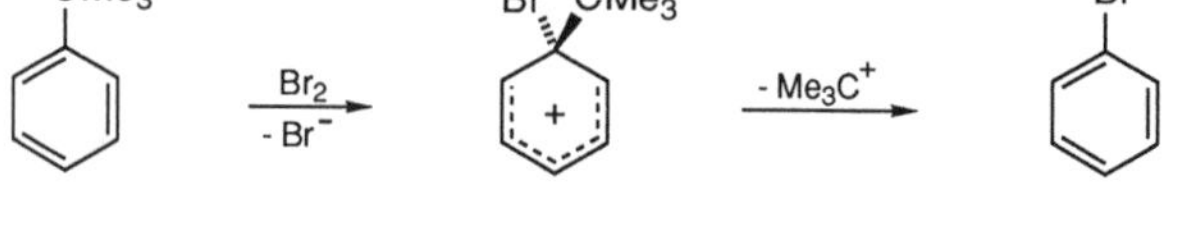

Nitration can also lead to dealkylation, e.g. the nitration of *p*-diisopropylbenzene gives comparable amounts of the deprotonated and dealkylated products:

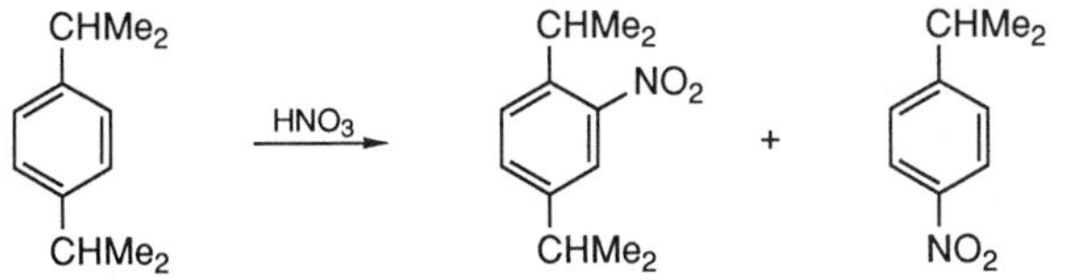

This side-reaction is only significant when highly branched alkyl groups are involved. In some cases it can be the dominant reaction, e.g. the nitration of 1,2,4,5-tetraisopropylbenzene gives only the triisopropylnitro product, evidently because of the steric hindrance to normal nitration.

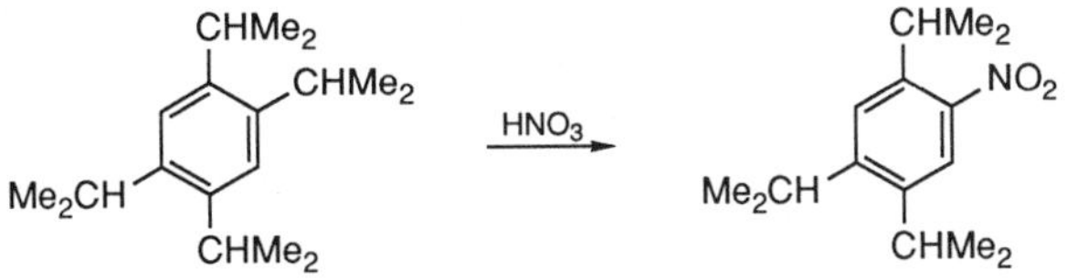

(*iv*) *Dehalogenation.* The reverse of halogenation occurs when the halogen atom is adjacent to two very large substituents. For example, 2,4,6-tri-t-butyl-bromobenzene is debrominated by strong acid. Evidently the reaction is facilitated by the release of steric strain between the halogen and the *ortho* substituents in passage from the eclipsed reactant to the staggered intermediate:

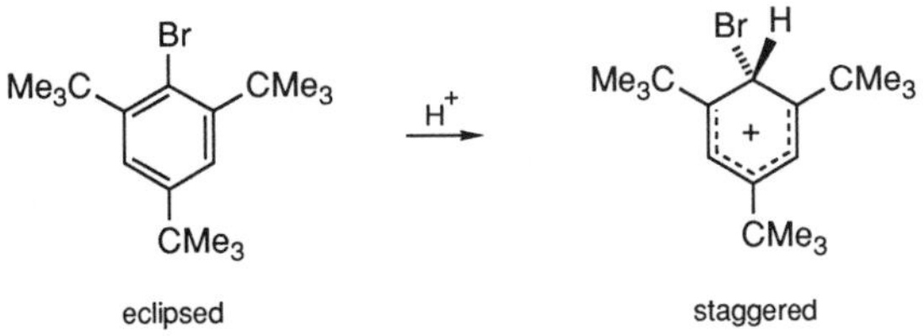

Dehalogenation as a side-reaction is more serious in nitration. For example, *p*-iodoanisole and nitric acid give *p*-nitroanisole:

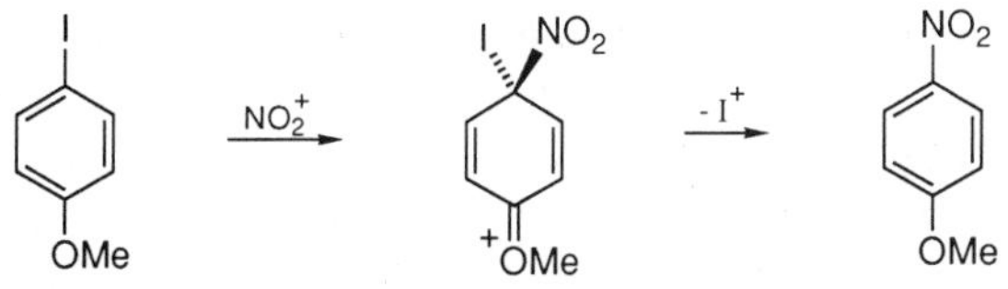

Reaction is here facilitated by the powerful activating effect of *p*-methoxyl, but deiodination occurs to some extent even in less activated environments, e.g. the nitration of iodobenzene gives a small amount of nitrobenzene. Debromination and dechlorination occur, but less readily.

Unless only one position in an aromatic compound is available for reaction, as in a symmetrically *para*-substituted benzene, electrophilic reactions invariably lead to mixtures of two or more products. This is usually a disadvantage in synthesis, as it is usually required to obtain only one product in as high a yield as possible. Although it is not possible to write down a series of principles, reference to which will indicate an efficient route to every possible aromatic derivative, some generalizations are relevant. These are given here, together with a summary of some of the special techniques mentioned earlier in the text, with respect to substitution in benzenoid rings.

(1)   The first consideration in devising a route to a disubstituted benzene, C_6H_4XY, is the directing character of each substituent. If X is *meta*-directing and Y is *ortho*, *para*-directing, it should be possible to obtain as main products both the *meta*-isomer, by starting with PhX, and the *ortho*- and *para*-isomers, by starting with PhY. For example, the chlorination of nitrobenzene in the presence of iron filings gives mainly *m*-chloronitrobenzene whereas the nitration of chlorobenzene in sulfuric acid gives about 30% of *o*-chloronitrobenzene and 60% of *p*-chloronitrobenzene.

(2)   This approach fails in many cases, a particularly common circumstance being that the *meta*-directing compound is too deactivated to undergo the necessary reaction, e.g. *o*- and *p*-nitrotoluene can be obtained by nitrating toluene, but *m*-nitrotoluene cannot be obtained by alkylating nitrobenzene. An indirect approach has then to be adopted and one commonly used is based on the versatility of the amino group, as in the synthesis of *m*-bromotoluene (p. 355).

(3)   In some cases the nature of a substituent may be modified to achieve the appropriate orientation. For example, both *m*- and *p*-nitropropylbenzene may be made from propiophenone: nitration, followed by selective reduction of the carbonyl group (p. 649), gives the *m*-nitro compound,

and reduction to propylbenzene followed by nitration gives about 50% of the *p*-nitro compound.

(4)   The main product of a substitution is more easily isolated and purified if it is a solid than if it is a liquid. Fractional distillation of a liquid (on a laboratory scale) is usually inefficient because the boiling points of *ortho*-, *meta*-, and *para*-isomers are very close together, whereas the purification of a solid by recrystallization is almost always practicable. It is therefore sometimes more satisfactory to rearrange a synthesis so that the aromatic substitution provides a solid that, after purification, can be transformed into the desired product. For example, although *m*-chloronitrobenzene can be obtained as the main product of the direct chlorination of nitrobenzene, it is more easily obtained in a pure state by the nitration of nitrobenzene, which gives a high yield of the solid *m*-dinitro-derivative, followed

by selective reduction of one nitro group (p. 668), diazotization, and Sandmeyer chlorination (p. 412).

(5)   It is more difficult to isolate a particular product from the substitution of *ortho*, *para*-directing compounds than from a *meta*-directing compound. Whereas the latter usually give almost entirely (*c*. 90%) the *meta*-derivative, the former often give comparable quantities of the *ortho* and *para* products. This not only limits the efficiency of the method for a particular derivative, since yields are likely to be less than 50%, but also makes purification difficult even if both products are solids, and losses result. In general, the *para* product is easier to obtain than the *ortho*-isomer because it is normally higher melting and therefore less soluble and can be obtained by recrystallization.

In a few instances a particular structural feature gives rise to a specific difference in physical properties of *ortho* and *para* isomers which can be employed for their separation. For example, the nitration of phenol with dilute nitric acid gives comparable amounts of *o*- and *p*-nitrophenol. The *ortho*-isomer is a chelated compound (p. 35) whereas the *para*-isomer is highly associated through intermolecular hydrogen-bonding. This results in the former being much more volatile and it may be cleanly separated from the latter by distillation in steam.

(6)   Alteration in the reaction conditions can change the proportions of products, as in the sulfonation of naphthalene (p. 383).

(7)   A more general method for modifying the directing character of a compound is to block the most reactive position with a group which can be readily removed after the appropriate substitution has been accomplished. Examples of the use of the t-butyl and sulfonic acid groups have been quoted (pp. 365, 382).

(8)   The amino group is particularly widely employed in aromatic substitutions by virtue both of its ease of conversion into other groupings *via* diazotization (chapter 13) and of its activating power coupled with the ease of its removal from the aromatic nucleus (p. 413). Both mild activation, through the acetyl derivative of the amino group, and powerful activation (e.g. the synthesis of 1,3,5-tribromobenzene, p. 414) can be achieved. The methyl group also has some applications as an activating group which is removable from the nucleus, as in the synthesis of 1,3,5-trinitrobenzene (p. 378).

1.   What do you expect to be the chief product(s) of the mononitration of the **Problems** following compounds?

(*a*)   (*b*)   (*c*)

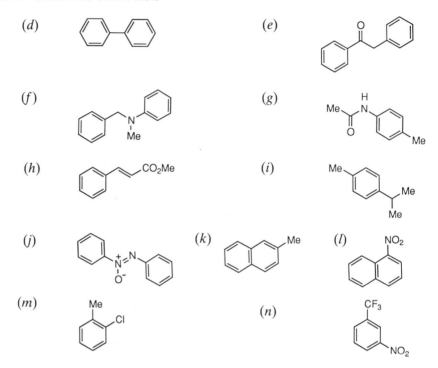

2. Arrange the following in order of decreasing reactivity towards an electrophilic reagent:

$$Ph-Cl \qquad Ph-Me \qquad Ph-OMe \qquad Ph-NMe_2$$

$$Ph-\overset{+}{N}Me_3 \qquad Ph-NO_2 \qquad Ph-CO_2Et \qquad Ph-OCOMe$$

3. Account for the following observations:

   (*i*)   Iodine is a catalyst for aromatic bromination.
   (*ii*) The product of the sulfonation of naphthalene depends on the temperature of the reaction.
   (*iii*) 2,6-Dimethylacetanilide is nitrated mainly at the 3-position.
   (*iv*) Pyrrole is more reactive to electrophiles at the 2- than at the 3-position, whereas the opposite holds for indole.
   (*v*)   Nitration of dimethylaniline gives mainly the *m*-nitro derivative when concentrated nitric and sulfuric acids are used but mainly the *o*- and *p*-nitro derivatives in less acidic conditions.
   (*vi*) Pyridine-1-oxide is more reactive than pyridine in nitration, and gives mainly the 4-nitro derivative.

4. Outline methods for the synthesis of the following:

(a) Me, Br (3-bromotoluene)

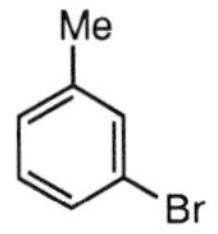

(b) NH$_2$, NO$_2$

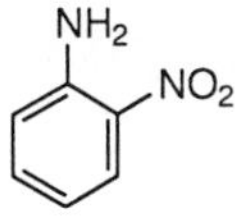

(c) NH$_2$, Br

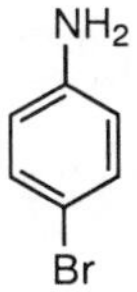

(d) NH$_2$, SO$_2$NH$_2$

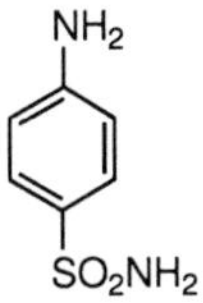

(e) NH$_2$, propyl

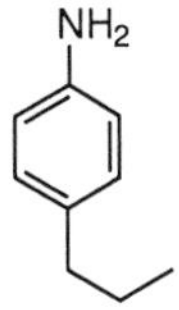

(f) NH$_2$, propyl

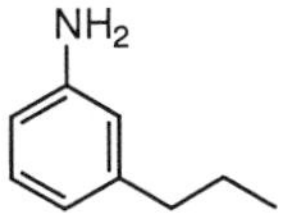

(g) OMe, CH$_2$OH

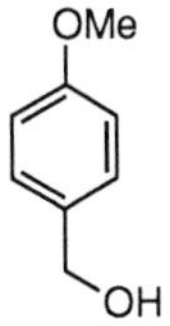

(h) Me, CHO

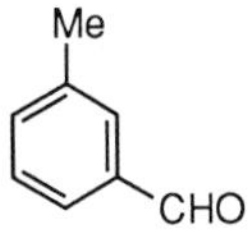

(i) NO$_2$, CHO

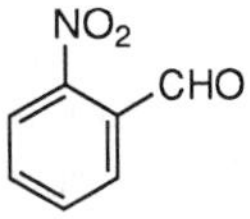

(j) OMe, CHO

(k) NO$_2$, O$_2$N, NO$_2$

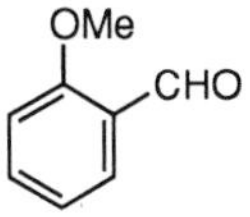

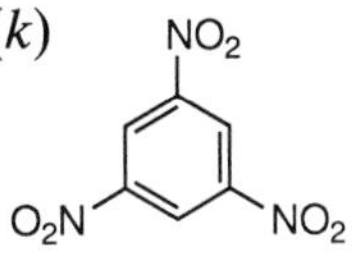

(l) furan, Cl

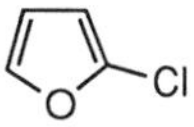

(m) pyrrole, N–H, CHO

(n) Ph, indene

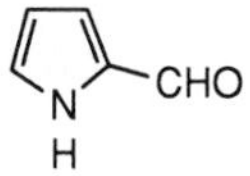

(o) phenanthrene, Me

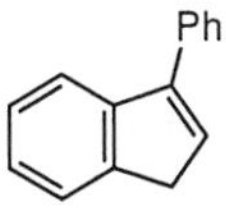

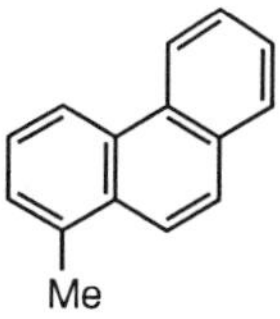

12 Nucleophilic aromatic substitution

12.1
Principles

Although benzene is moderately reactive towards electrophiles, it is inert to nucleophiles. However, just as the attachment of a group of $-M$ type to the C=C bond in an alkene activates that bond to nucleophiles, so the attachment of such a substituent to the benzene ring activates the ring to nucleophiles. For example, o- and p-nitrophenol are formed by heating nitrobenzene with powdered potassium hydroxide:

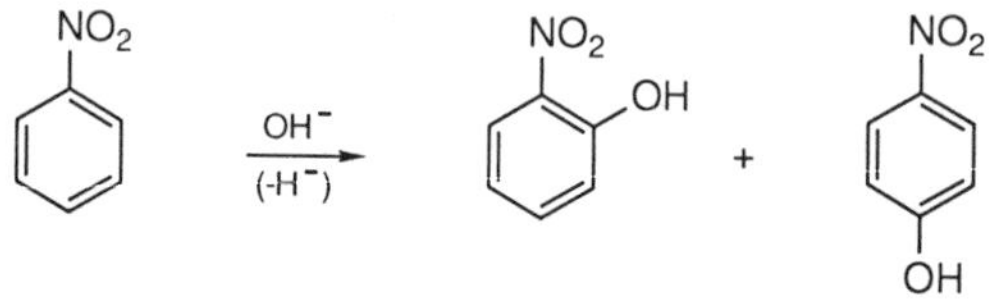

The mechanism of this, and most other, nucleophilic aromatic substitutions is similar to that of electrophilic aromatic substitutions except that an anionic rather than a cationic intermediate is involved. The nucleophile adds to the aromatic ring to give a delocalized anion from which a hydride ion is eliminated:

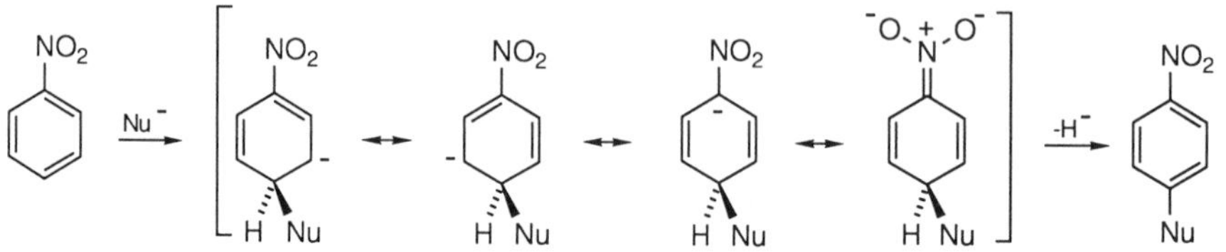

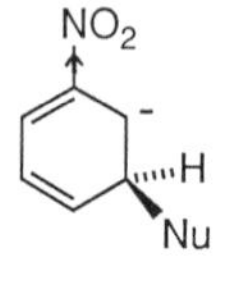

It is apparent that the reactivity towards nucleophiles of nitrobenzene compared with benzene stems from the ability of the nitro group to stabilize the anionic intermediate, and hence the preceding transition state, by accommodating the negative charge. Further, this delocalization can occur only when the reagent adds to the *ortho* or *para* position; addition to the *meta* position gives an adduct stabilized by the inductive effect of the nitro group, but not the mesomeric effect. Consequently, towards nucleophiles, electron-withdrawing substituents are activating and *ortho*, *para*-orienting, and conversely electron-releasing groups are deactivating and *meta*-directing. These principles are the reverse of those that apply to electrophilic substitution.

The example of nucleophilic substitution given above involves the displacement of hydride ion. This is a relatively high-energy species, and the reaction would be unfavourable but for the fact that an oxidizing agent, nitrobenzene itself, is present which removes the hydride ion. In doing so, a reduction product, azoxybenzene (p. 667), is formed, so that the yield of nitrophenols based on nitrobenzene is low.

It is therefore more satisfactory to carry out nucleophilic substitutions, where possible, on aromatic compounds from which a more stable anion than hydride can be displaced. Typical leaving groups are halide ions, as in the reaction of *p*-nitrochlorobenzene with hydroxide ion.

Four other mechanisms occur, though less widely, in nucleophilic aromatic substitution. The first is an S_N1 process, so far recognized only for reactions of aromatic diazonium ions. When aqueous benzenediazonium chloride is warmed, phenol is formed *via* the phenyl cation. Nuclear-substituted derivatives behave similarly.

The driving force for this reaction resides in the strength of the bonding in the nitrogen molecule which makes it a particularly good leaving group.

The three remaining mechanisms, which have closely defined structural requirements, are described later (sections 12.4, 12.5 and 12.6).

(*i*) *Substitution in heteroaromatic compounds.* The six-membered nitrogen-containing heteroaromatic compounds are activated towards nucleophiles. This is because the negative charge on the adduct formed by addition of the nucleophile to positions *ortho* or *para* to nitrogen is stabilized by delocalization onto the electronegative nitrogen atom:

Consequently pyridine resembles nitrobenzene in undergoing nucleophilic displacement at the 2- and 4-positions, just as it resembles nitrobenzene in its inertness towards electrophiles (p. 359). Similarly, anions such as halide ions can be readily displaced from the 2- and 4-positions of pyridine.

Pyrimidine, as expected, is even more strongly activated than pyridine, substitution being directed to the 2-, 4-, and 6-positions. Quinoline likewise reacts at the 2- and 4-positions, but in isoquinoline, in which both the 1- and 3-positions are *ortho* to the heterocyclic nitrogen, the 1-position is far more reactive than the 3-position.

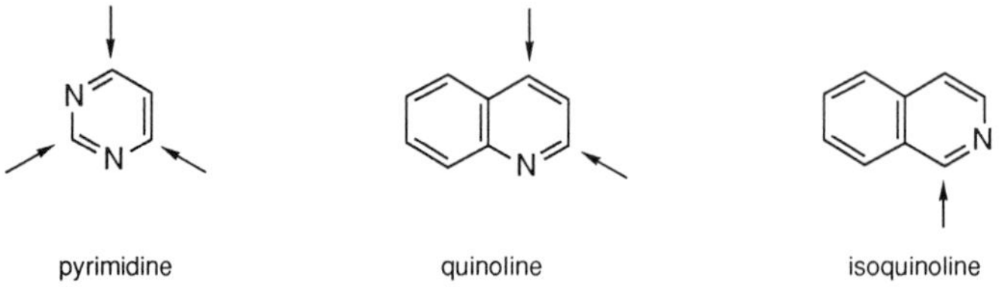

pyrimidine   quinoline   isoquinoline

The explanation of the last fact becomes apparent when the structures of the adducts formed at the 1- and 3-positions of isoquinoline are examined. The activating influence in each case is ascribed mainly to the resonance structures in which nitrogen accommodates the negative charge: that for addition to the 1-position is benzenoid and is therefore of lower energy than that for addition to the 3-position, which is not (cf. the discussion of why 2-naphthol directs electrophiles to the 1-position, p. 356).

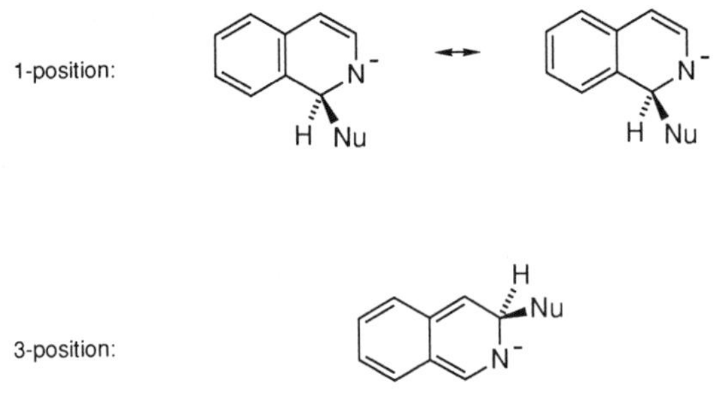

1-position:

3-position:

12.2
Displacement of
hydride ion

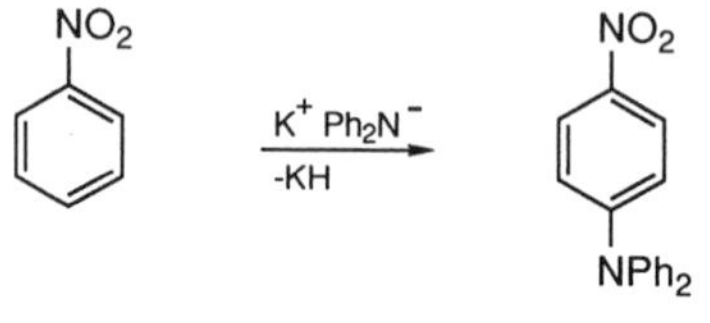

Benzene is not attacked by any nucleophile and nitrobenzene reacts only with the most reactive nucleophiles, such as amide or substituted amide ions, or less reactive nucleophiles such as hydroxide ion in very vigorous conditions (p. 396).

m-Dinitrobenzene is more strongly activated than nitrobenzene and reacts, for example, with cyanide ion:

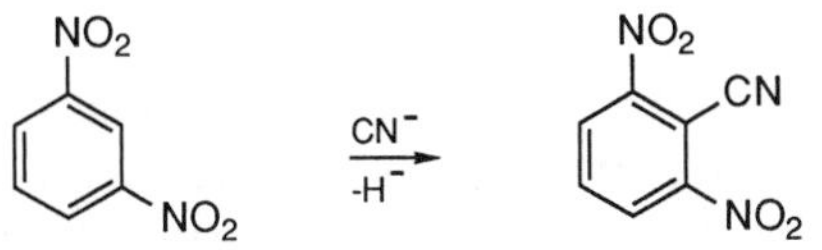

Pyridine in general resembles nitrobenzene save that organometallic reagents, which react with the nitro group of nitrobenzene, substitute in the nucleus of pyridine. The following are typical reactions, the predominant product being the 2-derivative.

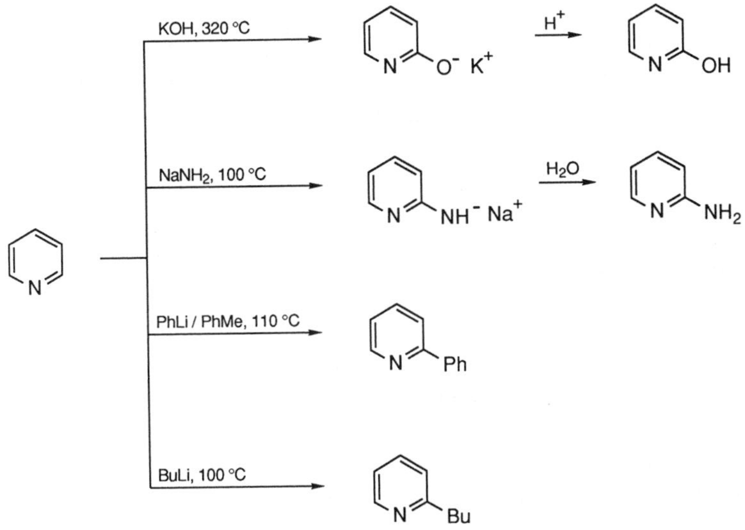

Similar reactions occur on quinoline, mainly at the 2-position, and on isoquinoline, at the 1-position.

**12.3
Displacement of
other anions**

(a) Halides

The four halobenzenes are, like alkenyl halides, very inert to nucleophiles in normal conditions. They do not react with methoxide ion in methanol or with boiling alcoholic silver nitrate, conditions in which alkyl halides react readily. Reaction can, however, be brought about in each of two ways. First, very vigorous conditions may be employed, as in the formation of phenol by heating chlorobenzene with 10% sodium hydroxide solution under pressure at 350°C: in these conditions reaction occurs mainly *via* the benzyne intermediate (section 12.4).

$$PhCl \quad \xrightarrow[-Cl^-]{OH^-} \quad [PhOH] \quad \xrightarrow{OH^-} \quad PhO^- \quad \xrightarrow{H^+} \quad PhOH$$

Second, reactions of halobenzenes with alkoxide ions occur many powers of ten faster in dimethyl sulfoxide than in hydroxylic media (p. 65). For example, bromobenzene and t-butoxide ion give phenyl t-butyl ether in about 45% yield [5]:

$$PhBr \quad \xrightarrow[-Br^-]{Me_3CO^- / DMSO} \quad Ph^{\diagdown O \diagdown}CMe_3$$

The introduction of substituents of $-M$ type into positions *ortho* or *para* with respect to the halogen atom considerably increases the ease of nucleophilic substitution. The activating effect increases with the number of such substituents present, so that a typical order of reactivity of chlorobenzenes is

$$p\text{-}NO_2 \quad < \quad 2,4\text{-dinitro} \quad < \quad 2,4,6\text{-trinitro}$$

In fact, 2,4,6-trinitrochlorobenzene is hydrolyzed by dilute alkali at room temperature.

These reactions are first-order with respect both to the halide and to the nucleophile, and in this sense they are analogous to those of aliphatic halides. Two differences should be noted. First, displacement on an aromatic halide cannot lead to inversion of configuration: the reagent must approach the aromatic carbon from the same side as the halide. Second, amongst the aromatic halides the usual order of reactivity is fluoride > chloride ~ bromide ~ iodide, whereas in the aliphatic series fluorides are the least reactive.

The reason for the latter difference is that the rate-determining step is the addition of the nucleophile to the ring:

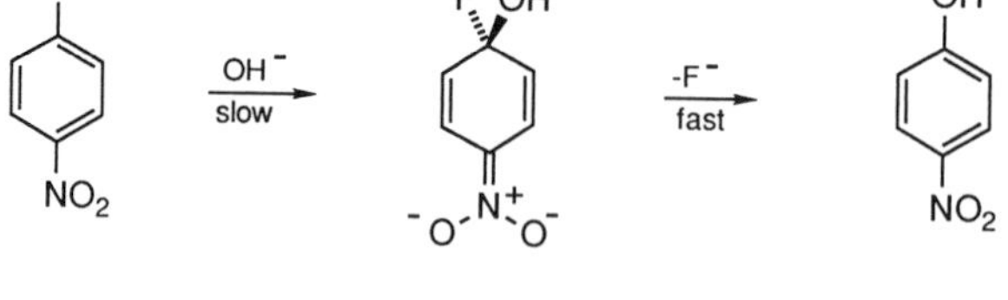

Fluorine, being the most electronegative of the halogens, is best able to stabilize the adduct and hence the preceding transition state. In S_N2 reactions on aliphatic halides, however, the rate-determining step involves the severance of the C–halogen bond and C—F, being the strongest of the four, is the hardest to break.

Nucleophiles other than hydroxide ion also react with activated aromatic halides. Typical examples are the formation of aromatic amines, e.g.

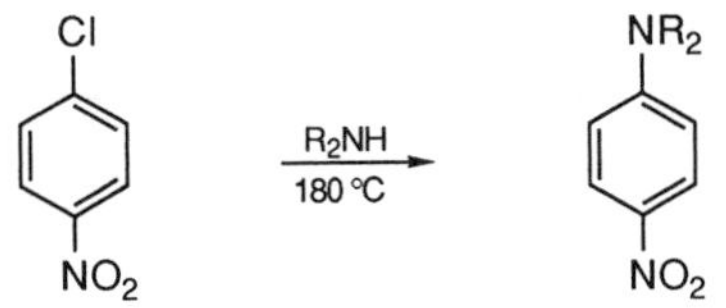

and 2,4-dinitrophenylhydrazine, from 2,4-dinitrochlorobenzene (itself obtained by the dinitration of chlorobenzene), which is used for the characterization of carbonyl compounds.

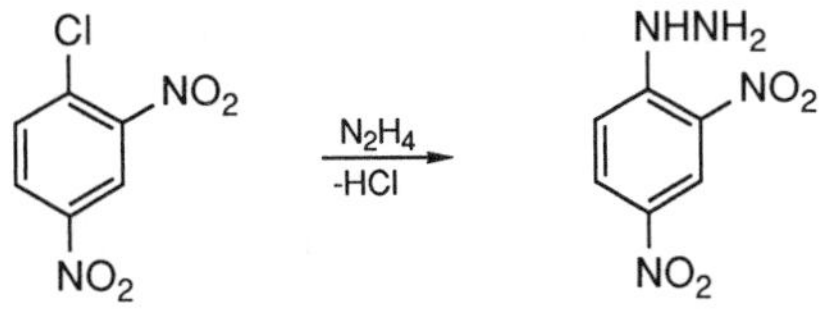

2,4-Dinitrofluorobenzene has a special use in 'marking' the terminal amino group(s) in a protein or polypeptide and was first used by Sanger in his elucidation of the structure of insulin. The nucleophilic amino group displaces fluoride, forming a secondary amine, and hydrolysis of the peptide links then leaves the terminal amino acid bonded to the 2,4-dinitrophenyl group. Isolation and identification of this derivative indicate the nature of the terminal amino acid.

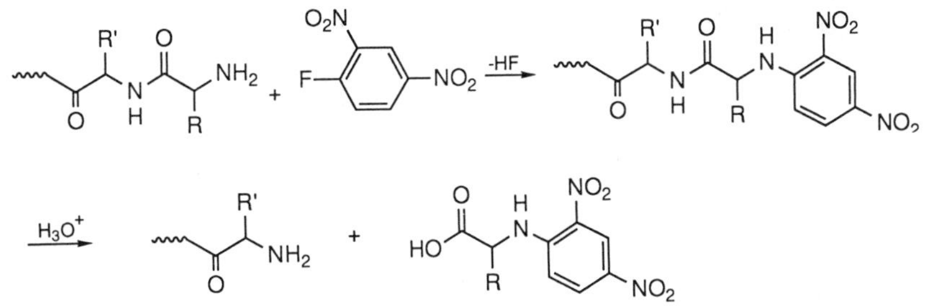

(*i*) *Copper-catalyzed reactions.* Aromatic halides which are not activated to nucleophilic displacement undergo copper(I)-catalyzed substitution on heating in an aprotic solvent such as DMF, e.g.

It is thought that reaction occurs by *oxidative addition* of the halide to Cu(I) followed by the reverse reaction but with cyanide transferring to the aromatic ring:

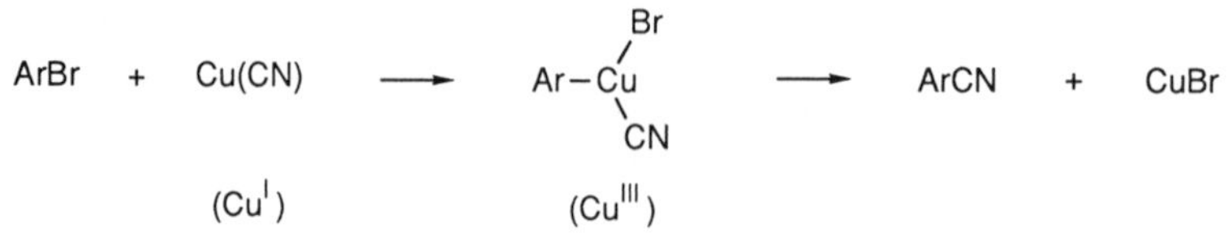

$$ArBr + Cu(CN) \longrightarrow Ar-\underset{\underset{CN}{|}}{\overset{\overset{Br}{|}}{Cu}} \longrightarrow ArCN + CuBr$$

(Cu^{I})   (Cu^{III})

(b)   Oxyanions

Although phenyl ethers are stable in basic conditions, the introduction of $-M$ substituents in the *ortho* and *para* positions induces hydrolysis, e.g.

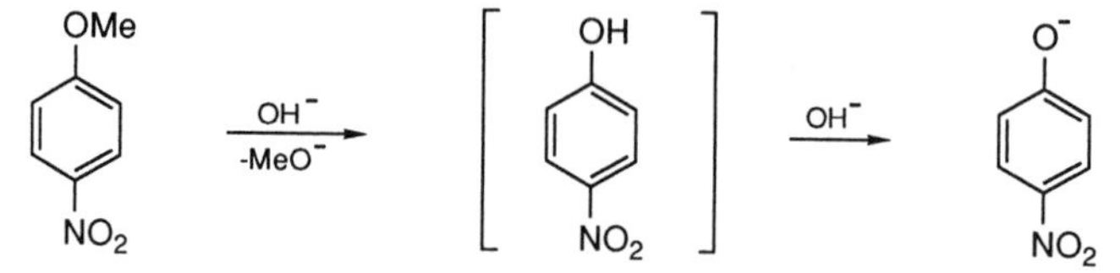

In this respect aromatic ethers resemble esters rather than aliphatic ethers. Indeed, the mechanism of reaction of aromatic ethers is entirely analogous to that of esters in that in each the addition of hydroxide ion is made possible by the presence of an electronegative atom which can accommodate the negative charge.
Compare

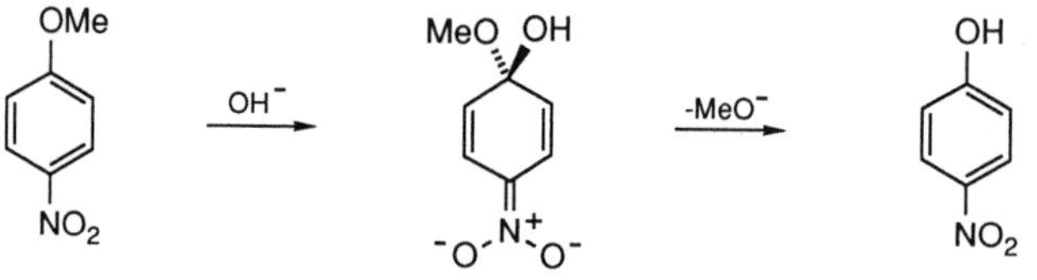

with

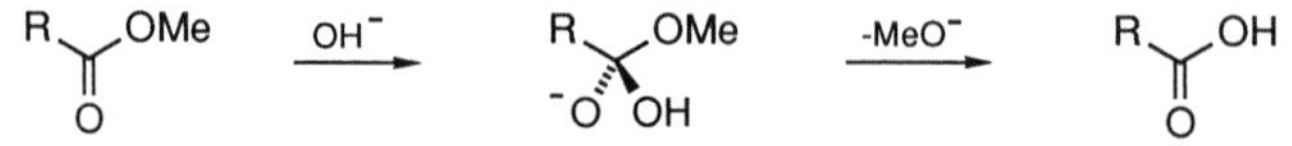

(c)   Sulfite ion

The fusion of aromatic sulfonates with caustic alkali at high temperatures gives phenols by displacement of sulfite ion. For example, sodium *p*-toluenesulfonate and a mixture of sodium hydroxide with a little potash (to aid fusion) give *p*-cresol in 65–70% yield at 250–300°C [1]:

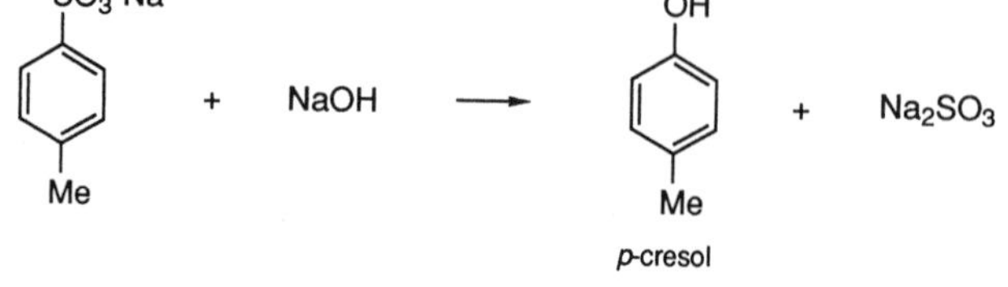

Because of the ready availability of aromatic sulfonic acids by direct sulfonation (section 11.5), this procedure is useful for the preparation of phenols. For example, 2-naphthol can be obtained by the high-temperature sulfonation of naphthalene (p. 383) followed by fusion with hydroxide and this provides a method for making 2-substituted naphthalenes *via* the Bucherer reaction (section 12.6).

There are, however, limitations to the method. At the high temperatures necessary for reaction, halogen substituents are themselves labile, so that halophenols cannot be prepared in this way. Again, a substituent of $-M$ type which is *meta* to the sulfonic acid group activates the *ortho* and *para* positions to hydride displacement, so that, for example, *m*-nitrophenol cannot be obtained from *m*-nitrobenzenesulfonic acid. In these instances the phenols are best obtained *via* the corresponding diazonium salt (section 13.3).

(d) Nitrogen anions

The nitrite ion may be displaced readily from aromatic nitro compounds if other nitro groups are present to activate the nucleus, e.g.

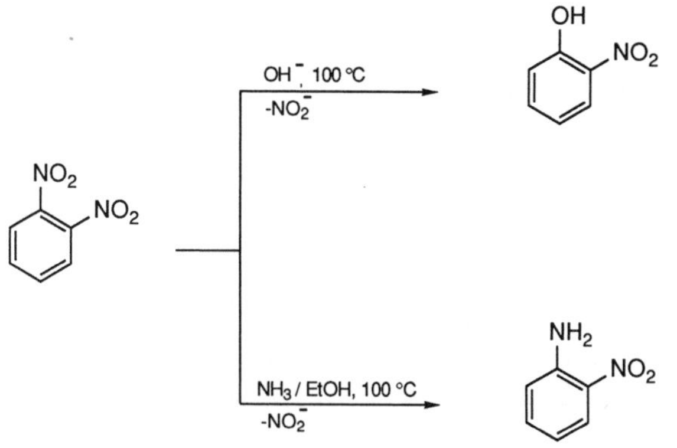

When 1,3,5-trinitrobenzene is refluxed in methanol containing methoxide ion, 3,5-dinitroanisole is formed in 70% yield [1]:

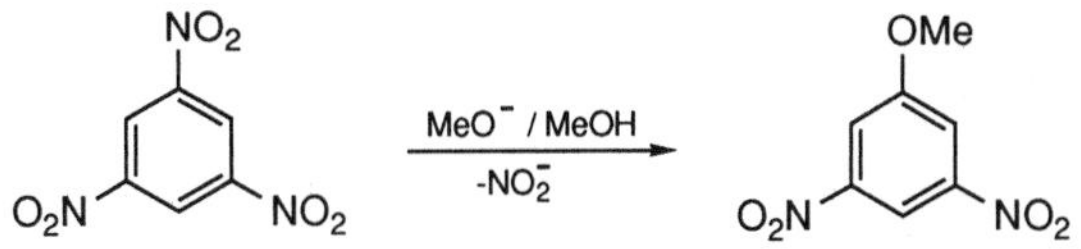

3,5-dinitroanisole

It should be noted that the three positions which are *ortho* and *para* to the nitro groups are more strongly activated to the addition of the nucleophile than a position containing a nitro group which is *meta* to the remaining nitro groups. The course taken in this reaction is evidently determined by the greater tendency of a nitro group to leave as nitrite ion than of a hydrogen to leave as hydride.

Amino substituents may be displaced as amide ions. A useful synthetic method for secondary amines consists of converting aniline into a tertiary amine, nitrosating (p. 380), and heating the nitroso-derivative with alkali. The secondary amines produced cannot be contaminated by primary and tertiary amines, as they are when prepared from alkyl halides (p. 302).

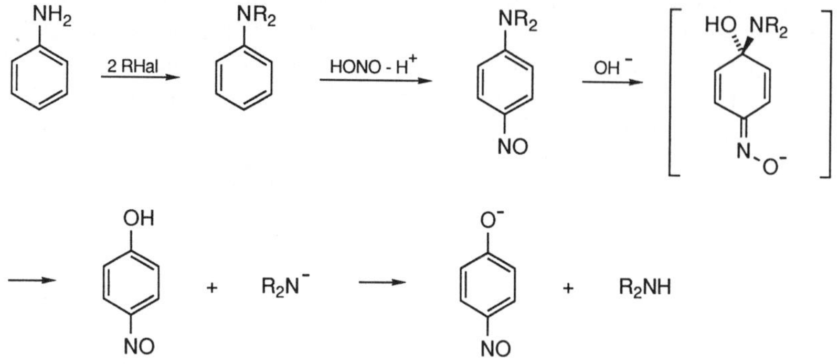

12.4 Substitution *via* benzynes

(a) Formation of benzynes

Treatment of chlorobenzene with sodamide in liquid ammonia at −33°C gives aniline. A mechanistic study has shown that this reaction is entirely different from those nucleophilic substitutions so far described. For example, chlorobenzene labelled with ^{14}C at the position bearing chlorine gives an equimolar mixture of unrearranged and rearranged products and this, together with other evidence, points to the mediation of a symmetrical species, benzyne, which is formed from chlorobenzene by an E2-type elimination and reacts with ammonia to give aniline:

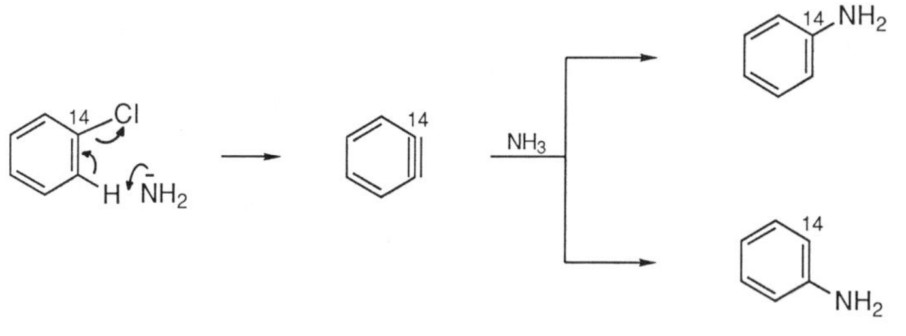

The ease of this reaction is due to the strong basicity of amide ion. Hydroxide ion, which is a much weaker base, reacts with chlorobenzene at about 340°C to give phenol *via* benzyne. Bromobenzene and iodobenzene react similarly to chlorobenzene, but fluorobenzene does not give benzyne directly with base, differing from the other halobenzenes because of the greater strength of the C—F bond.

There are other methods for generating benzynes. *o*-Halo-substituted organo-metallic compounds readily lose metal halide to give benzynes, e.g.

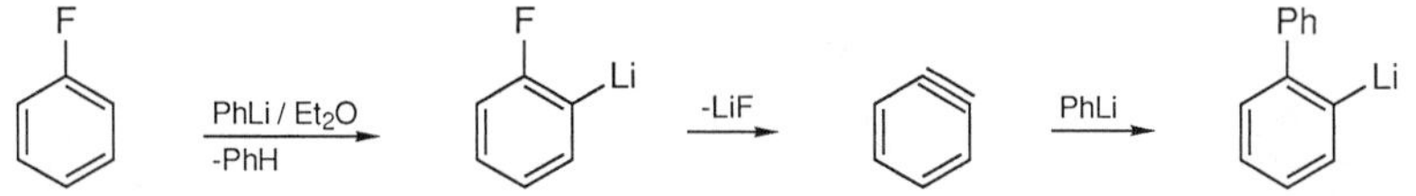

The resulting aryl-metal compound may be usefully employed in synthesis, e.g.

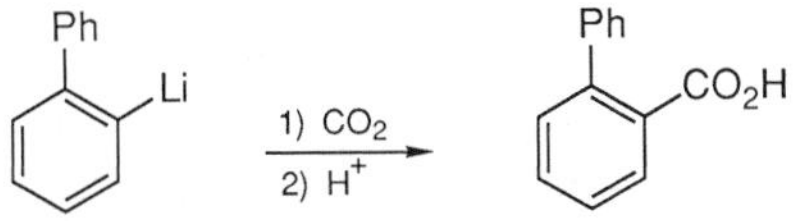

Second, *ortho* disubstituted benzenes from which two stable molecules can be formed by elimination give benzynes on thermolysis, e.g.

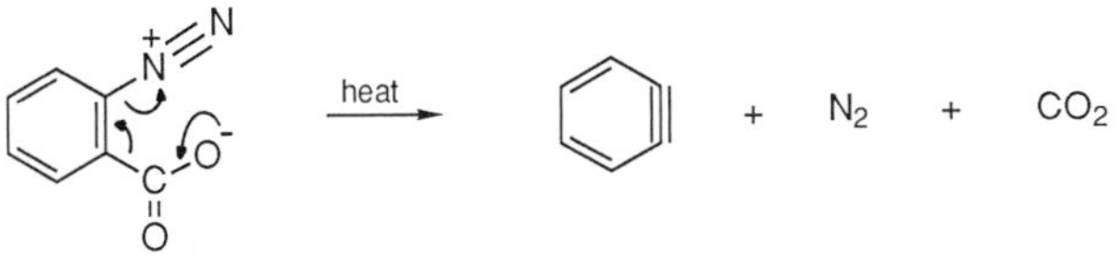

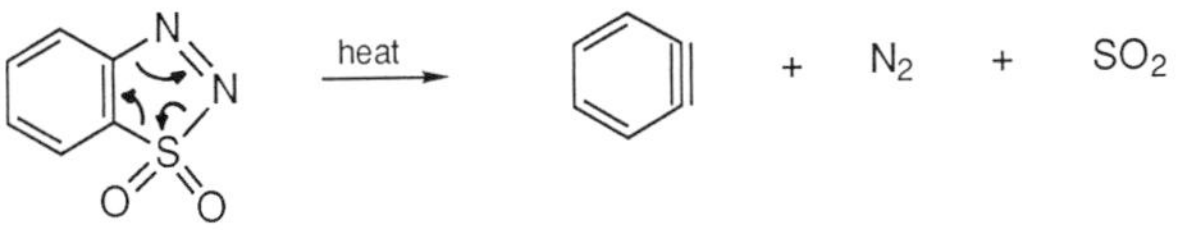

(b)   Reactions of benzynes

Benzynes are too unstable to be isolated and react with any nucleophile which is present, e.g. when generated in liquid ammonia they react to give amines and when formed from organometallic compounds they react with a second molecule of the organometallic compound.

Another nucleophilic species may be specially introduced into the system. For example, the addition of malonic ester to liquid ammonia containing amide ion gives the corresponding enolate which reacts with benzyne as it is generated:

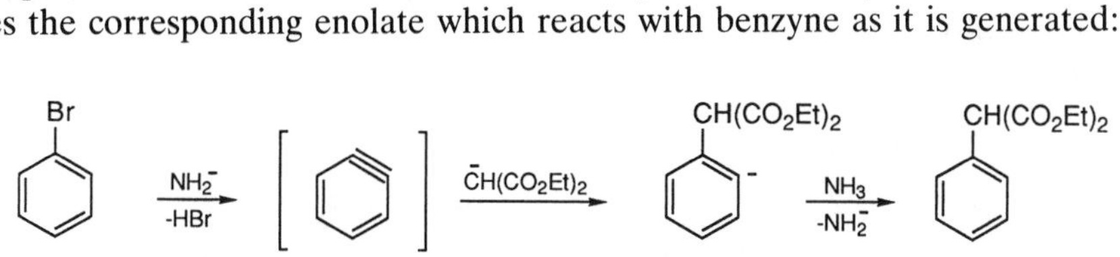

One disadvantage attendant upon the use of benzynes in synthesis is that the nucleophile may react at either end of the triple bond of the benzyne, giving a mixture of two products if the benzyne is monosubstituted. For example, *p*-

chlorotoluene with hydroxide ion at 340°C gives an approximately equimolar mixture of *m*- and *p*-cresol:

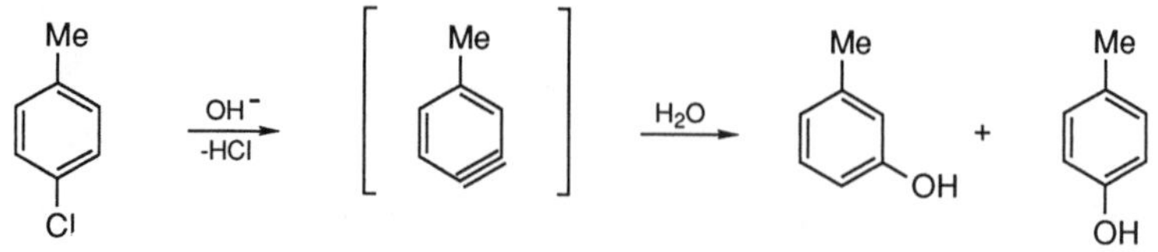

Fortunately, directive effects operate in certain cases, e.g. *m*-aminoanisole is the exclusive product from *o*-chloroanisole with sodamide in liquid ammonia. The direction of addition is such that the more stable of the two possible anions is formed.

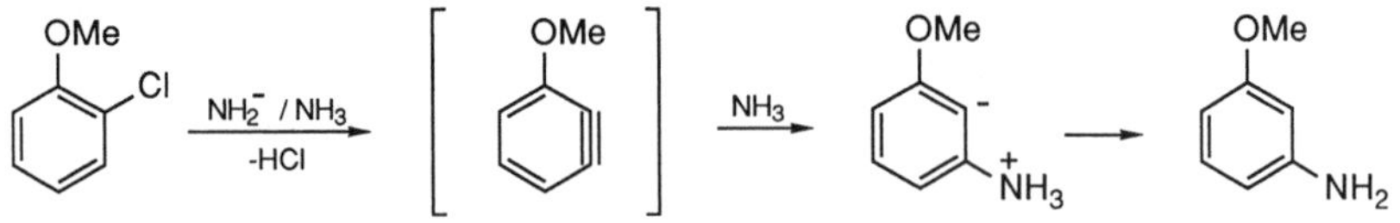

In the absence of nucleophiles, benzyne dimerizes to give biphenylene,

and in the presence of dienes it reacts as a dienophile to give Diels–Alder adducts, e.g.

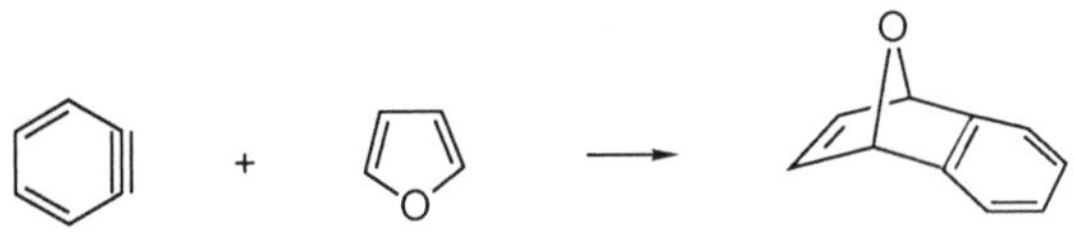

12.5
The S$_{RN}$1
reaction

Unactivated halides undergo nucleophilic substitution in a chain reaction involving anion radicals in which the initiation step is an electron transfer:

initiation $ArX \xrightarrow{e} ArX^{\overline{\cdot}}$

propagation
$$ArX^{\overline{\cdot}} \longrightarrow Ar\cdot + X^-$$
$$Ar\cdot + Nu^- \longrightarrow ArNu^{\overline{\cdot}}$$
$$ArNu^{\overline{\cdot}} + ArX \longrightarrow ArNu + ArX^{\overline{\cdot}}$$

Nucleophiles are typically ketone enolates, amide ion, and thiol anions. Initiation is usually either with the solvated electron (sodium in liquid ammonia) or by photochemical excitation in which the nucleophile is the electron donor. The similarity of the first two steps to those in the S$_N$1 reaction has led to the designation S$_{RN}$1 (R for radicals).

The solvent is usually liquid ammonia, a strong base being included if necessary (for example, for the formation of an enolate). Yields are variable, but the method can often be convenient. The third example shows a method for the conversion of phenols into amines:

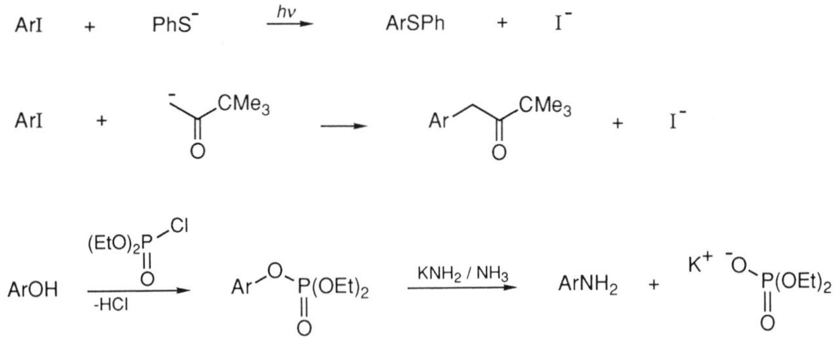

12.6 Bucherer reaction

Certain phenols react with aqueous ammonium sulfite to give aromatic amines. Reaction occurs through the sulfite adduct of the keto-tautomer of the phenol, e.g.

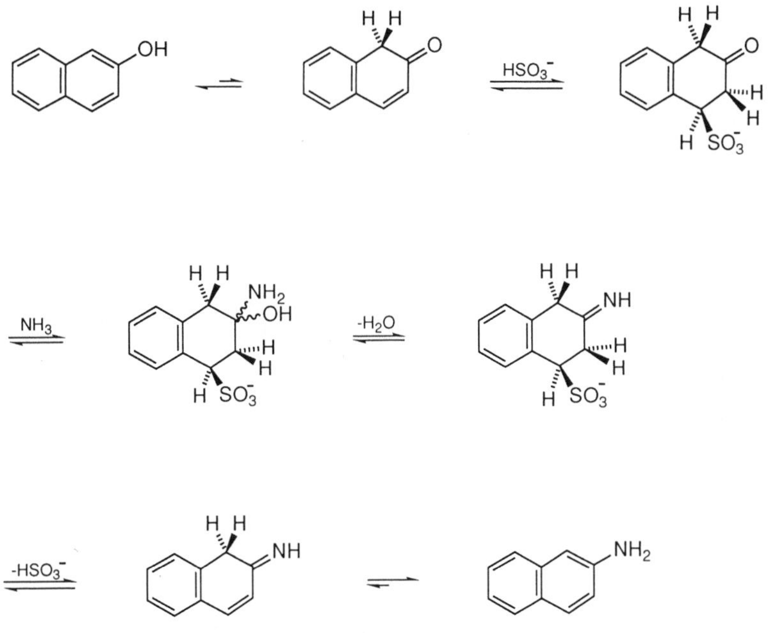

The reaction, named after Bucherer, is limited to those phenols which have a tendency to ketonize, such as 1- and 2-naphthol and *m*-dihydroxybenzene (p. 13). In the example above, 2-naphthylamine may be obtained in 95% yield by carrying out the reaction at 150°C under pressure and, since 2-naphthol may be readily obtained from naphthalene by high-temperature sulfonation (p. 383) followed by alkali-fusion (p. 402), this provides a method for obtaining 2-substituted naphthalenes *via* diazotization of 2-naphthylamine (p. 409).

The Bucherer reaction is reversible, and by careful control of the relative amounts of ammonia and water it may be used to convert amines into phenols.

Problems

1. Write the following compounds in order of their decreasing reactivity towards hydroxide ion: chlorobenzene; *m*-nitrochlorobenzene; *p*-nitrochlorobenzene; 2,4-dinitrochlorobenzene.

2. Account for the following:
   - (*i*) Whereas aliphatic fluorides are less easily hydrolyzed than the corresponding chlorides, 2,4-dinitrofluorobenzene is more rapidly hydrolyzed than 2,4-dinitrochlorobenzene.
   - (*ii*) Isoquinoline is more reactive towards nucleophiles at its 1-position than at its 3-position.
   - (*iii*) The treatment of *o*-bromoanisole with sodamide in liquid ammonia gives mainly *m*-aminoanisole.

3. Give methods for carrying out the following conversions:
   - (*i*) Naphthalene into 2-naphthol.
   - (*ii*) Chlorobenzene into 2,4-dinitrophenylhydrazine.
   - (*iii*) Chlorobenzene into *p*-chlorophenol.
   - (*iv*) Fluorobenzene into *o*-butylbenzoic acid.
   - (*v*) Anisole into *p*-EtO—C_6H_4—NO_2.
   - (*vi*) Pyridine into 2-aminopyridine.
   - (*vii*) Isoquinoline into 1-phenylisoquinoline.

Aromatic diazonium salts 13

Primary amines react with nitrous acid in acidic solution to give diazonium ions, R—$\overset{+}{N}$≡N. The mechanism of reaction may be summarized as follows:

$$HO-NO \ + \ H^+ \ \rightleftharpoons \ H_2\overset{+}{O}-NO \ \rightleftharpoons \ H_2O \ + \ NO^+$$

$$R-NH_2 \ \xrightarrow{NO^+} \ R\overset{+}{N}H_2{-}^{N\!\lessgtr}{}_O \ \xrightarrow{-H^+} \ RNH{-}^{N\!\lessgtr}{}_O \ \xrightarrow{H^+}$$

$$RNH{-}^{N\lessgtr}\!{}_{\overset{+}{O}H} \ \xrightarrow{-H^+} \ RN^{\lessgtr N}{-}^{OH} \ \xrightarrow{H^+} \ RN^{\lessgtr N}{-}^{\overset{+}{O}H_2} \ \xrightarrow{-H_2O} \ R-\overset{+}{N}{\equiv}N$$

The electrophilic nitrosonium ion reacts with the nucleophilic nitrogen of the amine, a series of prototropic shifts occurs, and finally water is eliminated.

This reaction scheme is common to both aliphatic and aromatic amines. However, the two groups differ in that aliphatic diazonium ions are too unstable to be isolated in the form of salts, decomposing into carbocations and nitrogen, whereas aromatic diazonium ions are moderately stable in aqueous solution at low temperatures when present with anions of low nucleophilic power and may also, in suitable cases, be isolated as solids. The difference in behaviour is related to the fact that singly bonded carbon more readily tolerates a positive charge than unsaturated carbon (p. 78).

The diazotization of primary aromatic amines is normally carried out by adding an aqueous solution of sodium nitrite to a solution (or suspension) of the amine hydrochloride in an excess of hydrochloric acid which is cooled by an ice-bath. The rate of addition is controlled so that the temperature of the reaction remains below about 5°C, and addition is continued until the solution just contains an excess of nitrous acid.

Amines which are substituted in the nucleus with electron-attracting groups are less easy to diazotize because the nucleophilicity of the amino-nitrogen is reduced by the partial withdrawal of the unshared electron pair into the nucleus, e.g.

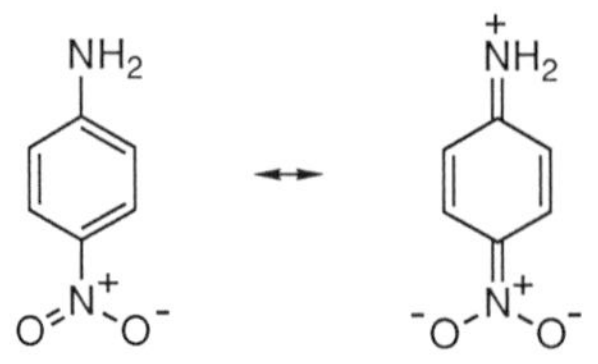

Acetic acid is usually a suitable reaction medium in these instances and even 2,4,6-trinitroaniline may then be diazotized.

Most reactions employing diazonium salts may be conducted in solution and it is not normally necessary to isolate the solid salt. If necessary, however, two procedures may be employed. First, an aqueous solution of the diazonium salt, usually the chloride or sulfate, is prepared and treated with fluoroboric acid; the insoluble diazonium fluoroborate, $ArN_2^+BF_4^-$, is precipitated. Second, the amine hydrochloride is treated with an organic nitrite and acetic acid in ether: the diazonium salt, being insoluble in ether, is precipitated.

**13.2
The reactions of
diazonium ions**   Three types of reaction may be distinguished.

(a)   *Reactions of nucleophiles at nitrogen*

Although diazonium ions coexist in solution with stable anions (i.e. those from strong acids) such as chloride, less stable anions (i.e. those from weaker acids) react to give covalent diazo-compounds, e.g.

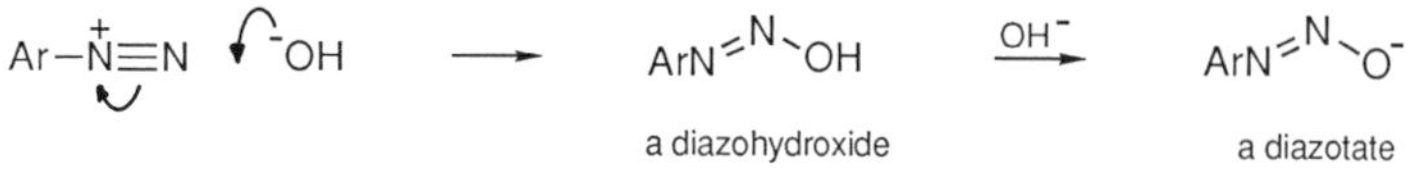

(Note that these products exist in *cis* and *trans* forms.) Many of these diazo-compounds are unstable and decompose in solution to give aryl radicals (p. 416).

An important group of nucleophiles that react with diazonium ions consists of aromatic nuclei containing powerfully electron-releasing substituents such as the —O⁻ group in a phenoxide ion (section 13.4):

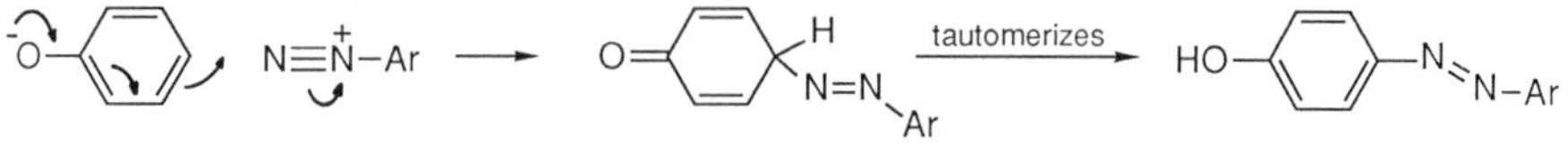

(b)   The S_N1 reaction

Diazonium ions decompose on warming into aryl cations and nitrogen. The aryl cation is highly reactive and relatively unselective, being attacked rapidly by any nucleophile in its vicinity. Reaction in aqueous solution therefore leads to the formation of phenols:

$$ArN_2^+ \xrightarrow{-N_2} Ar^+ \xrightarrow{H_2O} Ar\overset{+}{-}OH_2 \xrightarrow{-H^+} ArOH$$

The formation of aromatic fluorides by the Schiemann procedure (p. 412) also occurs by the S_N1 mechanism.

(c)   One-electron reductions

Diazonium ions are subject to one-electron reduction with the formation of an aryl radical and nitrogen.

$$ArN_2^+ \xrightarrow{e} Ar\cdot \;+\; N_2$$

Copper(I) ion is frequently used as the one-electron reducing agent. The aryl radical is highly reactive and is capable of abstracting a ligand from the transition metal ion (Sandmeyer reaction, p. 412) or a hydrogen atom from a covalent bond (p. 414).

The synthetic applications of aromatic diazonium salts are conveniently divided into those in which nitrogen is eliminated and those in which it is retained.

**13.3
Reactions in
which nitrogen
is eliminated**

(a)   Replacement by hydroxyl

When a diazonium salt is warmed in water the corresponding phenol is formed by the S_N1 mechanism. The reaction is normally carried out in acidified solution in order to preserve the phenol in its unionized form, for otherwise a further reaction may occur between diazonium salt and phenoxide ion.

This procedure for phenols is more elaborate and often gives lower yields than that in which the sulfonic acid is fused with alkali (section 12.3c): starting from an aromatic hydrocarbon, the former process requires nitration, reduction of nitro to amino, diazotization, and hydrolysis, whereas the latter requires only sulfonation and alkali-fusion. However, the diazonium method can be used in circumstances in which the sulfonate method fails: for example, m-nitrophenol, which cannot be obtained from m-nitrobenzenesulfonic acid (p. 403), may be prepared from m-nitroaniline in over 80% yield [1].

(b)   Replacement by halogens

The procedures used differ according to the halogen to be introduced.

Aromatic fluorides are prepared by the *Schiemann reaction*. An aqueous solution of the diazonium salt is treated with fluoroboric acid, precipitating the diazonium fluoroborate. This is dried and then heated gently until decomposition begins, after which reaction occurs spontaneously. In this way fluorobenzene may be obtained from aniline in about 55% yield [2]. The reaction involves the S_N1 mechanism:

$$ArN_2^+ \ BF_4^- \quad \xrightarrow[-N_2]{\text{heat}} \quad Ar^+ \ \ F\!\!-\!\!\bar{B}F_3 \quad \longrightarrow \quad ArF \ + \ BF_3$$

In a recent modification, fluorophosphoric is used instead of fluoroboric acid. The diazonium fluorophosphates are less soluble than the fluoroborates and give better yields of aromatic fluorides when heated at about 165°C:

$$ArN_2^+ \ PF_6^- \quad \longrightarrow \quad ArF \ + \ N_2 \ + \ PF_5$$

Aromatic chlorides and bromides may be obtained in an analogous manner from diazonium tetrachloroborates and tetrabromoborates, but, since these compounds often decompose violently when heated, it is in general safer to employ the *Sandmeyer reaction*. The procedure for chlorides consists of adding a cold aqueous solution of the diazonium chloride to a solution of copper(I) chloride in hydrochloric acid: a sparingly soluble complex separates which decomposes to the aryl chloride on being heated. The reaction involves one-electron reduction of the diazonium ion followed by ligand transfer:

$$ArN_2^+ \ + \ CuCl_2^- \quad \longrightarrow \quad Ar\cdot \ + \ N_2 \ + \ CuCl_2$$

$$Ar\cdot \ \ Cl\!\!-\!\!Cu\!\!-\!\!Cl \quad \longrightarrow \quad ArCl \ + \ CuCl$$

Aryl bromides may be prepared similarly from diazonium hydrogen sulfates and copper(I) bromide in hydrobromic acid. Typical conversions from the amines are between 70 and 80% for *p*-chlorotoluene and *p*-bromotoluene [1].

Aromatic iodides may be prepared without using a copper(I) salt. An aqueous solution of a diazonium salt is treated with potassium iodide and warmed, and the iodide is usually formed in good yield. For example, aniline gives about 70% of iodobenzene [2]. It was at one time thought that reaction occurs *via* the aryl cation and that the iodide ion is sufficiently nucleophilic, unlike bromide or chloride, to compete effectively for the cation with the much larger amount of water present. However, there is now evidence that iodide ion, which has a similar reducing potential to copper(I), likewise brings about a one-electron reduction of the diazonium ion,

$$ArN_2^+ \ + \ I^- \quad \longrightarrow \quad Ar\cdot \ + \ N_2 \ + \ I\cdot$$

thereby initiating a radical-chain sequence:

$$Ar\cdot \ + \ I_2 \ \longrightarrow \ ArI \ + \ I\cdot$$

$$I\cdot \ + \ I^- \ \longrightarrow \ I_2^{\bar{\cdot}}$$

$$ArN_2^+ \ + \ I_2^{\bar{\cdot}} \ \longrightarrow \ Ar\cdot \ + \ N_2 \ + \ I_2$$

(c)   Replacement by cyano

Aromatic nitriles are obtained in generally good yield by a Sandmeyer procedure using copper(I) cyanide in aqueous potassium cyanide solution. For example, *p*-toluidine gives 64–70% of *p*-tolunitrile [1]:

$$Me\!-\!\!\bigcirc\!\!-\!NH_2 \xrightarrow{HONO \cdot H^+} Me\!-\!\!\bigcirc\!\!-\!N_2^+ \xrightarrow{CuCN} Me\!-\!\!\bigcirc\!\!-\!CN$$

(d)   Replacement by nitro

Two general procedures are available for converting a diazonium salt into a nitro-compound.

(1)   A neutral solution of the diazonium salt is treated with sodium cobaltinitrite and the resulting diazonium cobaltinitrite is decomposed by aqueous sodium nitrite in the presence of copper(I) oxide and copper(II) sulfate.

$$3 \ ArN_2^+ + Co(NO_2)_6^{3-} \ \longrightarrow \ (ArN_2^+)_3Co(NO_2)_6^{3-} \xrightarrow[Cu_2O \cdot CuSO_4]{NaNO_2} 3 \ ArNO_2 + 3 \ N_2$$

(2)   A suspension of the diazonium fluoroborate in water is added to aqueous sodium nitrite solution in which copper powder is suspended. For example, *p*-dinitrobenzene may be obtained from *p*-nitroaniline in about 75% yield [2].

$$ArN_2^+ \ BF_4^- \ + \ NaNO_2 \ \xrightarrow{Cu} \ Ar \ NO_2 \ + \ N_2 \ + \ NaBF_4$$

(e)   Replacement by hydrogen

Two general methods may be used for the conversion $ArN_2^+ \rightarrow ArH$. In the first, the diazonium solution is warmed with ethanol:

$$ArN_2^+ \ Cl^- \ + \ C_2H_5OH \ \longrightarrow \ ArH \ + \ N_2 \ + \ HCl \ + \ CH_3CHO$$

The reaction almost certainly involves the aryl radical.

Yields are often low because of the competitive nucleophilic displacement,

$$ArN_2^+ \ \xrightarrow{-N_2} \ Ar^+ \ \xrightarrow[-H^+]{EtOH} \ ArOEt$$

although when electron-attracting groups are present in the aromatic nucleus the S_N1 heterolysis occurs less readily and the reduction pathway competes more effectively.

Because of this disadvantage, the second method is now more commonly chosen. The reducing agent is hypophosphorous acid and reaction occurs at room temperature when the competing S_N1 reaction is much slower. Copper(I) salts catalyze the process and a chain-reaction occurs:

initiation: ArN_2^+ + Cu^+ $\longrightarrow$ $Ar\cdot$ + N_2 + Cu^{2+}

propagation:

$$Ar\cdot \ + \ \underset{\substack{| \\ OH}}{\overset{\substack{H \\ |}}{H-P=O}} \ \longrightarrow \ ArH \ + \ \underset{\substack{| \\ OH}}{\overset{\substack{H \\ |}}{\cdot P=O}}$$

$$ArN_2^+ \ + \ O=PH(OH) \ \xrightarrow{H_2O} \ Ar\cdot \ + \ N_2 \ + \ H_3PO_3 \ + \ H^+$$

The ease of removal of an amino group from an aromatic ring by diazotization and reduction is of considerable value in synthesis, for the amino group and its derivatives are powerfully activating substituents with well-defined orienting properties in electrophilic reactions. The group may be introduced in order to promote the necessary reactivity and may later be removed. For example, 1,3,5-tribromobenzene may be made by the bromination of aniline followed by removal of the amino group:

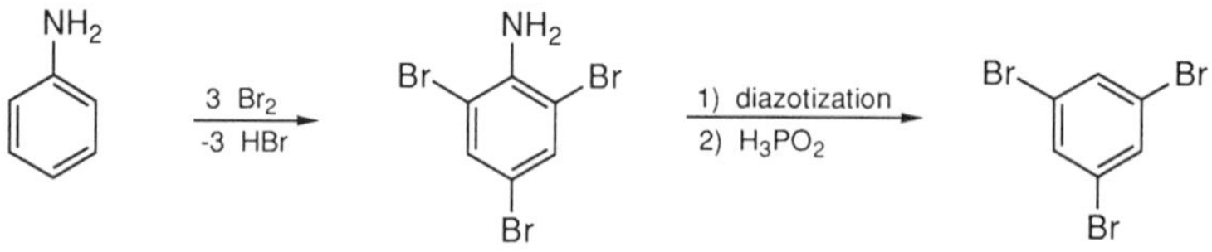

A more elaborate pathway is employed to make *m*-bromotoluene. The nitration of toluene gives the *p*-nitro-derivative in about 40% yield. Reduction of nitro to amino and the acetylation of amino are nearly quantitative reactions so that *p*-acetamidotoluene is easily available. Since the acetamido group is more strongly activating than the methyl group, despite the steric hindrance to reaction at its *ortho* position, bromination occurs largely *meta* to methyl. Hydrolysis and deamination give *m*-bromotoluene.

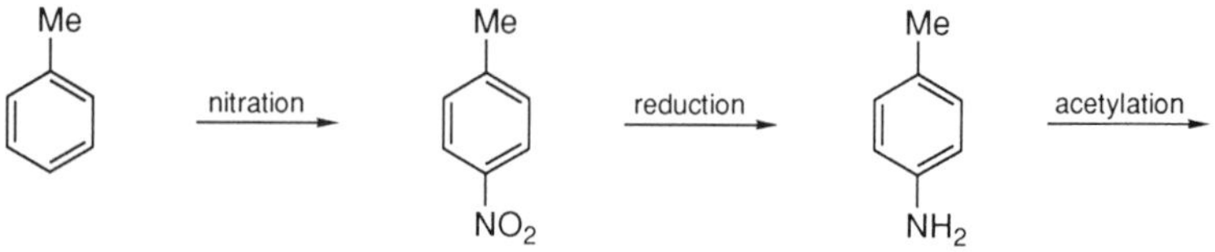

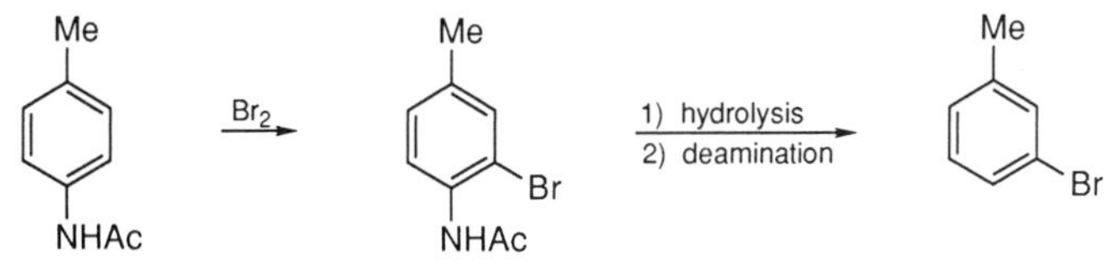

Despite the number of steps involved, this process is considerably more efficient than the direct bromination of toluene which gives less than 1% of the *meta*-derivative. Similar procedures may be used for preparing other *m*-haloalkylbenzenes (except fluoro compounds) and *m*-nitroalkylbenzenes.

(f) Replacement by aliphatic carbon

Alkenes in which C=C is conjugated to another unsaturated group (e.g. C=C, C=O, C≡N) react with diazonium salts in the presence of a catalytic amount of copper(I) chloride (*Meerwein reaction*), e.g.

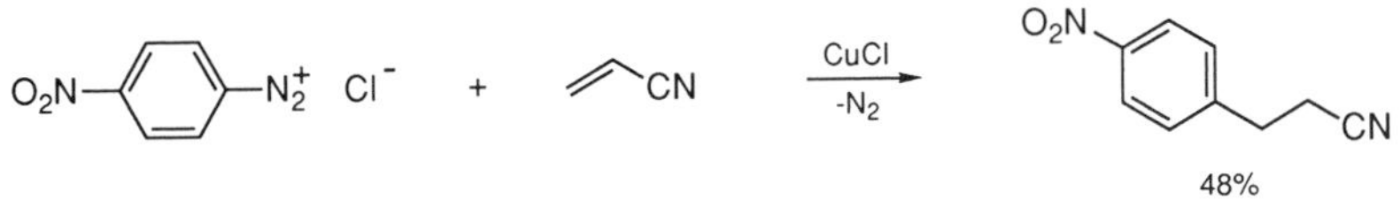

As in the Sandmeyer reaction, copper(I) reduces the diazonium ion to an aryl radical and copper(II) chloride acts as a ligand-transfer agent for the adduct of this radical and the alkene:

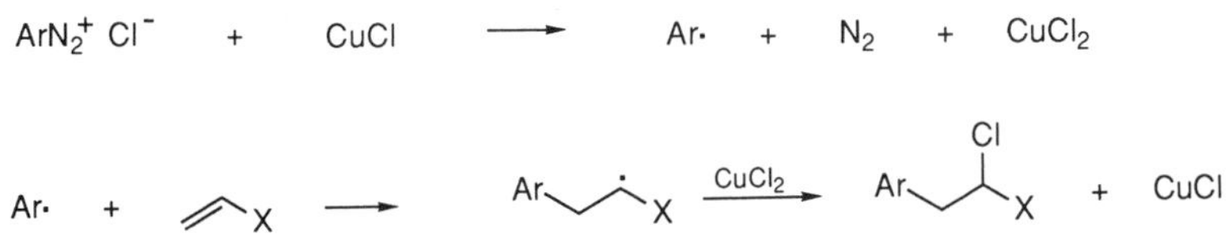

$$ArN_2^+ \; Cl^- \quad + \quad CuCl \quad \longrightarrow \quad Ar\cdot \quad + \quad N_2 \quad + \quad CuCl_2$$

In some cases, the final step is the loss of hydrogen, e.g.

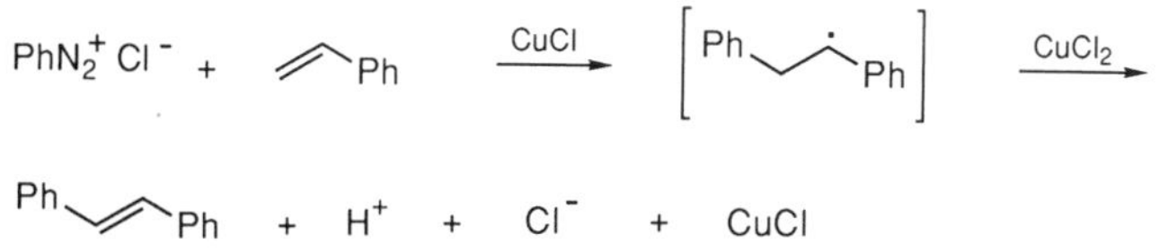

The conjugated unsaturated group serves to stabilize the transition state for the addition of the aryl radical (cf. the delocalization of the unpaired electron in the ensuing adduct, above), thereby increasing the rate of this reaction compared with the competing processes (Sandmeyer reaction and dimerization).

The aryl radical can also be formed from the diazonium ion with titanium(III) chloride. However, because titanium(IV) is a very weak oxidant, the final step consists of reduction by Ti(III) rather than oxidation,

$$\text{ArN}_2^+ \xrightarrow[-\text{N}_2]{\text{Ti}^{3+}} \text{Ar·} \xrightarrow{\qquad} \text{Ar}\diagup\diagdown_X$$

$$\xrightarrow{\text{Ti}^{3+}} \text{Ar}\diagup\diagdown_X^- \xrightarrow{\text{H}_2\text{O}} \text{Ar}\diagup\diagdown_X$$

(X = -*M* substituent)

for example [7],

70%

The yields in these processes are often low but the method can provide a short route to a compound that is not otherwise readily accessible. Yields are improved by the presence of electron-releasing groups in the diazonium ion.

(g) Replacement by aromatic carbon

A number of procedures lead to the arylation of aromatic carbon by diazonium salts.

(i) Gomberg reaction.   A two-phase liquid system consisting of an aqueous solution of the diazonium salt and an aromatic liquid, or a solution of an aromatic solid in an inert solvent, is treated with aqueous sodium hydroxide. The covalent diazohydroxide is formed ($\text{ArN}_2^+\text{Cl}^- + \text{NaOH} \rightarrow \text{Ar}-\text{N}=\text{N}-\text{OH} + \text{NaCl}$) and gives aryl radicals by a complex mechanism similar to that for reaction of diazoacetates, below; these radicals react with the aromatic liquid. Yields are never high and are usually less than 40%, e.g.

35%

When the aromatic compound is unsymmetrical, arylation gives a mixture of products. For example, benzenediazonium chloride reacts with nitrobenzene in Gomberg conditions to give a mixture containing all three nitrobiphenyls. The reaction is therefore more valuable when benzene itself or a symmetrical *para*-disubstituted benzene is involved. The mechanism of homolytic arylation and the factors governing the orientation of the substitution are discussed subsequently (section 17.3*c*).

(*ii*) *Diazoacetates.* Aromatic diazoacetates decompose to aryl radicals which bring about the homolytic arylation of aromatic compounds. The mechanism is complex:

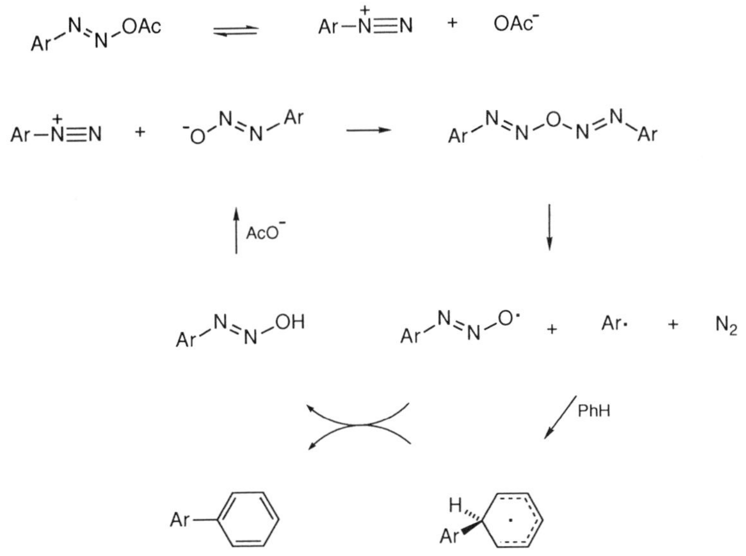

It is usually more convenient to use an *N*-nitrosoacylarylamine rather than a diazoacetate as the starting material; the former rearranges to a diazoester on being heated:

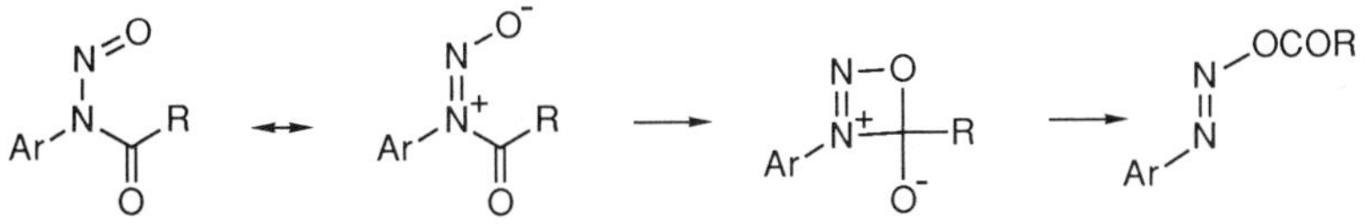

The use of either the Gomberg reaction or the diazoacetate method for the formation of biphenyls is normally less satisfactory than the aroyl peroxide method (p. 544). However, when the appropriate peroxide is difficult to obtain the other two methods provide practicable alternatives.

(*iii*) *Pschorr reaction.* A convenient synthesis of phenanthrene consists of reacting *o*-nitrobenzaldehyde with phenylacetic acid in the presence of acetic anhydride (Perkin reaction, p. 218), reducing the nitro group to amino, diazotizing, treating the diazonium salt with copper powder, and finally decarboxylating.

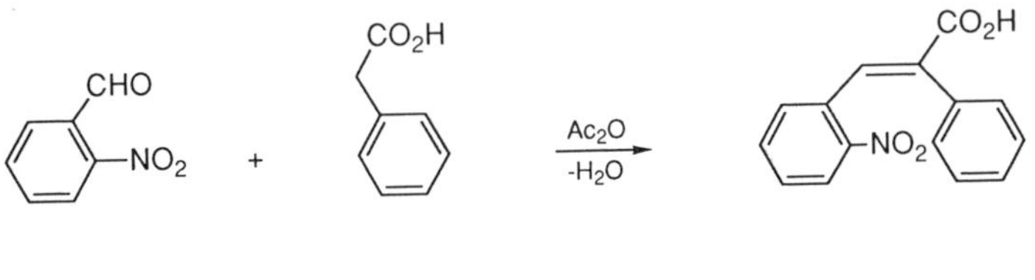

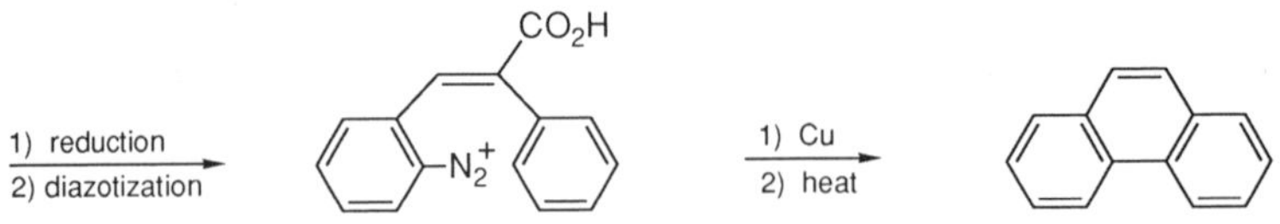

phenanthrene

This procedure, originally applied by Pschorr to the synthesis of phenanthrene and its derivatives, has been extended to forming other aromatic systems, e.g. fluorenone may be obtained from 2-aminobenzophenone:

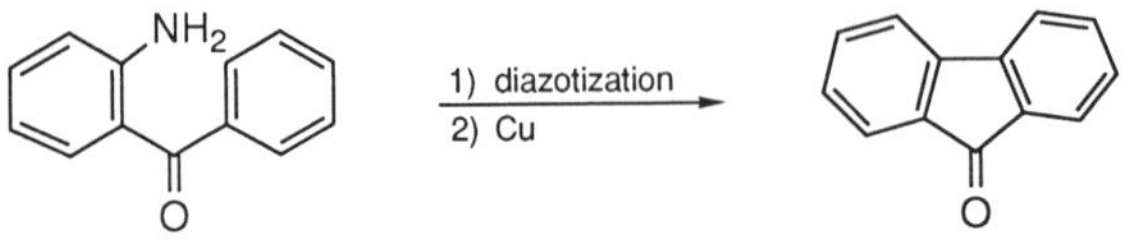

fluorenone

(iv)   Reduction by copper(i) ammonium ion. The addition of copper(i) ammonium hydroxide (obtained by treating copper(ii) sulfate in ammonia with hydroxylamine) to diazotized anthranilic acid gives diphenic acid in about 90% yield:

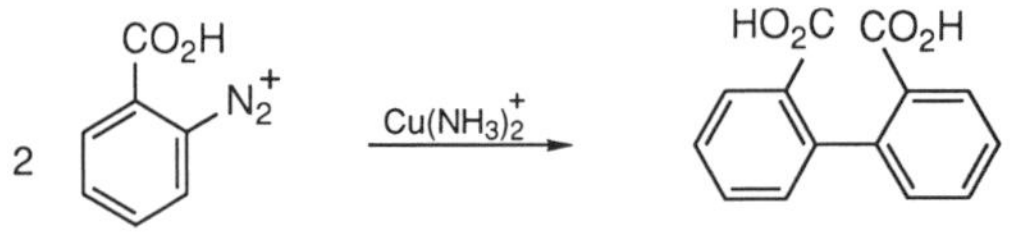

diphenic acid

Reaction probably occurs by one-electron reduction of the diazonium ion followed by dimerization of the resulting aryl radicals. Nucleophilic substitutions of the Sandmeyer type which occur in the presence of copper(ii) ion may here be inhibited because the copper(ii) ion formed is bound as copper(ii)-ammonium ion so that the ligand-transfer mechanism (p. 412) cannot operate.

Yields of biaryls are rarely as high as that for diphenic acid. Diazotized *o*-nitroaniline does not couple, and diazotized *m*-nitroaniline gives 3,3′-dinitrobiphenyl in 45% yield.

(h)   Replacement by other species

As well as iodide ion, azide ion and several types of sulfur-centred anion (RS^-, RSS^-, $EtOC(S)S^-$) react directly with diazonium salts. For example, diazotized anthranilic acid and sodium disulfide give dithiosalicylic acid [2] (Ar = *o*-carboxyphenyl),

$$2\ ArN_2^+\ Cl^-\ +\ Na_2S_2\ \longrightarrow\ ArSSAr\ +\ 2\,N_2\ +\ 2\,NaCl$$

and diazotized *m*-toluidine and potassium ethyl xanthate give *m*-tolyl ethyl xanthate, hydrolysis of which gives *m*-thiocresol in about 70% yield [3] (Ar = *m*-tolyl).

$$\text{ArN}_2^+ \text{ Cl}^- \xrightarrow[\text{-N}_2, \text{-KCl}]{\text{K}^+ \ {}^-\text{S}-\overset{\displaystyle S}{\underset{}{\text{C}}}-\text{OEt}} \text{Ar}-\text{S}-\overset{\displaystyle }{\underset{\displaystyle S}{\text{C}}}-\text{OEt} \xrightarrow[\text{2) H}^+]{\text{1) KOH}} \text{ArSH}$$

A radical-chain mechanism has been established for the reaction with PhS$^-$ (cf. reaction with I$^-$, p. 412),

initiation
$$\text{ArN}_2^+ + \text{PhS}^- \longrightarrow \text{Ar}-\text{N}{\overset{\nwarrow}{=}}\text{N}-\text{SPh}$$

$$\text{Ar}-\text{N}{\overset{\nwarrow}{=}}\text{N}-\text{SPh} \longrightarrow \text{Ar} \cdot + \text{N}_2 + \cdot\text{SPh}$$

propagation
$$\text{Ar} \cdot + \text{PhS}^- \longrightarrow \text{ArSPh}^{\overline{\cdot}}$$

$$\text{ArSPh}^{\overline{\cdot}} + \text{ArN}_2^+ \longrightarrow \text{ArSPh} + \text{Ar} \cdot + \text{N}_2$$

and the other nucleophiles may react by similar pathways.

**13.4
Reactions in
which nitrogen
is retained**

The reaction of a nucleophile with the terminal nitrogen of a diazonium ion gives a covalent azo-compound. In many cases the products are very unstable and readily lose nitrogen (e.g. diazohydroxides, p. 416), but two synthetic procedures lead to retention of the nitrogen atoms.

(a) Reduction to arylhydrazines

Aromatic diazonium ions are reduced by sodium sulfite to arylhydrazines. The probable mechanism consists of the reaction of the diazonium ion with a sulfite anion to give a covalent azo-sulfite which, having a double bond conjugated to an electron-accepting group, adds a second nucleophilic sulfite ion (cf. Michael addition, p. 240). Hydrolysis yields the hydrazine.

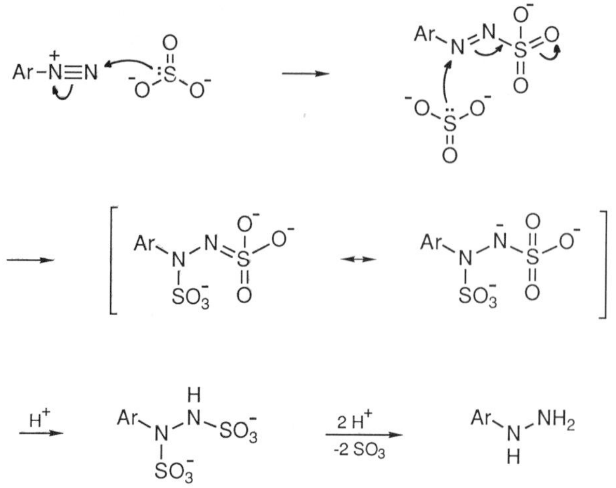

In this way phenylhydrazine may be obtained from aniline in over 80% yield [1].

Reduction to arylhydrazines is also brought about by tin(II) chloride and by electrolysis.

(b)   Coupling reactions

The reactions of diazonium ions with aromatic nuclei are known as coupling reactions. Diazonium ions are weakly electrophilic and react, with some exceptions (p. 423), only with those aromatic compounds which are very powerfully activated towards electrophiles (cf. the nitrosonium ion, p. 380). These compounds include amines, phenols, and heterocyclic systems such as pyrrole.

(i)   *Amines.*   Tertiary aromatic amines react with diazonium ions almost exclusively at the *para* position. The mechanism is that common to other electrophilic aromatic substitutions, i.e. the diazonium ion adds to the aromatic compound and a proton is then eliminated:

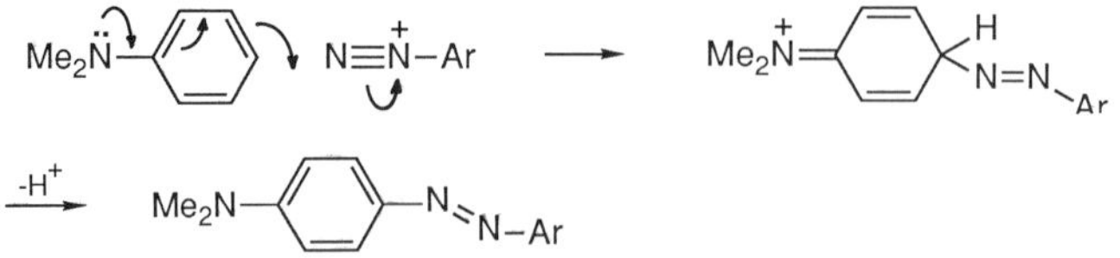

Careful control of the pH of the medium is necessary. In strongly alkaline conditions the diazonium ion is converted fairly rapidly into the covalent diazohydroxide, whereas in strongly acidic conditions the amine is converted largely into its unreactive conjugate acid. A pH in the region 4–10 is satisfactory and a sodium acetate buffer is commonly used to maintain suitable conditions.

Primary and secondary aromatic amines usually react preferentially with di-azonium ions at their nitrogen atoms, just as they do with the nitrosonium ion (p. 380). For example, when aniline is only partially diazotized in hydrochloric acid, the addition of sodium acetate liberates the remaining aniline from its conjugate acid and coupling then occurs to give diazoaminobenzene in about 70% yield [2]:

$$PhNH_3^+ \;+\; AcO^- \;\longrightarrow\; PhNH_2 \;+\; AcOH$$

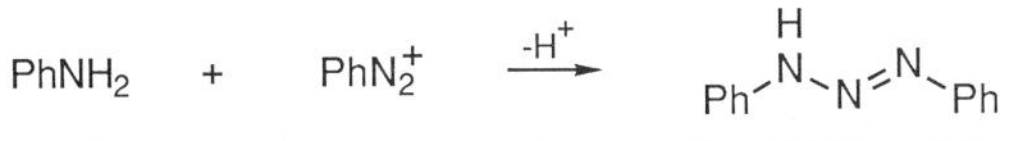

diazoaminobenzene

These *N*-coupled products can be rearranged to *C*-coupled products by treat-ment with mineral acid, e.g. diazoaminobenzene gives *p*-aminoazobenzene. Reac-tion occurs by uncoupling followed by *C*-coupling:

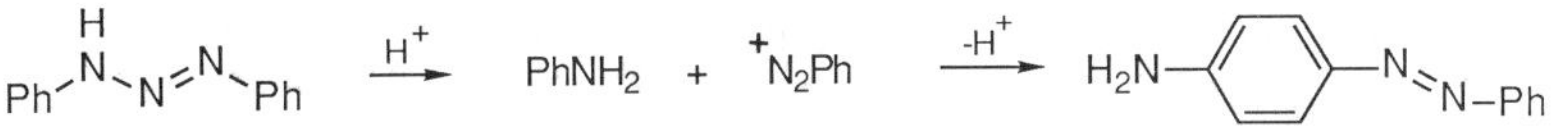

Direct *C*-coupling to primary and secondary amines occurs in two circumstances: first, when the diazonium ion is particularly reactive (e.g. *p*-nitrobenzenediazonium ion) and second, in aqueous formic acid solution, which is acidic enough for the *N*-coupled product to rearrange. When the structure and stereochemistry of the diazonium ion are suitable, *intra*molecular *N*-coupling can occur, for example the diazotization of *o*-phenylenediamine leads directly to benzotriazole in over 75% yield [3]:

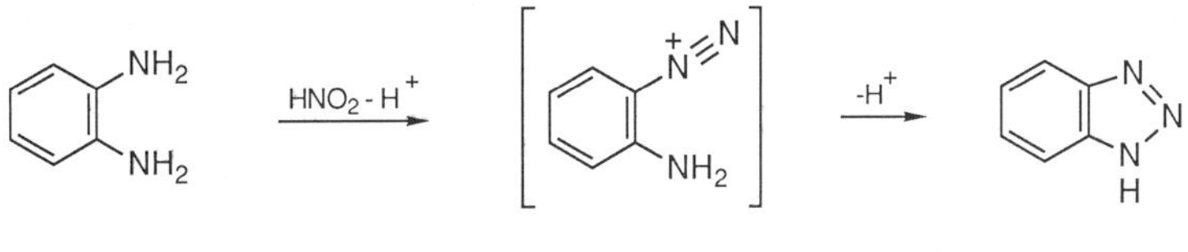

(*ii*) *Phenols.* The coupling of diazonium ions with phenols occurs *via* the phenoxide ions which are considerably more strongly activated to electrophiles than phenols themselves (p. 349). It is therefore necessary to carry out the reaction in alkaline solution and precautions must be taken to prevent the con-version of the diazonium ion into the diazohydroxide. The optimum pH for coupling is about 9–10 but it is normally satisfactory to add an acidic solution of the aqueous diazonium salt to a solution of the phenol in sufficient alkali to neutralize the acid formed and to maintain suitable alkalinity: coupling occurs rapidly enough to prevent significant destruction of the diazonium ion.

The mechanism of coupling with phenoxide ions is analogous to that with amines and reaction is similarly directed mainly to the *para* positon, e.g.

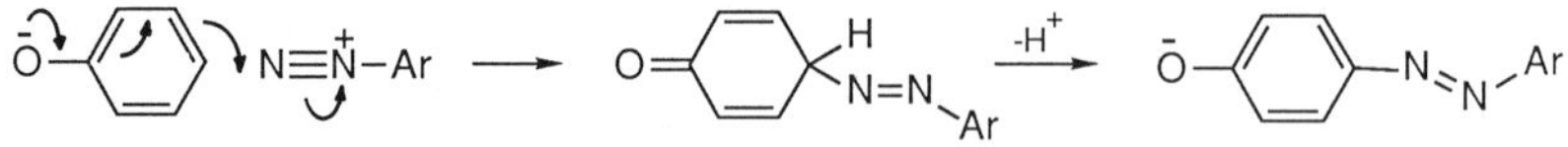

When the *para* position is already substituted, *ortho* coupling occurs, e.g.

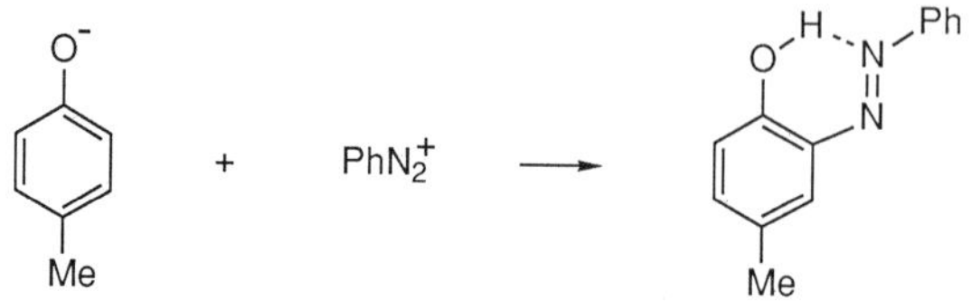

These *ortho*-azophenols are much weaker acids than their *para*-isomers because the unionized form is stabilized relative to the ionized form by chelation involving the hydroxylic proton. Use is made of this property in detecting primary aromatic amines: the compound is treated with hydrochloric acid and sodium nitrite and the resulting solution is poured into an alkaline solution of 2-naphthol. Primary aromatic amines form diazonium ions which couple in the 1-position of the 2-naphtholate (p. 356) to give red azophenols which are insoluble in the alkaline medium.

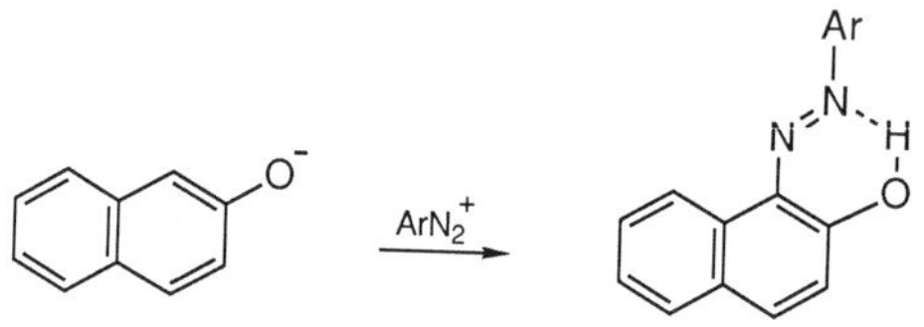

(*iii*) *Enols.*   Although alkenes do not react with diazonium ions, enols, in which the hydroxylic oxygen activates the double bond to electrophiles (p. 85), do so. Reaction occurs more readily in alkaline conditions since the enolate anion is more reactive than the enol.

A typical example is the reaction of acetoacetic ester with benzenediazonium chloride, in a sodium acetate buffer:

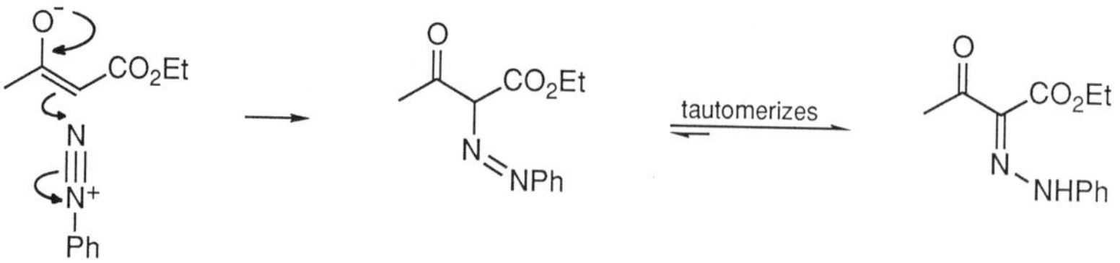

When the carbon atom at which coupling occurs does not possess a hydrogen atom so that the final prototropic shift cannot occur, the initial product readily undergoes C—C bond fission to give an arylhydrazone (*Japp-Klingemann reaction*), e.g.

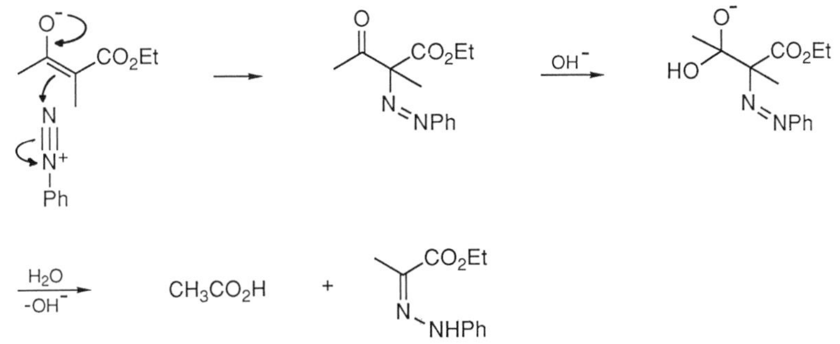

As expected for a reaction involving an enol, carbonyl-containing compounds react both in the normal way and in the Japp-Klingemann manner with decreasing ease as they become decreasingly enolic, for example acetophenone and its derivatives do not couple with diazonium ions. As usual, however, intramolecular reactions occur more readily than the intermolecular analogues if the stereochemistry is suitable, e.g. diazotization of *o*-aminoacetophenone leads to 4-hydroxycinnoline:

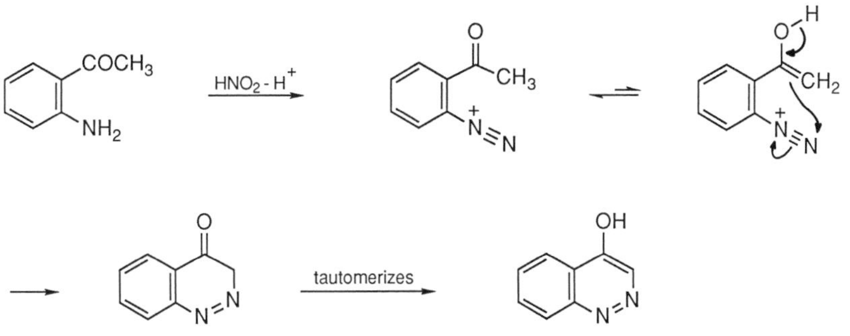

4-hydroxycinnoline

(*iv*) *Pyrroles*. Pyrrole, whose reactivity towards electrophiles is comparable with that of phenol (p. 358), couples with diazonium salts maiinly in the 2-position:

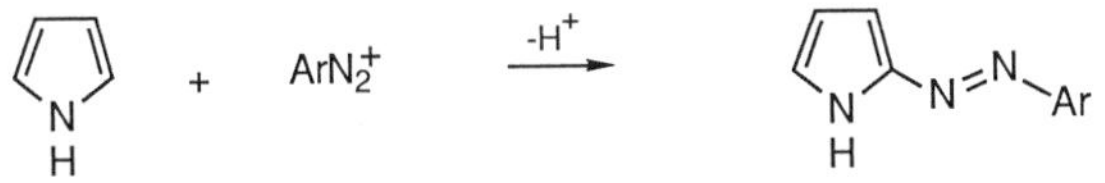

If the 2- and 5-positions are substituted, coupling occurs at the 3-position.

(*v*) *Less activated aromatic nuclei*. Nuclei which are less activated to electrophiles than those of phenoxide ions and aromatic amines do not couple with benzenediazonium ion. The introduction of an electron-attracting substituent into the diazonium ion increases the electrophilicity of this ion and less reactive nuclei then couple. For example, 2,4-dinitrobenzenediazonium ion is reactive enough to

couple with anisole and 2,4,6-trinitrobenzenediazonium ion couples even with
mesitylene:

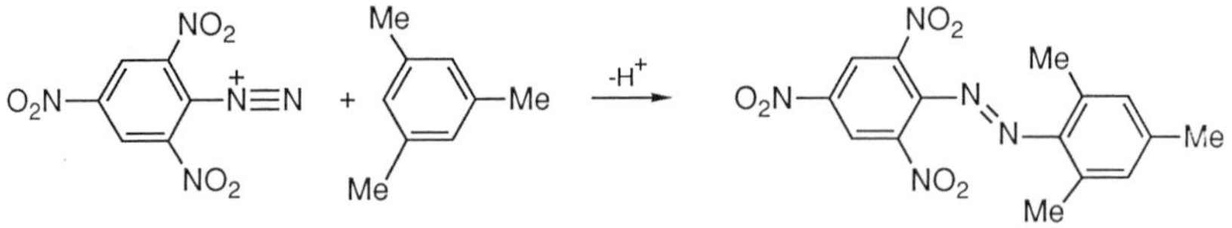

**13.5
The synthetic
value of
diazo-coupling** (a) *Dye-stuffs*

Aromatic azo-compounds are all strongly coloured (e.g. azobenzene is orange-
red, λ_{max} 448 nm) and many of those prepared by the coupling reaction are used
as dye-stuffs. Three classes of dye may be distinguished.

(1) Neutral azo-compounds are used as *azoic combination* (or *ingrain*) dyes,
that is, they are formed *in situ* in the fibre. An example is para red, from 2-
naphthol and *p*-nitrobenzenediazonium ion:

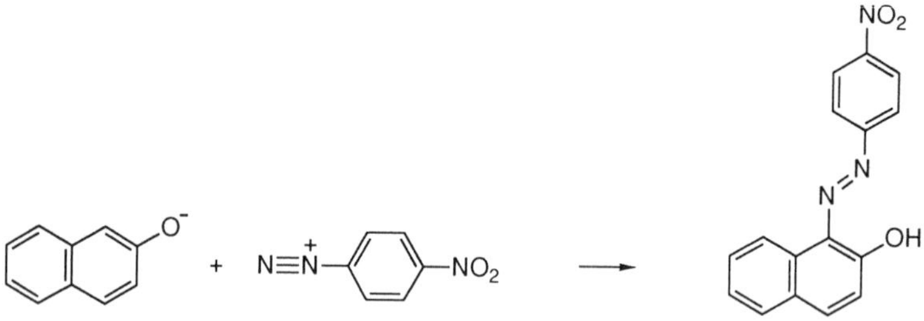

para red

(2) Azo-compounds which contain either a sulfonic acid group or an amino
group are adsorbed directly onto the fibre from aqueous solution. Examples are
orange II, an acidic dye, and Bismarck brown R, a basic dye:

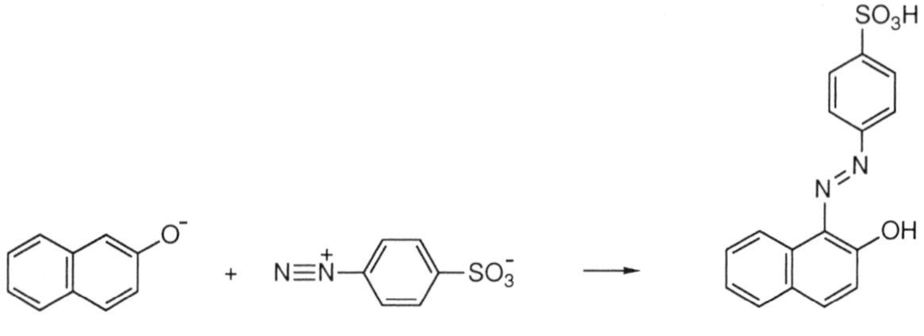

orange II

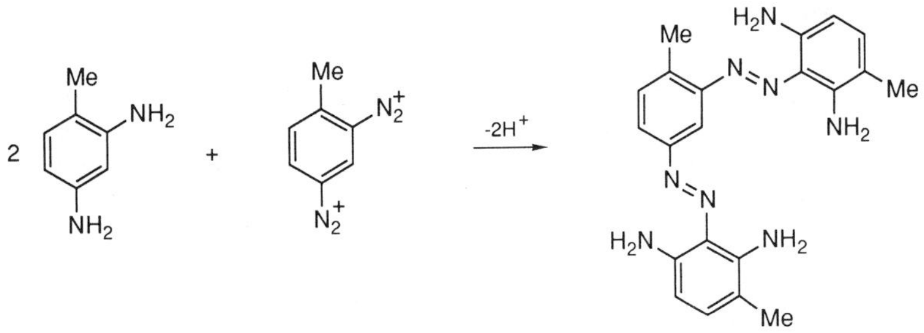

Bismarck brown R

(3) Azo-compounds which contain groups capable of chelation with a metal ion are used as *mordant* dyes: the metal ion is adsorbed on the fibre and binds the dye-stuff. An example is alizarin yellow R, which forms chelates by means of its phenolic and carboxyl groups. Aluminium and chromium oxides are commonly used as mordants.

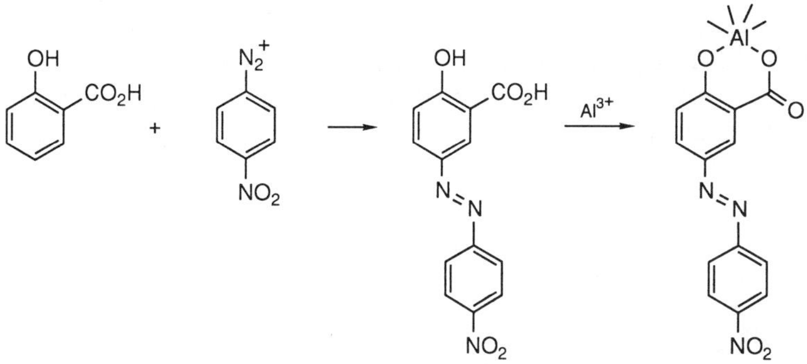

alizarin
yellow R

(b) Indicators

Azo-compounds which contain both an acidic and a basic group may be used as indicators since the colours of the conjugate base and the conjugate acid are invariably different. Examples are methyl red, made from diazotized anthranilic acid and dimethylaniline in about 64% yield [1], and methyl orange, from diazotized sulfanilic acid and dimethylaniline.

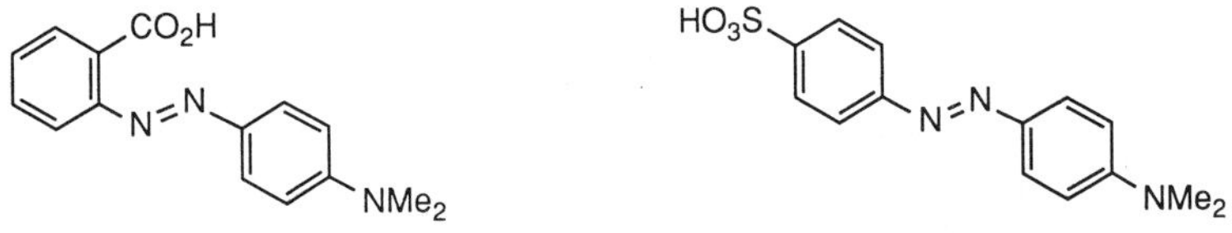

methyl red

methyl orange

(c) Synthesis of amines

Azo-compounds are susceptible to hydrogenolysis, giving amines. Sodium dithionite is usually employed as the reducing agent but catalytic methods may be used. For example, 4-amino-1-naphthol is obtained in about 70% yield from 1-naphthol by coupling with benzenediazonium chloride followed by reduction with dithionite [1]:

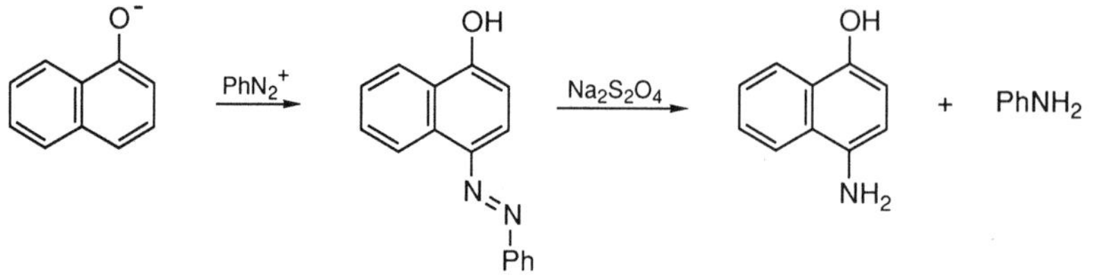

Aliphatic amines may be obtained similarly by coupling to an enol followed by reduction.

This method for preparing amines has the advantage of occurring in mild conditions and it is therefore of value in some cases as an alternative to the normal method for making aromatic amines, by nitration followed by reduction, which involves vigorous and strongly oxidizing conditions. An example occurs in a synthesis of adenine (p. 722).

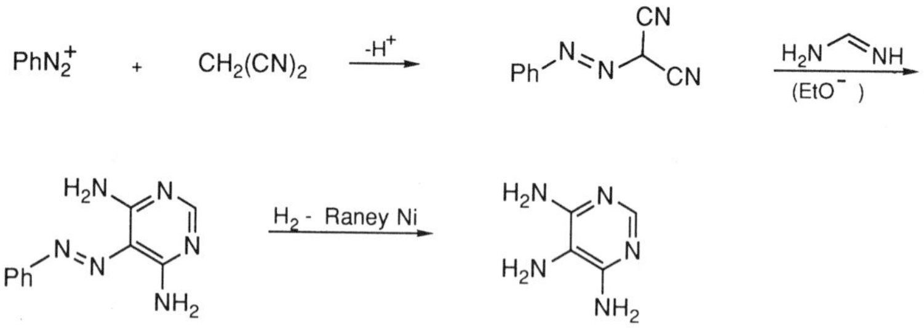

(d) Synthesis of quinones

The *ortho* and *para* diamines and aminophenols are readily oxidized to quinones. Both classes of compound are available by diazo-coupling followed by reduction, providing a useful route to quinones, e.g.

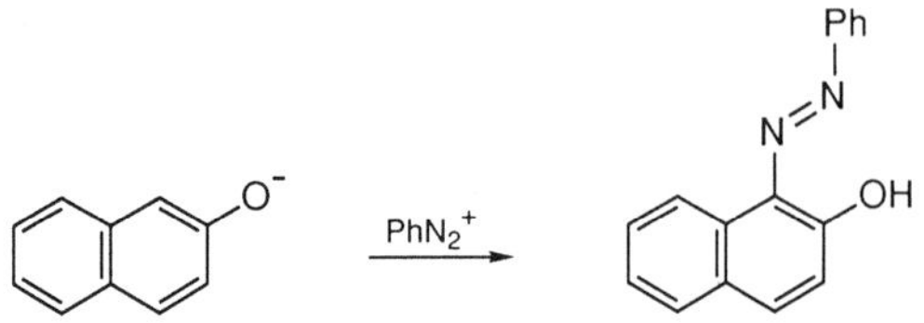

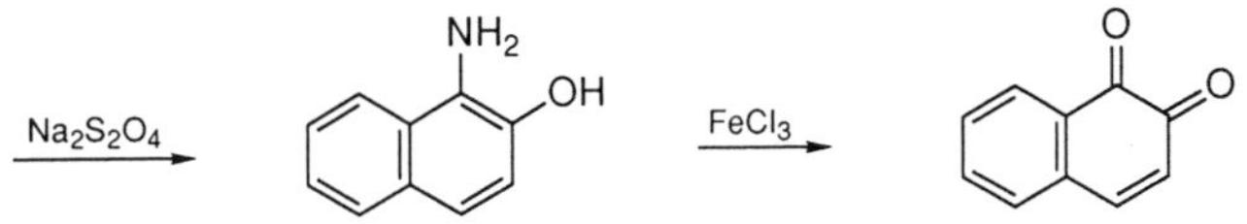

1. Outline routes to the following compounds which employ diazonium salts, **Problems** starting with any monosubstituted benzene:

(a)

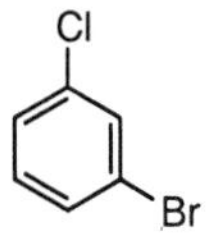

(b)

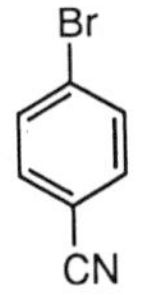

(c)

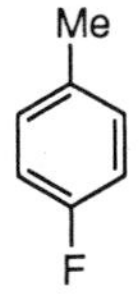

(d)

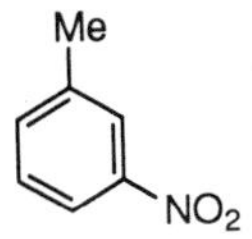

(e)

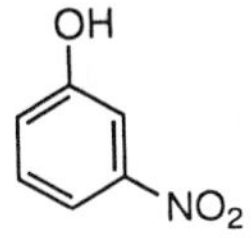

(f)

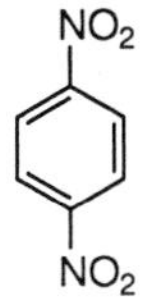

(g)

(h)

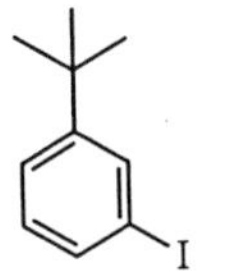

(i)

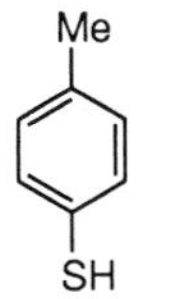

(j)

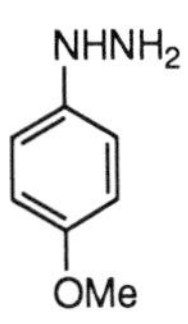

(k)

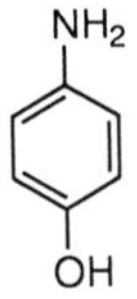

(l)

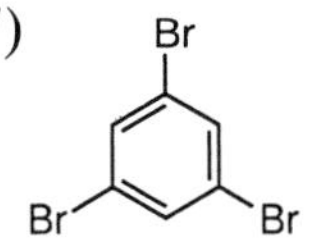

(m)

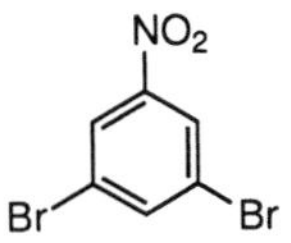

(n)

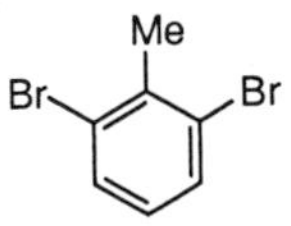

(o)

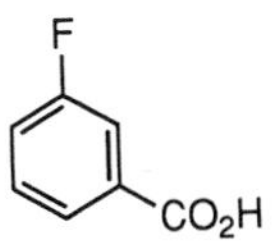

(p)

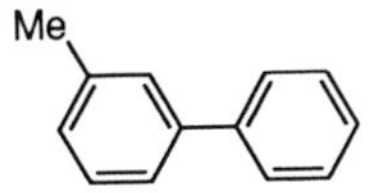

(q)

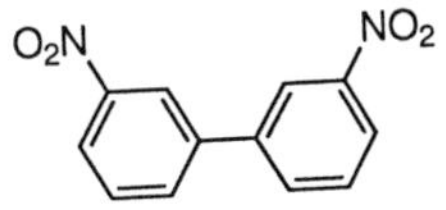

2. How would you obtain the following compounds?

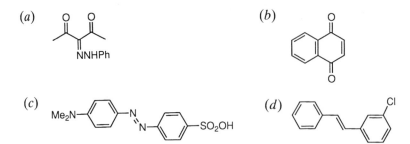

(a)   (b)

(c)   (d)

3. Account for the following observations:
    (*i*) Although *NN*-dimethylaniline couples with benzenediazonium chloride,
           its 2,6-dimethyl derivative does not.
    (*ii*) 2,4-Dinitrobenzenediazonium chloride couples with anisole, although
            benzenediazonium chloride does not.
    (*iii*) When *p*-chloroaniline is diazotized with sodium nitrite and hydrobromic
             acid, the resulting diazonium salt solution couples with *NN*-dimethyl-
             aniline to give largely 4-bromo-4′-dimethylaminoazobenzene.

Molecular rearrangements 14

**14.1
Types of
rearrangement**

Although in the majority of reactions of organic compounds the basic skeleton of the molecule remains intact, there are many in which a skeletal rearrangement occurs. Some of these processes are of value in synthesis and the main emphasis in this chapter will be on this aspect of rearrangements. In addition, however, rearrangements can be a disadvantage if they occur during the course of operations designed only to effect other changes of functionality, so that it is important to be familiar with the structural features and reaction conditions which lead to rearrangement.

Rearrangements are of two types: intramolecular processes in which the group which migrates does not become completely detached from the system in which rearrangement is occurring, and intermolecular processes, in which the migrating group is first detached and later re-attached at another site. The latter group may be regarded as elimination–addition processes and, except in the aromatic series, will not be discussed here; examples include the prototropic and anionotropic shifts which have been discussed earlier, e.g.

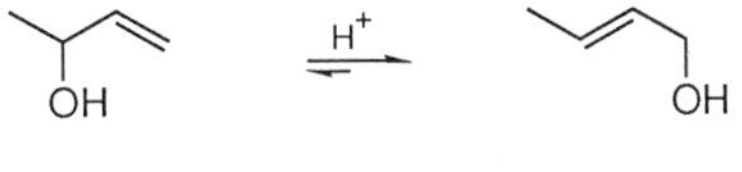

Intramolecular rearrangements are conveniently subdivided into those which occur in electron-deficient systems and those which occur in electron-rich systems. Rearrangements of free radicals are also known and are discussed separately (p. 528): they have little synthetic value. Rearrangements within aromatic nuclei are grouped together in the final section.

**14.2
Rearrangement
to electron-
deficient carbon**

These reactions are classified according to the nature of the group which migrates.

(a)   Carbon migration

(i)   Wagner–Meerwein rearrangement.   One of the simplest systems within which carbon migrates, with its bonding-pair, to an electron-deficient carbon atom is the neopentyl cation:

All reactions which give rise to this ion give products derived from the rearranged ion, e.g.

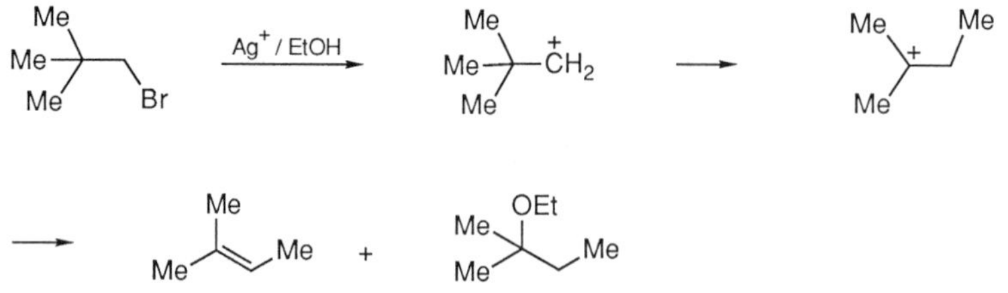

The driving force for the rearrangement resides in the greater stability of a tertiary than a primary carbocation.

In alicyclic systems, the relief of strain can provide a powerful driving force for rearrangement. For example, the addition of hydrogen chloride to α-pinene gives the rearranged product, isobornyl chloride: the strained four-membered ring in the carbocation expands to the less strained five-membered analogue, despite the fact that the former contains a tertiary and the latter a secondary carbocation.

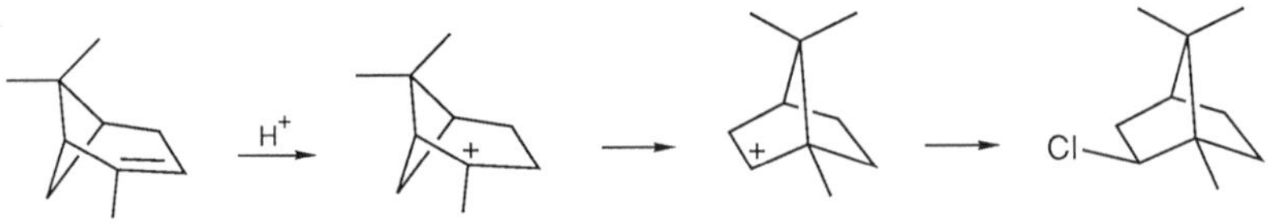

α-pinene

isobornyl chloride

The other principal features of these migrations are as follows.

1.   *The carbocation may be generated in a variety of ways.*

(1)   From a halide, by using a strongly ionizing solvent or by adding a Lewis acid such as silver ion, which aids carbocation formation by abstracting the halide:

(2)   From an alcohol, by treatment with acid to promote heterolysis:

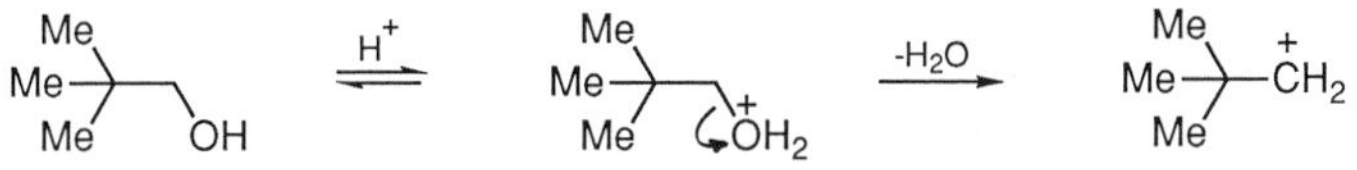

(3) From an amine, by treatment with nitrous acid: reaction occurs *via* the aliphatic diazonium ion from which molecular nitrogen is rapidly lost,

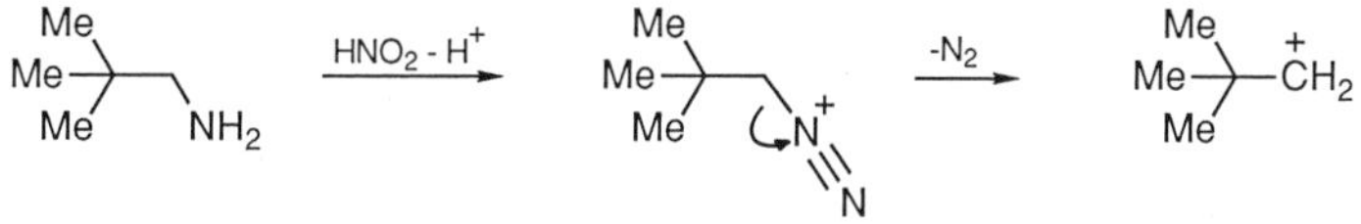

(4) From an alkene by protonation, e.g.

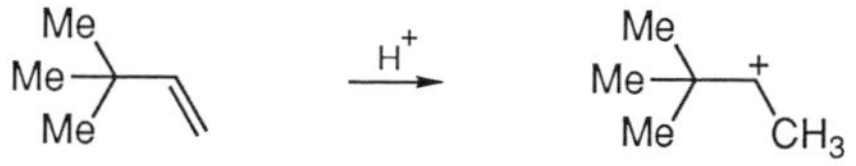

2. *Hydrogen can also migrate in these systems.* For example, reactions which occur through the isobutyl cation yield mainly products derived from the t-butyl cation:

A typical example of a hydride shift occurs in the reaction of a primary aliphatic amine with nitrous acid, e.g.

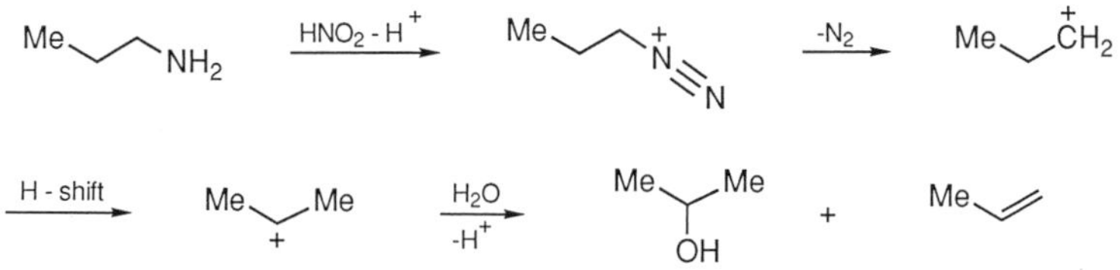

3. *Aryl groups have a far greater migratory aptitude than alkyl groups or hydrogen.* For example, the chloride below undergoes solvolysis with rearrangement many thousands of times faster than neopentyl chloride in the same conditions. This is ascribed to the fact that, whereas the rate-determining step in the reaction of neopentyl chloride is the formation of the high-energy primary carbocation, that in the reaction of the phenyl-substituted chloride is the formation of a lower-energy bridged phenonium ion:

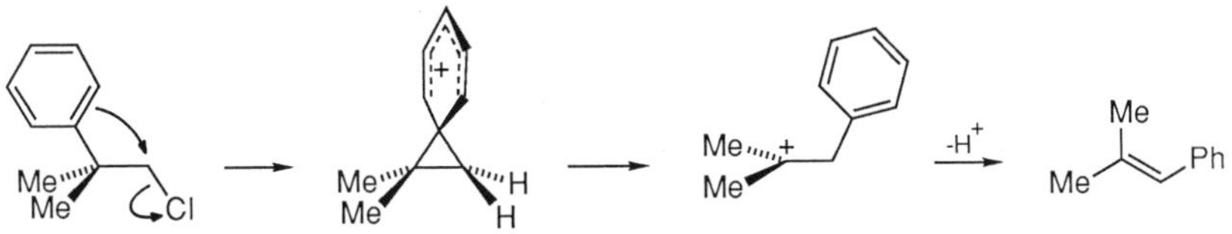

The aryl group is said to provide *anchimeric assistance* to the reaction.

The phenonium ion is similar in structure to the intermediate in an electrophilic aromatic substitution and understandably, therefore, electron-releasing groups in the aromatic ring (e.g. p-OCH$_3$) give rise to greater rates of migration and electron-attracting groups (e.g. p-Cl) to lower rates.

It should be noted, however, that the enormously greater tendency for rearrangement of aryl groups than of alkyl groups or hydrogen does not apply to deaminations. For example, the treatment of 3-phenyl-2-butylamine with nitrous acid in acetic acid gives comparable amounts of the acetates derived from the migration of phenyl, methyl, and hydrogen,

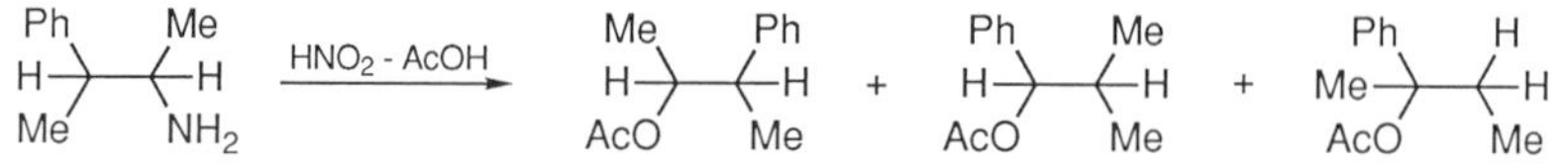

whereas the solvolysis of the corresponding tosylate in acetic acid gives only the product derived from the migration of phenyl, CH$_3$CH(OAc)—CHMePh. This is because the tendency for nitrogen to depart from the diazonium ion is so great that anchimeric assistance is not required in order to help to 'push off' the leaving group.

4. *The rearrangement is stereospecific*: the migrating group approaches the electron-deficient carbon atom from the direction opposite to that in which the departing group is moving, just as in the S$_N$2 reaction. Inversion of configuration therefore occurs at the electron-deficient carbon. This stereospecific requirement of the reaction, together with the *anti*-stereospecificity of E2 eliminations (p. 160), has significant consequences in alicyclic chemistry. For example, the treatment of cyclohexanol in acidic dehydrating conditions results in elimination; reaction occurs on the conformational isomer in which hydrogen is *anti* to the axial hydroxyl:

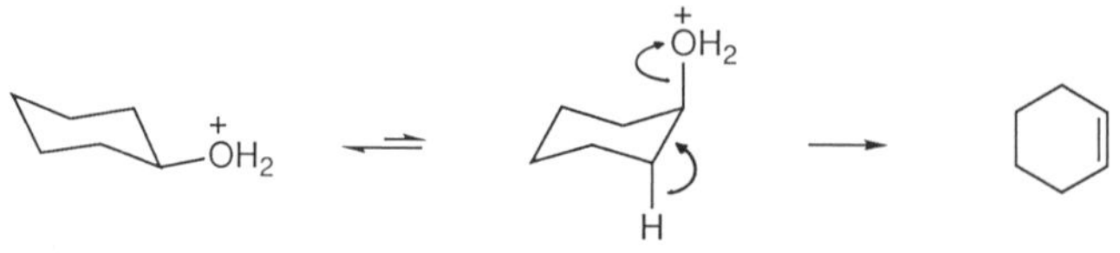

However, in the *trans*-decalin derivative below, the hydroxyl group is held in the equatorial position because the ring system cannot flap (p. 158). In this situation there is no hydrogen *anti* to the hydroxyl, but two of the ring carbon atoms are in the appropriate *anti* position for rearrangement and, in the presence of acid, ring contraction takes place:*

---

* Of the two carbon atoms which might migrate, that one does so which leaves behind the more stable carbocation.

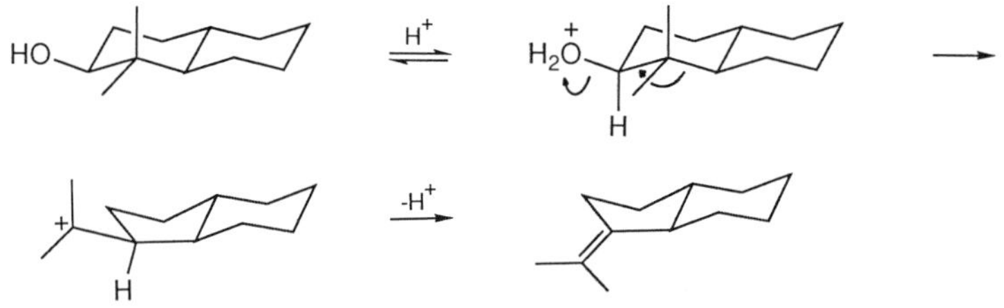

Particular attention must be paid to the possibility of rearrangements such as this in transformations involving steroids and terpenes which contain decalin systems.

5. *Rearrangements in bicyclic systems are particularly common,* as in the conversion of camphene hydrochloride into isobornyl chloride, catalyzed by Lewis acids. Note that the final uptake of chloride ion also occurs with inversion, illustrating another general feature of these rearrangements.

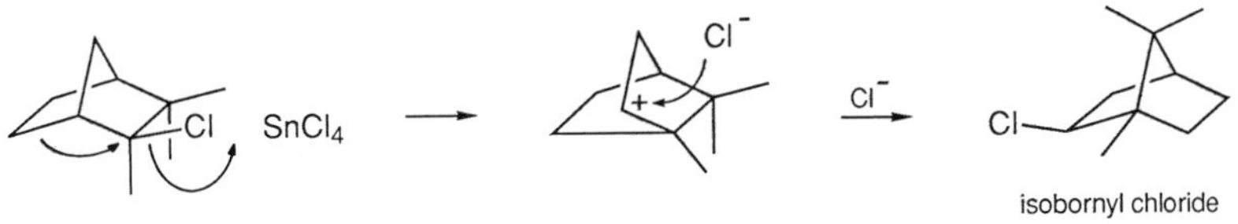

isobornyl chloride

6. *Two or more rearrangements may occur successively.* A spectacular illustration of this is the conversion,

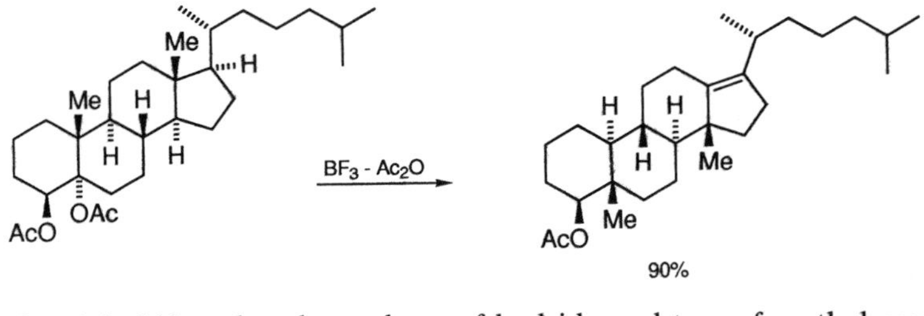

90%

in which five 1,2-shifts take place, three of hydride and two of methyl groups.

(*ii*) *Pinacol rearrangement.* The treatment of 1,2-diols (pinacols) with acid leads to rearrangement, e.g.

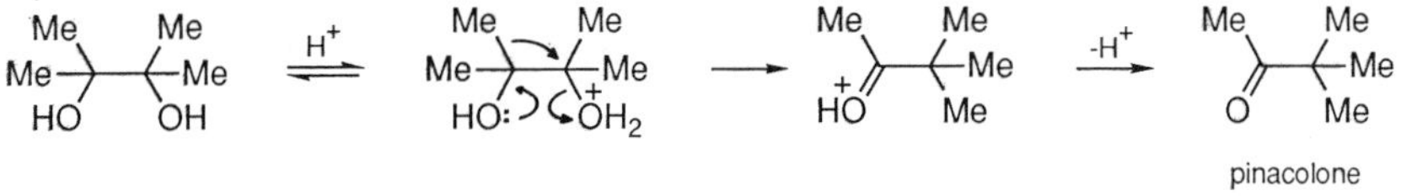

pinacolone

The pinacol rearrangement, although fundamentally similar to the Wagner–Meerwein rearrangement, differs in that the rearranged ion, the conjugate acid of a ketone, is relatively more stable than the rearranged carbocations formed in the Wagner–Meerwein reaction. Consequently, the driving force for the rearrangement of pinacols is much greater: whereas alcohols, other than those with the special structural features discussed above, can usually be dehydrated by acids without the occurrence of rearrangement, pinacols normally rearrange in preference to undergoing simple dehydration.

Pinacols are readily obtained by the one-electron reduction of carbonyl-containing compounds (p. 656) so that the pinacol rearrangement is synthetically useful. For example, pinacolone (methyl t-butyl ketone) can be obtained in about 70% yield by the distillation of a mixture of pinacol hydrate and sulfuric acid [1]. However, the distillation of a mixture of pinacol and aqueous hydrobromic acid gives mainly the dehydration product, 2,3-dimethyl-1,3-butadiene [3]:

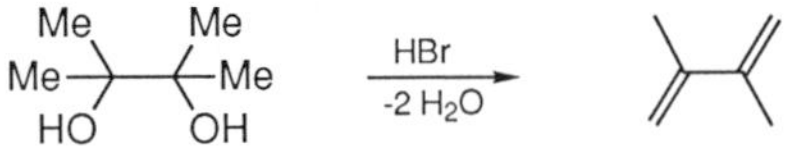

The characteristics of the Wagner–Meerwein rearrangement apply also to the pinacol rearrangement:

(1)   Alkyl and aryl groups and hydrogen migrate.
(2)   The migratory aptitude of an aryl group is much greater than that of alkyl or hydrogen, and amongst aryl groups the migratory aptitude increases as the aromatic nucleus is made increasingly electron-rich (e.g. *p*-chlorophenyl < phenyl < *p*-tolyl < *p*-methoxyphenyl), e.g.

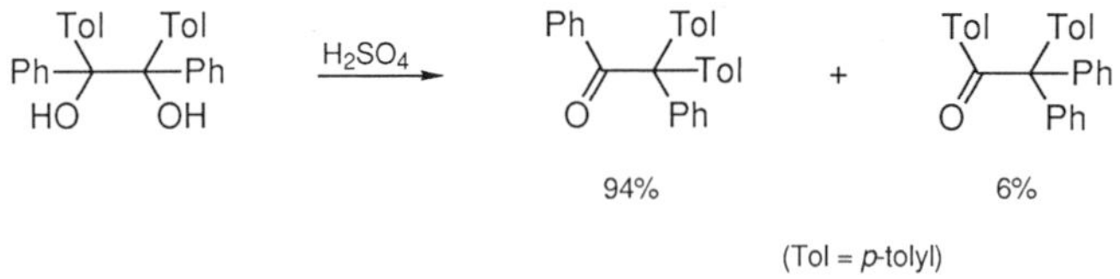

94%   6%

(Tol = *p*-tolyl)

(3)   The reaction occurs in an *anti* manner.

There is a further factor which does not apply to the Wagner–Meerwein reaction: of the two possible hydroxyl groups which are available as leaving groups, that one departs which leaves behind the more stable carbocation. This factor takes precedence over the migratory-aptitude factor, e.g. rearrangement of 1,1-dimethyl-2,2-diphenyl glycol leads to 3,3-diphenyl-2-butanone, by migration of methyl rather than phenyl:

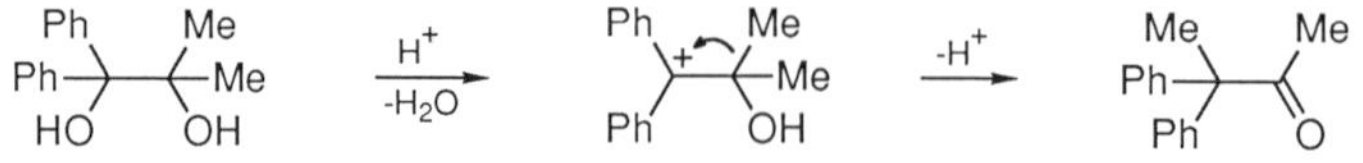

The requirement that the migrating group be *anti* to the leaving group has important consequences in alicyclic systems. For example, *cis*-1,2-dimethyl-cyclohexane-1,2-diol undergoes a methyl shift to give 2,2-dimethylcyclohexanone, whereas the *trans*-isomer undergoes ring contraction to give a cyclopentane derivative:

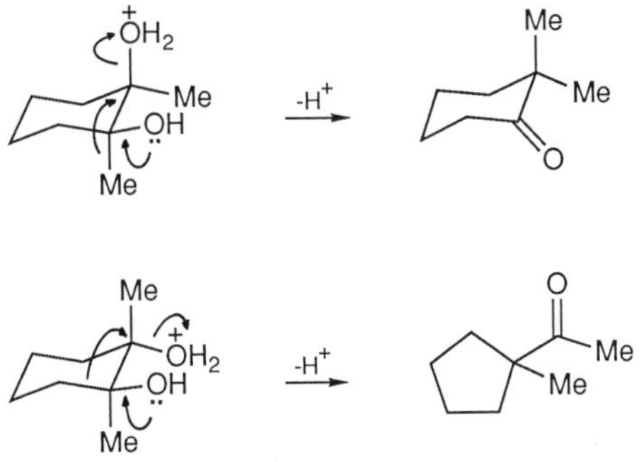

In addition to 1,2-diols, β-hydroxyhalides undergo rearrangement in the presence of Lewis acids, and β-aminoalcohols undergo rearrangement, *via* the diazonium ion, on treatment with nitrous acid.

A synthesis of cycloheptanone illustrates the use of the latter process. The appropriate β-aminoalcohol is obtained from cyclohexanone by base-catalyzed reaction with nitromethane (p. 222), followed by reduction of the nitro group (p. 666), and treatment with sodium nitrite in acetic acid gives cycloheptanone in 40% overall yield [4],

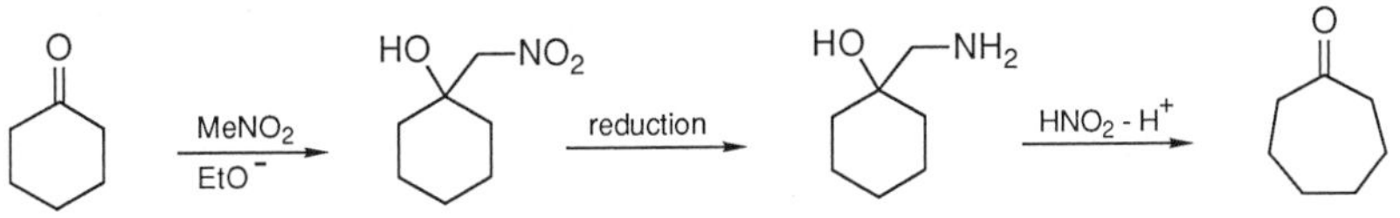

By converting a 1,2-diol into its monotosylate, base-catalyzed rearrangement can be brought about. Since the ease of formation of tosylates decreases in the order primary > secondary > tertiary, this provides a means of effecting rearrangement in the opposite direction from the acid-catalyzed reaction. For example, in the diol below, acid-catalyzed rearrangement occurs with migration of R', since the tertiary carbocation is formed more readily, whereas in the base-catalyzed process it is R that migrates:

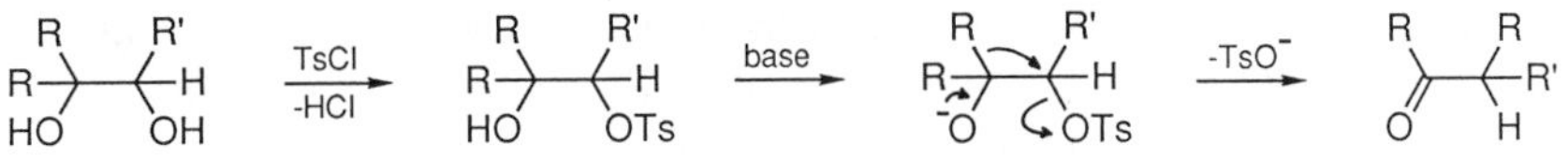

(*iii*) *Benzilic acid rearrangement.* α-Diketones undergo a rearrangement when treated with hydroxide ion, giving α-hydroxyacids. The best known example is the conversion of benzil into benzilic acid; the driving force for the reaction lies in the removal of the product by ionization of the carboxyl group.

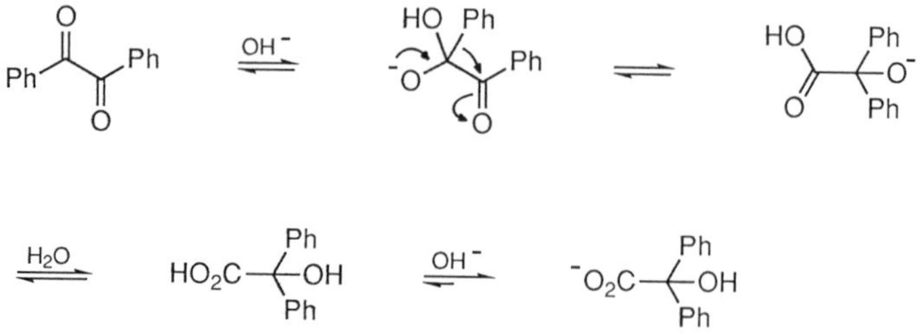

Ketones which contain α-C—H bonds usually undergo base-catalyzed reactions in preference to rearrangement, although ketipic acid gives citric acid:

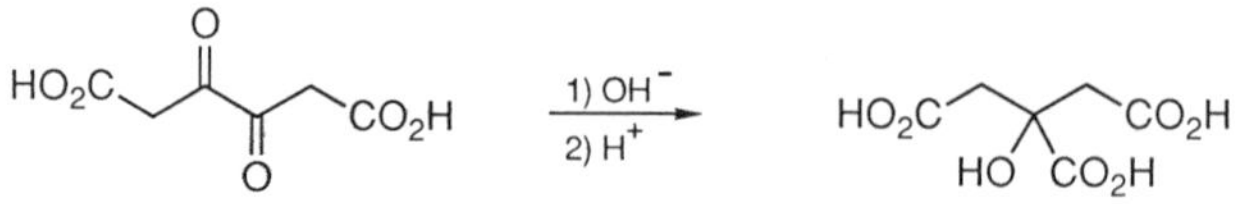

ketipic acid   citric acid

Aromatic α-diketones are normally prepared by the oxidation of the α-hydroxyketones obtained by the benzoin reaction (p. 246), and benzilic acids may be synthesized directly from benzoins by combining the oxidation and rearrangement reactions. For example, the treatment of benzoin itself with sodium bromate and sodium hydroxide gives benzilic acid in up to 90% yield [1].

(iv) Rearrangements involving diazomethane.   Diazomethane takes part in two types of reaction which lead, as the result of rearrangement, to the insertion of a methylene group into a chain of carbon atoms. In each, it acts first as a carbon nucleophile ($CH_2=\overset{+}{N}=\overset{-}{N} \leftrightarrow \overset{-}{C}H_2-\overset{+}{N}\equiv N$), giving a derivative from which nitrogen is readily lost.

1.   *Aldehydes and ketones are converted into the next highest homologue:*

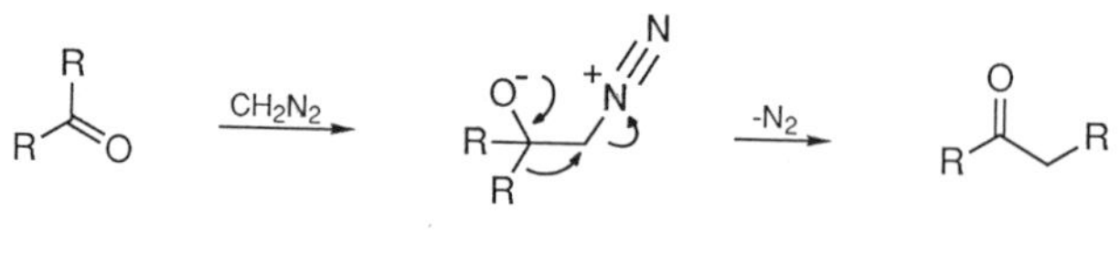

The migration step is similar to that in the pinacol reaction: the movement of R may be visualized as being brought about by the combination of the 'pull' from nitrogen and the 'push' from the oxyanion. Two disadvantages attend the use of this procedure in synthesis. First, unsymmetrical ketones give a mixture of two products and second, an epoxide is formed as a by-product and in some cases as the main product:

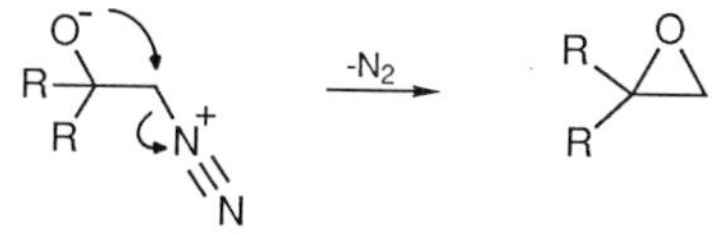

Nevertheless, the reaction in some cases gives practicable yields, in one step, of difficultly accessible compounds, e.g. cyclohexanone gives a 33–36% yield of cycloheptanone [4]. The diazomethane may be generated *in situ* by treating *N*-methyl-*N*-nitrosotoluene-*p*-sulfonamide with base.* A special example is the reaction of diazomethane with ketene to give cyclopropanone:

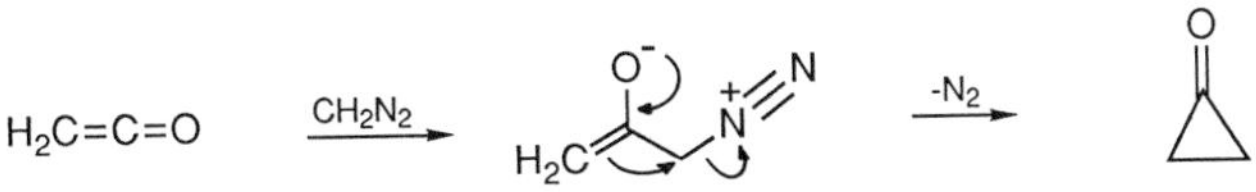

The reaction must be carried out at very low temperatures because of the high reactivity of cyclopropanone. An excess of ketene must also be used, for otherwise the cyclopropanone reacts with diazomethane to form cyclobutanone.

2. *The reaction of diazomethane with an acid chloride gives a diazoketone*[†] which, on being heated in the presence of silver oxide, undergoes the *Wolff rearrangement* to give a ketene:

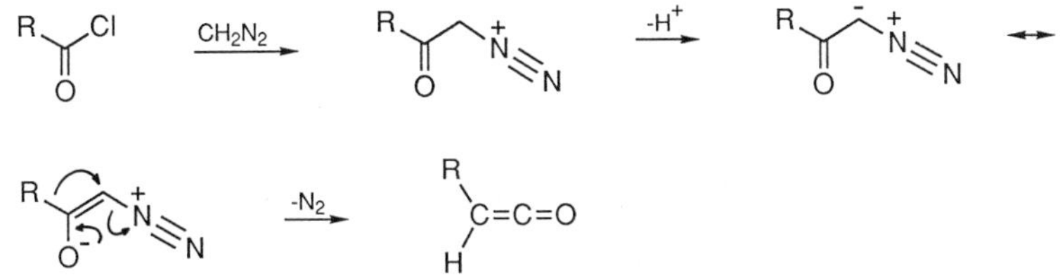

When the rearrangement is carried out in the presence of water or an alcohol, the ketene is converted directly into an acid or ester:

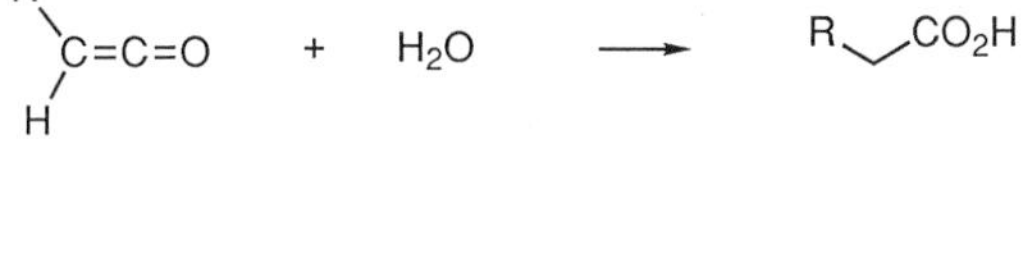

* A probable reaction path is:

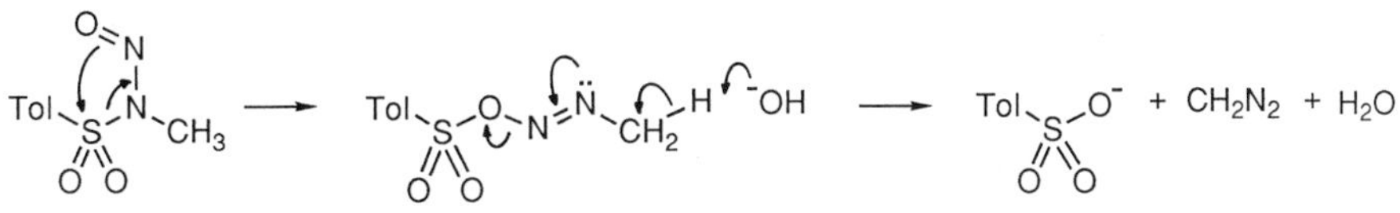

[†] The acid chloride is added to an excess of diazomethane to minimize the formation of the alternative product, $RCOCH_2Cl$.

$$\underset{H}{\overset{R}{\diagdown}}C=C=O \quad + \quad R'OH \quad \longrightarrow \quad R\diagup\diagdown CO_2R'$$

The overall process (*Arndt–Eistert* synthesis) provides a method for the conversion of an acid RCO_2H into the homologue RCH_2CO_2H in three stages. Total yields are normally good (*c.* 50–80%).

(*v*) *Rearrangement of alkanes.*   Saturated carbon chains undergo skeletal rearrangements when treated with a Lewis acid in the presence of a catalytic quantity of an organic halide. The products are equilibrium mixtures of all the possible isomeric compounds, e.g.

$$\diagup\diagdown\diagup \quad \underset{150\,^{\circ}C}{\overset{AlCl_3\ (RCl)}{\rightleftarrows}} \quad \diagdown\!\!\diagup\!\!\diagdown$$

ca. 20%   ca. 80%

The rearrangement occurs by way of a carbocation, formed from the halide, which abstracts hydride ion from the alkane:

$$RCl \quad + \quad AlCl_3 \quad \rightleftarrows \quad R^+ \quad + \quad AlCl_4^-$$

$$\overset{R^+}{\underset{H}{\diagup\!\!\diagdown}} \quad \underset{-RH}{\rightleftarrows} \quad \overset{+}{\diagdown}\!\!\diagup\!\!\diagdown Me \quad \rightleftarrows \quad \overset{}{\diagdown}\!\!\underset{+}{\diagup}\!\!\diagdown \quad \underset{RH}{\rightleftarrows} \quad \diagdown\!\!\diagup\!\!\diagdown$$

It is to be noted that the predominant product may be derived from the least stable carbocation: in the above example, isobutane is formed following the rearrangement of a secondary to a primary carbocation. This is the converse of the direction of rearrangement in the reactions previously discussed. The underlying basis is that whereas the latter reactions are kinetically controlled (the carbocation formed by rearrangement reacts almost immediately with a nucleophile), the former are thermodynamically controlled. Virtually no nucleophile is present, for only a trace of organic halide is added so that the concentration of nucleophilic $AlCl_4^-$ is negligible, and the reactions are freely reversible. Therefore the relative proportions of the alkanes alter gradually until they reach values determined by the relative free energies of the compounds.

Since the relative free energies of isomeric acyclic hydrocarbons differ little, a complex mixture of products is likely to be obtained by this process. Amongst alicyclic compounds, however, there are greater differences in relative free energies because of the occurrence of strain in ring systems. For example, the isomerization of methylcyclopentane gives cyclohexane but no ethyl- or dimethylcyclobutane.

$$\diagup\!\!\triangle\!\!\diagdown \quad \overset{AlCl_3\ (RCl)}{\rightleftarrows} \quad \bigcirc$$

12.5%   87.5%   (25 $^{\circ}$C)

A novel application of this rearrangement is the formation of adamantane in 15% yield by treatment of the reduction product of the readily available dimer of cyclopentadiene (p. 273) with aluminium trichloride at 150–180°C:

adamantane

Adamantane is the most stable of the saturated hydrocarbons of molecular formula $C_{10}H_{16}$.

(b)   Halogen, oxygen, sulfur, and nitrogen migration

An atom, X, with an unshared pair of electrons, in the system X—C—C—Y, can assist the heterolysis of the C—Y bond in the same way as a phenyl group:

In a symmetrical case, such as the solvolysis of Et—S—CH₂CH₂Cl, no rearrangement occurs because nucleophilic attack at either carbon atom of the bridged ion leads to the same product as would be formed in the absence of neighbouring-group participation. In unsymmetrical cases, however, nucleophilic attack at the less highly substituted carbon of the bridged ion predominates (cf. the opening of epoxides, p. 590), and a rearranged skeleton can result:

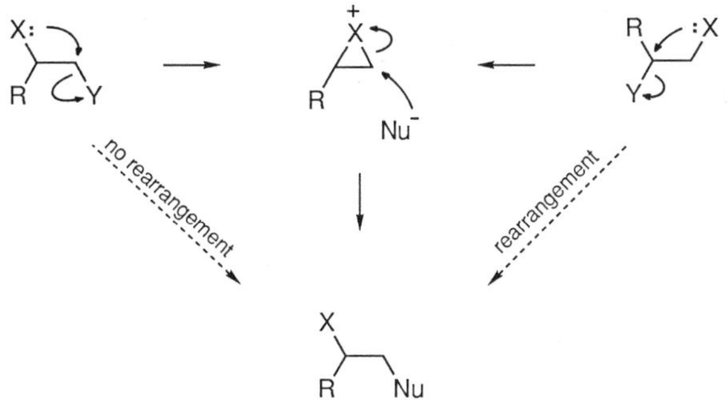

The following are typical examples:

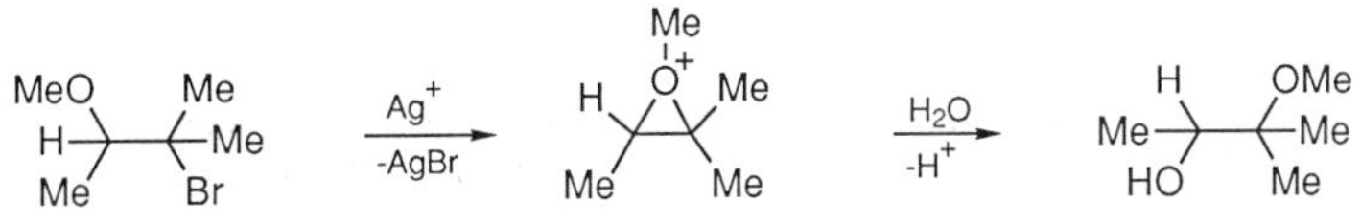

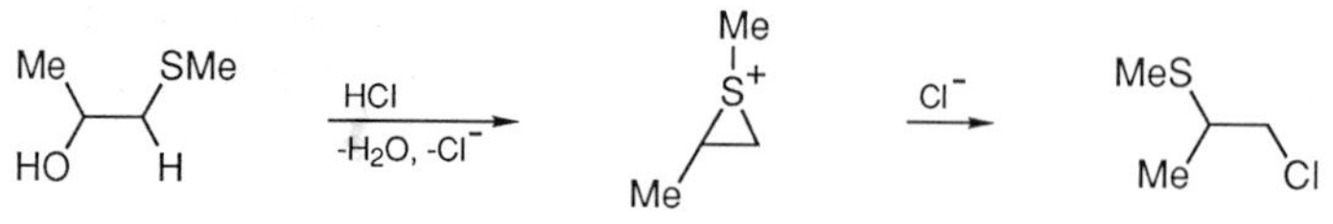

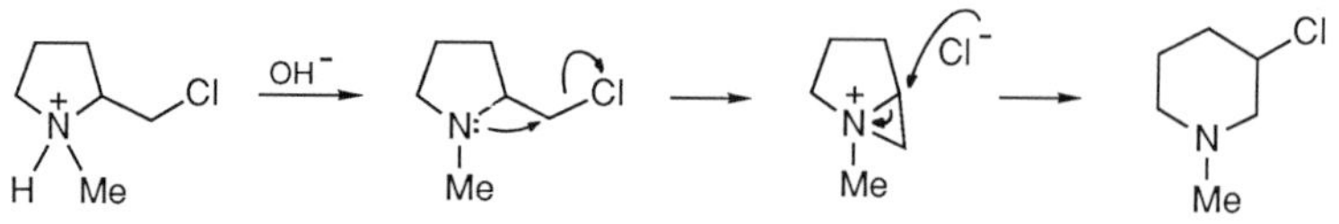

The bridged cation may be generated by protonation of an unsaturated bond as in the *Rupe rearrangement* of α-acetylenic alcohols, e.g.

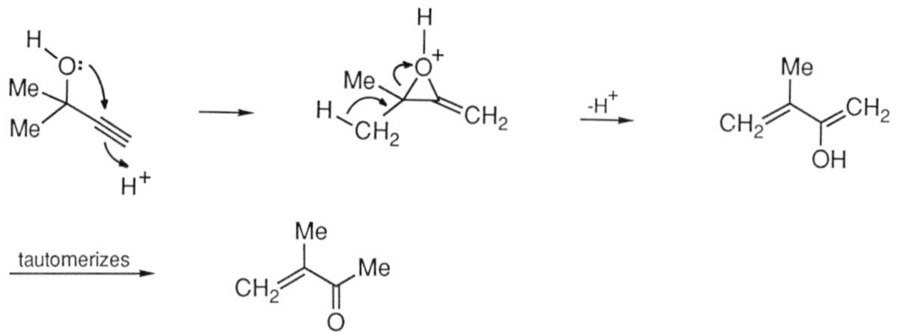

A neighbouring acetoxy group assists solvolysis by forming a *five*-membered acetoxonium ion:

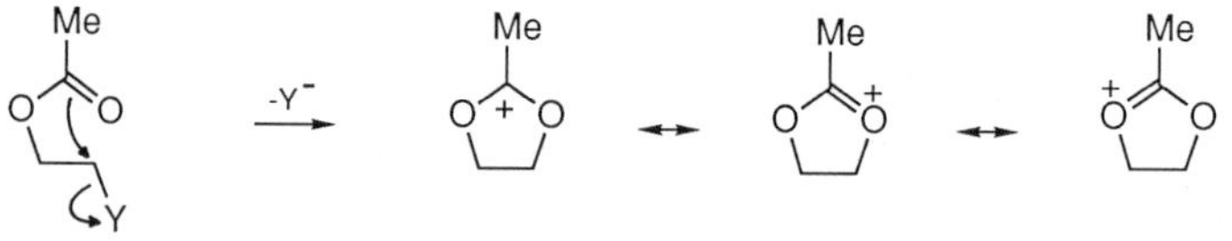

The cyclic ion is then opened by reaction with a nucleophile. Reaction with water occurs at the acetoxy-carbon atom to yield ultimately a *cis*-hydroxyacetate, whereas reaction with acetate ion occurs at alkyl carbon in the S_N2 manner to yield a *trans*-diacetate*:

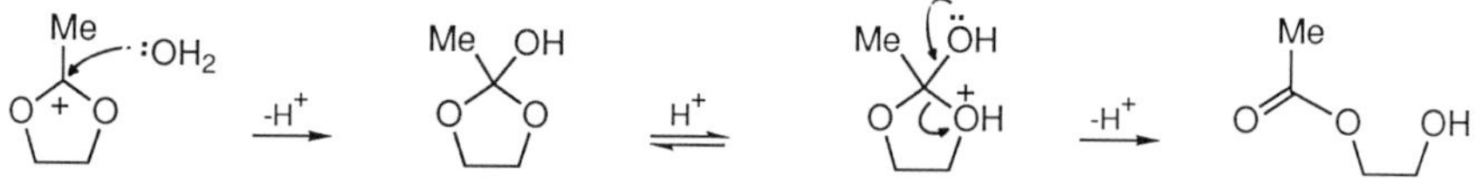

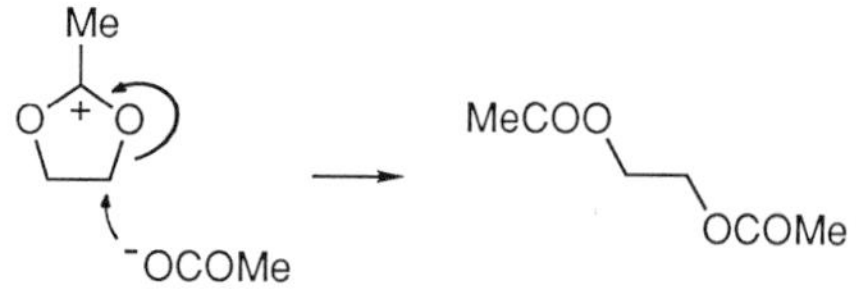

---

* It is probable that acetate ion reacts faster at the cationic centre, but reversibly, whereas reaction of water there is rendered irreversible by the subsequent proton loss and bond cleavage.

Procedures for making *cis*- and *trans*-diols are based on the principle of acetoxonium-ion participation (p. 592).

Acetoxonium-ion participation can affect both rates and stereochemistry. For example, *trans*-2-acetoxycyclohexyl tosylate solvolyzes in acetic acid about one-thousand times faster than its *cis*-isomer because its acetoxy group is ideally placed in the diaxial conformer to assist the departure of the tosylate: reaction of the cyclic ion with acetic acid gives product with retained stereochemistry:

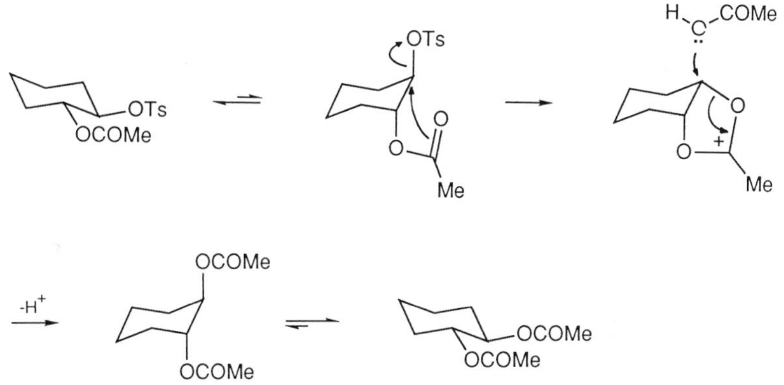

Solvolysis of the *cis*-compound is not kinetically assisted, but after departure of the tosylate the same acetoxonium ion can be formed and the product is therefore again the *trans*-diacetate.

14.3 Rearrangement to electron-deficient nitrogen

(a)   The Hofmann, Curtius, Schmidt, and Lossen rearrangements

There is a group of closely related rearrangements in which carbon migrates from carbon to nitrogen. They may be formulated generally as,

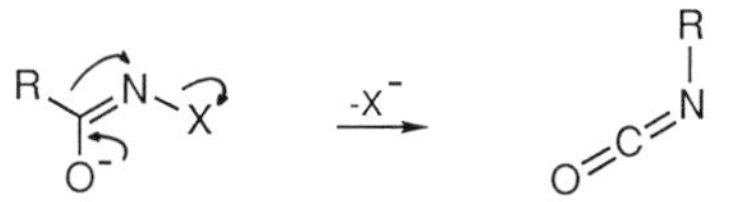

where R is an alkyl or aryl group and —X is a leaving group which may be —Br (*Hofmann rearrangement*), —$\overset{+}{N}$≡N (*Curtius and Schmidt rearrangements*), and —OCOR (*Lossen rearrangement*). In each case, if the alkyl carbon which migrates is asymmetric, it retains its configuration.

(i)   Hofmann rearrangement.   A carboxylic acid amide is treated with sodium hypobromite, or bromine in alkali. The *N*-bromo-amide is formed and reacts with

the base to give its conjugate base within which rearrangement occurs. The isocyanate which is produced may be isolated in anhydrous conditions, but reaction is normally carried out in aqueous or alcoholic solution in which the isocyanate is converted into an amine or a urethane, respectively.

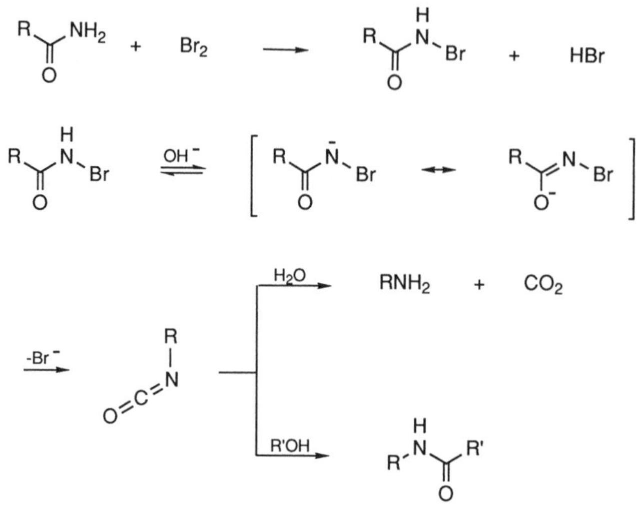

This rearrangement provides an efficient route for making both aliphatic and aromatic primary amines. For example, β-alanine can be obtained in about 45% yield by treating succinimide with bromine and aqueous potassium hydroxide [2]; reaction occurs through the half-amide of succinic acid.

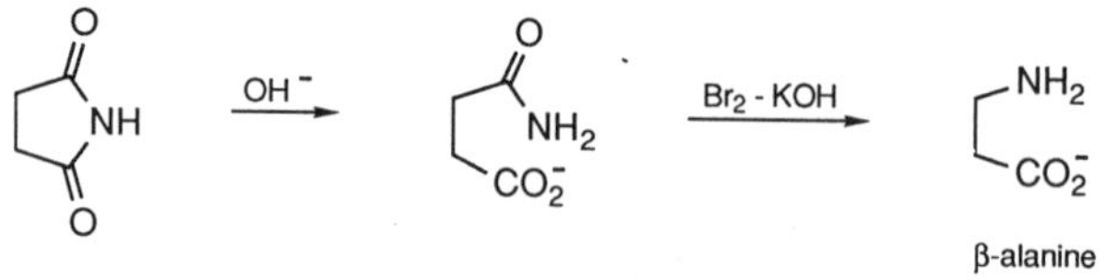

β-alanine

Anthranilic acid (c. 85%) may be obtained in a similar way from phthalimide:

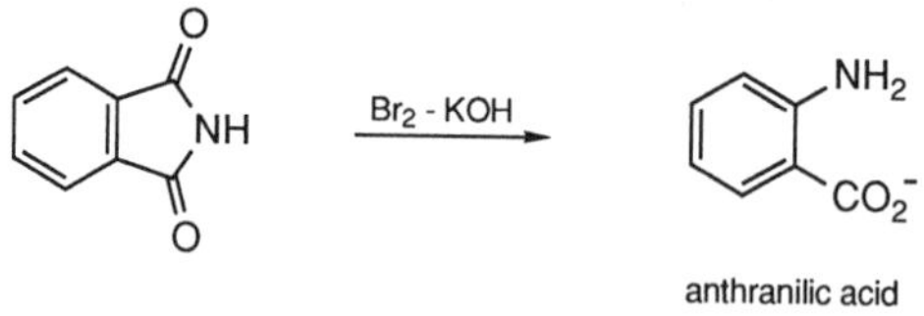

anthranilic acid

A useful example of the reaction is the preparation of β-aminopyridine (65–70%) [4] from nicotinamide (available from natural sources), for this cannot be obtained in good yield via the nitration of pyridine (p. 379).

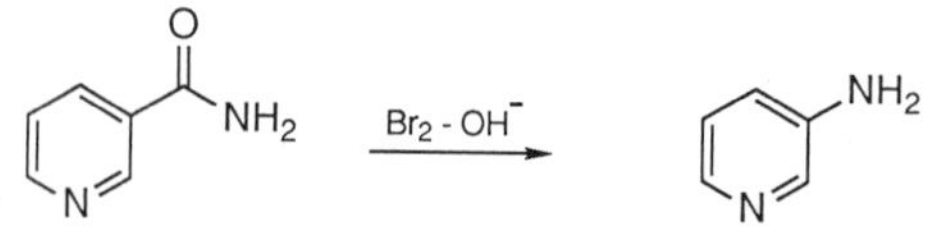

(ii) Curtius rearrangement. Acid azides* decompose on being heated to give isocyanates:

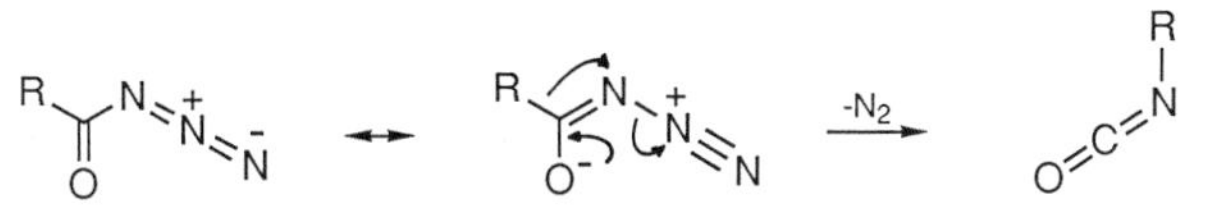

The isocyanate may be isolated by carrying out the reaction in an aprotic solvent such as chloroform but it is customary to use an alcoholic solvent with which the isocyanate reacts to form a urethane. Acid hydrolysis gives the corresponding amine.

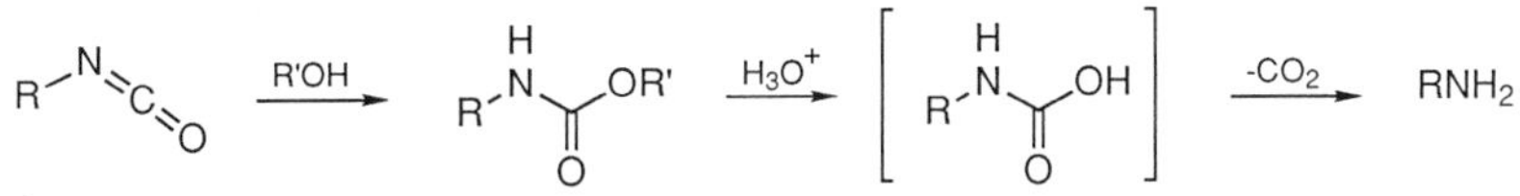

The Curtius rearrangement has been applied to the synthesis of α-amino acids, e.g.

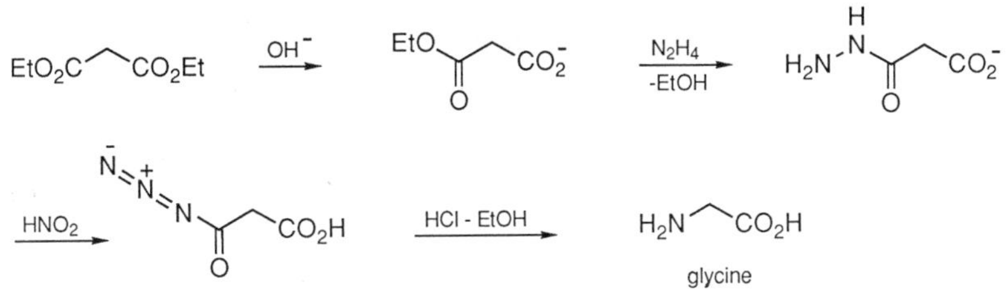

(iii) Schmidt reaction. Carboxylic acids react with hydrazoic acid in the presence of concentrated sulfuric acid to give isocyanates directly. Reaction occurs through the acid azide, but in the strongly acid conditions this is present as its conjugate acid from which nitrogen is lost without heating.

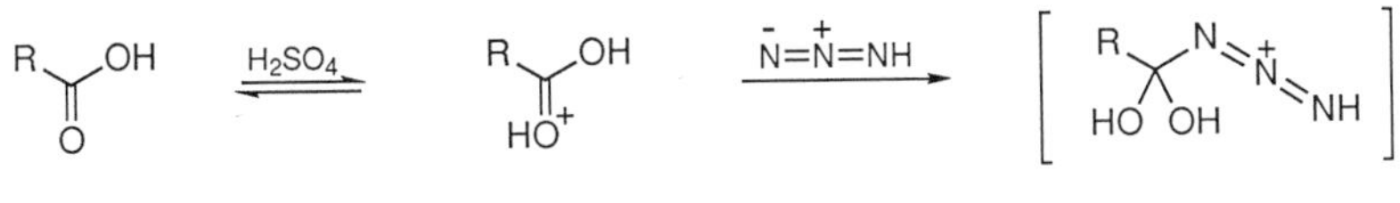

* Acid azides are readily prepared from acid chlorides by treatment with sodium azide,

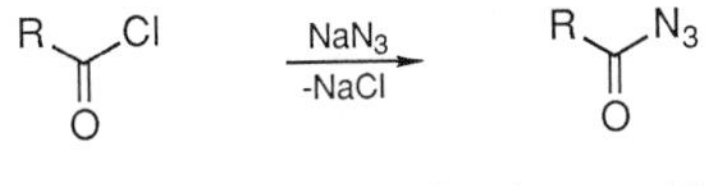

and from esters by treatment with hydrazine followed by nitrous acid,

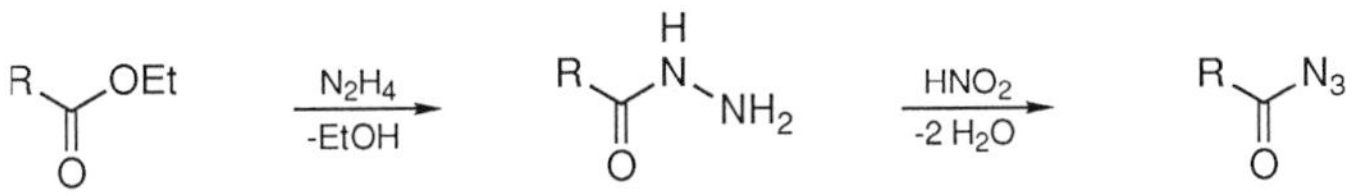

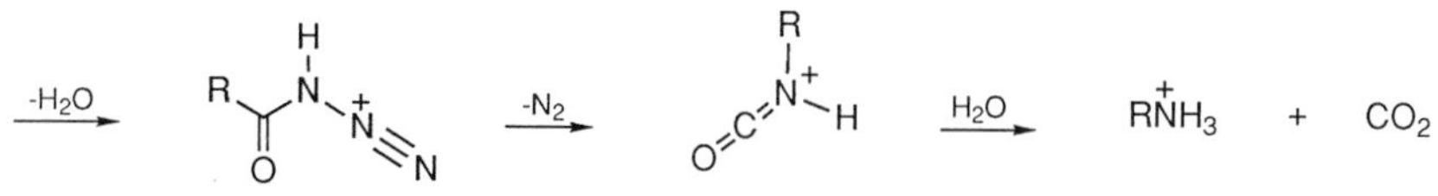

$$\xrightarrow{-H_2O} \quad \text{[structure]} \quad \xrightarrow{-N_2} \quad \text{[structure]} \quad \xrightarrow{H_2O} \quad R\overset{+}{N}H_3 \quad + \quad CO_2$$

(*iv*) *Lossen rearrangement.* This differs from the Hofmann rearrangement only in that the leaving group is a carboxylate anion rather than bromide ion. The starting material is the ester of a hydroxamic acid:

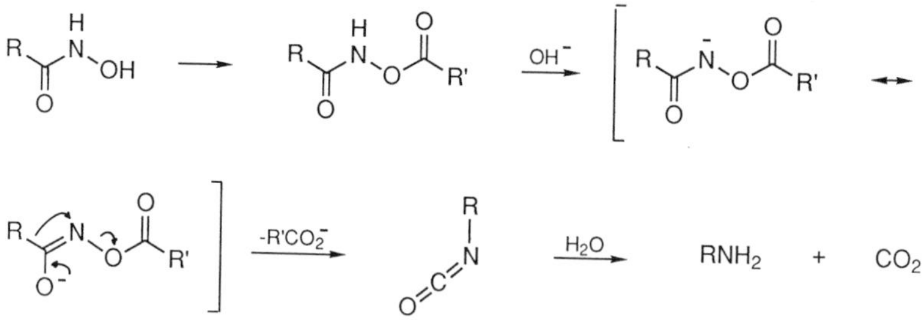

Of these four related processes the Lossen rearrangement is the least useful in the synthesis of amines because hydroxamic acids are not readily available. The Schmidt reaction is the most direct method but is only applicable if the acid does not contain groups which are sensitive to concentrated sulfuric acid. The Curtius rearrangement involves the mildest conditions but requires the preparation of the azide. The Hofmann rearrangement is convenient providing that other functional groups in the molecule do not react with bromine and alkali.

(*b*) *The Beckmann rearrangement*

Oximes undergo a rearrangement in acidic conditions to give substituted amides:

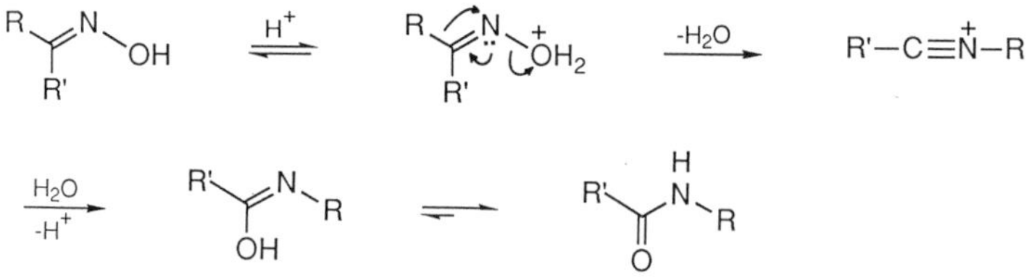

The Beckmann rearrangement is stereospecific: the group *trans* to the leaving group migrates. For example, acetophenone oxime, which has the stereochemistry shown, gives only acetanilide.

As in other intramolecular rearrangements, if the carbon atom which migrates is asymmetric, it retains its configuration during the reaction.

The rearrangement is also induced by reagents other than proton acids. Boron trifluoride removes the hydroxyl group as $HO-BF_3$, toluene-*p*-sulfonyl chloride forms the oxime tosylate which eliminates the stable tosylate anion,

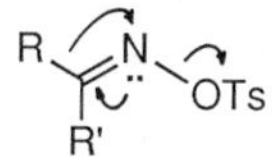

and phosphorus pentachloride, normally used in ether, induces rearrangement by providing a phosphate as leaving group,

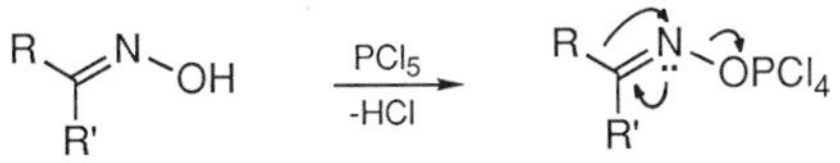

An interesting application of the rearrangement is the synthesis of caprolactam (70%) from cyclohexanone oxime and concentrated sulfuric acid [2], a ring expansion of analogous type to the formation of cycloheptanone from cyclohexanone (p. 435). Caprolactam gives a polymer of the nylon group when heated:

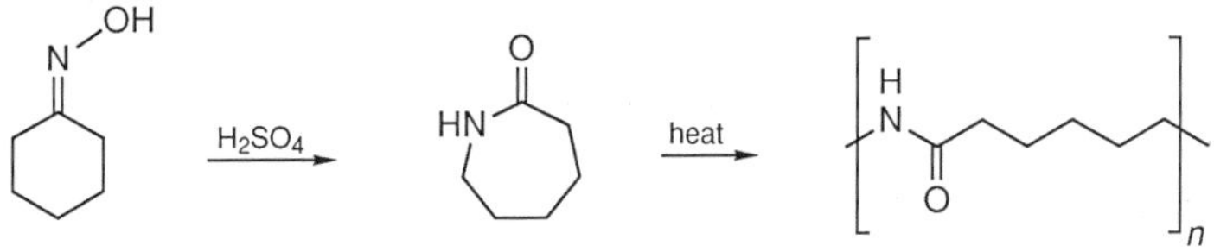

caprolactam

14.4
Rearrangement to electron-deficient oxygen

The most general rearrangement of this type is the *Baeyer–Villiger reaction* in which ketones are converted into esters and cyclic ketones into lactones, by treatment with a peroxyacid. The mechanism is closely related to that of the pinacol rearrangement: nucleophilic attack by the peroxyacid on the carbonyl group gives an intermediate which rearranges with the expulsion of the anion of an acid.

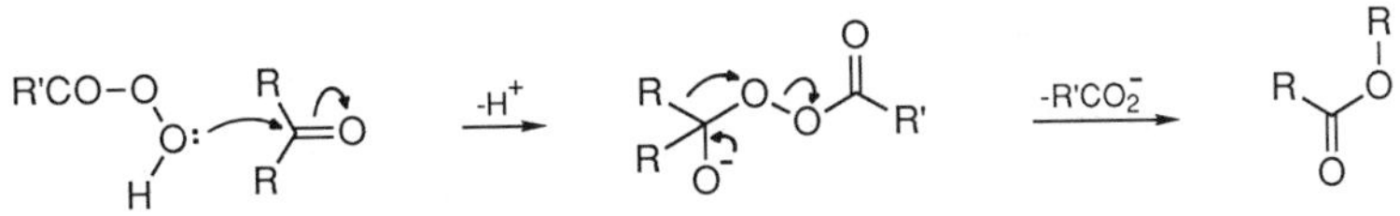

Acids catalyze the reaction by facilitating both the addition to carbonyl and the expulsion of the carboxylate.

A number of peroxyacids, including peroxyacetic, monoperoxyphthalic, and trifluoroperoxyacetic acid, have been successfully employed in the reaction. Trifluoroperoxyacetic acid is the most reactive of the peroxyacids, probably because the trifluoroacetate ion, derived from a strong acid, is a very good leaving group; it is necessary to buffer the solution for otherwise transesterification occurs to give the trifluoroacetate ester.

In an unsymmetrical ketone, that group migrates which is the better able to supply electrons, as in the Wagner–Meerwein and related rearrangements. Amongst alkyl groups, the ease of migration is tertiary > secondary > primary > methyl, e.g. pinacolone gives t-butyl acetate.

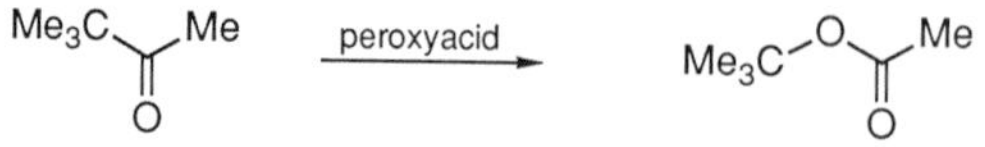

Amongst aryl groups, the order is, p-methoxyphenyl > p-tolyl > phenyl > p-chlorophenyl, etc. (cf. p. 432). Aryl groups migrate in preference to primary alkyl groups, e.g.

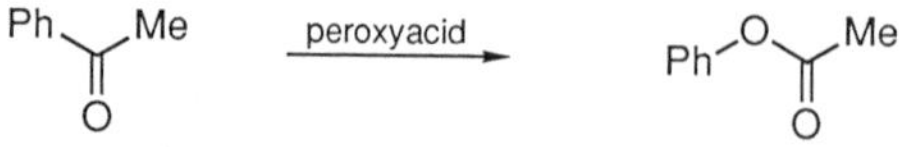

but only if they contain electron-releasing substituents do they migrate in preference to secondary and tertiary alkyl groups.

Cyclic ketones undergo ring expansion with peroxyacids. For example, cyclohexanone gives caprolactone (cf. the formation from cyclohexanone of cycloheptanone, p. 435, and caprolactam, p. 445):

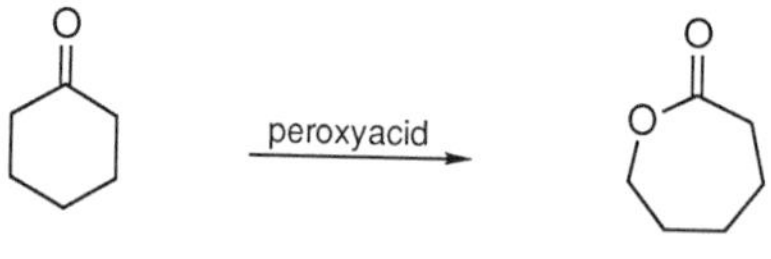

The lactone is hydrolyzed under the reaction conditions. In an aqueous medium, 6-hydroxycaproic acid is formed and undergoes condensation polymerization. Polymerization is prevented by carrying out the reaction in ethanol; ethyl 6-hydroxycaproate is then obtained.

The acid-catalyzed rearrangement of tertiary hydroperoxides is similar to the Baeyer–Villiger reaction. The product from the migration is a hemiacetal which is hydrolyzed in the reaction conditions (p. 86), e.g.

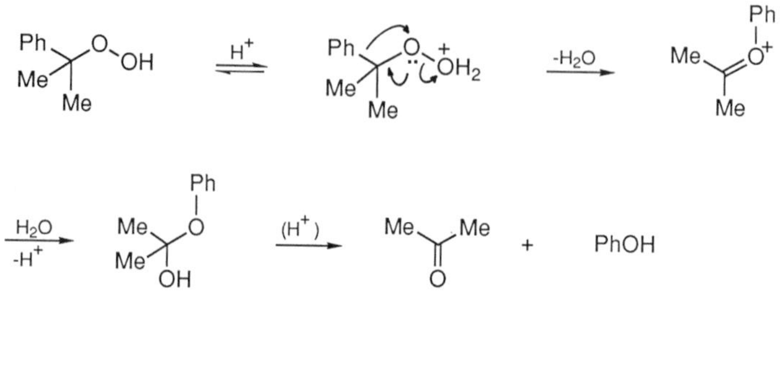

The above example is of industrial importance, for cumene is cheaply available from the Friedel–Crafts alkylation of benzene with propylene and, like other tertiary hydrocarbons, it readily forms the hydroperoxide by autoxidation (p. 557).

(*i*) *Dakin reaction.* Benzaldehydes containing *ortho*- and *para*-hydroxyl groups are converted into catechols and quinols, respectively, by alkaline hydrogen peroxide. For example, catechol itself may be obtained in 70% yield from salicylaldehyde [1].

The mechanism is similar to that of the Baeyer-Villiger reaction:*

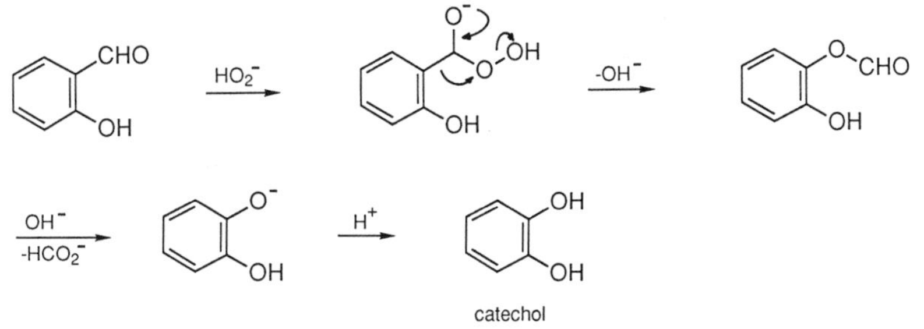

catechol

**14.5
Rearrangement
to electron-rich
carbon**

This group of rearrangements has been less extensively studied, and is of less synthetic importance, than rearrangements to electron-deficient carbon. The known examples are of the type,

$$(X = N^+, S^+, O)$$

in which the group R migrates from X to C.

(*i*) *Stevens rearrangement.* Quaternary ammonium ions which contain β-hydrogen atoms undergo E2 (Hofmann) elimination with base, e.g.

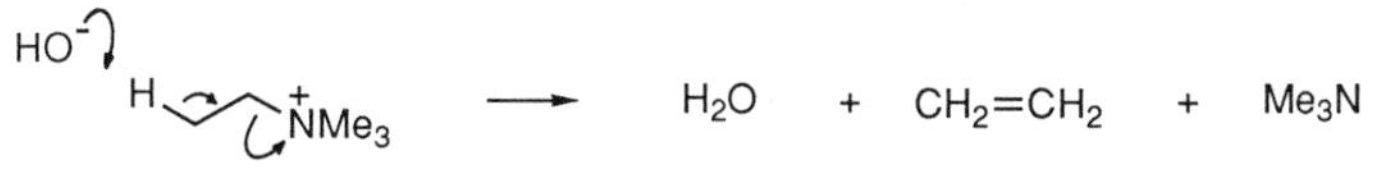

$$H_2O + CH_2{=}CH_2 + Me_3N$$

---

* The hydroxide ion is not a good leaving group and the function of the aromatic hydroxyl substituent may be to provide powerful anchimeric assistance (*o*-hydroxyphenyl ~ *p*-hydroxyphenyl ≫ phenyl) to effect heterolysis of the O—O bond.

If, however, none of the alkyl groups possesses a β-hydrogen atom but one has a β-carbonyl group, an α-hydrogen is removed by base to give an *ylide* (a species in which adjacent atoms bear formal opposite charges). The role of the carbonyl group is to assist formation of the ylide by stabilizing the negative charge, e.g.

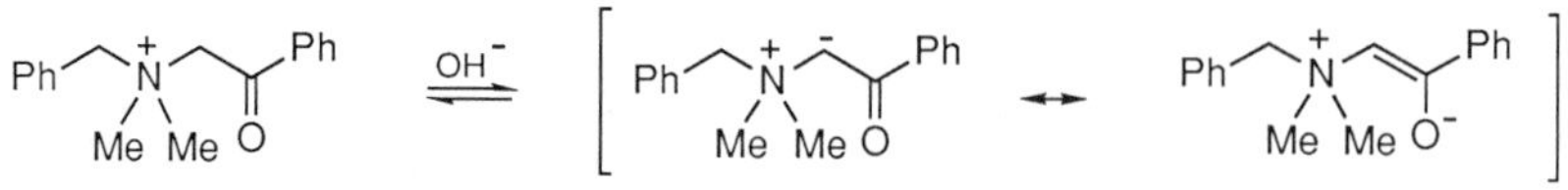

Rearrangement then occurs:

Sulfonium salts behave analogously (p. 470).

It used to be thought that these rearrangements occurred in a concerted manner, represented by:

However, the recognition that this would not be an allowed reaction according to orbital-symmetry rules (p. 293) prompted further study which has shown that reaction probably occurs *via* formation and combination of radical-pairs:

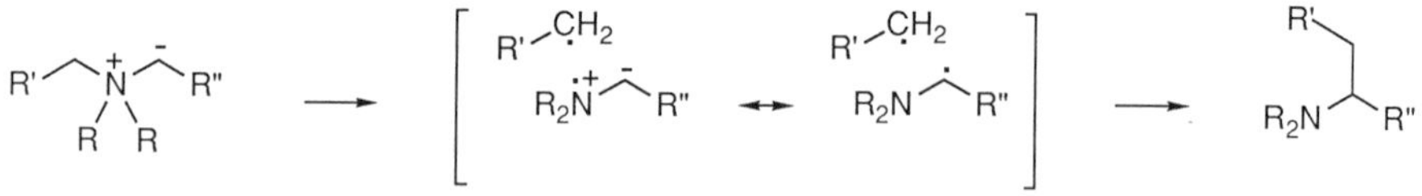

In the absence of a β-carbonyl group the α-hydrogen is too weakly acidic for rearrangement to be induced by hydroxide ion. A stronger base, such as amide ion in liquid ammonia, is effective, but the rearrangement takes a different course: instead of the [1,2]-shift, a [3,2]-sigmatropic rearrangement (*Sommelet rearrangement*) occurs:*

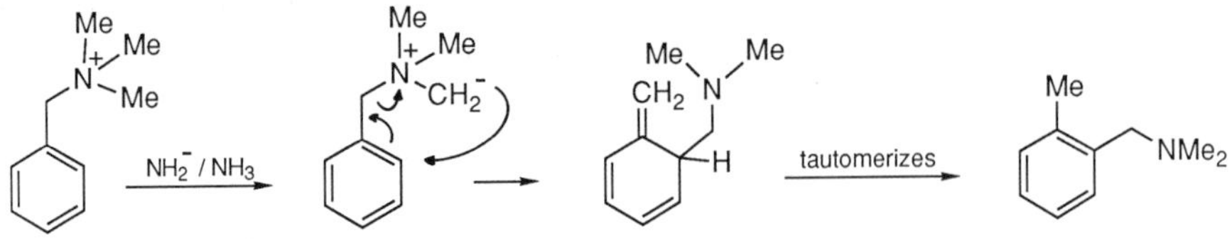

---

*The benzylic protons are more acidic than the methyl protons because the negative charge of the ylide is delocalized over the benzene ring. However, formation of this ylide is not followed by rearrangement.

High yields can be obtained by this procedure; in the example above the yield is over 90% [4].

(*ii*) *Wittig rearrangement.* Benzyl and allyl ethers undergo a base-catalyzed rearrangement analogous to the Stevens rearrangement. A benzylic or allylic carbanion* is generated by the action of a powerful base such as amide ion or phenyllithium and migration of carbon then leads to the more stable oxyanion, e.g.

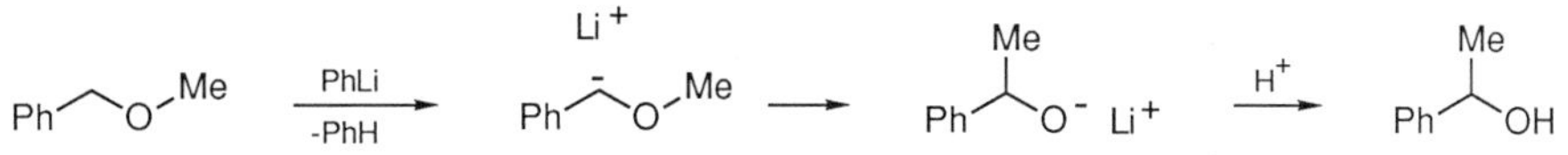

As with the Stevens rearrangement, there is evidence for a radical-pair mechanism:

(*iii*) *Favorskii rearrangement.* α-Halo-ketones react with base to give enolates which rearrange to esters *via* cyclopropanones:

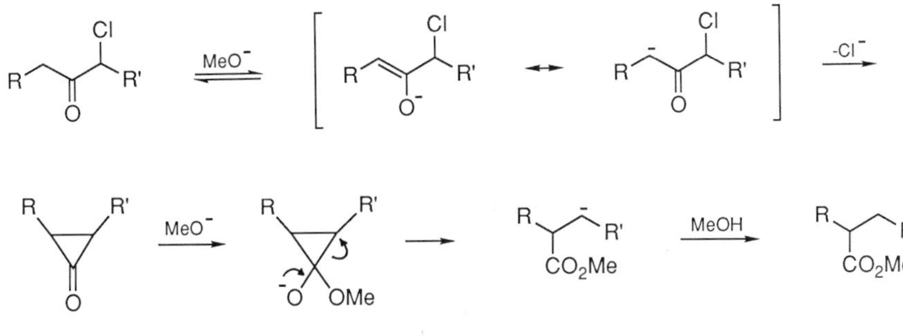

The direction of ring opening of the cyclopropanone is determined by which is the more stable of the two possible carbanions that can be formed. Alkyl groups destabilize carbanions, whereas aryl groups stabilize them by delocalization of the negative charge. For example, $PhCH_2$—$COCH_2Cl$ and $PhCHCl$—$COMe$ give the same product:

---

* The negative charge is delocalized over the benzylic or allylic system. Alkyl ethers are not sufficiently acidic to react.

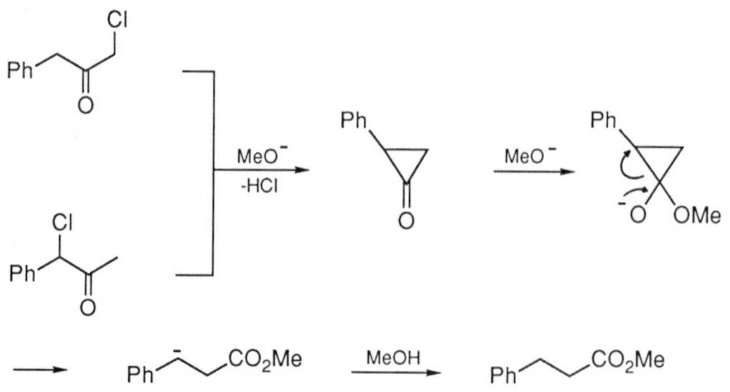

The rearrangement can be employed for bringing about ring contraction in cyclic systems, e.g. 2-chlorocyclohexanone and methoxide ion give methyl cyclo-pentanecarboxylate in 60% yield [4]:

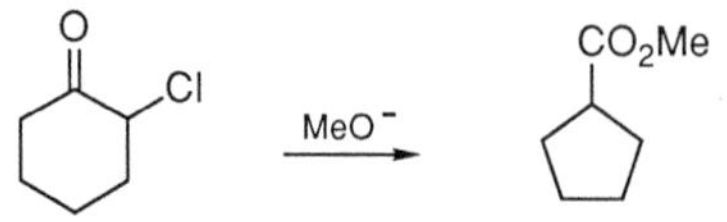

An interesting example is the conversion of the cyclobutyl into the cyclopropyl system, for this must apparently involve the highly strained bicyclobutane system:

14.6 Aromatic rearrangements

A number of rearrangements occur in aromatic compounds of the type:

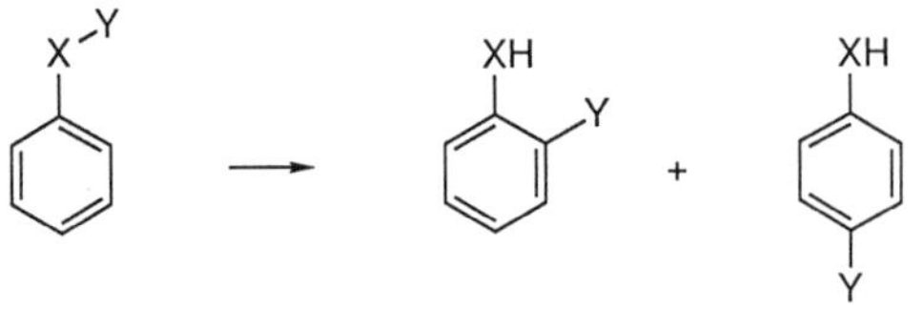

The element X is most commonly nitrogen and in some cases oxygen. Both intermolecular and intramolecular migrations are known.

(a) Intermolecular migration from nitrogen to carbon

Several derivatives of aniline undergo rearrangement on treatment with acid. In the following examples, the conjugate acid of the amine eliminates an electrophilic species which then reacts at the activated *ortho* and *para* positions of the amine.

(*i*) N-*Haloanilides*. For example, N-chloroacetanilide and hydrochloric acid give a mixture of o- and p-chloroacetanilde in the same proportions as in the direct chlorination of acetanilide:

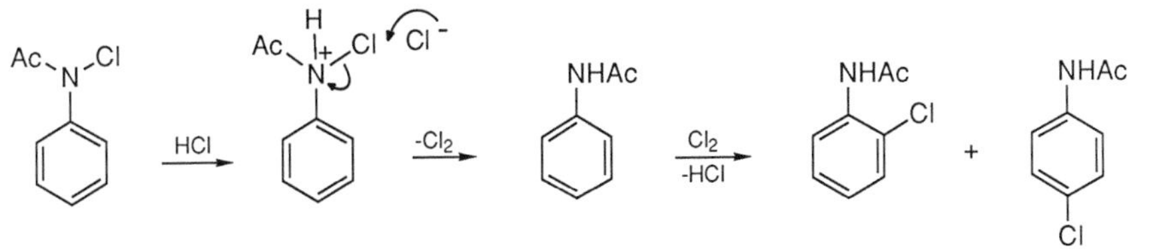

(*ii*) N-*Alkyl-N-nitrosoanilines*. The nitrosonium ion is released from the conjugate acid of the amine and nitrosates the nuclear carbon atoms, giving mainly the p-nitroso product:

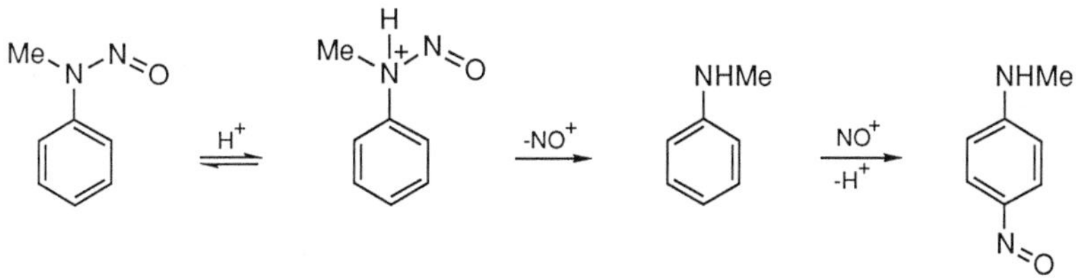

(*iii*) N-*Arylazoanilines*. The aryldiazonium cation is formed and couples, essentially completely, at the *para* carbon atom (p. 420).

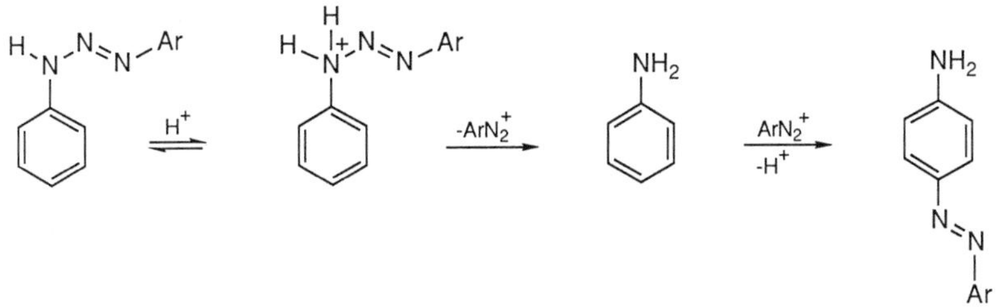

(*iv*) N-*Alkylanilines*. The mechanism of rearrangement of the salts of these amines is the same as those above, although higher temperatures (250–300°C) are required, e.g.

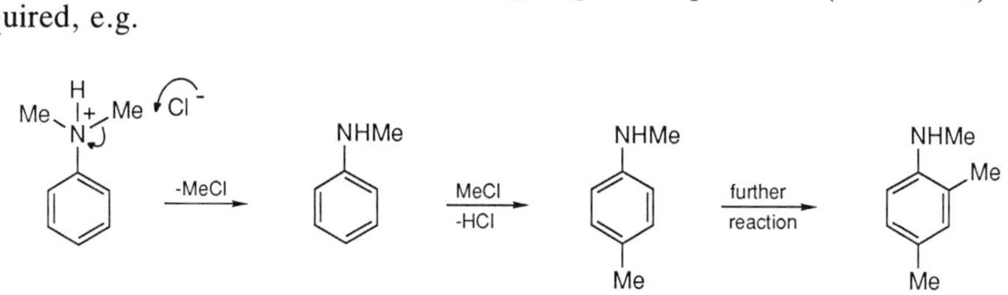

(v) N-*Arylhydroxylamines.* The rearrangement of arylhydroxylamines to aminophenols is mechanistically different: the conjugate acid of the hydroxylamine undergoes *nucleophilic* attack by the solvent, e.g.

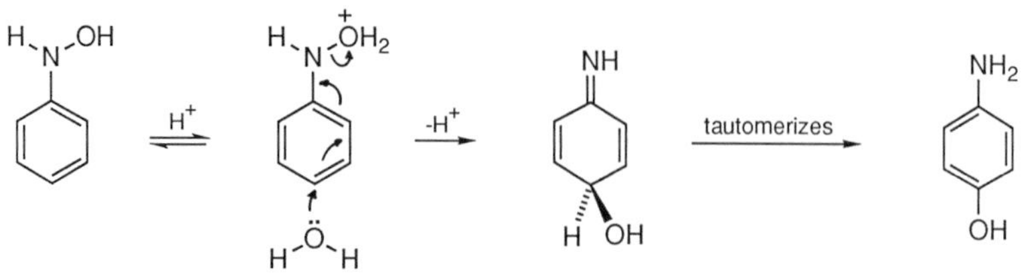

When an alcohol is used as solvent, the corresponding *p*-alkoxy compound is formed.

This rearrangement is useful in synthesis, especially since the hydroxylamine, normally made by reduction of the corresponding nitro compound, need not be isolated. For example, the electrolytic reduction of nitro compounds in the presence of sulfuric acid leads directly to the aminophenol, e.g. *o*-chloronitrobenzene gives 2-chloro-4-hydroxyaniline [4]:

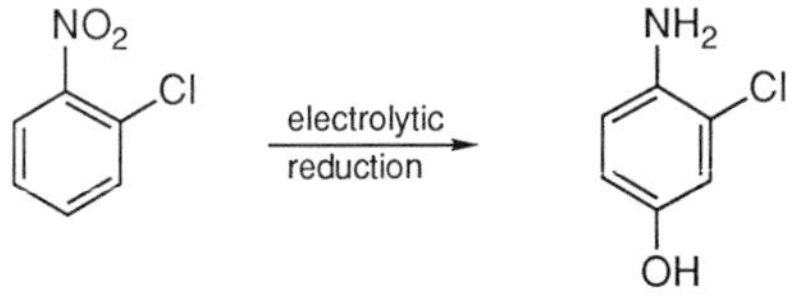

(b)   *Intermolecular migration from oxygen to carbon*

The only common rearrangement of this type is the *Fries rearrangement* in which aryl esters are treated with Lewis acids to give *ortho* and *para* hydroxy-ketones. The complex between the ester and the Lewis acid eliminates an acylium ion which substitutes at the *ortho* and *para* positions, as in Friedel–Crafts acylation (p. 365), e.g.

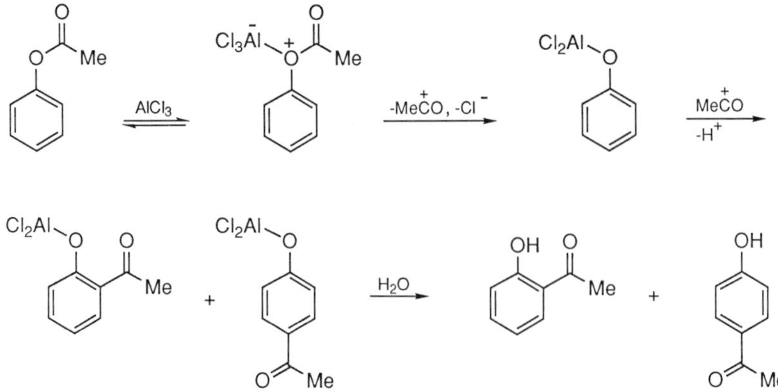

For example, the treatment of phenyl propionate with aluminium chloride at 130°C gives about 35% of *o*-propiophenol and 45–50% of the *para*-isomer [2].

In general, low temperatures favour formation of the *para*-substituted product and high temperatures favour the *ortho*-substituted product. This appears to be because the *para*-derivative is formed the faster (kinetic control) but the *ortho*-derivative is thermodynamically the more stable by virtue of possessing a chelate system:

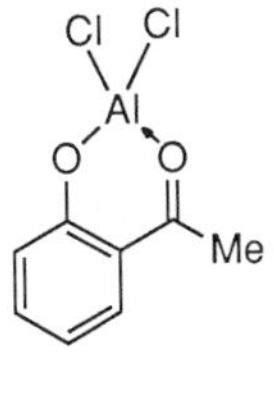

(c)   Intramolecular migration from nitrogen to carbon

The mechanisms of these reactions are not fully understood, although it is known in each case that the migrating group does not become detached from the aromatic system during rearrangement.

(*i*)   *Phenylnitramines*.   These compounds rearrange on being heated with acid, giving mainly the *o*-nitro-derivative, e.g.

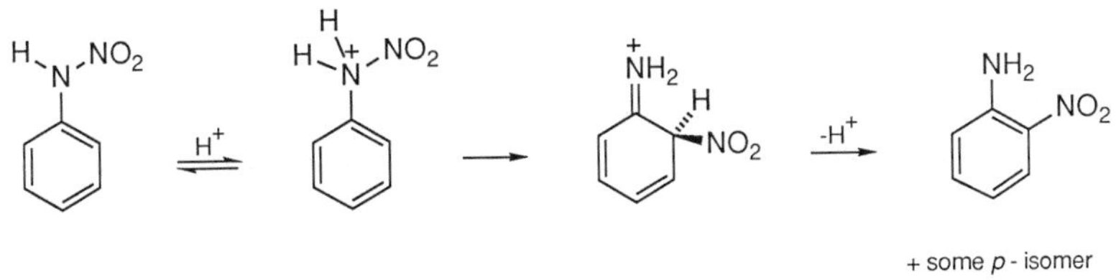

+ some *p* - isomer

(*ii*)   *Phenylsulfamic acids*.   These compounds rearrange on being heated, giving mainly the *o*-sulfonic acid derivatives which, at higher temperatures, are converted into the *p*-sulfonic acids, e.g.

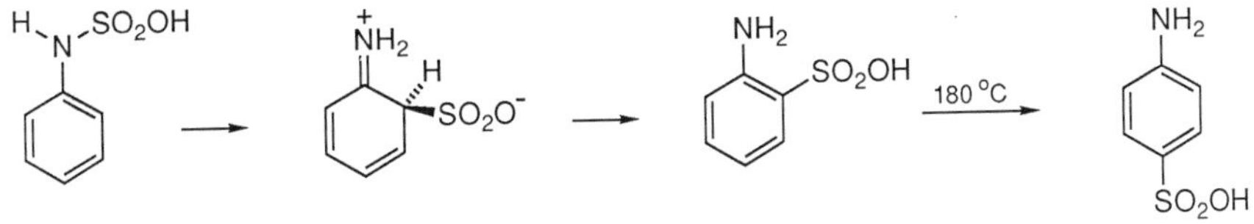

(*iii*)   *Hydrazobenzenes*.   These give benzidines on treatment with acid, e.g.

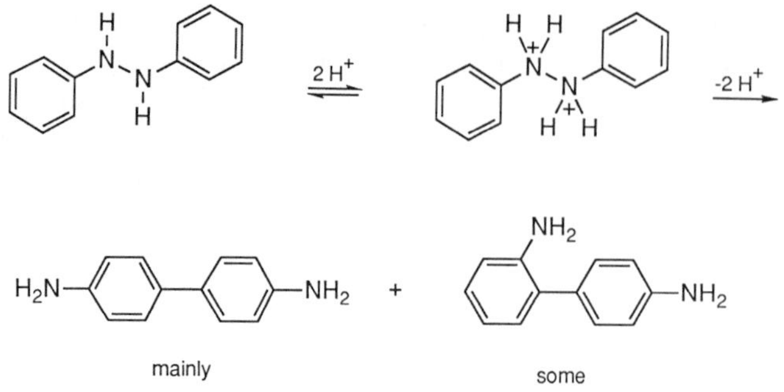

There is evidence that the 4,4'-compound is formed by a [5,5]-sigmatropic rearrangement:

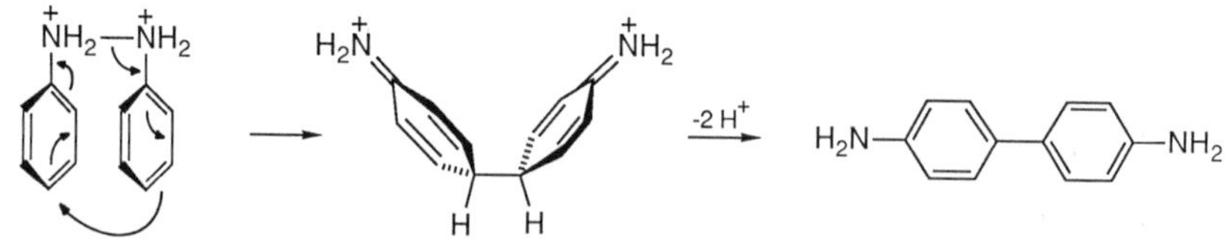

but the 2,4'-compound is formed by a different, unknown mechanism.

Since hydrazobenzenes are readily obtained by the reduction of aromatic nitro compounds (p. 666), this rearrangement gives access to 4,4'-disubstituted biphenyls.

(d) Intramolecular migration from oxygen to carbon

The Claisen rearrangement of aryl allyl ethers to allylphenols is a [3,3]-sigmatropic reaction, which was described earlier (p. 295). An example is:

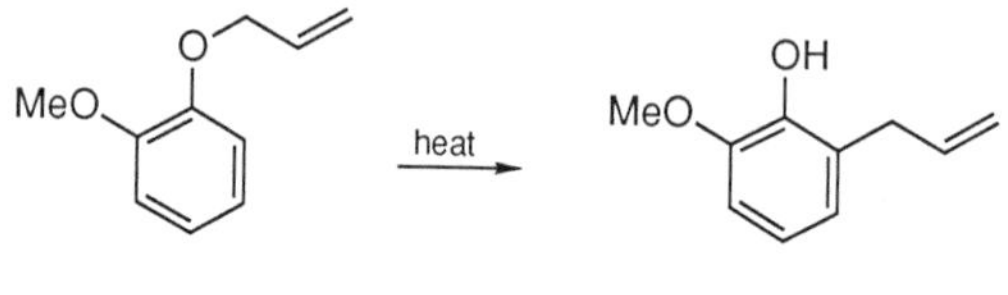

Problems   1.   What products would you expect from the following reactions?

(a)

(b)

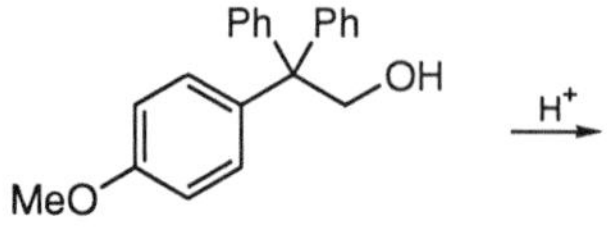

(c)

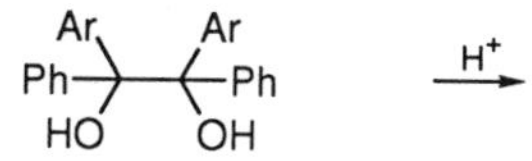

Ph—⟨Ar⟩(HO)—(Ar)(OH)—Ph $\xrightarrow{H^+}$

(Ar = p-methoxyphenyl)

(d)

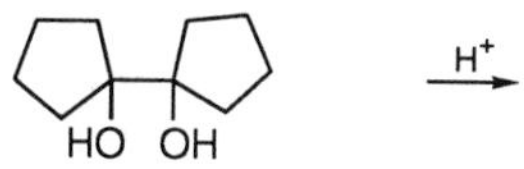

$\xrightarrow{H^+}$

(e)

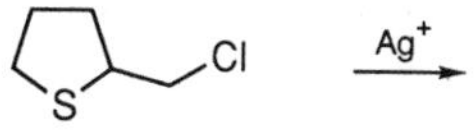

$\xrightarrow{Ag^+}$

(f)

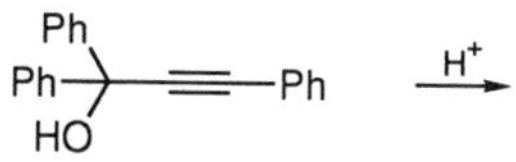

Ph—⟨Ph⟩(HO)—C≡C—Ph $\xrightarrow{H^+}$

(g)

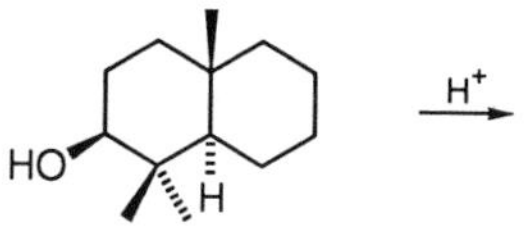

$\xrightarrow{H^+}$

(h)

$\xrightarrow{H^+}$

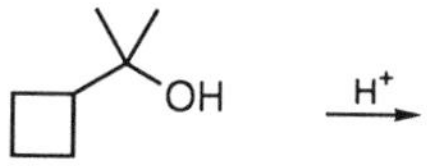

(i)

OHC—CHO $\xrightarrow{OH^-}$

(j)

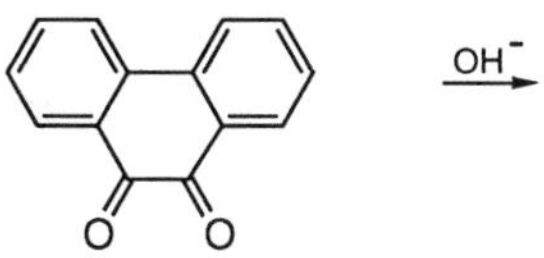

$\xrightarrow{OH^-}$

(k)

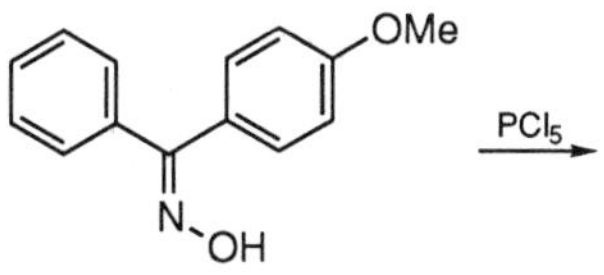

$\xrightarrow{PCl_5}$

(l)

Ph⟨⟩=O   + CH_2N_2 ⟶

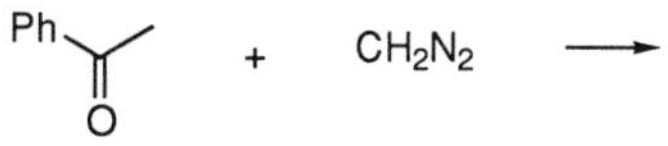

2. Account for the following:
    (i) *cis*- and *trans*-1,2-dimethylcyclohexane-1,2-diol give different
        products on treatment with concentrated sulfuric acid.

(*ii*) The dehydration of $MeCH(OH)CMe_3$ gives tetramethylethylene.

(*iii*)

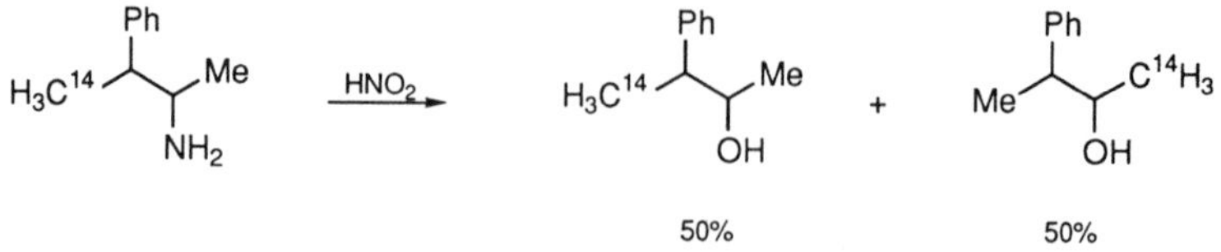

50%   50%

(*iv*) In the pinacol rearrangements of $PhMeC(OH)$—$C(OH)PhMe$ and $Ph_2C(OH)$—$C(OH)Me_2$, a phenyl group migrates in the former case but a methyl group migrates in the latter.

3. Outline routes involving rearrangement reactions to the following:

(*a*)

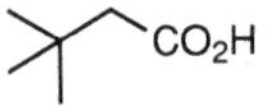

(*b*)

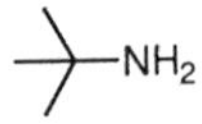

(*c*)

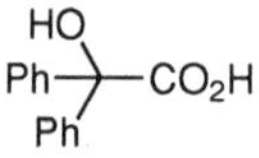

(*d*)

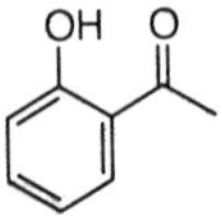

(*e*)

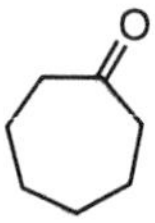

(*f*)

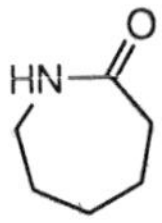

(*g*)

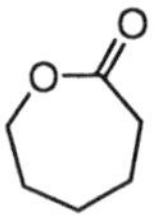

(*h*)

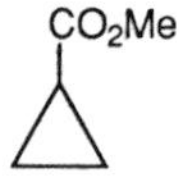

(*i*)

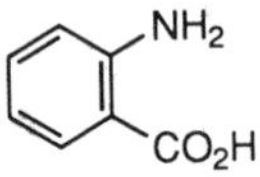

(*j*)

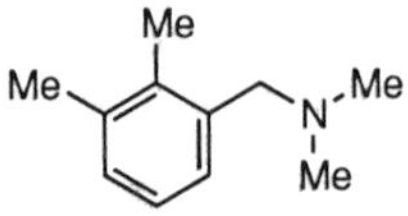

4. Suggest mechanisms for the following:

(*a*)

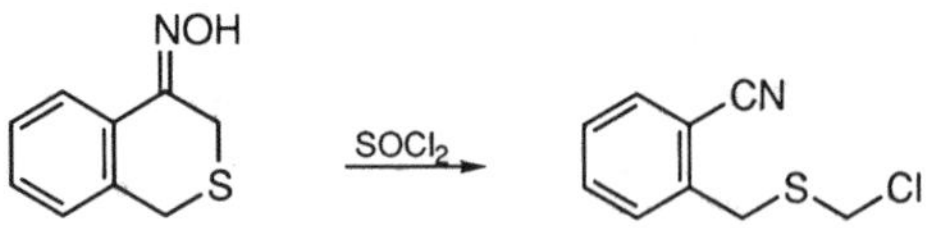

(*b*)

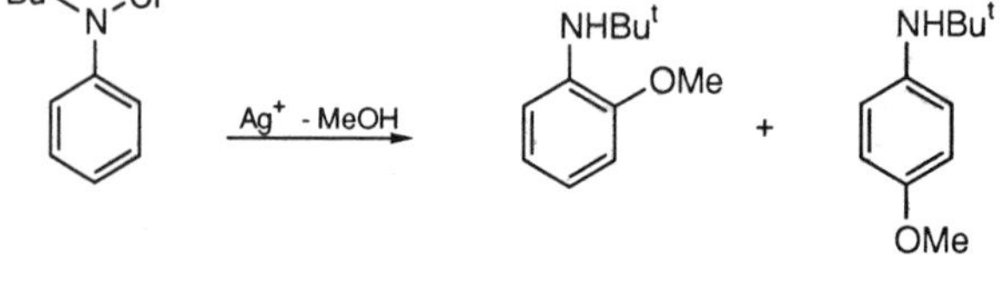

(*c*)

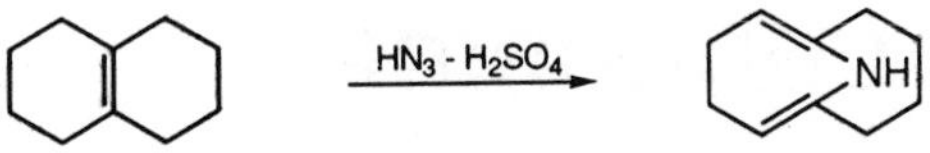
HN3 - H2SO4
NH

(*d*)

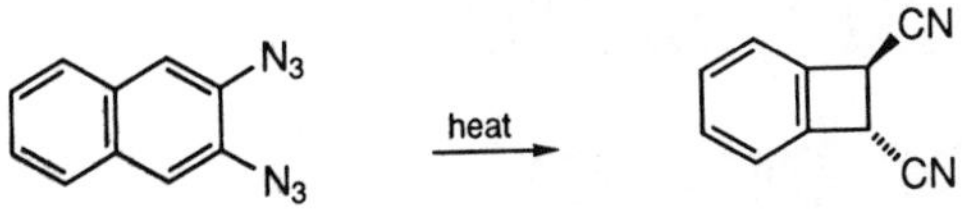
N3
N3
heat
CN
CN

(*e*)

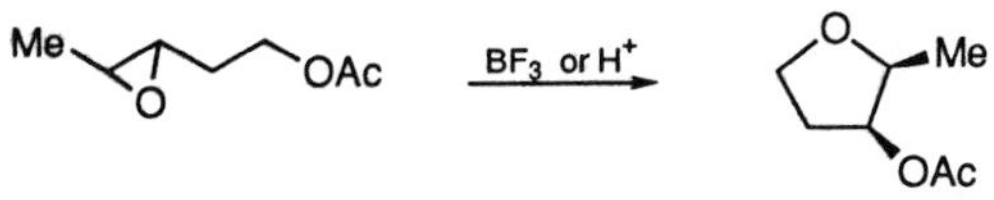
Me
O
OAc
BF3 or H+
O
Me
OAc

(*f*)

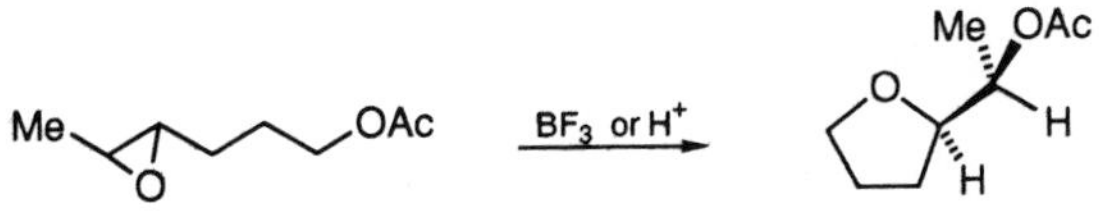
Me
O
OAc
BF3 or H+
Me OAc
O
H
H

(*g*)

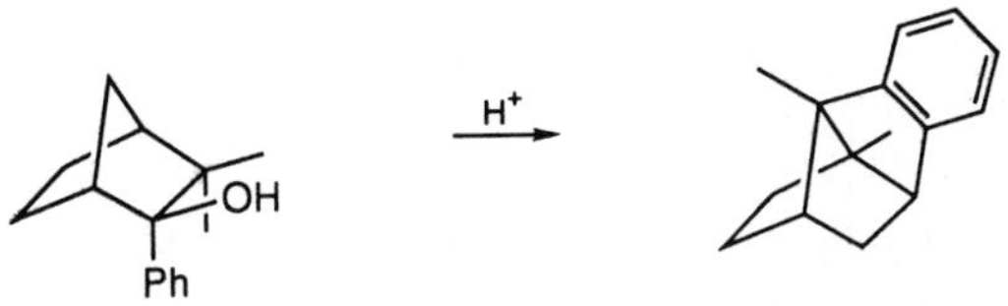
OH
Ph
H+

(*h*)

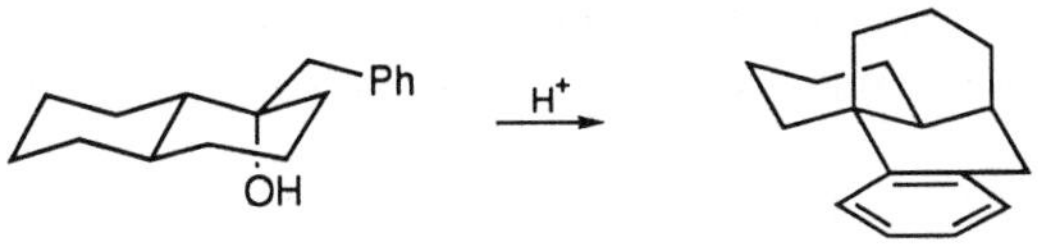
Ph
OH
H+

15 Reagents containing phosphorus, sulfur, silicon, or boron

15.1
Introduction

Compared with most of the synthetic methods discussed so far, those based on reagents which contain phosphorus, sulfur, silicon, or boron have been introduced relatively recently, almost all since 1950. There is still very active research towards further developments.

Phosphorus-containing reagents owe their usefulness to three characteristics of phosphorus chemistry: the ease with which phosphorus(III) is converted into phosphorus(V); the relatively strong bonds formed by phosphorus to oxygen and to sulfur; and the availability of $3d$ orbitals for bonding. In each of these respects phosphorus differs from nitrogen.

For example, hydroxide ion reacts with a phosphonium salt at phosphorus to form a P—O bond and gives ultimately a phosphine oxide, for example:

$$Me_3\overset{+}{P}\diagup\!\!\!\diagdown Ph \ + \ OH^- \ \longrightarrow \ \underset{\underset{OH}{|}}{Me_3P}\diagdown\!\!\!\diagup Ph \ \xrightarrow{OH^-} \ \underset{\underset{O^-}{|}}{Me_3P}\diagup\!\!\!\diagdown Ph$$

$$\xrightarrow{-Me_3P=O} \ Ph\bar{C}H_2 \ \xrightarrow{H_2O} \ PhMe \ + \ OH^-$$

In contrast, reaction with a quaternary ammonium salt occurs at carbon (S_N2 reaction),

$$HO^- \diagdown \underset{Ph}{\overset{\overset{H}{\underset{|}{\diagup}H}}{C}}\!\!\overset{+}{-}NMe_3 \ \longrightarrow \ HO\!\!-\!\!\underset{Ph}{\overset{\overset{H}{\diagup}H}{C}} \ + \ NMe_3$$

or, if the quaternary ion has a β-hydrogen atom, by the E2 process (p. 98).

The strong affinity of phosphorus for oxygen is shown, too, in the *Arbuzov reaction* of triethyl phosphite with an alkyl bromide:

$$(EtO)_3P: \diagup \underset{R}{\overset{\overset{H}{\diagup}H}{C}}\!\!-Br \ \longrightarrow \ (EtO)_3\overset{+}{P}\!\!-\underset{R}{\overset{\overset{H}{\diagup}H}{C}} \ + \ Br^-$$

Sulfur-containing reagents also owe their usefulness in part to the capacity of sulfur to utilise $3d$ orbitals for bonding and to occur in valence states higher than 2. However, although the S—O bond is fairly strong, the tendency for its formation does not dominate sulfur chemistry in the way that P—O bond formation dominates phosphorus chemistry. This generates some useful differences between the two groups of reagents.

Silicon-containing reagents differ from those containing phosphorus or sulfur in that silicon displays only one valency (four) and that, although the Si—O bond is readily formed, it is also readily cleaved (whence derives the value of silyl protecting groups for alcohols). The other features of silicon chemistry which are exploited in its applications to synthesis are: a silyl substituent stabilizes both a negative charge on the carbon to which silicon is bonded and a positive charge on a carbon atom β to it; silyl groups are usually displaced from carbon more readily than is a proton; and silyl groups remote from the functional centre can be relied on to survive all typical reactions except those involving strongly nucleophilic or electrophilic reagents.

Two main factors underlie the usefulness of boron-containing reagents. The first is the ability of diborane and of organic boranes to add to C=C bonds, for example,

The second is the ease of displacement of the boron atom from these adducts by a variety of reagents, often by a reaction which makes use of the ability of the boron atom to accept an electron-pair.

(a) *Reactions of phosphorus ylides*

**15.2
Phosphorus-
containing
reagents**

Phosphorus ylides are prepared by quaternizing a tervalent phosphorus compound with an alkyl halide and treating the salt with butyllithium, phenyllithium, or sodium hydride. The powerfully basic reagent abstracts a proton from the carbon atom adjacent to quaternary phosphorus, e.g.

$$Ph_3P: \overset{H}{\underset{R}{\overset{R}{\diagup}}}\!\!\!\diagup Br \longrightarrow Ph_3\overset{+}{P}\!\!-\!\!\overset{Br^-}{\underset{R}{\overset{H}{\diagup}}}\!\!\!R \xrightarrow{PhLi} Ph_3\overset{+}{P}\!\!-\!\!\overset{R}{\underset{R}{\diagup}} + PhH + LiBr$$

These compounds, known as alkylidenephosphoranes, are appropriately represented as hybrids of two structures,

$$Ph_3\overset{+}{P}\!\!-\!\!\overset{R}{\underset{R}{\diagup}} \quad\longleftrightarrow\quad Ph_3P\!\!=\!\!\overset{R}{\underset{R}{\diagup}}$$

The second structure, which represents phosphorus's ability to incorporate an empty $3d$ orbital in bonding, is thought to be much the less important contributor to the hybrid. Nonetheless, it accounts in part for the much readier formation of phosphorus ylides compared with their nitrogen analogues.

The importance of alkylidenephosphoranes lies in their reactions with aldehydes and ketones to form alkenes:

$$\overset{R}{\underset{R}{\diagup}}\!\!=\!\!O \;+\; \overset{R'}{\underset{R'}{\diagup}}\!\!-\!\!\overset{+}{PR''_3} \;\longrightarrow\; \overset{R}{\underset{R}{\diagup}}\!\!=\!\!\overset{R'}{\underset{R'}{\diagdown}} \;+\; R''_3P\!\!=\!\!O$$

The reaction, known after *Wittig*, is thought to occur in two stages: the phosphorane adds as a carbon nucleophile to the carbonyl group and the resulting intermediate reacts *via* a cyclic intermediate to form the products:

$$\underset{R\;\;\;R}{\overset{O}{\diagup}\!\!\diagdown}\;\;\underset{R'\;\;\;R'}{\overset{+}{PR''_3}} \longrightarrow \underset{R\;\;\;\;\;R}{\overset{O^-\;\;\;\;\overset{+}{PR''_3}}{\diagup\!\!\diagdown}}\;\overset{}{\underset{R'}{R'}} \longrightarrow \underset{R\;\;\;\;\;\;\;R}{\overset{O\!\!-\!\!PR''_3}{\diagup\!\!\diagdown\;\;R'}}\;\;R' \longrightarrow \overset{O=PR''_3}{\underset{\overset{R}{\underset{R}{\diagup}}=\overset{R'}{\underset{R'}{\diagdown}}}{+}}$$

Simple phosphoranes are very reactive and are unstable in the presence of air or moisture. They are therefore prepared in a scrupulously dry solvent (usually tetrahydrofuran) under nitrogen or argon and the carbonyl compound is added as soon as the phosphorane has been formed.

More stable phosphoranes are obtained when a $-M$ substituent is adjacent to the anionic carbon, for example,

$$\overset{+}{Ph_3P}\!\!\underset{H}{\diagdown}\!\!\overset{-}{\diagup}\!\!-\!\!\equiv N \quad\longleftrightarrow\quad \overset{+}{Ph_3P}\!\!\underset{H}{\diagdown}\!\!\diagup\!\!=\!\!C\!\!=\!\!\overset{-}{N}$$

They are formed more readily, requiring treatment of the phosphonium salt only with sodium hydroxide, and they are usually isolable, crystalline compounds. However, although they react with aldehydes, they do not do so effectively with ketones and for this purpose a modification has been introduced in which triphenylphosphine is replaced by triethyl phosphite (*Wadsworth–Emmons reaction*), e.g.

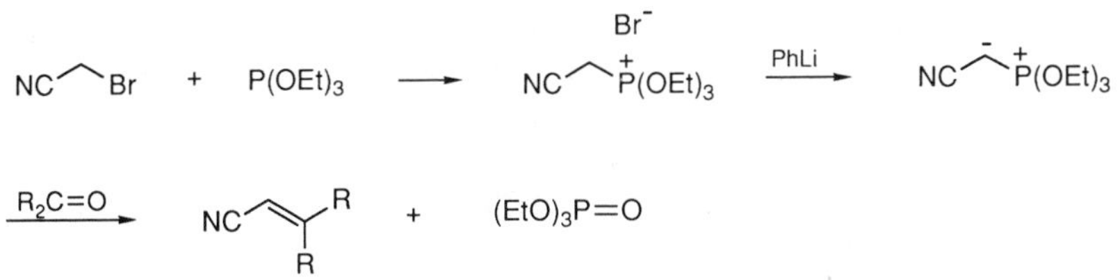

The importance of the Wittig reaction for the synthesis of alkenes stems from two properties. First, it is specific for the conversion $\diagdown C{=}O \rightarrow \diagdown C{=}C\diagup$. For example, the obvious alternative for the conversion of an aldehyde or ketone into an alkene involves the use of a Grignard reagent and can give a mixture of products, e.g.

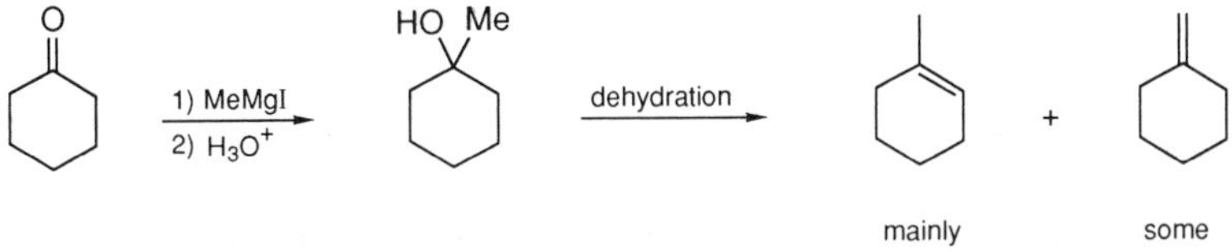

In contrast, the methylenephosphorane from triphenylphosphine and methyl bromide followed by butyllithium reacts with cyclohexanone to give methylene-cyclohexane as the only alkene product, in up to 40% overall yield [5]:

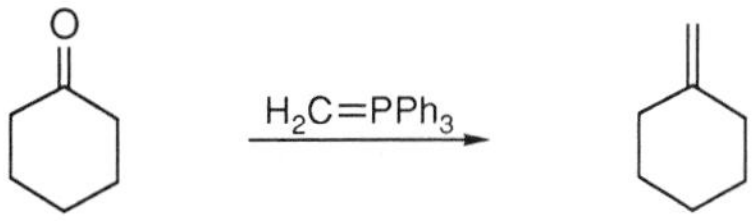

Second, the Wittig reaction can be carried out on aldehydes or ketones which contain a variety of other functional groups (e.g. hydroxyl, ester, C≡C) since these are unaffected. This is again in contrast to the use of Grignard reagents.

The reaction has been used in recent years in a number of syntheses of naturally occurring compounds. It is especially suitable for making rather sensitive compounds, such as polyunsaturated ones. An example is in a commercial synthesis of vitamin A₁:

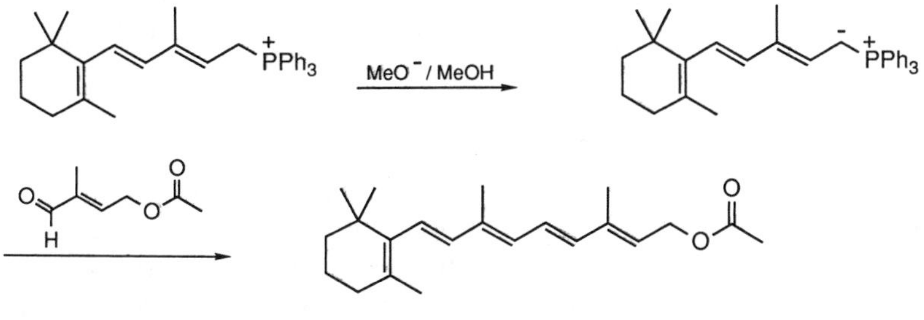

vitamin A₁ acetate

(i) Stereoselectivity in the Wittig and Wadsworth–Emmons reactions. In general, stablilized phosphoranes, including Wadsworth–Emmons reagents, give mainly or exclusively the thermodynamically more stable alkene, especially when a non-polar solvent like benzene is used. An example is in the vitamin A_1 synthesis above: the phosphorane is a stabilized one, as shown by its method of formation (the negative charge is delocalized over several unsaturated carbon atoms), and reacts to form the more stable (*E*)-alkene.

Unstabilized phosphoranes react less selectively, although high yields of (*Z*)-alkenes are often formed in polar, aprotic solvents such as DMF or DMSO, particularly in the presence of a salt, e.g.

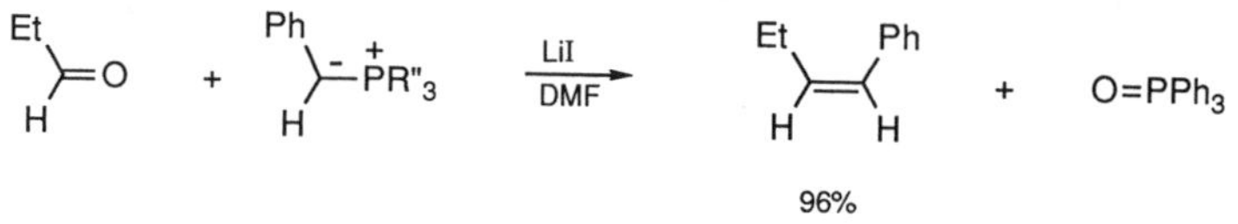

The factors that govern these stereoselectivities are not fully understood and the following theories which have been put forward have not been proved. With stabilized phosphoranes, it is suggested that formation of the stereoisomeric betaines is reversible, so that equilibrium is established between them: the more stable corresponds to the less congested alkene, with the opposite poles close together owing to electrostatic attraction: *syn*-elimination then gives mainly (*E*)-alkene from the more abundant betaine,

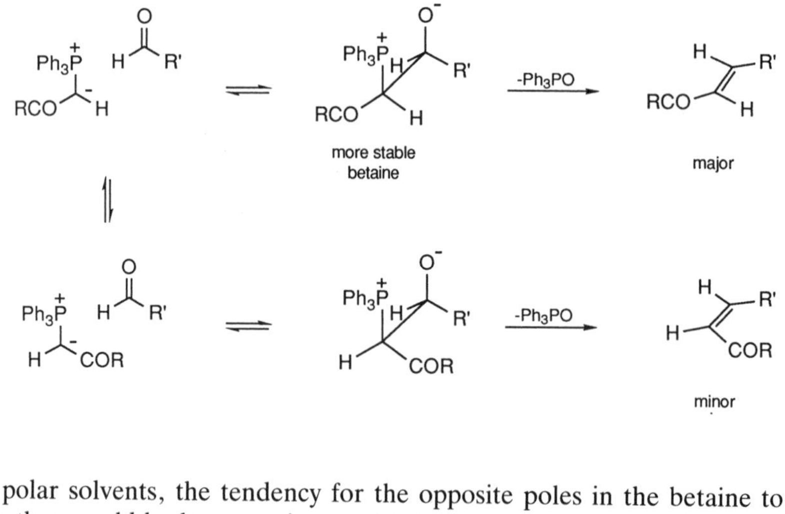

In polar solvents, the tendency for the opposite poles in the betaine to be close together would be less, consistent with the reduced stereoselectivity.

With unstabilized phosphoranes, the betaines are likely to be formed irreversibly, so that the one formed faster determines the major product. In a polar solvent and with salts present, not only should there be a smaller tendency for the opposite poles in the betaine to be close together, but their solvation by salt ions could so enlarge them that they would preferably be formed in the *anti*-conformation; coupled with the steric requirements of the other substituents and the *syn* nature of the elimination, this would lead to the less stable (*Z*)-alkene:

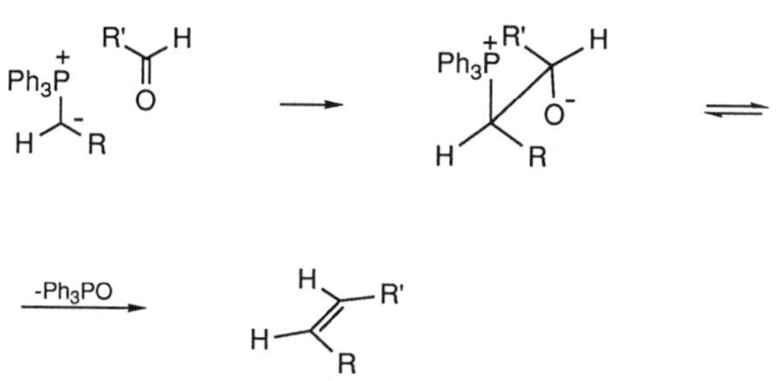

It is possible to form mainly (*E*)-alkenes from unstabilized phosphoranes by the *Schlosser modification* of the Wittig reaction. The reaction between the ylide, generated as a lithium bromide complex, and the aldehyde is carried out at a sufficiently low temperature that the betaine is stable. Phenyllithium is added, giving a new ylide that, on treatment with t-butanol, is protonated stereoselectively to give the more stable betaine. Warming the solution then gives mainly (*E*)-alkene:

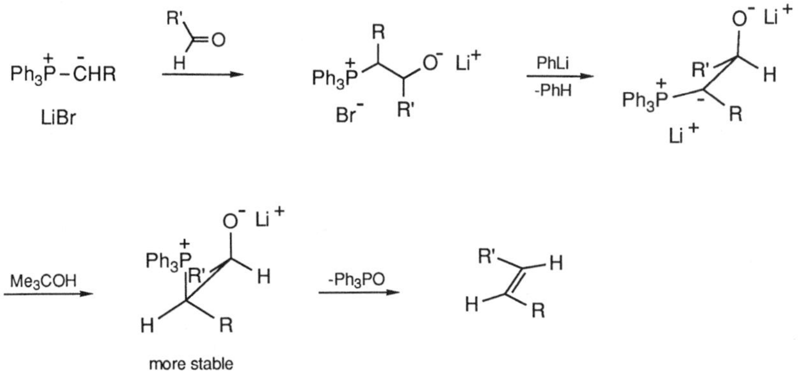

(*ii*)   *Developments of the Wittig reaction.*   These include the following:

(1)   The use of $Ph_3P=CH-OCH_3$ as the phosphorane, to introduce an aldehydic group, e.g.

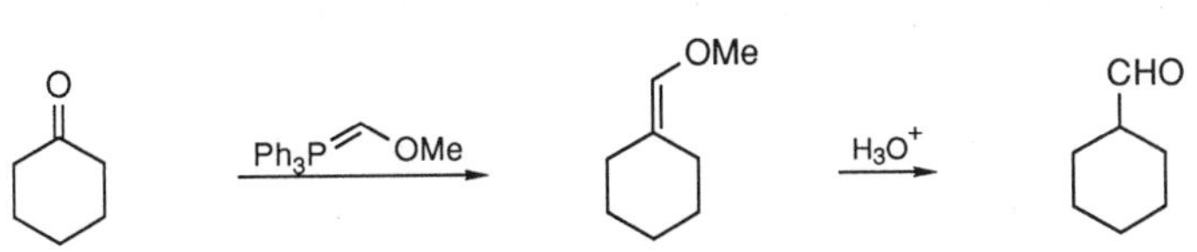

(2)   The use of bifunctional Wittig reagents, e.g.

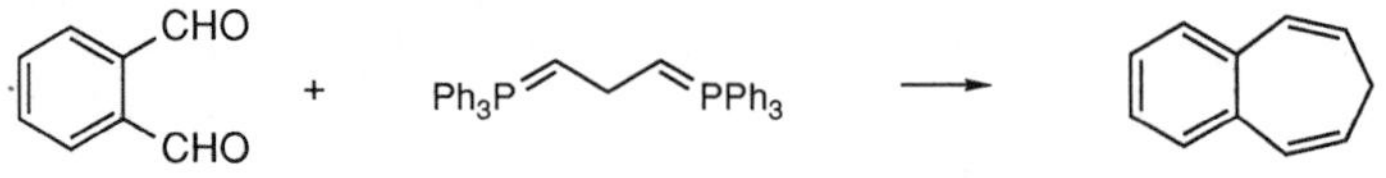

(3)   The formation of carbodiimides (used in the formation of the CO—NH bond in peptide synthesis, p. 338):

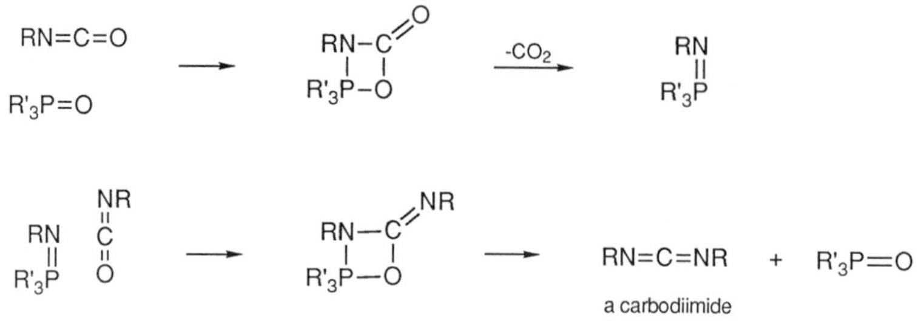

(b)   Reductive cyclization of nitro-compounds

When an aromatic nitro-compound is heated with triethyl phosphite in an inert solvent such as t-butylbenzene, the oxygen atoms of the nitro group are transferred to phosphorus and the nitrogen atom is inserted into a double bond or aromatic ring which is stereochemically suited for the reaction, for example:

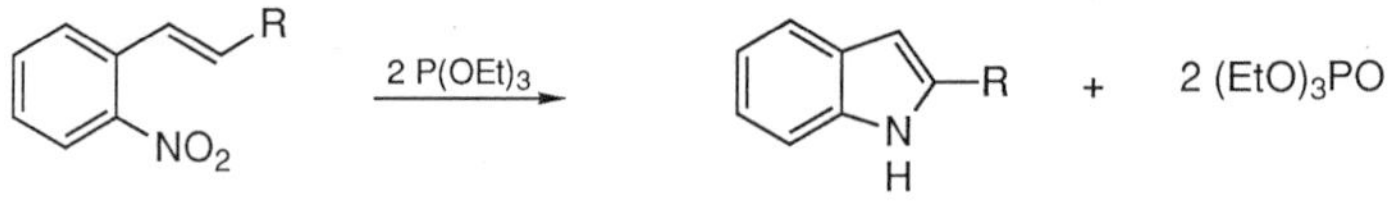

The process probably involves a nitrene as the reactive intermediate and is a further example of phosphorus's strong affinity for oxygen:

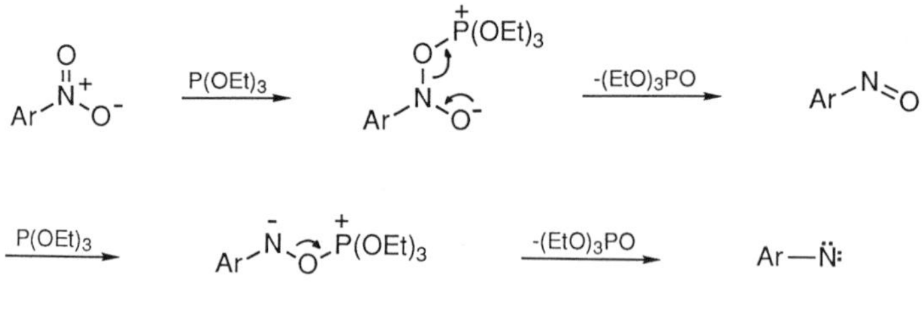

The reaction is capable of wide variation, for example:

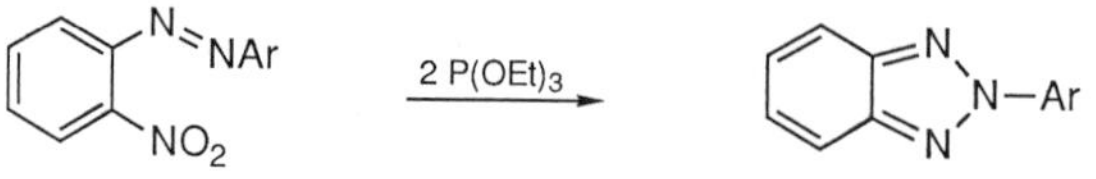

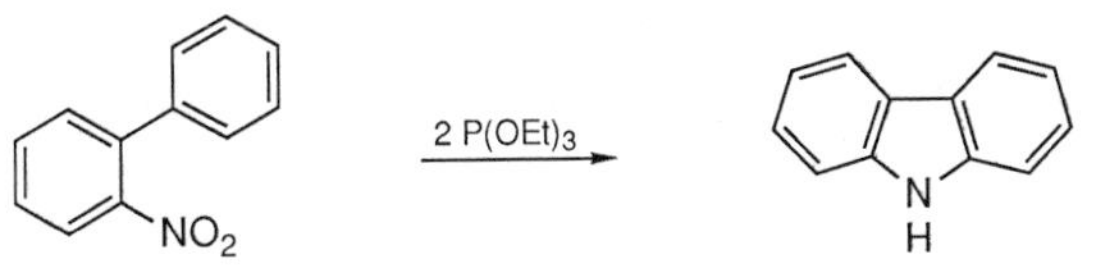

(c)   Synthesis of alkenes from 1,2-diols

The affinity of phosphorus for sulfur is exploited in a synthesis of alkenes from 1,2-diols, *via* thionocarbonate intermediates:

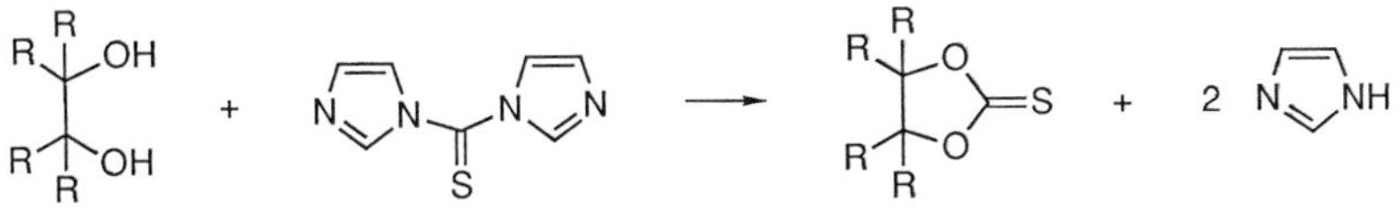

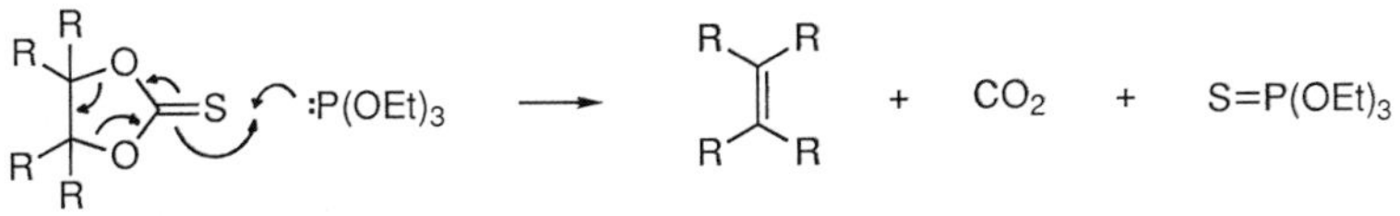

The especial value of the reaction lies in its stereospecificity. For example, the highly strained *trans*-cyclooctene has been made in 75% yield from the *cis*-isomer by conversion into the *trans*-diol (p. 590) and then elimination. The essence of the procedure is that, in order to form the thionocarbonate, the diol has to adopt a conformation which ensures that the *trans*-alkene is formed:

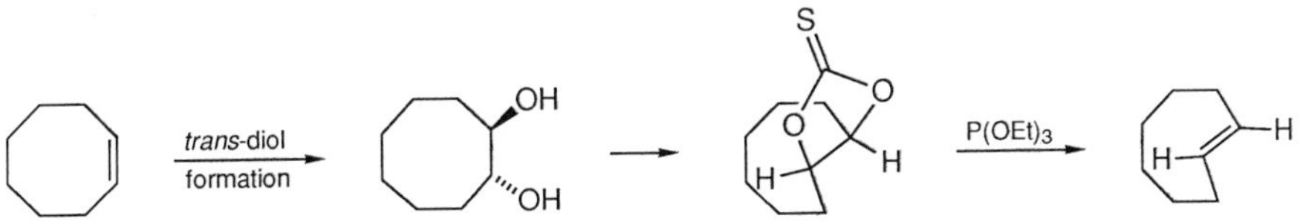

(d)   Conversion of alcohols into halides

Aliphatic halides are generally made from alcohols with either a strong acid, e.g.

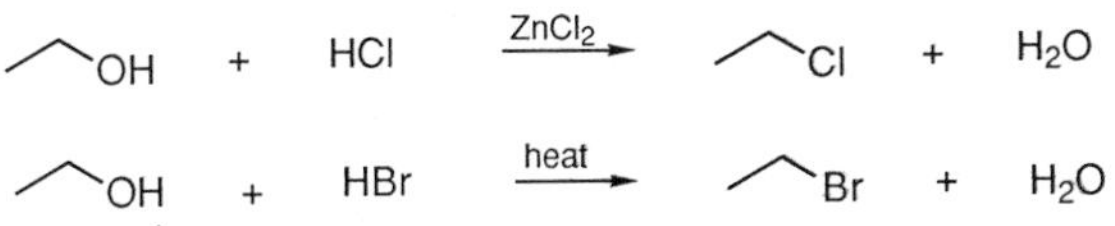

or with thionyl chloride (p. 108) or phosphorus tribromide,

$$3\ ROH\quad +\quad PBr_3\quad \longrightarrow\quad 3\ RBr\quad +\quad O{=}PH(OH)_2$$

which also provide strongly acidic media. These conditions are too vigorous for many organic compounds. Moreover, yields can be low because of side reactions.

For example, Et_2CHOH and PBr_3 give 10–15% of a rearranged bromide, $CH_3CH_2CH_2CHBrCH_3$, as well as Et_2CHBr, and optically active alcohols, while giving mainly bromides of opposite configuration, usually give some of retained configuration.*

These disadvantages have led to the development of methods which are more highly selective under milder conditions.

(i) Bromides. Triphenylphosphine and bromine form the adduct Ph_3PBr_2: the alcohol displaces one bromine as bromide and the new adduct ionizes (presumably helped by the $+M$ effect of alkoxy). Bromide ion then forms the organic bromide by a clean S_N2 reaction on the alkoxyphosphonium ion:

$$Ph_3PBr_2 \ + \ R''{-}\underset{R'}{\overset{H}{\mathstrut}}{-}OH \ \xrightarrow{-HBr} \ R''{-}\underset{R'}{\overset{H}{\mathstrut}}{-}O{-}PBrPh_3 \ \longrightarrow$$

$$Br^- \ \ R''{-}\underset{R'}{\overset{H}{\mathstrut}}{-}\overset{+}{O}{-}PPh_3 \ \longrightarrow \ Br{-}\underset{R'}{\overset{H}{\mathstrut}}{-}R \ + \ Ph_3PO$$

(ii) Chlorides. The corresponding reaction occurs when chlorine is used instead of bromine. Alternatively, the alkoxyphosphonium salt can be generated *in situ* from a compound that can transfer a chlorine substituent without its bonding-pair to form a relatively stable carbanion, e.g.

$$Ph_3P{:} \quad Cl{-}C(Cl)(Cl)_{Cl} \ \longrightarrow \ Ph_3\overset{+}{P}{-}Cl \ + \ {}^-C(Cl)(Cl)_{Cl}$$

---

* The reaction is mainly of S_N2 type on the trialkyl phosphite,

$$3 \ R''{-}\underset{R'}{\overset{H}{\mathstrut}}{-}OH \ + \ PBr_3 \ \longrightarrow \ R''{-}\underset{R'}{\overset{H}{\mathstrut}}{-}O{-}\underset{OCHRR'}{\overset{OCHRR'}{P}} \ + \ 3\,HBr$$

$$Br^- \ \ R''{-}\underset{R'}{\overset{H}{\mathstrut}}{-}O{-}\underset{OCHRR'}{\overset{OCHRR'}{P}} \ \xrightarrow{H^+} \ Br{-}\underset{R'}{\overset{H}{\mathstrut}}{-}R \ + \ O{=}PH(OCHRR')_2$$

The occurrence of skeletal rearrangement and some retention of configuration implies some contribution from the alternative S_N1 mechanism, presumably catalyzed by acid.

$$\text{ROH} + \text{Ph}_3\overset{+}{\text{P}}-\text{Cl} \longrightarrow \text{RO}-\overset{+}{\text{P}}\text{Ph}_3 + \text{HCl}$$

$$\text{Cl}^- + \text{RO}-\overset{+}{\text{P}}\text{Ph}_3 \xrightarrow{\text{S}_\text{N}2} \text{RCl} + \text{OPPh}_3$$

(*iii*) *Iodides.* The required alkoxyphosphonium ion cannot be formed in a corresponding way. It is generated instead from diethyl azodicarboxylate, triphenylphosphine, and methyl iodide (*Mitsunobu reaction*). The reaction occurs in an ether solvent at room temperature or below, and exploits, first, the reactivity of the azo compound as an electrophile in the formation of the first phosphonium ion (helped by the $-M$ effect of one ester group) and, second, the good leaving-group property of the reduced azo compound (helped by the $-M$ effect of the other ester group). The role of methyl iodide is to capture the enolate anion and, in so doing, to provide the iodide ion for the $\text{S}_\text{N}2$ formation of the product:*

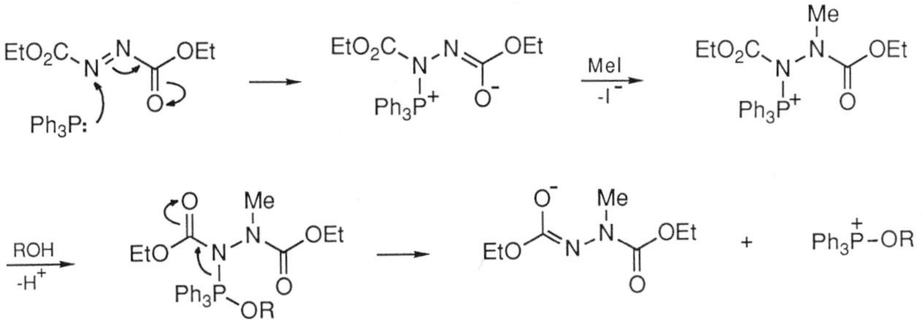

* A variant of the Mitsunobu reaction exploits its $\text{S}_\text{N}2$ character to invert the configuration of an optically active alcohol. A carboxylic acid is used instead of methyl iodide, giving the enantiomeric alcohol as its carboxylate ester from which it is obtained by hydrolysis:

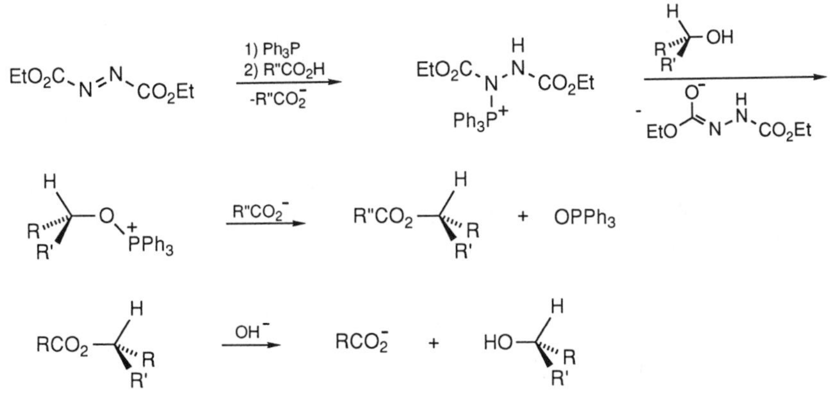

15.3
Sulfur-
containing
reagents

(a) *Reactions of sulfur ylides*

Sulfur ylides are prepared, analogously to phosphorus ylides, from sulfonium salts and an exceptionally strong base. The latter is usually the 'dimsyl' anion, $CH_3-SO-CH_2^-$, formed from dimethyl sulfoxide and sodium hydride:

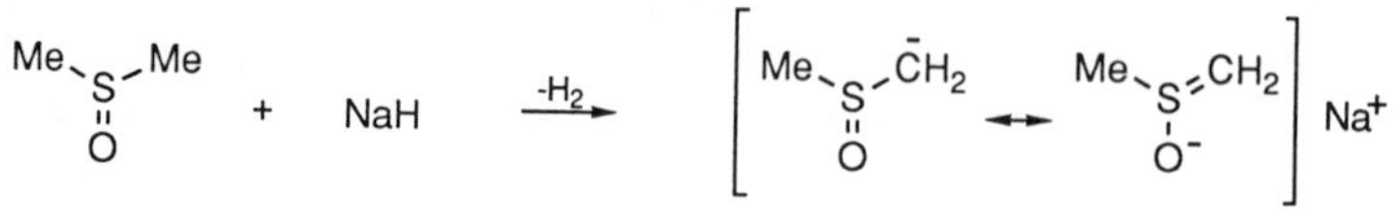

Ylide formation then occurs as follows, e.g.

Sulfoxonium salts can also be used, e.g.

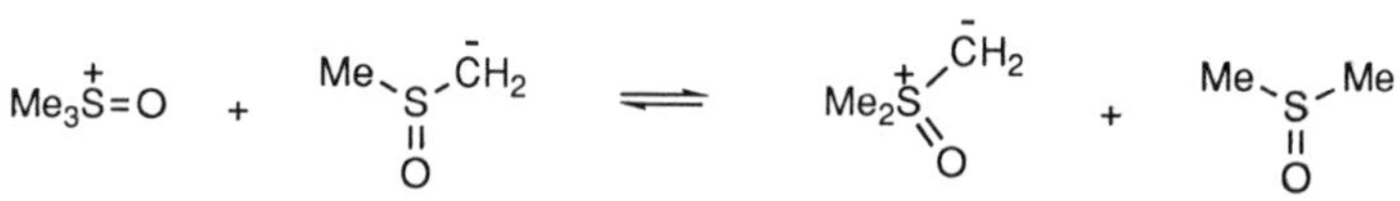

The ylides are thought to be stabilized by bonding which involves a sulfur $3d$ orbital, as represented by

(i) *Reactions with carbonyl compounds.* As with phosphorus ylides, the first stage of the reaction of a sulfur ylide with an aldehyde or ketone consists of nucleophilic addition:

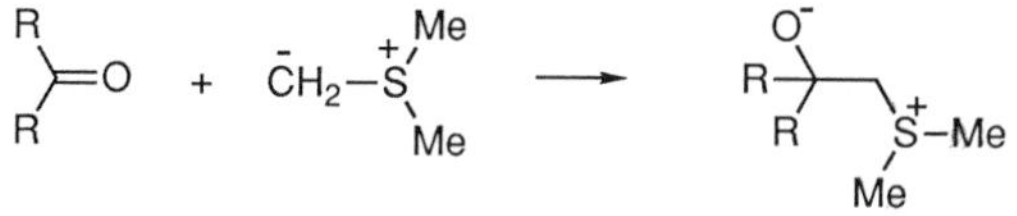

However, whereas the corresponding adduct from a phosphorus ylide forms an alkene and the phosphine oxide (p. 460), the sulfur-containing adduct undergoes intramolecular nucleophilic substitution to give an epoxide:

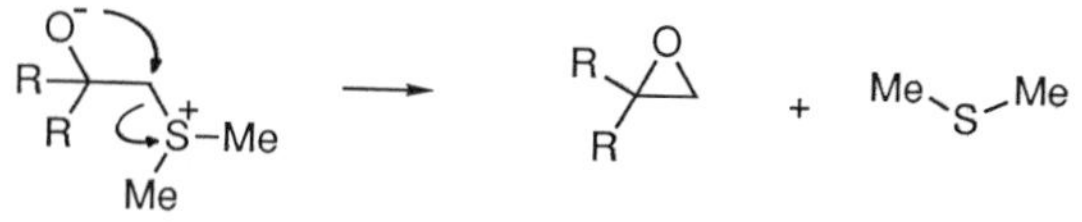

This illustrates the relatively weaker affinity of sulfur for oxygen compared with that of phosphorus.

Reaction may be brought about with an ylide derived from either a sulfonium salt or a sulfoxonium salt. Yields are generally good; for example, cyclohexanone and the ylide from trimethylsulfoxonium iodide give 70% of the epoxide [5]:

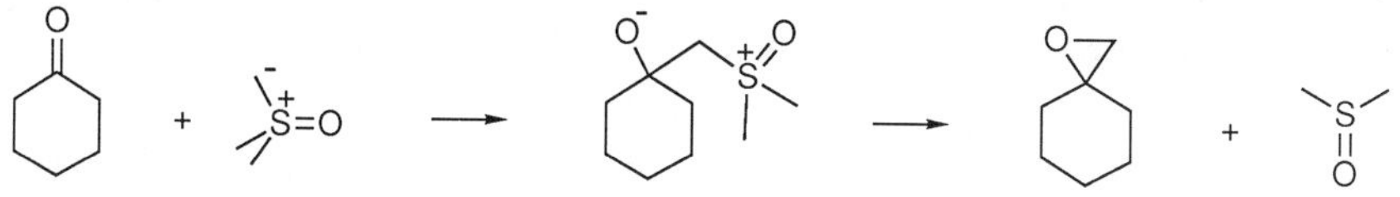

Since epoxides undergo rearrangement to carbonyl compounds when treated with Lewis acids (p. 590), the overall process can be used for the conversion $\diagdown C{=}O \rightarrow \diagdown CH{-}CHO$; for example,

A particularly useful sulfur ylide is the cyclopropyl-substituted one whose synthesis is shown:

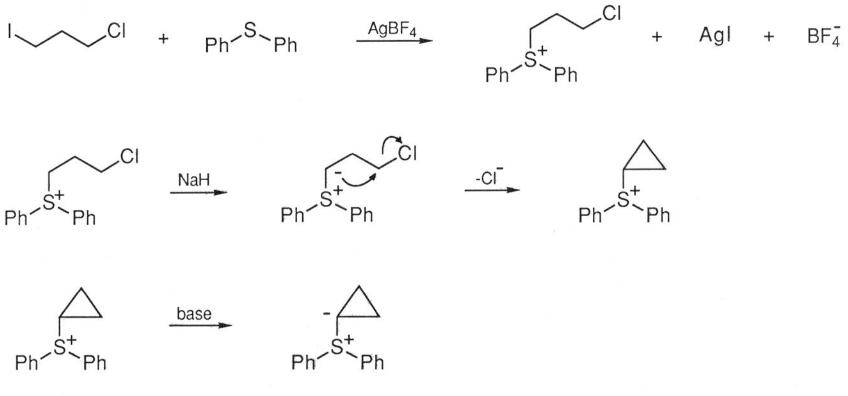

It is not necessary to make the ylide separately; the precursor sulfonium salt is treated with potassium hydroxide in the presence of a carbonyl compound, and, as the ylide is formed, it reacts with the carbonyl compound:

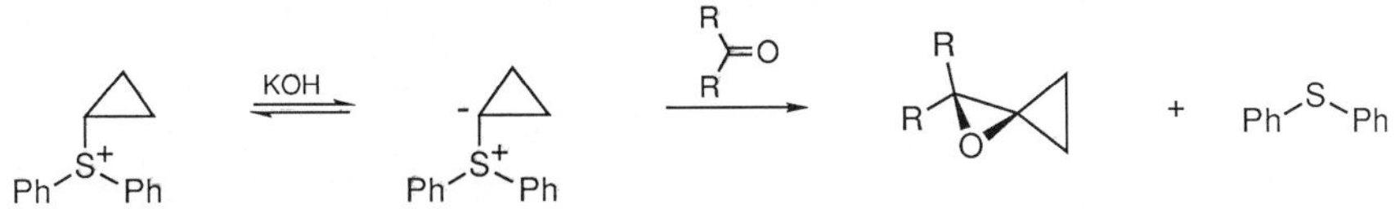

The resulting epoxides – oxaspiropentanes – are particularly strained and readily subject to ring-opening reactions. For example, treatment with an acid gives cyclobutanone derivatives:

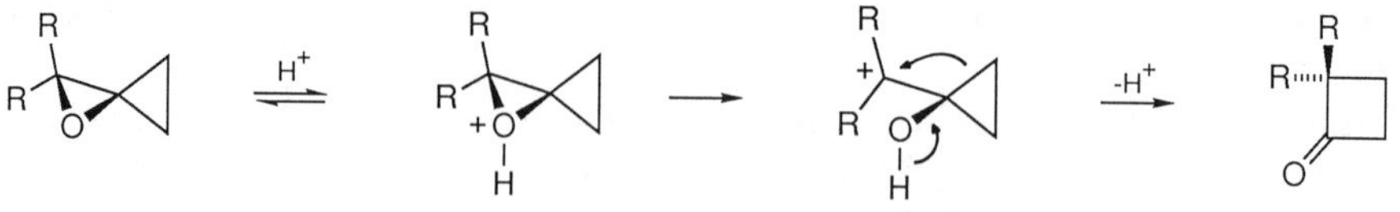

These in turn can undergo useful transformations, such as ring expansion to γ-lactones on treatment with alkaline hydrogen peroxide:*

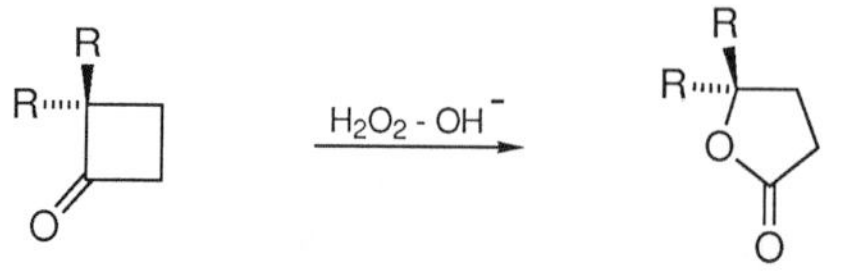

(*ii*) *Rearrangements.* Sulfur ylides rearrange in the same way as nitrogen ylides (p. 448),

When one of the substituents on the sulfur atom is allylic, a [3,2]-sigmatropic rearrangement occurs instead, e.g.

This can be usefully employed in synthesis because, first, a sulfur substituent can be readily introduced by making use of the nucleophilic property of sulfides and, second, a sulfur group can be removed after the rearrangement. In the following example, the [3,2]-shift occurs in preference to the [1,2]-shift, and the removal of sulfur at the end is made possible by the fact that allylic systems are susceptible to hydrogenolysis (p. 645):

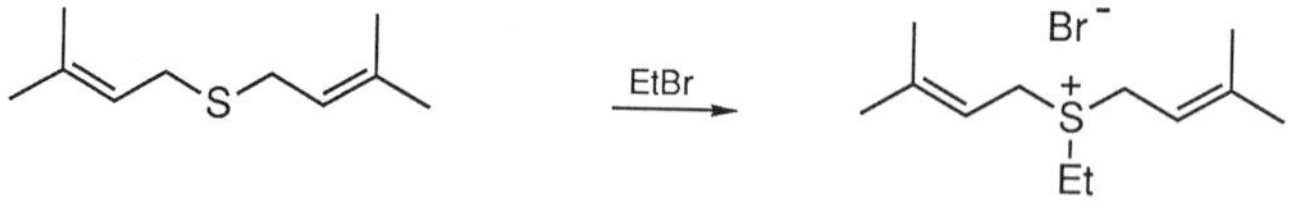

---

* This is an example of the Baeyer–Villiger reaction (p. 445). However, the Baeyer–Villiger reaction usually requires a peroxyacid: in this case, the milder reagent no doubt succeeds because of the relief of ring strain during rearrangement.

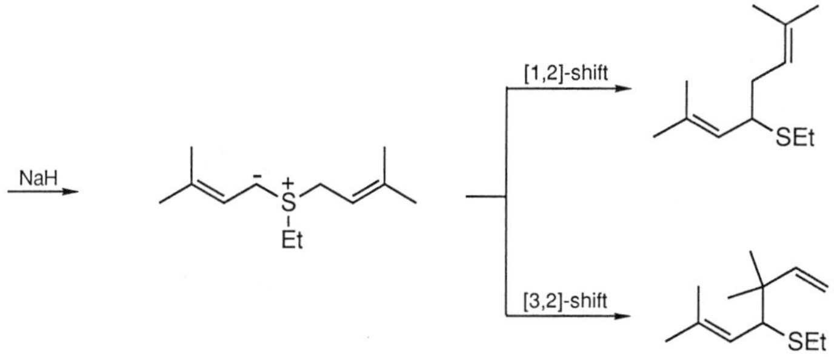

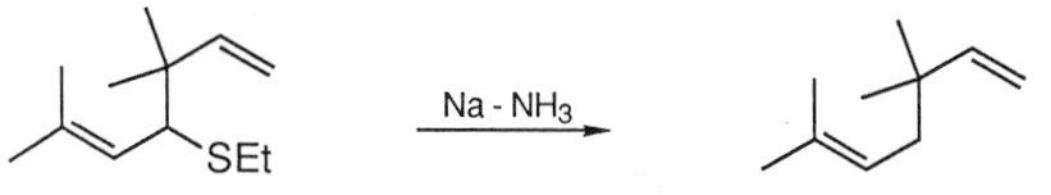

Aza- and oxa-sulfonium salts also give ylides which can rearrange and special use can be made of this in the aromatic series. The crucial reactions are of the types,

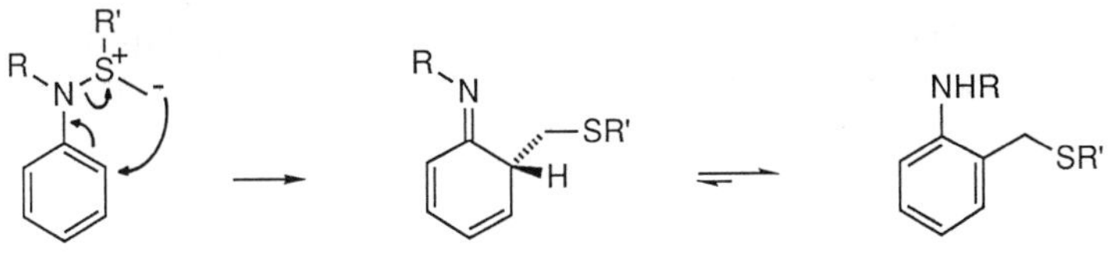

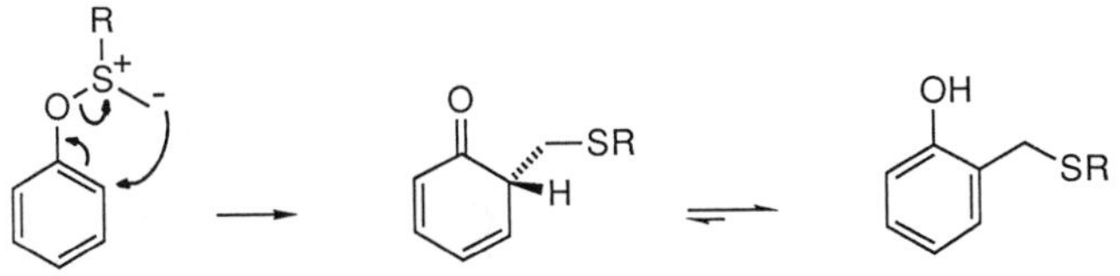

These [3,2]-sigmatropic rearrangements are analogues of the Sommelet rearrangement (p. 448).

In the nitrogen series, treatment of an aniline with t-butyl hypochlorite gives the N-chloro-compound which, with an organic sulfide, gives the azasulfonium salt, e.g.

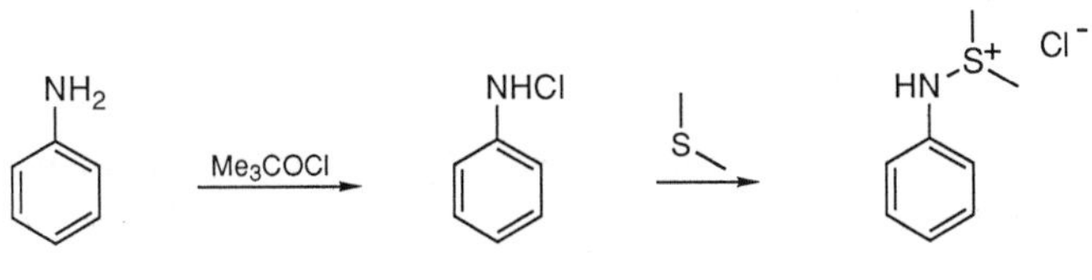

The ylide is generated with sodium methoxide and spontaneously rearranges; reduction over Raney nickel then yields an o-alkylaniline, e.g.

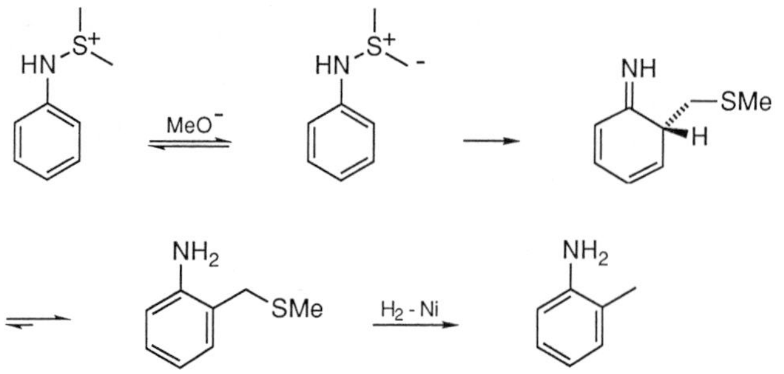

A simple variant enables an *o*-butyl group to be introduced, e.g.

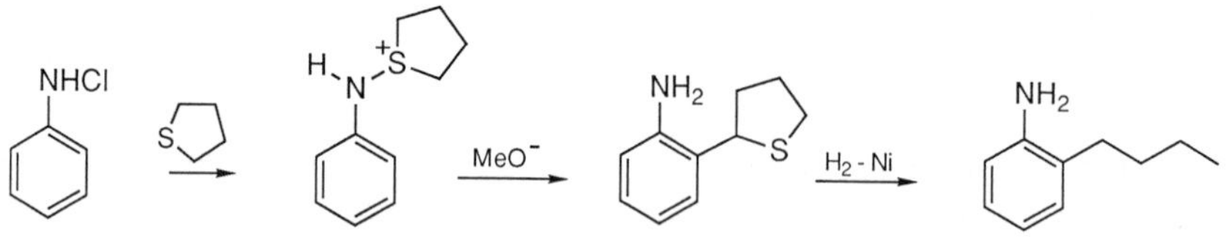

A further variant is to use a β-keto-sulfide. After rearrangement of the ylide,* intramolecular nucleophilic displacement on the carbonyl group occurs spontaneously to form the indole ring and the sulfur substituent can be removed by reduction, e.g.

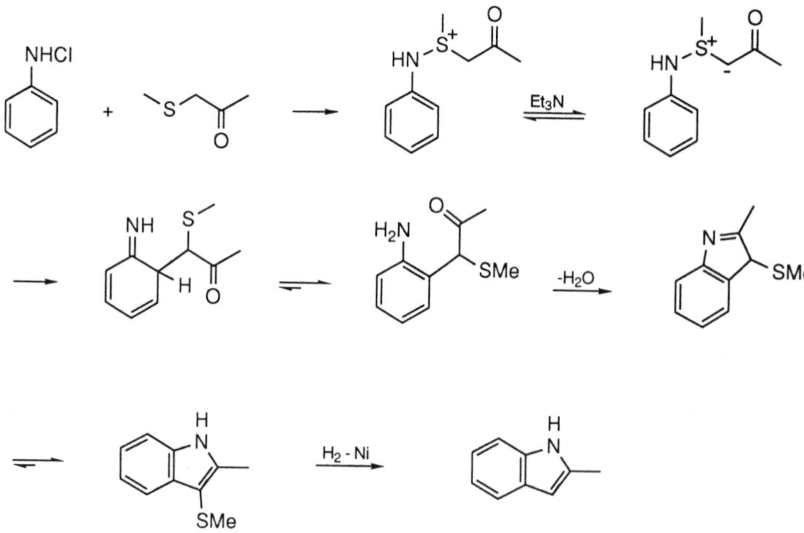

In the oxygen series, *o*-alkylation of phenols can be effected in a similar manner to that of anilines, e.g.

---

* The −*M* effect of the carbonyl group stabilizes the ylide formed by loss of a proton from the adjacent carbon atom and allows the use of a weaker base than methoxide ion.

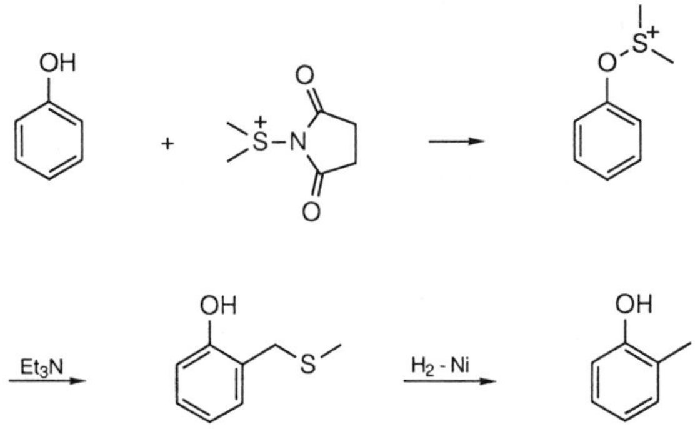

(b) Reactions of the dimsyl anion

As well as its use as a very strong base in the formation of ylides, the dimsyl anion is very reactive as a nucleophile, and several synthetic applications are based on this property together with the ease of removal of the sulfur substituent either by reduction or thermally.* For example,

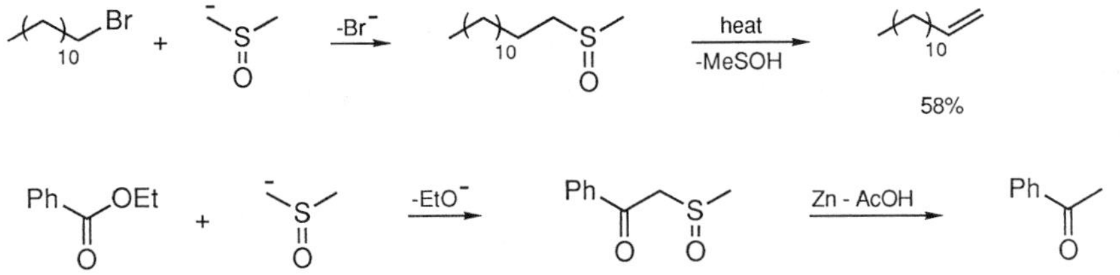

Since compounds of the type $RCOCH_2SOCH_3$ contain strongly activated methylene groups, the reaction of the dimsyl anion with an ester can be used as the basis of the synthesis of more complex ketones, as for example from $PhCOCH_2SOCH_3$:

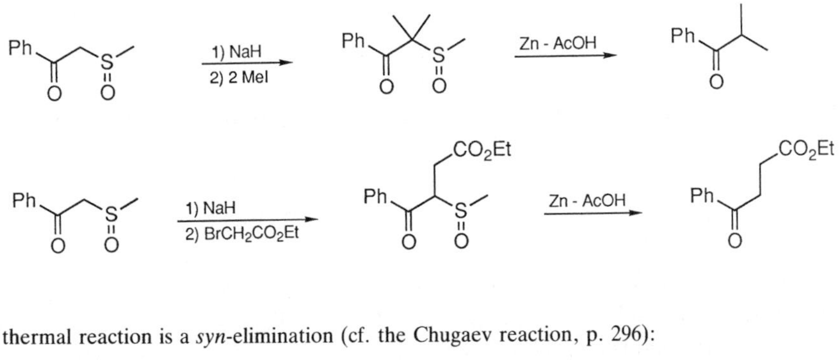

* The thermal reaction is a *syn*-elimination (cf. the Chugaev reaction, p. 296):

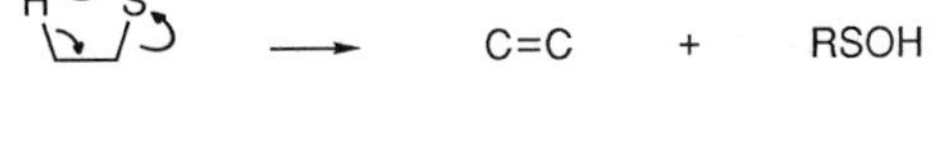

$$C=C \quad + \quad RSOH$$

(c)   Sulfoxide elimination

A route to sulfoxides which is an alternative to that involving the dimsyl anion is to treat a compound containing an activated C—H bond with diphenyl or dimethyl disulfide in the presence of base, followed by oxidation of the resulting sulfide (p. 626), e.g.

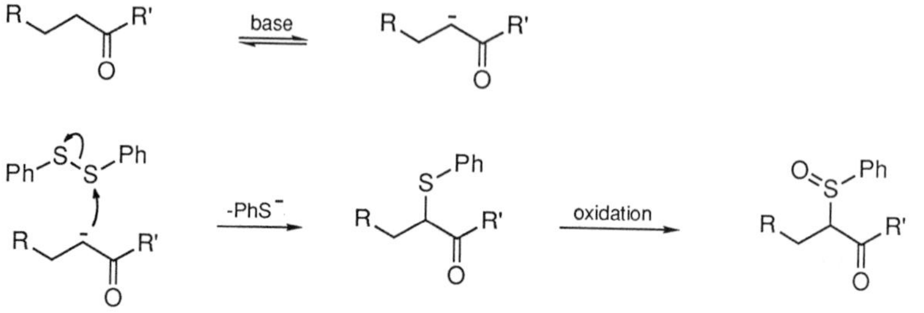

Since sulfoxides undergo elimination on heating,

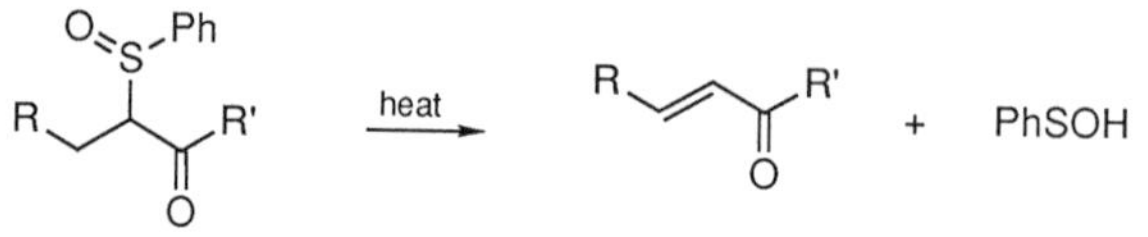

this provides a method for introducing a double bond next to a group of $-M$ type, for example:

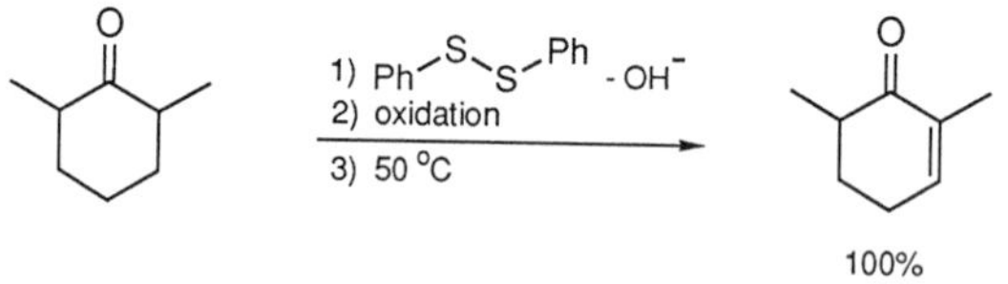

100%

If dimethyl disulfide is used, the final step requires a higher temperature (110°C).

(d)   Use of dithioacetals: 'reversed polarity' of carbonyl compounds

The characteristic of the carbonyl group is its susceptibility to attack by nucleophiles, e.g.

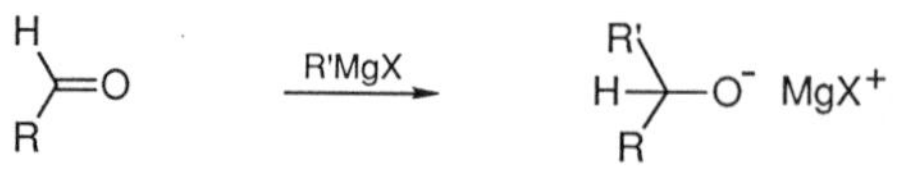

Its versatility in synthesis would be increased if it were possible to render the group itself nucleophilic, i.e. to generate RCO^-. This is not practicable, but an alternative approach is available: to attach groups X and Y to the carbonyl carbon which stabilize an adjacent negative charge, carry out reaction with the corresponding carbanion, and then replace X and Y by oxygen:

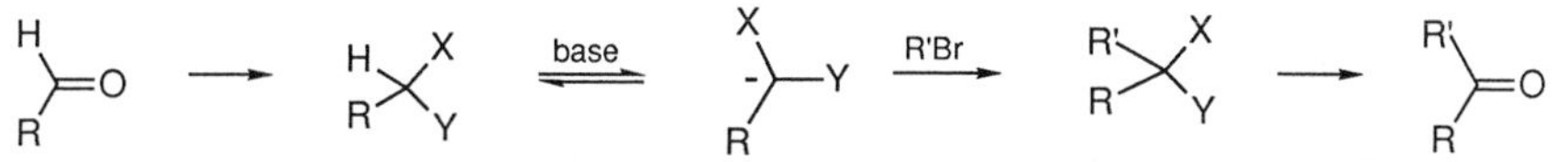

It has been found that the most satisfactory forms of X and Y are RS-groups: they can be readily introduced by an acid-catalyzed reaction between the aldehyde and a thiol, e.g.

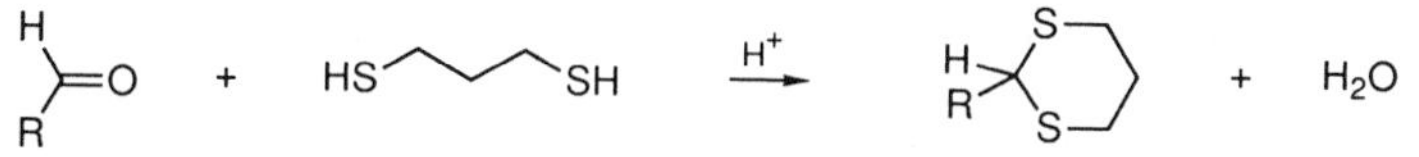

the sulfur atoms stabilize carbanions (p. 468), so making possible alkylation,

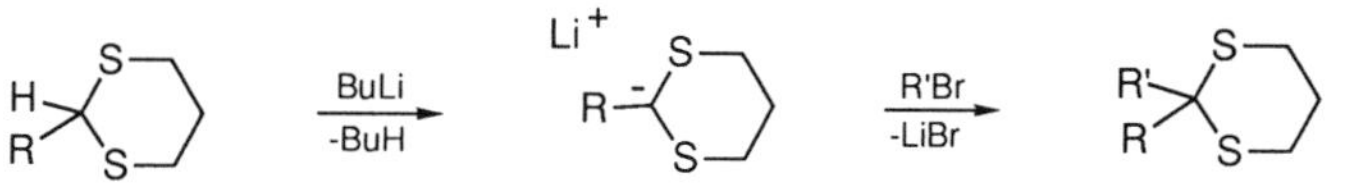

and the dithioketal can be hydrolytically cleaved in the presence of mercury(II) ion,

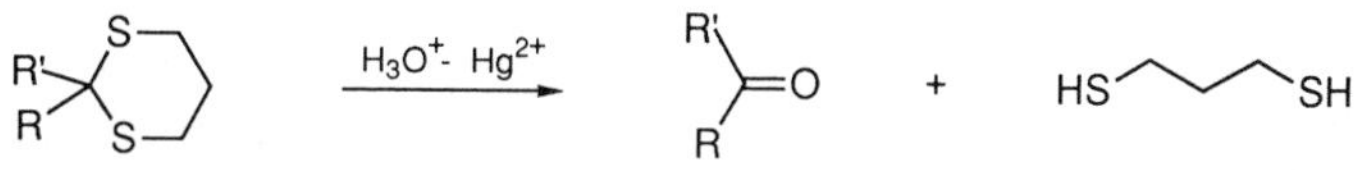

The overall process corresponds to the conversion, RCHO → RR'CO.

The use of the 1,3-dithiane derivative of formaldehyde enables two alkyl groups to be introduced sequentially,

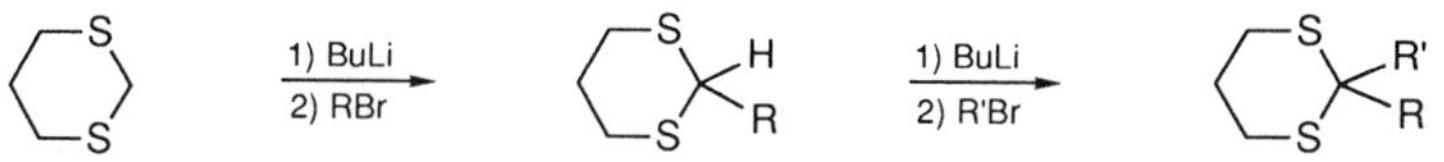

and, if the appropriate dihalide is used, 3- to 7-membered cyclic ketones can be obtained, e.g. [6]

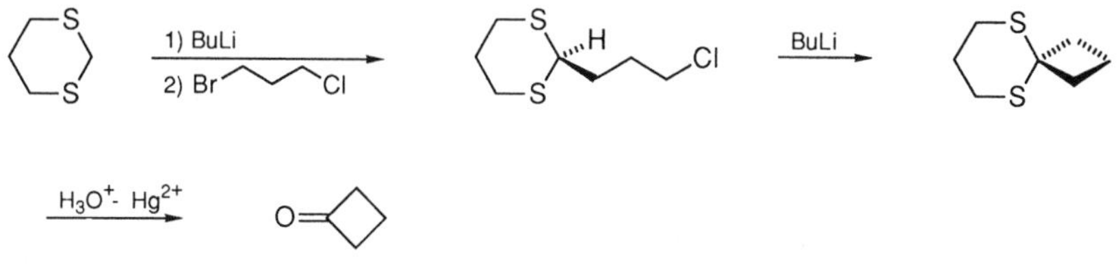

As usual in alkylations of carbanions, primary and secondary aliphatic halides react successfully but aromatic halides are unreactive and tertiary halides undergo elimination. The halide can be replaced by another source of electrophilic carbon in the form of an aldehyde, ketone, epoxide, or aromatic nitrile, e.g.

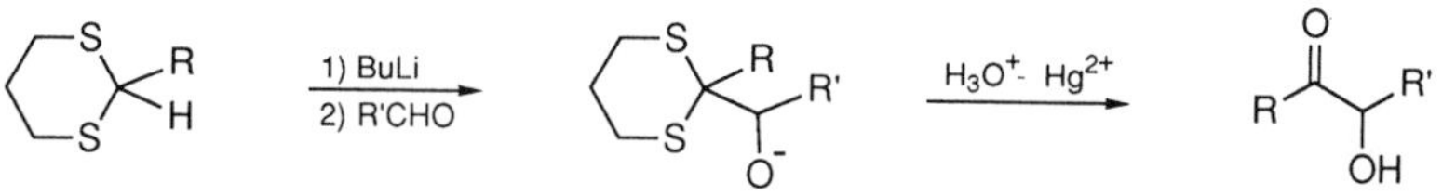

(e) Julia reaction

This is the reaction of an aldehyde or ketone, $RR'CO$, with a sulfone, $R''R'''CHSO_2Ph$, to give the alkene, $RR'C{=}CR''R'''$. A CH bond adjacent to sulfonyl is significantly acidic and can be metallated with lithium or magnesium. The anion reacts readily with the carbonyl compound to give an adduct whose acetyl derivative undergoes reductive elimination with sodium amalgam to form the alkene. Yields are usually at least 80%.

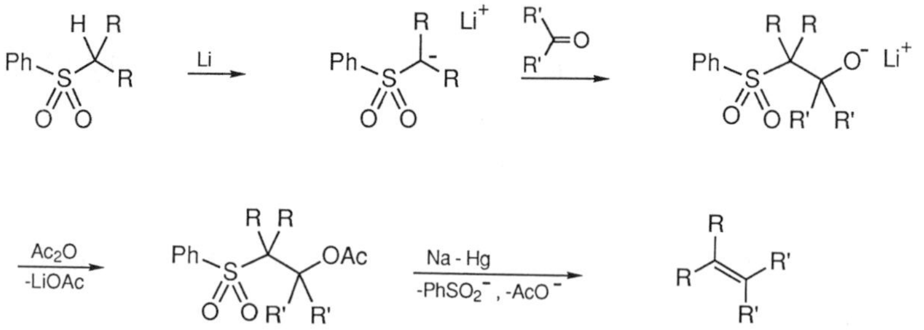

One of the important characteristics of the reaction is its high stereoselectivity for (E)-disubstituted alkenes. This is thought to result from mediation of a carbanion in the reductive elimination which can rotate into the most stable configuration, corresponds to (E)-alkene, before eliminating acetate:

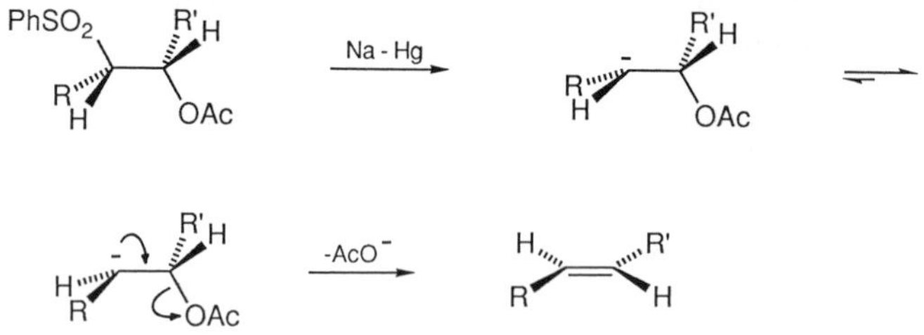

15.4 Silicon-containing reagents

(a) Synthesis of alkenes: Peterson reaction

This process, which is closely related to the Wittig reaction, can be carried out in two ways. In one, a CH group that is adjacent to both a silicon-containing substituent (usually $SiMe_3$) and a $-M$ group undergoes a base-induced reaction with an aldehyde or ketone to give the alkene directly, e.g.

Yields are generally high, e.g.

95%

In the second method, a carbanion-equivalent (as a C-metal bond) is generated adjacent to the $SiMe_3$ group in the absence of a $-M$ substituent. This reduces the tendency for the final step to occur since the alkene formed is not conjugated; acid work-up yields the β-hydroxysilane which is converted into the alkene in a second step.

The carbanion-equivalent is usually formed by one of the following routes:

(1) Addition of an organolithium compound to the commercially available trimethylvinylsilane:

The regioselectivity follows from the anion-stabilizing effect of silicon.

(2) Abstraction of a proton from an allylsilane, e.g.

(3) Formation of a silyl-containing Grignard reagent, e.g.

The β-hydroxysilanes, derived from these reagents with aldehydes and ketones followed by acid work-up, can be formed as diastereomers (each of which is a racemate). After separation, either diastereomer can be converted, usually with over 90% stereoselectivity, into either the (E)- or the (Z)-alkene, depending on whether basic (KH or NaH in THF) or acidic (BF₃·Et₂O) conditions are employed. This results from the *syn* nature of the former elimination and the *anti* nature of the latter, e.g.*

---

*Elimination of Me_3SiO^- from the oxanion occurs in the base-induced process whereas it does not occur when β-hydroxysilanes are formed *via* the oxanion derivatives, in the processes involving lithium or magnesium reagents described above. This is because the oxanions in the latter reactions form essentially covalent bonds with lithium or magnesium, whereas they are 'bare' and more reactive when associated with the more electropositive sodium or potassium.

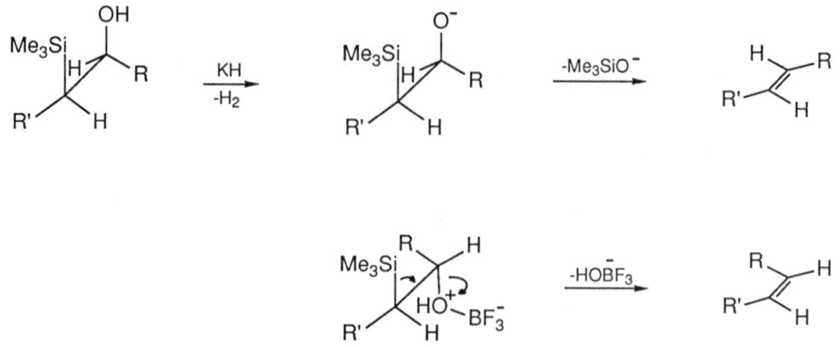

As well as this control of stereoselectivity, the Peterson reaction has the advantage over the Wittig reaction that it is more effective with hindered ketones.

(*i*) *Inversion of configuration of an alkene.* An inversion of the type

can be accomplished by conversion of the (*E*)-alkene into the epoxide, followed by treatment with Me₃Si—SiMe₃ and NaOMe (which form Me₃Si⁻ + Me₃SiOMe); S_N2 opening of the epoxide by Me₃Si⁻ is followed by *syn*-elimination:

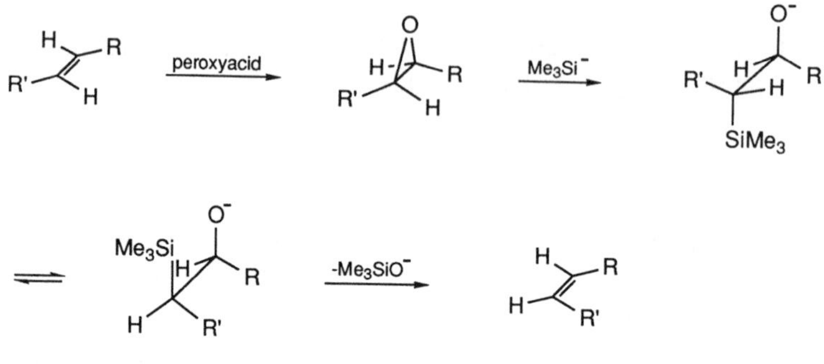

(*b*)   *Protection of alcohols*

Alcohols react rapidly with chlorosilanes to give silyl ethers. Reaction is usually carried out in the presence of a weak base such as triethylamine in order to neutralize the hydrogen chloride that is released:

$$ROH \quad + \quad Me_3SiCl \quad \xrightarrow{Et_3N} \quad RO-SiMe_3 \quad + \quad Et_3NH^+ \; Cl^-$$

This provides a method for protecting an alcoholic function while other reactions are carried out. The alcoholic group can be released with dilute acid,

$$RO-SiMe_3 \xrightarrow{H_3O^+} ROH + Me_3SiOH$$

A particular value of this method of protection depends on the fact that, for steric reasons, the order of reactivity to the chlorosilane amongst alcohols is primary > secondary > tertiary. This makes it possible, for example, to carry out reactions on a secondary or tertiary alcohol in the presence of a (protected) primary one.

Diethylaminotrimethylsilane is a less reactive but correspondingly more selective protecting reagent. For example, it is possible to protect the alcoholic group *trans* to R without affecting the more hindered one *cis* to R' in compounds of the following type,

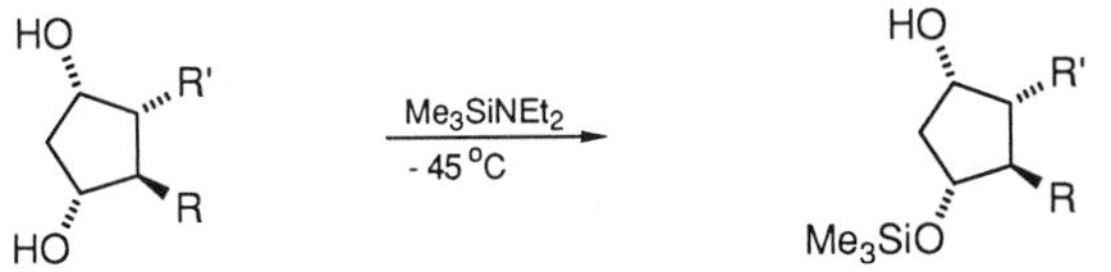

Bulkier silyl groups confer stronger protection. For example, $RO-SiMe_2(CMe_3)$ is stable to dilute acid, enabling acid-catalyzed processes to be carried out elsewhere in the molecule. This is the result of steric hindrance and the group is correspondingly harder to introduce. Reaction with the chlorosilane is too slow to be practicable but imidazole – a much stronger nucleophile than an alcohol and, in its protonated form, a good leaving group – is an effective catalyst:

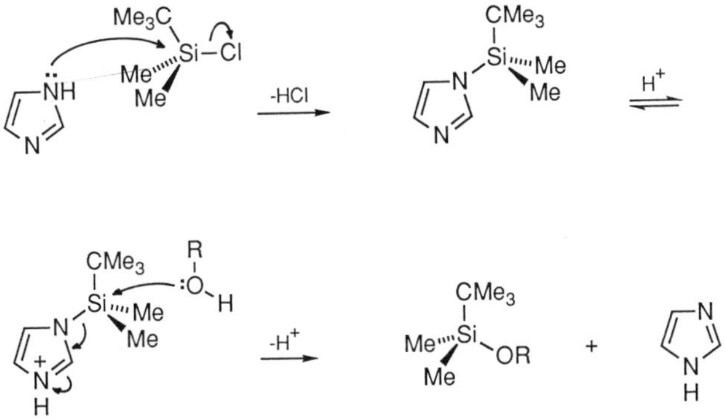

The group can be removed by either a strong base (e.g. MeO^-) or by nucleophilic displacement by fluoride ion, in the form of a quaternary ammonium fluoride such as $Bu_4N^+F^-$ or KF.18-crown-6 (p. 107), making use of the very strong Si—F bond (594 kJ mol^{-1}):

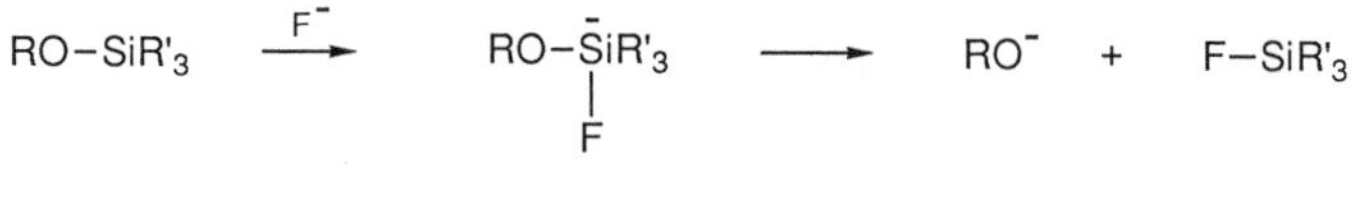

(c) Directive influence of —SiR₃ in electrophilic reactions

A trialkylsilyl group has a strongly stabilizing effect on a cationic centre which is beta to it, consistent with effective carbon–silicon hyperconjugation, that is the C—Si bonding orbital and the empty p orbital on the β-carbon atom overlap laterally to provide extra bonding, as represented by the hybrid,

A measure of the effect comes from the rate of the reaction,

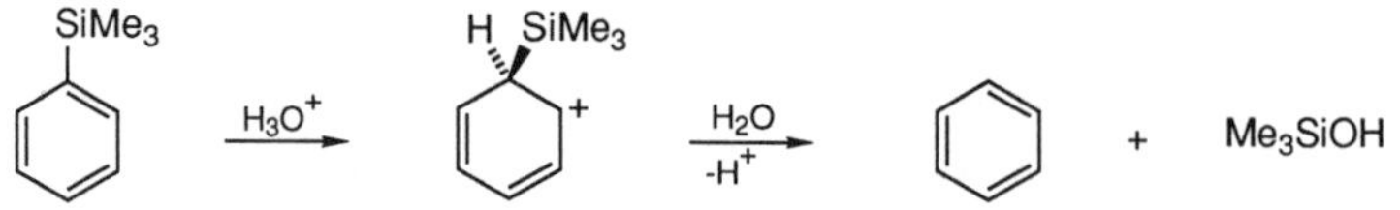

which is about ten-thousand times greater than that in which a proton is displaced from benzene.*

(i) Arylsilanes: ipso substitution.   The reactivity of silicon-substituted aromatic carbon described above provides a strong directing effect for electrophiles, e.g.

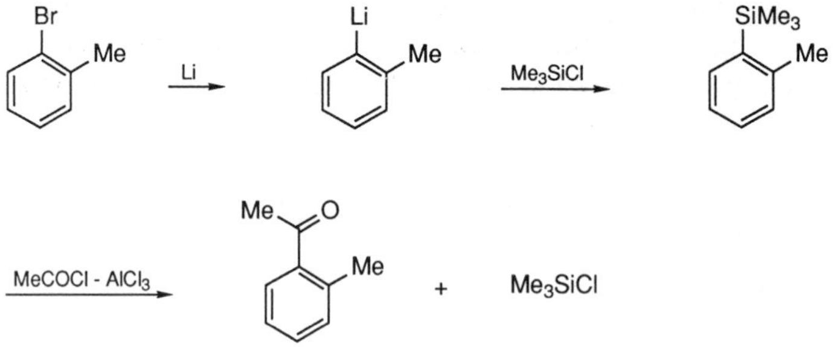

Any of the usual aromatic electrophilic reagents can be used in this way.

(ii) Alkenylsilanes can be prepared from alkynes by hydrosilylation (cf. hydroboration), catalyzed by hexachloroplatinic acid, e.g.

---

* Addition of a proton at the *meta* position also gives a cation with a β-silicon substituent,

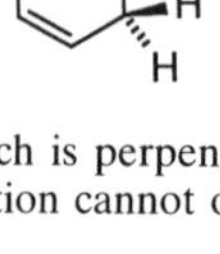

However, the C—Si bond is in a plane which is perpendicular to the p orbitals associated with the positive charge, so that C—Si hyperconjugation cannot occur. Consequently the *meta* position is not activated.

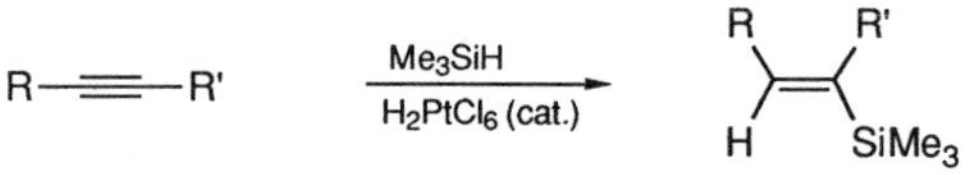

and from Grignard reagents, e.g.

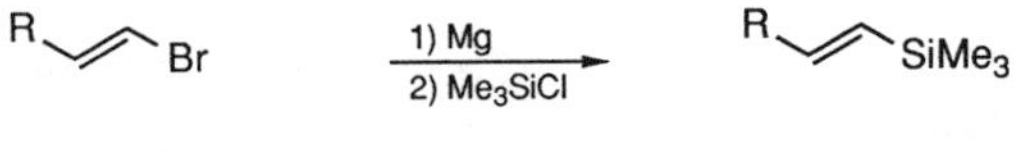

They undergo the aliphatic counterpart of aromatic *ipso* substitution, e.g.

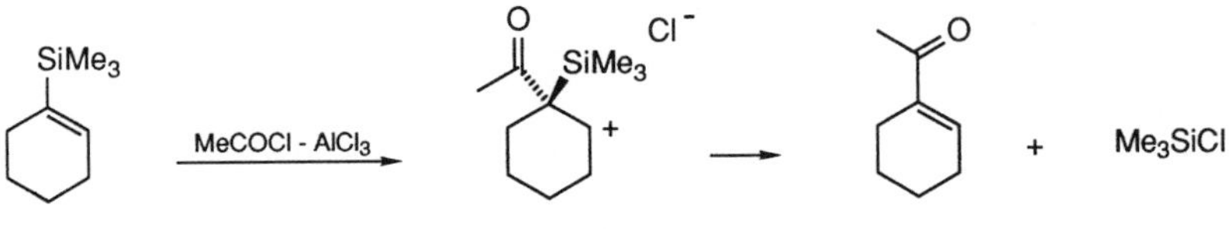

Reaction normally occurs with retention of configuration, e.g.

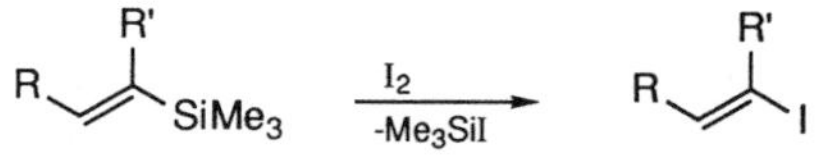

It is believed that this is because the carbocation formed with an electrophile E^+ undergoes rotation about the new C—C single bond in the direction shown as a result of the increase in the stabilizing interaction between the unoccupied $2p$ orbital and the C—Si bonding orbital:

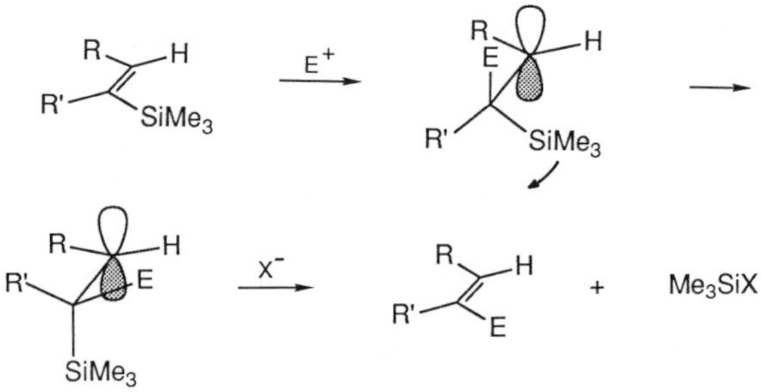

Inversion is observed in some cases (e.g. with bromine). This is probably because the initial carbocation rapidly completes an *anti*-addition; rotation about C—C in the adduct (e.g. the dibromo compound) then occurs to give the required antiperiplanar conformation for elimination:

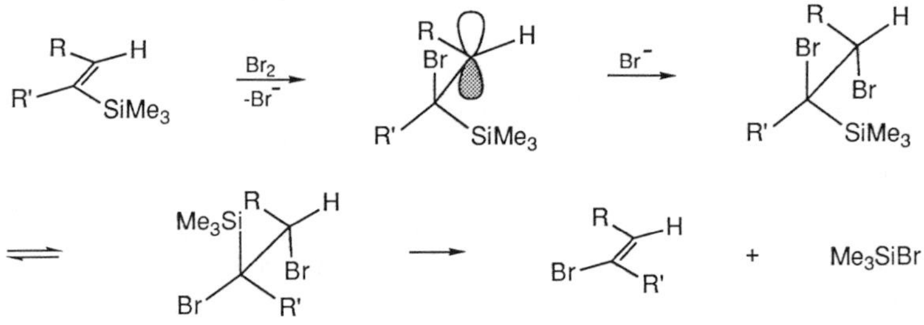

(*iii*)   *Allylsilanes* are readily available from allyl halides *via* Grignard reagents, e.g.

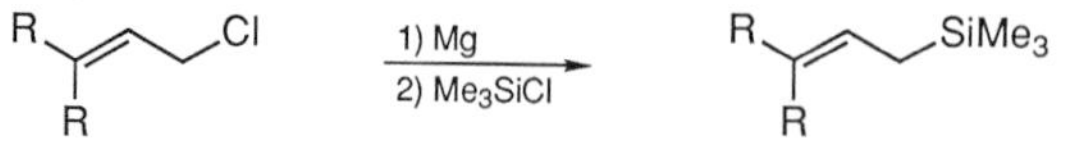

Their characteristic reaction is the addition of an electrophile at the further of the unsaturated carbon atoms, reflecting again the stabilizing influence of a C—Si bond on a β-cationic centre. Elimination of the silyl group then results in regeneration of the C=C bond in a new position, e.g.

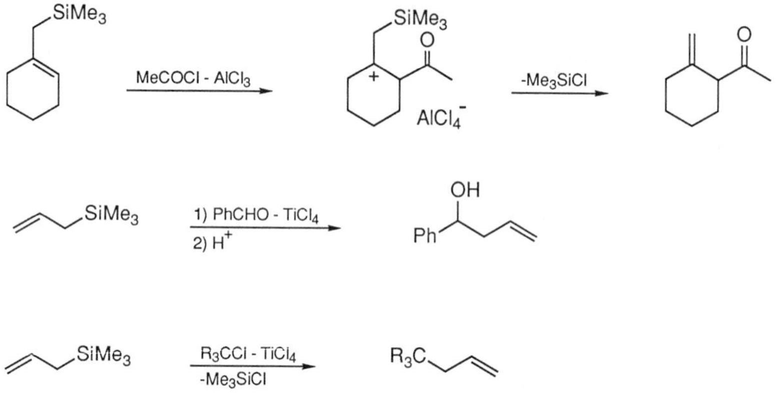

With a proton as reagent, double-bond migrations can be directed, e.g.

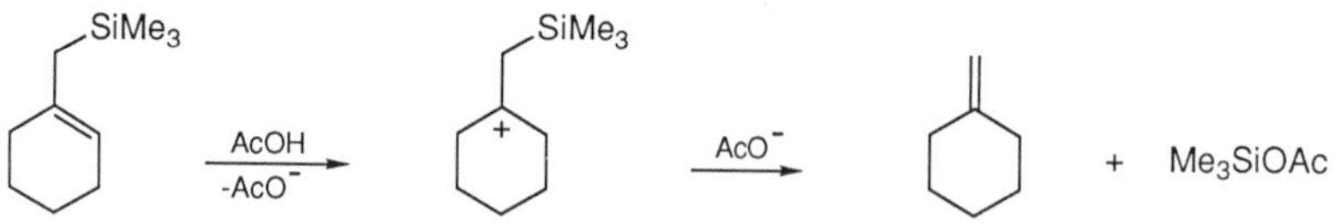

(*iv*)   *Silyl epoxides* can be obtained from vinylsilanes with peroxyacids or from ketones, e.g.

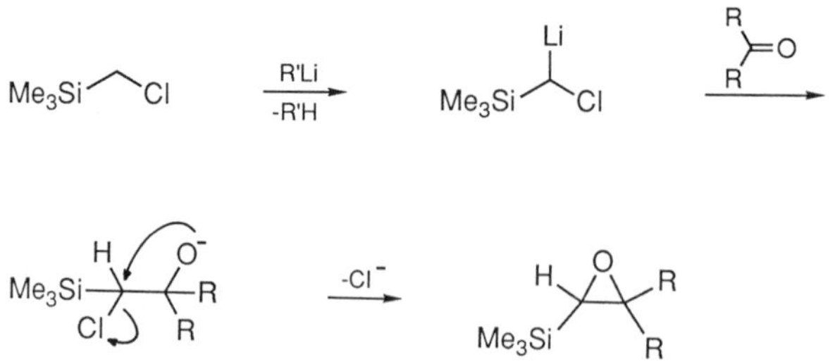

Their main uses are in the formation of ketones by acid-catalyzed ring opening, the regioselectivity of which is directed by the stabilizing effect of a β-Si—C bond on a carbocation, and in the formation of β-hydroxysilanes required for the Peterson reaction, e.g.

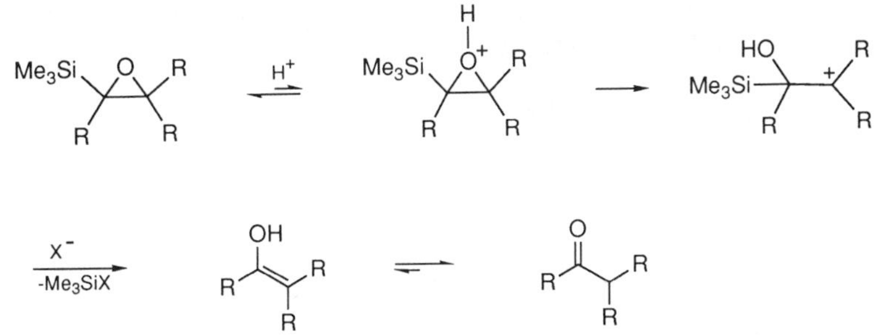

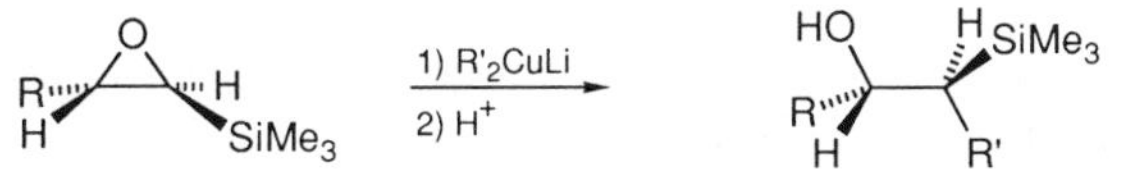

The regioselectivity is probably directed by the stabilization by α-SiMe$_3$ of the partial negative charge developed in the transition state at the carbon undergoing S$_N$2 substitution.

15.5 Boron-containing reagents

Diborane, B$_2$H$_6$, adds readily to alkenes to give organoboranes which have numerous synthetic uses. Diborane is available commercially in the form of the complexes which it forms with ethers, e.g.

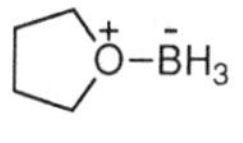

and dimethyl sulfide, Me$_2$S—BH$_3$. Alternatively, it can be generated *in situ* by the reaction of sodium borohydride with the boron trifluoride-ether complex:

$$3\,NaBH_4 \ + \ 4\,F_3B{-}OEt_2 \ \longrightarrow \ 3\,NaBF_4 \ + \ 4\,H_3B{-}OEt_2$$

One procedure is to mix the unsaturated compound and the borohydride in an ether solvent such as (MeOCH$_2$CH$_2$)$_2$O ('diglyme') and then to add the boron trifluoride complex slowly at room temperature. After the rapid reaction, the next reagent is introduced. However, if the borohydride would react with another group in the alkene or alkyne, gaseous diborane is passed into the solution.

(a) *Formation of organoboranes: hydroboration*

The characteristics of the reaction of diborane with an alkene are:

(1) If the alkene is either monosubstituted or disubstituted with groups which are not bulky, trialkylboranes are formed, e.g.

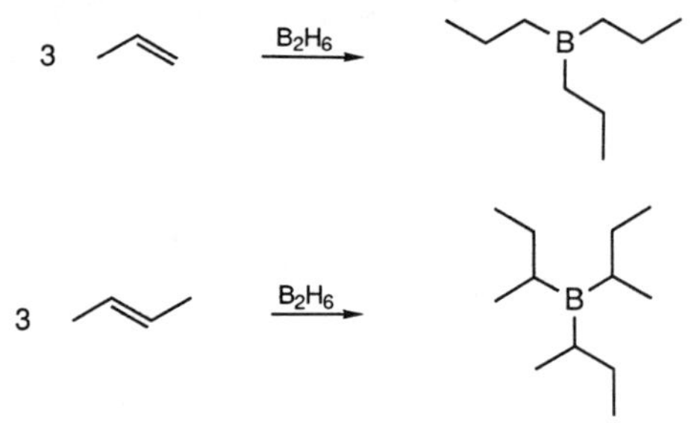

However, with sterically hindered alkenes such as trisubstituted ones, reaction can be controlled to give dialkylboranes, e.g.

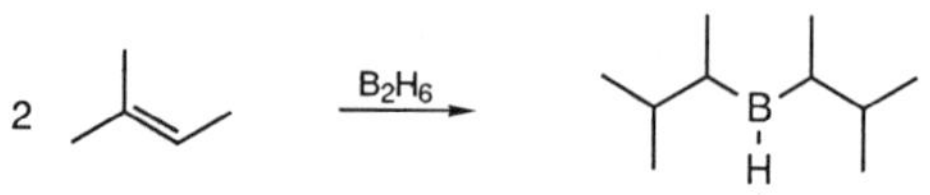

or even monoalkylboranes, e.g.

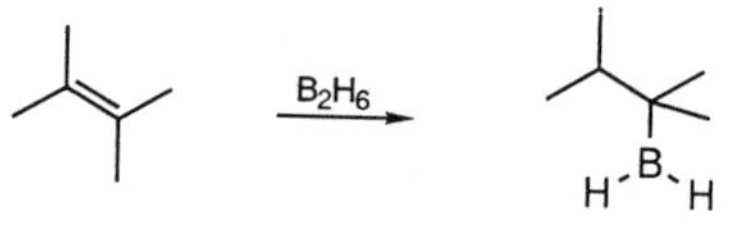

thexylborane

Monoalkylboranes can be made by reaction of an alkene with benzo-1,3-dioxa-2-borole ('catecholborane'), followed by reduction with aluminium hydride:

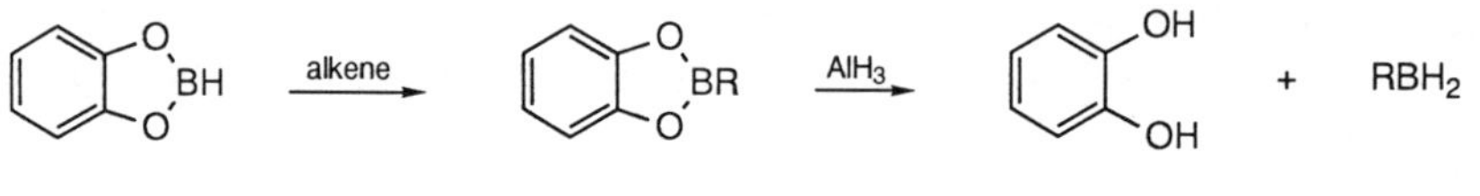

(2) The boron atom adds predominantly to the less substituted carbon atom, as shown above for propylene and trimethylethylene. For example, the preference for attack by boron at the two unsaturated carbon atoms of 1-butene is

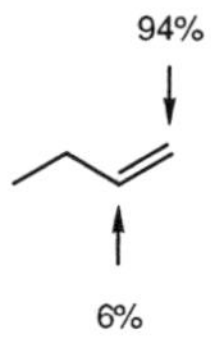

94%

6%

With 1,2-disubstituted alkenes, the selectivity is less marked, e.g.

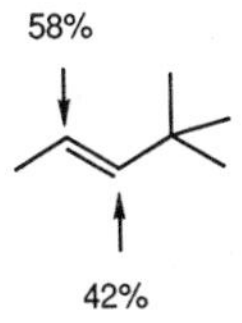

58%

42%

(3)   Addition occurs with *syn* stereospecificity, e.g.

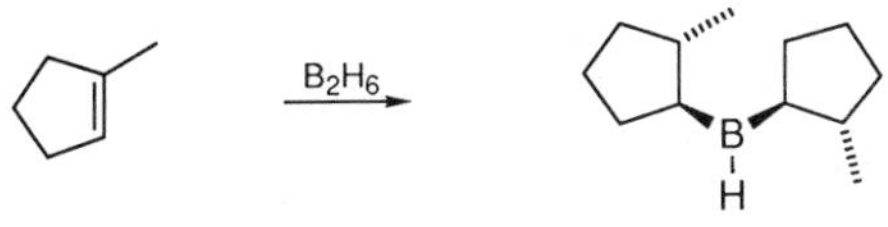

(4)   The alkene reacts at the less hindered face, e.g.

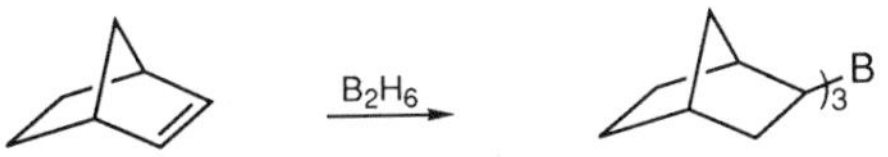

It is generally considered that reaction occurs *via* a complex in which the π-bond of the alkene donates to boron's vacant $2p$ orbital, followed by formation of a four-centred cyclic transition state:*

(*i*)   *Alkynes* also form organoboranes. The reaction with diborane is efficient with monosubstituted alkynes, but disubstituted alkynes tend to form polymers and an alternative boronating agent is used (p. 491).

(*ii*)   *Other boronating agents.*   Several organoboranes which are formed from alkenes with diborane and contain one or two B—H bonds are themselves of value as boronating agents. They are of two general types.

(1)   Sterically congested boranes are much more strongly regioselective than diborane. For example, the boron atom in the adduct of trimethylethylene and diborane (shown above) which is known as disiamylborane, Sia_2BH, reacts almost exclusively at the methyl-substituted unsaturated carbon atom in $MeCH{=}CHCMe_3$, in contrast to the 58:42 ratio given by diborane. Another such selective borane is 9-borabicyclo[3.3.1]nonane (9-BBN), formed from 1,5-cyclooctadiene,

* A concerted cycloaddition is symmetry-forbidden unless the vacant orbital on boron is involved.

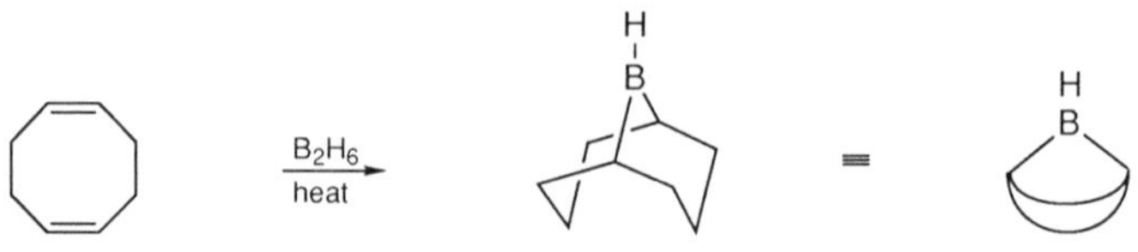

9-BBN

9-BBN and Sia$_2$BH are also effective boronating agents for disubstituted alkynes. These boranes have a second useful function. In some reactions of an organoborane R$_3$B, only one of the R groups is used; the other two become part of a boron-containing byproduct. If the alkene to which R$_3$B corresponds is hard to obtain or expensive, it is best to use Sia$_2$BH or 9-BBN to form the mixed organoborane R$_2$R'B: providing that (as is frequently the case) the R' group reacts the more readily, none is 'wasted' (e.g. p. 489). Thexylborane, from tetramethylethylene (shown above), can fulfil a similar role when two groups in an organoborane are utilized in a synthesis (p. 489).

(2) Asymmetric boranes can be made which are of value in enantioselective synthesis. The reagents of choice are prepared from (+)- and (−)-α-pinene: (+)-α-pinene and diborane form (−)-diisopinocampheylborane ('(−)-Ipc$_2$BH'),

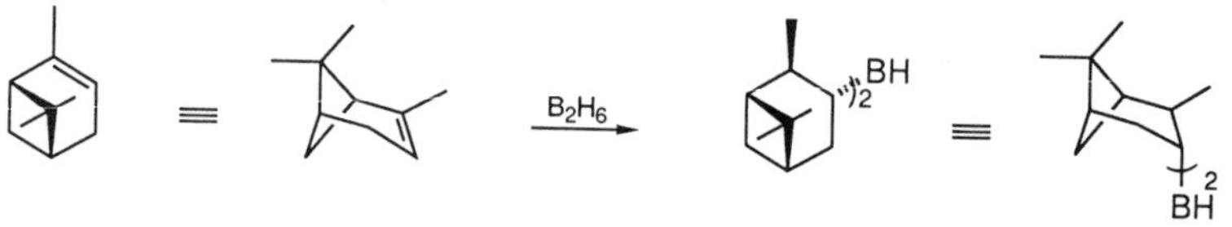

(+)-α-pinene   (-)-Ipc$_2$BH

and, with less alkene, (+)-monoisopinocampheylborane ('(+)-IpcBH$_2$'), and likewise the (−)-α-pinene forms (+)-Ipc$_2$BH and (−)-IpcBH$_2$.

(iii) *Isomerization of organoboranes.* On heating at about 150°C, organoboranes are isomerized to a mixture in which the major component has boron attached to the terminal carbon atom of the alkyl group, e.g.

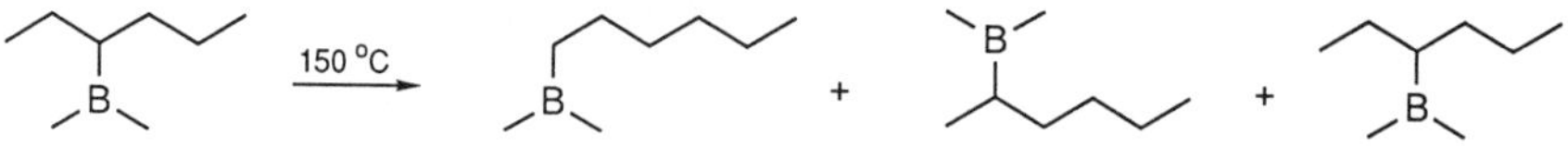

90%   6%   4%

The process is catalyzed by diborane and occurs by sequential elimination and addition steps. It provides a method by which a readily available internal alkene can be converted into, for example, a primary alcohol, e.g.

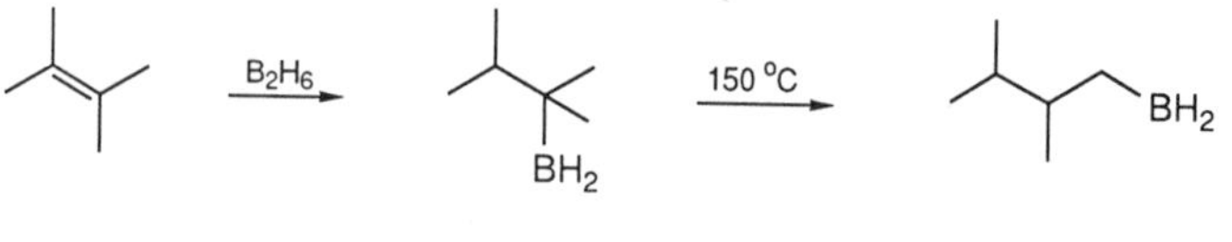

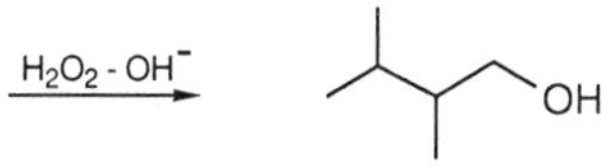

$$\xrightarrow{\text{H}_2\text{O}_2 \text{ - OH}^-}$$

(isopropyl carbinol structure with OH)

(b) Reactions of organoboranes from alkenes

The formation of alkanes is discussed later (p. 635). The other synthetic uses of organoboranes are as follows:

(i) Oxidation to alcohols.

Reaction with alkaline hydrogen peroxide solution gives alcohols, e.g.

$$\text{R}_3\text{B} \; + \; 3\,\text{H}_2\text{O}_2 \; + \; \text{NaOH} \longrightarrow 3\,\text{ROH} \; + \; \text{NaH}_2\text{BO}_3 \; + \; \text{H}_2\text{O}$$

The reaction occurs by addition of HO_2^- to boron, making use of boron's vacant $2p$ orbital, followed by rearrangement,

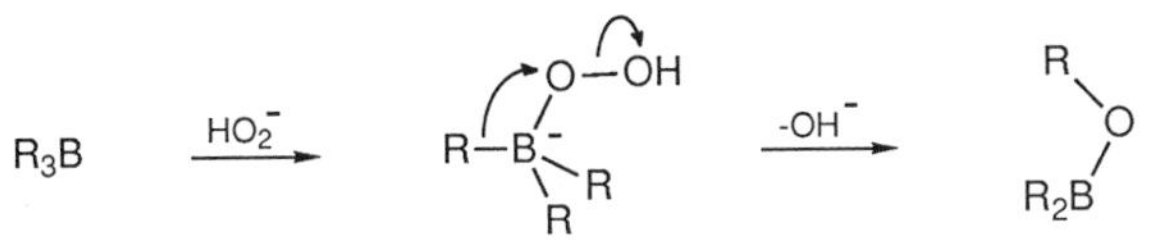

$$\text{R}_3\text{B} \quad \xrightarrow{\text{HO}_2^-} \quad \begin{array}{c} \text{R} - \underset{\overset{|}{\text{R}}}{\overset{}{\text{B}}} \overset{\text{R}}{\underset{}{}} - \text{O} - \overline{\text{O}}\text{H} \end{array} \quad \xrightarrow{-\text{OH}^-} \quad \begin{array}{c} \text{R} \\ \diagdown \\ \text{O} \\ \diagup \\ \text{R}_2\text{B} \end{array}$$

These steps are repeated to give, after all three alkyl groups have migrated, a borate ester which is hydrolyzed:

$$3\,(\text{RO})_3\text{B} \; + \; 3\,\text{H}_2\text{O} \quad \xrightarrow{\text{OH}^-} \quad 3\,\text{ROH} \; + \; \text{B(OH)}_3$$

The migrating group retains its stereochemistry and, if different groups are attached to boron, the order of ease of migration is normally primary > secondary > tertiary.

This provides a method for converting an alkene $\text{RCH}{=}\text{CH}_2$ into the alcohol $\text{RCH}_2\text{CH}_2\text{OH}$, complementing the method for the conversion $\text{RCH}{=}\text{CH}_2 \rightarrow \text{RCH(OH)-CH}_3$ by oxymercuration followed by reduction (p. 203). Because stereochemical integrity is retained in the rearrangement step, the overall reaction has *syn* stereospecificity as well as strong regioselectivity, e.g.

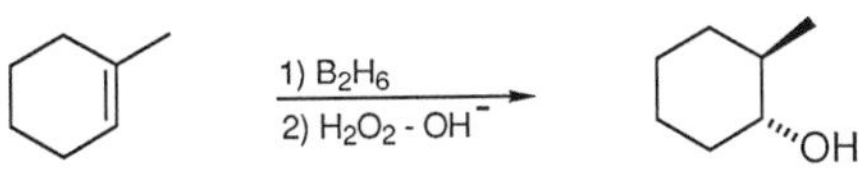

(methylcyclohexene structure) $\xrightarrow[\text{2) H}_2\text{O}_2 \text{ - OH}^-]{\text{1) B}_2\text{H}_6}$ (trans-2-methylcyclohexanol structure with "OH)

Enantioselectivity. The reaction of (+)- or (−)-Ipc$_2$BH with (Z)-1,2-disubstituted alkenes followed by oxidation gives a predominance (often >10:1) of the (S)- or (R)-alcohol, respectively, e.g.

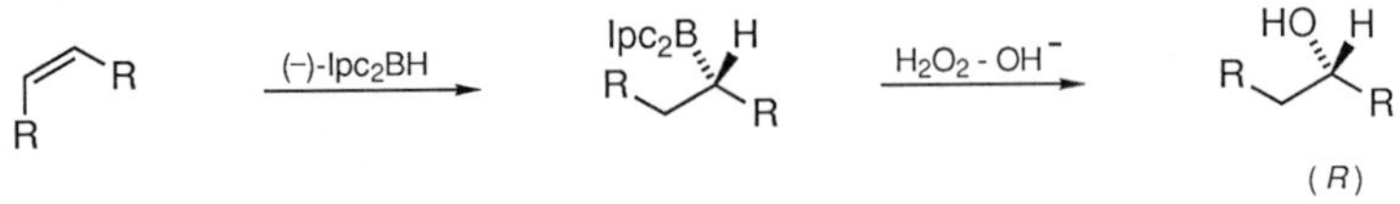

Ipc$_2$BH reacts very slowly with (*E*)-alkenes, but the less hindered (+)- or (−)-monoisopinocampheylborane can be used effectively, e.g.

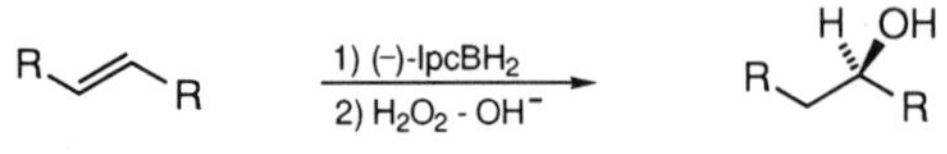

(*ii*)   *Conversion into primary amines.*   Chloramine and hydroxylamine-*O*-sulfonic acid react in a similar way to HO$_2^-$, e.g. (X = Cl or OSO$_2$OH),

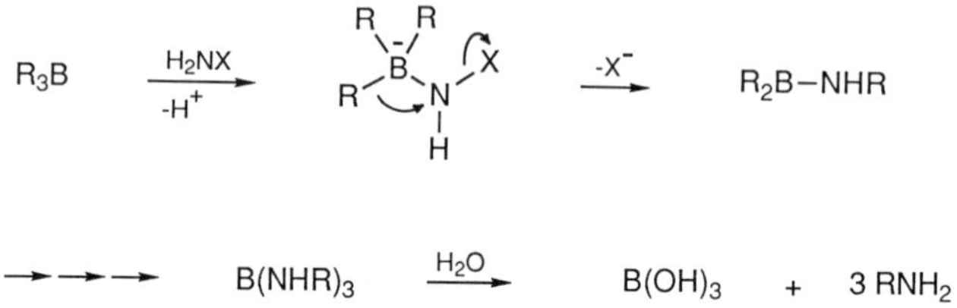

(*iii*)   *Carbonylation.*   The reaction of organoboranes with carbon monoxide opens up a variety of synthetic pathways.

1.   *Tertiary alcohols.*   Reaction of the borane with carbon monoxide in ethylene glycol at 100–125°C, followed by oxidation with alkaline hydrogen peroxide, gives a tertiary alcohol:

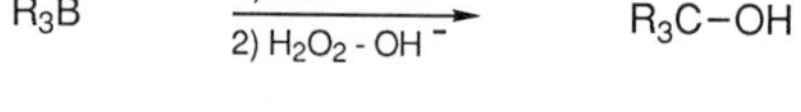

The mechanism is thought to be as follows:

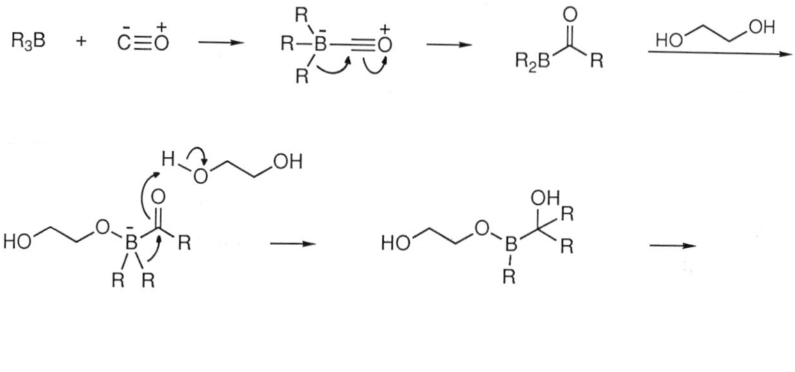

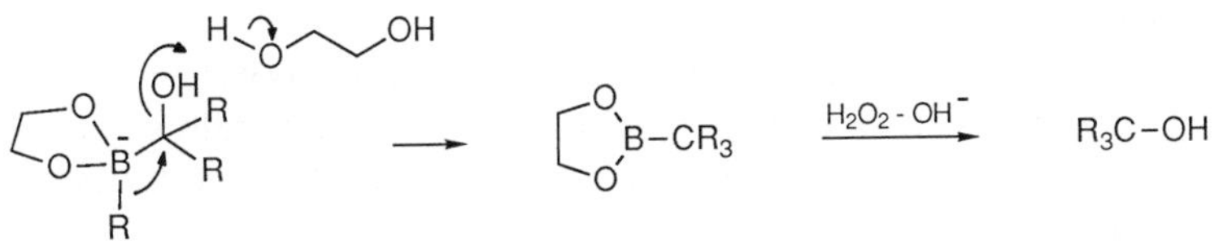

2. *Ketones and secondary alcohols.* By carrying out the reaction in the presence of water, the intermediate formed after the second rearrangement step (which contains a hydroxyl and not a glycol substituent) is intercepted. It can be oxidized (HO_2^-) to a ketone or hydrolyzed to a secondary alcohol:

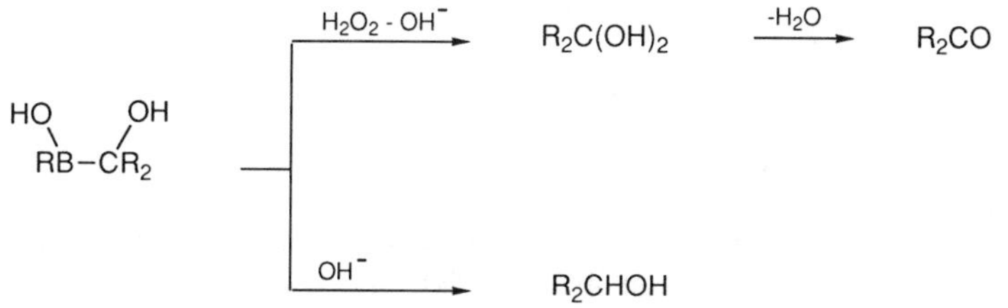

Thexylborane is normally used to provide one of the alkyl groups in the borane: since it is attached to boron by a tertiary alkyl group, it rearranges less readily than the other groups if they are primary or secondary, so avoiding waste. The sequential introduction of two alkenes enables unsymmetrical products to be formed, e.g.

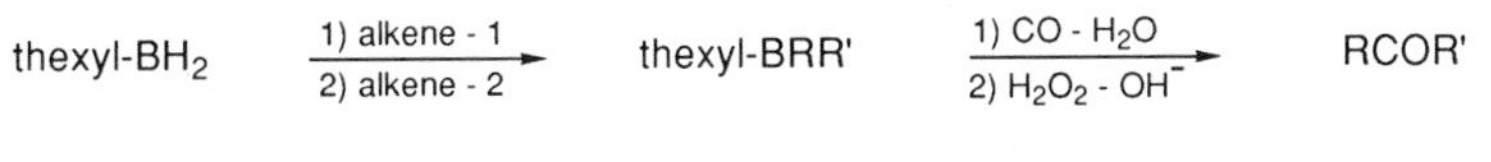

$$\text{thexyl-BH}_2 \xrightarrow[\text{2) alkene - 2}]{\text{1) alkene - 1}} \text{thexyl-BRR'} \xrightarrow[\text{2) H}_2\text{O}_2\text{ - OH}^-]{\text{1) CO - H}_2\text{O}} \text{RCOR'}$$

and cyclic compounds can be formed from dienes, e.g.

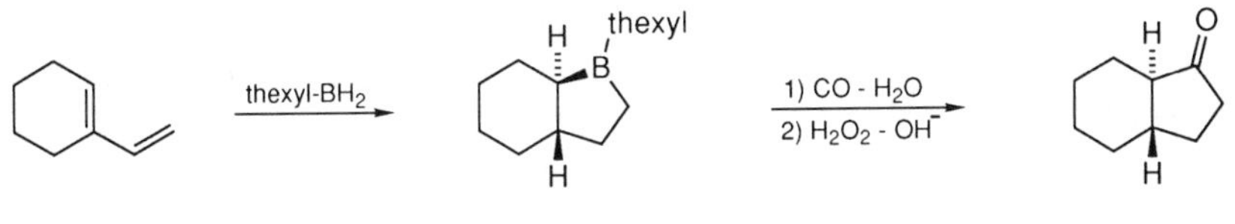

3. *Aldehydes.* By carrying out the carbonylation in the presence of a reducing agent such as lithium trimethoxyaluminium hydride, the intermediate following the first rearrangement step is intercepted. Oxidation yields aldehydes:

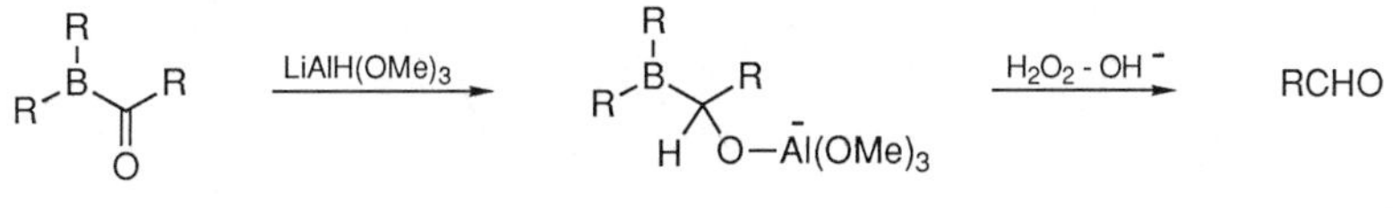

9-BBN is normally used to provide two of the alkyl groups and the process constitutes a method for the conversion $RCH{=}CH_2 \rightarrow RCH_2{-}CH_2{-}CHO$.

(*iv*) *Cyanidation.* Cyanide ion, which is isoelectronic with carbon monoxide, reacts similarly. The nitrogen atom in the initial adduct is not sufficiently electron-attracting to induce rearrangement but is made so by trifluoroacetylation: two alkyl groups are transferred at low temperatures, enabling ketones to be formed by oxidation,

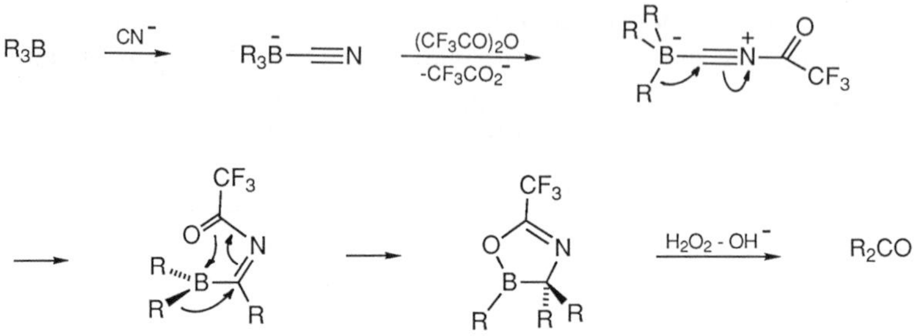

and the third group is transferred on heating with an excess of trifluoroacetic anhydride, to give tertiary alcohols after oxidation,

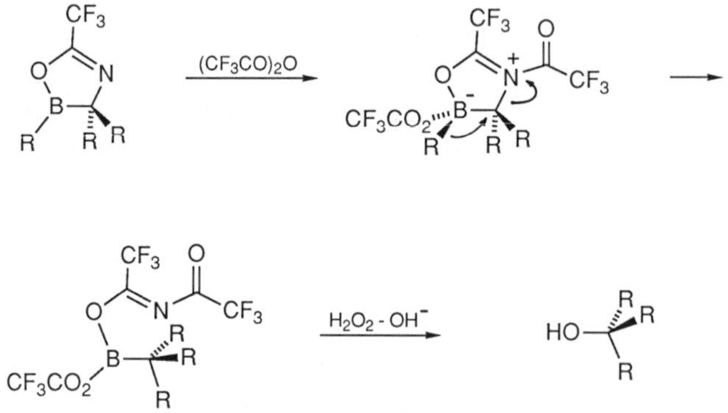

This method is superior to that based on carbonylation if lower temperatures are needed to avoid reactions of other groups.

(*v*) *Synthesis of esters.*   The enolate formed from an α-haloester with a strong, hindered base adds to an organoborane to give an anion that rearranges, e.g.

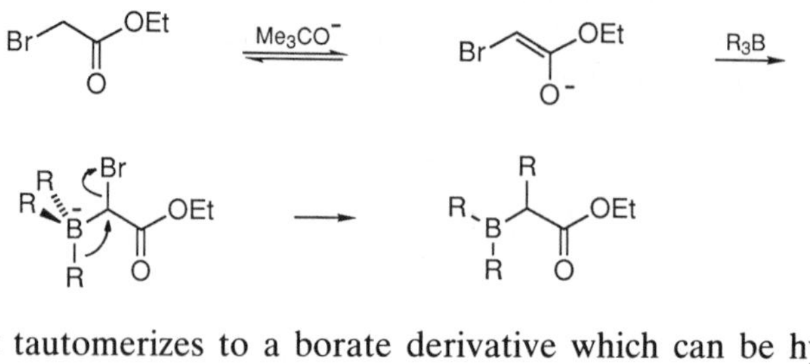

The product tautomerizes to a borate derivative which can be hydrolyzed to an ester,

9-BBN is often used to provide two of the alkyl groups in the borane.

(c)   Reactions of organoboranes from alkynes

(i)   Synthesis of ketones and aldehydes.   The alkyne is boronated, usually with 9-BBN or Sia$_2$BH, and the adduct is oxidized with alkaline hydrogen peroxide:

With a terminal alkyne, the product is an aldehyde,

and this provides a complementary method to the direct hydration catalyzed by mercury(II) sulfate, which gives methyl ketones (p. 91).

(ii)   Synthesis of alkenes.   Both (*E*)- and (*Z*)-alkenes can be made from a mono-substituted alkyne. Thexylborane is normally used to provide one of the alkyl groups in a borane RR′BH.

For (*E*)-alkenes, a l-haloalkyne, e.g.

is hydroborated with *syn* specificity. Methoxide ion induces rearrangement with elimination of the halide, and protonolysis gives the alkene:

For (Z)-alkenes, hydroboration of the alkyne itself is followed by treatment with iodine and alkali. It is believed that rearrangement occurs within an iodonium ion, from which (Z)-alkene is formed by elimination *via* an antiperiplanar transition state:

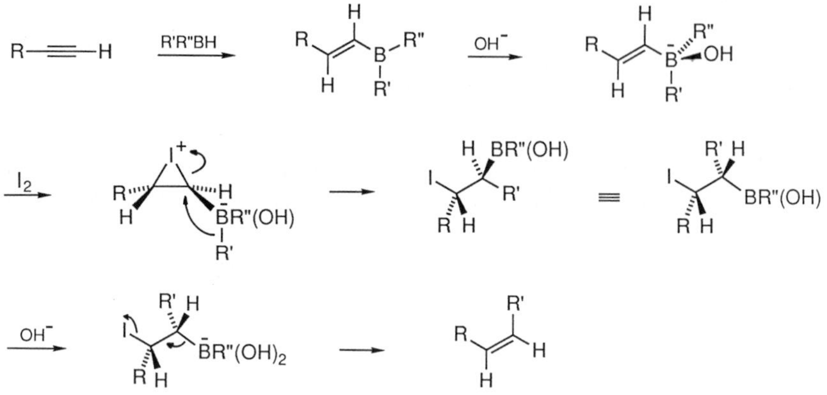

In some instances, thexyl groups migrate in this reaction.

(*iii*) *Synthesis of conjugated dienes.* The principle in the formation of (*E*)-alkenes can be applied to making (*E*,*E*)-dienes; the rearranging group is alkenyl:

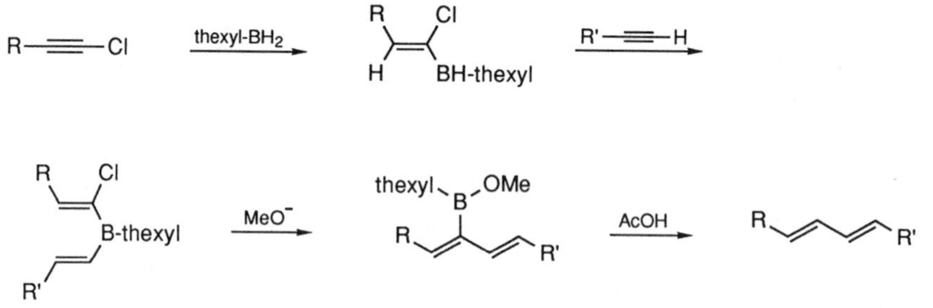

(*E*,*Z*)-dienes can be made *via* enynes by a route corresponding to that for (*Z*)-alkenes:

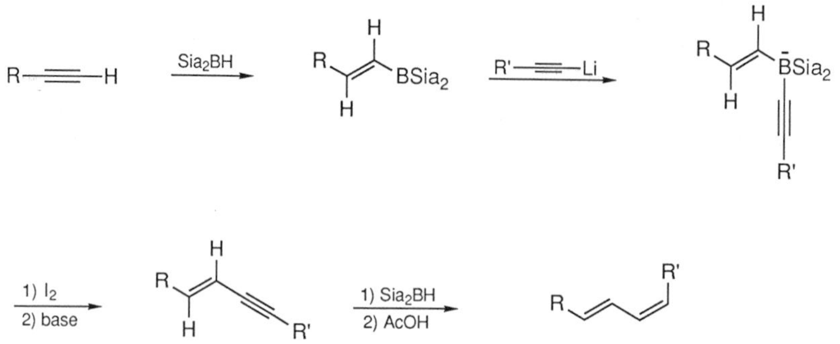

Finally, (Z,Z)-dienes can be made from diynes,

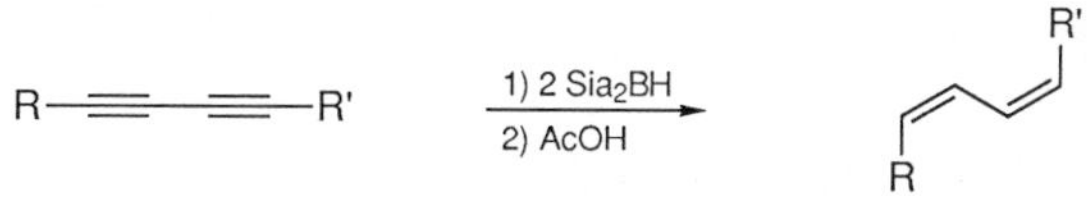

(iv) *Synthesis of bromoalkenes.* When catecholborane is used to hydroborate an alkyne, treatment of the adduct with bromine and base does not result in rearrangement as above. Instead, a (Z)-bromoalkene is formed, probably by *anti*-addition of bromine followed by an *anti*-elimination:

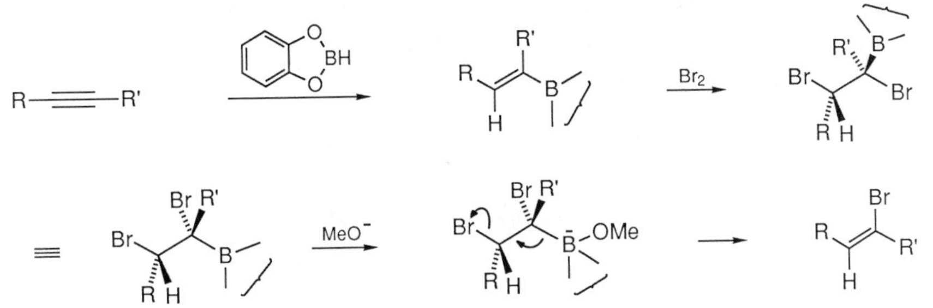

1. How would you employ phosphorus-, sulfur-, or boron-containing reagents in the synthesis of the following?: **Problems**

(a)

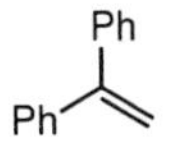

(b)

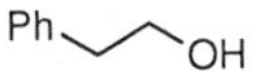

(c)

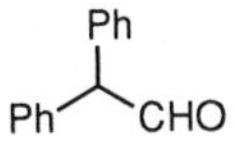

(d)

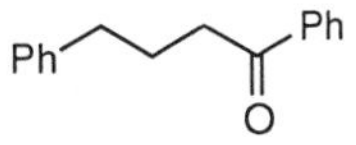

(e)

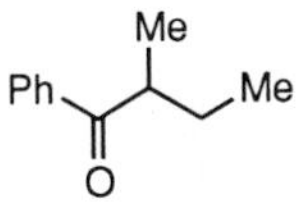

(f)

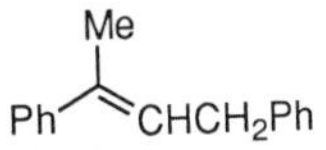

(g)

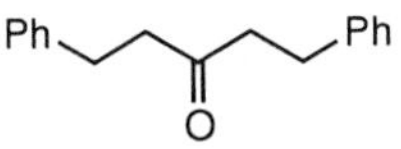

(h)

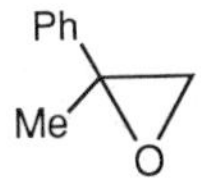

(i)

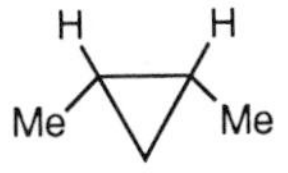

(j)

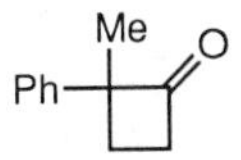

(k)

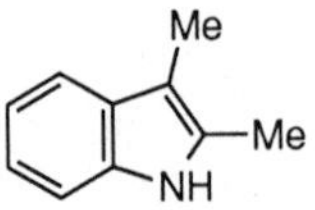

(l)

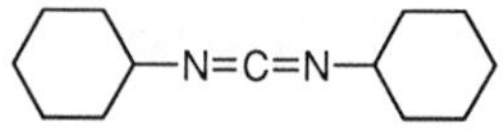

2. Rationalize the following reactions:

(a)

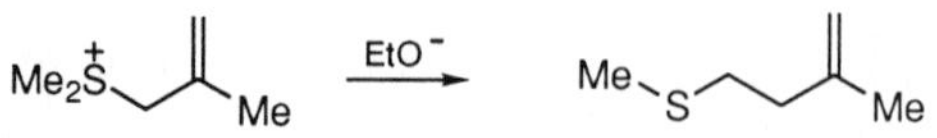

(b)

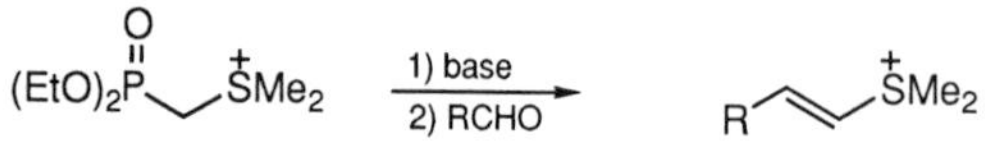

(c)

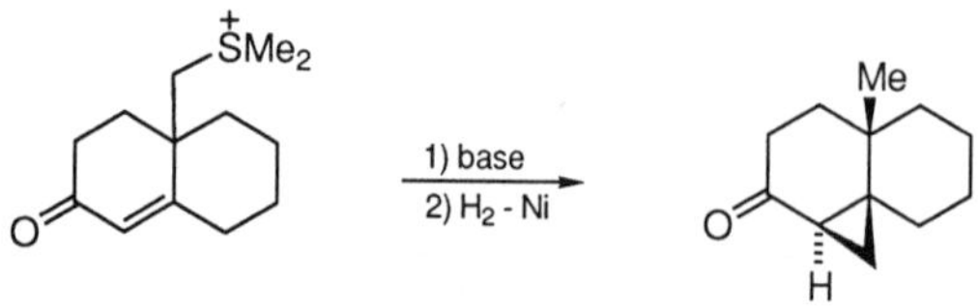

(d)

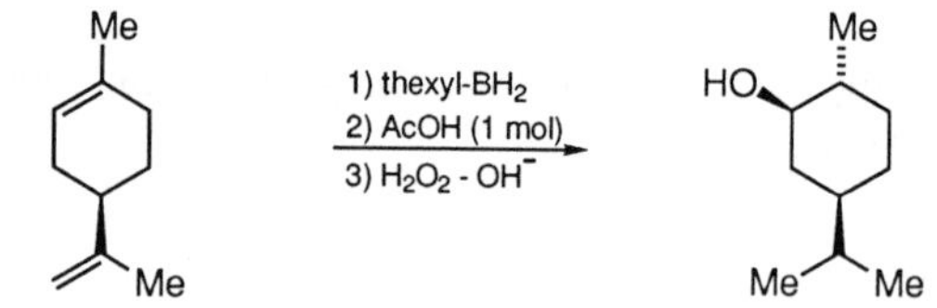

(e)

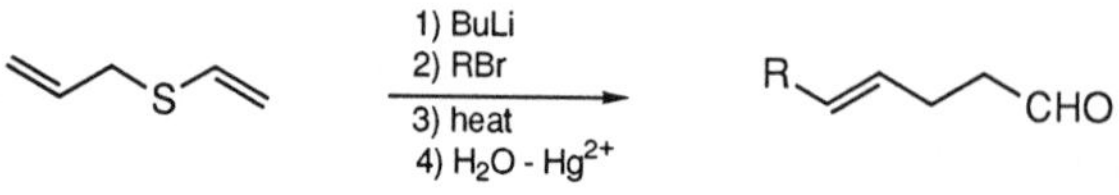

(f)

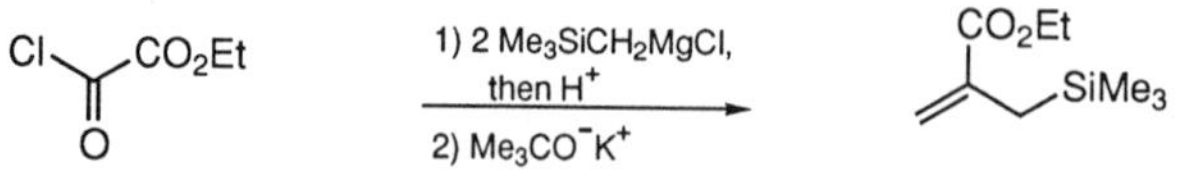

(g)

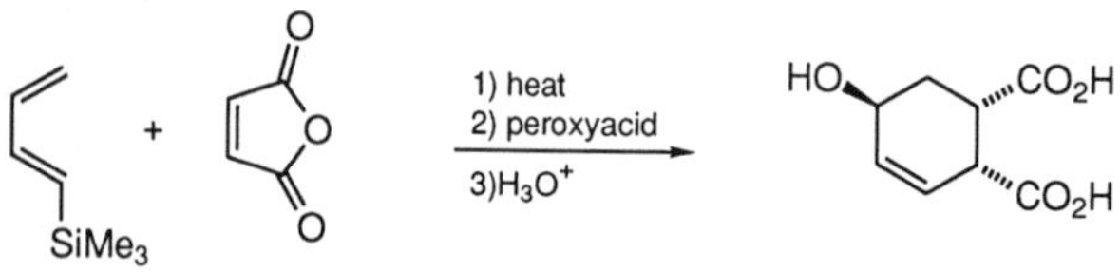

3. What products would you expect from the following reactions?

(a)

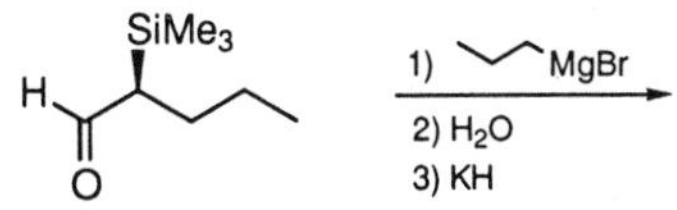

1) ⌇MgBr
2) H$_2$O
3) KH

(b)

SiMe$_3$

MeCOCl - AlCl$_3$

(c)

SiMe$_3$

DCl / CH$_2$Cl$_2$

(d)

SiMe$_3$

Br$_2$ / CCl$_4$

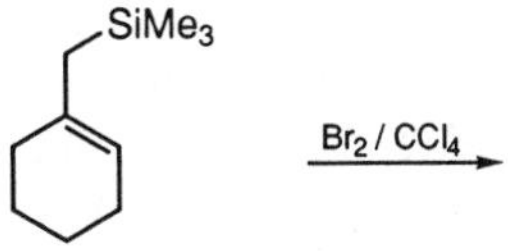

4. Explain why the ratios of (E):(Z)-alkene are different in the following Wittig reactions:

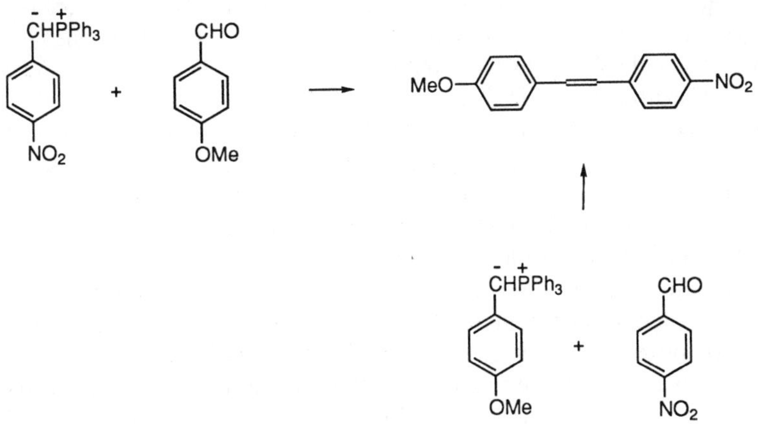

16 Photochemical reactions

**16.1
Principles**
The electronic excitations of compounds require quanta of energy which correspond to wavelengths of electromagnetic radiation lying in the ultraviolet or,
less frequently, the visible region of the spectrum. When a molecule absorbs
a photon of the appropriate energy, one of its electrons is raised to an orbital
of higher energy. The resulting molecule may then take part in a reaction while
in this, or an alternative, excited state. The reaction so induced may be either
intramolecular – for example, rearrangement or dissociation – or intermolecular –
for example, addition. Alternatively, the excited molecule may transfer its energy
to another molecule which in turn undergoes reaction. These variants open up a
wide range of synthetic possibilities.

The first requirement, then, is that the reactant should absorb light at a
wavelength which is experimentally accessible. In practice, the ultraviolet region
below 200 nm is inconvenient because the radiation is absorbed by air and vacuum
techniques are required; moreover, normal solvents absorb in the range up to
220 nm. The useful spectrum therefore begins at 220 nm. For the ultraviolet
region, a low-pressure mercury arc lamp has its strongest emission line at 254 nm
which is a wavelength at which aromatic molecules absorb. Broad-band emission
is in general more useful: it is obtained with a higher pressure mercury lamp,
which also gives more intense radiation, and wavelengths which might bring about
unwanted reactions can to some extent be removed by appropriate filters. For the
visible region, a tungsten lamp or the more powerful xenon arc are suitable.
Saturated organic compounds absorb only at wavelengths well below 200 nm, but
many unsaturated ones absorb above 230 nm, so that photochemical reactions are
essentially limited to those in which at least one of the reactants is unsaturated or
aromatic.

Consider a compound containing a C=C group. Light of the appropriate
wavelength – in the region 180–200 nm for compounds with isolated double
bonds, but of longer wavelength for conjugated systems – causes a π-electron to
be excited to the lowest antibonding π orbital (i.e. a π^* orbital): the transition is
described as $\pi \rightarrow \pi^*$. The new state of the molecule is described as the first
excited singlet, S_1 (since the promoted electron and its original partner still have
opposite spins); this particular excited singlet is of (π,π^*) type. For compounds
containing a carbonyl group, two types of excitation are commonly effected: one,

which requires the more energy and therefore corresponds to shorter wavelengths (e.g. $\lambda_{max.}$ 187 nm for acetone), is a $\pi \rightarrow \pi^*$ transition. The other, which corresponds to longer wavelengths (e.g. $\lambda_{max.}$ 270 nm for acetone), is an excitation of one of the non-bonding electrons on the oxygen atom into an antibonding π orbital; it is described as an $n \rightarrow \pi^*$ transition and the excited state, S_1, is of (n,π^*) type.

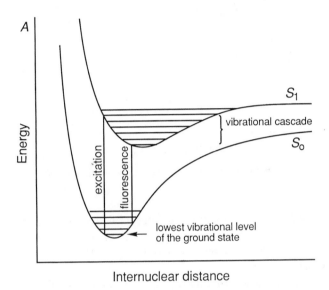

Figure 16.1 Excitation states. The *x*-axis is labelled internuclear distance for simplicity. The diagram is strictly appropriate for a diatomic molecule. For others a full description would require a multi-dimensional representation.

Figure 16.1 shows a typical situation. Initially, almost all molecules are in the lowest vibrational level of the ground state, S_0. The excitation occurs so rapidly that the nuclei do not have time to alter their relative positions; it is therefore represented by a vertical line. Consequently, the S_1 state can be reached at a point which corresponds to an excess of vibrational energy. However, this energy is rapidly dissipated through molecular collisions (vibrational cascade) and thereafter one of five processes takes place:

● The molecule returns to its ground state either by giving up its energy as heat, undergoing *internal conversion* to a high vibrational level of the ground state, followed by release of energy in a vibrational cascade, or by emitting radiation (fluorescence). As shown in Figure 16.1 the fluorescent light is of slightly lower energy (longer wavelength) than the absorbed light.
● A chemical reaction occurs.
● The energy of excitation is transferred to another molecule, which is thereby raised to an excited singlet level.

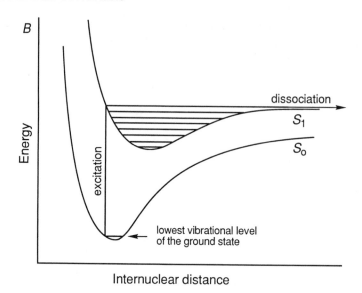

Figure 16.2   Excitation leading to a point where the excited molecule dissociates.

- One of the unpaired electrons in the excited molecule undergoes an inversion of its spin, leading to a lower-energy state described as a triplet, T_1, which contains two unpaired electrons of parallel spin; this is described as inter-system crossing.
- In the special case shown in Figure 16.2, excitation leads to a point in S_1 which is *above* that of the energy curve on the right-hand side; consequently the excited molecule, instead of vibrating, dissociates.

The triplet state may return to the ground state, with a further spin inversion, either by emitting radiation (*phosphorescence*) or by giving up energy as heat; it may take part in a chemical reaction; it may transfer its energy to another molecule which is thereby raised to the triplet state; or it may dissociate.

These processes are summarized as follows:

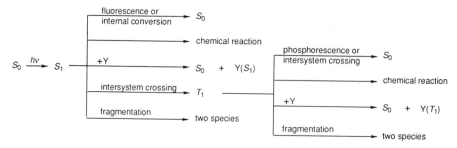

The life-times of excited singlets (i.e. the time-lag before radiative return to the ground state) are very short (10^{-6}–10^{-9} seconds), so that chemical reactions occurring *via* these species, despite their high energy, must be exceptionally fast and are comparatively uncommon. Conversion into the triplet state occurs with an efficiency which varies widely with the structure of the molecule: for benzophenone, for example, it is a highly efficient process whereas for alkenes it is not. The life-times of triplets are relatively long ($>c.$ 10^{-4} seconds for species containing relatively light atoms) so that the probability of the triplet taking part in a chemical reaction is much higher than for a singlet.

There are two key features of reactions which occur through excited states which underlie their special synthetic importance. First, since the excited state usually has a large excess of energy compared with the ground state,* it is often possible to effect reactions which, insofar as the *ground*-state reactant is concerned, are thermodynamically unfavourable. Second, reaction can usually be carried out, if required, at low temperature, so that the product can be formed 'cold'. For these reasons, it is often possible to make, for example, highly strained ring systems; energy is pumped in as light to overcome the activation-energy barrier in their formation, and they are produced in conditions which allow them to survive.

The reactions of excited states, including both singlets and triplets, can be classified under the following headings: reduction, addition, rearrangement, oxidation, aromatic substitution, and fragmentation.

**16.2
Photoreduction**

Carbonyl compounds can be converted into 1,2-diols and alcohols by irradiation in the presence of a hydrogen-donating compound. Reaction occurs through the triplet state of the carbonyl compound which abstracts a hydrogen atom from the second reactant,

$$R_2C=O \xrightarrow{h\nu} R_2\overset{1}{C}\overset{\cdot}{=}O^{\cdot} \xrightarrow{R'H} R_2\overset{\cdot}{C}-OH + R'\cdot$$

triplet

The resulting radicals either combine,

$$2\ R_2\overset{\cdot}{C}-OH \longrightarrow R\underset{HO}{\overset{R}{>}}\underset{OH}{\overset{R}{<}}R$$

or abstract hydrogen from the solvent,

$$R_2\overset{\cdot}{C}-OH + R'H \longrightarrow R_2CH-OH + R'\cdot$$

The latter reaction is favoured when $R'\cdot$ is a stabilized radical (e.g. tertiary) and the coupling reaction is favoured when R is a radical-stabilizing group (e.g. aryl). The coupling reaction is usually most efficient when the hydrogen donor is

---

* For example, light of wavelength 300 nm is equivalent to energy of $c.$ 400 kJ mol^{-1}.

itself the alcoholic reduction product of the carbonyl compound, for then only one dimeric product can be formed, e.g.

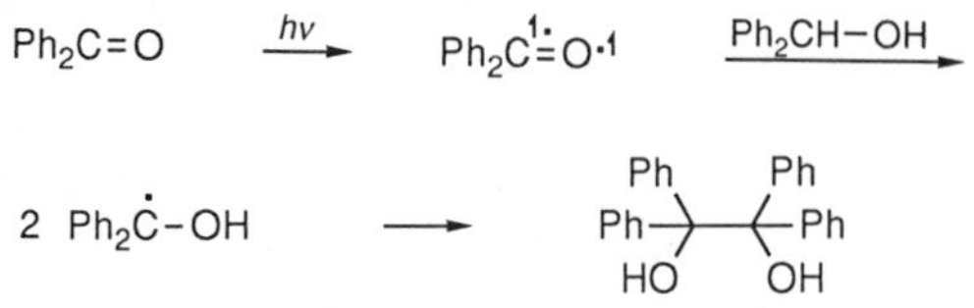

$$Ph_2C=O \xrightarrow{hv} Ph_2\overset{1}{\overset{.}{C}}=\overset{.1}{O} \xrightarrow{Ph_2CH-OH}$$

$$2\ Ph_2\overset{.}{C}-OH \longrightarrow \underset{\underset{HO}{}{}}{Ph}\overset{Ph}{\underset{}{}}\underset{\underset{OH}{}{}}{\overset{Ph}{}}{}-Ph$$

Reaction occurs in good yield only when the triplet is of the (n,π^*) type; for example, it is efficient for benzophenone and acetone, where this is so, but not for p-phenylbenzophenone whose lowest triplet state is of (π,π^*) type.

o-Alkyl-substituted aromatic ketones, even though giving (n,π^*) triplets, are not reduced in this way. In preference, intramolecular hydrogen-atom abstraction occurs to give the enol tautomer:

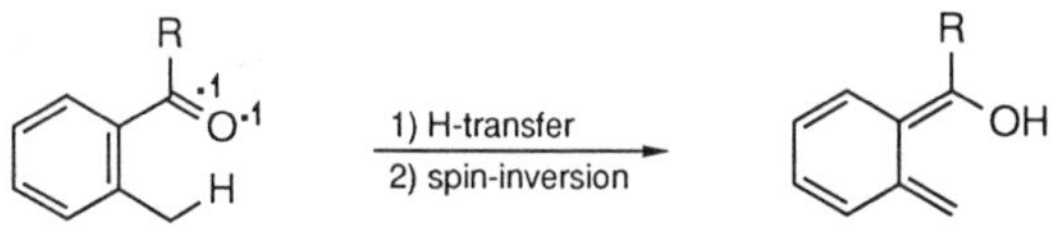

triplet

The enol then rearranges back to the more stable ketone, though it can be trapped as a Diels–Alder adduct with, for example dimethyl acetylenedicarboxylate:

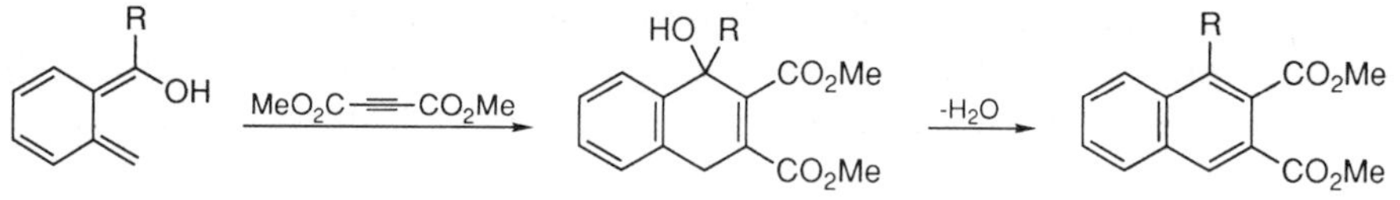

16.3 Photoaddition

Photoaddition is the formation of a 1:1 adduct by reaction of an excited state of one molecule with the ground state of another. The molecule in the ground state is commonly an alkene; the reactant which is excited can be a carbonyl compound, quinone, aromatic compound, or another molecule of the same alkene. The majority of the reactions lead to the formation of a ring.

(a) Photoaddition of alkenes to carbonyl compounds

This photoaddition – the *Paterno–Büchi reaction* – normally occurs by reaction of the triplet state of the carbonyl compound with the ground state of the alkene. As in photoreduction, it is more efficient when the triplet is of (n,π^*) rather than (π,π^*) type. Typical examples, which also show the marked variations in the yields of products, are the addition of benzophenone to propylene and isobutylene:

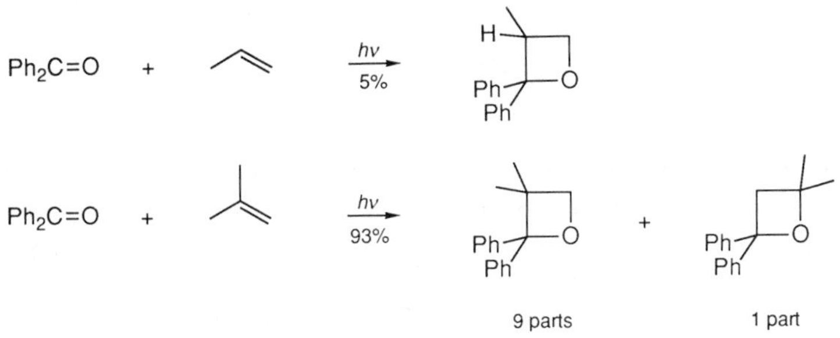

The ring is formed in two stages. The excited carbonyl compound (triplet) first adds through its oxygen atom to the alkene so as to give mainly the more stable of the two possible diradicals. A spin-inversion then occurs and the second bond is formed.

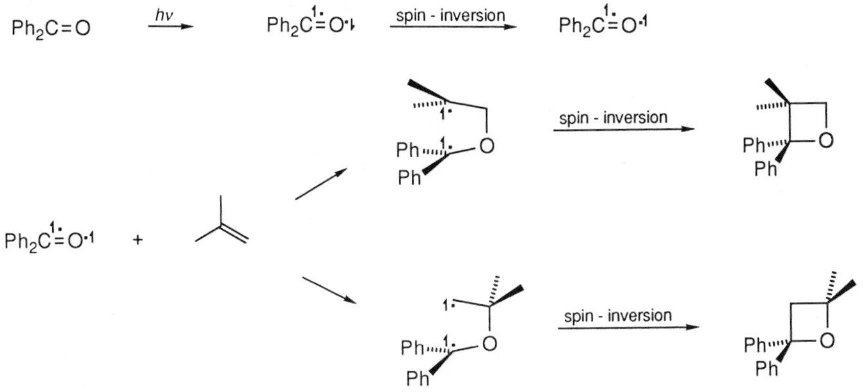

The reaction is not stereospecific, because the time-lag before the final spin-inversion is more than enough for rotation to occur about single bonds; for example, *cis*- and *trans*-2-butene give the same mixture of *cis*- and *trans*-oxetanes with benzophenone:

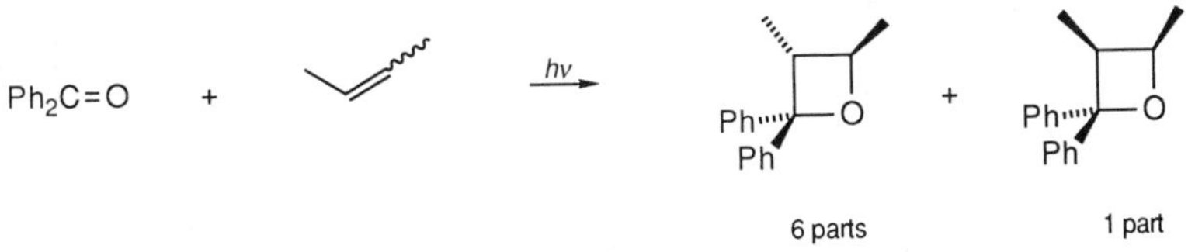

There is one general circumstance in which photoaddition of alkenes to carbonyl compounds fails: namely, when the energy difference between the triplet and ground state of the carbonyl compound is greater than that between the corresponding states of the alkene. In that event, the excited carbonyl compound can transfer its energy to the alkene so returning to its ground state and giving triplet-state alkene. This is then followed by alkene dimerization. For example, the energy of the triplet from benzophenone is less than that from norbornene, so that photoaddition occurs:

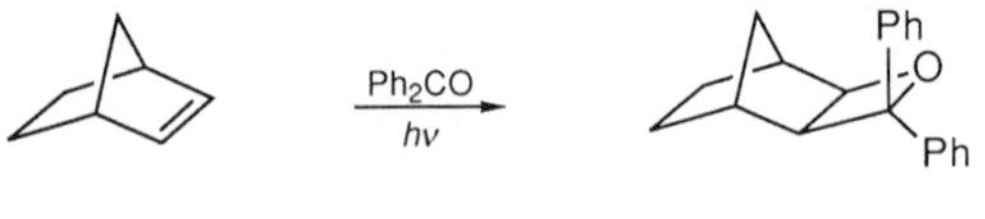

In contrast, the energy of the triplet from acetophenone is greater than that from norbornene, so that irradiation of acetophenone in the presence of norbornene yields mainly norbornene dimers. Acetophenone is here described as a *sensitizer*.

(b)   *Photoaddition of alkenes and alkynes to aromatic compounds*

Both alkenes and alkynes undergo photochemical addition to benzene. Cyclo-addition occurs to give 1,2-, 1,3-, or (less commonly) 1,4-adducts. Examples of 1,2-addition are:

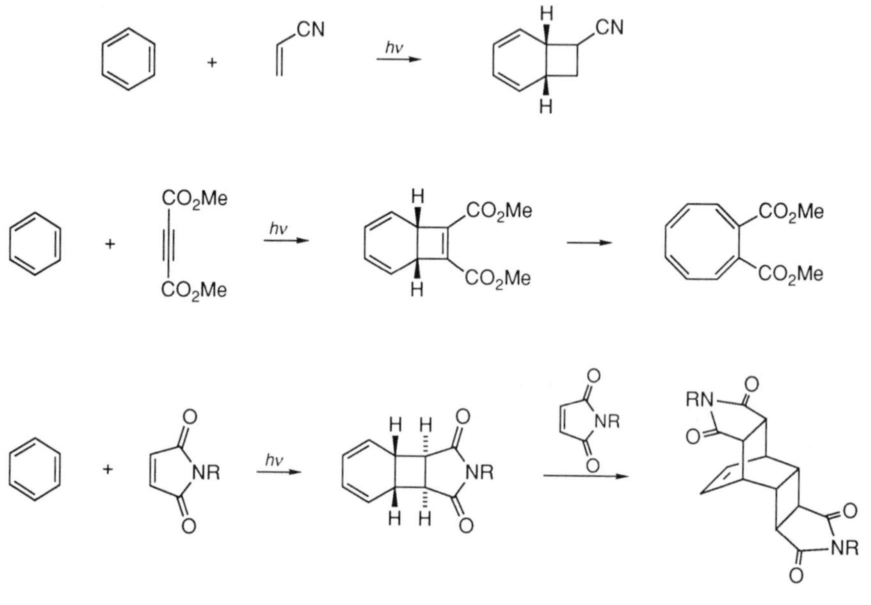

In the second example, the very strained cyclobutene undergoes a spontaneous electrocyclic ring-opening reaction. In the last, the photochemical 1,2-addition to the benzene ring, which is initiated by $n \rightarrow \pi^*$ excitation of the maleimide, is followed by a spontaneous 1,4-addition of Diels–Alder type to the resulting 1,3-diene.

Some alkenes react with benzene by 1,3-addition. It is thought that the key intermediate is prefulvene, which is also considered to be involved in the photoisomerization of benzene (p. 510) and is believed to arise from a high vibrational level of the first excited singlet of benzene, e.g.

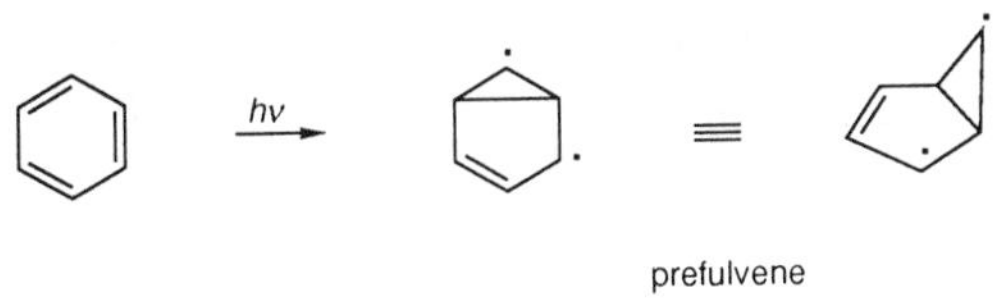

prefulvene

This reaction is of value in the synthesis of complex structures. The following examples show also its stereospecificity.

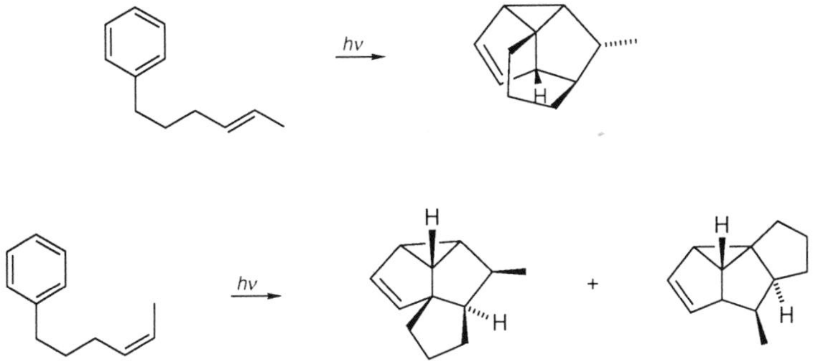

1,4-Addition to benzene is also known; it occurs when a mixture of benzene and a primary or secondary amine is irradiated, and involves the T_1 state of benzene, e.g.

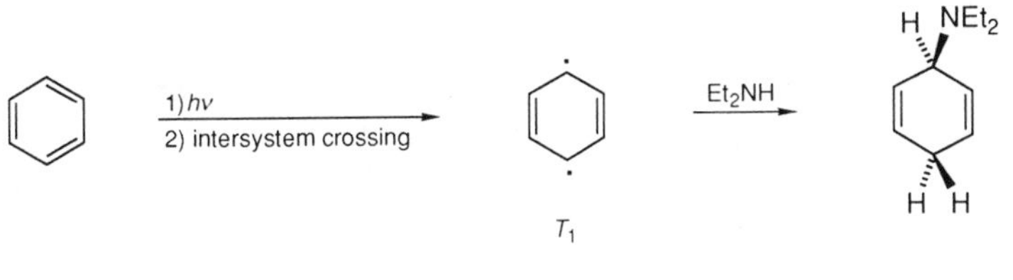

(c) Photodimerization of alkenes, conjugated dienes, and aromatic compounds

A special case of photoaddition is the formation of a 1:1 adduct by reaction of an excited state of one molecule of a reactant with another molecule of the same reactant in its ground state. This process – photodimerization – occurs with both alkenes and aromatic compounds. Depending on the structure of the reactant and the conditions of excitation, the excited state may be a singlet or a triplet.

Since non-conjugated alkenes absorb only in the difficultly accessible region below 200 nm, their unsensitized photodimerizations are comparatively little used. Reaction in this case occurs through the excited singlet S_1 and is concerted.

Reference to the earlier discussion of such reactions (p. 281) shows that the photochemical $[_\pi 2 + _\pi 2]$ cycloaddition is symmetry-allowed, e.g.

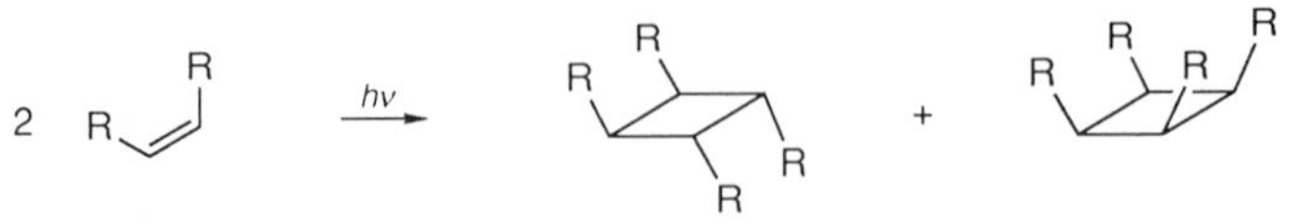

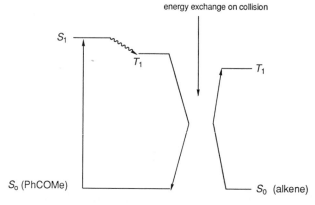

Because of the experimental difficulty, alkene photodimerization is better accomplished through the triplet state, this being generated with a photosensitizer. For example, acetophenone has $\lambda_{max.}$ 270 nm and its S_1 state undergoes intersystem crossing to T_1 efficiently; interaction of this species with the alkene then gives singlet sensitizer and triplet alkene:

The molecule in its triplet state adds through one carbon atom to a carbon atom of a second molecule of alkene and, after a spin-inversion, formation of the ring is completed.

Conjugated dienes do not undergo photodimerization through the excited singlet but instead undergo photocyclization (p. 506). However, photodimerization can be effected through triplet sensitization. Mixtures of products are often formed, as in the dimerization of cyclopentadiene in the presence of benzophenone:

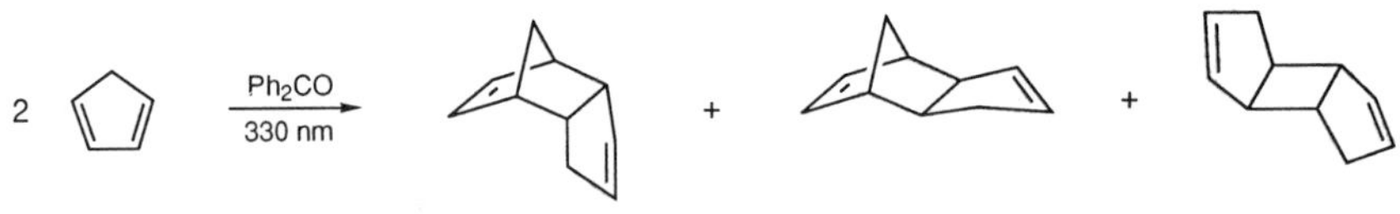

Here, the first two products represent $[\pi^4 + \pi^2]$ addition and the third represents a $[\pi^2 + \pi^2]$ addition. The three are formed in approximately equal proportions. whereas the concerted thermal dimerization of cyclopentadiene gives entirely *endo*-dicyclopentadiene (p. 273).

Benzene does not form a photodimer, but polycyclic aromatic compounds can do so. For example, anthracene dimerizes through its excited singlet (π,π^*) state:

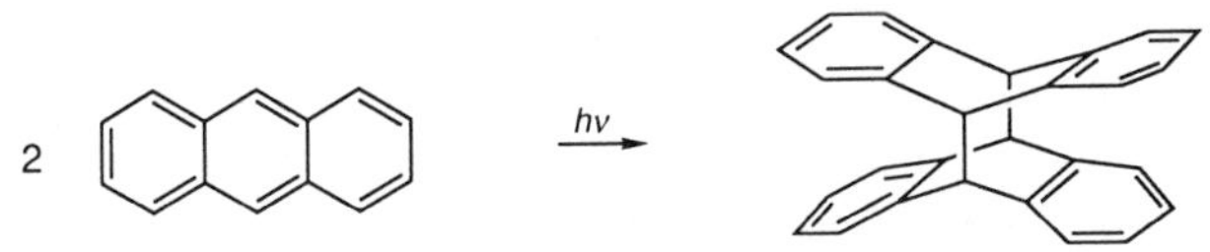

Light can effect rearrangements which lead either to structural isomers (where groups or atoms in reactant and product occupy entirely different positions) or to valence-bond isomers (where groups or atoms essentially maintain their relative positions but the bonding framework changes). These reactions provide the commonest general methods, and certainly often the shortest, for synthesizing highly strained compounds.

(a) Cis–trans isomerization

This provides the simplest case of light-induced structural isomerism. Examples are:

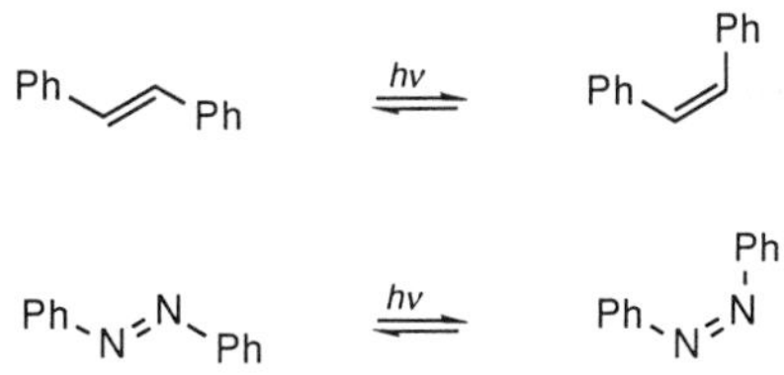

Isomerization can occur because the π-bond, which normally prevents it, is lost in passage to the excited state. Here, the two sets of substituents tend to occupy mutually perpendicular planes, e.g.

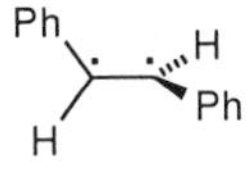

so as to minimize the repulsive forces between them. Consequently, when the molecule returns to the ground state, it can do so by twisting in either of two directions, to give both *cis*- and *trans*-isomers.

The usefulness of the reaction resides in the fact that the final mixture is usually richer in the less stable *cis* form. For example, *trans*-stilbene gives the *cis*- and *trans*-isomers in relative amounts of *c.* 10:1. This is because, at most wavelengths that are used, the *trans*-isomer has the greater molar decadic absorbance, i.e. the *trans*-isomer is excited more frequently than the *cis* and, even though de-excitation may give *trans*-isomer faster, the overall rate of formation of the *cis*-isomer is usually greater than that of the *trans* so that the equilibrium constant (the ratio of the two rate constants) >1.*

Even the highly strained *trans*-2-cycloheptenone (the smallest ring made that contains a *trans*-C=C bond) can be isolated at low temperatures by the irradiation of the *cis*-isomer:

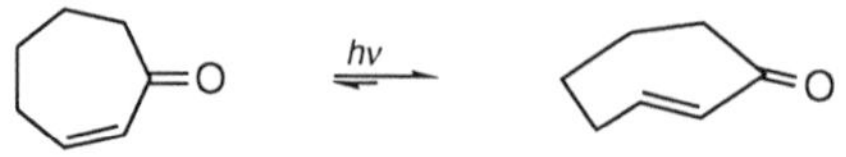

In this case, the non-planar C=C—C=O chromophore in the *trans*-isomer has the smaller molar decadic absorbance so that it is the *cis*-isomer that reacts the faster.

(b) Intramolecular photocyclization

Many dienes and polyenes are converted photochemically into cyclic isomers.

(i) Conjugated dienes. When the excited singlet is involved, reaction is of the concerted, electrocyclic type and is stereospecific; its course can be rationalized by the orbital-symmetry principles outlined earlier (p. 285).

For example, irradiation of *trans,trans*-1,4-dimethyl-1,3-butadiene gives entirely *cis*-3,4-dimethylcyclobutene by disrotatory ring closure,

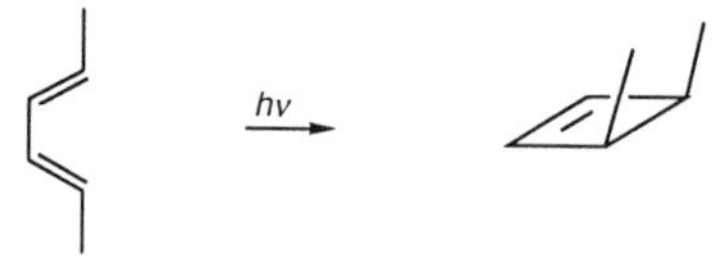

and *trans,trans*-1,6-dimethyl-2,4,6-hexatriene cyclizes with conrotatory motion:

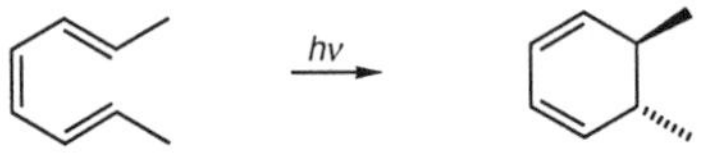

These reactions are reversible, and an early example of their synthetic utility was the electrocyclic ring opening which occurs on the irradiation of ergosterol, to give precalciferol:

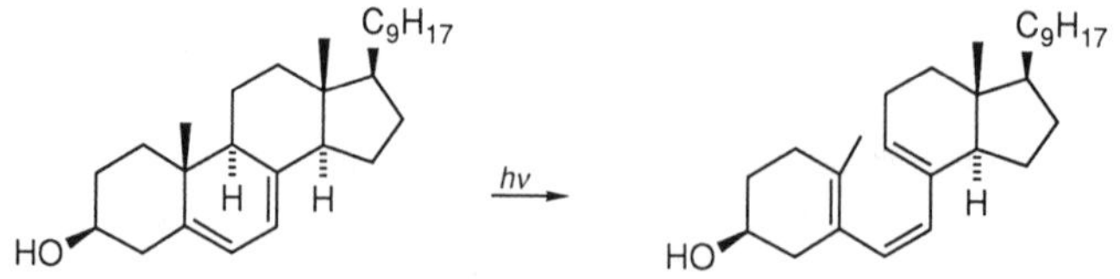

ergosterol   precalciferol

---

*The condition for this is that the ratio of the molar decadic absorbances > the ratio of the rates at which the excited state forms *trans*- and *cis*-isomers.

Highly strained compounds can also be made in this way; for example, irradiation of cyclopentadiene gives bicyclo[2.1.0]pent-2-ene [6]:

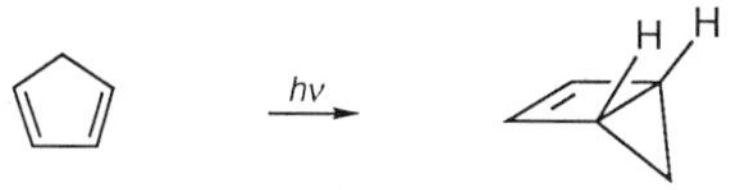

(*ii*) *1,4-Dienes.* Unless a conjugating substituent such as an aryl group is present, triplet sensitization is required. The reaction (*Zimmerman rearrangement*) results in formation of a vinylcyclopropane,

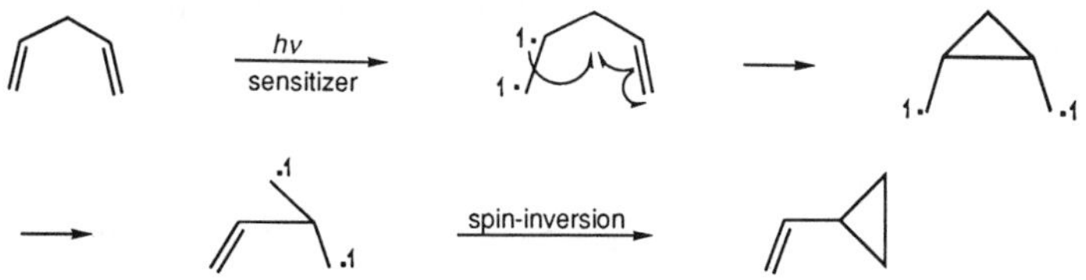

For example, norbornadiene gives quadricyclane in 70–80% yield with acetophenone as sensitizer [6],

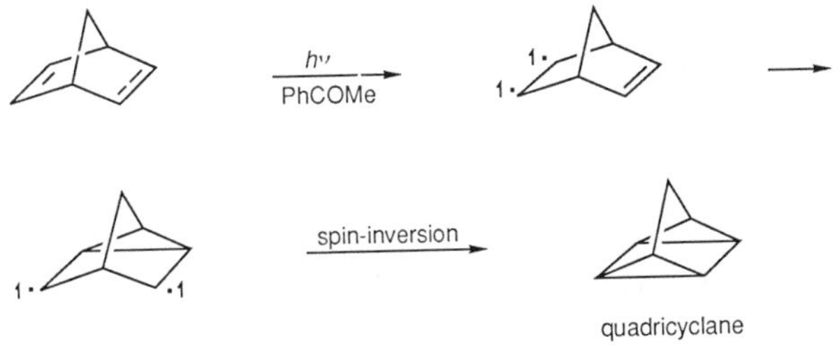

and barrelene likewise gives semibullvalene:

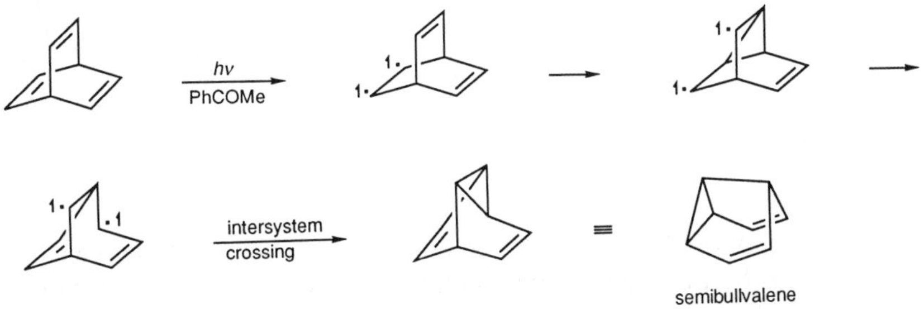

An analogous rearrangement occurs with βγ-unsaturated aldehydes and ketones,

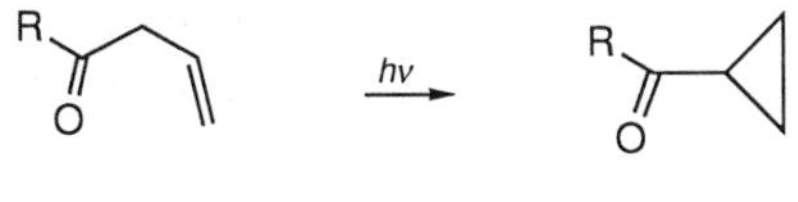

Reaction of this type on an enantiomer of bicyclo[2.2.2]oct-5-en-2-one gives an enantiomerically pure tricyclic compound which has been used as an asymmetric building block in the synthesis of natural products:

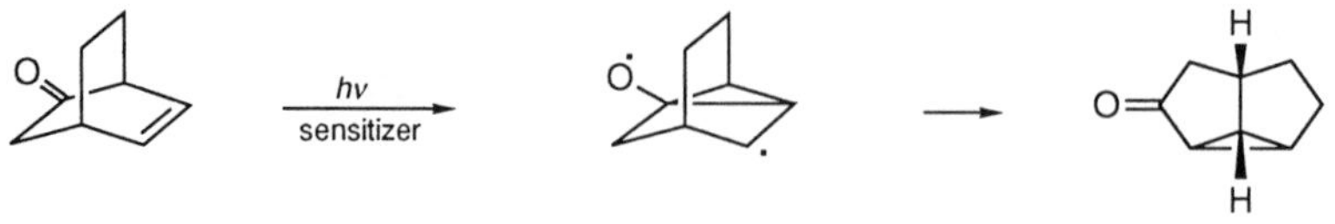

(iii) *Cage compounds.* Sequences which include photocyclization are used in assembling cage compounds. For example, cyclopentadiene adds readily to *p*-benzoquinone in the thermal [$_\pi4 + _\pi2$] manner, and irradiation of the product then effects ring closure:

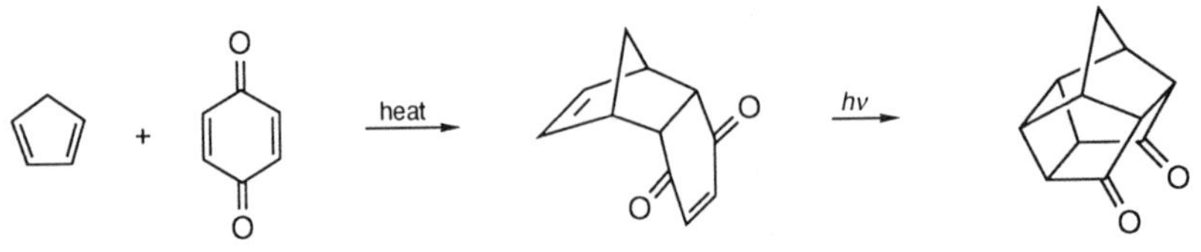

Again, irradiation of dicyclopentadiene yields a cage-like compound:

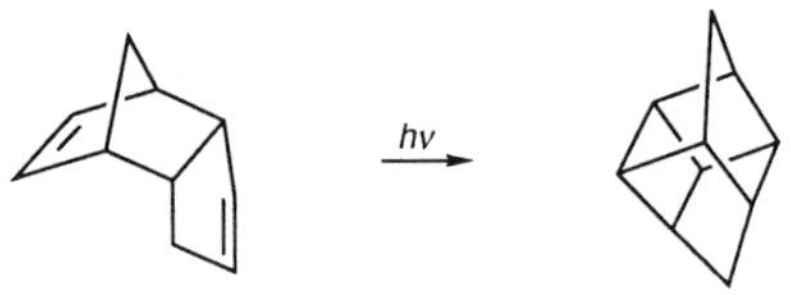

A synthesis of cubane is based on the latter reaction:

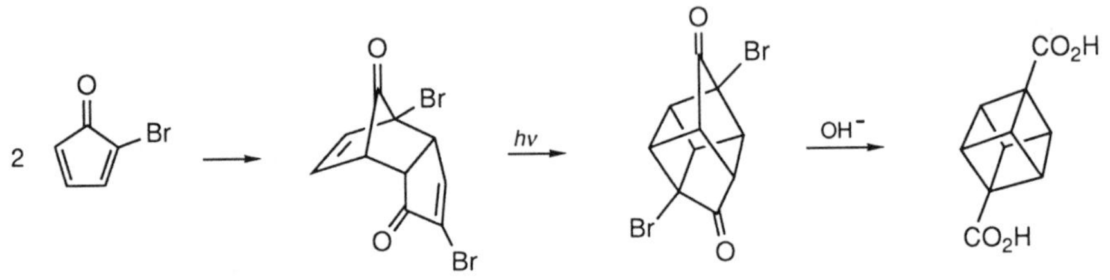

The dimerization of 2-bromocyclopentadienone occurs spontaneously. The irradiation step is best carried out on a ketal derivative of the dimer from which the keto group is subsequently regenerated by hydrolysis. The successive base-induced ring contractions are reactions of Favorskii type (p. 449). The cubanedicarboxylic acid has been converted into cubane itself by heating the t-butyl peroxyester (from the acid chloride with t-butyl hydroperoxide) in cumene; homolysis of the peroxide bond is followed by decarboxylation and uptake of a hydrogen atom from the solvent:

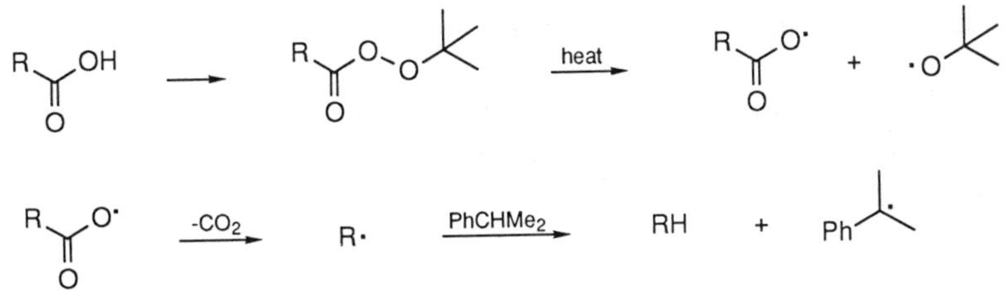

(c) Sigmatropic rearrangements

A suprafacial 1,3-alkyl or 1,3-acyl shift in an allylic system is symmetry-allowed as shown by examination of frontier orbitals, e.g.

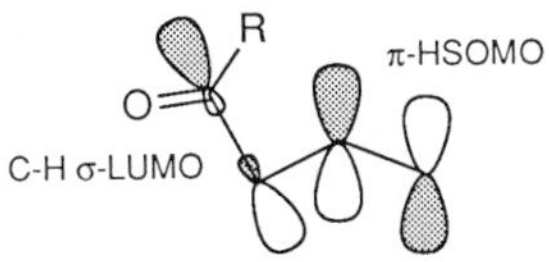

For example, eucarvone undergoes photochemically induced disrotatory ring closure followed by a 1,3-acyl shift:

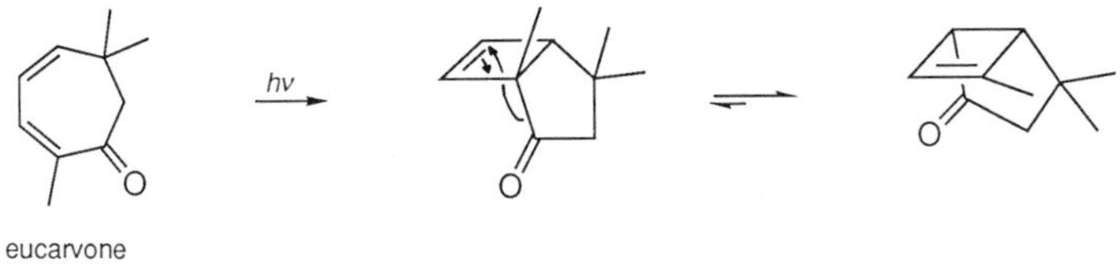

eucarvone

(d) Cyclohexadienones

The photochemical rearrangement of a cyclohexadienone was first observed with α-santonin over 150 years ago:

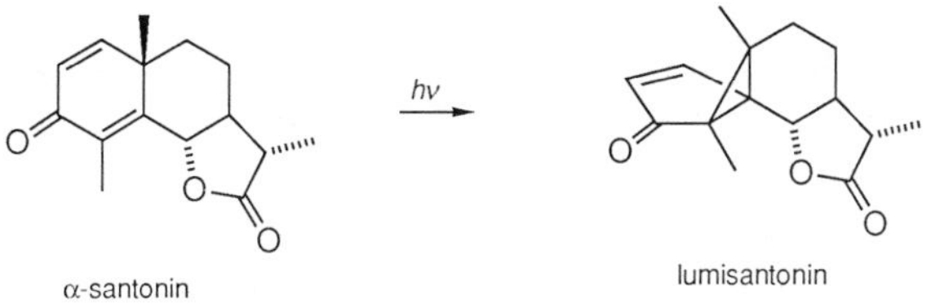

α-santonin

lumisantonin

The essential features of the rearrangements are:

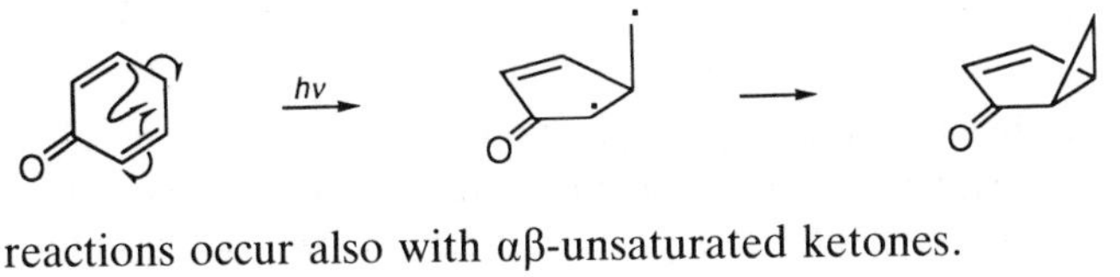

Analogous reactions occur also with αβ-unsaturated ketones.

(e)   Photoisomerization of benzenoid compounds

The irradiation of benzene itself brings about a low conversion into fulvene and benzvalene; the latter slowly reverts to benzene. Prefulvene is thought to be an intermediate:

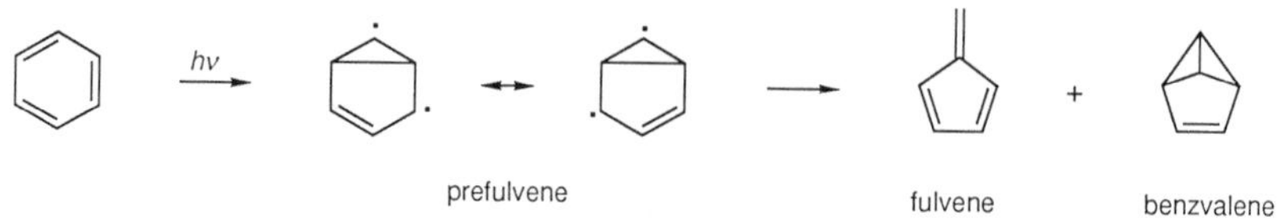

Substituted benzenes give other strained systems. For example, a Dewar benzene can be obtained from 1,2,4-tri-t-butylbenzene,

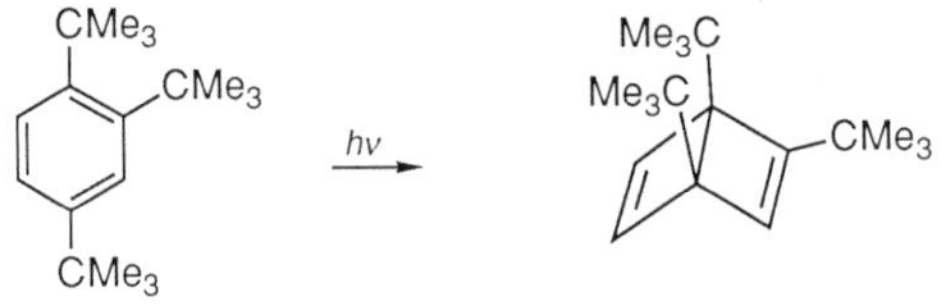

and a prismane from 1,3,5-tri-t-butylbenzene:

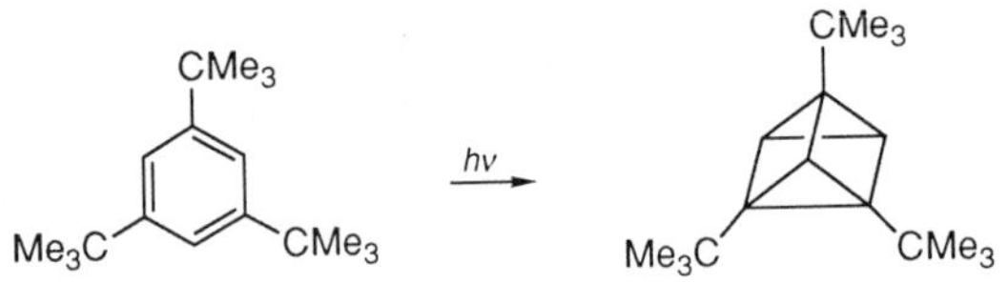

In these cases, the products are thought to arise from a 'bent' triplet state, represented for simplicity for benzene:

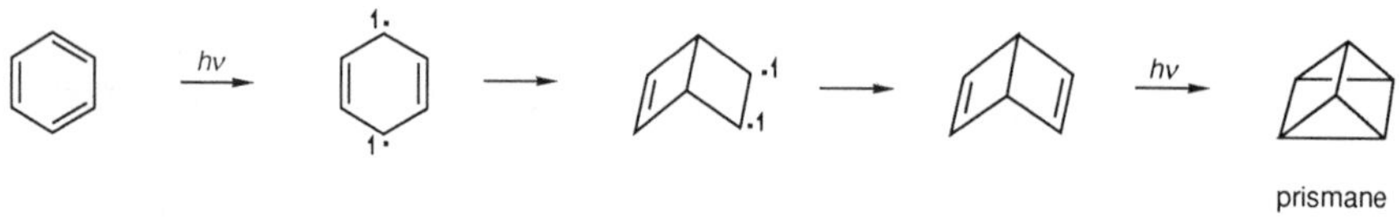

Dewar benzene itself has not been made directly from benzene, but it has been prepared *via* a disrotatory electrocyclic ring closure:

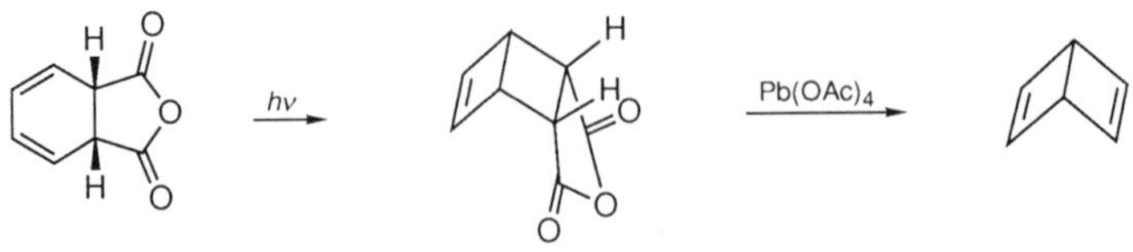

On irradiation, Dewar benzene gives prismane.

Benzvalene, Dewar benzene, prismane, and their derivatives are all thermally labile to varying extents, ultimately reverting to benzenoid compounds. For example, Dewar benzene itself has a half-life of about 2 days at room temperature.*

Some of these species are also thought to be intermediates in photochemical reactions which lead to benzenoid isomers. For example, the irradiation of *o*-xylene gives mixtures containing the *meta*- and *para*-isomers, probably as follows:

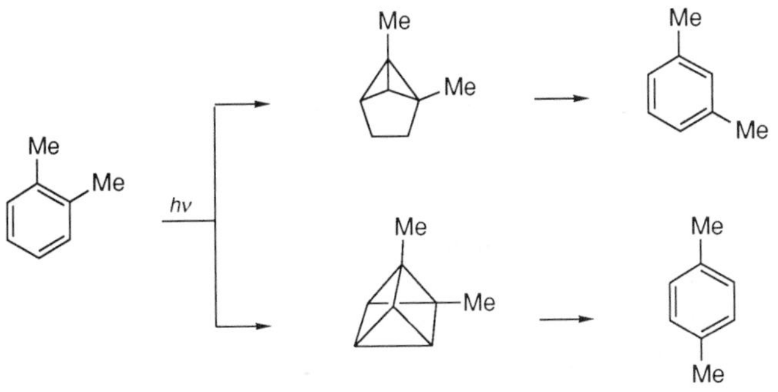

**16.5
Photooxidation**

(a)   Formation of peroxy-compounds

Certain types of peroxy-compound can be formed by irradiating the parent organic compound in the presence of oxygen and a sensitizer. Reaction occurs by the excitation of the sensitizer to its triplet state, and thereafter in one of two ways: either the triplet abstracts a hydrogen atom from the substrate to form a radical which then reacts with oxygen, or it interacts with the oxygen molecule so as to activate it.

The best-known example of the first type of reaction is the oxidation of a secondary alcohol to give a hydroxy-hydroperoxide, with benzophenone as the sensitizer. The triplet benzophenone reacts with the alcohol to give a carbon radical (cf. photoreduction, p. 499) which adds oxygen:

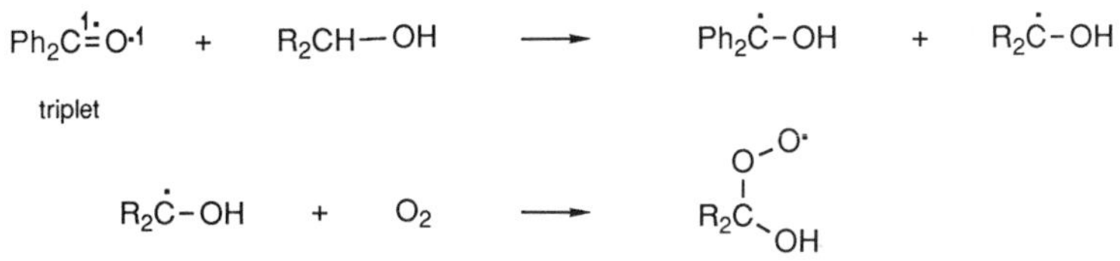

A chain reaction is now propagated:

---

* The existence of Dewar benzene even for this time is remarkable, considering the strain inherent in the compound on the one hand and the aromatic stabilization energy to be gained from its conversion into benzene on the other. The significant fact is that the thermal concerted conversion into benzene is symmetry-forbidden.

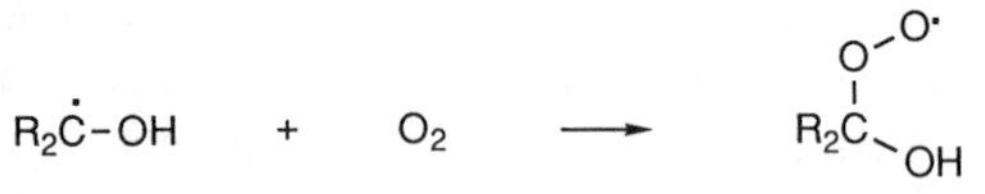

$$R_2\overset{\cdot}{C}-OH \quad + \quad O_2 \quad \longrightarrow \quad R_2C\overset{\displaystyle O-O^{\cdot}}{\underset{OH}{|}}$$

The usefulness of the reaction lies in the fact that hydroxy-hydroperoxides readily eliminate hydrogen peroxide to form carbonyl compounds:

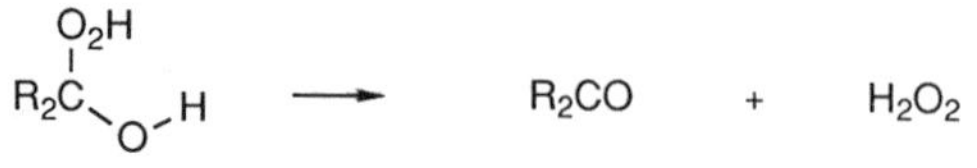

$$R_2\overset{\displaystyle O_2H}{\underset{O-H}{\overset{|}{C}}} \quad \longrightarrow \quad R_2CO \quad + \quad H_2O_2$$

In the second type of sensitized oxidation, the sensitizer in its triplet state interacts with the triplet oxygen molecule, i.e. the ground state of oxygen*, to give singlet oxygen, $^1O_2$, i.e. an *excited* oxygen molecule, while the sensitizer returns to its ground state, S_0. Singlet oxygen then interacts directly with the organic substrate. The common sensitizers are dyes such as fluorescein and Rose Bengal (a halogenated fluorescein).

Three types of oxidation can be brought about in this way. First, conjugated dienes yield cyclic peroxides in a reaction of Diels–Alder type. Examples are:

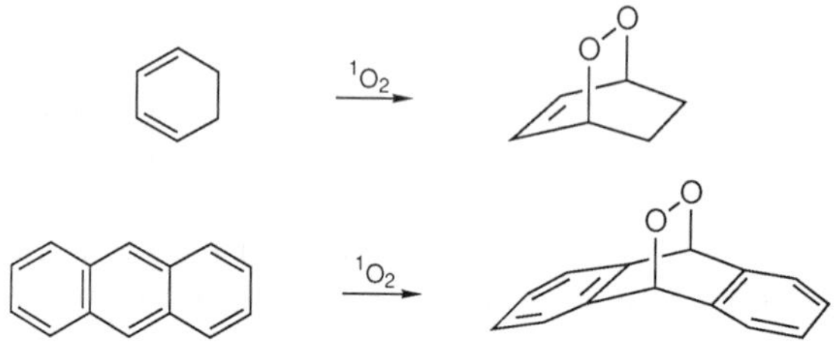

The usefulness of the reaction lies in the reducibility of the peroxides to diols:

---

* That the ground state of O_2 is a triplet can be readily understood by constructing molecular orbitals. Its highest occupied orbitals are the π^*2p; each contains one electron, and the two electrons have parallel spins. The lowest singlet state is only $92.4\,kJ\,mol^{-1}$ higher in energy. Singlet oxygen can also be made by non-photochemical reactions, e.g.

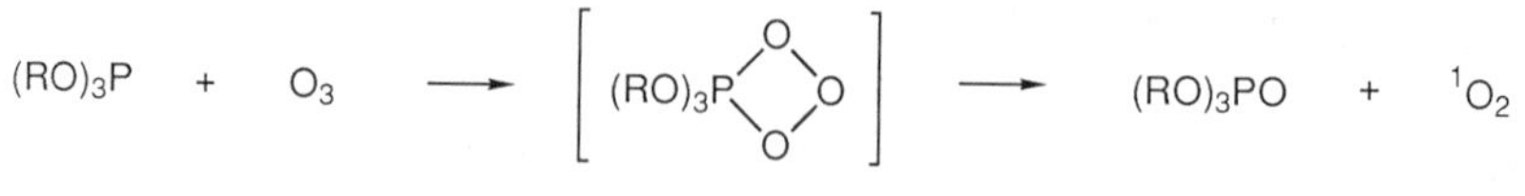

Second, alkenes with an allylic hydrogen atom form hydroperoxides, probably as follows (cf. the ene-reaction, p. 295):

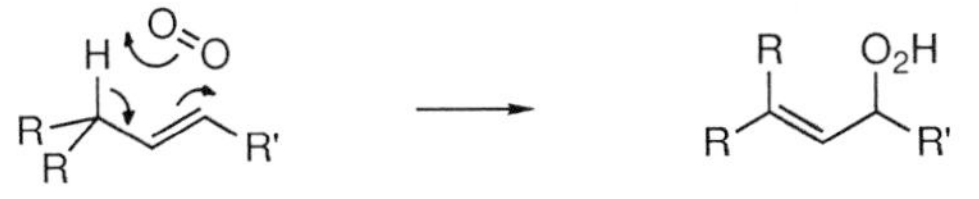

The products can be reduced to allyl alcohols.

Third, alkenes with no allylic hydrogen form dioxetans which, on warming, yield carbonyl compounds:

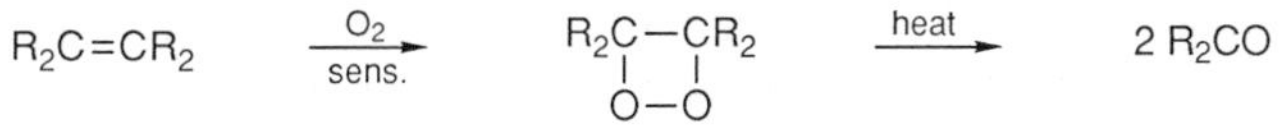

$$R_2C=CR_2 \xrightarrow[\text{sens.}]{O_2} \underset{\substack{| \quad | \\ O-O}}{R_2C-CR_2} \xrightarrow{\text{heat}} 2\,R_2CO$$

(b) Oxidative coupling of aromatic compounds

Although the irradiation of *cis*- or *trans*-stilbene in the absence of oxygen gives simply a mixture of the two (p. 505), in the presence of oxygen phenanthrene is formed. It has been inferred that *cis*-stilbene photocyclizes reversibly to give a small proportion of dihydrophenanthrene which, in the presence of oxygen, is oxidized irreversibly to phenanthrene:

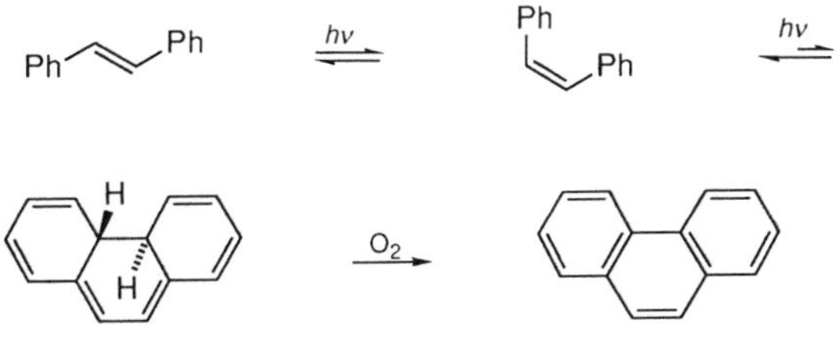

Compounds related to stilbene react analogously, providing a useful method for obtaining polycyclic systems, e.g.

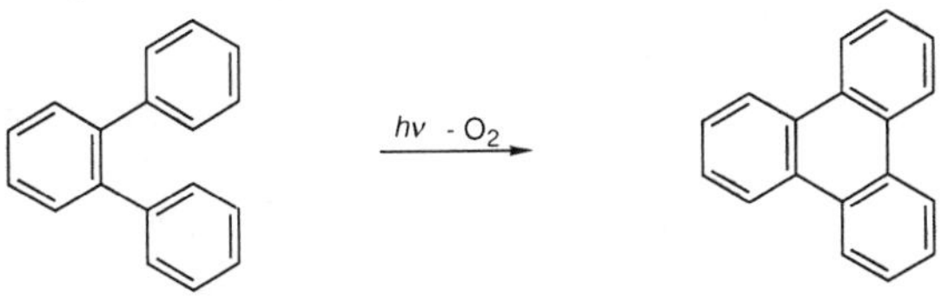

16.6 Photochemical aromatic substitution

The distribution of a charge in a species in an excited state can be wholly different from that in the ground state and especial use has been made of this difference in controlling the positional selectivity in nucleophilic aromatic substitutions. For example, when 3,4-dimethoxynitrobenzene is heated with hydroxide ion, the 4-methoxy substituent is replaced, whereas when the reaction is carried out at room temperature under ultraviolet irradiation it is the 3-methoxy substituent which is replaced:

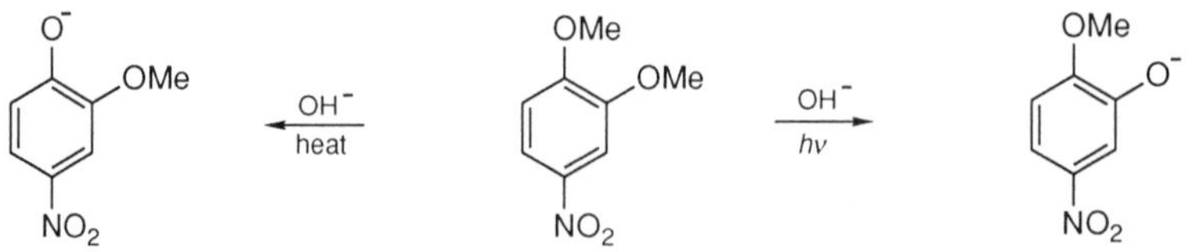

In general, an electron-attracting group such as nitro renders the *ortho* and *meta* positions positive relative to the *para* position in the excited state, whereas the *ortho* and *para* positions are positive relative to the *meta* position in the ground state. With an electron-releasing substituent, the converse obtains: in the ground state, the *ortho* and *para* positions are negatively charged compared with the *meta*, whereas in the excited state the *ortho* and *meta* positions are negatively charged relative to the *para* position.

Other examples which illustrate these effects in photoactivated nucleophilic substitution are:

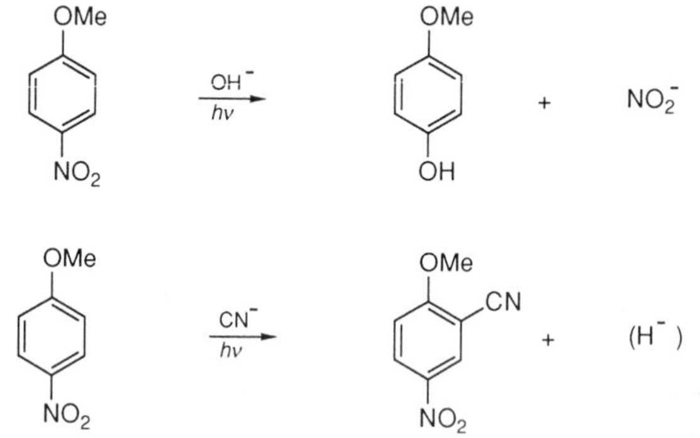

Compared with the first example, the thermal reaction causes displacement of the methoxy group; compared with the second, it leads to cyanation *ortho* to the nitro group.

There are so far hardly any examples of photoactivated electrophilic substitutions in which the substituents have the opposite effects to those in thermal reactions. However, in electrophilic deuteration with CH_3CO_2D, methyl and methoxyl groups direct mainly *meta* and *ortho*, and nitro directs mainly *para*, e.g.

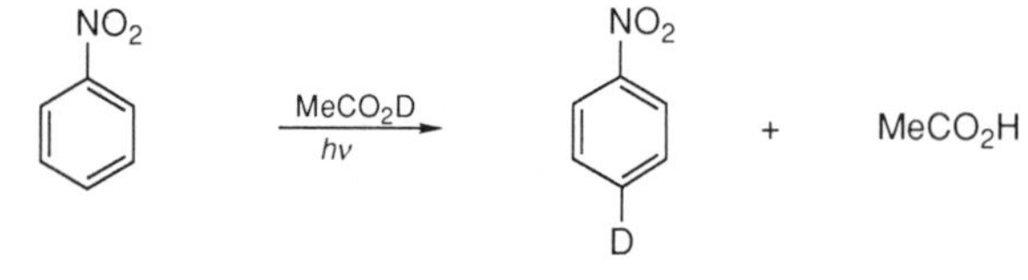

In order for light to bring about the fragmentation of a compound, two conditions must be met: the wavelength of the light must correspond both to an energy greater than that of the bond to be cleaved, and to an electronic excitation of the molecule concerned. This effectively eliminates saturated organic compounds since, although they have weaker bonds than correspond to typical, accessible ultraviolet wavelengths, they do not absorb in this region. For example, ethane is unaffected by light of wavelength 300 nm even though this is equivalent to a larger energy (c. $400 \, \text{kJ mol}^{-1}$) than the C—C dissociation energy ($347 \, \text{kJ mol}^{-1}$); and methyl iodide is essentially unaffected by light whereas iodobenzene, having an aromatic chromophore, dissociates into phenyl radicals and iodine atoms.

In general, then, fragmentation is characteristic of certain unsaturated organic compounds together with inorganic compounds such as the halogens. In many instances, the same fragmentation can be brought about by heat and these are considered in the following chapter: examples are the fragmentation of peroxides, e.g.

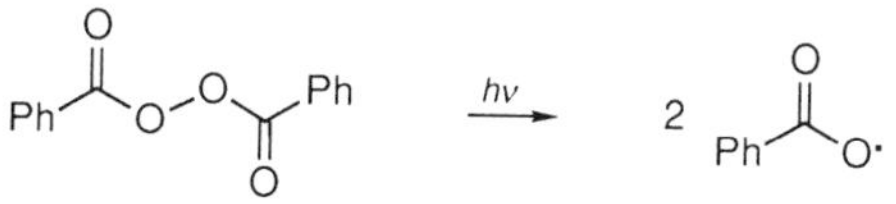

and nitrites,

In this section attention is restricted to carbonyl compounds – where the photochemistry is not paralleled thermally – and to aliphatic diazo-compounds.

(a) Photolysis of carbonyl compounds

The $n \to \pi^*$ excitation of aldehydes and ketones can bring about two types of fragmentation.

First, fragmentation can occur of the C—C bond adjacent to carbonyl:

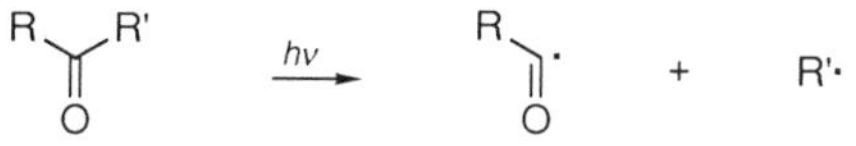

This is termed the *Norrish Type I* process, and is followed by one or more of three reactions: disproportionation to give a ketene,

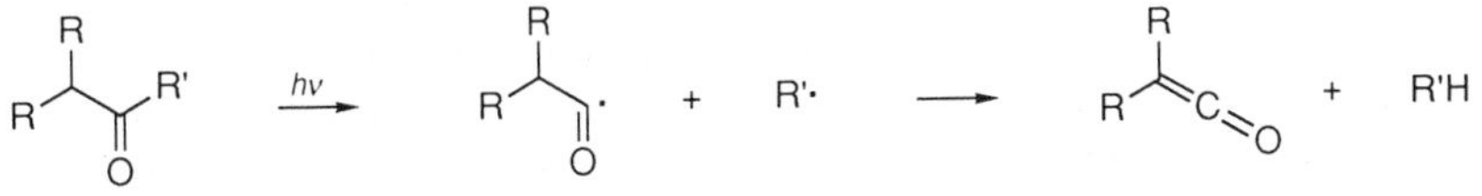

decarbonylation and then dimerization and/or disproportionation, e.g.

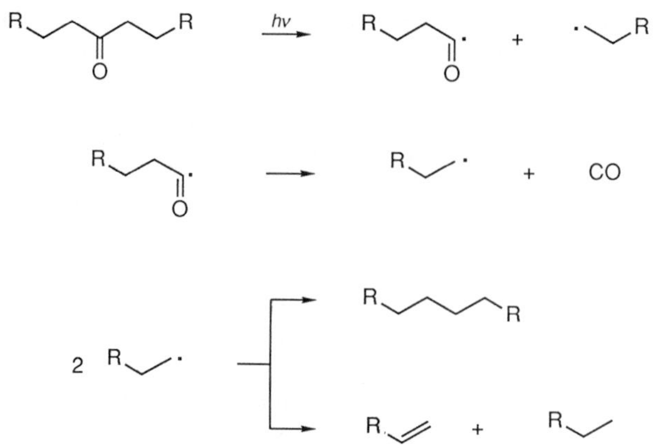

or intermolecular hydrogen-atom abstraction by the acyl radical,

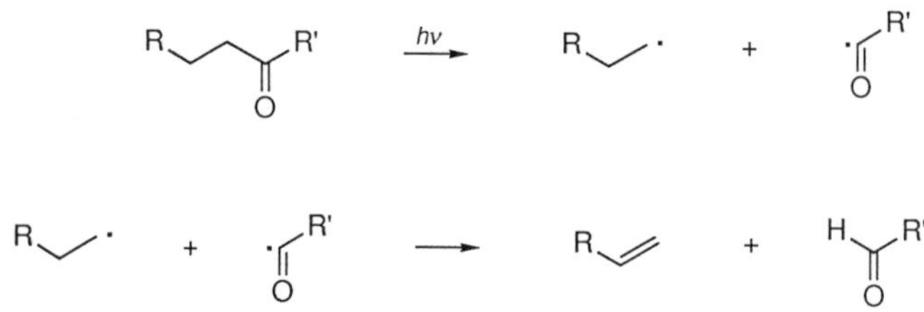

Second, if the carbonyl compound contains a γ-CH group, intramolecular hydrogen-atom abstraction can occur:

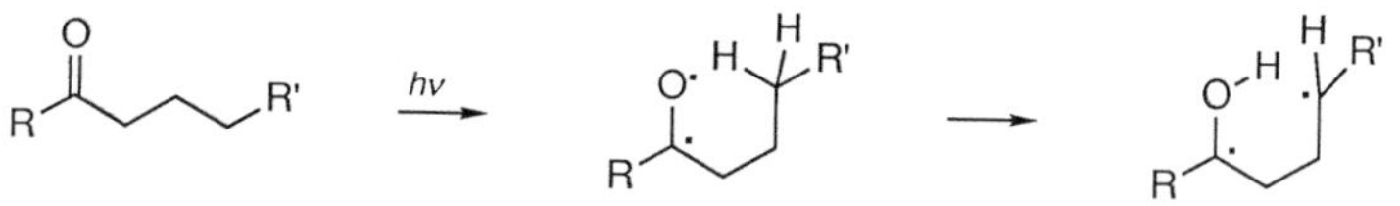

This – the *Norrish Type II* process – is followed by fragmentation,

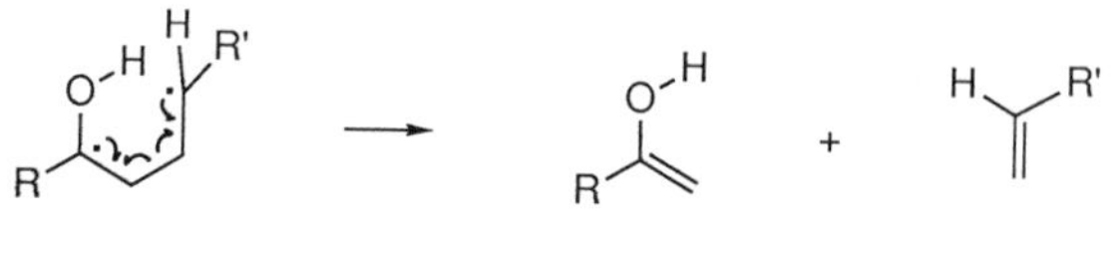

or by ring closure,

The Type I process occurs more readily in the vapour phase than in solution; under the latter conditions, it is thought that the two radicals usually recombine within the solvent cage in which they are formed, and the process is usually significant only when the alkyl radical formed is of the relatively stable tertiary or benzylic type.

Which of the two types of product is formed by the Type II process depends markedly on the substitution at the carbon atom adjacent to carbonyl. For example, in the following case,

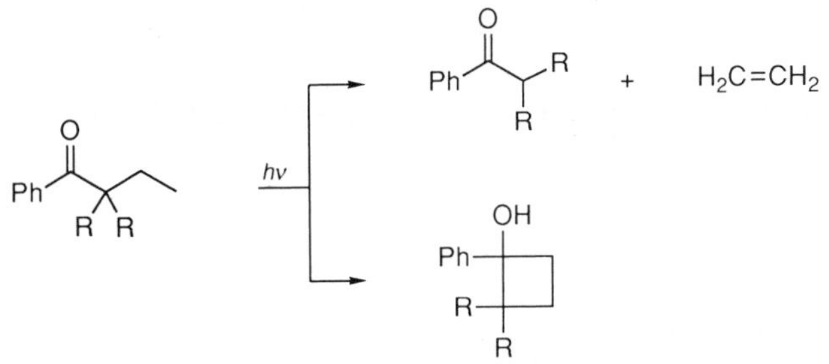

the extent of cyclization relative to fragmentation is 10% when R = H and 89% when R = CH$_3$.

The photochemistry of the smaller cyclic ketones is dominated by the Type I process, followed by intramolecular reactions:

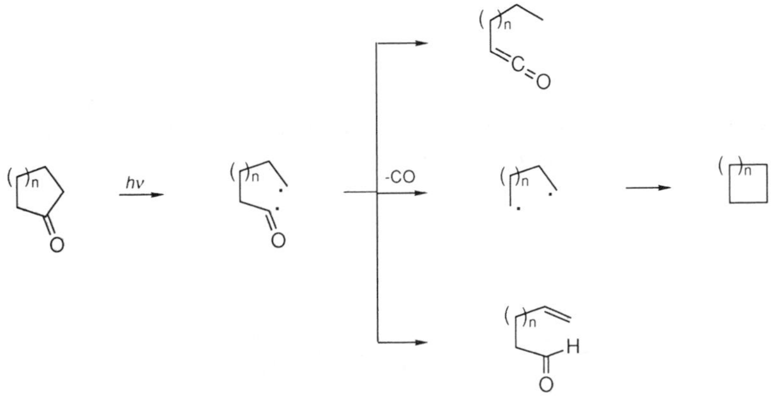

However, in larger rings, cyclization related to the Type II process can occur, e.g.

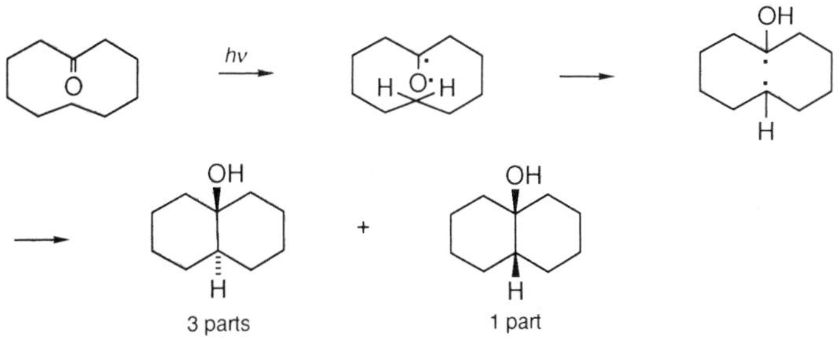

3 parts 1 part

(b) *Photolysis of compounds containing the* $\overset{+}{=}N\overset{-}{=}N$ *group*

(i) *Diazoalkanes.* The irradiation of diazomethane in ether gives methylene in its singlet state:

$$CH_2N_2 \xrightarrow{hv} :CH_2 \ + \ N_2$$

Methylene is highly reactive, inserting into both unsaturated and saturated bonds; its synthetic usefulness is therefore limited. However, suitably substituted methylenes – carbenes* – are relatively stable as a result of delocalization, e.g.

and are consequently much more selective. For example, whereas methylene inserts into both primary and tertiary C—H bonds in 2,3-dimethylbutane, carbomethoxymethylene reacts almost solely at the tertiary position:

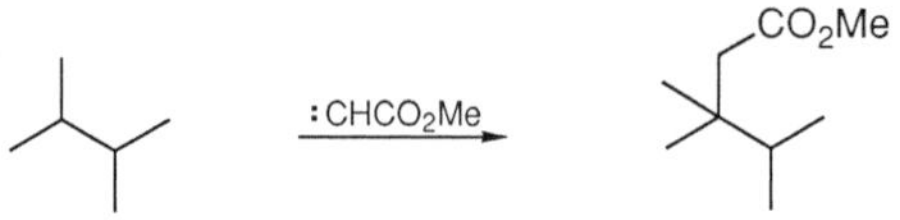

The most useful reaction of carbenes is with alkenes to form cyclopropane derivatives. When the carbene is generated in its singlet state, by direct photo-lysis, its insertion into the double bond is essentially stereospecific. For example, cis-2-butene and dicarbomethoxymethylene give almost entirely the cis-dimethyl-cyclopropane:

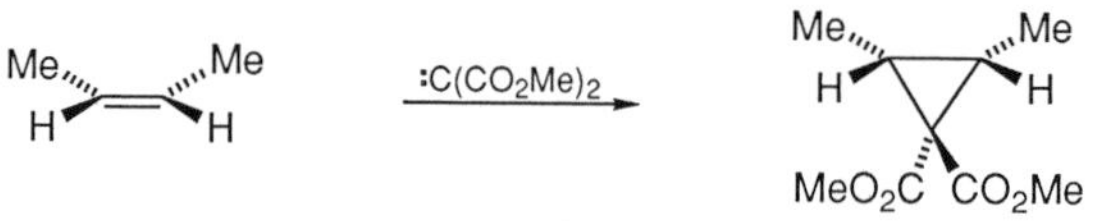

However, when the carbene is generated by photolysis in the presence of a triplet sensitizer, it is formed in its triplet state and the stereospecificity is lost; for example, cis-2-butene gives largely trans-dimethylcyclopropane. This is because

---

* Carbenes can be generated from hydrazones and arenesulfonylhydrazones, e.g.

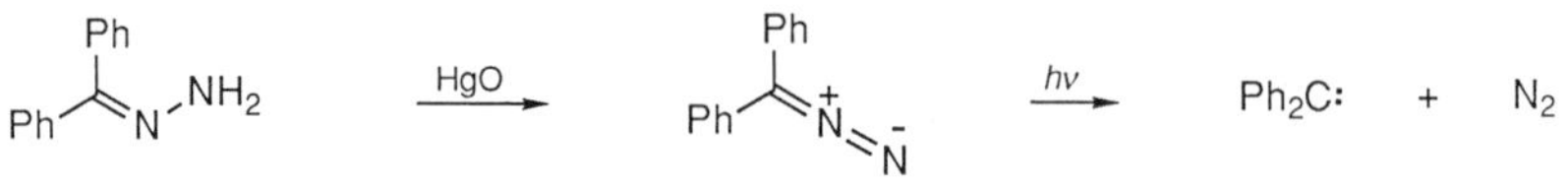

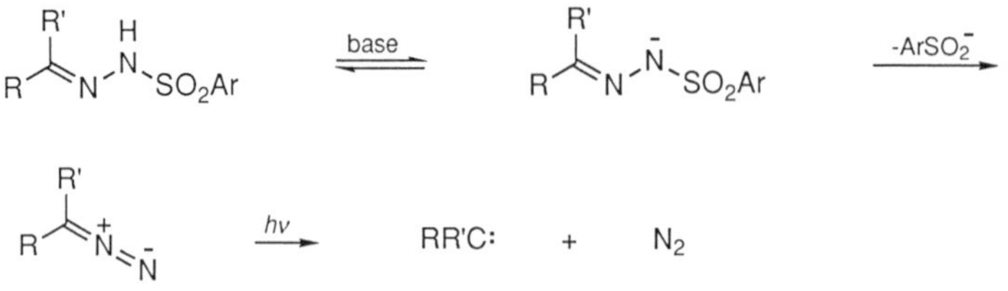

there is a time lag between formation of the two C—C bonds during which spin-inversion occurs and this allows rotation about the original C—C bond:

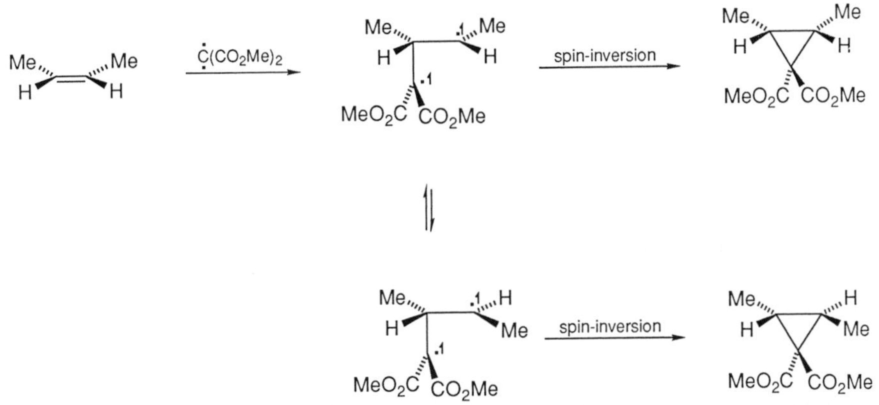

Carbenes also react with aromatic compounds. For example, irradiation of diazomethane in benzene gives about 30% of cycloheptatriene; addition across C=C is followed by a spontaneous Cope rearrangement:

However, this is not nearly so efficient a reaction as a non-photochemical process in which benzene and diazomethane react in the presence of copper(I) ion. It is thought that a copper-carbene complex is involved.

The carbenes from α-diazoketones have the special property that they undergo rearrangement faster than they are trapped by, for example, alkenes. Rearrangement gives a ketene which can be trapped by an alkene and this forms the basis of a method for synthesizing four-membered rings, e.g.

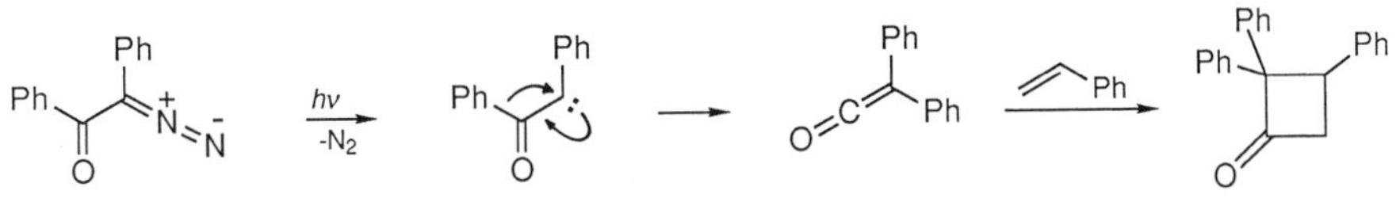

A variant of this reaction is the ring contraction of carbenes from cyclic α-diazoketones, e.g.

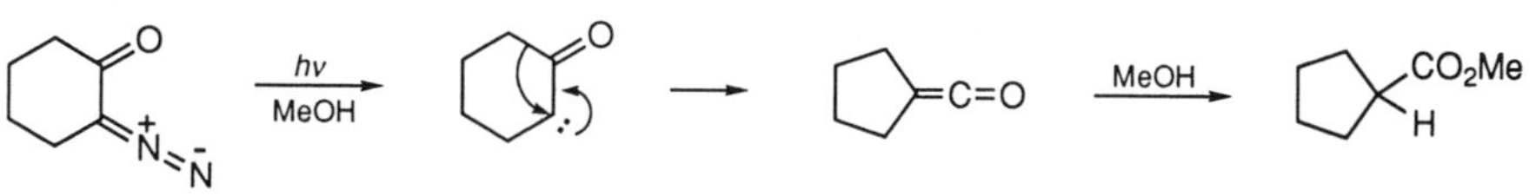

Some highly strained small-ring compounds have been made in this way, e.g.

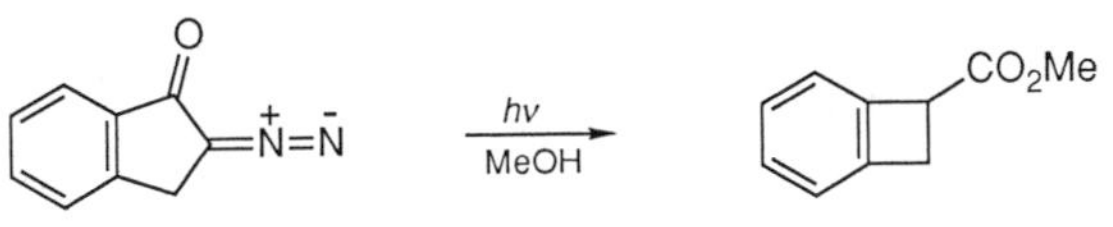

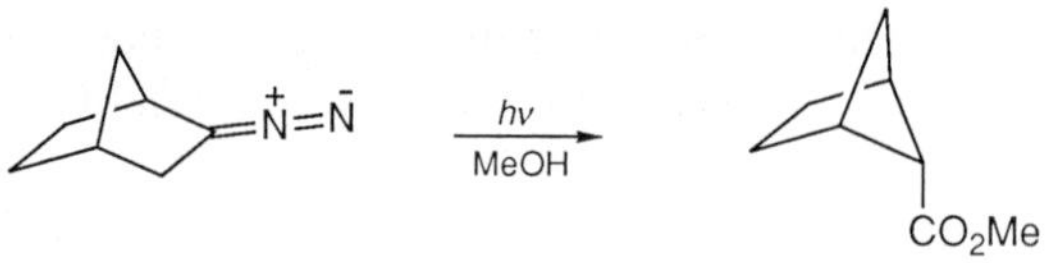

$$\text{(norbornene-CH=}\overset{+}{N}=\overset{-}{N}) \quad \xrightarrow[\text{MeOH}]{h\nu} \quad \text{(product with CO}_2\text{Me)}$$

(*ii*) *Alkyl azides.* The photolysis of an azide gives a nitrene,

$$RN=\overset{+}{N}=\overset{-}{N} \quad \xrightarrow{h\nu} \quad R\overset{..}{N}: \quad + \quad N_2$$

which can undergo a variety of intra- and intermolecular reactions: intermolecular hydrogen-atom abstraction,

$$R\overset{..}{N}: \quad + \quad R'H \quad \longrightarrow \quad R\overset{.}{N}H \quad + \quad R'\cdot$$

$$R\overset{.}{N}H \quad + \quad R'H \quad \longrightarrow \quad RNH_2 \quad + \quad R'\cdot$$

intramolecular abstraction of δ-H,

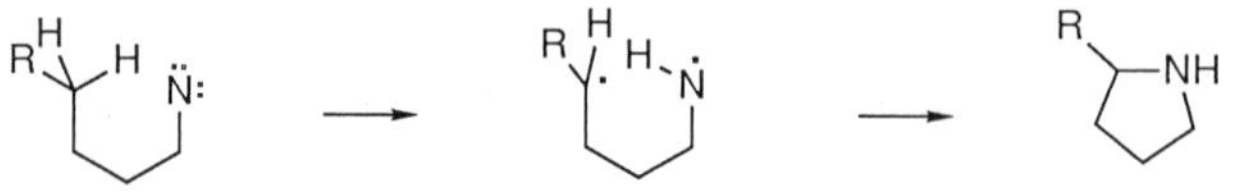

addition to an alkene

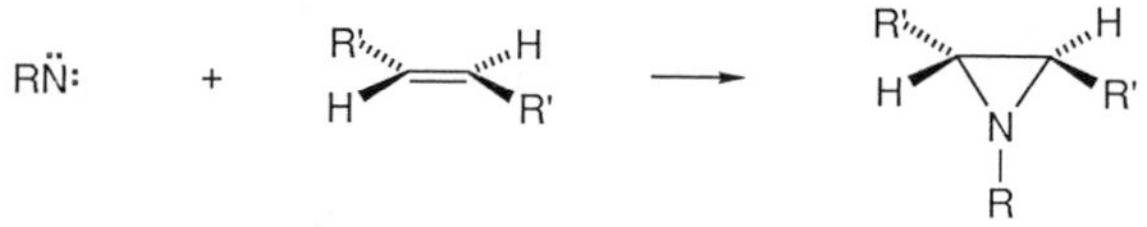

and coupling,

$$2 \ R\overset{..}{N}: \quad \longrightarrow \quad RN=NR$$

Synthetically, the most useful process is that involving abstraction of δ-H followed by cyclization, for this enables functionality to be introduced at an unactivated CH group which is stereochemically suitably placed (cf. p. 553).

Alkenyl and allyl azides give nitrenes which undergo intramolecular addition to give highly strained compounds, e.g.

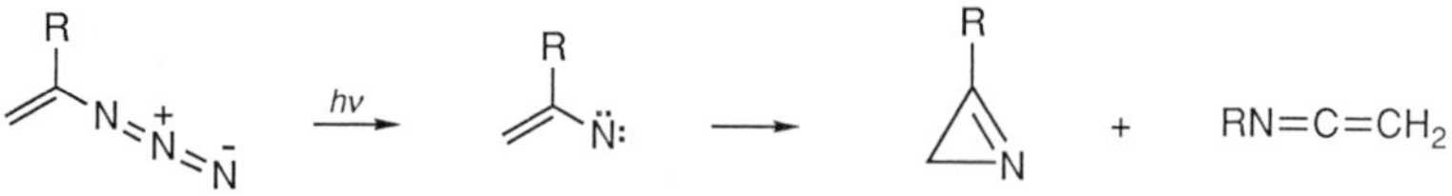

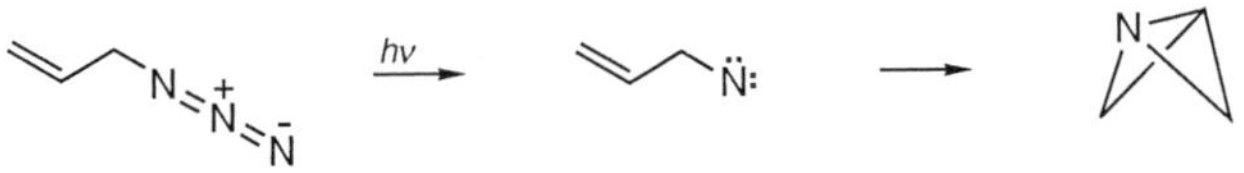

[1,2]-hydrogen shift,

$$R_2CH\overset{..}{-}\overset{..}{N}: \quad \longrightarrow \quad R_2C=NH$$

1. Draw the structure of the chief product formed by ultraviolet irradiation of   **Problems**
    each of the following:

(a)

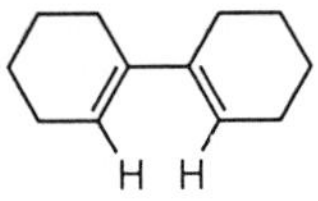

(b)

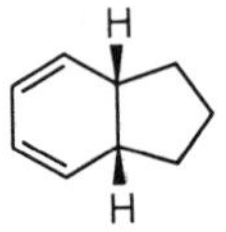

(c)

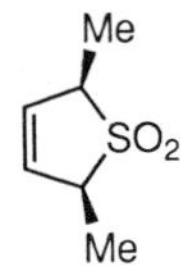

(d)

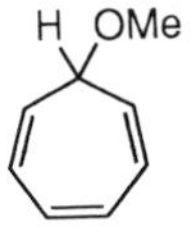

(e)

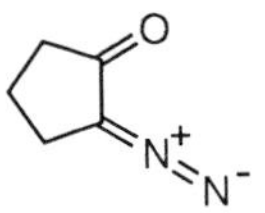

(f) HO$_2$C⌒⌒CO$_2$H

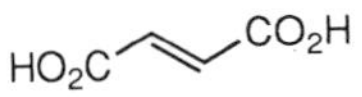

(g)

(h) 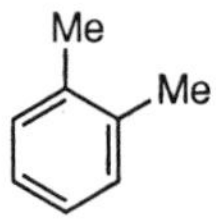 Me ... Me

(i) OMe ... NO$_2$ +   NH$_3$

(j)

(k) Ph

(l) Ph ... Ph   +   Ph

2. Rationalize the following reactions:

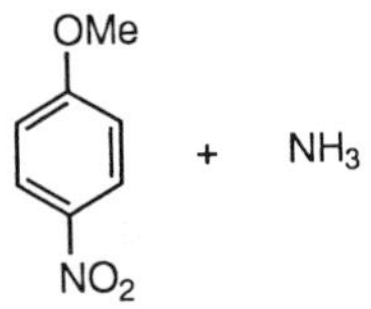

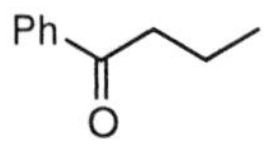

(i)

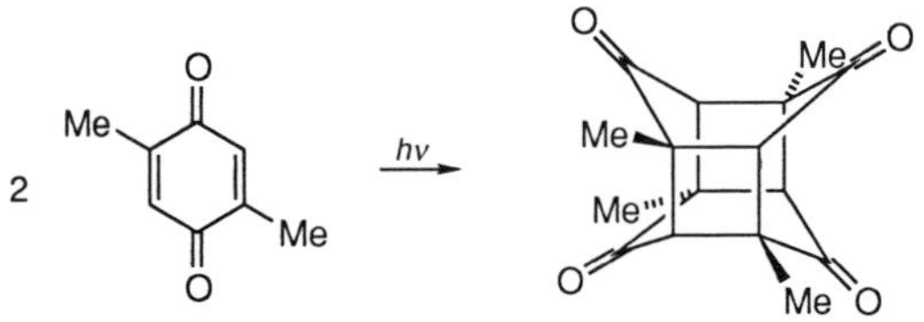

2   Me ... Me $\xrightarrow{h\nu}$ Me ... Me ... Me ... Me

(ii)

$\xrightarrow[-CO]{h\nu}$

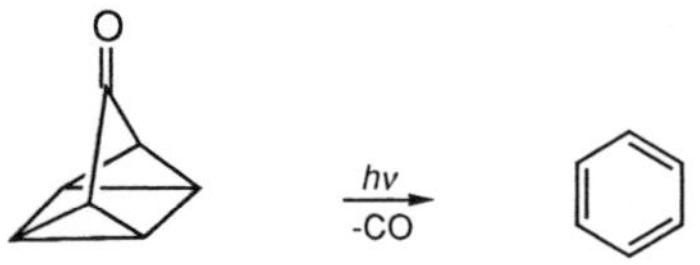

(*iii*)

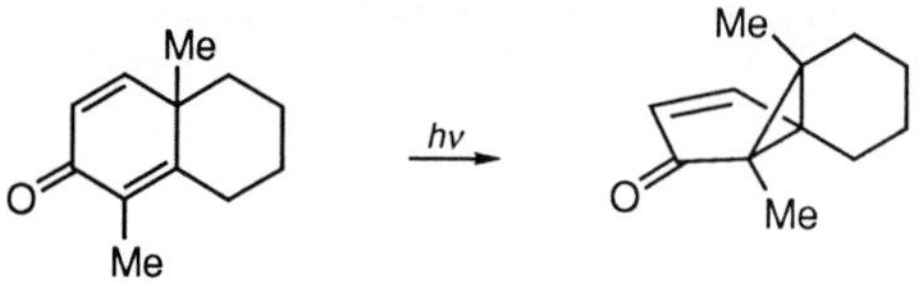

(*iv*)

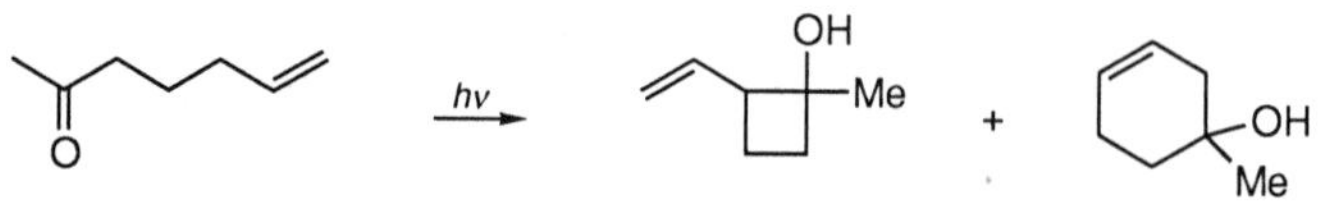

(*v*)

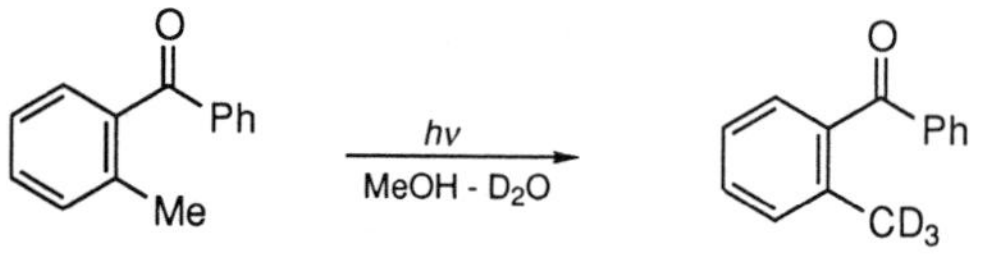

Free-radical reactions 17

A few organic free radicals are relatively stable. For example, the triphenylmethyl radical, generated from bromotriphenylmethane by reduction with silver, lives indefinitely in benzene solution in equilibrium with a dimer:

$$2\ Ph_3C\cdot \ \rightleftharpoons$$

Its stability is accounted for in part by the delocalization of the unpaired electron and in part by the steric forces inhibiting its reactions, for the conversion of the planar tervalent carbon atom into a tetravalent atom involves the compression of the three bulky phenyl groups. Indeed, dimerization gives a compound in which the aromatic stabilization energy of one ring has been sacrificed to retain one of two tervalent carbon atoms, in preference to yielding the fully aromatic but more sterically congested dimer, Ph_3C-CPh_3.

In the absence of significant stabilizing influences of a steric and/or electronic type, free radicals are very unstable, living for only a fraction of a second before reacting with another species. Nevertheless, they are intermediates in a wide variety of reactions, of which many have important applications in synthesis.

(a) The generation of free radicals

Three general methods are used to form free radicals.

(i) *Thermal generation.* The principles have already been described (p. 72) and will be summarized briefly.

Two types of compound undergo dissociation into free radicals at moderate temperatures: (1) compounds which possess an intrinsically weak bond, such as dialkyl peroxides ($D_{O-O} = 155\ kJ\ mol^{-1}$), chlorine ($D_{Cl-Cl} = 238\ kJ\ mol^{-1}$), and bromine ($D_{Br-Br} = 188\ kJ\ mol^{-1}$); (2) compounds which, on fragmentation, form especially strongly bonded products, such as azobisisobutyronitrile (AIBN) which yields N_2:

and peroxyoxalates which yield CO_2, e.g.

In each case, the presence of a substituent which stabilizes a resulting radical increases the rate of decomposition. For example, azomethane is stable up to 200°C whereas azobisisobutyronitrile decomposes with a half-life of 5 minutes at 100°C, to give a delocalized radical,

Again, $t_{1/2}$ for dibenzoyl peroxide (30 minutes at 100°C) is far less than for di-t-butyl peroxide (200 hours at 100°C) since the former gives a delocalized radical,

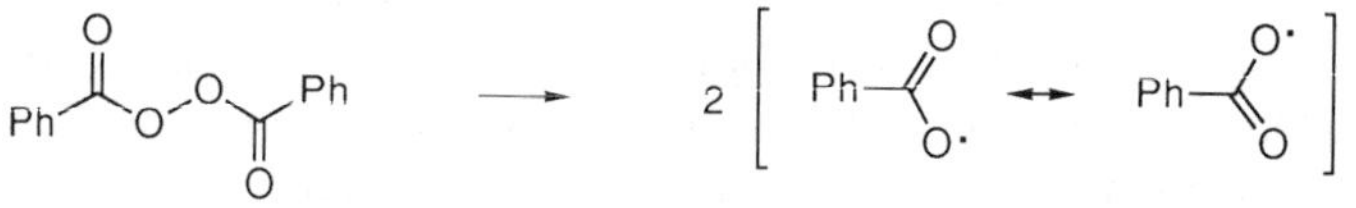

which subsequently decarboxylates to give the phenyl radical:

$$PhCO_2\cdot \longrightarrow Ph\cdot + CO_2$$

(*ii*) *Photochemical generation.* The principles have been described (p. 515). The method is suitable, for example, for generating chlorine or bromine atoms from the corresponding halogen molecules, alkoxy radicals from alkyl nitrites or hypochlorites,

$$RO-N{=}O \xrightarrow{h\nu} RO\cdot + NO$$

$$RO-Cl \xrightarrow{h\nu} RO\cdot + Cl\cdot$$

and nitrogen cation-radicals from protonated *N*-chloroamines:

$$R_2\overset{+}{N}H-Cl \xrightarrow{h\nu} R_2\overset{+\cdot}{N}H + Cl\cdot$$

(*iii*) *Redox generation.* Some covalent bonds may be broken by the acceptance of an electron from a species of powerful one-electron-donating type, others may be broken by the donation of an electron to a powerful one-electron acceptor.

In the first group, the donor may be a strongly electropositive element such as sodium, as in the acyloin reaction (p. 550),

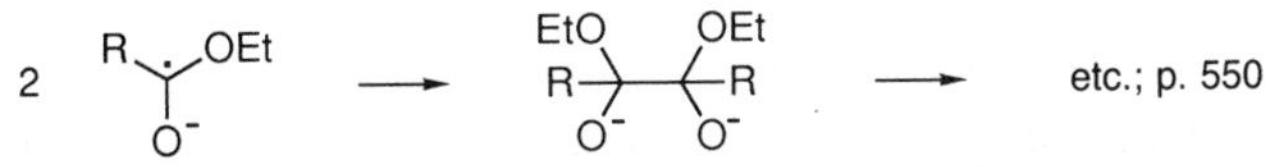

and in other metal-reducing systems (p. 636), or a low-valence transition-metal ion, as in *Fenton's reaction* in which hydroxyl radicals are formed,

$$Fe^{2+} \; + \; HO\text{--}OH \; \longrightarrow \; Fe^{3+} \; + \; HO\cdot \; + \; OH^-$$

In the second group, the acceptor is usually a transition-metal ion in a high-valence state, such as hexacyanoferrate(III) ion in the oxidation of certain phenols, e.g.

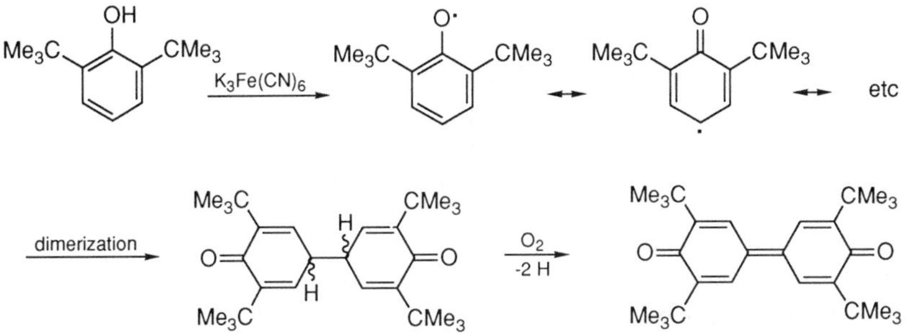

(b) The reactions of free radicals

Radicals normally react in one of three ways: by abstracting an atom from a saturated bond; by adding to an unsaturated bond; and by reacting with other radicals, leading to either combination or disproportionation. A fourth class of reaction, rearrangement, is relatively rare. There is no analogue of the S_N2 reaction, that is, one radical does not displace another from saturated carbon.

(i) Abstraction.

Free radicals react with saturated organic compounds by abstracting an atom, usually hydrogen, from carbon:

$$R\cdot \; + \; H\text{--}R' \; \longrightarrow \; R\text{--}H \; + \; R'\cdot$$

The selectivity of a free radical towards C—H bonds of different types is determined principally by two factors: bond dissociation energies and polar effects. First, the rate of abstraction increases as the bond dissociation energy decreases. The bond dissociation energies for C—H bonds in four simple alkanes are as follows:

Bond:	H–CH$_3$	H–CH$_2$Me	H–CHMe$_2$	H–CMe$_3$
Bond dissociation energy (kJ mol^{-1})	426	401	385	372

Consequently, the order of reactivity is, isobutane > propane > ethane > methane, and in general the order is, tertiary C—H > secondary C—H > primary C—H*. Allylic and benzylic C—H bonds (c. 322 kJ mol^{-1}) are significantly weaker than those in saturated systems because the unpaired electron in the resulting radicals is delocalized:

As a result, compounds containing these systems not only react readily with free radicals but also react selectively, e.g. ethylbenzene reacts with bromine atoms essentially entirely at its methylene group:

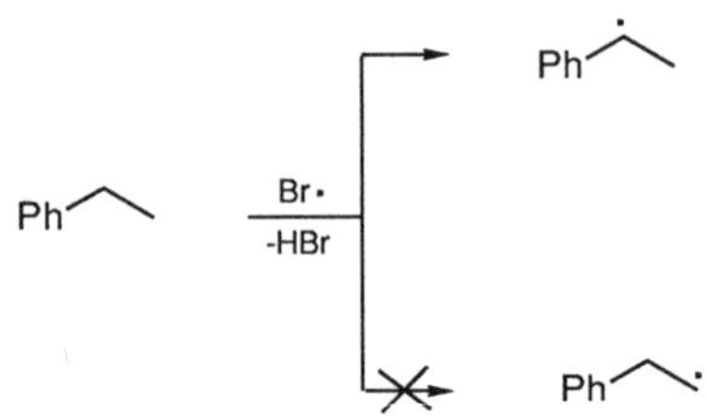

Halogenated compounds undergo abstraction of a halogen atom. For example, bromotrichloromethane reacts by loss of the bromine atom, C—Br rather than C—Cl cleavage occurring because the former bond is the weaker.

Secondly, polar factors are operative in many radical reactions. For example, the relative ease of abstraction by chlorine atoms from the C—H bonds in butyl chloride is:

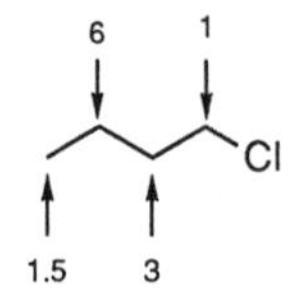

This results from the fact that the chlorine atom is strongly electronegative and preferentially reacts at C—H bonds of relatively high electron density. The effect of the chloro-substituent is to polarize the carbon atoms so that the electron density is relatively low at the carbon to which chlorine is attached, somewhat greater at the next carbon, and greater still at the third, the order of reactivities reflecting the fall-off in the inductive effect of chlorine. The lower reactivity of the methyl group than the adjacent methylene reflects the bond dissociation energy factor discussed above.

---

* Expressed differently, the stability of radicals falls in the order, tertiary > secondary > primary. This order is possibly attributable to hyperconjugative stabilization (p. 527).

Alkyl radicals, in contrast, react preferentially at sites where the electron density is low; they are described as *nucleophilic* radicals.

(*ii*) *Addition.*   Free radicals add to the common unsaturated groupings. The most important of the unsaturated groups in free-radical synthesis is the C=C bond, addition to which is markedly selective. In particular, addition to CH_2=CHX occurs almost exclusively at the methylene group, irrespective of the nature of X:

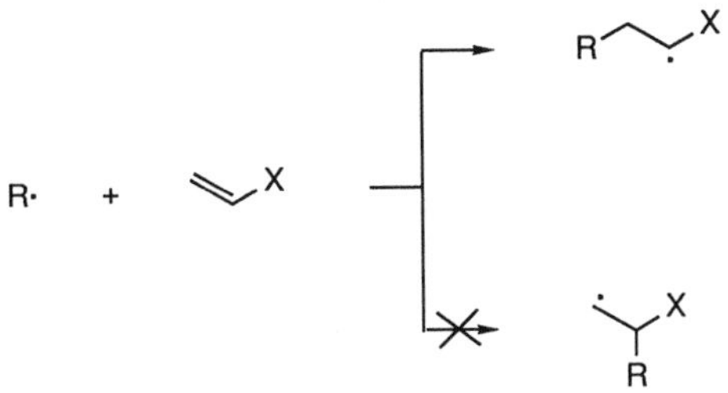

Two factors are responsible. First, steric hindrance between the radical and X slows reaction at the substituted carbon atom. Secondly, X can stabilize the radical RCH_2—$\dot{C}HX$ relative to the alternative $\cdot CH_2$—CHXR*, and this is reflected in the preceding transition state. The following relative rates of addition of the l-hexyl radical

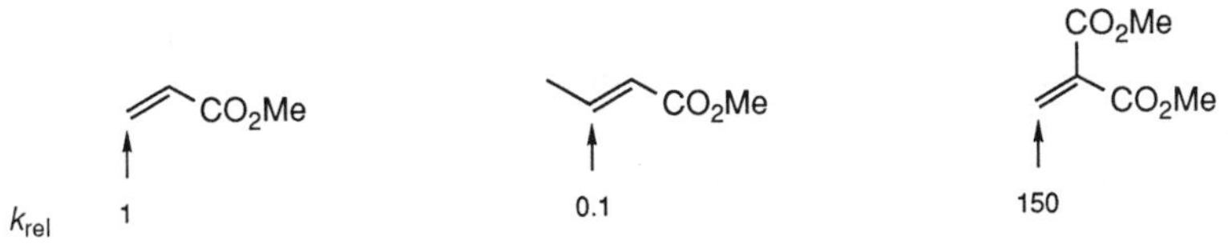

show the effect of steric hindrance at the site of addition and, in the di-ester, the stabilizing influence of two delocalizing substituents.

Alkynes react similarly to alkenes, monosubstituted alkynes reacting preferentially at the unsubstituted carbon. Addition to carbonyl groups is also known, but radicals normally react with carbonyl-containing compounds by abstracting

---

* If X = alkyl, stabilization may be the result of (hyperconjugative) delocalization, e.g.

If X is an unsaturated substituent or a substituent with one or more pairs of *p*-electrons, stabilization results from (conventional) delocalization, e.g.

hydrogen from a saturated carbon atom or from —CHO in an aldehyde. The lower reactivity in addition of C=O than C=C probably results from the fact that more energy is required to convert C=O into C—O ($c.$ 350 kJ mol^{-1}) than to convert C=C into C—C ($c.$ 260 kJ mol^{-1}).

(iii) *Combination and disproportionation.* Two free radicals can combine by dimerization, e.g.

$$2\ Br\cdot \longrightarrow Br_2$$

$$2\ \cdot CH_3 \longrightarrow CH_3 - CH_3$$

by the closely related reaction in which different radicals bond, e.g.

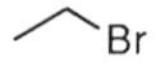

and by disproportionation, e.g.

$$2\ Et\cdot \longrightarrow EtH + CH_2{=}CH_2$$

These reactions are mostly very rapid, some having negligible activation energies.

(iv) *Rearrangement.* Free radicals, unlike carbocations, seldom rearrange. For example, whereas the neopentyl cation, $(CH_3)_3C{-}\overset{+}{C}H_2$, rearranges to $(CH_3)_2\overset{+}{C}{-}CH_2CH_3$, the neopentyl radical does not rearrange. However, the phenyl group migrates in certain circumstances, e.g.

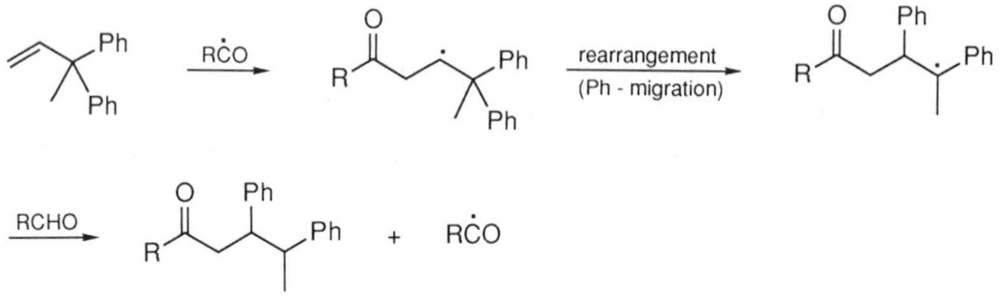

Rearrangements can also occur when they result in the relief of strain in a cyclic system, as in the radical-catalyzed addition of carbon tetrachloride to β-pinene:

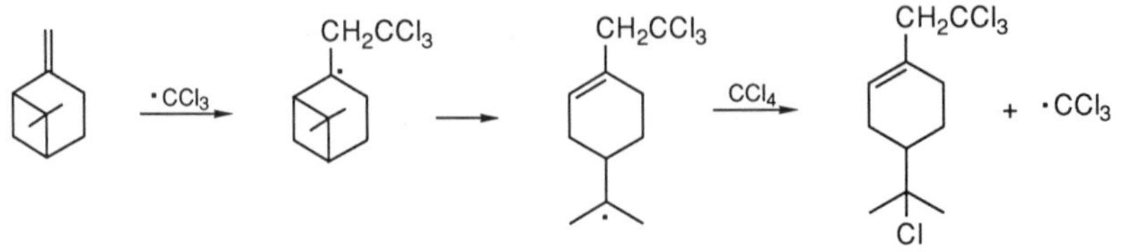

(c)   The characteristics of free-radical reactions

Free-radical reactions may be divided into two classes. In the first, the product results from the combination of two radicals, as in the Kolbe synthesis (p. 545),

$$RCO_2^- \xrightarrow{-e} RCO_2^\cdot \xrightarrow{-CO_2} R\cdot$$

$$2R\cdot \longrightarrow R-R$$

In the second class, the product results from the reaction of a radical with a molecule, as in the photochemical chlorination of methane (p. 530):

$$Cl_2 \xrightarrow{h\nu} 2Cl\cdot$$

$$Cl\cdot + CH_4 \longrightarrow HCl + \cdot CH_3$$

$$\cdot CH_3 + Cl_2 \longrightarrow CH_3Cl + Cl\cdot$$

The fundamental difference between the two types of reaction is that reactions in the latter class are *chain reactions*: that is, the step in which the product is formed results in the production of a new free radical which can bring about further reaction. Consequently, were it not for *chain-terminating* reactions, one radical could effect the complete conversion of reactants into products. In practice, chain-terminating processes inevitably occur: in the example above of the chlorination of methane both methyl radicals and chlorine atoms are destroyed by the unions,

$$\cdot CH_3 + \cdot CH_3 \longrightarrow C_2H_6$$

$$\cdot CH_3 + Cl\cdot \longrightarrow CH_3Cl$$

$$Cl\cdot + Cl\cdot \longrightarrow Cl_2$$

Nevertheless, these terminating steps, although having large rate constants, are of relatively low frequency because the concentration of radicals is low so that the probability of two of them colliding is small compared with the probability of a radical colliding with a molecule of a reactant. Hence the generation of a comparatively small concentration of radicals can suffice for extensive reaction providing that the rate constants for the radical-molecule reactions are large enough. Such reactions are sometimes described as being *radical-catalyzed* (although the use of the term catalyzed is misleading because the catalyst, e.g. a peroxide, is consumed).

Radical-catalyzed reactions are susceptible to *inhibitors*: certain compounds, present in low concentration, which are very reactive towards radicals can react with the radicals as they are generated to give inactive products. Inhibitors may be stable free radicals, such as nitric oxide, which combines with organic radicals to give nitroso compounds $(R \cdot + NO \rightarrow R—NO)$, or, more commonly in organic synthesis, compounds which react with organic radicals to generate radicals of such stability that they do not perpetuate the chain. An example of the latter type is quinol which reacts to give a relatively stable (delocalized) semiquinone radical:

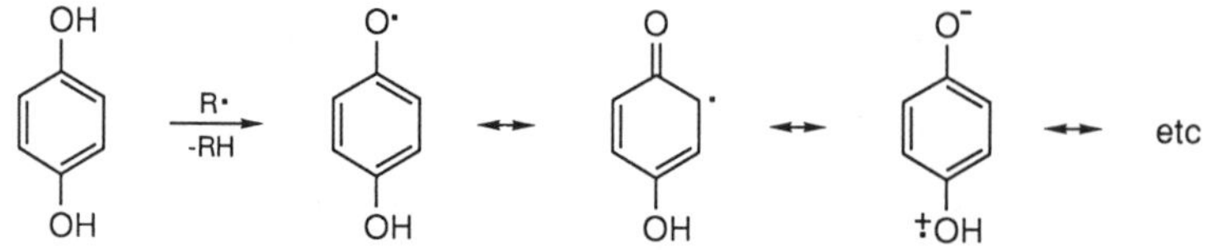

The chain-propagating reaction cannot begin until the inhibitor has been consumed. Since it frequently happens that alternative, ionic reactions can occur between the reactants which are unaffected by the presence of an inhibitor, it is necessary to rid the reactants of impurities which might act as inhibitors in order to achieve maximum efficiency in a radical-catalyzed process. On the other hand, some organic compounds are so susceptible to radical-catalyzed polymerization (p. 542) that it is necessary to store them in the presence of an inhibitor which can remove stray radicals generated by the action of light or oxygen.

17.2
Formation of
carbon–
halogen bonds

(a) *Substitution in saturated compounds*

(i) *Energetics.* Radical-catalyzed chlorination and bromination can be carried out both in the gas phase and in solution. For a typical chlorination, that of methane, the energetics are as follows:

$$\Delta H^\circ / \text{kJ mol}^{-1}$$

$$Cl\cdot \;+\; CH_4 \;\longrightarrow\; HCl \;+\; \cdot CH_3 \qquad\qquad -2$$
$$\cdot CH_3 \;+\; Cl_2 \;\longrightarrow\; CH_3Cl \;+\; Cl\cdot \qquad\qquad -101$$

That is, both propagating steps are exothermic. They are rapid reactions, so that the chain reaction competes very effectively with the terminating steps and the chains are long. In light-induced reactions, the quantum yield is high (i.e. one photon can lead to the conversion of many molecules of reactants) and in radical-initiated reactions only a small quantity of initiator is necessary.

In the corresponding bromination, however, the first of the propagating steps is endothermic:

$$\Delta H^\circ / \text{kJ mol}^{-1}$$

$$Br\cdot \;+\; CH_4 \;\longrightarrow\; HBr \;+\; \cdot CH_3 \qquad\qquad +62$$
$$\cdot CH_3 \;+\; Br_2 \;\longrightarrow\; CH_3Br \;+\; Br\cdot \qquad\qquad -92$$

Its activation energy is necessarily at least as great as $62\,\text{kJ mol}^{-1}$ so that reaction is much slower than in the case of chlorine. As a result, the terminating processes compete more effectively with the propagating step and the chains are short.

The corresponding reaction with iodine atoms,

$$I\cdot \ + \ CH_4 \longrightarrow \ HI \ + \ \cdot CH_3$$

is so strongly endothermic ($129\,kJ\,mol^{-1}$) that the reaction is ineffective; indeed, alkyl iodides can be reduced by hydrogen iodide to alkanes and iodine. For fluorine, both propagating steps are strongly exothermic; reaction is violent and is accompanied by the fragmentation of alkyl groups. It is therefore usually better to introduce fluorine by indirect means, although several fluorinations have been conducted successfully in the gas phase in the presence of nitrogen as a dilutent and with metal packing to dissipate the heat.

(*ii*) *Applications.*   Radical-catalyzed chlorination and bromination occur readily on alkanes and substituted alkanes, both in the gas phase and in solution. Both thermal and photochemical generation of the halogen atoms are employed, and, for chlorinations, sulfuryl chloride in the presence of an initiator such as dibenzoyl peroxide is also used.

$$Initiator \longrightarrow \ R\cdot$$

$$SO_2Cl_2 \ + \ R\cdot \longrightarrow \ RCl \ + \ \cdot SO_2Cl$$

propagation $\begin{cases} \\ \\ \\ \\ \end{cases}$

$$\cdot SO_2Cl \longrightarrow \ SO_2 \ + \ Cl\cdot$$

$$Cl\cdot \ + \ RH \longrightarrow \ HCl \ + \ R\cdot$$

$$R\cdot \ + \ SO_2Cl_2 \longrightarrow \ RCl \ + \ \cdot SO_2Cl$$

Although both the bond dissociation energy factor and the polar factor provide some selectivity in chlorinations, a mixture of products is normally obtained. For example, the chlorination of isobutane in the gas phase at 100°C gives comparable quantities of isobutyl chloride and t-butyl chloride,

and at 300°C 2-methylbutane gives the following products:

| 33.5% | 22% | 28% | 16.5% |

In addition, further chlorination occurs, e.g. methane gives a mixture of methyl chloride, dichloromethane, chloroform, and carbon tetrachloride. Nevertheless, these reactions are of great importance industrially because the products can be separated by efficient fractional distillation. Their use on the laboratory scale is limited, except when only one monochlorinated product can be formed, for this is usually readily separated from any unchanged starting material (lower boiling) and any dichlorinated product (higher boiling), e.g. cyclohexyl chloride can be isolated in 89% yield by treating cyclohexane with sulfuryl chloride in the presence of dibenzoyl peroxide:

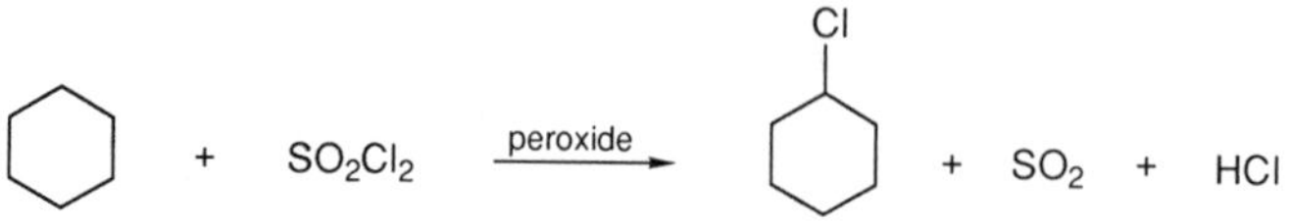

Bromine atoms are significantly more selective than chlorine atoms in their reactions at primary, secondary, and tertiary C—H.* For example, the gas phase bromination of isobutane gives t-butyl bromide essentially exclusively:

(b)   Substitution in allylic and benzylic compounds

Allylic and benzylic compounds are both more reactive than saturated compounds in radical-catalyzed halogenation and react selectively (p. 526). Allylic compounds of the type RCH=CH—CH$_2$R' give mixtures of two monohalogenated products because the allylic radical can react at each of two carbon atoms:

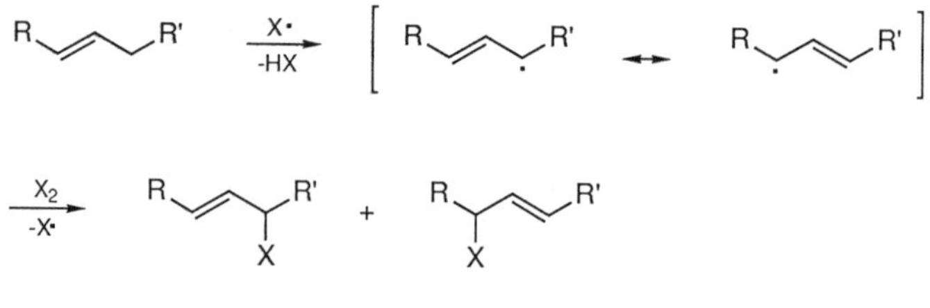

Benzylic compounds, on the other hand, give only one product because the possible isomers, being non-aromatic, are of much higher energy content.

---

* This can be understood by reference to Hammond's postulate (p. 61). Whereas abstraction by Cl· from C—H is exothermic, abstraction by Br· is endothermic. Consequently, the transition state in the latter reaction more closely resembles the product (alkyl radical) (i.e. bond-breaking has made more progress), so that the factors which determine the relative stabilities of the radicals (tertiary > secondary > primary) are of greater significance in determining the relative stabilities of the transition states. The relative reactivities of tertiary and secondary C—H are about 2:1 towards Cl· and about 30:1 towards Br·.

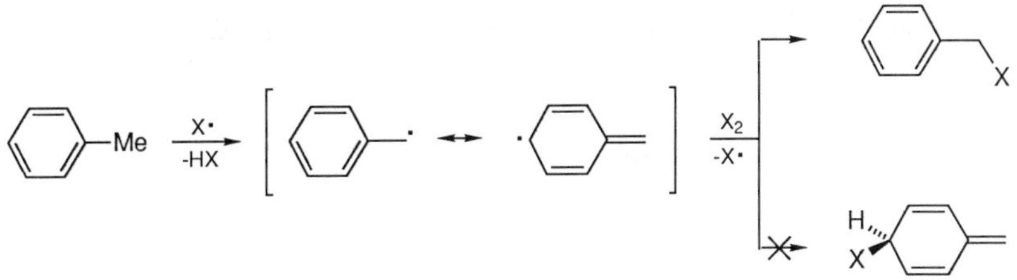

In addition to the thermal and photochemical* methods for the chlorination and bromination of saturated compounds, N-bromosuccinimide is used for allylic and benzylic bromination. It is prepared by treating succinimide in alkaline solution with bromine,

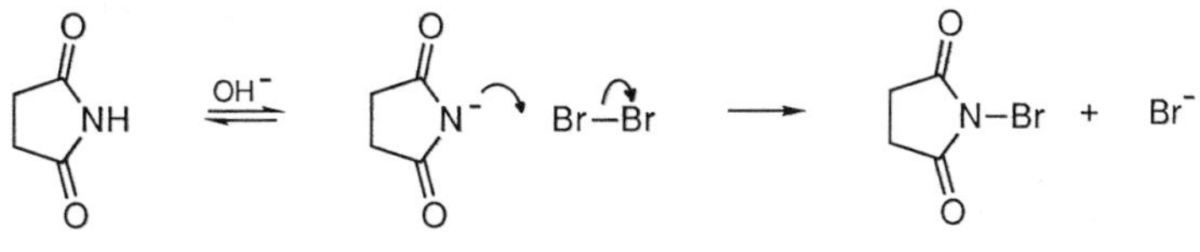

N-bromosuccinimide

and reacts, usually in the presence of a peroxide as initiator, by a chain mechanism; a trace of hydrogen bromide is necessary for initiation:

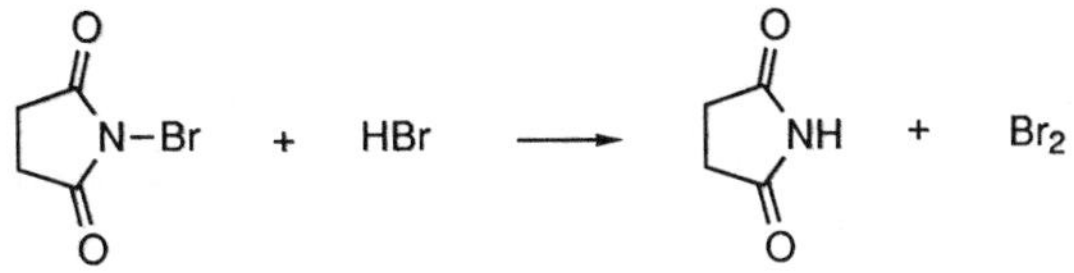

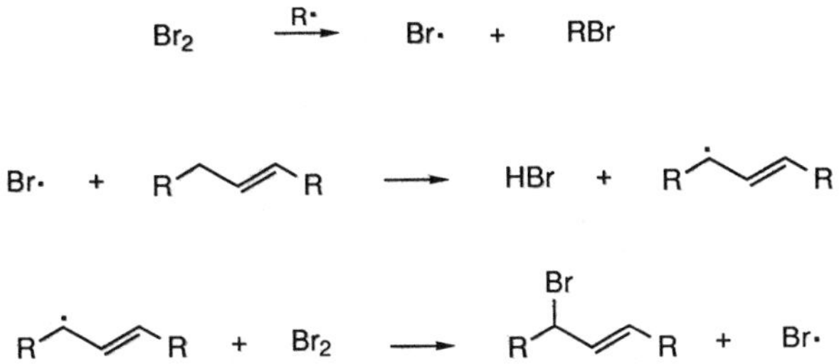

* Benzenoid compounds can also react by addition with chlorine atoms. For example, the irradiation of benzene and chlorine gives a mixture of stereoisomeric hexachlorocyclohexanes, one of which (shown) has powerful insecticidal activity (Gammexane):

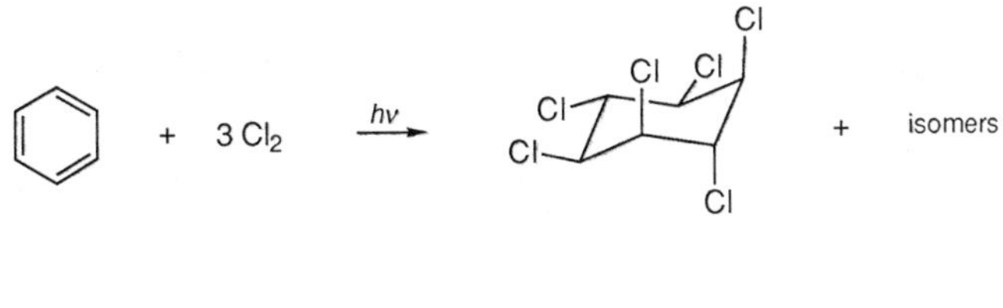

Examples of substitution in allylic and benzylic compounds are:

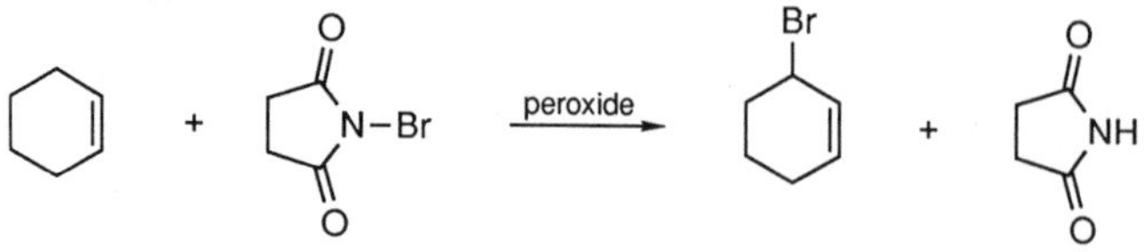

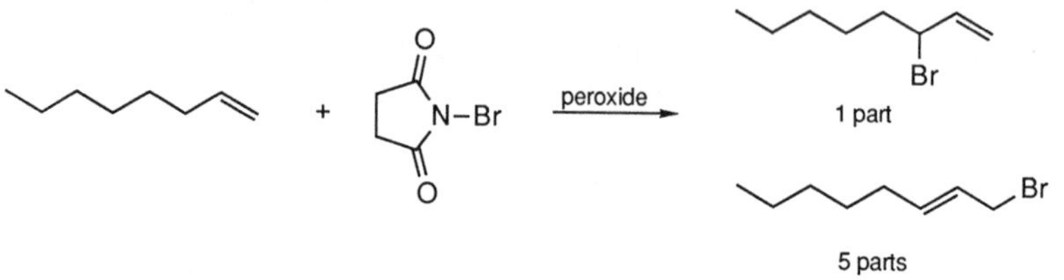

$$PhCH_3 + SO_2Cl_2 \xrightarrow{\text{peroxide}} PhCH_2Cl + SO_2 + HCl$$

80%

$$PhCH_3 + Br_2 \xrightarrow{\text{peroxide}} PhCH_2Br + HBr$$

98%

The bromination of allylic compounds is frequently used in the conversion of alkenes into conjugated dienes, which are obtained from the allylic bromides by base-catalyzed elimination. For example, 3-bromocyclohexene gives 1,3-cyclohexadiene:

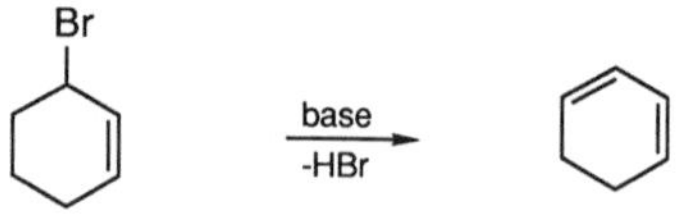

(c) *Addition to C═C and C≡C*

Hydrogen bromide adds to C═C and C≡C bonds in radical-catalyzed reactions, illustrated for propylene:

$$\text{Initiator} \longrightarrow R\cdot$$

$$R\cdot + HBr \longrightarrow RH + Br\cdot$$

propagation

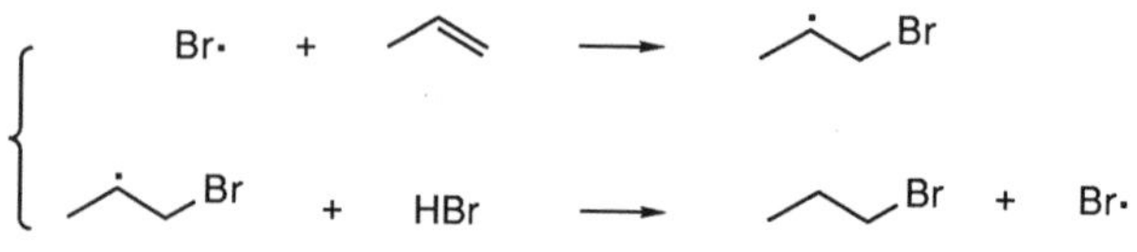

Yields are usually good, as in the preparation of propyl bromide (87%) and in the following examples:

88%

80%

The bromine atom always adds to the terminal carbon atom (p. 527), so that the process is complementary to the *ionic* addition of hydrogen bromide to alkyl- and halosubstituted alkenes in which hydrogen adds to the terminal carbon (Markovnikov's rule, p. 83). The ionic and radical-catalyzed processes are competitive and the latter usually occurs the more rapidly. In order to ensure that the ionic addition shall occur when required, it is necessary to free the alkene from the peroxidic impurities formed when alkenes are exposed to air (autoxidation; p. 556) which can initiate the radical reaction. The reaction of propyne (above) illustrates another feature of the process, namely, its *trans*-stereoselectivity.

The other hydrogen halides do not react at C=C bonds in this way, although hydrogen chloride is partially effective. The reason is apparent from inspection of the energetics of the two propagating steps which, for reaction on ethylene, are as follows:

$$\Delta H^\circ / \text{kJ mol}^{-1}$$

X in HX	X· + (ethylene) → X(radical)	X(radical) + HX → X(product) + X·
F	−209	+159
Cl	−101	+27
Br	−42	−37
I	+12	−104

Only for hydrogen bromide are both these steps exothermic. The reaction of hydrogen chloride is slow because of the endothermicity of the reaction between the carbon radical and hydrogen chloride, and hydrogen iodide fails to add because the reaction of iodine atoms with ethylene is endothermic.

(d)   Bromo-decarboxylation (Hunsdiecker reaction)

Treatment of the silver salt of a carboxylic acid with bromine in refluxing carbon tetrachloride gives a bromo-compound with elimination of carbon dioxide:

$$RCO_2Ag \ + \ Br_2 \ \longrightarrow \ RBr \ + \ AgBr \ + \ CO_2$$

The silver salt is prepared by treating an aqueous solution of the sodium salt of the acid with silver nitrate. Yields are variable but in many cases are high, e.g. that of neopentyl bromide is 62%:

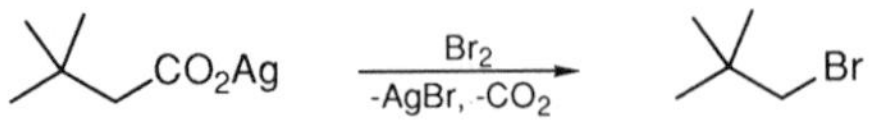

Reaction occurs by formation of the acyl hypobromite and its subsequent homolysis. The acyloxy radical loses carbon dioxide and the resulting radical abstracts bromine from a second molecule of the hypobromite:

$$RCO_2Ag \ + \ Br_2 \ \longrightarrow \ RCO_2Br \ + \ AgBr$$

$$RCO_2Br \ \longrightarrow \ RCO_2{}^{\cdot} \ + \ Br{\cdot}$$

$$RCO_2{}^{\cdot} \ \xrightarrow{-CO_2} \ R{\cdot} \ \xrightarrow{RCO_2Br} \ RBr \ + \ RCO_2{}^{\cdot}$$

Mercury(II) and lead(IV) compounds react similarly. For example, treatment of cyclopropanecarboxylic acid with mercury(II) oxide and bromine gives bromocyclopropane in about 45% yield [5]:

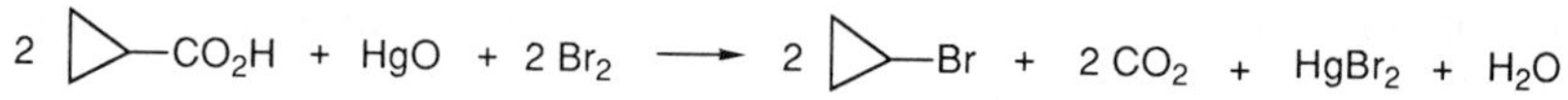

$$2 \ \triangleright\!-CO_2H \ + \ HgO \ + \ 2\,Br_2 \ \longrightarrow \ 2 \ \triangleright\!-Br \ + \ 2\,CO_2 \ + \ HgBr_2 \ + \ H_2O$$

(e) Iodo-decarboxylation

Iodides may be prepared by a process analogous to the Hunsdiecker reaction. The carboxylic acid is heated in solution, under illumination, with lead tetraacetate and iodine. An exchange equilibrium between lead-bonded acetate and carboxylate groups is established and the acyl hypoiodite is formed and photolyzed:

$$4\,RCO_2H \ + \ Pb(OAc)_4 \ \rightleftharpoons \ Pb(O_2CR)_4 \ + \ 4\,AcOH$$

$$Pb(O_2CR)_4 \ + \ I_2 \ \longrightarrow \ Pb(O_2CR)_2 \ + \ 2\,RCO_2I$$

$$RCO_2I \ \xrightarrow{h\nu} \ RCO_2{}^{\cdot} \ + \ I{\cdot}$$

$$RCO_2{}^{\cdot} \ \xrightarrow{-CO_2} \ R{\cdot} \ \xrightarrow{RCO_2I} \ RI \ + \ RCO_2{}^{\cdot}$$

The yields are in some cases excellent, e.g. hexanoic acid gives 1-iodopentane quantitatively.

(a)   *Addition to C=C*

A large group of synthetically useful radical-catalyzed reactions is based on the addition of aliphatic carbon radicals to C=C bonds. The reaction of bromoform with 1-butene is illustrative:

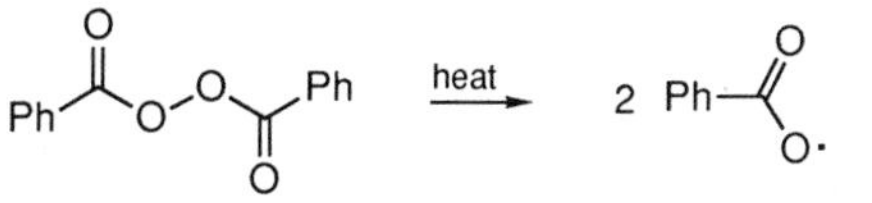

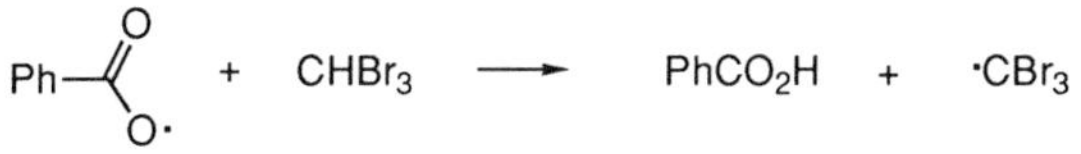

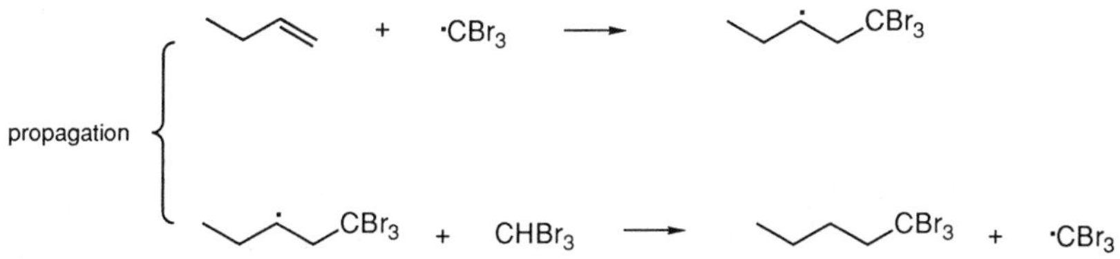

The following examples show that moderate to good yields can be obtained with a variety of types of starting material:

Aldehydes: addition of RĊO, e.g.

64%

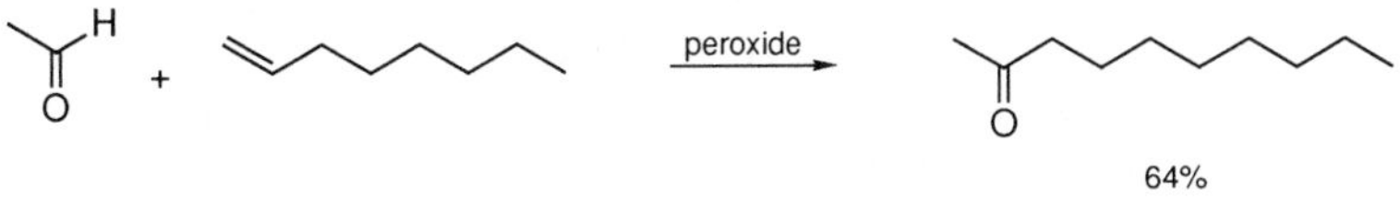

Ketones: addition of RCO—ĊH—R', e.g.

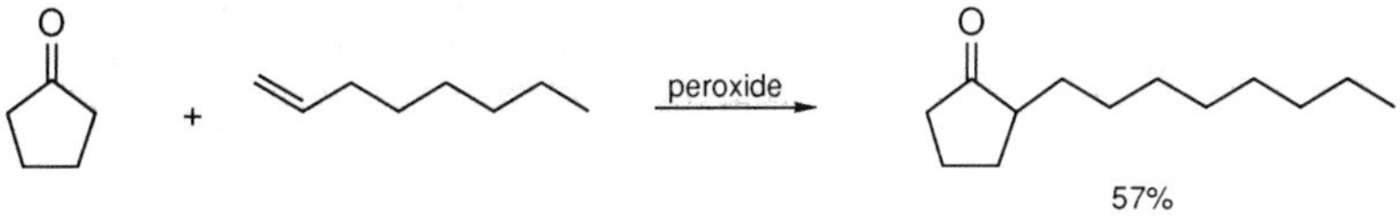

57%

Alcohols: addition of R—Ċ(OH)—R', e.g.

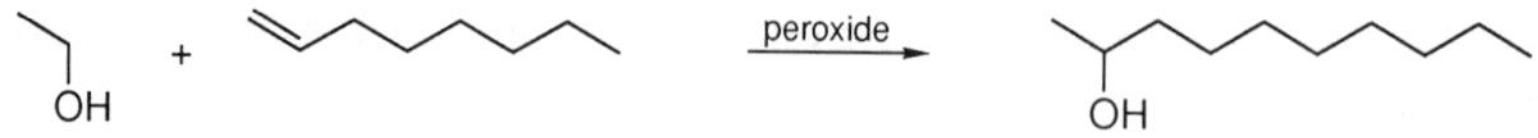

28%

Amines: addition of $R-\dot{C}(NHR')-R''$, e.g.

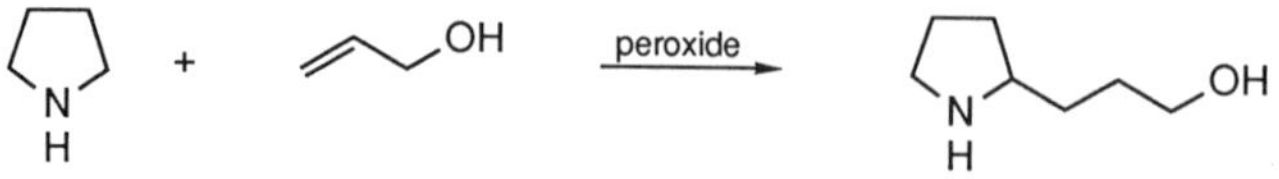

Nevertheless, this approach is limited in scope. In each of the examples above, one of the two types of C—H bond is significantly more reactive towards the abstracting radical than the other: in the case of acetaldehyde, the aldehydic C—H bond is considerably weaker than the bonds in the CH$_3$ group and in the other instances delocalization in the resulting radicals,

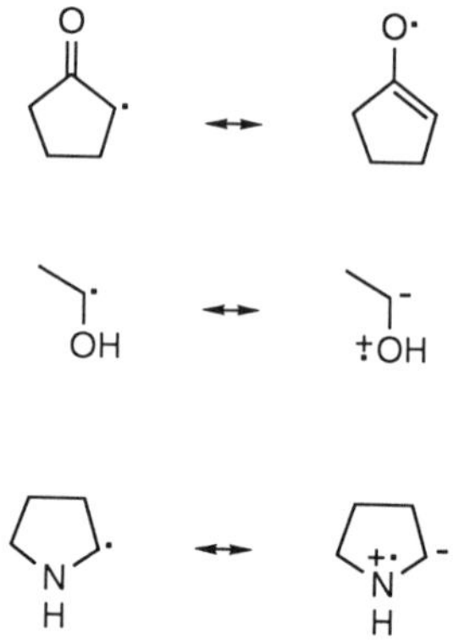

is reflected in the stability of the transition state for hydrogen-atom abstraction. In other cases, there may be little selectivity between the various C—H bonds and the yields of individual products are then low. It is therefore in general better to use an attacking radical which abstracts a group or an atom other than hydrogen in preference to a hydrogen atom and to locate that group or atom in the reactant in the required position. Such a radical is tributylstannyl, Bu$_3$Sn·, and the atoms and groups which it abstracts efficiently are the halogens (except fluorine), PhS, and PhSe. In contrast, since the Sn—H bond is weak, it does not abstract hydrogen from C—H. This method, while of limited application in inter-molecular C—C bond formation, is of very considerable value in the intra-molecular analogue.

Reaction is usually initiated with AIBN, either thermally or photochemically. The radical ·CMe$_2$CN readily abstracts a hydrogen atom from the weak Sn—H bond in tributylstannane, and the chain consists of the following steps (X = Cl, Br, I, SPh, or SePh):

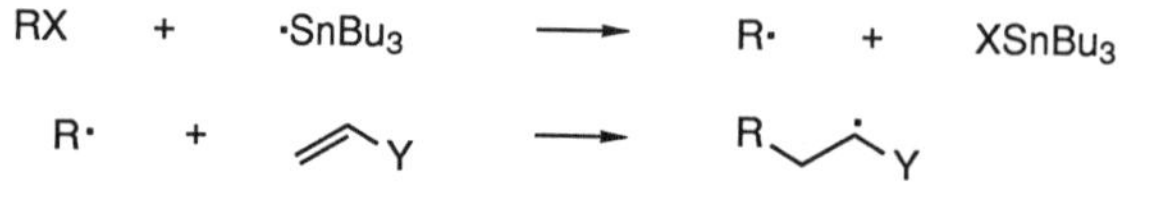

$$RX \quad + \quad \cdot SnBu_3 \quad \longrightarrow \quad R\cdot \quad + \quad XSnBu_3 \qquad (1)$$

(2)

(3)

There are also unwanted reactions:

(4)

(5)

However, these have little effect providing that Y is a group of $-M$ type and that the ratio [alkene]:[Bu$_3$SnH] is about 100:1, for the following reasons. First, alkyl radicals are nucleophilic and add to alkenes which contain $-M$ substituents with rate constants of $c.$ $10^6 \mathrm{M}^{-1}\mathrm{s}^{-1}$ which is about 1000 times greater than for other alkenes and is approximately the same as that for reaction (4). Therefore with a high [alkene]:[Bu$_3$SnH] ratio, reaction (4) is insignificant compared with reaction (2). Second, the radical RCH$_2$—ĊHY is electrophilic owing to the presence of the $-M$ substituent and the rate constant of reaction (5) is about 10000 times smaller than that of (2).

(i) Intramolecular addition.   Internal addition of a carbon radical to C=C does not occur to give three- or four-membered rings; indeed, these strained cyclic radicals rapidly undergo the reverse of addition:

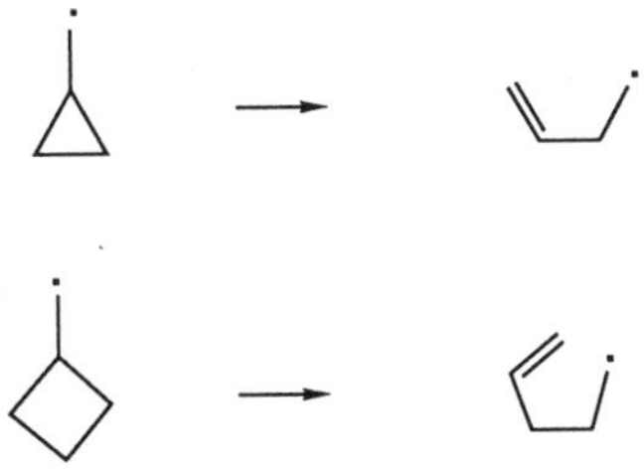

However, five-membered rings are readily formed by the tributylstannane method, e.g.

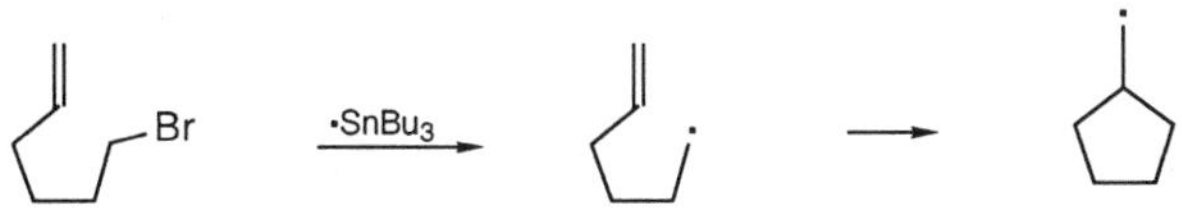

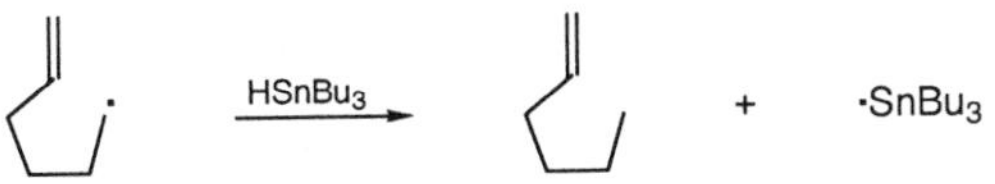

In contrast to the intermolecular reaction, a $-M$ substituent is not required on the alkene terminus. This is because the entropy of activation for the cyclization is much more favourable than for the intermolecular reaction: even without an activating substituent on the alkene, the rate constant for the cyclization step is $c.$ $10^5 \mathrm{s}^{-1}$. Since the rate constant for the competing hydrogen-atom abstraction, is *c.* $10^6 \text{M}^{-1}\text{s}^{-1}$, cyclization predominates providing that [Bu₃SnH] is less than about 0.05 M.

There is a further competing reaction: formation of the six-membered ring,

However, in spite of the fact that the five-membered cyclic radical is somewhat more strained and that it is a primary radical whereas the six-membered isomer is a secondary radical, six-membered cyclization occurs at only about one-fiftieth the rate of five-membered cyclization. This is an example of a general phenomenon and is attributed to a preference for a transition state in which the attacking radical approaches the double bond along a plane perpendicular to that of the $\overset{\diagdown}{\underset{\diagup}{C}}=\overset{\diagup}{\underset{\diagdown}{C}}$ framework and with an angle between the developing single bond and the C=C bond close to the tetrahedral value:

This is achieved for five-membered cyclization through the chair-like transition state (a)

(a)

with less strain than it is for six-membered cyclization (cf. Baldwin's rules, p. 678).

There are stereochemical consequences of this: substitutents preferentially occupy equatorial-type positions consistent with product distributions such as

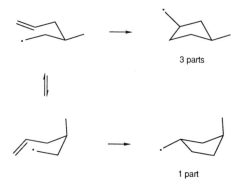

3 parts

1 part

Ring closure of the heptenyl radical also gives mainly the *exo*-cyclic radical:

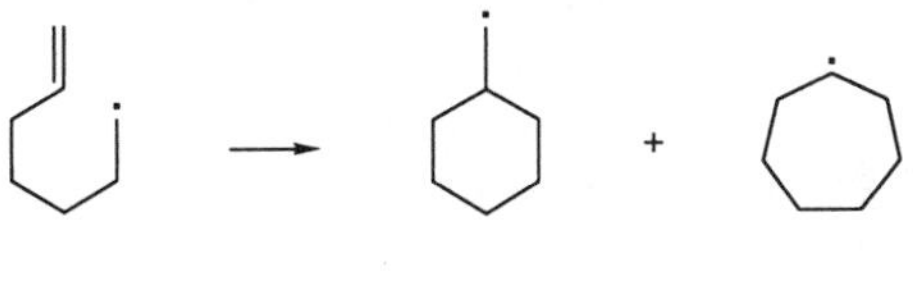

6 parts   1 part

However, the rate constant is about ten-times smaller than that for formation of the five-membered *exo*-cyclic radical so that reduction competes more effectively and the method is correspondingly less valuable in synthesis.

The use of PhS and PhSe substitutents, e.g.

$$R-SePh \quad + \quad \cdot SnBu_3 \quad \longrightarrow \quad R\cdot \quad + \quad PhSe-SnBu_3$$

instead of a halogen is desirable when the halogen would be sensitive to the reaction conditions in a step preceding cyclization, whereas PhS and PhSe are not.

Two examples illustrate the power of the tributylstannane method. In the first, reaction is induced at a bridgehead, where radical reactions are normally difficult because the radical cannot achieve planar geometry:

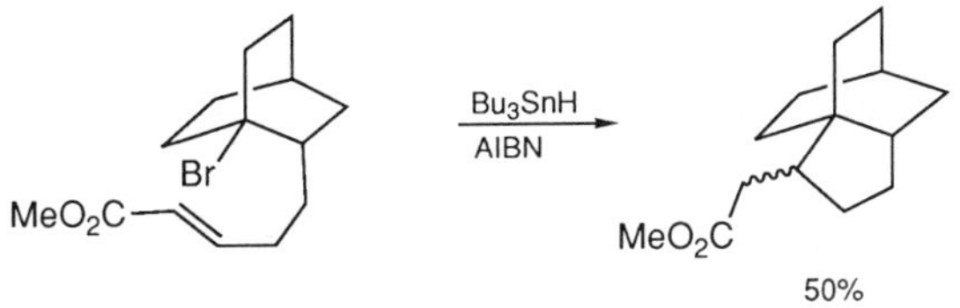

50%

The other is a tandem cyclization, the second cyclization of which involves an alkyne:

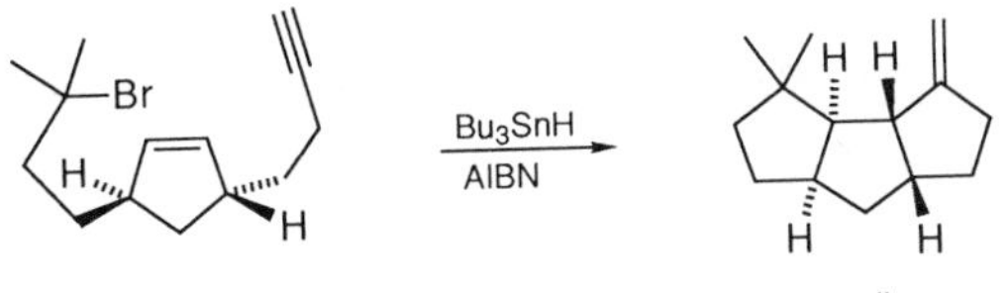

capnellene

(ii) Alkenyl and aryl radicals. Both alkenyl and aryl radicals can be generated by the tributylstannane method. Each forms *exo*-cyclic, five-membered rings with rate constants of *c.* $10^8 \, s^{-1}$ and, since the rate constants for their abstraction of hydrogen from the stannane are *c.* $10^8 \, M^{-1} s^{-1}$, ring closure competes effectively with reduction when $[Bu_3SnH] < c.$ 0.2 M, e.g.

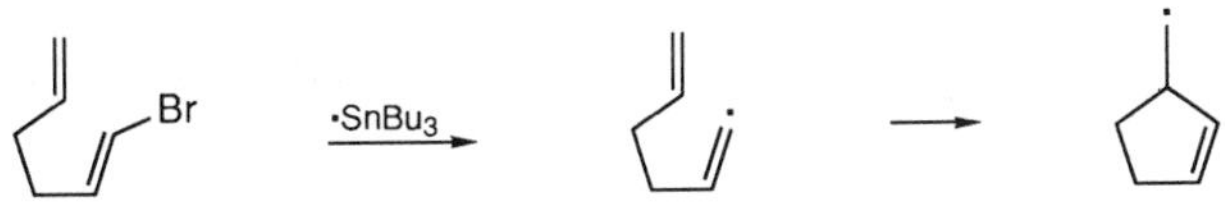

(iii)   The Sm(II) method for cyclization.   An alternative to the tributylstannane method for five-membered ring formation is to form the first carbon radical in the sequence as a ketyl by one-electron reduction of a carbonyl group with samarium(II) iodide. After cyclization on to a carbon–carbon double bond which is attached to a −*M* group, a second one-electron reduction yields the product as its enolate, e.g.

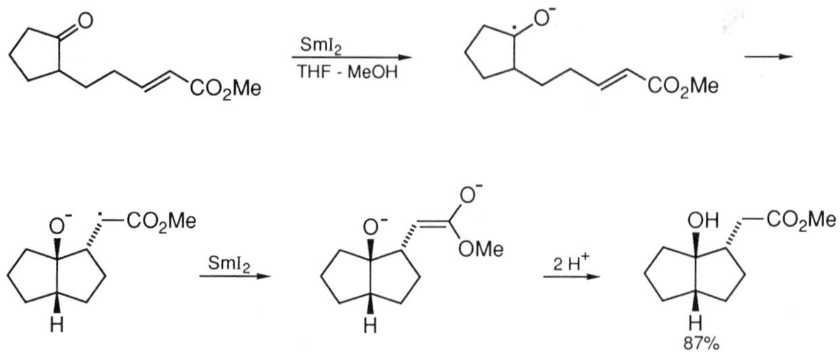

(b)   The polymerization of alkenes

Many alkenes are polymerized by free-radical initiators. The reaction of a monosubstituted alkene occurs as follows:

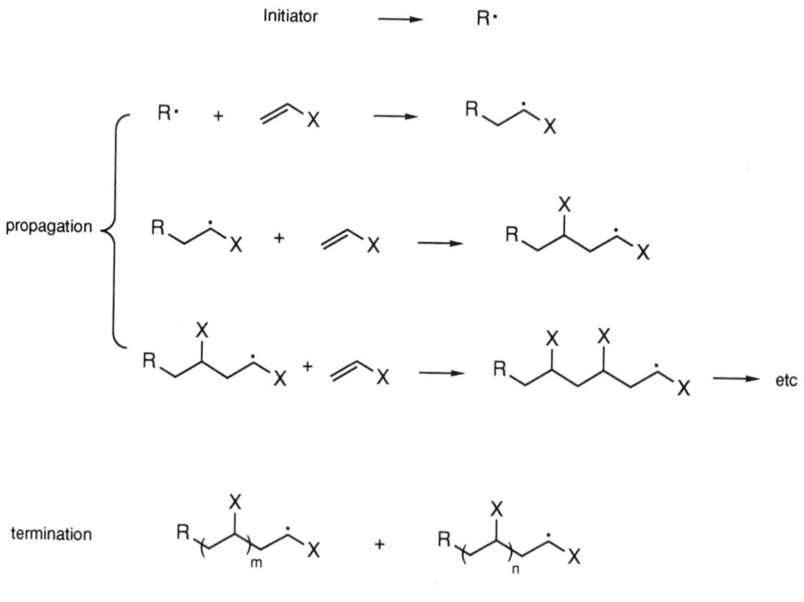

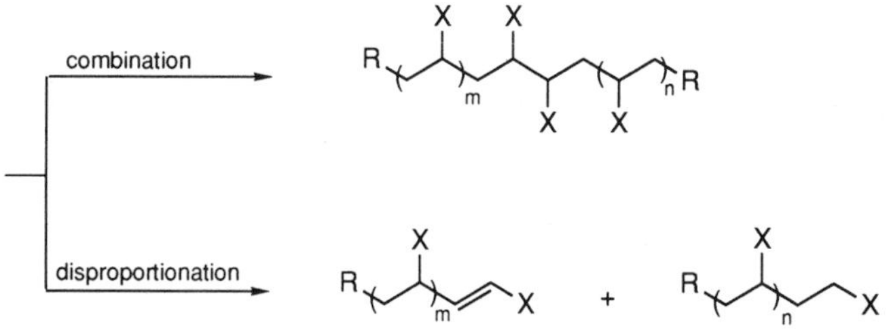

Initiation is normally effected by a peroxide or an azo-compound and the chains grow until two meet and the reaction terminates by dimerization or disproportionation. As the concentration of initiator is increased the number of growing chains increases and consequently termination becomes more probable relative to polymerization. Hence the average molecular weight of the polymer can be determined by adjusting the concentration of initiator. The molecular weight can also be controlled by the addition of *chain-transfer* agents, i.e. compounds which react with the growing-chain to terminate it and at the same time themselves give radicals which initiate new chains. Thiols are often chosen for the purpose.

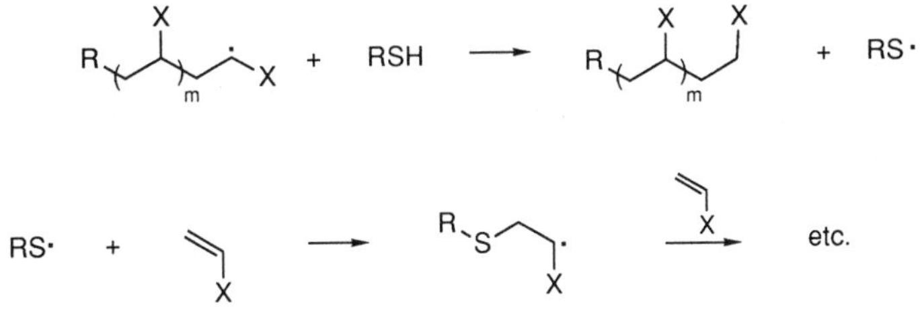

Polymerization is subject to inhibition (p. 529) and it is advisable to add a small quantity of an inhibitor to the more readily polymerized alkenes (e.g. styrene) to prevent polymerization during storage.

Free-radical polymerization is of immense industrial importance. Some of the widely used monomers are vinyl chloride (for polyvinyl chloride, PVC), styrene (for polystyrene), and methyl methacrylate ($CH_2{=}C(CH_3)CO_2Me$) (for Perspex). The polymerization of a mixture of two monomers (copolymerization) can also give polymers with useful properties, e.g. butadiene is copolymerized with both acrylonitrile and styrene for the manufacture of synthetic rubbers.

Ethylene and propylene are not polymerized in this way; Ziegler's catalysts are used (p. 568).

(c) Homolytic aromatic substitution

Both alkyl and aryl radicals substitute in aromatic nuclei by an addition–elimination sequence:

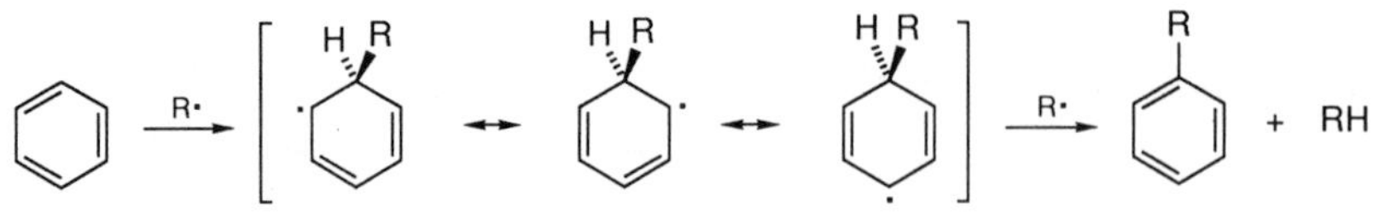

Alkylation can be effected by heating the aromatic compound with a diacyl peroxide or a lead tetracarboxylate, e.g.

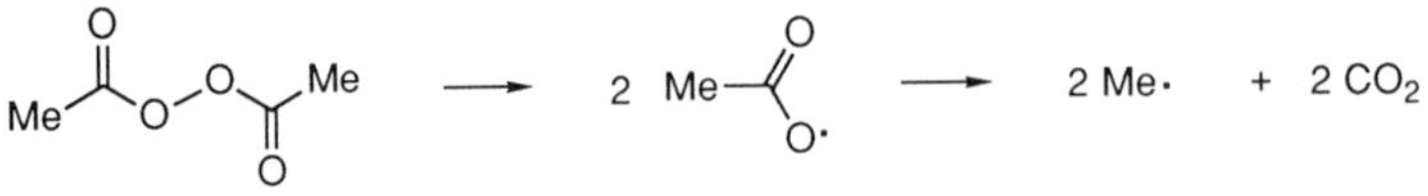

$$Pb(OCOMe)_4 \longrightarrow Pb(OCOMe)_2 + 2\,Me\cdot + 2\,CO_2$$

However, the yields are low, partly because the alkylbenzenes are very reactive towards free radicals at their benzylic carbon atoms, e.g. the methylation of benzene gives toluene which reacts readily with alkyl radicals to give benzyl radicals and thence bibenzyl. Since alkylated aromatic compounds can normally be obtained through Friedel–Crafts reactions (section 11.3), homolytic alkylation is of little synthetic utility.

Arylation is normally brought about in one of three ways: by the Gomberg reaction (p. 416), using *N*-nitrosoacylarylamines (p. 417), and using diacyl peroxides such as dibenzoyl peroxide,

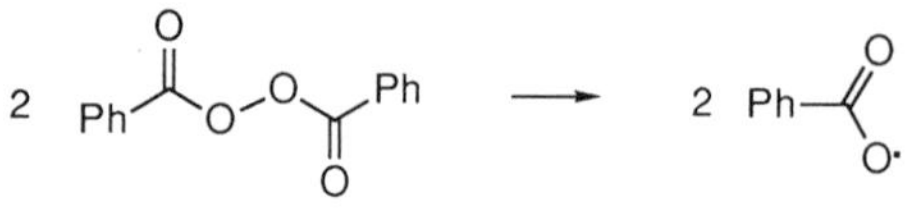

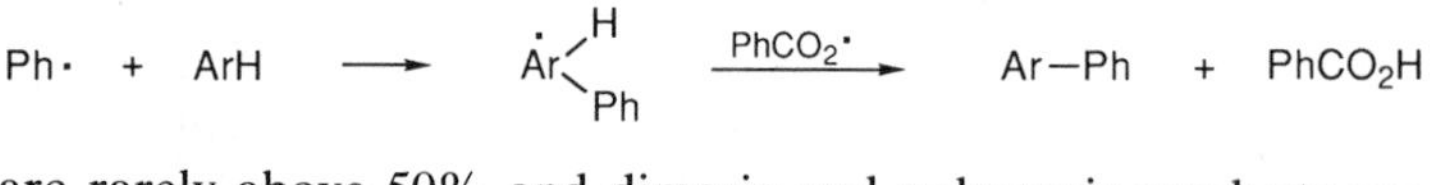

Yields are rarely above 50% and dimeric and polymeric products are formed by processes such as

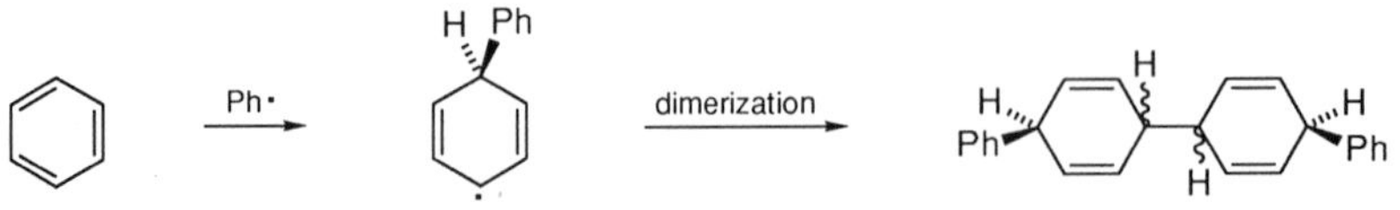

However, the addition of nitrobenzene and related oxidants increases the yield of the biaryl from peroxide reactions: for example, biphenyl can be obtained in about 80% yield from dibenzoyl peroxide and benzene. The mechanism of action of the nitro-compound is complex; effectively it, and species derived from it, such as diphenyl nitroxide, Ph_2N—$O\cdot$, oxidize the intermediate cyclohexadienyl radical and so divert it from dimerization.

Free radicals are not nearly so selective between different positions in aromatic compounds as they are between the two carbon atoms of CH_2=CHX. Reaction tends to occur preferentially at the *ortho* position of monosubstituted benzenes unless the substituent is particularly large (e.g. t-butyl), and it is understandable that *ortho* should predominate over *meta* substitution since the former reaction leads to a more effectively stabilized radical, e.g.

This cannot be the sole governing factor, however, for then *para* substitution should always predominate over *meta* substitution since similar conjugating influences are present for *para* and *ortho* substitution. In practice, more than twice as much of the *meta* as the *para* derivative is formed in some cases. Some typical data for phenylation are shown; the total spread in the relative reactivities of monosubstituted benzenes is less than a factor of ten, in marked contrast to the spread in electrophilic and nucleophilic reactions where powerful polar effects operate.

X in PhX	*o*	*m*	*p* (%)	Reactivity of PhX relative to benzene
NO_2	62	10	28	3.0
CH_3	67	19	14	1.2
F	55	31	14	1.1
$C(CH_3)_3$	24	49	27	0.6

Despite the fact that homolytic arylation usually gives a mixture of products in low yields, it is nevertheless the only readily applicable method for obtaining many biphenyls, for Friedel–Crafts reactions are unsuccessful and electrophilic substitution in biphenyl is specific only for certain derivatives. For example, although 4-nitrobiphenyl can be obtained in about 50% yield by the nitration of biphenyl, the 3-nitro-derivative is formed in negligible yield and is best prepared from the reaction between benzene and di-*m*-nitrobenzoyl peroxide or *m*-nitrobenzenediazonium chloride in Gomberg conditions. (The alternative route, reaction between phenyl radicals and nitrobenzene, would give all three nitrobiphenyls.)

(d)   Dimerization of alkyl radicals: the Kolbe electrolytic reaction

Electrolysis of the alkali-metal salts of aliphatic carboxylic acids results in the liberation of alkyl radicals at the anode and their subsequent dimerization:

$$RCO_2^- \xrightarrow{-e} RCO_2 \cdot \longrightarrow R \cdot + CO_2$$

$$2 R \cdot \longrightarrow R-R$$

The reaction is usually carried out by dissolving the acid in methanol containing enough sodium methoxide to neutralize about 2% of the acid and electrolyzing between platinum-foil electrodes. The sodium liberated at the cathode reacts with the solvent to generate further carboxylate anion until the acid has been consumed and the solution is just alkaline. Yields are variable but are sometimes nearly quantitative, e.g.

$$2 CH_3CO_2H \longrightarrow CH_3-CH_3 \quad (93\%)$$

$$Me-(CH_2)_{12}-CO_2H \longrightarrow Me-(CH_2)_{24}-Me \quad (60\%)$$

Electrolysis of the half-esters of dibasic acids gives long-chain dibasic esters, e.g.

$$2 EtO_2C-(CH_2)_n-CO_2H \xrightarrow[\substack{n = 2:76\% \\ n = 8:70\% [5]}]{} EtO_2C-(CH_2)_{2n}-CO_2Et$$

Partial hydrolysis and further electrolysis enable still longer chains to be built up.

(e) Dimerization of aryl radicals: the Ullmann reaction

Symmetrical biphenyls may be prepared by heating aryl halides (other than fluorides) with copper powder or copper bronze. For example, o-nitrochlorobenzene, when heated to 215–225°C with a specially prepared copper bronze in the presence of sand to moderate the reaction, gives 2,2'-dinitrobiphenyl in 50–60% yield [3].

The reaction, which probably involves aryl radicals, is more efficient when the aromatic nucleus contains nitro-substituents, as in the example above, and it can be carried out at lower temperatures by irradiating with ultrasonic waves, which act by disrupting the surface of the copper thereby increasing its reactivity.

(f) The coupling of alkynes

Acetylene and monosubstituted acetylenes react with copper(II) acetate in pyridine solution to give diynes. The reaction probably involves one-electron oxidation of the acetylide anion by copper(II) ion followed by dimerization of the resulting acetylide radicals:

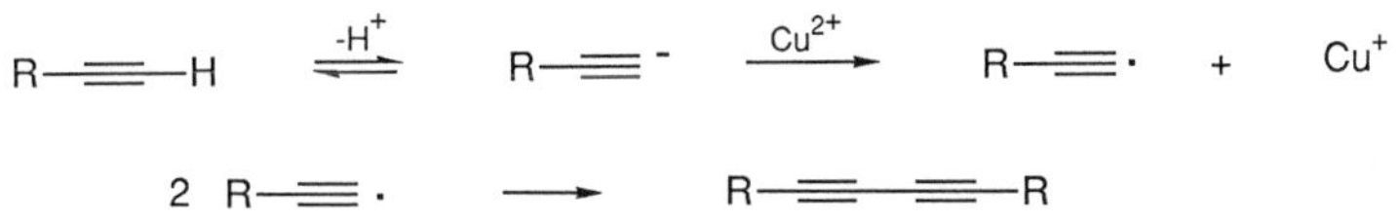

$$R\!-\!\!\equiv\!\!-H \quad \xrightarrow{-H^+} \quad R\!-\!\!\equiv^- \quad \xrightarrow{Cu^{2+}} \quad R\!-\!\!\equiv\!\cdot \quad + \quad Cu^+$$

$$2\ R\!-\!\!\equiv\!\cdot \quad \longrightarrow \quad R\!-\!\!\equiv\!\!-\!\!\equiv\!\!-R$$

For example, phenylacetylene gives 1,4-diphenylbutadiyne in 70–80% yield [5]:

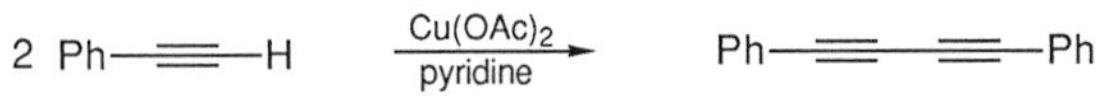

$$2\ Ph\!-\!\!\equiv\!\!-H \quad \xrightarrow[\text{pyridine}]{Cu(OAc)_2} \quad Ph\!-\!\!\equiv\!\!-\!\!\equiv\!\!-Ph$$

An alternative method for carrying out the reaction is to treat the alkyne with an aqueous mixture of copper(I) chloride and ammonium chloride in air. In these acidic conditions, removal of the proton may be facilitated by complexing between the C≡C bond and copper(I) ion,

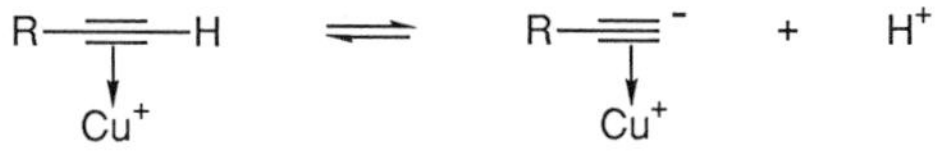

$$\underset{\underset{Cu^+}{|}}{R\!-\!\!\equiv\!\!-H} \quad \rightleftharpoons \quad \underset{\underset{Cu^+}{|}}{R\!-\!\!\equiv^-} \quad + \quad H^+$$

Some of the copper(I) ion is oxidized by air to copper(II) ion which in turn oxidizes the acetylide ion to the radical while returning to the copper(I) state. Yields can be high, although for acetylene itself the yield is less than 10%, the main product being vinylacetylene (CH_2=CH—C≡CH).*

The following are examples of the many applications of the coupling reaction.

(*i*) *Extension of carbon chains.* The coupling reaction on 3-hydroxy-1-butyne (from acetaldehyde and acetylene with amide ion in liquid ammonia, p. 242) gives a diyne used in a synthesis of β-carotene:

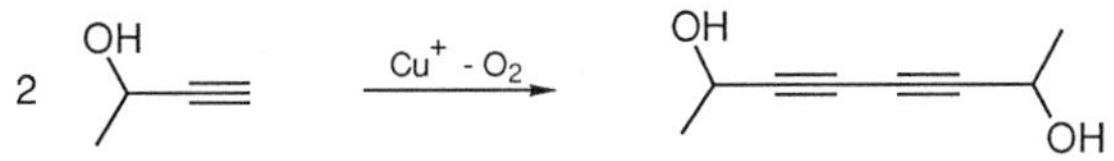

Phellogenic acid has been synthesized from undecylenic acid (obtained from castor oil):

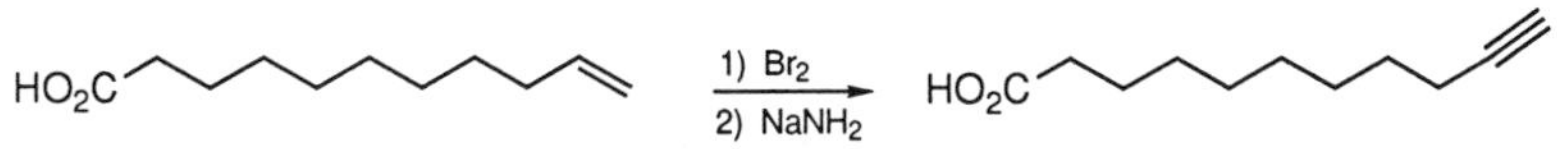

* Butadiyne may be prepared from 1,4-butynediol (p. 243):

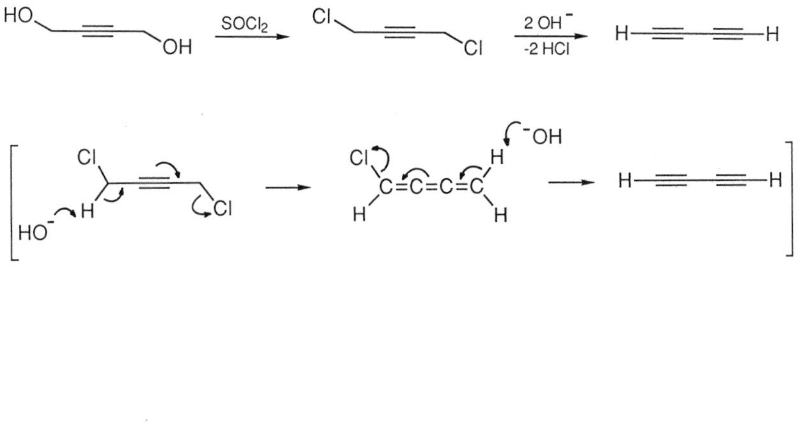

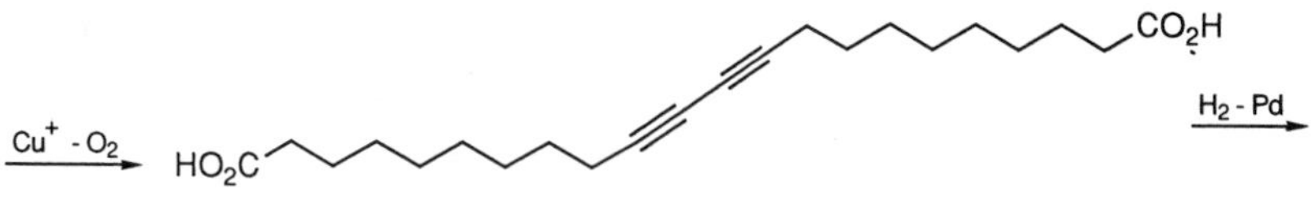

$$HO_2C-(CH_2)_{20}-CO_2H$$

phellogenic acid

Most of the polyynes of the type $R-(C{\equiv}C)_n-R$, where R is CH_3, $C(CH_3)_3$, and Ph, and $n = 2$–10, have been synthesized, e.g.

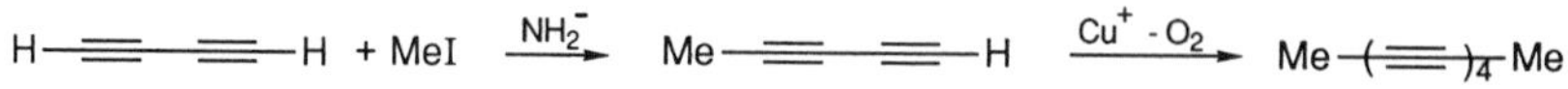

(*ii*) *Synthesis of cyclic compounds.* Terminal diynes can couple both inter-molecularly and, if the resulting ring is not highly strained, intramolecularly: the latter reaction is favoured by high dilutions. For example, the macrocyclic lactone, exaltolide, which is partly responsible for the sweet odour of the angelica root and is used in scents, has been synthesized in high yield by this means:

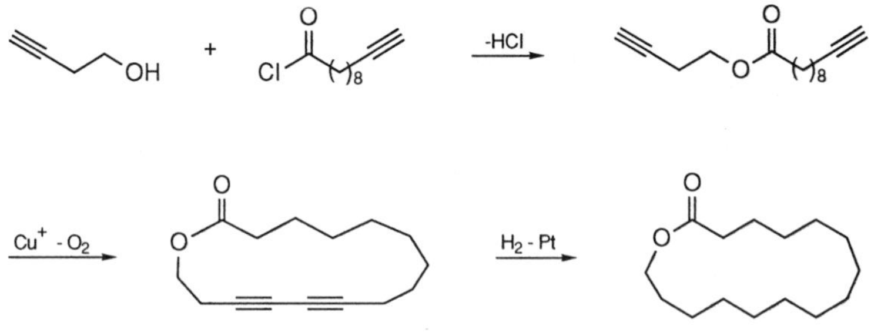

exaltolide

[18]Annulene* has been synthesized from 1,5-hexadiyne by coupling to give a mixture of cyclic trimer, tetramer, pentamer, etc., treating the trimer with base to effect conversion into the fully conjugated system *via* prototropic shifts and partially hydrogenating the remaining triple bonds (p. 638) [6]:

---

*[18]Annulene was the first compound to be isolated which is cyclic, conjugated, contains $(4n + 2)$ π-electrons where n is greater than 1, and in which the inside hydrogen atoms are sufficiently far apart for the molecule to adopt a planar or near-planar configuration. It should therefore be aromatic (p. 40), and the experimental evidence indicates that this is so. In particular, the NMR spectrum shows six protons at unusually high field (δ -3.0 ppm) and 12 at low field (δ 9.3 ppm), indicating that the molecule sustains an aromatic ring current such that the induced magnetic field causes shielding of the six inward-pointing protons and deshielding of the 12 outward-pointing ones.

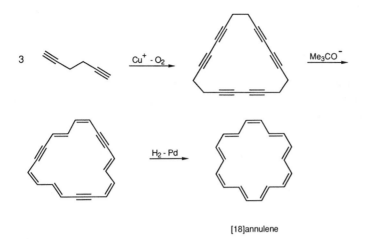

[18]annulene

(iii)   Synthesis of unsymmetrical di- and poly-ynes.   A coupling reaction between a mixture of two monosubstituted acetylenes normally gives a mixture of three products. Although this procedure has been used to obtain several unsymmetrical di- and poly-ynes in low yield, it is more satisfactory to employ a related coupling process which can give only one product. This procedure (*Chadiot–Chodkiewicz coupling*) consists of treating a monosubstituted acetylene with a 1-bromoalkyne in the presence of copper(I) ion:

$$R\!-\!\!\equiv\!\!-H \;+\; Br\!-\!\!\equiv\!\!-R' \xrightarrow{\text{Cu}^+} R\!-\!\!\equiv\!\!-\!\!\equiv\!\!-R' \;+\; HBr$$

The copper(I) ion acts catalytically, probably *via* the complexes which it forms with C≡C bonds.

Examples, the second of which gives an essential oil constituent, are:

$$H\!-\!\!\equiv\!\!-\!\!\equiv\!\!-H \;+\; Br\!-\!\!\equiv\!\!-Ph \xrightarrow[\text{-HBr}]{\text{Cu}^+} H\!-\!\!\equiv\!\!-\!\!\equiv\!\!-\!\!\equiv\!\!-Ph$$

dehydromatricarianol

(g)   The acyloin synthesis

The treatment of aliphatic esters with molten sodium in hot xylene (an inert, fairly high-boiling solvent) gives the disodium derivatives of acyloins from which the acyloin is liberated with acid. The reaction is preferably carried out under nitrogen because acyloins and their anions are readily oxidized. For example, ethyl butyrate gives butyroin in 65–70% yield [2]:

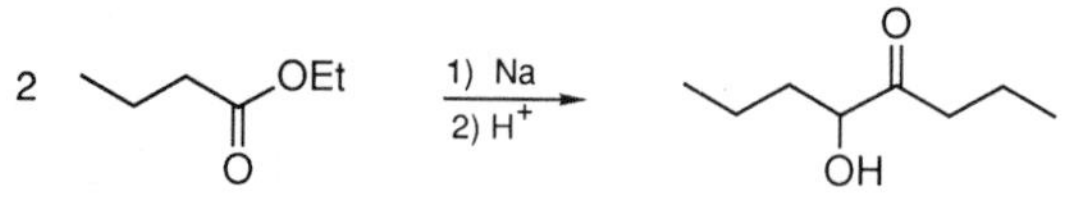

Reaction is initiated by electron transfer to the carbonyl group of the ester, the resulting radicals dimerize, alkoxide groups are eliminated, and further electron transfers give the disodium derivative of the acyloin:

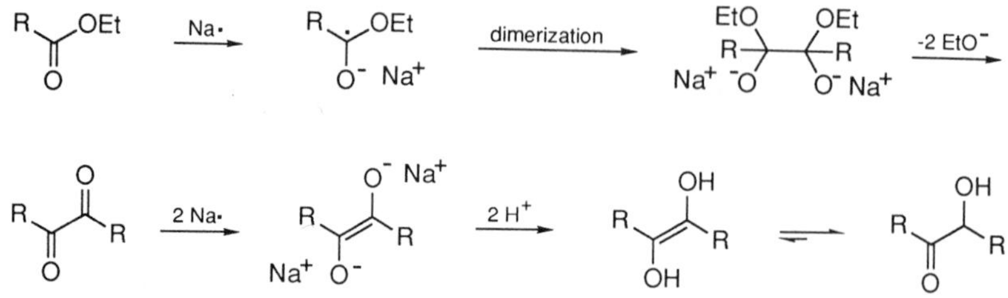

Since ethoxide ion is generated in the acyloin reaction, base-catalyzed ester condensations can occur in competition. This is especially disadvantageous in the formation of six-, seven-, and eight-membered acyloins, where the Dieckmann reaction to give, respectively, five-, six-, and seven-membered cyclic β-keto-esters is very efficient (p. 228). To prevent this competitive reaction, chloro-trimethylsilane is included to trap ethoxide ion as it is formed,

$$Me_3SiCl + EtO^- \longrightarrow Me_3Si{-}OEt + Cl^-$$

The acyloin dianion is also trapped, and the acyloin is released with acid:

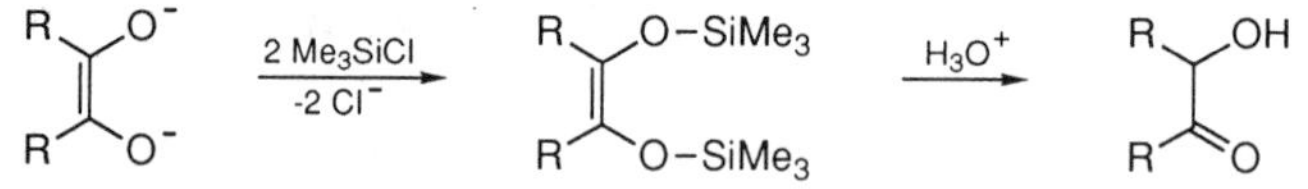

It is even possible to make four-membered cyclic acyloins in this way, e.g.

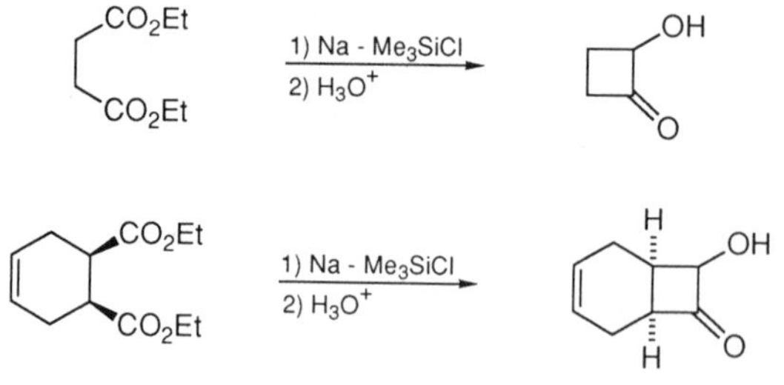

However, the *trans*-isomer of the ester above yields a particularly strained silyl derivative which undergoes a spontaneous electrocyclic reaction,

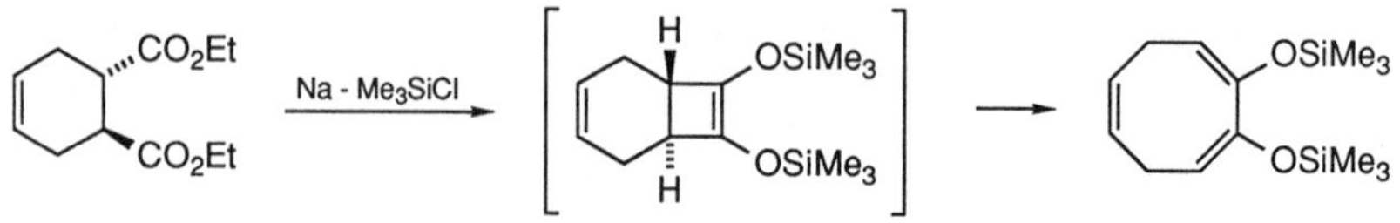

Alternatively, the silyl compound can be oxidized with bromine to the α-diketone, e.g. [7],

OSiMe3
OSiMe3
Br2
O
O
+
2 Me3SiBr

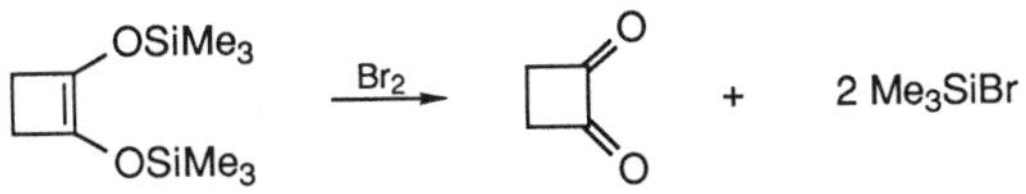

The acyloin reaction is particularly valuable for the synthesis of large rings from dibasic esters. Even rings with 10–13 members (cf. p. 229) are formed in fair to good yield and for these the method is superior to those of Ziegler (p. 229) and Blomquist (p. 283). (The intramolecular reaction is possibly facilitated by the bringing together of the polar ends of the molecule on the surface of the sodium.) Typical yields are:

Ring size	10	12	13	14	15	16	17	18	20
Yield (%)	45	76	67	79	77	84	85	96	96

17.4 Formation of carbon–nitrogen bonds

(a) *Nitration at saturated carbon*

Saturated carbon can be nitrated by nitric acid in both the gas and the liquid phases. The mechanism of reaction is not certain but is probably as follows:

initiation $\quad HO{-}NO_2 \longrightarrow HO\cdot + NO_2$

propagation
$$R{-}H + NO_2 \longrightarrow R\cdot + HONO$$
$$R\cdot + NO_2 \longrightarrow R{-}NO_2$$
$$HONO + HNO_3 \longrightarrow 2\,NO_2 + H_2O$$

The fragmentation of alkyl groups commonly occurs, but one or more products can often be isolated in good yield and the gas-phase reaction at about 450°C is of considerable industrial importance. Some examples are:

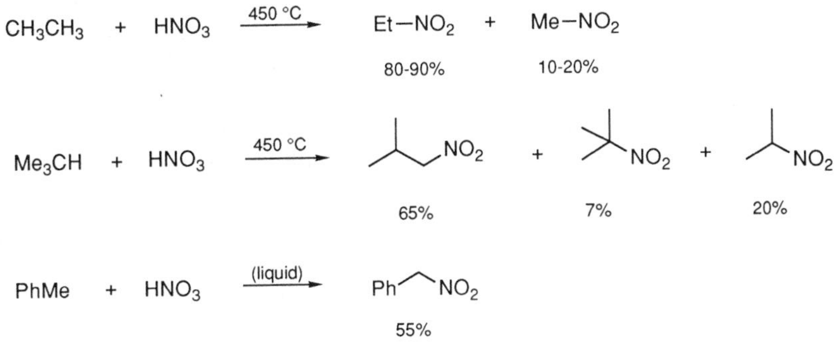

CH3CH3 + HNO3
450 °C
Et—NO2 + Me—NO2
80-90%
10-20%
Me3CH + HNO3
450 °C
NO2
+
NO2
+
NO2
65%
7%
20%
PhMe + HNO3
(liquid)
Ph
NO2
55%

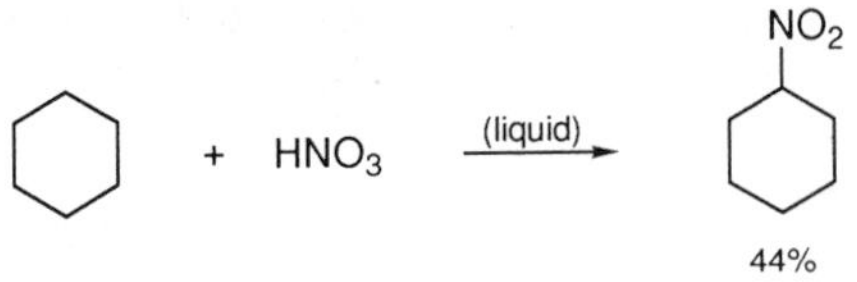

cyclohexane $+$ HNO$_3$ $\xrightarrow{\text{(liquid)}}$ nitrocyclohexane (NO$_2$)

44%

(b) Addition to C=C and C≡C

Dinitrogen tetroxide adds to C=C and C≡C to give a mixture of dinitro-compounds, nitroalcohols, and nitronitrates:

$$O_2N{-}NO_2 \rightleftharpoons 2\,NO_2{}^*$$

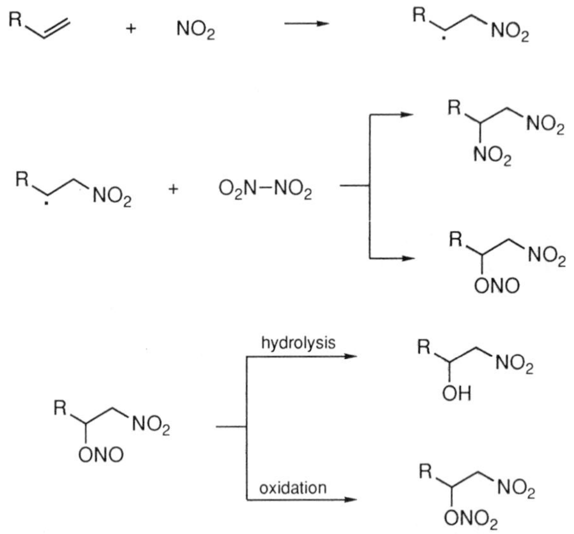

$$R{-}CH{=}CH_2 + NO_2 \longrightarrow R{-}CH(\cdot){-}CH_2{-}NO_2$$

$$R{-}CH(\cdot){-}CH_2{-}NO_2 + O_2N{-}NO_2 \longrightarrow$$

$R{-}CH(NO_2){-}CH_2{-}NO_2$

and

$R{-}CH(ONO){-}CH_2{-}NO_2$

$R{-}CH(ONO){-}CH_2{-}NO_2$:

$\xrightarrow{\text{hydrolysis}}$ $R{-}CH(OH){-}CH_2{-}NO_2$

$\xrightarrow{\text{oxidation}}$ $R{-}CH(ONO_2){-}CH_2{-}NO_2$

Yields of particular products tend to be low, e.g.

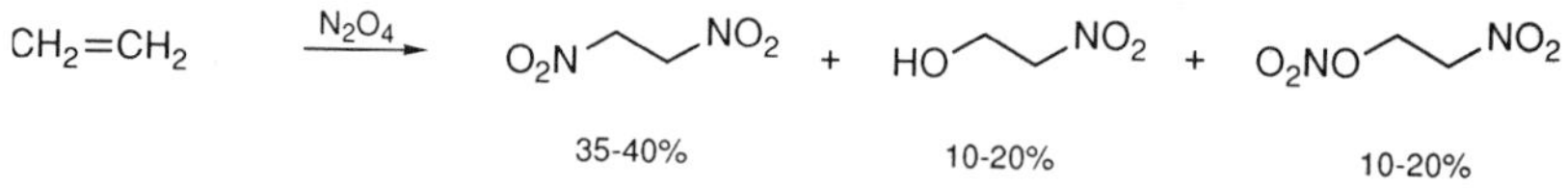

$$CH_2{=}CH_2 \xrightarrow{N_2O_4} O_2N{-}CH_2CH_2{-}NO_2 + HO{-}CH_2CH_2{-}NO_2 + O_2NO{-}CH_2CH_2{-}NO_2$$

35-40%   10-20%   10-20%

Nitryl and nitrosyl halides add in an analogous manner; the products from nitrosyl halides are often the β-halonitro-compounds which result from oxidation of nitroso-compounds.

---

* N$_2$O$_4$ is 20% dissociated at 27 °C.

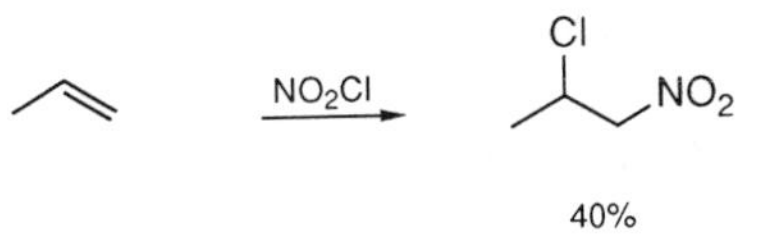

40%

(c) The Hofmann-Löffler-Freytag reaction

Irradiation of *N*-chlorodibutylamine in 85% sulfuric acid, followed by basification, gives *N*-butylpyrrolidine:

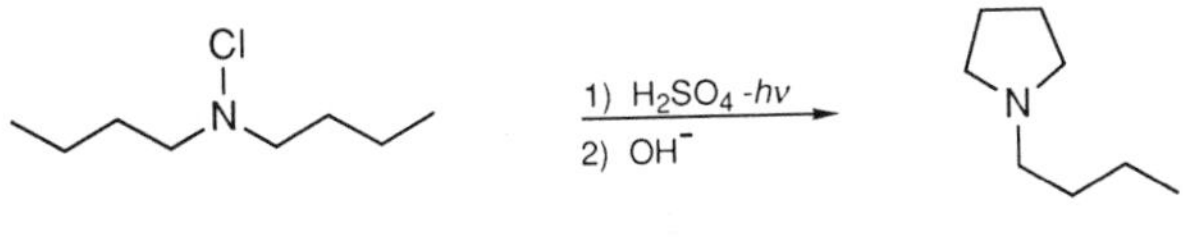

This is a general reaction of those *N*-chloroamines which possess a δ-CH group. Photolysis of the protonated amine gives a nitrogen cation-radical which abstracts a hydrogen atom from the δ-carbon *via* a six-membered cyclic transition state. The resulting alkyl radical abstracts a chlorine atom from more of the chloroamine and when the amino group is released with base it displaces intramolecularly on the halide:

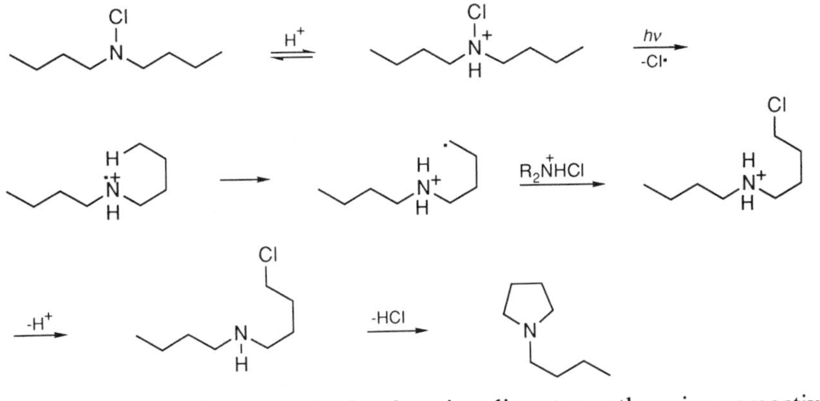

This provides a method for introducing functionality at an otherwise unreactive aliphatic carbon atom and the reaction has been used in the steroid series to modify angular methyl groups, as in a synthesis of dihydroconessine:

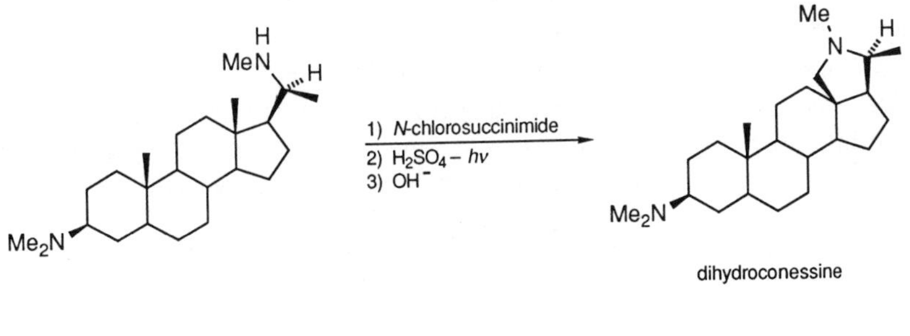

1) *N*-chlorosuccinimide
2) H$_2$SO$_4$ – *hv*
3) OH$^-$

dihydroconessine

(d)   Photolysis of nitrites: the Barton reaction

The irradiation of organic nitrites may also be used to introduce functionality at an otherwise unreactive aliphatic carbon. The oxy-radical produced by photolysis abstracts hydrogen from a δ-CH bond and the resulting alkyl radical combines with the nitric oxide liberated in the photolysis to give a nitroso-compound and thence, when primary and secondary CH groups are involved, the tautomeric oxime.

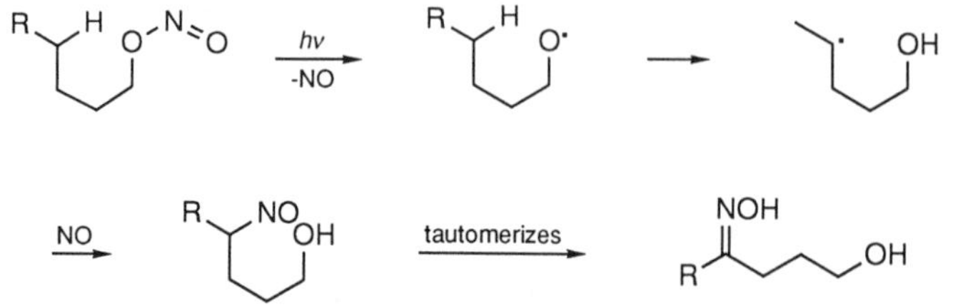

This method has been used in a synthesis of aldosterone 21-acetate from corticosterone acetate. The nitrite ester of the 11β-hydroxyl group was formed from the alcohol with nitrosyl chloride and, after photolysis, the oxime was hydrolyzed in mild conditions (nitrous acid). (Note that aldosterone 21-acetate exists as a hemiacetal; cf. the ring-chain tautomerism of glucose, p. 169.)

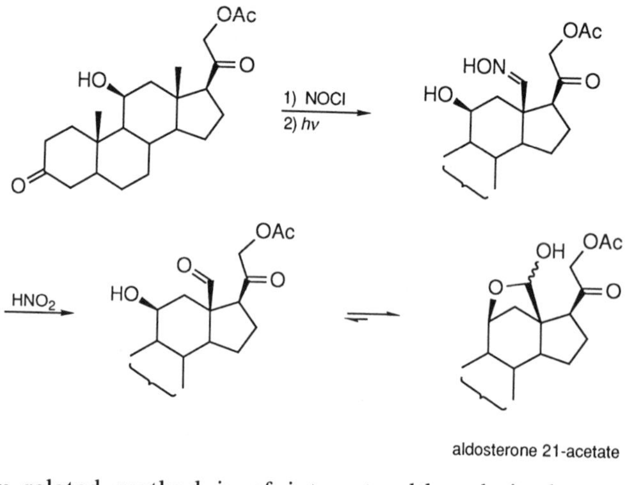

aldosterone 21-acetate

A closely related method is of interest, although it does not involve the generation of a C—N bond. The photolysis of tertiary hypochlorites (readily prepared from the corresponding alcohols with chlorine in alkali or with chlorine monoxide) gives alkoxy radicals which can abstract hydrogen from δ-CH bonds; the resulting alkyl radical abstracts chlorine from a second molecule of the hypochlorite, giving a δ-chloro-alcohol which may be cyclized with base, e.g.*

---

* The fragmentation products, mainly butyl chloride and acetone, are also formed in 13% yield (see below).

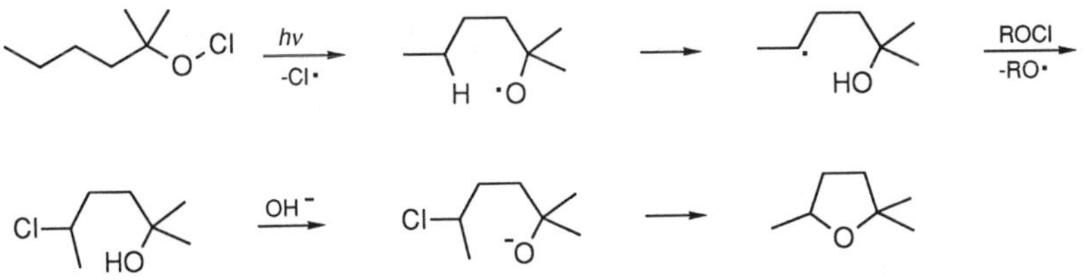

This method has also been used in the steroid series for introducing functionality at angular methyl groups,

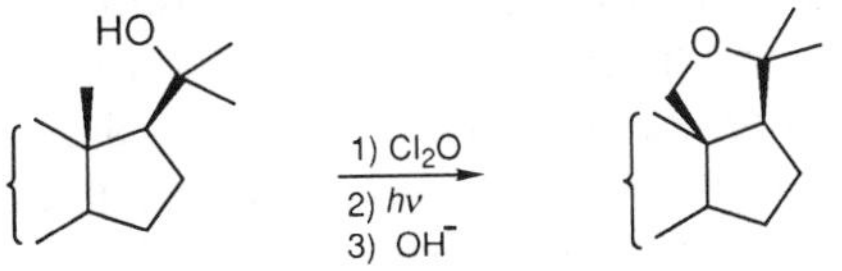

It should be emphasized that in both the Barton and the hypochlorite reactions, intramolecular hydrogen-abstraction has to compete with other reactions of alkoxy radicals. In particular, these radicals undergo fragmentation: for example, the t-butoxy radical gives acetone and the methyl radical. The reaction,

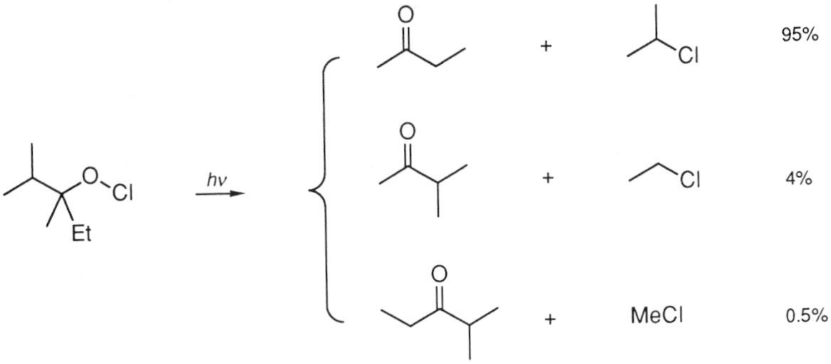

indicates that the ease of fragmentation of a particular group increases with the stability of the radical formed. However, in the examples involving steroid systems cited above, the intramolecular reaction competes favourably with fragmentation, for not only is the transition state of the necessary six-membered type but also the angular methyl group and oxy-radical are suitably placed with respect to each other.

17.5
Formation of carbon–oxygen bonds

C—H bonds in a wide variety of environments are oxidized on standing in air to hydroperoxide groups. Reaction is apparently initiated by the appearance of stray radicals produced, for example, by sunlight photolysis, and thereafter a chain process operates:

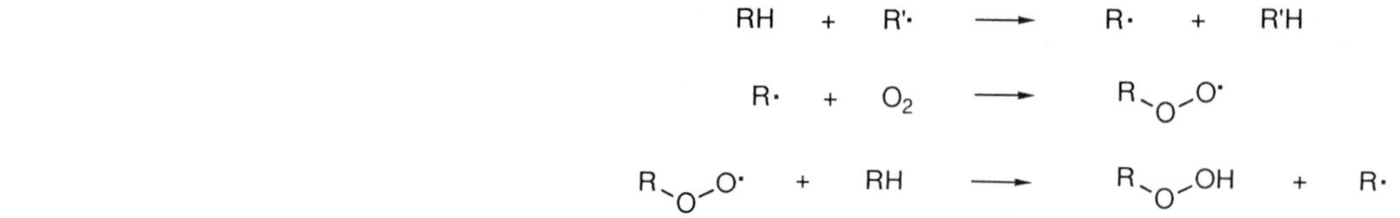

$$RH + R'\cdot \longrightarrow R\cdot + R'H$$

$$R\cdot + O_2 \longrightarrow R\text{-O-O}\cdot$$

$$R\text{-O-O}\cdot + RH \longrightarrow R\text{-O-OH} + R\cdot$$

The rates of these reactions (*autoxidations*) vary markedly with structure: for example, alkanes react at a negligible rate at room temperature whereas allylic compounds react at a significant rate. Autoxidation is catalyzed by the usual initiators and retarded by inhibitors and it is advisable to store the more readily autoxidized compounds in the presence of an inhibitor.

In the alkane series the order of reactivity is, tertiary > secondary > primary C—H, as usual in free-radical reactions. For example, isobutane can be converted in good yield into t-butyl hydroperoxide in the presence of an initiator:

Allylic compounds owe their greater reactivity to the greater stability of allyl than alkyl radicals. Unsymmetrical compounds give mixtures of products,

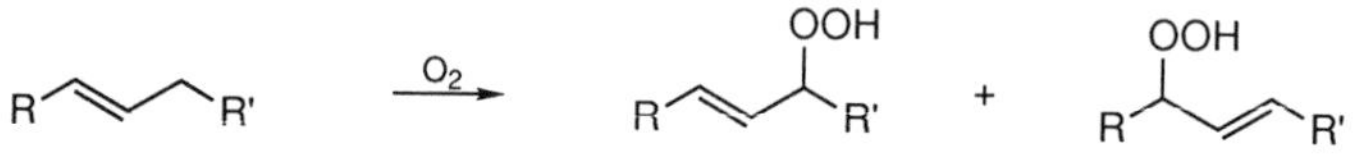

Those allylic compounds which, as alkenes, are readily polymerized, react differently, for the hydroperoxy radical initiates polymerization:

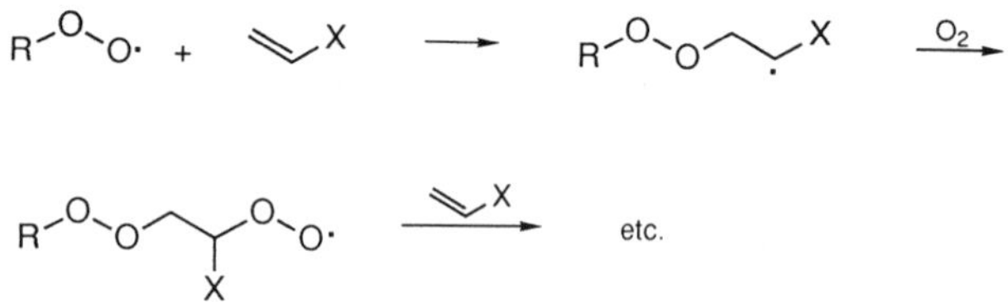

Polymerization of a complex type is also responsible for the formation of a hard skin on drying oils, which are allylic compounds.

Ethers are particularly prone to autoxidation, e.g. tetrahydrofuran gives the α-hydroperoxide,

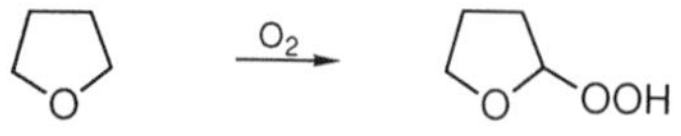

Since hydroperoxides can explode on heating it is essential to remove them from ethers (e.g. by reduction with aqueous iron(II) sulfate) before using the ethers as solvents for reactions which require heat.

Aldehydes also autoxidize readily but the initial product, a peroxyacid, reacts with more of the aldehyde to give the carboxylic acid. For example, benzaldehyde gives benzoic acid on standing in air:

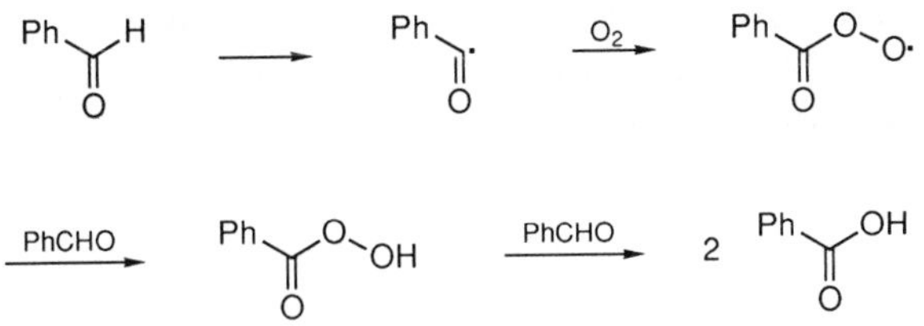

The final step is a Baeyer–Villiger oxidation involving a hydride shift (p. 445). Ketones are unreactive towards oxygen, but their enolate ions react:

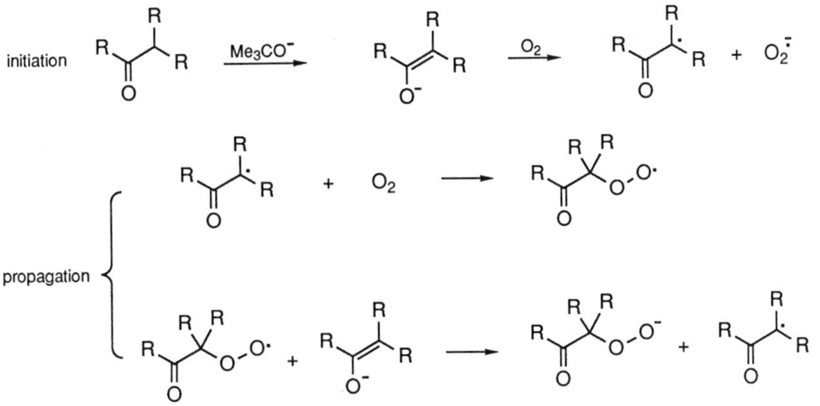

By carrying out the reaction in the presence of a trialkyl phosphite, α-ketols can be obtained:

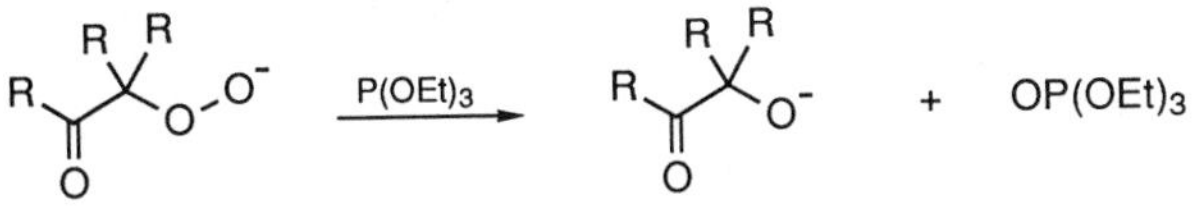

Hydroperoxides are not themselves of much synthetic value but they are employed as intermediates in certain reactions. For example, cumene is converted industrially into phenol and acetone *via* cumene hydroperoxide (p. 446),

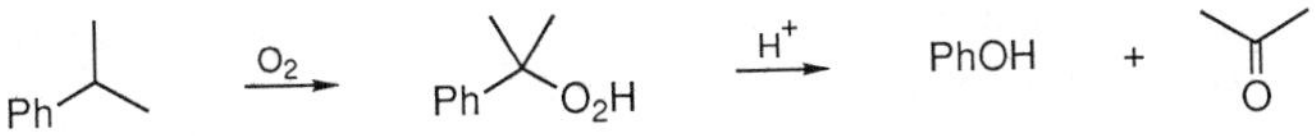

and tetralin can be converted into α-tetralone by E2 elimination on its hydroperoxide,

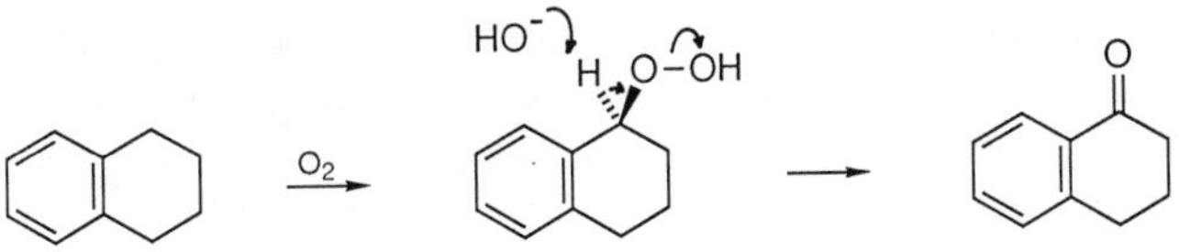

**17.6
Formation of
bonds to other
elements**

Elements such as sulfur, phosphorus, and silicon can be bonded to alkenes and alkynes by radical-catalyzed reactions which have the characteristics of those already described. Typical examples are:

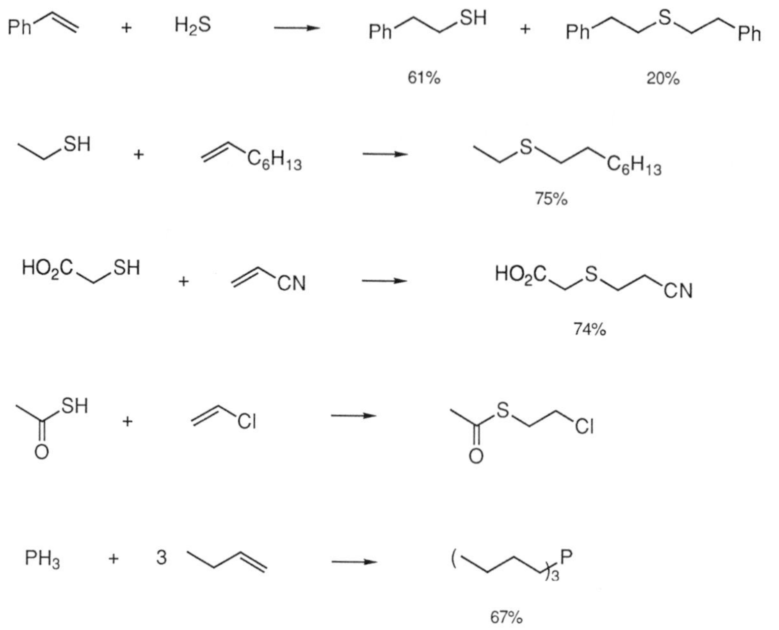

61%   20%

75%

74%

67%

(the intermediate compounds, butylphosphine and dibutylphosphine, may be obtained by using smaller proportions of 1-butene)

95%

Problems   1.   Account for the following:

    (*i*) The orientation in the addition of hydrogen bromide to allyl bromide depends on whether or not the reactants are contaminated with peroxide impurities.

    (*ii*) The radical-catalyzed chlorination of optically active CH_3CH_2—$CH(CH_3)$—CH_2Cl gives mainly racemic 1,2-dichloro-2-methylbutane.

    (*iii*) The peroxyester $PhCH{=}CHCH_2$—CO—O—O—CMe_3 decomposes several thousand times faster than the peroxyester CH_3—CO—O—O—CMe_3 at the same temperature.

    (*iv*) The radical-catalyzed chlorination, $ArCH_3 \rightarrow ArCH_2Cl$, occurs faster when Ar = phenyl than when Ar = *p*-nitrophenyl.

    (*v*) The aldehyde Ph—$C(CH_3)_2$—CH_2—CHO undergoes a radical-catalyzed decarbonylation to give a mixture of Ph—$C(CH_3)_2$—CH_3

and Ph—CH$_2$—CH(CH$_3$)$_2$. The proportion of the latter product decreases as the concentration of the reactant is increased.

(*vi*) When *p*-cresol is oxidized by potassium hexacyanoferrate(III), the compound (I) is one of the products.

(*vii*) When the compound (II) (formed from cyclohexanone and hydrogen peroxide) is treated with iron(II) sulfate, 1,12-dodecanedioic acid is formed.

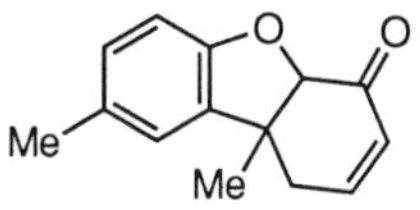

(I) 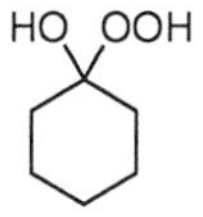   (II)

2. Summarize the free-radical reactions which may be used for:
    (*i*) the extension of carbon chains;
    (*ii*) the introduction of functionality at unactivated methyl groups;
    (*iii*) the formation of medium- and large-sized rings.
    In the last case, compare these methods with those which do not involve radical reactions.

3. What products would you expect from the following reactions?

(*i*)   Me$_3$CH + Br$_2$ $\xrightarrow{\text{light}}$

(*ii*)

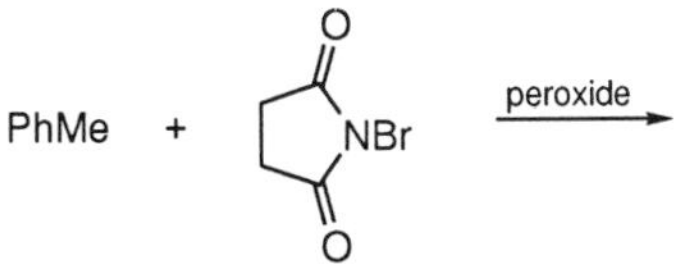

PhMe + [N-bromosuccinimide] $\xrightarrow{\text{peroxide}}$

(*iii*)

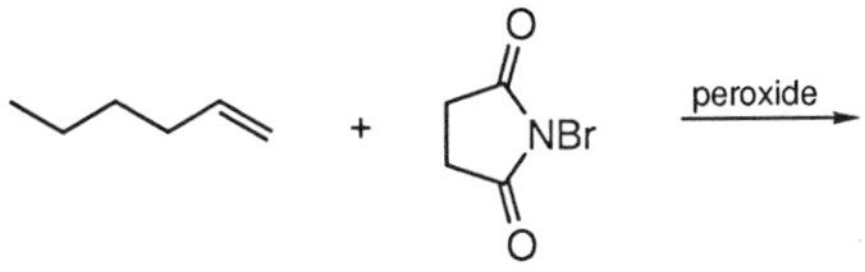

+ [N-bromosuccinimide] $\xrightarrow{\text{peroxide}}$

(*iv*)

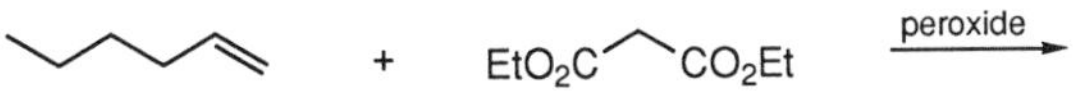

+ EtO$_2$C⌒CO$_2$Et $\xrightarrow{\text{peroxide}}$

(*v*)

+ CHCl$_3$ $\xrightarrow{\text{peroxide}}$

4. How would you employ radical reactions in the synthesis of the following compounds?

(*a*)

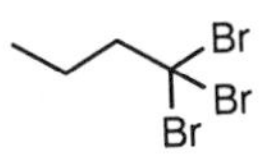

(*b*)

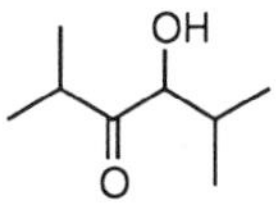

(c)

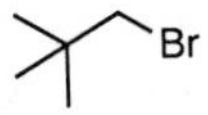

(d)

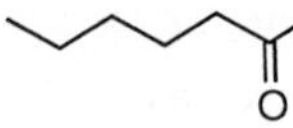

(e)

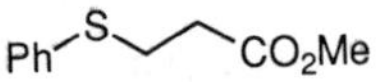

Ph—S—CO₂Me

(f)   Ph—Ph

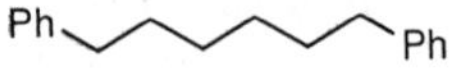

(g)

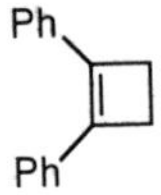

Ph
Ph

(h)

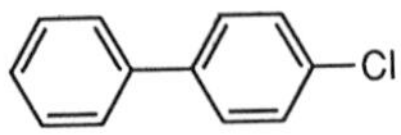

(i)

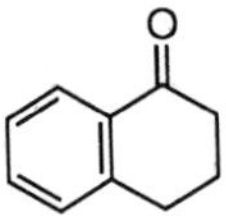

(j)

OOH

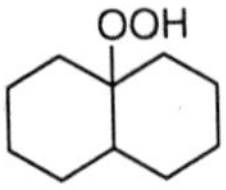

Organotransition metal reagents 18

The application to organic synthesis of complexes formed by transition-metal ions and organic ligands is a rapidly developing field; its scope is far from fully defined. In this chapter, only those processes that have proved their value in synthesis are described.

(a) Organic ligands

The organic ligand can bind to the metal ion either through an *s*-containing orbital, giving a σ-organometallic compound, or with π orbitals which bind from one face of the organic group, as in a π-alkene complex, written

In the latter case, bonding is provided not only by donation from a filled π orbital of the ligand into an empty metal *d* orbital but also by interaction of an unfilled π^* orbital of the ligand with a filled metal *d* orbital:

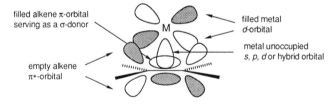

Carbon monoxide and cyanide ion are ligands whose donor orbitals are of *s*-containing type but which can accept electrons into π^* orbitals (X = O or N$^-$):

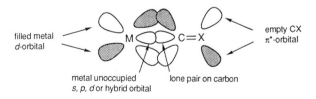

The number of atoms in the ligand within bonding distance of the metal is described as its *hapto* number, η. For a σ-bonding ligand the hapto number is therefore one, regardless of whether the ligand provides one electron in the formation of the bond to the metal (e.g. alkyl, H) or two (e.g. O=C:, H_2C:, Ph_3P:). For a π-bonding ligand it is equal to the number of carbon atoms whose *p* orbitals are involved. Commonly occurring ones are:

2: alkenes

3: π-allyl

4: butadiene, cyclobutadiene
5: π-cyclopentadienyl
6: benzene

where L_n indicates that *n* other ligands are bonded.

(b)   The 18-electron rule

The chemistry of organometallic transition-metal compounds is marked by the tendency of the metal to achieve an electronic configuration corresponding to that of the next highest noble gas. This equates to a total of 18 electrons in the outermost *d*, *s*, and *p* orbitals (e.g. $3d^{10}$, $4s^2$, $4p^6$ for Kr). An exception referred to in this chapter is titanium which is low in available electrons and would be overcrowded if sufficient ligands were present to give a total of 18.

The numbers of outer-shell electrons in the metals whose organometallic chemistry is described in this chapter are:

Outer-shell electrons:	4	6	8	9	10
Metal:	Ti	Cr	Fe	Co, Rh	Ni, Pd

from which it follows, for example, that chromium can bond to ligands that provide a total of 12 electrons.

In checking on the validity of a structure, it must be remembered that, where a metal is positively charged, each charge corresponds to one less electron in the outermost shell. The following are examples of complexes which obey the 18-electron rule:

Examples of 16-electron titanium complexes are:[*]

(c) *Commonly encountered reaction mechanisms*

The reactions described in this chapter mostly include one or more of the following steps (M = metal, L = an unspecified ligand).

(i) *Coordination–dissociation*

$$M \quad + \quad :L \quad \underset{\substack{\text{coordinative} \\ \text{dissociation}}}{\overset{\substack{\text{coordinative} \\ \text{association}}}{\rightleftharpoons}} \quad M{-}L$$

A complex which fulfils the 18-electron rule must necessarily dissociate by loss of a ligand before it can coordinate, e.g.

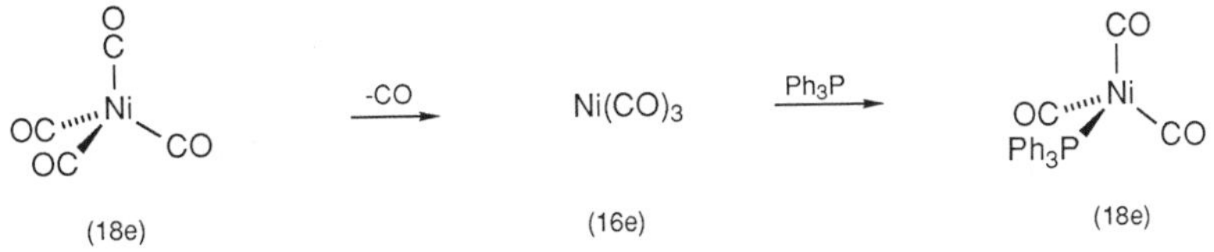

but this is not necessary with a complex containing less than 18 electrons, e.g.

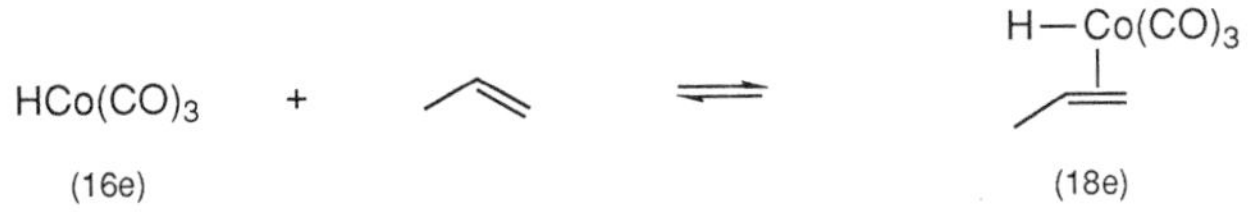

(ii) *Oxidative addition–reductive elimination*

$$L_nM \quad + \quad Z{-}Y \quad \underset{\text{reductive elimination}}{\overset{\text{oxidative addition}}{\rightleftharpoons}} \quad L_nM \overset{Z}{\underset{Y}{\diagdown}}$$

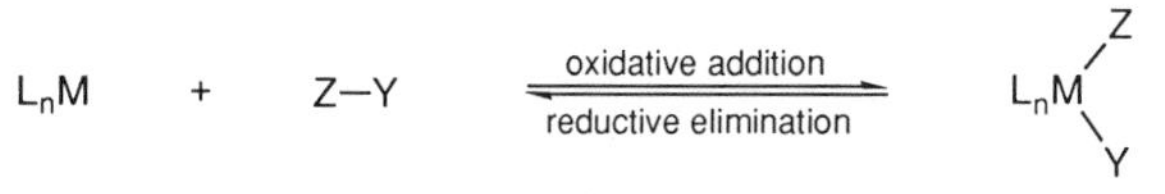

[*] The reactions by which ligands are introduced often involve the ligand as an anion, such as the cyclopentadienyl anion (6e donor) and chloride ion (2e donor) in the formation of the first of these titanium complexes. However, since these reactions involve displacement on titanium which is formally Ti^{4+}, the electron count is the same as that described above in which titanium has four electrons in its outer shell and cyclopentadienyl and chloride ligands are five and one electron-donors, respectively.

where ZY is typically RX, RCOX, or H_2. The oxidation state of the metal formally increases by two in the forward reaction, since the atoms or groups Z and Y are considered less electropositive than the metal. This is so even for the addition of H_2, e.g.

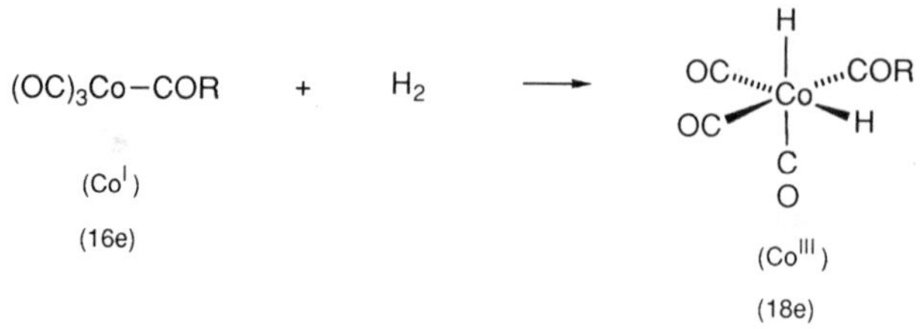

$$(OC)_3Co-COR \quad + \quad H_2 \quad \longrightarrow$$

(Co^I)

(16e)

(Co^{III})

(18e)

The product of oxidative addition can often undergo reductive elimination with the release of an organic product, e.g.

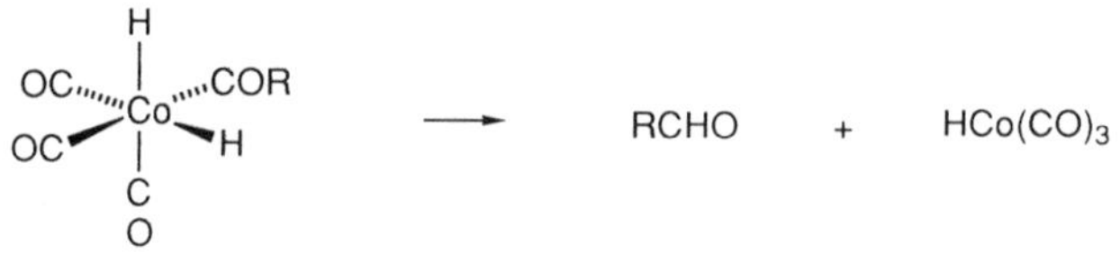

$$\longrightarrow \quad RCHO \quad + \quad HCo(CO)_3$$

(iii)   Ligand insertion–deinsertion

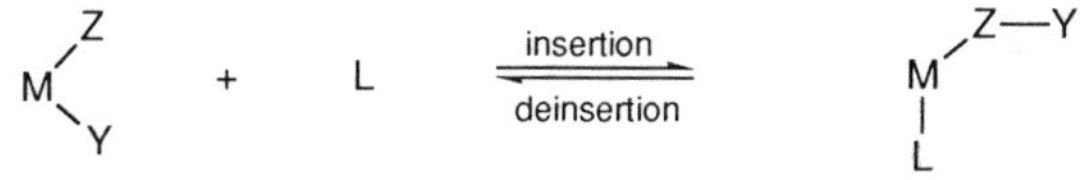

$$M\begin{smallmatrix}Z\\Y\end{smallmatrix} \quad + \quad L \quad \underset{\text{deinsertion}}{\overset{\text{insertion}}{\rightleftharpoons}} \quad M\begin{smallmatrix}Z-Y\\L\end{smallmatrix}$$

Z is typically CO or an alkene and Y is typically H, R, or COR, e.g.

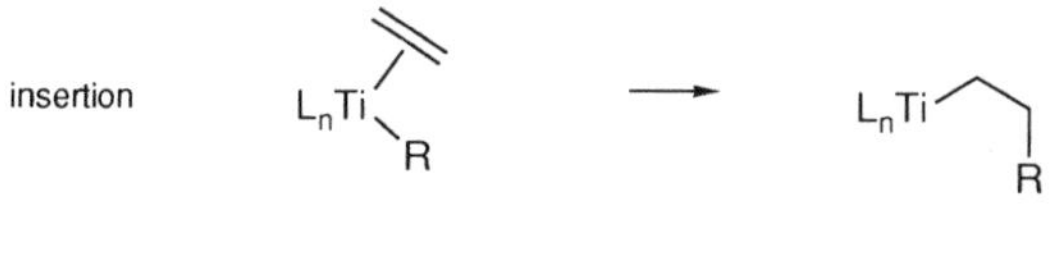

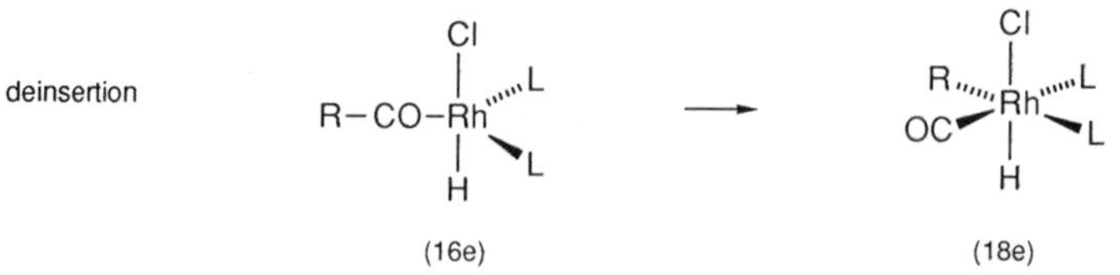

insertion $L_n Ti$ $\longrightarrow$ $L_n Ti$

deinsertion $R-CO-Rh$ $\longrightarrow$ Rh

(16e)   (18e)

β-Hydride elimination is an important example of ligand deinsertion:

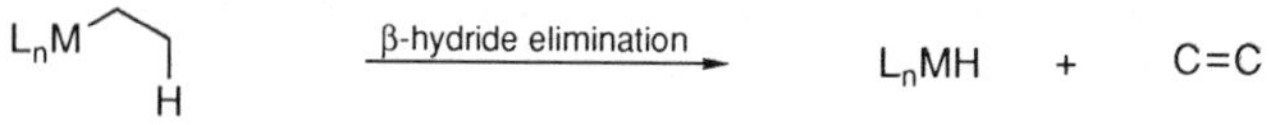

$$L_n M \quad \xrightarrow{\text{β-hydride elimination}} \quad L_n MH \quad + \quad C=C$$

(a)   *σ-Organotitanium compounds*

These are prepared from titanium tetraalkoxides:

$$3 \ Ti(OR')_4 \xrightarrow{\ TiCl_4\ } 4 \ ClTi(OR')_3$$

$$ClTi(OR')_3 \xrightarrow[\text{or RLi}]{RMgX} R-Ti(OR')_3$$

The carbon atom bonded to titanium is strongly negatively polarized and, as in a Grignard reagent, the organic group behaves as a nucleophile. However, the reagents are significantly more selective than Grignard reagents in two important respects. First, they react only with the carbonyl group in aldehydes and ketones even when other functions are present which react with Grignard reagents (esters, nitriles, epoxides, halides), e.g.

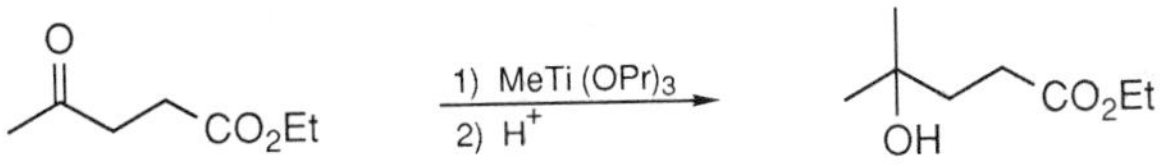

Moreover, an aldehyde group reacts selectively in the presence of a keto group.

Second, they show a much higher degree of diastereoselectivity with asymmetric aldehydes (cf. p. 191), e.g.

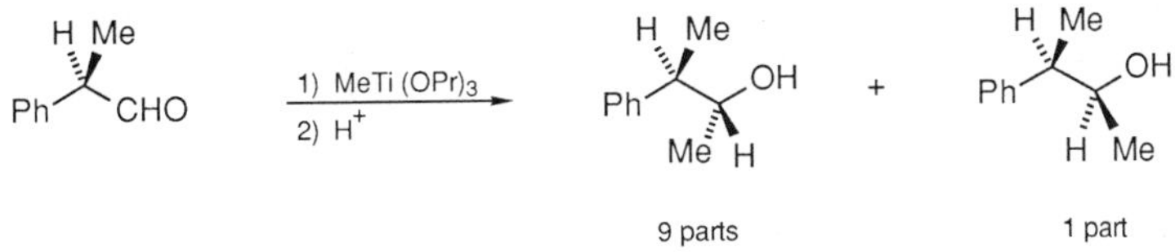

(b)   *The Tebbe reagent*

This is prepared from Cp_2TiCl_2 (Cp = cyclopentadienyl):*

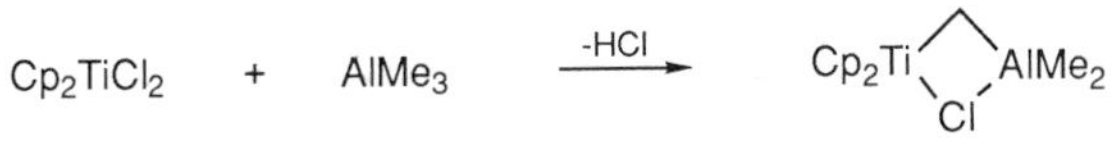

(i)   *Reaction with carbonyl compounds*   Tebbe's reagent reacts with the carbonyl group in aldehydes, ketones, esters, lactones, and amides to replace oxygen by

---

*Tebbe's reagent, and other complexes with bridging halogens, should formally be written with charges, e.g.

The charges are conventionally omitted.

CH_2. A weak base is necessary: in some cases the carbonyl oxygen atom itself is effective but otherwise pyridine is added. It is believed that the reactive intermediate is $Cp_2Ti{=}CH_2$, which reacts analogously to Wittig reagents:

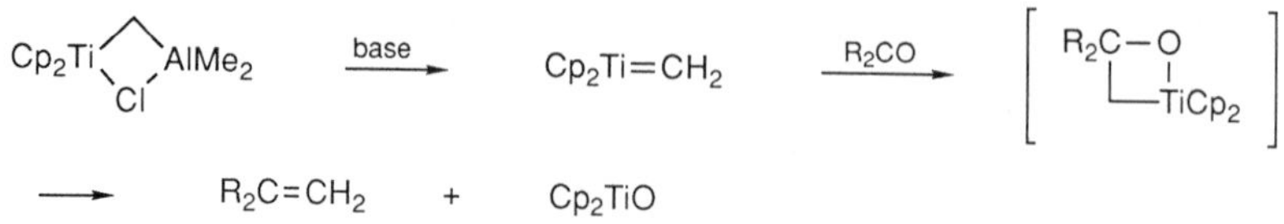

The process is superior to the Wittig reaction in two circumstances: first, if an asymmetric carbon atom with an enolizable proton is adjacent to the carbonyl group, it retains its stereochemical integrity; second, highly hindered carbonyl compounds react more efficiently, e.g.

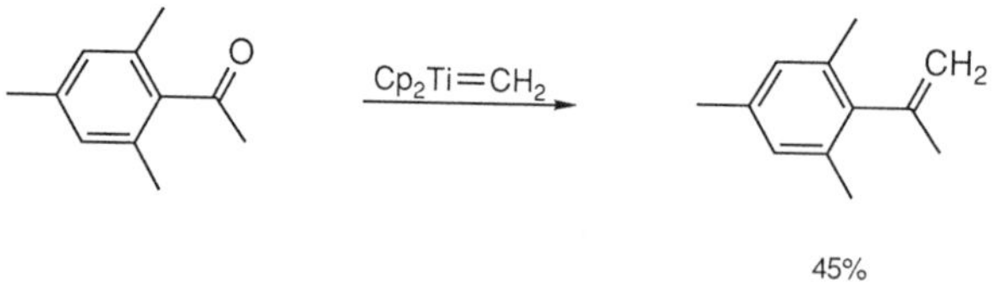

45%

(ii) *Reaction with alkenes* The Tebbe reagent reacts with alkenes to form metallocycles, e.g.

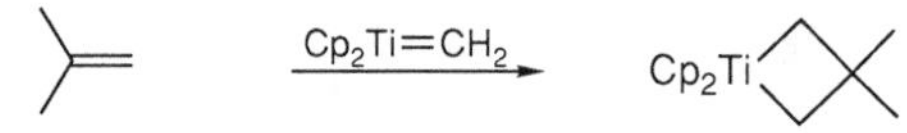

On heating, these normally regenerate the reactants. However, if the $C{=}C$ bond is part of a strained ring, cleavage can occur to give a new titanium-carbene complex. The following example, showing the formation of an intermediate required for the synthesis of capnellene, also demonstrates the value of the method in forming strained rings:

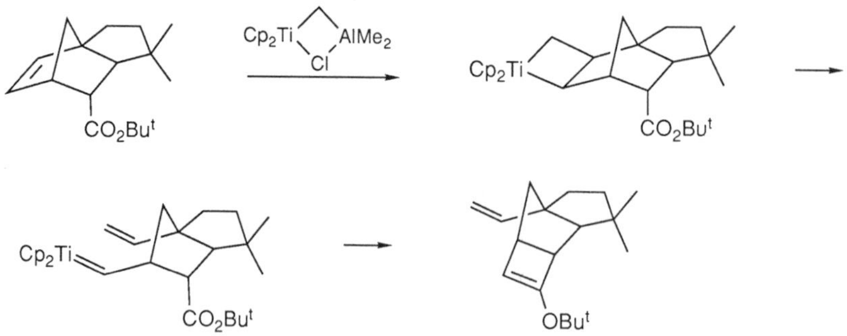

Alkene metathesis.   Alkenes undergo disproportionation reactions when heated with a catalytic quantity of the Tebbe reagent, e.g.

Thermodynamic equilibrium is established which, in this case, favours 2-butene (mainly *trans*) and ethylene.

The process makes use of the reversibility of the formation of metallocycles and the fact that, whereas the preferred regiochemical reaction in their formation may be unproductive, the less preferred one may establish a desired reaction pathway. In the above case, the preferred reaction simply establishes the dynamic equilibrium (M = Cp$_2$Ti)

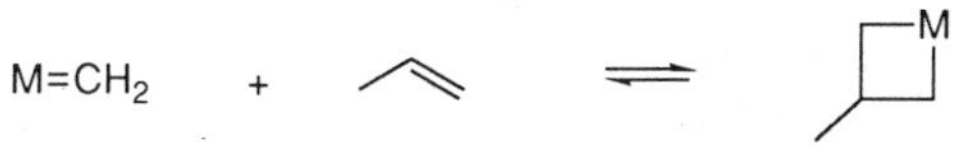

$$M{=}CH_2 \quad + \quad \diagup\!\!\!\diagdown \quad \rightleftharpoons \quad \text{[metallocycle]}$$

whereas the alternative can give ethylene and a new carbene complex,

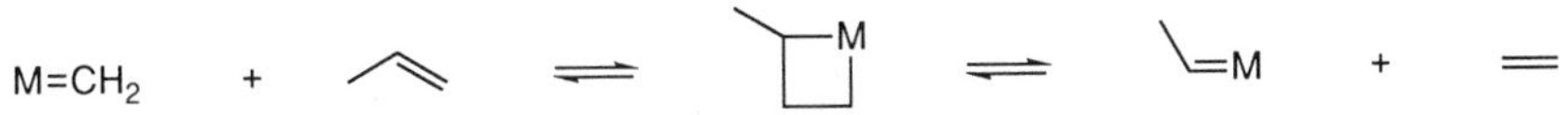

$$M{=}CH_2 \quad + \quad \diagup\!\!\!\diagdown \quad \rightleftharpoons \quad \text{[metallocycle]} \quad \rightleftharpoons \quad {=}M \quad + \quad {=}$$

The latter can react with more propylene to give 2-butene:

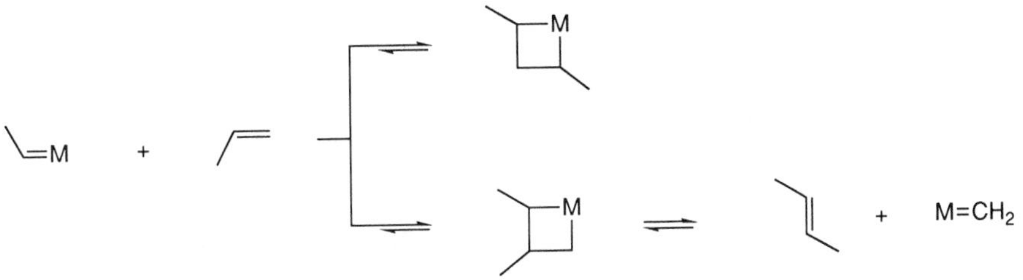

$${=}M \quad + \quad \diagup\!\!\!= \quad \text{[metallocycles]} \quad \rightleftharpoons \quad + \quad M{=}CH_2$$

This was the first commercial application of alkene metathesis. Two alkenes can also be employed as reactants. The reaction of ethylene with the readily available dimer of isobutylene (p. 252) is used in a commercial process for making neohexene:

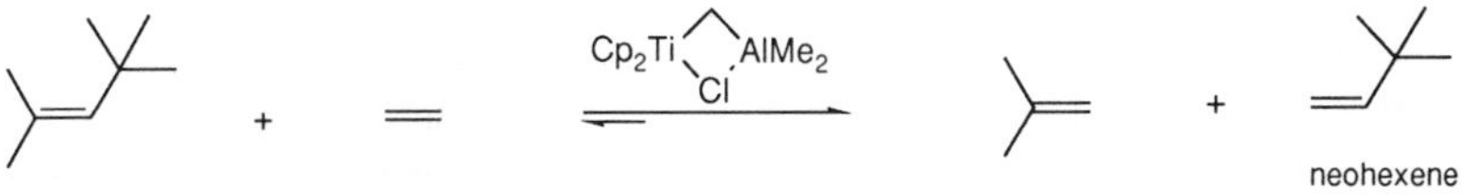

$$\text{(isobutylene dimer)} \quad + \quad {=} \quad \xrightarrow{\underset{Cl}{Cp_2Ti{\diagup}\!\!{\diagdown}AlMe_2}} \quad + \quad \text{neohexene}$$

Alkene polymerization. Application of the above method to cycloalkenes generally gives high molecular weight polymers in a process known as ROMP (*ring-opening metathesis polymerization*), e.g.

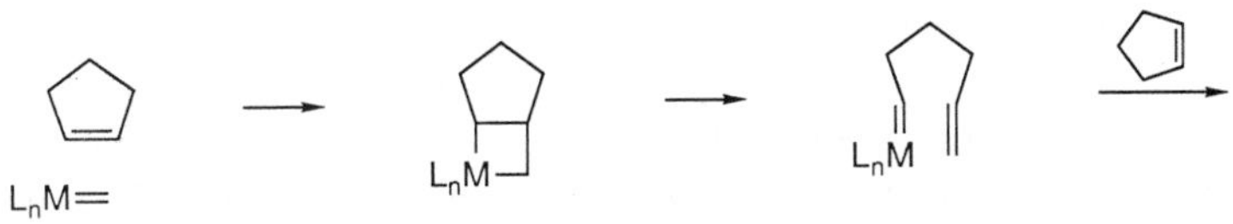

$$L_nM{=} \quad \rightarrow \quad L_nM{-} \quad \rightarrow \quad L_nM \quad \rightarrow$$

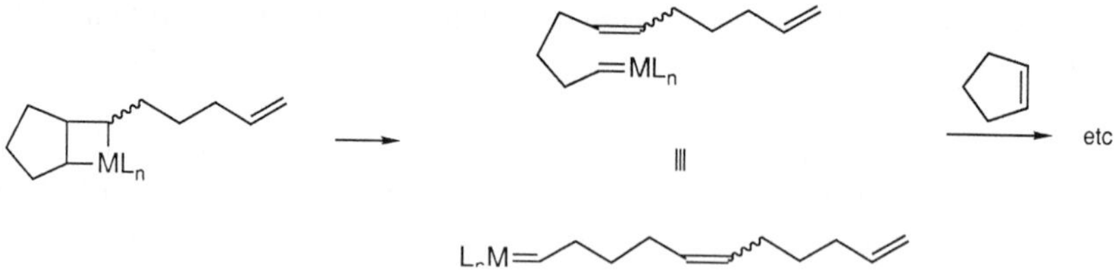

Various carbene complexes are used, some containing tungsten in place of titanium, and in this way products with different bond geometries and therefore different properties are obtained. For example, the all-*trans* polymer is an elastomer which is an excellent material for tyre manufacture, whereas the all-*cis* isomer is a non-elastomeric solid.

Ziegler's catalysts. Ethylene and propylene are polymerized by Ziegler's catalysts, which consist of titanium trichloride and an aluminium alkyl derivative. The essence of the method is the insertion of the alkene into the C—Ti bond of the growing polymer, e.g.

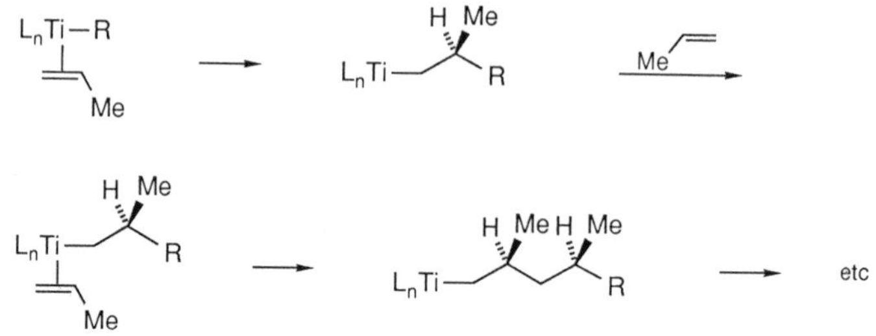

Chain termination is by β-elimination or hydrogenation, e.g.

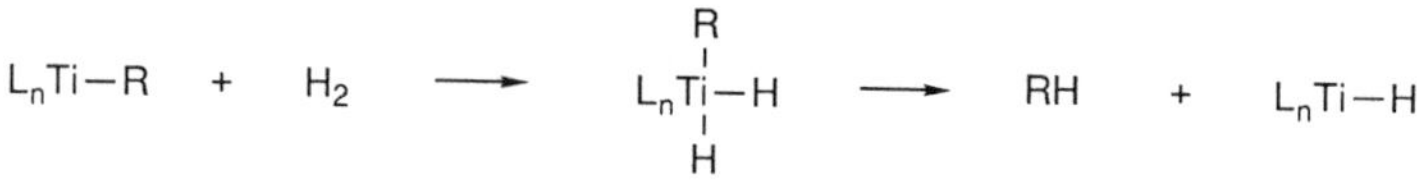

As shown above, the method gives *isotactic* polypropylene, in which each carbon atom has the same configuration. The material has a higher density and is better suited to making films and moulded articles than the randomly oriented *atactic* polymer made by older methods. By modifying the catalyst, *syndiotactic* polypropylene, in which alternate carbon atoms are of opposite configuration, can be made.

18.3
Chromium

(a) σ-Organochromium(III) compounds

These react with aldehydes and ketones in the same way as Grignard reagents. They are prepared *in situ* by the reaction of an alkyl, aryl, alkenyl, or allyl halide with chromium(II) chloride in an anhydrous solvent such as DMF or THF at 0°C (cf. p. 647):

$$Cr^{II} \ + \ R{-}X \ \longrightarrow \ Cr^{III} \ + \ R{\cdot} \ + \ X^-$$

$$R{\cdot} \ + \ Cr^{II} \ \longrightarrow \ R{-}Cr^{III}$$

They have two advantages over Grignard reagents. First, an alkenyl halide reacts with retention of configuration, e.g.

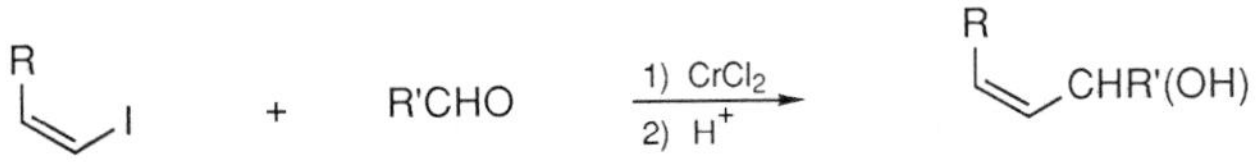

Second, either stereoisomer of an allyl halide reacts mainly at the more substituted carbon atom to give wholly or mainly *threo* product, e.g.

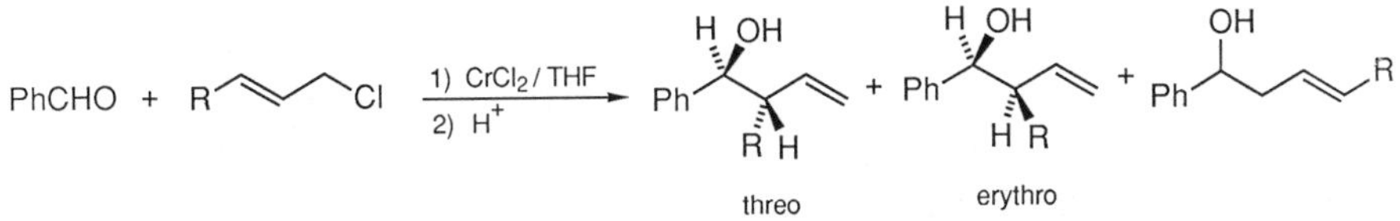

	threo	erythro	
R = Me	70	0	30 %
R = Ph	83	17	0 %

It is thought that this is because the new C—C bond is formed *via* a chair-like transition state in which R and Ph preferentially occupy equatorial positions:*

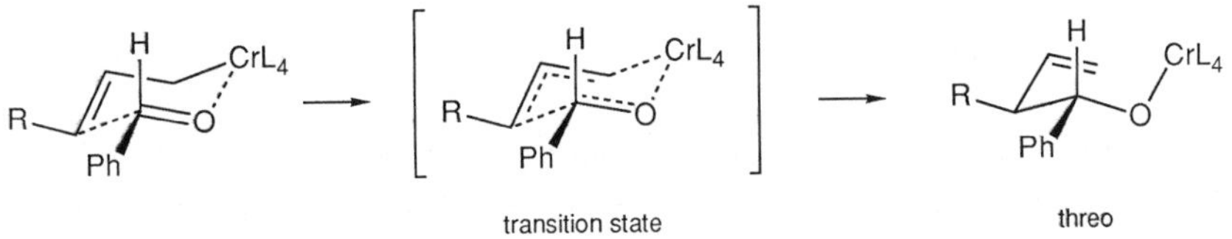

transition state   threo

*This is analogous to both aldol and Diels–Alder reactions:

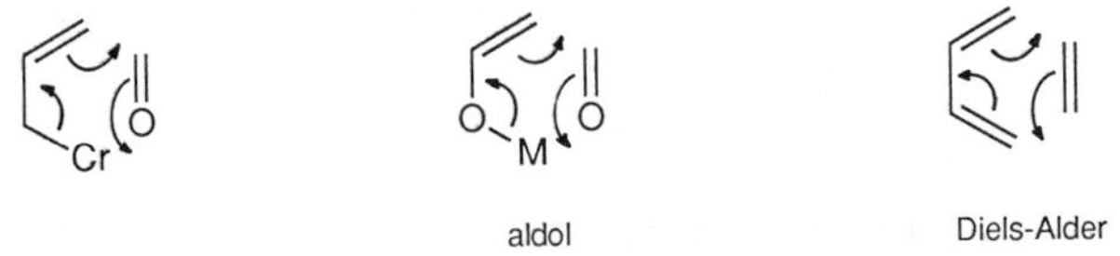

aldol   Diels-Alder

(b) Chromium-arene complexes

These are formed by heating the arene with chromium hexacarbonyl and have three principal applications.

(i) *Nucleophilic reactions.* The $Cr(CO)_3$ fragment is electron-withdrawing and activates the aromatic compound to nucleophiles. With a monosubstituted benzene, addition of the nucleophile is fastest at the *meta* position, and with a powerful nucleophile such as a carbanion, for which k_1 is large, the adduct can be efficiently trapped by the addition of acid, e.g.

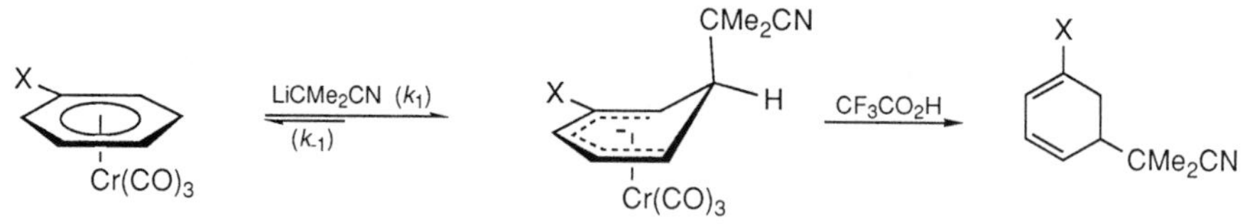

If X = OMe, acid-catalyzed hydrolysis follows, to give the enone:

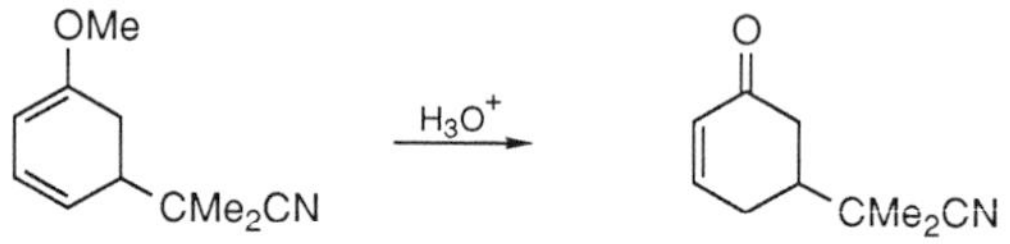

However, with less reactive nucleophiles and when X is a good leaving group (halide), the reversibility of the fast addition to the *meta* position means that the slower addition at the substituted position followed by irreversible elimination of the halide becomes the preferred course, e.g.

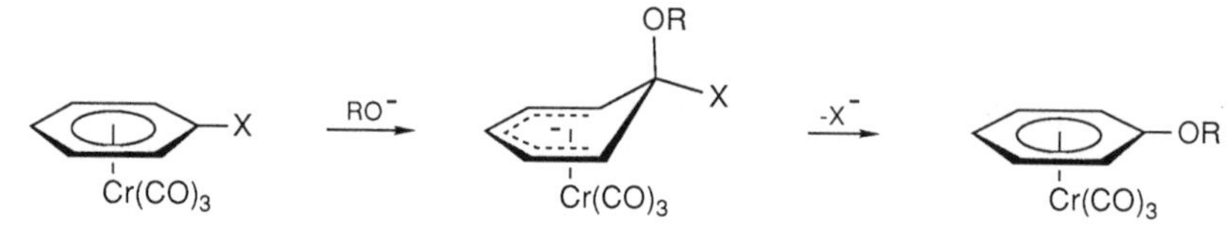

The product is released from the complex by mild oxidation, usually with cerium(IV).

The method has been applied to the synthesis of benzocrown ethers, using a quaternary ammonium salt as a phase-transfer catalyst (p. 290), e.g.

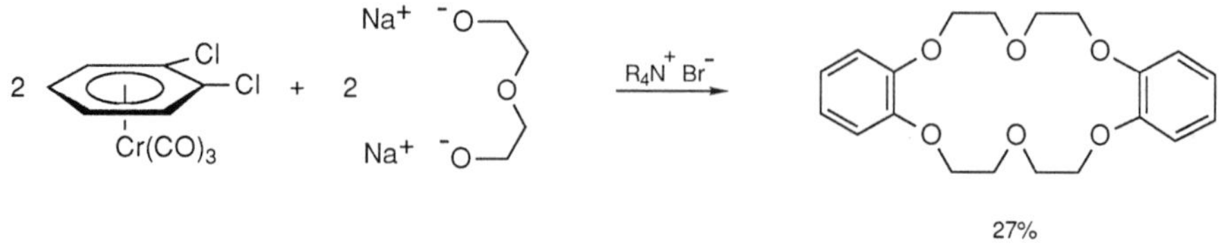

27%

(ii) *Lithiation.* Aromatic C—H bonds in η^6-arene-$Cr(CO)_3$ complexes are rendered sufficiently acidic by the electron-withdrawing effect of $Cr(CO)_3$ to undergo ready lithiation. This is of value in making *ortho*-substituted derivatives from PhX (X = OMe, F, Cl), since lithiation preferentially occurs *ortho* to the substituent:

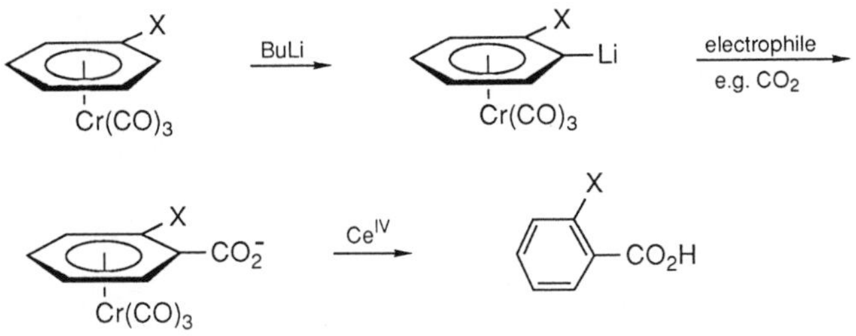

(iii) Activation of benzylic carbon. Benzylic anions are stabilized by the electron-withdrawing effect of $Cr(CO)_3$, so that, for example, Michael-type addition to styrene and its derivatives is possible, e.g.

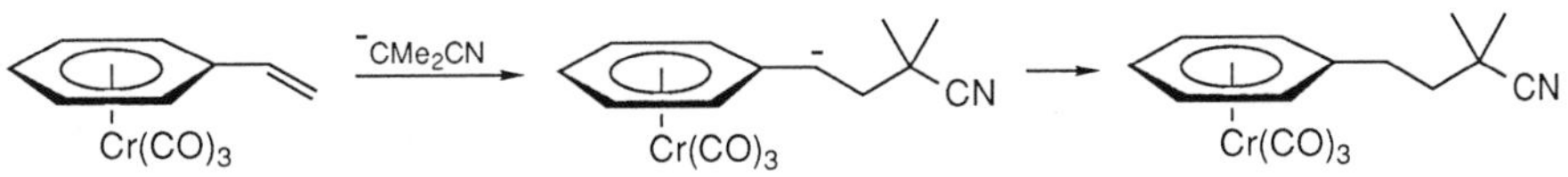

18.4 Iron

(a) Acyl-iron complexes

(i) Synthesis of ketones. Alkyl-iron bonds are readily formed by nucleophilic displacement on an alkyl halide or tosylate by an anionic iron complex,

$$2Na^+ [Fe(CO)_4]^{2-} \quad + \quad RX \quad \xrightarrow{-NaX} \quad R-\overset{-}{Fe}(CO)_4$$

Collman's reagent Na^+

In the presence of carbon monoxide and triphenylphosphine, insertion and ligand exchange reactions give a complex which reacts with alkenes to give, after isomerization and hydrolysis, ketones, e.g.

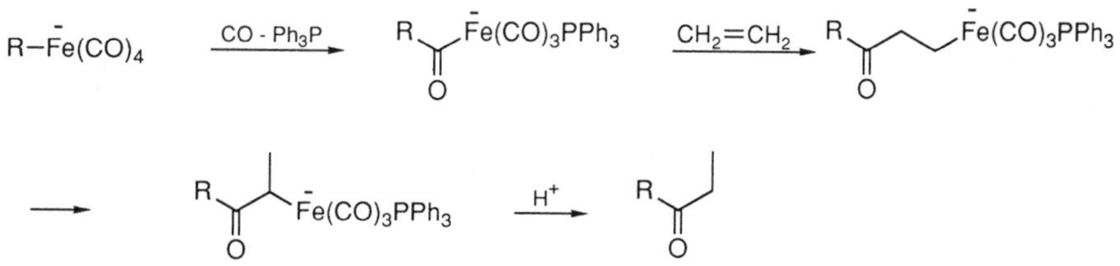

The method is of value in the synthesis of cyclopentanones,

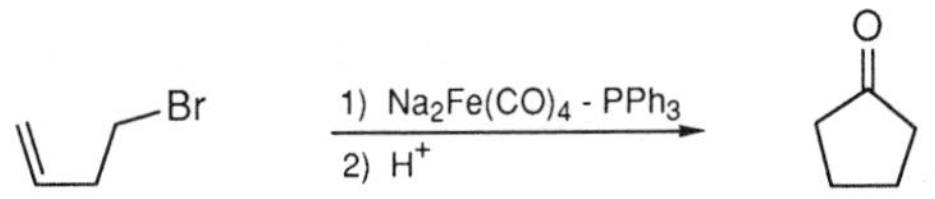

and was used in making a key compound in a synthesis of aphidicolin:

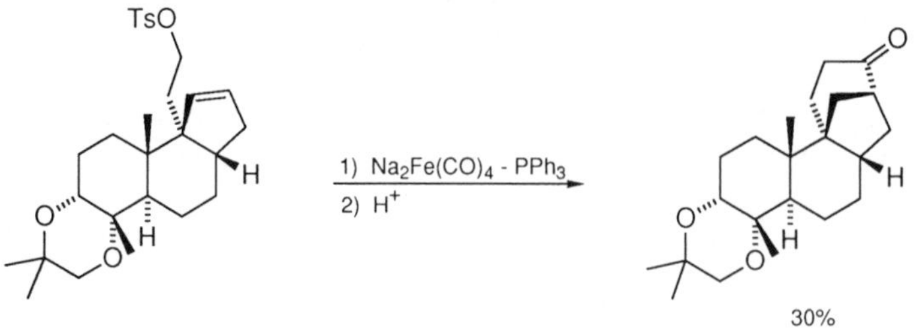

30%

In addition, acyl-iron anions can be converted into aldehydes by protonolysis and into ketones by alkylation:

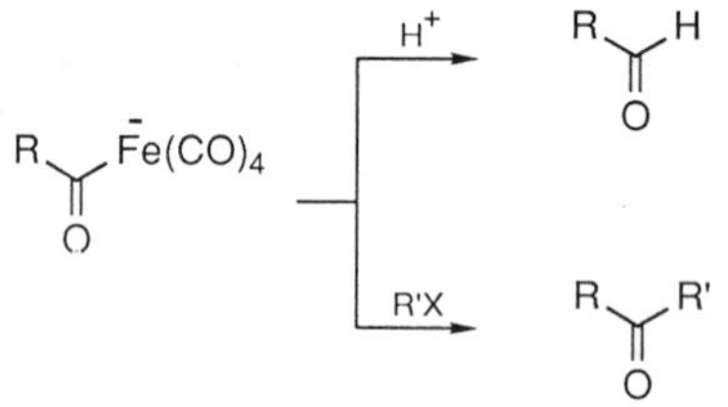

This provides an alternative to the 'reversed polarity' procedure using dithiane for the conversions RX → RCHO and RCOR′ (p. 474).

(*ii*) *Enantioselective synthesis.* An enolizable acyl group bound to iron in the asymmetric complex below can be alkylated with a very high degree of enantio-selectivity. The enolate is formed with butyllithium and, owing to hindrance to attack by a phenyl group on one face, it is alkylated, often exclusively, on the other. One-electron oxidation by bromine in water, an alcohol, or an amine can then yield acid, ester, or amide, often in enantiomerically pure form, e.g.

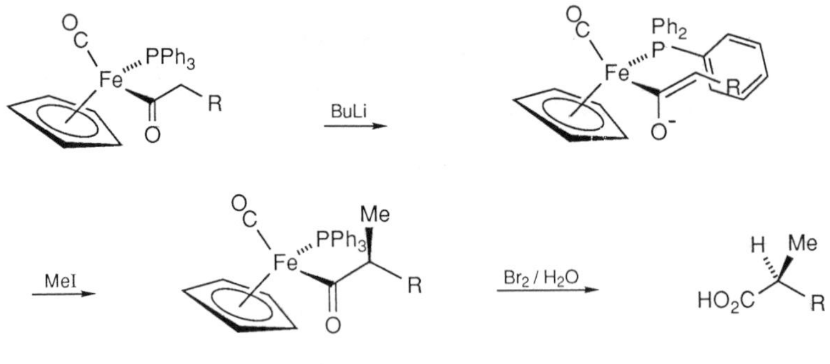

(*b*)   *Stabilization of cyclobutadiene*

The highly strained cyclobutadiene is unstable and has not been isolated. However, it can be obtained as a complex with $Fe(CO)_3$, and this enables it to be used in synthesis. In the form of its complex, it reacts with electrophiles, e.g.

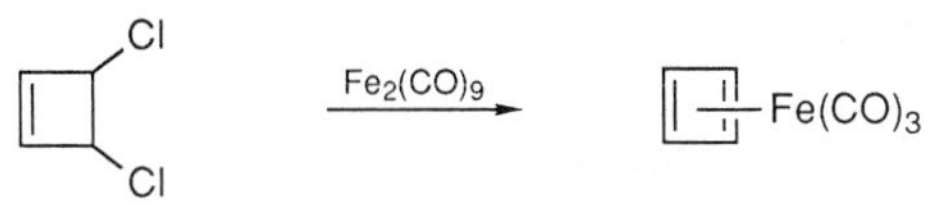

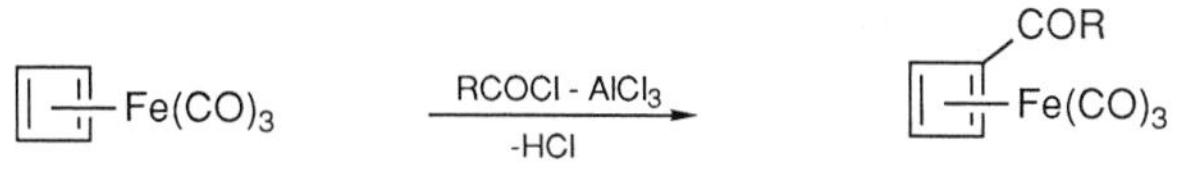

and, when released from the complex by oxidation with cerium(IV) in the presence of a dienophile, it undergoes Diels–Alder reactions, e.g.

18.5 Cobalt

(a)   Pauson-Khand reaction

Alkynes react with octacarbonyldicobalt to form complexes in which all four π-electrons are involved in bonding, e.g.

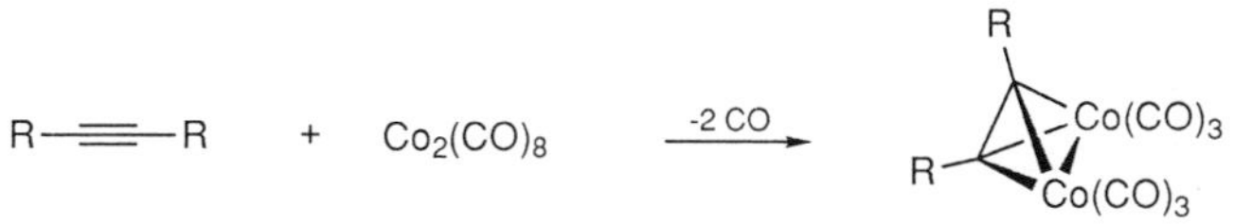

The most useful reactions of the complexes are with alkenes and carbon monoxide to form cyclopentenones. It is not necessary to isolate the cobalt complex: a mixture of octacarbonyldicobalt, the alkyne, and the alkene are heated at 60–110°C under carbon monoxide. An alkene $RCH=CH_2$ reacts with high regioselectivity:

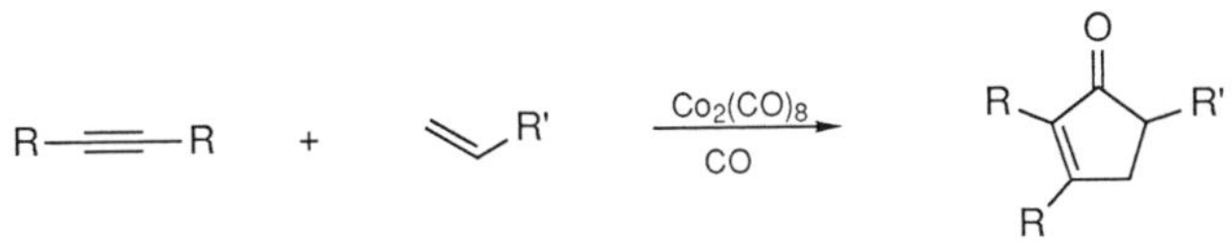

Bicyclic compounds can be constructed by using enynes, e.g.

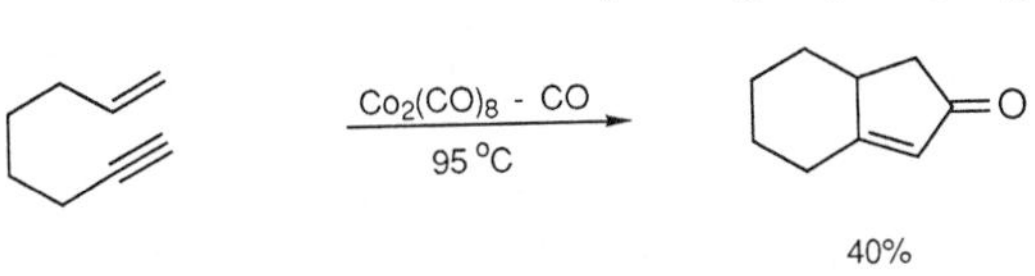

40%

(b) *Trimerization of alkynes*

$CpCo(CO)_2$ catalyzes the trimerization of alkynes to give aromatic compounds, e.g.

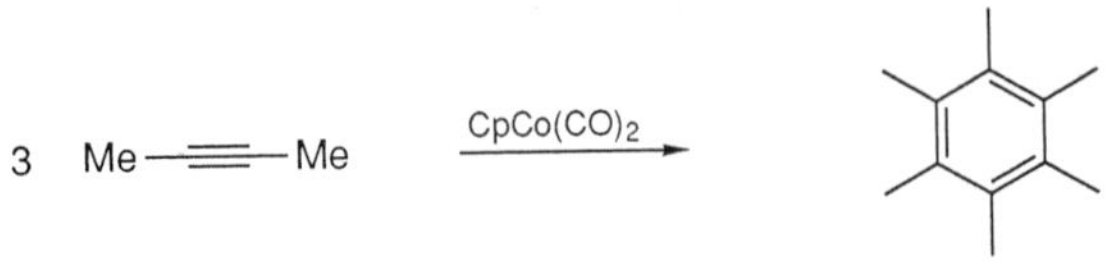

This principle was applied in a synthesis of a derivative of the hormone oestrone in which $Me_3Si—C\equiv C—SiMe_3$, which does not trimerize, was used in a mixed alkyne cyclization; formation of the aromatic ring was followed by an electrocyclic ring opening and Diels–Alder ring closure:*

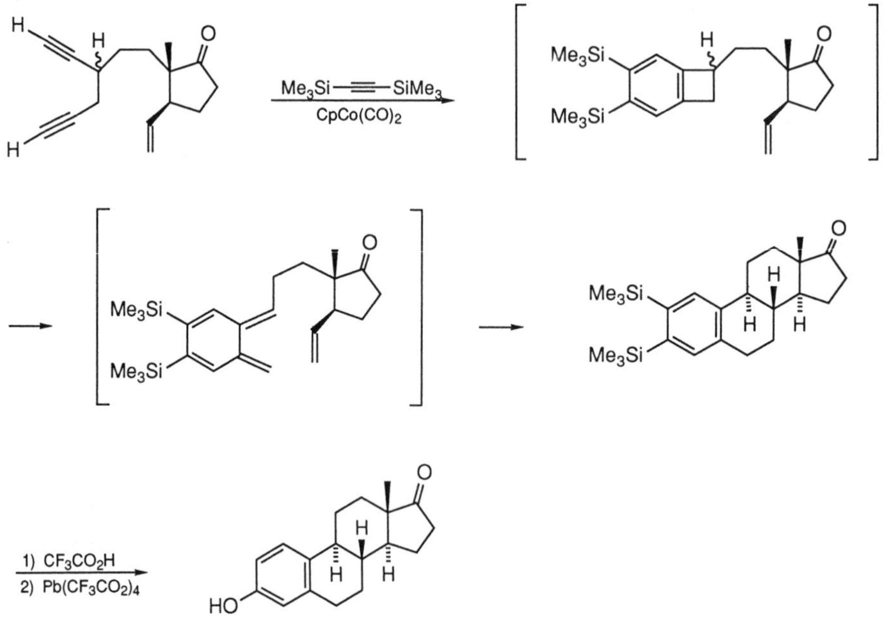

(c) *Hydroformylation*

The conversion

$$C=C + CO + H_2 \longrightarrow \overset{\displaystyle C}{\underset{\displaystyle H}{C}}{\sim}CHO$$

---

* It was found that one — $SiMe_3$ group could be removed selectively from the last intermediate by protonolysis with CF_3CO_2H (p. 345), and the second was replaced by — OH with a lead(IV) oxidant.

which is especially of industrial importance, takes place at high temperatures (*c.* 150°C) and pressures (300 atmospheres) in the presence of octacarbonyldicobalt. The key cobalt intermediate is $HCo(CO)_3$:

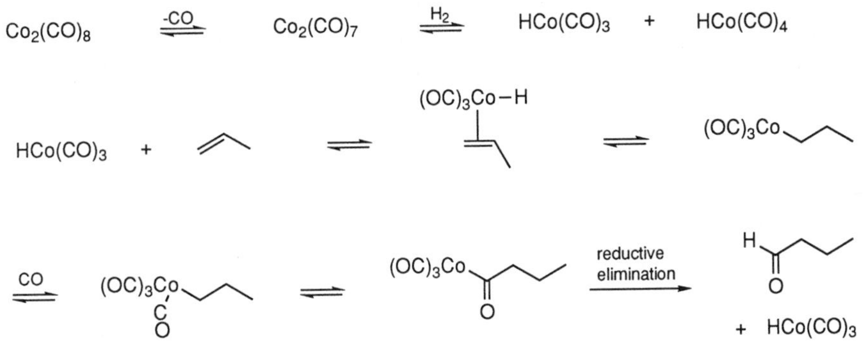

Steric effects control the reactivity: straight-chain terminal alkenes > straight-chain internal alkenes > branched-chain alkenes.

18.6
Rhodium

The most useful rhodium complex is *Wilkinson's catalyst*, $(Ph_3P)_3RhCl$, which is prepared by heating rhodium chloride, $RhCl_3 \cdot 3H_2O$, with triphenylphosphine in boiling ethanol. As well as its value in homogeneous hydrogenation (p. 633), it is used in the decarbonylation of aldehydes (the reverse of the carbonyl insertions described earlier):

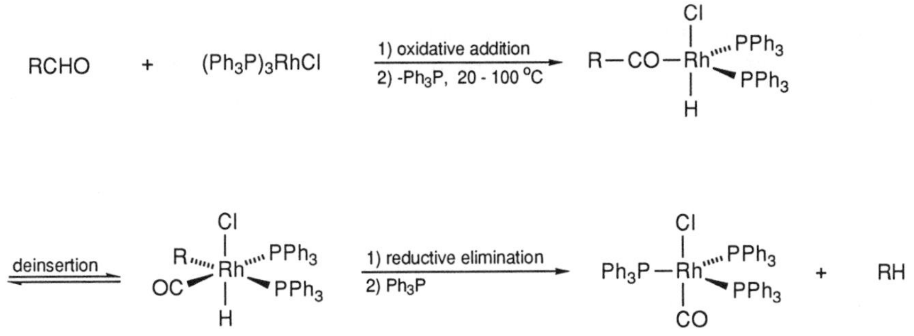

This is of value when the carbonyl group has been required for activation of a step earlier in the synthesis and has subsequently to be removed. If R is asymmetric, it retains its stereochemistry.

18.7
Nickel

Nickel tetracarbonyl effects the reductive coupling of allyl halides in aprotic solvents such as DMF:

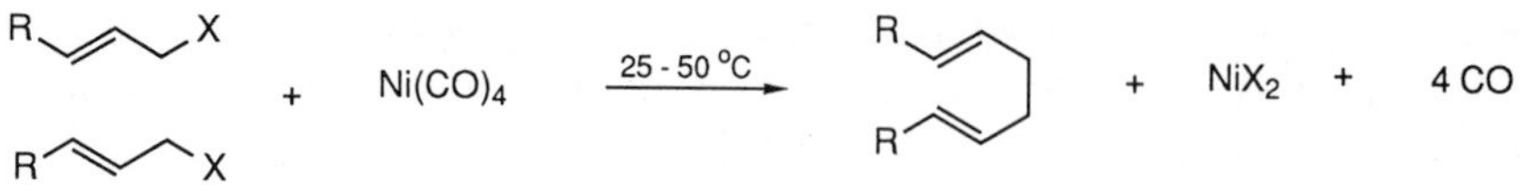

Reaction occurs by reversible dissociation of nickel tetracarbonyl, oxidative addition of the allyl compound, formation of the η^3-allyl complex, dimerization of the complex through halogen bridging, and finally C—C coupling:

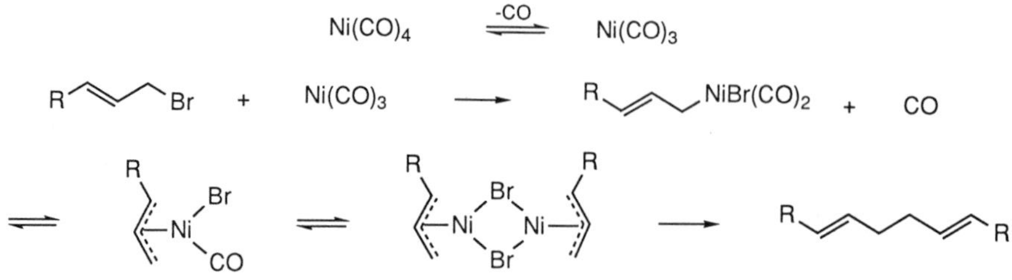

The coupling takes place at the less substituted carbon of the allyl group and gives mainly *trans*-products, as above, regardless of the stereochemistry of the allyl halide.

Since the dimeric nickel complex brings the coupling termini into proximity, the method is particularly suitable for macrocyclization, e.g.

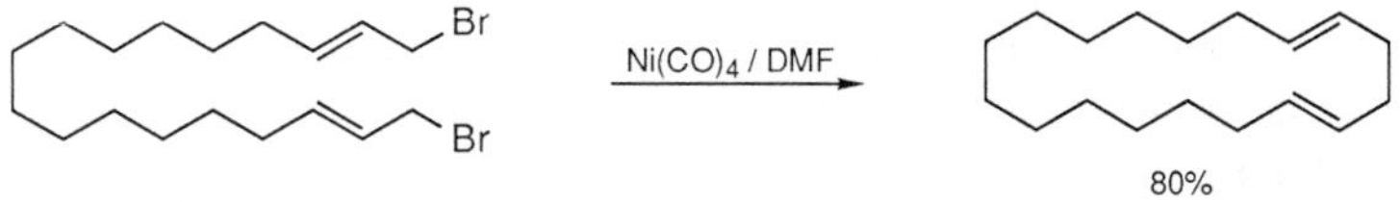

80%

When the reaction is carried out in a hydrocarbon solvent such as benzene, the η^3-allyl complex is formed but does not couple; it can be isolated by crystallization at very low temperatures. This makes it possible to effect crossed coupling by treating the complex with an alkyl halide in a polar aprotic solvent such as DMF. For example, α-santalene has been made in 88% yield in this way:

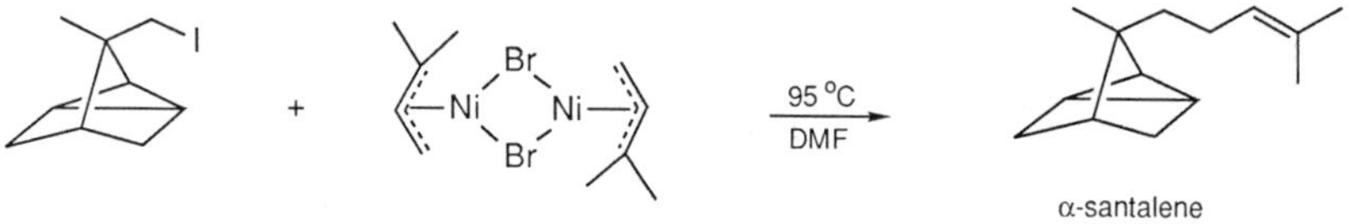

α-santalene

Terpenes and dihydrocoumarins can also be constructed, e.g.

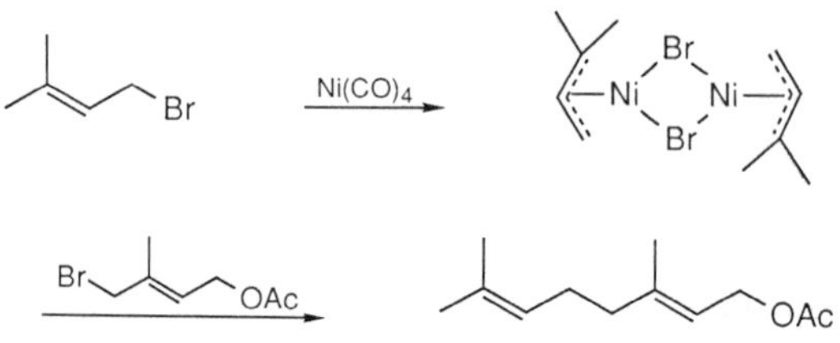

geranyl acetate

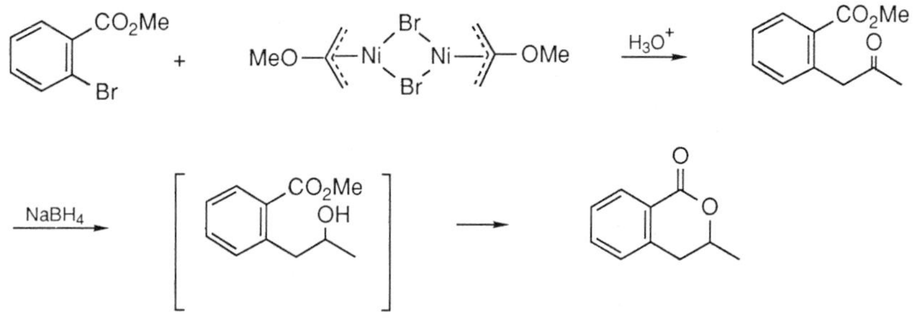

(i) *Ni(COD)₂.* The complex of nickel(o) with 1,5-cyclooctadiene, Ni(COD)₂,

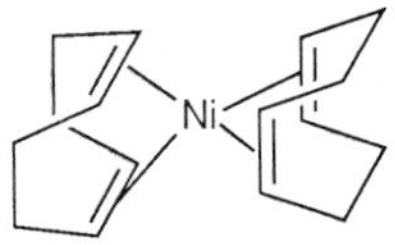

can generally be used in place of nickel tetracarbonyl in coupling reactions and, since it is much less toxic, is preferred.

**18.8
Palladium**

(a) *(η²-Alkene)palladium(ii) complexes*

The π-complexes formed from palladium(ii) and monosubstituted or 1,2-disubstituted alkenes are activated to nucleophiles, enabling reactions of the type

$$RCH=CHR' \ + \ X^- \ + \ Pd^{2+} \ \longrightarrow \ RC\underset{X}{=}CHR' \ + \ H^+ \ + \ Pd$$

to be brought about; XH is usually water, an alcohol, a carboxylic acid, or a primary or secondary amine.

The sequence of reactions, illustrated for ethylene and with water as the nucleophile, is:

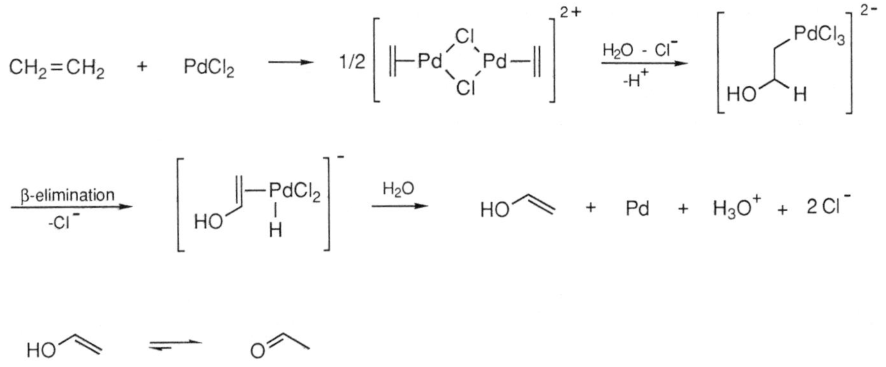

By carrying out the reaction in air and in the presence of a catalytic amount of $CuCl_2$, $Pd(o)$ is reoxidized *in situ*,

$$Pd(0) \quad + \quad 2\,Cu^{2+} \quad \longrightarrow \quad Pd^{2+} \quad + \quad 2\,Cu^{+}$$

$$Cu^{+} \quad \xrightarrow{O_2} \quad Cu^{2+}$$

so that only a catalytic quantity of $Pd(\text{II})$ is needed.

This is the basis of an important industrial method (*Wacker process*) for manufacturing acetaldehyde.

Monosubstituted alkenes bond to palladium through the unsubstituted carbon atom so that, for example, reaction with water gives methyl ketones, e.g. [7],

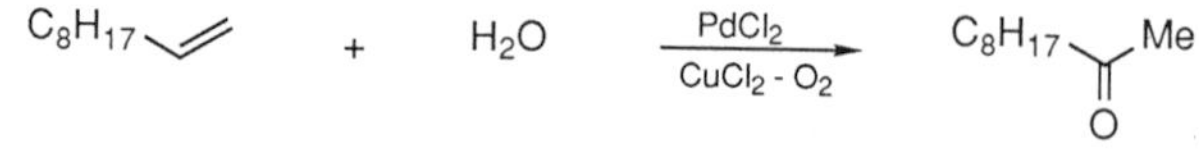

70%

1,2-Disubstituted alkenes react less rapidly (*cis* faster than *trans*), so that selective reactions can be carried out, e.g.

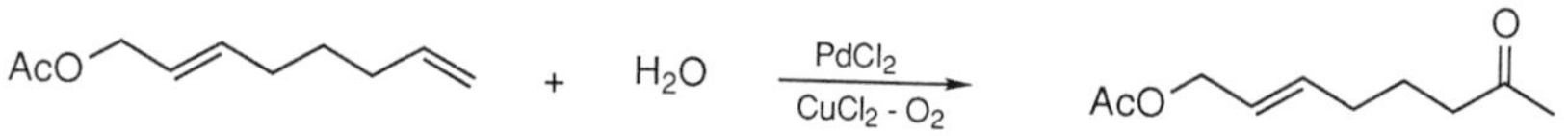

Cyclization of suitable hydroxyalkenes can be brought about and can be combined, in the presence of carbon monoxide, with an insertion reaction of the σ-alkylpalladium species, e.g.

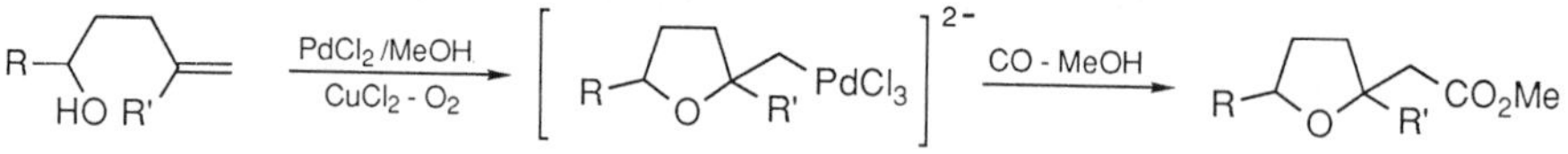

(b) (η³-Allyl)palladium complexes

(i) Formation. The complexes can be made from either alkenes or allyl compounds which possess reasonable leaving groups.

From alkenes. Reaction with $PdCl_2$ gives, *via* the π-alkene complex, the allylpalladium complex as a chlorine-bridged dimer:

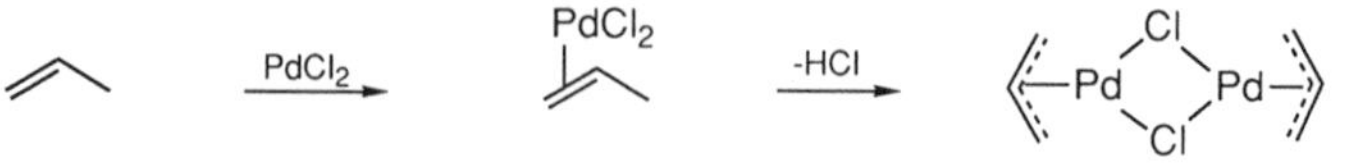

From allyl compounds: with Pd(II). Allyl acetates and other allyl compounds with leaving groups react with $PdCl_2$ by either the π-alkene route or formation of a σ-allylpalladium bond,

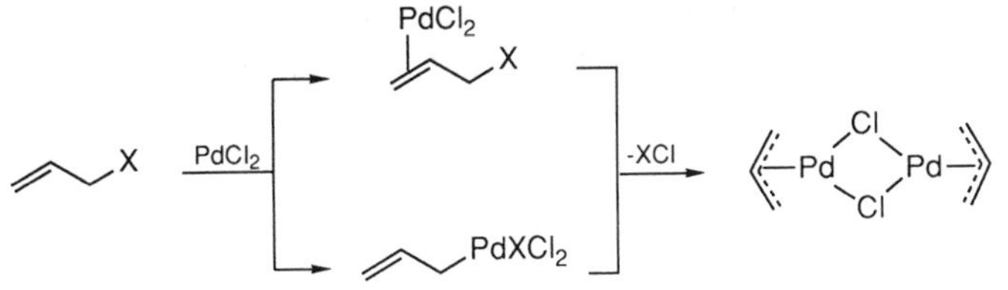

From allyl compounds: with Pd(o). Reactions corresponding to those of allyl compounds with $PdCl_2$ take place with $Pd(PPh_3)_4$ to give an 18-electron allyl complex which dissociates,

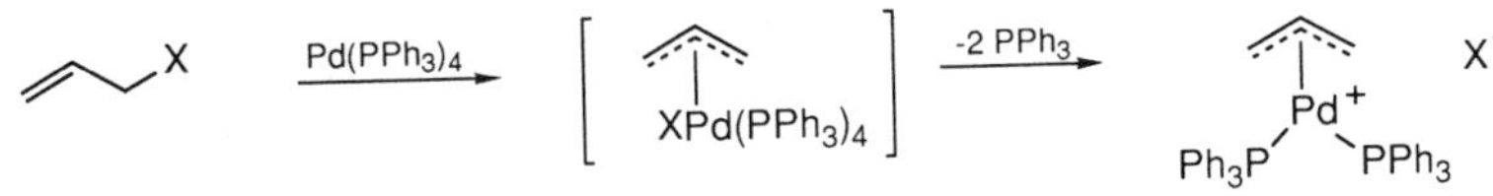

Since Pd(o) is regenerated on reaction with a nucleophile, only a catalytic quantity is required.

(*ii*) *Reactions of the complexes.* In order to activate the chlorine-bridged π-allyl dimers to nucleophiles, they are treated with triphenylphosphine to give the same type of allyl complex as is formed by the Pd(o) method. The reaction of this complex with a nucleophile then gives product by reductive elimination:

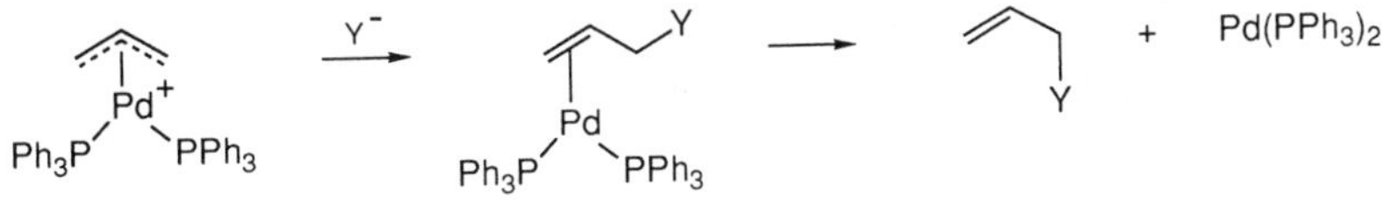

If the allyl group contains terminal substituents, the nucleophile reacts at the less hindered carbon atom.

The following examples illustrate the usefulness of the process.

(1) *Prenylation.* Isoprene units can be introduced by reaction with a readily formed enolate from a diactivated CH group, as in a synthesis of (*E,E*)-farnesol:

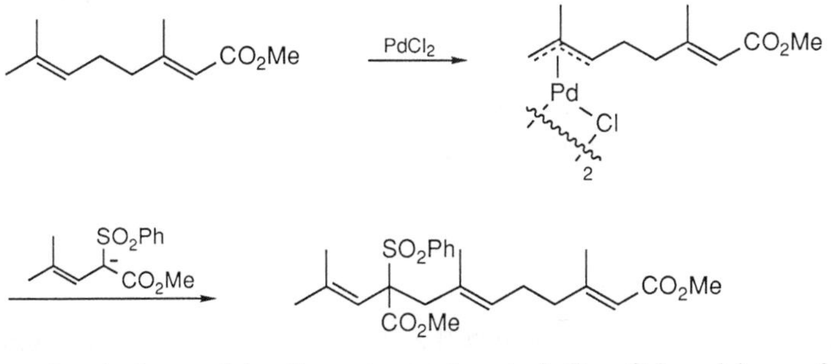

The allyl complex is formed by the non-conjugated C=C bond in preference to the conjugated, stabilized one. Hydrolysis of the ester group, monodecarboxylation (assisted by the sulfone group, cf. β-keto-acids, p. 296), reduction of the remaining carboxyl by DIBAL (p. 660), and finally hydrogenolysis to remove the sulfone gave (*E,E*)-farnesol:

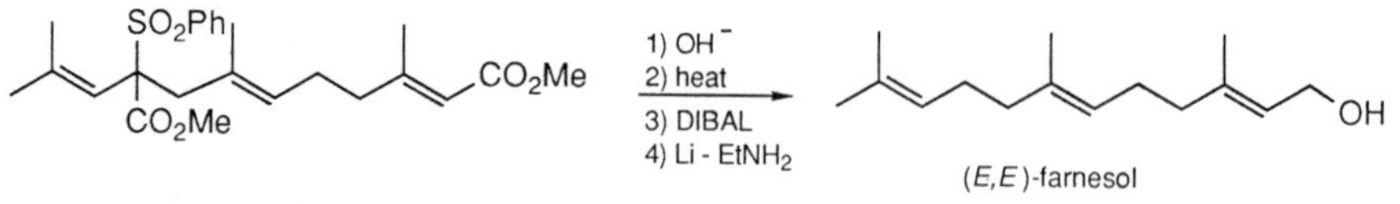

(2) *Formation of conjugated dienes.* A doubly allylic acetate acts as a masked conjugated diene, e.g.

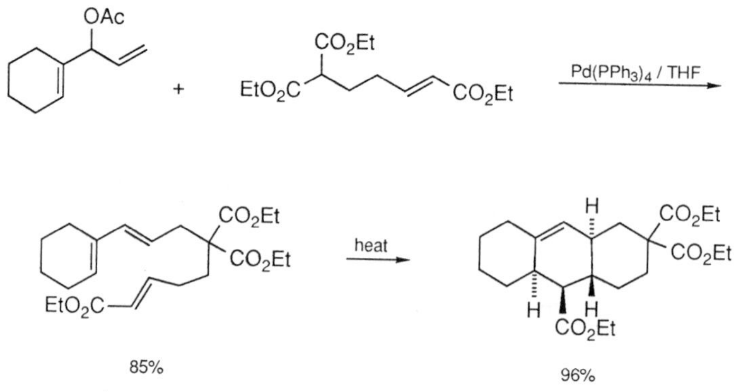

(3) *Synthesis of macrocyclic nitrogen heterocycles.* The following example shows that very large rings can be formed efficiently:

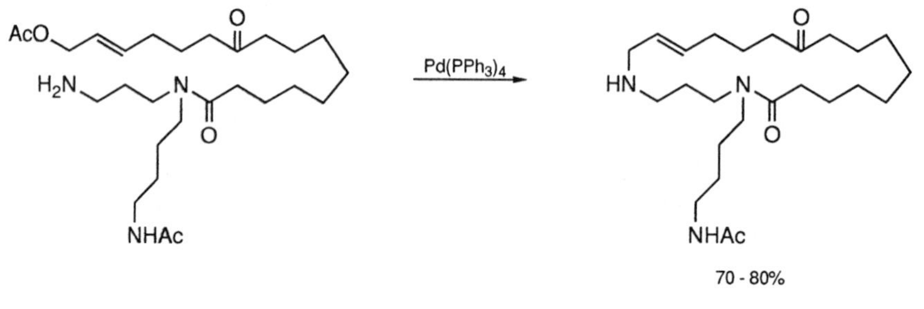

Hydrogenation of the C=C bond and hydrolysis of the NHAc group gave indandenin-12-one (a spermidine alkaloid).

(c) *Heck reaction*

This is a coupling reaction in which the R group in RPdX (X = halide or acetate) replaces hydrogen at the less hindered carbon atom of an alkene. It is believed to occur by *syn*-addition of RPdX followed by *syn*-elimination of HPdX, so that there is inversion of configuration at the carbon atom where substitution takes place, e.g.

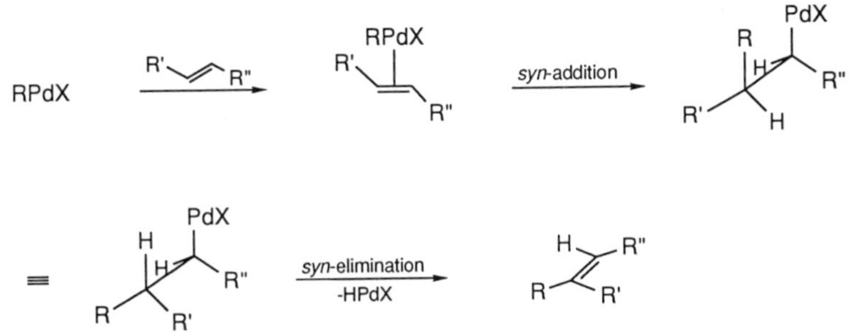

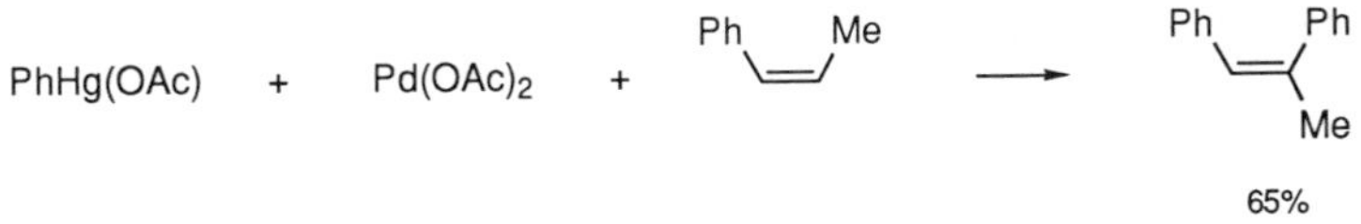

65%

The species RPdX can be formed *in situ* in several ways. The method using PhHg(OAc) as above is especially suitable for R = aryl, since ArHg(OAc) is available by mercuration of ArH (p. 388). Alternatively, RPdI can be generated from the iodide and Pd(OAc)$_2$ in the presence of a weak base such as Bu$_3$N. A particularly useful method is to employ palladium(o), since the process is then catalytic:

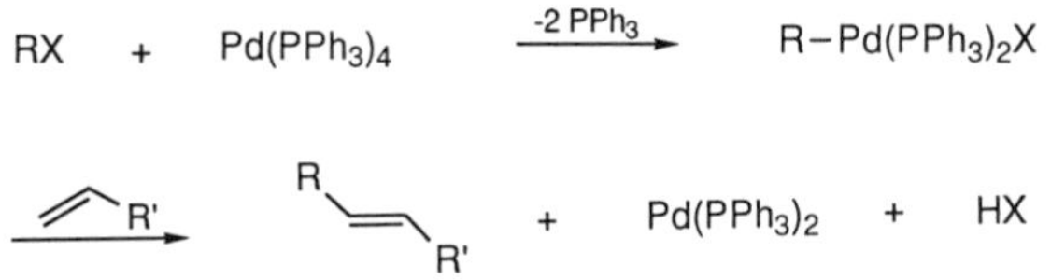

$$RX + Pd(PPh_3)_4 \xrightarrow{-2\,PPh_3} R-Pd(PPh_3)_2X$$

As well as aryl groups, alkenyl groups and alkyl groups that do not contain a β-hydrogen (e.g. CH$_3$, PhCH$_2$, Me$_3$C—CH$_2$) can be employed. The alkene substituent can be varied widely (e.g. alkyl, aryl, CO$_2$Me, CN, CH(OEt)$_2$), with yields usually >80% with a monosubstituted alkene and some 1,1- and 1,2-disubstituted alkenes, for example in coupling with phenyl,

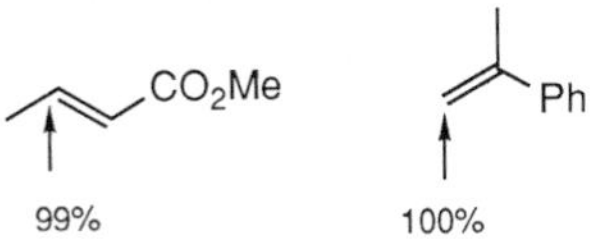

99%   100%

The rate of coupling is strongly dependent on steric effects: for example, in the reactivity sequence

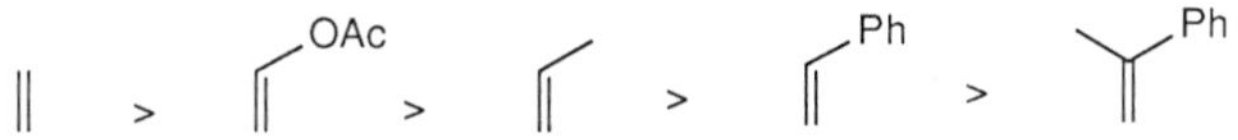

ethylene is over ten-thousand times as reactive as α-methylstyrene in phenylation.

(d)   Aromatic palladation

Aromatic compounds which contain a benzylic-type nitrogen substituent undergo *ortho*-metallation with palladium(II):

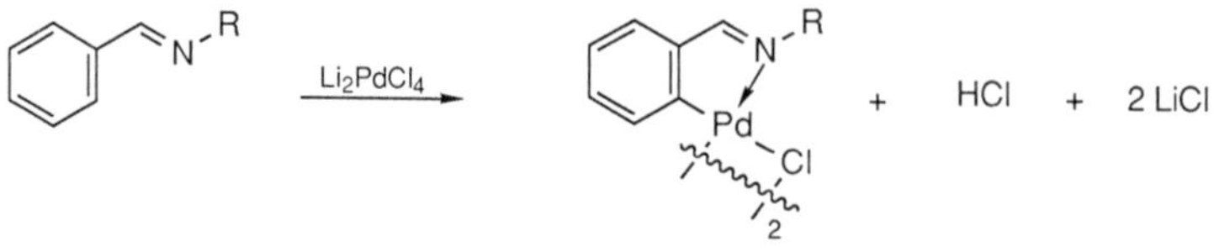

This is the basis of a method for generating *ortho* C—C bonds: the carbon atom bonded to palladium has nucleophilic character and reacts with acid chlorides and activated C=C bonds, e.g.

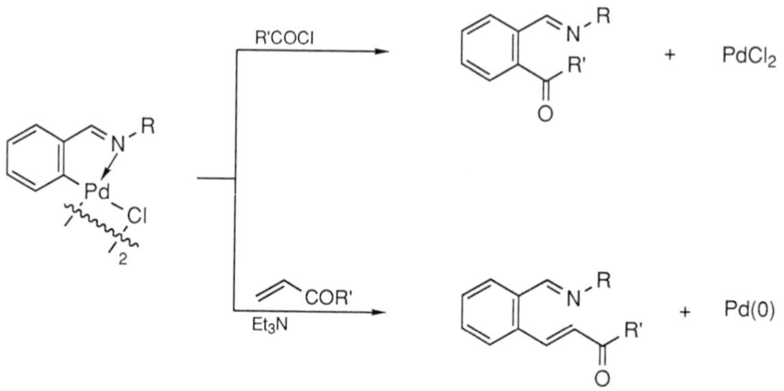

(e)   Bis-η³-allyl complexes from dienes

Conjugated dienes react with palladium(0) in the presence of a wide range of nucleophiles, such as water, alcohols, carboxylic acids, and carbanions, in a process known as *telomerization*. It involves a Pd(0)-catalyzed reductive dimerization to give a bis-π-allyl complex, nucleophilic attack, and reductive elimination, e.g.

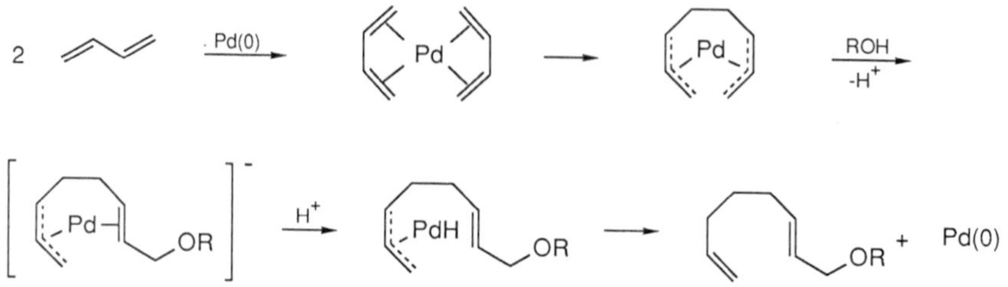

1. Indicate the formal oxidation state and total number of electrons in the   **Problems**
   valence shell of the transition metal in each of the following compounds:

(a)

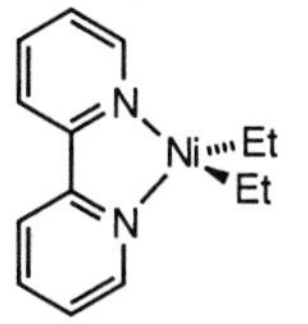

(b)

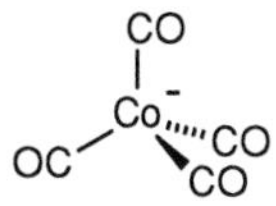

(c)

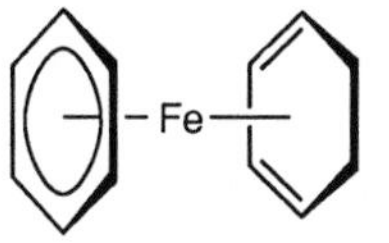

(d)

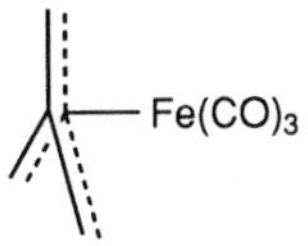

(e)

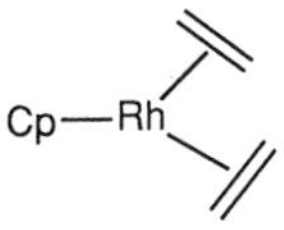

(f)

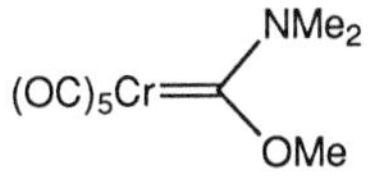

2. Predict structures for the following compounds using the 18-electron rule as a
   guide:

   $C_7H_7Co(CO)_3$ $(C_5H_5)_2Fe(CO)_2$ $(C_5H_5)Fe(CO)_2(allyl)$

3. How would you employ organotransition metal compounds in the synthesis of
   the following?

(a)

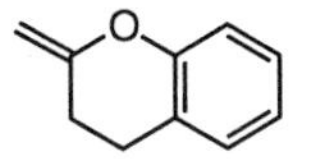

from

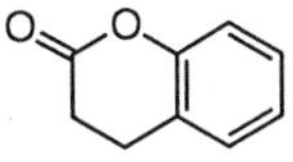

(b)

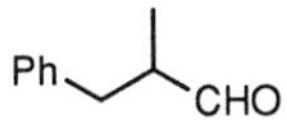

from

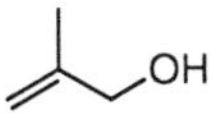

(c)

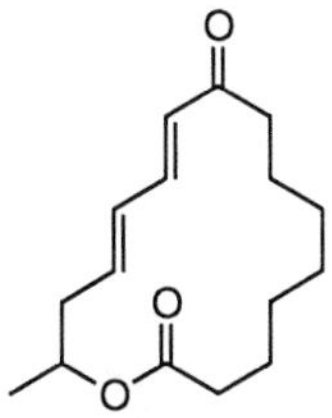

from

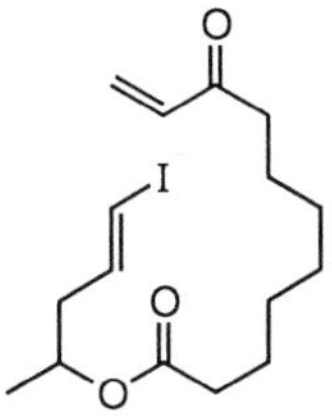

(d)

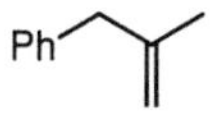

from

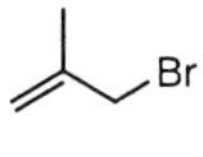

(e)

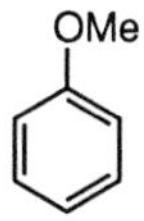

from

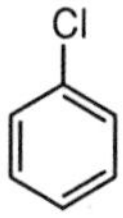

4.   Rationalize the following reactions:

(a)

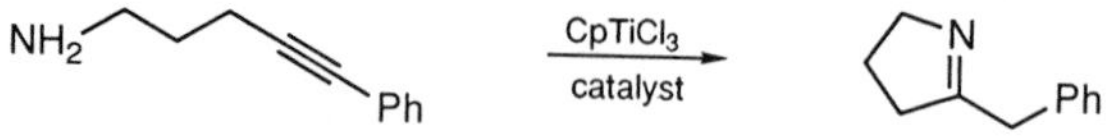

(b)

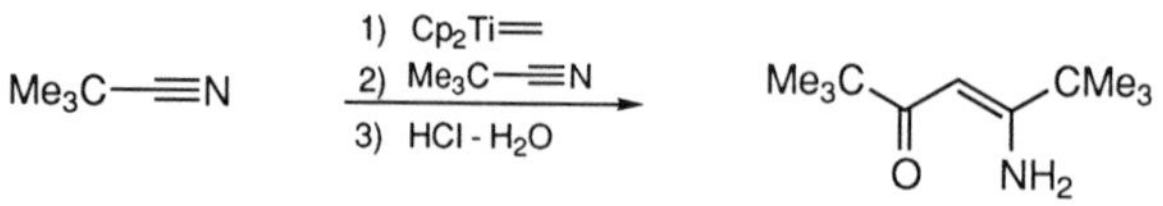

(c)

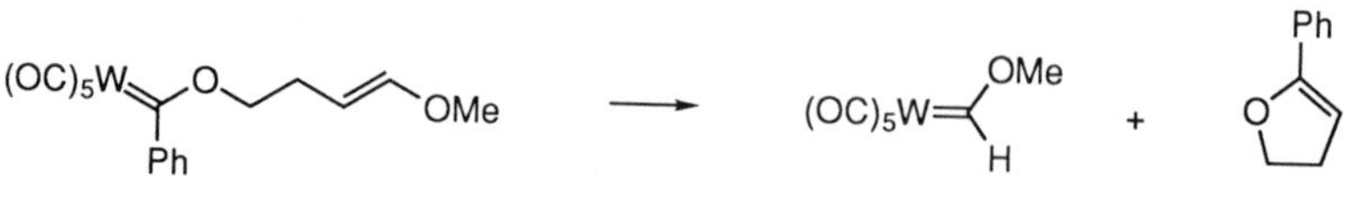

(d)

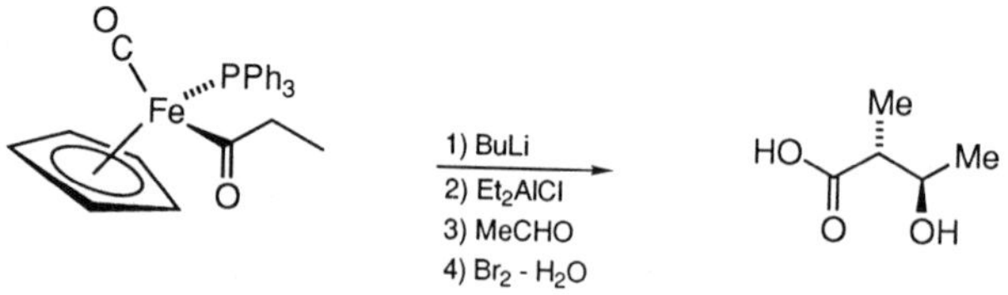

(e)

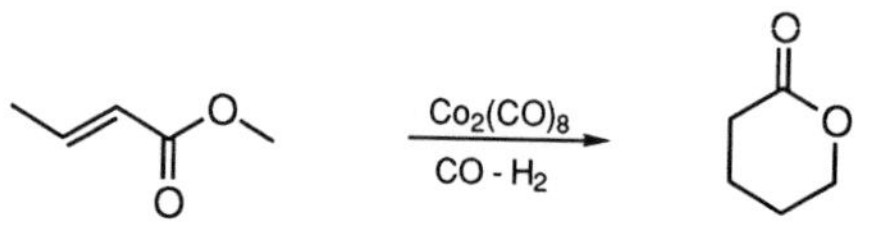

(f)

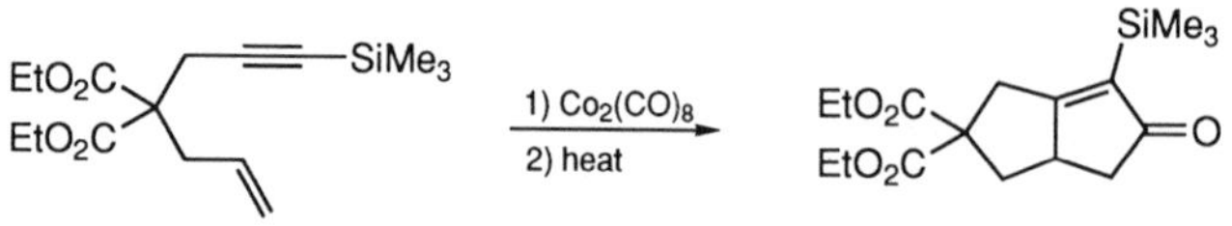

(g)

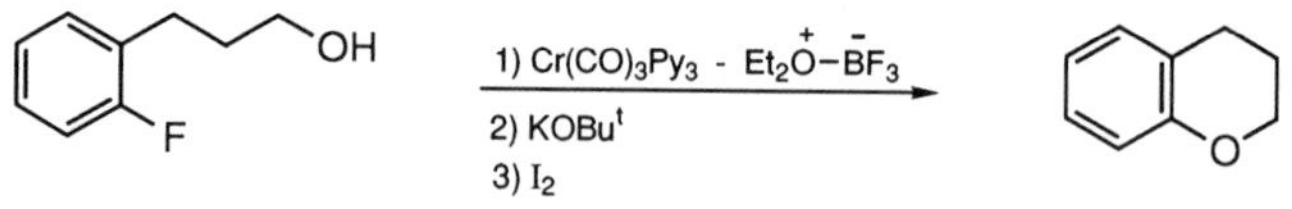

(h)

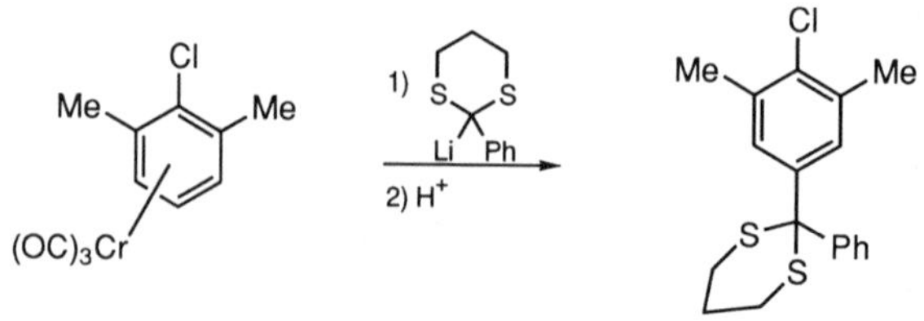

(*i*)

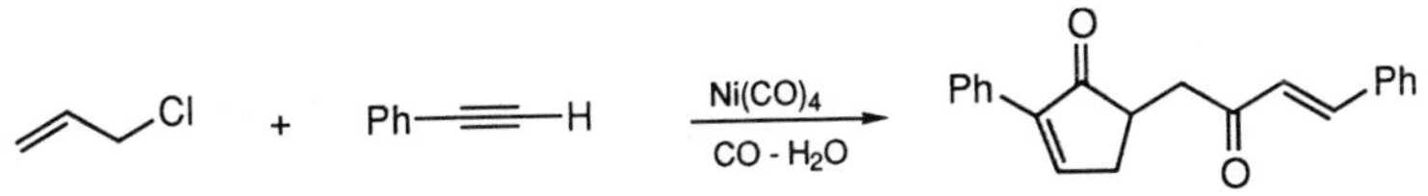
Cl
+
Ph
H
Ni(CO)4
CO - H2O
Ph
O
O
Ph

(*j*)

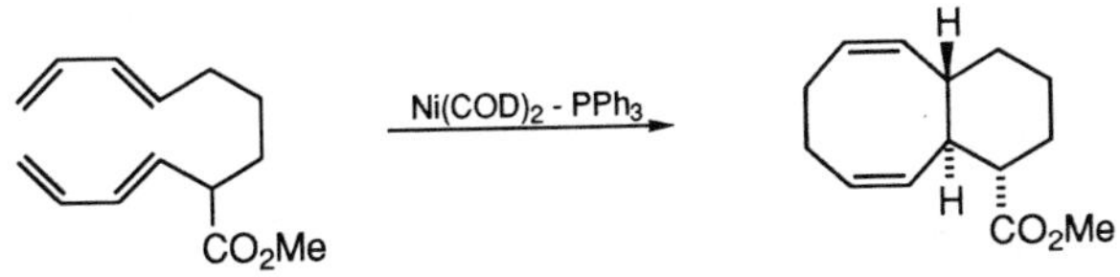
CO2Me
Ni(COD)2 - PPh3
H
H
CO2Me

(*k*)

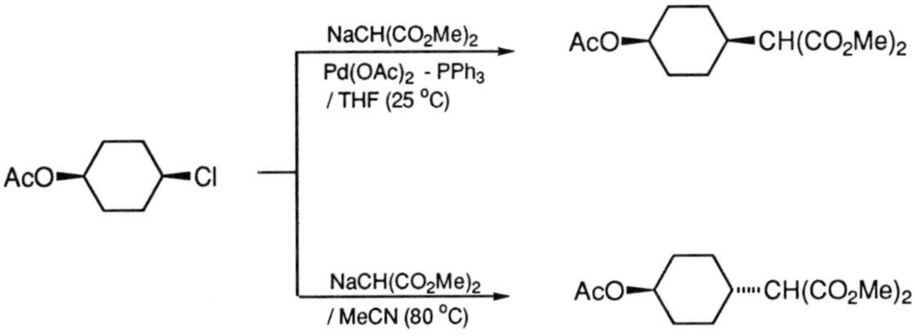
NaCH(CO2Me)2
Pd(OAc)2 - PPh3
/ THF (25 °C)
AcO
CH(CO2Me)2
AcO
Cl
NaCH(CO2Me)2
/ MeCN (80 °C)
AcO
CH(CO2Me)2

(*l*)

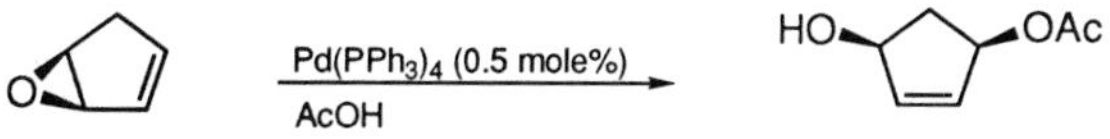
O
Pd(PPh3)4 (0.5 mole%)
AcOH
HO
OAc

(*m*)

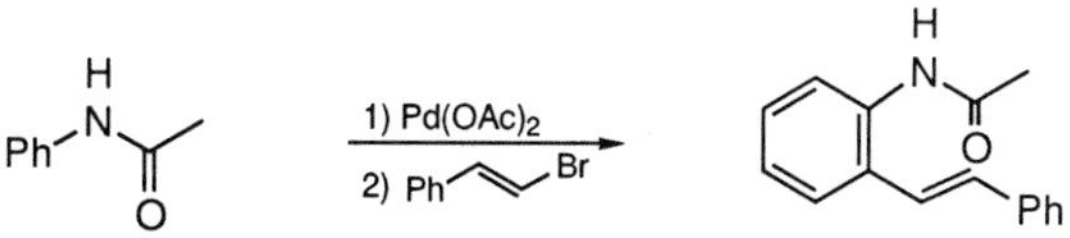
H
Ph
N
O
1) Pd(OAc)2
2) Ph
Br
H
N
O
Ph

(*n*)

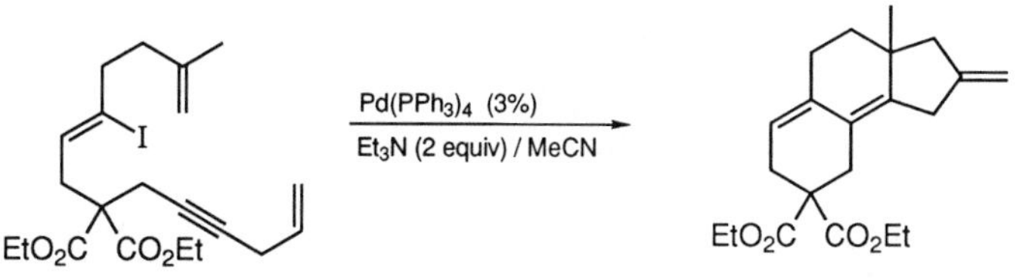
I
EtO2C
CO2Et
Pd(PPh3)4 (3%)
Et3N (2 equiv) / MeCN
EtO2C
CO2Et

(*o*)

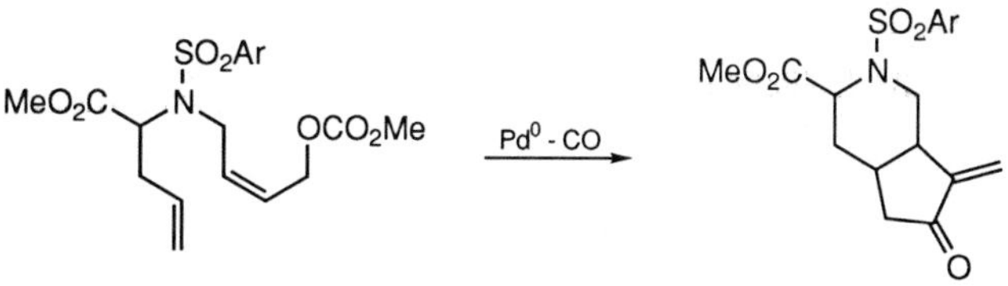
SO2Ar
MeO2C
N
OCO2Me
Pd0 - CO
SO2Ar
MeO2C
N
O

(*p*)

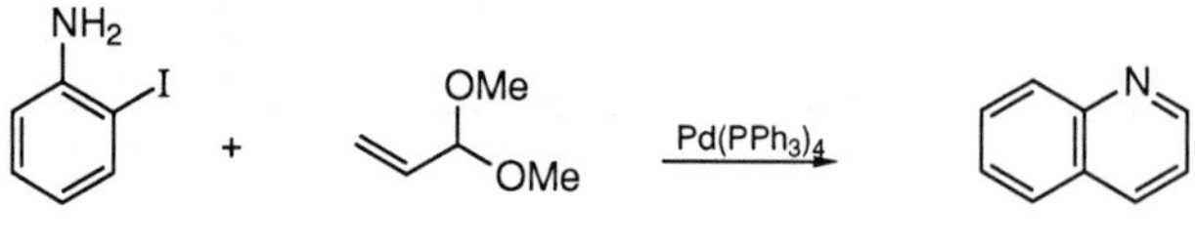
NH₂
I
+
OMe
OMe
Pd(PPh₃)₄
N

(*q*)

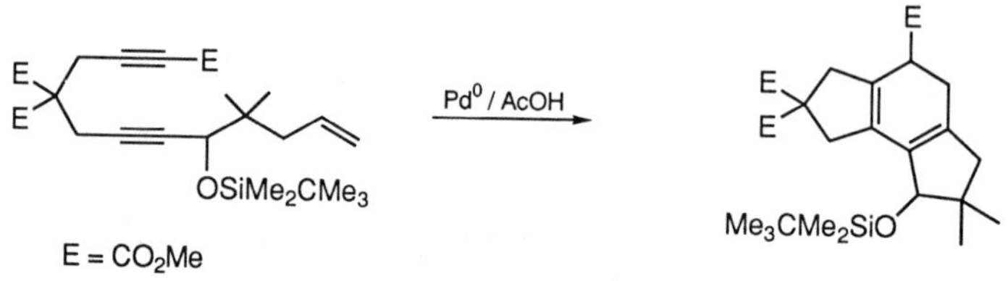
E
E
E
OSiMe₂CMe₃
Pd⁰ / AcOH
E
E
E
Me₃CMe₂SiO
E = CO₂Me

(*r*)

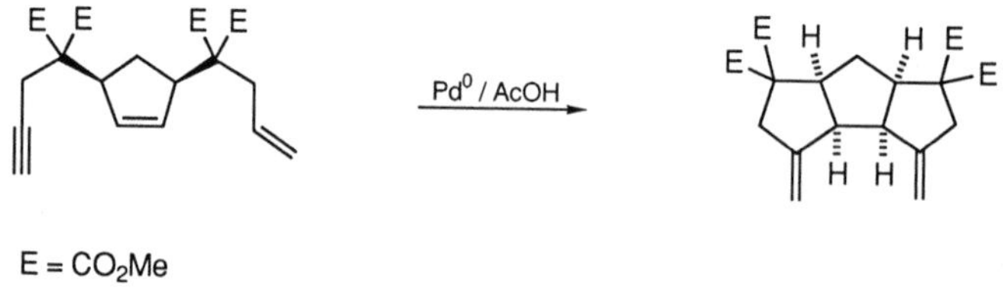
E E
E E
Pd⁰ / AcOH
E
E
H
H
E
E
H
H
E = CO₂Me

Oxidation 19

The scope of the term oxidation has been discussed earlier (section 4.9). The processes described in this chapter are those which lead to the incorporation of oxygen or the removal of hydrogen from a molecule; those which result in the partial removal of electrons (e.g. the bromination of methane) are included only where they are employed *en route* to the introduction of oxygen or the removal of hydrogen, as in the conversion of toluene into benzaldehyde *via* benzal chloride (p. 599).

The different techniques used in the laboratory and in large-scale industrial processes are stressed. In the former, complex polyfunctional molecules are often involved; regioselective and/or stereoselective reagents are required, many of which are expensive, and high temperatures have usually to be avoided. In industrial high-tonnage processes, relatively simple compounds are often involved which can withstand vigorous conditions; the first essential is cheapness. Air is the natural industrial oxidizing agent, as in the conversions of isopropanol into acetone, propylene into propenal, and naphthalene into phthalic anhydride, each over metal oxide catalysts at high temperatures.

Bacterial oxidation is also employed industrially. For example, *Acetobacter suboxydans* is the best reagent for the specific oxidation of the C_2-carbon of D-glucitol in the production of vitamin C (L-ascorbic acid) from glucose:

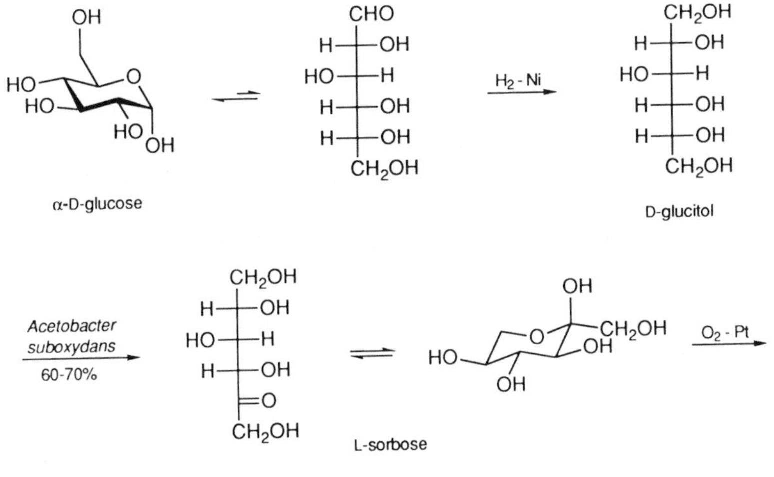

α-D-glucose

D-glucitol

L-sorbose

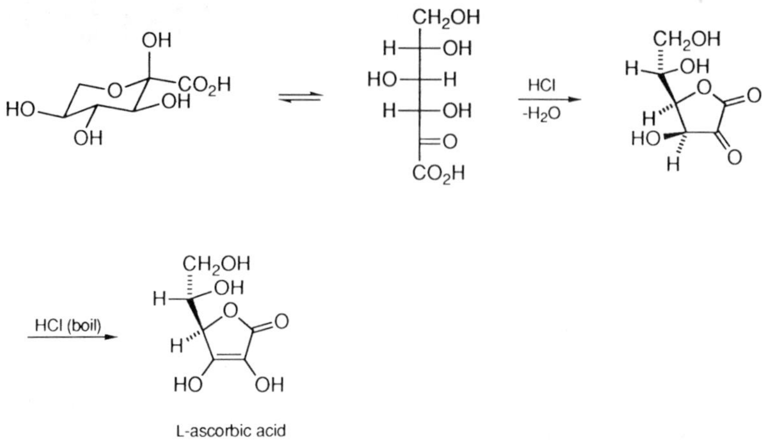

L-ascorbic acid

Both bacterial and enzymatic methods may eventually prove to be powerful tools for selective oxidation in the laboratory.

This chapter is arranged according to the type of system to be oxidized, so that whereas the appropriate method for a particular oxidation can readily be found, the properties of any one oxidizing agent are not collected together.

19.2 Hydrocarbons

(a) *Alkene double bonds*

(i) *Epoxidation.* Alkenes react with peroxyacids to give epoxides. The reaction is usually expressed as

but this does not reflect the fact that electron-releasing groups in the alkene increase the reactivity (e.g. $RCH{=}CHR \sim R_2C{=}CH_2 > RCH{=}CH_2 > CH_2{=}CH_2$), as do electron-attracting groups in the peroxyacid: that is, there is a partial transfer of electrons from alkene to peroxyacid at the transition state.

The important characteristics of the reaction are:

- It is regioselective for the more electron-rich double bond when more than one is present, e.g.

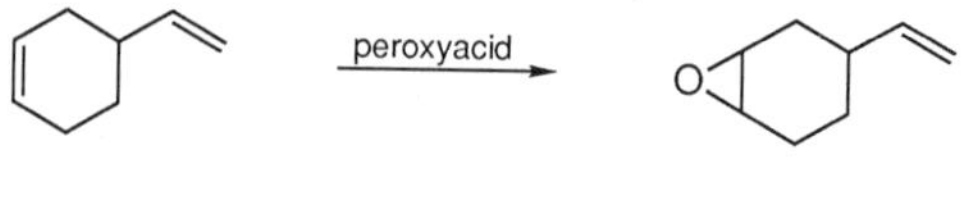

- It is stereospecific, as expected from the above mechanism; for example, *cis*-2-butene gives only the *cis* product:

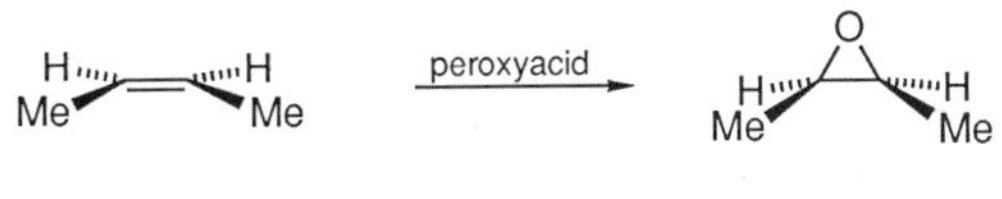

- Cyclic alkenes are attacked predominantly from the less hindered side, e.g.

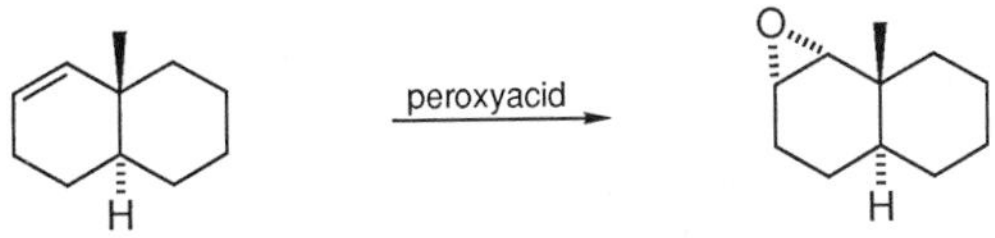

The peroxyacid can be prepared *in situ* from the parent acid and hydrogen peroxide. However, it is generally more convenient to use a commercially available compound. One which has been widely used and gives good yields is *m*-chloroperoxybenzoic acid but safety considerations have led to its replacement by the magnesium salt of monoperoxyphthalic acid. A peroxyacid from a relatively strong acid such as CF_3CO_2H is unsuitable since the acid released during epoxidation is sufficiently strong to bring about ring opening of the epoxide (see p. 591).

αβ-Unsaturated carbonyl compounds. These can be expoxidized by an alkaline solution of hydrogen peroxide. Nucleophilic addition of HO_2^- to C=C is facilitated by the C=O group (cf. Michael reaction, p. 240):

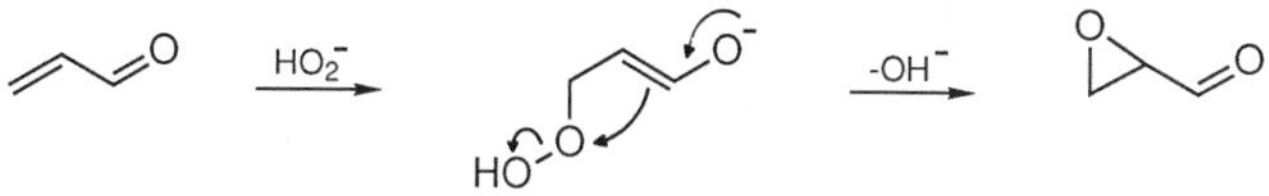

Enantioselective epoxidation: Sharpless reaction. An alkene which is achiral inevitably yields a non-optically active product with an achiral oxidant. However, allylic alcohols give mainly one enantiomer of a pair when epoxidized with t-butyl hydroperoxide in the presence of $Ti(OCHMe_2)_4$ and a chiral dihydroxy compound such as (2*R*,3*R*)-diethyl tartrate, e.g.

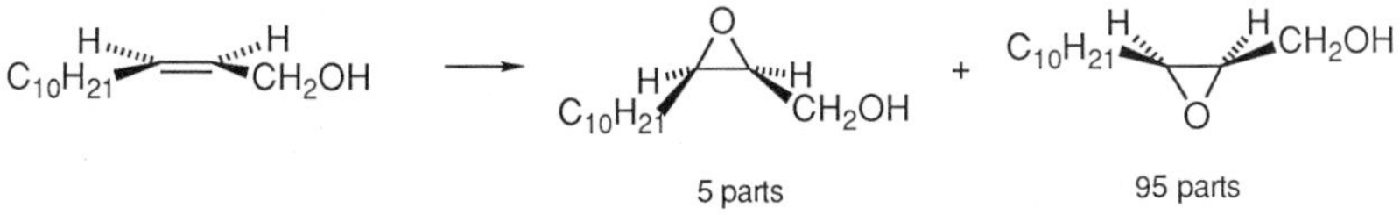

The (2*S*,3*S*)-tartrate gives the same enantiomers but in reversed proportions. It is thought that oxygen is transferred from the hydroperoxide to the alkene within a complex in which the configurations of the two ester groups determine whether the oxygen is delivered mainly from above the plane of the alkene or from below it.

Reactions of epoxides. The value of epoxidation lies in the variety of synthetic pathways readily available by cleavage of the strained ring:

(1)   Reduction (lithium aluminium hydride) gives an alcohol (p. 656)

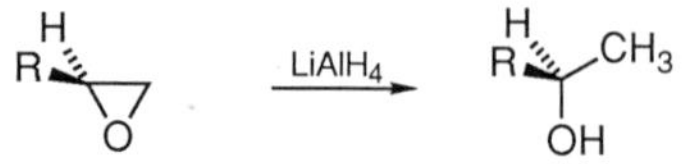

(2)   Treatment with a Lewis acid gives a carbonyl compound, e.g.

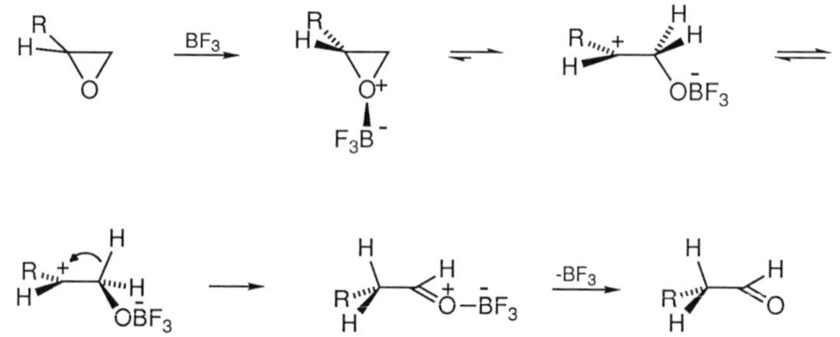

(3)   Treatment with dimethyl sulfoxide gives an α-ketol (cf. p. 610):

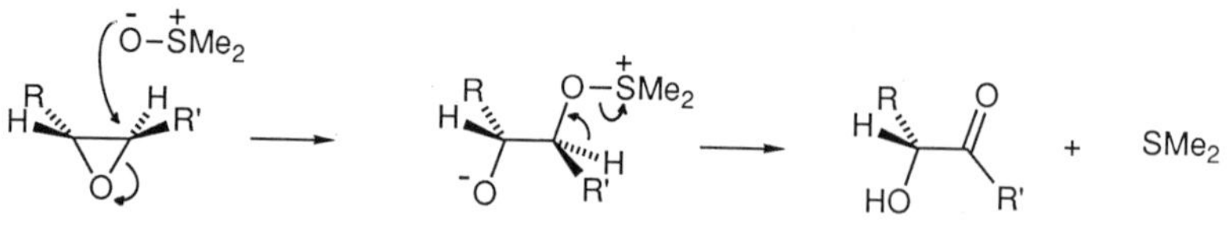

(4)   Hydrolysis gives a 1,2-diol. Either acidic or basic conditions can be used; in the former, the reactivity of the epoxide to a weak nucleophile (water) is enhanced by protonation of the ring oxygen. The product has *trans* stereochemistry as a result of the stereospecificity of S_N2 displacements (p. 160):

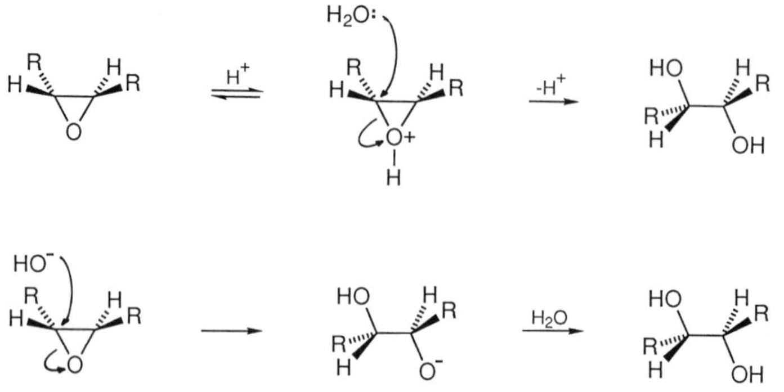

Other weak nucleophiles such as alcohols and bromide ion are also effective when the epoxide's reactivity is increased by protonation. As in other S_N2 reactions, the less alkylated carbon atom is the more reactive, e.g.

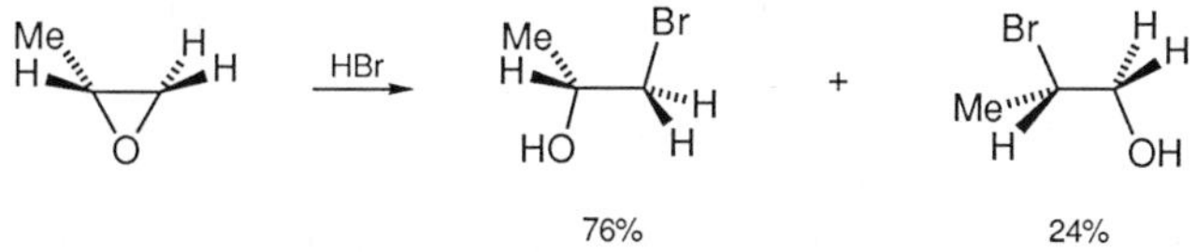

However, if ring opening of the protonated epoxide gives a carbocation that is strongly stabilized (e.g. by a phenyl substituent), the S_N1 mechanism is favoured. Consequently the stereospecificity is lost, and the nucleophile attacks at the substituted carbon.

The ring opening of epoxides of rigid cyclohexene systems gives *trans*-diaxial products:

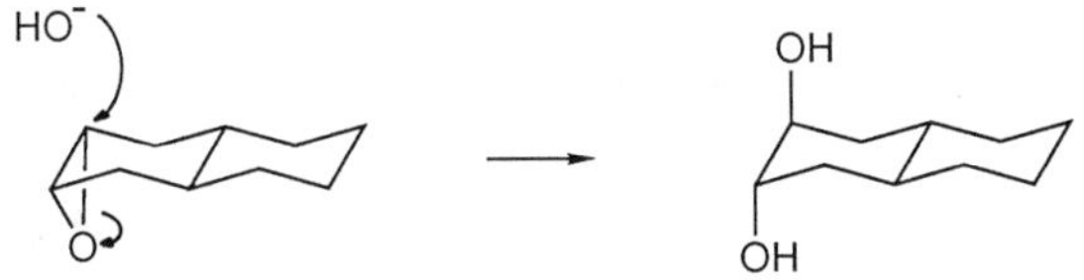

Industrially, ethylene oxide is made by the oxidation of ethylene with atmospheric oxygen at pressures up to 300 p.s.i. over a silver catalyst at 200–300°C. Its most important use is in the production of ethylene glycol, required, for example, for the manufacture of terylene.

(*ii*) *Diol formation.* As well as the method for obtaining *trans*-1,2-diols *via* epoxides, there are three methods for converting alkenes directly into diols.

1. Osmium tetroxide. The addition of an alkene to osmium tetroxide in ether causes the rapid precipitation of a cyclic osmate ester. Pyridine, which complexes the osmium atom in the ester, is often added as a catalyst. The ester is then hydrolyzed, commonly with aqueous sodium sulfite, to give a *cis*-1,2-diol.

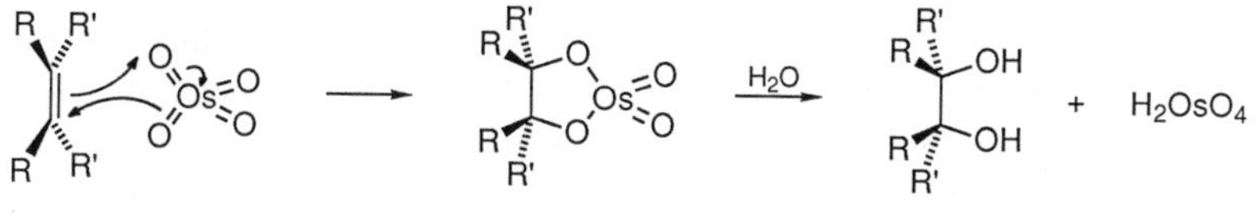

As well as its *syn*-stereospecificity, its other characteristics are the same as those of epoxidation:

- It reacts with the most electron-rich double bond when more than one is present, so that it can be used regioselectively.
- It attacks rigid cyclic systems from the less hindered side, thereby yielding the more stable of the two possible *cis*-diols (compare the iodine-silver acetate method below), e.g.

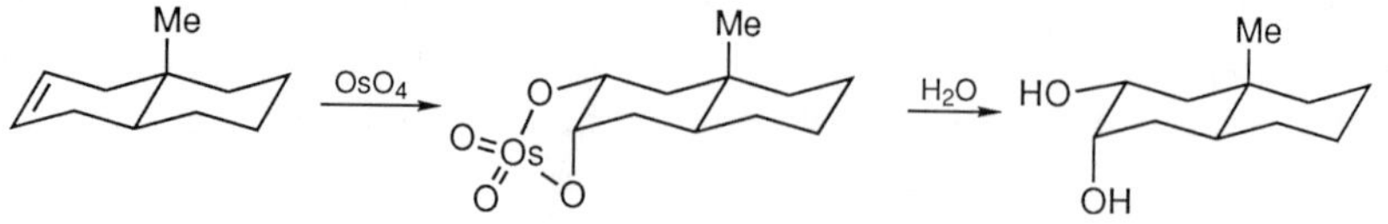

Because of both its expense and its toxicity it is best to use osmium tetroxide in catalytic quantities by carrying out the reaction in the presence of a tertiary amine oxide or alkaline t-butyl hydroperoxide; the intermediate osmate ester regenerates the tetroxide, e.g.

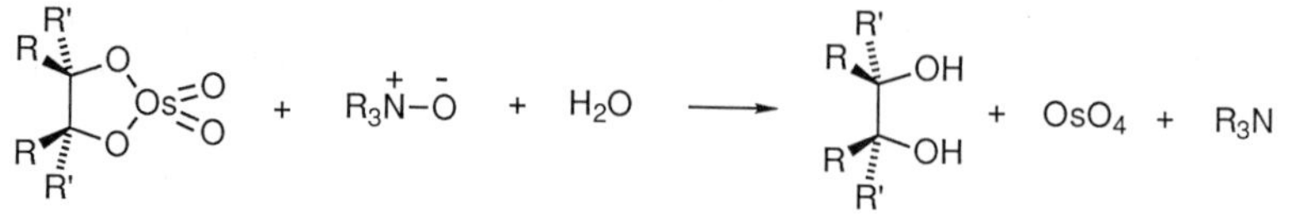

2. *Potassium permanganate acts similarly to osmium tetroxide:*

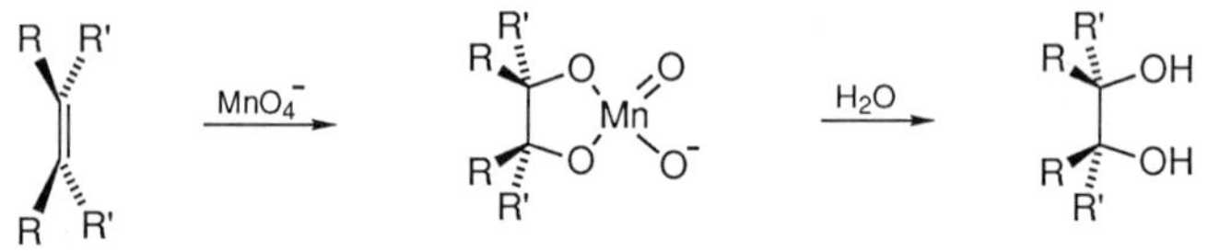

It is normally used in an aqueous solution in which the organic compound is dissolved or suspended; a co-solvent (usually t-butanol or acetic acid) is sometimes employed. In order to prevent further oxidation it is necessary to work in alkaline conditions, for otherwise the *cis*-diol is oxidized further with formation of an α-hydroxyketone or by cleavage of the C—C bond. In general, permanganate is less selective, and therefore less satisfactory, than osmium tetroxide, but it has the advantage of being less hazardous to use and very much cheaper.

3. *Iodine-silver acetate ('wet')*. This method also yields *cis*-diols. The alkene is treated with iodine in aqueous acetic acid in the presence of silver acetate. Iodine reacts with the double bond to give an iodonium ion which undergoes displacement by acetate in the S_N2 manner, giving a *trans*-iodoacetate. Anchimeric assistance by the acetate group, together with the powerful bonding capacity of silver ion for iodide, then lead to the formation of a cyclic acetoxonium ion which in turn reacts with water to give a *cis*-hydroxyacetate. Final hydrolysis gives the *cis*-diol:

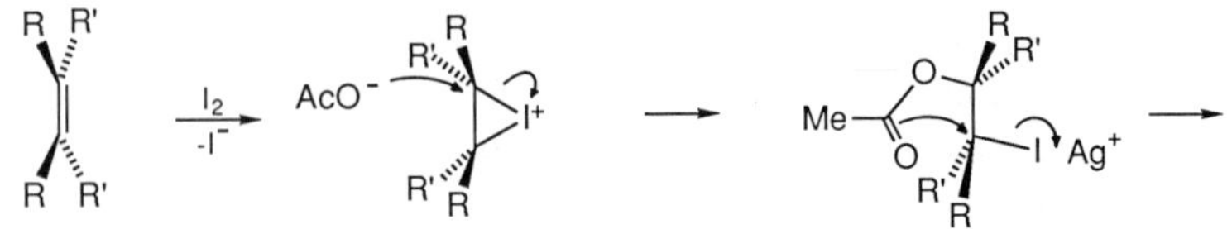

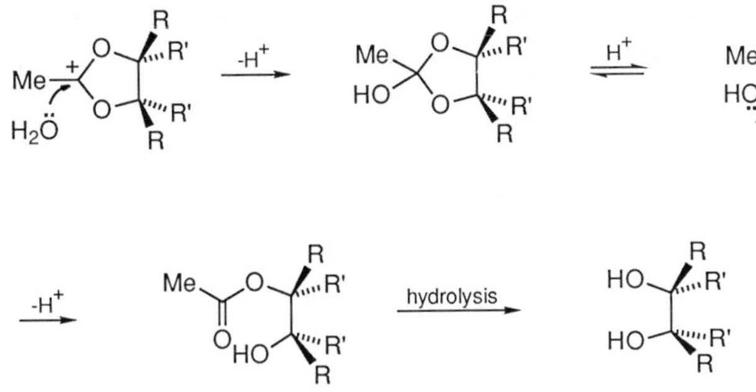

In a rigid cyclic alkene, approach of the iodine from the less hindered side leads ultimately to the *less* stable *cis*-diol, e.g.

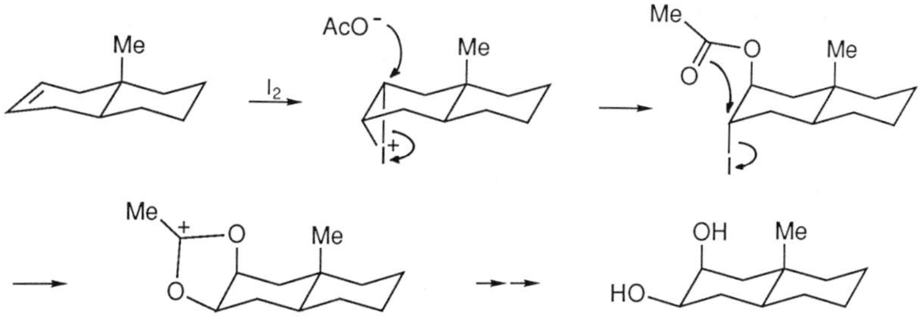

4.   Iodine-silver acetate ('dry').   The same reaction carried out in the absence of water leads to a *trans*-1,2-diacetate from which the *trans*-diol is obtained by hydrolysis (*Prévost reaction*):

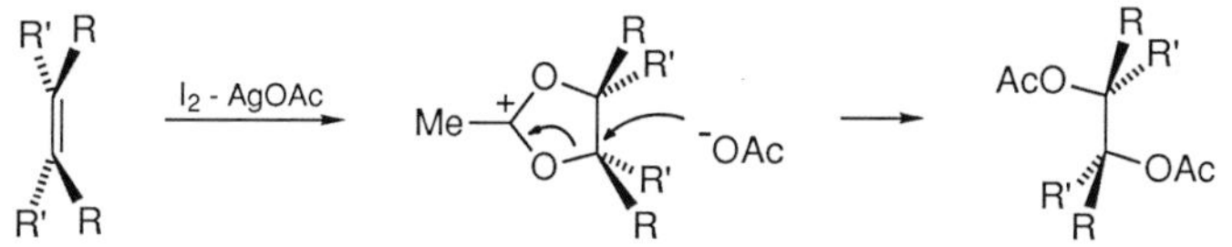

(iii)   Cleavage of the double bond.

1.   Ozonolysis.   Ozone reacts as an electrophile with alkenes forming a primary ozonide which rearranges through a zwitterionic intermediate to an isolable ozonide:

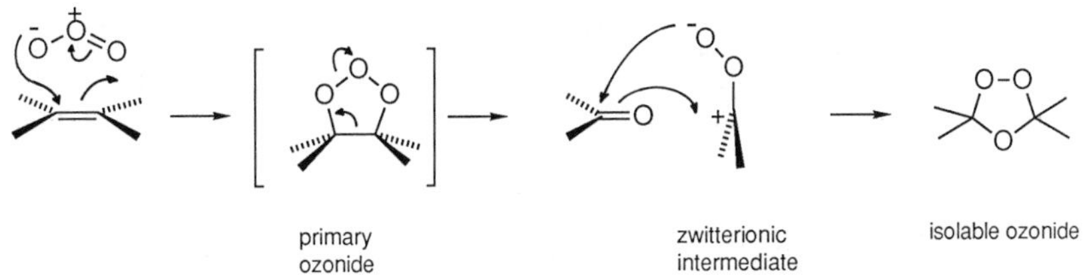

primary
ozonide

zwitterionic
intermediate

isolable ozonide

Ozonides are dangerously explosive, however, and are usually converted at once into products. Direct solvolysis gives ketones and/or acids, depending on the structure of the alkene e.g.

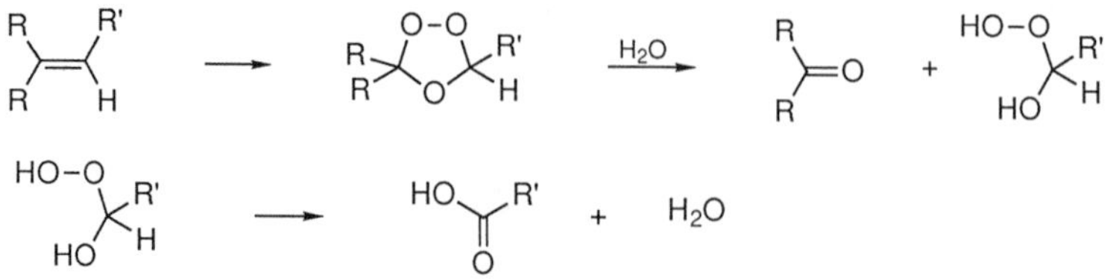

It is often more useful to carry out the solvolysis in the presence of a reducing agent. With zinc and acetic acid or dimethyl sulfide, only carbonyl compounds are formed; dimethyl sulfide has the advantage that other reducible groups in the compound are not affected and its suitability stems from its ability to reduce hydroperoxy compounds:

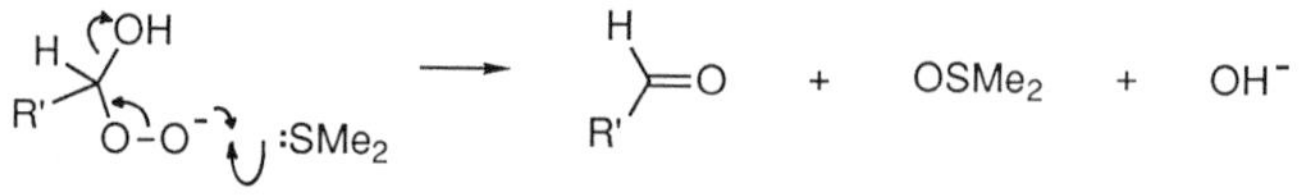

For example,

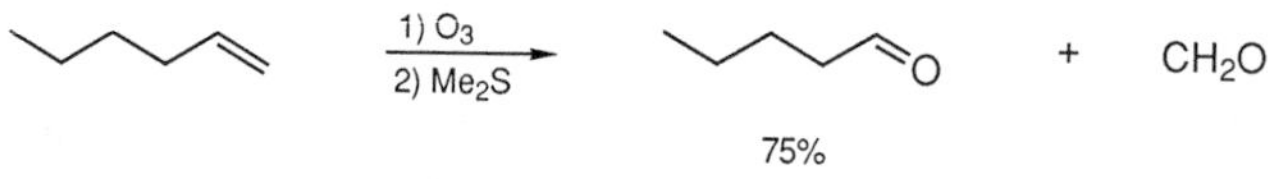

With lithium aluminium hydride, further reduction occurs to give alcohols, e.g.

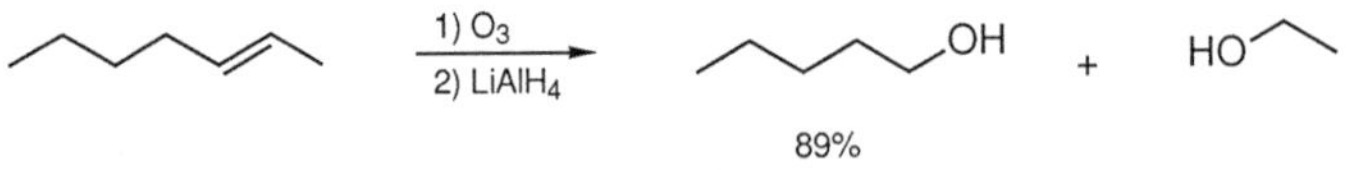

Alkynes are also oxidized by ozone but they generally react at only about one-thousandth the rate of alkenes just as they are less reactive than alkenes towards other electrophiles (p. 87). Selective oxidation of C=C in the presence of C≡C can therefore be achieved. Reaction with alkynes gives carboxylic acids, together with small amounts of α-dicarbonyl compounds, probably by a mechanism analogous to that for alkenes:

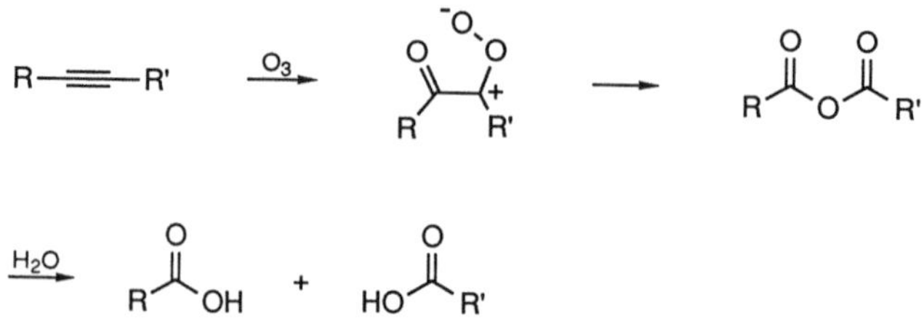

Aromatic rings also undergo ozonolysis (p. 596).

2. *The Lemieux reagents.* Ozone is unpleasant to handle and is not selective for alkenes, e.g. secondary alcohols are oxidized to ketones and tertiary C—H bonds to alcohols. Ozone has therefore largely been displaced by the Lemieux reagents which consist of dilute aqueous solutions of sodium periodate with a catalytic quantity of potassium permanganate and of osmium tetroxide, respectively. In each case, the alkene is first oxidized to the *cis*-diol which is then cleaved by the periodate (p. 616) to give aldehydes and/or ketones. The permanganate reagent then oxidizes aldehydic products to carboxylic acids. The low-valent states of manganese and osmium generated during the reaction are re-oxidized by the periodate to their original state so that only catalytic quantities are required. The reactions are rapid at room temperature and are selective for alkenes, e.g.

$$\text{(alkene)} \xrightarrow{\text{NaIO}_4 - \text{KMnO}_4} \text{(ketone)} + \text{MeCO}_2\text{H}$$

$$\text{(alkene)} \xrightarrow{\text{NaIO}_4 - \text{OsO}_4} \text{(ketone)} + \text{MeCHO}$$

3. *Chromium(VI) oxide.* The cleavage of C=C bonds with chromic oxide is competitive with oxidation of allylic C—H bonds, e.g. cyclohexene gives a mixture of 3-cyclohexenone and adipic acid:

$$\text{(cyclohexene)} \xrightarrow{\text{CrO}_3} \text{(cyclohexenone)} + \text{(adipic acid)}$$

The use of a partially aqueous medium favours the cleavage process, whereas an anhydrous medium such as glacial acetic acid favours allylic oxidation. In addition, cleavage is promoted by the presence of phenyl substituents, evidently because the first step involves the formation of a carbocation which is stabilized by adjacent aromatic rings:

$$\text{Ph}\diagup\diagdown + \text{O=Cr-OH} \longrightarrow \text{Ph}\diagup\diagdown\text{O-Cr-OH}$$

(b) *Aromatic rings*

The oxidation of unsubstituted aromatic rings which results in the loss of the associated stabilization energy requires vigorous conditions. Reaction can result either in the cleavage of the ring or in the formation of quinones.

Ozone effects cleavage, e.g.

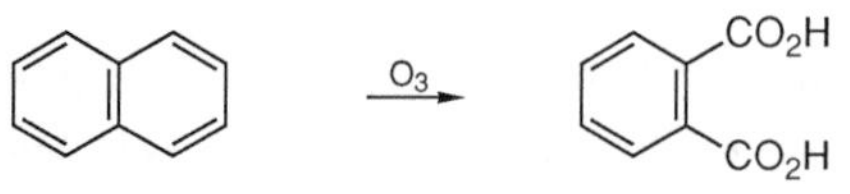

phthalic acid (88%)

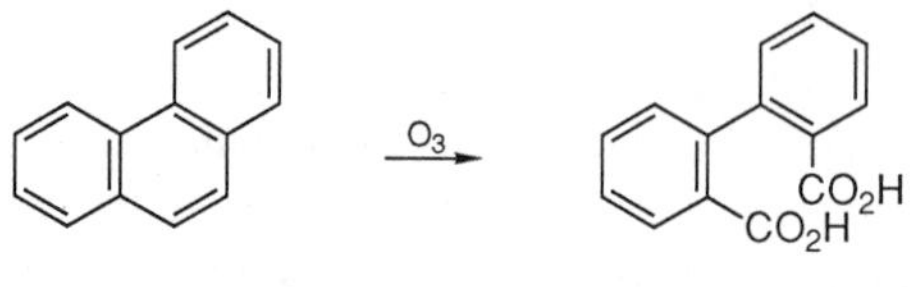

diphenic acid (65%)

An industrial method, which is far cheaper than ozonolysis, employs aerial oxidation over a vanadium pentoxide catalyst at 400–500°C. For example, maleic anhydride and phthalic anhydride are cheaply available from benzene and naphthalene:

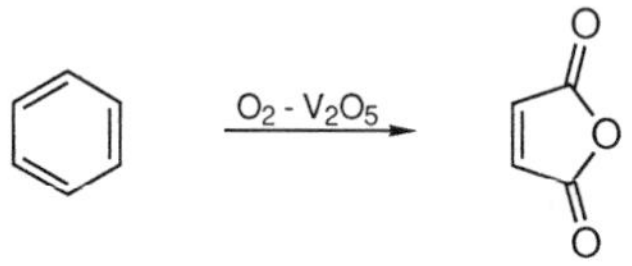

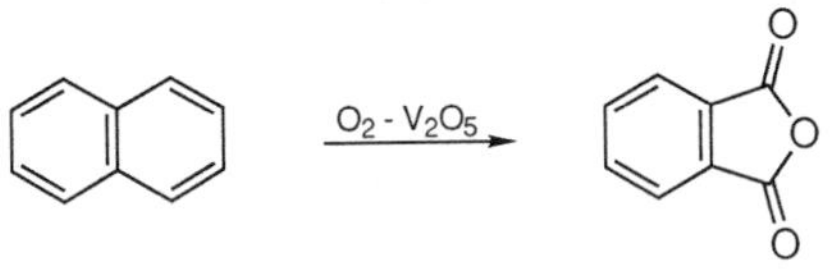

The latter process is particularly important since some of the derived diesters (e.g. dinonyl phthalate) are used as plasticizers for cellulose acetate and polyvinyl chloride. Polyester resins and phthalocyanine dyes are also manufactured from phthalic anhydride.

Chromium(VI) oxide can also be used: for example, quinoline is oxidized to pyridine-2,3-dicarboxylic acid which readily loses the 2-carboxy-substituent on being heated (p. 389), providing an easy route to nicotinic acid:

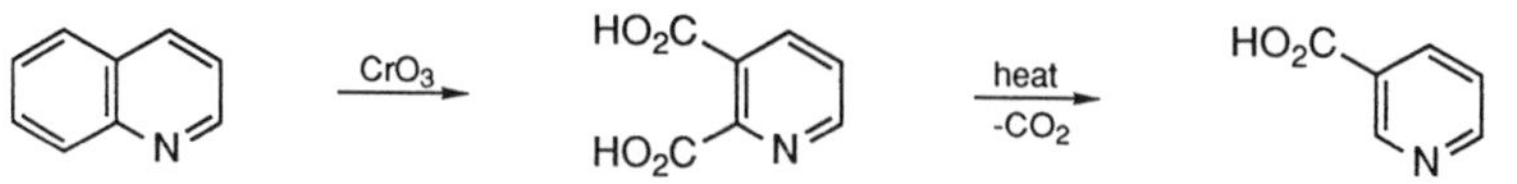

nicotinic acid

Chromium(VI) oxide does not invariably lead to ring fission. For example, it reacts with naphthalene in acetic acid solution at room temperature to give 1,4-naphthoquinone in about 20% yield [4]:

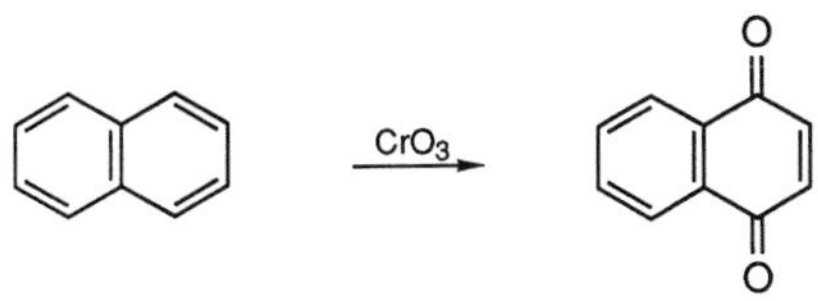

The aromatic rings of phenols are very susceptible to oxidation by one-electron oxidants, for the removal of a hydrogen atom gives a delocalized aryloxy radical,

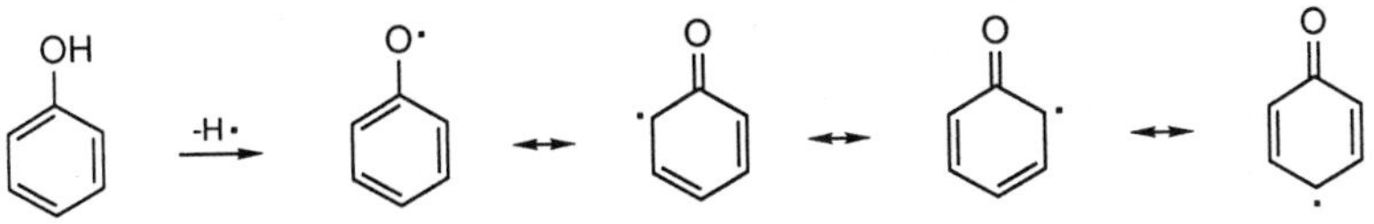

The fate of the radical depends on the structure of the phenol. Diphenols can be formed, as in the oxidation of 2-naphthol by iron(III) ion,

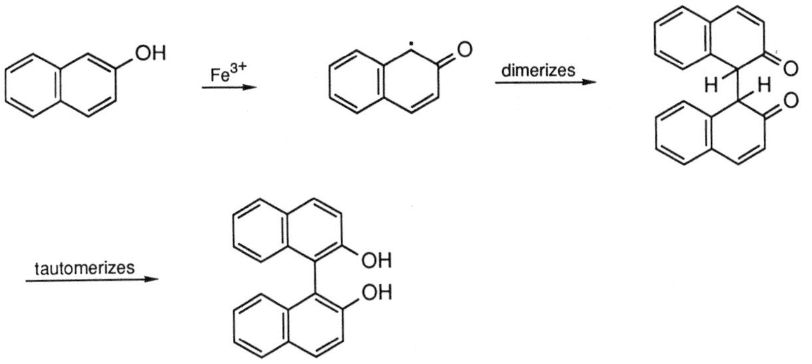

further oxidation to quinones can occur, e.g.

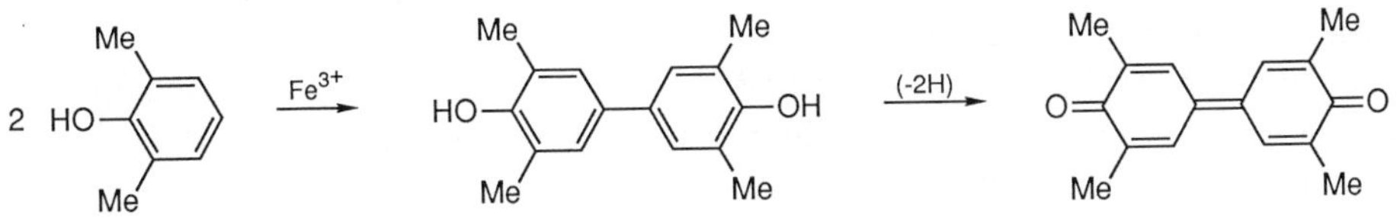

and cyclization to derivatives of dibenzofuran can arise through intramolecular nucleophilic attack on an intermediate quinone. For example, the oxidation of *p*-cresol by hexacyanoferrate(III) ion gives, in addition to carbon–carbon dimers, Pummerer's ketone:

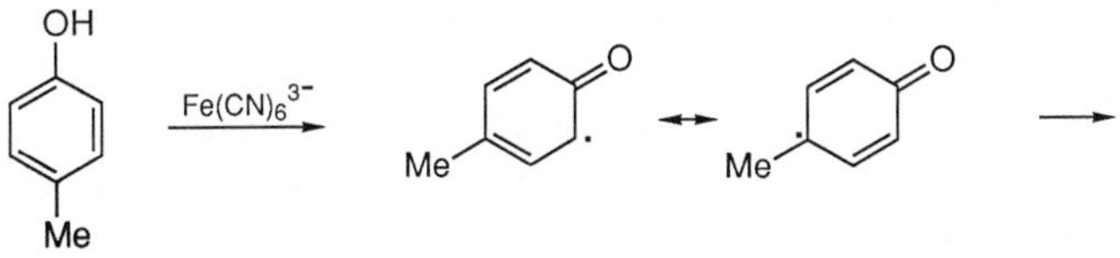

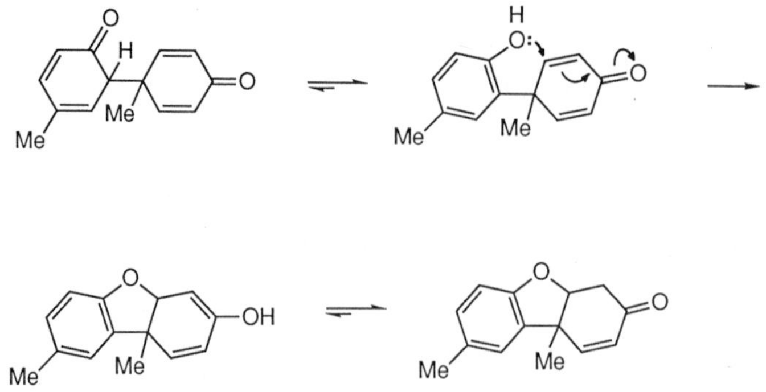

Pummerer's ketone

A synthesis of usnic acid is based on this mode of coupling. Hexacyanoferrate(III) oxidation of methylphloroacetophenone gives a dimer which is dehydrated by concentrated sulfuric acid to (±)-usnic acid:

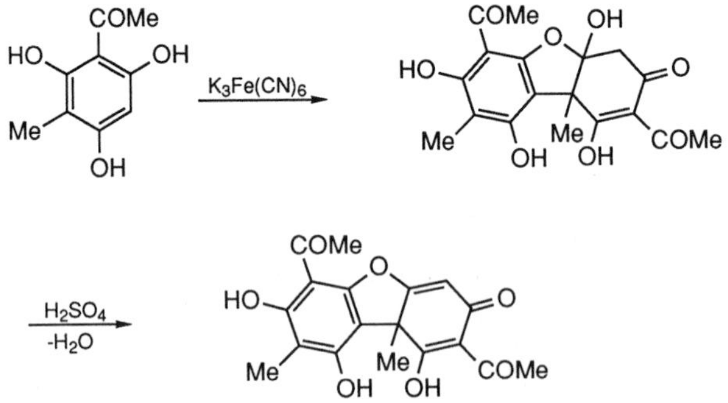

(±)-usnic acid

Aromatic amines are also sensitive to oxidation and discolour in air. The products are complex and the processes are of little use except in two cases: first, specific methods for oxidizing the amino group rather than the aromatic ring are available (p. 621) and second, aniline is oxidized to *p*-benzoquinone in 60% yield by dichromate:

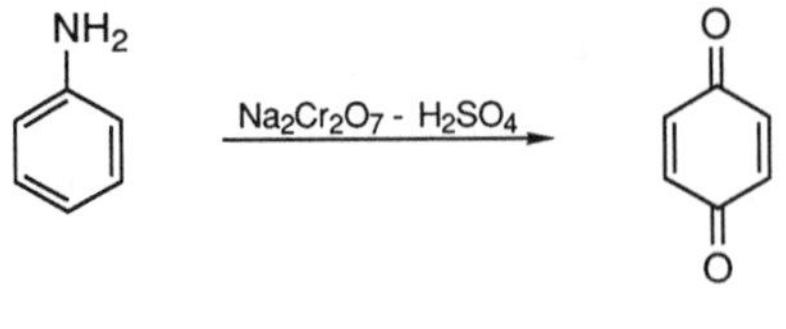

Industrially, aniline is oxidized to *p*-benzoquinone with manganese dioxide and sulfuric acid.

The *ortho* and *para* dihydric phenols are readily oxidized to the corresponding quinones by one-electron oxidants. Reaction occurs through delocalized semi-quinone radicals, e.g.

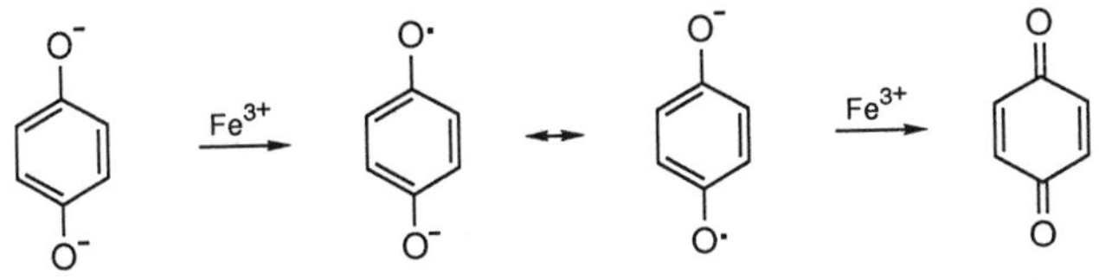

o-Benzoquinone may be prepared by the oxidation of catechol (from sali-cylaldehyde by Dakin's reaction; p. 447) with silver oxide suspended in ether; sodium sulfate is added as a dehydrating agent since the quinone is rapidly attacked by water (addition to the αβ-unsaturated carbonyl system):

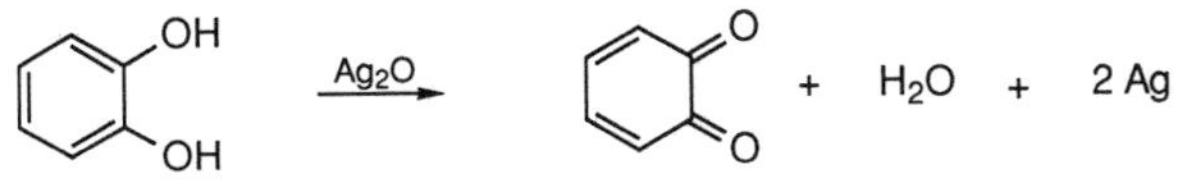

The corresponding aminophenols react similarly. For example, 1-amino-2-naphthol (readily available from 2-naphthol by nitrosation or diazo-coupling fol-lowed by reduction, p. 380) is oxidized by iron(III) chloride to 1,2-naphthoquinone in over 90% yield [2]:

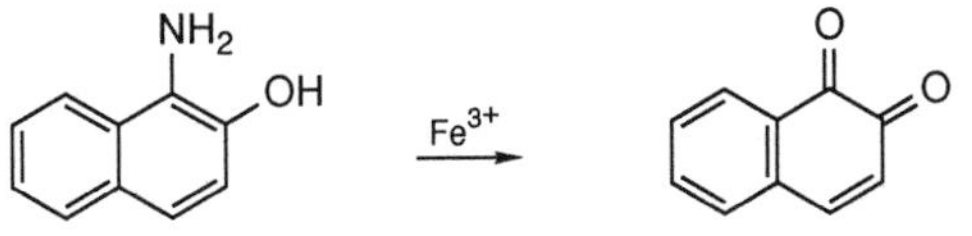

1,2-naphthoquinone

(c)   Saturated C—H groups

(i)   Allylic and benzylic systems.   The comparative stability of allyl and benzyl radicals, e.g.

$$CH_2{=}CH{-}\dot{C}H_2 \quad \longleftrightarrow \quad \dot{C}H_2{-}CH{=}CH_2$$

renders allylic and benzylic systems susceptible to oxidation *via* free-radical reac-tions. Autoxidation has been described (section 17.5). An alternative approach is through radical-catalyzed halogenation (p. 531). For example, the chlorination of toluene in the vapour phase or under reflux gives benzyl chloride, benzal chloride, and benzotrichloride, from which benzyl alcohol, benzaldehyde, and benzoic acid, respectively, are available by hydrolysis: the alcohol and aldehyde are produced in this way on an industrial scale. However, the difficulty of separating the halogenated products efficiently in the laboratory necessitates the use of more specific methods.

Oxidation to alcohols.   Toluene can be oxidized to benzyl alcohol with sulfuryl chloride in the presence of a radical initiator such as a peroxide followed by hydrolysis of benzyl chloride,

$$PhMe \quad + \quad SO_2Cl_2 \quad \xrightarrow[\text{-HCl, -SO}_2]{\text{peroxide}} \quad PhCH_2Cl \quad \xrightarrow{\text{hydrolysis}} \quad PhCH_2OH$$

A method of wide application for allyl systems is the use of selenium dioxide. It is thought to act by an 'ene' reaction (p. 295) followed by a [3,2]-sigmatropic rearrangement and hydrolysis of a selenium ester:

$$\text{hydrolysis} \longrightarrow \quad \diagdown\!\!\diagup^{OH} \quad + \quad Se(OH)_2$$

By including t-butyl hydroperoxide to reoxidize $Se(OH)_2$, selenium dioxide can be used in catalytic amount.

Oxidation to aldehydes.   Commercially, a particularly important example is the oxidation of propylene to propenal $CH_2{=}CH{-}CHO$, over copper(II) oxide at 300–400°C. The product is largely converted into glycerol (required for nitroglycerine), by reduction to allyl alcohol followed by hydroxylation with hydrogen peroxide. A development involves the air oxidation of propylene in the presence of ammonia over a catalyst (e.g. bismuth molybdate); propenal is probably formed first and reacts further:

$$\diagup\!\!\diagdown\!\!\diagup\!\!\diagdown_O \quad + \quad 1/2\,O_2 \quad + \quad NH_3 \quad \longrightarrow \quad \text{(acrylonitrile)} \quad + \quad 2H_2O$$

This process for acrylonitrile has superseded that from acetylene and hydrogen cyanide.

Three methods are available for the selective oxidation of $ArCH_3$ to $ArCHO$.

(1) *With chromyl chloride (Etard reaction)*. A solution of chromyl chloride in carbon disulfide is added cautiously to the benzylic compound at 25–45°C and the brown complex which separates is decomposed by water to give the aldehyde and chromic(VI) acid, e.g.

1) CrO₂Cl₂ / CS₂
2) H₂O

75%

The aldehyde must be removed rapidly by distillation or extraction to prevent its further oxidation.

(2) *With chromium(VI) oxide in acetic anhydride*. Oxidation is carried out with chromium(VI) oxide in a mixture of acetic anhydride, acetic acid, and sulfuric acid

at low temperature. As it is formed, the aldehyde is converted into its 1,1-diacetate which is stable to oxidation; this is isolated and reconverted into the aldehyde by acid hydrolysis. For example, *p*-nitrotoluene gives *p*-nitrobenzaldehyde in about 45% overall yield [2]:

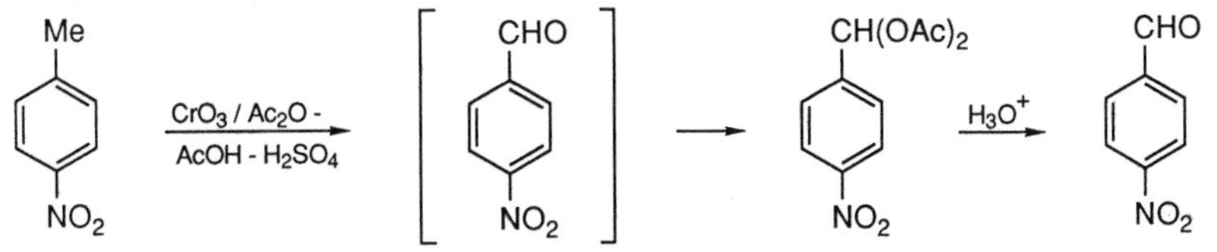

(3)   *With p-nitrosodimethylaniline.* The benzylic compound is treated with *p*-nitrosodimethylaniline and the resulting imine is hydrolyzed. The method is applicable only to those compounds whose methyl groups are strongly activated, e.g. 2,4-dinitrotoluene

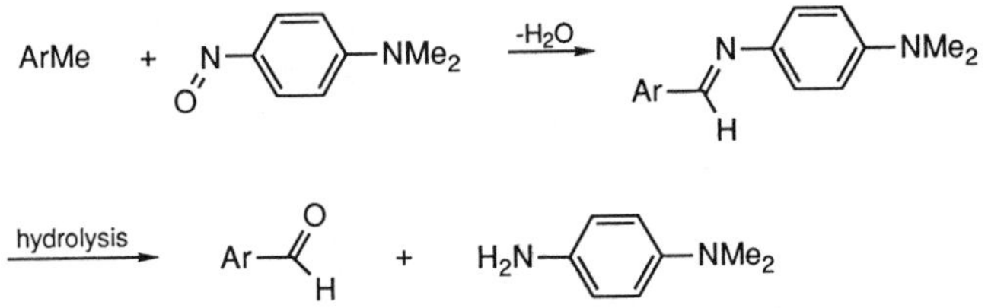

Oxidation to carboxylic acids.   Methyl groups on aromatic rings can be oxidized with chromium(VI) oxide, permanganate, or nitric acid. For example, dilute permanganate oxidizes *o*-chlorotoluene to *o*-chlorobenzoic acid in about 65% yield [2],

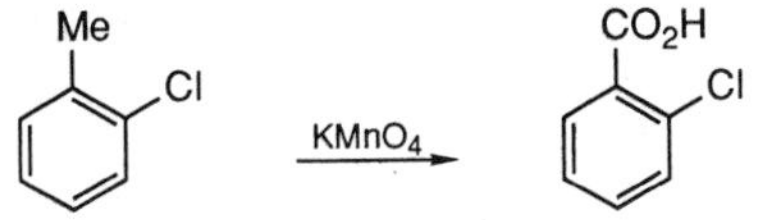

and concentrated nitric acid oxidizes *o*-xylene to *o*-toluic acid in 54% yield [3]:

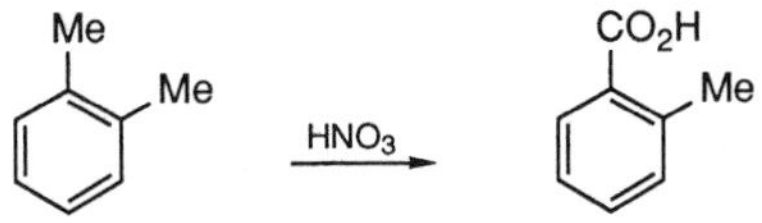

Substituents on the methyl group are also removed in these conditions, e.g. nicotine is oxidized to nicotinic acid by concentrated nitric acid at 70°C:

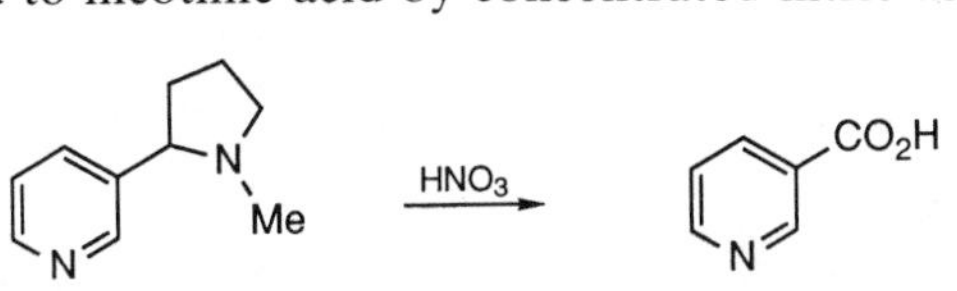

nicotinic acid

Industrially, benzoic acid is manufactured by the catalytic air oxidation of toluene. Its main use is in the production of sodium benzoate which is employed as a preservative for foodstuffs and pharmaceuticals. Benzoyl chloride is used to prepare dibenzoyl peroxide, used for initiating addition polymerization (and, on the laboratory scale, for other radical-catalyzed reactions) and as a bleaching agent.

$$2\ PhCOCl\ +\ H_2O_2\ \xrightarrow{\text{base}}\ \underset{\text{Ph}}{\overset{O}{\|}}\!\!-\!\!O\!-\!O\!-\!\underset{\text{Ph}}{\overset{O}{\|}}\ +\ 2\ HCl$$

(ii) The —CH₂—CO— system. Methylene groups adjacent to carbonyl may themselves be oxidized to carbonyl in two ways.

1. Through the oxime. The methylene group is activated by the carbonyl group towards reaction with organic nitrites in the presence of acid or base (p. 319). The resulting nitroso compound tautomerizes to the oxime which may be hydrolyzed to the α-dicarbonyl compound:

$$R\!-\!CH_2\!-\!CO\!-\!R'\ \xrightarrow{\text{RONO - H}^+\ (or\ OH^-)}\ \underset{\text{(nitroso)}}{R\!-\!CO\!-\!CH(N\!=\!O)\!-\!R'}\ \rightleftharpoons\ \underset{\text{(oxime)}}{R\!-\!CO\!-\!C(NOH)\!-\!R'}$$

$$\xrightarrow{\text{hydrolysis}}\ R\!-\!CO\!-\!CO\!-\!R'$$

Methylene groups are oxidized in preference to methyl. For example, butanone gives $CH_3COCOCH_3$.

2. By selenium dioxide (Riley reaction). The reaction is thought to occur as follows:

$$R\!-\!CO\!-\!CH_2\!-\!R'\ \rightleftharpoons\ R\!-\!C(OH)\!=\!CH\!-\!R'\ \xrightarrow{\text{SeO}_2}\ R\!-\!CO\!-\!CH(Se(O)OH)\!-\!R'$$

$$\xrightarrow{-H_2O}\ R\!-\!CO\!-\!C(=Se=O)\!-\!R'\ \xrightarrow{H_2O}\ R\!-\!CO\!-\!CO\!-\!R'\ +\ H_2SeO$$

For example, camphor reacts in refluxing acetic anhydride to give camphorquinone in 95% yield,

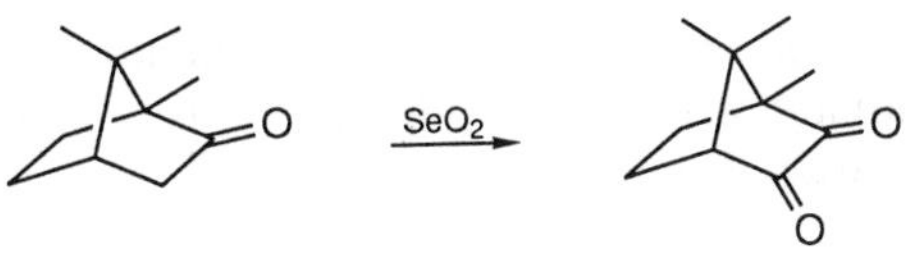

camphorquinone

and acetophenone reacts in dioxan to give phenylglyoxal in about 70% yield [2]:

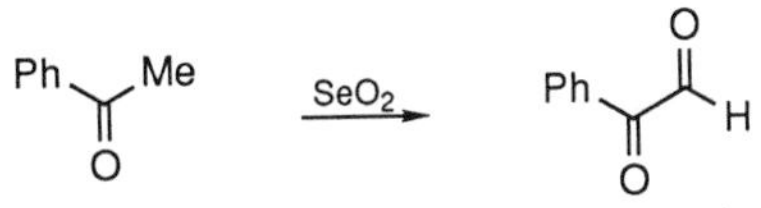

phenylglyoxal

In contrast to the oxime method, selenium dioxide preferentially oxidizes methyl rather than methylene groups. For example, butanone gives mainly CH_3CH_2COCHO. Selenium dioxide is of limited scope because it is unselective: it can bring about the following oxidations:

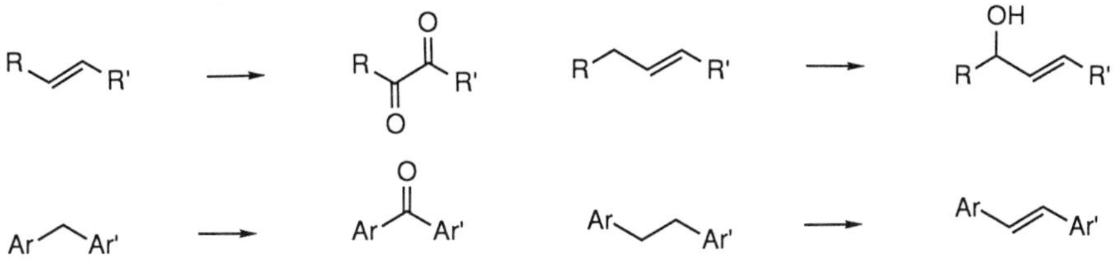

Oxidation of $-CH_2-CO-$ and $\rangle CH-CO-$ to $-CH(OH)-CO-$ and

$\rangle C(OH)-CO-$ can also be effected. The enolate is formed in the presence of a strong base such as potassium t-butoxide and reacts with oxygen to give a hydroperoxide (p. 557). This is reduced to the corresponding alcohol with a trialkyl phosphite.

(*iii*)   *The* $-CH_2-CH_2-CO-$ *and related systems.*   Dehydrogenation to give an αβ-unsaturated carbonyl compound can be brought about by acid-catalyzed bromination followed by base-catalyzed dehydrobromination:

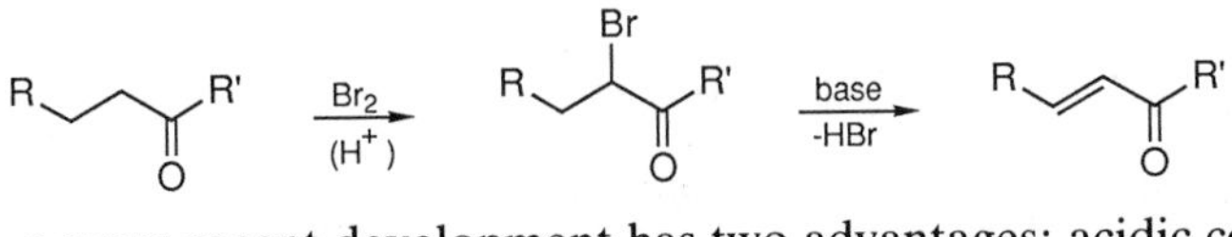

However, a more recent development has two advantages: acidic conditions are not required, and no base is present when the product is formed, so that possible base-catalyzed reactions are avoided. It consists of making either a sulfide or a selenide derivative and oxidizing with hydrogen peroxide to the sulfoxide or selenoxide which undergoes a ready thermal *syn*-elimination. The procedure *via* a sulfide has been described (p. 474). That *via* a selenide is usually carried out by generation of the enolate with LDA followed by treatment with benzeneselenyl bromide and then treatment with hydrogen peroxide:

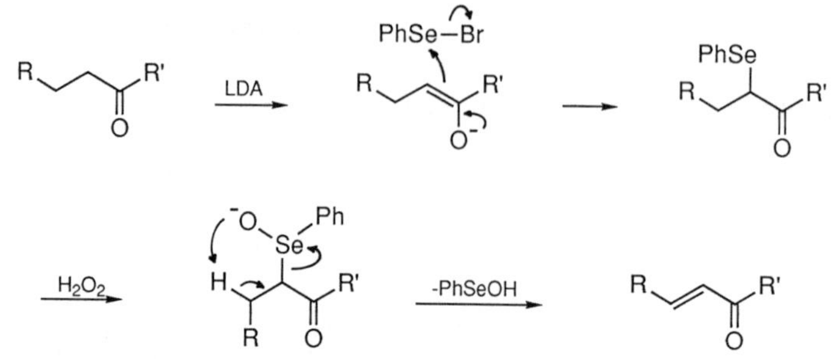

The carbonyl group may be ketonic or part of an ester function, and αβ-unsaturated nitriles also react. Yields are mostly very good, e.g.

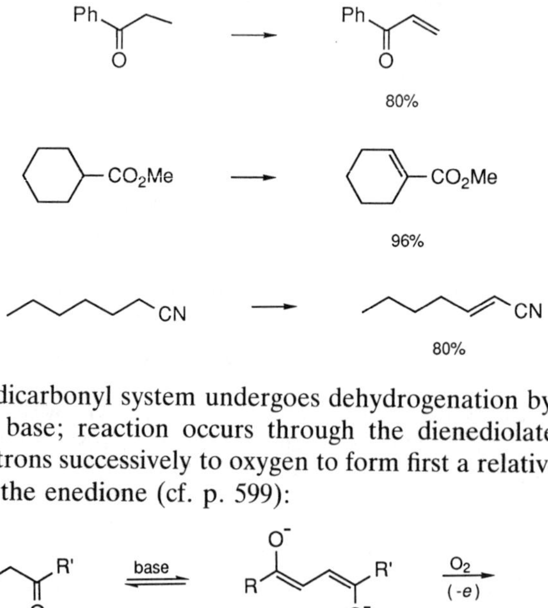

Finally, the γ-dicarbonyl system undergoes dehydrogenation by autoxidation in the presence of base; reaction occurs through the dienediolate dianion which donates two electrons successively to oxygen to form first a relatively stable anion-radical and then the enedione (cf. p. 599):

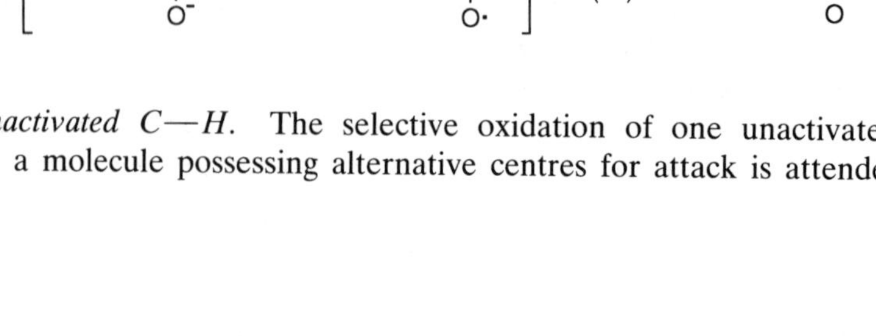

(*iv*) *Unactivated C—H.* The selective oxidation of one unactivated C—H group in a molecule possessing alternative centres for attack is attended by the difficulty that the reagents with which C—H bonds react – essentially only free radicals – are relatively unspecific. However, selectivity is obtained in two circumstances. First, a relatively unreactive free radical, such as the bromine atom (p. 532), discriminates quite sharply in favour of tertiary C—H compared with primary or secondary C—H, and the oxidation of tertiary C—H to C—OH can sometimes be achieved directly with alkaline permanganate; reaction occurs with retention of configuration:

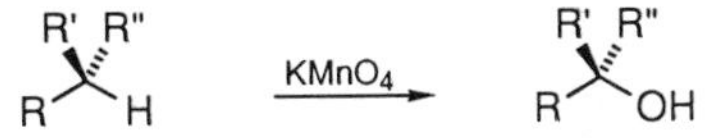

Second, intramolecular free-radical abstraction reactions occur specifically through six-membered cyclic transition states, so that it is possible selectively to oxidize δ-CH bonds, e.g.

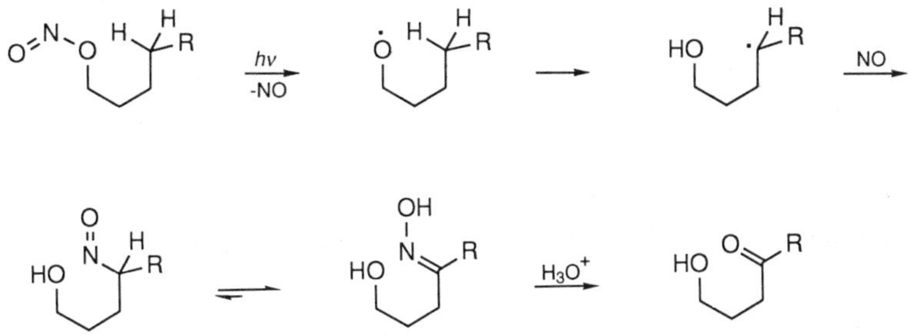

Examples of the application of this principle have been described earlier (p. 554).

The very vigorous oxidation of hydrocarbon chains by chromium(VI) oxide in concentrated sulfuric acid oxidizes most substances to carbon dioxide and water, but C-methyl groups give mainly acetic acid. This procedure has often been applied in determining the number of C-methyl groups in a compound of unknown structure (Kuhn–Roth method).

Regioselective dehydrogenation. A method for introducing a C=C bond in a steroid, with good regioselectivity, can be illustrated by the following conversion of 3α-cholestanol:

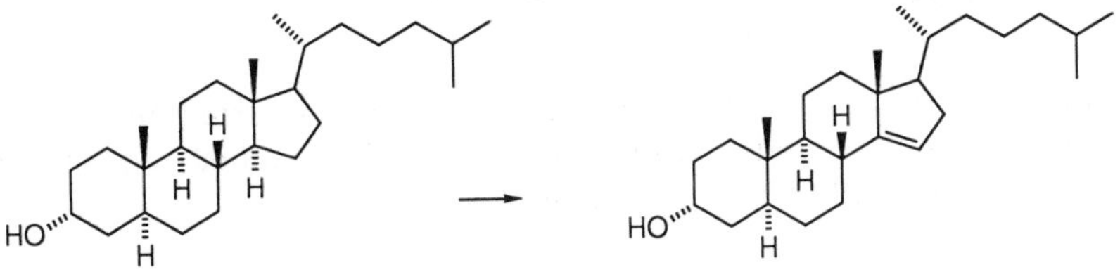

It is accomplished by irradiation of the ester shown; the geometry is such that the ketone's triplet can come to within bonding distance only of the hydrogen atom at C_{14}, and abstraction of that hydrogen followed by a second hydrogen gives the unsaturated ester from which the product is obtained by hydrolysis.

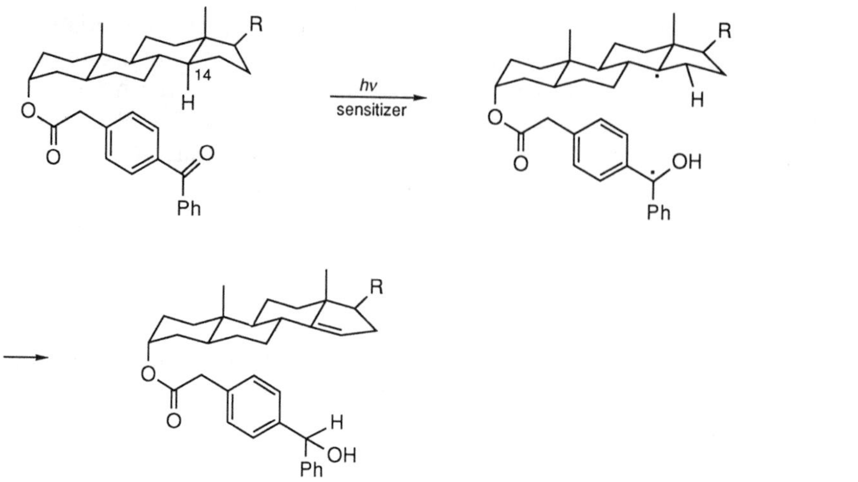

The method is capable of adaptation to other regioselectivities by varying the nature of the esterifying acid.

(v) *Aromatization.* Alicyclic compounds which are reduction products of aromatic systems can be dehydrogenated to the respective aromatics in several ways.

1. With sulfur or selenium. Reaction occurs with sulfur at about 200°C and with selenium at about 250°C, the hydrogen being removed as hydrogen sulfide or hydrogen selenide. Skeletal rearrangements may occur and carbon atoms may also be removed; in particular, angular methyl groups are degraded, e.g.

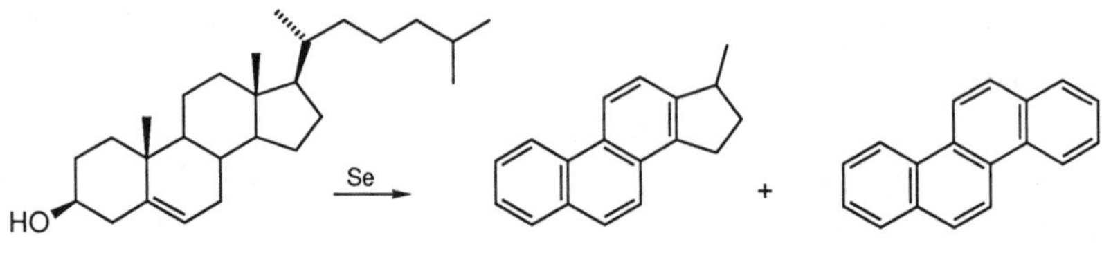

cholesterol   Diels' hydrocarbon   chrysene

Selenium is less destructive than sulfur and is usually preferred but the methods are of more value in degradation than in synthesis. Distillation with zinc dust has essentially the same effect but is more destructive.

2. Catalytically. Alicyclic rings which contain some unsaturation can be dehydrogenated on those catalysts which are successful for hydrogenation (section 19.2): palladium on charcoal or asbestos is the most commonly used. The conditions are far milder than those using selenium and the procedure is widely applied, e.g. the reduced isoquinolines obtained by the Bischler–Napieralski synthesis are usually oxidized in this way (p. 706):

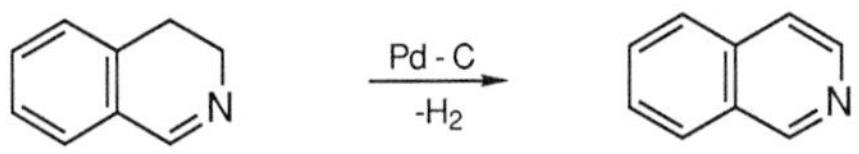

3. With quinones.   Partially unsaturated alicyclic rings are oxidized by quinones through hydride-ion transfer, e.g.

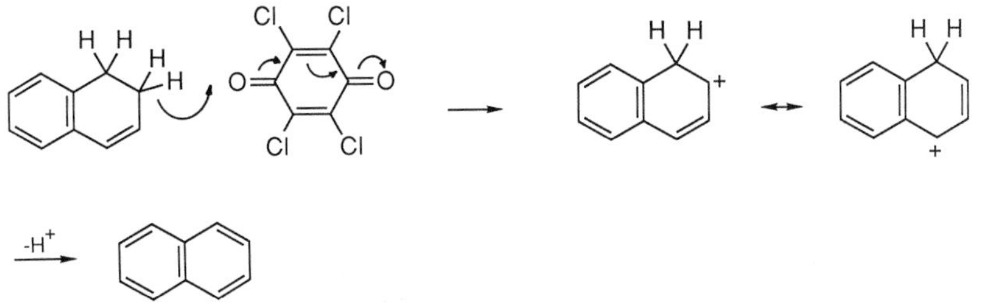

The driving force results from the conversion of both the quinone and the alicyclic system into aromatic compounds. Quinones which contain electron-releasing substituents are stabilized relative to their quinols, e.g.

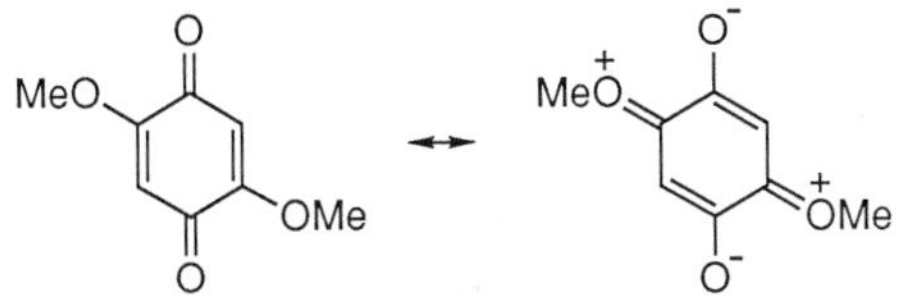

and are less powerful oxidants than those containing electron-attracting substituents. Chloranil* is frequently used, as in a synthesis of *p*-terphenyl,

---

* Chloranil is obtained by heating benzoquinone with potassium perchlorate and hydrochloric acid. Nucleophilic attack by chloride ion on the quinone gives a chloroquinol which is oxidized by the perchlorate to the chloroquinone. Successive reactions of this type give chloranil:

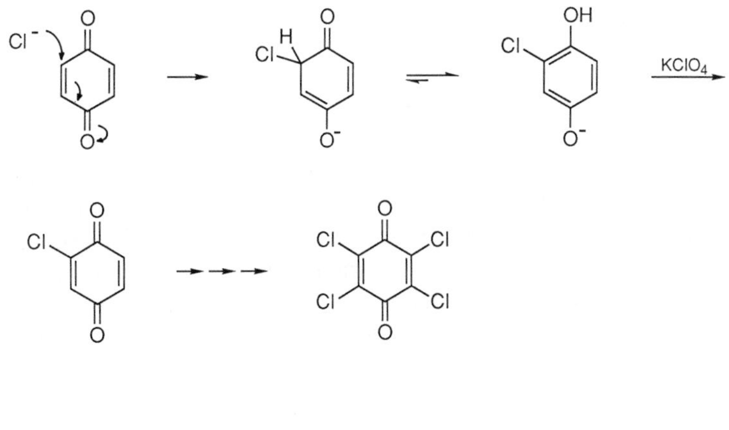

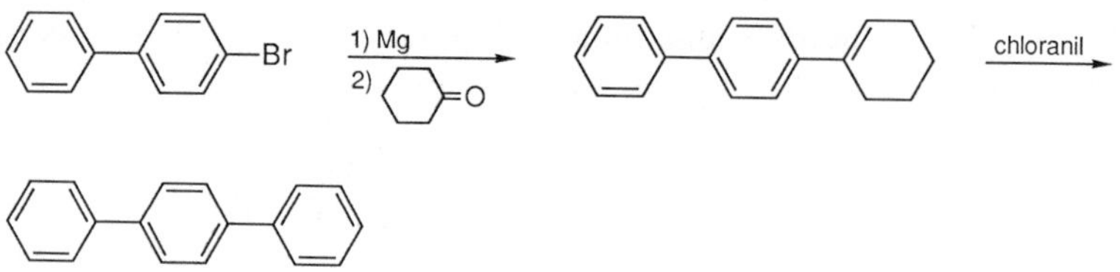

and 2,3-dichloro-5,6-dicyanobenzoquinone is a still more powerful reagent.

**19.3
Systems
containing
oxygen**   (*a*)   *Primary alcohols*

(*i*)   *To aldehydes.*   The oxidation of a primary alcohol to an aldehyde is attended by the difficulties that, first, aldehydes are readily oxidized to acids by
many reagents and, second, under acidic conditions the aldehyde reacts with
unchanged alcohol to give, in an equilibrium mixture, some of the hemiacetal
which is readily oxidized to an ester:

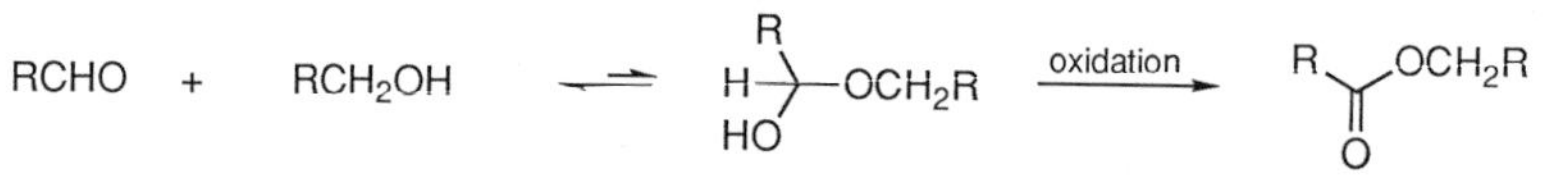

Mechanistically, one commonly employed principle is to attach a strongly
electrophilic group, X, to the oxygen of the alcohol, so that H^+ and X^- can be
eliminated,

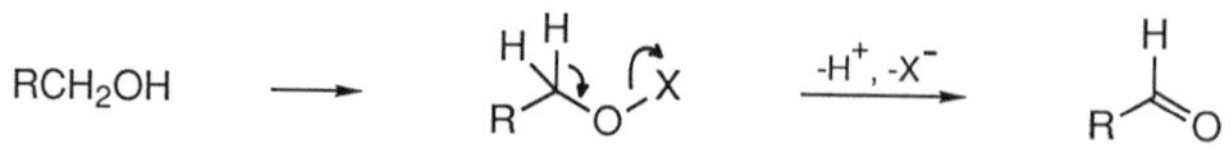

A second approach is to employ a hydride-accepting oxidant,

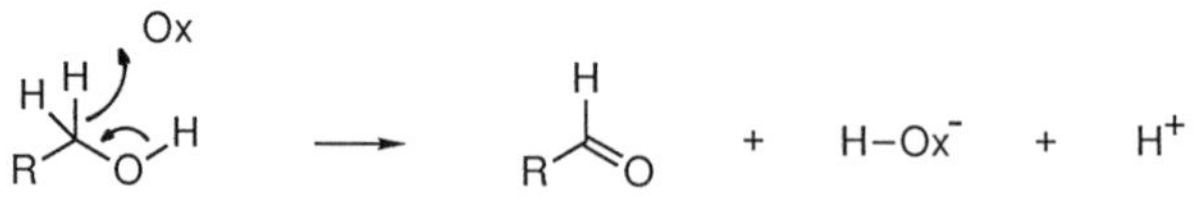

1.   Chromium(VI) oxide.   For the lower-boiling aldehydes, the simplest oxidation procedure is to add an acid solution of potassium dichromate slowly to the
alcohol. Use is made of the fact that aldehydes boil at lower temperatures than
the corresponding alcohols so that by maintaining the temperature above the
boiling-point of the aldehyde but below that of the alcohol, the aldehyde distils as
it is formed, e.g. propionaldehyde can be obtained in about 47% yield [2]:

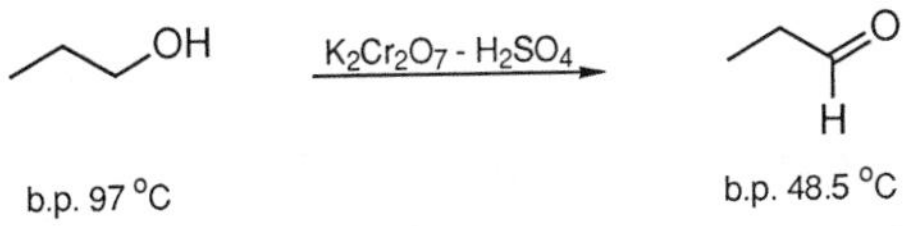

b.p. 97 °C

b.p. 48.5 °C

Reaction occurs through the chromate ester, *via* a cyclic transition state:

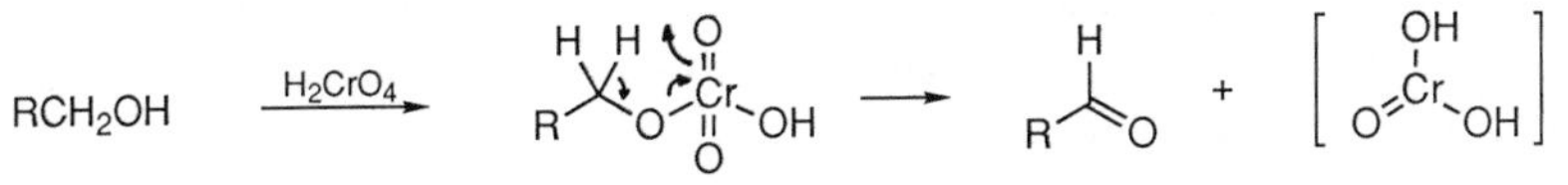

The chromium(IV) then disproportionates to give chromium(III) and chromium(VI).

The yields of aldehydes can often be increased by use of the complex, $CrO_3 \cdot 2C_5H_5N$, formed by chromium(VI) oxide with pyridine (cf. the use of pyridine-sulfur trioxide as a mild sulfonating agent, p. 381). Addition of the chromium(VI) oxide to pyridine is followed either by addition of the alcohol or, better, by isolation of the complex followed by oxidation in dichloromethane. For example, with the latter variant, 1-heptanol gives heptanal in 70–80% yield [6]:

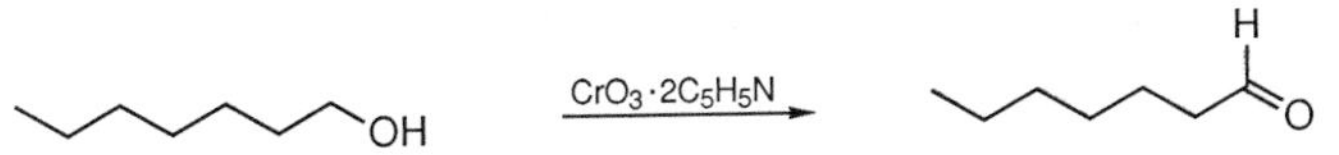

This method is particularly useful for compounds which contain acid-sensitive groups (e.g. acetals) or other easily oxidized groups (e.g. C=C bonds).

In a variant of this method, pyridinium chlorochromate, $C_5H_5NH^+ \, ClCrO_3^-$, which is precipitated when pyridine is added to a solution of chromic oxide in hydrochloric acid, is used in dichloromethane solution.

2. *Chlorine.* Chlorine oxidizes by the acceptance of hydride ion from the alcohol:

The hydrogen chloride produced can catalyze chlorination of the C—H bonds adjacent to the carbonyl group, so that the method is not in general suitable for the formation of simple aldehydes. However, a number of important compounds are available through the oxidation of ethanol with chlorine:

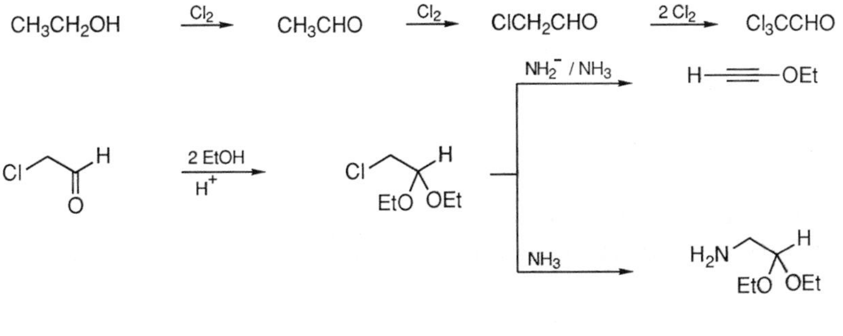

$$Cl_3CCHO \xrightarrow{OH^-} CHCl_3 \ + \ HCO_2^-$$

By passing chlorine into ethanol until the specific gravity reaches 1·025, mono-chloroacetaldehyde is obtained in good yield. Addition of more ethanol to the now acidic solution gives the corresponding acetal from which ethoxyacetylene and the amino-acetal (for the Pomeranz–Fritsch synthesis of isoquinolines, see pp. 707, 708) can be obtained. Further chlorination of monochloroacetaldehyde gives chloral and thence chloroform.

3. Dimethyl sulfoxide. There are two methods of using dimethyl sulfoxide. Either the sulfoxide is treated with an electrophile to form a species which is activated towards addition of the alcohol to the sulfur atom and which also possesses a good leaving group, or the alcohol is activated by conversion into its tosylate which reacts readily with the sulfoxide. In each case an alkoxysulfonium ion is formed which undergoes base-catalyzed elimination to give the carbonyl compound.

The first method can be illustrated by reaction with an acid chloride:

$$\xrightarrow{-RCO_2^-} \ Me_2\overset{+}{S}-OCH_2R' \xrightarrow{Et_3N} \ Me_2S \ + \ R'CHO \ + \ Et_3\overset{+}{N}H$$

Oxalyl chloride, ClCO—COCl, is a particularly effective electrophile (*Swern's method*); reaction occurs at temperatures as low as −60°C.

Dicyclohexylcarbodiimide is also a very effective reagent. An acid is added to catalyze the first step, and the addition of base to complete the reaction is unnecessary because an intramolecular hydrogen shift occurs spontaneously:

$$\longrightarrow \ R'CHO \ + \ Me_2S$$

In the second method (*Kornblum's method*), the tosylate is treated with dimethyl sulfoxide in the presence of sodium hydrogen carbonate for a few minutes at 150°C:

$$RCH_2-OTs \xrightarrow[-TsO^-]{Me_2SO} RCH_2-\overset{+}{O}SMe_2 \xrightarrow[-H^+]{base} RCHO + Me_2S$$

Since the first, and key, step is an S_N2 substitution, benzylic compounds, which are more reactive than alkyl derivatives towards nucleophiles, react at a lower temperature (100°C) and the very reactive α-bromo-ketones undergo oxidation directly, even though bromide ion is not as good a leaving group as toluene-*p*-sulfonate, e.g.

Amine oxides act similarly to dimethyl sulfoxide, and pyridine *N*-oxide in particular has been used to prepare aldehydes from halides and sulfonates:

4. *Catalytic dehydrogenation.* Dehydrogenation over copper or copper chromite occurs at about 300°C. Industrially, silver is employed as the catalyst for the production of formaldehyde and acetaldehyde. The dehydrogenation step, e.g.

$$CH_3OH \xrightarrow{Ag} CH_2O + H_2$$

is endothermic ($\Delta H > +100\,kJ\,mol^{-1}$), but in the presence of air a second type of oxidation occurs,

$$CH_3OH + 1/2\,O_2 \xrightarrow{Ag} CH_2O + H_2O$$

This is strongly exothermic ($\Delta H = -150\,kJ\,mol^{-1}$), so that by carefully controlling the air supply, a steady temperature is maintained and the reaction becomes self-supporting. Acetaldehyde is also manufactured from ethylene (p. 578).

(*ii*) *To carboxylic acids.* Primary alcohols can be oxidized directly to acids by reagents such as chromic(VI) acid, nitric acid, and potassium permanganate but in each case side reactions occur and yields are usually not high. For example, chromic(VI) acid degrades carboxylic acids to smaller molecules, ultimately giving acetic acid (from *C*-methyl groups) and carbon dioxide (see Kuhn–Roth oxidation; p. 605).

A selective method is, however, available: namely, the use of molecular oxygen on a platinum catalyst. For example, pentaerythritol gives trihydroxymethylacetic acid in 50% yield in carefully controlled conditions (sodium hydrogen carbonate buffer, 35°C):

An example of the selectivity of this system as between primary and secondary alcohols occurs in a synthesis of ascorbic acid (p. 587). The reagent has the added advantage that double bonds are not attacked, but halides and amines are degraded.

(b)   Secondary alcohols

Since ketones are only oxidized (with C—C bond-breakage) under vigorous conditions, the oxidation of secondary alcohols to ketones is not attended by the difficulty which applies to the oxidation of primary alcohols to aldehydes. Three methods are widely employed.

(1)   *Chromic oxide.* Large-scale oxidations of alcohols which can withstand vigorous conditions can be carried out with an inexpensive reagent such as sodium dichromate in hot aqueous acetic acid. However, milder conditions, at room temperature or below, are often necessary: either pyridinium chlorochromate in dichloromethane or chromium trioxide in a solution of sulfuric acid in aqueous acetone (*Jones' reagent*) are used. For the latter, it is usual to add a solution of chromic oxide in aqueous sulfuric acid from a burette to a cooled solution of the secondary alcohol in acetone. Addition is stopped when a permanent yellow colour, indicating a slight excess of Cr(VI), is obtained. In this way, over-oxidation is prevented and yields of over 90% can be obtained.

Alternative techniques are available. For example, the oxidant power is increased by using acetic acid as the solvent and milder conditions can be obtained by using chromium(VI) oxide in pyridine (see also primary alcohols).

(2)   *Oppenauer method.* The secondary alcohol and a ketone are equilibrated with the corresponding ketone and secondary alcohol by heating in the presence of aluminium t-butoxide. By using the added ketone, usually acetone, in large excess, the equilibrium is forced to the right. The reaction, which is the reverse of Meerwein–Ponndorf–Verley reduction (p. 654), involves the transfer of hydrogen within a cyclic complex:

Benzene or toluene is often added as a co-solvent to raise the temperature of the reaction, or alternatively cyclohexanone is used instead of acetone.

The method is specific for alcohols and is therefore suitable, for example, for compounds containing C=C bonds or phenolic groups, e.g.

80%

One disadvantage of the method is that the aluminium compounds are basic and can bring about prototropic shifts within the product, e.g. the oxidation of cholesterol is accompanied by migration of the C=C bond to give the αβ-unsaturated ketone:

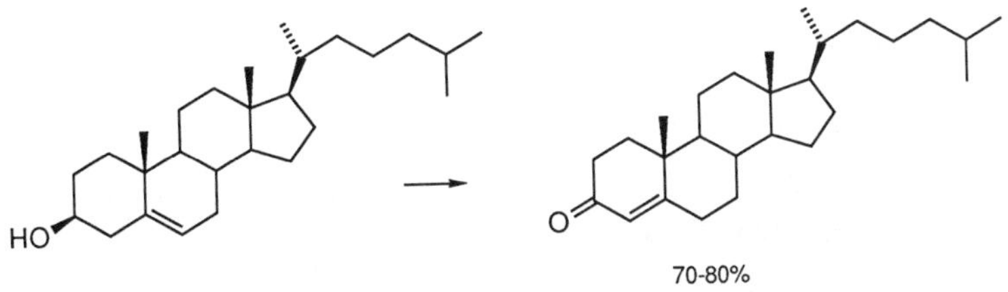

70-80%

Aldehydes are not satisfactorily prepared by Oppenauer oxidation because the basic medium induces reactions between the aldehyde and the ketone.

(3) *Catalytic dehydrogenation*. Like primary alcohols, secondary alcohols are dehydrogenated when passed over certain heated catalysts. For example, acetone is obtained industrially by the dehydrogenation of isopropanol over copper or zinc oxide at about 350°C. (Acetone is also produced by the cumene hydroperoxide process, p. 446).

(c) Allylic alcohols

In addition to the methods described above, allylic alcohols are oxidized to αβ-unsaturated carbonyl compounds on the surface of manganese dioxide suspended in an inert solvent such as dichloromethane. The nature of the manganese dioxide affects the yields: it is best prepared as a non-stoichiometric compound by reducing permanganate ion with manganese(II) ion in alkaline solution. Yields are then high, e.g.

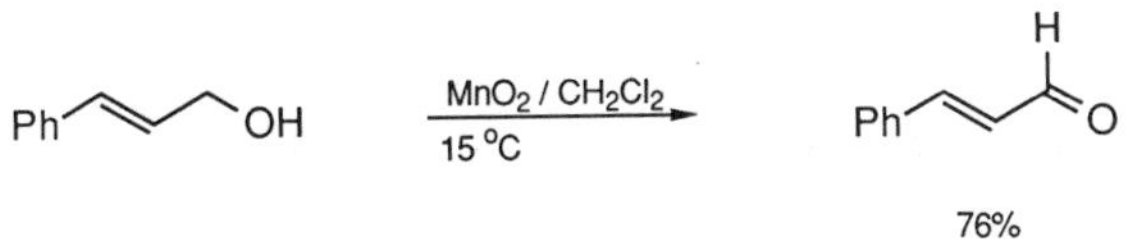

76%

The method is suitable also for benzylic alcohols but not for primary and secondary alcohols which are oxidized only very slowly.

The high potential quinones (p. 607) are suitable for the oxidation of allylic, benzylic, and propargylic (—C≡C—CH(OH)—) alcohols: reaction occurs through the relatively stable carbocations formed by the loss of hydride ion:

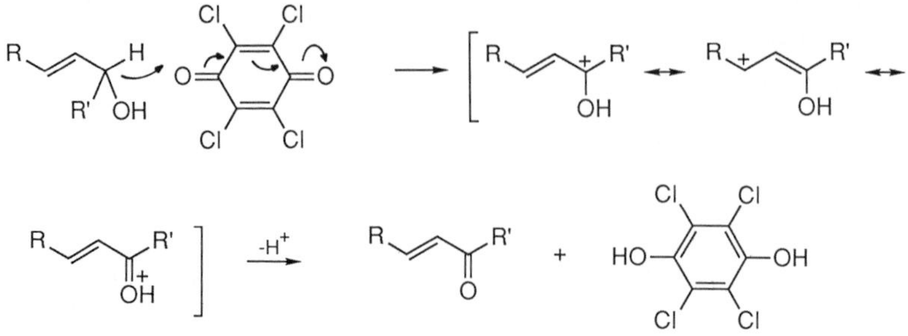

(d) Benzylic alcohols

Several specific processes have been developed.

(1) *Nitrogen dioxide (Field's method)*. Primary and secondary benzyl alcohols react with dinitrogen tetroxide in chloroform at 0°C, e.g.

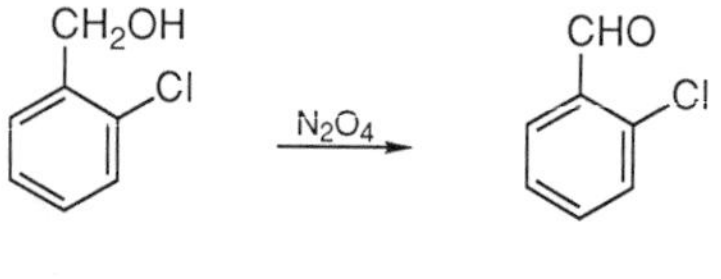

96%

Reaction occurs through (the radical) nitrogen dioxide with the formation and decomposition of a hydroxynitro compound:

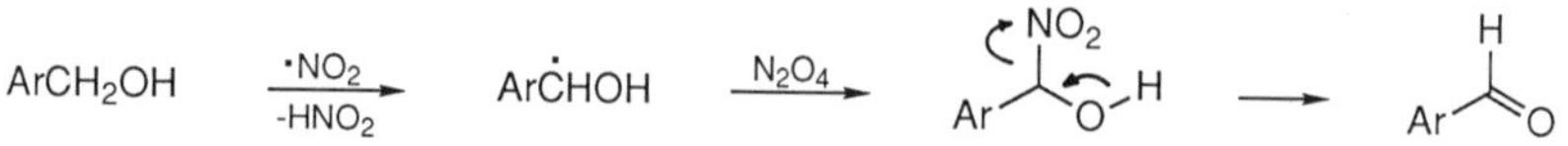

(2) *Hexamethylenetetramine (Sommelet reaction)*. The halide from the benzyl alcohol is treated with hexamethylenetetramine (p. 307) and the resulting salt is hydrolyzed in the presence of more hexamethylenetetramine, usually with aqueous acetic acid, to the aldehyde, e.g.

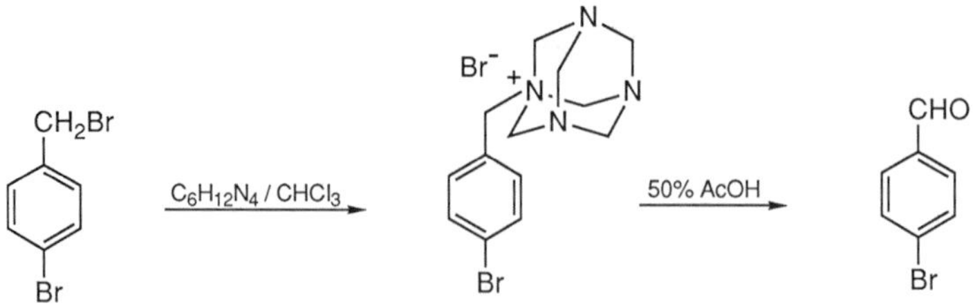

70%

It is not necessary to isolate the intermediate salt.

Electron-withdrawing substituents decrease the yield and *ortho*-substituents hinder the reaction, e.g. neither 2,4-dinitro- nor 2,6-dimethylbenzaldehyde can be prepared in this way.

The reaction involves hydride-ion transfer. At the acidity employed, the quaternary benzyl salt is hydrolyzed to the benzylamine, ammonia, and formaldehyde. The benzylamine transfers hydride ion to methyleneimine (from formaldehyde and ammonia) giving an imine which is hydrolyzed to the aromatic aldehyde:

$$Ar\text{-}CH_2\text{-}N^+(\text{hexamine}) + 6H_2O \longrightarrow ArCH_2NH_2 + 6\,CH_2O + 2\,NH_3 + NH_4Br$$

$$\begin{array}{c} Ar\text{-}CH(\cdot\cdot NH_2)(H) \\ CH_2{=}\overset{+}{N}H_2 \end{array} \longrightarrow \begin{array}{c} Ar\text{-}CH{=}\overset{+}{N}H_2 \\ + \\ MeNH_2 \end{array} \xrightarrow{H_2O} ArCHO + NH_4^+$$

(3) *Kröhnke reaction.* A benzyl halide is converted into its pyridinium salt and thence, with *p*-nitrosodimethylaniline, into a nitrone. Acid hydrolysis gives the aromatic aldehyde.

$$Ar\text{-}CH_2X + \text{(pyridine)} \xrightarrow{-X^-} Ar\text{-}CH_2\text{-}\overset{+}{N}(\text{pyridinium}) \cdots O{=}N\text{-}Ar' \longrightarrow Ar\text{-}CH(H)\text{-}\overset{+}{N}(\text{=}O)\text{-}Ar'$$

$$\xrightarrow{OH^-} Ar\text{-}CH{=}\overset{+}{N}(O^-)\text{-}Ar' \xrightarrow{H_3O^+} ArCHO$$

Reaction occurs under mild conditions so that it is suitable for the preparation of sensitive aldehydes. In addition, it is facilitated by electron-attracting substituents, e.g. 2,4-dinitrobenzaldehyde can be prepared in this way (cf. the Sommelet reaction).

(e) 1,2-diols

1,2-Diols are cleaved by lead tetraacetate, phenyliodoso acetate, and periodic acid or sodium metaperiodate:

$$\begin{array}{c} HO \quad R \\ R\text{-}C\text{-}C\text{-}R \\ R \quad OH \end{array} \longrightarrow \begin{array}{c} R \\ R \end{array}{=}O + O{=}\begin{array}{c} R \\ R \end{array}$$

The first two reagents are used in an organic medium, commonly glacial acetic acid, whereas periodate is used in aqueous solution. Phenyliodoso acetate is the least powerful and is not often employed, but all three are entirely specific: the aldehydes formed are not oxidized further. Yields are usually excellent, e.g. dibutyl tartrate and lead tetraacetate give butyl glyoxylate in up to 87% yield [4],

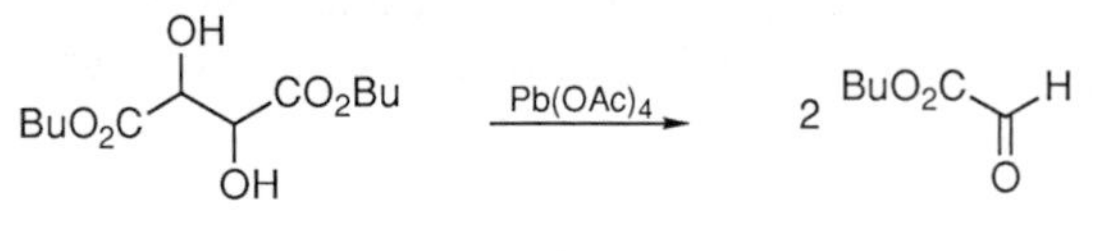

and 2,3-butane diol and sodium periodate give acetaldehyde quantitatively,

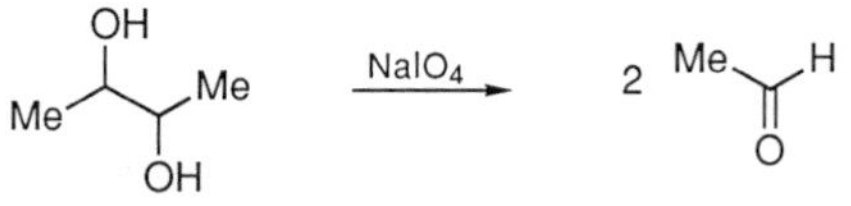

Reaction normally occurs by two-electron oxidation (Pb(IV) to Pb(II) and I(VII) to I(V)) within a cyclic intermediate:

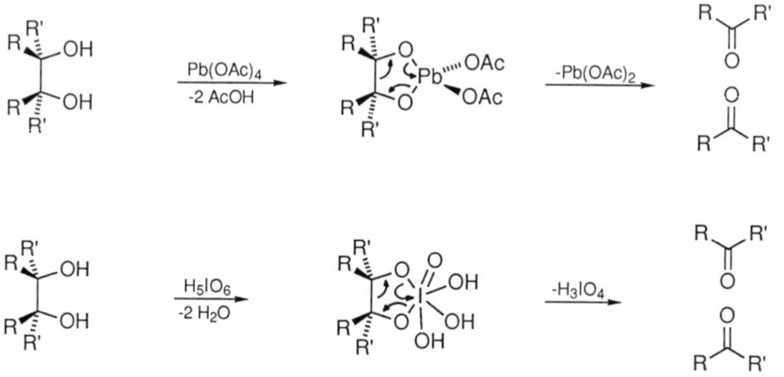

so that *cis*-diols are oxidized faster than *trans*-diols, e.g. *cis*-cyclohexane-1,2-diol reacts about 25 times faster than its *trans*-isomer. However, an acyclic path is also followed, e.g.

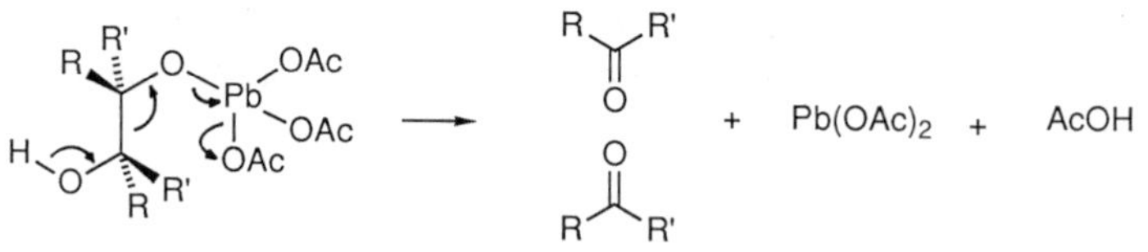

In addition to diols, α-amino-alcohols, α-ketols, and α-dicarbonyl compounds are cleaved:

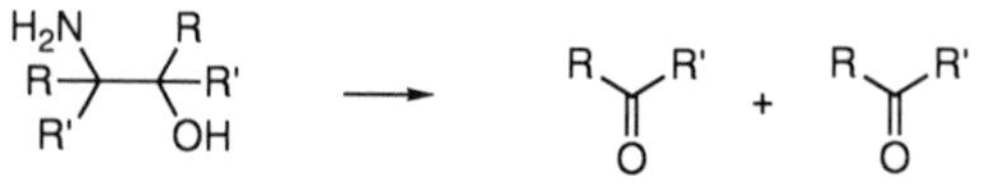

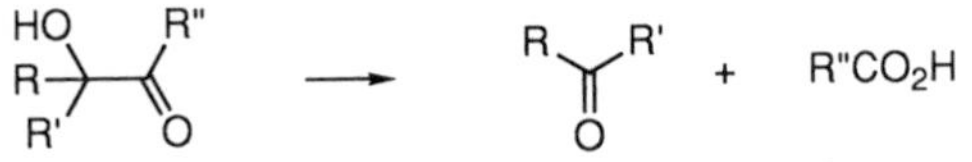

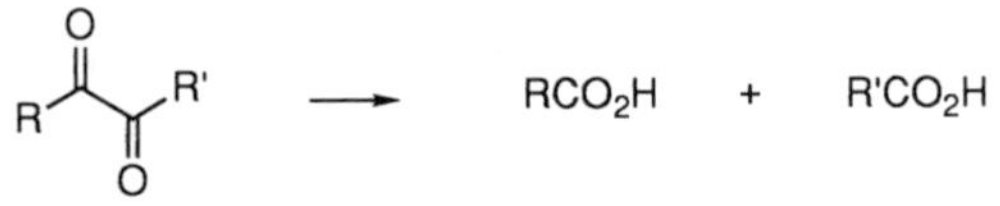

The grouping —CH(OH)—CH(OH)—CH(OH)— is oxidized to a mixture of aldehydes and formic acid,

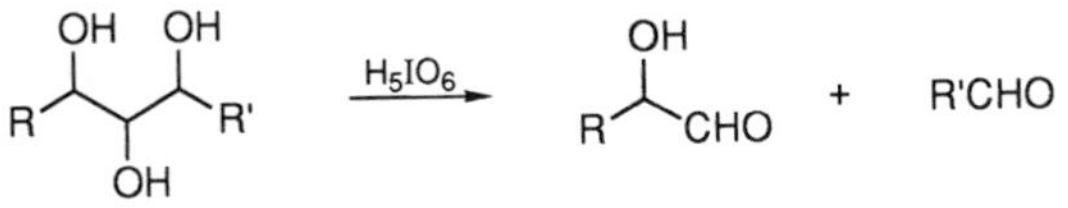

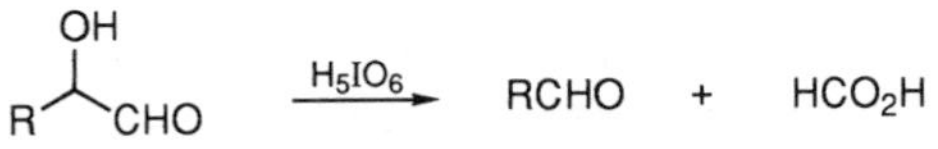

Periodate has proved a valuable degradative agent in carbohydrate chemistry, for it reacts nearly quantitatively. Since the grouping —CH(OH)—CH$_2$OH is oxidized to formaldehyde and the grouping —CH(OH)—CH(OH)—CH(OH)— gives one equivalent of formic acid, the estimation of formaldehyde and formic acid provides information about the occurrence of these groups and the total periodate consumed indicates the total number of 1,2-diol groups present.

An interesting example of the synthetic use of periodate, in the oxidation of the system —CO—CH(OH)—CH(OH)—, occurs in the synthesis of reserpine (section 22.1). Other uses are in the Lemieux reagents (p. 595).

Lead tetraacetate also effects the bisdecarboxylation of succinic acids:

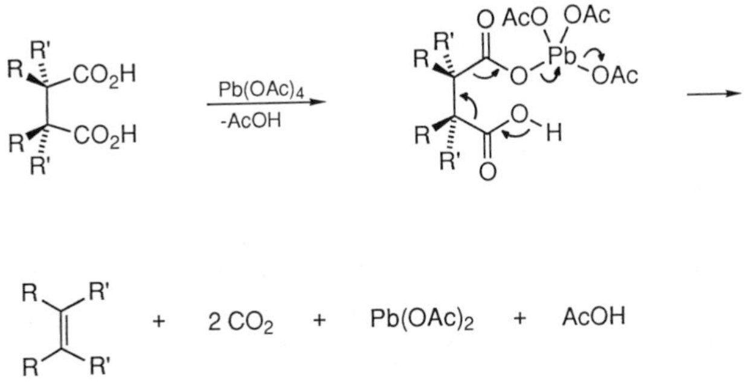

(f) Aldehydes

Aldehydes can be oxidized to acids by vigorous reagents such as chromic(VI) acid and permanganate. This can be satisfactory for compounds which do not possess sensitive groups, e.g. heptanal gives heptanoic acid in 76–78% yield with permanganate in sulfuric acid at 20°C [2],

$$\text{(heptanal)} \xrightarrow{\text{KMnO}_4 - \text{H}_2\text{SO}_4} \text{(heptanoic acid, CO}_2\text{H)}$$

In most cases, however, milder conditions are necessary. Three methods are available.

(1) *Jones' reagent* (p. 612). Reaction is rapid at room temperature. It is thought to occur *via* the 1,1-diol:

$$\text{RCHO} + \text{H}_2\text{O} \rightleftharpoons \underset{R}{\overset{HO\ H}{\diagdown}}\text{OH} \xrightarrow{\text{CrO}_3} \cdots$$

$$\longrightarrow \text{RCO}_2\text{H} + \text{CrO(OH)}_2$$

(2) *Baeyer–Villiger oxidation*. In the intermediate formed with a peroxyacid (cf. ketones, p. 445), hydrogen migrates in preference to alkyl and phenyl groups:

$$\text{RCHO} + \text{R'CO}_2\text{OH} \longrightarrow \cdots \xrightarrow{-\text{R'CO}_2^-} \text{RCO}_2\text{H}$$

These conditions are less strongly acidic than with Jones' reagent.

(3) *Silver oxide*. If acidic conditions have to be avoided, silver oxide is used. For example, thiophen-3-aldehyde gives the 3-carboxylic acid nearly quantitatively in 5 minutes at 0°C [4]:

$$\text{(thiophene-3-CHO)} \xrightarrow{\text{Ag}_2\text{O} - \text{H}_2\text{O}} \text{(thiophene-3-CO}_2\text{H)}$$

In general, it is usually more convenient to synthesize acids by routes which do not involve the aldehyde, such as by the carbonation of organometallic compounds or from malonic ester.

Those aldehydes which do not possess α-hydrogen atoms undergo the Cannizzaro reaction with base (p. 654), e.g. benzaldehyde gives benzyl alcohol and benzoate ion. Finally, *ortho* and *para* hydroxybenzaldehydes undergo oxidative rearrangement with alkaline hydrogen peroxide (Dakin reaction, p. 447).

(g) Ketones

The C—CO bond in ketones can be oxidized in three ways.

(1) *By nitric acid or alkaline potassium permanganate*. These powerful con-

ditions give carboxylic acids, reaction occurring through the enol (acid solution) or the enolate (basic solution), e.g.

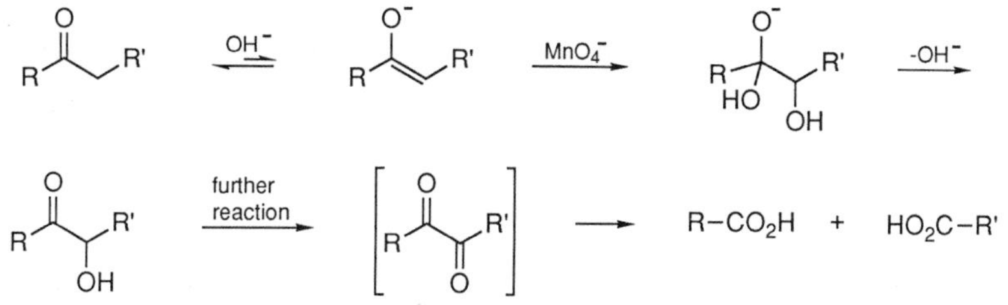

Attack can occur on both sides of the carbonyl group so that, unless the ketone is cyclic, a mixture of products is obtained. With cyclic ketones, however, reasonable yields can be achieved, e.g. cyclohexanol is oxidized by hot 50% nitric acid, *via* cyclohexanone, to give adipic acid in 60% yield [1]:

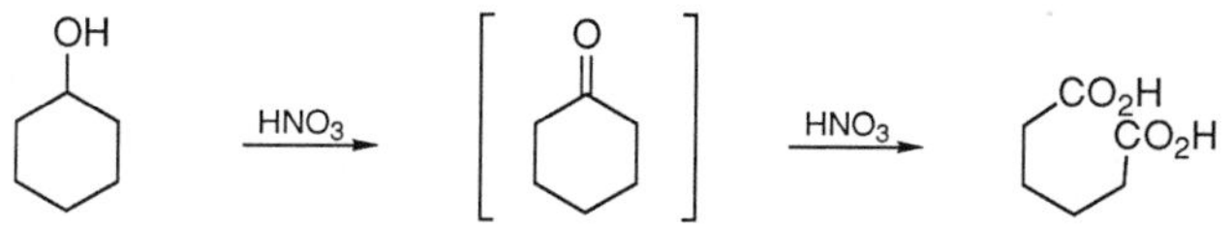

(2) *By the halogens in alkali.* Methyl ketones are oxidized by chlorine, bromine, or iodine in alkaline solution to give acids and the corresponding haloform. Reaction occurs by base-catalyzed halogenation followed by elimination of the conjugate base of the haloform:

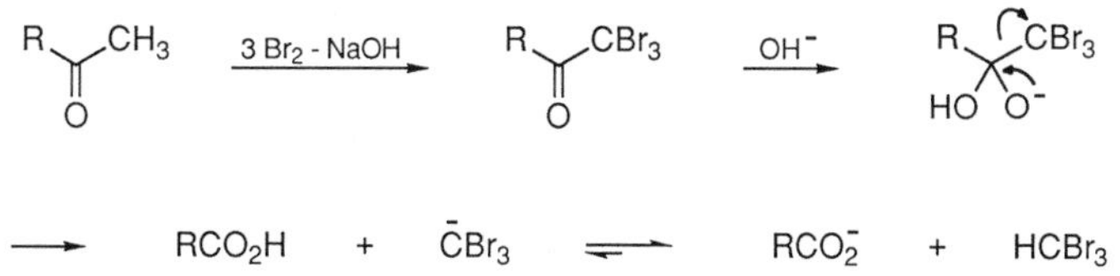

This provides a particularly useful method for the synthesis of aromatic acids, for the corresponding methyl ketones are often readily available through the Friedel–Crafts reaction (p. 365). For example, the acetylation of naphthalene in nitrobenzene solution gives 2-acetylnaphthalene (p. 368) from which 2-naphthoic acid can be obtained in 97% yield with chlorine in sodium hydroxide solution at 55°C:

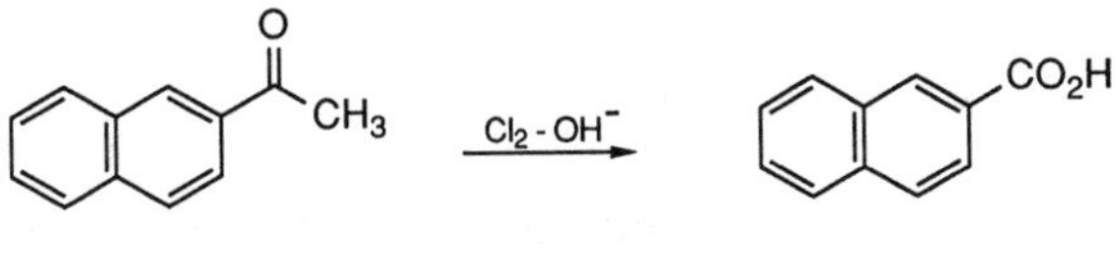

There are also applications in the aliphatic series. For example, pinacolone, available from acetone by reduction and rearrangement (p. 433), reacts with bromine in sodium hydroxide solution at below 10°C to give trimethylacetic acid in over 70% yield [1]:

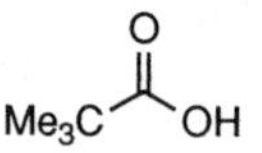

$$Me_3C\overset{O}{\underset{}{\parallel}}CH_3 \xrightarrow{Br_2 \cdot OH^-} Me_3C\overset{O}{\underset{}{\parallel}}OH$$

(3) *By peroxyacids.* Ketones undergo oxidative rearrangement with peroxy-acids to give esters or lactones, $RCOR \rightarrow RCO_2R$ (Baeyer–Villiger reaction, p. 445).

(h)   α-Ketols

These systems are very readily oxidized to α-dicarbonyl compounds. One-electron oxidants in basic solution are effective, for the enediolate formed by base can donate one electron to the oxidant to give a delocalized radical (cf. semiquinone radicals, p. 598). Loss of a second electron completes the oxidation.

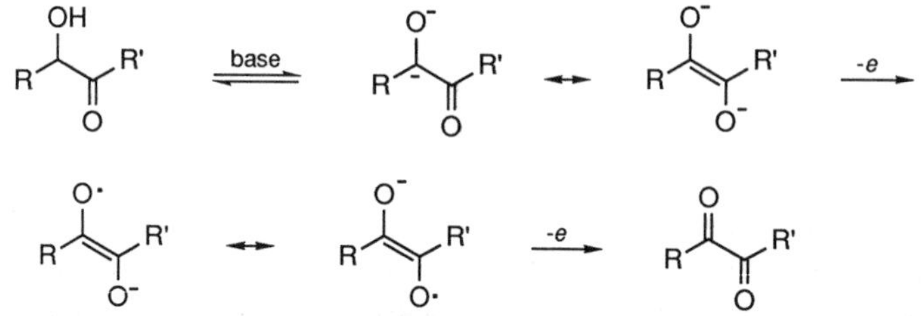

For example, copper sulfate in pyridine at 95°C oxidizes benzoin to benzil in 86% yield [1]:

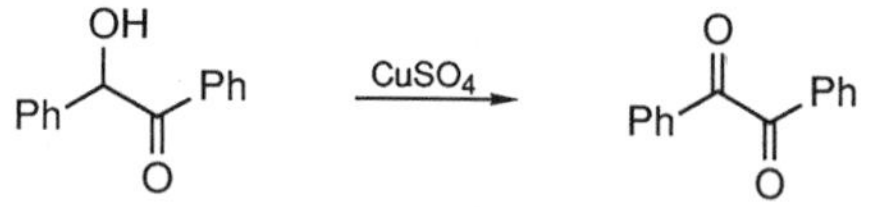

(i)   α-Dicarbonyl compounds

Oxidation can be brought about in three ways: peroxyacids in inert solvents give anhydrides, $RCO\!-\!COR \rightarrow RCO\!-\!O\!-\!COR$ (Baeyer-Villiger reaction, p. 445); and both warm hydrogen peroxide in acetic acid and the glycol-cleavage reagents give carboxylic acids, $RCO\!-\!COR \rightarrow 2RCO_2H$.

(j)   Acids: oxidative decarboxylation

Carboxylic acids undergo oxidative decarboxylation when heated with lead tetra-acetate in the presence of a catalytic amount of a copper(II) salt:

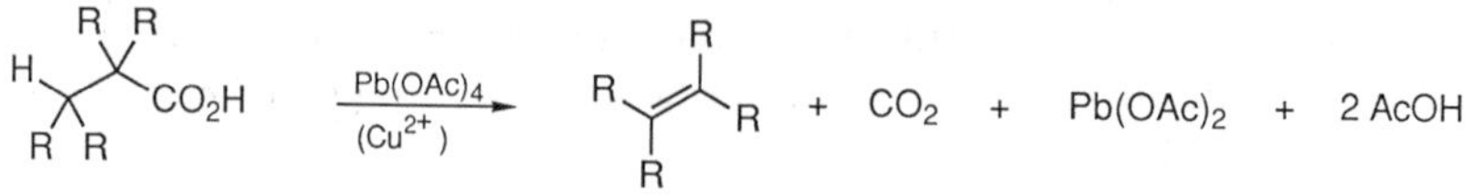

Reaction is thought to occur by homolysis of the lead carboxylate, followed by oxidation of the resulting radical by copper(II), e.g.

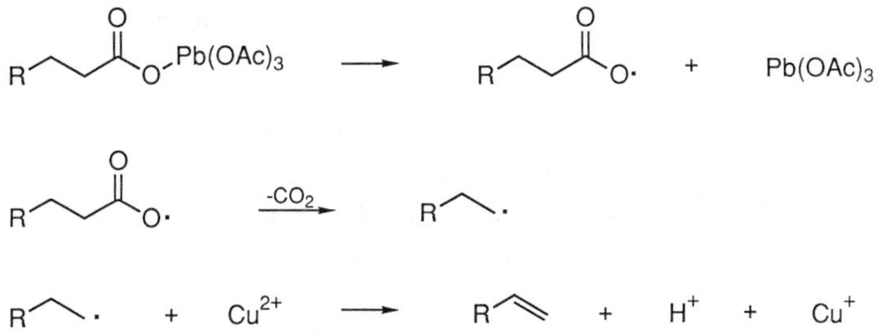

The copper(I) ion is then oxidized by lead(IV) to regenerate copper(II).

**19.4
Systems
containing
nitrogen**

(a) Primary amines

These are very sensitive to oxidation and generally darken on exposure to air, through autoxidation at the surface, to give mixtures of complex products. This is particularly true of aromatic amines: a number of products arise by oxidation to nitroso and nitro compounds followed by condensations, e.g. $ArNH_2 \rightarrow ArNO$; $ArNH_2 + ArNO \rightarrow ArN=NAr$. Repeated reactions give aniline blacks.

Synthetically useful methods have therefore to be highly selective. The most successful reagents are hydrogen peroxide and peroxyacids.

(1) *Hydrogen peroxide* converts primary aliphatic amines into aldoximes, e.g.

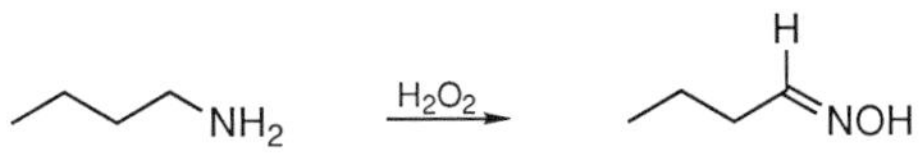

57%

by nucleophilic displacement by the amine on the peroxide:

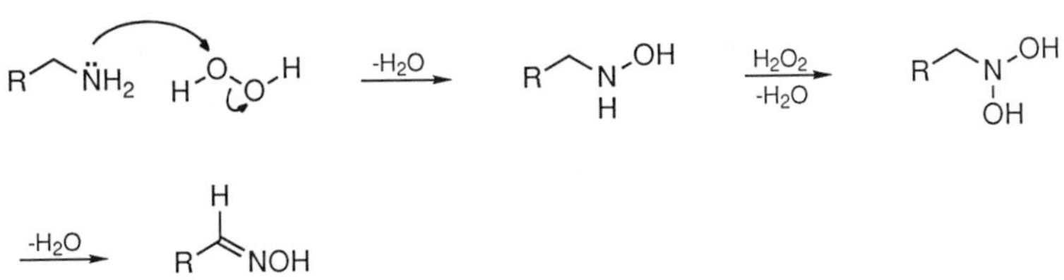

Oxidation of aromatic amines to the corresponding level (nitroso compounds) is generally brought about with peroxydisulfuric acid, $HO_3S—O—O—SO_3H$, e.g.

$$\text{(structure: benzene ring with NH}_2\text{ and NO}_2) \xrightarrow{(NH_4)_2S_2O_8 \,/\, H_2SO_4} \text{(structure: benzene ring with NO and NO}_2)$$

75%

Aromatic nitroso compounds can also be prepared from the corresponding hydroxylamines by oxidation with dichromate at low temperatures (to prevent further oxidation). For example, phenylhydroxylamine, sodium dichromate, and 50% sulfuric acid at 0°C give nitrosobenzene in about 50% yield [3]:

$$\text{PhNHOH} \xrightarrow{Na_2Cr_2O_7 \,/\, H_2SO_4} \text{PhNO}$$

Since arylhydroxylamines are readily obtained by the reduction of nitro compounds in neutral solution (p. 667), this method is convenient for introducing the nitroso group *via* nitration.

(2) *Trifluoroperoxyacetic acid*, a more powerful oxidant than hydrogen peroxide, converts primary amines directly into nitro-compounds. The yields with aromatic amines are generally high, e.g. *o*-nitroaniline is oxidized in refluxing dichloromethane to *o*-dinitrobenzene in 92% yield:

$$\text{(structure: benzene ring with NH}_2\text{ and NO}_2) \xrightarrow{CF_3CO_2OH} \text{(structure: benzene ring with NO}_2\text{ and NO}_2)$$

With aliphatic amines, however, the yields are low. Other oxidants have been used with moderate success. For example, 1-hexylamine is oxidized in 33% yield to the nitro compound by peroxyacetic acid.

Aromatic nitro-compounds can also be obtained by the oxidation of nitroso-compounds with dichromate or dilute nitric acid, but it is normally more convenient to oxidize the amine.

(b) Secondary amines

Oxidation with hydrogen peroxide gives hydroxylamines (cf. primary amines):

$$R_2NH \xrightarrow{H_2O_2} R_2N-OH \;+\; H_2O$$

(c) Tertiary amines

Hydrogen peroxide converts tertiary amines into their *N*-oxide hydrates, by nucleophilic displacement analogous to the reaction of primary amines. The *N*-oxide is obtained by warming the hydrate *in vacuo*.

$$R_3N \xrightarrow{H_2O_2} R_3\overset{+}{N}-OH \ OH^- \xrightarrow[-H_2O]{warm} R_3\overset{+}{N}-O^-$$

(d) Hydrazines

Hydrazine itself is readily oxidized to nitrogen. Some at least of the oxidants, e.g. copper(II) ion, yield diimide, HN=NH, as an intermediate and this is employed for the reduction of alkenes (p. 634).

Monosubstituted hydrazines react with one-electron oxidants such as copper(II) and iron(III) ion to give unstable azo compounds which decompose with loss of nitrogen to hydrocarbons, e.g.

$$PhNH-NH_2 \xrightarrow{Cu^{2+}} \left[PhN{=}NH \right] \longrightarrow PhH + N_2$$

Arylhydrazines are oxidized by the two-electron oxidants chlorine and bromine giving diazonium salts, e.g.

$$PhNHNH_2 + 2\,Cl_2 \longrightarrow PhN_2^+\,Cl^- + 3\,HCl$$

$$PhNHNH_2 + 3\,Br_2 \longrightarrow PhN_2^+\,Br_3^- + 3\,HBr$$

N,N'-Disubstituted hydrazines give azo-compounds readily. For example, azobisisobutyronitrile (a useful reagent for initiating free-radical reactions) can be obtained from acetone, hydrazine, and cyanide (compare the Strecker synthesis of α-amino acids, p. 329) followed by oxidation, e.g. with mercury(II) oxide:

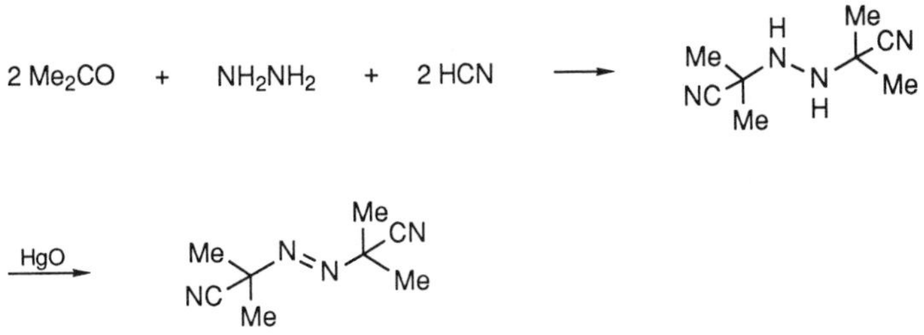

$$2\,Me_2CO + NH_2NH_2 + 2\,HCN \longrightarrow$$

Diarylhydrazines oxidize with exceptional ease since a conjugated system is formed. For example, hydrazobenzene gives azobenzene even on standing in air for some time.

Azo-compounds may be further oxidized to azoxy-compounds in more vigorous conditions, e.g.

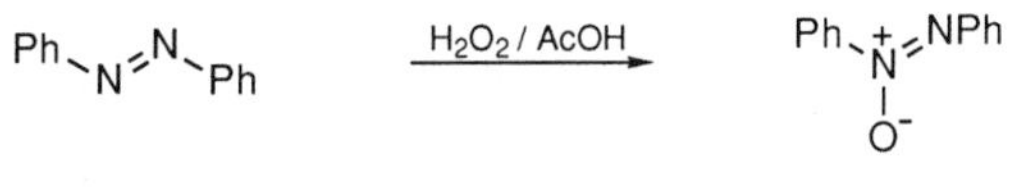

azoxybenzene

(e) Hydrazones

Hydrazones are oxidized to diazoalkanes by mercury(II) oxide. Those containing aryl-substituents give moderately stable (conjugated) products which can be isolated, e.g. benzophenone hydrazone gives diphenyldiazomethane nearly quantitatively [3]:

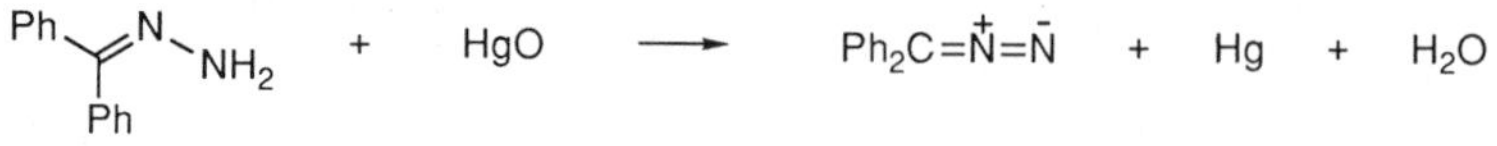

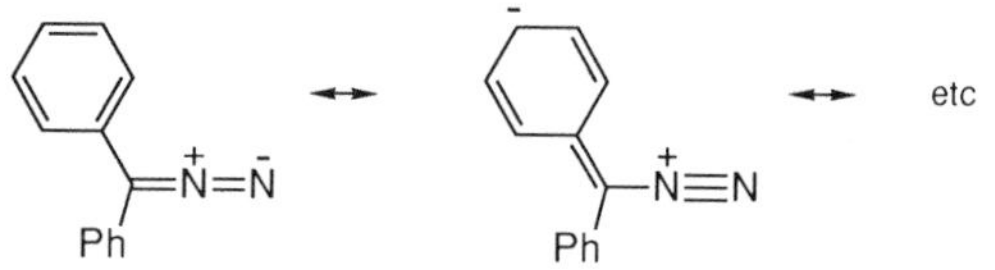

$$Ph_2C=\overset{+}{N}=\overset{-}{N} \quad + \quad Hg \quad + \quad H_2O$$

Diazoalkanes containing only saturated groups decompose rapidly to nitrogen and products derived from carbenes ($R_2CN_2 \rightarrow R_2C$: $+ N_2$). Use is made of this in a synthetic procedure for alkynes: the bishydrazone of an α-diketone is oxidized with mercury(II) oxide and the unstable bisdiazo-compound decomposes to the alkyne:

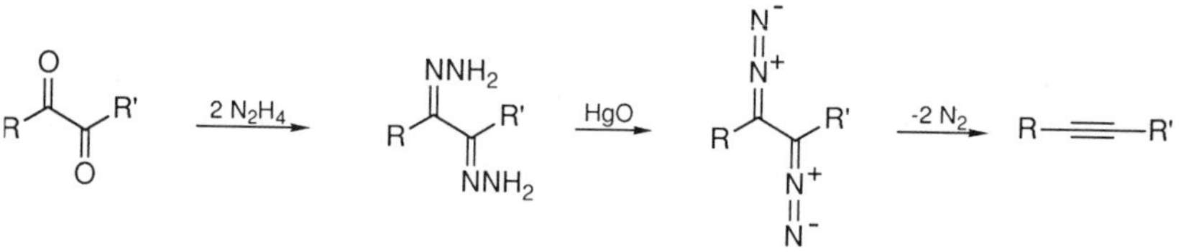

For example, benzil bishydrazone gives diphenylacetylene in about 70% yield [4] and the bishydrazone from 1,2-cyclooctanedione gives cyclooctyne in 9% yield, the low yield doubtless resulting from the extreme ring strain in the product (which is the smallest cycloalkyne to have been made).

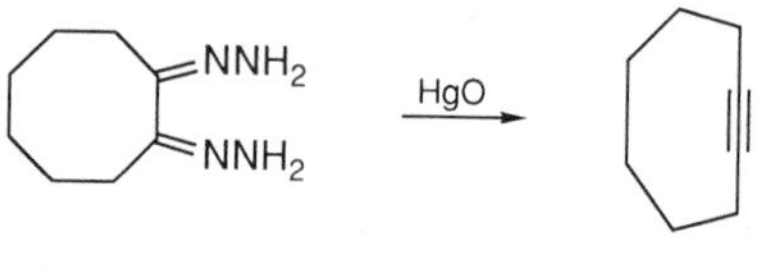

The monohydrazones of α-diketones give ketenes in these conditions, e.g. the treatment of benzil monohydrazone with mercury(II) oxide gives a diazo-compound which rearranges on distillation, with loss of nitrogen, to diphenyl-ketene (58%) [3]:

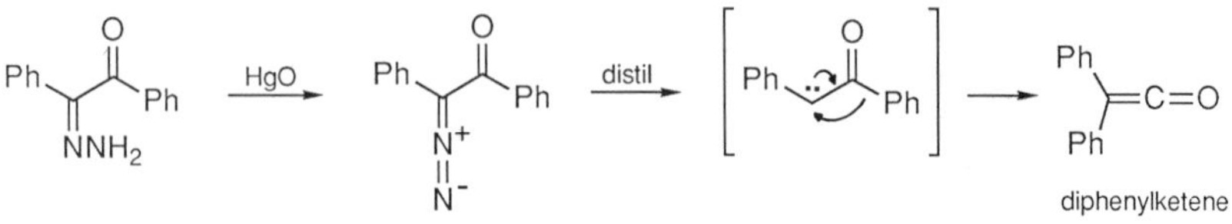

diphenylketene

(a) Thiols

Whereas hydroxyl-containing compounds are oxidized at carbon, thiol-containing compounds are oxidized at sulfur, largely because of the comparatively low bond energy of S—H compared with O—H (p. 46).

(1) *Oxidation to disulfides.* A variety of relatively weak oxidants, such as hydrogen peroxide, iron(III) ion, and iodine, oxidize thiols to disulfides,

$$2\ RSH \xrightarrow{\ -H_2\ } RS-SR$$

For example, cysteine is readily oxidized to cystine, and the disulfide ring in thioctic acid is readily formed from the dithiol precursor (p. 255).

(2) *Oxidation to sulfenyl chlorides.* Thiols are oxidized by chlorine, through disulfides, to sulfenyl chlorides,

$$2\ RSH \xrightarrow{\ Cl_2\ } R_{\diagdown S}{\diagup}^{S}{\diagdown}_R \xrightarrow{\ Cl_2\ } 2\ RSCl$$

from which derivatives of sulfenic acids may be obtained, e.g.

$$RSCl\ +\ R'OH \longrightarrow RS-OR'$$

a sulfenic ester

$$RSCl\ +\ NH_3 \longrightarrow RS-NH_2$$

a sulfenamide

(3) *Oxidation to sulfonic acids.* Vigorous reagents such as nitric acid and permanganate give sulfonic acids, probably through sulfenic and sulfinic acids, which are too easily oxidized to be isolated:

$$RSH \longrightarrow \left[RS-OH \right] \longrightarrow \left[\underset{R}{\overset{O}{\underset{\|}{S}}}{}_{OH} \right] \longrightarrow \underset{R}{\overset{O\ \ O}{\underset{\diagup}{S}}}{}_{OH}$$

Sulfonic acids are also formed by the treatment of lead thiolates with nitric acid:

$$2\ RSH \xrightarrow{\ Pb(NO_3)_2\ } Pb(SR)_2 \xrightarrow{\ HNO_3\ } 2\ RSO_2OH$$

(b) Sulfides

Sulfides can be oxidized to both sulfoxides and sulfones. The former are obtained in high yield (*c.* 90%) with liquid dinitrogen tetroxide in ethanol cooled with solid carbon dioxide. Further oxidation to the sulfone does not occur, and the product is anhydrous (wet sulfoxides are very difficult to dry).

$$R\text{---}S\text{---}R' \quad \xrightarrow{\;N_2O_4\;} \quad R\overset{\displaystyle O}{\underset{\displaystyle}{\overset{\|}{\text{---}S\text{---}}}}R'$$

The oxidation can also be carried out with a slight excess of sodium metaperiodate at 0°C and with hydrogen peroxide but these methods cause some further oxidation to the sulfone (see below).

Very sensitive sulfides require more delicate treatment. For example, manganese dioxide is the only reagent which oxidizes diallyl sulfide to diallyl sulfoxide in reasonable yield.

Sulfones are normally obtained from sulfides by oxidation with hydrogen peroxide in aqueous or acetic acid solution:

$$R\text{---}S\text{---}R' \quad \xrightarrow{\;2\,H_2O_2\;} \quad R\overset{\displaystyle O\;\;O}{\underset{\displaystyle}{\overset{\|\;\;\|}{\text{---}S\text{---}}}}R'$$

19.6 Systems containing phosphorus

The characteristics of phosphorus chemistry compared with that of nitrogen are that three-valent phosphorus is readily oxidized to the five-valent state and that P—O bonds are more stable than N—O bonds. Oxidations at phosphorus occur under mild conditions, e.g.

$$R_3P \quad \xrightarrow{\;air\;} \quad R_3P{=}O$$

$$R\overset{\displaystyle O}{\underset{\displaystyle R}{\overset{\|}{\text{---}P\text{---}}}}H \quad \xrightarrow{\;H_2O_2\;} \quad R\overset{\displaystyle O}{\underset{\displaystyle R}{\overset{\|}{\text{---}P\text{---}}}}OH$$

19.7 Systems containing iodine

The iodine atom in aryl iodides can be oxidized to both the three- and the five-valent states. For example, iodobenzene reacts with chlorine in dry chloroform to give dichloroiodobenzene in about 90% yield [3]; this may be hydrolyzed to iodosylbenzene in about 60% yield [3] and iodylbenzene may be obtained in over 90% yield by steam-distilling iodosobenzene to remove the iodobenzene formed by disproportionation.

$$PhI \quad + \quad Cl_2 \quad \longrightarrow \quad PhICl_2$$

dichloroiodobenzene

PhICl$_2$ +   2 NaOH $\longrightarrow$ PhIO   +   2 NaCl   +   H$_2$O

iodosylbenzene

2 PhIO $\xrightarrow[\text{-PhI}]{\text{steam distil}}$ PhIO$_2$

iodylbenzene

The aliphatic analogues of these higher-valent iodine compounds are unstable.

Problems

1. How would you carry out the following transformations?
    (*i*) R—CH=CH$_2$ into (*a*) R—CHO, (*b*) R—CH$_2$OH,
    (*c*) R—CH$_2$—CHO, (*d*) R—CH(OH)—CHO,
    (*e*) R—CH(OH)—CH$_3$.
    (*ii*) R—CH=CH—CH$_2$OH into (*a*) R—CH(OH)—CH(OH)—
    CH$_2$OH, (*b*) R—CH=CH—CHO.
    (*iii*) R—CH$_2$—CO—CH$_3$ into (*a*) R—CH$_2$—CO—CHO, (*b*) R—
    CO—CO—CH$_3$.
    (*iv*) PhCHO into (*a*) PhCH(OH)—COPh, (*b*) PhCO—COPh.
    (*v*) PhCO—COPh into (*a*) Ph$_2$C=C=O, (*b*) PhC≡CPh.
    (*vi*) PhCH$_2$Br into PhCHO.
    (*vii*) PhCOCH$_2$Br into PhCOCHO.
    (*viii*) PhNH$_2$ into (*a*) *p*-benzoquinone, (*b*) azoxybenzene, (*c*) *o*-
    dinitrobenzene.
    (*ix*) PhOH into (*a*) *o*-benzoquinone, (*b*) chloranil (tetrachloro-*p*-
    benzoquinone).
    (*x*) PhSH into (*a*) PhS—SPh, (*b*) PhSCl, (*c*) PhSO$_2$OH.
    (*xi*) *trans*-2-Butene into (*a*) (±)-2,3-butanediol, (*b*) *meso*-2,3-butanediol.
    (*xii*)

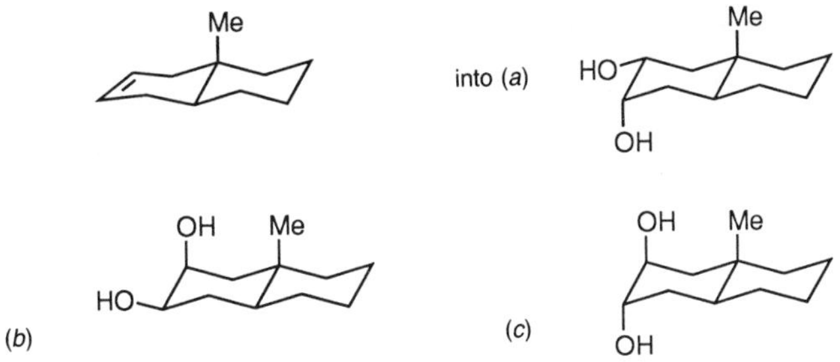

    (*xiii*)

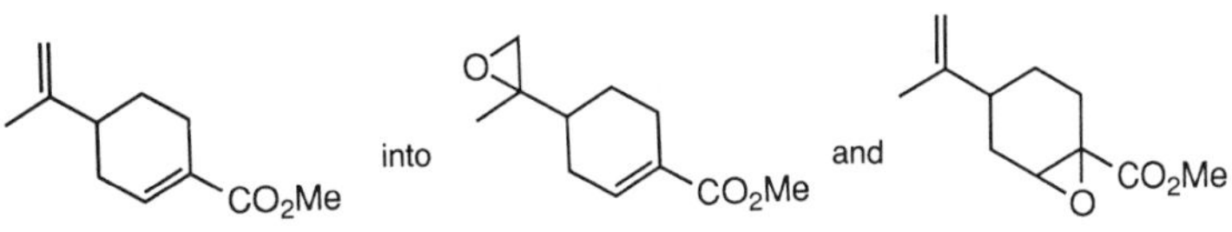

    (*xiv*)

(xv)

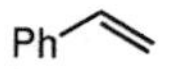

Ph⌒═   into   Ph—CH(OH)—CH₂—OMe   and   Ph—CH(OMe)—CH₂—OH

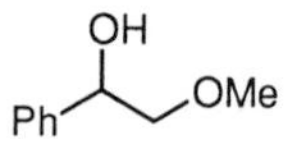

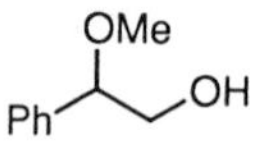

2. Summarize the reagents which may be used to oxidize the methyl group in a compound X—CH₃ according to the nature of the group X (i.e. X = acetyl, phenyl, etc.).

3. A methylated derivative of D-glucose is thought to have the structure (I). How could an oxidative method be used to provide evidence that the ring is six-membered?

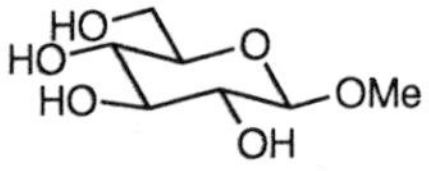

(I)

4. Rationalize the following reactions:

(i)

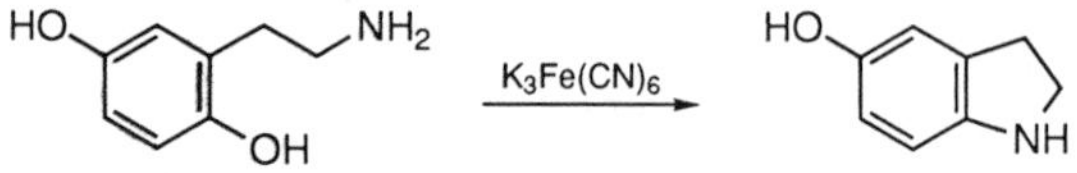

K₃Fe(CN)₆

(ii)

MeCO₂OH   CHO

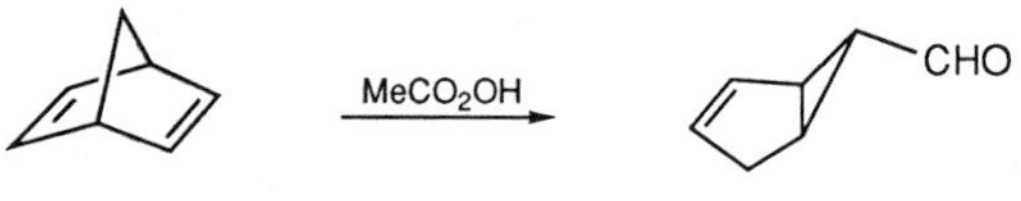

5. Griseofulvin (a fungal metabolite which is an important antibiotic) has been made by the following sequence:

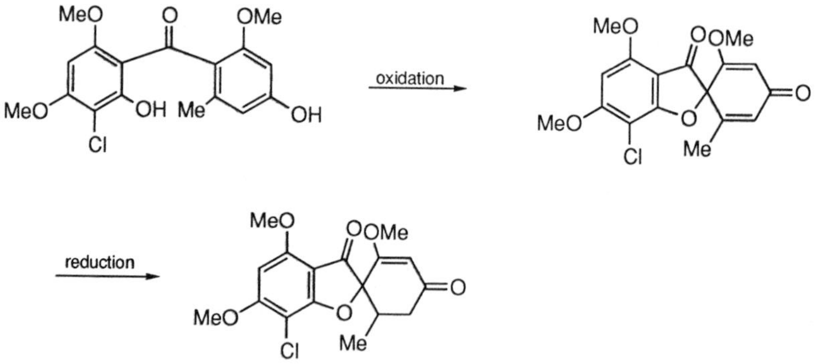

oxidation

reduction

How would you attempt to effect the oxidative step?

Reduction 20

The reductive processes described in this chapter fall into three categories: the removal of oxygen, the addition of hydrogen, and the gain of electrons. The addition of hydrogen may be subdivided into *hydrogenation*, the addition of hydrogen to an unsaturated system, e.g.

$$CH_2=CH_2 \ + \ H_2 \ \xrightarrow{\text{catalyst}} \ CH_3-CH_3$$

and *hydrogenolysis*, the addition of hydrogen with concomitant bond rupture, e.g.

$$ArCH_2-NMe_2 \ + \ H_2 \ \xrightarrow{\text{catalyst}} \ Ar-CH_3 \ + \ HNMe_2$$

Mechanistically, there are three main pathways for reduction:

(1) *By the addition of electrons*, followed either by the uptake of protons, as in the reduction of an alkyne by sodium in liquid ammonia (p. 639),

$$R\!\!=\!\!=\!\!R \ \xrightarrow{e} \ [RC\!=\!CR]^{\overline{\cdot}} \ \xrightarrow[-NH_2^-]{NH_3} \ [R\overset{\cdot}{C}\!=\!CHR]$$

$$\xrightarrow{e} \ [R\overset{-}{C}\!=\!CHR] \ \xrightarrow[-NH_2^-]{NH_3} \ \overset{R}{\underset{H}{\diagdown}}\!\!=\!\!\overset{H}{\underset{R}{\diagup}}$$

or by coupling, as in the reduction of ketones to pinacols (p. 656),

$$R_2C\!=\!O \ \xrightarrow{e} \ R_2\overset{\cdot}{C}\!-\!O^- \ \xrightarrow{\text{dimerizes}} \ \overset{R}{\underset{O^-}{R\!-\!}}\!\!\diagdown\!\!\overset{R}{\underset{O^-}{\!-\!R}} \ \xrightarrow{2\,H^+} \ \overset{R}{\underset{OH}{R\!-\!}}\!\!\diagdown\!\!\overset{R}{\underset{OH}{\!-\!R}}$$

(2) *By the transfer of hydride ion*, as in the reduction of the carbonyl group by lithium aluminium hydride (p. 652):

$$H_3\overset{-}{Al}\!\!-\!\!H \quad \overset{R}{\underset{R}{\diagup}}\!\!=\!\!O \ \longrightarrow \ H\!\!-\!\!\overset{R}{\underset{R}{\diagup}}\!\!-\!O\overset{-}{Al}H_3$$

Such transference may also occur intramolecularly, as in Meerwein–Ponndorf–Verley reduction (p. 654).

(3) *By the catalyzed addition of molecular hydrogen*, as in the reduction of alkenes on metals (p. 631).

Subsidiary reactions, such as the breaking of C—C bonds, are relatively uncommon in reductive processes, but some are usefully applied. For example, pimelic acid (heptanedioic acid) may be obtained in 45% yield from salicylic acid by refluxing for 8 hours with sodium in isoamyl alcohol (3-methyl-1-butanol) [2]; reaction occurs through a β-keto-acid which undergoes a reverse Claisen reaction

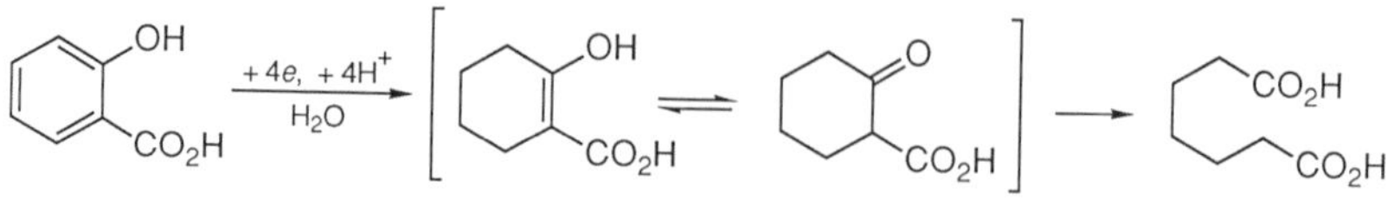

There have been enormous developments in reductive methods since the 1950s, with respect both to the types of bond which may be reduced and to the selectivity of the processes. The older methods involving electron-transfer, such as sodium and alcohol, and zinc and acetic acid (in which the metal acts as the electron-source and the hydroxylic-compound as the proton-donor), are now supplemented by the metal-ammonia and metal-amine systems which have increased the scope of these reductions. Hydride-transfer agents such as formic acid (Leuckart reduction, p. 670) are supplemented by the complex hydrides some of which are of remarkable selectivity. Catalytic methods have been improved by procedures for obtaining more active catalysts.

Enzymic systems are as yet of little general importance but are likely to achieve wider use as methods for isolating enzymes are developed. Their particular merit is that many (being themselves optically active) are enantioselective so that optically active compounds can be obtained from inactive reactants. For example, hydroxyacetone is reduced specifically to (R)-propylene glycol (50%) by incubation with yeast reductase at 32°C for 3 days [2]:

This chapter, like that on oxidation, is classified according to the type of group to be reduced. Each major class of reducing agent is described in the context of the system for which it has been most used.

20.2 Hydrocarbons

(a) Alkanes

Alkanes can be reduced only by rupturing carbon–carbon bonds. Reducing agents are not normally sufficiently powerful to bring this about, in contrast to oxidizing agents of which the strongest cleave aliphatic chains ultimately to carbon dioxide. There is, however, one circumstance in which reduction can be effected catalytically, namely, with strained cyclic compounds, since C—C cleavage relieves the strain, e.g.

$$\triangle \quad \xrightarrow[120\,°C]{H_2 - Ni} \quad \diagup\!\!\diagdown\!\!\diagup$$

$$\square \quad \xrightarrow[200\,°C]{H_2 - Ni} \quad \diagup\!\!\diagdown\!\!\diagup\!\!\diagdown$$

The nature of the catalyst for the reduction is described in the following section.

(b) Alkenes

(i) *Catalytic hydrogenation.* Almost all alkenes can be saturated in very high yield by treatment with hydrogen and a metal catalyst. The most active catalysts are specially prepared platinum and palladium.

1. *Palladium.* An active form of palladium can be obtained from palladium chloride in a similar way to that used for the Adam's catalyst, but more commonly the palladium chloride is reduced in the presence of a suspension of charcoal or other solid support on which the metal is deposited in a very finely divided state.

2. *Adams' catalyst.* Chloroplatinic acid is fused with sodium nitrate to give a brown platinum oxide (PtO_2) which can be stored. When required, it is treated with hydrogen to give a very finely divided black suspension of the metal. This is the reagent usually chosen in the laboratory, reaction being carried out in solvents such as acetic acid, ethyl acetate, and ethanol. About $0·2$ g of platinum oxide is usually employed per 10 g of reactant.

A particularly reactive catalyst is obtained by the reduction of platinum oxide *in situ* with sodium borohydride in the presence of carbon. The molecular hydrogen required for hydrogenation is then generated by the addition of acid to the excess of borohydride.

Most $C{=}C$ bonds are reduced on these catalysts at temperatures below 100°C and at atmospheric or slightly increased pressure. For example, maleic acid gives succinic acid in 98% yield in 30 minutes on platinum at 20°C and 1 atmosphere [1],

$$\underset{\text{CO}_2\text{H}}{\overset{\text{CO}_2\text{H}}{\bigg\|}} \quad \xrightarrow{H_2 - Pt} \quad HO_2C\diagup\!\!\diagdown\!\!\diagup CO_2H$$

and this particular reaction is often used for testing the apparatus and assessing the catalyst.

3. *Raney nickel.* This is a slightly less active catalyst than platinum or palladium. It is prepared by treating a nickel-aluminium alloy with sodium hydroxide,

$$Ni-Al \ + \ NaOH \ + \ H_2O \ \longrightarrow \ Ni \ + \ Na^+AlO_2^- \ + \ 3/2\,H_2$$

and washing away the sodium aluminate to leave the nickel as a black suspension saturated with hydrogen (c. 50–100 cm³ per g) which is pyrophoric when dry. Most alkenes are hydrogenated over Raney nickel at about 100°C and pressures up to 3 atmospheres. Since the Raney alloy is fairly cheap, it is usually convenient to employ relatively large quantities of the catalyst (c. 2 g per 10 g) so as to minimize the effects of catalyst-poisons. This is particularly important with sulfur-containing compounds (p. 672).

4. 'Copper chromite'. The early catalytic methods developed by Sabatier and Senderens employed metals such as iron, cobalt, nickel, and copper at temperatures of about 300°C. This approach has been considerably improved by the use of high pressures (up to 300 atmospheres of hydrogen) and more active catalysts. The best, *Adkins' catalyst*, the so-called copper chromite, is made from copper nitrate and sodium dichromate and corresponds to $CuO \cdot CuCr_2O_4$. This is a far cheaper catalyst than palladium or platinum and is therefore more attractive industrially.

5. Transfer hydrogenation. The hydrogen is supplied by a donor such as cyclohexene or hydrazine, e.g.

The driving force in the case of cyclohexene is the gain in aromatic stabilization energy when benzene is formed; with hydrazine, the strongly bonded N_2 molecule is formed. The advantage of this method is that no special apparatus for handling and measuring gaseous hydrogen is required.

Selectivity. The C≡C bond is reduced more readily than C=C but other unsaturated groupings, with the exception of nitro groups and acid chlorides, are reduced less readily. Catalytic hydrogenation can therefore be used for the selective reduction of C=C in the presence of aromatic rings and carbonyl groups, whether or not the unsaturated functions are conjugated. For example, benzylideneacetophenone is reduced over platinum at 20°C in 90% yield [1]:

Stereochemistry. Reduction is highly stereoselective, giving predominantly *cis*-alkene,* e.g.

* With palladium, however, the *trans*-product predominates. This is because Pd catalyzes the isomerization of the alkene:

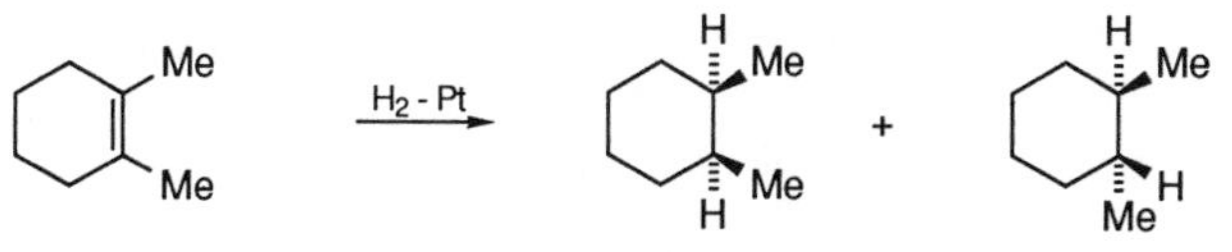

80%   20%

This can be understood in terms of the following simplified representation:

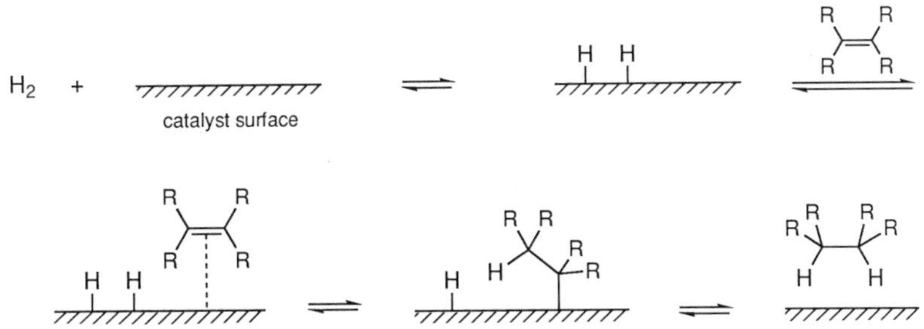

In rigid ring systems, the direction of addition is from the less hindered side. For example, octalins of the type shown below (e.g. cholesterol; X=OH) give mainly *trans* ring-junctions on reduction, since the angular methyl group hinders the fit of the catalyst on the opposite side of the double bond:

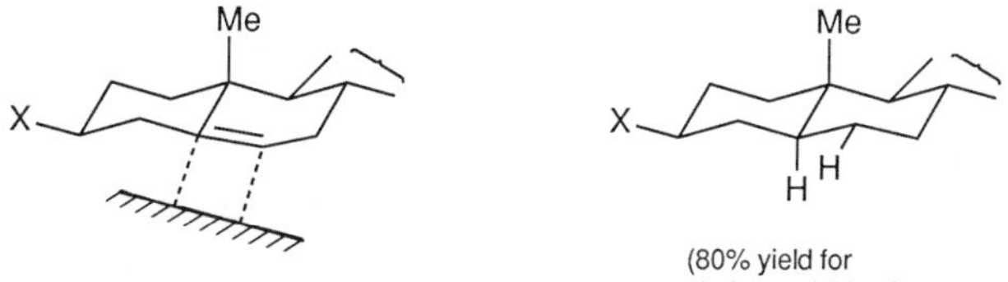

(80% yield for cholesterol (X = OH) [2])

On the other hand, when the substituent X is axial, the fit to the catalyst is hindered on both sides and reduction gives a mixture of *cis* and *trans* decalins in comparable amounts.

6.   *Homogeneous catalytic hydrogenation.*   The use of complexes of rhodium or ruthenium as catalysts enables hydrogenation to be carried out in homogeneous solution.

The rhodium reagent is Wilkinson's catalyst, $(Ph_3P)_3RhCl$ (p. 575). Alkenes and alkynes are reduced at room temperature and atmospheric pressure, usually in benzene solution. The mechanism, illustrated for ethylene ($L = PPh_3$), is:

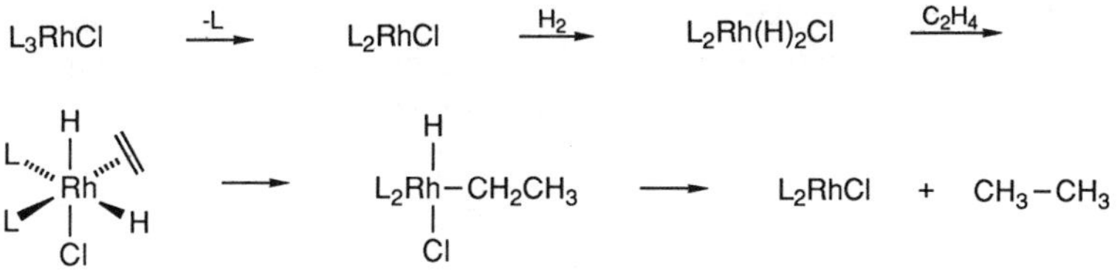

The method has several advantages. First, only alkenes and alkynes are reduced; other common groups such as C=O, C≡N, and NO$_2$ are unaffected. This makes possible selective reductions such as

Second, mono- and disubstituted alkenes are reduced much more rapidly than tri- or tetrasubstituted ones. This provides a further degree of selectivity, illustrated by the reduction of carvone to dihydrocarvone – leaving the trisubstituted double bond intact – in over 90% yield [6]:

Third, hydrogenolysis does not occur. For example, benzyl cinnamate gives the dihydro derivative,

whereas hydrogenation on a metal catalyst results also in cleavage of the O-benzyl bond to give PhCH$_2$CH$_2$CO$_2$H (cf. p. 643).

The major disadvantage of the method is that, because of the strong affinity of the rhodium complex for carbon monoxide, aldehydes are degraded, for example, PhCH=CHCHO → PhCH=CH$_2$. The ruthenium complex used is (Ph$_3$P)$_3$RuClH, which is formed *in situ* from (Ph$_3$P)$_3$RuCl$_2$ and molecular hydrogen in the presence of a base such as triethylamine. It is specific for the hydrogenation of monosubstituted alkenes and disubstituted acetylenes, which yield cis-alkenes.

(ii) The diimide method.   Diimide is an unstable compound which is usually made by the copper(ɪɪ)-catalyzed oxidation of hydrazine with air or hydrogen peroxide. In the absence of a species which traps it, it decomposes to nitrogen and hydrogen but when it is generated in the presence of an alkene and a small quantity of acetic acid, rapid *syn*-stereospecific reduction occurs, the driving force being the great stability of the nitrogen molecule compared with the —N=N— system. The role of the acetic acid is almost certainly to catalyze the formation of the *cis*-isomer of diimide from the more stable *trans*-isomer, the former then reacting with the alkene *via* a cyclic transition state:

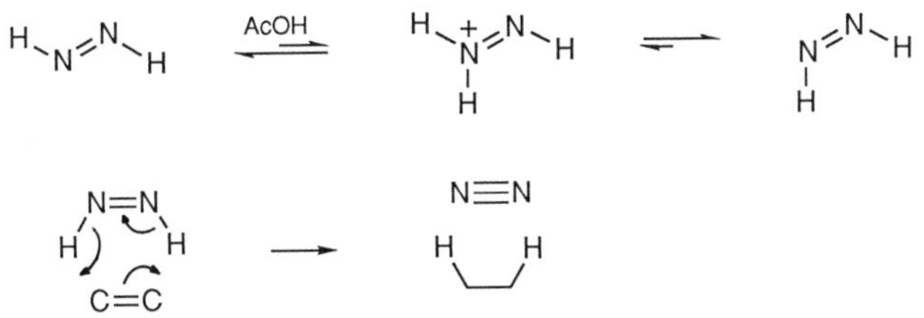

Alkynes and azo-compounds are also reduced, but carbonyl-containing groups, nitro groups, sulfoxides, and S—S bonds are not affected.

(iii) Hydroboration. The organoboranes formed by alkenes with diborane or other boronating agents (p. 483) are solvolyzed by organic acids (but not by mineral acids) to alkanes:

$$R_3B \quad + \quad 3\,R'CO_2H \quad \longrightarrow \quad 3\,RH \quad + \quad (R'CO_2)_3B$$

It is thought that reaction occurs as follows:

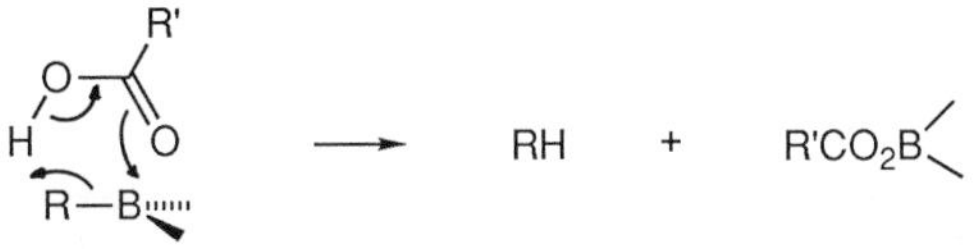

$$\longrightarrow \quad RH \quad + \quad R'CO_2B\big\langle$$

In the case of dienes, it is possible to reduce selectively the less substituted alkene by using a hindered borane such as that from trimethylethylene, Sia_2BH (cf. p. 485), e.g.

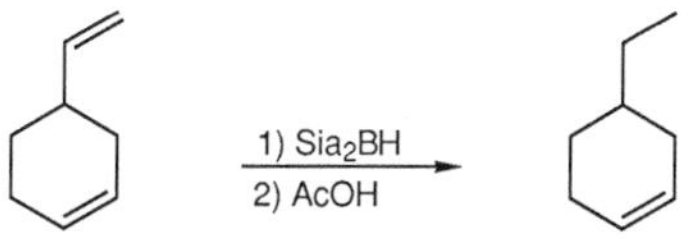

$$\xrightarrow[\text{2) AcOH}]{\text{1) Sia}_2\text{BH}}$$

Aldehydes, ketones, acids, esters, nitriles, and epoxides are also reduced.

(c) Conjugated alkenes

(i) Electron transfer. In addition to being reducible by catalytic methods, conjugated dienes and polyenes, and compounds containing C=C bonds conjugated with carbonyl-containing groups, are reducible with electron-transfer agents. The reason is that the uptake of electrons by these systems gives delocalized intermediates whereas the addition of electrons to alkenes does not:

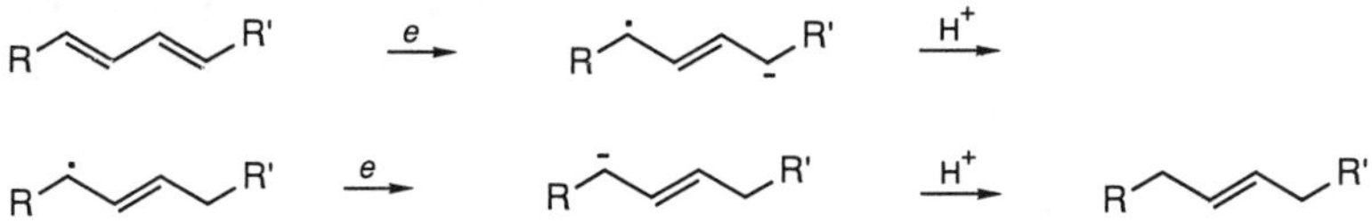

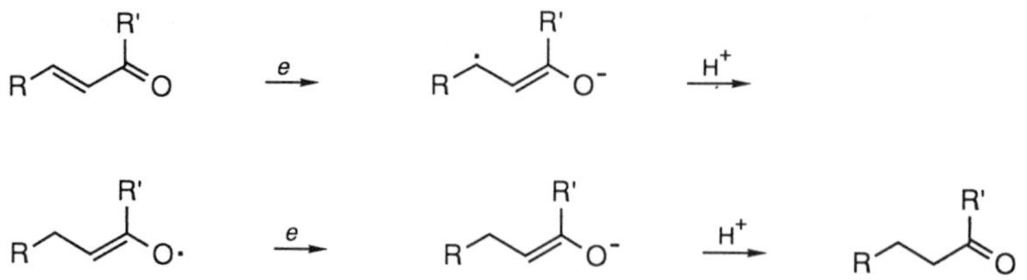

The common reducing agents in this category are sodium and an alcohol, sodium amalgam, zinc and acetic acid, and the metal-ammonia and metal-amine systems. The last two groups are discussed more fully below.

The reduction of a conjugated diene occurs by 1,4-addition, e.g.

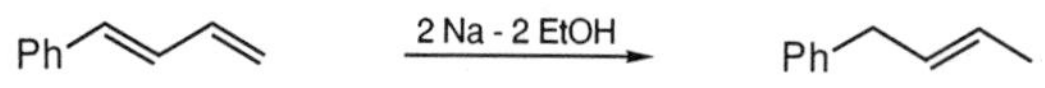

A C=C bond conjugated with an aromatic ring is also reduced, for the aromatic π-system is available to delocalize the charge on the anion, but the 1,2-adduct is formed since formation of the 1,4-adduct would be accompanied by the loss of aromatic stabilization energy. For example, stilbene is reduced to bibenzyl,

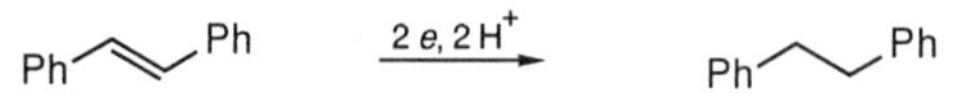

The reduction of an αβ-unsaturated carbonyl compound gives an enolate ion as the intermediate. Although this is protonated more rapidly on oxygen than carbon, the enol rapidly tautomerizes to the more stable carbonyl compound, e.g.

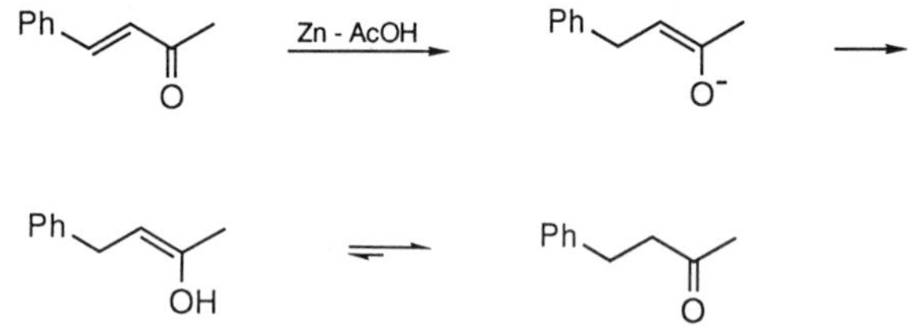

When only one equivalent of the proton-donor is used, the enolate remains as such and can be used for generating new C—C bonds, e.g.

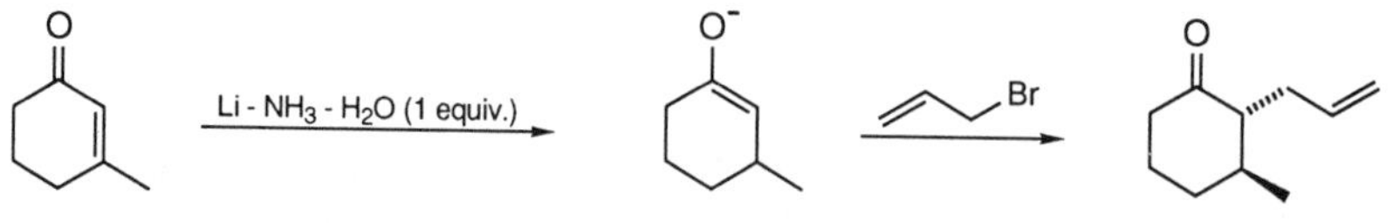

Metal–ammonia and metal–amine solutions. Strongly electropositive metals dissolve in liquid ammonia and in some amines to give characteristic blue solutions which contain metal cations and solvated electrons. Reduction to give amide ions is slow, but can be accelerated by the addition of catalytic amounts of transition-metal ions, e.g.

$$Na \xrightarrow{NH_3} Na^+ + e^- \text{(solvated)} \xrightarrow{(Fe^{3+})} Na^+ + NH_2^- + 1/2 H_2$$

The use of sodamide in liquid ammonia as a powerful catalyst for condensations has already been described (p. 242). The solution before amide formation acts as a powerful source of electrons for the reduction of organic compounds. A proton source is usually added to complete the reaction, for ammonia itself is a very weak proton donor. The donor should not be a strong enough acid to liberate hydrogen rapidly with sodium; ethanol is commonly used, for it reacts with sodium only very slowly at the boiling point of liquid ammonia. The reducing power of the system increases with the strength of the proton donor: the stronger this is, the more rapidly is the initial equilibrium displaced to the right:

$$R \; + \; e \; \rightleftharpoons \; R^{\bar{\cdot}} \; \xrightarrow{\;XH\;} \; RH\cdot \; + \; X^-$$

As the affinity of R for electrons is reduced, increasingly powerful proton donors are needed to effect reduction.

The use of ammonia is limited by the low solubility in it of most organic compounds at its boiling-point ($-33°C$). Co-solvents such as ethers can be added, but they depress the solubility of the metals: high mutual solubility is rare. Primary amines such as ethylamine are more powerful solvents and lithium is the most soluble of the alkali metals in such solvents. Since the reducing power of the system increases with the concentration of the metal, lithium and ethylamine provide about the most vigorous available conditions.

In general, these systems are of less value for the reduction of C=C bonds than in the aromatic series (p. 641) partly because so many effective alternatives are available for the former and partly because the conditions are often too vigorous. However, yields in simple cases are high, e.g.

98%

and in certain situations these methods succeed where catalytic methods fail. For example, the C_8—C_9 and especially the C_8—C_{14} double bonds in steroid systems are almost impossible to reduce catalytically but are reduced efficiently with sodium in liquid ammonia, e.g.

The main product is usually the thermodynamically most stable stereoisomer, as in the above example. The first protonation, at C-8, preferentially occurs from above the molecular plane, giving an axial C—H bond and therefore equatorial C—C bonds and protonation of the ensuing enolate gives the more stable *trans* adduct.

(ii) Hydride-transfer.   Compounds in which C=C is conjugated to substituents of −*M* type are susceptible to nucleophiles (p. 91), so that reductions of the type,

might be expected with αβ-unsaturated carbonyl compounds. When the −*M* group is —CHO, this does not normally occur to a significant extent and reduction of the carbonyl occurs instead, e.g.

When conjugation is to a keto group (which is less reactive than —CHO to nucleophiles), mixtures of carbonyl-reduced and fully reduced products are formed, e.g.

Lithium aluminium hydride has a relatively stronger tendency for reaction at the carbonyl group: in the above case, it gives 94% of 2-cyclohexenol and only 2% of cyclohexanol.

However, if the −*M* substituent is not itself reducible by the hydride, selective reduction of the C=C bond can be effected. For example, reduction of an αβ-unsaturated nitro-compound by sodium borohydride,

was employed in the synthesis of chlorophyll (section 22.4). The more powerful hydride-donor lithium aluminium hydride reduces both the C=C bond and the nitro group in this situation and the sequence, $ArCH=CHNO_2 \rightarrow ArCH_2CH_2NH_2$, is employed in the building up of isoquinoline rings from aromatic aldehydes (p. 706).

(d) Alkynes

(i) Catalytic reduction.   The C≡C bond can be fully reduced on the catalysts used for the reduction of alkenes, but this is seldom needed in synthesis. It is much more useful to exploit the fact that the catalytic reduction of C≡C to C=C is faster than the reduction of C=C to C—C. However, it is not satisfactory simply to use the conventional catalysts and to inject the amount of hydrogen needed for reduction to the alkene level, since this gives mixtures of

partially and fully reduced compounds. The technique is to deactivate ('poison') the catalyst so that reduction of C=C to C—C is inhibited, and this is usually done by using *Lindlar's catalyst*, in which palladium is deposited on a calcium carbonate support to which lead acetate and a small quantity of quinoline are added as the 'poison'. The reduction product is the *cis*-alkene.

(*ii*) *Hydroboration.* The hydroboration of alkynes is carried out in the same way as that of alkenes. With terminal alkynes, dihydroboration occurs to some extent,

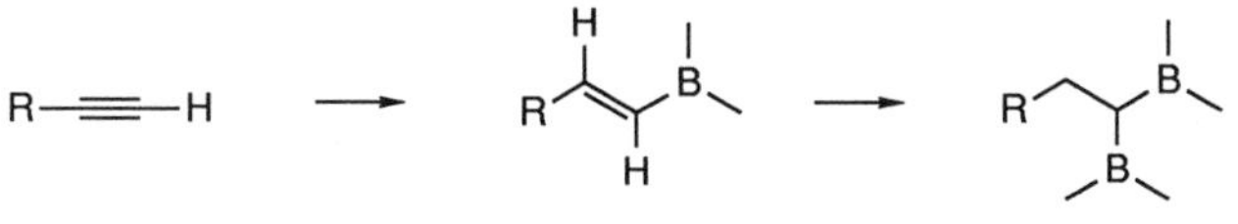

so that, on hydrolysis, complete reduction occurs. To prevent this, it is best to use the hindered and more selective disiamyl borane (p. 485) which allows quantitative monohydroboration of both mono- and disubstituted acetylenes, e.g.

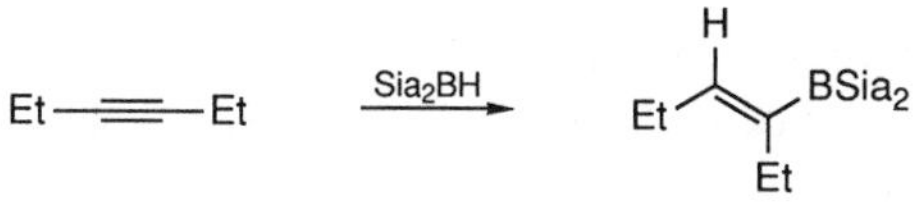

The products can be hydrolyzed with acetic acid at room temperature to give specifically *cis*-alkenes.

(*iii*) *Electron transfer.* Electrons are transferred to alkynes more readily than to alkenes (cf. the greater reactivity of alkynes towards nucleophiles). The best reagents are the metal–amine and metal–ammonia systems; no added proton donor is needed. Reaction is thought to occur by alternating electron and proton transfers:

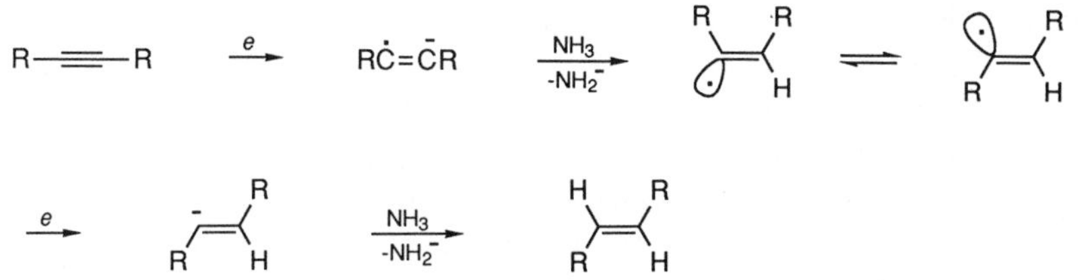

The formation of the *trans*-product is thought to be dictated by the occurrence of a rapid equilibrium between the *cis* and *trans* vinyl radicals in which the latter, being the less congested, predominates; the *trans* anion is thereby formed the faster. This method is therefore complementary to the catalytic methods. For example, *trans*-cyclononene can be obtained from cyclononyne:

71%

(*iv*) *Hydride-transfer.* Diisobutyl aluminium hydride (p. 659) reduces alkynes to alkenes, but lithium aluminium hydride is effective only when the alkyne has a hydroxyl group in the α-position: the hydroxyl group facilitates reaction through the formation of a complex with aluminium, and the product is the *trans*-alkene.

(*e*) *Aromatic rings*

(*i*) *Catalytic hydrogenation.* Aromatic rings can be reduced on the catalysts suitable for alkenes but more vigorous conditions are required because aromatic stabilization energy is lost in the process. The conditions vary according to the amount of stabilization energy which is sacrificed, e.g. naphthalene is reduced to tetralin more easily than benzene is reduced to cyclohexane (cf. p. 38). Typical conditions on Adams' catalyst are 10 hours at 100–150°C and 100–150 atmospheres (i.e. in a steel bomb), compared with 1 hour at 20°C and atmospheric pressure for the reduction of simple alkenes.

If there is a danger of the catalyst becoming poisoned, Raney nickel is normally selected because it is cheaper than palladium or platinum and can be used in excess. This is particularly important with sulfur-containing molecules but it must be remembered that sulfur is removed from the system during reduction (p. 672). Pyridine also poisons catalysts, but its hydrochloride can be reduced efficiently over platinum in ethanol.

Reduction in mild conditions gives mainly the *cis* product, but if the *trans* product is the more stable it can be formed in vigorous conditions:* e.g. the reduction of naphthalene in solution over platinum gives mainly *cis*-decalin whereas over copper chromite in the gas phase *trans*-decalin predominates.

Those aromatic compounds which are in equilibrium with significant proportions of aliphatic tautomers can be reduced in conditions appropriate to alkenes. For example, *m*-dihydroxybenzene is reduced, *via* its diketo-tautomer (p. 13), at 50°C and 80 atmospheres over Raney nickel:

---

* This is an example of thermodynamic control (p. 68). In vigorous conditions, equilibrium is established with the *cis* product while the slower *trans* reduction gradually removes the reactant to give the more stable product.

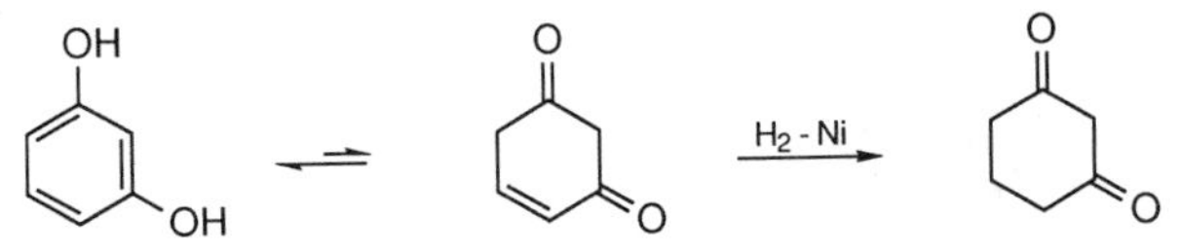

(*ii*) *Electron transfer*. Whereas catalytic methods normally fully reduce aromatic rings, electron-transfer reagents can be highly selective.

Reaction occurs by successive electron and proton transfers, e.g.

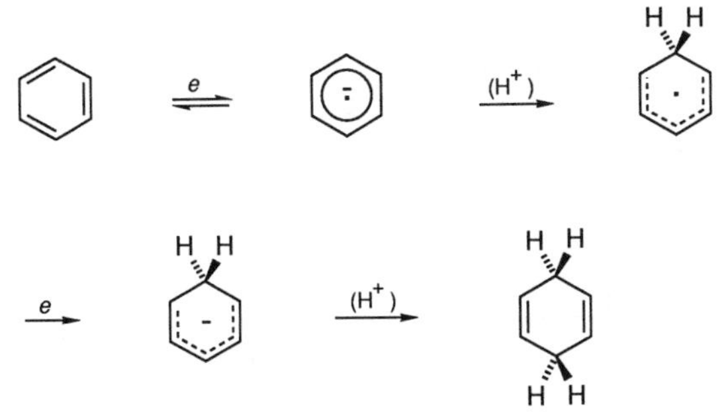

Sodium in liquid ammonia is commonly used as the reducing agent. Ammonia can act as the source of protons, but when an electron-releasing substituent such as alkyl, OMe, or NMe_2 is present, the first equilibrium lies well to the left and, because ammonia is a weak proton-transfer agent, reaction is very slow. It is therefore usual to include an alcohol to trap the anion radical rapidly.

The structure of the product is determined by the site of the first protonation: the anion that is then formed reacts at the central carbon of the delocalized system to complete the 1,4-addition. Electron-releasing substituents direct the first protonation to the *ortho* position, e.g.

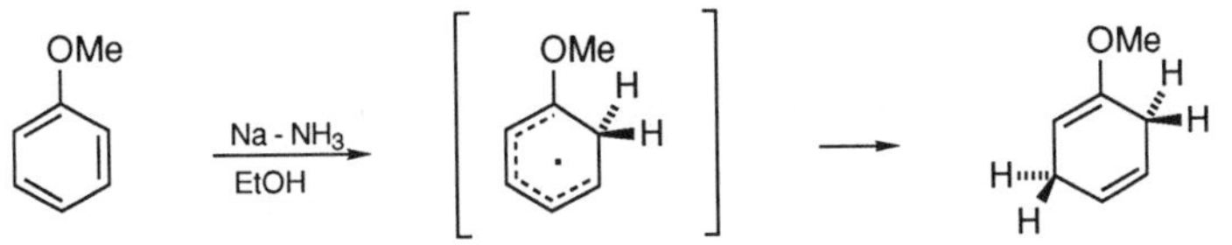

Acid-catalyzed hydrolysis (pH 2–3) of the product (an enol ether) gives an unsaturated ketone which, in slightly more acidic conditions (pH 1), tautomerizes to the conjugated isomer:

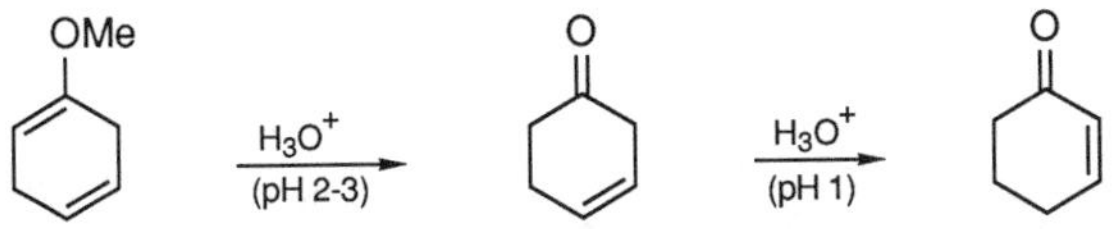

In contrast, the $—CO_2^-$ substituent directs protonation to the *para* position, e.g.

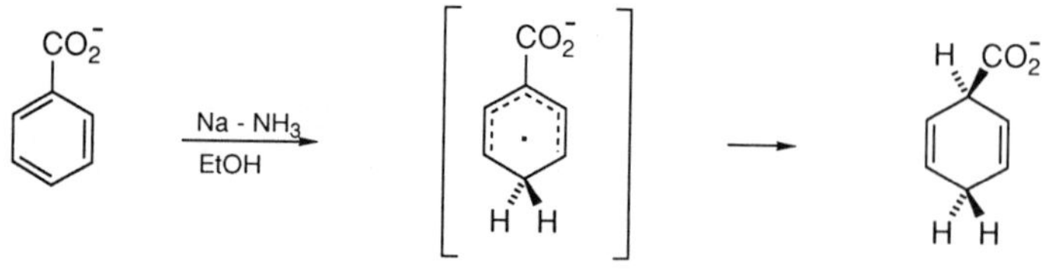

95% [5]

With naphthalene, a variety of products can be obtained, depending on the conditions:

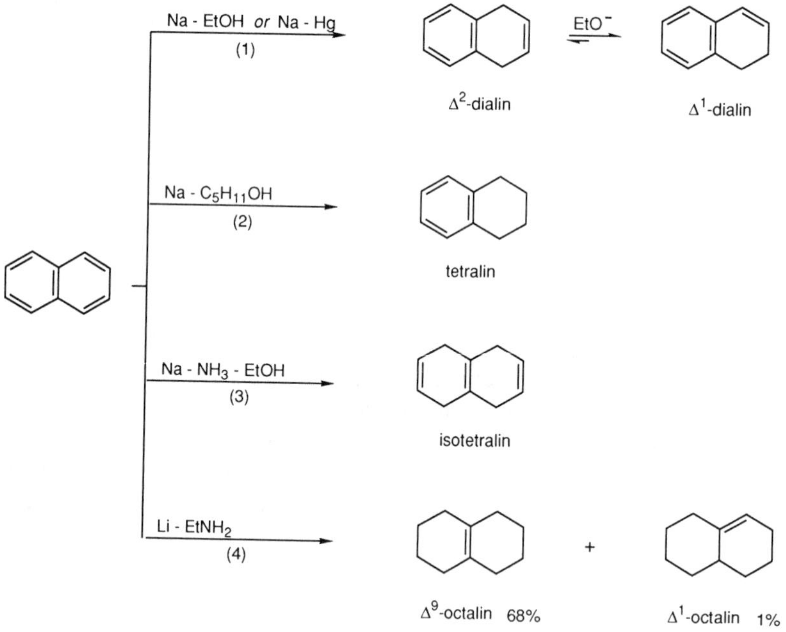

(1) The mildest conditions give a dihydro compound:

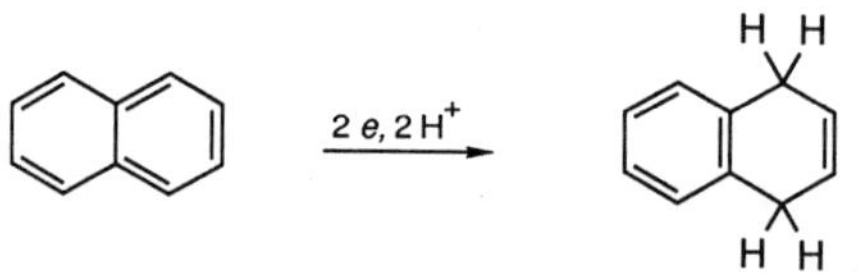

The product is isomerized with base to the more stable (conjugated) Δ^1-dialin, *via* a delocalized anion,

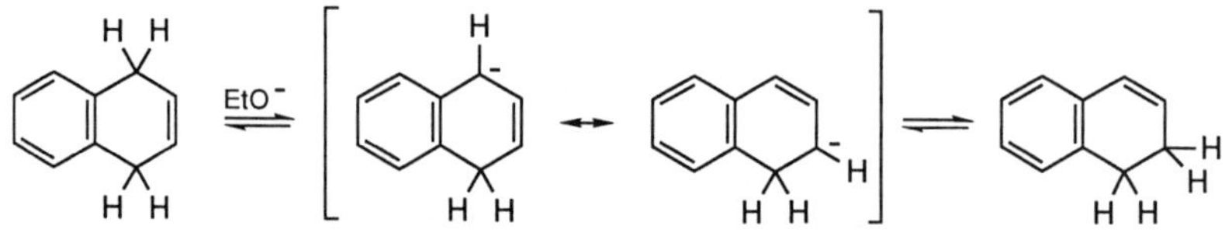

(2)   Further reduction occurs in the higher-boiling 1-pentanol probably through formation of Δ^2-dialin, isomerization as in (1), and reduction of the conjugated Δ^1-dialin.

(3)   The use of liquid ammonia with an added proton-donor enables both benzenoid rings to be reduced in the 1,4-manner.

(4)   Lithium in ethylamine provides the most vigorous conditions (p. 637): initial reduction as in (3) is followed by isomerization and further reduction.

The reductions of 1-naphthol and ethyl 2-naphthyl ether provide an interesting contrast. The former is reduced by lithium in liquid ammonia containing ethanol in the *unsubstituted* ring in 98% yield [4],

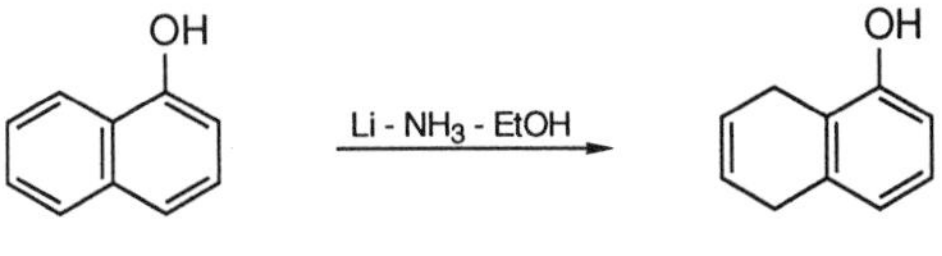

whereas the latter is reduced by sodium and ethanol in the *substituted* ring to give 2-tetralone in 45% yield after hydrolysis [4]:

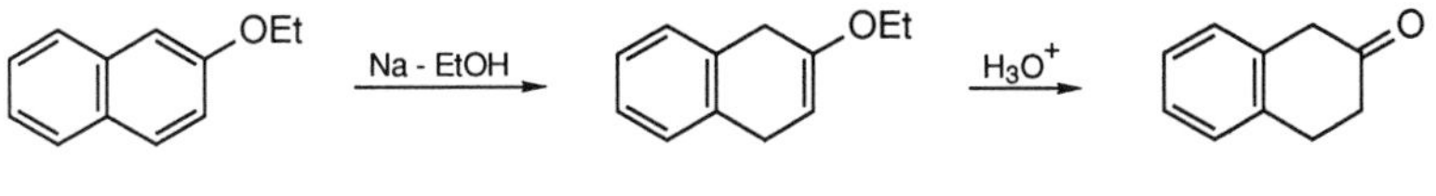

The difference is consistent with the mechanism described for the reduction of anisole. With 1-naphthol, protonation of the anion-radical at the 2-position and subsequent protonation at the 5-position would result in the complete loss of the aromatic stabilization energy and is less favourable than 1,4-addition in the unsubstituted ring. In contrast, protonation of the anion-radical from ethyl 2-naphthyl ether at the 1-position and subsequent protonation at the 4-position leaves a benzenoid system.

20.3 Hydrogenolysis

(a)   *Benzylic systems*

The benzylic group, when attached to OH, OR, OCOR, NR_2, SR, or halogen, is particularly susceptible to nucleophilic reducing agents, catalytic reduction, and electron-transfer reagents.

Catalytic reduction is widely employed, as in the following examples:

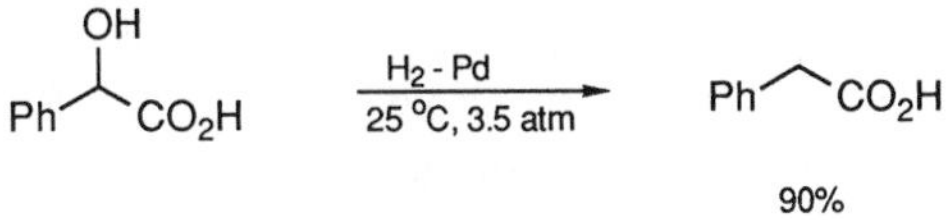

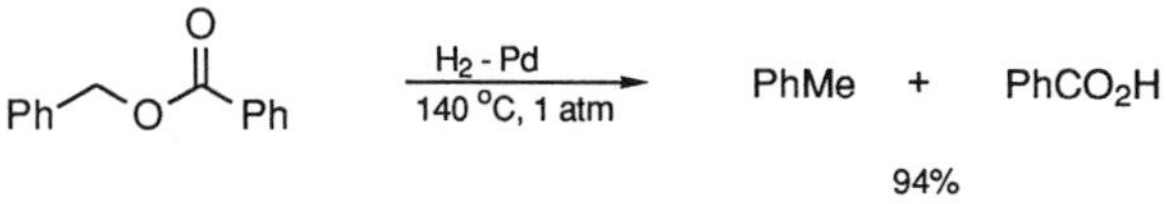

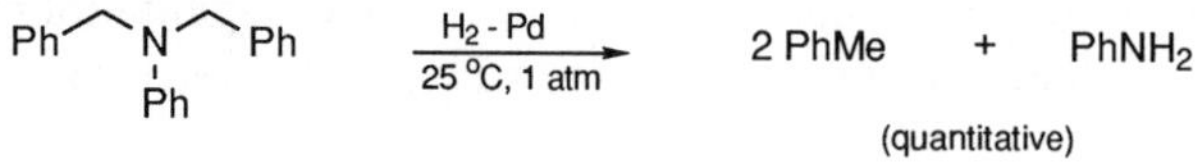

$$Ph \diagdown N(Ph) \diagdown Ph \xrightarrow[25\,°C,\,1\,atm]{H_2 - Pd} 2\,PhMe \quad + \quad PhNH_2$$

(quantitative)

Electron-transfer reduction is probably facilitated by the ability of the benzenoid ring to delocalize the charge of an anion, i.e.

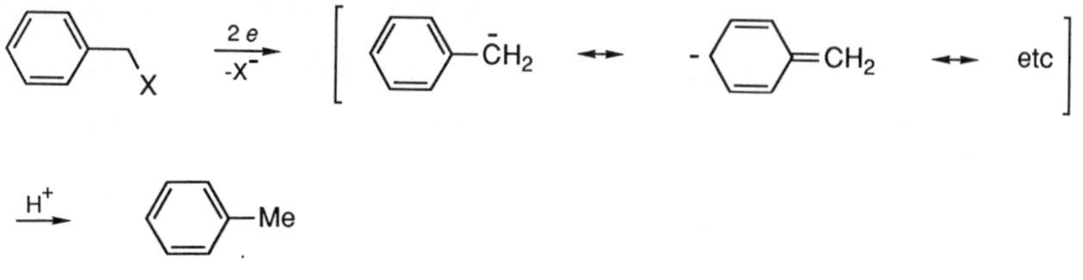

$$\xrightarrow{H^+} \text{(ring)}\!-\!Me$$

The following are examples:

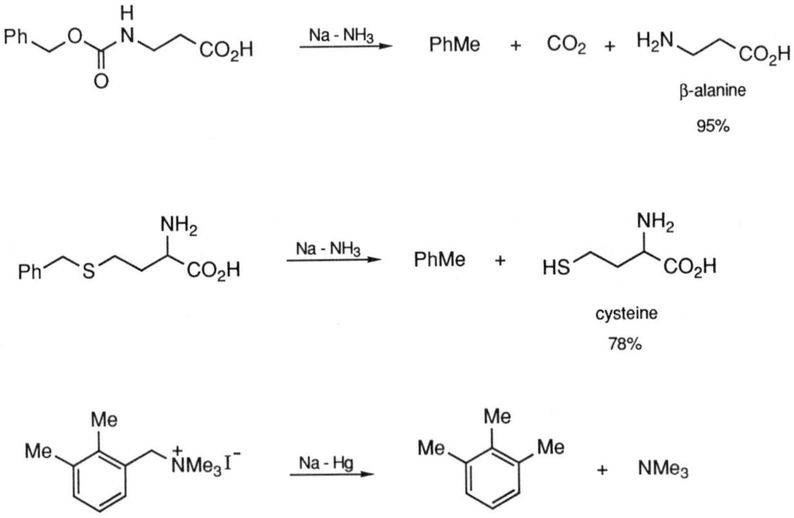

β-alanine

95%

cysteine

78%

The first example illustrates the use of the benzyl group as part of a protecting agent in amino acid chemistry: the benzyloxycarbonyl group is used to protect the amino group in peptide synthesis and is afterwards removed as above or catalytically (p. 330). The second example represents a standard procedure for the protection of thiol groups: the catalytic method is unsuitable for the removal of the benzyl residue because sulfur is present. The final example illustrates a synthetic method for introducing methyl groups *ortho* to existing methyl groups in aromatic rings: the reactant is obtained from *o*-xylene *via* the Stevens rearrangement (p. 447) and is converted into 1,2,3-trimethylbenzene by sodium amalgam at 80°C in 87% yield [4].

Diphenylmethyl systems are more easily hydrogenolyzed than benzyl systems. For example, benzilic acid is reduced in 95% yield to diphenylacetic acid by treatment with red phosphorus and iodine in refluxing acetic acid [1]:

Ph—C(OH)(Ph)—CO$_2$H →(red P - I$_2$)→ Ph—C(Ph)(Ph)—CO$_2$H

Triphenylmethyl (trityl) systems are still more easily reduced. Advantage is taken of this, together with the selectivity of triphenylmethyl chloride for primary alcoholic groups as compared with secondary and tertiary, in protecting primary alcohols (e.g. in the synthesis of adenosine triphosphate, section 22.2):

$$Ph_3C-Cl \quad + \quad HOCH_2R \quad \longrightarrow \quad Ph_3COCH_2R \quad \xrightarrow{H_2 - Pd} \quad Ph_3CH \quad + \quad HOCH_2R$$

Triphenylmethyl systems readily form the trityl cation and this is particularly susceptible to reduction by hydride-transfer agents. For example, triphenylmethane can be obtained in 70–80% yield by Friedel–Crafts reaction between benzene and carbon tetrachloride in the presence of aluminium trichloride (p. 361) followed by the addition of diethyl ether [1]:

$$3\,PhH \quad + \quad CCl_4 \quad \xrightarrow[-3\,HCl]{AlCl_3} \quad Ph_3C-Cl \quad \xrightarrow[-Cl^-]{AlCl_3} \quad Ph_3C^+$$

$$Ph_3C^+ \quad \text{H—CH(Me)—OEt} \quad \longrightarrow \quad Ph_3CH \quad (+ \quad \text{=CH—OEt}^+ \quad \xrightarrow{Cl^-} \quad EtCl \quad + \quad MeCHO\,)$$

(b)   Allylic systems

Allylic systems behave in most reactions analogously to benzylic systems. However, whereas the latter are normally reduced by hydrogenolysis rather than by hydrogenation of the aromatic ring, the former more readily undergo hydrogenation with those reducing agents which react with the C=C bond, so that catalytic reduction is in general unsuitable for the hydrogenolysis of allylic systems.

Fortunately, lithium aluminium hydride and sodium in an amine are suitable reagents since neither reacts with isolated C=C bonds. For example, allyl ethers are cleaved by lithium aluminium hydride in tetrahydrofuran,

$$R\diagup\!\!\!\diagdown\!\!\!\diagup OR' \quad \xrightarrow{LiAlH_4} \quad R\diagup\!\!\!\diagdown\!\!\!\diagup Me \quad + \quad R'OH$$

and allyl esters are cleaved by sodium in ethylamine,

$$R\diagup\!\!\!\diagdown\!\!\!\diagup O\!-\!CO\!-\!R' \quad \xrightarrow{Na - EtNH_2} \quad R\diagup\!\!\!\diagdown\!\!\!\diagup Me \quad + \quad R'CO_2H$$

The corresponding alkyl compounds are unaffected in these conditions.

(c)   Alkyl systems

Alkyl systems are generally much less easy to hydrogenolyze than the corresponding benzyl systems.

(*i*)   *Alcohols* cannot be hydrogenolyzed directly but can be converted into their toluene-*p*-sulfonates which are reducible with sodium cyanoborohydride or with lithium aluminium hydride (i.e. by taking advantage of the fact that the toluene-*p*-sulfonate anion is so good a leaving group in S_N2 displacements, p. 103):

$$R \diagdown OH \xrightarrow{\text{TsCl}} R \diagdown OTs \xrightarrow{\quad} RMe \ + \ TsO^-$$

with $H-\bar{A}lH_3$ adding across.

Alternatively, the thioester formed with PhCSCl can be reduced with tributylstannane in a radical-chain reaction that is usually initiated with AIBN (cf. p. 538):

$$ROH \ + \ \underset{Cl}{\overset{S}{\diagup\!\!\diagdown}}Ph \quad\longrightarrow\quad \underset{R\diagdown O}{\overset{S}{\diagup\!\!\diagdown}}Ph \ + \ HCl$$

propagation

$$RO\diagdown\overset{S}{\diagup}Ph \ + \ \cdot SnBu_3 \quad\longrightarrow\quad RO\diagdown\overset{S-SnBu_3}{\diagup}Ph$$

$$\longrightarrow \quad R\cdot \ + \ O\diagup\overset{S-SnBu_3}{\diagdown}Ph$$

$$R\cdot \ + \ H-SnBu_3 \quad\longrightarrow\quad RH \ + \ \cdot SnBu_3$$

This method is of particular value compared with the tosylate method when the organic compound contains other groups, such as carbonyl, which would also be reduced by LiAlH$_4$. However, although good yields are obtained from secondary alcohols, they are generally low with primary alcohols.

(*ii*)   *Halides* are much more reactive than alcohols in S_N2 displacements and are correspondingly more susceptible to hydride-transfer from lithium aluminium hydride or, better, sodium cyanoborohydride. For example, with the latter reagent in hexamethylphosphoramide, 1-iododecane gives decane in about 90% yield [6]:

$$\diagup\!\!\diagdown\!\!\diagup\!\!\diagdown\!\!\diagup\!\!\diagdown\!\!\diagup\!\!\diagdown I \xrightarrow{\text{NaBH}_3\text{CN}} \diagup\!\!\diagdown\!\!\diagup\!\!\diagdown\!\!\diagup\!\!\diagdown\!\!\diagup\!\!\diagdown$$

Electron-transfer agents are also efficient. Since the highly electropositive metals like sodium induce Wurtz coupling, the metal–amine systems are unsuitable, but magnesium amalgam in water and a zinc-copper couple in ethanol can be used, e.g.

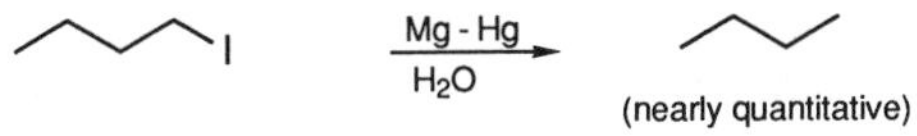

$$\text{(propyl iodide structure)} \xrightarrow[\text{H}_2\text{O}]{\text{Mg - Hg}} \text{(butene structure)}$$

(nearly quantitative)

Alternatively, magnesium in ether is used to form the Grignard reagent which is subsequently decomposed by water,

$$\text{RX} + \text{Mg} \longrightarrow \text{RMgX} \xrightarrow{\text{H}_2\text{O}} \text{RH}$$

Two radical-mediated processes are available. In one, the reducing agent is chromium(II) ion, usually complexed with ethylenediamine. It is thought to act by first removing the halide to form an alkyl radical,

$$\text{R--Br} + \text{Cr}^{\text{II}} \longrightarrow \text{R·} + \text{Cr}^{\text{III}} + \text{Br}^-$$

and then to complete reduction *via* an organochromium species:

$$\text{R·} + \text{Cr}^{\text{II}} \longrightarrow \text{R--Cr}^{\text{III}} \xrightarrow{\text{H}^+} \text{RH} + \text{Cr}^{\text{III}}$$

Alkenyl halides can also be reduced in this way.

The second is a radical-chain reaction, initiated by AIBN, in which the reducing agent is tributylstannane:

propagation $\left\{ \begin{array}{l} \text{R--Br} + \text{·SnBu}_3 \longrightarrow \text{R·} + \text{Br--SnBu}_3 \\ \\ \\ \text{R·} + \text{H--SnBu}_3 \longrightarrow \text{RH} + \text{·SnBu}_3 \end{array} \right.$

Electron-transfer agents react with 1,2-dihalides in a different manner: both halogen atoms are eliminated and a C=C bond is introduced. Zinc is normally used as the reducing agent:

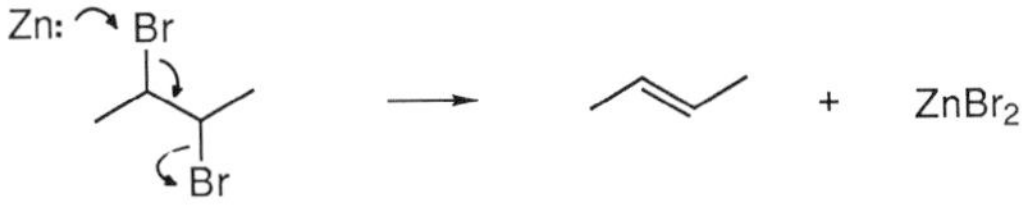

$$\text{Zn:} \overset{\text{Br}}{\underset{\text{Br}}{\text{(structure)}}} \longrightarrow \text{(alkene)} + \text{ZnBr}_2$$

For example, allene may be obtained in 80% yield by the treatment of 2,3-dichloropropylene with zinc in aqueous ethanol:

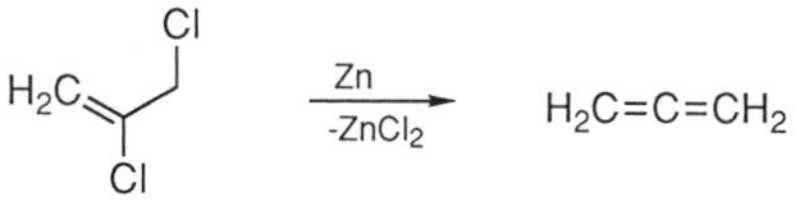

$$\text{H}_2\text{C} \overset{\text{Cl}}{\underset{\text{Cl}}{=}} \xrightarrow[\text{-ZnCl}_2]{\text{Zn}} \text{H}_2\text{C=C=CH}_2$$

(*iii*) *Acetals and ketals* are hydrogenolyzed to ethers in high yield by lithium aluminium hydride in the presence of aluminium trichloride:

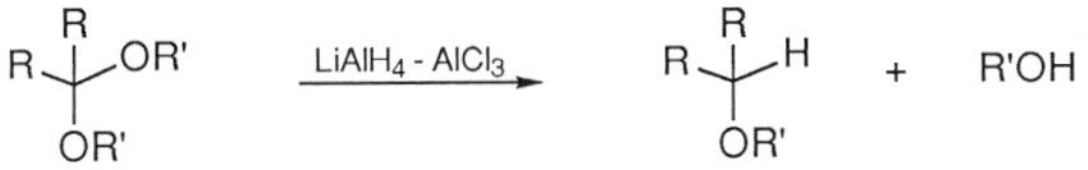

$$\text{R} \overset{\text{R}}{\underset{\text{OR'}}{\overset{|}{\underset{|}{\text{C}}}}} \text{OR'} \xrightarrow{\text{LiAlH}_4 \text{ - AlCl}_3} \text{R} \overset{\text{R}}{\underset{\text{OR'}}{\overset{|}{\underset{|}{\text{C}}}}} \text{H} + \text{R'OH}$$

(*iv*) *Primary amines* can be reduced by treating the derived toluene-*p*-sulfonamide with a large excess of hydroxylamine-*O*-sulfonic acid in alkaline ethanol. Use is made of the strong leaving-group potential of sulfates and sulfinates and the instability of the —N=NH grouping (cf. p. 650).

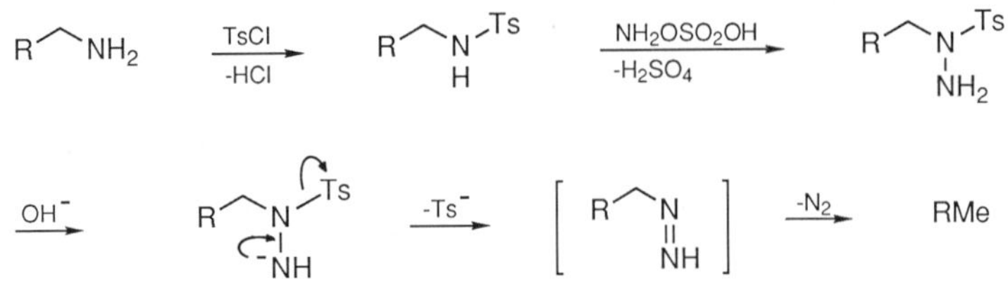

For example, 1-hexylamine gives hexane in 41% yield.

(*d*) Aromatic systems

Aryl halides can be reduced by the formation of the Grignard reagent (from bromides and iodides) or the lithium compound (from chlorides) followed by treatment with water, or with hydrazine over a palladium catalyst. For example, 2-bromonaphthalene is reduced to naphthalene in 95% yield in 5 minutes in boiling ethanol:

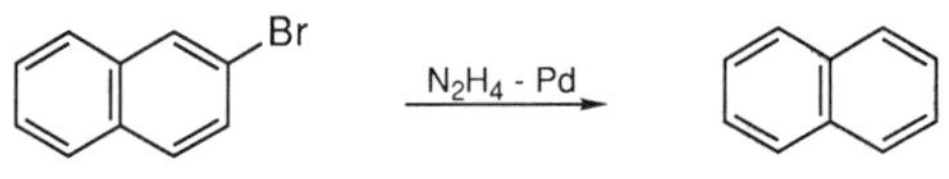

Reduction with chromium(II) ion is also efficient. For example, 1-bromonaphthalene with chromium(II) complexed with ethylenediamine in aqueous dimethylformamide gives 93–98% of naphthalene [6].

The halogen atoms at the 2- and 4-positions of pyridine and its derivatives can be efficiently removed both catalytically by hydrogen and by electron-transfer reducing agents (cf. the ease of nucleophilic substitution at these positions, p. 399). Since hydroxyl groups at these positions are convertible into chlorosubstituents with phosphorus oxychloride, this provides a method for reducing phenols, as in the Conrad–Limpach synthesis (p. 705), e.g.

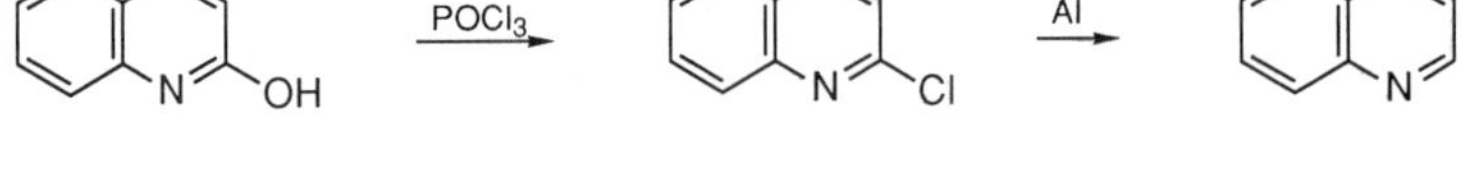

Phenols can be hydrogenolyzed *via* their phosphate esters by reduction with sodium in liquid ammonia, but yields are not usually above 50%.

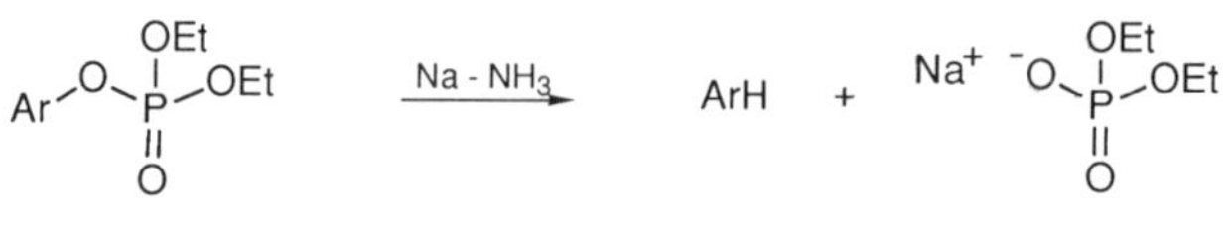

Aromatic amines cannot be hydrogenolyzed directly; they are deaminated by reduction of the corresponding diazonium ions (p. 413).

Aldehydes and ketones can be reduced to hydrocarbons, alcohols, and 1,2-diols.

(a) To hydrocarbons

Five methods are available. The choice between them is based on the sensitivity of other functional centres in the reactant in the reducing conditions.

(1) *Clemmensen method*. The reducing agent is amalgamated zinc and concentrated hydrochloric acid. The mechanism is not certain but may be as follows:

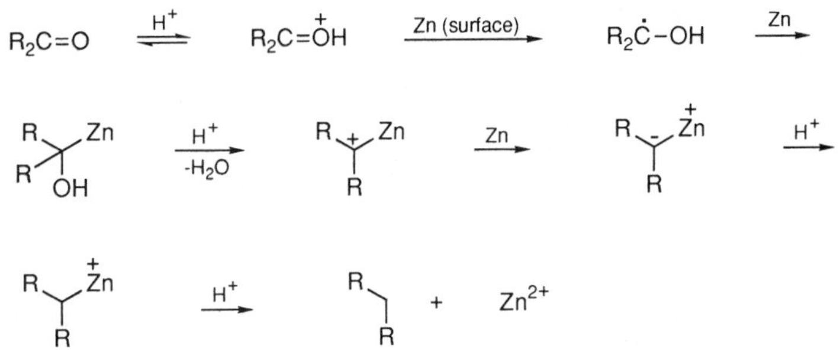

The method is particularly useful for ketones which contain phenolic or carboxylic groups. For example, the reduction of 3-benzoylpropionic acid in toluene gives 4-phenylbutyric acid in 85% yield [2]:

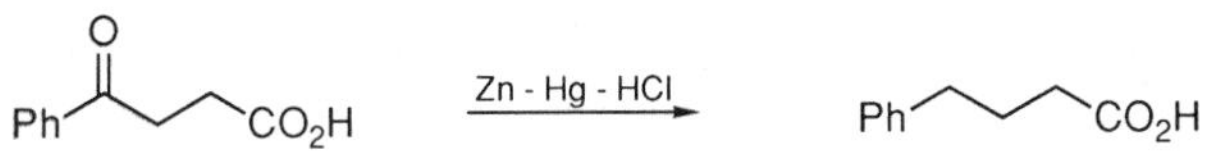

This type of reduction forms one step in the extension of benzenoid systems *via* Friedel–Crafts acylations (p. 368).

The reagent also reduces the $C=C$ bond in $\alpha\beta$-unsaturated ketones, acids, and esters. Benzyl halides and alcohols are hydrogenolyzed. Strongly hindered ketones give low yields and sometimes rearrangement products (e.g. $Ph_3C-CO-Ph \rightarrow Ph_2C=CPh_2$).

(2) *Wolff–Kishner method*. The hydrazones of aldehydes and ketones are reduced in vigorously basic conditions with the evolution of nitrogen, probably as follows:

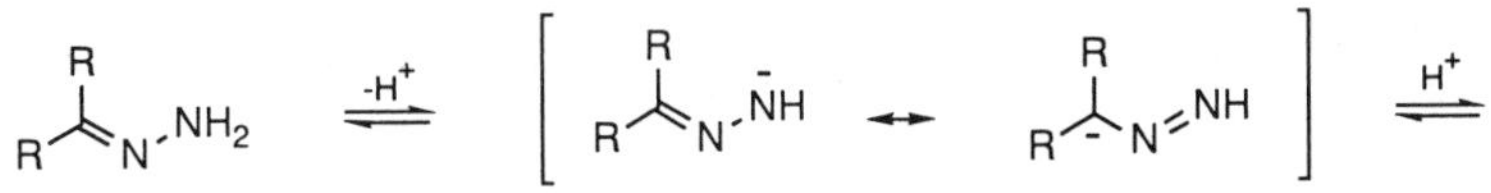

$$\left[\begin{array}{c} \text{H} \\ \text{R} \end{array} \begin{array}{c} \text{R} \\ \text{N=NH} \end{array} \right] \longrightarrow \begin{array}{c} \text{H} \\ \text{R} \end{array} \begin{array}{c} \text{R} \\ \text{H} \end{array} + N_2$$

One procedure is the *Huang–Minlon* modification. The hydrazone is formed by heating the carbonyl compound with hydrazine hydrate and potassium hydroxide in $(HOCH_2CH_2)_2O$ under a water condenser. After completion of the formation of the hydrazone, the water condenser is removed so that the water liberated in the first reaction is distilled and the temperature rises to over 200°C, so bringing about decomposition of the hydrazone.

A newer modification employs potassium t-butoxide as the base and dimethyl sulfoxide as solvent. Alkoxide bases are very much more powerful in this solvent than in hydroxylic solvents (p. 64) and reaction occurs at room temperature in high yield. For example, benzophenone gives about 90% of diphenylmethane:

$$\begin{array}{c} \text{Ph} \\ \text{Ph} \end{array}=O \xrightarrow{N_2H_4} \begin{array}{c} \text{Ph} \\ \text{Ph} \end{array}=N^{NH_2} \xrightarrow{Me_3CO^-/Me_2SO} Ph\diagdown Ph$$

(3) *Mozingo method*. The carbonyl compound is converted with ethanedithiol in the presence of a Lewis acid into its dithio-acetal or ketal and this is hydrogenolyzed over Raney nickel:

$$\begin{array}{c} \text{R} \\ \text{R} \end{array}=O + HS\diagdown\diagup SH \xrightarrow[20\,°C]{Et_2\overset{+}{O}-\overset{-}{B}F_3} \begin{array}{c} \text{R} \\ \text{R} \end{array}\diagup_S^S \xrightarrow{H_2-Ni} \begin{array}{c} \text{R} \\ \text{R} \end{array}\diagup_H^H$$

Alternatively, the cyclic dithio-compound is reduced by hydrogen-transfer from hydrazine at 100–200°C.

The Mozingo reaction is useful for reducing carbonyl compounds which are sensitive to mineral acid and bases, for the Clemmensen and Wolff–Kishner methods are then unsuitable.

(4) *Tosylhydrazone method*. Reaction of the carbonyl compound with toluene-*p*-sulfonylhydrazine gives the tosylhydrazone which is efficiently reduced by sodium borohydride or cyanoborohydride (Ar = *p*-tolyl):

$$\begin{array}{c} \text{R} \\ \text{R} \end{array}=O + H_2N_{\backslash N}\overset{O\ O}{\underset{H}{S}}{}_{Ar} \xrightarrow{-H_2O} \begin{array}{c} \text{R} \\ \text{R} \end{array}=N_{\backslash N}\overset{O\ O}{\underset{H}{S}}{}_{Ar}$$

$$\xrightarrow{NaBH_4} \begin{array}{c} \text{H} \\ \text{R}\overset{|}{\underset{R}{\,}}H \end{array} + N_2 + Na^+\ {}^-O_2SAr$$

Reaction probably occurs as follows:

$$H_3\overset{-}{B}-H \quad \begin{array}{c} \text{R} \\ \text{R} \end{array}=N_{\backslash N}\overset{O\ O}{\underset{H}{S}}{}_{Ar} \xrightarrow{-ArSO_2^-} \left[\begin{array}{c} \text{R} \\ \text{H}\overset{|}{\underset{R}{\,}} \end{array} N=NH \right] \longrightarrow \begin{array}{c} \text{R} \\ \text{H}\overset{|}{\underset{R}{\,}}H \end{array} + N_2$$

(5) *Lithium aluminium hydride.* Aromatic ketones are reduced by lithium aluminium hydride in the presence of aluminium trichloride. Reaction occurs by reduction to the alcohol (p. 652) followed by hydrogenolysis of the benzylic system, aided by the Lewis acid:

(b) *To alcohols*

Carbonyl compounds are reduced to alcohols by a variety of reagents. Of the three general classes of reductive process, catalytic hydrogenation is not normally chosen because it is slow, but both hydride-transfer and electron-transfer reagents are employed.

(i) *Hydride transfer.* The alkali-metal hydrides such as sodium hydride, NaH, are unsuitable as reducing agents because of their insolubility in organic solvents and, particularly, because of their powerful effects as catalysts for the base-catalyzed reactions which aldehydes and ketones undergo. Instead, lithium aluminium hydride, $LiAlH_4$, or sodium borohydride, $NaBH_4$, or derivatives in which some of the hydrogens have been replaced by other groups, are used; they are effective at transferring a hydride ion to a carbonyl group, but not as bases in condensations.

1. *Lithium aluminium hydride* was the first such compound to be developed. It is prepared by treating lithium hydride with aluminium trichloride,

$$4\ LiH\ +\ AlCl_3 \longrightarrow LiAlH_4\ +\ 3\ LiCl$$

and is soluble in ethers such as diethyl ether and tetrahydrofuran, in which it is normally used. The solvent must be scrupulously dry, since the hydride is destroyed by water,

$$LiAlH_4\ +\ 4\ H_2O \longrightarrow LiOH\ +\ Al(OH)_3\ +\ 4\ H_2$$

and it must also be handled as a solid with great care since it may inflame spontaneously while being crushed in a pestle and mortar and it explodes violently on strong heating. All hydroxyl-, amino-, and thiol-containing compounds liberate hydrogen quantitatively from it, e.g.

$$4\ EtOH\ +\ LiAlH_4 \longrightarrow (EtO)_4Al^-\ Li^+\ +\ 4\ H_2$$

Each of the four hydrogen atoms in lithium aluminium hydride is available for transfer to carbonyl groups, e.g.

each step occurring less rapidly than the preceding one. (As a result, the replacement of two or three hydrogen atoms of the aluminium hydride anion by alkoxy groups gives less reactive and more selective reducing agents: examples are described later.) Finally, hydrolysis of the aluminium alkoxide gives the alcohol.

2. *Sodium borohydride* is much less reactive. It can be used in alcoholic solvents and even in water, for it decomposes only enough to make the solution alkaline, after which it is stable.

$$NaBH_4 \quad + \quad 4\,H_2O \quad \longrightarrow \quad NaOH \quad + \quad B(OH)_3 \quad + \quad 4\,H_2$$

Chemoselectivity. The two reagents differ considerably in their reducing power. Lithium aluminium hydride reduces not only aldehydes and ketones but also acids, acid chlorides, esters, nitriles, imines, and nitro groups, whereas sodium borohydride reduces only aldehydes, ketones, acid chlorides, and imines. Neither reagent normally reduces C=C, C≡C, or N=N, although lithium aluminium hydride reduces alkynes containing α-hydroxy-substituents (p. 640) and reduces azo-compounds in the presence of a Lewis acid (p. 672). For the reduction of aldehydes and ketones, sodium borohydride is normally preferred, both because of its greater chemoselectivity and because of its greater ease of handling. Sodium borohydride is also selective between carbonyl groups in different environments. For example, with only one equivalent of reducing agent, the compound

is reduced only at the non-conjugated carbonyl position; the other carbonyl group is less reactive, owing to the delocalization energy in the conjugated system,

which is lost on addition of a nucleophile. Sodium cyanoborohydride, $NaBH_3CN$, is less reactive, and more selective, than the borohydride because the electron-attracting cyano-group stabilizes the negative charge on boron. It reduces aldehyde groups and imines but not the less reactive keto-groups and acid chlorides.

1. Stereoselectivity with cyclohexanones. With relatively small hydride-donors, such as the AlH_4^- and BH_4^- ions, steric hindrance to axial approach is less important than the stereoelectronic effect that favours it (p. 175), e.g.

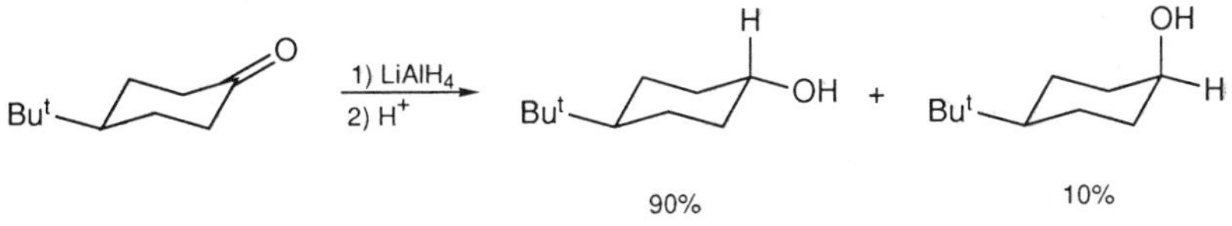

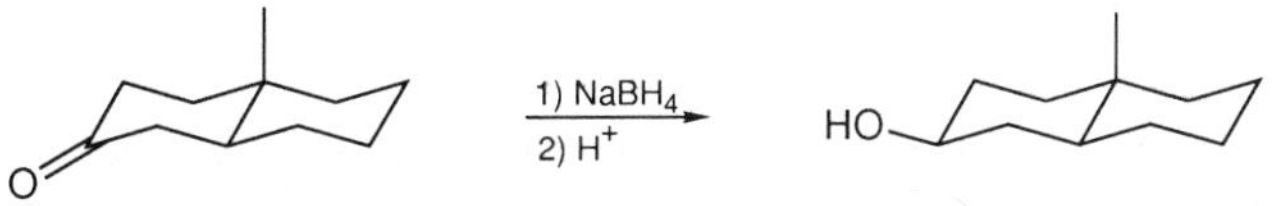

90%   10%

However, the opposite is the case with bulky hydrides, e.g.

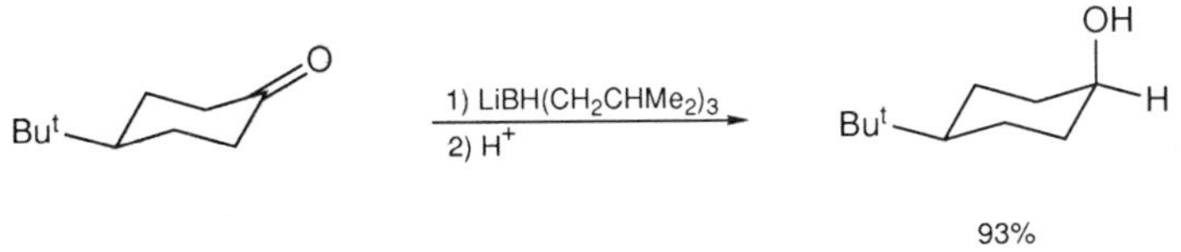

93%

2. Stereoselectivity with acyclic asymmetric ketones. The major product can be predicted by Cram's rule (p. 150), e.g.

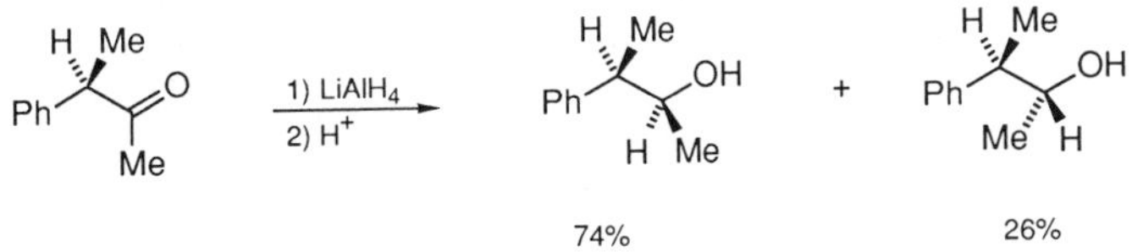

74%   26%

3. Stereoselectivity with symmetric ketones and an asymmetric hydride donor. (+)-α-Pinene reacts with the boron derivative 9-BBN (p. 485) to give an asymmetric adduct,

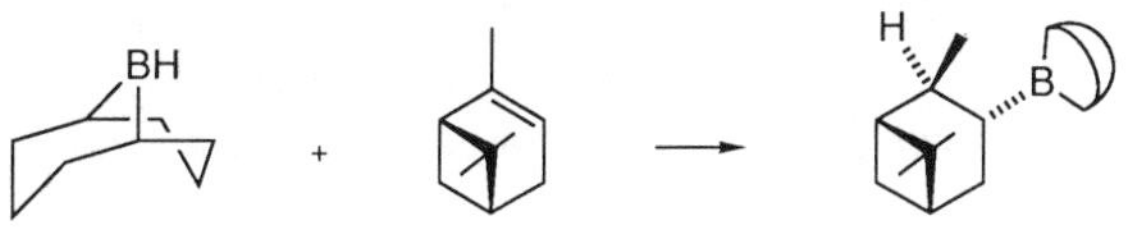

which reduces ketones with a very high degree of enantioselectivity, in some cases 100%. It is thought that the adduct transfers hydride to the ketone *via* a six-membered, boat-shaped cyclic transition state in which the larger group on the ketone preferentially lies away from the α-pinene residue to minimize steric congestion, e.g.

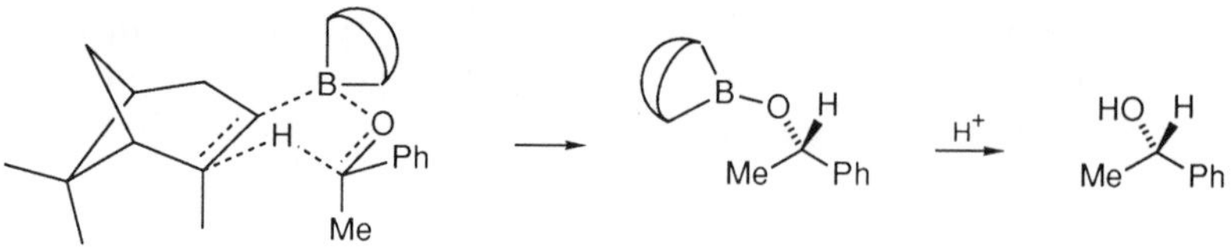

The adducts of 9-BBN with both (+)- and (−)-α-pinene are available commercially, so that either enantiomeric alcohol can be synthesized.

(ii)   Other hydride-transfer systems.

(1)   *Cannizzaro reaction.* Aldehydes which do not have α-CH groups cannot undergo base-catalyzed reactions. Instead they react with bases by disproportionation involving the transfer of hydride ion, e.g.

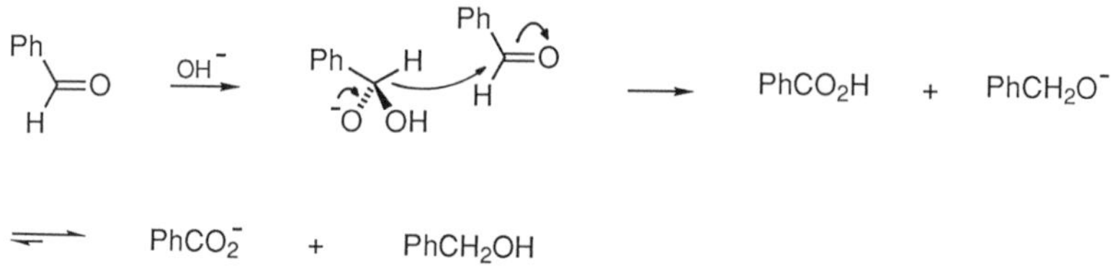

Crossed Cannizzaro reactions between one such aldehyde and formaldehyde result in the reduction of the former and the oxidation of the latter, for formaldehyde is more reactive than other aldehydes towards nucleophiles and rapidly gives a high concentration of the donor anion:

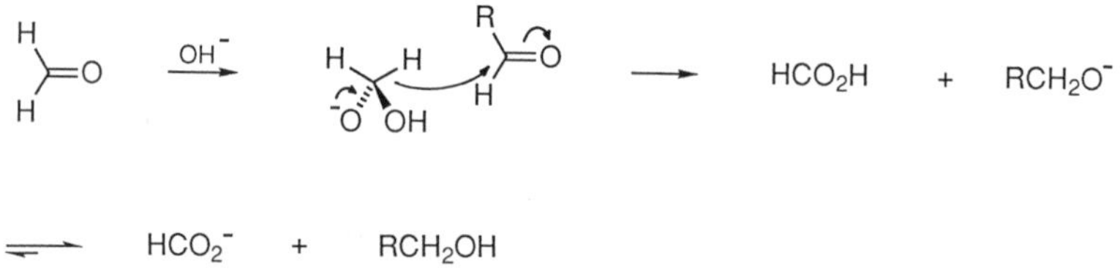

This fact can be exploited for reductions. For example, benzaldehyde is reduced by formaldehyde in the presence of potassium hydroxide in refluxing methanol to give 80% of benzyl alcohol [2]. The preparation of pentaerythritol from acetaldehyde and formaldehyde is also dependent on a crossed Cannizzaro reaction (p. 212).

(2)   *Meerwein–Ponndorf–Verley reaction.* This is the reverse of Oppenauer oxidation (p. 612): equilibrium is established between the carbonyl group to be reduced and isopropanol on the one hand, and the required alcohol and acetone on the other, in the presence of aluminium isopropoxide. Since acetone is the lowest boiling constituent of the mixture, it can be continuously distilled so that the equilibrium is displaced to the right.

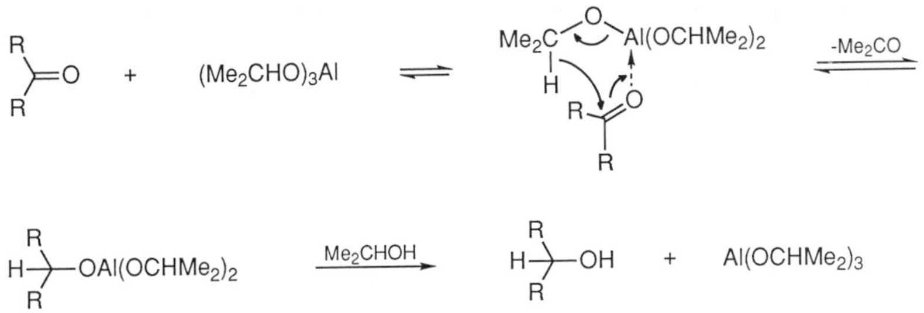

For example, trichloroacetaldehyde is reduced to trichloroethanol in about 80% yield. The reaction is specific to aldehydes and ketones; in particular, the C=C bond in αβ-unsaturated aldehydes or ketones is not reduced (compare lithium aluminium hydride, p. 638). However, the basic conditions may bring about side reactions (as in the synthesis of reserpine, section 22.1). Very hindered Grignard reagents effect reduction in a similar way (p. 190).

(ii) *Electron-transfer reagents.* These reagents are less selective than sodium borohydride and the Meerwein–Ponndorf–Verley reagent, e.g. they also reduce the C=C bond in αβ-unsaturated carbonyl compounds. Nevertheless, in simple cases they are rapid and efficient, e.g. 2-heptanone is reduced by sodium in ethanol to 2-heptanol in over 60% yield [2],

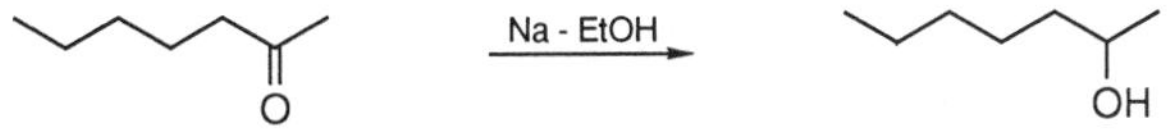

and heptanal is reduced by iron in aqueous acetic acid to 1-heptanol in 80% yield.
    If the ketone contains at the α-position a substituent which is a good leaving group, the intermediate anion-radical undergoes elimination:

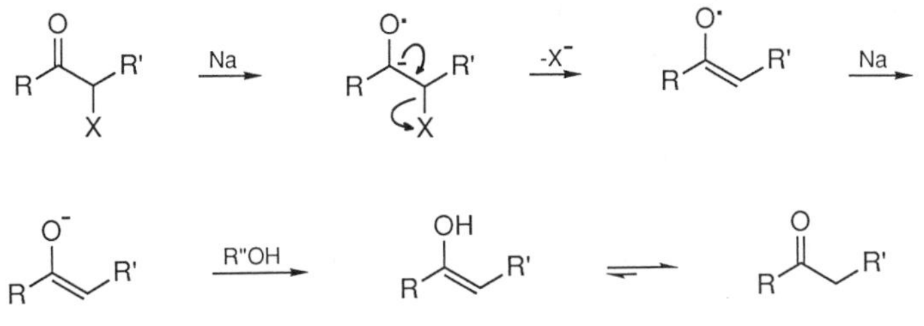

Zinc is often used as the electron source in these reductions. For example, α-hydroxycyclodecanone gives cyclodecanone in about 75% yield when treated with zinc in a mixture of hydrochloric and acetic acids at 75–80°C [4]:

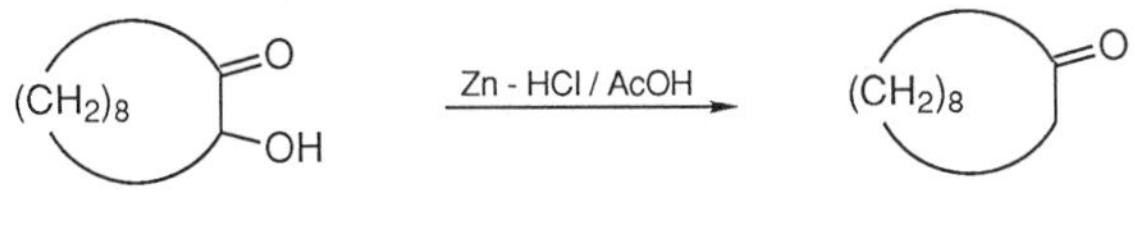

(iii) The borane 9-BBN. Whereas lithium aluminium hydride and sodium borohydride reduce αβ-unsaturated aldehydes and ketones by both 1,2- and 1,4-addition, 9-BBN (p. 485) reduces only the carbonyl group.

(c) To 1,2-diols

In the absence of a proton-donor, electropositive metals reduce ketones to 1,2-diols *via* the dimerization of anion-radicals:

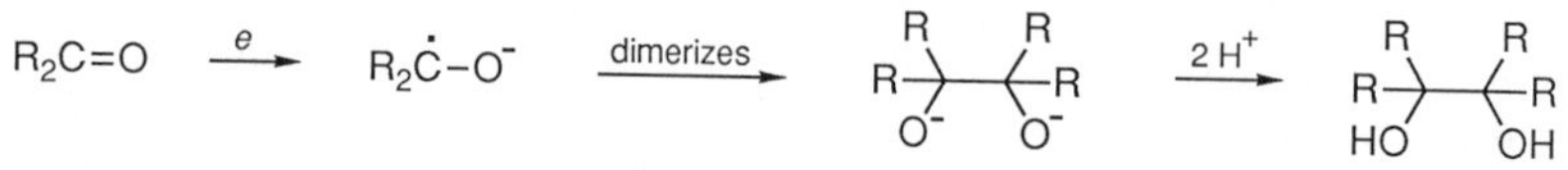

The standard procedure employs amalgamated magnesium with benzene as solvent: the solid magnesium salt of the diol is formed and is hydrolyzed to the diol itself. For example, acetone gives pinacol in 45% yield after 2 hours in refluxing benzene [1]:

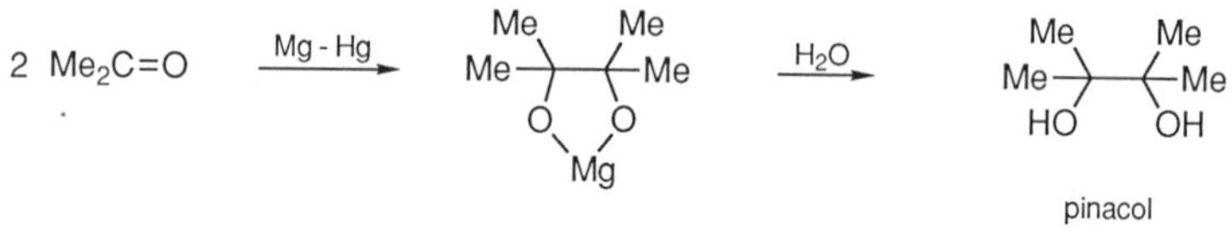

In an improved procedure, titanium tetrachloride is included; reaction is probably brought about by titanium metal formed *in situ*, e.g.

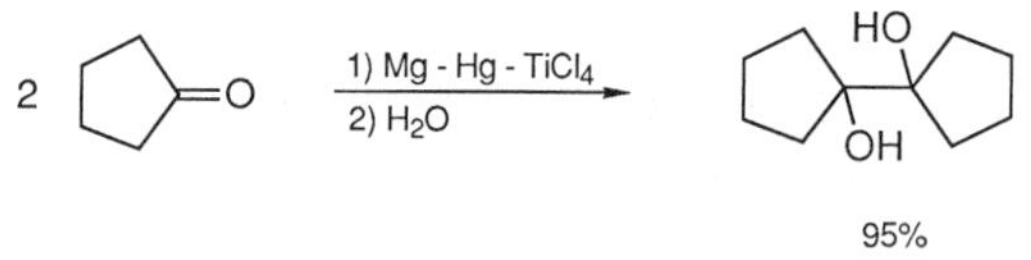

95%

The reaction is generally ineffective for aldehydes because they are too readily reduced to alcohols. 1,2-Diols may also be formed by photochemical dimerization (p. 499).

**20.5
Epoxides** *(a) Lithium aluminium hydride*

Epoxides are reduced to alcohols by lithium aluminium hydride. Since epoxides are readily obtained from alkenes (p. 588), the overall reaction serves to hydrate the alkene. The procedure is complementary to the hydroboration method (p. 657) since the hydride selectively attacks the less alkylated carbon of the epoxide ring, so giving the more highly alkylated alcohol (p. 590), e.g.

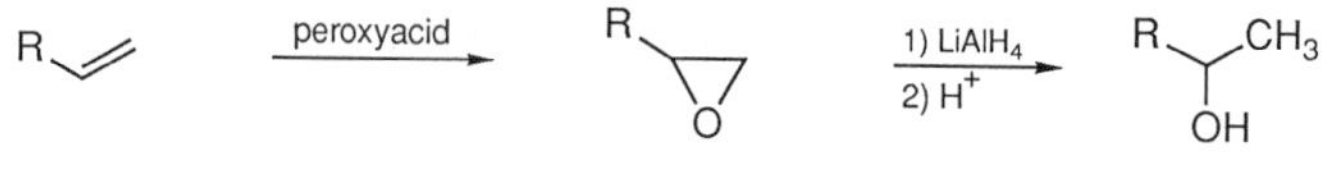

The reaction has the stereochemistry characteristic of S_N2 processes, for example in a rigid cyclic system the *axial* alcohol is formed,

The strained ring in four-membered cyclic ethers is also cleaved by lithium aluminium hydride, e.g.

but the near-strainless five-membered ethers are resistant. They are, however, opened in the more vigorous conditions obtained by using lithium aluminium hydride in the presence of aluminium trichloride (cf. p. 651), e.g. tetrahydrofuran gives 1-butanol:

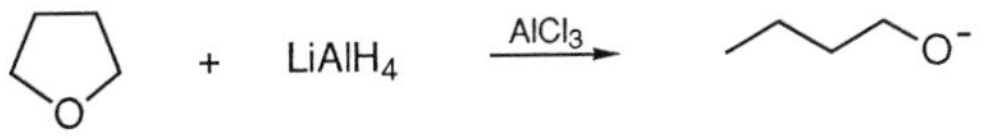

(b)   Hydroboration

Epoxides are reduced by diborane to give mainly the *less* substituted alcohol, e.g.

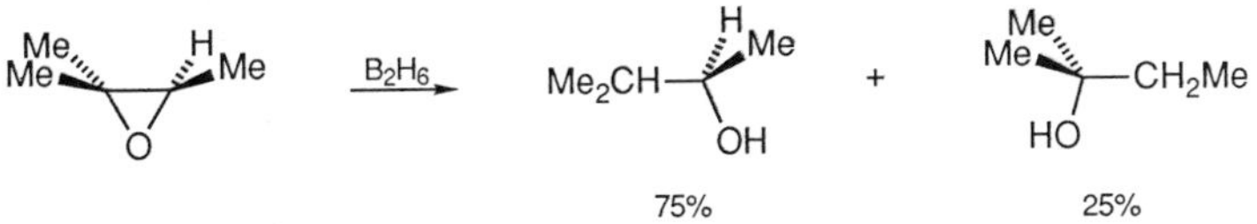

The method is therefore complementary to the use of lithium aluminium hydride.

20.6
Acids and their derivatives

Until the advent of the complex metal hydrides (1947), the only methods for reducing acids and related compounds were relatively unselective. For example, esters could be reduced to alcohols over copper chromite in vigorous conditions, although not over other catalysts, e.g.

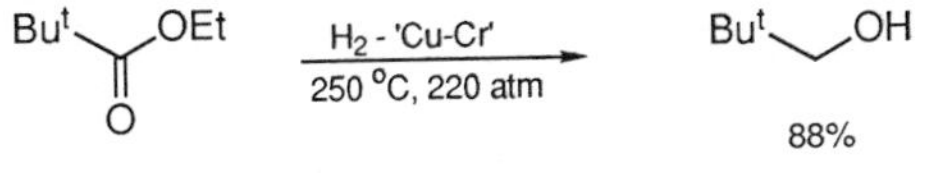

but other reducible groups in the molecule would also react (e.g. $C{=}C$, $C{\equiv}C$, $C{=}O$).

Many methods, some of unique and others of moderate selectivity, are now available.

(a) Reduction to the alcohol or amine

Lithium aluminium hydride reduces acids and all their derivatives to the alcoholic or amino level.* The general mechanisms are:

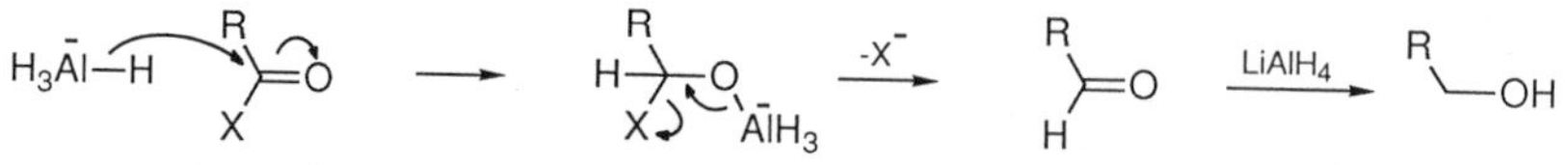

$$H_3\bar{Al}-H \quad R_2C{=}O(X) \longrightarrow H-C(R)(X)-O^-\bar{Al}H_3 \xrightarrow{-X^-} R\,CH{=}O \xrightarrow{LiAlH_4} R-CH_2-OH$$

(X = OH, OR, OCOR, halogen)

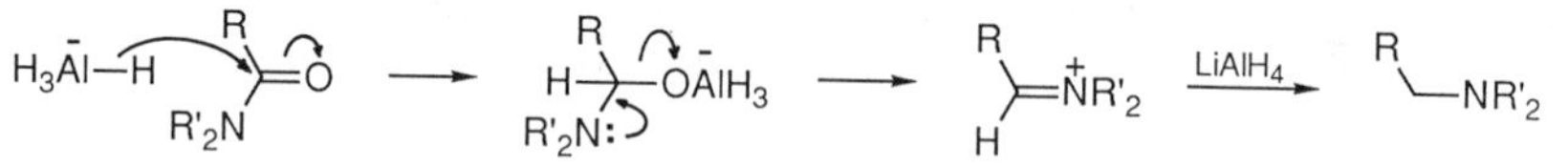

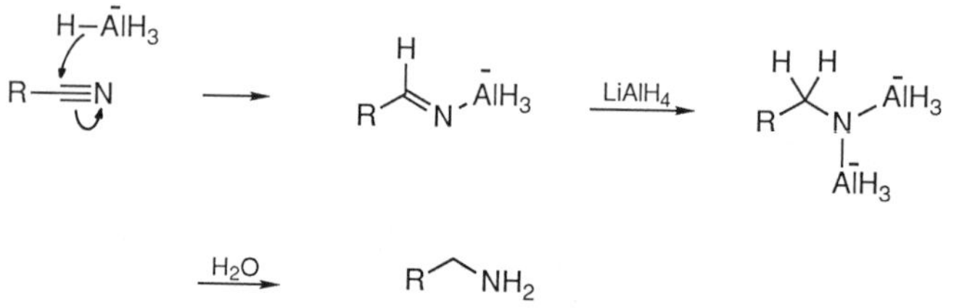

$$H_3\bar{Al}-H \quad R(R'_2N)C{=}O \longrightarrow H-C(R)(NR'_2)-O\bar{Al}H_3 \longrightarrow R\,CH{=}\overset{+}{N}R'_2 \xrightarrow{LiAlH_4} R-CH_2-NR'_2$$

$$H-\bar{Al}H_3 \quad R-C{\equiv}N \longrightarrow R-CH{=}N-\bar{Al}H_3 \xrightarrow{LiAlH_4} R-CH_2-N(\bar{Al}H_3)(\bar{Al}H_3)$$

$$\xrightarrow{H_2O} R-CH_2-NH_2$$

For example:

Acids:

$$Bu^t\,C(=O)OH \longrightarrow Bu^t\,CH_2OH \quad 92\%$$

Esters:

$$Ph\,C(=O)OEt \longrightarrow Ph\,CH_2OH \quad 90\%$$

Acid halides:

$$Cl_2CH\,C(=O)Cl \longrightarrow Cl_2CH\,CH_2OH \quad 65\% \quad [4]$$

* Note the difference between the reaction with amides and those with other acid derivatives. It arises because —NR$_2$ is a not as good a leaving group as, e.g. —Cl.

Anhydrides:

87%

Amides:

88%   [4]

Nitriles:

Ph—CN   ⟶   Ph⌒NH₂

72%

The order of reactivity is the same as that in other nucleophilic displacements on acid derivatives: $RCOCl > RCO_2R > RCONR_2 > RCN > RCO_2H$. In contrast, with a *neutral* aluminium hydride (AlH_3) – an *alane* (cf. boranes) – the order of reactivity is almost the exact opposite: $RCO_2H > RCONR_2 > RCN > RCO_2R > RCOCl$. This is because the crucial step is the formation of a complex in which the aluminium hydride acts as an electrophile, so that it is the capacity for donation of an electron pair by the organic compound that is important, e.g.

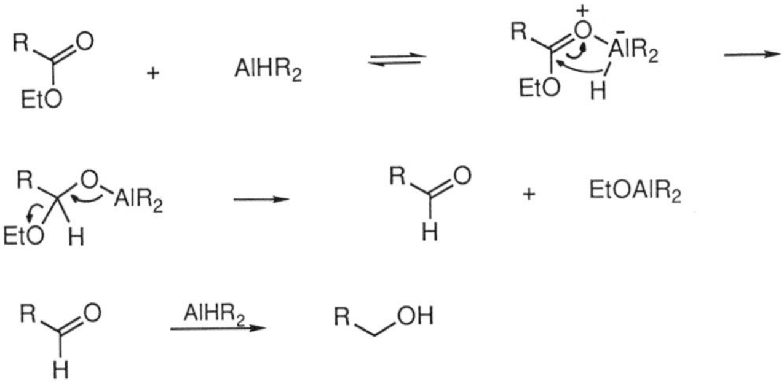

A suitable aluminium compound for this purpose is prepared by the reduction of isobutylene in the presence of aluminium:

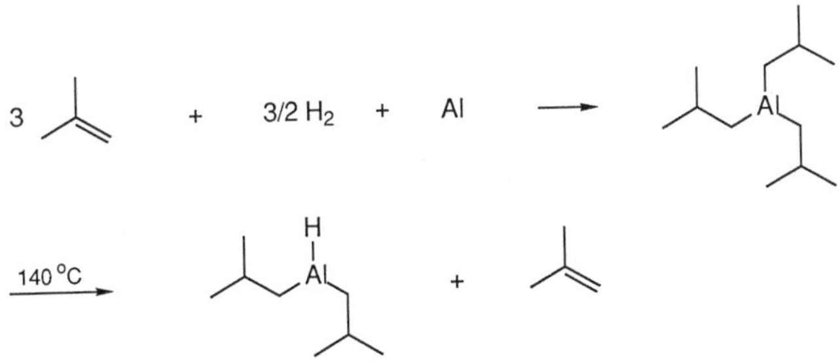

diisobutylaluminium hydride

(DIBAL)

DIBAL is a convenient reagent for reducing carboxylic acids to alcohols. It can also be used for reducing esters to alcohols *via* aldehydes as above, but it is possible to stop reaction at the aldehyde stage (p. 662). Amides and nitriles also yield aldehydes (pp. 665, 666).

Diborane reduces carboxylic acids to alcohols and nitriles and amides to amines, again through complexes:

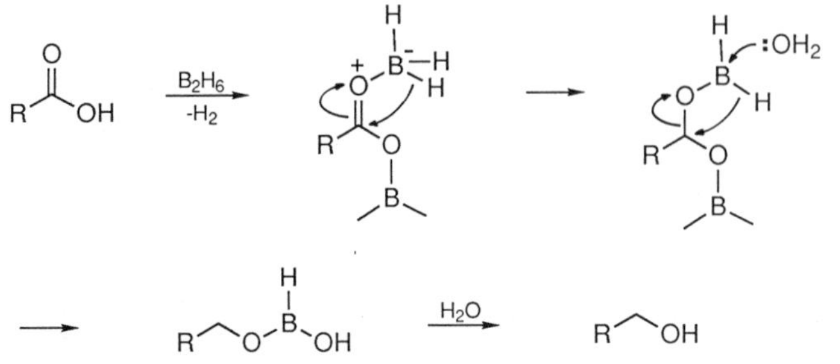

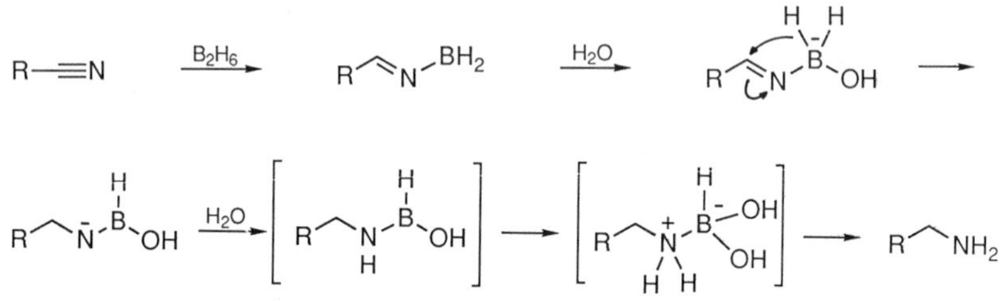

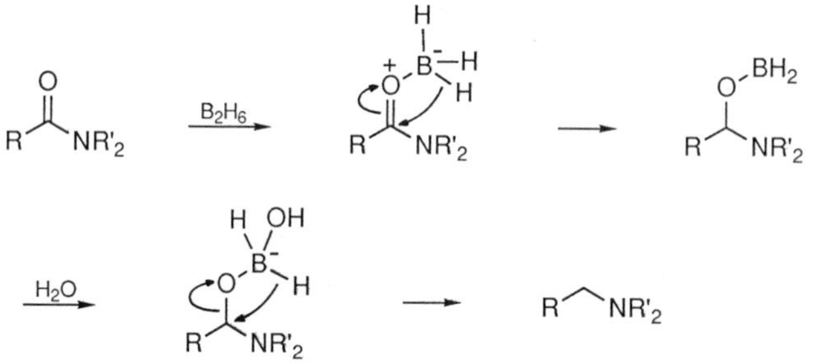

This method has the advantage, compared with lithium aluminium hydride, that ester and nitro groups are not reduced. For example, *p*-nitrobenzoic acid gives *p*-nitrobenzyl alcohol in about 80% yield.

Esters can be reduced by the acyloin method (p. 549) and by the *Bouveault–Blanc method* (sodium in refluxing ethanol), e.g.

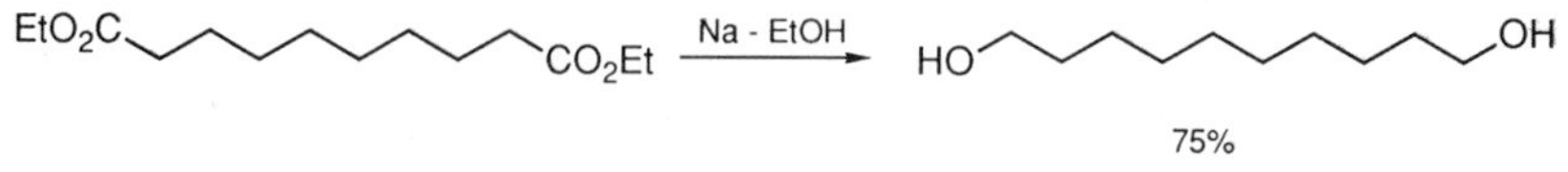

75%

Acid chlorides can be reduced with sodium borohydride in diglyme solution. Acid, ester, amide, C=C, and C≡C groups in the molecule are not reduced but aldehydic and ketonic groups are. Sodium trimethoxyborohydride, $NaBH(OCH_3)_3$, also reduces —COCl to —CH_2OH without affecting ester groups.

Nitriles can be reduced catalytically. In the absence of added ammonia, considerable quantities of secondary amines are formed since the primary amine can add to the imino intermediate (i.e. the half-reduced stage):

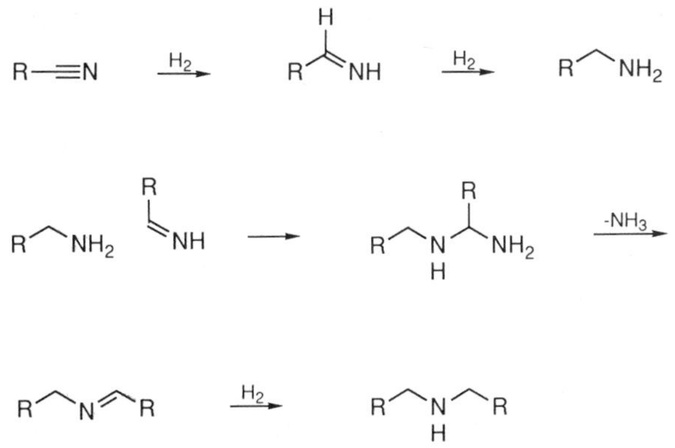

This is circumvented by the addition of an excess of ammonia which assumes the role of nucleophile in place of the primary amine and yields are then high, e.g. over Raney nickel, benzyl cyanide gives phenylethylamine in about 85% yield in liquid ammonia under pressure [3] and 1,8-octanedinitrile gives decamethylenediamine (79–80%) in ammoniacal ethanol [3]:

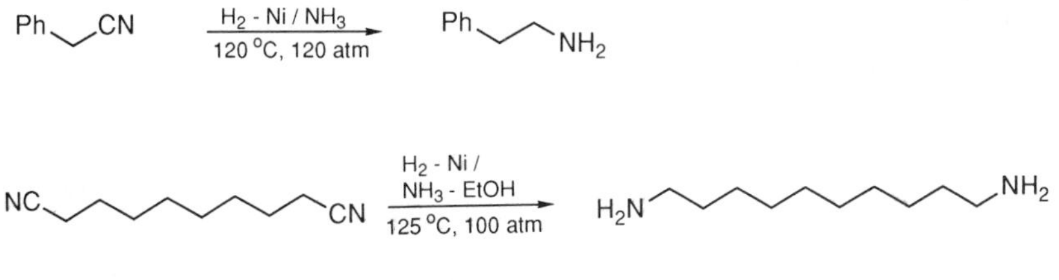

The reduction of nitriles by sodium in ethanol suffers from the disadvantage that secondary amines are formed, as in the ammonia-free catalytic method.

It should be noted that none of the reagents described above is uniquely selective, but in some cases where two methods are available one complements the other. For instance, diborane is the reagent of choice for acids when ester groups are present but not when C=C or C≡C bonds are, whereas lithium aluminium hydride is suitable in the latter case but not in the former. Benzylic systems are liable to hydrogenolysis, so that reductions of $ArCO_2H$, $ArCO_2R$, ArCOCl, $ArCONR_2$, $ArCH_2OCOR$, and $ArCH_2NHCOR$ are likely to give $ArCH_3$.

(b)   Reduction to aldehydes

It is often necessary to convert acids and their derivatives into aldehydes, for aldehydes are not readily made by other methods whereas acids are readily available. Selective procedures are required since aldehydes are easily reduced to alcohols, and a number of older, lengthy procedures have been superseded by techniques using modified complex metal hydrides.

(i)   Acids.   Lithium in ethylamine can be used but yields are in general low, e.g.

26%

and sometimes zero (e.g. for Me_3CCO_2H), but in isolated instances yields of up to 80% have been recorded.

Satisfactory yields can be obtained by reducing the imidazolide of the acid with lithium aluminium hydride, e.g.

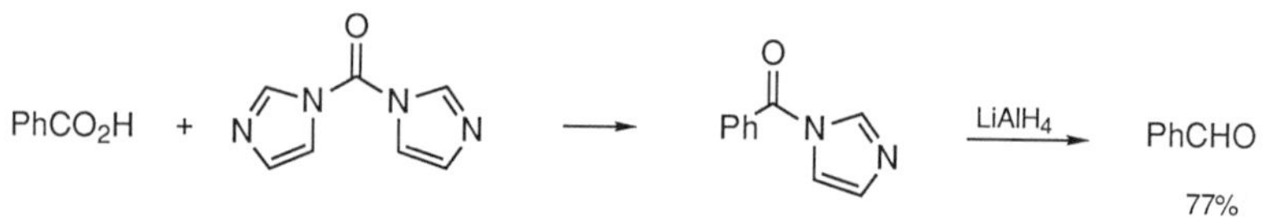

77%

(ii)   Esters

(1)   *McFadyen and Stevens' method* consists in converting the ester into its hydrazide, treating this with benzenesulfonyl chloride, and hydrolyzing the product with base. The principle of the method is analogous to that of the Wolff–Kishner reduction (p. 649) and to the method for hydrogenolyzing primary aliphatic amines (p. 648):

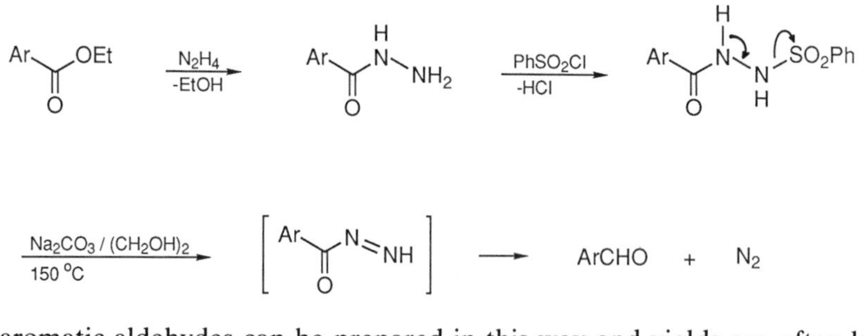

Only aromatic aldehydes can be prepared in this way and yields are often low.

(2)   *DIBAL* reduces esters to alcohols at room temperature (p. 660). Reaction occurs by way of an aluminium alkoxide intermediate which spontaneously forms aldehyde which, being more readily reduced than the original ester, yields the alcohol. However, at −78°C, the alkoxide is stable enough to accumulate and can

be trapped by added water which also destroys the residual reducing agent. The resulting adduct forms aldehyde:

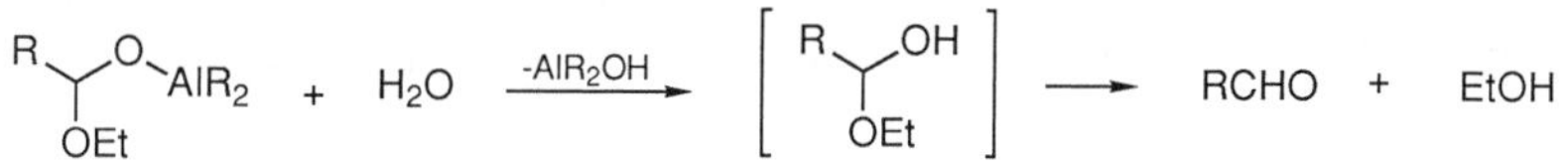

Amide, nitrile, and C≡C groups are also reduced.

(iii) Acid halides

(1) *Rosenmund's method* consists of reducing the acid chloride with hydrogen on a palladium catalyst supported on barium sulfate or, less commonly, on calcium carbonate or charcoal.

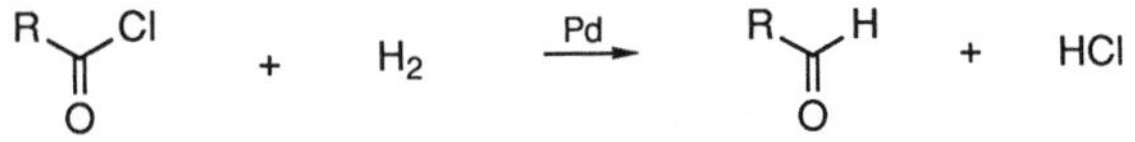

Temperature control is important: the reaction should be conducted at the lowest temperature possible in order to prevent further reduction. Refluxing toluene or xylene (i.e. 110–140°C) is usually suitable and the reaction can be followed by passing the hydrogen chloride into standard base. It is sometimes helpful partially to poison the catalyst: a commonly used poison is obtained by refluxing sulfur in quinoline.

Yields are normally high, even with sterically hindered compounds, e.g.

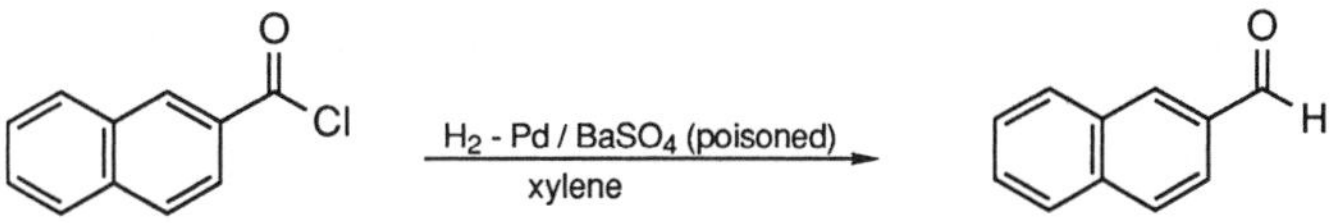

74-81% [3]

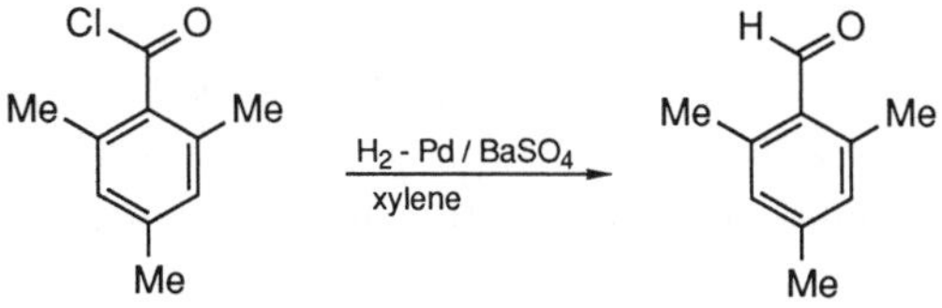

70-80% [3]

(2) *Desulfurization* of thiol esters on Raney nickel which has been partially deactivated by boiling with acetone is effective, e.g.

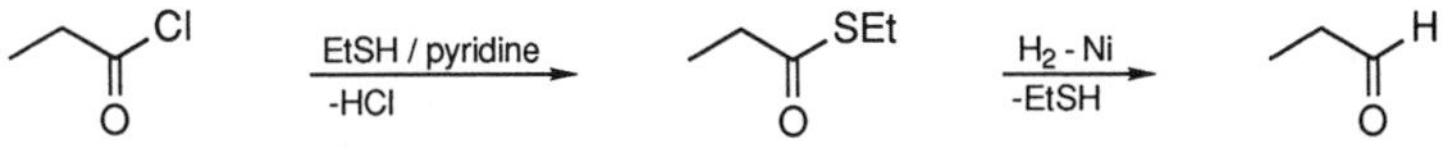

73%

(3)   *Lithium tri-t-butoxyaluminium hydride*, prepared from lithium aluminium hydride and t-butanol in ether, is considerably less reactive than $LiAlH_4$ since the electron-attracting alkoxy-group stabilizes the negatively charged aluminium ion. It reduces acid chlorides only to aldehydes at $-78°C$, and does not affect nitro, cyano, and ester groups, e.g.

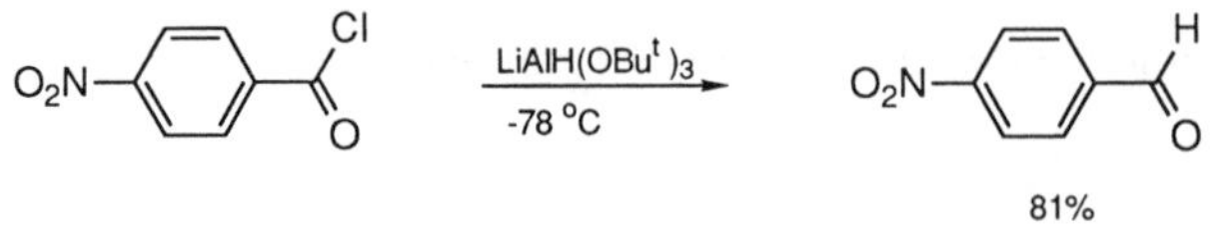

81%

(iv)   *Amides*

(1)   The *Sonn–Müller* reaction involves conversion of an anilide or toluidide into the imino-chloride which is reduced to the imine with tin(II) chloride; hydrolysis gives the aldehyde.

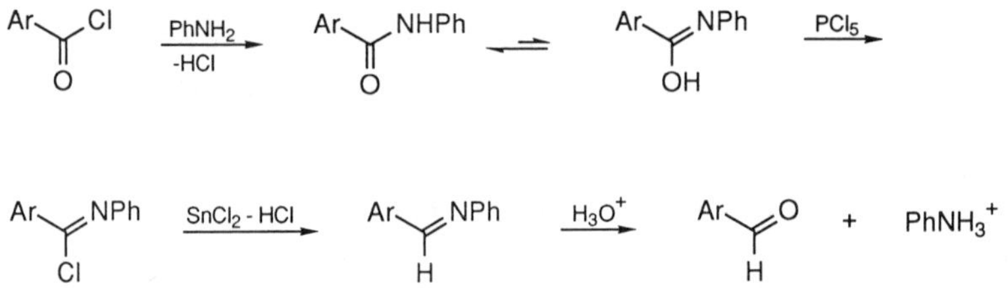

For example, *o*-tolualdehyde can be obtained in up to 70% yield [3]:

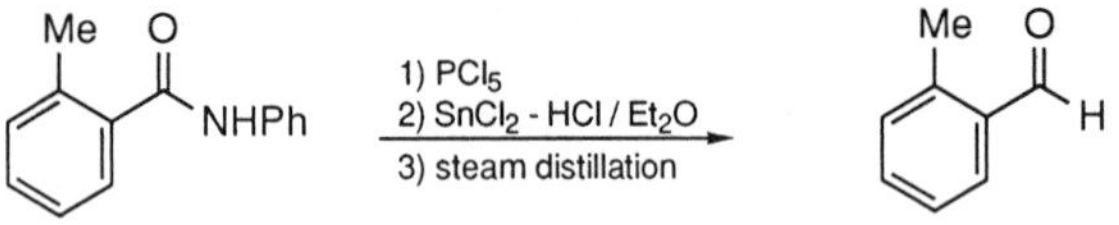

This method is not applicable to aliphatic aldehydes because the intermediate imino-chlorides are unstable.

(2)   *The hydrolysis of Reissert compounds* has proved suitable for the preparation of some aromatic aldehydes. The acid chloride is added to quinoline in aqueous potassium cyanide and the resulting Reissert adduct is hydrolyzed with mineral acid:

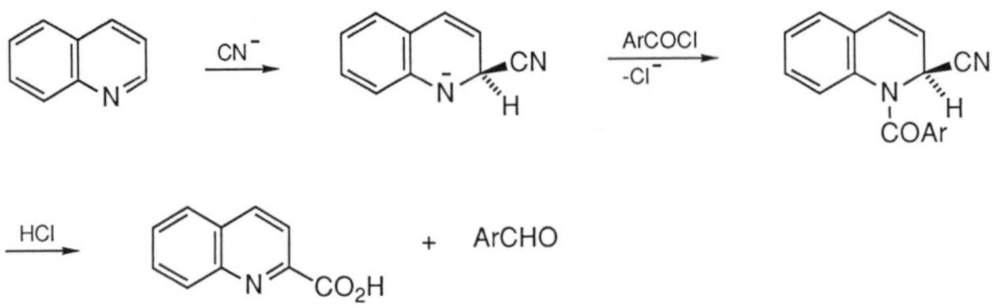

For example, *o*-nitrobenzoyl chloride gives *o*-nitrobenzaldehyde in about 65% yield. The method has also been used for introducing the carboxyl group into quinolines.

(3) *Lithium aluminium hydride* reduces disubstituted amides to aldehydes provided that the temperature is kept low:

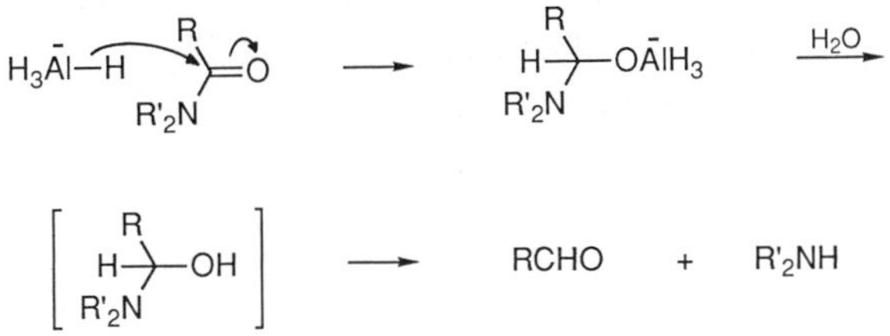

At higher temperatures the first intermediate undergoes elimination, and further reduction to the amine ensues (p. 659). The *N*-methylanilide is usually employed, e.g.

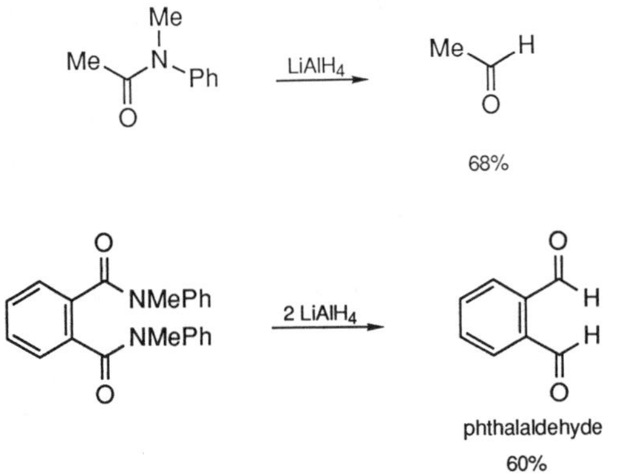

68%

phthalaldehyde
60%

(4) *DIBAL* and *lithium diethoxyaluminium hydride*, $LiAlH_2(OEt)_2$, from lithium aluminium hydride with two equivalents of ethanol, can be used for the conversion $R—CONR'_2 \rightarrow R—CHO$.

(v) Nitriles

(1) The *Stephen reaction* involves the addition of the nitrile to a suspension of anhydrous tin(II) chloride in diethyl ether saturated with hydrogen chloride, followed by hydrolysis. The reduction occurs at room temperature and is equivalent to that in the Sonn–Müller reaction (p. 664). For example, 2-naphthaldehyde can be prepared in about 75% yield [3]:

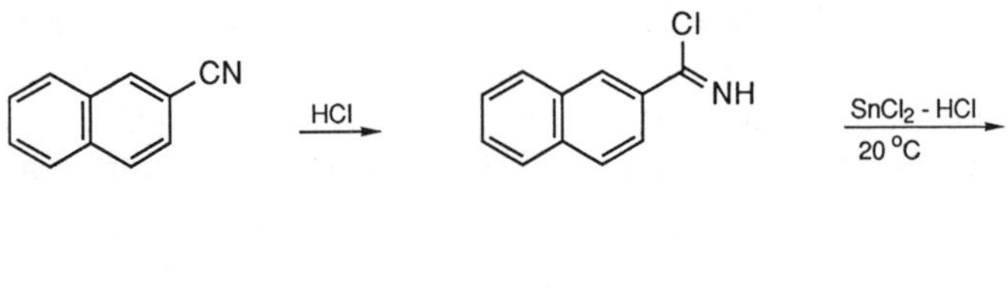

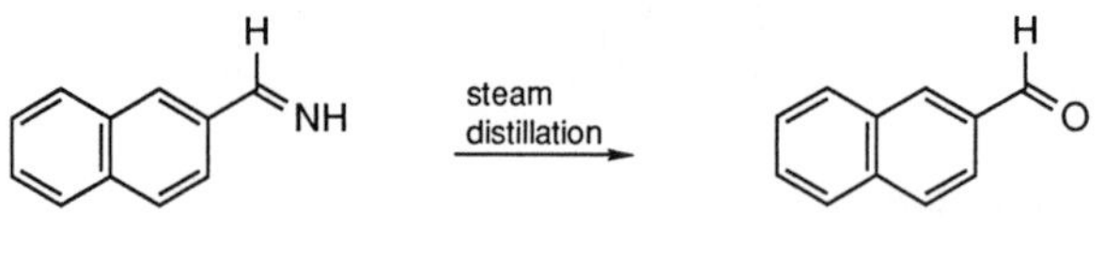

(2) *Hydrogen and Raney nickel* in the presence of semicarbazide and water gives the semicarbazone of the aldehyde from which the aldehyde is liberated by an exchange reaction with formaldehyde. Over-reduction is here prevented by the trapping of the aldehyde.

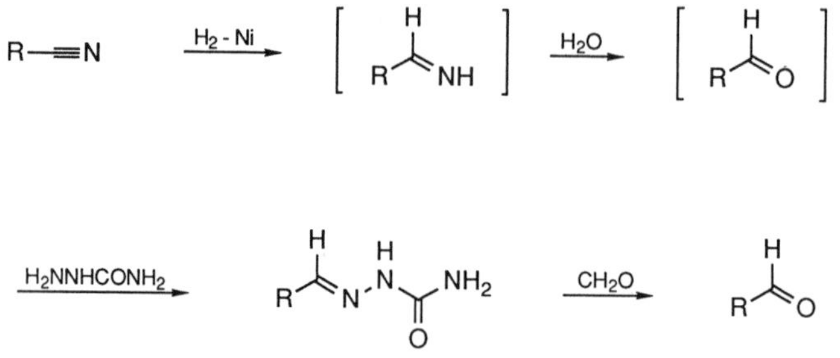

(3) *DIBAL* and *lithium triethoxyaluminium hydride*, LiAlH(OEt)$_3$ (from lithium aluminium hydride and three equivalents of ethanol) can be used, e.g.

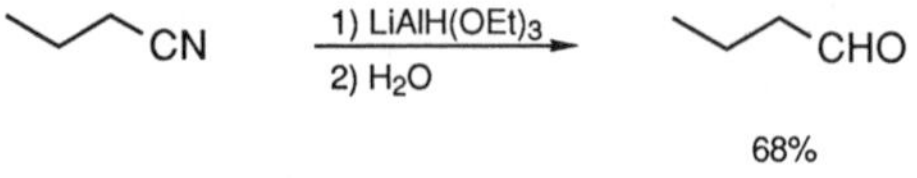

68%

20.7 Systems containing nitrogen

(a) Nitro-compounds

The reduction of aliphatic nitro-compounds to amines is of little importance because the amino group is easily introduced in a number of other ways (e.g. from the halide and an amine, p. 302). Lithium aluminium hydride and catalytic reduction on Raney nickel are both suitable methods, e.g. 2-nitrobutane is reduced to 2-aminobutane in 85% yield by the former reagent (compare the mode of reduction of aromatic nitro-compounds by LiAlH$_4$: below).

The reduction of aromatic nitro-compounds is much more important because the nitro group can be introduced into a wide variety of aromatic systems by nitration whereas the amino group can only be introduced directly into those aromatic compounds which are strongly activated to nucleophiles (p. 397). Several types of reduction product are obtainable, as illustrated for nitrobenzene:

Zn - NH$_4$Cl / H$_2$O
50-55 °C

PhNHOH

phenylhydroxylamine   (55%) [1]

dextrose - NaOH
100 °C

$$Ph-\overset{\overset{\displaystyle O^-}{|}}{\overset{+}{N}}=N-Ph$$

azoxybenzene   (80%) [2]

PhNO$_2$

Zn (2 moles) - NaOH / MeOH / H$_2$O

$$Ph-N=N-Ph$$

azobenzene   (85%) [3]

Zn (3 moles) - NaOH / MeOH

$$Ph-\overset{\overset{\displaystyle H}{|}}{N}-\overset{\overset{\displaystyle }{|}}{\underset{\underset{\displaystyle H}{|}}{N}}-Ph$$

hydrazobenzene   (88%)

Sn (or Fe) - HCl

PhNH$_2$

aniline

In neutral solution buffered with ammonium chloride, phenylhydroxylamine is the main product. In basic solution, using a weak reducing agent such as dextrose or glucose, the intermediate nitrosobenzene reacts with phenylhydroxylamine to give azoxybenzene, which the more powerful reducing agents convert into azobenzene and hydrazobenzene. Lithium aluminium hydride gives azobenzene quantitatively and, in the presence of Lewis acids, brings about further reduction to hydrazobenzene. Metals in acid solution bring about reduction to aniline; industrially, iron and dilute hydrochloric acid are used, the product is neutralized with lime, and aniline is isolated by distillation in steam.

The only member of the series which cannot be isolated is nitrosobenzene, but this is obtainable by the oxidation of phenylhydroxylamine (p. 622).

Controlled electrolytic reduction occurs in exact stages:

$$Ar-\overset{\overset{\displaystyle +}{N}}{\underset{\underset{\displaystyle O}{\|}}{}}-O^- \xrightarrow{2\,e,\;H^+} Ar-\overset{\overset{\displaystyle }{N}}{\underset{\underset{\displaystyle OH}{|}}{}}-O^- \xrightarrow{-OH^-} Ar-N=O$$

$$\xrightarrow{2\,e,\,2H^+} Ar-\overset{\overset{\displaystyle }{N}}{\underset{\underset{\displaystyle H}{|}}{}}-OH \xrightarrow[-OH^-]{2\,e,\;H^+} ArNH_2$$

It is usually the amino group which is required from the reduction and the problem is to find a selective reagent which leaves other reducible groups intact, so that the standard reagent for reducing nitrobenzene to aniline, tin and hydrochloric acid, is unsuitable for many substituted nitro-compounds.

Catalytic methods are suitable provided that C=C and C≡C bonds are not present: e.g.

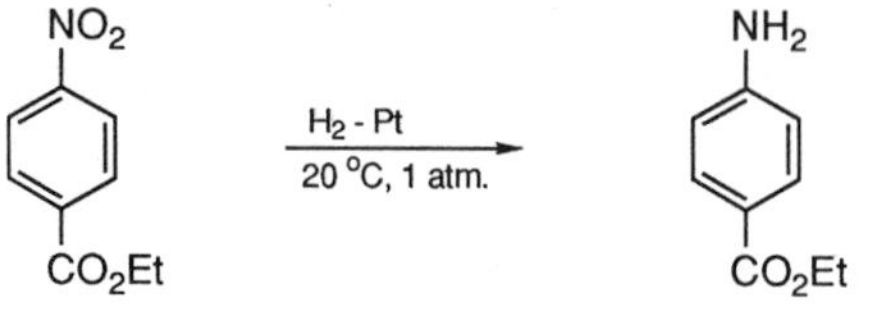

90-100% [1]

and transfer-hydrogenation by hydrazine on palladium is also very effective. Electron-transfer reagents of mild type can be used, e.g.

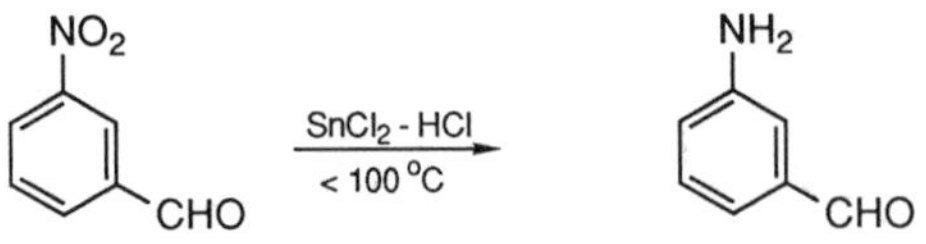

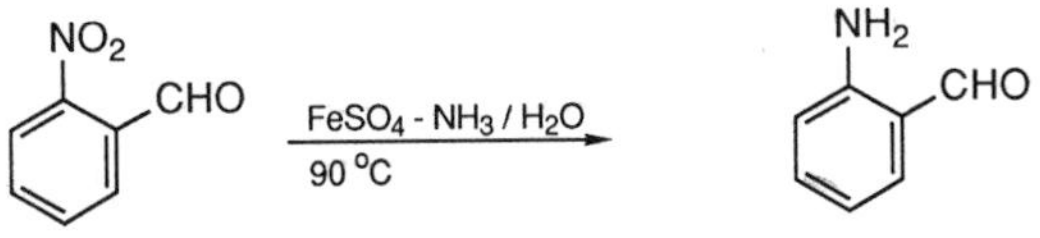

70-75% [3]

The complement to the last two reactions, reduction of the aldehydic group without reduction of the nitro group, can be effected with sodium borohydride.

It is also possible to reduce one nitro group in the presence of another. The older method employed ammonium sulfide, e.g.

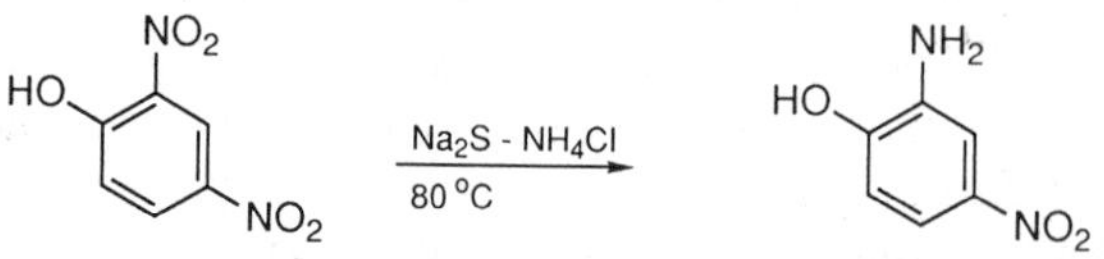

65%   [3]

but catalytic methods are usually more efficient. For example, *m*-dinitrobenzene gives *m*-nitroaniline almost quantitatively when reduced on palladium with cyclohexene as the hydrogen donor (p. 632):

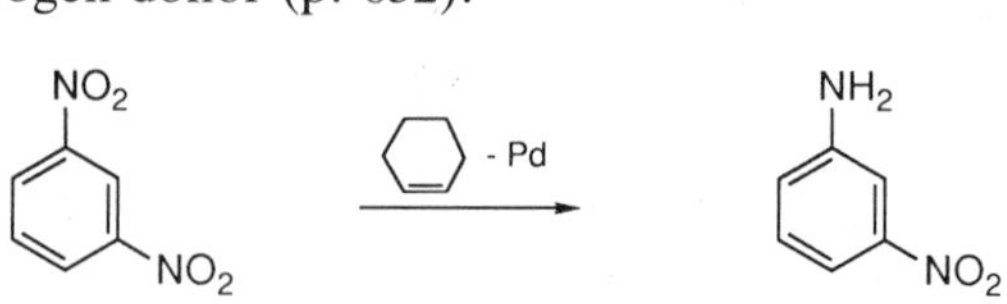

(b) Imines

Imines of the type $RCH=NR'$ and $RCH=\overset{+}{N}R'_2$ can be selectively reduced with sodium borohydride:

$$RCH=NR' \longrightarrow RCH_2-NHR'$$

An alternative reagent is dimethylamine borane, which reduces $C=N$ without reducing acids, esters, or nitro groups. Yields are high, e.g.

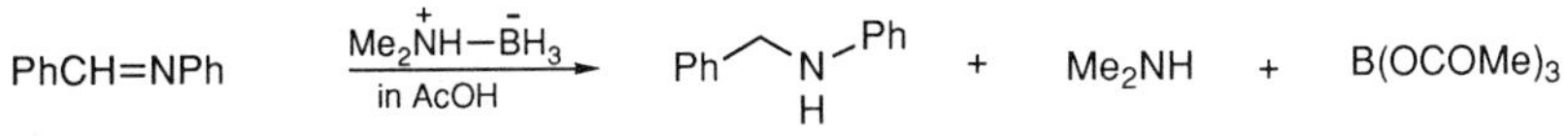

84%

(i) Reductive amination.

When the reaction between an aldehyde or ketone and ammonia or a primary or secondary amine is carried out in the presence of a suitable reducing agent, the imine or immonium ion which is formed is reduced *in situ* to an amine. The usual reducing agents are hydrogen on Raney nickel, e.g.

$$PhCHO \; + \; NH_3 \; \underset{-H_2O}{\rightleftharpoons} \; \left[PhCH=NH\right] \; \xrightarrow{H_2 \text{-} Ni} \; PhCH_2NH_2$$

89%

or sodium cyanoborohydride, e.g.

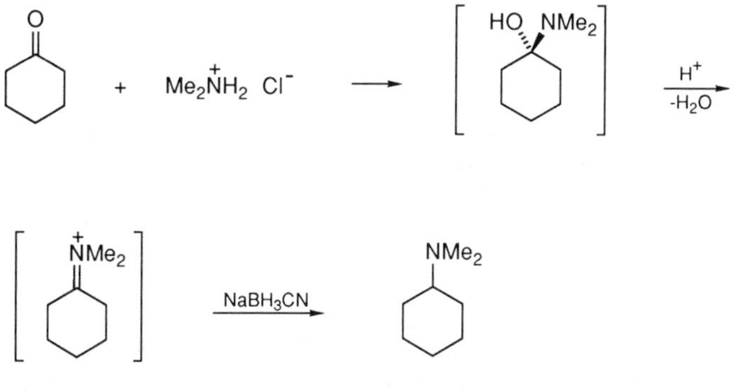

85%

A side reaction can be disadvantageous: as the amine is formed, it can react with more of the carbonyl compound to give a new imine and, by reduction, a new amine, e.g.

$$RCHO \; \underset{NH_3}{\rightleftharpoons} \; RCH=NH \; \xrightarrow{reduction} \; RCH_2NH_2 \; \xrightarrow{RCHO}$$

$$RCH_2N=CHR \xrightarrow{\text{reduction}} RCH_2NHCH_2R \longrightarrow \longrightarrow (RCH_2)_3N$$

This can be minimized by using a large excess of ammonia or primary amine.

A variation of the method suitable for making tertiary amines in which there is at least one *N*-methyl group is to heat an amine with formaldehyde and formic acid under reflux (*c.* 100°C) (*Eschweiler–Clarke reaction*). With a secondary amine, the intermediate immonium ion is reduced by formate ion:

$$R_2NH \quad + \quad CH_2O \quad \rightleftharpoons \quad R_2N-CH_2OH$$

$$R_2N-CH_2OH \quad + \quad HCO_2H \quad \rightleftharpoons \quad R_2\overset{+}{N}=CH_2 \quad + \quad HCO_2^- \quad + \quad H_2O$$

$$R_2\overset{+}{N}=CH_2 \longrightarrow R_2N-Me \quad + \quad CO_2$$

When the starting amine is primary, successive condensations and reductions occur:

$$RNH_2 \xrightarrow{CH_2O \cdot HCO_2H} RNHMe \xrightarrow{CH_2O \cdot HCO_2H} RNMe_2$$

For example, the conversion $PhCH_2CH_2NH_2 \rightarrow PhCH_2CH_2NMe_2$ occurs in about 80% yield [3].

The *Leuckart reaction* is a variant of the Eschweiler–Clarke reaction in which aldehydes or ketones other than formaldehyde are heated with ammonium formate or a formamide at 180–200°C, e.g.

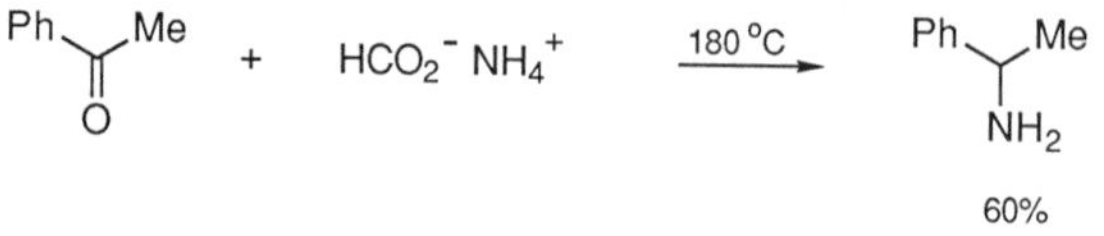

60%

However, while it is a more general procedure than the Eschweiler–Clarke process, yields are usually lower.

(c) Oximes

Electron-transfer reagents are normally employed, as in the formation of 1-heptylamine in 60% yield from heptanal oxime with sodium in ethanol [2],

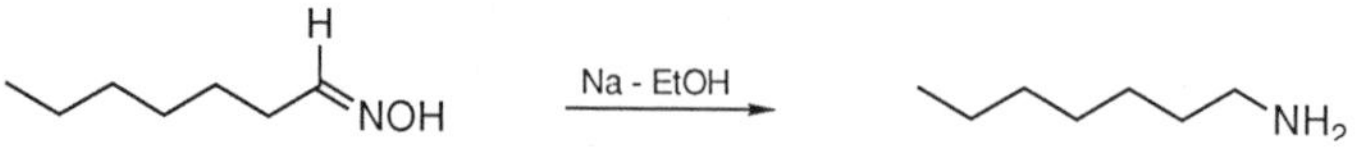

and in the Knorr synthesis of pyrroles (p. 686), where sodium dithionite is preferably used:

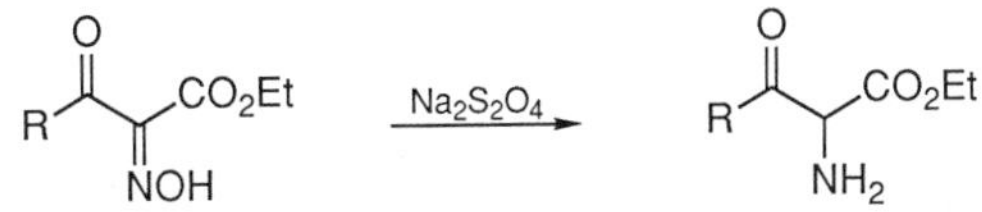

Catalytic methods are reasonably efficient but tend to produce some of the secondary amine, e.g.

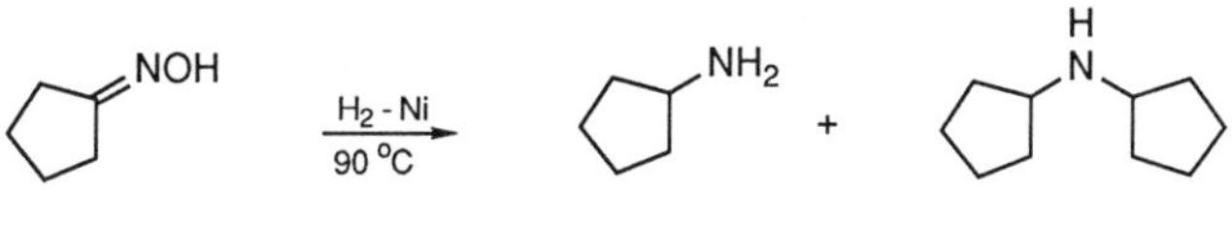

80%   10%

(d) Nitroso-compounds

C-Nitroso-compounds can be converted into hydroxylamines by controlled electrolytic reduction (p. 667). Reduction to amines is normally brought about by electron-transfer reagents, e.g.

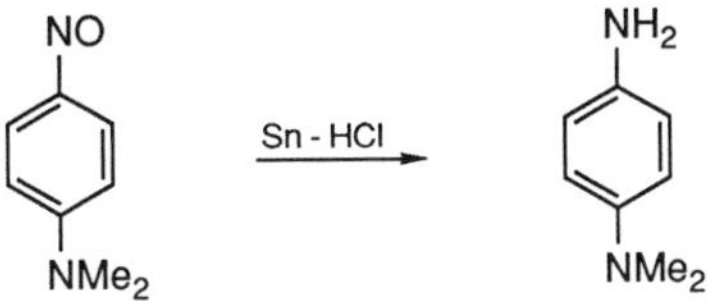

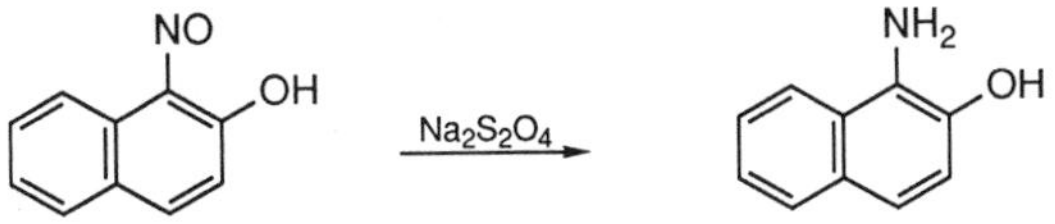

70% [2]

These reactions are commonly used for obtaining amino groups in aromatic nuclei which are strongly activated towards electrophiles, since it is often easier to nitrosate these compounds than to nitrate them (see the synthesis of pteridine, p. 719).

N-Nitroso-compounds are reduced by mild electron-transfer reagents to substituted hydrazines, e.g. N-nitroso-N-methylaniline with zinc in acetic acid gives 55% of N-methyl-N-phenylhydrazine [2]:

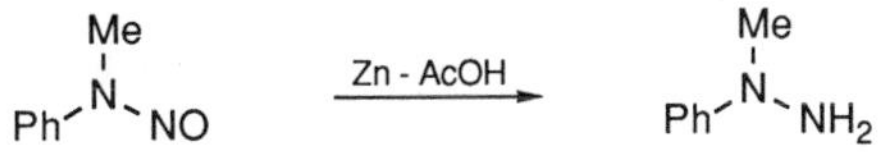

Stronger reducing agents rupture the N—N bond, e.g.

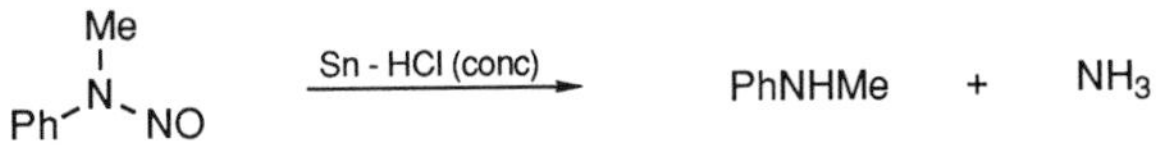

(e) Azo-compounds

Azo-compounds are reduced to hydrazo-compounds by lithium aluminium hydride in the presence of a Lewis acid, as already described (p. 667). A more useful type of reduction is brought about by sodium dithionite which cleaves the $N{=}N$ bond to give amines.* Since azo-compounds can be obtained by diazonium-coupling both to strongly activated aromatic compounds and to activated methylene compounds (section 13.4), this provides a useful route for the introduction of the amino group, e.g.

An alternative method, nitrosation followed by reduction, is also available.

The reduction of diazonium salts both to arylhydrazines and to hydrocarbons has been described earlier (pp. 420, 413).

20.8 Systems containing sulfur

Catalytic reduction of sulfur-containing compounds removes the sulfur as hydrogen sulfide. Reduction with retention of the sulfur atom requires selective methods; the less active electron-transfer reagents are usually employed.

(a) Disulfides

Zinc in refluxing acetic acid is normally used. For example, thiosalicylic acid can be obtained from anthranilic acid in 75–80% overall yield in this way [2]:

---

* The dithionite ion is in equilibrium in solution with the sulfur dioxide anion-radical,

$$^-O_2S{-}SO_2^- \quad \rightleftharpoons \quad 2\,SO_2^{\cdot -}$$

and it is the latter which is the reducing agent, acting as a one-electron donor, probably as follows:

$$RN{=}NR \xrightarrow{\;SO_2^{\cdot -}\;} R\overset{-}{N}{-}\overset{\cdot}{N}R \xrightarrow{\;H^+\;} RNH{-}\overset{\cdot}{N}R$$

$$\xrightarrow{\;SO_2^{\cdot -}\; -\,H^+\;} RNH{-}NHR \xrightarrow{\;SO_2^{\cdot -}\;} RNH^- + R\overset{\cdot}{N}H$$

$$\xrightarrow{\;SO_2^{\cdot -}\; -\,2\,H^+\;} 2\,RNH_2$$

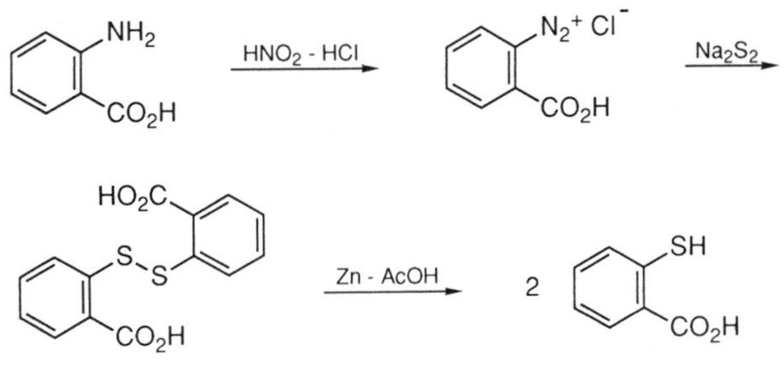

(b) Sulfonyl chlorides

Sulfonic acids are not easily reduced, but the readily derived chlorides can be reduced to sulfinic acids and to thiols. There is a resemblance to reduction of the nitro group (p. 667): zinc or tin in mineral acid gives the thiol, whereas in neutral or alkaline solution intermediate products are obtainable. For example, benzene-sulfonyl chloride with zinc in sulfuric acid gives thiophenol in 96% yield [1],

$$PhSO_2Cl \xrightarrow{\text{Zn - H}_2\text{SO}_4} PhSH$$

and toluene-*p*-sulfonyl chloride with zinc in aqueous sodium hydroxide gives sodium toluene-*p*-sulfinate in 64% yield [1],

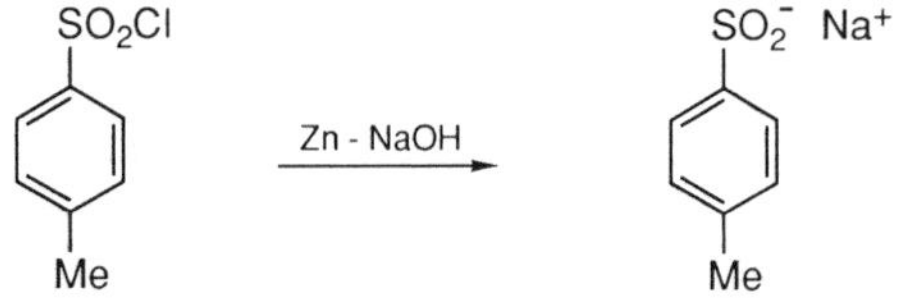

Problems

1. How would you carry out the following transformations?
    - (*i*) R—CH=CH$_2$ into (*a*) R—CH$_2$—CH$_2$OH,
      (*b*) R—CH(OH)—CH$_3$, (*c*) R—CH$_2$—CH$_2$—NH$_2$.
    - (*ii*) R—C≡CH into (*a*) R—CO—CH$_3$, (*b*) R—CH$_2$—CHO.
    - (*iii*) R—C≡C—R into the corresponding (*a*) cis- and (*b*) *trans*-alkenes.
    - (*iv*) EtO$_2$C—CH$_2$—CH$_2$—CO$_2$H into (*a*) EtO$_2$C—CH$_2$—CH$_2$—CH$_2$OH and (*b*) HOCH$_2$—CH$_2$—CH$_2$—CO$_2$H.
    - (*v*) Ph—CH=CH—CO$_2$H into (*a*) Ph—CH$_2$—CH$_2$—CO$_2$H, (*b*) Ph—CH=CH—CH$_2$OH, (*c*) Ph—CH=CH—CHO, (*d*) Ph—CH$_2$—CH$_2$—CH$_2$OH, (*e*) γ-cyclohexylpropanol.
    - (*vi*) *m*-Nitrobenzoyl chloride into *m*-nitrobenzaldehyde.
    - (*vii*) *m*-Nitrobenzaldehyde into (*a*) *m*-nitrobenzyl alcohol, (*b*) *m*-aminobenzaldehyde, (*c*) *m*-nitrotoluene.
    - (*viii*) 1-Methylcyclohexene into *trans*-2-methylcyclohexanol.

(ix) $EtO_2C—(CH_2)_8—CO_2Et$ into   (a)   decamethylenediol,   (b)
cyclodecanone, (e) *trans*-cyclodecene.

(x)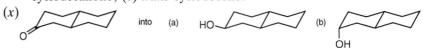

(xi)

(xii)

2. Outline a synthesis, which includes a reductive method, of each of the
    following, from readily available compounds:

(a)

(b) CMe_4

(c)

(d)

(e)

(f)

(g)

(h)

(i)

(j)

3. How would you attempt to convert acrylonitrile into adiponitrile
    $(NC—(CH_2)_4—CN)$ (which is required for the manufacture of nylon)?

4. An attempt to make the *N*-dimethyl derivative of the amine (I) by treatment
    with aqueous formaldehyde in the presence of formic acid gave the compound
    (II). Discuss.

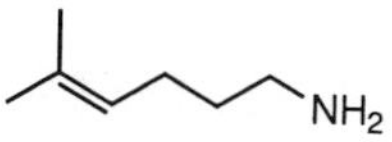

(I)

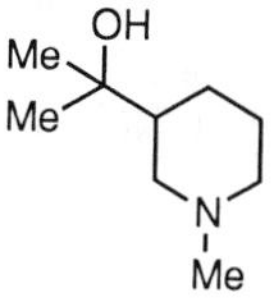

(II)

5. Summarize the uses of hydrazine and its derivatives in effecting the reduction
    of various types of organic groupings.

21 The synthesis of five- and six-membered heterocyclic compounds

**21.1
Introduction**

Heterocyclic systems are of widespread occurrence in nature, particularly in such natural products as nucleic acids, plant alkaloids, anthocyanins and flavones, and the haem pigments and chlorophyll. In addition, some of the vitamins contain aromatic heterocyclic systems and proteins contain the imidazole and indole rings (p. 324). In this chapter, the general methods for the synthesis of the more important heterocycles are discussed and, in the following chapter, applications of some of these methods in the synthesis of natural products are described.

The reactions employed for making heterocyclic compounds involve only those principles and procedures which have been discussed in earlier chapters. For example, C—N bonds are usually formed by reaction between amino groups and esters, aldehydes, ketones, halides, or activated alkenes (Michael-type addition); aliphatic C—C bonds are usually formed by acid- or base-catalyzed reactions involving activated methylene groups and carbonyl groups; and ring closure on to benzene rings is usually effected by Friedel–Crafts reactions. Systems with two heteroatoms are usually constructed from two compounds of which one contains both heteroatoms as nucleophiles and the other contains two electrophilic groups (usually C=O or C—halogen), e.g.

pyrazoles   isoxazoles   thiazoles   pyrimidines

Reactions between compounds which each contain both a nucleophilic and an electrophilic group are much less commonly used because of the likelihood of the self-condensation of each component.

Pericyclic reactions, especially the addition of 1,3-dipolar compounds to C=C,

$C\equiv C$, and $C\equiv N$, are also widely used to make heterocyclic compounds with one, two, or three heteroatoms (chapter 9).

(a) Ease of ring closure

In general, a reaction that forms a strainless or near-strainless five- or six-membered ring occurs more readily than the corresponding intermolecular reaction because the entropy of activation is much more favourable (p. 66). For example, whereas two molecules of an alcohol form an acyclic ether by dehydration only under very vigorous acidic conditions, diethanolamine gives morpholine (a useful basic solvent) in much milder conditions:

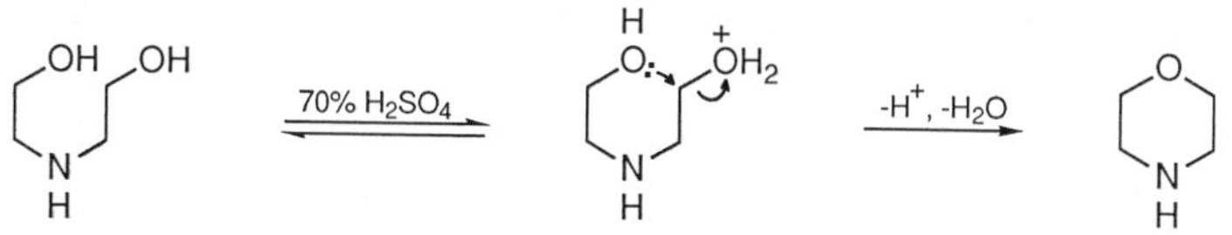

There is one important exception to this generalization in the formation of five-membered rings: ring closure is disfavoured when it involves reaction of a nucleophile at a trigonal (i.e. sp^2) carbon atom which forms a double bond to an atom *inside* the incipient ring (the *5-endo-trig* system):

5-endo-trig

For example, whereas the reaction of the *5-endo-dig* system (*dig* = digonal carbon),

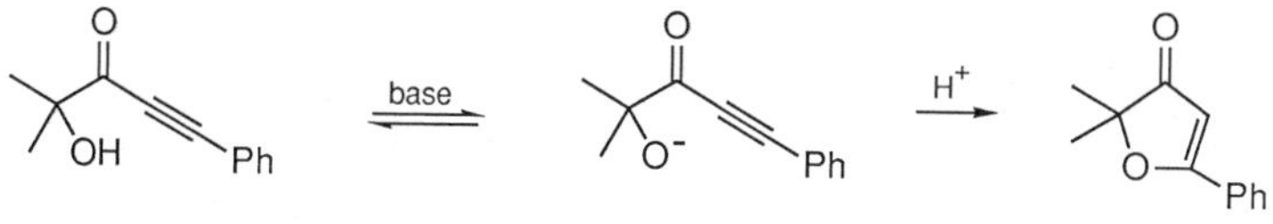

5-endo-dig

occurs, the reaction

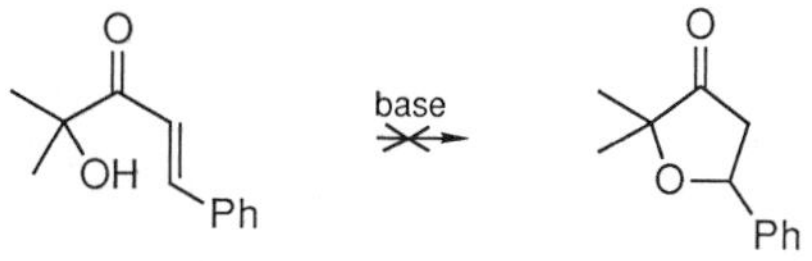

5-endo-trig

fails. On the other hand, *5-exo-trig* reactions in which reaction is at a double bond which is *outside* the incipient ring,

5-exo-trig

are successful.

The differences can be understood by considering the transition states for the processes. In that for each of the *endo* examples, the nucleophile's electron-pair orbital interacts with one on the phenyl-substituted carbon atom which was originally a *p* orbital and is undergoing rehybridization to *sp²* and *sp³* in the *dig* and *trig* cases respectively. In the *5-endo-dig* system this interaction can be achieved without appreciable strain when the original *p* orbital is the one *in the plane* of the molecular framework, corresponding to the relatively strainless five-membered ring of the product:

5-endo-dig   transition state

However, in the *5-endo-trig* system the only *p* orbital available is perpendicular to this plane, so that the molecule must adopt a non-planar framework in which the nucleophilic centre is above or below the C=C planar fragment. This is a highly strained arrangement and the transition state is consequently of high energy.

In contrast, in the *5-exo-trig* case, rotation about the single bond between the carbon of C=Y and its neighbour allows the *p* orbital on the former carbon atom to move into an approximately coplanar position for a relatively strainless interaction with the nucleophilic centre, the C—Y bond moving above or below the developing ring.*

In some cases, intramolecular reactions take a different course from their intermolecular analogues as a result of this steric requirement. For example, the *5-endo-trig* cyclization corresponding to the intermolecular Michael-type reaction

*These stereochemical considerations underlie *Baldwin's rules* for ring closure (cf. also the preference for reaction of the 6-hexenyl radical to give a five-membered primary rather than a six-membered secondary cyclic radical, p. 540). Common examples of these rules (F, favoured; D, disfavoured) are:

Ring size	exo-dig	exo-trig	endo-dig	endo-trig
3	D	F	F	D
4	D	F	F	D
5	F	F	F	D
6	F	F	F	F

does not occur; instead, *5-exo-trig* cyclization gives the amide, e.g.

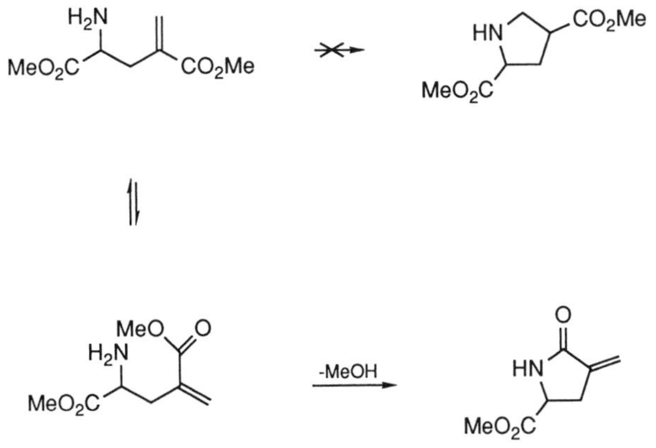

(*b*) *Retrosynthetic analysis*

The processes described in earlier chapters have mostly been concerned with single *functional group interconversions* in which one function is converted into another. In the synthesis of heterocyclic compounds, usually at least two such reactions occur, e.g.

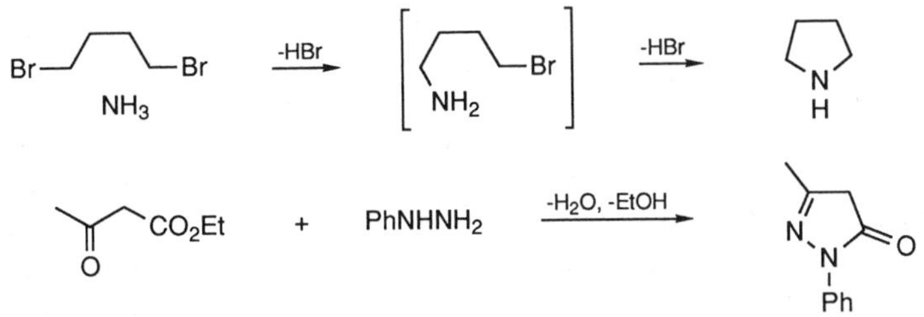

In planning such syntheses, it is helpful to employ *retrosynthetic analysis*: the target molecule is (mentally) disconnected one bond at a time and, after each disconnection, a check is made that a reaction is available to carry out the reverse, connecting, process. This is continued until available compounds and reagents are identified.

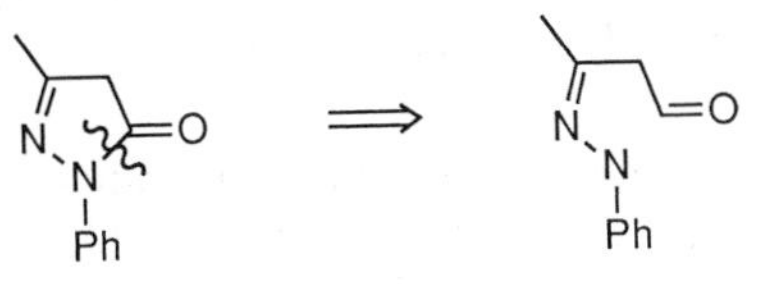

Consider the synthesis of the pyrazolone above. An obvious point of disconnection is the N—CO bond,

[⟹ denotes disconnection]

since this type of bond is readily formed by the reaction of an NH-containing group and an acid derivative, -COX, where X is a good leaving group; that is, the reaction,

should be successful if X is, for example, Cl. The next disconnection might be made at the C=N bond,

since -NH$_2$ and $>$C=O groups react together readily. Available compounds are now recognized: phenylhydrazine and a derivative of acetoacetic acid. A further check is necessary, however, since the former compound possesses two nucleophilic centres and the latter possesses two electrophilic centres: would the reactions occur with the required positional selectivity?

In phenylhydrazine, the nitrogen atom in the NH$_2$ group is more strongly nucleophilic than that in the PhNH group, since the unshared electron-pair on the latter nitrogen takes part in a delocalized system with the aromatic ring. The NH$_2$ group will therefore be the first to react. Now, if the leaving group, X, in the acetoacetic acid derivative is chlorine, then, since acid chlorides are more reactive than ketones towards nucleophiles, the first reaction would occur with the wrong chemoselectivity. However, esters are less reactive than ketones towards nucleophiles, so that the desired sequence of reactions would occur if X were OEt. Acetoacetic ester is therefore a suitable starting material and the sequence of reactions is

Other disconnections can, and should, be explored. However, as readers can discover for themselves, no alternative emerges which has the simplicity of the synthesis above.

Retrosynthetic analysis is aided by applying the concept of *synthetic equivalents*. For example, ethylene oxide is a synthetic equivalent for the structural unit —CH$_2$CH$_2$OH, since it can be transformed into this unit by reaction with a Grignard reagent,

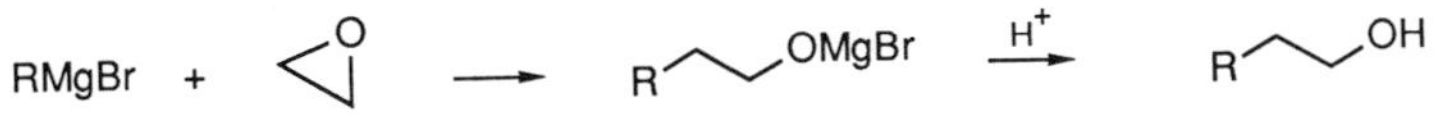

Cyanide ion is a synthetic equivalent for $-CO_2H$ and $-CH_2NH_2$

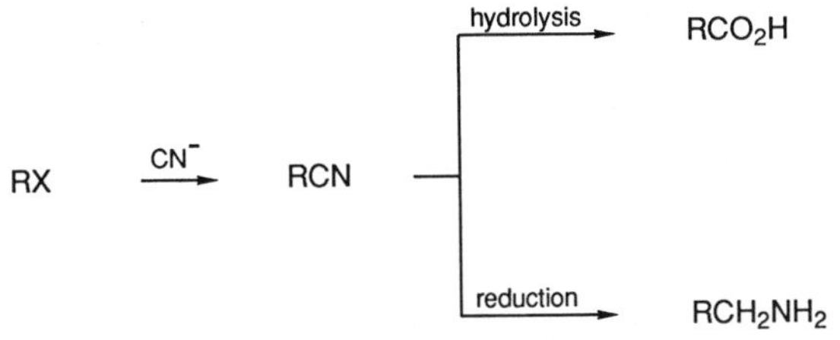

and, especially in heterocyclic synthesis, for aromatic amino groups, as the following disconnections show:

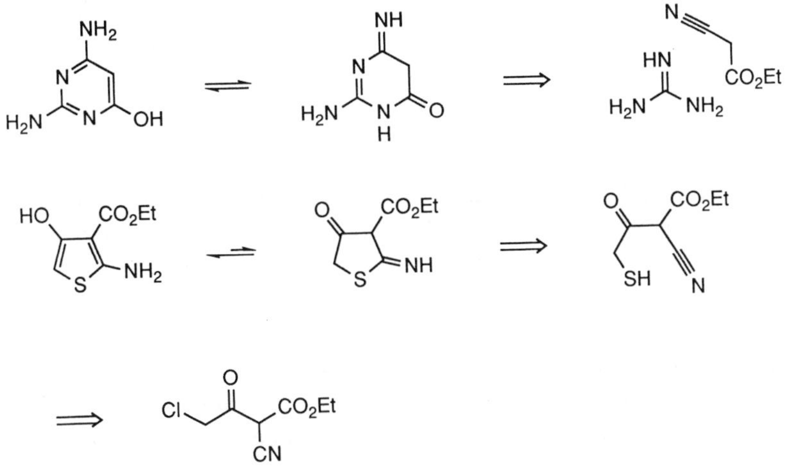

(c) Properties of heterocyclic compounds

The properties of the heterocyclic compounds whose syntheses are outlined in this chapter will not be described in detail. The non-aromatic compounds have properties closely related to their acyclic analogues, e.g. piperidine behaves like an acyclic secondary amine, being a moderately strong base (pK 11.2; cf. diethylamine, 11.1) and a reactive nucleophile; pyrrolidine is similar. The properties of those aromatic compounds which contain two heteroatoms can be generally inferred from those of benzenoid and simpler heterocyclic compounds (discussed in chapters 11 and 12). For example, pyrazole is derived from pyrrole by substitution of $-N=$ for $-CH=$, just as pyridine is derived from benzene; and the chemistries of pyrazole and pyrrole are related in very much the same way as those of pyridine and benzene. The following are broad outlines of the relevant chemistry.

(*i*)   *Basicity of heterocyclic nitrogen.*

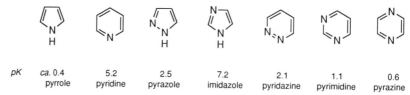

pK	*ca.* 0.4	5.2	2.5	7.2	2.1	1.1	0.6
	pyrrole	pyridine	pyrazole	imidazole	pyridazine	pyrimidine	pyrazine

Pyrrole is an extremely weak base because protonation destroys the aromatic system and is accompanied by a significant loss of aromatic stabilization energy (p. 39). This is not so for pyridine, which is nevertheless a weaker base than aliphatic amines because the nitrogen atom is sp^2-hybridized (p. 49). Both pyrazole and imidazole are more strongly basic at the pyridine-like nitrogen than at the pyrrole-like nitrogen, imidazole being a stronger base than pyridine for the same reason that an amidine is a stronger base than an amine: namely, that the conjugate acid is a delocalized cation (p. 49).

Both pyrazole and imidazole, like pyrrole (p. 49), are weakly acidic.

The weaker basicity of each of the three diazines compared with pyridine derives from the electron-attracting capacity of nitrogen relative to carbon.

(*ii*)   *Reactions at the nucleus.*   Pyridine is much less reactive towards electrophiles than benzene and in the same way the electron-attracting —N= reduces the reactivity of pyrazole and imidazole compared with that of pyrrole. Both compounds readily undergo reactions such as halogenation but only imidazole reacts (as its conjugate base) with the weaker electrophile benzenediazonium ion. Pyrazole reacts predominantly at the 4-position, imidazole at the 2-position.

Just as furan and thiophen are less reactive than pyrrole towards electrophiles, so oxazole, isoxazole, thiazole, and isothiazole are less reactive than pyrazole and imidazole, being slightly less reactive than benzene.

The three six-membered diazines are related to pyridine as pyridine is to benzene. Each is even more deactivated than pyridine towards electrophiles, being essentially inert. The reactivity is increased by the incorporation of electron-releasing groups, e.g. a hydroxyl or amino group in the 2-, 4-, or 6-position of pyrimidine enables reaction to occur at the 5-position (e.g. p. 719)

Conversely, the diazines are strongly activated towards nucleophiles, at each carbon in pyridazine and pyrazine and at all save the 5-position in pyrimidine.

Methyl substituents in these positions are powerfully activated towards acid- or base-catalyzed reaction and halogen atoms are activated towards nucleophilic displacement (p. 399).

Pyrrole and indole give Grignard reagents with an alkylmagnesium halide which can be used to introduce electrophiles at the 2- and 3-positions, respectively (p. 189). N-Substituted pyrroles and indoles, and also compounds containing pyridine-type nitrogen, oxygen, or sulfur can be metallated at the 2-position with butyllithium, again enabling electrophiles to be introduced, e.g.

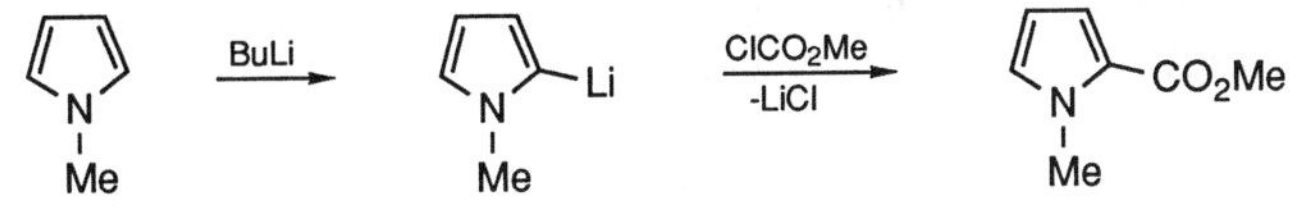

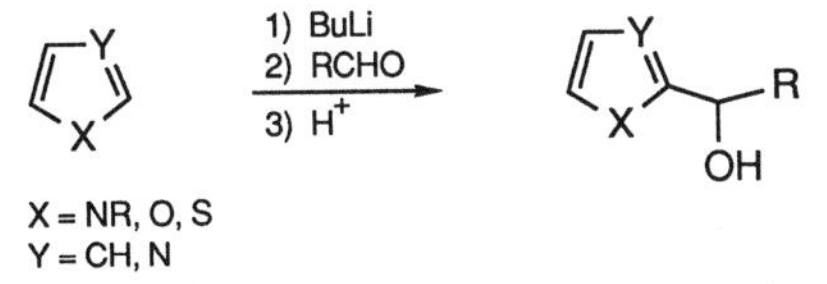

X = NR, O, S
Y = CH, N

(*iii*) *Tautomeric equilibria.*   The hydroxy derivatives of the six-membered aromatic nitrogen heterocycles exist in equilibrium with keto tautomers, except for those compounds in which the hydroxyl group is in a 3-position. For 2- and 4-hydroxypyridine, equilibrium favours the keto form ($K \sim 10^3$):

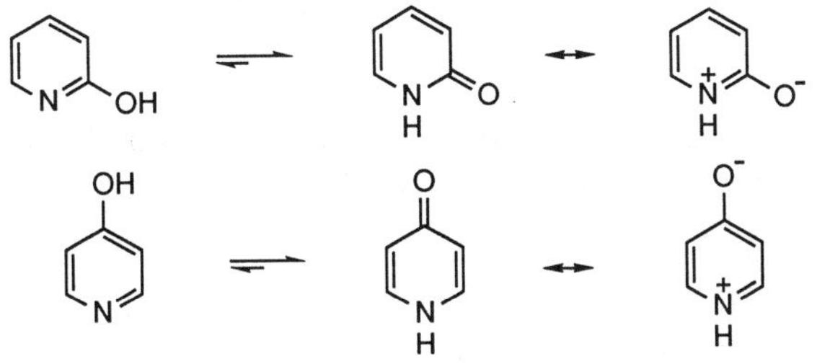

This situation is in strong contrast to that for phenol, where the keto tautomer is undetectable, the reason being that in this case the keto tautomers are themselves aromatic, as shown by the Kekulé representations above, and the resulting stabilization energy serves to make the pyridone the more strongly bonded tautomer.

This is also the case in the 2- and 4-hydroxypyrimidines, e.g.

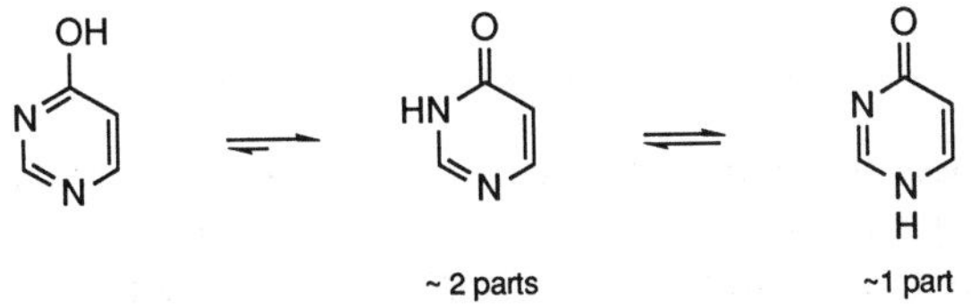

~ 2 parts   ~1 part

The same tautomeric possibilities apply to the corresponding amino derivatives, but equilibrium favours the amino over the imino form, possibly because the involvement of the amino group's unshared pair of electrons in the π-system,

stabilizes the amino form relative to its tautomer more strongly than the corresponding interaction in the hydroxy compounds (nitrogen having a greater $+M$ effect than oxygen).

21.2 Five-membered rings containing one heteroatom

(a) Pyrroles

Pyrrole itself may be extracted from coal tar and bone oil by distillation, so that its syntheses are relatively unimportant. It has been prepared industrially by passing acetylene and ammonia through a red-hot tube and by passing furan, ammonia, and steam over hot alumina. It has been made in the laboratory by distillation of the ammonium salt of mucic acid (HO_2C—$(CHOH)_4$—CO_2H). It is also produced by the distillation of succinimide over zinc dust:

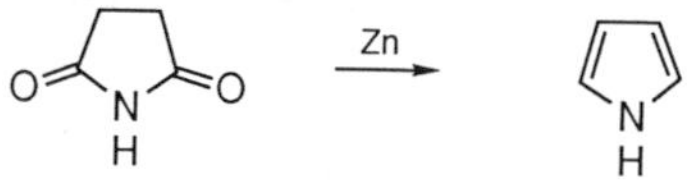

Substituted pyrroles are usually made by combining two aliphatic fragments in one of two ways, represented skeletally as follows:

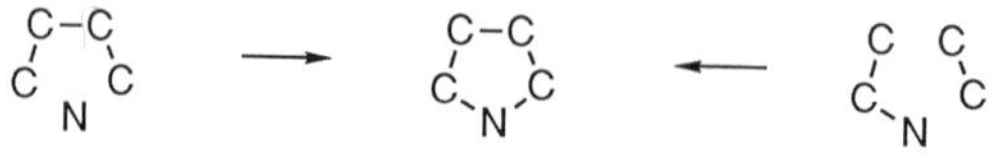

Retrosynthetic analysis can be helpfully applied to planning individual syntheses. For example, the recognition that the pyrrole (1) could be derived from its non-aromatic tautomer (2) by a 1,5-prototropic shift leads to the disconnection:

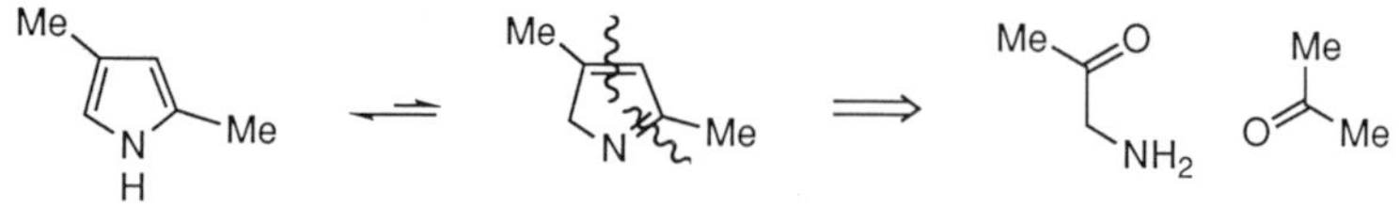

(1)   (2)   (3)   (4)

A scheme to connect the two components is clear: reaction of the amino group in (3) with the keto group in (4), and aldol-type reaction of the keto group in (3) with the activated CH_3 group in (4), either acid-catalyzed *via* the enol or base-catalyzed *via* the enolate. However, further thought reveals a difficulty: (4) is only feebly enolic and weakly activated to base, and self-condensation of the amino-ketone might occur in preference to the mixed reaction (as it does). A better possibility might be to employ acetoacetic ester in place of (4), relying on the much more strongly enolic character of the keto-ester to promote the mixed reaction (as it does, with mild acid catalysis). After reaction, the ester group can be removed by hydrolysis and – making use of the strong reactivity of pyrroles to electrophiles – protodecarboxylation:

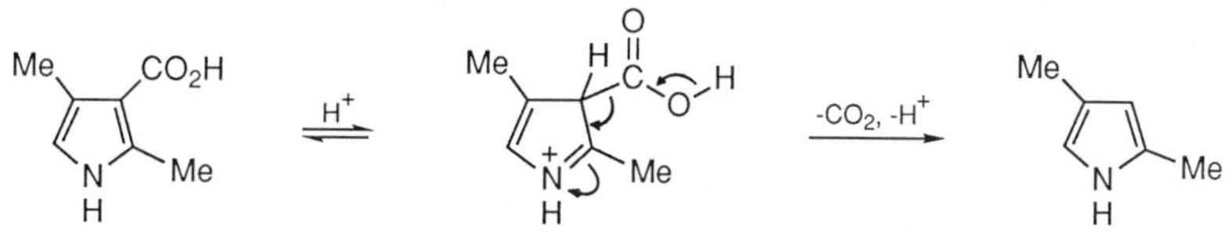

One further inference comes from this analysis: since the amino-ketone might self-condense before introduction of the β-keto-ester, it might be wisest to form it *in situ* (as is usually done).

(1)   *The Paal–Knorr synthesis.* A 1,4-dicarbonyl compound is treated with ammonia or a primary amine. Successive reactions of the nucleophilic nitrogen at the carbonyl groups are followed by dehydration, which occurs readily because the product is aromatic:

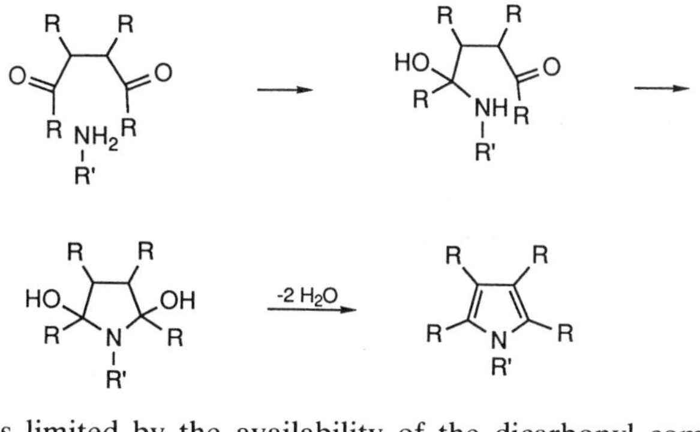

The reaction is limited by the availability of the dicarbonyl compounds. Those which are symmetrical may be obtained by reaction between the appropriate β-keto-ester and iodine (p. 239) and others can be made by treatment of furans with lead tetraacetate followed by hydrolysis and reduction:

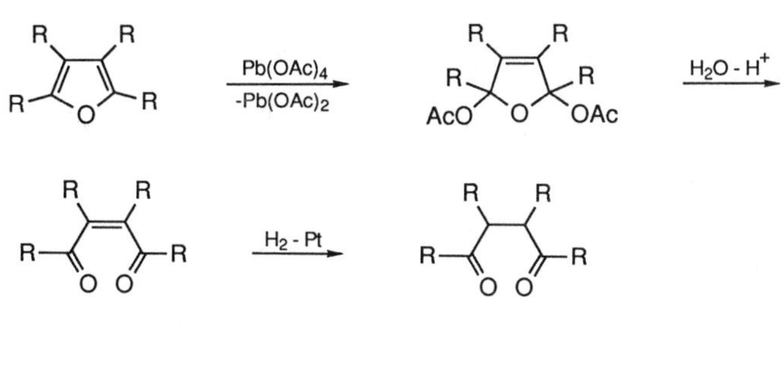

Analogous methods are used in the synthesis of furans (p. 692) and thiophens (p. 693).

(2) *The Knorr synthesis.* This is the most important route to pyrroles. The principle was outlined above: an α-amino-ketone is reacted with a ketone containing an activated α-methylene group in acetic acid solution,

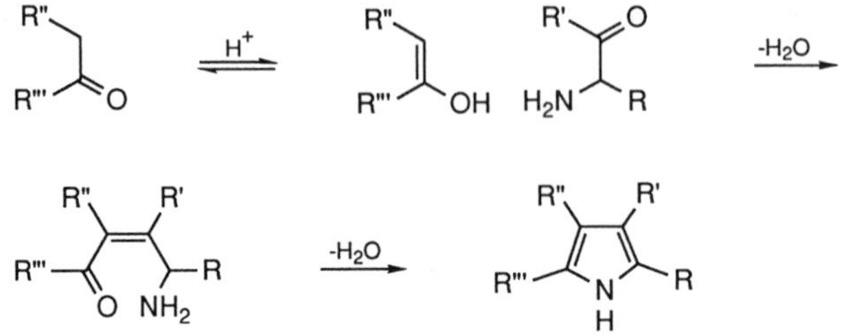

where R″ is a group such as ester that promotes rapid acid-catalyzed enolization. The α-amino-ketone is usually prepared from a β-keto-ester and an alkyl nitrite (p. 320), followed by reduction, usually *in situ*, with sodium dithionite or zinc and acetic acid, e.g.

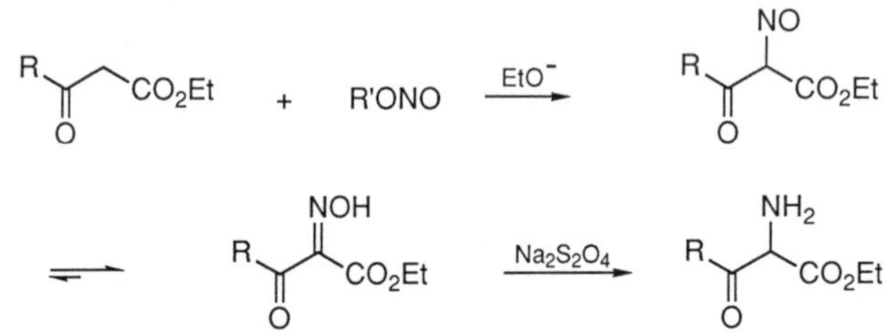

If the ester group is not required in the final product, the benzyl ester is employed, hydrogenolysis and decarboxylation then occurring in the reduction step:

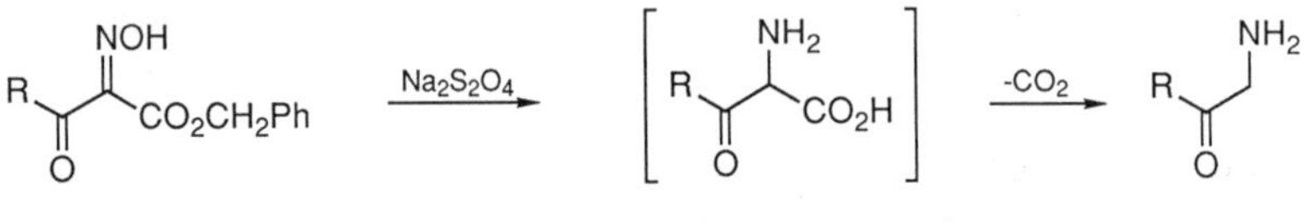

(3) *The Hantzsch synthesis.* A β-keto-ester is treated with an α-chloro-ketone in the presence of ammonia:

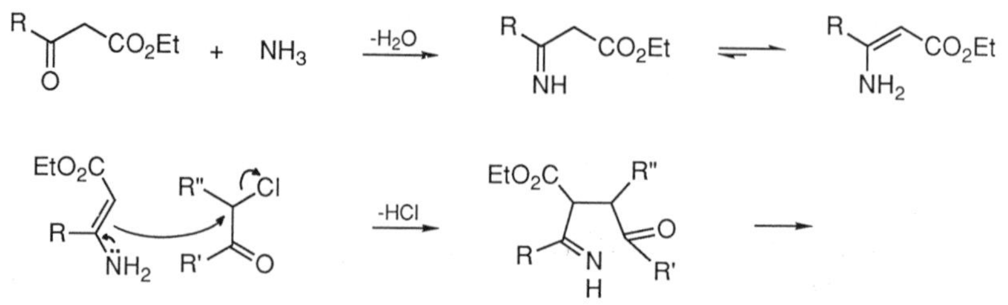

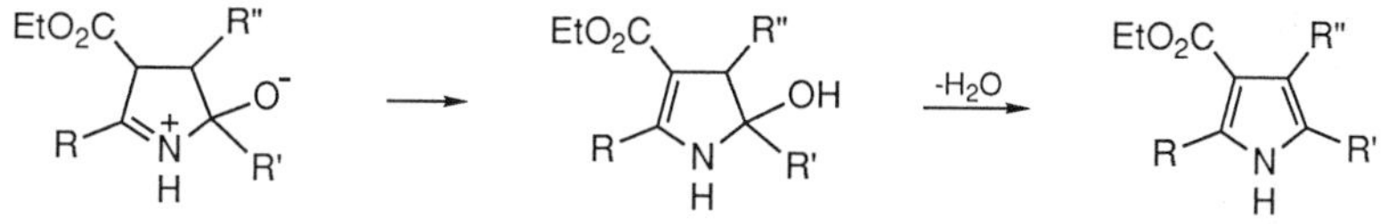

Precautions must be taken to prevent the chloro-ketone combining with the β-keto-ester to give a furan (p. 692).

Porphyrins.   Porphin, which contains four pyrrole units joined across their 2-positions by —CH= fragments,

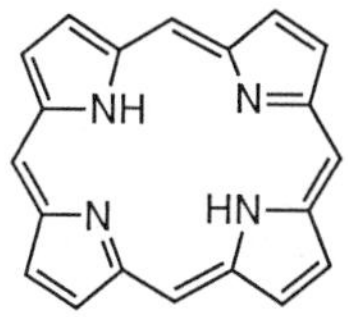

porphin

is the parent of the porphyrins, a group of naturally occurring compounds which includes haemin (in, e.g. haemoglobin) and chlorophyll. The porphyrin skeleton is usually constructed from four individual pyrroles by first bringing these units together in pairs to give two dipyrrylmethenes and then joining the pairs. The following are three of the methods for synthesizing dipyrrylmethenes.

(1)   A pyrrole 2-aldehyde is treated with a second pyrrole which must possess a free 2-position, in the presence of hydrogen bromide. The acid increases the reactivity of the aldehydic group towards nucleophiles (i.e. its reactivity as an electrophile), and the success of the reaction depends on the great reactivity of pyrroles at the 2-position towards electrophiles. The final dehydration occurs readily because the product is an effectively delocalized cation.*

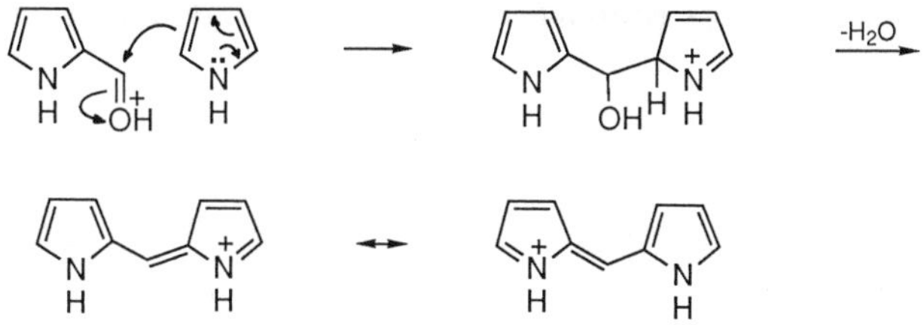

The required aldehydic group is readily introduced into the 2-position by Gattermann formylation (p. 371).

(2)   Preparation of the 2-aldehyde as in method (1) may be by-passed if a symmetrical dipyrrylmethene is required. A pyrrole is treated with formic acid

---

* The delocalization in the cation makes dipyrrylmethenes quite strong bases, analogous to amidines (p. 49) and in contrast to pyrroles themselves. For simplicity, other substituents are not shown in this and the two following general reaction schemes.

in the presence of hydrogen bromide, leading to successive reactions of Friedel–Crafts type:

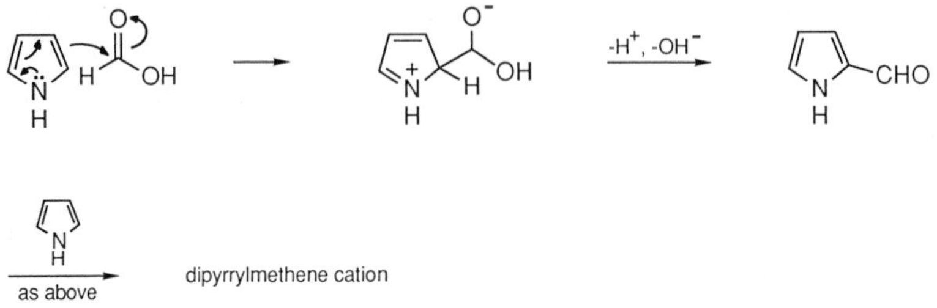

(3)   A 2-methylpyrrole containing a free 5-position is treated with bromine. A benzylic-type bromide is formed by one of the pyrrole units and this reacts at the 5-position of the second pyrrole in the Friedel–Crafts manner. Nuclear-bromination also occurs.

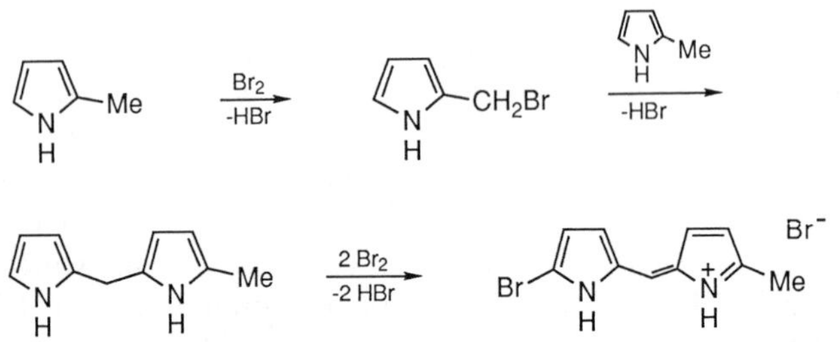

The coupling of two dipyrrylmethenes to give a porphyrin is usually accomplished by heating a 2-methyl-derivative with a 2-bromo-derivative in sulfuric acid at about 200°C. Although the yields are low (less than 5%), the route is an attractive one to such complex products because the starting materials are readily available.

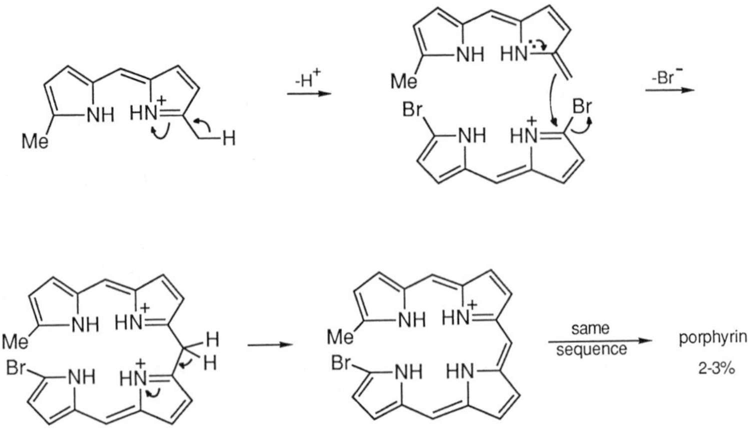

(b) *Indoles*

The indole (2:3-benzopyrrole) system occurs in the essential α-amino acid trypto-phan, in the plant alkaloids which owe their biogenetic origin to tryptophan (e.g. reserpine, section 22.1), and in indigo and Tyrian purple, two dyestuffs obtained from natural sources. Of the many methods of synthesis which have been developed, six are of particular importance.

(1) *The Fischer synthesis.* The phenylhydrazone of an aldehyde or ketone is heated with an acid catalyst such as boron trifluoride, zinc chloride, or poly-phosphoric acid. The reaction is analogous to the benzidine rearrangement (p. 453); the fourth step is an example of a [3,3]-sigmatropic rearrangement.

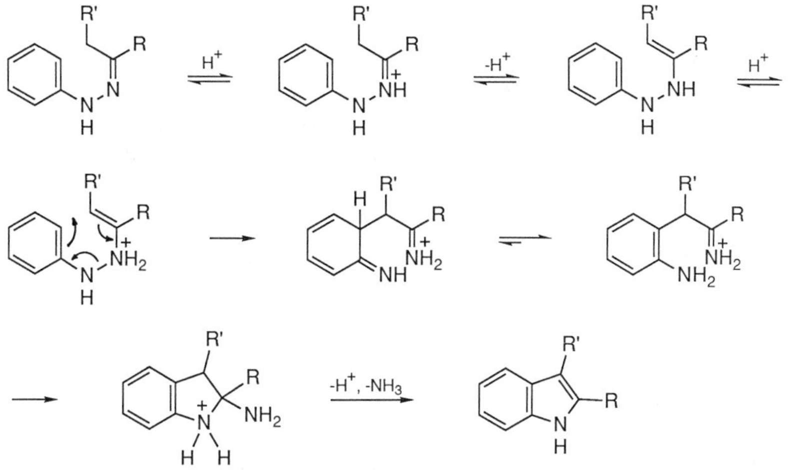

For example, acetophenone phenylhydrazone (R = Ph, R' = H), treated with zinc chloride at 170°C, gives 2-phenylindole in up to 80% yield [3] and yields are usually of this order. However, the reaction fails with acetaldehyde phenyl-hydrazone (R,R' = H), so that indole itself must be made indirectly. A simple method is to carry out the Fischer reaction on the phenylhydrazone of pyruvic acid (R = CO$_2$H, R' = H) and to decarboxylate the resulting indole-2-carboxylic acid thermally.

Substituents on the phenyl ring of the hydrazone influence the regiochemistry of the [3,3]-rearrangement. For example, electron-releasing *meta*-substituents give mainly 6-substituted indoles (i.e. *para* ring closure) whereas electron-attracting substituents give mainly 4-substituted indoles (i.e. *ortho* ring closure).

(2) *The Madelung synthesis.* An *o*-acylaminotoluene is treated with a base such as potassium t-butoxide or LDA. The probable mechanism is as follows:

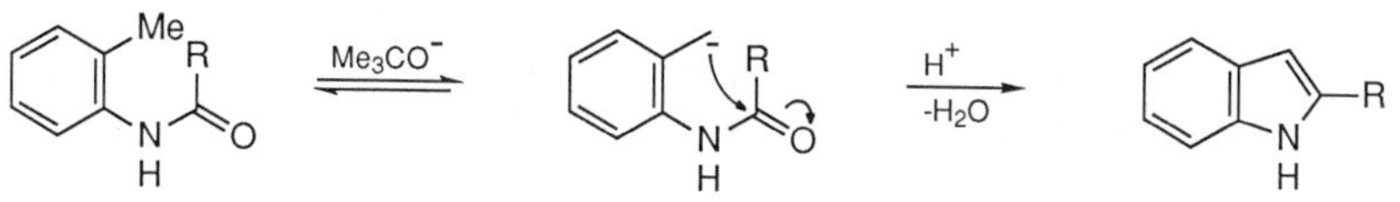

Yields are low, for decarbonylation of the starting material also occurs.

(3) *The Bischler synthesis.* An α-hydroxy- or α-halo-ketone is treated with an arylamine in the presence of acid. The probable mechanism is:

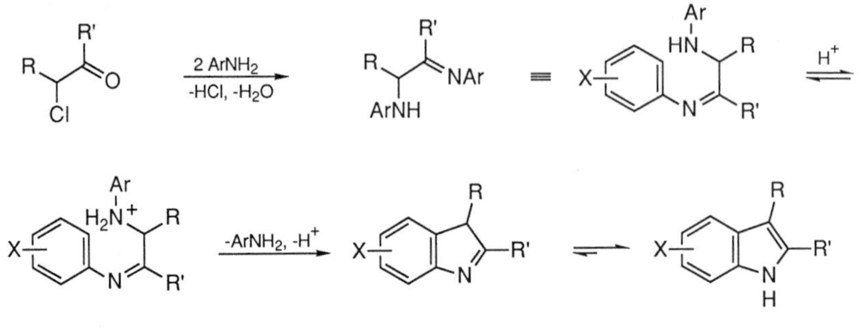

(4) *The Reissert synthesis.* An *o*-nitrotoluene is reacted with diethyl oxalate in the presence of a base. The nitro group of the resulting α-keto-ester is reduced to amino and cyclization then occurs spontaneously. The ester substituent in the indole may be removed, if required, by hydrolysis and thermal decarboxylation.

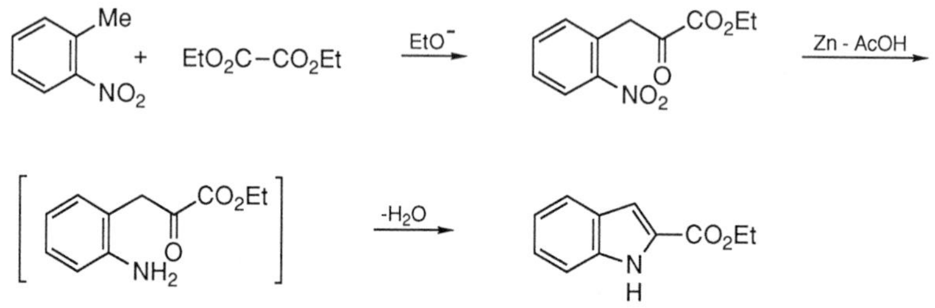

(5) *The Gassman synthesis.* An *N*-chloroaniline and a β-keto-sulfide form a sulfonium salt the ylide from which undergoes a [3,3]-sigmatropic rearrangement followed by ring closure. Desulfurization yields the indole:

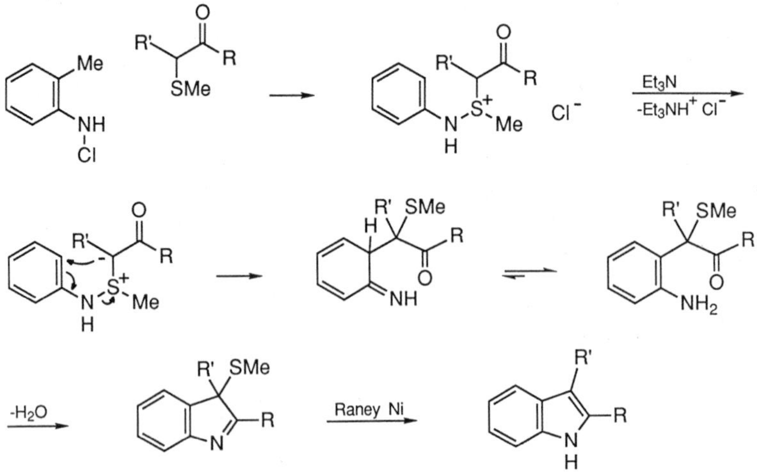

(6) *The Nenitzescu synthesis*. 5-Hydroxyindoles can be obtained from *p*-benzoquinones. The first intermediate is oxidized by further benzoquinone to one which cyclizes; the quinol formed in this reaction reduces the cyclized product to the indole, e.g.

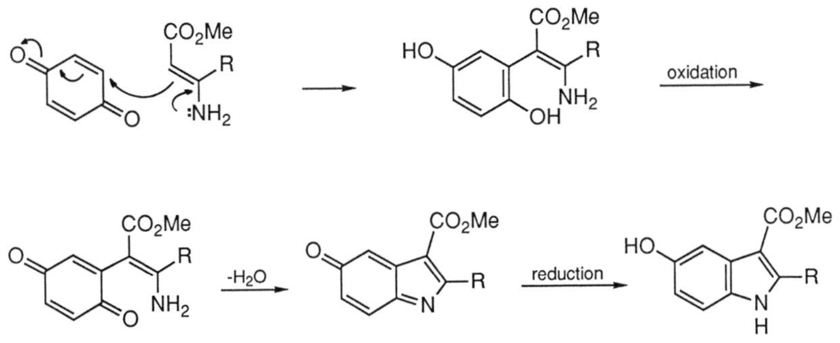

(c) *Isoindoles*

Isoindole is formed by low-pressure sublimation of the *N*-acetoxy-compound shown at 50–70°C (cf. pyrolysis of esters; p. 296) and is not very stable.

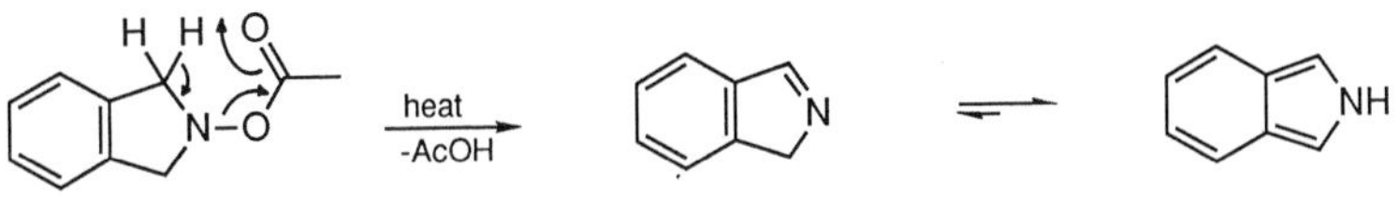

The most important derivatives of isoindole are the metal complexes of the phthalocyanines, used as dyes and pigments. They are prepared by passing ammonia into molten phthalic anhydride (cheaply available by the oxidation of naphthalene, p. 596) in the presence of a metal, e.g.

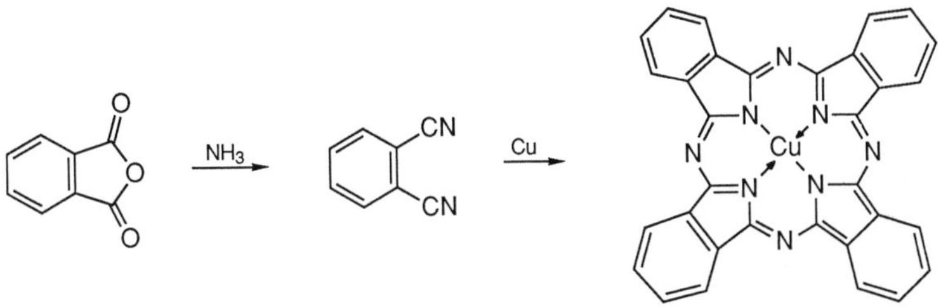

monastrol fast blue

Phthalocyanine itself is obtained by acidification of its metal complexes.

(d)   Furans

Furan itself is obtained industrially by the catalytic decomposition on a nickel catalyst of its 2-aldehyde, furfural. Furfural is readily available from oat husks or maize cobs (which contain pentoses), by digestion in dilute mineral acid followed by distillation in steam.

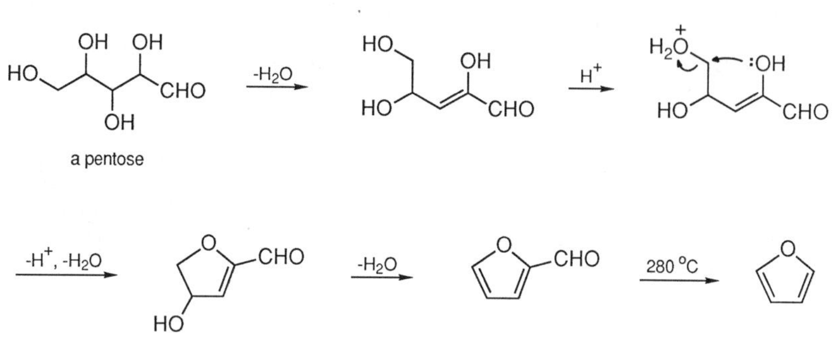

The following are the general methods for synthesizing furans.

(1)   *From 1,4-dicarbonyl compounds*. Treatment of 1,4-dicarbonyl compounds with dehydrating agents such as sulfuric acid, zinc chloride, acetic anhydride, and phosphorus pentoxide brings about cyclization and dehydration (cf. the Paal–Knorr procedure for pyrroles, p. 685). The reaction is limited by the ease of obtaining the dicarbonyl compounds.

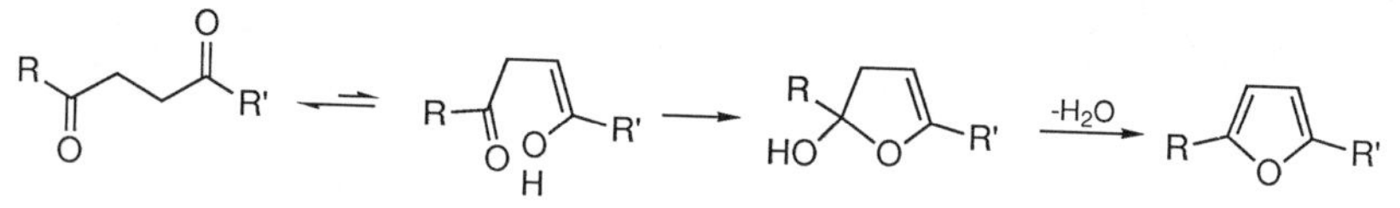

(2)   *The Feist–Benary synthesis*. An α-chloro-ketone is reacted with a β-keto-ester in pyridine solution (cf. the Hantzsch synthesis of pyrroles, p. 686).

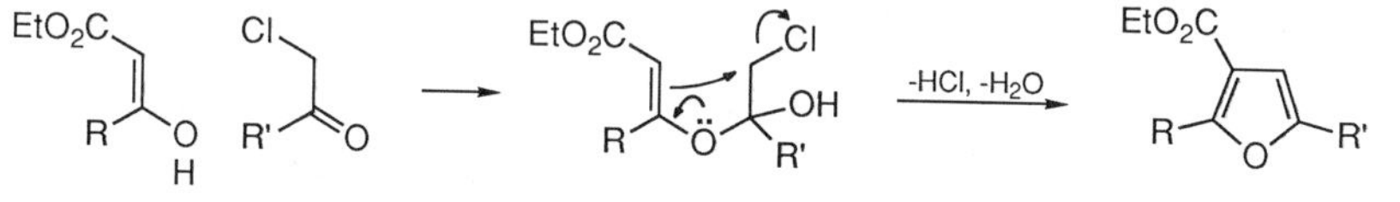

(e)   Thiophens

Thiophen itself occurs in the same fraction as benzene in coal-tar distillates. The methods for its removal in the purification of benzene are based on its greater reactivity towards electrophiles (p. 358): treatment with concentrated sulfuric acid gives the water-soluble thiophensulfonic acids, and mercury(II) acetate gives the mercury-derivative. It is obtained industrially by the continuous cyclization of hydrocarbons such as butane, the butenes, and butadiene over sulfur or sulfur-containing compounds such as pyrites, at temperatures of 500–600°C. Furans and pyrroles are converted in moderate yield into their thiophen analogues on treatment with hydrogen sulfide on an alumina catalyst at high temperatures.

The general methods for thiophens are as follows.

(1)   *Hinsberg's procedure.* An α-dicarbonyl compound is reacted with a thio-ether in which the sulfur atom is adjacent to two activated methylene groups, e.g.

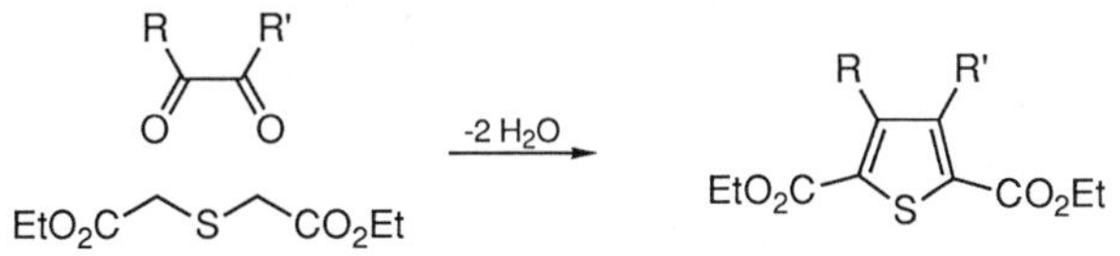

(2)   *From 1,4-dicarbonyl compounds.* 1,4-Dicarbonyl compounds react with phosphorus tri- or penta-sulfide, the former usually being the more satisfactory. The reaction is analogous to the Paal–Knorr procedure for pyrroles (p. 685).

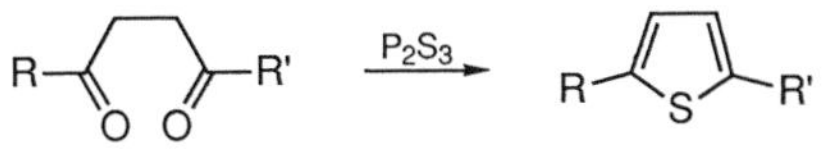

There are many variants of this method, of which the following is one example:

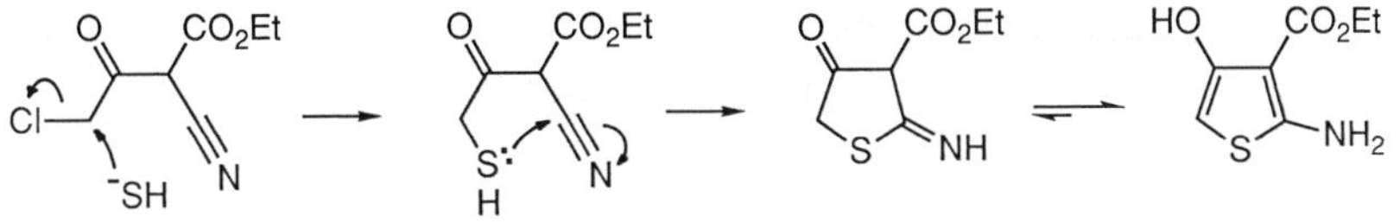

(f)   *Reduced pyrroles, furans, and thiophens*

These may be made either by the cyclization of an aliphatic compound or, in most cases, by reduction of the corresponding aromatic compound.

(1)   Pyrrolidine can be made by reaction between 1,4-butanediol and ammonia or by catalytic reduction. Reduction with zinc and acetic acid, however, results in 1,4-addition of hydrogen:

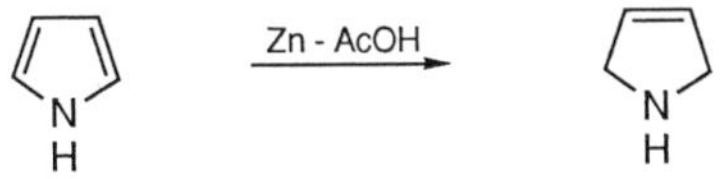

N-Substituted pyrroles can be made from a 1,4-dibromobutane with a primary amine or by the Hofmann–Löffler–Freytag reaction (p. 553).

(2)   The furan ring is fairly resistant to reduction unless it is substituted with electron-withdrawing groups such as carboxyl; furan itself requires Raney nickel at over 80°C for reduction to tetrahydrofuran. Tetrahydrofuran is, however, obtained cheaply by reaction between acetylene and formaldehyde followed by reduction and dehydrative cyclization (p. 243).

(3)   Reduction of thiophen cannot be brought about catalytically because

the sulfur atom is removed in the process and poisons the catalyst. Reduction with sodium in liquid ammonia containing methanol gives a mixture of di- and tetrahydrothiophens:

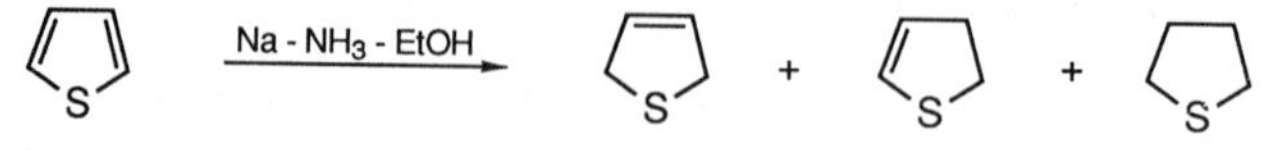

Tetrahydrothiophen is best prepared by reaction between 1,4-dichlorobutane and sodium sulfide or between 1,4-butanediol and hydrogen sulfide on alumina at 400°C.

21.3
Five-membered rings containing two heteroatoms*

(a)   Pyrazoles

Few pyrazoles occur naturally, but some of the synthetic derivatives have useful properties (e.g. antipyrine). They are prepared from 1,3-dicarbonyl compounds and hydrazine, e.g. pyrazole itself can be obtained by treating a malondialdehyde diacetal[†] with hydrazine in the presence of acid:

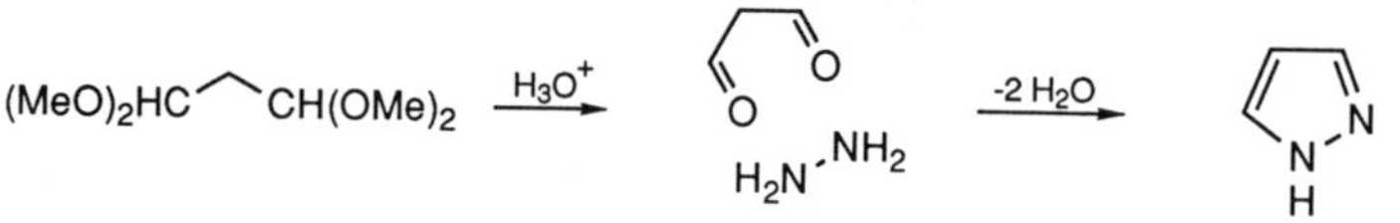

Pyrazolines (partially reduced pyrazoles) are prepared from alkenes and diazoalkanes (p. 278) or from $\alpha\beta$-unsaturated aldehydes or ketones and hydrazine, e.g.

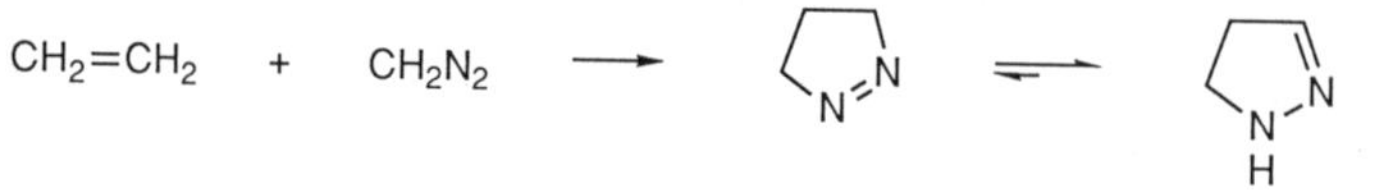

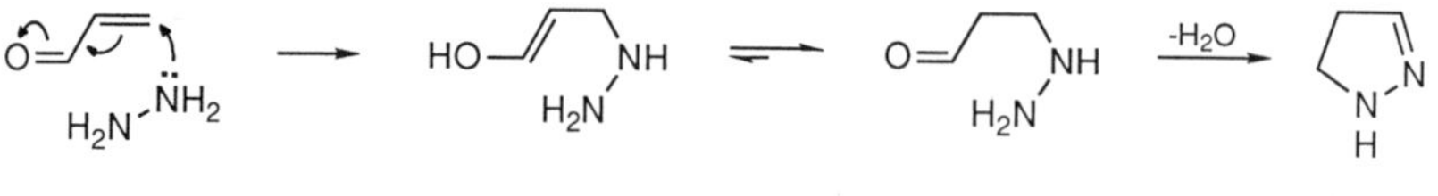

---

* Nomenclature: oxa, thia, and aza denote the presence of oxygen, sulfur, and nitrogen respectively, and, in compounds containing more than one of these substituents, are used in that order. The aromatic compounds have the suffix -ole; the partially reduced compounds, -oline, and the fully reduced compounds, -olidine. Common names are also used (e.g. imidazole for the isomer of pyrazole).

[†] Malondialdehyde readily undergoes self-condensation and is used in synthesis in the form of its diacetal from which the aldehyde is generated with acid.

β-Keto-esters give pyrazolones similarly:

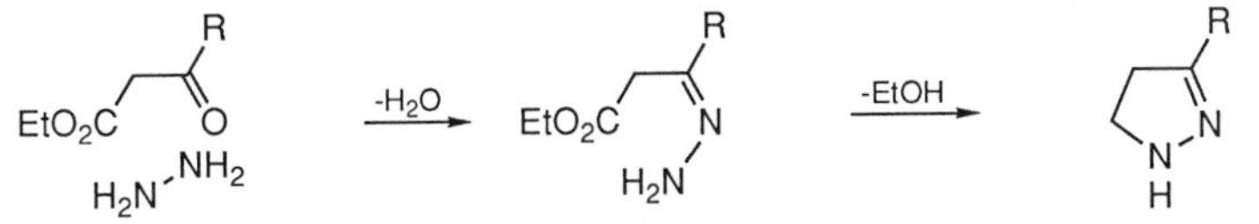

Pyrazolidines (fully reduced pyrazoles) are prepared from 1,3-dihalides and hydrazine, e.g.

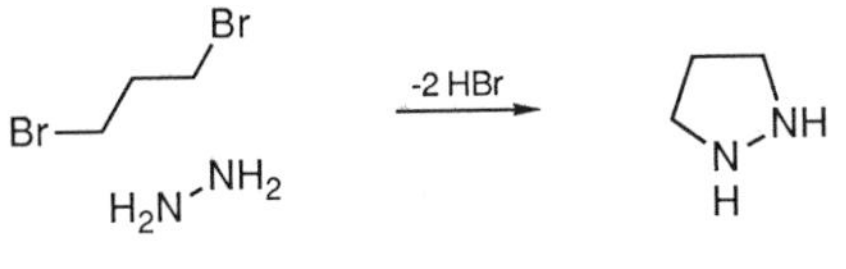

(b) Imidazoles

Imidazoles occur naturally, the best known being histidine, an α-amino acid contained in proteins (p. 324). Imidazole itself is prepared by treating acetaldehyde with bromine in ethylene glycol, giving the cyclic acetal of bromoacetaldehyde, and treating this with formamide and ammonia:

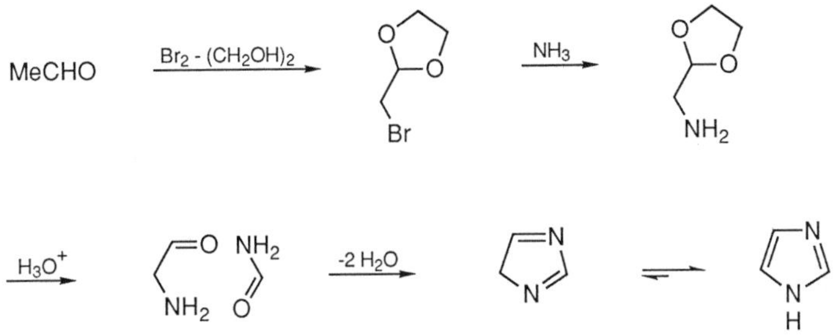

α-Hydroxyaldehydes and ketones may replace (the intermediate) aminoacetaldehyde in general applications.

In a second general procedure, a salt of an α-aminoaldehyde or ketone is treated with a hot aqueous solution of potassium thiocyanate, giving a thione from which sulfur is removed with Raney nickel:

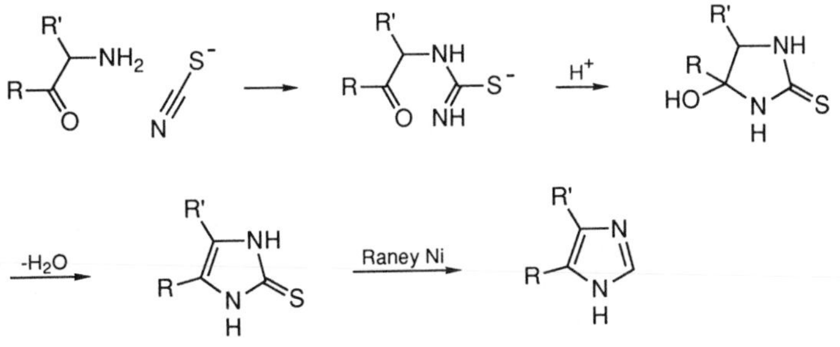

1,2-Diamines react with carboxylic acids and aldehydes or ketones to give, respectively, imidazolines and imidazolidines:

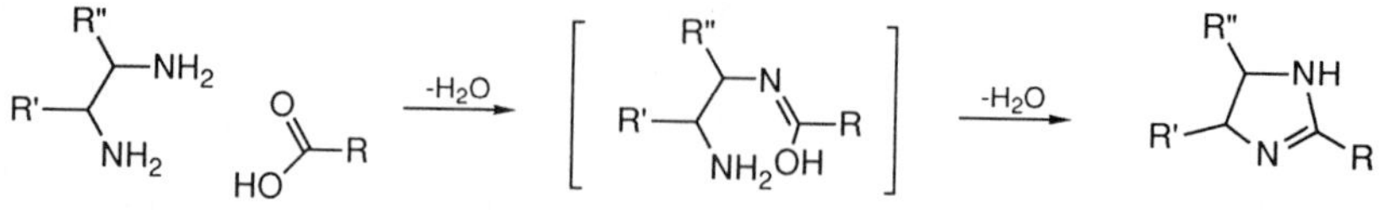

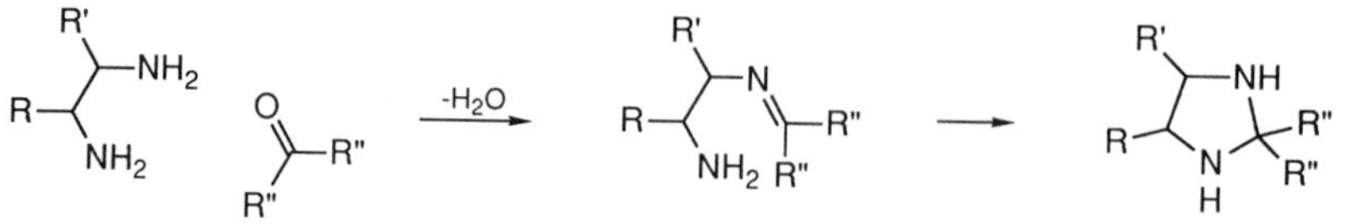

(c) Isoxazoles

Isoxazoles and their reduction products are prepared by methods analogous to those for pyrazoles and their reduction products, hydrazine being replaced by hydroxylamine. 1,3-Dicarbonyl compounds give isoxazoles,

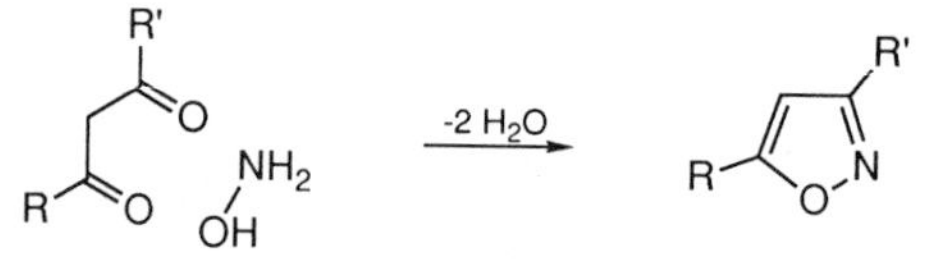

αβ-unsaturated aldehydes or ketones give isoxazolines,

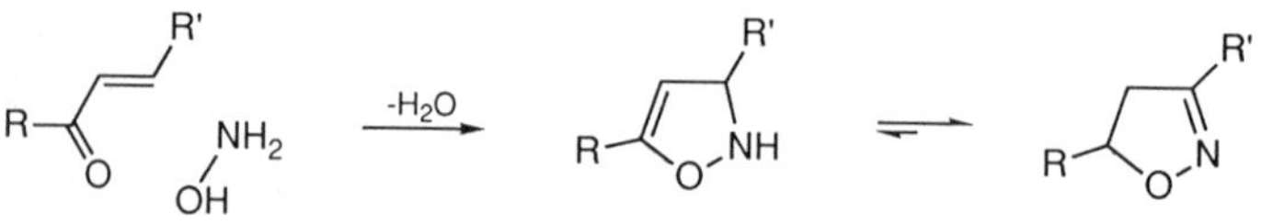

β-keto-esters give isoxazolones,

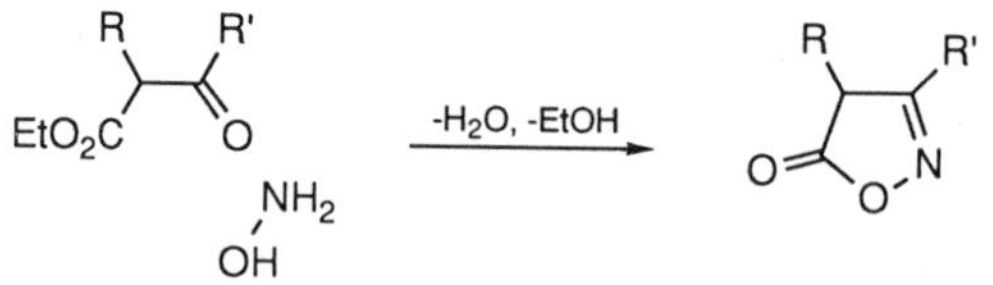

and 1,3-dihalides give isoxazolidines, e.g.

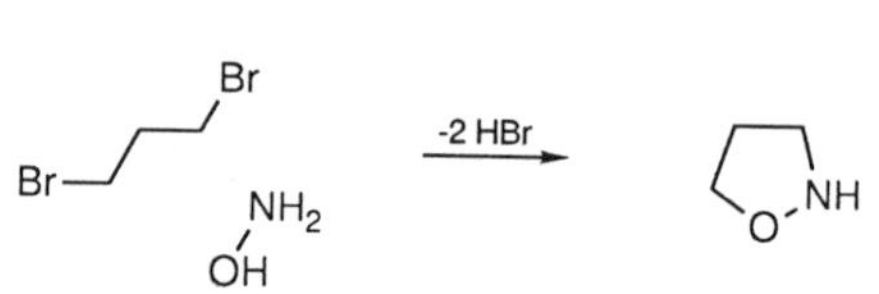

(d) Oxazoles

The oxazole ring is constructed by heating an α-haloketone with an amide,

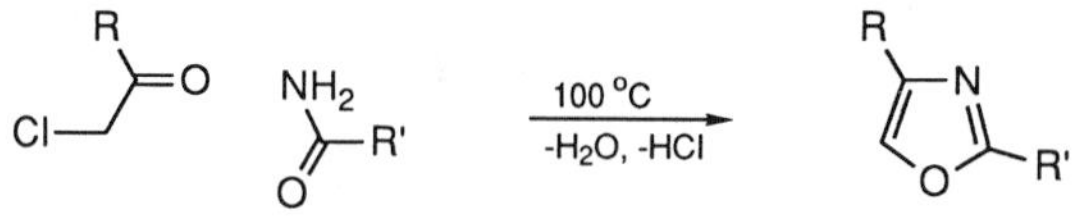

or by reductive acetylation of the α-oximino-derivative of a ketone followed by dehydrative cyclization in the manner in which furans are formed from 1,4-dicarbonyl compounds (p. 691):

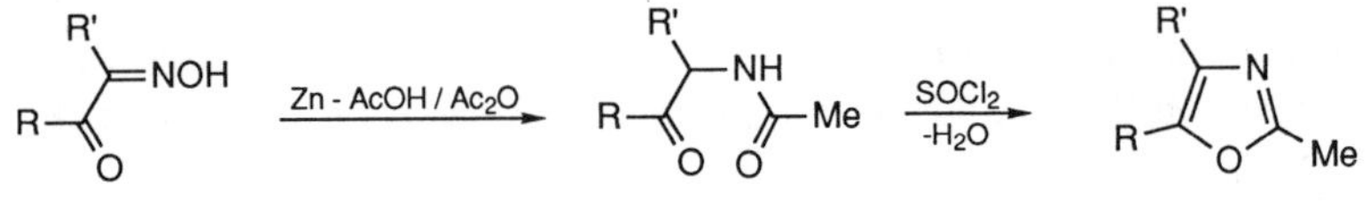

Oxazolines may be made from derivatives of β-chloroethylformamide, e.g.

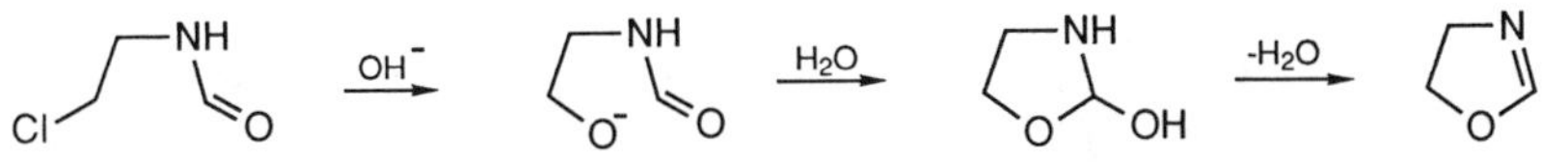

Their 5-keto-derivatives, the oxazolones or azlactones, prepared by the dehydration of N-acyl-α-amino acids, are employed in Erlenmeyer's synthesis of α-amino acids (p. 219). Oxazolidines are made from β-hydroxyamines and carbonyl compounds, analogously to the preparation of imidazolidines (p. 696).

(e) Isothiazoles

Isothiazole has been prepared as follows:

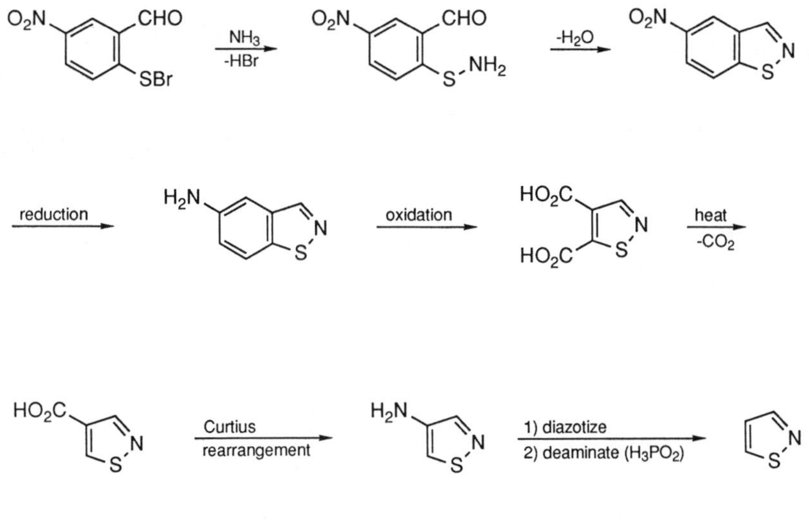

The sugar-substitute, saccharin, may be regarded as a derivative of isothiazole:

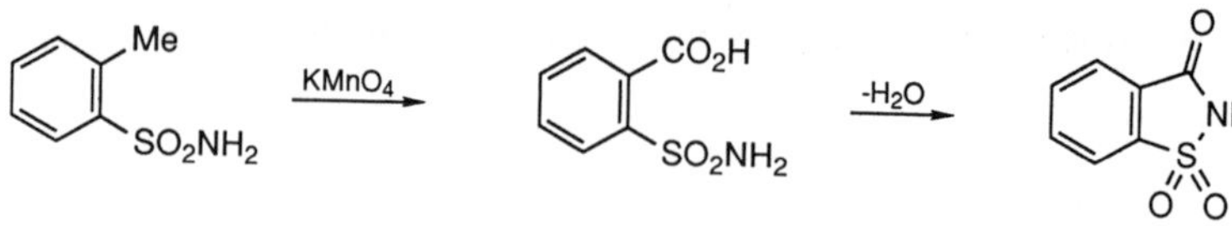

(f) Thiazoles

The thiazole nucleus occurs in vitamin B_1 (thiamine), the penicillins and cephalo-sporins (section 22.3), and the synthetic sulfathiazole (a member of the group of 'Sulfa' drugs).

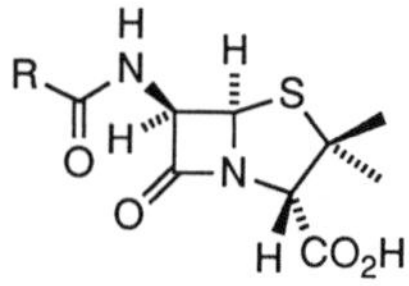

penicillins 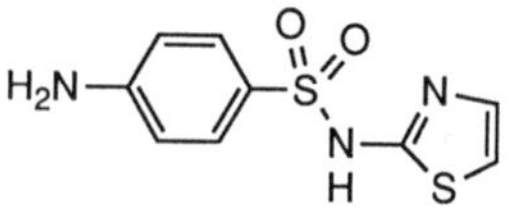   sulfathiazole

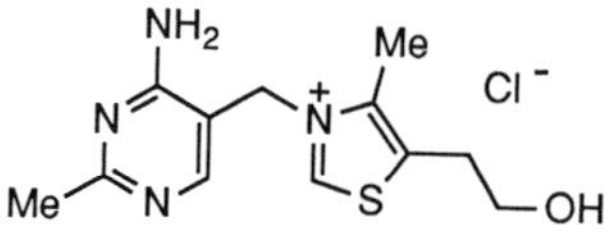

thiamine chloride

The important physiological properties of these compounds have stimulated interest in thiazole chemistry and three general synthetic routes have been developed.

(1) The most important method is from thioamides and α-halo-carbonyl com-pounds:

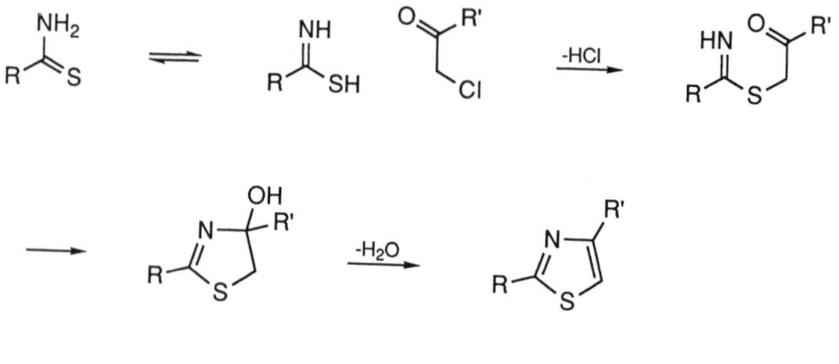

(2) From ammonium dithiocarbamates and α-halo-carbonyl compounds:

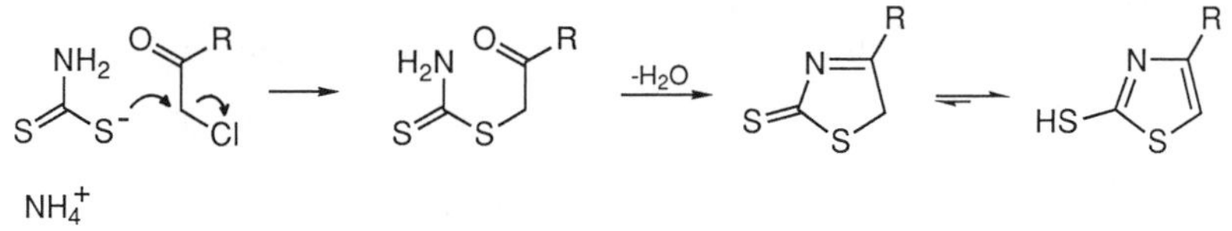

(3) From an α-acylamino-carbonyl compound and phosphorus pentasulfide (analogously to the formation of thiophens from 1,4-dicarbonyl compounds, p. 693):

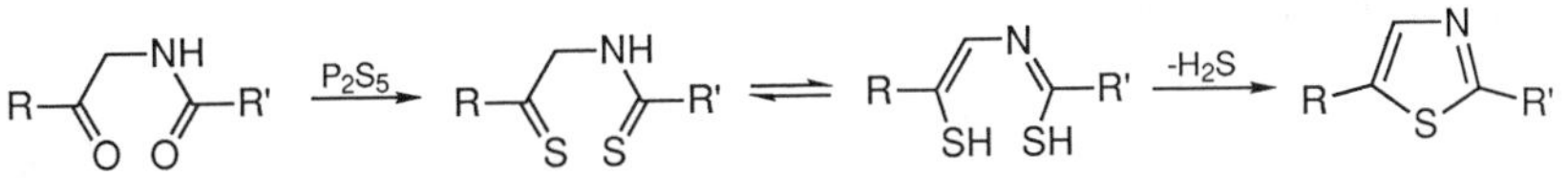

Δ^2-Thiazolines have been prepared from β-amino-thiols, e.g.

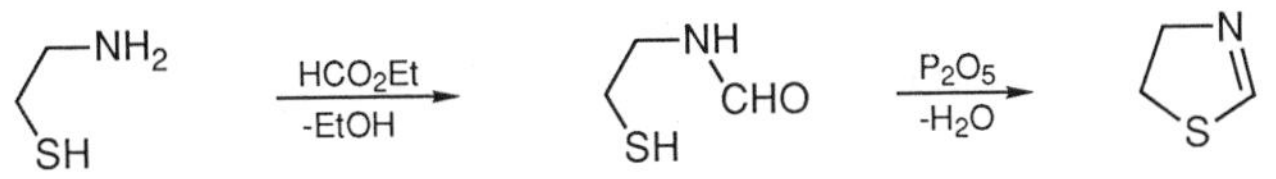

and are readily reduced by aluminium amalgam to thiazolidines. Thiazolidine itself is usually prepared from 2-aminoethanethiol and formaldehyde (cf. the formation of pyrazolidines, p. 695).

**21.4
Six-membered
rings containing
one heteroatom**

(a) Pyridines

The pyridine nucleus occurs in a number of plant alkaloids (e.g. nicotine) and in two members of the vitamin B group. One of these, pyridoxol, is a component of enzymes responsible for transamination and decarboxylation and the other, nicotinamide, is a component of enzymes in the electron-transport chain (respiratory enzymes concerned with the utilization of oxygen, see p. 721). Pyridines are also products of the distillation of coal-tar and bone oil.

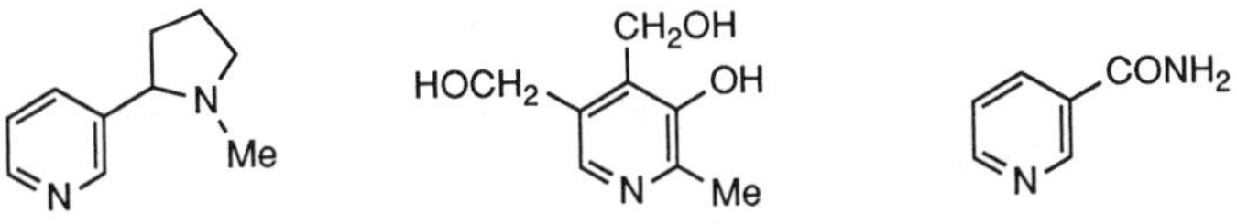

nicotine   pyridoxol   nicotinamide

Whereas the five-membered heterocycles are comparatively reactive towards a wide variety of electrophilic reagents (p. 357), so that substituents may be introduced into accessible pre-formed rings, pyridines are very unreactive towards electrophiles (p. 359) and few nucleophilic reagents are available for introducing substituents. It is therefore usual to synthesize substituted pyridines from aliphatic compounds containing the appropriate groups. A number of general routes are available, the following being the skeletal types from which the ring is constructed:

In addition, pyridines can be made from furans and pyrones.

(1)   *From five-carbon units and ammonia.* Pyridine itself is formed from gluta-conic aldehyde and ammonia:

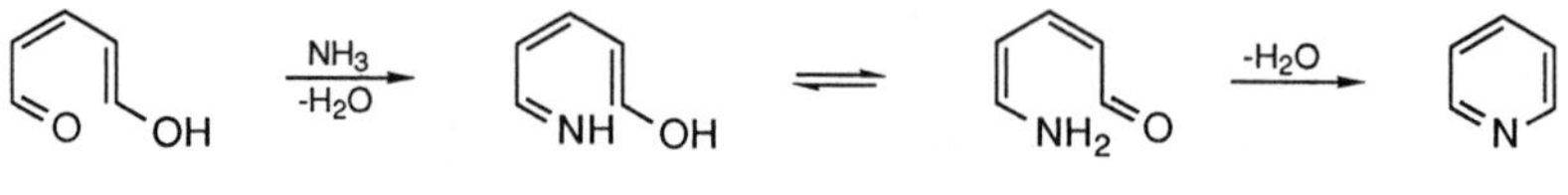

Similarly, primary amines give pyridinium salts and hydroxylamine gives pyridine-*N*-oxide:

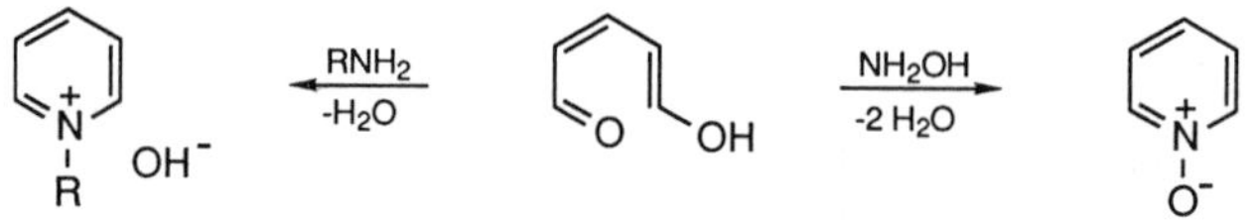

More generally, 1,5-dicarbonyl compounds and hydroxylamine form pyridines:

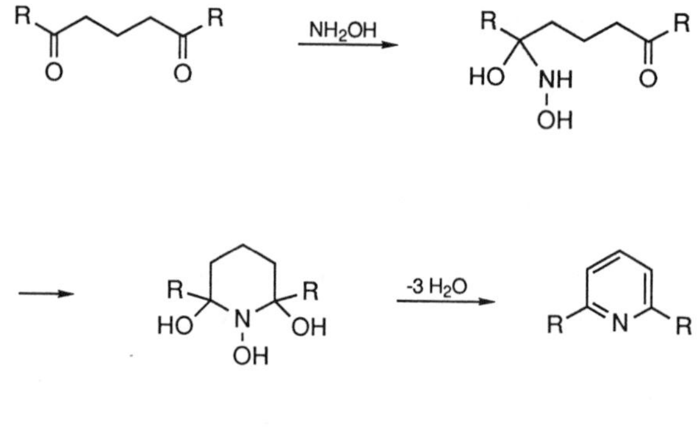

(2)   *From aldehydes or ketones and ammonia.* Aldehydes and ketones react with ammonia at high temperatures, under pressure, by ammonia-catalyzed aldol reactions together with the incorporation of the nitrogen atom of ammonia by Michael-type addition, e.g.

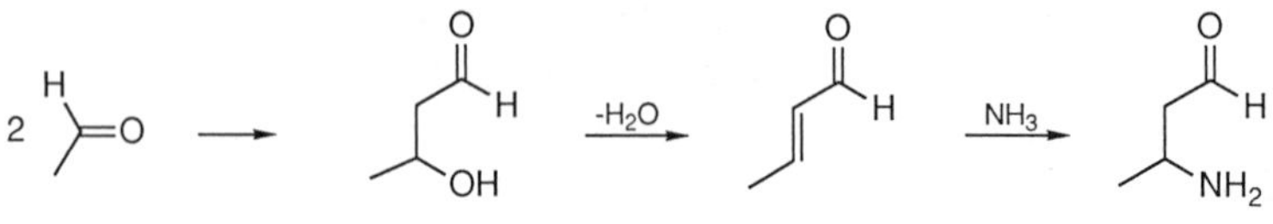

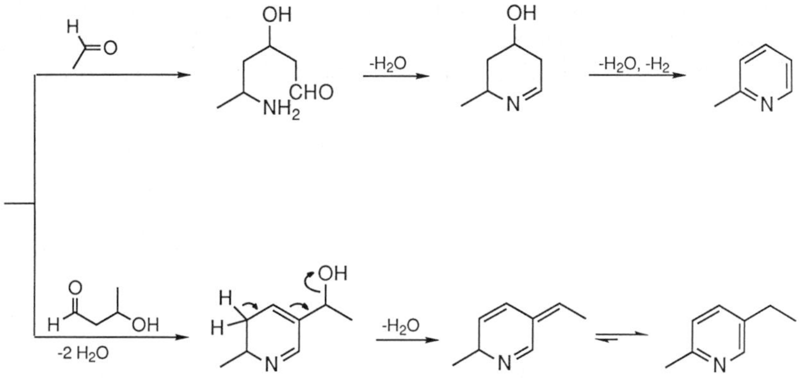

The products are usually mixtures which require separation by chromatography or distillation; in the above example, 2-methyl-5-ethylpyridine (53%) is the major product and 2-methylpyridine the minor one.

(3) *From β-keto-esters, aldehydes, and ammonia: the Hantzsch synthesis.* Two molecules of a β-keto-ester and one of an aldehyde react in the presence of ammonia to give a dihydropyridine which is then dehydrogenated, usually with nitric acid. The aldehyde reacts with one molecule of the β-keto-ester, under the influence of ammonia or an added base, and ammonia itself reacts with the second molecule of the β-keto-ester; the two resulting units are then joined by a Michael-type addition followed by ring closure:

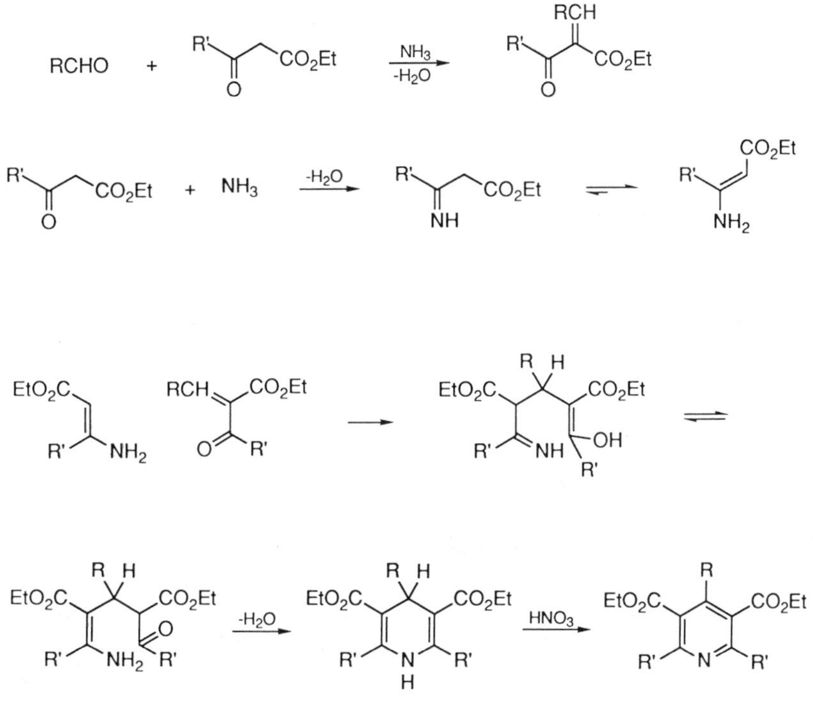

For example, acetoacetic ester, formaldehyde, and ammonia, in the presence of diethylamine, give a dihydropyridine which, after oxidation with nitric acid, hydrolysis, and decarboxylation, gives 2,6-lutidine (2,6-dimethylpyridine) in 65% yield [2]:

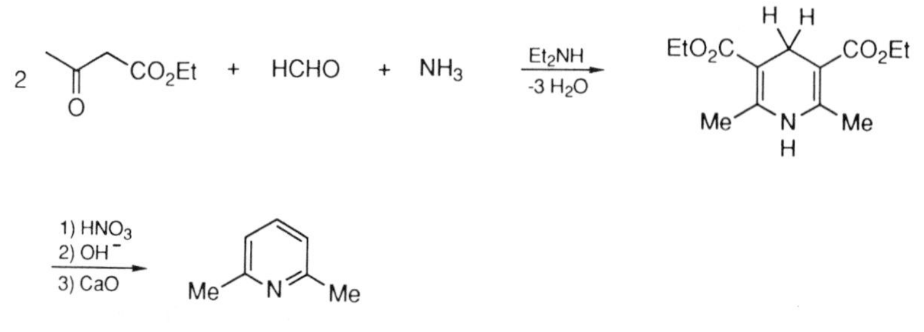

(4) *From 1,3-dienes*. Some pyridines are synthesized industrially by four-centre reactions between a 1,3-diene and a nitrile in the gas phase, e.g.

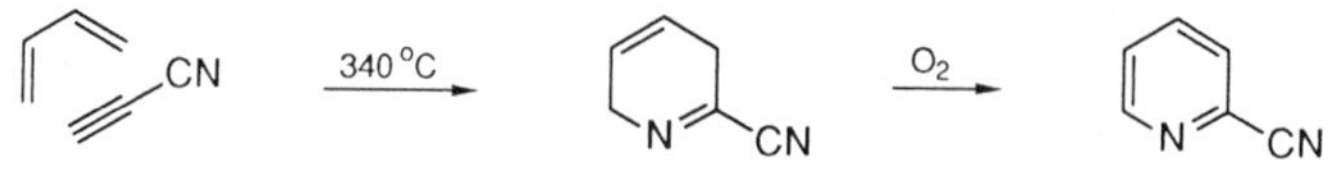

Cycloadducts of cyclopentadienones and nitriles spontaneously eliminate carbon monoxide, e.g.

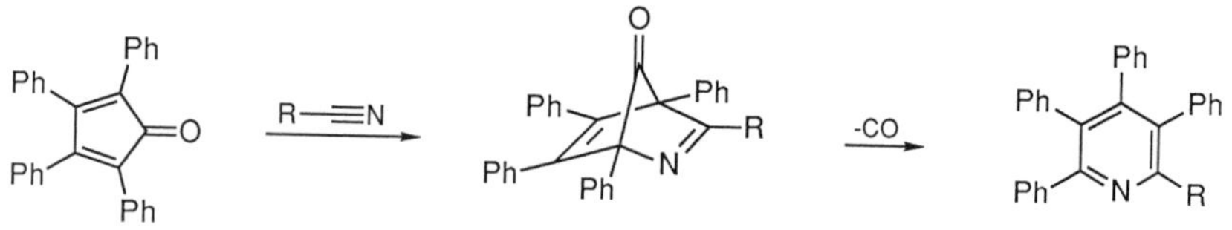

(5) *From furans and pyrones*. Furans containing 2-carbonyl substituents react with ammonia to give pyridines, e.g.

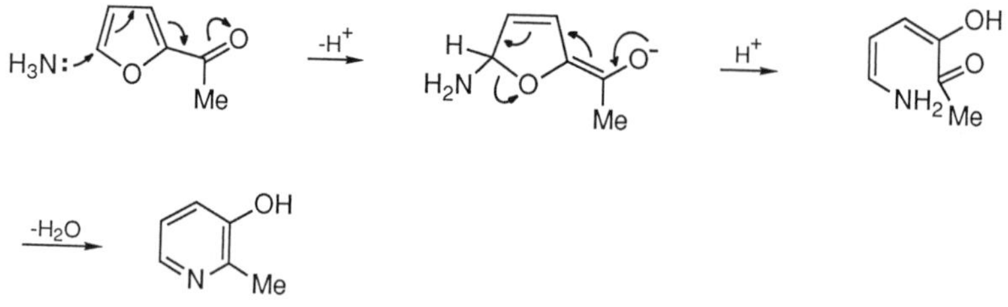

2-Pyrones also give pyridines with ammonia, e.g.

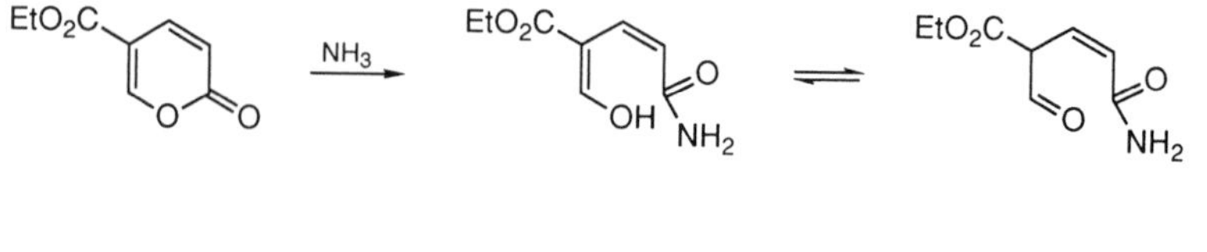

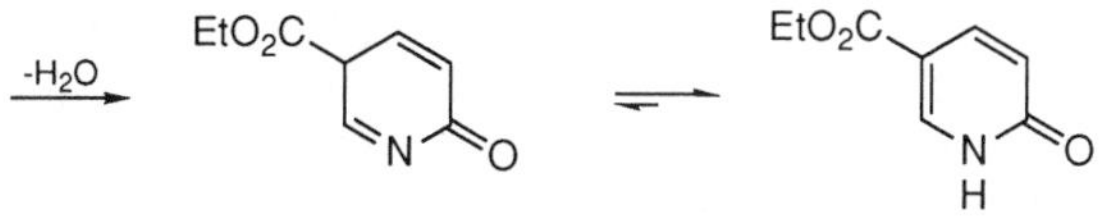

Pyridines are converted into their *N*-oxides with hydrogen peroxide in acetic acid and are catalytically reduced to piperidines over nickel at 200°C. Piperidines are also available by the cyclization of pentamethylene-containing compounds, e.g. pentamethylenediamine dihydrochloride gives piperidine on being heated (p. 304) and pentamethylene dichloride and a primary amine give an *N*-substituted piperidine.

(b)   Quinolines

The quinoline nucleus is contained in several groups of alkaloids (e.g. quinine), in a number of synthetic materials with important physiological properties such as the antimalarial plasmoquin, in the cyanine dyes (p. 261), and in the analytical reagent, oxine (8-hydroxyquinoline).

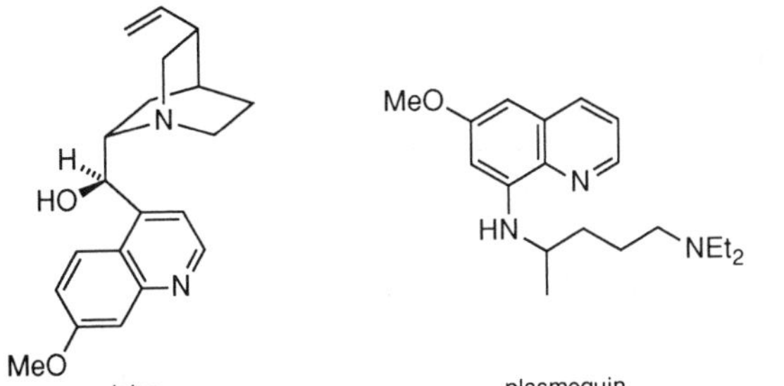

quinine   plasmoquin   oxine

The nucleus is usually formed in one of two ways:

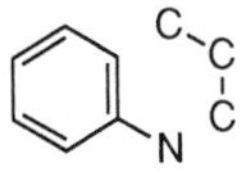

Skraup, Döbner-von Miller, and Conrad-Limpach syntheses

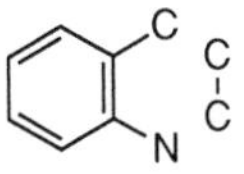

Friedlaender and Pfitzinger syntheses

(1)   *The Skraup synthesis.* In the simplest case, aniline, glycerol, nitrobenzene, iron(II) sulfate, and sulfuric acid are mixed and gently heated. A vigorous exothermic reaction occurs which is completed by further heating. After removal of the residual nitrobenzene by distillation in steam, quinoline is liberated from the reaction mixture by basification and is isolated by steam-distillation. The yield of purified product is 84–91% [1].

The reaction involves dehydration of glycerol to propenal, Michael-type addition of aniline to propenal, acid-catalyzed cyclization, and dehydrogenation by nitrobenzene. Iron(II) sulfate acts to moderate the reaction. Propenal itself is not employed because of its tendency to polymerize; the success of the reaction depends on the rapid addition of aniline to the propenal as it is formed.

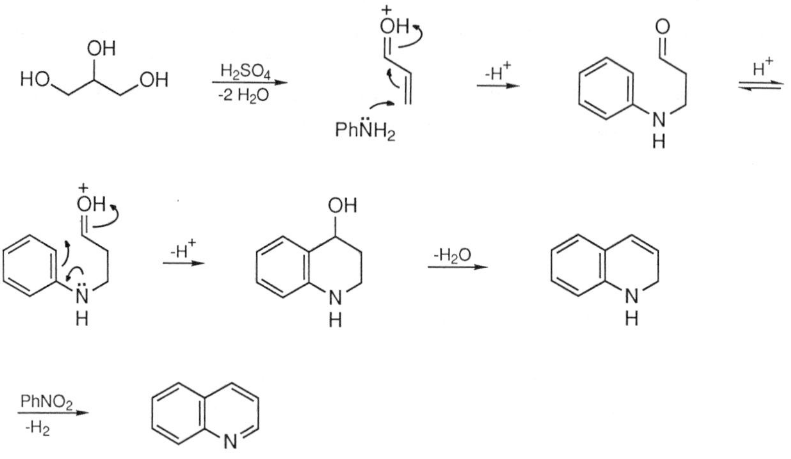

The synthesis is of wide application. The aniline may contain nuclear substituents: *ortho*- and *para*-substituted anilines give 8- and 6-substituted quinolines, respectively, and *meta*-substituted anilines give mixtures of 5- and 7-substituted quinolines, the latter predominating with activating substituents such as methoxyl and the former predominating with deactivating substituents such as nitro. Propenal (as glycerol) may be replaced by 2-butenal giving 2-methylquinolines, and by butenone, giving 4-methylquinolines.

(2)   *The Döbner–von Miller synthesis.* This is very similar to the Skraup method, the differences being that the three-carbon fragment is formed *in situ* by an acid-catalyzed aldol reaction and that no dehydrogenating agent is added. The oxidative step is thought to be brought about by hydride-transfer to the imine which is formed by reaction between the aniline and the aldehyde or ketone, e.g.

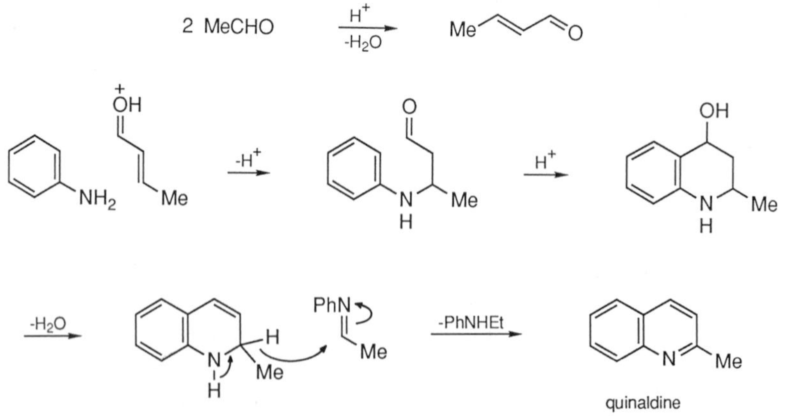

(3) *The Conrad-Limpach synthesis.* β-Keto-esters react with anilines in each of two ways: at low temperatures reaction occurs at the keto-group to give an imine, and at high temperatures reaction occurs at the ester group to give an amide. The resulting compounds can each be cyclized to quinolines, e.g.

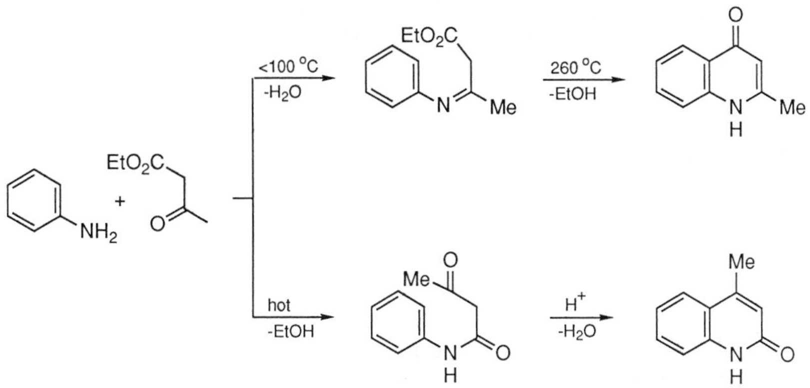

(4) *The Friedlaender synthesis.* An *o*-aminobenzaldehyde is treated with an aldehyde or ketone in a basic medium; formation of the imine is followed by dehydrative cyclization:

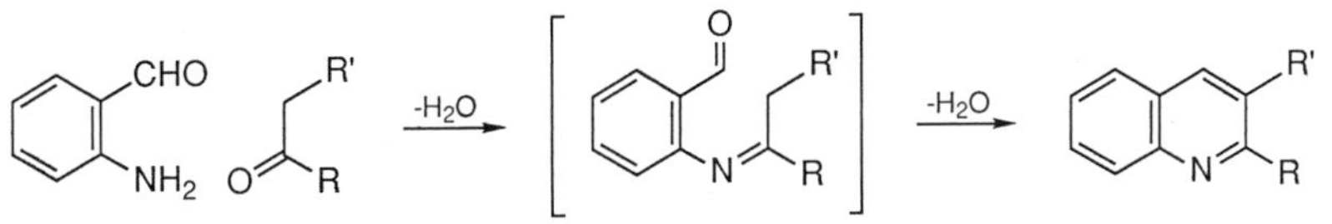

The main problem in this synthesis is that *o*-aminobenzaldehydes are very unstable, readily undergoing self-condensation. One way of overcoming this is to start instead with an *o*-nitrobenzaldehyde: acid- or base-catalyzed reaction gives an intermediate which cyclizes spontaneously on reduction (cf. the Reissert indole synthesis, p. 690), e.g.*

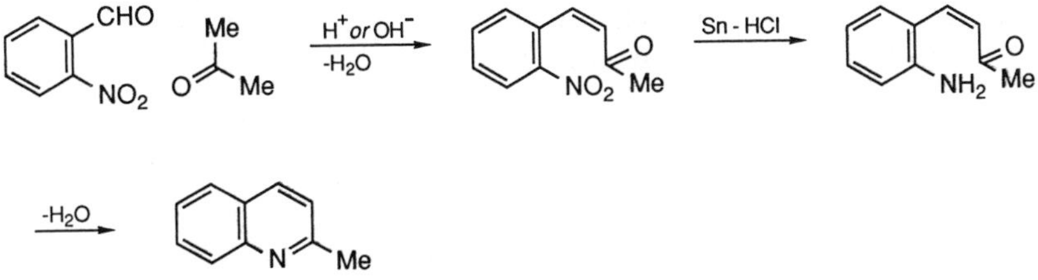

(5) *The Pfitzinger synthesis.* This modification of Friedlaender's method employs isatin in place of *o*-aminobenzaldehyde. Reaction with base gives *o*-

---

* With *hot* alkali, the first-formed aldol product undergoes a series of transformations leading to indigo.

aminobenzoylformate ion which reacts with an aldehyde or ketone to give a 4-carboxyquinoline from which the carboxyl group can be removed thermally:

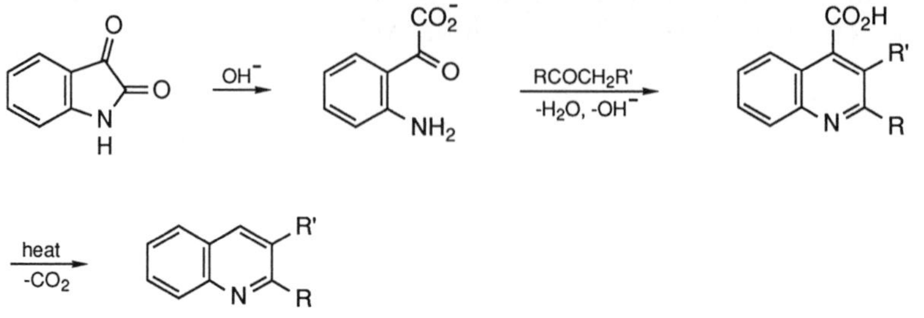

(c) Isoquinolines

The isoquinoline ring occurs in a large number of alkaloids, the synthesis of one of which, papaverine, is outlined below (p. 707). The general synthetic routes to isoquinolines involve the following skeletal types:

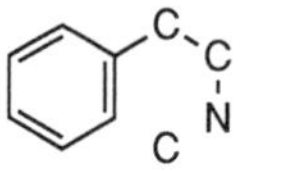

Bischler-Napieralski and Pictet-Spengler syntheses

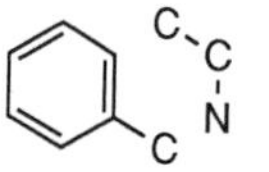

Pomeranz-Fritsch synthesis

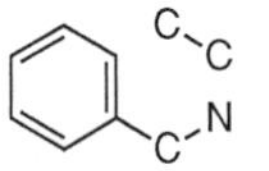

Schlittler-Müller synthesis

(1) *The Bischler–Napieralski synthesis.* A β-arylethylamine* is converted into an amide which is cyclized with an acidic reagent such as phosphorus oxychloride. Dehydrogenation of the product, usually with palladium on charcoal, gives the quinoline. Yields are normally of the order of 90%.

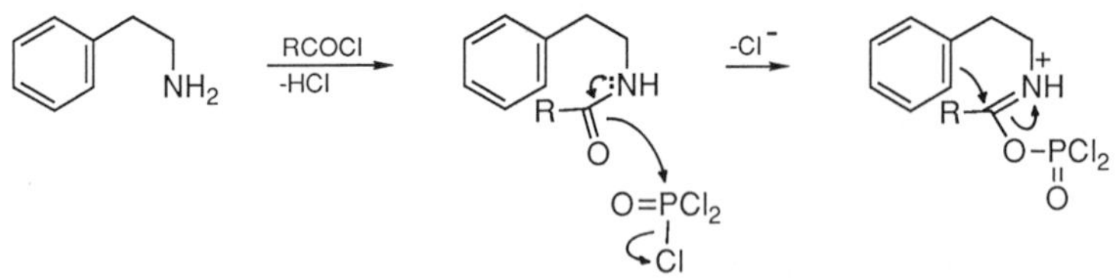

---

*These compounds are readily obtained from aromatic aldehydes by base-catalyzed reaction with nitromethane followed by reduction: $ArCHO \rightarrow ArCH{=}CHNO_2 \rightarrow ArCH_2CH_2NH_2$.

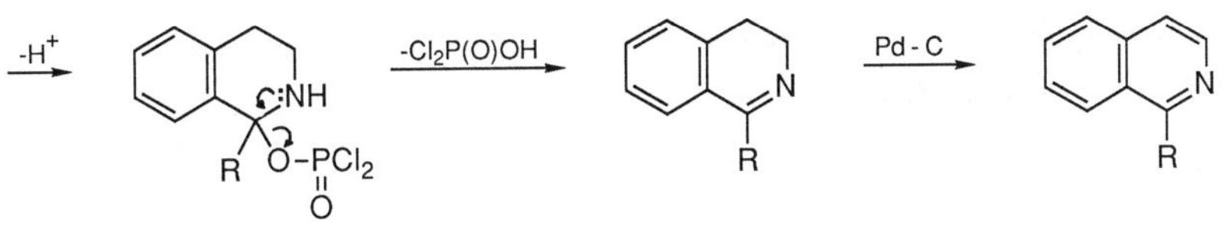

A synthesis of papaverine is based on the Bischler–Napieralski procedure:

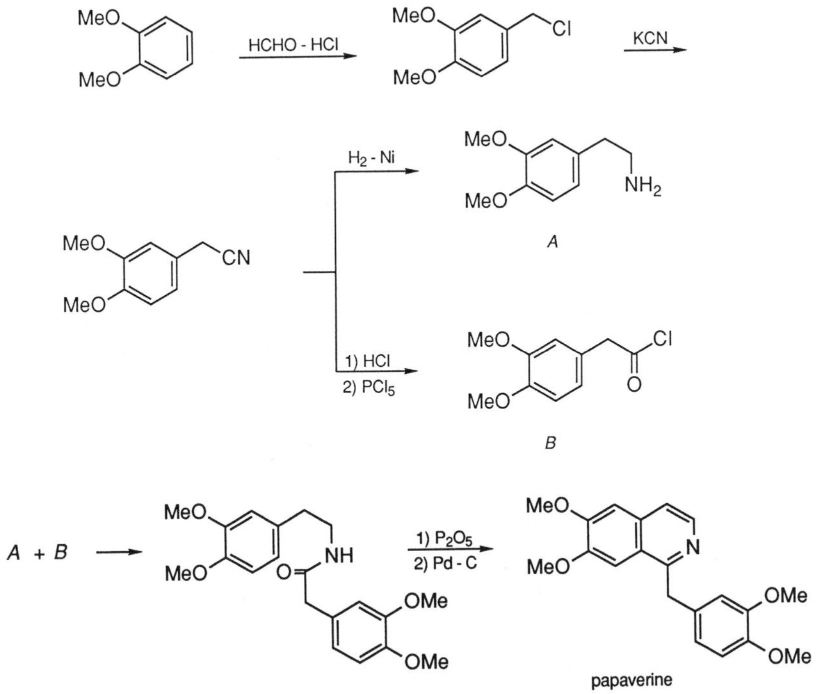

papaverine

(2) *The Pictet–Spengler synthesis.* A β-arylethylamine is treated with an aldehyde in the presence of dilute acid; ring closure occurs by a reaction of Mannich type and the tetrahydroisoquinoline is dehydrogenated on palladium.

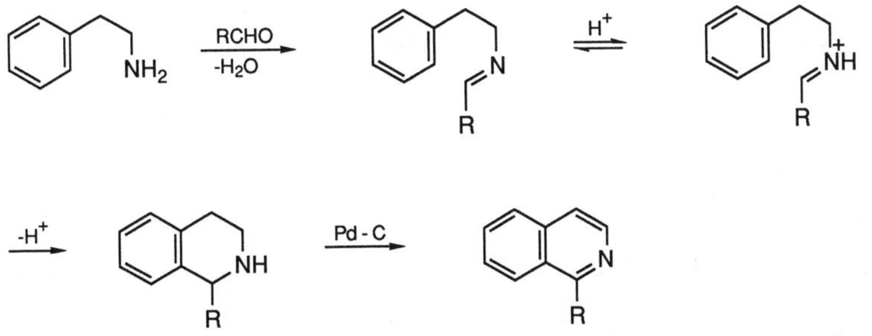

(3) *The Pomeranz–Fritsch synthesis.* The acetal of an α-aminoaldehyde (prepared from the corresponding chloro-compound with ammonia at 0°C) is treated with an aromatic aldehyde or ketone to give the imine which is then cyclized with

acid. (The acetal is employed because of the instability of aminoaldehydes, which undergo self-condensation.)

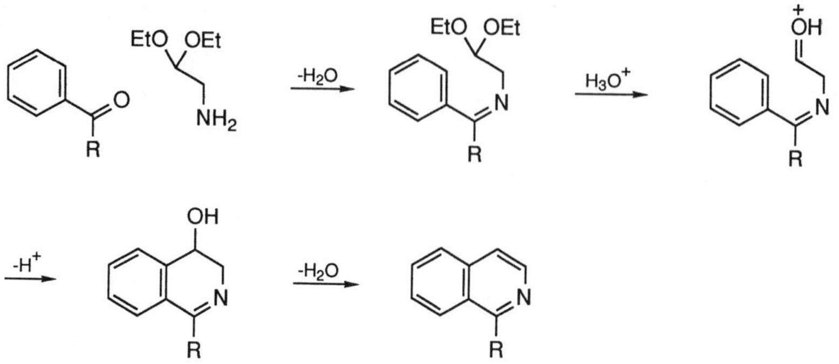

Yields are usually not greater than 50%, and are lower with aromatic ketones than with aromatic aldehydes.

(4) *The Schlittler–Müller synthesis.* A similar approach is adopted but, by avoiding the inefficient step involving reaction between an aromatic ketone and an amine, yields are improved.

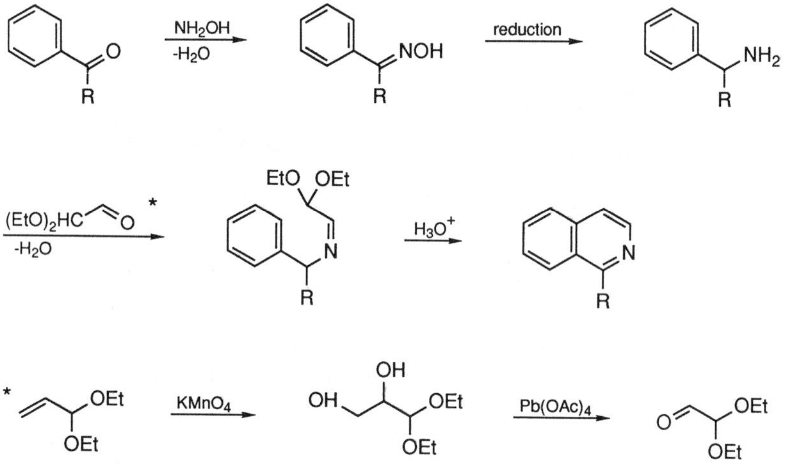

(d) Pyrans, pyrones, and pyrylium salts

Although there can be no uncharged oxygen analogue of pyridine, six-membered heterocyclic compounds containing oxygen are well known and many occur naturally. The simplest compounds, 2- and 4-pyran, have not themselves been prepared, although simple derivatives can be synthesized, e.g.

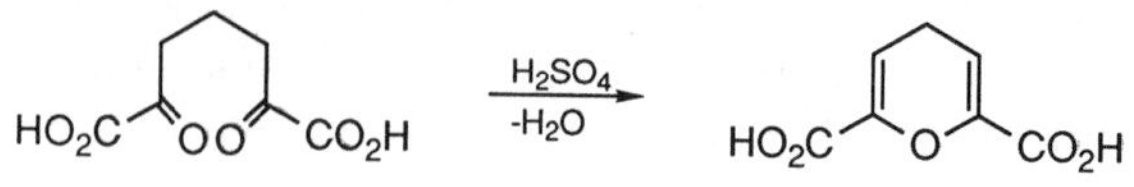

(i) Pyrones. 2-Pyrones can be obtained by treating malic acid and its derivatives with concentrated sulfuric acid; decarbonylation of the α-hydroxyacid* is followed by cyclization, e.g.

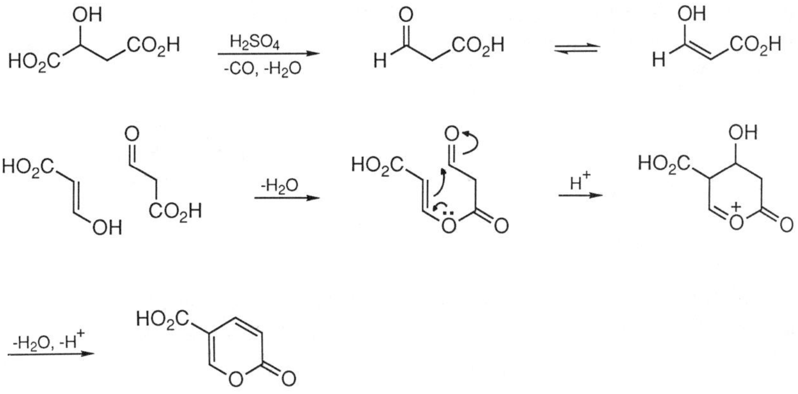

4-Pyrone can be obtained by way of chelidonic acid, which also occurs naturally:

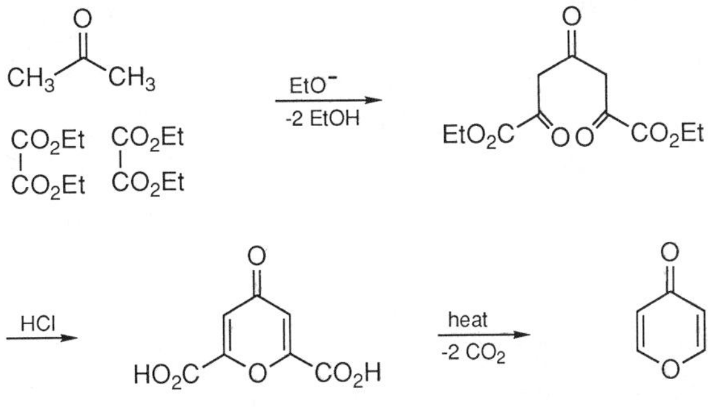

(ii) Dihydropyrans. Δ²-Dihydropyran† can be obtained by the acid-catalyzed rearrangement of tetrahydrofurfuryl alcohol over alumina at 270°C:

* This is a general reaction of α-hydroxy-acids,

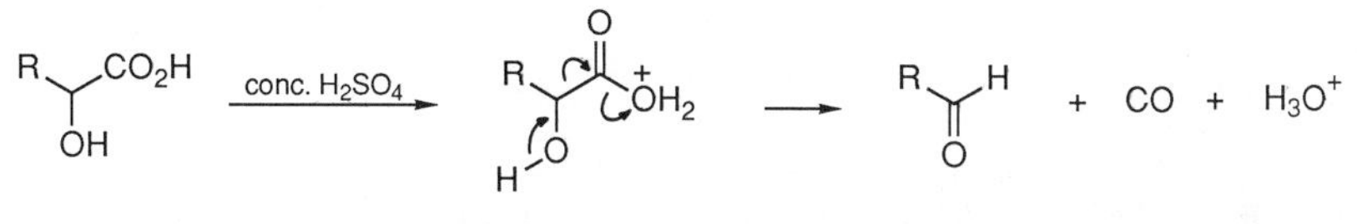

† This compound is used as a protecting agent for alcohols. Reaction with the alcohol in the presence of anhydrous acid gives an acetal from which the alcohol may be regenerated with mineral acid (see p. 89):

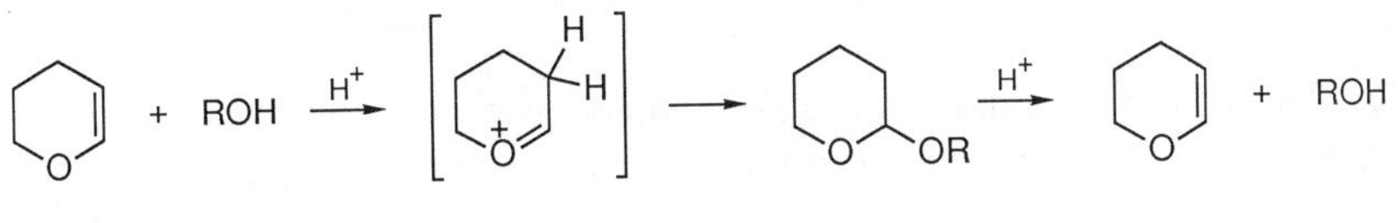

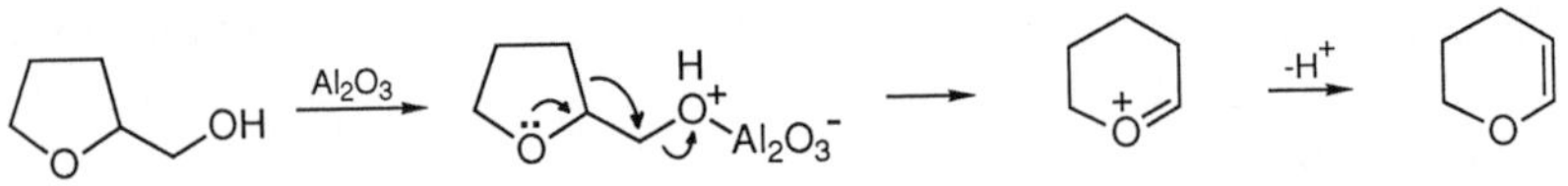

Reduction of the dihydropyran gives the tetrahydro-derivative, used as a solvent. The sugars, in their six-membered cyclic form, are derivatives of tetrahydropyran (pyranoses):

tetrahydropyran

β-D-glucose

The formation and transformations of pyrones are controlled by subtle effects which are illustrated by reactions of the dehydroacetic acid system:

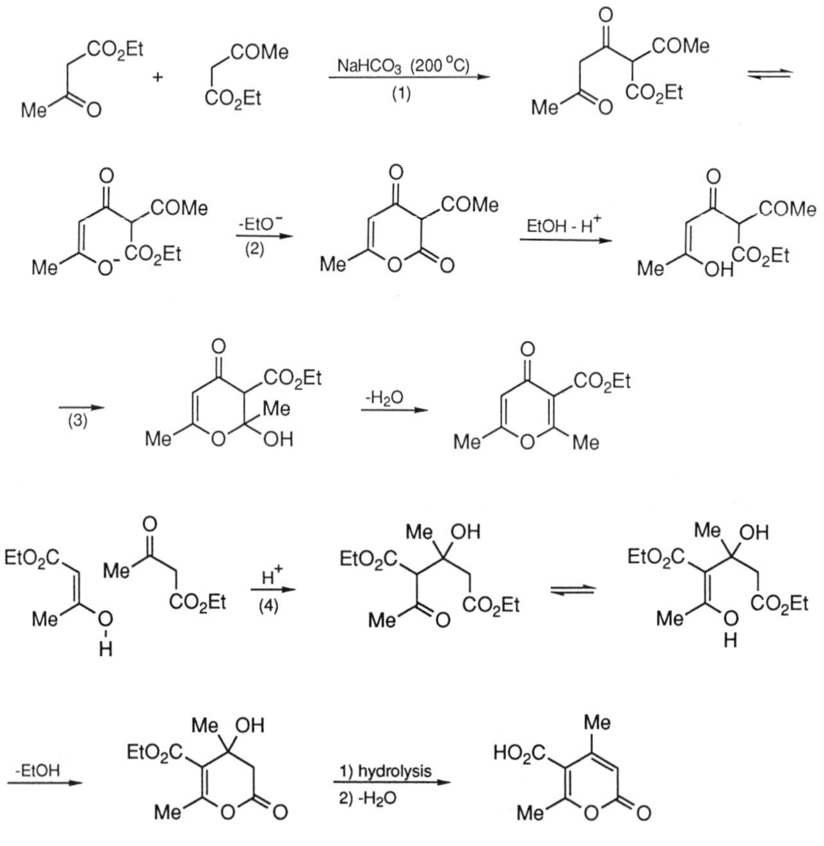

isodehydroacetic acid

(1) The Claisen condensation takes precedence over the aldol reaction (on to the keto group) because the product of the former reaction is removed from

equilibrium as a strongly resonance-stabilized anion; cf. the reaction between acetone and ethyl acetate, p. 229.

(2)   This anion reacts intramolecularly with the ester group through oxygen rather than through (the more usual) carbon because the latter would lead to a strained four-membered ring, cf. the reaction of acetoacetic ester with 1,3-dibromopropane, p. 239.

(3)   Ethanolysis of the lactone is followed by acid-catalyzed formation of a hemiacetal and dehydration.

(4)   Acid-catalyzed self-condensation of acetoacetic ester occurs on the carbonyl group because this is more reactive towards nucleophiles than the ester group and the driving force mentioned in (1) is absent.

(*iii*)   *Benzopyrones.*   Two of the three benzopyrones, coumarin and chromone, are notable. Coumarin, which occurs naturally (e.g. in clover), has been synthesized by the Perkin reaction,

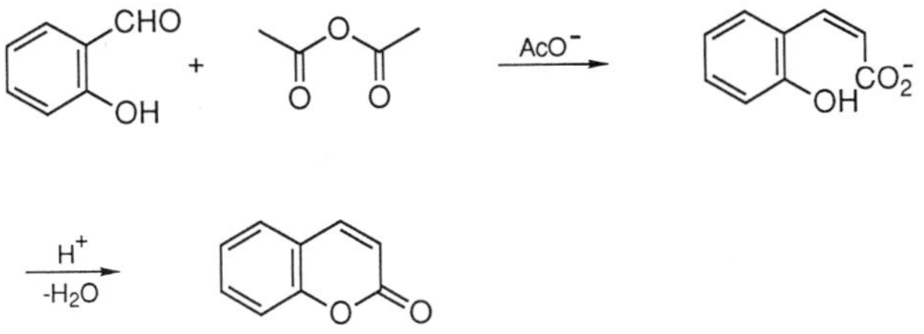

coumarin

and by the von Pechmann reaction from phenol, malic acid, and concentrated sulfuric acid,

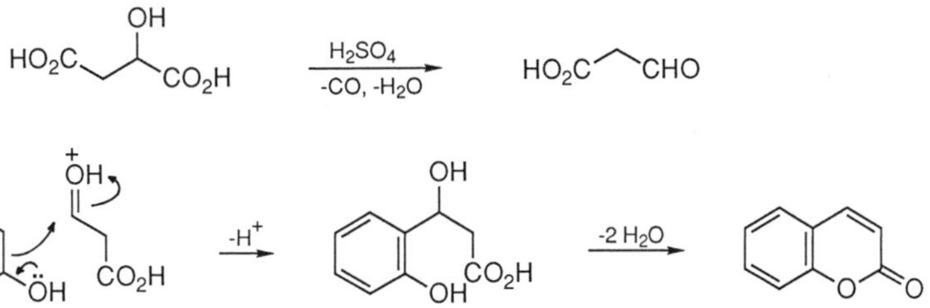

The latter reaction may also be applied to β-keto-esters.

Chromone is the parent member of a group of plant colouring pigments, the flavones (2-arylchromones), which are found both free and as glycosides. Three general synthetic routes are available.

(1)   From salicylic acid derivatives:

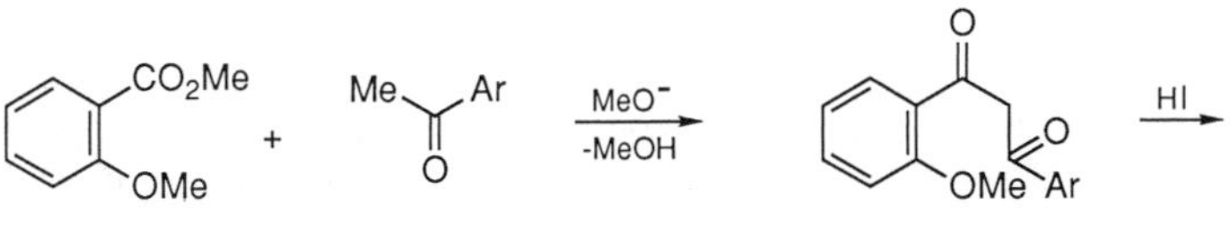

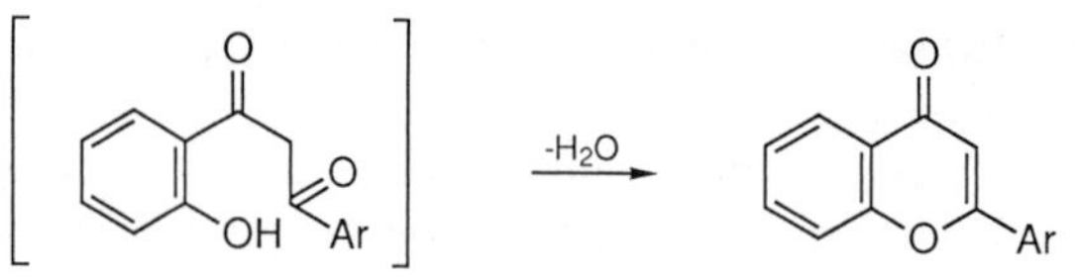

(2) From *o*-hydroxyacetophenones and aromatic acid anhydrides:

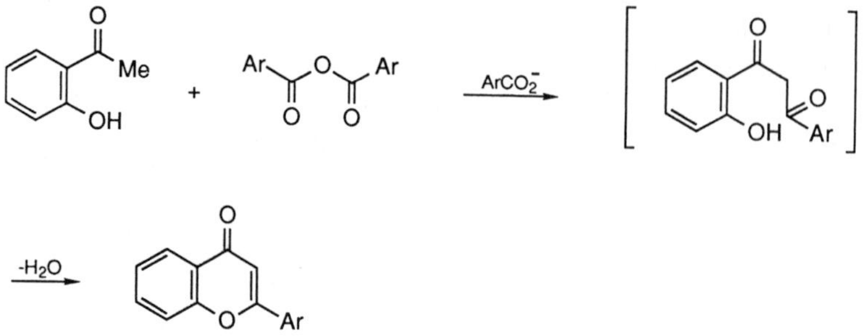

(3) From *o*-hydroxyacetophenones and aromatic aldehydes. The synthesis of quercetin is illustrative:

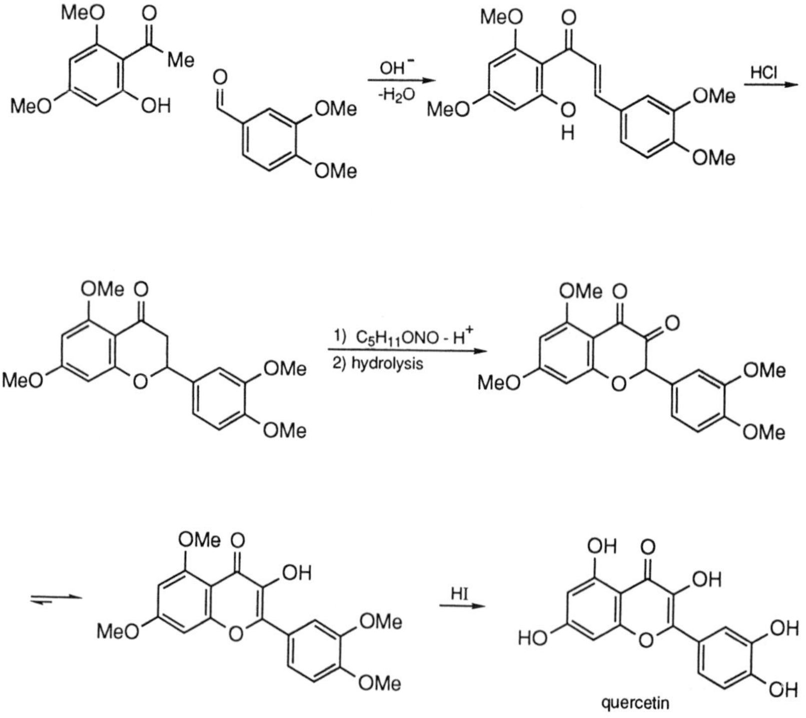

(*iv*) *Pyrylium salts.* Salts of the type $R_3O^+X^-$ are normally unstable and cannot be isolated unless X^- is an especially stable (non-nucleophilic) anion such as

BF_4^-.* However, the oxonium salts derived by oxidation of the pyran nucleus are comparatively stable because the cation possesses aromatic stabilization energy which is lost when an anion bonds covalently to it. The parent member of the series, the pyrylium ion, is formed by acidification of the sodium salt of glutaconic aldehyde at low temperatures:

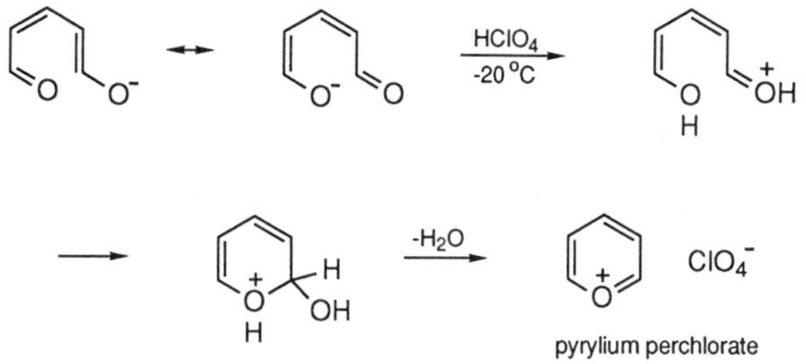

pyrylium perchlorate

The glycosides of a number of 2-arylbenzopyrylium salts occur naturally as plant colouring matters, the anthocyanins.

The parent salts, the anthocyanidins, are synthesized from salicylaldehydes and acetophenones by an acid-catalyzed aldol reaction followed by ring closure in ethyl acetate solution:

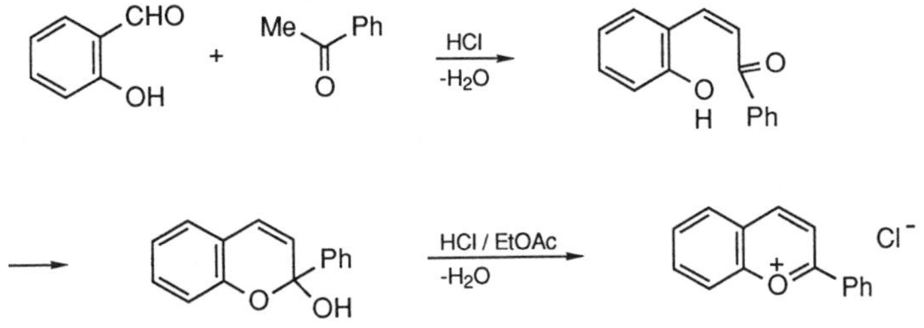

For example, cyanidin chloride has been prepared as follows:

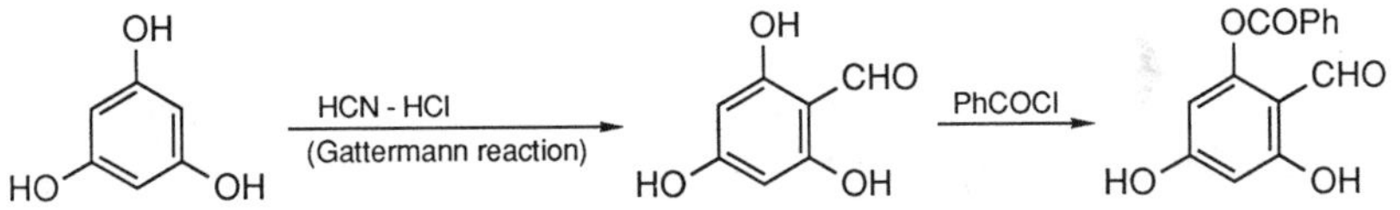

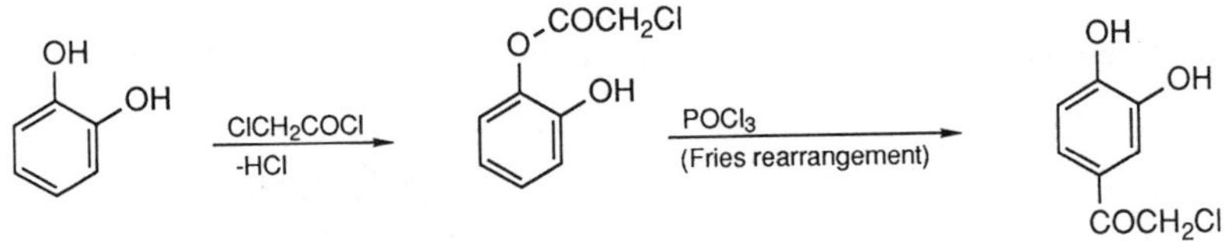

---

* For example, $(CH_3)_3O^+BF_4^-$ is a crystalline solid [6] which is a powerful alkylating agent; it can be used to methylate weakly nucleophilic species such as amides.

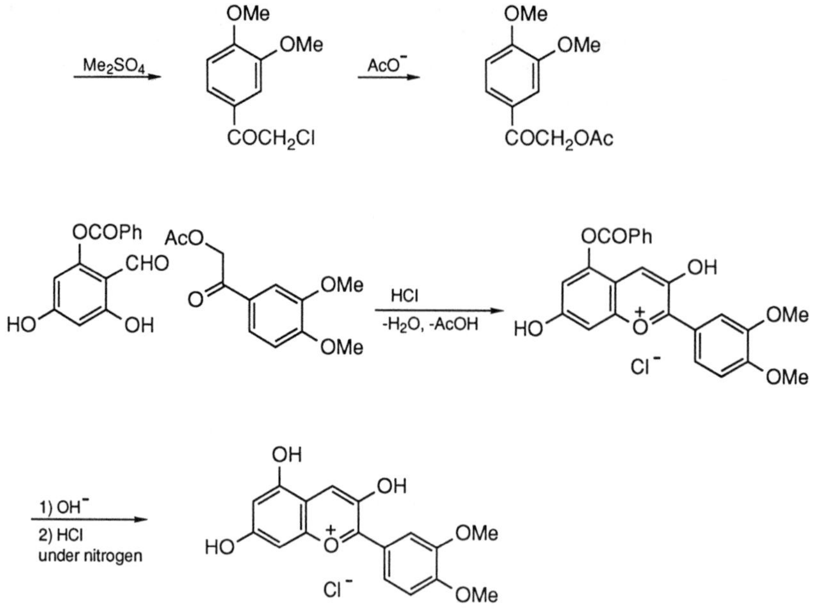

In the formation of the anthocyanins themselves, it is necessary to introduce the sugar units before the acid-catalyzed reaction. Glucose is introduced by the reaction of tetraacetyl-α-bromoglucose with the appropriate hydroxyl group:

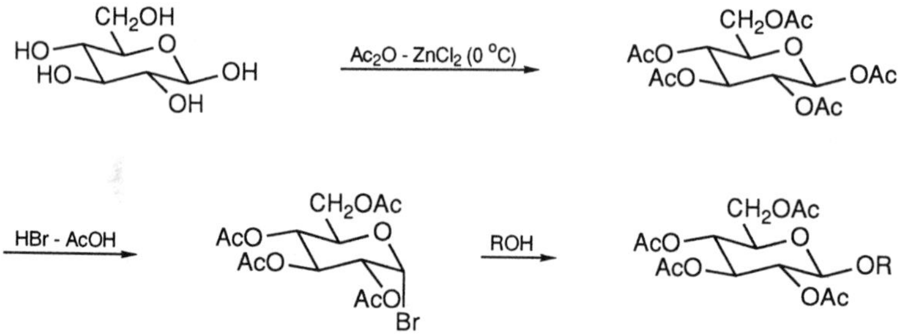

The remainder of the synthesis is an adaptation of that for the anthocyanidins, e.g.

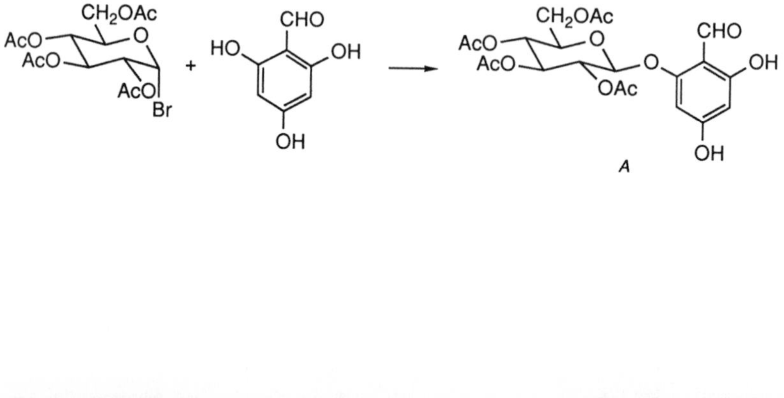

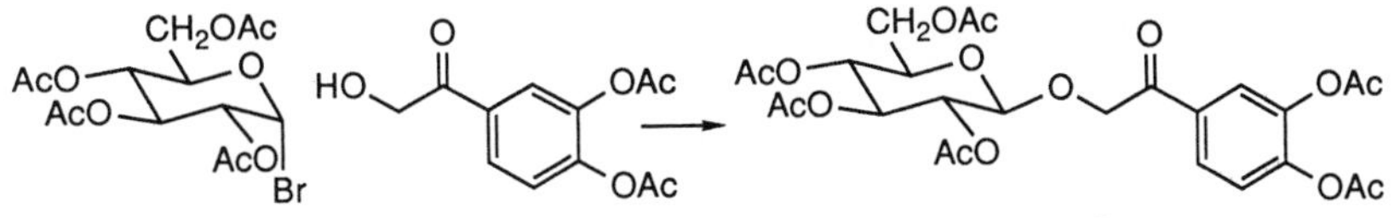

A + B $\xrightarrow{\text{HCl / EtOAc}}$

$\xrightarrow[\text{2) HCl}]{\text{1) KOH}}$

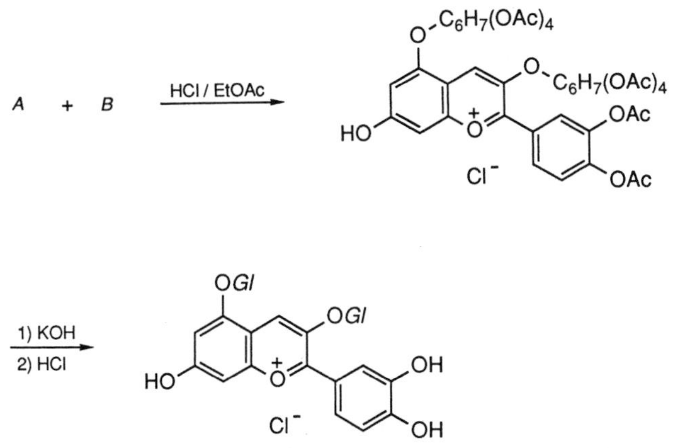

cyanin chloride
($Gl \equiv \beta$-glycosidyl)

(a) Pyrimidines

Of the three isomeric analogues of benzene which possess two nitrogen atoms (pyridazine, pyrimidine, and pyrazine), pyrimidine and its derivatives are by far the most important. Both the ribonucleic acids and the deoxyribonucleic acids contain cytosine (2-hydroxy-4-aminopyrimidine), the former group contains uracil (2,4-dihydroxypyrimidine), and the latter group contains thymine (2,4-dihydroxy-5-methylpyrimidine). The heterocycles are linked through their 1-nitrogen atoms to ribose and deoxyribose, the hydroxypyrimidines being in their tautomeric -one forms.

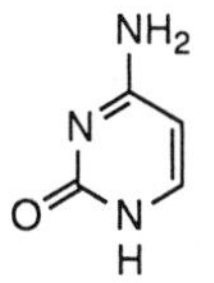

cytosine 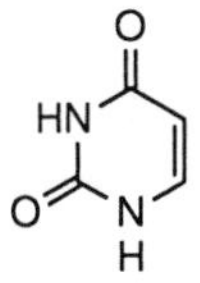   uracil 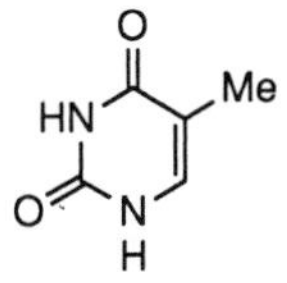   thymine

The full structure of the nucleic acids is as follows:

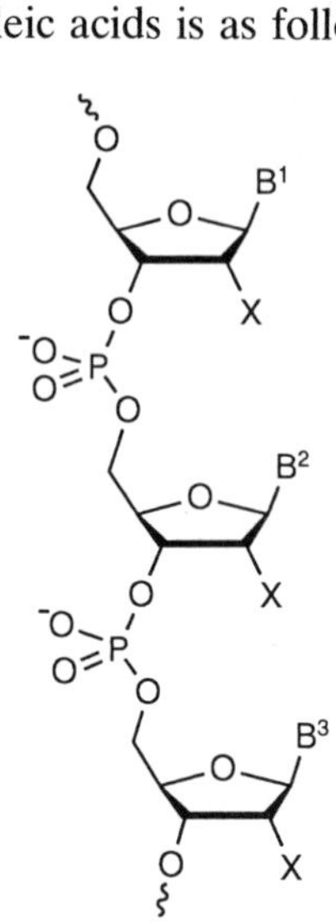

where X═OH (ribonucleic acids) or X═H (deoxyribonucleic acids) and B^1, B^2, etc. are heterocyclic bases which comprise the three pyrimidines mentioned above and two purines (adenine and guanine, p. 721).

In addition, vitamin B_1 (p. 717) contains a pyrimidine ring as do barbituric acid (2,4,6-trihydroxypyrimidine) and several of its derivatives (e.g. veronal) which are used as hypnotics.

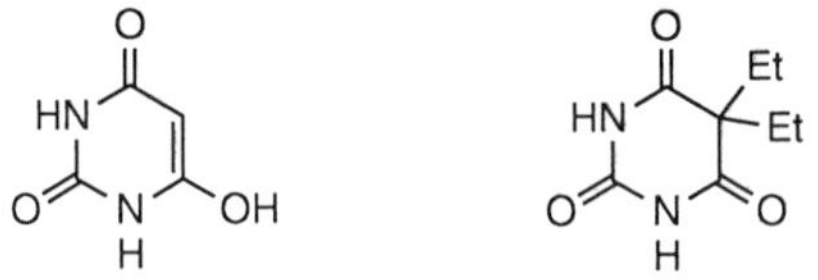

barbituric acid   veronal

The pyrimidine ring is usually constructed by a base-catalyzed reaction between a 1,3-dicarbonyl or related compound and an amidine or related compound:

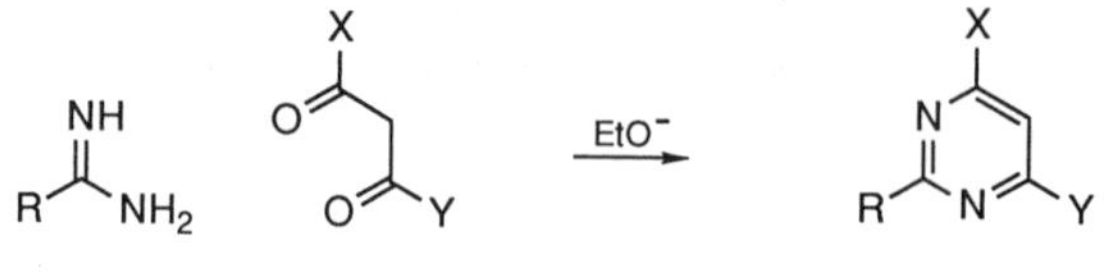

There is a wide range of possibilities: the 2-substituent, R, can be H, NH_2, OH, or SH, by the use of formamidine, guanidine, urea, or thiourea, respectively; and X and Y can be H, OH, or NH_2 by the use of aldehyde, ester, or nitrile substituents. For example, 2,4-diamino-6-hydroxypyrimidine is obtained in about 80% yield from guanidine and ethyl cyanoacetate [4]

and barbituric acid is obtained in 79% yield from urea and malonic ester [2]

A variant of this approach utilizes Michael-type addition to an αβ-unsaturated ester, e.g.

Vitamin B$_1$ (thiamine) is a pyrimidine derivative which has been synthesized as follows (for the synthesis of the thiazole unit, see p. 698):

thiamine bromide

* The 2-, 4-, and 6-positions of pyrimidines are strongly activated towards nucleophiles, even more so than those in pyridines, see p. 682.

(b)   Pyridazines

The ring system is built up from an unsaturated 1,4-dicarbonyl compound and hydrazine, e.g.

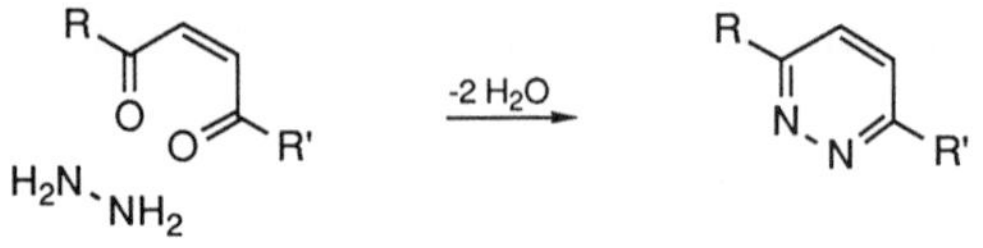

(c)   Pyrazines

The pyrazine system occurs naturally (e.g. in aspergillic acid, an antibiotic) and has been incorporated in a number of synthetic drugs, such as sulfapyrazine (a Sulfa drug) and chlorocyclizine (an antihistamine).

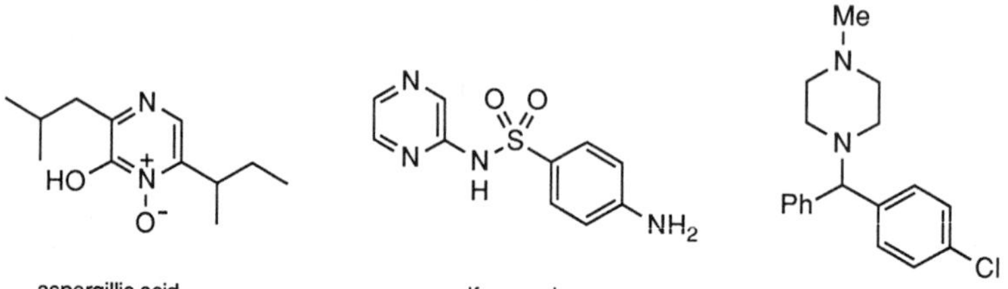

aspergillic acid   sulfapyrazine   chlorocyclizine

The ring is usually constructed by the spontaneous self-condensation of α-aminocarbonyl compounds followed by aerial oxidation:

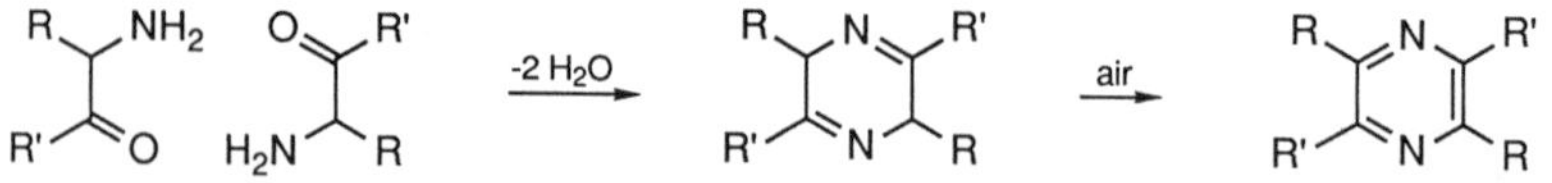

The α-aminoketones are usually prepared *in situ* by the reduction of α-ketooximes formed by nitrosation of ketones (p. 320).

This method is not suitable for pyrazine itself, for aminoacetaldehyde readily polymerizes. A convenient synthesis is from ethylene oxide and ethylenediamine:

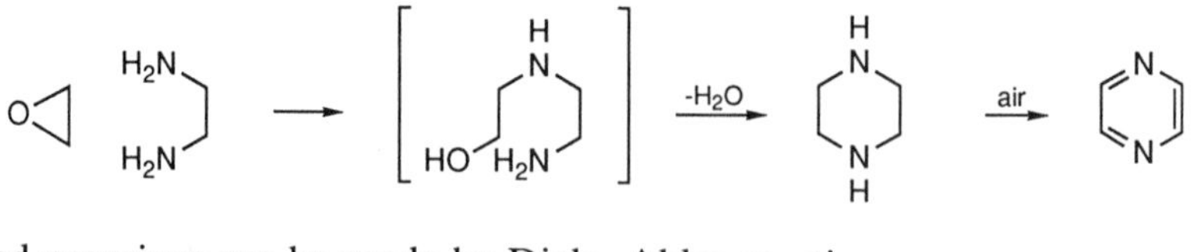

Reduced pyrazines can be made by Diels–Alder reactions, e.g.

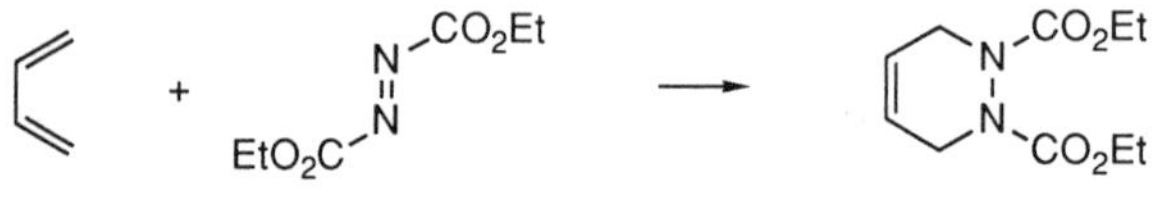

(d) Pteridines

The pteridine nucleus occurs in the pterins (the pigments of butterfly wings) such as xanthopterin, and in folic acid (a growth factor) and vitamin B_2 (riboflavin) (see also flavin adenine dinucleotide, p. 721).

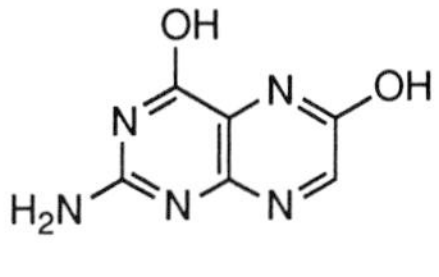

xanthopterin 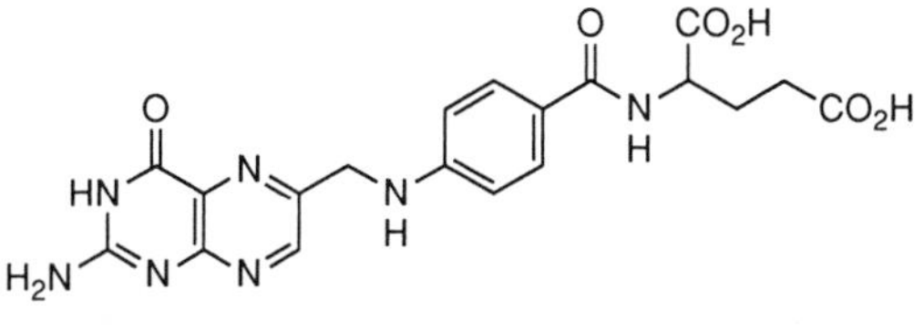   folic acid

A general synthesis of pteridines, illustrated by that of the parent member, is from a 4,5-diaminopyrimidine and an α-dicarbonyl compound. The former compounds are readily available from 4-aminopyrimidines by nitrosation followed by reduction*.

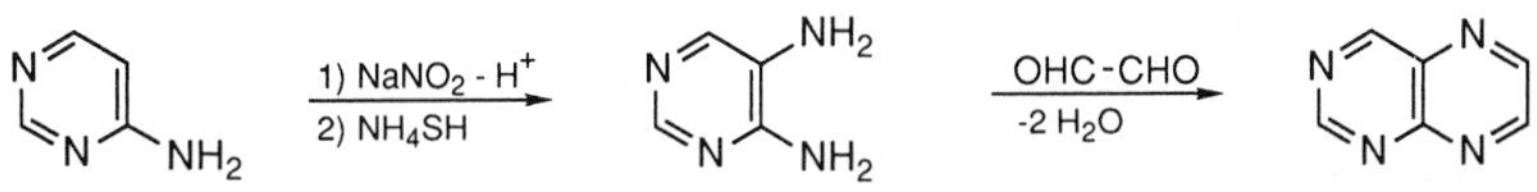

pteridine

A modification of this method gives tetrahydropteridines, as in a synthesis of folic acid:

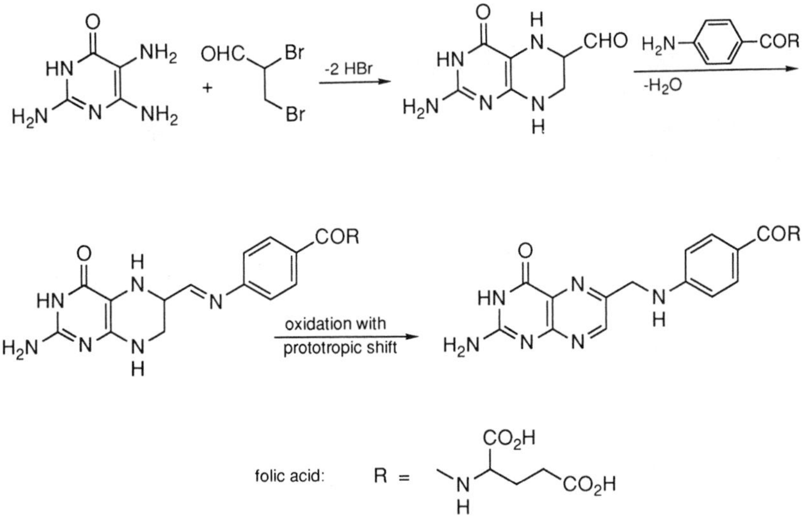

folic acid: R =

* The unshared electron-pair on the amino-nitrogen of a 2- or 4-aminopyrimidine is delocalized over the nucleus and particularly on to the nuclear nitrogens. Diazotization consequently does not occur but the nucleus is activated towards electrophiles, in this case the nitrosating species.

Riboflavin has been synthesized as follows:

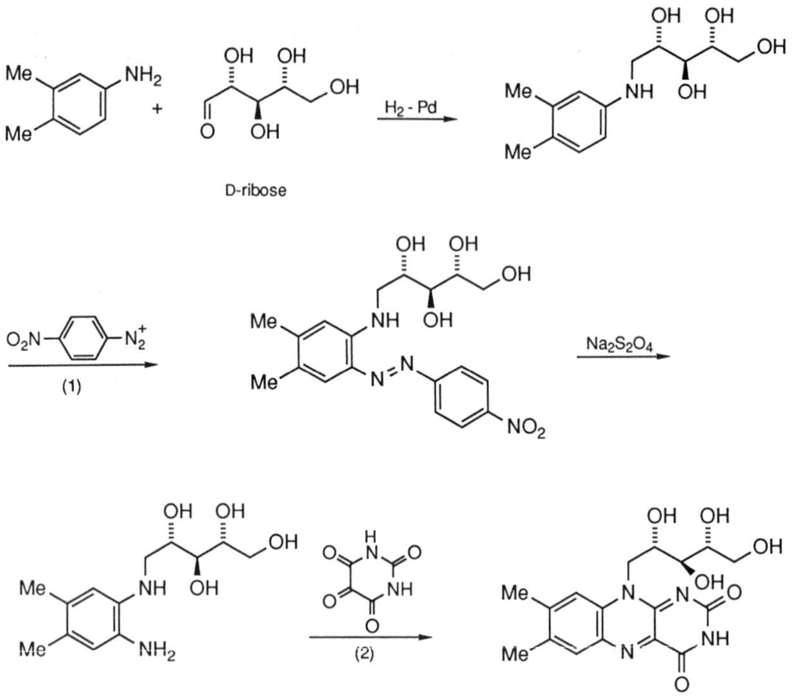

D-ribose

(1)

(1) The amino group activates the *ortho* position to diazonium coupling; p. 682.

(2) The pyrazine ring could be formed in either of two ways. Only this path enables two molecules of water to be eliminated and this may provide the necessary driving force. The reagent is alloxan:

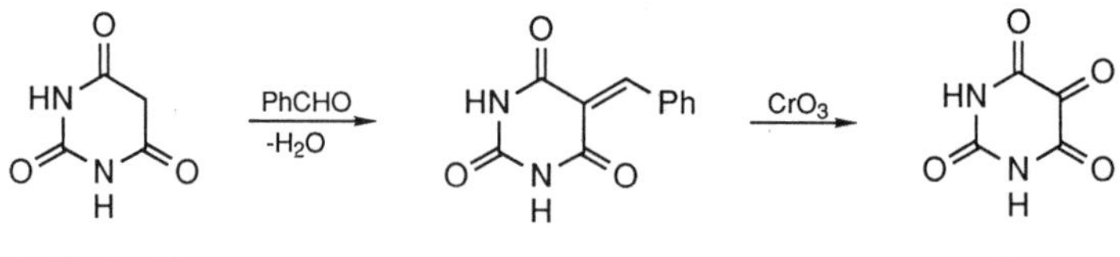

barbituric acid   alloxan

(e) Purines

The purine system occurs widely in nature. Two purines, adenine and guanine, are constituents of the nucleic acids (p. 716); adenine is a component of coenzymes I and II (the so-called phosphopyridine nucleotides), of flavin adenine dinucleotide and of the adenosine phosphates (section 22.2). Caffeine, theophylline, and theobromine are physiologically active constituents of coffee, cocoa, and tea and uric acid is the end-product of nitrogen metabolism in many animals.

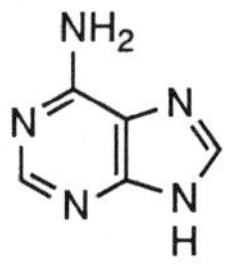

adenine

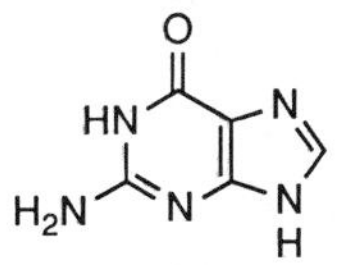

guanine

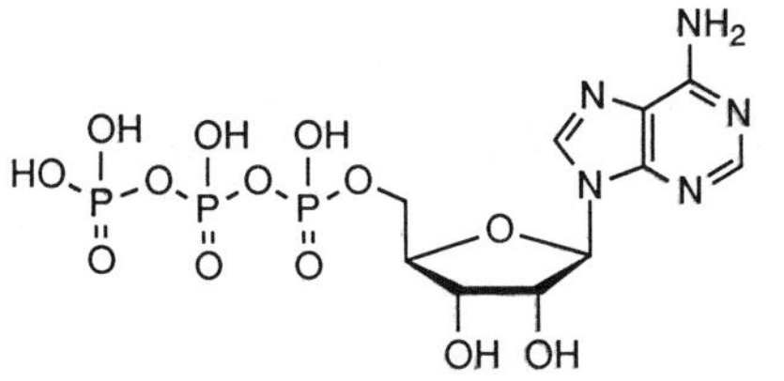

adenosine triphosphate

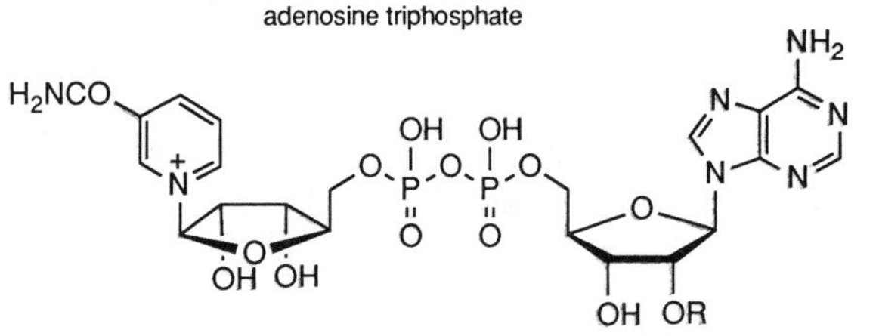

R = H: coenzyme I   (diphosphopyridine nucleotide)
R = P(OH)$_2$: coenzyme II (triphosphopyridine nucleotide)

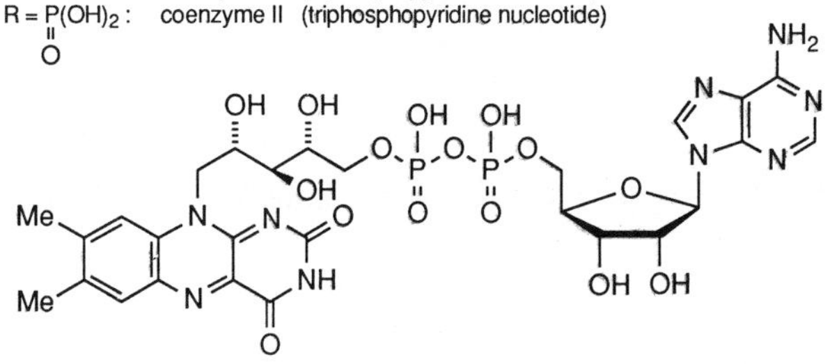

flavin adenine dinucleotide

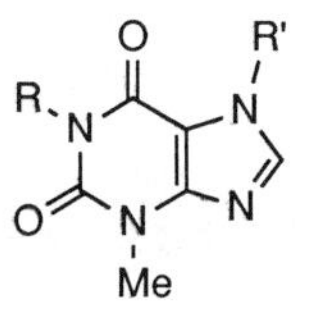

R = R' = Me: caffeine
R = Me, R' = H: theophylline
R = H, R' = Me: theobromine

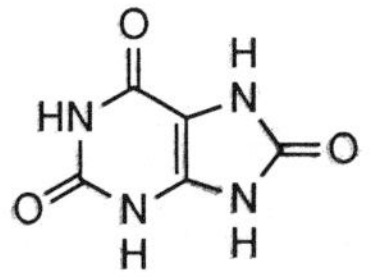

uric acid

Purines are usually synthesized by Traube's method in which a 4,5-diaminopyrimidine is treated with formic acid or, better, sodium dithioformate. 4,5-Diaminopyrimidines are themselves obtained from 4-aminopyrimidines by nitrosation followed by reduction (p. 671) or *via* diazonium coupling of activated methylene compounds (p. 422) followed by cyclization and reduction. Typical examples are:

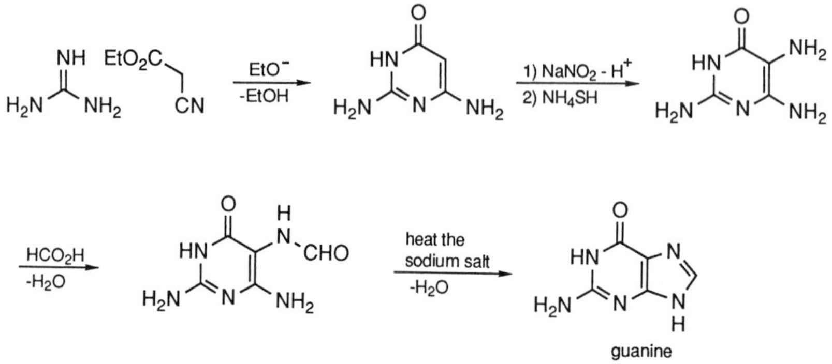

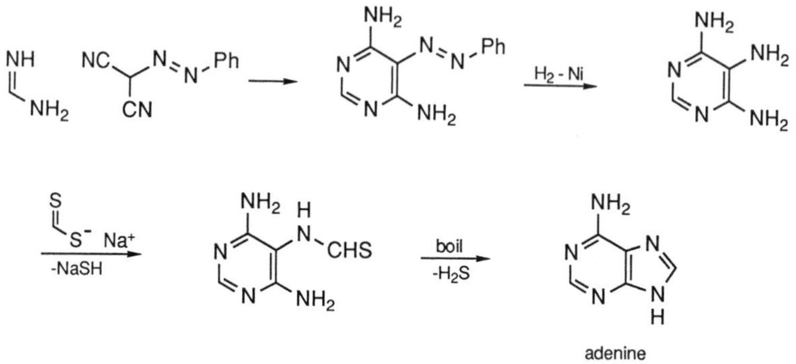

guanine

adenine

8-Hydroxypurines such as uric acid are made by using ethyl chloroformate in place of formic acid:

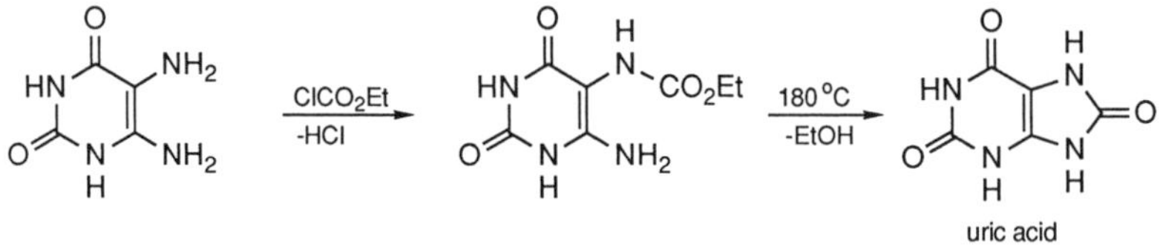

uric acid

Uric acid is used as the starting material for other purines, e.g.

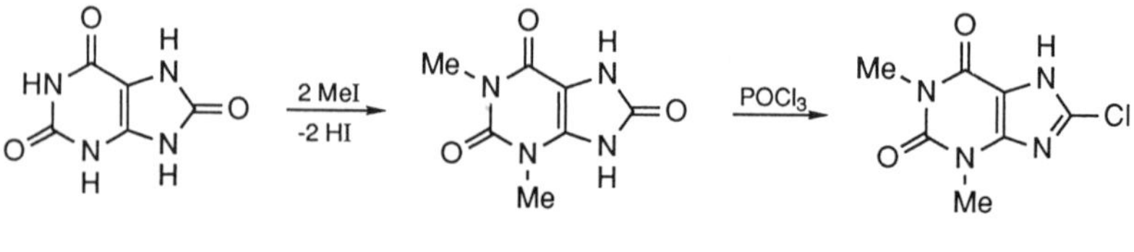

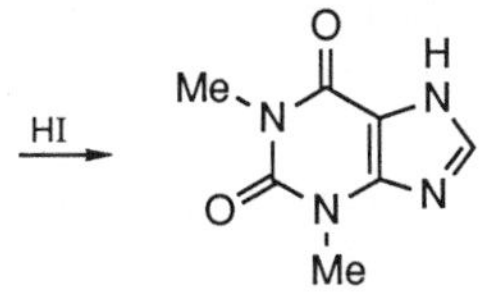

theophylline

21.6 Six-membered rings containing three heteroatoms

1,3,5-Triazine is obtained by the thermal or base-catalyzed polymerization of formamidine hydrochloride, and is very easily hydrolyzed by acids to formic acid and ammonia. Its best-known derivative, cyanuric chloride, obtained by the vapour-phase polymerization of cyanogen chloride, is an important intermediate in the dyestuffs industry. Its chlorine atoms are very easily displaced by nucleophiles (cf. p. 400), so that a chromophoric system can be attached to the molecule and the whole can then be attached to the cellulose of the fibre by reaction at the latter's hydroxyl groups, e.g.

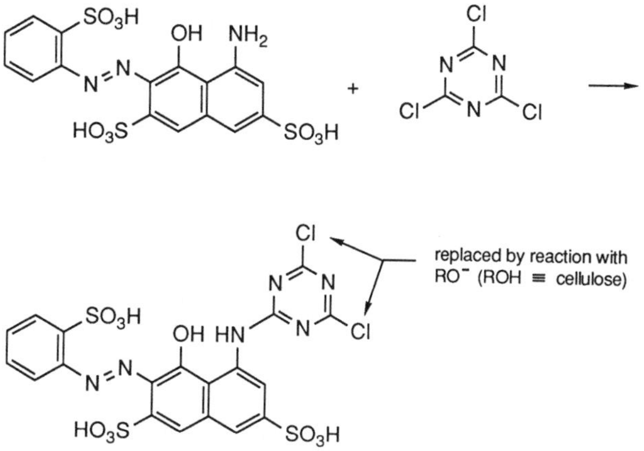

The dyestuffs (Procion dyes) are consequently tightly bound to the fibre and, in particular, have considerable wet-fastness.

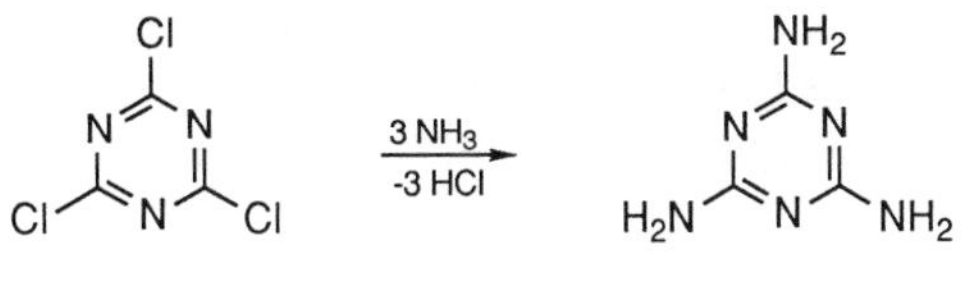

Melamine (2,4,6-triamino-1,3,5-triazine) can be made by treating cyanuric chloride with ammonia, but a cheaper industrial method involves the thermal decomposition of urea. The polymers derived from melamine and formaldehyde are used, for example, for plastic vessels and tableware and as electrical insulators.

1,3,5-Trinitrohexahydro-1,3,5-triazine, made by nitrating hexamethylenetetramine (p. 307), is an important high explosive (Cyclonite, or RDX).

1,2,3-Triazines can be made by the rearrangement of cyclopropenyl azides,

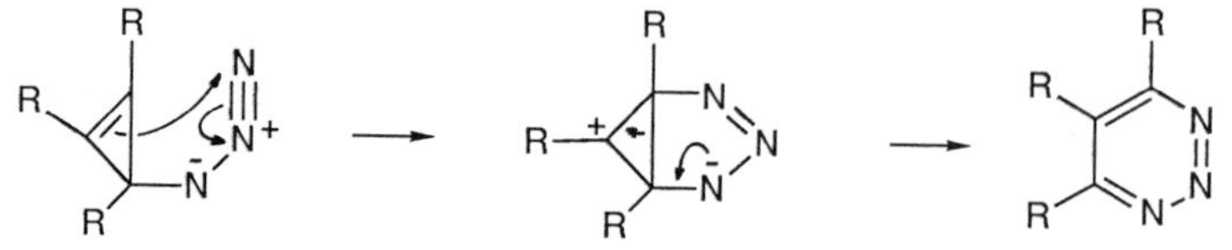

or by the oxidation of *N*-aminopyrazoles with nickel peroxide in a mixture of dichloromethane and acetic acid,

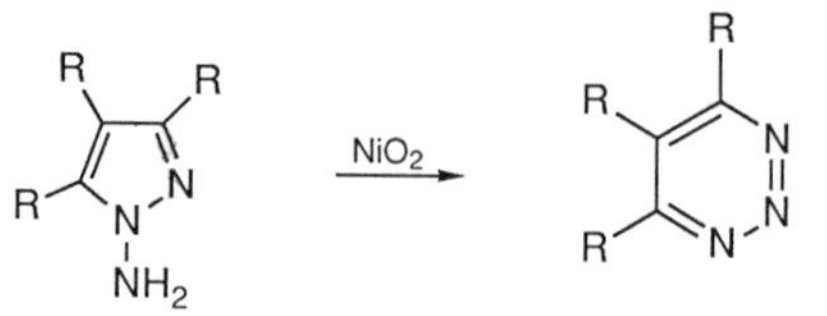

Only the latter method can be used for the parent compound (R=H). The mechanism of oxidation is unknown.

1,2,4-Triazines. Amidrazones, from nitriles with sodium hydrazide followed by hydrolysis, react with α-dicarbonyl compounds to give 1,2,4-triazines:

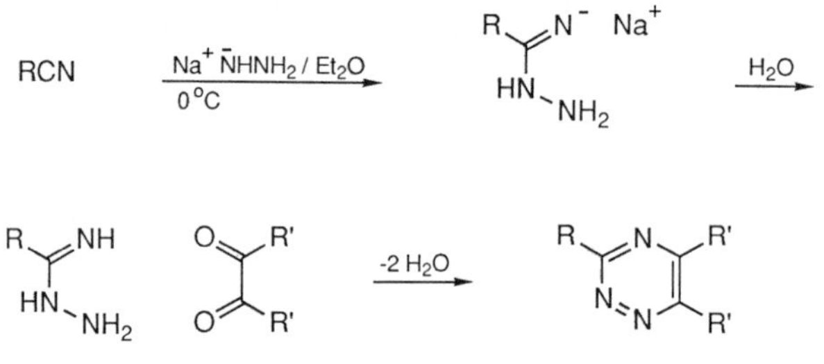

Problems

1. Classify the types of heteroaromatic compound which may be synthesized from acetoacetic ester. What structural variant would be introduced in those reactions in which an αβ-unsaturated ester can replace acetoacetic ester?

2. Annotate the steps in the syntheses of (*a*) papaverine (p. 707), (*b*) quercetin (p. 712), (*c*) cyanidin chloride (p. 714).

3. Account for the following:
    (*i*)   Pyrrole is both more acidic and less basic than pyrrolidine.
    (*ii*) Pyridine is less basic than piperidine.
    (*iii*) Imidazole is both more acidic than pyrrole and more basic than pyridine.

4. How would you carry out the following transformations?

(i)

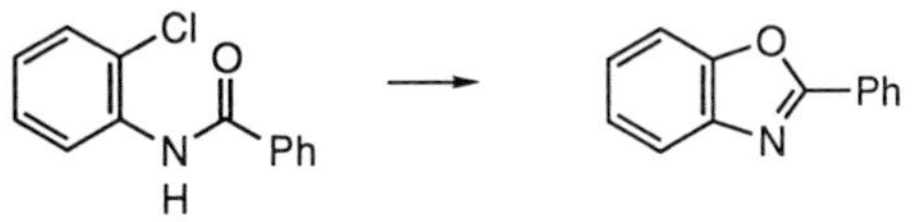

(ii)

(iii)

harman
a *Harmala* alkaloid

(iv)

(±)-hygrine
(a coca alkaloid)

(v)

(±)-coniine
(a hemlock alkaloid)

5. Complete the schemes outlined below by inserting the reagents and the intermediates which have been omitted:

(i) Synthesis of plasmoquin (an antimalarial):

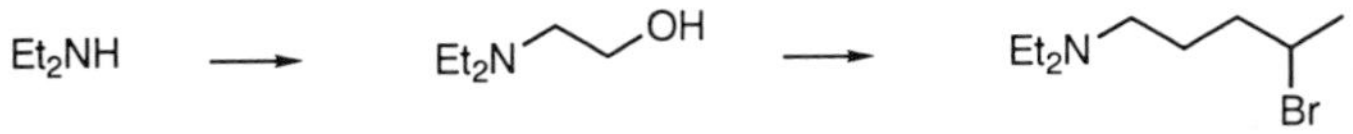

A

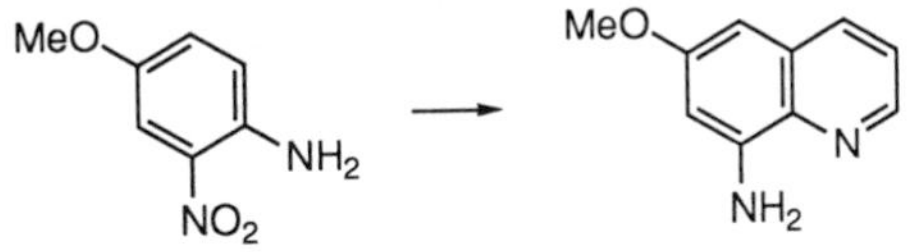

B

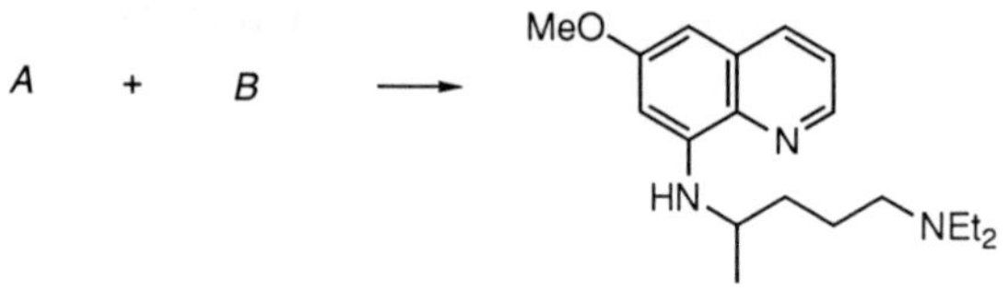

A + B ⟶

plasmoquin

(*ii*) Synthesis of nicotine (a tobacco alkaloid):

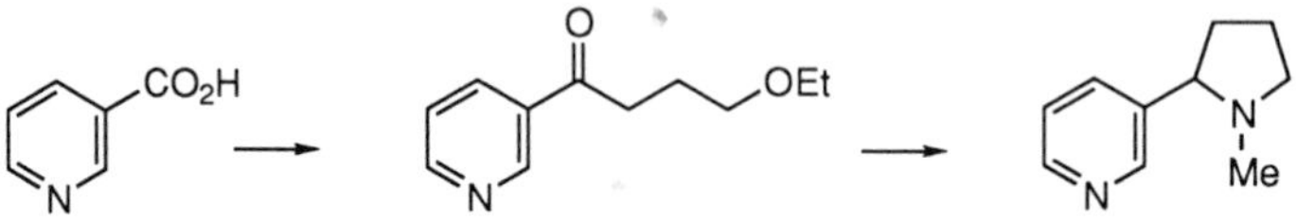

(±)-nicotine

(*iii*) Synthesis of biotin (vitamin H, a growth substance):

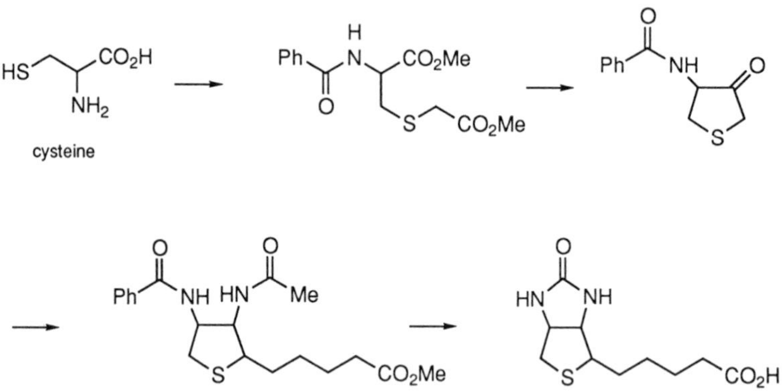

cysteine

biotin

(*iv*) Synthesis of pyridoxol (vitamin B₆):

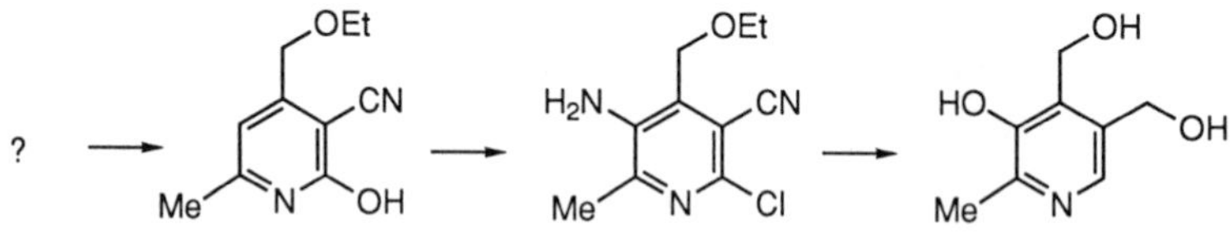

pyridoxol

6. Outline syntheses of the following compounds from readily available materials:

(*a*)

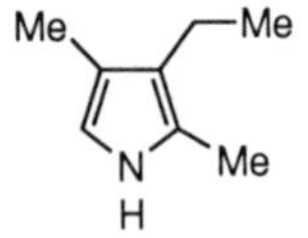

(*b*)

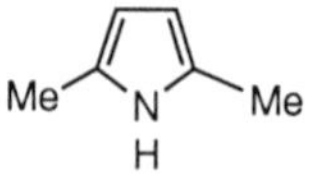

(*c*)

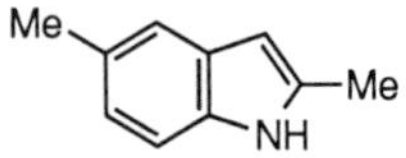

(*d*)

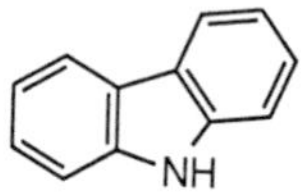

(e)

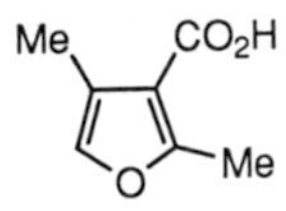

(f)

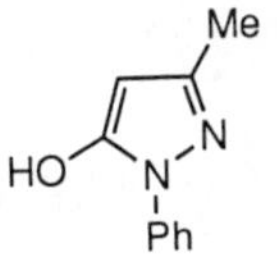

(g)

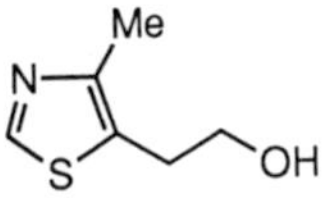

(h)

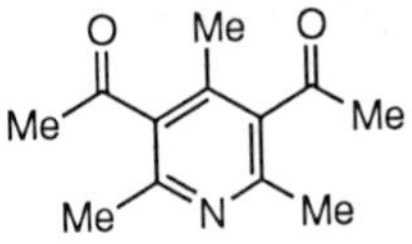

(i)

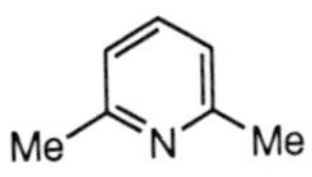

(j)

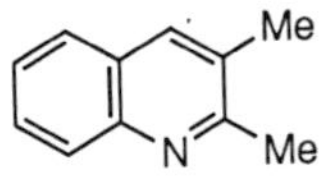

(k)

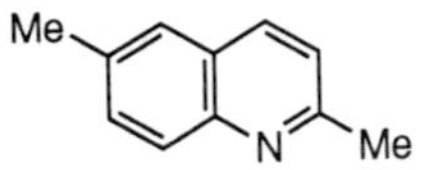

(l)

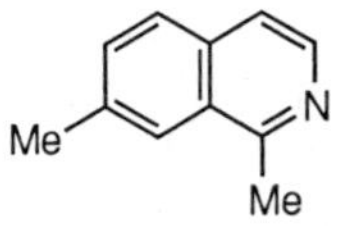

(m)

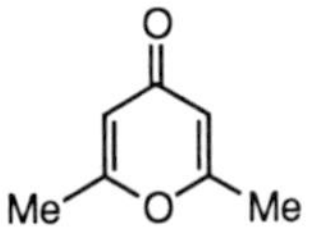

(n)

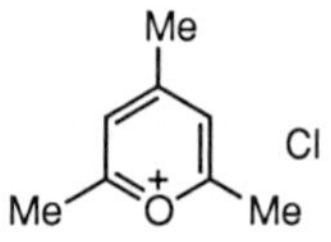

(o)

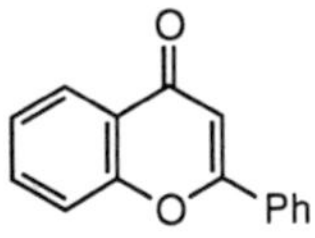

(p)

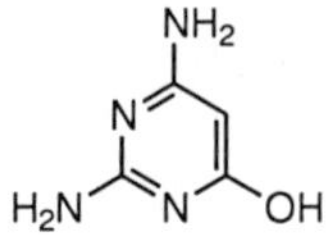

(q)

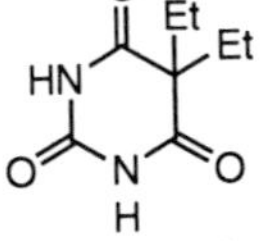

(r)

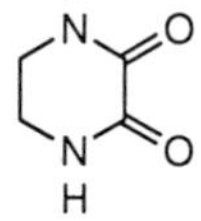

(s)

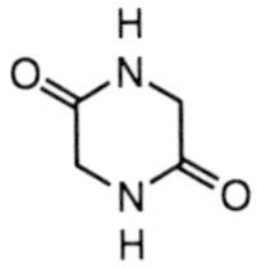

(t)

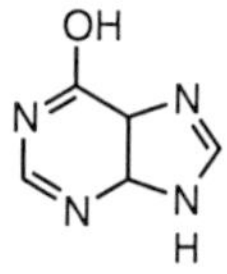

22 The syntheses of some naturally occurring compounds

At one time, the total synthesis of a compound represented the essential ultimate proof of its structure. This is no longer so; physical techniques such as nuclear magnetic resonance spectroscopy and X-ray crystallography can supply proof of structure in a much shorter time. Nevertheless, the synthesis of naturally occurring compounds remains a vigorously pursued field, for two reasons. First, some materials can be obtained from natural sources only in such small quantity or with such difficulty that synthesis can provide cheaper and more abundant material. This is of especial importance if the compound has valuable pharmaceutical properties. Moreover, the total synthesis will usually give precursors of the natural compound from which related structures can be synthesized and this also can be of value in the pharmaceutical field since the structural variants may be associated with different or more powerful biological properties. Second, natural product synthesis is a most fruitful source of new synthetic methods: a particular step in the sequence may lead to the development of a new and general method for its successful completion.

The following syntheses have been selected on the basis that each product is of biological importance and that they illustrate the application of a wide variety of regio- and stereoselective methods, many of which are still undergoing development.

22.1 Reserpine

Reserpine is found, with other alkaloids, in the roots of the plant genus *Rauwolfia*. It has useful tranquillizing properties and is used in the treatment of some mental disorders and for the reduction of hypertension. It was synthesized by R. B.

Woodward *et al.* (*J. Amer. Chem. Soc.*, 1956, **78**, 2023; *Tetrahedron*, 1958, **2**, 1). The strategy was based on building five contiguous stereocentres into a decalin derivative which could be opened to a monocyclic compound that would form ring E, as the following retrosynthetic analysis shows:

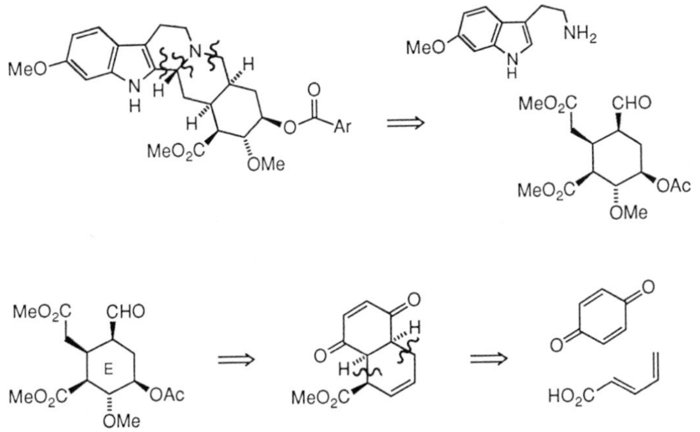

The synthesis is as follows:

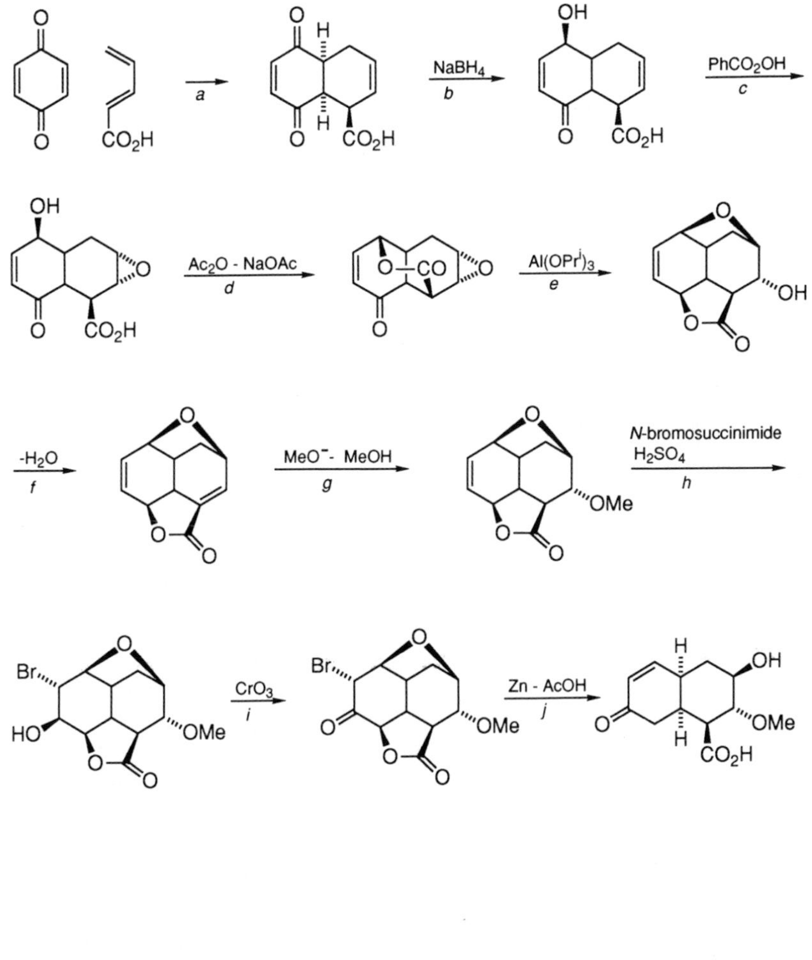

reserpine

a A Diels–Alder reaction. The stereochemical principles governing this reaction (p. 272) lead to the ring junction having *cis* stereochemistry and the
carboxyl group lying on the same side as the rings with respect to the ring
junction (i.e. reaction is represented as in (*i*) rather than as in (*ii*)).

(*i*)

(*ii*)

This step fixes the stereochemistry at C_{15}, C_{16} and C_{20} of reserpine. The
remainder of the synthesis concerns the stereochemistry at C_{17}, C_{18} and
C_3. (For clarity, the *cis*-hydrogens at the ring junction are omitted in the
remaining diagrams.)

b Sodium borohydride reduces the less hindered of the two carbonyl groups.
The nucleophile attacks from the less hindered α side.

c The peroxyacid reacts more rapidly with the isolated double bond than with
the carbonyl-conjugated double bond. Attack is from the less hindered α side.
If the epoxide were opened at this stage by reaction with, e.g. acetate ion, the
product would be a diaxial diol-derivative having the wrong stereochemistry
for ring E in reserpine:

Several transformations are therefore effected to produce the diol-derivative
with the required stereochemistry.

d Almost all the reactant in this step has the conformation in which both the
hydroxyl and the carboxyl groups are equatorial since the conformer in which
these groups are axial contains strong steric compressions, but only the latter
conformer can undergo dehydration to form a lactone. This reaction therefore
has the effect of changing the conformation of the system; in particular, the
epoxide becomes axial at C_{17} and equatorial at C_{18}[*]:

---

[*] This transformation is more readily understood with the aid of molecular models.

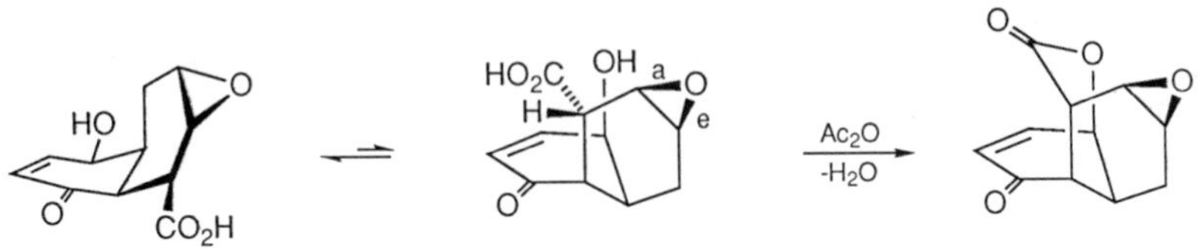

e   Meerwein–Ponndorf–Verley reduction converts the keto group into hydroxyl (p. 654). The resulting nucleophilic oxygen displaces on the carbonyl of the six-membered lactone ring, giving a five-membered lactone, and the hydroxyl group so released brings about opening of the epoxide ring:

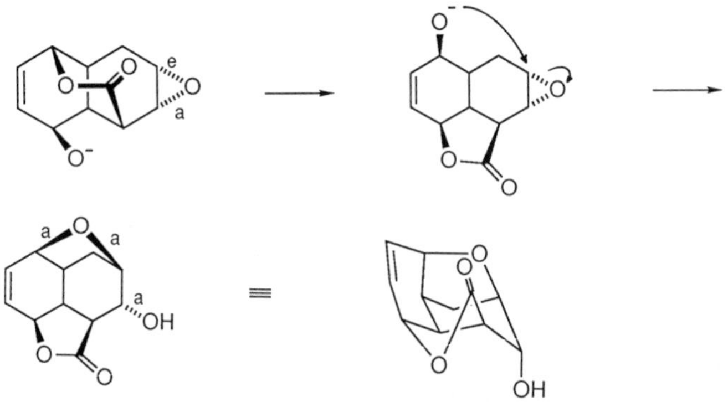

f   Dehydration, involving elimination of the axial hydroxyl group and formation of an $\alpha\beta$-unsaturated carbonyl-derivative, occurs readily.

g   The C=C bond is activated towards nucleophiles by the conjugated carbonyl group. Methoxide ion approaches from the less hindered α-side, so that the resulting —OCH$_3$ substituent is in the axial position at C$_{17}$. Ring E now has the correct stereochemistry.

h   *N*-Bromosuccinimide in acid is an ionic brominating agent. Approach from the α-side as usual gives a bromonium ion which is opened by water to give the diaxial bromo-alcohol:

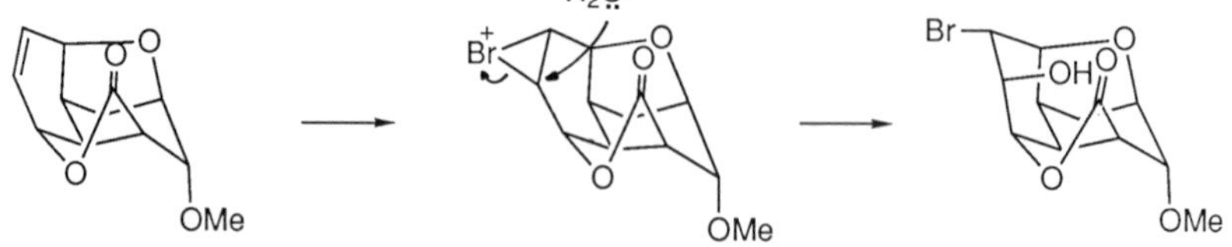

i   Mild oxidation of the secondary alcohol.

j   Zinc in acetic acid brings about the reductive opening of both the lactone and the strained ether:

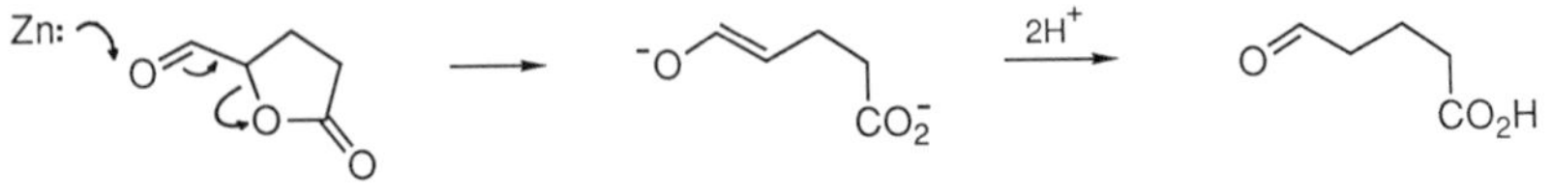

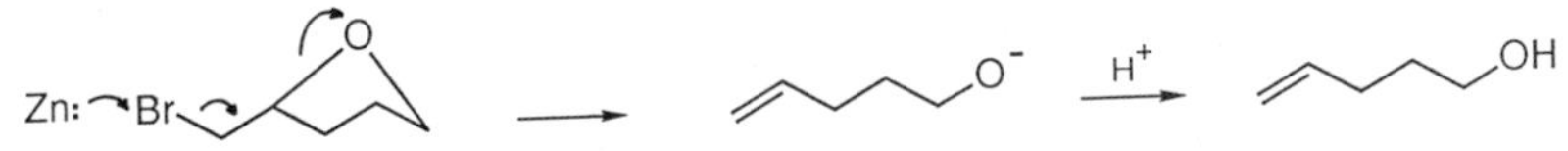

All the substituents in ring E are now able to adopt equatorial conformations:

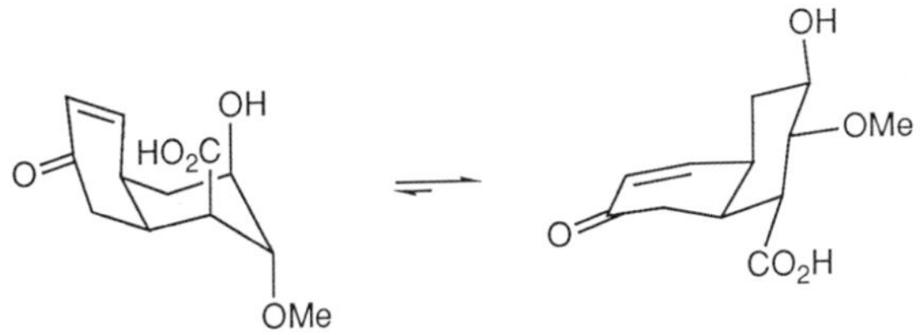

k The carboxyl group is esterified by diazomethane (which does not react with the alcoholic group), the alcoholic group is acetylated, and the 1,2-diol is formed with osmium tetroxide (p. 591).

*l** Treatment with periodate cleaves ring D and one carbon atom is lost as formic acid (p. 617) (later to be replaced, with ring closure, by nitrogen in step *n*). The new carboxyl group is then esterified with diazomethane.

m 6-Methoxytryptamine may be prepared as follows:

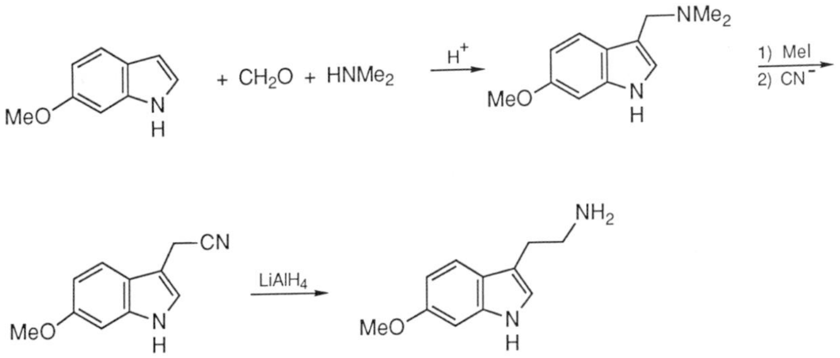

Indoles are very reactive towards electrophiles, particularly at the 3-position, so that the Mannich reaction can be effected (p. 261). The quaternized benzyl-type system is susceptible to nucleophilic displacement by cyanide ion (p. 265) and the final reduction is a standard procedure for nitriles (p. 659). The tryptamine reacts readily with the aldehydic group in benzene solution.

n Reduction of the imine gives an amine and the resulting nucleophilic nitrogen displaces intramolecularly on the ester to close ring D.

o Phosphorus oxychloride brings about ring closure as in the Bischler–Napieralski synthesis of isoquinolines (p. 706), giving an immonium salt which is then reduced with sodium borohydride:

---

*The intermediates resulting from steps *l–n* were not isolated.

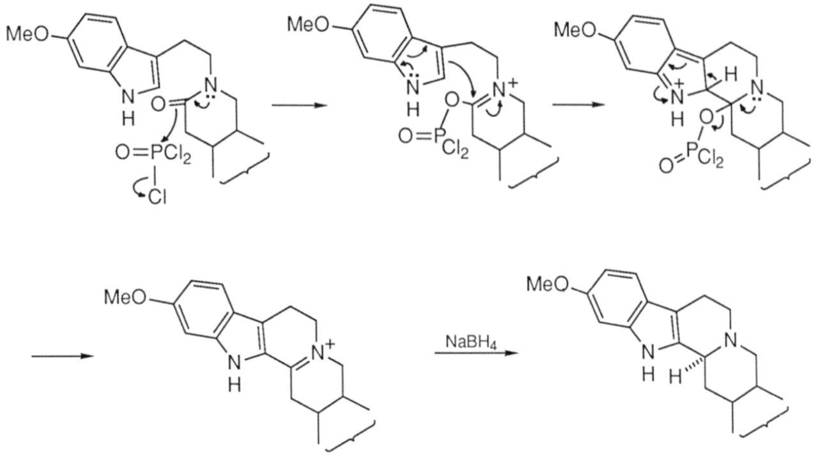

The borohydride attacks from the α-side, so that the product has the wrong stereochemistry at C_3 for reserpine. The remaining steps are concerned with correcting this and introducing the aroyl substituent at C_{18}.

p   Treatment with base followed by acid removes the acetyl group (C_{18}) and hydrolyzes the methyl ester (C_{16}). The freed hydroxyl and carboxyl groups are then joined to give a lactone, using dicyclohexyl carbodiimide, which increases the susceptibility of the acid to nucleophilic attack by the alcohol (p. 338). Lactonization occurs through the conformer in which the carboxyl and hydroxyl groups are axial:

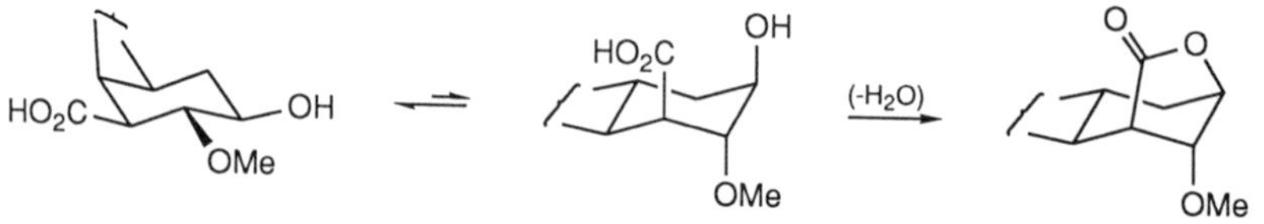

q   The result of the preceding step is that the molecule is now less stable than the stereoisomer obtained by altering the configuration of the hydrogen at C_3.* Equilibration of the two isomers therefore yields almost entirely the product with the stereochemistry required for reserpine at this centre. The equilibrating agent chosen was trimethylacetic acid because of its suitable boiling-point and the fact that it is too weak a nucleophile to open the lactone ring. Interconversion presumably occurs through protonation (activated by the pyrrole nitrogen) and deprotonation:

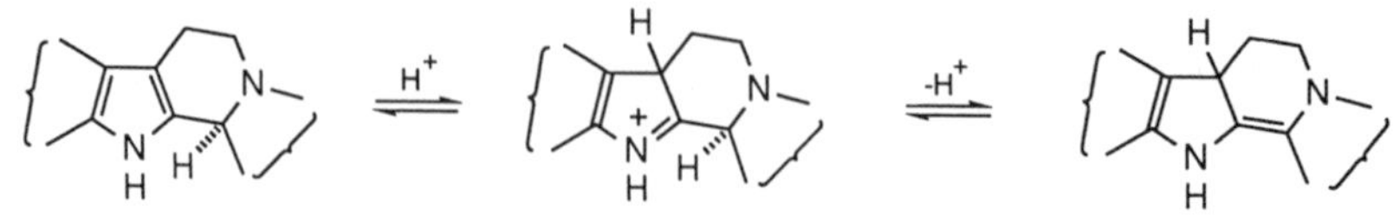

* The steric compressions which are responsible for this can best be seen by constructing molecular models.

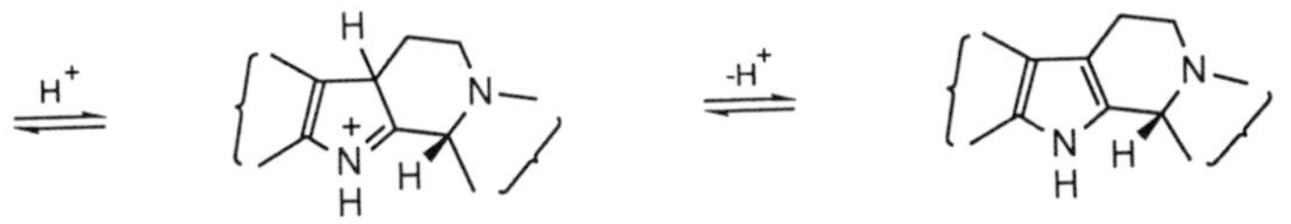

r   The lactone ring is opened with methanol, and the molecule then reverts to the more stable conformation in which the three substituents on ring E are equatorial. Finally, the aroyl residue is introduced at C_{18} with 3,4,5-trimethoxybenzoyl chloride.

22.2 Adenosine triphosphate

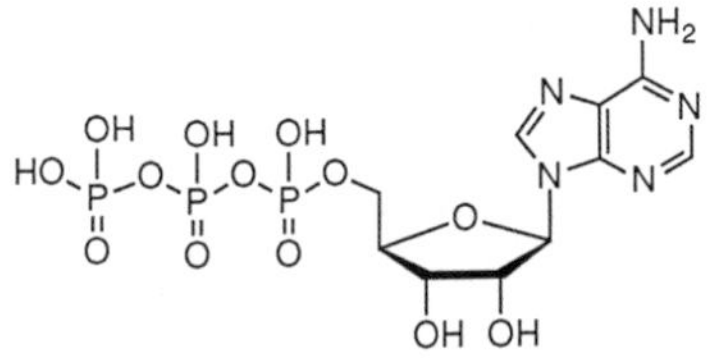

adenosine triphosphate

Adenosine occurs naturally in a variety of derived forms each of which performs important functions.

(1)   Adenosine triphosphate, ATP (above), acts in conjunction with its diphosphate, ADP, as a reversible phosphorylating couple and energy store. For example, it is concerned with the supply of energy for muscular contraction.

(2)   Coenzyme A, a derivative of the vitamin pantothenic acid, is involved, *inter alia*, in the biological formation and oxidation of fatty acids and in the decarboxylation of α-keto-acids.

(3)   *S*-Adenosyl-L-methionine is concerned with the biological transfer of methyl groups.

(4)   Adenosine is a component of the ribonucleic acids (p. 716), di- and triphosphopyridine nucleotides (p. 721), and flavin adenine dinucleotide (p. 721).

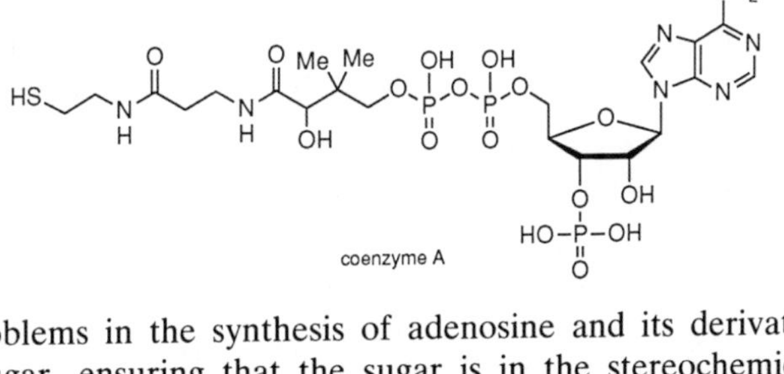

coenzyme A

The main problems in the synthesis of adenosine and its derivatives involve attaching the sugar, ensuring that the sugar is in the stereochemically correct form, and selectively phosphorylating a sensitive molecule. The first approach was to attach the sugar before closing the five-membered ring of the purine but it was

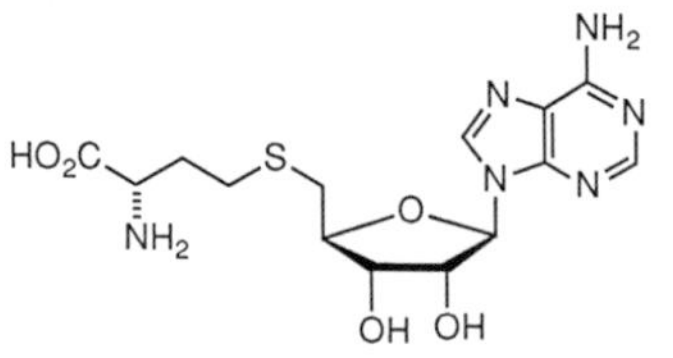

S-adenosyl-L-methionine

later found to be far more efficient to attach the sugar to the completed purine. The latter synthesis is described.

(1) *Preparation of the appropriate sugar derivative from* D-*ribose* (A. R. Todd et al., *J. Chem. Soc.*, 1947, 1052).

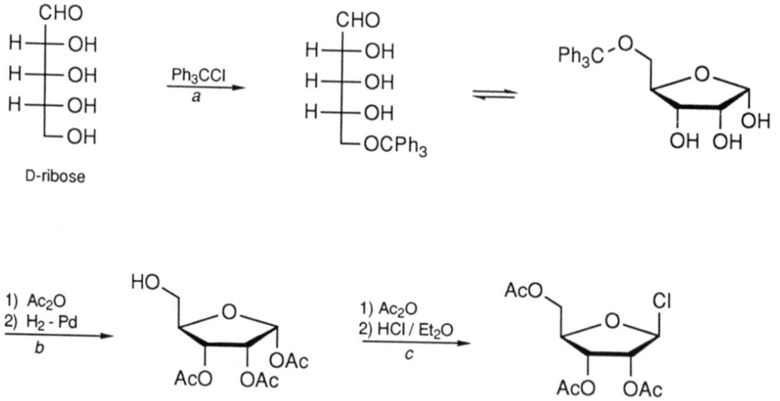

D-ribose

β-chloro-2,3,5-triacetyl-
D-ribofuranose

a   Triphenylmethyl chloride reacts only with the *primary* alcoholic group (p. 645). This ensures that the sugar adopts the furanose ring-system rather than the pyranose system.

b   After acetylation of the remaining hydroxyl groups, the triphenylmethyl group is removed by hydrogenolysis (p. 645).

c   S$_N$2 reaction, with inversion of configuration, occurs readily at the carbon next to the ring-oxygen (cf. the reactivity of α-chloro-ethers, p. 108).

(2) *Complete synthesis of adenosine from uric acid* (p. 722) (A. R. Todd et al., *J. Chem. Soc.*, 1948, 967).

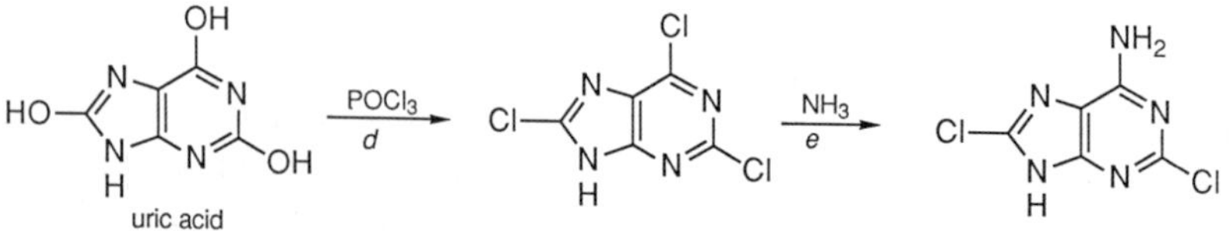

uric acid

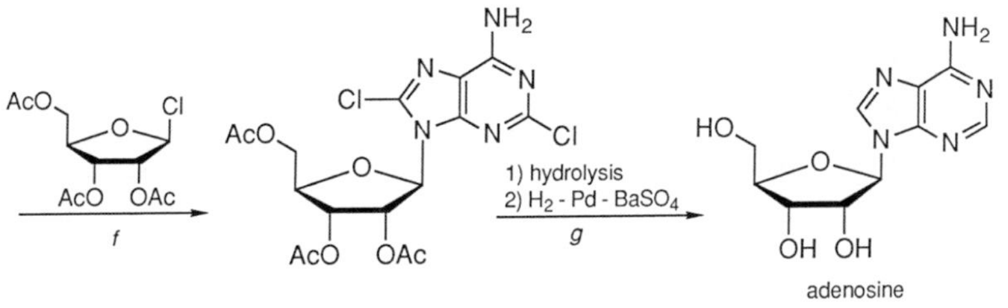

d   The hydroxyl groups behave differently from those in phenol in being re-placed by chlorine with phosphorus oxychloride (p. 648).

e   The chlorine atoms are all activated towards nucleophilic displacement by the heterocyclic nitrogen atoms (p. 400) but careful treatment leads only to the replacement of the 6-substituent.

f   The chloro-furanoside gives the β-furanosidyl-derivative of the purine, pre-sumably as a result of the formation of an acetoxonium ion (p. 440) followed by an S_N2 reaction with inversion:

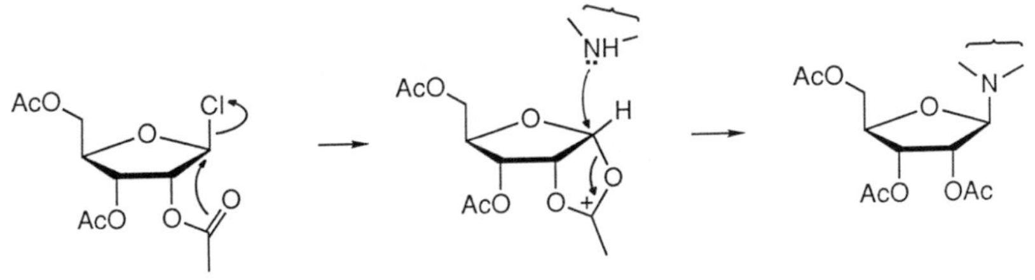

g   Hydrolysis of the acetyl groups, followed by hydrogenolysis of the C—Cl bonds on palladium, gives adenosine.

(3) *Adenosine triphosphate* (A. R. Todd *et al.*, *J. Chem. Soc.*, 1947, 648; 1949, 582). The phosphorylating agent, dibenzyl chlorophosphonate, is prepared by the chlorination of dibenzyl phosphite in carbon tetrachloride:

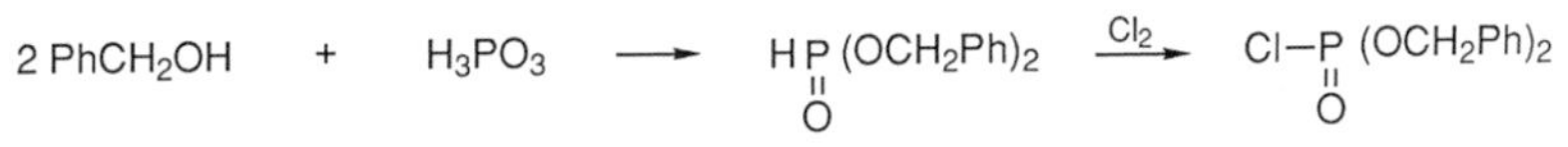

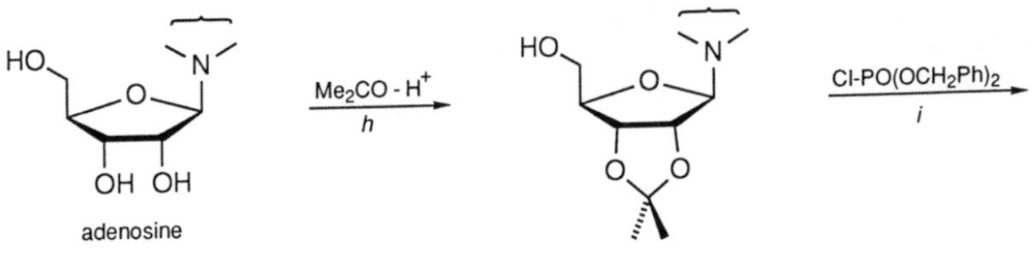

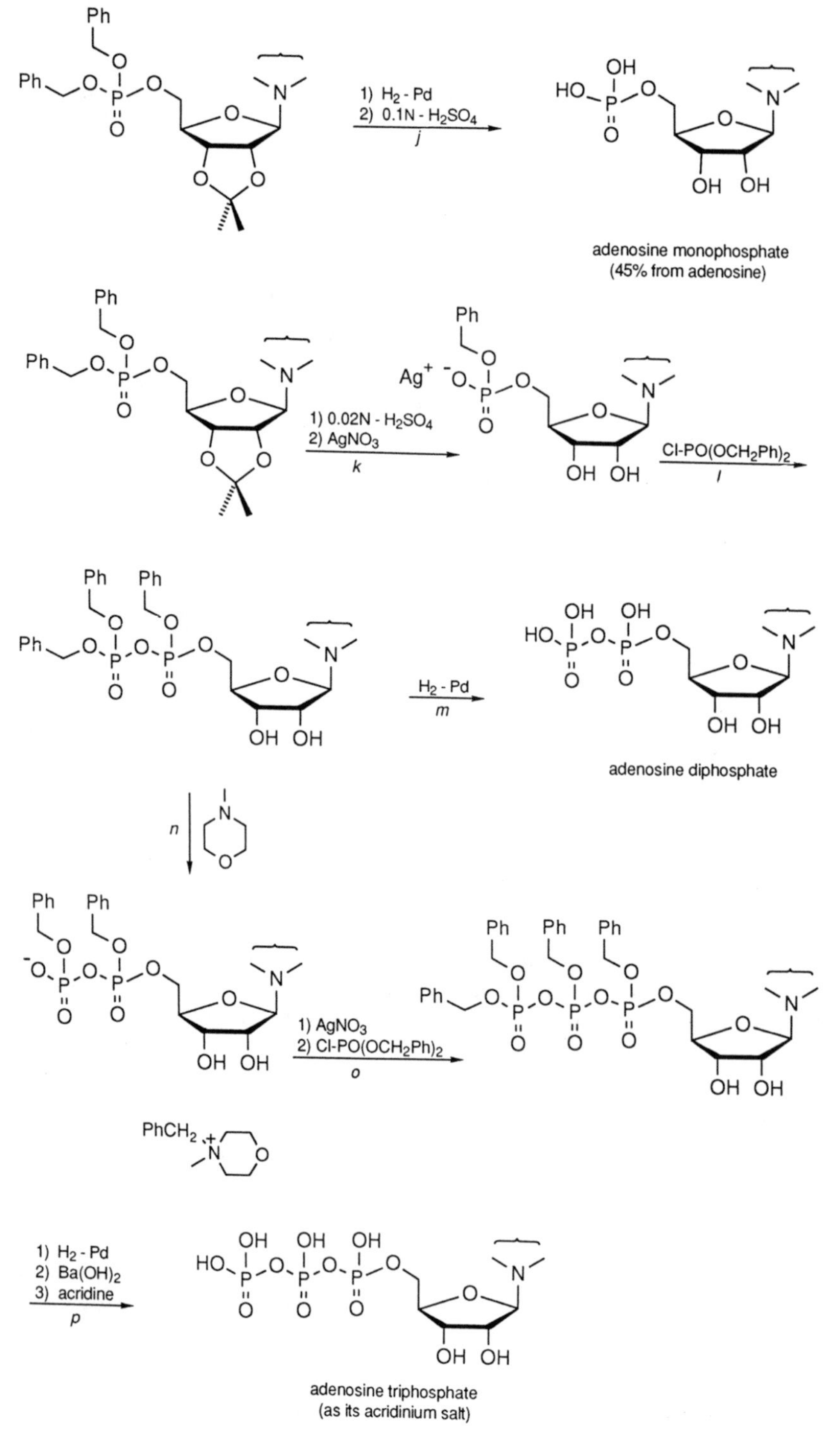

1) H_2 - Pd
2) 0.1N - H_2SO_4
j
adenosine monophosphate
(45% from adenosine)
1) 0.02N - H_2SO_4
2) $AgNO_3$
k
Cl-PO(OCH$_2$Ph)$_2$
l
H_2 - Pd
m
adenosine diphosphate
n
1) $AgNO_3$
2) Cl-PO(OCH$_2$Ph)$_2$
o
1) H_2 - Pd
2) Ba(OH)$_2$
3) acridine
p
adenosine triphosphate
(as its acridinium salt)

h The 2- and 3-hydroxyl groups of the furanose ring are protected against phosphorylation by formation of the ketal, the establishment of a cyclic ketal from acetone being selective for *cis*-1,2-diols (p. 90).

i Phosphorylation is effected at a low temperature; pyridine is used as solvent to remove the hydrogen chloride.

j The benzyl groups are removed by catalytic hydrogenolysis (p. 644) and the ketal is hydrolyzed with dilute sulfuric acid. Removal of the acid as barium sulfate allows the monophosphate to crystallize out.

k The dibenzyl phosphate is hydrolyzed under very mild conditions to remove the isopropylidene group and, at the same time, one of the two benzyl groups. After removal of the acid as barium sulfate, the product is dissolved in alkali and precipitated as its silver salt.

l Phosphorylation in anhydrous acetic acid gives a resinous product which may be converted into both the di- and triphosphates.

m Hydrogenolysis, followed by purification of adenosine diphosphate as its acridinium salt.

n It was found that the tribenzyl ester can be selectively debenzylated with *N*-methylmorpholine.

o The product, isolated as its silver salt, is phosphorylated in a mixture of acetonitrile and phenol.

p The four benzyl groups are removed by hydrogenolysis and the product is precipitated as its barium salt, liberated with sulfuric acid, and isolated as its acridinium salt.

**22.3
Penicillins and
cephalosporins**

(*a*) *Penicillins*

There are at least five naturally occurring penicillins, each based on the same structural nucleus (a β-lactam fused to a thiazolidine). They are produced from the mould *Penicillium notatum*, different strains of which produce different penicillins. The synthesis of penicillin V (R is $PhOCH_2$) is described below. Other penicillins contain R as $CH_3CH_2CH{=}CHCH_2$(F), $PhCH_2$(G), and 1-heptyl (K). The penicillins owe their importance to their powerful effect on various pathogenic organisms.

The main difficulty in synthesizing the penicillins is to close the β-lactam ring in conditions which do not disrupt the rest of the sensitive molecule. This was overcome by the carbodiimide method (also widely used for making peptide bonds, p. 338). The synthetic strategy employed by J. C. Sheehan and K. R.

Henery-Logan for penicillin V (*J. Amer. Chem. Soc.*, 1959, **81**, 3089) is shown by the following disconnections:

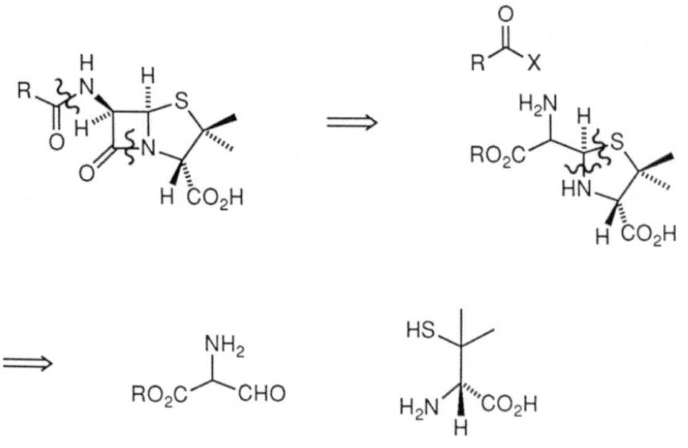

The synthesis is as follows:

(1) (+)-*Penicillamine*

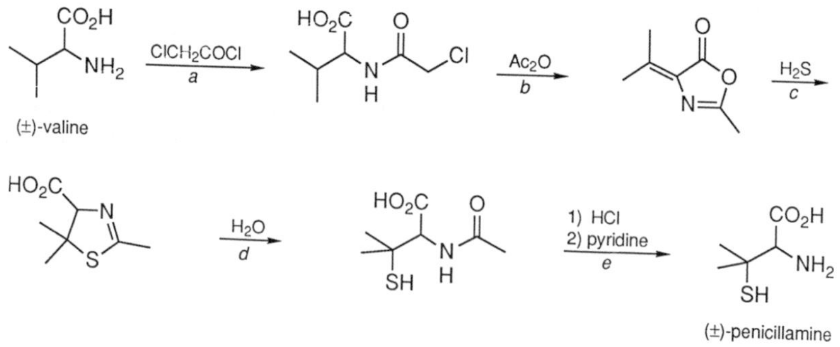

a   Acylation. The —COCl group is more reactive towards nucleophiles than —CH₂Cl.

b   Acetic anhydride induces dehydrative cyclization and this is accompanied by the elimination of hydrogen chloride and a prototropic shift:

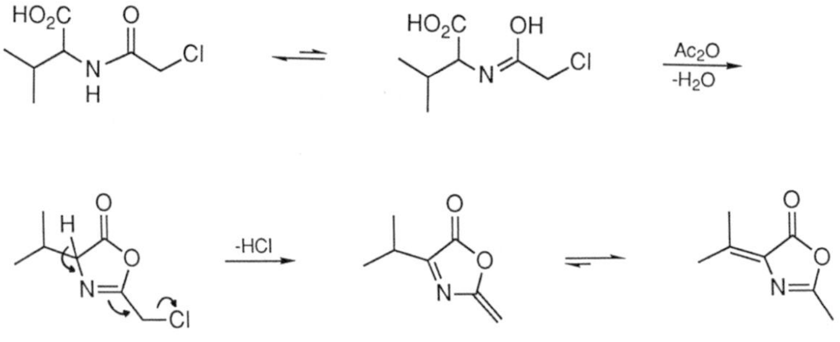

c   The azlactone is cleaved by hydrogen sulfide and the resulting thiol cyclizes *via* Michael-type addition to the αβ-unsaturated acid:

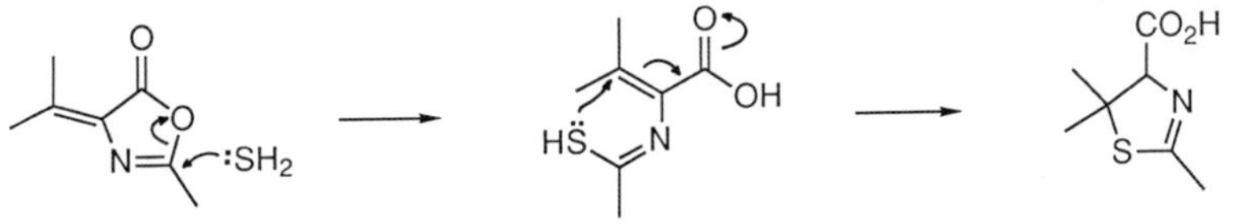

d The thiazoline ring is opened with boiling water.
e The *N*-acetyl group is removed by acid hydrolysis followed by basification. (±)-Penicillamine may be resolved by formylation (HCO$_2$H), separation of the diastereomers formed with brucine, and hydrolysis (HCl followed by pyridine).

(2) *Penicillin V*

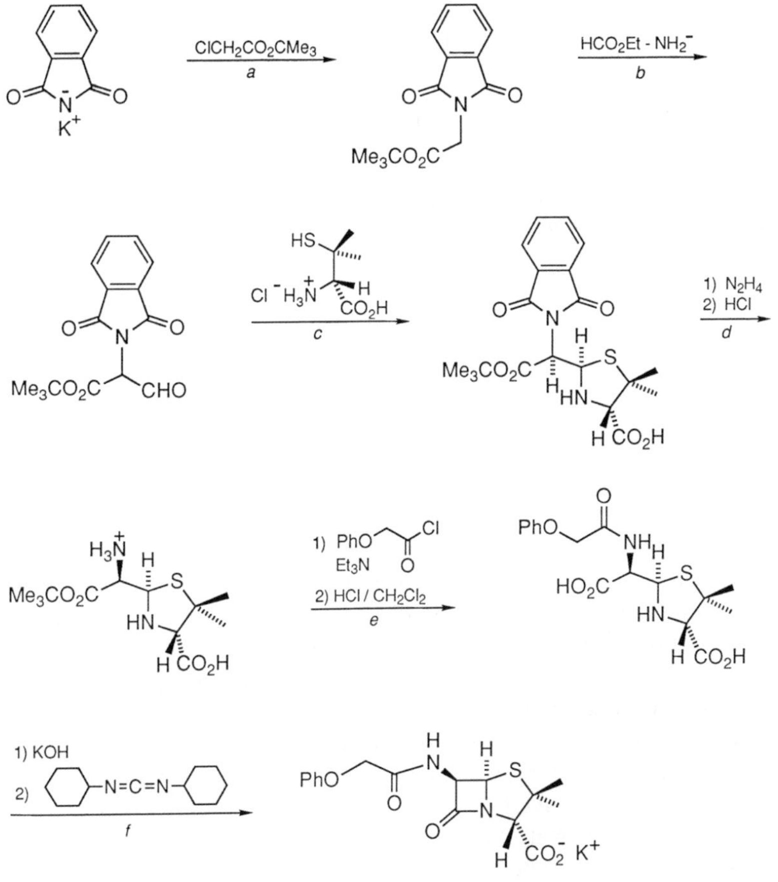

penicillin V

a Gabriel's synthesis (p. 303).
b A crossed Claisen condensation, standard procedure for the formylation of activated methylene groups (p. 230).

c   Reaction with penicillamine hydrochloride occurs at room temperature in sodium acetate buffer:

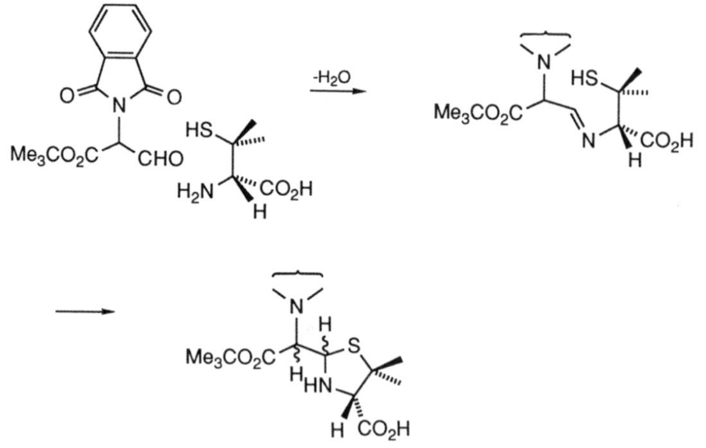

As expected, a mixture of diastereomers is obtained, since two asymmetric carbon atoms are generated in steps b and c. However, only two of the four possible isomers are formed in significant amounts, of which the one required,

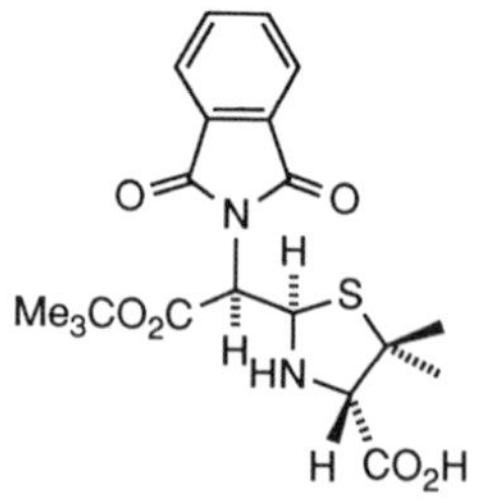

is the minor component. The major isomer is its epimer with the alternative configuration at the ester-substituted carbon and, since the C—H bond there is weakly acidic, the two can be equilibrated by boiling in pyridine under hydrogen. The required isomer is still the minor component but, after its removal from the mixture, equilibration and isolation can be repeated on the remainder until an acceptable yield is obtained.

d   The phthalimido group must be removed in conditions which do not cleave the t-butyl ester. This is achieved with hydrazine at 13°C for 3 hours and then at room temperature for 1 day, followed by treatment with HCl in acetic acid.

e   Acylation of the amino group in the presence of triethylamine is carried out at 25°C so that the thiazolidine ring is not opened. The t-butyl ester is then hydrolyzed by treatment with hydrogen chloride in dichloromethane at 0°C.

f   The potassium salt is cyclized with dicyclohexyl carbodiimide (p. 338), giving the potassium penicillinate. Penicillin V can be extracted after acidification with phosphoric acid, and it crystallizes from aqueous solution at pH 6·8. It may be purified by counter-current extraction.

(b) Cephalosporins

The success of the penicillins as antibiotics led to a search for related compounds, especially ones with a broader range of activity and which would be less easily defeated by the organisms which had developed resistance mechanisms to penicillins. The key to the next advance was the isolation of cephalosporin C from *Cephalosporium* sp. The cephalosporin skeleton, which differs from that of the penicillins only in having a 1,3-thiazine ring instead of a thiazolidine fused to a β-lactam, is now the basis of a wide variety of antibiotics.

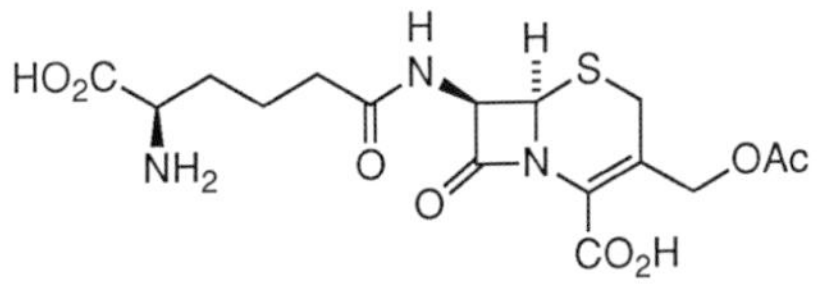

cephalosporin C

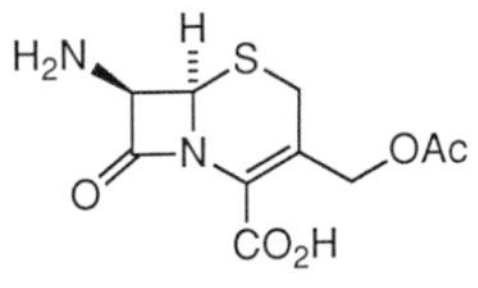

7-aminocephalosporanic acid

(7-ACA)

The cephalosporins that are used as drugs are mostly variants of cephalosporin C in which the carboxylic acid that is attached to the β-lactam ring as an amide and/or the acetoxy group is changed. Industrially, syntheses are usually from 7-aminocephalosporanic acid, which is readily available by hydrolysis of cephalosporin C. However, it is also possible to convert a penicillin into a cephalosporin by a ring expansion effected by acetic anhydride on the penicillin S-oxide:

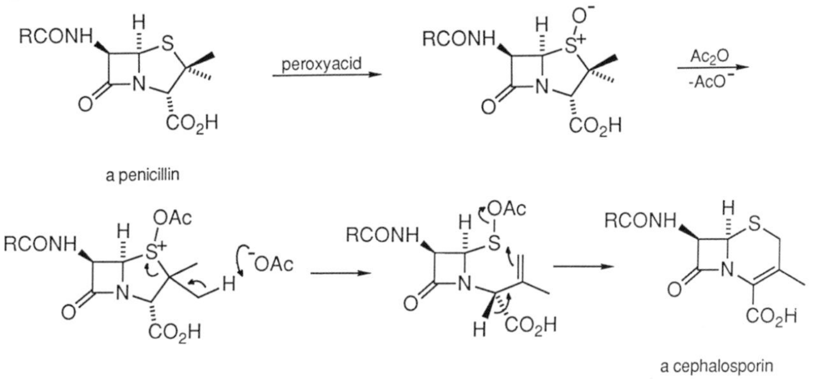

a penicillin

a cephalosporin

The synthesis of ceftazidime is illustrative. Retrosynthetic analysis of the carboxylic acid required to form the product with 7-aminocephalosporanic acid suggests the following strategy:

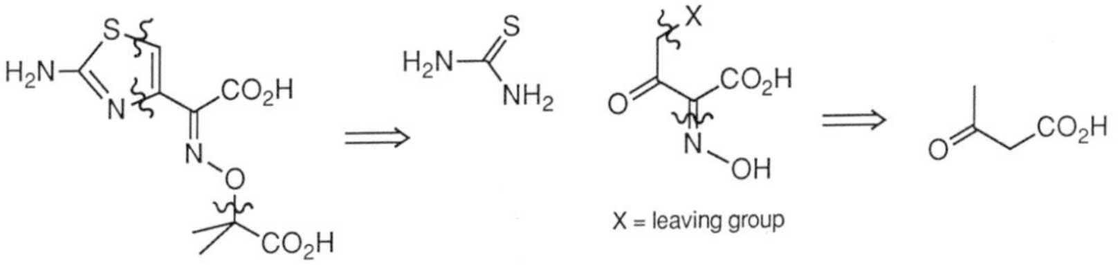

X = leaving group

The synthesis is as follows:

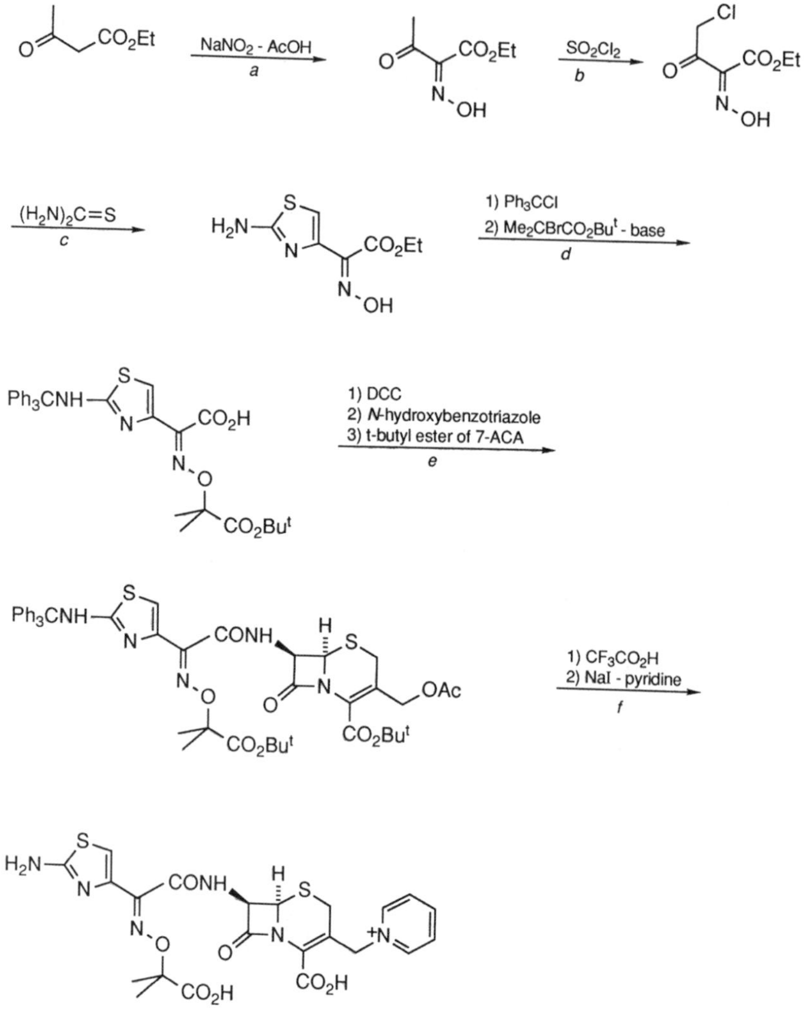

ceftazidime

a Nitrosation (NO$^+$) of the enol is followed by the nitroso-oxime tautomeric shift (p. 320).

b This is the equivalent of acid-catalyzed chlorination of a ketone *via* the enol.

c cf. The synthesis of thiazoles (p. 698).

d The amino group is protected with triphenylmethyl chloride, so that the oxime oxygen becomes the most nucleophilic centre for reaction with the bromo-ester. The base hydrolyzes the ethyl ester but does not affect the sterically hindered t-butyl ester.

e The acid reacts with dicyclohexyl carbodiimide to give an intermediate which, with *N*-hydroxybenzotriazole, gives an activated ester that reacts readily with the amino group of 7-ACA (whose carboxyl group is protected) (cf. peptide synthesis, p. 339).

f   The three protecting groups are removed (as carbocations) by trifluoroacetic acid, to which the β-lactam ring is stable. Finally, although an acetoxy group is usually not a good leaving group in S_N2 reactions, it is in a reactive allylic position (p. 108). Iodide ion – a better nucleophile than pyridine – displaces it, and is then itself displaced by pyridine:

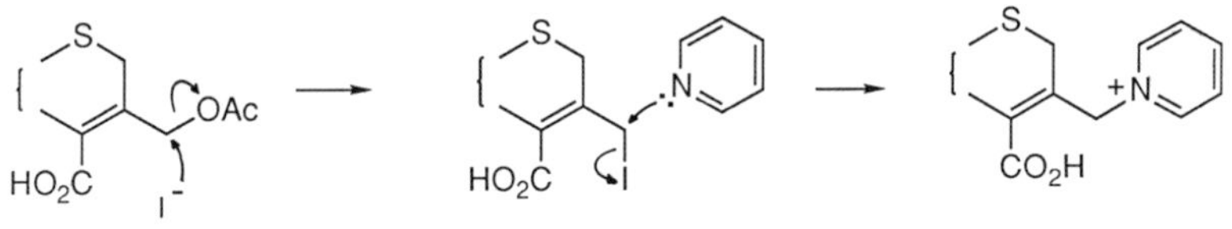

22.4
**22.4
Chlorophyll***

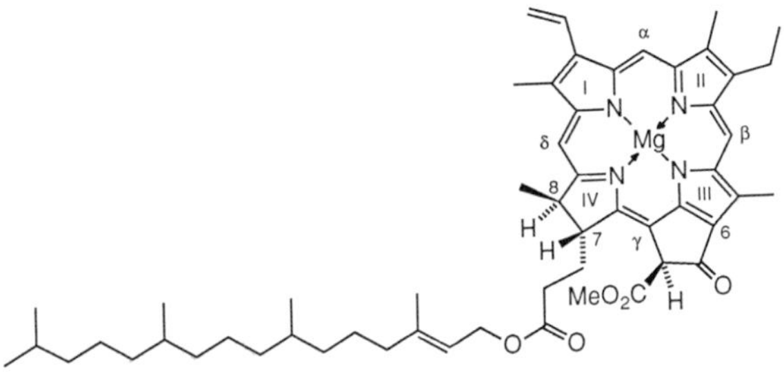

Chlorophyll is a dihydroporphyrin contained in plants and concerned in the photochemical process. There are in fact two chlorophylls of which that shown above is chlorophyll *a*; chlorophyll *b* differs in having an aldehyde group in place of methyl in ring II.

The synthesis described was carried out by R. B. Woodward *et al.* (*J. Amer. Chem. Soc.*, 1960, **82**, 3800; *The Chemistry of Natural Products*, I.U.P.A.C. Symposium, 1960, p. 383). They based their approach on observations relating to the structure and properties of chlorophyll. In particular, dihydroporphyrins are normally very easily oxidized to porphyrins but chlorophyll is not. The inference is that resistance to removal of the hydrogen atoms at C_7 and C_8 stems from the fact that in the oxidized system the alkyl groups at these positions would eclipse and strongly repel each other; in the reduced form (*trans*) the staggered arrangement reduces the steric strain. In addition, it is apparent from a consideration of bond lengths that the unsaturated ketone ring connecting C_6 and C_γ must be very strained, whereas the ring is closed with ease, the inference being that ring closure relieves the strain between the C_γ and C_7 substituents. The application of these observations in the synthesis is clear from the discussion below.

---

* At the time of the synthesis, the absolute configuration of chlorophyll was not known. It has since been determined as the enantiomer of the structure given here. For consistency with the papers by Woodward *et al.* which are referred to, their structures are retained in this section – the relative configurations are of course correct.

The Woodward synthesis was completed with the formation of chlorin e_6 trimethyl ester in step s, for this had previously been converted into chlorophyll a in three steps. One of these involves the introduction of phytol which has been synthesized from (+)-citronellol as follows:

1. *Synthesis of phytol*

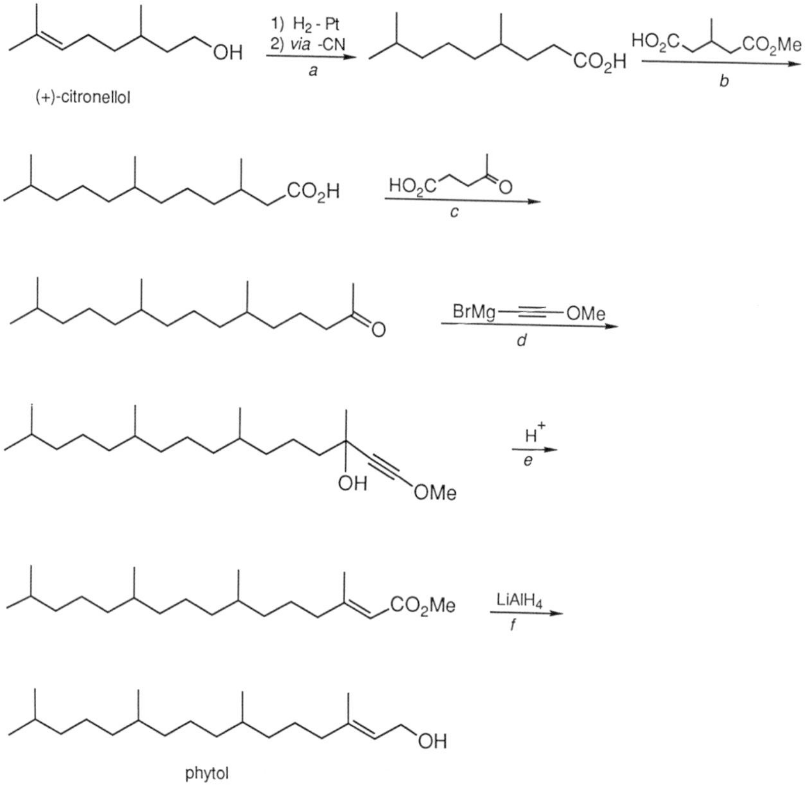

(+)-citronellol

a Reduction of C=C, followed by ascent of the homologous series through the nitrile (p. 244).

b Crossed anodic coupling with (+)-methyl hydrogen β-methylglutarate (p. 546), followed by hydrolysis of the ester.

c A similar reaction with the keto-acid.

d Standard procedure using the Grignard reagent from methoxyacetylene.

e Acid-catalyzed rearrangement:

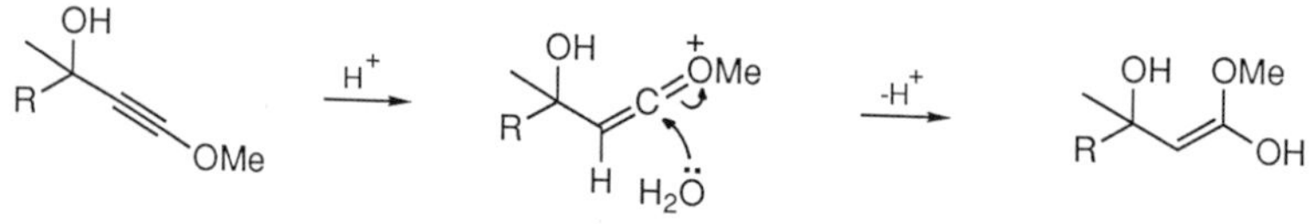

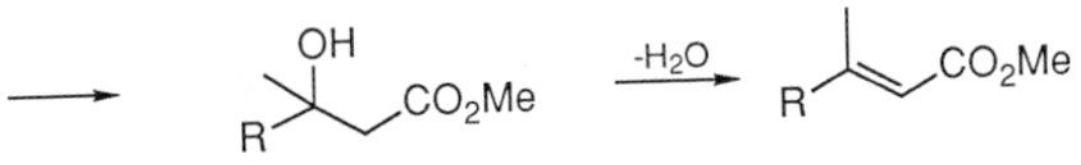

$$\longrightarrow \quad \underset{R}{\overset{OH}{|}}CO_2Me \quad \xrightarrow{-H_2O} \quad R \diagup CO_2Me$$

f   The *cis*- and *trans*-isomers are separated after reduction; the (−)-*trans*-isomer is the naturally occurring compound.

2.   *Synthesis of chlorophyll a.*

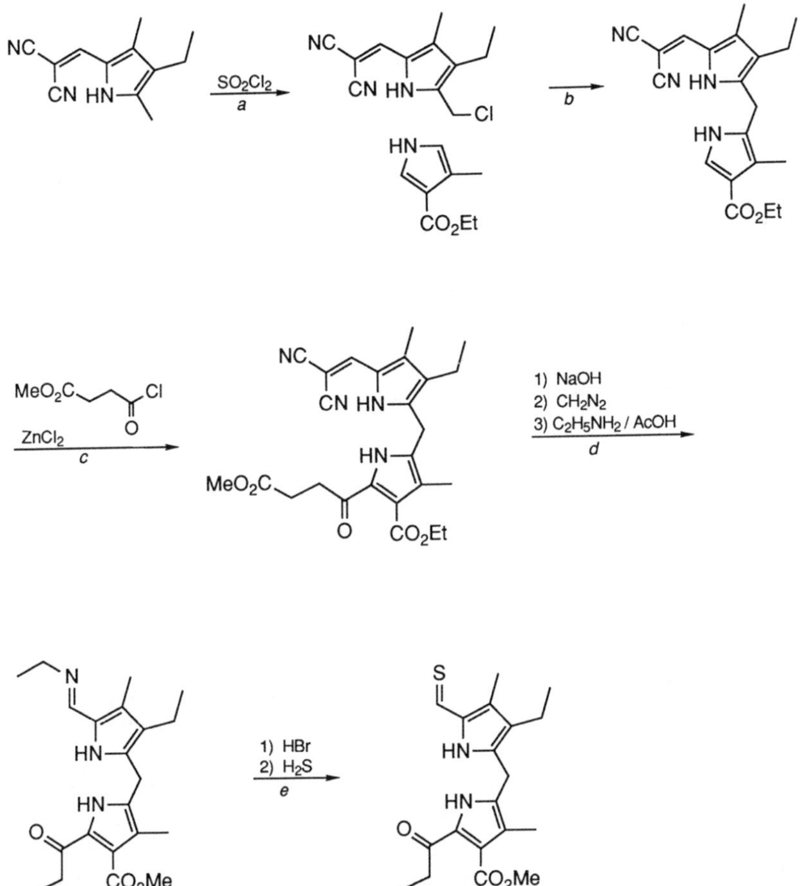

a   The free-radical chlorination of the benzylic-type methyl group (p. 531) occurs rapidly in acetic acid at 55°C; the 2- is more activated than the 3-position.

b   The Friedel–Crafts alkylation takes place readily on the activated pyrrole nucleus (p. 357). Of the two 2-positions available, that adjacent to the methyl-substituent is the more reactive, first, because the electron-releasing methyl group stabilizes the positive charge in the transition state whereas the electron-attracting ester group destabilizes it, and second, because the ester group probably provides more hindrance to the reagent than the methyl group.

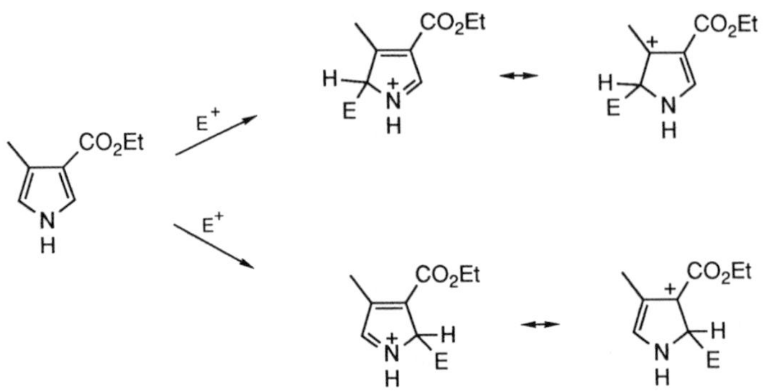

c   This Friedel–Crafts acylation, catalyzed by a mild Lewis acid, takes place at the only unsubstituted nuclear position.

d   The dicyanovinyl group has acted until this point as a protective group for the aldehydic function required at this carbon atom. The group is removed with 33% sodium hydroxide solution by the reverse-aldol reaction (p. 210), use being made of the anion-stabilizing influence of the cyano groups.

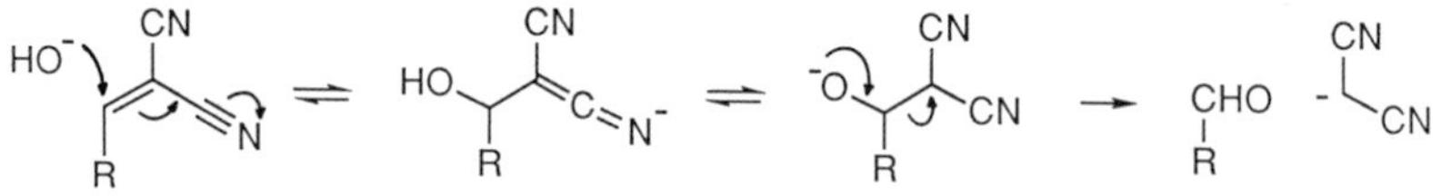

This procedure also hydrolyzes the ester groups which are then re-esterified with diazomethane. The aldehyde is converted with ethylamine into an imine, preparatory to conversion into the thioaldehyde in the next step.

e   The imine is converted into the thioaldehyde for the following reason. Two dipyrrylmethanes are to be combined (step *j*). Reaction must be specific: that is, the molecules must be oriented as in the representation preceding step *i* and not in the alternative manner. To ensure this, it was hoped to form an imine as in step *i* before coupling the dipyrrylmethanes. The first attempt used an aldehyde group for formation of the imine, but pyrrole aldehydes are unreactive towards nucleophiles such as amines as a result of delocalization,

The reactivity of the carbonyl group is increased by acids, but in this case even the mild acid conditions required (triethylammonium acetate) degraded the dipyrrylmethane. However, the thioaldehyde is much more reactive and proved suitable in this instance.

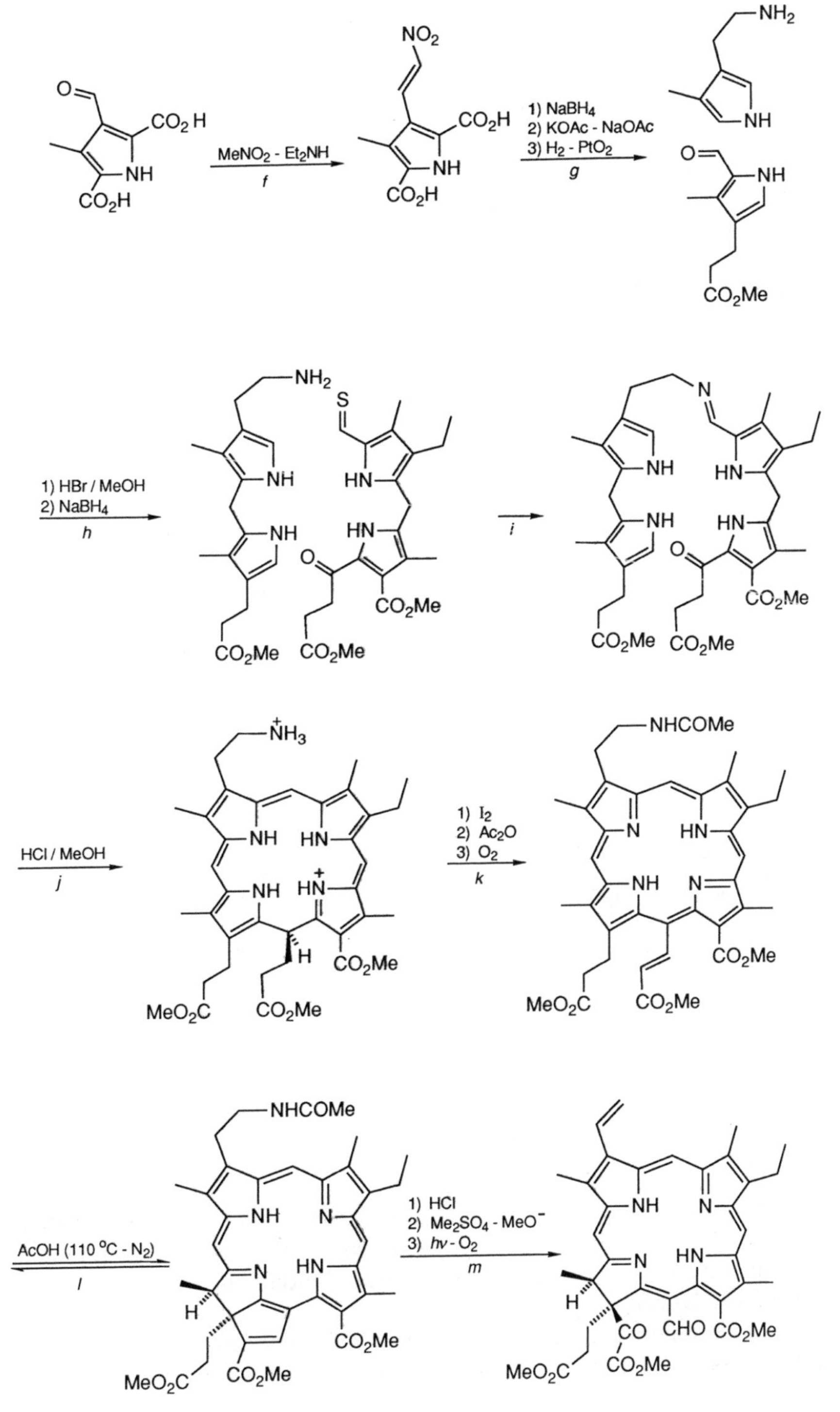
MeNO2 - Et2NH
f
1) NaBH4
2) KOAc - NaOAc
3) H2 - PtO2
g
1) HBr / MeOH
2) NaBH4
h
i
HCl / MeOH
j
1) I2
2) Ac2O
3) O2
k
AcOH (110 °C - N2)
l
1) HCl
2) Me2SO4 - MeO⁻
3) hν - O2
m
NHCOMe
CO2Me
MeO2C
CHO

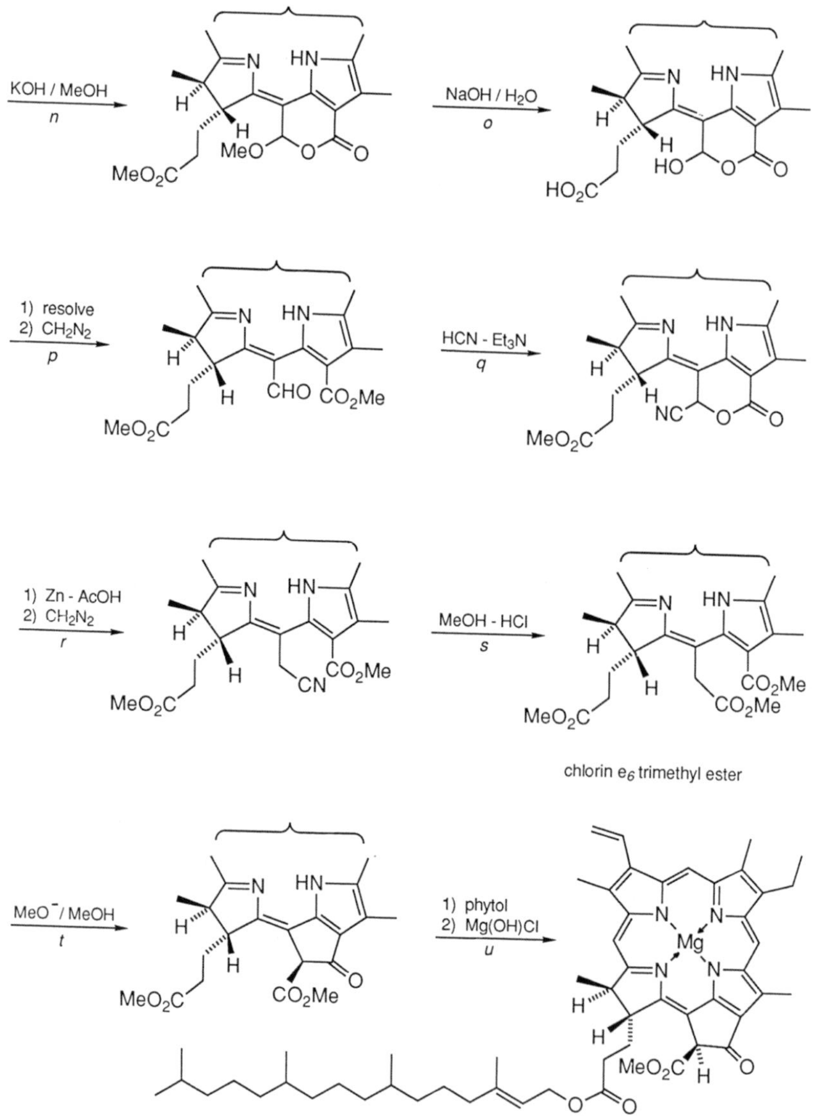

chlorin e_6 trimethyl ester

chlorophyll-a

f   Base-catalyzed reaction involving the activated methyl group of nitromethane (p. 222).

g   The C=C bond is activated by the nitro group to nucleophiles and is reduced by borohydride (p. 638). Fusion of the product with a mixture of sodium and potassium acetates at 120°C effects decarboxylation. The nitro group is then reduced catalytically.

h   Hydrogen bromide catalyzes the Friedel–Crafts reaction which joins the two pyrrole rings; reaction occurs at the 2-position shown rather than at the alternative 2-position as a result of the steric effect. The dipyrrylmethanol is readily reduced by borohydride.

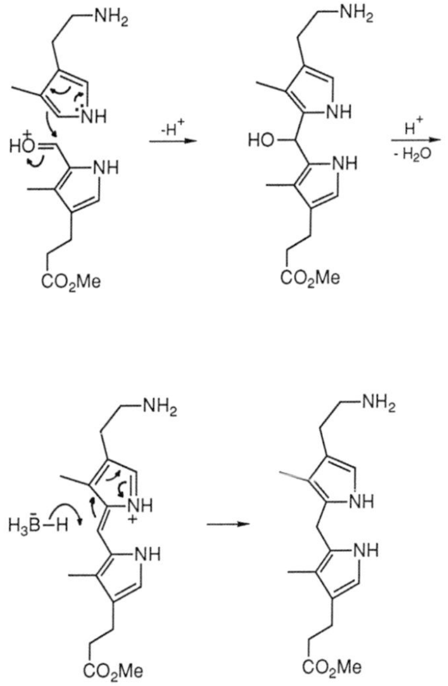

The dipyrrylmethane formed is a labile compound which is extracted into dichloromethane and used immediately in step *i*.

i   Formation of the imine (see *e*).

j   12M Hydrogen chloride in methanol brings about a series of transformations. The bridges are formed as follows:

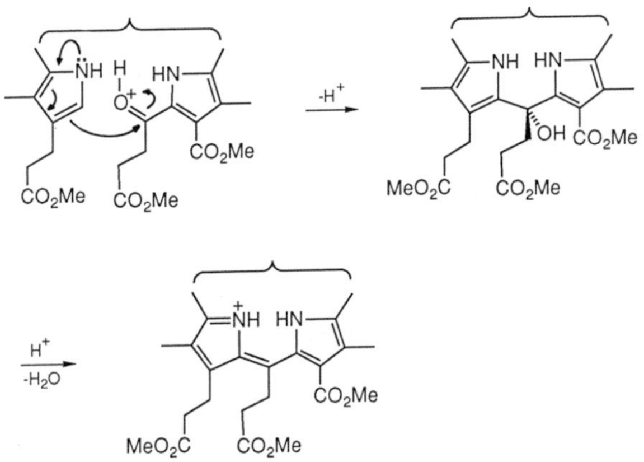

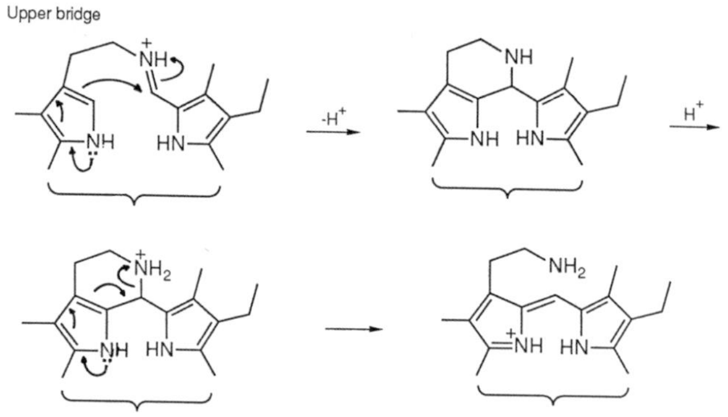

Upper bridge

There is spectroscopic evidence for the occurrence of the resulting intermediate and for those which succeed it:

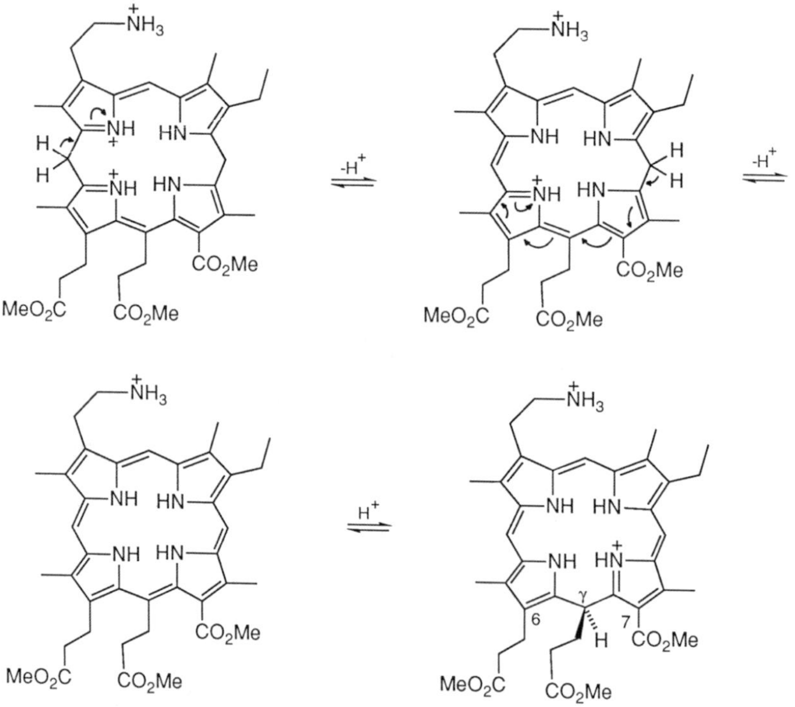

Of the various cations which exist in equilibrium, all save the last contain strong steric compression forces owing to the eclipsing of the three substituents at C_6, C_γ and C_7. In the last cation, however (only one enantiomer is shown), the tetrahedral nature of C_γ allows these substituents to assume a staggered relationship to one another, and as a result this is the most stable of the cations. It is not necessary to separate the cations.

k   An excess of iodine is added to the reaction mixture to remove by oxidation two nuclear hydrogen atoms; the amino group is acetylated by the standard procedure (acetic anhydride in pyridine solution), and the C_γ-side-chain is dehydrogenated in warm acetic acid in the presence of air.

l   The temperature of the acetic acid solution is raised to 110°C and air is excluded to prevent further oxidation. These conditions establish equilibrium with the dihydroporphyrin (a purpurin), for which $K \sim 2$; the product is isolated and the reactant is recycled to increase the yield. The reaction involves electrophilic attack by the activated C=C bond at the 3-position of the pyrrole nucleus:

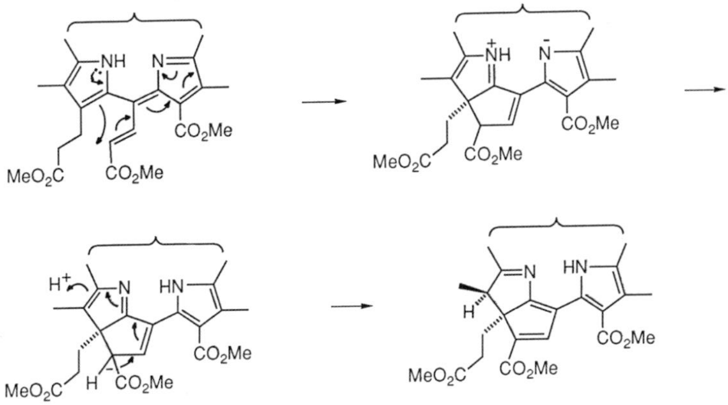

A further important aspect of the reaction is that the uptake of the proton at the last stage is from that direction which results in the methyl and methoxycarbonylethyl groups in ring IV being *trans* to one other, for the alternative product is more severely strained.

m   Hydrogen chloride in methanol removes the protective acetyl group and the resulting amine is submitted to Hofmann elimination yielding the required vinyl-substituent. Strong illumination with visible light in the presence of air then oxidatively cleaves the five-membered ring formed in *l*, the driving force for the reaction being derived partly from the release of the ring strain.

n   The substituent at C_7 is removed with base. This reaction is no doubt facilitated by the presence of the electron-accepting nitrogen atom in ring II, i.e.

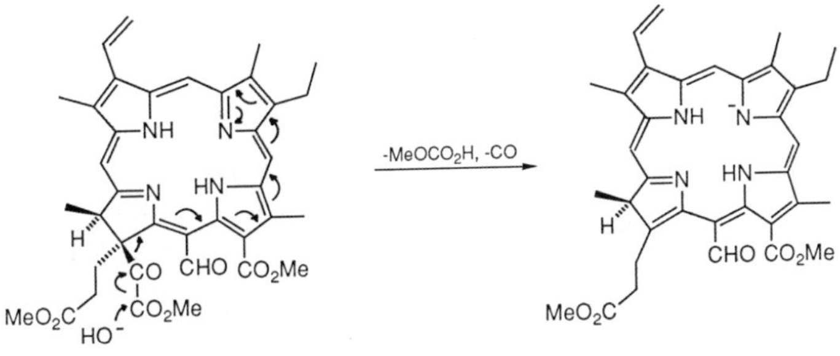

Again, the final protonation occurs so as to give the more stable stereoisomer (see *l*).

In addition, the base induces lactonization across C_6 and C_γ:

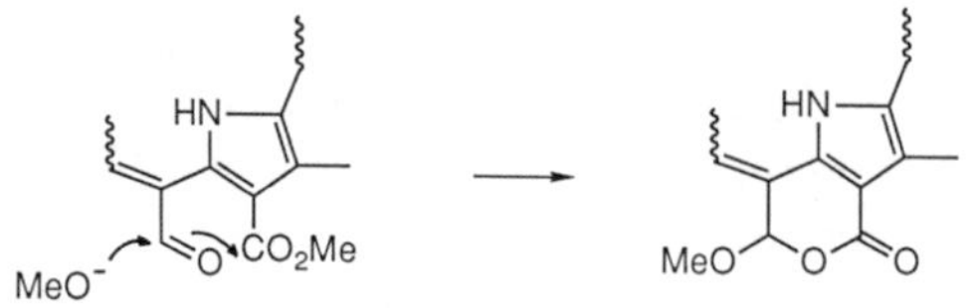

o   Very dilute sodium hydroxide solution hydrolyzes the methyl ester and, by the reverse of the lactonization in step *n* and re-lactonization, gives the hydroxy-containing lactone.

p   The optical isomers are resolved (the diastereomers formed with quinine were used). Diazomethane methylates the carboxylic acid and also opens the lactone by reacting with the open-chain tautomer:

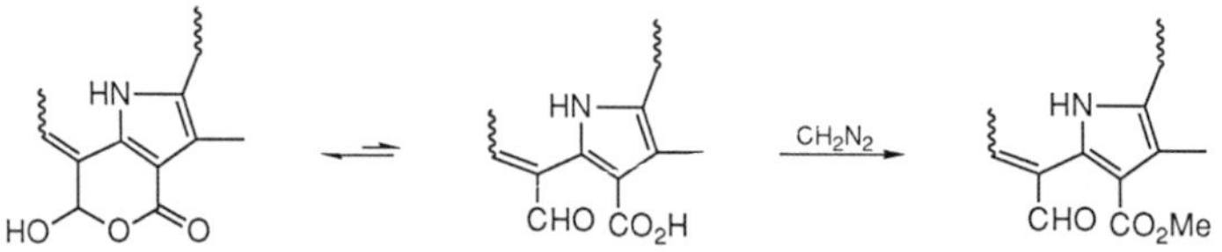

q   Cyanide ion forms a cyano-lactone in the manner of the methoxy-lactonization in step *n*.

r   The lactone is opened reductively and the resulting acid is methylated.

s   Methanolic hydrogen chloride converts the nitrile into the methyl ester.

t   Dieckmann cyclization to give a five-membered cyclic β-keto-ester (p. 228).

u   The phytyl residue is introduced in place of methyl (alcoholysis) at the less hindered of the two ester groups and the magnesium atom is inserted.

22.5 Prostaglandins E_2 and $F_{2\alpha}$

The prostaglandins are a series of closely related hormones which are derivatives of 'prostanoic acid':

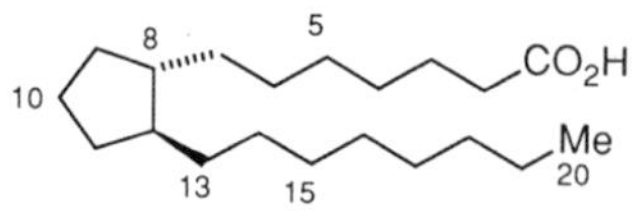

They are present in many mammalian tissues at very low concentrations: one of the richest known sources is human seminal plasma which contains at least 13 different prostaglandins at a total concentration of about $300\,\mu g\,cm^{-3}$. Prostaglandins have potent effects on various kinds of smooth muscle and they are of considerable medical interest for the control of hypertension, for gynaecological purposes, and as abortifacients.

Prostaglandins E_2 and $F_{2\alpha}$ are two of the six primary prostaglandins. In general, the E series of prostaglandins have a β-hydroxy-ketone structure in the ring and differ in the degree of unsaturation in the side-chains, whereas F series prostaglandins have a β-dihydroxy function in the ring and likewise differ in the extent of unsaturation in the side-chains.

PGE$_2$ PGF$_{2\alpha}$

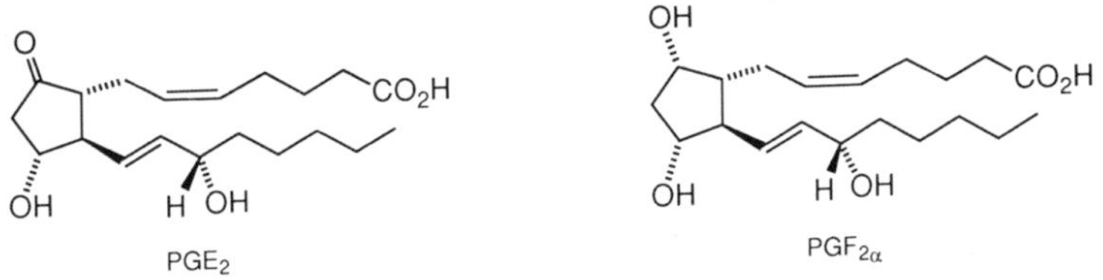

The synthesis of E series prostaglandins poses a particular problem in that the β-hydroxy-ketone function is especially prone to dehydration (p. 96). (Such dehydration gives rise to series A prostaglandins which readily rearrange to series B prostaglandins which are fully conjugated and have the ring double bond between carbon atoms 8 and 12.) Further challenges in the syntheses include the control of C=C diastereoisomerism in the side-chains and of the configuration at C$_{15}$. The synthesis given below is that of E. J. Corey and co-workers; the evolution of the synthesis from its original form to that of 1975 is discussed (*J. Amer. Chem. Soc.*, 1969, **91**, 5675; *ibid.*, 1971, **93**, 1489; 1491; *ibid.*, 1972, **94**, 8616; *ibid.*, 1975, **97**, 6908).

Corey's strategy was based on the retrosynthetic disconnections:

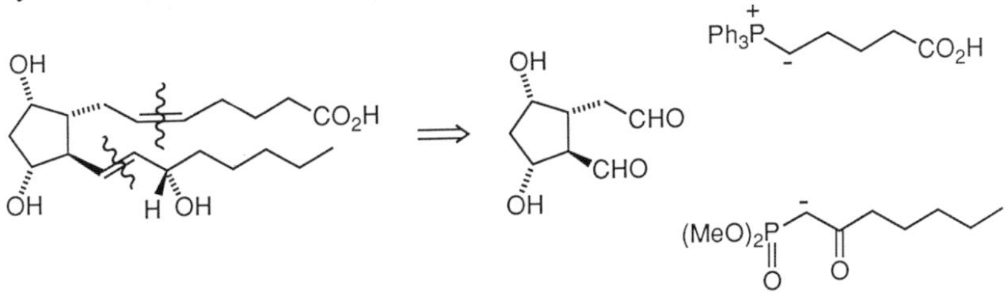

Both the C=C bonds would be introduced by Wittig-type reactions, using a non-stabilized phosphorane to generate the (Z)-alkene and a stabilized phosphorane (Wadsworth–Emmons variant) to generate the (E)-alkene (p. 462). The carbonyl group in the latter would be the source of the required hydroxyl group at C$_{15}$. However, it would be necessary to protect one of the aldehyde groups to ensure that the first Wittig reagent reacted at the appropriate position; release of the protected aldehyde would then be followed by introduction of the second Wittig reagent. The protective method chosen was to introduce the aldehydic group at C$_7$ as an acid which would form a lactone with the C$_9$ hydroxyl; reduction would then yield a hemiacetal which would be in equilibrium with the required aldehyde:

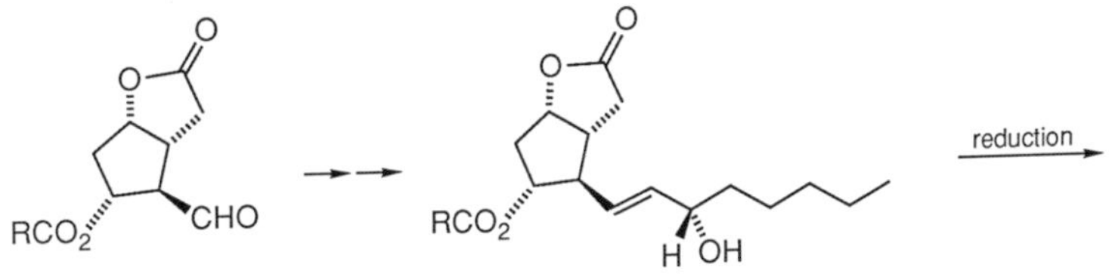

The synthesis is as follows:

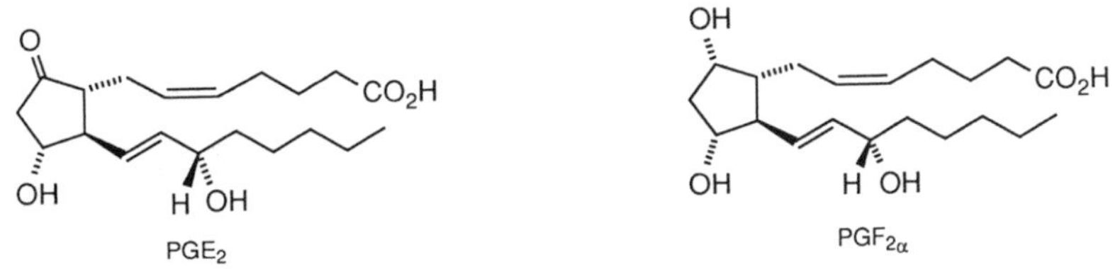

PGE$_2$

PGF$_{2\alpha}$

a   The cyclopentadienyl anion is alkylated at low temperature with a chloro-methyl ether. The original version of the synthesis used the sodium salt of cyclopentadiene and chloromethyl methyl ether. However, although the reaction worked adequately on a small scale, a considerable disadvantage was that unchanged sodium cyclopentadienide was sufficiently basic to catalyze isomerization of the product even at 0°C and the longer manipulation times necessary for work on a larger scale caused this isomerization to be troublesome:

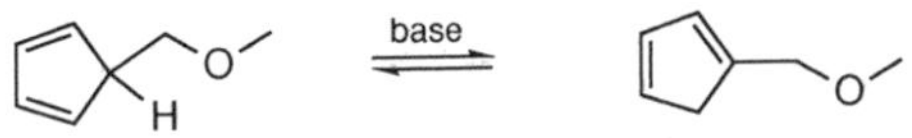

The problem was overcome by use of the thallium(I) derivative of cyclopen-tadiene, prepared from the diene with Tl$_2$SO$_4$ and KOH in water. An additional advantage was that the thallium(I) cyclopentadienide was capable of manipulation in air. The success deriving from the use of thallium is presumably attributable to the thallium derivative's having more covalent character than the sodium derivative, so that the cyclopentadienyl moiety is correspondingly less basic. Subsequent syntheses used benzyl chloromethyl ether as the alkylating agent, which had the advantage that subsequent debenzylation is more easily accomplished than demethylation (see step *e*).

b   This step was one which underwent the most change during development of the synthesis. Essentially it is a cycloaddition reaction. However, the apparent dienophile in the reaction is ketene. As was seen earlier (p. 282), this participates more readily in $[_\pi 2 + _\pi 2]$ than in $[_\pi 4 + _\pi 2]$ cycloadditions, so that a synthetic equivalent had to be found. The first dienophile used was 2-chloroacrylonitrile:

This dienophile was found to be insufficiently reactive in the $[_\pi 4 + _\pi 2]$ mode at temperatures where the isomerization of the precursor (this time by 1,5-sigmatropic rearrangement) did not occur but the elegant device of catalyzing the Diels–Alder reaction to give it advantage over the competitive pericyclic reaction was successful. The best catalyst was Cu(BF$_4$)$_2$; it brought about the desired coupling in tetrahydrofuran at −55°C in yields in excess of 80%.

The product of the reaction is a mixture of stereoisomers (both Cl and CN *endo*, and the side-chain mainly *syn* to the double bond) which is efficiently hydrolyzed to the required bicyclic ketone by KOH in aqueous dimethyl sulfoxide.

The second variant in accomplishing step *b* involved a change of dienophile, for 2-chloroacrylonitrile is an unusually hazardous material, undesirable for large-scale work. A preferable alternative was found to be 2-chloroacryloyl chloride. This dienophile is of higher reactivity than its predecessor, so that no catalyst is necessary:

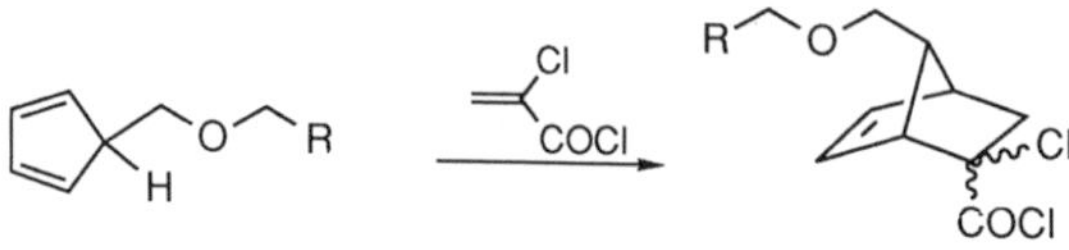

Again the product is a mixture of stereoisomers in respect of the *endo/exo* positions of Cl and COCl, but the side-chain is now exclusively *syn* to the double bond. The conversion into the bicyclic ketone is brought about by Curtius rearrangement (p. 443) and hydrolysis:

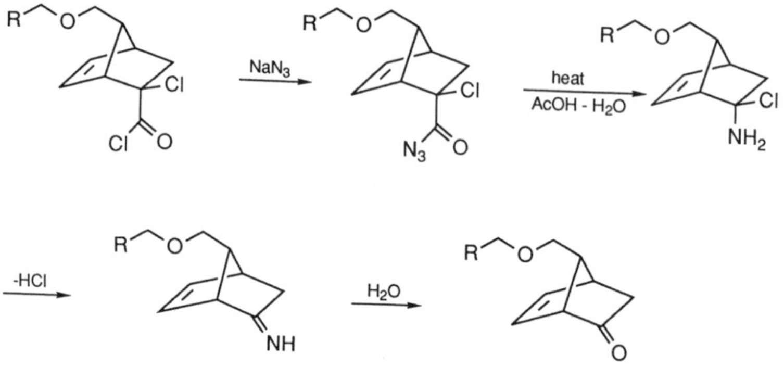

Although this variant of the reaction involves more formal steps than its predecessor, the conversion of cyclopentadiene into the bicyclic ketone was accomplished in greater than 90% yield, without isolation of intermediates.

The bicyclic ketone contains three asymmetric centres and the relative stereochemistry at these is governed by the suprafacial character of the Diels–Alder reaction on the less hindered side of the diene so that the product is a single diastereomer at this stage; it is, however, racemic. The early variants of the synthesis included a resolution step later in the sequence (step *d*); however, this 'wastes' half the material so far synthesized. The final variant of the synthesis overcame this disadvantage and involved an asymmetric synthesis of the bicyclic ketone. In order to achieve this, the dienophile for reaction with 5-benzyloxymethylcyclopentadiene was changed again, this time to an acrylate ester:

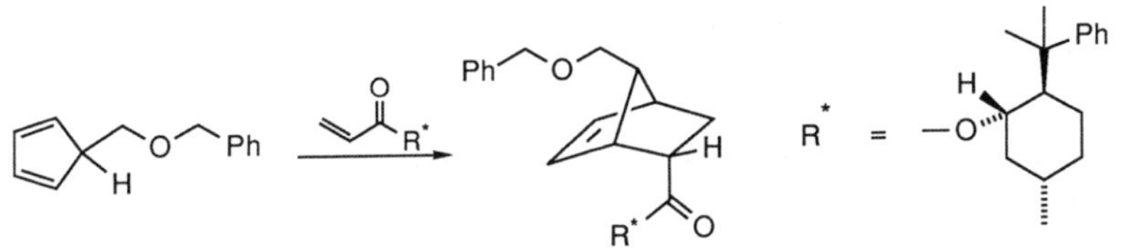

The optically pure asymmetric alcohol R*H was converted into its acrylate ester by reaction with acryloyl chloride in the presence of triethylamine. Diels–Alder reaction of the optically active acrylate, in the presence of aluminium chloride as catalyst, in dichloromethane at −55°C gave the bicyclic *endo* ester in 89% yield. This adduct contains four stereocentres in the acyl moiety. As before, the ring junction diastereoisomerism is determined by the stereochemistry of cycloaddition, whilst the asymmetry induced in the product depends upon that present in R*. The (+)-acrylate gave the stereochemistry required for the synthesis.

Conversion of the bicyclic ester into the bicyclic ketone now required new conditions:

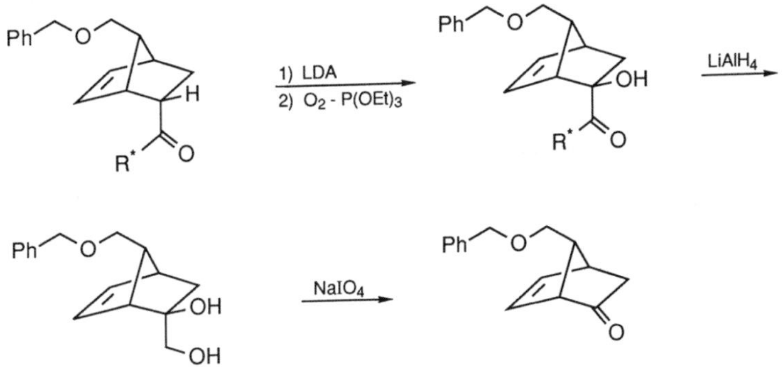

Treatment of the optically active bicyclic ester with lithium diisopropylamide generates its enolate anion which is autoxidized by molecular oxygen. The hydroperoxide so formed is reduced by triethyl phosphite, *in situ*, to the hydroxy-ester. Note that, although the product at this stage is a mixture of diastereomers, only one enantiomer of each is present. Reduction of the mixture of hydroxy-esters gives the asymmetric alcohol, R*H (optically pure and suitable for recycling), and a bicyclic 1,2-diol which is cleaved to give the bicyclic ketone as its required laevorotatory enantiomer.

In this synthesis, therefore, the key prostaglandin intermediate, the bicyclic ketone, is synthesized optically pure without resolution and the chiral agent used to achieve the asymmetric synthesis can be recovered, unaffected, suitable for re-use.

c   Baeyer–Villiger oxidation (p. 445) of the bicyclic ketone gives a single ring-expanded lactone resulting from migration of secondary carbon in preference to the alternative primary carbon.

d   The lactone is hydrolyzed with aqueous sodium hydroxide and the free acid is obtained by neutralization with carbon dioxide. (In syntheses which had not been asymmetric a resolution step was then included by crystallization of the (+)-amphetamine salt.) Finally, iodolactonization is achieved by treatment of the acid with iodine in aqueous KI solution (cf. Prévost reaction, p. 593):

Preference for γ- over δ-lactone formation ensures a unique product now containing five asymmetric centres.

e   The hydroxy group is protected by acylation. Originally acetylation was used but, later, acylation with 4-phenylbenzoyl chloride was found advantageous in that materials were more consistently crystalline and the aromatic chromophore facilitated location of the products during chromatography. A further important benefit was to accrue later (step h).

    Deiodination is effected with tributyltin hydride, the radical-mediated reaction being initiated with azobisisobutyronitrile (p. 538). The protecting benzyl group is removed by hydrogenolysis (p. 643) (where methyl was present in place of benzyl, it was removed with BBr$_3$).

f   The hydroxymethyl group is oxidized to aldehyde with the pyridine complex of chromium(VI) oxide in dichloromethane (p. 609).

g   The aldehyde couples with the anion of dimethyl 2-oxoheptyl phosphonate (a variant of the Wadsworth–Emmons reaction, p. 460) to give the trans C=C of the C$_{12}$—C$_{20}$ side-chain.

h   Prostaglandins E$_2$ and F$_{2\alpha}$ have the (S)-configuration at C$_{15}$. The evolution of synthetic procedures to build in this configuration enantioselectively is one of the most interesting aspects of this synthesis. The first synthetic sequence used Zn(BH$_4$)$_2$ as reducing agent. This hydride-transfer reducing agent was preferable to more common alternatives in that the amount of concomitant reduction of the double bond conjugated with the ketone grouping was minimized. However, there was no stereoselectivity in reduction of the carbonyl group, since other stereocentres in the molecule were too distant to influence the course of reaction when the new stereocentre at C$_{15}$ was created. In an attempt to introduce stereoselectivity in this reduction, various hindered borohydrides were synthesized and used. The most effective one was obtained by reduction of the trialkylborane derived from thexylborane (p. 484) and limonene, and used in tetrahydrofuran at $-120°C$ in the presence of HMPA as catalytic Lewis base. With this combination, the desired α-alcohol predominated over the β-isomer to the extent of 4.5:1. It was noted, however, that the $\alpha : \beta$ ratio varies with the nature of the acyl group used to protect the ring-hydroxyl group: the ratio was greater in the desired sense when an elongated rigid function such as 4-phenylbenzoyl was used. This was rationalized as follows. Molecular models showed that the 4-phenylbenzoyl group could lie alongside the -enone group in the aliphatic side-chain, possibly being weakly bonded to it by $\pi-\pi$ interaction,

Moreover, the models showed also that the interaction could be more extensive when the -enone function is in the s-cis conformation, i.e.

When the -enone is reduced in the s-cis conformation from the side opposite to the biphenyl moiety, alcohol of the desired stereochemistry results.

Structural variations were made in the protecting function to improve further the stereoselectivity of reduction. Eventually, by using biphenyl-4-isocyanate in place of the acyl halide as the protecting agent, the resultant urethane could be reduced with a stereoselectivity of 92% α-alcohol and 8% β-alcohol at C_{15}.

The biphenyl moiety is therefore a stereochemical control element as well as having the other advantages listed under step e.

i   Alkaline hydrolysis removes the protecting group, whether ester or urethane, and the resultant diol is treated with dihydropyran in dichloromethane containing toluene-p-sulfonic acid. This protects both hydroxyl groups as 2-tetrahydropyranyl derivatives (p. 709). The lactone ring is then reduced to a hemiacetal with diisobutylaluminium hydride in toluene. A hemiacetal is a masked aldehyde:

j   Wittig reaction on the masked aldehyde in DMSO, with the dimsyl anion as base to generate the phosophorus ylide, gives the *cis*-alkene (p. 462).

k,l   The prostanoid material so formed is the immediate precursor of both prostaglandins E_2 and $F_{2\alpha}$. Treatment with aqueous acetic acid hydrolyzes the pyranyl protecting groups to give prostaglandin $F_{2\alpha}$ (step *k*), whilst oxidation of the unprotected hydroxyl group by chromium(VI) oxide followed by similar deprotection (step *l*) yields prostaglandin E_2.

22.6
Ibogamine

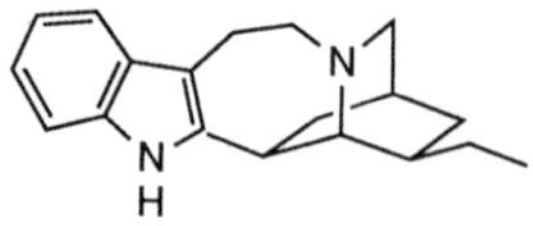

Ibogamine is an alkaloid in the oboga family which contains the clinically import-ant antitumour alkaloid vinblastine. This provides a stimulus for efficient synthetic approaches to this family of compounds. The following is a short, stereocontrolled synthesis of ibogamine reported by B. M. Trost *et al.*, *J. Amer. Chem. Soc.*, 1978, **100**, 3930, which makes use of a silver-palladium catalyzed alkene cyclization.

A Diels–Alder transformation provides the obvious approach to establishing the bicyclic ring with its required stereochemistry. The one chosen is depicted in the final product by the thickened bonds,

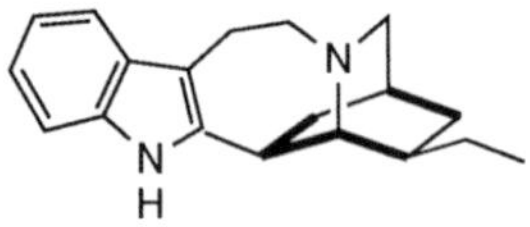

Five other such approaches are revealed by disconnection, some of which have been used in other syntheses.

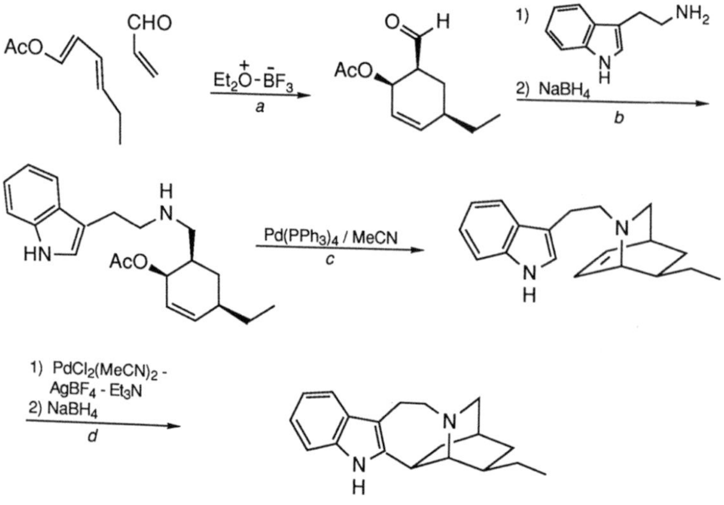

a   The regio- and stereochemical principles governing the Diels–Alder reaction (p. 272) result in all three substituents in the six-membered ring being *cis*. This is the required relationship for the ethyl and aldehyde groups (the latter will eventually form part of the nitrogen bridge), but is immaterial for the acetoxy group which is later to be removed. Complexation of the aldehyde oxygen (the more basic of the two) with BF_3, thereby lowering the energy of the dienophile's LUMO (p. 274), enables reaction to be carried out at $-10°C$, increasing the regio- and stereoselectivity.

b   Reaction to give the imine, followed by its reduction to the amine (p. 669).

c   The Pd(o) compound gives a π-allyl complex, with loss of acetate ion, which reacts with the nucleophilic nitrogen atom of the amino group (p. 580).

d   It is believed that $PdCl_2$ effects electrophilic substitution of the indole ring, its reactivity being enhanced by silver ion, and that intramolecular addition to the alkene follows:

where M is either a silver-palladium salt complex or a partially ionized palladium salt. The palladium substituent is removed by reduction ($NaBH_4$).

   In an enantioselective elaboration of the synthesis, the optically active diene

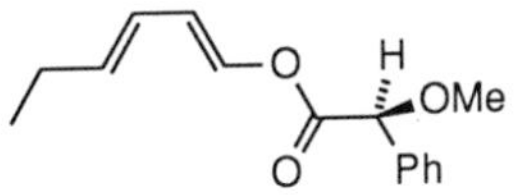

was used as starting material. This resulted ultimately in the formation of (+)- and (−)-ibogamine in the ratio 4:1 (i.e. an enantiomeric excess of $[(4 - 1)/5 \times 100] = 60\%$).

**22.7
11-Oxygenated
progesterones**

(*a*)   *11-Oxoprogesterone:*

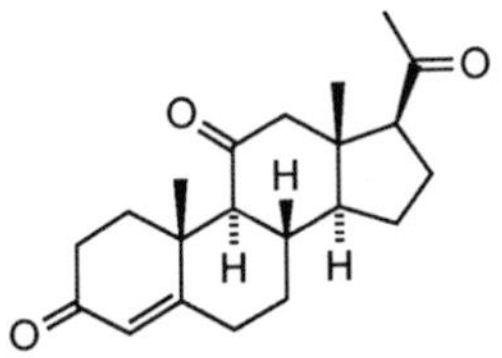

Steroids have played a central role in understanding the effect of conformation and configuration on reactivity and, because of the importance of these compounds as hormones and pharmaceutical products, it is not surprising that this framework has a central place in the development of regio- and stereo-chemically controlled synthetic methods. The importance of corticosteroids has prompted many attempts to synthesize 11-oxygenated steroids and two contrasting approaches to give closely related compounds (11-oxo- and 11α-hydroxyprogesterone) will be described. The first is an elegant methodology pioneered by G. Stork *et al.*, *J. Amer. Chem. Soc.*, 1981, **103**, 4949 and 1982, **104**, 3758, for the preparation of 11-keto steroids:

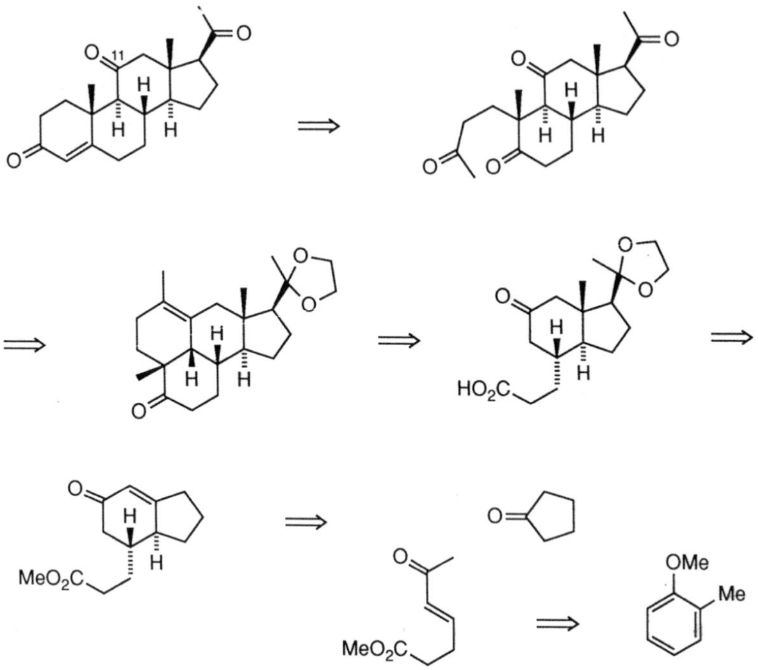

The following is the route:

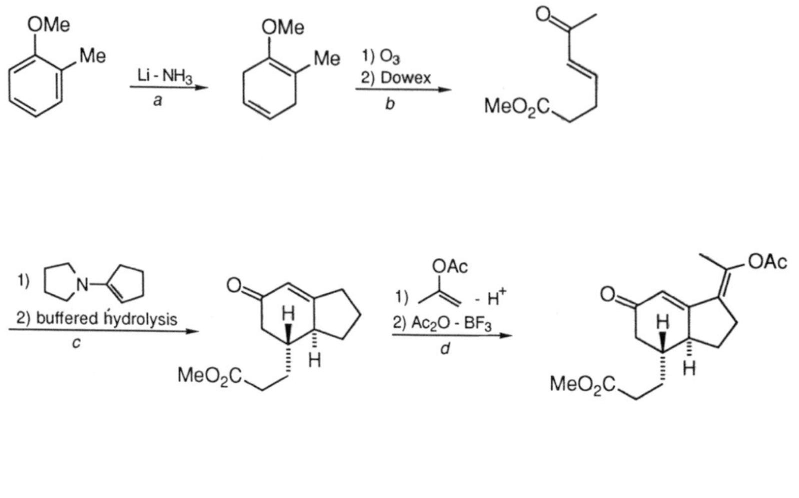

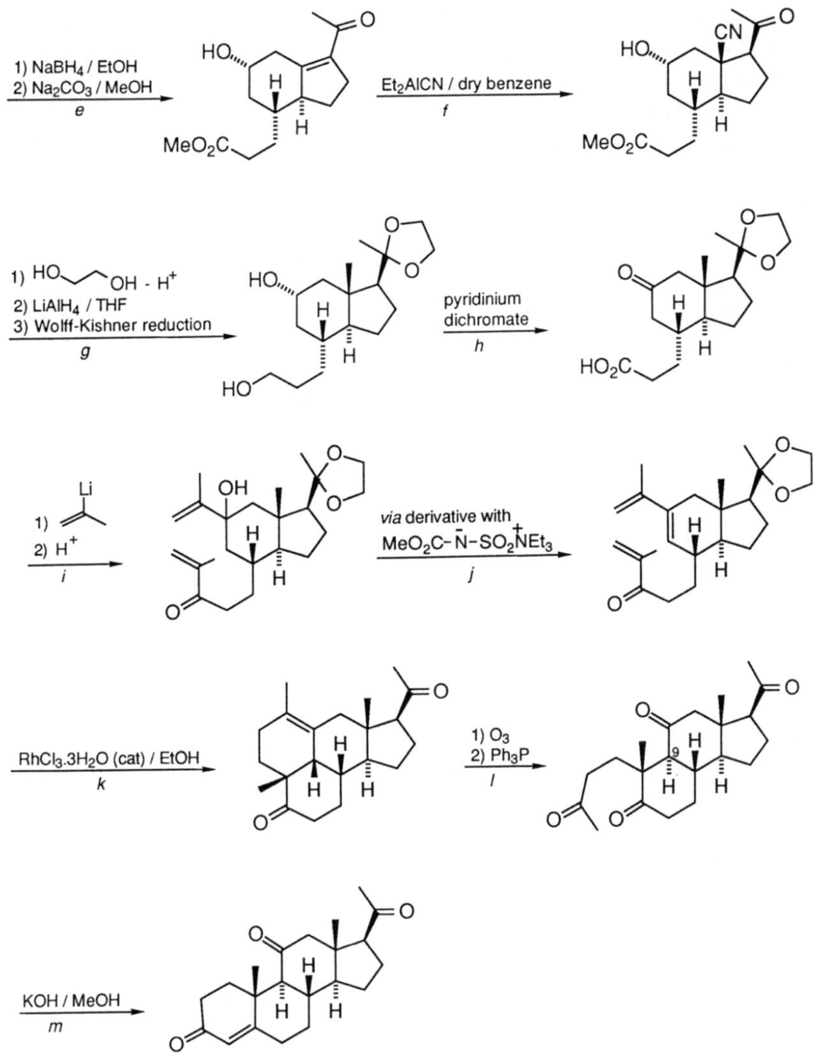

a   Reduction with lithium in liquid ammonia gives the 1,4-adduct and is directed by the strongly electron-releasing methoxyl group (p. 641).

b   Ozonolysis occurs more readily at the electron-rich, more substituted, double bond (p. 593), and refluxing over Dowex (a resin that is available in both acidic and basic forms) converts the product into the more stable *trans*-αβ-unsaturated ketone:

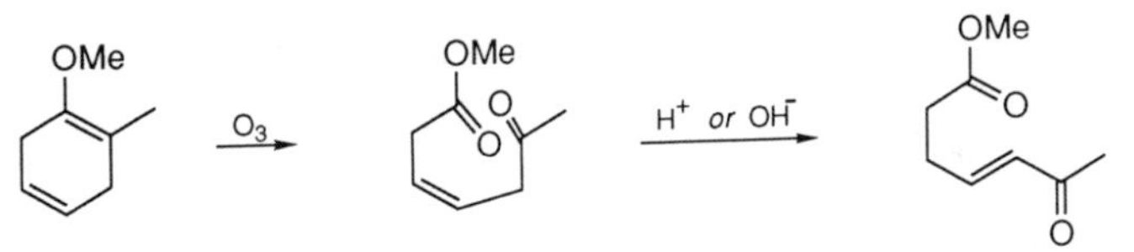

c   Michael addition by the enamine from cyclopentanone and pyrrolidine occurs through the least crowded transition state to give *trans* product as a racemate (only one enantiomer is shown). A Mannich electrophile is thereby formed which reacts with the enol tautomer of the ketone to give the bicyclic product. Careful hydrolysis (buffered solution) removes pyrrolidine but leaves the ester intact:

transition state

d   Acid-catalyzed ester-exchange with isopropenyl acetate gives the dienol acetate,

which is activated to acetylation by acetic anhydride–boron trifluoride (p. 259). The acetoxy group is removed during this process, probably as follows, and the enol formed by acid-catalyzed tautomerism of the methyl ketone is acetylated.

e   Borohydride ion approaches from the upper, less hindered, side of the molecule. Hydrolysis of the enol acetate with base regenerates the methyl ketone and the base also catalyzes a prototropic shift to give the conjugated ketone.

f   Michael-type addition of cyanide ion (p. 245) is also sterically favoured from the upper side.

g   The keto group is protected so that only the nitrile group is reduced by LiAlH$_4$. This yields the imine which, under Wolff–Kishner conditions (p. 649), is reduced to methyl.

h   Pyridinium dichromate in DMF oxidizes primary alcohols to carboxylic acids and secondary alcohols to ketones.

i   Addition of an excess of lithiumisopropenyl (from 2-bromopropene with a suspension of lithium and 2% sodium in diethyl ether at −78 °C), followed by quenching with acetic acid, introduces isopropenyl substituents both by addition to C=O and by substitution at —CO$_2$H (p. 199).

j   The sequence

provides very mild conditions for *syn*-elimination from an alcohol; in this case the alcohol derivative yields alkene at room temperature in THF. Some of the alternative conjugated diene is also formed, but it is isomerized to the desired diene in the next step.

k   RhCl$_3 \cdot$ 3H$_2$O catalyzes the isomerization of alkenes and, by carrying out the Diels–Alder cycloaddition in refluxing ethanol containing a trace of the rhodium salt, the diene mixture from step *j* is all converted into the cyclic product; the protecting group is hydrolyzed under these conditions. Two stereocentres are generated in the cycloaddition, so that four stereoisomers are possible. However, only two are likely, because approach of the enone side-chain is sterically easier from the underside of the molecule than from the top, where the angular methyl group impedes it. Formation of the required product is consistent with the preference for *endo*-addition in Diels–Alder reactions with conjugated dienophiles (p. 273).

l   Reaction with ozone in dichloromethane and methanol at −78°C and treatment of the ozonide with triphenylphosphine give the tetraketone. Epimerization at C$_9$ occurs *via* the enol, catalyzed by traces of acid, to give the required, thermodynamically more stable, product.

m There are two possible aldol adducts: that which gives the required αβ-unsaturated ketone on dehydration and a cyclopentanol derivative formed by reaction of the enolate at C_{11}. The latter is formed the faster, and can be obtained quantitatively by treating the triketone with 4% methanolic potassium hydroxide at a low temperature. However, its base-induced dehydration occurs much less rapidly than that of the alternative adduct. Advantage is taken of this, coupled with the reversibility of the aldol addition: by using more concentrated base at 40°C, the required six-membered product is obtained.

(*b*) *11α-Hydroxyprogesterone:*

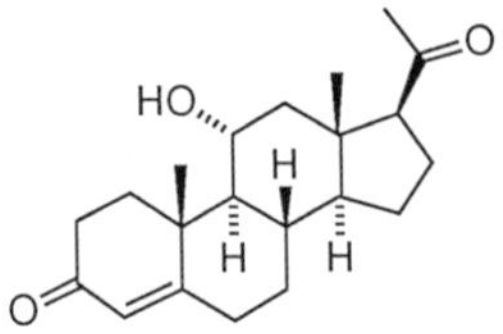

The (+)-enantiomer of 11α-hydroxyprogesterone is a key intermediate in the commercial production of hydrocortisone acetate. The following stereospecific polycyclization of a polyenic substrate to give racemic 11α-hydroxyprogesterone was carried out by W. S. Johnson *et al.* (*J. Amer. Chem. Soc.*, 1976, **98**, 1039). The development of polyene cyclizations, now recognised as biomimetic, introduced a new approach for the synthesis of multi-ring substrates. In this sequence the polyenic substrate was produced by convergent synthesis, the connecting step being the addition of the lithium salt of an alkyne to an aldehyde to give an alkynic alcohol. A retrosynthetic analysis of the synthesis follows:

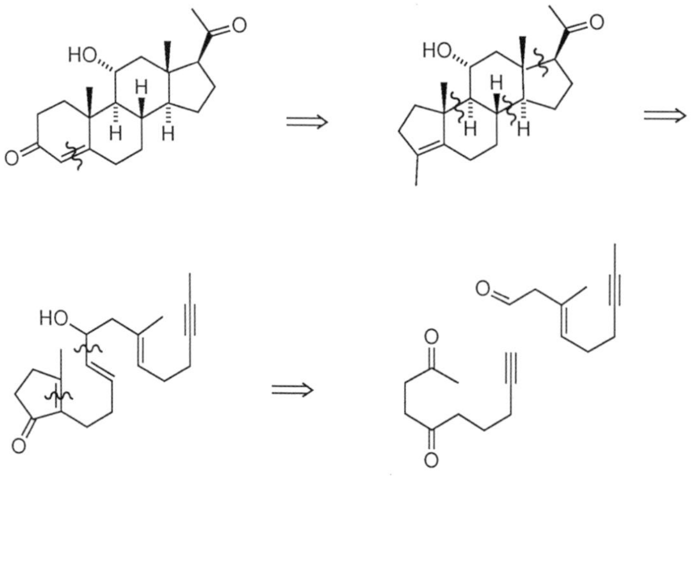

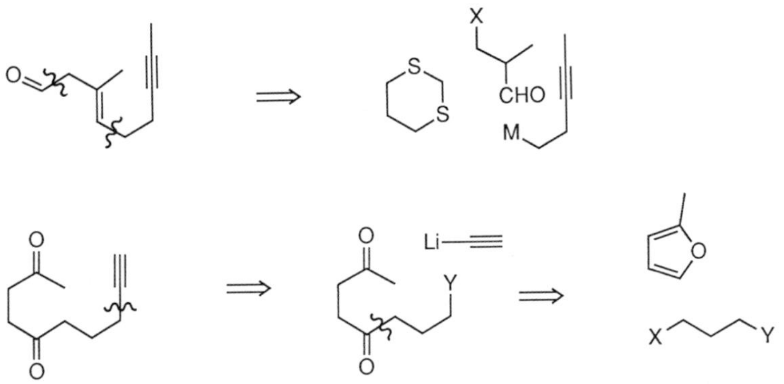

The synthesis gave a 15% overall yield of 11α-hydroxyprogesterone in 16 steps from simple compounds, corresponding to a 7% overall (22 steps) yield of racemic hydrocortisone acetate. The stereoselectivity of the cyclization portends well for this methodology.

The synthesis is as follows:

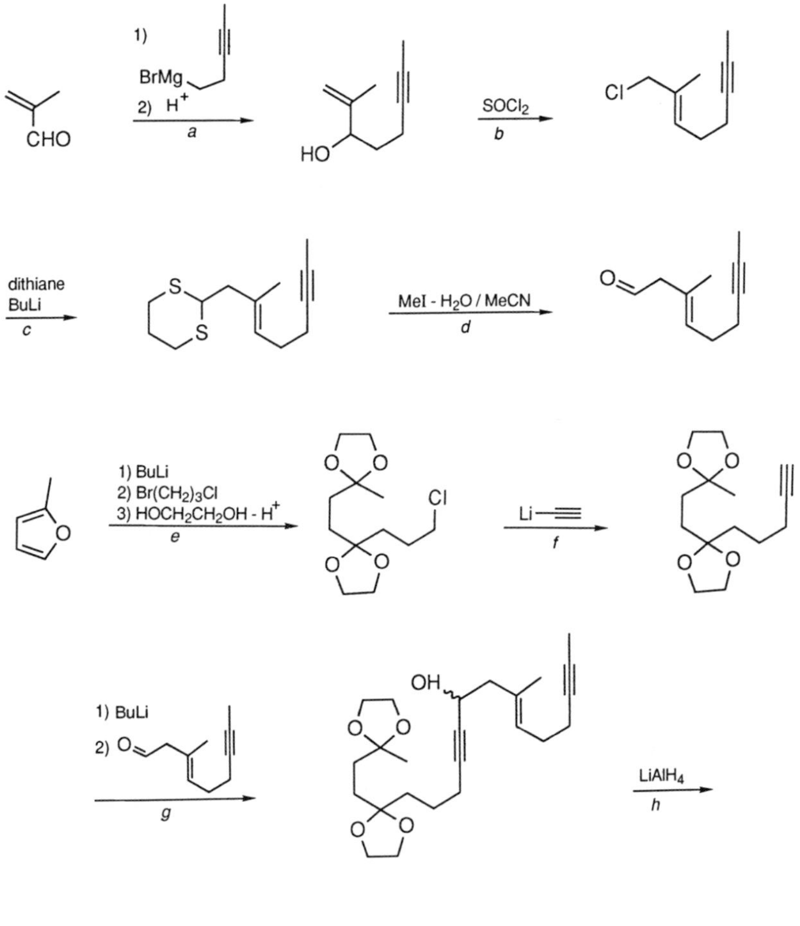

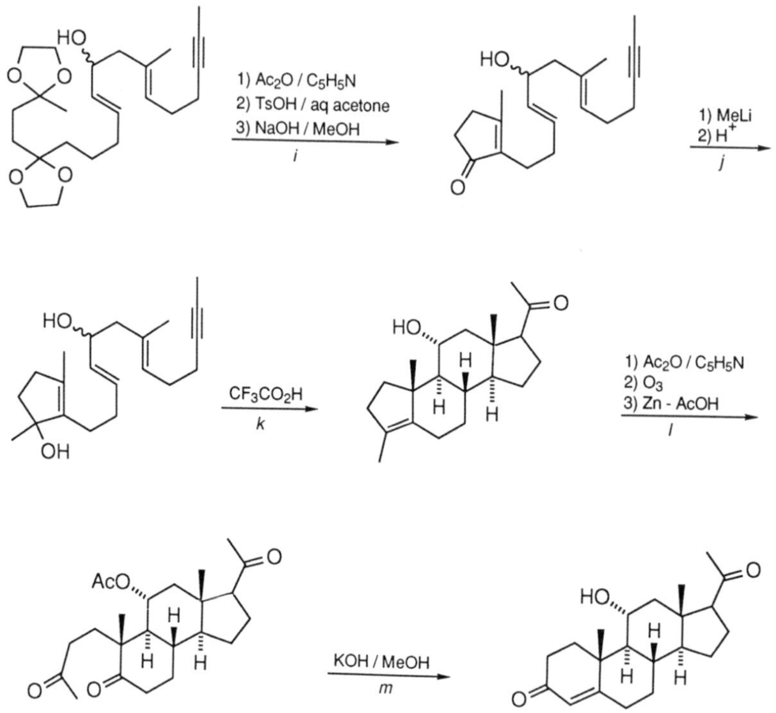

a   A standard Grignard reaction.

b   Conversion of —OH to —Cl is accompanied by a shift of the double bond (S$_N$2′ reaction; p. 108).

c   Displacement of chloride by the carbanion from dithiane (p. 475).

d   The usual method for hydrolysis of a thioacetal to the aldehyde involves mercury(II) ion as catalyst (p. 475), but this also catalyzes the hydration of alkynes (p. 91). A newer, promising method was chosen: methyl iodide forms a sulfonium cation to provide a good leaving group in a nucleophilic displacement by water,

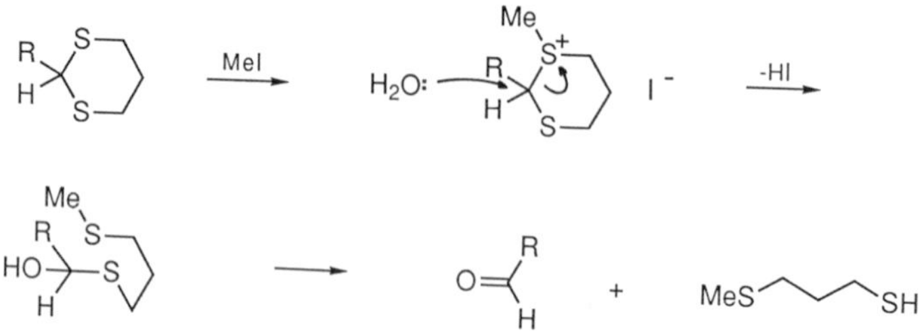

Calcium carbonate is included to absorb the HI formed.

e   Metallation of the furan occurs at the 5-position (p. 683) and the organometallic derivative displaces bromide in preference to chloride (p. 103). Acid-catalyzed ring opening of the furan, with formation of the diketal, follows essentially the reverse pathway to that in the formation of furans by dehydration of 1,4-dicarbonyl compounds (p. 692).

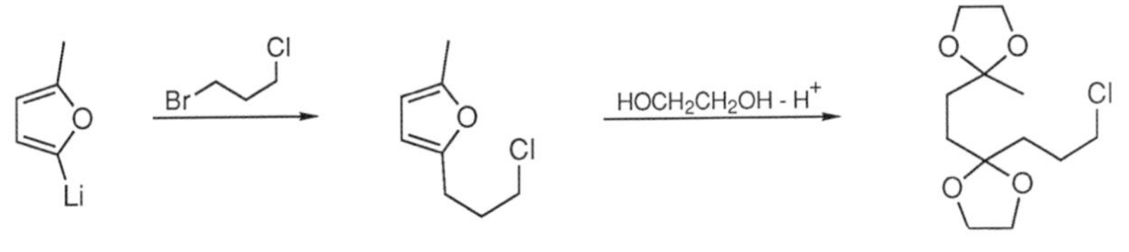

f Lithium acetylide was used as its complex with ethylenediamine in DMSO.

g Formation of the alkynyllithium (p. 185) is followed by its reaction at the carbonyl group.

h Only the α-hydroxyalkyne is reduced by LiAlH$_4$ (p. 640).

i After acetylation of —OH, the ketal residues are removed with acid, and aldol reaction and hydrolysis of the acetate are induced with base:

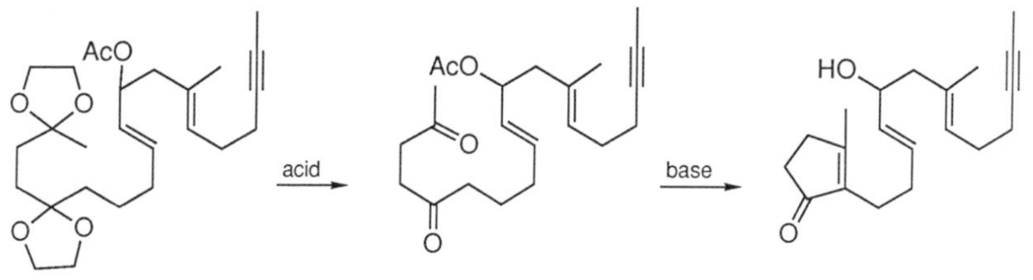

j The product is not stable and is therefore submitted to cyclization without purification.

k Acid-induced cyclization,

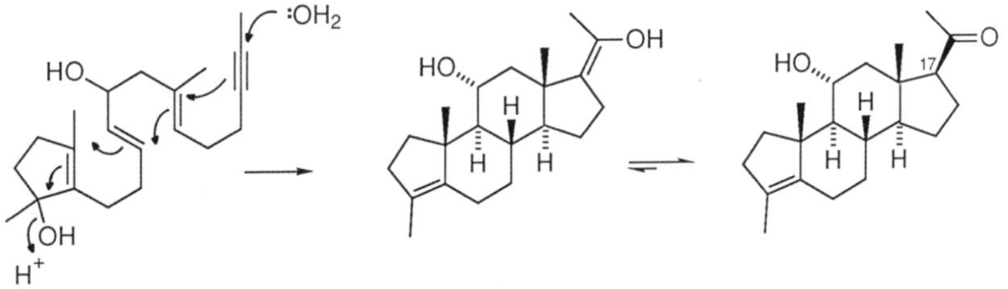

gives only the product with the required configuration of the hydroxyl group at C$_{11}$. Both epimers are formed at C$_{17}$ but each can be used in the later steps (epimerization occurs in step *m*; only one enantiomer is shown).

l The —OH group is protected by acetylation, and treatment with ozone followed by reduction of the ozonide gives the trione.

m Base-induced aldol reaction and hydrolysis of the acetate give a mixture of three parts of racemic 11α-hydroxyprogesterone and one part of the racemic 17α-epimer.

22.8
Fredericamycin
A

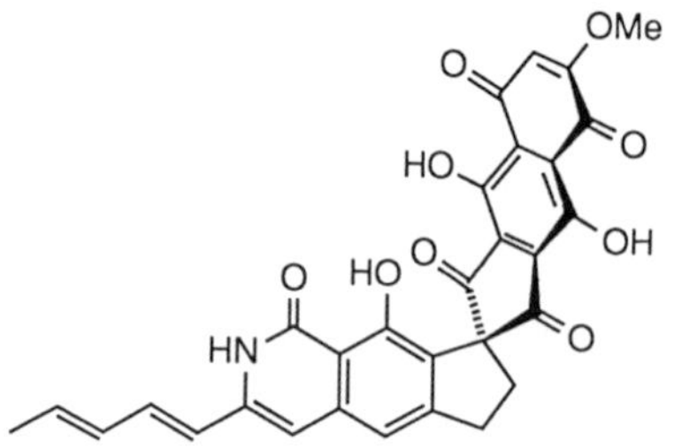

Fredericamycin A was isolated from a strain of *Streptomyces griseus* at the Frederick Cancer Research center in Frederick, Maryland and named accordingly. The compound shows anti-cancer activity and this, coupled with the novel structure, makes its synthesis and that of analogues of importance. The strategy employed by T. Ross Kelly *et al.* (*J. Amer. Chem. Soc.*, 1988, **110**, 6471) in their synthesis of (±)-fredericamycin A was to construct synthetic equivalents for the top and bottom units and then to effect their coupling in conjunction with elaboration of the spiro centre. For the bottom half of the molecule a double bond was temporarily introduced in the five-membered ring to direct the regio-chemistry for spiro formation.

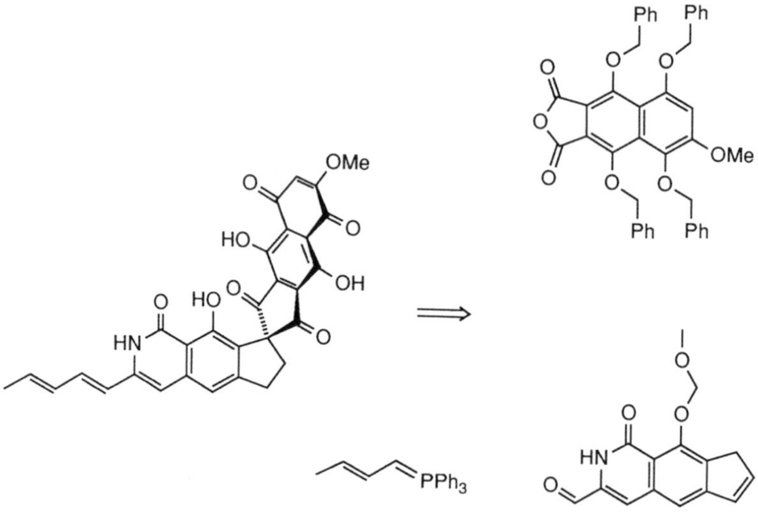

1. *Synthesis of top half*:

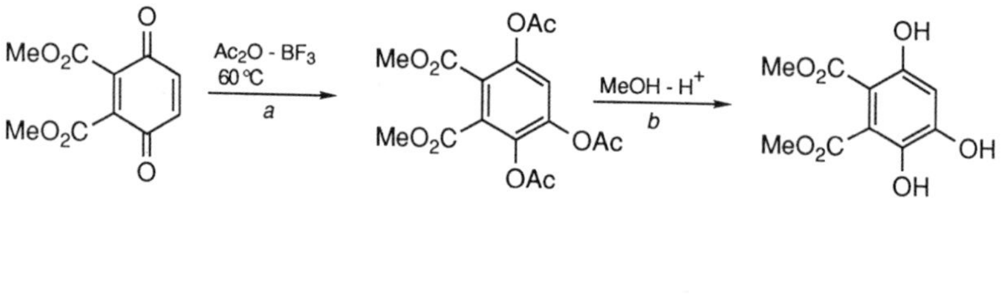

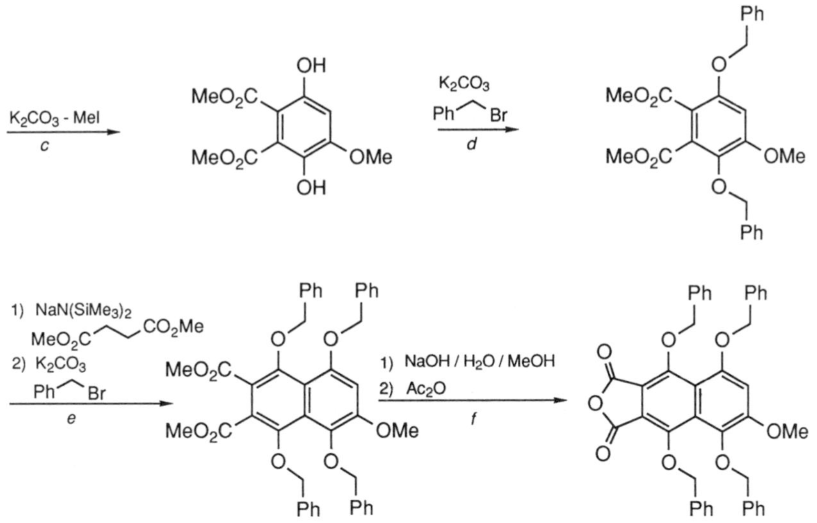

a Although this type of reaction (*Thiele–Winter acetylation*) has been known for about 100 years, its mechanism is not well established. Acetic anhydride and boron trifluoride are known to generate the acetyl cation (p. 254),

$$Ac_2O + BF_3 \rightleftharpoons MeCO^+ + AcO{-}\bar{B}F_3$$

$$AcO{-}\bar{B}F_3 \rightleftharpoons AcO^- + BF_3$$

and a possible mechanism, shown for *p*-benzoquinone, is:

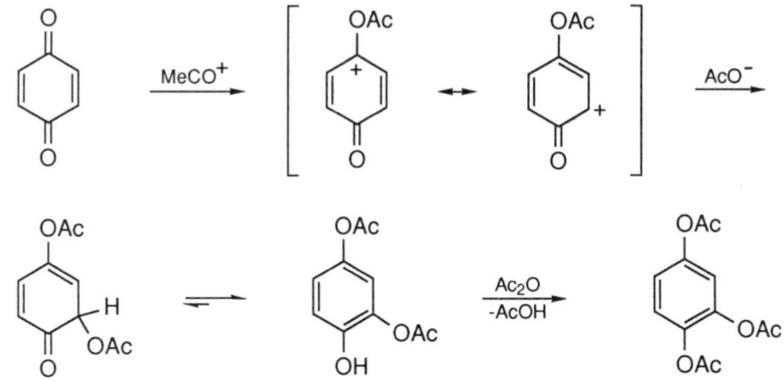

Alternatively, the reactivity of the quinone to Michael-type addition of acetate ion is enhanced by BF_3:

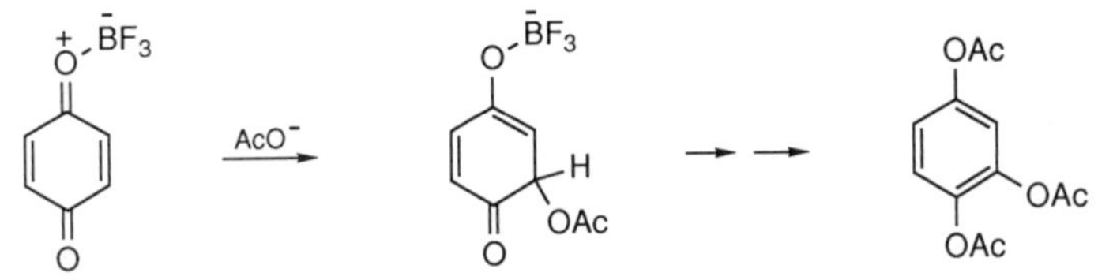

b   Transesterification with methanol, catalyzed by hydrogen chloride, gives the trihydroxy-compound.

c   Two of the hydroxyl groups are hydrogen-bonded to ester substituents and are less acidic than the third. Consequently, with only just over one equivalent of K_2CO_3, reaction through the phenoxide ion gives mainly the required methyl ether.

d   The remaining hydroxyls are benzylated with use of an excess of K_2CO_3.

e   Two crossed Claisen condensations catalyzed by $(Me_3Si)_2N^-$. Like LDA, this is a very strong base but is hindered as a nucleophile so that, for example, substitution at the benzyl groups does not occur. The product, in its phenolic form, is dibenzylated.

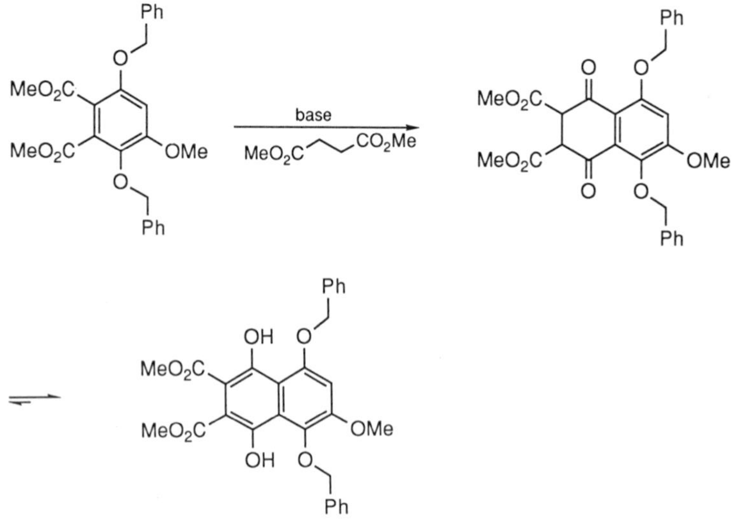

f   Hydrolysis of the ester groups followed by dehydration of the dibasic acid to give the anhydride.

2.   *Synthesis of bottom half and of fredericamycin A*:

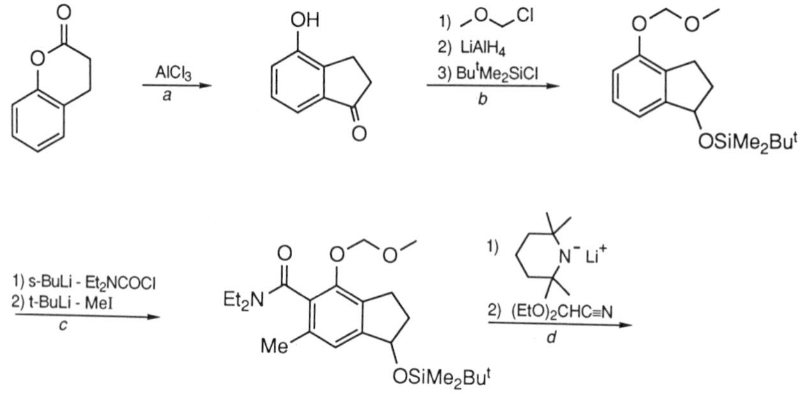

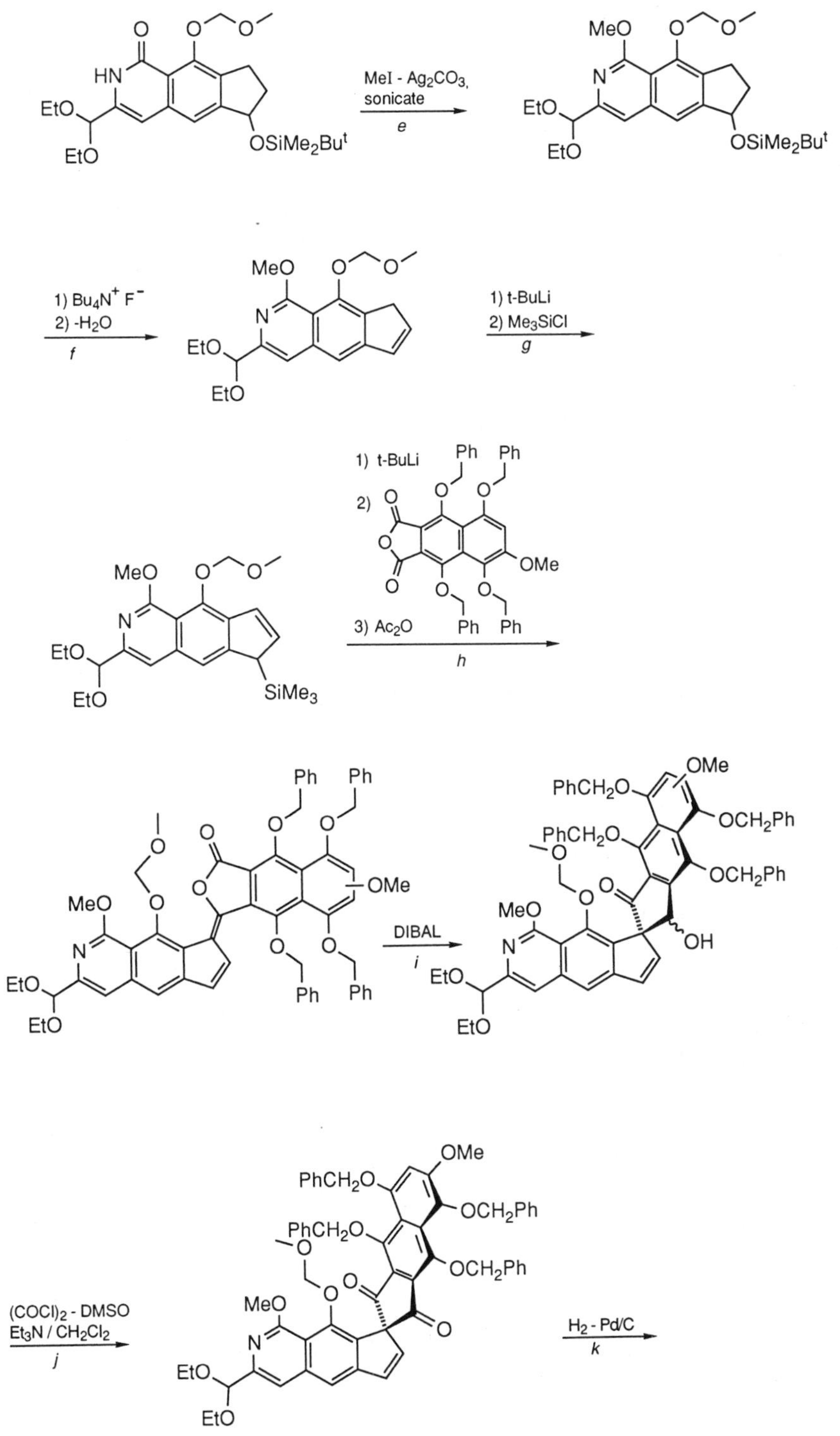
MeI - Ag2CO3, sonicate
e
1) Bu4N+ F−
2) -H2O
f
1) t-BuLi
2) Me3SiCl
g
1) t-BuLi
2)
3) Ac2O
h
DIBAL
i
(COCl)2 - DMSO
Et3N / CH2Cl2
j
H2 - Pd/C
k
OSiMe2But
SiMe3
OMe
OCH2Ph
PhCH2O
OH
EtO
MeO

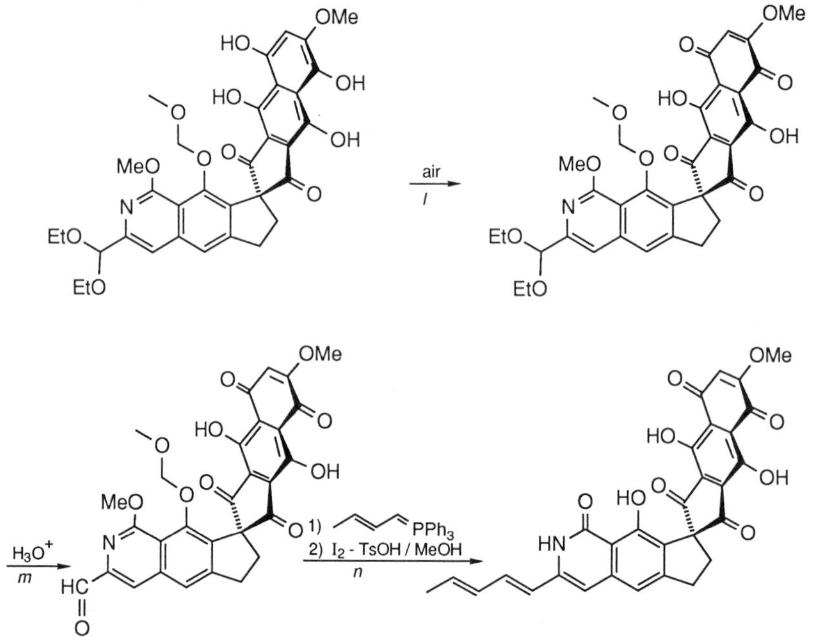

a   An intramolecular Fries rearrangement (p. 452) of dihydrocoumarin.

b   Protection of phenolic —OH as an acetal, reduction of C=O, and protection of the new —OH as a silyl derivative, the introduction of which is catalyzed by imidazole (p. 479): the latter protecting group can later be removed under conditions (F⁻) which do not affect the former.

c   Metallations are followed by reactions with electrophiles. The first metallation takes place *ortho* to the oxygen substituent (cf. p. 683), and the second *ortho* to the carbonyl group.

d   The combination of the aromatic ring and the *ortho*-carbonyl-substituent renders the methyl group weakly acidic. A very strong base is needed to generate the carbanion for the ensuing reaction on the nitrile, but one which is severely hindered as a nucleophile. The lithium derivative of 2,2,6,6-tetramethylpiperidine, generated from the amine with butyllithium, was used. Addition to the nitrile,

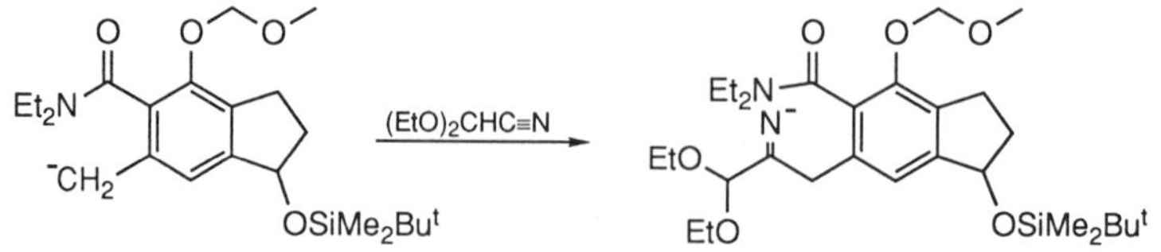

is followed by cyclization.

e   Under the conditions used, the ambident amido nucleophile reacts mainly at oxygen. Ultrasonic irradiation, by activating the surface of the silver carbonate, facilitates the reaction (p. 546).

f The silyl protecting group is removed with fluoride ion. Dehydration of the alcohol so formed cannot be effected in the usual acid conditions since the acetal would be hydrolyzed and there would be the possibility of isomerization of the new C=C bond to the alternative conjugated position. The method chosen was a variant of one used for dehydrogenation *via* a selenide derivative (p. 604). The alcohol ($ArCH(OH)CH_2R$) was converted into the selenide derivative by treatment with *o*-nitrophenyl selenocyanate (Ar′SeCN) and tributylphosphine in THF* and this was oxidized by hydrogen peroxide at 0 °C. When the solution was allowed to warm, elimination occurred:

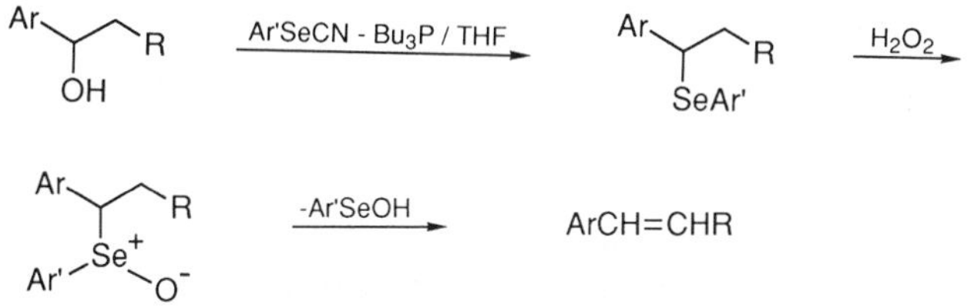

g Silylation of the delocalized allyl anion formed with t-butyllithium occurs at the less hindered position,

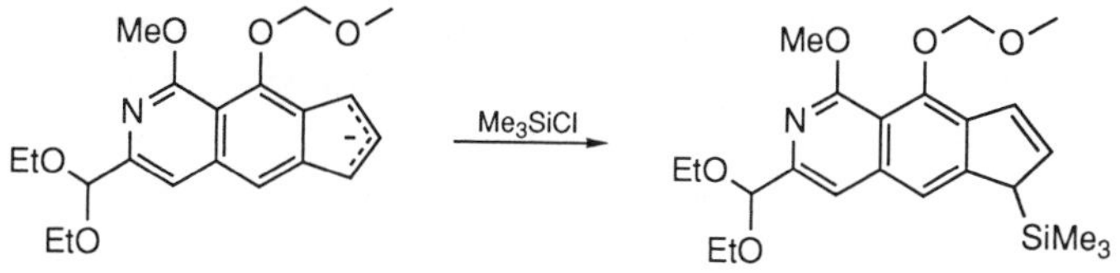

thereby blocking this position and ensuring the correct regiochemistry in the coupling reaction in the following step.

h The allylic anion, again formed with t-butyllithium,

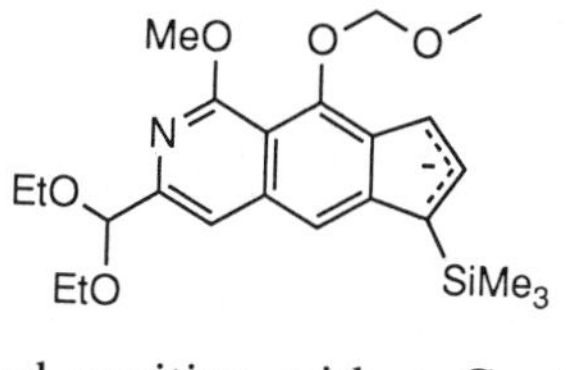

reacts at its unsubstituted position with a C=O group in the anhydride. The two such groups are regiochemically different because of the —OMe substituent, and a mixture is formed but this is immaterial (see the product of step *j*). Dehydration with acetic anhydride forms the C=C bond.

* The probable mechanism (where the alcohol is designated ROH) is:

$$Ar'SeCN + Bu_3P \longrightarrow Ar'Se\overset{+}{P}Bu_3 \quad CN^-$$

$$\xrightarrow{ROH} RO\text{–}\overset{+}{P}Bu_3 + HCN + Ar'Se^-$$

$$RO\text{–}\overset{+}{P}Bu_3 + Ar'Se^- \longrightarrow R\text{–}SeAr' + O{=}PBu_3$$

i   Reduction of the remaining C=O with DIBAL (p. 659) releases an aldehyde and an enolate which react to give spiro-compounds in the four isomeric forms associated with two asymmetric carbon atoms:

j   Swern oxidation (p. 610) removes one asymmetric carbon and the regio-isomerism associated with the —OMe group: the product consists of one pair of enantiomers.

k   Hydrogenation of the aliphatic C=C bond, together with removal of the four benzyl protecting groups by hydrogenolysis (p. 643).

l   Oxidation of the quinol to quinone occurs simply on the admission of air.

m   Acid-catalyzed hydrolysis of the more reactive —CH(OEt)$_2$ acetal occurs selectively without disturbing the ArOCH$_2$OMe acetal.

n   The Wittig reaction gives a mixture in which the *trans,trans-* and *cis,trans-* isomers predominate. However, treatment with iodine effects isomerization to give mainly the required *trans,trans* product and by carrying out this reaction in methanol in the presence of anhydrous NaBr and toluene-*p*-sulfonic acid, the acetal and ether groups are hydrolyzed.

22.9 Daphnilactone A and methyl homo-daphniphyllate

The Daphniphyllum alkaloids are a group of complex squalene-derived natural products which fall into two groups: those containing 30 skeletal carbons (e.g. daphniphylline) and those containing 22 skeletal carbons. Daphnilactone A is unique in that it contains 23 skeletal carbons, one of which is not derived from squalene. More than half of the 33 Daphniphyllum alkaloids have been isolated from methanol extracts of the bark and leaves of *D. macropodum*, a Japanese tree commonly named 'Yuzuriha' from the Japanese word 'Yuzuru' which means 'transfer from hand to hand'. The tree is unusual in that it puts forth new leaves before defoliation of the old leaves which gradually fall to the ground. Two other related species, *Daphniphyllum teijsmanni* and *Daphniphyllum humile*, have provided other alkaloids in this family. Their syntheses have led to the development by the Heathcock group of elegant methodology for complex ring skeletons: R. B. Ruggeri, K. F. McClure, and C. H. Heathcock, *J. Amer. Chem. Soc.*, 1989, **110**, 1530; R. B. Ruggeri and C. H. Heathcock, *J. Org. Chem.*, 1990, **55**, 3714; C. H. Heathcock, *Angew. Chem. Int. Ed. Engl.* 1992, **31**, 665.

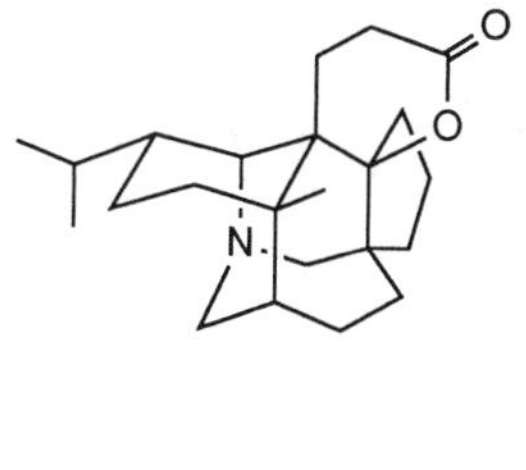

(±)-daphnilactone A

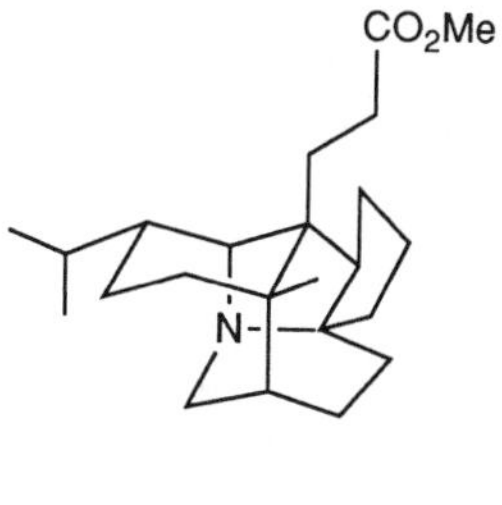

(±)-methyl homodaphniphyllate

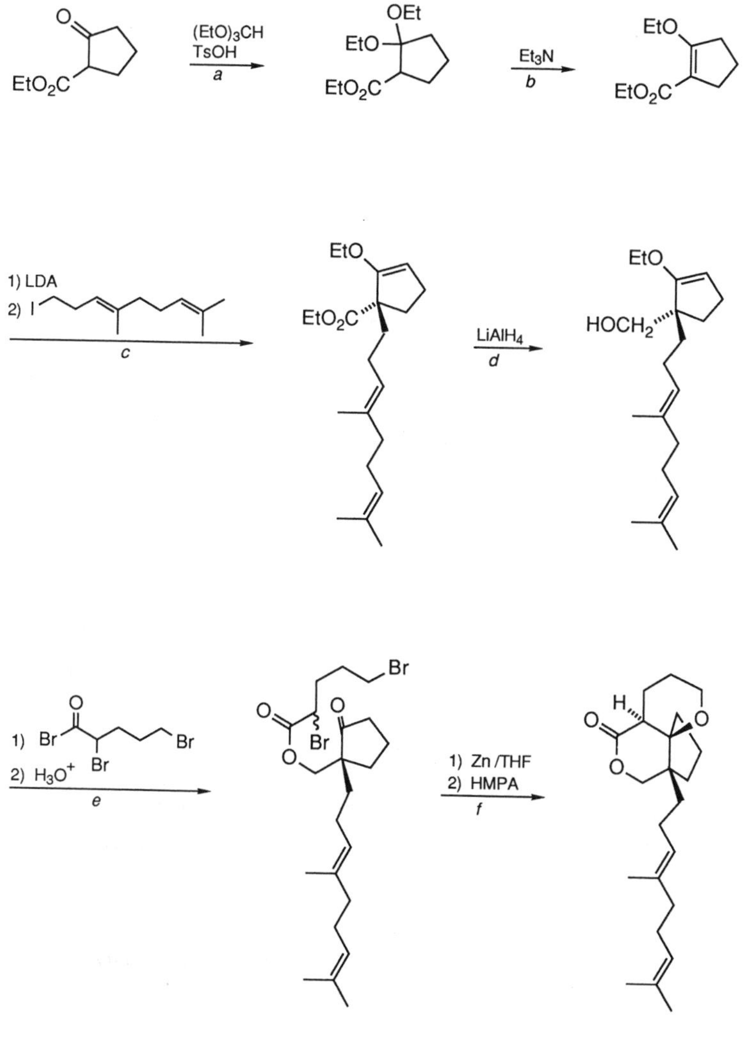

secodaphniphylline skeleton

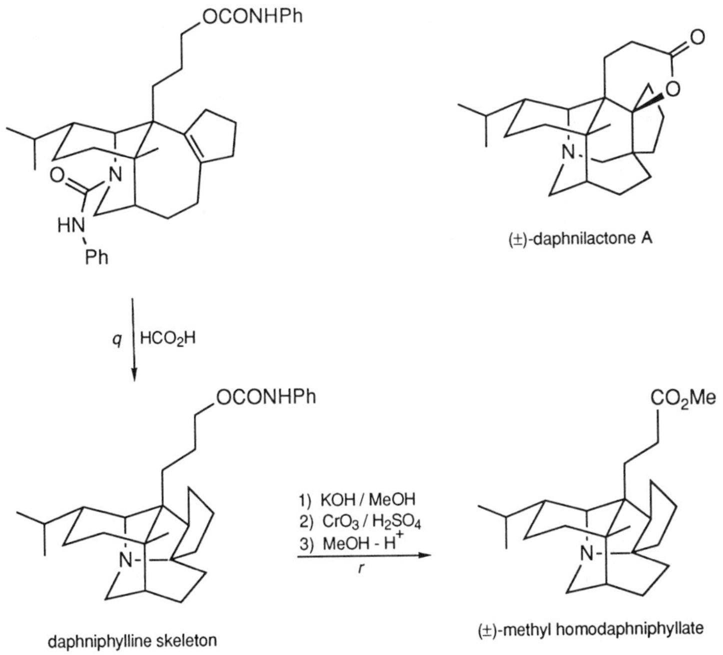

OCONHPh

(±)-daphnilactone A

q │ HCO₂H

OCONHPh

1) KOH / MeOH
2) CrO₃ / H₂SO₄
3) MeOH - H⁺
r

CO₂Me

daphniphylline skeleton

(±)-methyl homodaphniphyllate

a   Ketones do not gives ketals with monohydric alcohols but do so with ethyl orthoformate (p. 90).

b   Base-induced elimination on heating, aided by the development of conjugation.

c   The delocalized enolate formed with LDA could react with the iodide at either of two carbon atoms,

EtO

EtO₂C

EtO

EtO₂C

but does so predominantly at the ester-substituted carbon atom (p. 235). The product is asymmetric but for simplicity only one enantiomer is shown from now on.

d   The ester, but not C=C, is reduced by LiAlH₄.

e   The tribromo-compound was prepared from a readily available lactone,

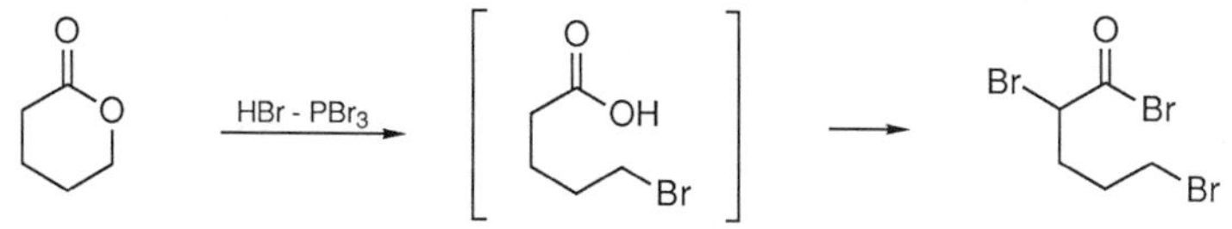

HBr - PBr₃

OH

Br

Br

Br

Br

Acid-catalyzed ring opening is followed by conversion of alcohol into bromide by PBr₃ (p. 465) and of the acid into the α-bromo-acid bromide (Hell–Volhard–Zelinsky reaction, p. 326). After reaction of the acid bromide with —CH₂OH, the enol ether is hydrolyzed.

f   An intramolecular Reformatsky reaction gives the thermodynamically favoured *cis* ring junction. Addition of HMPA coordinates the zinc, releasing
the oxyanion as a reactive nucleophile to form a second (pyran) ring by
intramolecular S_N2 displacement:

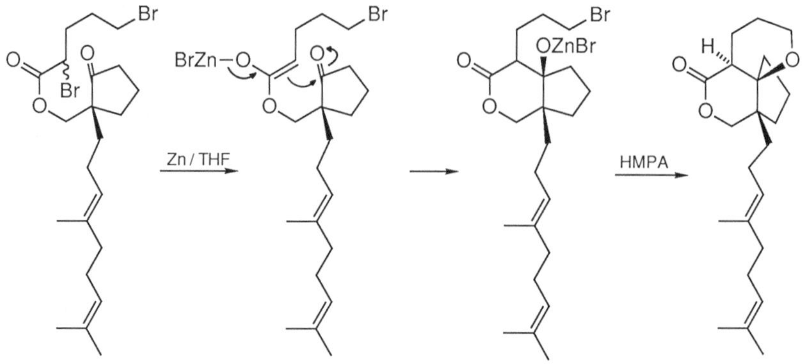

g   Reduction of the lactone.
h   Swern oxidation (p. 610).
i   Successive nucleophilic reactions by nitrogen at the —CHO groups and
elimination of water.
j   Protonation of the azadiene initiates cyclization:

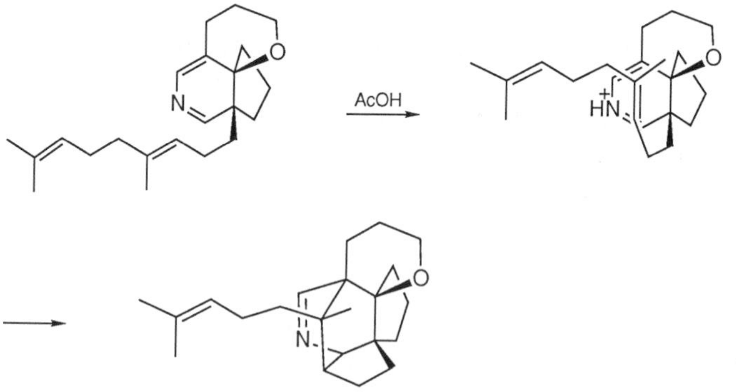

k   An intramolecular ene-type reaction occurs at 45°C:

Steps *j* and *k* represent the second occasion in the synthesis, step *f* being the
other, where two rings are formed sequentially in a one-pot reaction.
l   Hydrogenation of C=C completed the synthesis of the secodaphniphylline
skeleton.
m   Diisobutylaluminium hydride (DIBAL) complexes with the oxygen atom to

increase its leaving-group ability and bonds to the nitrogen atom; fragmentation is thereby induced:

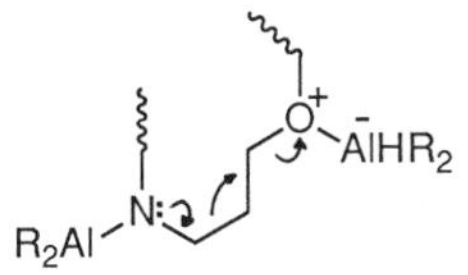

Finally, DIBAL reduces the resulting immonium ion:

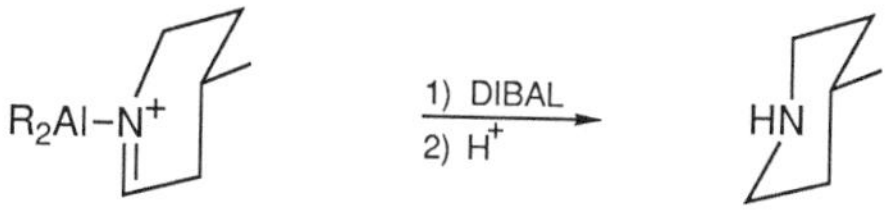

n Jones' oxidation (p. 612) gives an intermediate suitable for the synthesis of both ($\pm$)-daphnilactone A and methyl homodaphniphyllate.

o Reaction of formaldehyde at nitrogen leads to a Mannich electrophile which adds to C=C. The resulting carbocation reacts with —CO_2H to form the lactone ring and complete the synthesis of ($\pm$)-daphnilactone A:

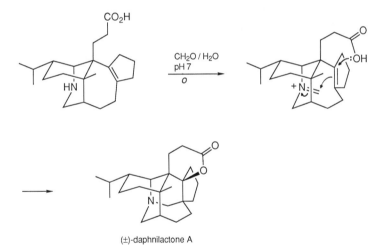

($\pm$)-daphnilactone A

p An excess of phenyl isocyanate protects the hydroxyl group as its carbamate derivative and, by reacting similarly at the amino group, reduces the nitrogen atom's basicity so that it is not significantly protonated in the following step; this enables it to act as a (weak) nucleophile in that step.

q In formic acid under reflux, the C=C bond is protonated and the carbocation reacts with the nucleophilic nitrogen to close the ring. Nucleophilic displacement on the resulting quaternary salt releases the amine.

r Hydrolysis of the carbamate followed by Jones' oxidation to the carboxylic acid and then esterification yield ($\pm$)-methyl homodaphniphyllate.

Problems   1.   Some of the products of the sequences outlined below have been used in the synthesis of naturally occurring compounds, and others are themselves important natural products. Complete the synthetic schemes in as much detail as you can: insert the formulae of intermediates which have been omitted and state the experimental conditions which you consider suitable for carrying out the individual steps. Comment, where relevant, on the stereochemistry of the intermediates.

(a)

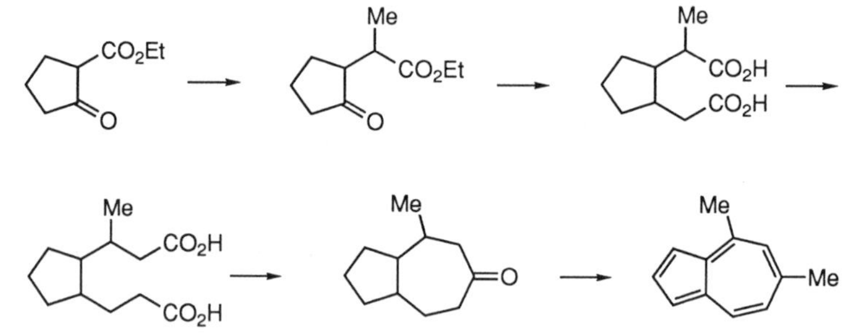

(b)

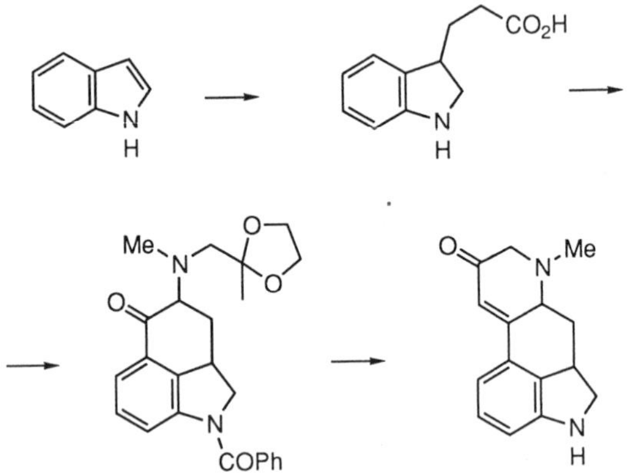

(c)

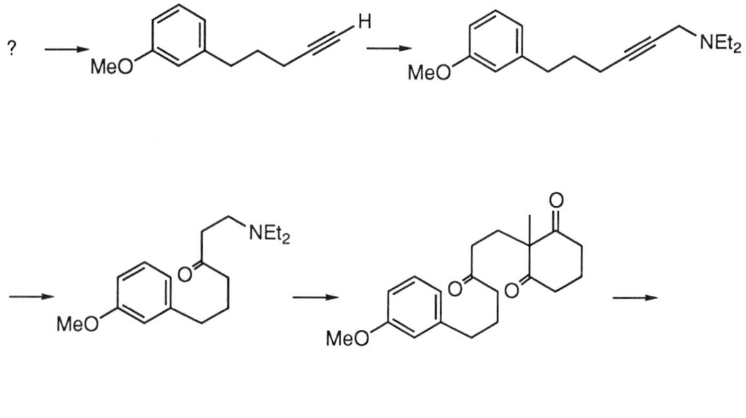

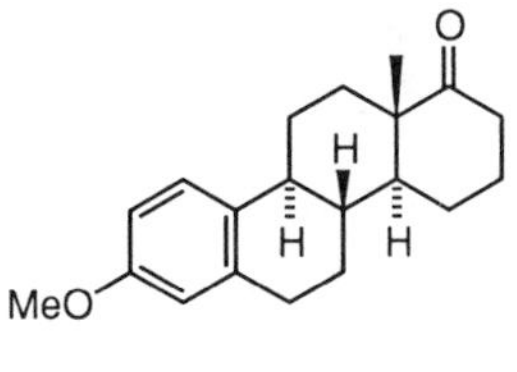

(d)

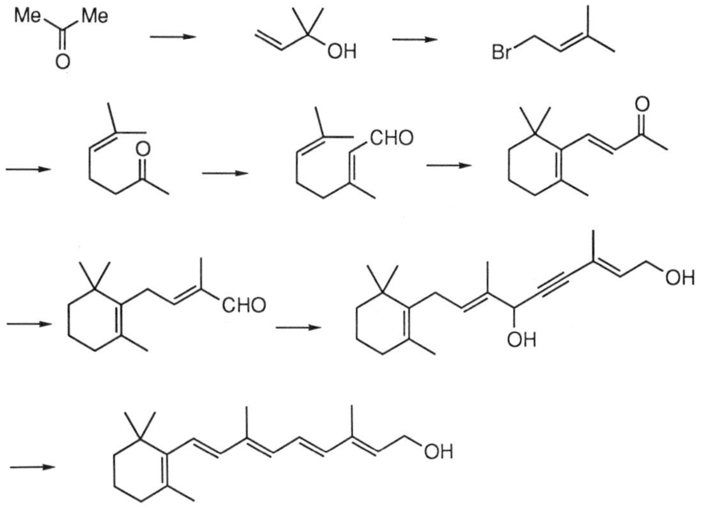

vitamin A₁

(e)

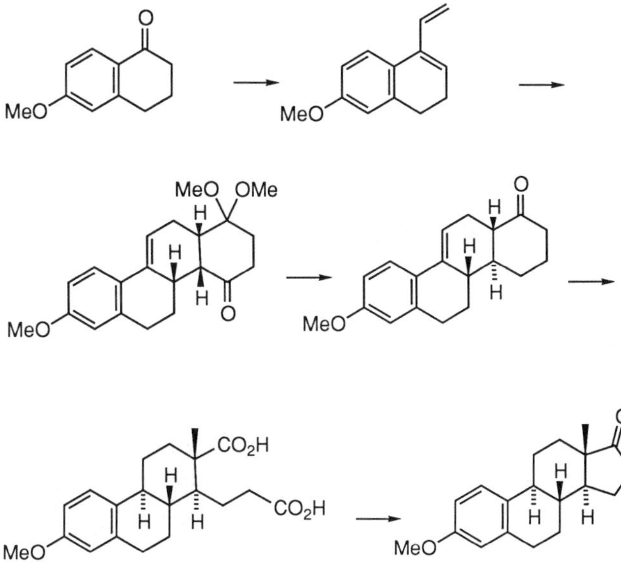

oestrone

(f)

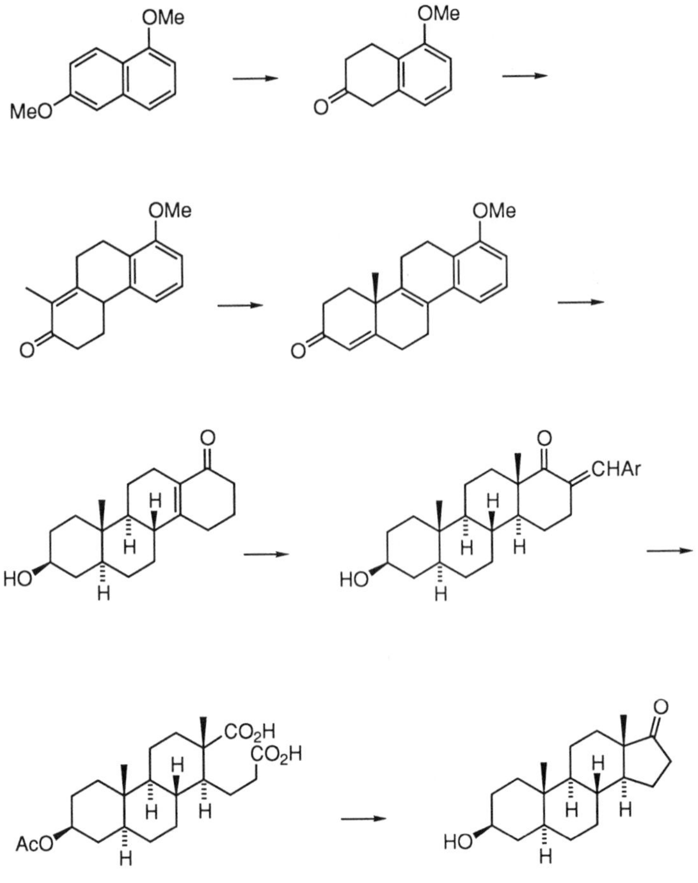

(±)-epiandrosterone

(g)

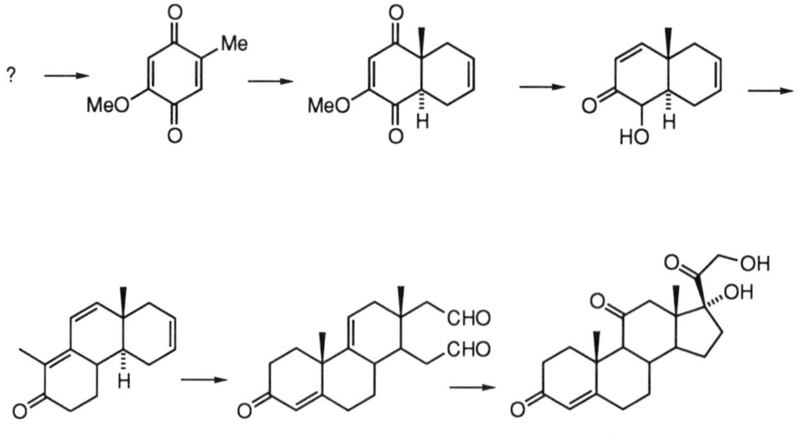

cortisone

(h)

quinine

(i)

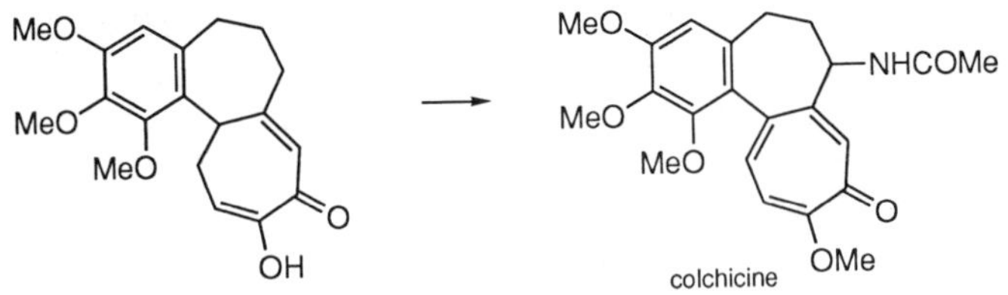

colchicine

(j)

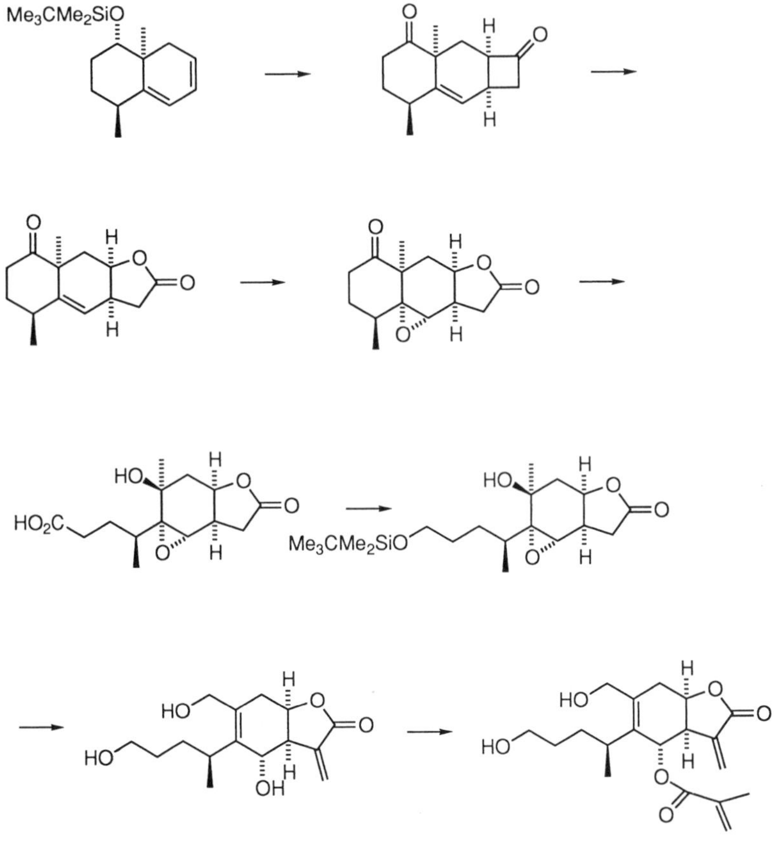

eriolanin

2. Outline methods for the synthesis of the following compounds from readily available materials.

(a) Me⌁⌁⌁⌁⌁⌁CO₂H   (b)

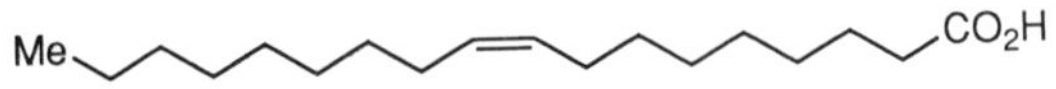

oleic acid

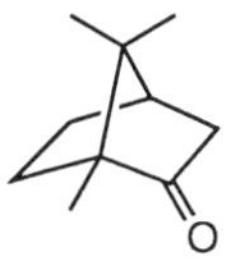

camphor

(c)

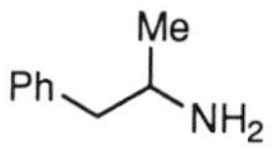

Me
Ph⏜⏜NH₂

benzedrine
(amphetamine)

(d)

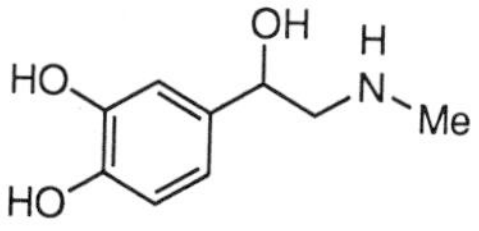

OH H
HO⏜⏜ ⏜ N⏜Me
HO

adrenaline
(epinephrine)

(e)

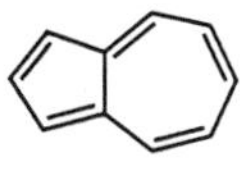

azulene

(f)

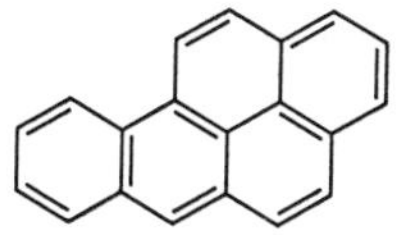

3:4-benzopyrene

(g)

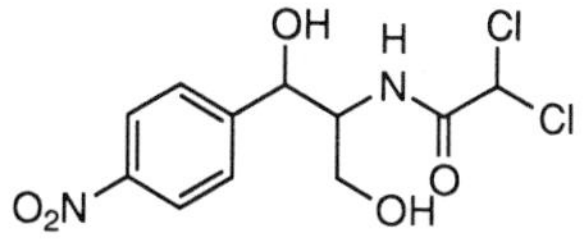

OH H   Cl
⏜⏜ N⏜Cl
O₂N   O   OH

chloramphenicol
(an antibiotic)

(h)

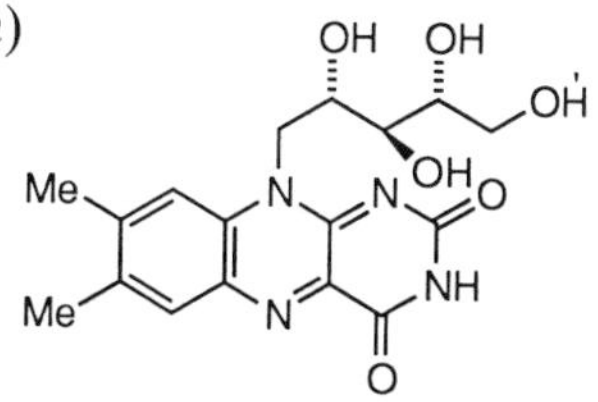

OH OH
⏜⏜⏜⏜OH
Me   N   N   OH O
Me   N   N   NH
O

riboflavin
(vitamin B₂)

3. Use retrosynthetic analysis to suggest syntheses of the following:

(a)

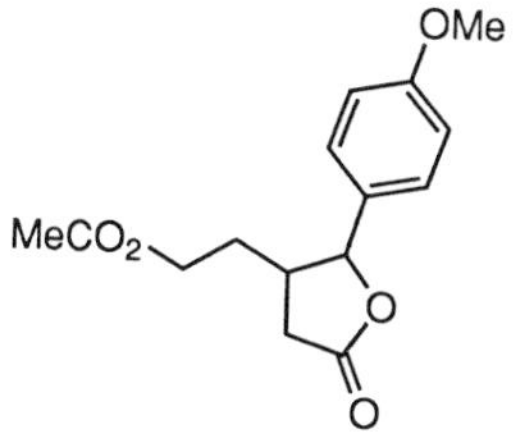

OMe

MeCO₂⏜

O
O

(b)

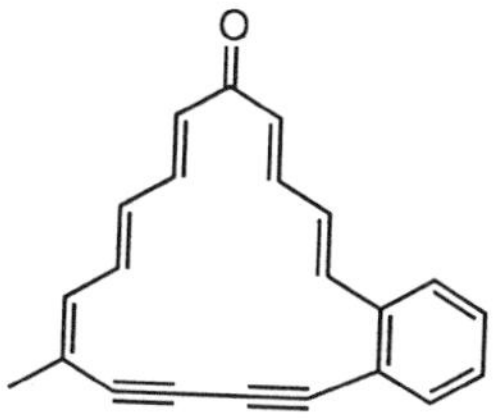

O

(c)

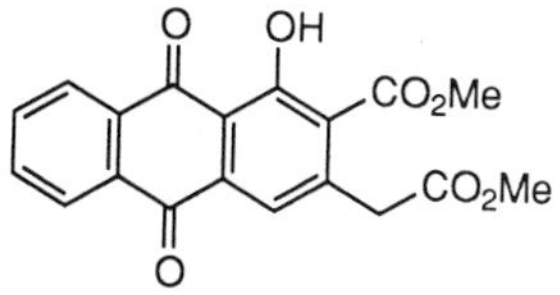

O   OH
CO₂Me
CO₂Me
O

(d)

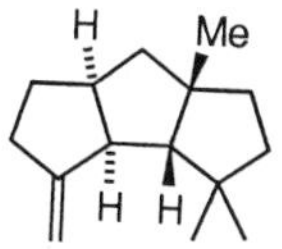

H   Me

H H

(e)

MeO

O
H
H   H   N⏜Me
HO⏜

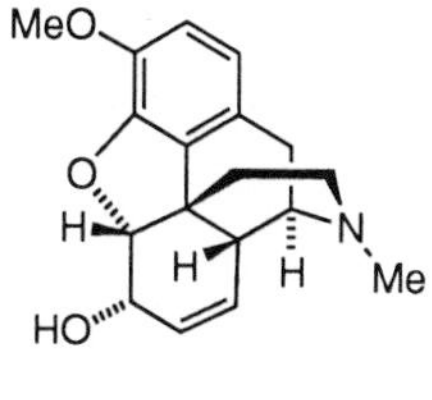

codeine

(f)

H
HO⏜⏜
H   CO₂H

OH

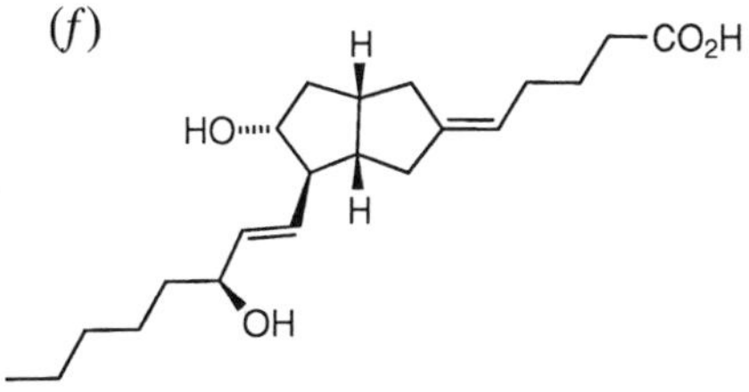

4. Rationalize the following reaction sequence that gives racemic dihydroprotodaphniphylline in 65% yield (one of the enantiomers is displayed on the
cover):

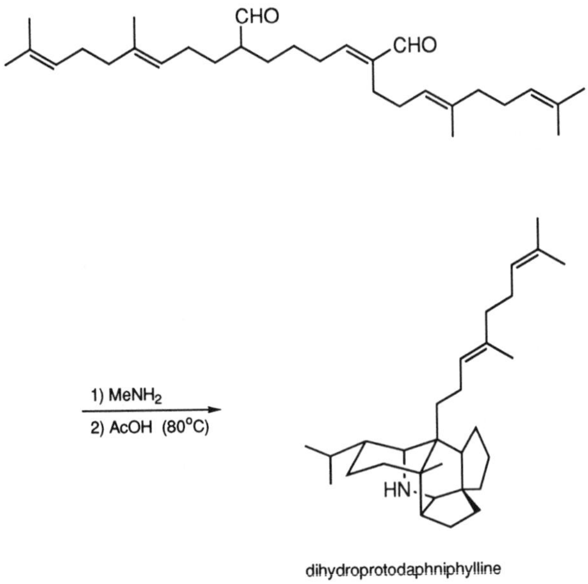

1) MeNH$_2$

2) AcOH (80°C)

dihydroprotodaphniphylline

Index